I0054735

Mechanical Design and Manufacturing of Electric Motors

Mechanical Design and Manufacturing of Electric Motors

Second Edition

Wei Tong

CRC Press
Taylor & Francis Group
Boca Raton London New York

CRC Press is an imprint of the
Taylor & Francis Group, an **informa** business

Second edition published 2022
by CRC Press
6000 Broken Sound Parkway NW, Suite 300, Boca Raton, FL 33487-2742

and by CRC Press
4 Park Square, Milton Park, Abingdon, Oxon, OX14 4RN

CRC Press is an imprint of Taylor & Francis Group, LLC

© 2022 Wei Tong

First edition published by CRC Press 2014

Reasonable efforts have been made to publish reliable data and information, but the author and publisher cannot assume responsibility for the validity of all materials or the consequences of their use. The authors and publishers have attempted to trace the copyright holders of all material reproduced in this publication and apologize to copyright holders if permission to publish in this form has not been obtained. If any copyright material has not been acknowledged, please write and let us know so we may rectify in any future reprint.

Except as permitted under U.S. Copyright Law, no part of this book may be reprinted, reproduced, transmitted, or utilized in any form by any electronic, mechanical, or other means, now known or hereafter invented, including photocopying, microfilming, and recording, or in any information storage or retrieval system, without written permission from the publishers.

For permission to photocopy or use material electronically from this work, access www.copyright.com or contact the Copyright Clearance Center, Inc. (CCC), 222 Rosewood Drive, Danvers, MA 01923, 978-750-8400. For works that are not available on CCC, please contact mpkbookspermissions@tandf.co.uk

Trademark notice: Product or corporate names may be trademarks or registered trademarks and are used only for identification and explanation without intent to infringe.

ISBN: 9780367564285 (hbk)
ISBN: 9780367564308 (pbk)
ISBN: 9781003097716 (ebk)

DOI: 10.1201/9781003097716

Typeset in Times
by codeMantra

To my wife, Zhangqing Zhuo, and daughter, Winnie Tong,
for their love, support, and patience.

Contents

Contents   xi

Preface (2nd Edition)

As the first edition of this book was published 8 years ago, many advanced technologies have emerged and developed rapidly, such as cloud computing, 3D metal printing, robotics, augmented reality, molecular motors, electrified and autonomous vehicles, quantum communication, artificial intelligence and machine learning, nanotechnology, and smartphones. These new technologies are constantly changing the landscape of almost all industries, affecting the design, analysis, manufacturing, and testing of electric machines in various degrees. Furthermore, new materials, such as graphene, silicon carbide, and enforced carbon fiber, have transited from the laboratory stage to industrial production. Servomotors and associated control systems are increasingly used in new fields, affecting modern society and human life with unprecedented depth and breadth.

Modern manufacturing has shifted the manufacturing model from mass production to mass customization. Trends in manufacturing today require greater flexibility, higher efficiency, lower cost, and minimized energy consumption. This occurs because of consumer demands such that more options and constant upgrades translate to shorter product cycles and increased stock keeping unit (SKU) proliferation. For instance, the manufacturing industry is able to benefit from collaborative robots that provide greater flexibility over industrial robotics. 3D printers are enabling manufacturers to handle greater demand for mass customization at lower costs. The digitization of manufacturing is bringing long tail economics to physical products. This alleviates inventory excesses and enables just-in-time manufacturing with the virtual guarantee.

In the global market of electric machines, the increasing competitive pressure, the requirement for reducing energy consumption, and a continuous push for decarbonization have provided an impetus for more efficient, cleaner (or greener), and quieter electric machines. Moreover, the emergence of relatively new applications such as electrified vehicles, industrial robots, 5G communication, drones, advanced medical devices, and new energy machinery has led to even more severe constraints.

To incorporate developments in motor design and manufacturing to date, this new edition of the book is significantly restructured to provide extensive information to engineers on various design and manufacturing aspects of electric motors and actuators. The change in the book title reflects this edition's focus on both design and manufacturing aspects of electric motors. The book remains focused on the combination of the theory/concepts of motor with practical approaches to benefit motor engineers in their design and manufacturing practices.

Nowadays, engineering has become more specialized as scientific and technological knowledge increased exponentially. It is important for engineers to keep abreast of the latest breakthroughs in technologies, in order to continuously innovate and design motors with a holistic perspective.

The main purpose of the second edition is to provide readers with comprehensive design, analysis, test, and manufacturing methods of electric machinery. The major additions in this second edition comprise three new chapters, which address the designs and selections of motor brakes, motor sensors and feedback devices, and power transmissions and gearing systems.

Today, brakes have an essential role to play in precision motion control systems. Chapter 7 is devoted to the design, development, and application of a variety of motor brakes. The purpose of this chapter is to introduce principles, operations, and maintenances of brake systems and provide information regarding advanced technologies, braking calculations, analyses, and selection considerations.

Chapter 8 focuses on fundamental sensing technologies, particularly various servo feedback devices and sensors used in electric machines. This chapter reviews the different physical principles and technologies that are applied in various types of measurement sensors. It starts with the measurement of rotational movement in the form of displacement, velocity, and acceleration motion using a variety of feedback devices such as encoders and resolvers. The operating features and performing characteristics of these feedback devices are evaluated and compared. The rapid progress of sensing technology in recent years allows smart sensors to gradually enter the various industrial sectors to partially replace conventional sensors. Following this, the various sensors in the form of variable inductance, capacitance, electromagnetic field intensity, piezoresistance, ultrasound, infra-red, eddy current, and Hall effect devices are reviewed in detail. Finally, the selection strategy of feedback devices and sensors for different applications and environmental conditions is provided.

Chapter 9 is concerned with mechanical power transmissions required by electric machines, including modern and conventional gearing systems, as well as newly developed innovative non-gearing, non-contact systems. To better meet or exceed the customers' requirements, considerations and factors that affect the selection of suitable transmission devices are addressed.

According to the suggestions of readers, more contents have been added in the original chapters from the first edition to comprehensively reflect new developments in the technology and manufac-turing process, including

- Chapter 1 has been expanded upon through a number of new topics, including specialty motor design, *e.g.*, explosive/flame proof motors, submersible motors, ultrahigh speed motors, motors operating under high vacuum and nuclear radiative environments, and piezoelectric motors. In addition, the section regarding the developments of new magnetic materials for replacing either partially or fully the rare earth permanent magnets (PMs) has been rewritten to reflect the updated progress.
- The magnetization and development of ring magnets are introduced in Chapter 2. This type of PM has been extensively applied in many automation and servo motion control systems for its superior magnetic properties and available magnetizing options for vari-ous applications. In order to show the latest progress in the research on the replacement of rare earth magnets, a review of the development of alternative rare earth magnets, as well as the reduction of reliance on rare earths, has been conducted in this chapter.
- A new section about non-contact coupling is introduced in Chapter 3. From a technical point of view, the type of non-contact coupling has numerous advantages over its contact competitor, including no wear, no noise, elimination of vibration transfer between a motor and its driven load, high tolerance of misalignment between shafts, reduced rotating equip-ment maintenance and operation costs, and increased seal and bearing life.
- Additive manufacturing has several advantages over traditional manufacturing technolo-gies, such as lower resource requirement, faster production cycle, flexible design, and sub-stantial savings on tooling. In Chapter 5, 3D printing and other additive manufacturing processes are discussed.
- Slewing ring bearings can improve load support and power transmission in all directions. They are typically employed to support heavy loads for slow speed applications and large diameter equipment such as CT and MRI scanners (Chapter 6).
- Hydrogen cooling is widely used for cooling large-size electric machines due to its supe-rior thermal properties. Compared to air cooling, hydrogen cooling can offer 150% higher heat transfer capability and a 20%–30% size deduction at similar power ratings. Moreover, direct liquid cooling through both the stationary and rotary hollow windings can signifi-cantly enhance heat-dissipating capabilities in electric machines (Chapter 11).
- Self-locking fasteners have been developed to virtually eliminate the possibility of bolted assemblies coming apart during operation because of vibration, shock, impact, and other dynamic loads. A variety of self-locking fastener types are addressed in the newly created section of Chapter 12.
- Heat transfer coefficients of two-phase flows are typically larger than those of single-phase flows due to the latent heat of vaporization in liquid-vapor flows or enhanced convec-tional heat transfer in liquid-solid particle flows (*e.g.*, nanofluids). Modeling of two-phase flow and heat transfer is addressed in Chapter 14 for its importance in cooling industrial machinery and medical equipment.

- Micro-, nano- and molecular-motors are miniaturized machines that are able to convert chemical or other energy into mechanical motion. This represents a fundamental step toward the realization of practical nanomachines. As an emerging technology, it can offer some unique functions in life science such as drug carriers. The related section in Chapter 15 has been rewritten to cover recent developments in the field.

I would like to express my sincere gratitude to the following reviewers for their insightful comments and suggestions. There is no doubt that with their dedication and efforts, the quality of the book is greatly enhanced.

Jeffrey J. Farrenkopf	Director of Engineering (Retired), Kollmorgen Corporation
Prof. Lih-Wu Hourng	Professor Emeritus, Department of Mechanical Engineering, National Central University (Taiwan)
Prof. Hakan Gürocak	School of Engineering and Computer Science, Washington State University Vancouver
Prof. Hanz Richter	Mechanical Engineering Department, Cleveland State University
Dr. Zhangqing Zhuo	Gas Turbine Application, General Electric Company
Dr. Winnie Tong	University of Chicago

Since the first edition of the book was published in 2014, I have received lots of feedback from readers with their comments, suggestions, and corrections. Accordingly, many revisions have been made to take all of them into account in this new edition. Errors appearing in the first edition have been tracked down and corrected as much as possible. Indeed, the inputs from the readers have helped shape the priorities for change and elaboration in the second edition. My particular gratitude goes to all of them. Moreover, I would like to express my thanks to Ms. Nicola Sharpe, the editor at CRC Press, for her kindly provided suggestions and supports throughout this project. Readers may contact me using the email address at: motor.mechanical.design@gmail.com, or via *CPC Press*.

Wei Tong, Ph.D., PE, MBA
Clifton Park, NY
January 2022

Preface (1ˢᵗ Edition)

Electric motors are extensively used in industrial, commercial, and military applications such as automobiles, elevators, electronic devices, robots, appliances, medical equipment, energy-conversion systems, machine tools, aircraft carriers, and satellites. Along with the rapid growth in energy demand all over the world in recent years, it has become a challenge for the motor industry to design and manufacture high-efficiency, high-reliability, low-cost, and quiet electric motors with superior performance.

Kollmorgen Corporation, a subsidiary of Danaher Corporation, is the global leader in the design, development, manufacture, and service of innovative and reliable products in the motion control industry. As the chief engineer of Kollmorgen, I have long been eagerly awaiting a modern book on the mechanical design of electric motors. However, most available books on electric motors in the market focus primarily on electromagnetic design rather than mechanical design. After a long unsuccessful search, one day I was suddenly struck with a thought: Why not write this book myself? This thought has inspired me to embark on the arduous task of compiling recent developments and advances in the mechanical design of electric motors into one book.

Despite significant advances in the last decades, the design, development, and implementation of modern electric motors still pose challenges for motor engineers and designers. This is because they involve a wide variety of disciplines, such as mechanics, electromagnetics, electronics, fluid and solid dynamics, heat transfer, material science, tribology, acoustics, control theory, manufacturing technology, and engineering economics, to name a few. In order to meet customer-specific requirements, today's motor manufacturers must effectively provide customized premium efficient products and services for accommodating multiple markets.

The ultimate goal of this book is to provide readers an in-depth knowledge base of designs, techniques, and developments of modern electric motors. The text is suitable for engineers, designers, and manufacturers in the motor industry, as well as maintenance personnel, graduate and undergraduate students, and academic researchers. This book has addressed in detail many aspects of motor design and performing characteristics, including motor classification, design of motor components, material selections, power losses, cooling, design integration, vibration, and acoustic noise. It also covers motor modeling, simulation, and engineering analysis. To reflect state-of-the-art electric motors, innovative and advanced motors developed in recent decades are reviewed in the book.

The book consists of a total of 12 chapters that cover electric motor fundamentals, practical design, engineering analysis, model setup, material and bearing selection, manufacturing processes, testing methods, and other design issues/practical considerations of electric motors. Chapter 1 presents the overview of the basic knowledge and fundamental principles of electric motors. Attempts have been made to introduce to readers motor operating characteristics, design parameters, manufacturing processes, and design methodologies.

Chapters 2 through 5 sequentially address design details of the most important components or subsystems of electric motors, including rotor, shaft, stator, and motor frame. In each of these chapters, design and manufacturing activities are addressed in detail to cover material selection, manufacturing process, assembly method, engineering analysis, *etc*. The potential applications of some newly developed manufacturing technologies on electric motors are also discussed.

Statistical data have revealed that the majority of electric motor failures are attributed to premature bearing failures. Thus, it is critical to understand bearing operating characteristics, determine bearing failure modes, identify bearing failure root causes, and evaluate the effects of some factors (*e.g.*, temperature, lubrication condition, contamination, bearing current, excessive loading, misalignment, *etc*.) on bearing life. All these are addressed in Chapter 6.

When a motor operates under load conditions, a certain portion of the input power does not convert into the mechanical power to do useful work. Most of the wasted energy is converted into heat, which must be effectively dissipated into the environment. This results in a decrease in motor efficiency. As a matter of fact, motor efficiency varies widely with the technology used. In order to improve motor efficiency, it is highly desirable to minimize these power losses. The analysis and calculation of each type of power loss are discussed in Chapter 7.

Motor cooling plays a major role in ensuring optimum performance, operating reliability, and cost-effectiveness of electric motors. The adoption of high-efficiency cooling techniques can significantly increase in the power density and overall motor efficiency. A variety of cooling techniques, including air cooling, liquid cooling, mist cooling, and phase change cooling, are reviewed in Chapter 8. The selection of cooling techniques depends on many factors, such as cooling load, cooling effectiveness, thermophysical properties of coolant, complexity of cooling system, type of motor application, and others. State-of-the-art cooling techniques that have been successfully used in electronics cooling are also reviewed, and these have been suggested for application in motor cooling.

Chapter 9 describes vibration and noise issues of electric motors. Motor vibration can not only cause high noise and high fatigue stress of motor, but also deteriorate the performance of motor and, in the worst-case scenario, lead to premature motor failure. High acoustic noise is prohibited by many legislative regulations. For some special motor applications, acoustic sound pressure levels are restricted by law.

Computer technology has become a key component in motor development and design processes over recent decades. Computer-aided design (CAD), computer-aided manufacturing (CAM), and computer-aided engineering (CAE) techniques today are widespread in their application to motor design. However, no matter how perfect the engineering design and how precise the theoretical calculations, motor testing is a necessary step to validate the conceptual design, ensure normal and safe operation of the motor, and further optimize the motor design. Chapter 10 presents the motor testing methods and procedures for both component and system levels. The important testing equipment is also addressed.

A variety of engineering modeling, simulation, and analysis that are frequently encountered during motor design processes are addressed in Chapter 11. The successful integration of the engineering analysis into the product design cycle can significantly enhance the design quality, improve design standards, and accelerate the design process. In this chapter, some important engineering analyses are discussed in detail, including CFD analysis, rotordynamic analysis, and stress/strain analysis.

With the rapid strides made in computer technology, material science, manufacturing process, control technology, and other fields, the motor industry has developed immensely in recent years. In fact, sustainable growth of the motor industry requires innovation. In the age of globalization, motor manufacturers tend to aim at high-efficiency, high-precision, durable, and energy-saving products. Chapter 12 reviews and discusses some innovative and advanced motors.

Electric motors and generators are commonly referred to as electric machines. While electric motors convert electrical energy to mechanical energy, electric generators convert mechanical energy into electrical energy. Due to the similarities between electric motors and generators in many aspects such as mechanical construction, electromagnetic operating principles, power losses, cooling methods, *etc.*, some contents in this book can also be applied in generator design and serve as a reference for generator design engineers. Similarly, a few generator design practices covered in this book can also benefit motor design engineers.

"Details determine success or failure" — this is true not only for industrial production but also for many aspects of human society. As a consequence, engineers and designers should pay special attention to details in designing and manufacturing electric motors.

I am greatly indebted to the excellent reviewers of this book. Their valuable and pertinent comments have greatly enhanced the quality of the book. The experts who have kindly read different parts of this book include the following:

Jeff Farrenkopf	Kollmorgen Corporation, USA
John Keesee	Kollmorgen Corporation, USA
Lih-Wu Hourng	Department of Mechanical Engineering, National Central University, Taiwan, Republic of China
Chen Fang	Heavy Oil Recovery Research Program, Exxon Mobil Corporation, USA
Igor V. Shevchuk	MBtech Group GmbH & Co. KGaA, Germany
Shanshan Conway	Remy International Inc., USA
Kamlesh Suthar	Argonne National Laboratory, USA
Shijun Ma	School of Mechanical Engineering, Southwest Jiaotong University, China
Zhangqing Zhuo	Power & Water Group, General Electric Company, USA

I wish to acknowledge my colleagues at Kollmorgen Corporation for their constructive discussions, suggestions, and comments during the preparation of the book, including, but not limited to, John Boyland, Jeff Farrenkopf, John Keesee, Todd Brewster, David Coulson, Lee Stephens, Gary Hodge, Jerry Brown, Mark Fields, Brad Trago, Denny Hu, Steve VanAken, James Davison, Tony Nozzi, Ron Bishop, David Guy, Stephen Funk, Brad Monday, Tommy Bunch, Ethan Filip, Amber Hollins, Kevin Garrison, and many others. It has been a great pleasure and honor working with them.

From November 2010 to May 2013, I was dispatched to Shanghai, China, for building up an engineering team. I wish to acknowledge all the engineers in the team for offering me some useful information. While providing engineering training to them, I have also learned a lot from them.

I am especially grateful to Jonathan Plant, Jennifer Ahringer, and Jennifer Stair of CRC Press for their valuable advice and guidance. Vijay Bose as the project manager at SPi Global has done an excellent job in editing and proofreading the manuscript with splendid efficiency.

I deeply miss my parents. Their imperceptible influence in my growing years helped me set my own life goals. At the most difficult time of my life (during a special historical period), they encouraged me to look forward and keep my faith and hope for the future. My biggest regret is that they cannot see this published book.

Finally, I wish to express my sincere gratitude to my lovely wife, Zhangqing Zhuo, and daughter, Winnie Tong, for their love and patience. Without their continuous support and encouragement, this book could have only been published in my dreams.

I greatly welcome and appreciate feedback, corrections, and suggestions from readers for improving and enhancing the technical contents of the book. Readers can directly reach me via e-mail at: motor.mechanical.design@gmail.com.

Wei Tong
Radford, Virginia
January 2014

Author

Dr. Wei Tong has 40 years of industrial experience in power systems, electric machines, energy conversion equipment, industrial robots, and power tools. He started his career as a mechanical technician at a factory producing enamel products and chemical equipment in the early 1970s. From 2003 to 2019, he served as Chief Engineer at Kollmorgen Corporation, responsible for the initiation, design, analysis, development, and implementation of electric motors, industrial robots, and automated systems. He made significant contributions in developing and promoting advanced servomotors, robot arms, cooling technologies, optimized mechanical structures, sealing systems, and manufacturing process improvements.

Prior to joining Kollmorgen, Dr. Tong worked for GE Energy at Schenectady, New York, where he conducted designs of high-efficiency, high-powered generators and steam turbines. He was the project leader of a number of design and R&D projects, such as the advanced cooling system, hydrogen sealing system, noise attenuation system, dynamic thrust control system, steam turbine axial clearance design, and thermodynamic operation design. All these projects had been led to great technical and commercial success. Prior to GE, he worked at Babcock & Wilcox, Lynchburg, Virgina, to carry out design and analysis for the superconducting magnetic energy storage system, solar-powered satellite, and medical isotope production reactor. Before entering industry, he spend two years each at the University of Colorado at Boulder and Rensselaer Polytechnic Institute (Troy, NY) as a research associate.

Dr. Tong received a Ph.D. degree from the University of Minnesota in 1989, an MBA degree from State University of New York (*SUNY*) at Albany in 1999, an M.E. degree from Howard University, and a B.E. degree from East China University (Shanghai). He served as an Associate Editor of the *ASME Journal of Heat Transfer* for two terms (2009–2012 and 2012–2015), Associate Editor of the *International Journal of Rotating Machinery* (2008–2016), and Associate Editor of *IEEE Transactions on Transportation Electrification* (2016–2018). He was invited as a key lecturer in *2016 IEEE Transportation Electrification Conference and Expo (ITEC'16)*.

Dr. Tong currently holds 29 US patents and 16 foreign patents. He has been a licensed Professional Engineer in Virginia since 1996. He is an internationally recognized expert on mechanical–electrical–thermal systems.

As a fellow (2006–now) of the American Society of Mechanical Engineering (*ASME*), Dr. Tong has served on several *ASME* committees, including Heat Transfer under Extreme Conditions (K-18), Heat Transfer in Energy Systems (K-6), Aerospace Heat Transfer (K-12), and Long Range Directions and Issues in Heat Transfer (K-2). He was the chair of K-18 Committee during 2006–2009. In 2014, he was elevated to Senior Member of the Institute of Electrical and Electronics Engineers (*IEEE*).

In 2006, Dr. Tong was appointed as an adjunct professor by the Department of Mechanical Engineering, Virginia Polytechnic Institute and State University (Virginia Tech), where he taught undergraduate core courses on mechanical engineering design.

List of Abbreviations

LIST OF SYMBOLS

A	area, m^2
A_c	contact area, m^2
a	acceleration, m/s^2
a_c	centrifugal (radial) acceleration, $a_c = v^2/r = r\omega^2$, m/s^2
$\mathbf{B}$	magnetic flux density (a vector quantity), T (Tesla, T = Wb/m^2), equivalently $N/(A \cdot m)$ (Newton per ampere-meter) (Chapters 2 and 8)
B_r	remanence magnetic flux density when H = 0 ($B_r = \mu_o M_r$), T (Chapter 2)
C_c	viscous damping of coupling, Ns·m/rad (Chapter 9)
C_e	electric time constant, s
C_g	center of gravity
C_H	Hall effect coefficient
C_l	viscous damping between load and ground, Ns·m/rad (Chapter 9)
C_m	viscous damping between motor and ground, Ns·m/rad (Chapter 9)
C_m	mechanical time constant, s (Chapter 1)
C_{th}	thermal conductance, W/°C or W/K
c	viscous damping coefficient, N·s/m
c	sound speed, m/s (Chapter 9)
c_p	specific heat, J/(kg·K) or J/(kg·°C)
D	outer diameter, m (Chapter 1)
d	diameter, m
d_e	effective diameter, m
d_p	pitch diameter of gear, mm or m (Chapter 9)
E	Young's modulus of elasticity $E = \sigma/\varepsilon$, Pa, MPa or GPa
E	energy, J (Chapters 2 and 12)
E_b	braking energy for a single operation, J
$E_{b,a}$	allowable braking energy in manufacturer's catalog, J
E_k	kinetic energy, J
E_p	potential energy, J
E_r	radiative energy, J
E_t	total energy, J
E_{th}	thermal energy, J
e	mass eccentricity, m (Chapter 2)
e	emissivity (Chapter 11)
e	specific internal energy, J/kg (Chapter 11)
F	force, N
F_a	axial force of gear, N
F_{bt}	bolt axial tensile force, N
F_c	centrifugal force, N
F_d	dynamic friction force, N
F_m	magnetomotive force, $F_m = IN$, A-turn (ampere-turn[1])
F_{mc}	fastener clamping force, N

---

[1] Since *turn* has no unit, the SI unit of F_m is ampere, but to avoid any possible confusion, *ampere-turn* is often used in the literature.

F_n	normal force, N
F_r	radial force of gear, N
F_s	static friction force, N
F_t	tangential force of gear, N
F_u	exciting force resulting from unbalanced mass, N
f	frequency, Hz
f_c	centrifugal force, N
f_d	damping force, N
f_e	input excitation frequency, Hz (Chapter 8)
f_e	external force, N
f_R	resonant frequency, Hz
f_r	frequency response of encoder, Hz
f_s	shape factor (Chapter 3)
f_{slip}	slip frequency, Hz (Chapter 1)
f_{tr}	torsional resonant frequency, Hz
G	shear modulus of elasticity, Pa, MPa, or GPa
Gy	a unit of ionizing radiation dose, defined as the absorption of one joule of radiation energy per kilogram, J/kg (Chapter 8)
g	acceleration due to gravity, m/s^2
H	magnetic field strength, (or magnetization force), a vector quantity measured in units of ampere per meter, A/m (Chapters 2 and 8)
H_c	intrinsic coercivity of magnetic material for resisting demagnetization, Oe (Chapter 2)
H_e	exchange bias shift, Oe (Chapter 2)
h	heat transfer coefficient, W/(m$^2 \cdot K$)
h	specific enthalpy, J/kg (Chapter 14)
I	electric current, A
I	moment of inertia, kg-m^2
I_a	second moment of area, m^4 (Chapter 3)
I	electric DC current, A
I_r	rated DC current, A
i	electric AC current, A
i_p	peak AC current, A
i_r	rated AC current. A
J	second polar moment of area, m^4
J	current density in conductor, A/m^2 (Chapter 2)
Ja	Jakob number
J_b	bearing inertia, kg·m^2
J_c	coupling inertia, kg·m^2
J_{gb}	gearbox inertia, kg·m^2
J_l	load inertia, kg·m^2
J_m	motor inertia, kg·m^2
J_o	current density at conductor surface, A/m^2 (Chapter 2)
J_{rl}	reflected load inertia, kg·m^2
J_t	total polar moment of inertia, kg·m^2
J_v	polar moment of inertia, kg·m^2
j	j-Colburn factor, $j = St/Pr^{2/3}$
K	tightening factor (Chapter 5)
K	gauge factor (Chapter 8)
K	thermal conductance, W/°C (Chapter 8)
K	loss coefficient (Chapter 14)
K_φ	ratio of electrical loading on rotor and stator

K_a	loss coefficient for actual winding, $K_a = 2\Delta p_a / \rho u_a^2$ (Chapter 14)
K_c	coupling elasticity, Nm/rad (Chapter 9)
KE	kinetic energy, J
K_e	back EMF constant, V/(rad/s) or V/rpm
K_i	current waveform factor
K_m	motor constant, Nm/W$^{1/2}$ (Chapter 1)
K_p	loss coefficient for porous media system, $K_p = 2\Delta p_p / \rho u_p^2$ (Chapter 14)
K_p	electrical power waveform factor (Chapter 1)
K_t	normal stress concentration factor (Chapter 2)
K_t	torque constant, Nm/A (Chapter 1)
K_{ts}	shear stress concentration factor
K_v	motor velocity constant, (rad/s)/V or rpm/V
K_{vd}	viscous damping, Nm/rpm or Nm/(rad/s) (Chapter 1)
k	stiffness, N/m
k	shape parameter (Chapter 6)
k	thermal conductivity, W/(m·K) (Chapters 11 and 14)
k	spring coefficient, $k = -F_s/x$, N/m (Chapter 12)
k	radius of gyration, m (Chapter 14)
k_b	bearing support stiffness, N/m
k_{eff}	effective thermal conductivity, W/(m·K) (Chapter 14)
k_l	linear stiffness, N/m
k_w	specific wear rate coefficient, m^3/Nm (Chapter 7)
L	inductance, $L = v(t)/(di/dt)$, H
L	the largest length scale in turbulent flow (Chapter 14)
L_e	effective stack length, m
L_p	sound pressure level, dB (Chapter 9)
L_s	sliding distance, m (Chapter 7)
L_w	sound power level, dB (Chapter 9)
l	length, m
l_g	grid length, m
l_u	a half of coupled arm (the distance between the two unbalance masses, m (Chapter 2)
M	bending moment, Nm
$\mathbf{M}$	magnetization (a vector quantity representing magnetic moment per unit volume), A/m (ampere per meter) (Chapter 2)
M_a	angular momentum, J·s or Nm·s
M_r	rotor mass, kg (Chapter 2)
M_r	remanence magnetization when $H = 0$, A/m (ampere per meter) (Chapter 2)
M_s	saturation magnetization, A/m (ampere per meter) (Chapter 2)
m	gear module, the ratio of the reference diameter of gear to the number of teeth, $m = d/z$, mm (Chapter 9)
m	mass, kg
m	number of motor phases (Chapter 1)
m_c	correction mass, kg
m_m	magnetic moment, Am^2
m_u	rotor unbalance mass, kg (Chapter 2)
	mass flow rate through actual winding, kg/s (Chapter 14)
	mass flow rate through porous media, kg/s (Chapter 14)
N	number of winding turns
N_{phase}	phase number
N_{slot}	number of stator slots

Nu	Nusselt number, $Nu = hl/k$, the ratio of convection heat transfer to conduction in a fluid slab of thickness l
n	revolutions per minute, rpm
n_s	factor of safety
n_s	synchronous speed, rpm (Chapter 1)
n_{slot}	slot number per pole
P	rated motor power, W
P_a	apparent power, W
P_b	bearing power loss, W
P_b	brake power loss, W (Chapters 7 and 13)
PE	potential energy, J
PF	power factor
P_g	gearbox power loss, W
P_{hp}	rated motor power, hp
P_{in}	motor input power, W
P_m	eddy current loss in PMs, W
P_{out}	motor output power, W
P_p	preload of fastener, N
P_r	rated power of motor, W (Chapter 1)
P_r	power loss due to electric resistance, W (Chapter 10)
P_{react}	reactive power, W
P_{real}	real power, W
P_s	stator winding I^2R loss, W
P_{sc}	stator iron core loss, W
P_s	sealing power loss, W
P_w	windage power loss, W
p	pressure, Pa (Chapter 14)
p	thread pitch, mm or m (Chapter 5)
p_c	circular pitch of gear, mm or m (Chapter 9)
	averaged contact pressure of brush seal acting on shaft (Chapter 10)
Pr	Prandtl number, $Pr = \nu/\alpha$
p_r	number of pole-pair per phase
Q_c	heat absorbed power, W (Chapter 8)
Q_h	heat rejected power, W (Chapter 8)
Q_θ	directivity factor (Chapter 9)
q	notch sensitivity of material (Chapter 3)
q''	heat flux, W/m^2 (Chapters 7 and 10)
q'''	volumetric heat generation, W/m^3
R	electric resistance, Ω (Chapter 8)
R	supporting force of shaft, N (Chapter 3)
R	displacement amplification ratio $R = dy/d(\Delta l)$ (Chapter 7)
Re	Reynolds number, $Re = \rho u l/\mu$, the ratio of inertia force to viscous force
R_m	magnetic flux path, A/Wb (ampere per Weber)
R_s	resultant friction force, N, $R_s = \sqrt{F_n^2 + F_s^2}$
$R(T)$	electrical resistance at temperature T, Ω (Chapter 8)
R_{th}	thermal resistance, $^\circ$C/W or K/W
R_o	electrical resistance at temperature $0^\circ C$, Ω (Chapter 8) or at the reference temperature (other Chapters)
r	radius, m
r_c	radius from the center of correction mass to rotation axis, m

r_u	radius from the center of unbalance mass to rotation axis, m (Chapter 2)
S	material strength, Pa or MPa
S	Seebeck coefficient, $\mu V/°C$ or $V/°C$ (Chapter 8)
S	Sommerfeld number (Chapter 6)
S_h	volumetric heat source, W/m^3
St	Stribeck number (Chapter 6)
St	Stanton number, the ratio of heat transfer at the surface to that transported by fluid by its thermal capacity (Chapter 11)
S_t	torsional stiffness, Nm/rad (Chapters 2 and 3)
S_{ut}	ultimate tensile strength, Pa or kPa
S_y	yield tensile strength, Pa or kPa
s	slip (%)
T	torque, Nm
T	oscillating period, s (Chapter 12)
T	temperature, $°C$ or K (Chapter 8)
T_a	ambient temperature, $°C$ or K (Chapter 1)
$T_{a,r}$	real ambient temperature, $°C$ or K (Chapter 1)
T_c	Curie temperature, $°C$ or K (Chapter 2)
T_d	damping torque, Nm
T_{dyno}	dynamometer torque, Nm
T_f	final temperature, $°C$ or K
T_f	friction torque, Nm
T_{hs}	hot spot allowance, $°C$ or K
T_J	inertia torque, Nm
T_l	load torque, Nm (Chapter 7)
T_{max}	maximum motor winding temperature, $°C$ or K (Chapter 1)
T_{mj}	temperature at the measurement junction (*hot* junction) of thermocouple, $°C$ or K (Chapter 8)
T_o	mean motor torque, Nm (Chapter 1)
T_p	peak torque, Nm
$T_p(t)$	periodic motor torque, as a function of time, Nm (Chapter 1)
T_r	continuous rated torque, Nm
T_{rj}	temperature at the reference junction (*cold* junction) of thermocouple, $°C$ or K (Chapter 8)
T_s	stall torque, Nm
t	time, s
t_f	lubricant film thickness, mm or m
t_L	the largest time scale in turbulent flow (Chapter 14)
t_l	time resulted from the effect of load, either increase or decrease the deceleration rate, s (Chapter 7)
t_r	brake response time, s (Chapter 7)
t_s	brake actual stopping time, s (Chapter 7)
t_t	total braking time, s (Chapter 7)
t_η	the smallest time scale in turbulent flow (Chapter 14)
U	total strain energy, J (Chapter 3)
U	unbalance, kg·m (Chapter 2)
u	velocity, m/s
u_L	the largest velocity scale in turbulent flow (Chapter 14)
u_η	the smallest velocity scale in turbulent flow (Chapter 14)
V	voltage, V
V_e	excitation voltage, V (Chapter 8)

V_H	Hall voltage, V
V_s	Seebeck voltage, µV or mV
V_w	wear volume, m^3 (Chapter 7)
v	phase velocity of wave, m/s
W_{def}	work done by friction of bolt head, J
W_f	work done by friction of nut, J
W_t	work done by fastener tightening torque, J
W_{th}	work done by fastener thread friction, J
Z	figure of merit, $1/°C$
Z_{po}	impedance of primary winding at rated frequency with second windings open circuited (Chapter 8)
Z_{ps}	impedance of primary winding at rated frequency with second windings short circuited (Chapter 8)
Z_{so}	impedance of second windings at rated frequency with primary windings open circuited (Chapter 8)
Z_{ss}	impedance of second windings at rated frequency with primary windings short circuited (Chapter 8)
ZT	dimensionless figure of merit
z	number of gear teeth or number of threads

GREEKS

α	angular acceleration, rad/s^2
α	coefficient of thermal expansion, K^{-1} or $°C^{-1}$ (Chapters 6, 10, and 13)
α	half thread angle, degree (°) (Chapter 5)
α	load angle (between the resultant load and the radial load), degree (°) (Chapter 6)
α	mounted contact angle, degree (°) (Chapter 6)
α	temperature coefficient of resistance, $1/°C$ or $1/K$ (Chapters 1, 10, and 13)
α	thermal diffusivity, $\alpha = k/\rho c_p$, m^2/s
α	pressure angle of gear, degree (°) (Chapter 9)
α	wear coefficient (Chapter 7)
α_h	high pressure angle at the concave side of spiroid or helicon gear, degree (°) (Chapter 9)
α_1	low pressure angle at the convex side of spiroid or helicon gear, degree (°) (Chapter 9)
α_n	normal pressure angle of helical gear, degree (°) (Chapter 9)
a_r	transverse pressure angle of helical gear, degree (°) (Chapter 9)
$\bar{\alpha}$	average sound absorption coefficient (Chapter 12)
β	bearing contact angle, degree (°) (Chapter 6)
β	coefficient of thermal expansion, K^{-1} or $°C^{-1}$ (Chapter 11)
β	frequency ratio, $\beta = \omega/\omega_n$ (Chapter 12)
β	thread lead angle, degree (°) (Chapter 5)
χ_m	magnetic susceptibility
Δp_a	pressure drop through actual winding, Pa (Chapter 14)
Δp_p	pressure drop through porous media, Pa (Chapter 14)
ΔV_s	differential voltage at thermocouple open terminals (also known as Seebeck voltage), µV or mV
δ	clearance, mm or m (Chapter 2)
δ	change in bolt length, m (Chapter 5)
δ	interference, mm or m (Chapter 6)
δ	logarithmic decrement (Chapter 11)
δ_d	thickness of friction disc, m (Chapter 7)
δ_p	thickness of friction pad, m (Chapter 7)

ε	load distribution factor (Chapter 6)
ε	material elongation (%)
ε	$\varepsilon \rightarrow$ strain; $\varepsilon = \Delta l / l$, mm/mm or μm/mm (Chapters 7 and 14)
ε_{cr}	gear contact ratio, the length of contact path divided by the base pitch (Chapter 9)
ε_h	hoop strain
Φ	magnetic flux, Wb (Weber)
ϕ	torsional deflection angle, degree (°) (Chapter 3)
φ	dissipation function, Pa·s^{-1} (Chapter 14)
φ	phase angle, degree (°) (Chapters 2 and 9)
φ	pitch core angle of bevel gear, degree (°) (Chapter 9)
γ	shear strain
γ_g	gear ratio, $\gamma_g = \omega_i / \omega_o$
η	efficiency (%), $\eta = P_{out} / P_{in}$
η	loss factor, defined as the ratio of vibration energy dissipated per cycle to the energy stored in all structural elements (Chapter 12)
η	porosity in porous media, $\eta = A_d / A_p$ (Chapter 14)
η	the smallest length scale in turbulent flow (Chapter 14)
η_g	efficiency of gearing system (%)
η_m	efficiency of electric motor (%)
η_s	overall system efficiency (%)
λ	sound wavelength, m (Chapter 12)
λ	coefficient of bulk viscosity (or expansion viscosity), Pa·s or N·s/m^2 (Chapter 14)
λ	scale parameter (Chapter 6)
μ	dynamic viscosity, Pa·s or N·s/m^2 (Chapter 14)
μ	magnetic permeability, Wb/(A·m) (Weber per ampere-meter), or H/m (Henry per meter), or N/A^2 (Newton per square ampere) (Chapters 2 and 12)
μ_d	coefficient of dynamic friction
μ_{eff}	effective viscosity, Pa·s or N·s/m^2
μ_h	friction coefficient under bolt head
μ_n	friction coefficient under nut
μ_o	magnetic permeability of material in vacuum, Wb/(A·m) (Weber per ampere-meter or N/A^2 (Newton per square ampere)^3 (Chapters 2 and 12)
μ_r	relative magnetic permeability, $\mu_r = \mu / \mu_o$
μ_s	coefficient of static friction
μ_{th}	fastener thread friction coefficient
θ	angle, radian or degree (°)
θ_d	angle of dynamic friction, degree (°) (Chapter 7)
θ_h	helix angle of gear tooth, degree (°) (Chapter 7)
θ_l	lead angle of gear tooth, degree (°) (Chapter 7)
θ_s	angle of static friction, degree (°) (Chapter 7)
ρ	density, kg/m^3
ρ	electrical resistivity, Ω·m (Chapters 8, 10, and 13)
Σ	shaft angle of gear pair, degree (°) (Chapter 9)
σ	normal stress, $\sigma = F/A$, N/m^2 or Pa
σ	pitch point angle, degree (°) (Chapter 9)
σ	Stefan's constant, $\sigma = 5.6703 \times 10^{-8}$ W/(m^2·K^4) (Chapter 11)
σ_h	hoop stress, $\sigma_h = E\varepsilon$, N/m^2
σ_{rad}	radiation efficiency (Chapter 12)
τ	shear stress, $\tau = F_s/A$, N/m^2

---

[2] As of May 20, 2019, the vacuum permeability μ_o is no longer a defined constant ($4\pi \times 10^{-7}$ N/A^2).

τ_e	electrical time constant, s
τ_m	mechanical time constant, s
τ_{th}	thermal time constant, s
ω	rotational speed, rad/s (Chapters 1 and 3)
ω	oscillating frequency, Hz (Chapter 12)
ω_b	rotor burst speed, rad/s (Chapter 2)
ω_{dyno}	dynamometer rotational speed, rad/s
ω_n	natural frequency, Hz
ω_p	peak rotational speed, rad/s
ω_r	rated rotational speed, rad/s
χ_m	magnetic susceptibility, $\chi_m = \mu_r - 1$
ξ	load-to-motor inertia ratio, $\xi = J_l/J_m$
ξ	porous media inertial resistance factor, $\xi = K_p/\Delta L$, m^{-1} (Chapter 14)
ζ	damping ratio, $\zeta = c/2(mk)^{1/2}$ (Chapter 12)
ζ_T	torque ratio, $\zeta_T = T_{out}/T_{in}$

ABBREVIATIONS

ADC	analog-to-digital conversion
AC	alternative current
AF	axial flux
AFM	antiferromagnetic
AFPM	axial flux permanent magnet
AISI	American Iron and Steel Institute
AJ	aerosol-jet
AGMA	American Gear Manufacturers Association
AM	additive manufacturing
AMP	additive manufacturing process
AMR	anisotropic magnetoresistive
ANSI	American National Standards Institute
ASIC	application-specific integrated circuit
BDRM	brushless double-rotor machine
BLDC	brushless DC motor
BN	boron nitride
BOM	bill of material
CAD	computer-aided design
CAGR	compound annual growth rate
CDF	cumulative distribution function
CE	carbon equivalent
CFD	computational fluid dynamics
CFM	cubic feet per minute
CFRP	carbon fiber reinforced polymer or carbon fiber reinforced plastic
CMM	coordinate measuring machine
CMOS	complementary metal-oxide semiconductor
CNC	computer numerical control
COP	coefficient of performance
CPI	cycles per inch (Chapter 8)
CPR	counts per revolution or cycles per revolution (both used by manufacturers, though there are big differences between the two)
CS-PMSM	compound-structure permanent magnet synchronous machine

CT	computed tomography
CTE	coefficient of thermal expansion
CVD	chemical vapor deposition
CVSWT	continuously variable strain wave transmission
CWD	circular wave drive
DC	direct current
DDR	direct drive rotary
DFA	design for assembly
DFM	design for manufacturing
DFMA	design for manufacturing and assembly
DI	diffusion interface
DIR	direction of rotation
DMLM	direct metal laser melting
DMLS	direct metal laser sintering
DNS	direct numerical simulation
DQMOM	direct quadrature method of moments (Chapter 14)
DSP	digital signal processor
DSPM	doubly salient permanent magnet
DVA	dynamic vibration absorber
EBM	electron beam melting
EDM	electrical discharge machining
ECP	electronically controlled pneumatic
EE	Eulerian-Eulerian (Chapter 14)
EMC	electromagnetic compatibility
EMF	electromotive force (denoted ε and measured in volts) or electromagnetic field (measured in Tesla)
EMI	electromagnetic interference
EP	electro-pneumatic
EPC	evaporative pattern casting
EPR	electron paramagnetic resonance
EVS	electric vehicle system
FCPM	flux-controllable permanent magnet
FC-TFM	flux concentrated transverse flux machine
FDM	finite difference method (Chapter 14)
FEA	finite element analysis
FEM	finite element method (Chapter 14)
FFT	fast Fourier transform
FIT	failure in time (defined as a failure rate of 1 per billion hours)
FNM	flow network modeling
FS	factor of safety
FVM	finite volume method
GD&T	geometric dimension and tolerance
GFRP	glass fiber reinforced polymer
GMR	giant magnetoresistive
HD	harmonic drive
HDD	hard disk drive
HESM	hybrid excitation synchronous machine
HNS	high nitrogen steel
HPC	high-performance computing
HSM	hysteresis synchronous motor
HTS	high-temperature superconductor

HTS	high-throughput screening (Chapter 2)
ICE	internal combustion engine
ID	inner diameter
IDC	insulation displacement contact
IEC	International Electrotechnical Commission
IGBT	insulated gate bipolar transistor
IIW	International Institute of Welding
ILC	iterative learning control
IM	induction motor
IoT	internet of things
IR	infrared
ISO	International Organization for Standardization
IP	ingress protection or international protection rating
IP	internet protocol (Chapter 8)
IPM	interior-buried permanent magnet
LDV	laser Doppler vibrometer
LE	Lagrangian-Eulerian (Chapter 14)
LED	light-emitting diode
LHM	linear hybrid motor
LIM	linear induction motor
LMR	multiple linear regression
LPI	line per inch (Chapter 8)
LPM	lumped parameter model
LPTN	lumped parameter thermal network
LRC	locked rotor current
LRT	locked rotor torque
LS	level-set method
LSM	linear synchronous motor
MCNT	multi-walled carbon nanotube
MDO	multidiscipline design optimization
MEMS	microelectromechanical system
MF	magnification factor
MG	magnetic gearing
MGPM	magnetic-geared permanent magnet
MI	magnetoinductive
MIG	metal inert gas (welding)
MLR	multiple linear regression (Chapter 14)
MMF	magnetomotive force
MO	magneto-optical
MPC	model predictive control
MPM	mixing plane model
MPM	multicell piezoelectric motor (Chapter 1)
MPU	microprocessor unit (Chapter 8)
MR	magnetorheological (Chapter 7)
MR	magnetoresistive (Chapter 8)
MRB	magnetorheological brake (Chapter 7)
MRE	magnetorheological elastomer
MRF	multiple reference frame
MRI	magnetic resonance imaging
MTBF	mean time between failures
Nd-Fe-B	neodymium-iron-boron

NEC	National Electrical Code
NEMS	nanoelectromechanical system
NIST	National Institute of Standard and Technology
NMR	nuclear magnetic resonance
NSM	negative stiffness mechanism
NTC	negative temperature coefficient (Chapter 8)
OD	outer diameter
ODP	open drip proof
OEM	original equipment manufacturer
PAI	polyamide-imide
PBF	powder bed fusion
PCB	printed circuit board
PCMM	pole-changing memory motor
PCR	part count reduction
PDF	probability density function
PEEK	polyetheretherketone
PF	power factor
PFM	pulse field magnetization
PI	polarization index
PM	permanent magnet
PMHS	permanent magnet hysteresis synchronous
PMM	permanent magnet motor
PMSM	permanent magnet synchronous motor
PPI	pulses per inch (Chapter 8)
PPI	pores in a linear inch (Chapter 11)
PPM	parts per million
PPM	pulses per millimeter (Chapter 8)
PPR	pulses per revolution
PTC	positive temperature coefficient (Chapter 8)
PUR	polyurethane
PVC	polyvinyl chloride
PVD	physical vapor deposition
PWM	pulse width modulation
QMOM	quadrature method of moments (Chapter 14)
RD	resolver differential
RDC	resolver-to-digital converter
RDT	rim-driven thruster
RF	radial flux
RFI	radio frequency interference
RM	refractory metal
RMS	root mean square
RNG	renormalization group
ROI	return on investment
RTD	resistance temperature detector
RV	rotate vector
RT	resolver transformer (also known as resolver receiver)
RX	resolver transmitters
rpm	revolutions per minute
rps	revolutions per second
SAE	Society of Automotive Engineers
SAW	surface acoustic wave

SCR	silicon controller rectifier
SFM	static field magnetization
SHS	selective heat sintering
SLS	selective laser sintering
SKU	stock keeping unit
Sm-Co	samarium-cobalt
SPL	sound pressure level
SPM	surface-mounted permanent magnet
SPM-TFM	surface permanent magnet transverse flux machine
SQUID	superconducting quantum interference device
SRF	single reference frame
SRM	switched reluctance motor
STL	sound transmission loss
SWG	strain wave gearing
TCP	transmission control protocol
TEAO	totally enclosed air over
TEFC	totally enclosed fan cooled
TEFV	totally enclosed forced ventilated
TENV	totally enclosed non-ventilated
TEWD	totally enclosed washdownww
TF	transverse flux
TFM	transverse flux machine
TFRM	transverse flux reluctance motor
TIG	tungsten inert gas (welding)
TIR	total indicated runout or total indicate reading
TMD	tuned mass damper
TMR	tunnel magnetoresistive
UL	Underwriters Laboratories
UMP	unbalanced magnetic pull
USM	ultrasonic motor
VCA	voice coil actuator
VCM	variable-capacitance micromotor
VFMM	variable flux memory motor
VOF	volume of fluid
VPI	vacuum pressure impregnation
VR	variable reluctance
WSN	wireless sensor network
VRM	variable reluctance motor
YASA	yokeless and segmented armature

1

Introduction to Electric Motors

Electric motors are devices that convert electrical energy into magnetic energy and finally into mechanical energy. Electromagnetism is the basis of electric motor operation by generating magnetic forces necessary to produce either rotational or linear motion. For rotating electric motors, it is the interaction between the stator and rotor magnetic fields that creates motor torque to drive external loads.

Today, electric motors come in a wide variety of types, sizes, styles, winding topologies, magnetic flux loops, operating characteristics, and structural configurations to suit different applications. They have been used almost everywhere in the world, including industrial drives, household appliances, medical devices, electronic products, robots, drones, electrified vehicles, machine tools, spacecrafts, and military equipment. As one of the fastest-growing industrial sectors, electric motor manufacturing represents a major industry worldwide. According to the study conducted by the International Energy Agency [1.1], electric-motor-driven systems account for approximately 53% of total global electricity consumption, or 10,500 terawatt hours (TWh) per year, and emit a total of 6,800 metric tons of carbon dioxide. By 2030, energy consumption from electric motors is expected to rise to 13,360 TWh per year, ticking up CO_2 emissions to 8,570 million metric tons per year. End users now spend USD 565 billion annually on electricity used in these motor-driven systems and by 2030, that could rise to close USD 900 billion [1.2]. In the United States, motor-driven equipment accounts for 64% of the electricity consumed in the manufacturing sector. That is approximately 290 billion kilowatt hours (kWh) of power per year [1.3]. There are more than 40 million electric motors used in manufacturing operation [1.4]. In addition, more than 95% of electric motor's life-cycle cost is the energy cost. In China, the total power consumption of electric motors is about 3 trillion kWh, accounting for about 64% of the total power consumption [1.5, 1.6]. It is estimated that improving the efficiency of all the world's motors by just 1% would reduce the motor power consumption by 94.5 TWh of electricity and shrink their CO_2 footprint by the equivalent of 60 million metric tons per year [1.7]. This would significantly reduce the need for new power plants and push down the total environmental cost of electricity generation. There is no doubt that the continuous push for decarbonization will greatly increase global demand for more efficient electric motors.

The report that provides information on the industrial motor market growth, trends, and forecast during the period of 2020–2025 by Motor Intelligence was released in 2019 [1.8]. According to the report, the global market of industrial motors is expected to register a compound annual growth rate of 6.11% during 2020–2025 and reaches a market value of USD 64.25 billion by 2025. However, it is worth noting that this report did not consider the consequence of the COVID-19 pandemic that has severely upended global economy. According to World Trade Organization, the volume of world merchandise trade declined by 5.3% in 2020 as the COVID-19 pandemic disrupts normal economic activity all over the world and is expected to increase by 8.0% in 2021 [1.9]. Nevertheless, in the long term, the market will generate remunerative prospects for producers post COVID-19 crisis. Substantial advances in technology have opened up opportunities to develop and manufacture electric motors for a wide range of applications in industries, such as automation, automotive, agriculture, aerospace, construction, power generation, and other industrial sectors.

DOI: 10.1201/9781003097716-1

1.1   HISTORY OF ELECTRIC MACHINES

Both electric motors and generators are integral parts of electric powered machines. They have a similar structure, current-carrying loop, and operating mechanisms for converting energy between the electrical form and the mechanical form. However, they have significant differences in terms of function, driving mode, rule followed (*e.g.*, the motor follows Fleming's left-hand rule, and the generator follows Fleming's right-hand rule), and power source.

The discoveries of phenomena of static electricity can be traced back to ancient Greece about 2,600 years ago. However, there was little real progress until the English scientist William Gilbert in 1600 described the electrification of many substances and coined the term *electricity*, the Greek word for amber.

In 1742, Andrew Gordon invented a simple electrostatic device known as *electric chimes* [1.10]. This is the first device that converts electrical energy into mechanical energy in the form of repeating mechanical motion [1.11]. This device was further developed by Benjamin Franklin in 1752 to detect approaching thunderstorms. Hans Oersted discovered electromagnetism in 1820 and additional works were made by a number of other scientists such as William Sturgeon, Joseph Henry, Ander Marie Ampere, Michael Faraday, Thomas Davenport, Moritz von Jacobi, and Antonio Pacinotti. In 1831, Michael Faraday discovered electromagnetic induction, the principle behind the electric motor and generator. This discovery was crucial in allowing electricity to be transformed from a curiosity into a powerful new technology.

Using a broad definition of *motion* as meaning any apparatus that converts electrical energy into motion, it is widely accepted that Michael Faraday invented the first direct current (DC) electric motor in 1821. This motor was basically used to confirm his concept of electric motor and had no actual value in application. He succeeded in building the practical electric motor 10 years later. Following his groundbreaking work, many scientists had contributed to the development of electric motors. William Sturgeon invented the first commutator DC electric motor capable of turning machinery in 1832 [1.12]. Moritz von Jacobi built an electric motor in 1834 that actually developed a remarkable output power (about a power of 15 W). The improved version of his motor was released in 1838, which was capable of 300 W output power and could drive a boat occupied with 14 people across a wide river, powered by zinc batteries [1.13]. The first US patent on electric motor was granted in 1837 to Thomas Davenport [1.14]. In 1887, Nikola Tesla introduced the world's first alternating current (AC) motor and obtained a US patent next year [1.15]. Three-phase *cage-rotor* induction motors (IMs) were invented by Mikhail Dolivo–Dobrovolsky during 1889–1890 [1.16]. The three-phase synchronous motor was first developed by Friedrich A. Haselwander in 1887 [1.13]. Even today, this type of motor is still in service for the vast majority of commercial applications.

Electromagnetic generators are based on the principle of electromagnetic induction, which was first discovered by Michael Faraday in 1831. By entailing a copper disc (*i.e.*, *Faraday disc*) that rotated between two magnets with their poles perpendicular to it, Faraday invented the first electromagnetic generator that was capable of converting mechanical power into electricity. Due to his remarkable contributions, electricity became available for use in a large variety of today's technologies.

Based on the principle of electromagnetic induction, the first magneto electric generator was built in 1832 by Hippolyte Pixii, a French instrument maker. The magneto generator converts mechanical energy (motion) into electrical current. In parallel, William Ritchie constructed an electromagnetic generator with four rotor windings, a commutator, and brushes, also in 1832 [1.13]. Thereafter, with the help of Zenobe Gramme, the dynamo generator was produced and used for producing enough power for commercial applications. Stimulated by the successful invention of new lighting systems of Thomas Edison in 1878, DC and AC generators had experienced an unprecedented development. In 1882, Edison installed a number of DC generators in the central station plant on Pearl Street in lower Manhattan, which was one of the earliest power generation stations [1.17]. In 1887, Nikola Tesla made significant changes to the generator progression by considerably improving AC generators. Instead of standard single rotating poles, Tesla introduced polyphaser AC that consisted of several outputs combined into one phase.

1.2   MOTOR DESIGN CHARACTERISTICS

Electric motors are manufactured in a variety of types and configurations. Typically, an electric motor assembly is formed from a collection of parts, including a stator, a rotor, a shaft, a pair of end bells, bearings, and a motor housing supporting and enclosing the various components. In addition to these primary motor components, some motors may include electronic components that are used to modify operating characteristics for particular applications.

Motor design characteristics are the essential elements in the design, analysis, prototype, manufacturing, and test processes of electric motors. In designing high efficiency and cost-effective motors or selecting appropriate motors for specific applications, these design characteristics must be well understood and addressed.

1.2.1   MOTOR TORQUE

Torque is a measurement of the turning force acting on an object to cause that object rotating or twisting about an axis or pivot. Torque, like *work*, is measured as Newton meter (Nm) in the International System of Units (SI system) or pound-foot (lb$_f$-ft) in the English system. However, unlike *work* that only occurs during displacement, torque may exist even though no displacement or no rotation occurs. A typical example is the static holding torque.

1.2.1.1   Static and Dynamic Torque

Torque can be divided into two major categories, either static or dynamic torque. From the standpoint of physics, the system is considered static if it has no angular or linear acceleration. Static torque refers to the amount of torque that an electric motor produces at zero speed (*i.e.*, the motor is in a real static state) with the power output $P_{out}=0$. By contrast, dynamic torque refers to the amount of torque that an electric motor produces at variable speeds of rotation with load applied (*i.e.*, the motor is in a dynamic state) with $P_{out}>0$. Simply, static torque is associated with forces that do not involve angular acceleration/deceleration, and dynamic torque is associated with dynamic forces that arise from acceleration/deceleration, following Newton's second law.

Some motor manufacturers may provide the information of continuous static torque rating for customers, indicating that the motor is capable of supplying that static torque at zero speed of rotation continuously. However, this information may be not very useful in motor selection because it does not define the continuous torque available from the motor at a specific speed (or in a range of speeds) to drive an external load.

As a continuous torque rating is provided at a specific rated speed (or in a range of speeds), the torque is thus the dynamic torque, indicating the capability of the motor to provide up to a corresponding rated torque continuously. Furthermore, the maximum torque appearing on the motor nameplate refers to the highest dynamic torque for the motor at a rated motor speed.

As shown in Figure 1.1, when a force vector **F** acts on a solid body at the point A to make the body rotate about its axis through the point O, the torque vector **T** around the point O is obtained by crossing the product of the radial displacement vector **r** and the force vector **F**:

$$\mathbf{T} = \mathbf{r} \times \mathbf{F} \tag{1.1}$$

The direction of the torque vector **T** is determined by the right-hand rule, that is, it is perpendicular to both **r** and **F**. Correspondingly, the magnitude of the torque acting on the body is

$$T = rF\sin(\theta) \tag{1.2}$$

where the moment arm $r = \overline{OA}$, defined as the distance from the axis to the point where the force is applied, and θ is the measure of the smaller angle between the displacement vector **r** and the force

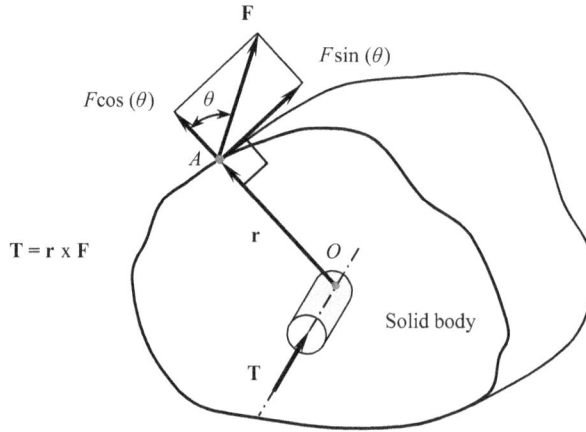

FIGURE 1.1   Torque vector **T** acting on a solid body to make it rotate with respect to its axis.

vector **F**. It is worth to note that torque calculated from Equation 1.2 can be either static or dynamic torque depending on whether F is a static or dynamic force.

For a rotating system with a fixed axis, the dynamic torque on the rotating system along the axis of rotation is determined by the rate of change of the system angular momentum:

$$T = \frac{dM_a}{dt} \tag{1.3}$$

where M_a is the angular momentum of the rotating system, measured in Nm·s. M_a can be expressed as the product of the polar moment of inertia of the rotating system J_p and the rotating speed ω, that is,

$$M_a = J_p \omega \tag{1.4}$$

As a result, the torque on the rotating system can be expressed as

$$T = \frac{d(J_p \omega)}{dt} = J_p \frac{d\omega}{dt} = J_p \alpha \tag{1.5}$$

where α is the angular acceleration of the rotating system, measured in rad/s². This equation indicates that for electric motors, the less the motor inertia, the less torque the motor needs to produce to meet the desired acceleration rate. As a result, it is advantageous to minimize motor inertia to the greatest extent to maximize acceleration.

Servomotors are typically expected to accelerate loads from a stop to a given velocity and then decelerate the loads once again to a stop at a precise position. Accordingly, to move or stop loads as fast as possible, the angular acceleration/deceleration α must be maintained high enough. As a result, the motor's polar moment of inertia J_p has to be kept at very low levels. For this reason, the motor inertia must be taken into account in the earlier stage of motor design.

1.2.1.2   Motor Torque in Motor-Load System

When a motor drives a load machine to perform work, it usually connects with some power transmission components such as coupling and gearbox. For demonstration purposes, a two-shaft gearbox is shown in Figure 1.2. The gear ratio γ_g is defined as the rotating speed of the gearbox input shaft that is directly coupled with the rotor shaft to the rotating speed of the gearbox output shaft, which is directly coupled with the load machine, that is,

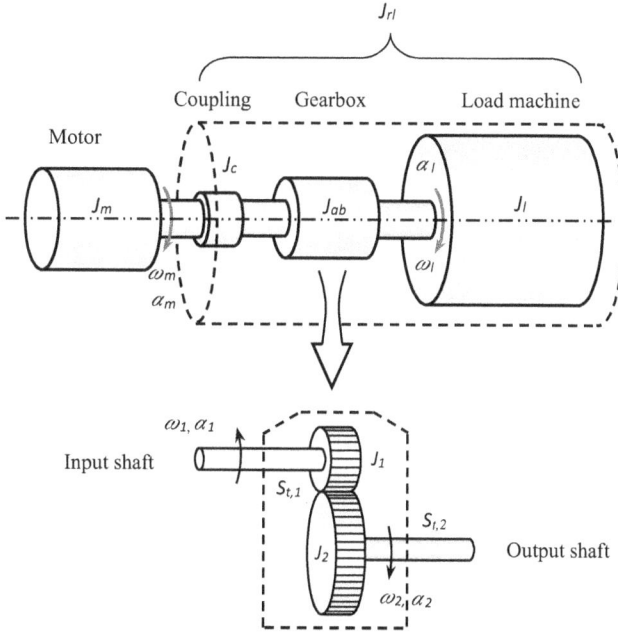

FIGURE 1.2 A motor is connected to a driven load machine via a coupling and a gearing system. The reflected load inertia is the equivalent inertia of the load seen by the motor.

$$\gamma_g = \frac{\omega_i}{\omega_o} = \frac{\omega_m}{\omega_l} \tag{1.6}$$

where

ω_i and ω_o are the rotating speeds of the input and output shaft of the gearbox
ω_m and ω_l are the rotating speeds of the motor and load, respectively

Unlike other components, the inertia of each shaft must be calculated separately because the input and output shafts of a gearbox have different rotating speeds.

To determine the motor torque required for the system, a concept of reflected load inertia is introduced as the equivalent inertia of the load seen by the motor:

$$J_{rl} = J_c + J_{i,gb} + \frac{J_{o,gb} + J_l}{\gamma_g^2} \tag{1.7}$$

where J_c and J_l are the inertias of the coupling and load machine, respectively. This equation indicates that load inertia is reduced by the square of the gear ratio, which can be derived from the conservation of energy or power in a set of rotating components. In this resultant equation, it is assumed that the inertias of the coupling and gearbox are considered part of the load inertia.

The torque required at the motor thus becomes

$$T_m = (J_m + J_{rl})\alpha_m = \left(J_m + J_c + J_{i,gb} + \frac{J_{o,gb} + J_l}{\gamma_g^2} \right)\alpha_m \tag{1.8}$$

The motor angular acceleration α_m has the relationship with the load angular acceleration α_l [1.18]:

$$\alpha_m = \alpha_l \, \gamma_g \tag{1.9}$$

Combining the aforementioned equations yields

$$\alpha_l = \frac{\alpha_m}{\gamma_g} = \frac{T_m}{\gamma_g}\left(J_m + J_c + J_{i,gb} + \frac{J_{o,gb} + J_l}{\gamma_g^2}\right)^{-1} \tag{1.10}$$

Taking the derivative of α_l with respect to γ_g, it follows that

$$\frac{d\alpha_l}{d\gamma_g} = T_m \frac{\left(J_l + J_{o,gb}\right) - \left(J_m + J_c + J_{i,gb}\right)\gamma_g^2}{\left[\left(J_l + J_{o,gb}\right) + \left(J_m + J_c + J_{i,gb}\right)\gamma_g^2\right]^2} \tag{1.11}$$

Let $d\alpha_l/d\gamma_g = 0$; thus,

$$\left(J_l + J_{o,gb}\right) = \left(J_m + J_c + J_{i,gb}\right)\gamma_g^2 \tag{1.12}$$

Thus, the optimum gear ratio is found at

$$\gamma_g = \sqrt{\frac{J_l + J_{o,gb}}{J_m + J_c + J_{i,gb}}} \tag{1.13}$$

The torque on a motor shaft can be measured using a number of strain gages to the shaft in a proper orientation. This allows directly measuring torsional shear strains that can be calibrated to output torque.

1.2.1.3 Continuous Torque

It is the interaction of the stator revolving field and the rotor induced field that produces the motor torque to drive load machines. Consequently, the motor torque is a function of both the field and armature currents and acts on both the rotor and stator simultaneously. The continuous rated torque T_r (in Nm) of a motor can be determined from its rated power P_r (in W) and rated rotor angular speed ω_r (in rad/s):

$$T_r = \frac{P_r}{\omega_r} \tag{1.14}$$

At normal operation, a motor provides a continuous torque, known as nominal torque, to drive an external load device smoothly. It can be seen from the previously mentioned equation that for a given motor power, the continuous torque is inversely proportional to the motor rotating speed. In practice, the continuous torque generated in a motor commonly has a cyclic variation as a result of the cyclic permeance variation that occurs as the rotating member moves with respect to the stationary member. The instantaneous torque of a motor can be expressed as (Figure 1.3)

$$T(t) = T_o + T_p(t) \tag{1.15}$$

where T_o and $T_p(t)$ are the mean component and the periodic component of the motor torque, respectively.

The rated continuous torque listed in manufacturers' catalogs is usually based on a specifically defined ambient temperature T_a, often 40°C. When the actual ambient temperature $T_{a,r}$ exceeds the predetermined value T_a, the motor performance deteriorates and the rated continuous torque $\tilde{T}_c$ is reduced to $\tilde{T}_{c,d}$ using the formula

$$\tilde{T}_{c,d} = \tilde{T}_c \sqrt{\frac{T_{max} - T_{a,r}}{T_{max} - T_a}} \tag{1.16}$$

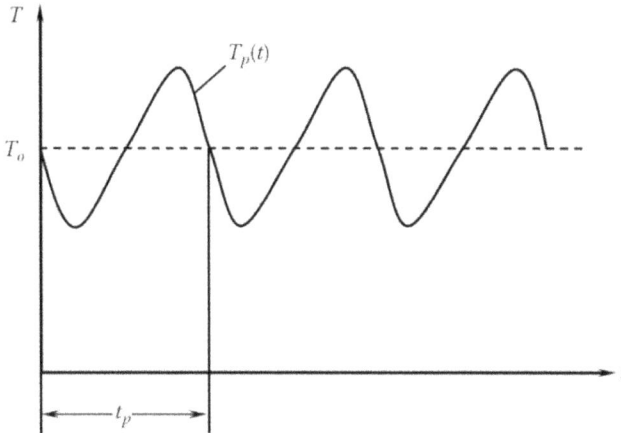

FIGURE 1.3 Mean component and periodic component of motor torque.

where

T_{max} is the maximum motor winding temperature, $°C$

T_a is the ambient temperature defined in a motor catalog (usually $T_a=40°C$), $°C$

$T_{a,r}$ is the real ambient temperature where $T_{a,r}>T_a$, $°C$

For instance, with $T_{max}=155°C$, $T_a=40°C$, and $T_{a,r}=58°C$, the derated torque becomes 91.8% of the rated continuous torque in the catalog.

1.2.1.4 Peak Torque

The motor peak torque, which is always associated with the peak current, is the maximum torque a motor can produce for short periods of time without exceeding the motor temperature limit or safe operating torque. A motor that operates under a peak torque condition is typically associated with a quick temperature rise. When the motor temperature exceeds the allowable value, it will cause degradation of insulation materials, irreversible demagnetization of permanent magnets (PMs) in PM motors (PMMs), and winding damage and, in turn, lead to the degradation of motor performance or even motor failure.

Peak torque contains two components [1.19]: (a) acceleration torque as inflicted by the inertia from the mechanical setup plus the maximum angular acceleration; and (b) constant torque due to all other non-inertial forces such as gravity, friction, preloads, and other push-pull forces.

The continuous torque and peak torque curves are usually determined through motor testing and provided by motor manufacturers.

1.2.1.5 RMS Torque

Root mean square (RMS), which is defined as the square root of the mean square, is a mathematical measure of the magnitude of a varying quantity. It can be applied to either a series of discrete values or a continuous time-varying function.

Many applications of modern servomotors today require higher power efficiency and more sophisticated motion control. The RMS torque calculation provides a simple method of assessing how the range of torque requirements during the full duty cycle will affect the motor's performance. For a series of discrete values, the characteristic motion profile is broken up into multiple segments; each segment has a constant torque with a specific amount of time. Thus, the RMS torque can be calculated as

$$T_{RMS} = \sqrt{\frac{\sum_i T_i^2 t_i}{\sum_i t_i}}$$

(1.17)

where

    T_i is the torque required at the ith segment (where $i \geq 1$), Nm

    t_i is the time interval $(t_i - t_{i-1})$ at the ith segment, s

This indicates that the RMS torque calculation takes into account not only the amount of torque but also the amount of time at each segment.

When a motor torque is a continuously varying function of time, *i.e.*, $T(t)$, the RMS torque over a time interval $(t_1 \leq t \leq t_2)$ can be determined as

$$T_{RMS} = \sqrt{\frac{1}{(t_2 - t_1)} \int_{t_1}^{t_2} [T(t)]^2 \, dt} \tag{1.18}$$

The maximum torque from a rotary motor is typically the sum of three components: the torque due to angular acceleration, the torque due to load, and the torque due to friction. In most applications, torque requirements vary throughout the duty cycle of the motor. The RMS torque reflects the heating of the motor during its duty cycle, *i.e.*, it is used to determine the torque that would result in the same amount of motor heating as all the various torques and durations that the motor actually experiences during its duty cycle.

1.2.1.6   Stall Torque

The stall torque is the torque that a motor produces at zero rotating speed where $P_{out} = 0$. The stall torque, also known as locked rotor torque (LRT), can be measured with the rotor being locked. For a DC motor, the motor torque has a linear relationship with the motor rotating speed. The maximum torque occurs at zero rotating speed and zero torque occurs at a maximum rotating speed. For an IM, the stall torque can be calculated as [1.20]

$$T_s = \frac{C_1 k_2 (P_{in,s} - P_s - P_{sc})}{n_s} \tag{1.19}$$

where

    C_1 is a reduction factor ($0.9 \leq C_1 \leq 1.0$) to account for non-fundamental losses

    $k_2 = 9.549$ for torque, Nm

    $P_{in,s}$ is the input power to stator, W

    P_s is the stator *PR* loss, W

    P_{sc} is the stator iron core loss, W

    n_s is the synchronous speed, rpm

1.2.1.7   Cogging Torque and Reduction Methods

The torque generated from a PMM usually consists of two components: the effective driving torque, which is basically proportional to the supplied electric current, and the *nocurrent torque*, which is independent of the current, such as cogging torque.

Cogging torque originates from the interaction between the rotor-mounted PMs and the stator teeth, which produces reluctance variations depending on the rotor position [1.21]. In a motor design, it is highly desirable to reduce cogging torque because it can result in vibration, speed ripple and noise, particularly at light loads and low speeds. As depicted in Figure 1.4, when the rotor rotates, the magnets attached to the rotor successively pass through the stator teeth and slot openings, resulting in the periodic variations of the magnetic field. During the process, the PM rotor tends to lock onto the position where the permeability reaches the largest. When the rotor deviates from this equilibrium position, tangential forces are produced between the magnets and stator teeth, either to return the rotor back to the old equilibrium position or to push the rotor to the next equilibrium position, leading to cogging torque.

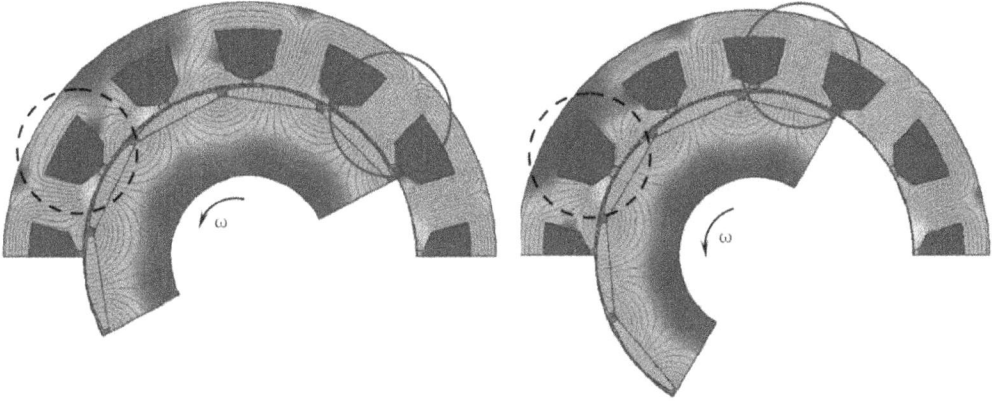

FIGURE 1.4 As a rotor rotates, magnets successively pass through stator teeth and slot, leading to different magnetic flux distributions, as shown earlier for magnet-teeth (solid circle) and magnet-slot (dashed circle) conditions.

Cogging torque is highly undesirable because it is the major cause of motor vibration and acoustic noise, particularly at light loads and low speeds. Even for high-speed applications, lower cogging torque always benefits smooth operation. For instance, in a servo system, the motor may come to a stop and then accelerate to another high speed after the stop. As the motor approaches zero speed, it is often to use settling time or settling error to measure how smoothly the motor approaches zero speed. This can definitely be impacted by cogging torque.

Because the rotor always tends to lock onto a position where it is aligned with the stator poles, it makes precise positioning of the rotor difficult. In addition, cogging torque is also an important source of torque ripples that have adverse effects in many demanding motion control applications.

Motor manufacturers often adopt the cogging torque ratio to quantitatively describe the level of cogging torque. This torque ratio is defined as the absolute cogging torque, which is characterized as peak-to-peak cogging torque, to the rated continuous torque. It is widely accepted that for regularly controlled servo systems, the maximum cogging torque ratio should keep under 5%. More ideally, for precisely controlled servo systems, the cogging torque ratio is 2% or less. Cogging torque below 1% really requires a special design. For example, in some high-precision film roller systems that need super smooth operation at very low speeds, an effective solution is to use DC torque motors for excellent operation. Three-phase brushless motors have been designed for very smooth telescope azimuth motion that rotates at extremely slow earth rates of rotation of one revolution in 24 h. It is worth to note that each application may allow for different extremes of cogging torque.

Though cogging torque can be calculated accurately using numerical approaches such as finite element methods (FEMs), with simplified motor models, analytical methods may provide greater physical insight into the mechanism of cogging torque production. This is especially useful at the initial stage of the motor design. One approach to predict cogging torque is represented by [1.22]

$$T_{cog} = -\frac{\phi_g^2}{2}\frac{dR}{d\theta} \tag{1.20}$$

where
ϕ_g is the airgap magnetic flux, Wb
R is the airgap reluctance, H^{-1}
θ is the position of the rotor, rad or degree (°)

This equation indicates that in order to reduce cogging torque, either the airgap magnetic flux ϕ_g or the rate of change of the airgap magnetic reluctance $dR/d\theta$ must be minimized. In practice, cogging torque is reduced by forcing the airgap reluctance R as close as possible to constant with respect to rotor position.

Another method for calculating cogging torque in PMMs was proposed by Lu *et al.* [1.23]:

$$T_{cog} = \frac{p_r B_r}{2} \frac{l_m}{\mu_o \mu_r} \frac{d\Phi}{d\theta} \qquad (1.21)$$

where

p_r is the number of magnet pole pairs

B_r is the remanent magnetic flux density at $H=0$

Φ is the magnetic flux calculated over a surface perpendicular to its direction of magnetization

l_m is the magnet length along the magnetization direction

μ_o and μ_r are the permeability of free space and the relative permeability of the magnet material, respectively

This equation shows that the magnitude of cogging torque is proportional to the number of magnet pole pairs, the remanent flux density value, the magnet length in the direction of magnetization, and the variation of the magnetic flux with respect to the rotor position. It has shown a good agreement of the predicted cogging torque with the cogging torque calculated using the Maxwell stress method and measured results.

There are a number of techniques available for reducing cogging torque:

- The most effective way in practice to reduce cogging torque is to skew either the stator teeth relative to the rotor centerline (Figure 1.5) or rotor magnets relative to the stator teeth, where Figure 1.6 is for step-skewing of segmented magnets and Figure 1.7 [1.24] is for skewing whole magnets. More detailed descriptions can be found in reference [1.25]. Skewing stator teeth need special care in stacking steel laminations. Skewing whole magnets may have smoother rotor performance than step-skewing segmented magnets. However, it is to be noted that rotor skewing may also decrease the rotor saliency and thus reduce the back electromotive force (EMF) and motor effective torque. All forms of skewing will reduce the back EMF and motor effective torque as skewing affects total flux linkage to the coils.

- Another technique is to use a fractional number of slots per pole. The use of this method not only reduces the amplitude of the cogging torque but also increases the fundamental order. This is because the stator slots are located at different relative circumferential positions with respect to the edges of the magnets [1.26–1.28]. To address the problem, the parameter q, which is defined as the slot number per pole n_{slot} divided by the phase number N_{phase}, is introduced as

$$q = \frac{N_{slot}}{2 p_r N_{phase}} = \frac{n_{slot}}{N_{phase}} \qquad (1.22)$$

where

p_r is the number of pole pair

N_{slot} is the number of the stator slots

The fractional slot winding arrangement with $q<1$ is attractive for lower cogging torque. The investigations have shown that the cogging torque of the fractional slot motors can be less than 1% of the rated torque. In the case of multipole machines, the cogging torque of 0.05% could be estimated [1.29].

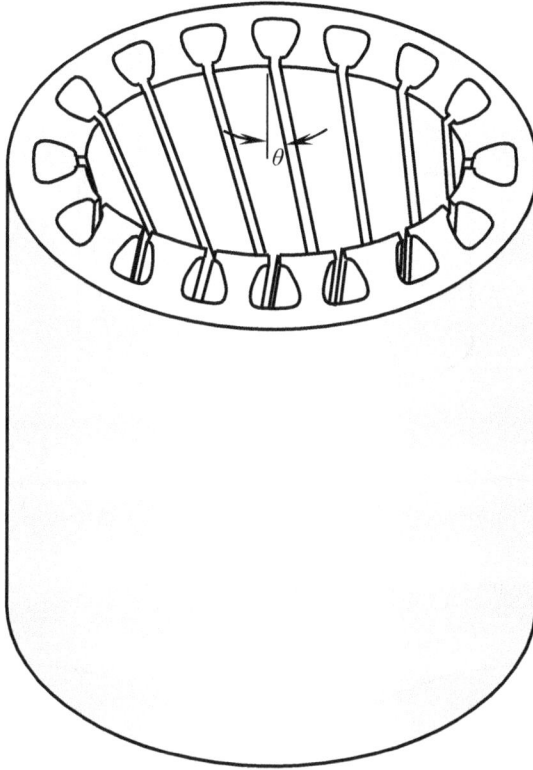

FIGURE 1.5 Skewed stator teeth with respect to its axial centerline for reducing cogging torque, where θ is the skew angle.

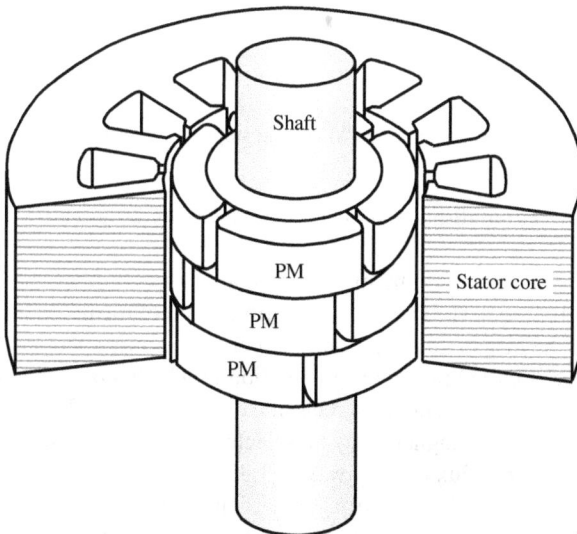

FIGURE 1.6 Step-skewed permanent magnets relative to stator teeth for reducing cogging torque.

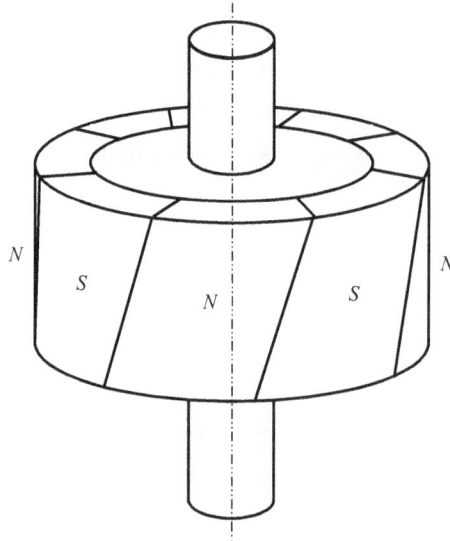

FIGURE 1.7   Skewed PMs on a rotor core for reducing cogging torque.

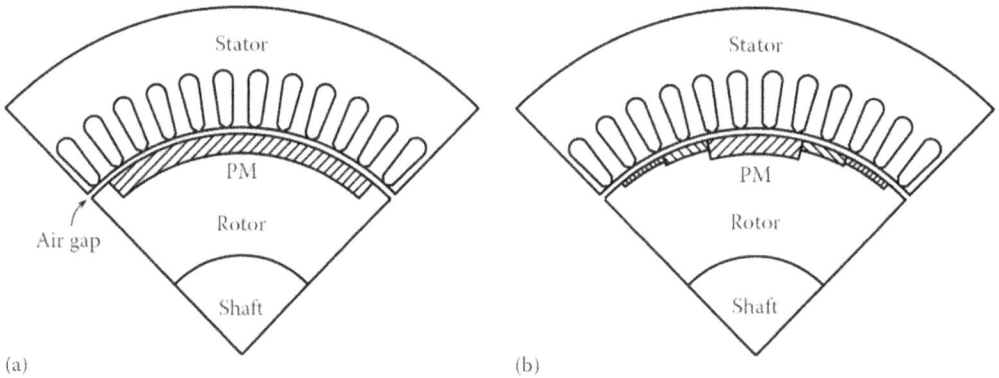

FIGURE 1.8   Rotors with surface-inset single PM (*a*) and stepped PMs (*b*).

- The reduction in cogging torque can also be achieved by modulating drive current waveform [1.30].
- It is a well-established fact that the magnet pole arc can have a large effect on the magnitude of the cogging torque [1.31, 1.32]. The optimization of the magnet pole arc can reduce the harmonics of the airgap flux wave and the permeance wave [1.33].
- Another technique is called magnet segmentation in which a pole magnet consists of several elementary magnet segments with the same polarities [1.34]. Furthermore, as shown in Figure 1.8, the segmented magnets may be selected with different thicknesses and lengths to obtain more sinusoidal flux density waves [1.35].
- The shape of the magnetic pole has a strong impact on the uniformity of the stator–rotor airgap. A comparison of uniform and nonuniform airgaps with different pole shapes of magnets is illustrated in Figure 1.9. It has been reported that the nonuniform pole shape of magnets can reduce cogging torque as high as 50% [1.35]. The normalized cogging torque to the peak torque between uniform airgap and nonuniform airgap is given in Figure 1.10. The simulation studies of the pole surface effect on cogging torque were performed by Lao *et al.* [1.36].

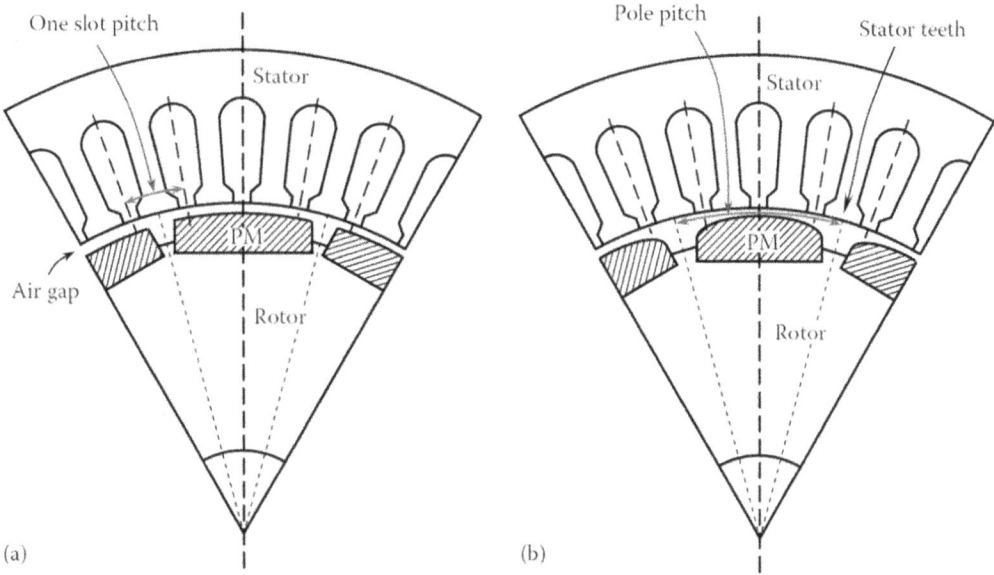

FIGURE 1.9 Surface-mounted PMs with different shapes for (a) uniform airgap and (b) nonuniform airgap.

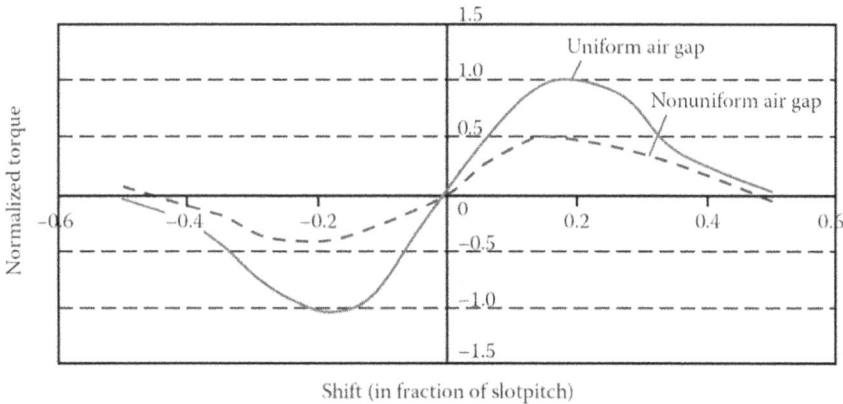

FIGURE 1.10 Comparison of cogging torque for uniform and nonuniform airgap (bread-loaf magnet shape).

- Other mechanical factors can affect cogging torque. Geometry factors such as rotor or magnet concentricity of the magnetic airgap can create an eccentric magnet to stator teeth flux linkage. These variations in flux linkage can cause variations in cogging torque and torque ripple.
- In order to achieve decreasing torque ripple, iron losses, and cogging torque for interior PM (IPM) synchronous machines, Soleimani *et al.* [1.37] proposed a novel structure of rotor. In their design, three layers of PM have been used and each layer has a fragmental trapezoid structure, as shown in Figure 1.11. With the optimized dimensions and shapes of the buried rotor magnets, the cogging torque ratio of 1.82% is achieved compared with a torque ratio of 5% in conventional IPM machines.

1.2.1.8 Torque Ripple

PM synchronous motors usually generate torque ripple during their normal operation. The studies of Hsu *et al.* [1.25] have revealed that torque ripple can be classified into four types depending on the nature of their origin:

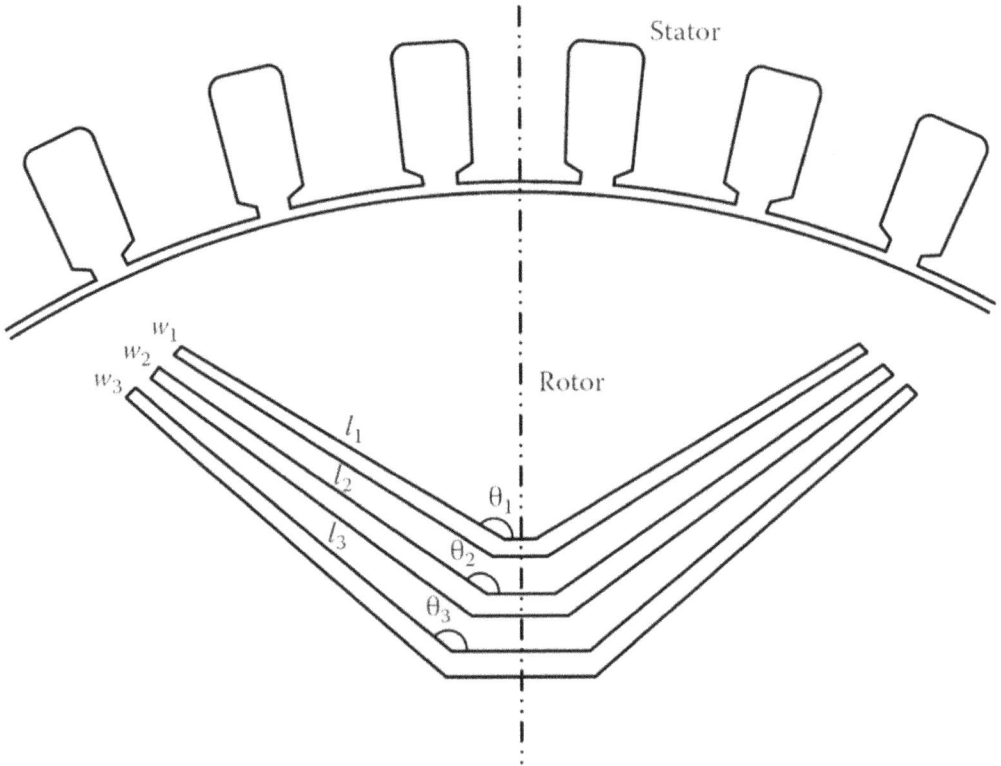

FIGURE 1.11 Novel structure of rotor for minimizing cogging torque for IPM synchronous motors.

- Pulsating torque, which is inherently produced by the trapezoidal back EMF. The torque ripples caused by pulsating torque may be reduced by purposely producing fluctuating counter torques.
- Fluctuating torque, which is produced by altering the magnitudes of phase current in the same ratio.
- Reluctance cogging torque, which is produced by nonuniformly distributed airgap permeance associated with the teeth and PMs. Actually, this type of torque exists even when the motor is not energized.
- Inertia and mechanical system torques, which are generated by the dynamic motions of the mechanical components of motor. They are affected by the driven device.

Later, Holtz and Springob [1.38] have investigated the different sources of torque ripple in PM machines, including the distortion of the stator flux linkage distribution, variable magnetic reluctance at the stator slots, and secondary phenomena. In addition, the feeding power converter also contributes to torque ripple due to the time harmonics in the current waveform's time-varying delays between the commanded and the actual current.

1.2.2 MOTOR SPEED

There are a few terms to describe rotational motion. Angular speed represents the change in angle per unit of time, typically measured in radians per second (rad/s) in the SI system and in degrees per second in the English system. The rotational speed is the measurement of revolutions per unit of time, expressed as either revolutions per second (Hz) or revolutions per minute (rpm). It is to be noted that speed is a scalar quantity and velocity is a vector. Thus, the difference between angular (or rotational) speed and angular (or rotational) velocity is that the latter contains the information of the rotational direction. Angular speed ω (in rad/s) and rotational speed n (in rpm) can be converted into each other:

$$\omega = \frac{\pi n}{30} \qquad (1.23)$$

1.2.2.1 Continuous Speed

The continuous speed of a motor is an important design parameter, defining the nominal speed at which the motor continuously operates at the rated voltage and current to drive the full motor load. It depends on the number of the motor pole, the frequency of AC, and the amount of torque.

1.2.2.2 Peak Speed

The peak speed is the maximum speed a motor can reach during operation. It must be noted that at the peak rotating speed, all rotating components are subject to high centrifugal forces. Even a tiny unbalance in a high rotating speed system can cause severe motor vibration or serious damage to rotating components. Therefore, the motor peak speed needs to be carefully determined during the motor design phase.

1.2.2.3 Speed Ripple

Speed ripple, also known as speed fluctuation, is characterized by undesirable periodic variation in speed as the motor shaft rotates. It is induced by parasitic torque pulsations that vary periodically with rotor position. Speed ripple values are usually expressed as a percentage (%) of deviation from the nominal value. In fact, periodic fluctuation in motor speed and torque eventually causes vibration at low speeds and acoustic noise at high speeds.

The effects of speed ripple are particularly undesirable in some precision motion control and machine tool applications because they can lead to speed oscillations that deteriorate electric machine performance, excite resonances, and produce acoustic noise, especially in low-speed operations. As an example, in a machine tool application, speed ripple leaves visible patterns in high-precision machined surfaces. In the case of escalators and elevators, speed ripple can introduce jerky movement, making passengers uncomfortable. Furthermore, speed ripples have an impact on the magnetic flux and back EMF waveform. They can significantly change the harmonic spectrum of the back EMF when compared to the constant speed situation.

There are a large number of factors that affect motor velocity ripple and torque ripple [1.39], including

- Load-to-rotor inertia ratio
- Motor pole count—higher pole count leads to lower ripple
- Encoder resolution
- Commanded velocity
- System bandwidth (different between an analog and a digital drive)
- Commutation type (sinusoidal vs. trapezoidal commutation)
- System trajectory update rate (when using a digital drive)
- Natural (harmonic) frequencies of the system as a whole
- Mechanical resonance of motor components
- Mechanical friction in the system
- Physical alignment of mechanical components
- Inherent performance variances between two like components
- Load damping
- Consistency of the AC supply voltage to the drive (*i.e.*, power supply integrity)
- Sampling rate/resolution of the tachometer (or velocity-measuring device)

To minimize speed and torque ripples, many techniques have been proposed and developed in the past several decades. These techniques can be briefly classified into two categories. The first category involves the design and improvement of the motor structure, such as skewing the slot or magnet, ensuring a fractional number of slots per pole, and improving the winding distribution [1.40, 1.41]. The second category involves the use of active control algorithms to improve the performance of motors. Fei *et al.* [1.42] propose a scheme that combines model predictive control and iterative learning control to speed up the response time of the system and effectively reduce the periodic speed ripples.

1.2.3 TORQUE DENSITY

Torque density is defined as the ratio of the nominal continuous torque T to the motor volume V, as the measure of the torque-carrying capability per unit volume. High torque density and high efficiency are two of the most desirable features for electrical motors. Torque density is a measure of the torque-carrying capability per unit volume of a motor, expressed in units of N/m^2 or lb_f/ft^2. Torque density is a system property since it depends on the design of motor components and their interconnections. One of the main design goals in motor design is to improve the torque density of motors.

1.2.4 MOTOR POWER AND POWER FACTOR

The output power P_{out} of rotary motors is expressed as the product of the motor torque and the angular rotating speed, that is,

$$P_{out} = T\omega \qquad (1.24)$$

Since the angular rotating speed ω (in rad/s) is related to the rotating speed n (in rpm) as $\omega = \pi n/30$, Equation 1.24 can also be expressed as

$$P_{out} = \frac{\pi n T}{30} \qquad (1.25)$$

When an IM is connected to a power supply but still at rest, it appears just like a short-circuited transformer, drawing a very high current known as the locked rotor current (LRC). The torque corresponding to the LRC is defined as the locked rotor torque (LRT). A motor that exhibits a high starting current will generally produce a low starting torque and vice versa. Both the torque and current of the locked rotor are a function of the terminal voltage of the motor. Under a constant voltage, the torque and current vary with the rotor speed during the motor acceleration/deceleration process.

The most important two parameters for motor performance are motor efficiency and power factor. In the electric power industry, power factor PF is defined as the ratio of real power P_{real} to apparent power P_a. As shown in Figure 1.12, power factor PF (where $0 \leq PF \leq 1$) can also be expressed as the cosine of the impedance phase angle ϕ:

$$PF = \cos\phi = \frac{P_{real}}{P_a} \qquad (1.26)$$

The power supply system provides both real and reactive power to operate the motor. Useful mechanical work is developed from real power P_{real} and is measured in watts (W) or kilowatts (kW). For AC motors, reactive power P_{react} is to develop magnetic fields. It is worth to note that reactive power does not provide any mechanical work. From the power triangle in Figure 1.12, apparent power P_a can be expressed as

$$P_a = \sqrt{P_{real}^2 + P_{react}^2} \qquad (1.27)$$

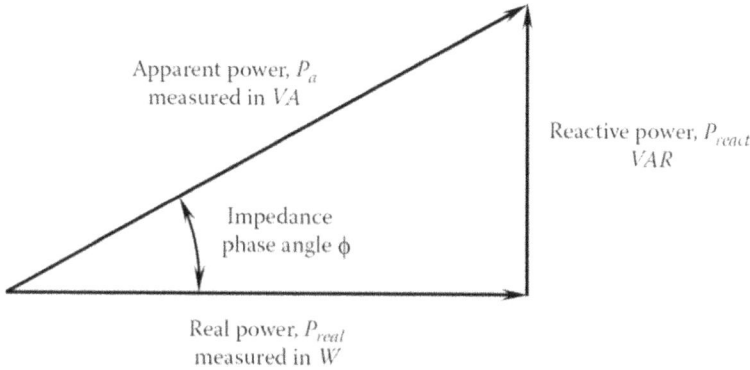

FIGURE 1.12 The power triangle: the relationship of real power, reactive power, and apparent power.

Power factor depends upon motor load. When a motor is in normal operation, the electric current drawn by the motor varies with the external load. Under a no-load condition such as the motor start, the power factor is minimum, typically 0.1–0.25. As the shaft load increases, the load current through the stator windings increases significantly, and consequently, the power factor increases until reaching the maximum at the full-load point. Then, the power factor falls again as the motor approaches full speed. In engineering practice, *PF* is determined at each load point using the following formula:

$$PF = \frac{P_{in}}{\sqrt{3}\,VI} \tag{1.28}$$

where

P_{in} is the motor input power

V and I are line-to-line voltage and current, respectively

The full-load power factor of an IM can vary from 0.5 for a small low-speed motor up to 0.9 for a large high-speed machine.

1.2.5 Torque–Speed Characteristics

In the power industry, a commonly used method of displaying motor performance characteristics graphically is to use motor torque–speed curves, as shown in Figure 1.13 for a typical DC motor. There are two pairs of parameters used to define the DC motor performance: one pair of torque, which includes peak torque T_p and rated torque T_r, and another pair of speed, which includes peak speed ω_p and rated speed ω_r. Since the peak torque occurs at $\omega = 0$, it is also called the stall torque. The peak rotating speed is also called no-load speed because the motor generates no torque at the point. As depicted in the figure, the linear relationship between the motor torque and the rotating speed can be expressed as

$$T = T_p\left(1 - \frac{\omega}{\omega_p}\right) \tag{1.29}$$

As shown in Equation 1.24, the motor output power is the product of torque and angular rotating speed. Thus, the power of a DC motor is given as

$$P_{out} = T_p\omega\left(1 - \frac{\omega}{\omega_p}\right) \tag{1.30}$$

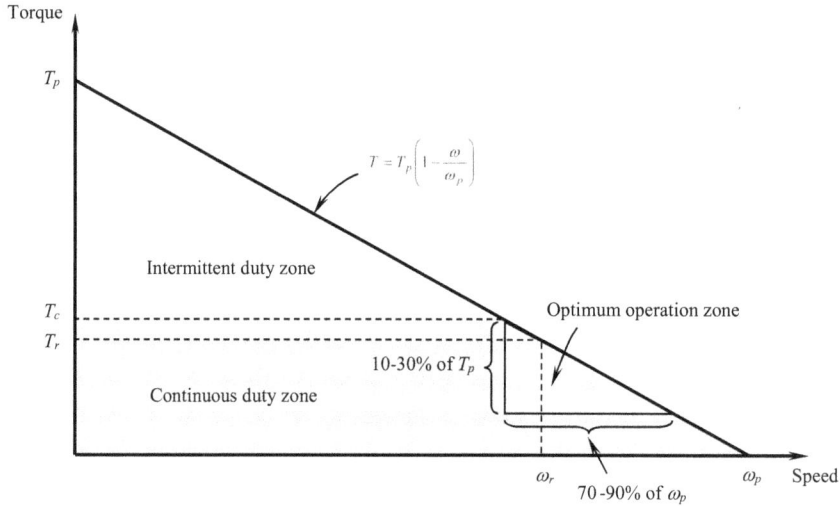

FIGURE 1.13   Torque–speed characteristics of a typical DC motor.

This indicates that $P_{out}=0$ at both $\omega=0$ and $\omega=\omega_p$. Differentiating P_{out} with respect to ω and letting $dP_{out}/d\omega=0$, the maximum power output is found to occur at $\omega=\omega_p/2$ where $T=T_p/2$. However, at this operating point, the motor efficiency is rather low, causing higher power losses and higher temperature rises. For optimal continuous performance, the operating speed may be set between 70% and 90% of the no-load speed and the operating torque between 10% and 30% of the stalled torque [1.43]. The optimum operating zone is shown in Figure 1.13 as a small triangle.

With the two pairs of parameters (torque and speed), two motor operation zones can be identified: the intermittent duty zone and the continuous duty zone. In the intermittent duty zone, the motor operates intermittently with a higher torque. This is especially useful when a rotor requires frequent starts and frequent reversals in the rotating direction. In such cases, extra torque is required to overcome the inertia of the load as well as the rotor itself. However, it is especially noted that in the intermittent duty zone, the motor can only produce torque and speed for a limited amount of time; otherwise, it will cause the motor to overheat. As the combination of torque and speed produced by the motor falls in the continuous operation zone, the motor can be loaded until the rated torque remains constant for a speed range up to the rated speed. In this zone, the motor can run as long as needed without any chance of overheating. The slope of the torque–speed curve represents inherent motor damping. For DC motors, damping is a constant.

The torque–speed characteristics of a typical AC IM are displayed in Figure 1.14. When the motor is initially started from standstill at the starting current I_s, the starting torque (also called LRT) is produced by the motor to overcome the inertia of the motor drive system. The starting torque depends on the terminal voltage and the stator and rotor design. As the motor accelerates, the torque generated by the motor may drop slightly to the local minimum point known as the pull-up torque. In the case that the pull-up torque of the motor is less than that required by its application load, the motor will overheat and eventually stall. Then, a further increase in motor speed will lead to an increase in torque until it reaches the breakdown torque, which is the highest torque the motor can attain without stalling. Starting from this point, the continuous increase in speed causes a sharp decrease in torque, as well as in motor current. When the motor reaches its full operation speed ω_r, it is loaded to its full-load torque T_r and the corresponding rated current I_r and slip s. At the synchronous speed, no torque can be developed as zero slip ($s=0$) implies no induced rotor current ($I=0$) and thus no torque ($T=0$). This situation only occurs for motors that run while not connected to a load. Therefore, in the strict sense, an IM can never reach the synchronous speed.

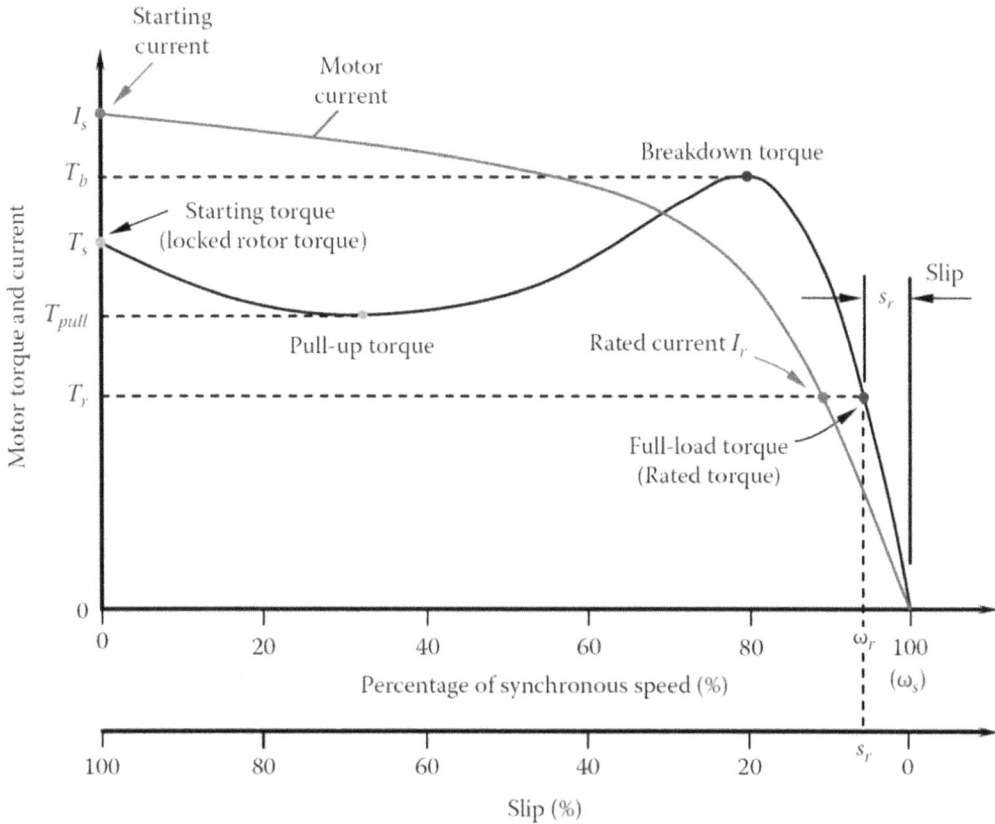

FIGURE 1.14 Torque–speed characteristics of a typical AC IM.

In order to clearly present the relationship between rotor torque and speed, an absolute value of torque (usually Nm) is used in Figure 1.14. More frequently, torque is expressed in terms of a percentage of full-load torque, together with speed in terms of a percentage of synchronous speed.

The torque–speed curve for a servomotor is given in Figure 1.15. The whole working zone is defined by the peak torque line $T = T_p$, the peak speed line $\omega = \omega_p$, and a diagonal line equation

$$T = T_p - \left(T_p - T_{ms}\right)\frac{\omega - \omega_k}{\omega_p - \omega_k} \tag{1.31}$$

where T_{ms} is the peak torque at the motor maximum speed ω_p and ω_k is the speed at the knee in the peak envelope.

The two duty zones, that is, the intermittent duty zone and the continuous duty zone, are separated by the maximum continuous torque line, which is expressed as

$$T = T_c - \left(T_c - T_r\right)\frac{\omega}{\omega_r} \tag{1.32}$$

At normal operation, a motor runs at the continuous duty zone to provide a continuous torque to drive external loads.

Unlike DC motors, motor damping in AC motors always varies along with the motor speed.

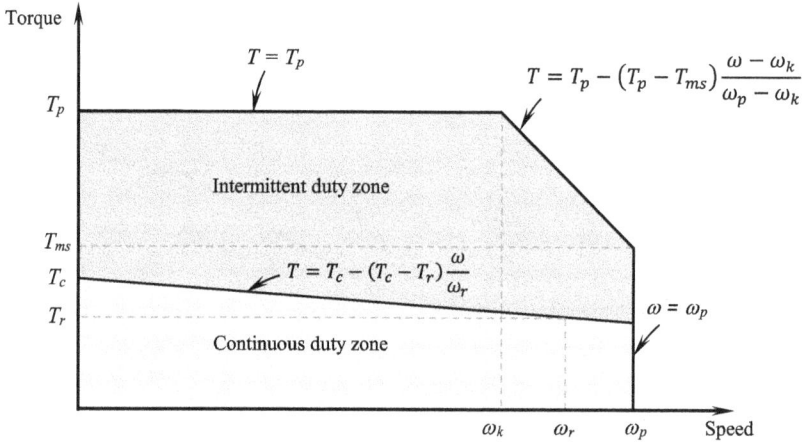

FIGURE 1.15 Torque–speed characteristics of a servomotor.

1.2.6 MECHANICAL RESONANCE AND RESONANT FREQUENCY

Mechanical resonance occurs when an external source amplifies the vibration level of a mass or structure at its natural frequency. For a rotating mass like a motor or a pump, this occurs at what is called the *critical speed*.

Every mechanical system can be resonated at a certain frequency, defined as resonant frequency. However, when two systems (*e.g.*, a motor and a driven machine) are connected together by some power transmission components such as shaft coupling, gearboxes, and belts, it forms a new resonant frequency, based on all components in the system.

As illustrated in Figure 1.16, in a regular motor drive system, the total system inertia J_t is the sum of the motor inertia J_m and the reflected load inertia J_{rl}:

$$J_t = J_m + J_{rl} = J_m + J_c + J_{i,gb} + \frac{J_{o,gb} + J_l}{\gamma_g^2} \tag{1.33}$$

The unit of inertia is kg·m² in the SI system and lb_m·ft² in the English system. It is worth to note that 1 kg·m²=1 Nm·s² and 1 lb_m·ft²=1/32.185 lb_fft·s².

Because the motor-load system is not rigid (*i.e.*, the stiffness is not infinitely large), as the motor torque is applied on the system, each of these connecting components twists slightly like a torsional spring.

To calculate the resonant frequency, motor and load inertias must be distinguished from each other. As discussed previously, load inertia is the inertia of the load reflected to the motor shaft J_{rl}. It is the lumped inertia of the shaft coupling, gearbox, and driven load machine:

$$J_{rl} = J_c + J_{i,gb} + \frac{J_{o,gb} + J_l}{\gamma_g^2} \tag{1.34}$$

The total motor inertia J_m is the lumped inertia of the rotor (including the rotor core and shaft), bearing, and other rotating components:

$$J_m = J_{rotor} + J_{bearing} + \sum J_{other} \tag{1.35}$$

The effects of interference fit on the shaft stiffness can vary depending on the actual interference fit. For a very tightly fitted rotor, the rotor assembly can be viewed as one body with variable outer diameters

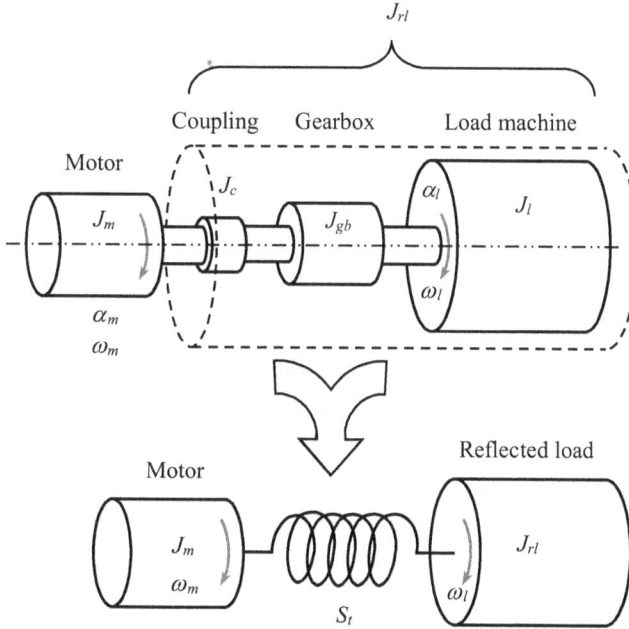

FIGURE 1.16 The equivalent motor-load model, consisting of two masses connected by a massless torsionally elastic spring.

(ODs) along its axis. Thus, the rotor core virtually increases the shaft stiffness. On the contrary, if the rotor is loosely coupled with the driven machine, its compliance can contribute to resonance.

For a rotor with a hollow shaft, the torsional stiffness (or spring constant) at each segment is calculated as

$$S_{t,rotor,i} = \frac{\pi\left(d_{out,i}^4 - d_{in,i}^4\right)G}{32 l_i} \tag{1.36}$$

where
$d_{out,i}$ and $d_{in,i}$ are the outer and inner diameter (ID) at i^{th} segment, respectively
l_i is the length of the ith segment
G is the shear modulus of elasticity

It is worth noting that the stiffness of the motor shaft can be influenced by the various types of construction. In general, the machined or welded webs on the motor shaft can add significant stiffness (typically 10%–40% over the base shaft diameter stiffness), while keyed on laminations typically add minimal stiffness. A finite element analysis (FEA) has confirmed that with six welded spider arms on an IM shaft, the equivalent diameter is 7% larger than the base shaft diameter, which corresponded to a 33% increase in torsional stiffness [1.44].

Similarly, the torsional stiffness of the load system at each segment becomes

$$S_{t,load,i} = \frac{\pi\left(d_{out,i}^4 - d_{in,i}^4\right)G}{32 l_i} \tag{1.37}$$

For a solid shaft, $d_{in,i} = 0$.

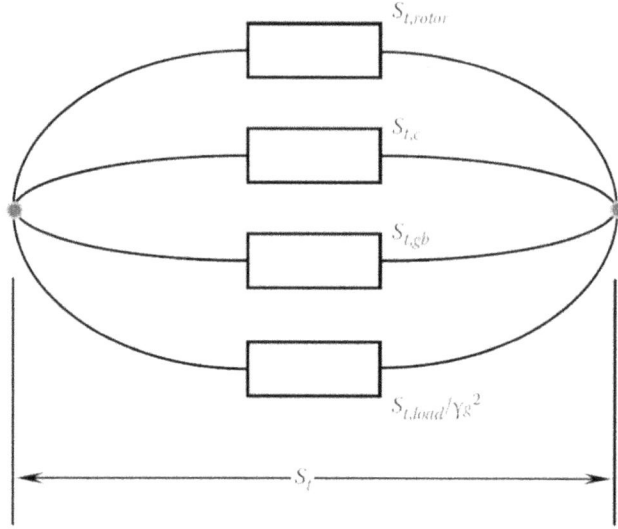

FIGURE 1.17 Taking advantage of the analogy between mechanical and electric systems to calculate the total torsional stiffness S_t.

Thus, the total torsional stiffness (spring constant) for the rotor and load machine is given as, respectively,

$$S_{t,rotor} = \frac{1}{\sum\left(1/S_{t,rotor,i}\right)} \tag{1.38a}$$

$$S_{t,load} = \frac{1}{\sum\left(1/S_{t,load,i}\right)} \tag{1.38b}$$

For two geared shafts with a gear ratio γ_g (where $\gamma_g = \omega_i/\omega_o$), the torsional stiffness is given as [1.45]

$$\frac{1}{S_{t,gb}} = \frac{1}{S_{t,i}} + \frac{1}{S_{t,o}/\gamma_g^2} \tag{1.39}$$

It follows that

$$S_{t,gb} = \frac{S_{t,i}S_{t,o}}{S_{t,i}\gamma_g^2 + S_{t,o}} \tag{1.40}$$

Utilizing the analogy between the mechanical and electric systems, an electric circuit is developed for the total torsional stiffness S_t, as shown in Figure 1.17. Thus, S_t, which is measured in Newton-meter per radians (Nm/rad), can be determined as

$$S_t = \frac{1}{\dfrac{1}{S_{t,rotor}} + \dfrac{1}{S_{t,c}} + \dfrac{1}{S_{t,gb}} + \dfrac{1}{S_{t,load}/\gamma_g^2}} \tag{1.41}$$

Resonant frequency (Hz) is thus calculated as

$$f_R = \frac{1}{2\pi}\sqrt{S_t\left(\frac{1}{J_m} + \frac{1}{J_{rl}}\right)} = \frac{1}{2\pi}\sqrt{\frac{S_t\left(J_m + J_{rl}\right)}{J_m J_{rl}}} \tag{1.42}$$

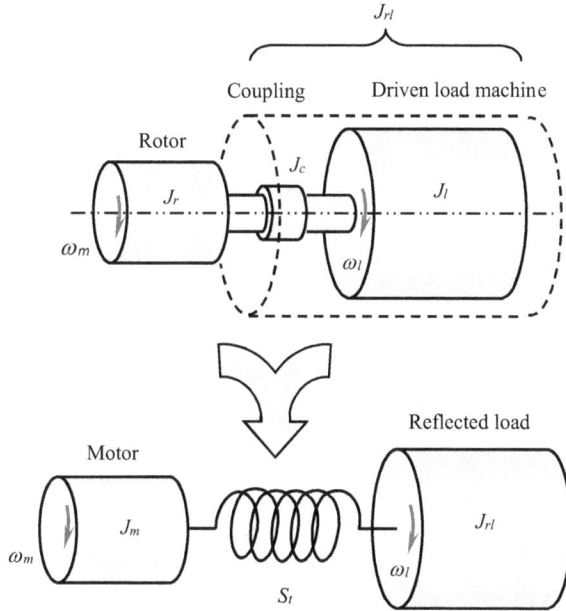

FIGURE 1.18 The equivalent motor-load model for direct drive systems.

This shows that J_m and J_{rl} have the same effect on the resonant frequency. From this equation, it can be seen that there are four ways to increase the system resonant frequency:

- Make the motor shaft more rigid—this increases the stiffness and, in turn, increases the overall torsional stiffness S_t.
- Reduce motor inertia—this actually increases the ratio of load-to-motor inertia, but it also results in higher performance of servo system.
- Reduce reflected load inertia.
- Increase stiffness of components attached/mounted to shafts.

In a direct drive system, a direct drive rotary (DDR) motor is connected with a load without using a gearbox (Figure 1.18). The coupling inertia is considered as a part of the load inertia. Therefore, the system model can be reduced to two lumped masses; the resonant frequency of the system is given by

$$f_R = \frac{1}{2\pi}\sqrt{\frac{S_t\left(J_m + J_{rl}\right)}{J_m J_{rl}}} = \frac{1}{2\pi}\sqrt{S_t\frac{J_m + \left(J_c + J_l\right)}{J_m\left(J_c + J_l\right)}} \tag{1.43}$$

The torsional stiffness is

$$S_t = \frac{1}{\dfrac{1}{S_{t,rotor}} + \dfrac{1}{S_{t,c}} + \dfrac{1}{S_{t,load}}} \tag{1.44}$$

For a three-mass system shown in Figure 1.19, the resonant frequencies are [1.46]

$$f_R = \frac{1}{2\pi}\sqrt{A \pm \left(A^2 - B\right)^{1/2}} \tag{1.45}$$

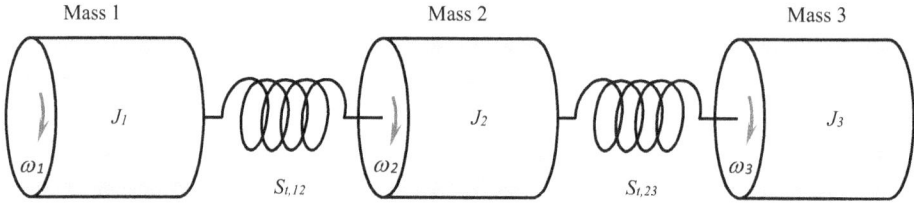

FIGURE 1.19 Three-mass system with massless torsionally elastic springs among them.

where

$$A = \frac{S_{t,12}(J_1 + J_2)}{2J_1J_2} + \frac{S_{t,23}(J_1 + J_2)}{2J_1J_2}$$

$$B = \frac{(J_1 + J_2 + J_3)S_{t,12}S_{t,23}}{J_1J_2J_3}$$

The notations $S_{t,12}$ and $S_{t,23}$ indicate that the torsional stiffness applies to the shaft between rotors 1 and 2 and rotors 2 and 3, respectively.

1.2.7 LOAD-TO-MOTOR INERTIA RATIO

The load-to-motor inertia ratio is defined as

$$\xi = \frac{J_l}{J_m} \tag{1.46}$$

The mismatch of the load-to-motor inertia is very important for the performance of electric motors. The importance of the inertia mismatch is relative to how responsive the system needs to be and how stiff the system is. In designing motion systems, the generally used rule of thumb is that the motor inertia should match the load inertia, that is, a ratio of 1:1 between load and motor inertia would be the ideal scenario. Regularly, the load-to-motor inertia ratio should stay within 10:1. Lower load-to-motor inertia ratios improve motor response, reduce mechanical resonance, and minimize power dissipation. An inertia mismatch of greater than 10:1 may cause motor speed oscillations (Figure 1.20), produce less than optimal response, waste power, and reduce system bandwidth.

In terms of efficiency, a 1:1 ratio between load and motor inertia provides the optimum power transfer. However, a 1:1 ratio is rarely useful in an actual application, because it requires an over-sized and unnecessarily expensive motor. Furthermore, it wastes energy and may not perform to specifications [1.18]. In a typical servo system with a *stiff* coupling methodology, a load-to-motor inertial mismatch of 5:1 is generally accomplished.

However, in some cases, the inertia ratio for servomotors can be much higher than 5:1. As an example, direct drive elevator PM traction machines can have 40:1 inertia mismatches. More recently, Kollmorgen has developed a new generation of drive with digital biquadratic filters, enabling servomotors to be successfully applied to medical imaging gantry applications with high inertia mismatches up to 1,000:1 [1.46]. To overcome the challenges of inertia mismatch, DDR motors, which directly couple to the load, have been developed in the last two decades. DDR motors do not require a high degree of responsiveness and thus significantly reduce inertia mismatch concerns. Directly coupled motors have been successfully tuned to ratios as high as 1,600:1 [1.47].

Some servomotor applications require larger inertia rotors to achieve good system controllability. This is due to mass or inertia acting as mechanical filters to load disturbances. The easiest

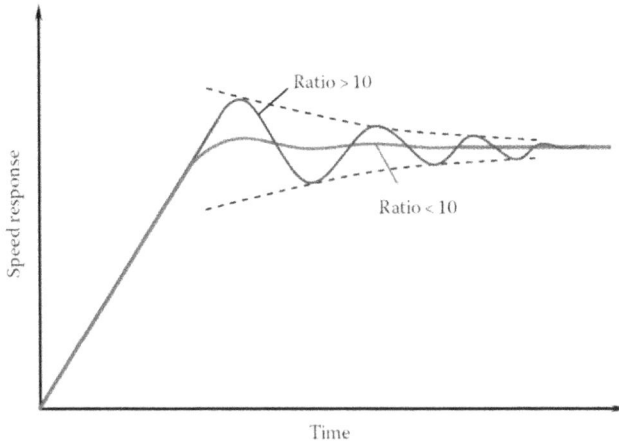

FIGURE 1.20 The effect of load-to-motor inertia ratio on motor speed response. For ratio >10, the speed response curve oscillates and asymptotically approaches its final value in a long period of time.

way is to attach an inertia disk or wheel to the shaft [1.48]. If possible, it is preferred to arrange the inertia disk inside the motor or have a larger diameter rotor in the same frame size. Otherwise, the inertia disk has to be arranged outside of the motor (Figure 1.21). As an example, by adding an inertia wheel or wheels to servomotors, Kollmorgen's engineers increase the motor inertia by a factor of 7. Thus, low-inertial motors that have a 14:1 inertia ratio can be changed to medium-inertia motors with a 2:1 inertia ratio. Correspondingly, the resonant frequency decreases from almost 4 times the antiresonant frequency to about 1.6 times the antiresonant frequency.

1.2.8 Duty Cycle

In the power industry, duty cycle is a measure of the fraction of time that a power device is in an active state, that is, $t_{on}/(t_{on}+t_{off})$. For electric motors, duty cycle is defined as the ratio the motor produces rated continuous power divided by the total elapsed time. In fact, duty cycle is a variation of load over a given period of time. The load variation may have a repetitive pattern or a fluctuating pattern.

Duty cycle is used to determine the acceptable level of running time so that the rated motor temperature is not exceeded. For a fixed repetitive load pattern, duty cycle is determined as the ratio of on-time to total cycle period. When the operating cycle is such that electric motors operate at idle or a reduced load for more than 25% of the time, duty cycle becomes a factor in sizing electric motors. Also, the energy required to start electric motors (*i.e.*, accelerating the inertia of the electric motor as well as the driven load) is much higher than for steady-state operation, so frequent starting could overheat the electric motor.

According to the applications of electric motors, a running load on an electric motor can be either steady or variable (*e.g.*, follows a repetitive cycle of load variation or has pulsating torque shocks). For instance, electric motors in ventilating fans or blowers run continuously over an extensive period of time with almost constant loads. By contrast, electric motors in electric vehicle systems have wide variations in running loads. Elevator machines have typical duties of 120, 180, and 240 starts per hour ratings but may only run at those rates for a few hours a day during heavy traffic time. The temperature variations of electric motors under different operation conditions are presented in Figure 1.22.

The International Electrotechnical Commission (*IEC*) defines ten duty cycle designations to describe operation conditions of electric motors, denoted S1–S10, as shown in Table 1.1 [1.49].

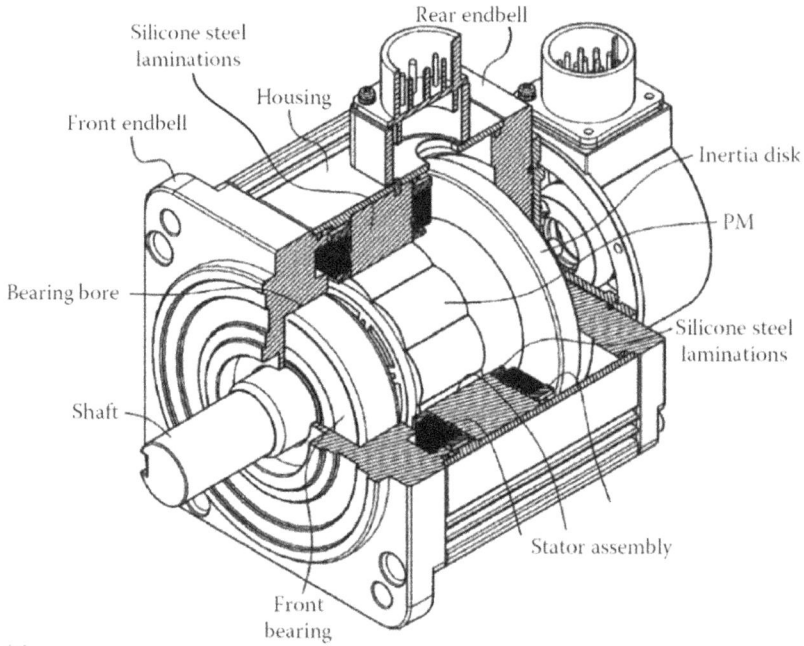

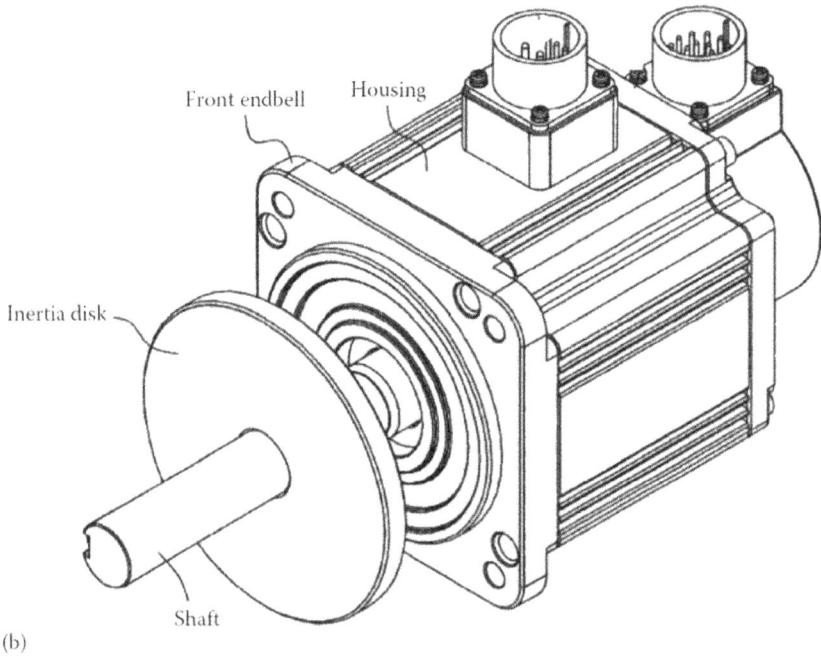

FIGURE 1.21 Addition of an inertia disk either inside (*a*) or outside (*b*) of a motor to increase the rotor inertia (US patent 7,911,095) [1.48]. (Courtesy of the U.S. Patent and Trademark Office, Alexandria, VA.)

1.2.9 MOTOR EFFICIENCY

The fast-rising energy demands all over the world and continuously increasing energy costs have created motivations for many developed and developing countries to focus on energy saving and consumption reduction. Improvements in energy efficiency are most often achieved by producing

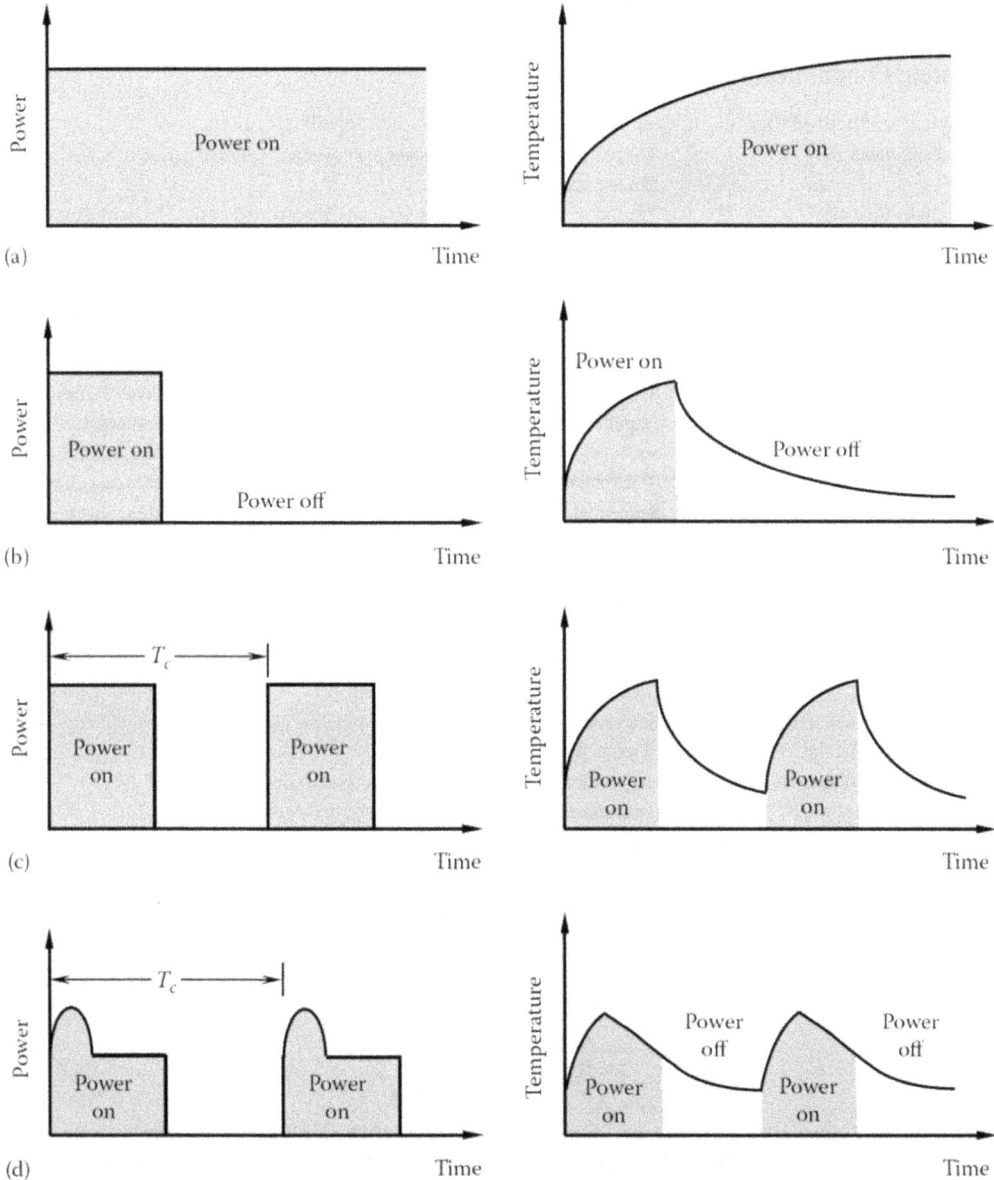

FIGURE 1.22 Temperature variations under different operation conditions: (*a*) SI—continuous operation; (*b*) S2—short time operation; (*c*) S3—intermittent periodic operation, where the start current has no impact on temperature rise; and (*d*) S4—intermittent periodic operation, where the start current has an impact on temperature rise.

more efficient machines or adopting more advanced technologies. Because electric motors consume a significant amount of electric energy motor manufacturers in recent years have committed to the development of more efficient electric motors to save energy.

1.2.9.1 Definition of Motor Efficiency

Motor efficiency is a measure of how effectively a motor converts electrical energy into mechanical energy. Given as a percentage, motor efficiency is defined as the ratio of the power output P_{out} to the power input P_{in}. Since the power output P_{out} represents the difference between the power input

TABLE 1.1

Operating Conditions of Electric Motors

Class	Motor Duty	Note
S1	Continuous duty	The motor operates at a constant load over an extensive period of time to reach temperature equilibrium.
S2	Short-time duty	The motor operates at a constant load over a period of time not sufficient to reach the thermal equilibrium. The rest periods are long enough for the motor to cool down.
S3	Intermittent periodic duty	The motor operates with repeated cycles consisting of a constant output power period followed by an off period. Temperature equilibrium is never reached. Starting current has little effect on temperature rise.
S4	Intermittent periodic duty with starting	Sequential, identical start, run, and rest cycles with constant load. Temperature equilibrium is not reached, but starting current affects temperature rise.
S5	Intermittent periodic duty with electric braking	Sequence of identical duty cycles—starting, operation, braking, and rest. Thermal equilibrium is not reached.
S6	Continuous operation with intermittent load	Sequential, identical cycles of running with constant load and running with no load. No rest periods.
S7	Continuous operation periodic duty with electric braking	Sequential identical cycles of starting, running at constant load, and electric braking. No rest periods.
S8	Continuous operation with periodic changes in load and speed	Series of identical repeating duty cycles, where within each cycle the motor operates at several different load levels and speed. There is no stopped time and thermal equilibrium is not reached.
S9	Duty with nonperiodic load and speed variations	Load and speed vary periodically within the permissible operating range. Frequent overloading may occur.
S10	Duty with discrete constant loads and speeds	Duty with a discrete number of load/speed combinations, with these maintained long enough to reach thermal equilibrium.

P_{in} and the variety of power losses $\sum P_{loss}$ and the power input P_{in} represents the summation of the power output P_{out} and the variety of power losses $\sum P_{loss}$, motor efficiency can be expressed in several different forms:

$$\eta = \frac{P_{out}}{P_{in}} = \frac{P_{in} - \sum P_{loss}}{P_{in}} = \frac{P_{out}}{P_{out} + \sum P_{loss}} \tag{1.47}$$

This equation shows that in order to increase motor efficiency it must minimize various power losses in motor, such as copper losses, iron losses, mechanical losses, and windage losses. Thus, the efficiency of electric motors can be boosted by the following points [1.50]:

- Reducing the copper losses in the motor windings. This can be done by increasing the cross-sectional area of the conductor or by improving the winding technique to reduce the winding length, especially at the end turns.
- Using better materials for lowering eddy-current-related power losses.
- Taking advantage of advanced nanotechnology to fabricate thinner laminations.
- Improving the aerodynamics of a motor ventilating system for reducing motor windage losses.
- Applying more efficient cooling methods in motor cooling.
- Improving manufacturing tolerances.

For electric motors, high efficiency and high torque density are two of the most desirable features. Today, motor efficiency is normally in the range of 80%–95%. Usually, larger motors with higher power output have higher efficiency than smaller motors. High-efficiency motors can provide significant benefits, including reductions in energy consumption and carbon emissions over their entire life cycles. Because operating costs comprise the majority of lifetime motor costs, even a 1% gain in efficiency can make a big difference to costs.

1.2.9.2   IEC Standards on Efficiency Classes of AC Electric Motors

In 2008, IEC published the standard IEC 60034-30: 2008 [1.51] on efficiency classes of AC electric motors. The scope of this standard covers almost all motors (*e.g.*, standard, hazardous area, marine, and brake motors) but excluded motors made solely for converter operation and completely integrated into a machine:

- Single speed, three phases, and 50 and 60 Hz
- 2, 4, or 6 pole
- Rated output from 0.75 to 375 kW (1–500 hp)
- Duty type S1 (continuous duty) or S3 (intermittent periodic duty) with a rated cyclic duration factor of 80% or higher

This standard defines the requirements for the efficiency classes and aims to create a basis for international consistency. The international efficiency (IE) classes IE1, IE2, and IE3 defined in this standard are based on the test methods specified in IEC 60034-2-1 [1.52]. It is noted that the methods with this standard determine efficiency values more accurately than the methods previously used.

In order to promote a competitive motor market transformation, a new international standard IEC 60034-31 [1.53] was released in 2010 for the addition of the IE4 motor efficiency level. Since no sufficient market and technological information is available to allow IE4 standardization, this efficiency class is intended to be informative. It is expected that advanced technologies will be developed in the near future that can enable manufacturers to design motors for the IE4 class efficiency levels, while maintaining motor dimensions compatible with the existing motors having lower efficiency classes. The four electric motor efficiency classes, testing standards, and regulation over time are listed in Table 1.2.

TABLE 1.2
Electric Motor Efficiency Level, Class, Testing Standard, and Regulation over Time

Efficiency Level	Efficiency Class Standard	Test Standard	Regulation over Time
Standard efficiency	IE1	IEC 60034-2-1	China
	IEC 60034-30	Medium uncertainty	Taiwan 2003
			Switzerland 2010
High efficiency	IE2	IEC 60034-2-1	Australia 2006
	IEC 60034-30		Brazil 2009
			Canada 1997
			China 2009
			Europe 2011
			Korea 2008
			Switzerland 2011
Premier efficiency	IE3	IEC 60034-2-1	Europe 2015–2017
	IEC 60034-30	Low uncertainty	United States 2011
Super premier efficiency	IE4		
	IEC 60034-31		

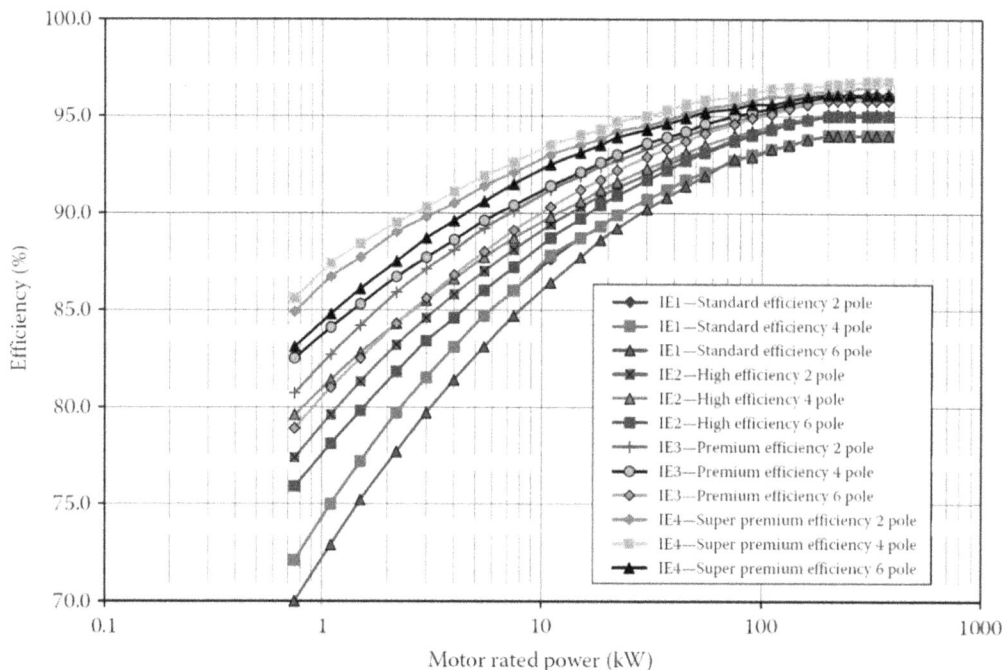

FIGURE 1.23   Comparison of four efficiency classes for electric AC motors with 50 Hz.

A rated efficiency of a motor is a function of its rated output. The comparisons of the efficiencies for all four efficiency classes (IE1, IE2, IE3, and IE4) are presented in Figures 1.23 and 1.24 for 50 Hz and 60 Hz electric motors, respectively.

Furthermore, the comparison of four efficiency classes for 50 Hz and 4-pole motors is given in Figure 1.25.

1.2.10   MOTOR INSULATION

The maximum operating temperature of a motor depends on the selection of motor insulation. As the maximum temperature is determined, the maximum motor load, typically specified as the amount of power the motor can deliver on a continuous basis, can also be determined. Thus, motor insulation needs to be defined at the motor's conceptual design stage.

According to the maximum allowable operating temperature that insulation systems can withstand, electric insulation can be categorized into several classes, characterizing the capability of the insulator to resist aging and failures due to overheating. During motor operation, an insulation material may gradually lose its insulating ability to perform the task. The lifetime of insulation materials depends on thermal, chemical, electrical, and mechanical factors. Among them, thermal and electrical stresses are the most important. In fact, dielectric strength of insulation is very sensitive to temperature aging. Average insulation lifetime decreases rapidly with the increase in motor internal temperatures. A generally accepted rule of thumb is that each $10°C$ rises above the rating temperature may reduce the motor lifetime by one-half.

There are generally five specialized insulation elements used in an electric motor, including turn-to-turn insulation between separate wires in each coil, phase-to-phase insulation between adjacent coils in different phase groups, phase-to-ground insulation between windings and the electrical ground, slot wedge to hold conductors firmly in the slot, and impregnation to bring all the other components together and fill in the air space.

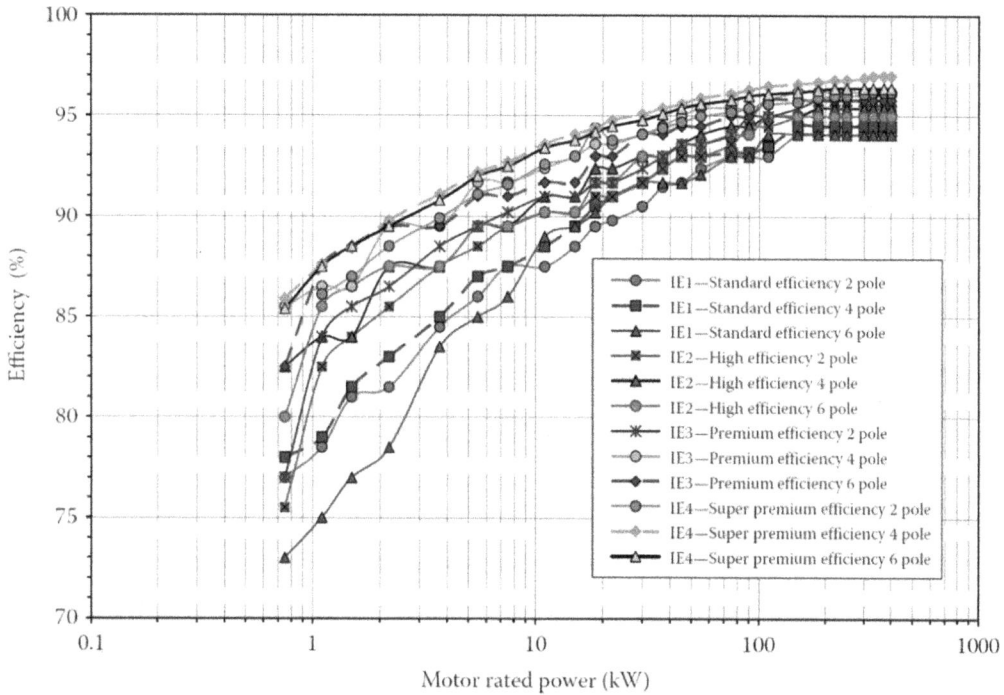

FIGURE 1.24 Comparison of four efficiency classes for electric AC motors with 60 Hz.

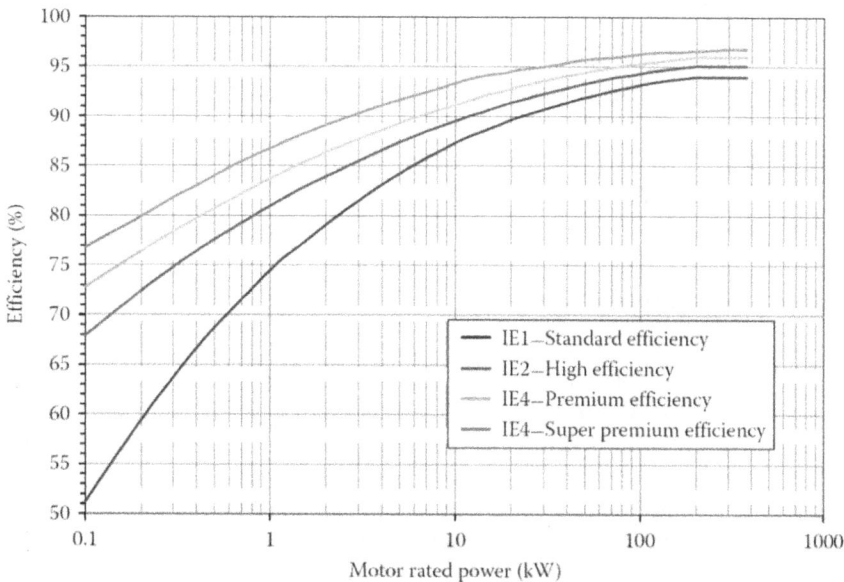

FIGURE 1.25 Comparison of four efficiency classes for 50 Hz, 4 pole electric motors.

Dielectric strength refers to the maximum electric field (in V/m in the SI system and V/mil in the English system, where 1 mil=0.001 in.) that a material can withstand without breaking down and losing its insulating capabilities. The dielectric strength of a material depends on the specimen thickness, the electrode shape, the rate of the applied voltage increase, the shape of the voltage–time curve, and the medium surrounding the sample (*e.g.*, air, gas, or liquid). It should be noted that in strong electric fields, Ohm's law does not hold for insulation materials that have extremely high electric resistance. The current density increases almost exponentially with the electric field, and at a certain value, the current jumps to very high magnitudes at which a specimen of the material is destroyed [1.54].

Dielectric strength values of some solid insulation materials at room temperature and normal atmospheric pressure are presented in Table 1.3.

Based on an average 20,000 h lifetime, the maximum allowable temperature and insulation materials for each insulation class are listed in Table 1.4. The motor classification is based on the temperature rating of the lowest-rated component in the motor.

Thermal aging is an irreversible, permanent reduction of material properties, defined as deterioration. Many of the insulation failures can be attributed to the fact that the operation temperature exceeds the temperature limits of insulation materials. In fact, temperature has a strong impact on the lifetime of insulations. As the temperature rises above the normal operating temperature, the lifetime of insulation can be quickly shortened (Figure 1.26). This phenomenon is called thermal aging. As discussed by Bonnett and Soukup [1.64], the increase in temperature may have resulted from various causes, which are as follows: (a) voltage variation or unbalance (per Bonnett and Soukup, 3.5% voltage unbalance per phase will lead to an increase in winding temperature of 25% in the phase with the highest current), (b) frequent motor starts and stops, (c) improper motor cooling, (d) severe environmental conditions such as high ambient temperature, and (e) motor operation under overloading conditions.

In many motor manufacturers, the reference ambient temperature T_a is assumed to be 40°C. The temperature rise of a motor is referred to as the difference between the measured temperature of the motor winding and the ambient temperature, that is, $\Delta T = T_w - T_a$. However, the standard method of measuring the winding temperature involves taking the ohmic resistance of the winding. This provides the average temperature of the whole winding, including the motor leads, end turns, and wires deep inside the stator slots. Therefore, to reflect the temperature difference within the winding, the so-called hot spot allowance must be added to adjust the allowable temperature rise. The hop spot allowance T_{hs} is usually assumed to be 5°C–15°C, depending on the insulation class. For a specific insulation class, the allowable temperature rise of a motor is determined for preventing motor overheating under all loading conditions (no load, full load, locked rotor, etc.):

$$\Delta T = T_{\max} - \left(T_a + T_{hs} \right) \tag{1.48}$$

1.2.11 MOTOR OPERATION RELIABILITY

Motor operation reliability and testing are two key components for design optimization and safe operation of motors. Motor reliability is the measure of the chance that the motor can operate normally over a period of time without failure. In the motor industry, one of the useful parameters is the mean time between failures (MTBF), defined as the reciprocal of the failure rate of the system.

Two failure modes dominate the life of motors: bearing failures and winding failures. In all motor failures, bearing failures account the majority. Winding failures generally occur in the early stage of motor running and are attributed to shorts and grounds resulting from assembly quality rather than the long-term insulation degradation. By utilizing more than 2,000 actual failure cases of fractional horsepower motors in the existed data failure bank, Wilson and Smith [1.65] developed a mathematical reliability model for each failure mode by means of Weibull cumulative distribution function and regression technique for use in predicting overall motor life and failure rates.

1.3 CLASSIFICATIONS OF ELECTRIC MOTORS

Electric motors can be classified in a variety of ways according to their operating characteristics, such as the source of electric power, type of rotor winding, type of motion, control pattern, the

TABLE 1.3

Dielectric Strength of Some Solid Insulation Materials

Insulation Material	Dielectric Strength (V/m$\times 10^6$)	Reference
Ceramics		
• Porcelain	35–160	[1.55]
• Titanates of Mg, Ca, Sr, Ba, and Pb	20–120	[1.56]
Glasses		
• Fused silica, SiO_2	470–670	[1.55]
• Alkali-silicate glass	200	
Insulating films and tapes		
• Low-density polyethylene film	300	[1.57]
• Poly-p-xylylene film	410–590	[1.58]
• Aromatic polymer films		[1.59]
• Kapton H (DuPont)	389–430	
• Ultem	437–565	
• Hostaphan	338–447	
• Amorphous Stabar K2000 (ICI film)	404–422	
• Stabar S100 (ICI film)	353–452	
• Polyetherimide film (26 µm)	486	[1.60]
• Parylene N/D (25 µm) film	275	[1.61]
• Cellulose acetate film	157	[1.61]
• Cellulose triacetate film	157	[1.61]
• Polytetrafluoroethylene film	87–173	[1.61]
• Fluorinated ethylene-propylene copolymer film	157–197	[1.61]
• Ethylene-tetrafluoroethylene film	197	[1.61]
• Ethylene-chlorotrifluoroethylene copolymer film	197	[1.61]
• Polychlorotrifluoroethylene film	197	[1.61]
	118–153.5	[1.61]
Micas		
• Muscovite, ruby, natural	118	
• Phlogopite, amber, natural	118	[1.61]
• Fluorophlogopite, synthetic	118	
Potassium bromide, KBr, crystalline	80	[1.55]
Sodium chloride, NaCl, crystalline	150	[1.55]
Varnish		
• Vacuum-pressure-impregnated baking-type solventless varnish	79.9	
• Epoxy baking-type varnish	90.6	
• Solventless, rigid, low viscosity, one part	82.7	
• Solventless, semiflexible, one part	106.3	[1.61]
• Solventless, semirigid, chemical resistant	181.1	
• Solvable, for hermetic electric motors		
• Polyurethane coating—clear conformal, fast curing	78.7	
• Standard conditions	47.2	
• ommersion conditions		
Various insulations		
• Natural rubber	100–215	[1.55]
• Silicon rubber	26–36	[1.56]
• Calendered Aramid paper	28.7	[1.56]
• Aramid with mica	39.4	[1.56]

TABLE 1.4
Standard Insulation Classes

Insulation NEMA Class [1.62]	Maximum Temperature Rating (°C)	Insulation Materials
Class A	105	Cotton, silk, paper, synthetic fibers, vinyl acetate
Class E[a]	120	Polyurethane, epoxy resins, polyethylene terephthalate, phenolics, alkyds, leatheroid
Class B	130	Shellac, bitumen, silk, mica, polyesters
Class F	155	Epoxy, polyamides, silicone, mica, glass
Class H	180	Silicone elastomers, epoxy, polymides, silicone, mica, glass
Class N	200	Glass fibers, mica, asbestos, Teflon
Class R	220	Glass, silicone, mica, Nomex, Teflon

[a] *IEC 60085 Thermal class [1.63].*

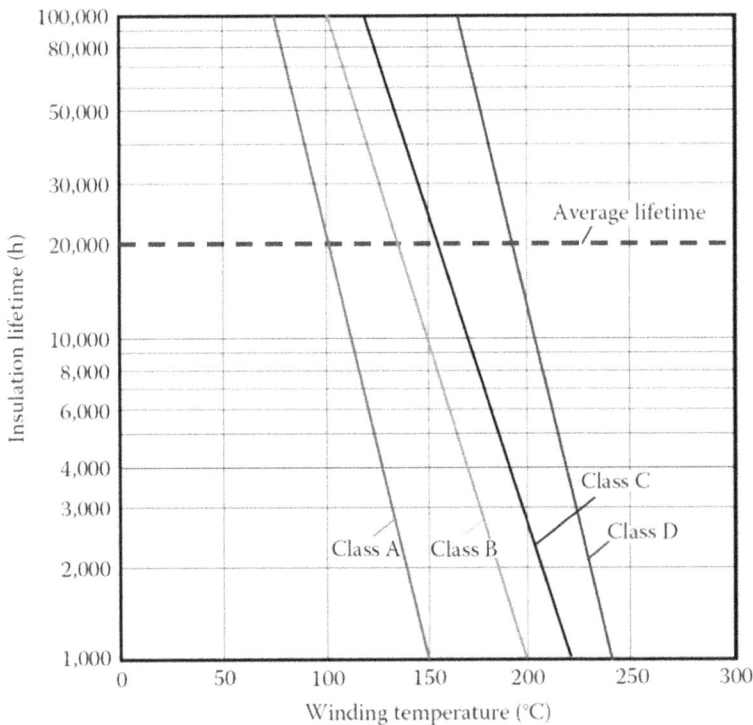

FIGURE 1.26 Thermal aging effect on lifetime of different insulation classes.

magnetic flux orientation, structure topology, power rating, and the cooling methods. A brief classification of rotary electric motors is given in Figure 1.27.

1.3.1 DC AND AC MOTORS

DC motors are designed to run on DC power. This type of motor was invented much earlier than the type of AC motors. They are often used when high torque at a low speed is required. The speed adjustments of DC motors can be as much as 20:1, and they can operate at 5%–7% of the motor's base speed (some can even operate at 0 rpm) [1.66]. There are a number of different types of DC motors:

```
                          ┌─────────────────┐
                          │ Rotary electric │
                          │     motor       │
                          └─────────────────┘
```

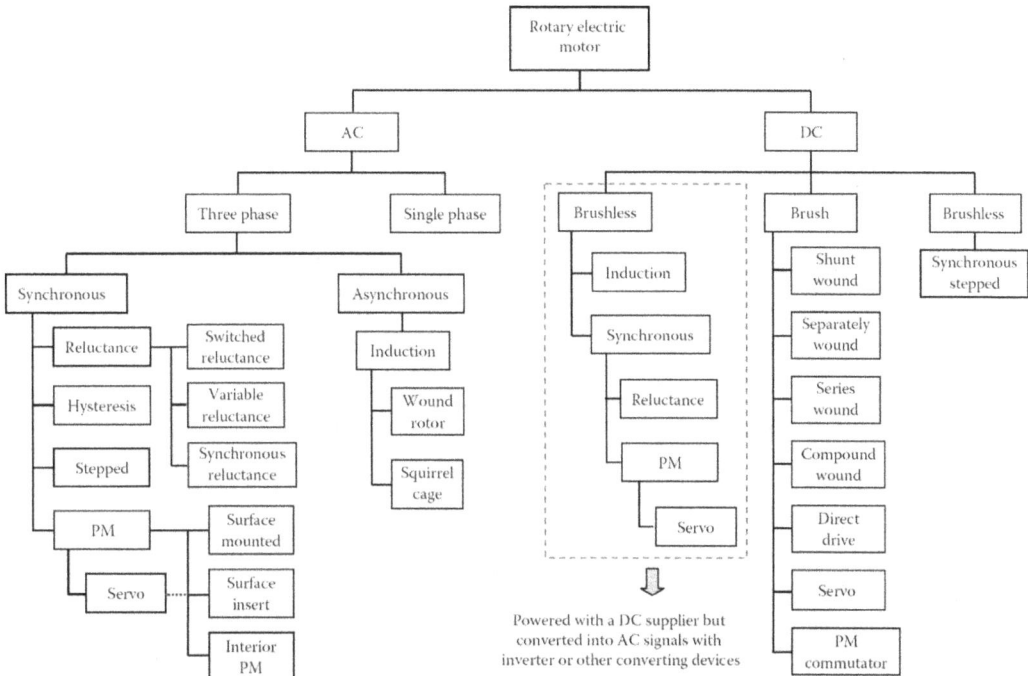

FIGURE 1.27 The classification of rotary electric motors.

- Shunt wound motor—In this type of motor, the rotor and stator windings are connected in parallel. Hence, the current in the rotor and stator windings is independent of one another. The characteristics of a shunt motor provide it very good speed regulation.
- Separately excited motor—In this type of motor, the rotor and stator windings are connected to different power suppliers. For a long time, it has been the most common configuration used in industrial applications for DC motors with electronic speed control. Due to their excellent controllability and simple operational performance, separately excited motors are extensively used in speed or position control systems.
- Series motor—As its name indicates, the rotor and stator windings in a series motor are connected in series. The torque is proportional to I^2 so it gives the highest torque per current ratio over all other DC wound field motors at economical costs. It is therefore used in starter motors of cars and some old generation elevator motors [1.67].
- Compound motor—As the combination of the series motor and the shunt motor, the motor has the torque characteristics of the series motor and the regulated speed characteristics of the shunt motor. A compound motor comprises an armature winding on the rotor and two field windings on the stator. The stator is connected to the rotor through a compound of shunt and series windings. Compound motors are often adopted to drive loads such as shears, presses, and reciprocating machines.
- Brushed DC PM field motors—Field windings are now replaced by PM materials and the winding has a series of commutator bars with carbon brushes arranged to provide mechanical commutation of the motor. Commutation is the switching of the current direction in the winding at the right timing to produce continuous torque vectors in one direction for continuous rotating speeds of the shaft.
- Universal motors with brushes and commutators have been designed to operate on either AC or DC power. For this type of motor, the stator's windings are connected in series with the rotor windings through a commutator. The advantages of universal motors are high

starting torque, compact size, and ability to run at a high operating speed. The negative aspects include the short lifetime caused by the commutator and brushes and high noise and vibration. Therefore, this type of motor is often used in applications where the motor only operates intermittently. Universal motors generally run at high speeds, making them suitable in home appliances (*e.g.*, vacuum cleaners and food mixers) and home power tools (*e.g.*, electric drills) where single-phase wall plug-in power is abundant.

- Brushless DC (BLDC) motor—This special type of motor has been developed to overcome the limitations and drawbacks of brushed DC motors. Practically, brushless motors are powered by a DC electric source via an integrated inverter/switching power supply, which produces an AC electric signal to drive/commutate the motor. It usually requires sensors and electronic control system for controlling the inverter output amplitude, waveform, and frequency. Though a number of brush-less motors are available (*e.g.*, induction, switched reluctance), typical brushless motors are made with PMs. With built-in PM, the motor does not need a separate excitation winding, resulting in reduced power losses and increased efficiency. The size of the PMDC motor is the smallest (high power density) among all other DC motors. Often, PMDC motors are served as servomotors.

Advantages of brushed DC motors include low initial cost, high reliability, and simple speed and torque control characteristics. Brushless DC motors have evolved from brush DC motors as power electronic devices and became available to provide electronic commutation in place of the mechanical commutation provided by brushes. Brushless motors have increased reliability, longer lifetime, higher efficiency, less maintenance, reduced noise, elimination of ionizing sparks from commutator, and overall reduction of electromagnetic interference.

Today, DC motors are still used in a wide range of applications, especially for those applications requiring precise speed control over a large range around the rated speed, such as steel mills, mines, and electric trains. However, DC motors usually require additional operational elements such as brushes and commutators that transfer electric power to motor armatures. However, commutators usually can cause power ripples and limit the rotor speed and brushes increase the frictional power lower and radiofrequency interference. Furthermore, as brushes wear and tear, carbon dust spreads throughout the motor, causing some operation and performance problems. As a consequence, the maintenance of the interface between the brush and commutator becomes critical for motor operation reliability. In addition, the use of brushed can enhance the motor acoustic noise.

AC motors are driven by AC power sources. By eliminating commutators and brushes, AC motors offer several advantages over DC motors, including increased operation reliability, longer lifetime, higher efficiency, less maintenance, reduced cost, shorter frame size, and simpler motor structure. AC motors are dominant in industrial motion control for cost/performance reasons. As the most common type, single-phase AC motors are mainly used for residential and commercial applications. Three-phase AC motors are especially suitable for high-power applications. In general, single-phase motors operate less efficiently than three-phase motors.

Brushless DC motors are similar to AC synchronous motors. The major difference is that the waveform used to drive AC motors is typically sinusoidal and could come directly from an AC source or could be using the pulse-width modulation technique. Therefore, AC synchronous motors develop a sinusoidal back EMF, as compared to a rectangular or trapezoidal back EMF for brushless DC motors.

Motors used in hospital equipment or other patient-care facilities are required to comply with low noise level standards to endorse patient comfort and reduce anxiety. Brushless DC motors are ideal for noise-sensitive environments due to the lack of brushes, which emit audible noise during rotation [1.68].

The disadvantages of AC motors include difficulty in speed control, high control complexity, less torque density, inability for operating at low speeds, induced eddy current and hysteresis power losses in the stator and rotor cores, high cost, and poor positioning control.

1.3.2 SINGLE-PHASE AND THREE-PHASE MOTORS

In most countries, household power is usually single phase due to the low cost of single-phase power distribution. In practice, single-phase power is best to use for low power units, usually less than 3.7 kW (5 HP). A single-phase motor is run from an AC single-phase power system. The most standard frequencies of single-phase power systems are either 50 or 60 Hz, although other frequencies may also be available. However, unlike three-phase motors, a single-phase motor is unable to produce the start torque itself; it must be started by some external means such as an auxiliary start winding or a start capacitor. Single-phase motors need some other forcing functions to set direction of rotation (DIR). The auxiliary start winding or start capacitor creates a simulated phase to set the DIR.

A three-phase system can transmit three times the power of a single-phase system by using only two additional wires. Consequently, three-phase power is a common form of electric power for its inherent benefits in high-powered transmission applications. In a three-phase system, AC is carried by three circuit conductors with the same frequency and amplitude but different phases. As sinusoidal functions of time, the current at each conductor has shifted 120° in phase from each other. Correspondingly, in a three-phase motor, there are also three windings (separated equally in space by 120°) per pole on stator to produce a rotating magnetic field.

Today, three-phase AC motors dominate in almost all industries and consume more than half of all the electricity used in the industry. Three-phase motors are more efficient and compact than single-phase motors with comparable power ratings. Three-phase motors have generally lower vibrations and noise, and last longer than single-phase motors under the same operating conditions. In fact, the effectiveness, low cost and operational flexibility are major reasons for three-phase motors to be extensively used in industry.

1.3.3 INDUCTION AND PM MOTORS

Depending upon the method of generation of the magnetic field in the rotor, an AC motor can be classified as either a PM AC motor, where the magnetic field of the rotor is directly produced by PMs, or an IM, where the magnetic field of the rotor is produced though the induction effect onto the rotor bars/winding.

A typical IM is presented in Figure 1.28. The primary components of an IM include a stator, a rotor, a shaft, one or two end bells located at the motor ends, bearings, and a motor housing. Both stator and rotor cores are made of thin silicon steel laminations to minimize eddy current losses. The field windings in the stator are normally made of copper wires with coated insulation films, distributed in slots around the stator for optimizing the electromagnetic field. It is common practice to bring three motor leads (*i.e.*, U, V, and W) to a terminal block. The motor housing is often made from diecasting aluminum with a machined finish inside. The shaft of the motor is press fit at the center of the rotor core. To enhance cooling efficiency, a cooling fan mounts on the shaft at its back end to ventilate cooling air flowing through either the interior or exterior of the motor. Cooling plate fins are fitted on both the motor housing and the front end bell for maximizing convective heat dissipation.

In an IM, the AC power supply is connected to the stator winding to generate a rotating magnetic field. Because of this rotating field, the rotor is powered by means of electromagnetic induction. The change in magnetic flux through the rotor induces AC in rotor windings and, in turn, creates its own magnetic field. The induced current in the rotor gives rise to magnetic forces, which cause the rotor to rotate in the direction defined by the stator rotating magnetic field. In actual operation, the rotor speed always lags the magnetic field speed, which is defined as the synchronous speed, allowing the rotor windings or conducting bars to cut magnetic lines of force and produce useful torque.

According to the rotor structure, IMs can be further divided into two subcategories: (*a*) squirrel cage motors in which the rotor is made of conductive aluminum/copper bars that are parallel (or have a small skew angle) to the rotor centerline and short-circuited by the end rings (Figure 1.29) and (*b*) wound rotor motors where windings are made on the rotors (Figure 1.30).

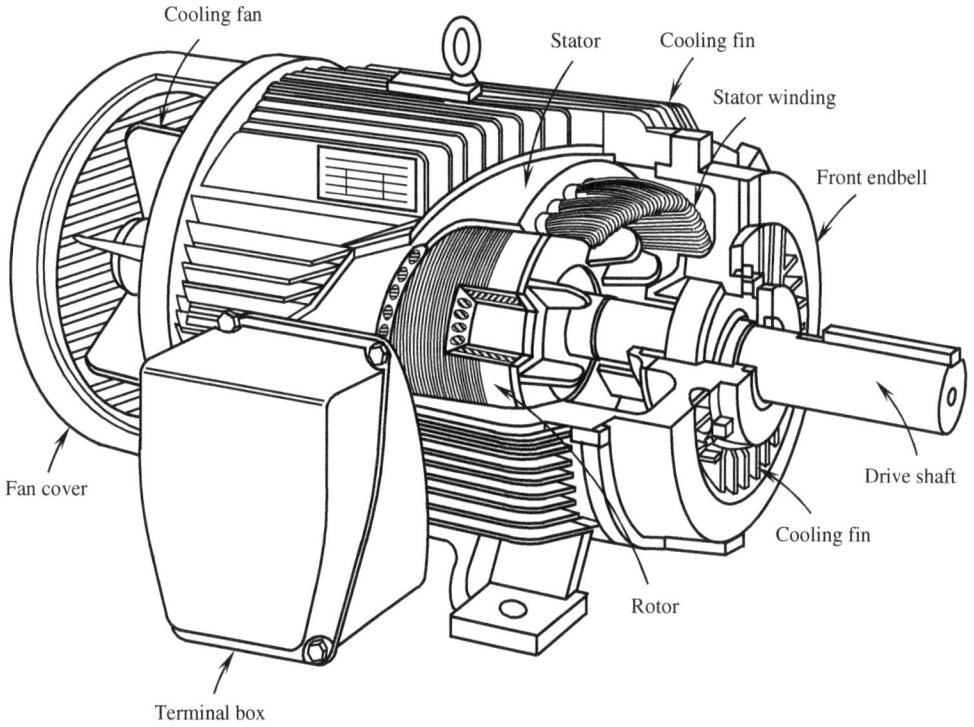

FIGURE 1.28 The structure of a typical three-phase IM motor. To cool the motor, a cooling fan is mounted on rotor shaft at the non-driving end and cooling fins are casted on the housing outer surface and the front end bell.

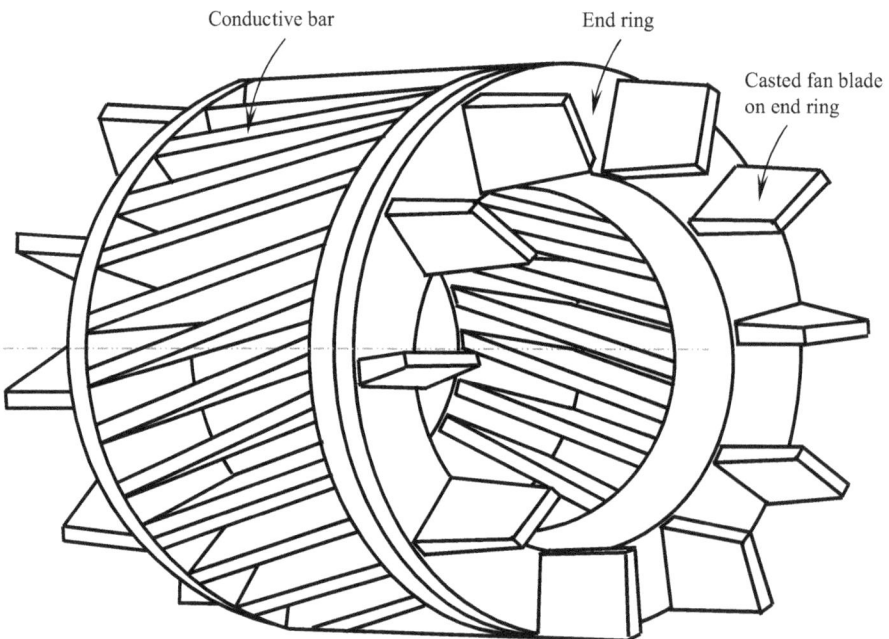

FIGURE 1.29 Conductive bars and end rings in a squirrel cage motor form a closed electrical circuit.

FIGURE 1.30 Rotor winding in a wound IM.

IMs are perhaps the simplest and most rugged motors, which have been extensively used in a variety of applications, from household appliances to heavy industrial equipment, due to their simple structures and low costs. In modern squirrel cage IMs, the conducting bars are formed in the skewed slots distributed axially along the rotor surface by casting. These conducting bars are connected at the ends of the rotor slots as the end rings to form the closed electric circuit.

For IMs, the stator operates with the power supply frequency f ($f=60\,Hz$ in North America and $f=50\,Hz$ in most countries of the world) and the rotor winding contains current and voltage with slip frequency f_{slip} or sf, where f_{slip} is defined as the frequency corresponding to the slip speed $n_{slip} = (n_s - n_r)$.

A low rotor resistance will result in the current being controlled by the inductive component of the circuit, yielding a high out-of-phase current and a low torque.

PMMs use PMs to generate the required magnetic field and thus standalone excitation windings are no longer required. A permanently excited synchronous motor has a sinusoidal back EMF and therefore operates with a sinusoidal voltage. PMMs are designed to provide a wide operation speed and load range with high motor efficiency, usually in excess of 90% from less than 50% speed driving a typical pump or compressor load to a peak range in excess of 97% for high-speed applications [1.69]. In many industrial applications, PMM is the preferred choice due to its high power and torque densities, along with its high efficiency. The structure of a typical PMM is shown in Figure 1.31.

PMMs can provide excellent performance in terms of torque density, energy efficiency, and controllability. They have numerous advantages over IMs. First, the built-in PMs in a PMM eliminate rotor windings that are required in an IM. Hence, it greatly reduces the rotor inertia and power losses and, consequently, increases operational reliability and improves dynamic load response. Second, since the stator current in a PMM is only for torque production and thus magnetizing current through the stator is no longer necessary, a PMM operates at a higher power factor and thus a higher efficiency over an IM for the same power output. Third, the application of rare earth PMs makes PMMs with high power density and high torque-to-inertia ratio, which lead to fast dynamic response capability. As a matter of fact, rare earth PMMs have the highest power density of any motor type. This feature is especially desirable for applications in which the motor size is the main consideration. Today, PMMs have become increasingly popular in various industrial and commercial sectors. They are especially ideal in high-accuracy, high-performance motion control applications, for instance, computer numerical control (CNC) machine tools, robots, embedded motion, engraving machines, packaging and printing machines, semiconductor fabrication facilities, medical equipment, and satellite servo systems.

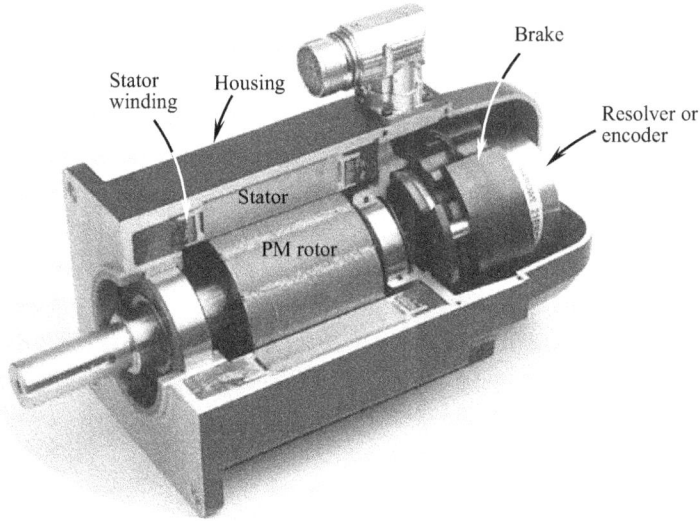

FIGURE 1.31   The typical structure of a PMM. (Courtesy of Kollmorgen Corporation, Radford, VA.)

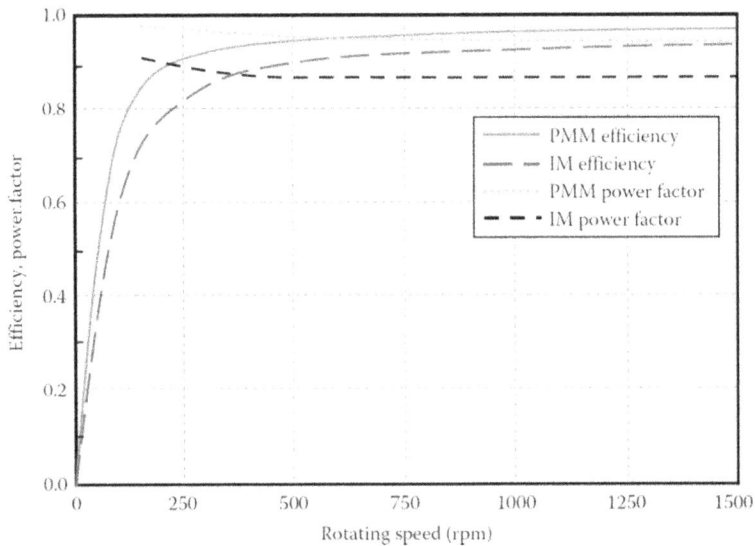

FIGURE 1.32   Comparison of calculated motor efficiency and power factor between a 55 kW, 1,500 rpm IM, and a PMM under an identical operation condition.

The comparison of motor efficiency and power factor between a 55 kW, 1,500 rpm IM and a comparable performing PMM under an identical operation condition and power rating is given in Figure 1.32 [1.70]. The efficiency of IM is inherently lower than that of PMM due to its high power losses. However, the efficiency of PMM at very high speeds may drop because of the risk of PM's demagnetization [1.71].

Engineers at Parker Hannifin did an excellent job for quantitatively investigating and comparing the performing characteristics and costs between IMs and PMMs [1.72]. Instead of examining a full operation cycle, three operating conditions are selected and examined: (*a*) low speed, high torque; (*b*) mid speed, mid torque; and (*c*) high speed, low torque. The results are displayed in Table 1.5, as a good reference for motor designers and end users.

TABLE 1.5

Comparison of Motor Performance between IM and PMM under Three Operating Conditions (The Data are Cited from Reference [1.72])

Parameter	PMM	IM
Low Speed, High Torque		
Torque, Nm	600	600
Speed, rpm	500	500
Total rotor losses, W	34	1,625 (+4,679%)
Stator iron losses, W	110	187 (+70%)
Stator I^2R losses, W	5,842	15,370 (+163%)
Total loss, W	5,986	17,182 (+187%)
Mid Speed, Mid Torque		
Torque, Nm	250	250
Speed, rpm	2,000	2,000
Total rotor losses, W	38	750 (+1,874%)
Stator iron losses, W	474	588 (+24%)
Stator I^2R losses, W	759	1,382 (+82%)
Total loss, W	1,271	2,770 (+118%)
High Speed, Low Torque		
Torque, Nm	50	50
Speed, rpm	4,500	4,500
Total rotor losses, W	49	332 (+578%)
Stator iron losses, W	1,137	411 (−64%)
Stator I^2R losses, W	102	198 (+94%)
Total loss, W	1,288	941 (−27%)

Note: The percentages in the parentheses are calculated using the formula $\dfrac{IM - PMM}{PMM} \times 100\%$.

Under high-speed and low torque conditions, the total power loss of IM is approximately 2.87 times higher than that of PMM. Breaking it down further into component power losses, a huge difference comes from the rotor losses (34 W for PMM and 1,625 W for IM). Most of the IM rotor losses are derived from the resistive losses (I^2R) in the copper bars. The resistive losses of IM are about 3 times higher than PMM. This is because IM needs enough current passing through the stator windings to create both the rotor and stator magnetic fields. Consequently, the IM can only produce its peak torque in a short time.

In the middle region of the speed–torque curve, while the IM's efficiency improves substantially, the IM has twice as many losses as the PMM. This indicates that the IM generates more heat that must be effectively dissipated into the surrounding environment for avoiding overheating.

At the region with high speed and low torque, the IM outperforms the PMM. The very low flux density in both stator and rotor keeps the total power loss to a minimum.

According to the law of conservation of energy, most of the power losses in electric machines are finally converted into thermal energy to either dissipate to the surroundings or cause the machine to heat up. In general, IMs generate more heat than PMM and therefore require a most efficient cooling system. However, while PMMs produce less heat during operation, some rare earth PMs cannot work at very high temperatures due to the limitation of their Curie temperatures (*e.g.*, 150°C for Nd-Fe-B and 300°C for Sm-Co magnets).

The cost of IM is perhaps the strongest benefit. Taking the total cost of the PMM as the benchmark (100%) of the comparison, the most expensive part of the IM is the stator winding, accounting for 27%. The following are the stator core (22%), the rotor core (12%), the rotor bar (10%), and the rotor end ring (2%). Thus, the total cost of IM is 73%, indicating that the IM is about 27% less expensive than the equivalent PMM. As to the PMM, the most expensive parts are PMs, which accounted for 61% of the total cost. The costs of other active components, *i.e.*, the state core, the stator winding, and the rotor core, account for 19%, 12%, and 8% of the total cost, respectively. The main reason for the lower cost of IM is due to the use of copper bars and end rings on the rotor instead of PMs.

Overall, PMMs are considered excellent candidates for applications that require high efficiency, high torque/power density, and a wide speed range. In comparison, as the most mature technology, IMs are characterized by low cost, good high-speed performance, high-temperature tolerance, and robust structure. This indicates that IMs are suitable for high-speed and high-temperature generation applications, as well as applications where the cost is a main decisive factor.

According to the mounting pattern of PMs, PMMs can be further divided as surface-magnet mounted and buried-magnet mounted. Surface-mounted PM (SPM) motors have dominated for decades. However, this mount method has relatively weak bounding strength, especially at high rotational speeds. In recent years, some emerging markets and perhaps the sharp price rise of rare earth magnets have boosted demand for interior-buried PM (IPM) motors. Since in this buried-magnet method, all PMs are buried inside the rotor core, it is more robust than the SPM and enables to achieve higher airgap flux density. With other advantages such as high-speed performance, low heat generation in rotor, and special designs that can support near-constant power over a broad range of speed, IPM motors provide an excellent solution for applications like traction motors in electrical vehicles, machine tools, and high-speed electric machines [1.73].

Torque produced by an IPM motor is based on two different mechanisms [1.74]: one is the same as an SPM motor; PM torque is generated by the magnetic flux linkage between the PM rotor magnetic field and the electromagnetic field of the stator. The greater number of magnet poles usually generates a higher torque for the same current level. Another is known as reluctance torque. PMs buried inside a rotor pole piece exhibit high reluctance directly along the magnetic axis due to the low permeability of the PMs and pole pieces among the magnetic poles or magnet barriers, creating inductance saliency and reluctance torque. Thus, IPM motor designs augment PM torque with reluctance torque. As a result, the magnets used in IPM motors can be thinner, achieving significant cost reductions.

1.3.4 SYNCHRONOUS AND ASYNCHRONOUS MOTORS

A synchronous motor refers to a three-phase AC motor in which the rotor runs at the same speed as the rotating magnetic field of the stator. The synchronous speed n_s can be expressed in different ways, depending on the unit it adopted:

$$n_s = \frac{f}{p} \quad \text{in rps (Hz)} \tag{1.49a}$$

$$n_s = \frac{60f}{p} \quad \text{in rpm} \tag{1.49b}$$

where
  f is the frequency of the AC supply current in Hz
  p is the number of magnetic pole pairs per phase

The synchronous speeds of 50 and 60 Hz machines are presented in Figure 1.33 for various magnetic pole pairs.

If N denotes the number of magnetic poles per phase, then it gives that $N=2p$ and yields

$$n_s = \frac{120f}{N} \quad \text{in rpm} \tag{1.50}$$

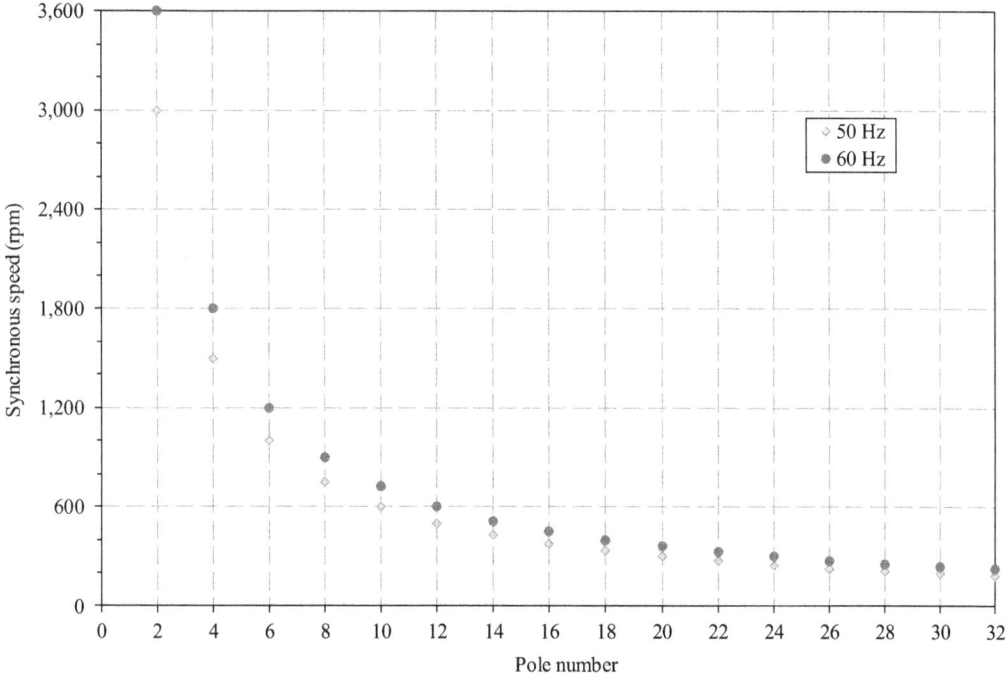

FIGURE 1.33 Synchronous speeds of 50 and 60 Hz machines vary with different pole numbers. For both units, an increase in the pole number will decrease motor rotating speeds and reduce the gap between them.

Therefore, synchronous speed can be altered by changing either the frequency applied to the motor or the number of magnetic poles. Some multispeed motors adopted external connections that enable to switch the stator poles, for example, from 4 to 6 poles. By using an adjustable frequency drive, the motor speed can vary in a large speed range under a constant voltage. In the unit of radians per second (rad/s), the angular synchronous speed ω_s is given as

$$\omega_s = \frac{\pi n_s}{30} = \frac{2\pi f}{p} = \frac{4\pi f}{N} \tag{1.51}$$

Substituting (1.51) into (1.24), the output power of the synchronous motor is given as

$$P_{out,s} = \frac{4\pi f T}{N} \tag{1.52}$$

This indicates that a change in AC power frequency can result in a change in motor output power. Thus, for a motor that remains a constant torque, when the frequency changes from f_1 to f_2, the corresponding output power $P_{out,2}$ can be determined as

$$P_{out,2} = \frac{f_2}{f_1} P_{out,1} \tag{1.53}$$

Similarly, the rotational speed becomes

$$\omega_2 = \frac{f_2}{f_1} \omega_1 \tag{1.54}$$

For instance, a motor with the output power of 12 kW at 60 Hz can produce only 10 kW at 50 Hz, while it reduces the rotational speed about 16.7%, calculated from the formula $(\omega_2 - \omega_1)/\omega_1 = f_2/f_1 - 1$.

In contrast, in an IM, the stator windings are wound around the rotor to produce a rotating magnetic field with a three-phase power supply. It is the varying magnetic field that induces currents in the rotor conductors/end rings (or rotor windings). Thus, the interaction between the magnetic fields of the stator and rotor causes a rotational motion of the rotor.

Because the current in the rotor conductors/end rings is induced by the stator windings, the rotating speed of the rotor must lag the rotating speed of the stator's magnetic field (*i.e.*, the synchronous speed). This speed difference is called slip, which is defined as

$$s = \frac{n_s - n_r}{n_s} = \frac{\omega_s - \omega_r}{\omega_s} \tag{1.55}$$

where
n_s and n_r are the synchronous speed and rotor speed in rpm
ω_s and ω_r are the angular synchronous speed and angular rotor speed in rad/s, respectively

Thus, the rotor rotating speed of an asynchronous motor becomes

$$n_r = n_s(1 - s) \quad \text{in rpm} \tag{1.56a}$$

$$\omega_r = \omega_s(1 - s) \quad \text{in rad/s} \tag{1.56b}$$

Slip increases with load. Usually, full-load slip ranges from less than 1% to 3% for large power motors and 4%–6% for small power motors.

The output power of the asynchronous motor can be determined by the following equation:

$$P_{out,r} = \frac{4\pi f T}{N}(1 - s) \tag{1.57}$$

Synchronous motors can be built with either salient pole rotor or nonsalient pole (cylindrical) rotors, depending on the customers' specifications and applications. Generally, salient pole motors are typically found in high-power, low-speed applications. Motors with cylindrical rotors are typically found in high-power, high-speed applications such as spindle motors in machine tools.

1.3.5 SERVO AND STEPPER MOTORS

There are two main types of motion control systems: closed loop and open loop. A closed-loop system applies a feedback system to verify whether or not the desired output has been reached. For instance, an encoder is commonly attached with a servomotor to measure the velocity and position of the servomotor for providing the information to the motion controller. Obviously, servomotor systems require the use of the closed-loop system. In contrast, an open-loop system does not need any feedback for verifying the output. In practice, most step motor systems use an open-loop system. The difference between the two control schemes is whether or not the use of the feedback system (Figure 1.34).

A closed motion control system usually consists of a motion controller, a feedback system, a motor drive, and an electric servomotor. The motion controller acts primarily as the brain of the motion control system. Its main function is to make sure the output of the system is as close as possible to the desired result. Based on the information from the feedback system (such as encoders, resolvers, and sensors), the motion controller sends the electronic signals to the motor drive to adjust the motion path or trajectory. The feedback system is to obtain the motor motion information (*e.g.*, displacement and speed), which are then converted into a set of digitized output signals and fed to the motion controller for making necessary adjustments. The motor drive takes the low-power

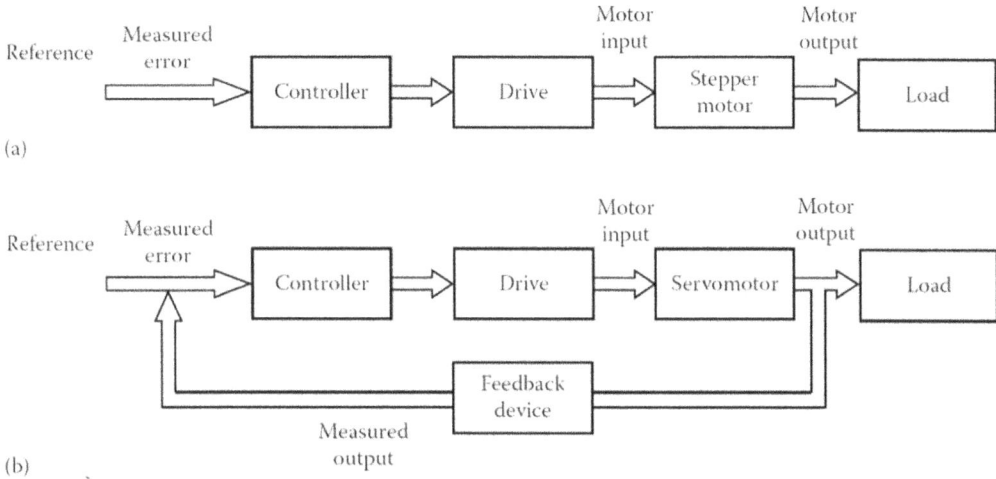

FIGURE 1.34 Comparison of two control schemes: (*a*) the open control loop and (*b*) the closed control loop.

electronic signal from the motion controller and converts it into high-power current/voltage to the motor, which executes the tasks from the drive as an actuator. All these components are integrated to form the complete motion control system to provide the desired movement for various applications.

The word *servo* originally comes from the Latin word *servus*, meaning slave or servant. National Electric Manufacturers Association (NEMA) [1.75] defined a servomotor as "an electric motor that employs feedback with the purpose of producing mechanical power to perform the desired motion of the servo mechanism." Simply speaking, any control element that employs feedback is a servo.

Today, servomotors are increasingly being used in a variety of applications due to their high power density, high efficiency, and excellent dynamic performance as compared with other motor drive technologies. Servomotors operate with closed-loop control systems. The ability of the servomotor to adjust to differences between the motion profile and feedback signals depends greatly upon the types of control systems and servomotors. There are two types of servomotors: one is classical DC servomotor and another is AC servomotor. Generally, AC servomotors can handle higher current surges compared to DC servomotors. It is to be noted that in some references, AC servomotors are referred to as brushless DC motors, causing confusion to some readers. An AC servomotor or the so-called brushless DC motor is essentially a three-phase AC synchronous motor. It has a position transducer inside the motor to transmit motor shaft position to the drive amplifier for the purpose of controlling current commutation in the three phases of the motor windings.

Stepper motors are special motors used in motion control systems. This type of motor has high torque at low speeds, high reliability, low cost, and simple rugged construction that operates in almost any environment. Typical applications include printers, image scanners, CNC machines, and volumetric pumps. Unlike common AC and DC motors rotating continuously, a stepper motor moves in fixed angular increments, called the step angles. In a stepper motor, a full revolution is divided into a large number of discrete steps. In fact, a stepper motor is an actuator that converts electrical pulse signals into angular displacements. Once the stepper motor receives an electrical pulse signal, it moves a fixed step angle in a predefined direction. Thus, by controlling the pulse number, the angular displacement (*i.e.*, the motor positioning) can be controlled precisely. By altering the pulse frequency, the accurate control of the motor rotating speed and acceleration can be achieved. It is important to note that in this control mechanism, no feedback systems are required. This allows the stepper motor to operate in an open-loop system, making the motion control easier and less expensive. In addition, stepper motors can operate at a set speed regardless of load as long as the applied load is less than the limited torque rating.

As an open-loop control system, the stepper motor requires only the input of the current state. The number of steps per revolution varies by model and manufacturer. Stepper motors were introduced in the early 1960s as an economical replacement to closed-loop DC servo systems. Unlike conventional motors that rotate continuously, stepper motors rotate in fixed angular increments. One revolution of a stepper motor involves taking a number of steps, depending on the number of rotor teeth, motor construction, and type of drive scheme used in the motor control.

Stepper motor technology does have some disadvantages. The most critical drawback is the loss of synchronization and torque if a large load exceeds the motor's capacity. A high-inertia load can cause the rotor to slip or not advance when the step pulse is given. Consequently, the user typically selects a stepper motor's capability with 2:1 factors of safety to torque to minimize or eliminate loss of synchronization. Stepper motors also tend to run hot because phase current is independent of the load. In some applications, if the motor needs to be overdriven by the load, it may be undesirable to feel the poles of the stepper motor as the rotor is being pulled by the load [1.76].

The first three-phase variable reluctance (VR) stepper motor was introduced as a position control device for British warships. With a mechanical rotary switch, this stepper motor could move in steps of 15°. The design and operation of the motor were presented in an article published in 1927 [1.77]. The first hybrid stepper motor, the combination of VR and PM, was invented, patented, and manufactured by General Electric in 1952 [1.78]. This type of motor was originally used in low-speed smooth motion applications and thereafter has been used increasingly as position control devices. In the same year, Sigma Instruments introduced an industrial cyclonome stepper motor, which is regarded as the first practical two-wire stepper motor [1.79].

Today, stepper motors are available in many types, topologies, and step sizes for various industrial and commercial applications due to their advantages of low cost, high reliability, no cumulative error, high torque at low speeds, simple and rugged construction, and excellent response to start-up, stopping, and reverse operation.

There are primarily three types of stepper motors: PM, VR, and hybrid [1.80]. The PMM, as the name implies, has PMs added to the motor structure. The rotor is magnetized with alternating north and south poles in a straight line with the rotor shaft to increase magnetic flux intensity. This type of stepper motor has relatively low torque and low speed with large step angles. Its simple construction and low cost make it an ideal choice for nonindustrial applications. The variable reluctance motor (VRM) usually has three-phase stator windings. This type of motor is characterized by soft iron multi-toothed rotor and wound stator for achieving large torque outputs. However, due to the large vibration and high noise emission, the VRMs gradually withdraw from the main industrial market. The hybrid stepper motor combines the best features of both the VRM and PMMs by constructing multi-stator poles and PMs inside the rotor. The typical step angle for this type of motor ranges from 0.72° to 3.6° (correspondingly, 500–100 steps per revolution). Therefore, the hybrid stepper motors are the most widely used stepper motors in various industries today. The comparison of three types of stepper motors is listed in Table 1.6.

TABLE 1.6
Comparison of Three Types of Stepper Motors

Stepper Motor Type	Phase	Step Angle	Torque	Vibration and Noise	Note
Permanent magnet	2 or 4	7.5° or 15°	Low	Low vibration at low frequencies, low noise	High efficiency, low current, low heat generation
Variable reluctance	3	1.5°	High	Large	Out of main industrial market
Hybrid	2	1.8°	High	Low	With the most wide applications
	5	0.72°			

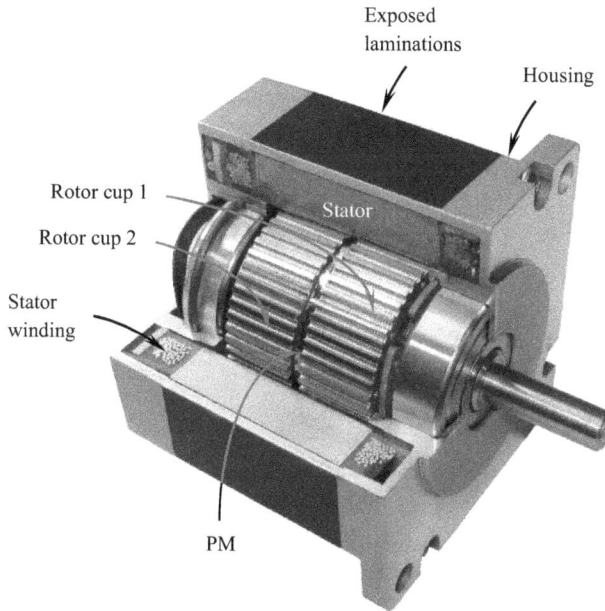

FIGURE 1.35 Structures of hybrid stepper motors, which combine features of the PM stepper and the variable reluctance stepper motors.

A cutaway diagram of hybrid stepper motors is demonstrated in Figure 1.35. The stator coils are wound on stator poles. The rotor is a cylindrical PM, magnetized along the axis with radial soft iron teeth. Thus, one pole of the rotor may align with the stator in distinct positions.

The polarity of rotor laminations is shown in Figure 1.36. Unlike conventional motors, the magnet in a stepper motor is magnetized axially rather than radially. As shown in Figure 1.36, two axially spaced sections each formed with radially projecting and angularly spaced teeth. In addition, the teeth on the north end of the rotor are displaced by a half of a pole pitch from the teeth on the south end. Both of these two features make stepper motors unique from other motors.

For standard reluctance stepper motors (*e.g.*, without PM), the rotors look similar to the aforementioned rotor except there is no PM and the magnetic field is not permanently present in the rotor lamination. The coils in the stator create a magnetic field in the stator that attracts or repels the soft iron rotor laminations into position. Then, the coil current is reversed to reverse the coil magnetic field and keeps the rotation moving. A stepper motor drive is required to switch the currents.

1.3.6 GEAR DRIVE AND DIRECT DRIVE MOTORS

A power transmission is a way of manipulating the output power of a motor to suit the need of the system. There are several motor power transmission patterns in motion control applications: (*a*) gear drive, (*b*) direct drive, (*c*) tangential drive (*e.g.*, belt, chain), (*d*) ball/lead screw drive, (*e*) worm drive, and (*f*) pulse drive [1.81] and others.

Gear drive is a conventional mechanism in connecting electric motors to external mechanical loads. The shaft of motor connects to the input shaft of gear drive and thus the desired output torque and speed are obtained at the output shaft of the gear drive. In this way, the motor is coupled with the transmission device to reduce the rotating speed and increase the output torque. The most important design parameter of the gearing system is the gear ratio γ_g, which is defined as the ratio of the input speed to the output speed. In the gearing system, the ratio of the load torque T_l (*i.e.*, the

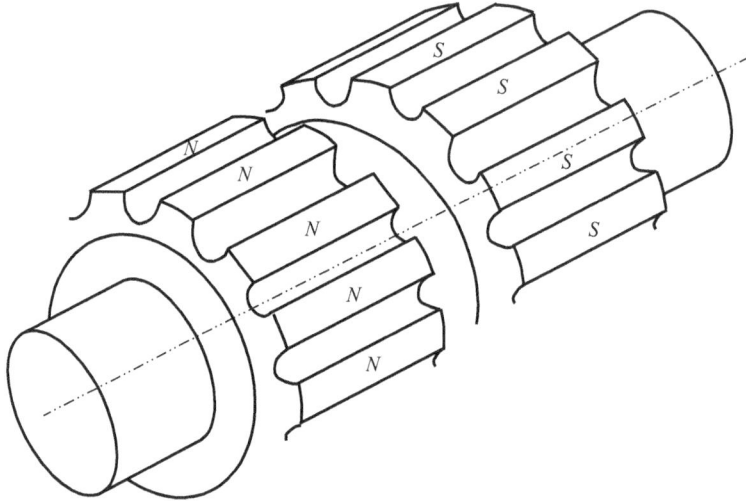

FIGURE 1.36 Magnetic polarity of rotor laminations in a hybrid step motor. The PMs are magnetized axially and placed in the center of lamination stacks. The rotor teeth on the north end of the magnet are magnetized north and the rotor teeth on the south end of the magnet are magnetized south. The north and south halves are displaced by a half of a pole pitch from the teeth on the south end.

gearing system output torque) to the motor torque T_m (*i.e.*, the gearing system input torque) is determined by the gear ratio γ_g and the gearing system efficiency η_g:

$$\frac{T_l}{T_m} = \gamma_g \eta_g \tag{1.58}$$

Gearing systems are available in many different types, sizes, gear ratios, structural configurations, efficiencies, and backlash characteristics. They can be generally categorized into several different types: parallel shaft gearing (*e.g.*, spur and helical gearing), perpendicular shaft gearing (*e.g.*, bevel gearing), planetary gearing, cycloidal gearing, and worm gearing, just to name a few. Usually, the type of parallel shaft gearing system takes more space than other gearing types. Planetary and cycloidal gearing systems are the most commonly used in the motion control industry due to their high torsional rigidity high torque ratio, high transmission efficiency, compact size, lightweight, low inertia, and low backlash.

However, there are some disadvantages of using gearing systems. The main problem is backlash introduced by gearing systems. Backlash is the gap between the teeth of two adjacent gears. Thus, the rotational backlash of a gearing system is the accumulated backlash from all paired gears. It can be measured as the free rotational angle at the gearing system output shaft when the input shaft is locked or vice versa. In precise motion control applications involving frequent load reversals (*e.g.*, CNC machines, elevators, wind pitch control systems), backlash plays a crucial role in determining the repetitive positioning accuracy. For this type of application, gearing systems must be made with low backlash and high stiffness. For demonstration purposes, the theoretical backlash of the gearing system and torsional stiffness is shown in Figure 1.37.

Second, the use of separate motor and gearing system can greatly increase the system volume, introduce extra backlash, and lower torque density and system reliability. To properly address these issues, servomotor manufacturers have introduced gearmotors in which a motor is integrated with a built-in gearing system for achieving better performing characteristics such as high torque capacity, smooth operation, compact structure, high torque density, minimal maintenance, and high operation reliability, as shown in Figure 1.38.

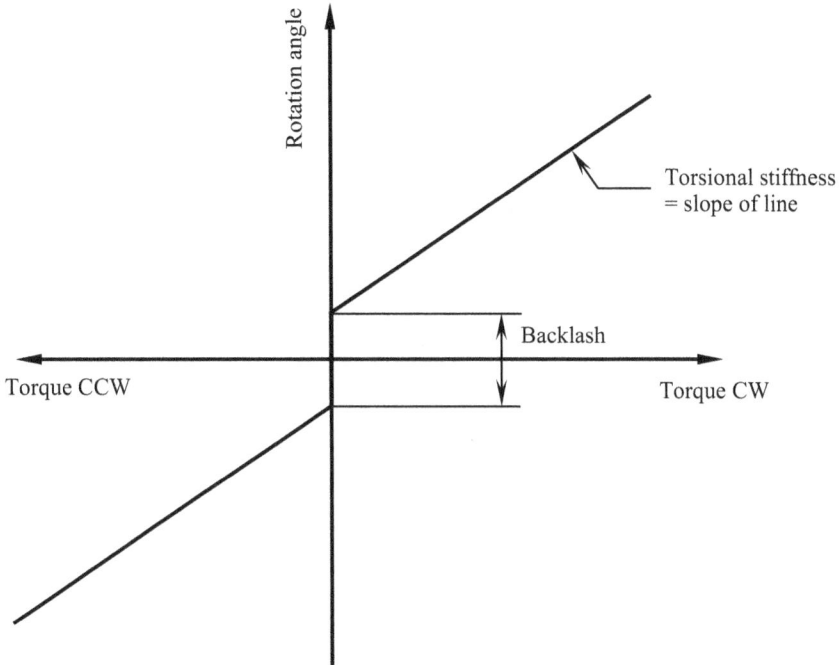

FIGURE 1.37   Theoretical backlash of a gearbox and stiffness curve.

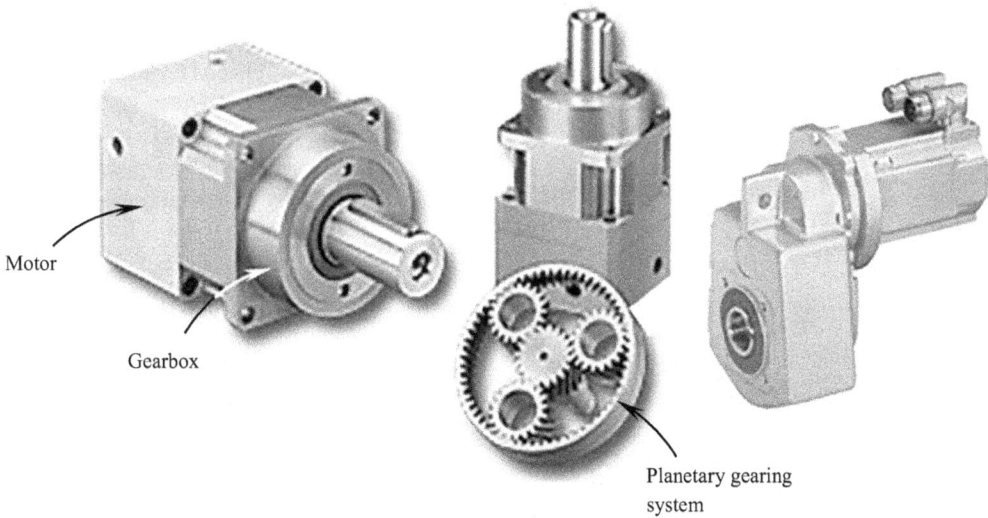

FIGURE 1.38   Integration of electric motor and gearing system as gearmotor.

FIGURE 1.39 Frameless DDR motors. (Courtesy of Kollmorgen Corporation, Radford, VA.)

The direct drive technology was pioneered by Kollmorgen in the 1960s [1.82]. DDR motors are designed to directly drive external loaded machines and thus eliminate the need for power transmission elements. They can be housed or frameless depending on customers' needs. A housed DDR motor is commonly delivered as an assembled device in a standardized frame size. By contrast, a frameless motor (sometimes also known as a built-in motor or kit motor) is delivered as a separate rotor and stator (Figure 1.39). This enables the frameless motor to be directly built into the mechanical structure of machine, using the machine's own bearings, shaft, and housing. As a result, frameless motors offer the most compact and lightweight solution to meet a variety of industrial automation needs. Generally, this technological solution has been beneficially used in applications that require high-power and high-performance motor systems with an optimized power density. To help integrate frameless motors, motor manufacturers often provide a reusable assembly tool called bridge to maintain the alignment of the rotor and stator during assembly.

By eliminating the need for additional mechanical transmission devices, the direct drive technology allows improved motor operating reliability, efficiency, dynamic performance, torque density, motion accuracy, and disturbance response. This also leads to large mass savings in the mechanical structure and enables further cost reductions. There are other advantages such as better load inertia matching, ease of control, low noise emission, and streamlined machine design. All these benefits make direct drive motors the ideal solution for applications requiring precision motion control and high bandwidth.

Typical DDR motors are shown in Figure 1.40, which offer many high-performance features and zero maintenance in a precision servo solution. Among them, the cartridge DDR motor, developed by Kollmorgen in 2005, is the first in the motor industry to combine the space-saving and performance advantages of frameless DDR technology. The cartridge DDR motors are supported by customers' bearings, thus providing up to 50% more torque density than comparably sized conventional servomotors and the ability to remove the motor without dissembling the machine. By making direct drive benefits available to simple mechanisms, this type of motor has been successfully used in many applications such as packaging, printing, medical equipment, converting, and others.

For a DDR motor, torque is proportional to rotor diameter squared. Thus, in order to increase motor torque, DDR motors are typically designed with high diameter-to-length ratios, that is, large diameters and short axial lengths. In addition, a DDR motor can simultaneously have both a very large OD and ID. This implies that the DDR motor looks more like a thin ring. In fact, the large OD and ID help develop the large motor torque. This thin-ring shape may be particularly suitable for some applications requiring large diameters and short lengths such as computed tomography scanners, magnetic resonance imaging scanners, and wind turbines. As an example, a DDR motor for a telescope drive has a diameter of 2.5 m and a length of less than 50 mm. This motor can produce a continuous torque exceeding 10,000 Nm [1.83]. In large machine tool applications, DDR diameters of 1 m or so are commonly encountered.

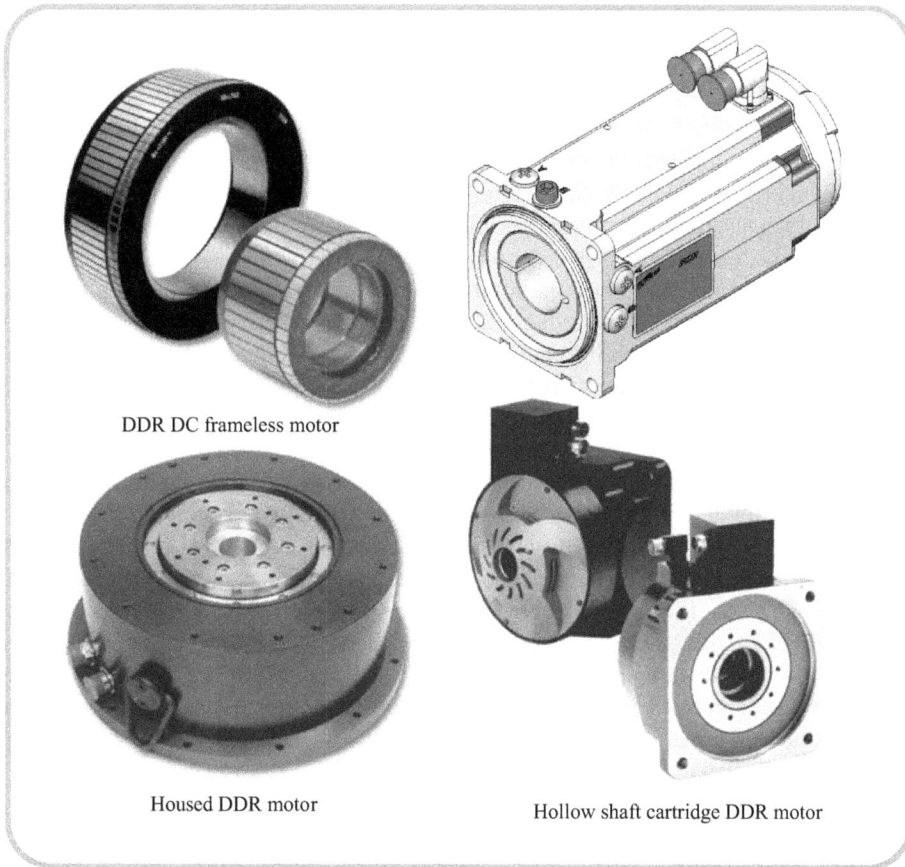

DDR DC frameless motor

Housed DDR motor

Hollow shaft cartridge DDR motor

FIGURE 1.40 Various DDR motors. (Courtesy of Kollmorgen Corporation, Radford, VA.)

The design of DDR motors is mainly influenced by their rotating speed, torque, and power. The achievable power of a motor is approximately proportional to the square of the airgap diameter and the axial length of the motor at the airgap. This indicates that with a doubling of the airgap diameter, the motor power can increase four times. As a matter of fact, direct motors with low rotating speeds usually have a large diameter and a short length. Compared with conventional servo systems, the advantages of direct drive systems are as follows:

- It significantly increases system efficiency by eliminating inefficient gearboxes.
- Because the gearing system in a motor is a primary noise source, the elimination of the gearbox greatly reduces the noise emission from the motor.
- Motor reliability increases due to the reduction in the number of rotating parts.
- The positioning inaccuracy from transmission compliance no longer existed.
- Lower inertia enables fast and accurate positioning.
- A direct drive servomotor is capable of precise revolution control with high resolutions.
- Without speed reduction devices, mechanical backlash, hysteresis, and elasticity in gear transmission systems are eliminated.
- There is no need for gearbox lubrication, and thus it simplifies maintenance.

However, direct drive motors require more precise control systems. Since the direct drive motors need special design, their costs are much higher than other servomotors, but their system costs could be lower than the total installed cost.

1.3.7 Brush and Brushless Motors

Brush motors have a long history of applications since the first electric motor was invented in the late nineteenth century. For a classical DC motor, an electrical power source is connected to the rotor winding through a commutator and brushes. A commutator usually consists of a set of contact bars attached to the rotating shaft of a motor. As the stator windings are sequentially energized to generate a rotating magnetic field, the rotor aligns itself with the magnetic field of the stator. A number of stationary brushes come into contact with the corresponding contact bars of the commutator for reversing the flow of current in the rotor winding. In order to reduce the contact electric resistance at the contacting surface, brushes are pressed against the commutator and thus result in some friction between the brushes and commutator. In addition, at high rotating speeds, brushes become increasingly difficult to maintain reliably in contact with the commutator. Brushes may bounce off from the commutator surface to produce sparks and consequently break the circuit in a very short period of time. Continuous sparking can cause overheat, erode, or even melt the commutator. For this reason, brush motors are restricted to relatively low speeds to avoid brushes excessively bouncing and sparking.

To overcome these drawbacks, brushless motors have been developed to replace brushed motors in some motion control applications. Brushless DC motors typically operate with PM rotors, offering distinct mechanical and electrical advantages over conventional systems. Attachment of PMs on rotor and elimination of rotor coil windings allow significant reductions in rotor inertia and increase in motor acceleration and efficiency. Winding heat can be dissipated directly from the stator into the environment and rotor heating is much lower than that in a brush motor. Generally, compared to brush motors, brushless motors offer higher operation reliability, higher dynamic accuracy, higher efficiency, less product variation, and smaller size but require complex and expensive electronic control systems.

With the widespread use of brushless AC motors in industrial, commercial, and military applications, brushes are gradually exiting the electrical machine market.

1.3.8 Reluctance Motors

Reluctance motors are special motors that use salient pole rotors but without windings and PM on the rotors. There are several types of reluctance motors, with different construction and slightly different functions [1.84]:

- Switched reluctance motor (SRM)—By eliminating windings and PMs from the rotor, this type of motor has inherent advantages over some DC and AC motors, such as simple structure, structure robustness, low cost, and operation in high temperatures. SRMs have been designed mainly for applications in high-power, high-efficiency, variable speed drives, enabling to deliver a wide range of torques. This type of motor requires the closed-loop position control.
- VRM—This type of motor is an evolution of the stepper motor but with less salient poles. The motors are generally designed for use in low-power, open-loop position and speed control systems where efficiency is not of prime importance. In fact, there are similarities between switched and VRMs such as the operating principles, mechanical structures, performing characteristics, and power losses. Both of the types are similar to the brushless PMMs except that the rotors are made from laminated *soft* magnetic materials.
- Synchronous reluctance motor—This type of motor is similar to synchronous AC motors. The rotor has salient poles but the stator has smooth, distributed poles, whereas both the switched and variable motors have salient poles for both the rotor and the stator.

SRMs can be traced back to the invention of Robert Davidson in 1938 [1.85]. However, this type of motor did not find widespread use until the late 1970s due to the difficulty in controlling the machine. For an SRM, there are no windings or PMs on the rotor. When the stator windings are

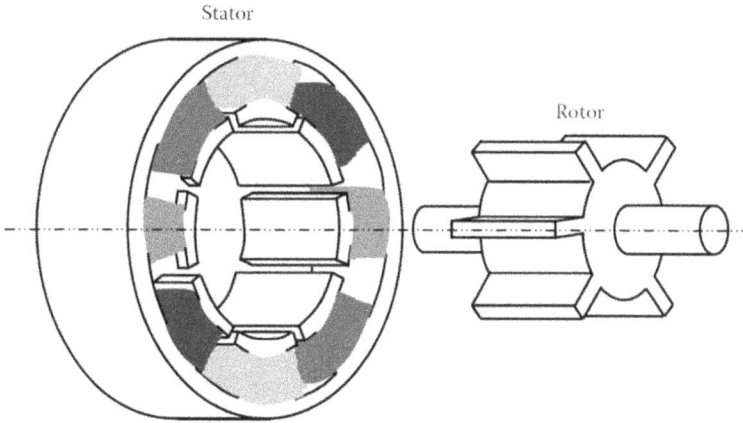

FIGURE 1.41　The salient rotor in an SRM.

powered, the rotor's magnetic reluctance produces a force that attempts to align the rotor with the exited stator poles. The coils in the adjacent slots are powered successively so that the rotor can rotate continuously. This operation principle is based on the difference in magnetic reluctance for magnetic field lines between aligned and unaligned rotor positions when a stator coil is excited; the rotor experiences a force that will pull the rotor to the aligned position [1.86].

Due to the elimination of magnets and rotor windings, SRMs offer several performance, efficiency, and cost advantages. In an SRM, torque is produced by the tendency of its move-able part to move to a position of least reluctance. The SRM has salient poles on both the rotor and the stator, but only the stator poles carry windings. When the stator windings are energized, they create a magnetic field that pulls the nearest pole on the rotor toward it. Consequently, the performance of SRMs is largely a function of the power electronics that control the sequencing of pole energizations. Besides, the SRMs have characteristically high power-to-weight ratios and are well suited for vehicle applications [1.87].

SRMs have been found to offer important advantages over conventional motors in many industrial and commercial applications. In recent years, SRMs have been receiving increasing attention because they can provide similar performance to brushless DC motors without using expensive rare earth magnets. In addition, their robust construction and efficiency over a wide speed range make them well suited to a variety of applications including hybrid and electric vehicles. However, the main drawbacks are its high-torque ripple and consequently high audible noise. As shown in Figure 1.41, the stator of an SRM is similar to a brushless DC motor but the rotor has no magnets or windings attached.

1.3.9　RADIAL FLUX AND AXIAL FLUX MOTORS

The configuration of the electromagnetic field in a motor determines the motor's geometry and structural topology. Electric motors in which the magnetic flux travels in the radial direction are classified as radial flux (RF) motors. This type of motor is cylindrical in shape, and the rotor is typically located inside a stator but can also be placed outside the stator in some special applications, sometimes referred to as *inside out* motors or *outer rotating rotor* motors. In contrast, motors in which the magnetic flux travels in the axial direction are classified as axial flux (AF) motors. This type of motor usually has multiple disk- or pancake-shaped rotors and stators. A typical brushless AF motor is demonstrated in Figure 1.42, showing a stator-rotor-stator structural topology. AF motors offer low axial length and are common in automotive radiator cooling fan applications, table-top medical or lab devices, and specialized elevator machines to name but a few. The structure

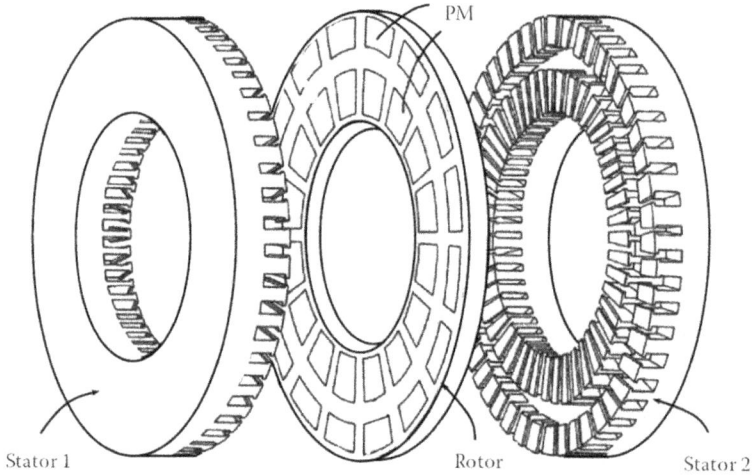

FIGURE 1.42 Brushless AF motor with the stator-rotor-stator structural topology, where PMs mounting on the rotor side surfaces and rotating with the rotor.

FIGURE 1.43 DC brush AF motor, where PMs are mounted on the stators to create an axial magnetic field. (Courtesy of Kollmorgen Corporation, Radford, VA.)

of a DC brush AF motor is shown in Figure 1.43. As can be seen from the figure, the armature is constructed from several layers of copper conductors in a unique flat-disk configuration. Because it adopts ironless design, the thin, low-inertia armature design leads to exceptional torque-to-inertia ratios. The ironless armature enables the motors to deliver more torque over their entire speed range. In fact, the torque is almost constant from 0 to 3,000 rpm, which is not attainable with conventional iron core motor designs. In addition, this type of motor can accelerate from 0 to 3,000 rpm in only 60° of rotation [1.88].

1.3.10 ROTARY AND LINEAR MOTORS

According to the motion pattern of motor, an electric motor can be classified as either rotary or linear motor. As indicated in their names, the rotor in a rotary motor always rotates with respect to

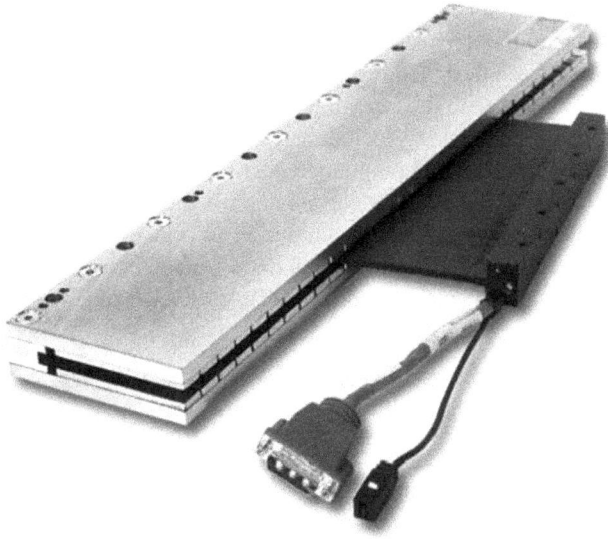

FIGURE 1.44 A linear motor. (Courtesy of Kollmorgen Corporation, Radford, VA.)

its rotating axis and the rotor in a linear motor moves linearly along the flat stator. Actually, a linear motor has the motion of a rotary motor if it was laid out in a flat state.

Linear motors are typically used in the applications that require linear or reciprocating motions (Figure 1.44). A famous example is the high-speed maglev train that was built in 2004 in Shanghai, China, and has been in normal commercial operation thereafter. Similar to rotating motors, linear motors can be grouped as induction or PM linear motors. When very fast linear acceleration or high linear forces are required, linear motors are often employed. A disadvantage is that the mechanical structure to support them is more expensive than rotary motor structures.

1.3.11 OPEN AND ENCLOSED MOTORS

An open motor refers to the motor that has ventilating openings that permit passage of external cooling air to get in the motor for flowing over and around the windings. With the openings on the motor housing, it makes possible for the external objects such as dusts and moisture to enter into the motor, causing motor damage even failure. Therefore, this type of motor cannot work under harsh environments.

An enclosed motor refers to the motor that is completely enclosed to prevent the free exchange of air between inside and outside of the motor housing. In such a configuration, heat is usually dissipated relying on conduction from the heat source (*i.e.*, stator windings) to the motor housing and then from the housing to the environment by convection. This cooling mode is applicable only to small motors with relatively low heat loads. For high-powered motors, a build-in cooling system such as a liquid cooling system or using heat pipes is necessary. This type of motor usually has high International Protection Rating or Ingress Protection (IP) Rating values and thus can be used in harsh and dangerous environments.

1.3.12 HOUSED AND FRAMELESS MOTORS

Housed motors, as a standard type of motor, are most commonly used in industrial applications. They provide customers with an array of options to develop their own automated systems. A housed motor is often delivered as an assembled device, containing all motor components in a single housing. In today's motor market, a wide range of housed motors is commercially available. In addition, modern motor manufactures also provide a variety of customized motors to customers.

Frameless motors have been designed to fit applications where size, weight, and response time are the main concerns. Compared to the type of housed motor, this type of motor offers many benefits to customers in terms of technology, including excellent torque density, enhanced heat dissipation capability, eliminated mechanical compliance, increased bandwidth capability, and flexibility to end users. Integration of a frameless kit motor (stator and rotor only) directly into the customer's bearing allows minimal assembly size and elimination of redundant components. For example, using an integrated frameless motor kit eliminates extra bearings to support the rotor, a separate shaft, and a coupling to engage the motor shaft to the input of a gearing system.

1.3.13 Internal Rotor Motor and External Rotor Motor

A conventional motor constructs a rotor located inside a wound stator. By comparison, an external rotor motor has the opposite construction. The cowl-shaped rotor is placed around the stator and has PMs mounted on the inner surface of the rotor housing. This configuration is especially useful for some applications, such as elevator motors and power tool motors. The external rotor design provides several advantages over the conventional design, including the following points:

- The large airgap radius of an external rotor motor helps maximize the output torque.
- An external rotor has a larger diameter and thus a higher inertia. This would help reduce torque ripple, minimize cogging torque, and stabilize motor operation, especially at low speeds. Furthermore, this type of motor produces lower acoustic noise, which makes them ideal for applications requiring a quiet environment.
- The axial length of an external rotor motor is shorter than that of an internal rotor motor with similar performing characteristics, enabling space-saving and reduced power losses.
- This type of motor is usually compact and lightweight.
- In some applications, this design can greatly simplify the system structure. For example, in an elevator driving system, the integration of the external rotor into a traction sheave to directly drive the elevator cage (also known as *car*) can increase the operation reliability of the driving system.

In an external rotor motor, the shaft may be stationary or rotating, depending on the application. The shaft in an elevator motor is usually stationary and has press fit with the stator. In practice, the elevator often uses a hollow shaft to allow cooling air passing through it and provide a path for motor cables. In a cordless power tool, the rotor is connected to the motor shaft through a rotor cover so that the shaft and the rotor rotate at the same speed. A gear is mounted on the front side of the shaft to drive a gearing system for increasing the output torque and reducing the output speed. For this configuration, bearings are usually placed between the shaft and the stator, allowing free rotation of the shaft.

However, the main drawbacks of an external rotor motor are cooling and mechanical constraints. In this external rotor configuration, the stator winding is located the inner portion of the motor, making it difficult to dissipate heat from the winding to the surrounding environment. In addition, another problem is mechanical constraints. Compared to an internal rotor motor, an external rotor has a larger diameter and thus the peripheral speed becomes much higher with the same motor speed, leading to higher centrifugal force acting on magnets. To avoid magnets from being broken, the rotor speed must be limited within a certain range.

1.3.14 Specialty Electric Motors

Specialty electric motors refer to the motors that are designed for specific applications under certain operational conditions or designed based on different working principles. Some specialty motors operate under very special conditions, such as hazardous environment, vacuum, and deep undersea. Obviously, each specialty motor has its own specific design requirements and applications.

1.3.14.1 Explosion Proof Motor and Flame Proof Motor

In some hazardous environments, special care must be taken in designing and selecting suitable motors for ensuring their reliable and safe operation. For instance, in chemical process, coal mining, textile, metal powder, foundry, papermaking, and petrochemical industries, as well as some fields such as waste disposal and landfills, explosion-proof or flameproof motors must be employed to prevent the ignition of flammable gas/vapor, fibers, flour, or metal powders surrounding them.

The terms explosion-proof and flameproof refer to very similar products that adhere to two different sets of standards. While flameproof is the preferred international term, adhering to IEC regulations, explosion-proof is the term that is commonly used in North America, subjected to American National Standards Institute and Underwriters Laboratories (UL) regulations. Generally, explosion-proof motors have more heavily constructed and undergo more intense testing than flameproof motors.

The most effective approach for preventing an explosion or fire is to design a robust motor enclosure to isolate the motor from the environment. This protection method allows the explosion to occur but confines it to a well-defined motor enclosure, thus avoiding the propagation to the surrounding environment The explosion-proof enclosure must be strong enough to withstand the maximum internal explosion pressure created by an explosion of flammable gases/vapors or combustible powders without rupture or burst. Furthermore, the explosion proof motor should operate at relatively low temperatures. Its frame construction must be designed to maintain the external surface temperature below the minimum ignition temperature.

Explosion-proof motors are rated for use in hazardous locations as defined by UL [1.89] and NEC [1.90]. A detailed overview of hazardous area designations and the motor features required for these areas was presented by von Amerom [1.91]. Flameproof motors are rated by IEC 60079 [1.92].

1.3.14.2 Submersible Motor

Industrial processes often require motors to operate in the presence of water or aqueous fluids. Typical applications include food processing, pharmaceutical manufacturing, underwater construction, wastewater treatment, and municipal systems. In these applications, specially designed and manufactured submersible motors are used to prohibit liquid from entering the motor housing to contact any part of electrical circuit, bearings, and other mechanical parts while providing long life in these harsh applications.

In general, submersible motors often use totally enclosed non-ventilated enclosures or totally enclosed washdown enclosures for isolating motor internal components from the external liquid. Motor cooling depends a great extent upon conduction to transfer heat from the stator windings to the enclosure, and then on natural convection to the liquid in which it is submerged. Considering the convective heat transfer coefficient of water (or other liquids) is much higher than that of air, a submersible motor has a much better heat dissipation capability than standard motors exposed to air.

Hydrostatic pressure increases in proportion to the water depth measured from the surface with an approximate rate of 10 MPa per kilometer. For submersible motors such as sump pumps that submerge in water less than tens of meters, casted aluminum enclosures are fully competent. However, for submersible motors such as undersea robots that dive hundreds or even thousands of meters in deep sea, heavy-duty cast iron or high-strength aluminum alloy enclosures are often adopted for their high compressive strength, deformation resistance, wear resistance, and low damping capacity. These enclosures are precisely machined to tightly fit the stator and other components such as seals. Depending upon the operating condition, the submersible motors can be filled with either air or mineral oil. Air-filled motors are usually used when the hydrostatic pressure is low or moderate. The advantages of air-filled motors include their higher efficiency, easier maintenance, and environmentally friendly design. However, they suffer poor cooling and require frequent maintenance.

For a submersible motor that operates in deep water, mineral oil or synthetic oil is normally filled in the interior of the motor for a number of purposes: (*a*) using the incompressibility of oil to minimize the pressure difference between the interior and exterior of the motor; (*b*) providing continuous lubrication for bearings and other motor components; (*c*) facilitating cooling of the motor during operation; (*d*) offering electrical protection to the motor windings; and (*e*) increasing corrosion resistance to the internal motor components.

To avoid the influence of air trapped in a motor or dissolved gas in oil, vacuum degassing may be necessary during the oil filling process. A simple and cost-effective method to remove air from oil is to place the motor in a vacuum chamber. When the container reaches a certain level of vacuum, oil is filled into the motor by means of gravity. By maintaining a vacuum state for a certain period of time, dissolved gases are removed from oil. To achieve the best results, a bumping technique is recommended, which involves a few cycles of pulling and releasing the vacuum.

However, by comparing with air-filled motors, oil-filled motors may have some disadvantages. They are often accused of being less efficient because of a purported increase in the airgap between the stator and rotor and have higher windage loss due to the higher fluid density (where $P_w \propto \rho$ and $\rho_l \gg \rho_{air}$).

The key factor affecting the life and operation of a submersible motor is sealing. When a motor is immersed in deep water, the motor construction must be designed to withstand high hydrostatic pressure. In addition, the power cable that connects a power supply to a motor needs to be integrated into the motor system robustly. Many manufacturers utilize double seal design to incorporate two separate seals for providing dual protection for submersible motors. The double seal system usually employs a seal chamber or rubber diaphragm for equalizing the possible differential pressure across the seals, minimizing leakage, and providing maintenance-free for years.

In normal operation, oil-filled submersible machines may experience lubricant leakage to some extent. In many cases, oil leakage is quite minor, often without obvious signs. But in other cases, significant oil leakage could cause some severe problems, including oil consumption, environmental pollution, penetration of liquid or moisture into motor, increased operation cost, and in worse cases, motor failure. It is worth noting that in practice a minimal oil leakage to some controlled degree may help reduce shaft friction and improve heat dissipation, leading to a considerable reduction of seal wear.

In general, sealing tends to improve as the fluid pressures increase. This is because the fluid pressure acting on the seal surface compresses the seal axially; the seal is forced more tightly into the gland, improving conformability of the seal with the metal surfaces around it. If the seal is correctly designed, as the fluid pressure increases, sealing force and effectiveness also increase [1.93].

In addition, seals in dynamic systems must resist shear forces resulting from the differential between the pressurized and unpressurized sides of the seal. Shear forces tend to push the seal into the gap between adjacent metal surfaces (known as the clearance gap), and thus the seal material must be sufficiently strong and stiff to prevent them from being broken, damaged, or destroyed.

Submersible motor components (*e.g.*, housing and shaft) usually use corrosion-resistant materials such as stainless steel for dealing with corrosive acidic fluids as well as seawater. Moreover, the submersible motors often use specifically designed gaskets to seal the mounting surfaces. To prevent electrical short circuits, they must utilize heavy winding insulation.

In normal operation, a submersible motor relies on the liquid in which it is submerged to dissipate heat from the frame. Except high-temperature designs, a standard submersible motor operates at the rated ambient temperature (usually, 40°*C*). Under high temperature operating conditions such as in food processing applications, it is necessary to apply an active cooling system, which is integrated into the motor system.

1.3.14.3 Ultrahigh-Speed Motor

Some motors operate at ultrahigh speeds, from hundreds of thousands, up to millions rpm [1.94]. This type of motor is commonly used in gyroscopes, dental drills, micro turbines, air compressors

[1.95], and other applications. Because this type of motor operates at ultrahigh speeds, gearing systems are no longer needed and thus high power density and high efficiency of the motor system can be achieved.

However, the surge in motor speed will inevitably incur a number of design challenges, namely, high stressed rotor caused by great centrifugal forces, rigorous requirements for motor bearings, temperature increase in rotor due to induced high eddy current and friction, and possible thermal instability due to high-frequency losses, among others. In order to ensure secure and reliable operation of motors, all these key design issues must be carefully addressed, as in the following discussions:

- Under an ultrahigh rotational speed, all rotating components of motor are subjected to great centrifugal loading. From Newton's law of motion, the magnitude of the centrifugal force on an object of mass m moving at tangential speed v along a path with a radius of curvature r becomes

$$f_c = ma_c = m\frac{v^2}{r} = mr\omega^2 \tag{1.59}$$

  where a_c is the centrifugal (radial) acceleration of an object in an inertial frame and ω is the angular velocity of the object. This equation indicates that doubling the rotational speed causes four times the centrifugal force. Thus, the rotor and bearings, as well as other rotating components, must be specifically designed with robust constructions to withstand high loads/stresses. Usually, FEA is the most direct and effective way to investigate the stress distribution in an ultrahigh-speed machine.
- Rotor resonance is a serious problem in ultrahigh-speed machines and thus must be carefully considered and avoided. The analysis of rotordynamics is a critical step for ensuring motor stable and safe operation. A rotor in an ultrahigh-speed motor regularly operates in the supercritical region; therefore, issues such as unbalance response and stability of rotation need to be addressed in the earliest design phase to prevent motor vibration and failure. It is worth to note that not only mass unbalance of the rotor can cause resonant vibrations, circulating fluids in bearings or harmonics of unbalanced magnetic pull of motor can also excite resonant vibrations [1.96].
- The selection of suitable bearing type for ultrahigh-speed machines depends not only on their rotational speeds but also on applications, as well as the environmental requirement. Generally, specially designed ball bearings (e.g., precision ceramic ball bearings) are commonly found for rotational speeds up to 100,000 rpm. For dental drill spindles that are subjected to large axial loads but small radial loads, ball bearings can support speeds up to 500,000 rpm. As to turbomachinery that subject to high radial loads, the speed of ball bearing is limited to 160,000 rpm [1.97]. Beyond this limit, contactless bearings such as magnetic bearings and air bearings may be used for eliminating contact friction and bearing wear. Using magnetic bearing technology, the bearing forces are only generated electromagnetically and thus bearing lubrication is no longer needed. Thus, as a unique technology it enables vacuum operation with high rotational speeds. One of the remarkable advantages is that the windage losses are eliminated. Theoretically, it increases lifetime to unlimited runtime, irrespective of the operating speed and the number of start/stop cycles. With this technology, some motors can reach 500,000 rpm and more. However, magnetic bearings require additional volume and lower motor efficiency due to the increased reactive power and energy losses. In addition, the active control of the magnetic bearing is needed for control of the rotordynamics at critical speeds as well as to damp vibrations. Bearing forces are only generated electromagnetically and neither a lubricant nor an atmosphere is needed.

Air bearings utilize a thin film of pressurized air or gas to support loads. There are numinous technical advantages of air bearings over other types of bearings, including near zero friction and wear, high rotating speed, high precision capability, and without need for lubrication. Due to the small airgap (usually in the order of tens of μm), air bearings require flawless geometry with very low tolerances. In general, air bearings have high dynamic stiffness and low viscous drag; both benefit the bearings operating stability.

- Ultrahigh-speed machines often suffer from the high windage power losses because the power losses are proportional to the cube of the rotor speed (*i.e.*, $P_w \propto \omega^3$), leading to the high-temperature rise and possible hot sports occurring inside the machines. It is widely recognized that the temperature rise can increase winding resistance and cause extra copper loss, degrade PM properties, and lower iron core permeability [1.98]. Therefore, thermal analysis must be conducted using CFD software to ensure the motor temperature not exceeding the allowable value.

- An additional challenge lies in the control of high-speed motors. Due to limitations of the processing powers of microcontrollers, ultrahigh-speed machines are usually controlled in open loop [1.99], which is considered unstable after exceeding a certain speed [1.100]. Thus, special control algorithms are required that would ensure stability without a great computational burden.

1.3.14.4 Motor Operating under Vacuum Environment

Motors that operate under vacuum environmental conditions need to be specially designed. This type of motor has been developed to avoid problems of outgassing and contamination in many material processing applications, for instance, semiconductor wafer fabrication processes. Other applications of vacuum motors are found for space-related systems such as spacecrafts, space vehicles, and satellites.

The key design challenge for vacuum motors is how to cool them effectively. In a regular ambient environment, a motor is typically cooled by natural or forced convection of air. However, in a vacuum environment, convective heat transfer is no longer possible due to the lack of medium such as air or water. The motor can dissipate heat only through radiation. Given that thermal radiation is the relatively weak mode of heat transfer for most motor applications, it is necessary to adopt indirect cooling techniques in a vacuum. One example is to use liquid-cooled cold plates or vapor-cooled cooling jackets to attach to the motor housing.

Moreover, under vacuum conditions, motor components are subjected to increased outgassing, causing the emitted material to deposit on the surrounding component surfaces. Therefore, it is important to use low outgassing motor materials and lubricants in vacuum. The low outgassing metal materials include stainless steel, aluminum alloys, and oxygen-free copper. Several specially developed vacuum greases with low outgassing properties are the common choice for vacuum systems.

1.3.14.5 Motor Operating under Nuclear Radiation Environment

Nuclear radiation, also known as ionizing radiation, is defined as energy transmission through certain kinds of ionizing particles and photons during nuclear reactions. Nuclear radiation includes α-rays, β-rays, x-rays, and the more energetic portion of the electromagnetic spectrum. Electric servomotors and drives used in nuclear power plants are subjected to nuclear radiation, high temperature, and high humidity. These severe environmental conditions can degrade the performance of the servo system in different ways: (*a*) The radiation with high-energy particles can demagnetize PMs in the motor [1.101, 1.102]. (*b*) Ionizing radiation can break down materials within electrical equipment (*e.g.*, motors switches, incandescent lights, wiring, and solenoids). For instance, when wiring is exposed to γ-rays, no change is noticed until the wiring is flexed or bent. The wire's insulation becomes brittle and may cause short circuits in the equipment [1.103]. (*c*) The gamma and neutron radiation can cause extensive damage to integrated circuit devices [1.104]. (*d*) Aging mechanisms could significantly affect electric motors/components. Under a nuclear radiation condition, aging effects are most commonly due to radiation exposure and heat, as well as other phenomena such as mechanical vibration and chemical degradation.

For safe operation, motors in a nuclear reactor must be carefully designed and selected. For instance, the motor insulation material and bearing lubrication needs to be selected carefully to withstand the nuclear radiation without aging. In order to create an environment protected from radiation hazards, specifically designed radiation shield systems may be required and applied to electric motors, drives, and cables.

1.3.14.6   Piezoelectric Motor

The piezoelectric effect was first demonstrated by French physicists Jacques Curie and Pierre Curie in 1880 [1.105] that electrical charge could accumulate in certain materials (*e.g.*, crystals and poly-crystalline ceramics, known as piezoelectric materials) in response to applied mechanical stress (squeezed or pressed). Oppositely, when the same material is subject to mechanical stress, it generates an electrical voltage, which is known as the inverse piezoelectric effect, discovered by Gabriel Lippmann in 1881 [1.106]. This suggests that piezoelectric materials allow direct energy conversion between the mechanical form and the electrical form. This feature is especially useful in developing various piezoelectric sensors, transducers, actuators, and other piezoelectric devices. Today, due to their peculiar properties, piezoelectric materials have found extensive applications in the fields of energy conversion, robotics, medicine, biomechanics, aerospace, electrical vehicles, and many others.

Piezoelectric motors and actuators are electro-mechanical drive systems using the inverse piezo-electric effect, where microscopically small oscillatory motions are converted into continuous or stepping rotary or linear motions. They can produce constant force in a static condition without energy consumption due to the capacitive energy transformation principle. Compared to regular PMMs, they have high torque density, high holding torque at zero input power, low rotor inertia, rapid start and stop characteristics, and high bandwidth sufficient for most machines. Moreover, electromagnetic motors typically have high speed but low torque. To produce useful torque, they are often coupled with gearing systems to reduce the speed and increase the output torque, thus adding more weight and reducing motor efficiency. As a comparison, piezoelectric motors can be designed with high torque and low speed without the need for gears.

However, the displacement generated by most piezoelectric materials is rather small, usually on the order of 0.1%–0.2% to its original length [1.107]. To overcome this limitation, it is often to employ stacked piezoelectric elements with certain compliment mechanisms and/or displacement amplification mechanisms to multiple the displacements.

Piezoelectric motors are especially suited for direct drive in small or miniature sizes. By comparison, standard electromagnetic motors smaller than 1 cm^3 are very difficult to fabricate using the conventional manufacturing methods. Moreover, the superior properties of piezoelectric motors, such as high precision, fast responding, compact structure, magnetic insensitive, and vacuum compatible, make them excellent candidates in precise and miniature instruments (*e.g.*, digital cameras, micro-robots, medical equipment, and ultra-precision stages) and those operating under severe environments (strong magnetic field, vacuum, cryogenic temperature, *etc.*).

According to the different drives and functional principles for generating unlimited rotary or linear movement, piezoelectric motors can be briefly categorized into three groups: resonance-drive (ultrasonic motors, USMs), inertia-drive, and piezo-walk drive. Among them, USMs are mostly used in many industrial, military, and medical applications for their attractive properties, such as micro-positioning capability, fast response, high torque at low speed, compact structure, quiet operation, absence of electromagnetic noise, and high integration capability into application. This type of piezoelectric motor converts oscillatory piezoelectric motion at their resonant frequencies in the ultrasonic range (usually 20 kHz – 10 MHz) into continuous rotary or linear motion. In addition, another advantage of USMs over conventional electromagnetic motors, with expensive copper coils, is the improved availability of piezoelectric ceramics at a reasonable cost [1.108].

USMs can be categorized in a number of different ways based on (*a*) the number of mechanical resonance modes (single-mode or multi-mode) that are excited at the operating frequency of the vibration element; (*b*) the number of vibrating elements; (*c*) the number of driving signals applied to a vibrating element; (*d*) the type of generated motion (oblique or elliptical motion); (*e*) the type of wave generation (standing wave or propagating/traveling wave); (*f*) full or partial drive of a piezoelectric element (applied to both single-mode and multi-mode excitation types); and (*g*) unidirectional or bidirectional motion.

On a stator, if only one mechanical resonance mode at the motor operating frequency is excited, then this motor is referred to as a single-mode excitation type. If more than one mechanical resonance mode is excited, then the motor is referred to as a multi-excitation type. An oblique motion is generated from single-mode and an elliptical motion is generated by exciting two orthogonal mechanical resonance modes simultaneously at the same or similar operating frequencies [1.109]. The generation of standing or traveling wave depends on the contact mechanism at the interface between the stator and the rotor. In standing wave motors, the contact between the stator and rotor is intermittent (sometimes the standing-wave type is referred to as a woodpecker type). Although this type of motor is subject to intermittent torque, due to the inertia of rotor and high excitation frequency, the speed ripple is very small. In general, the use of standing waves leads to a simpler mechanical structure of motor and more basic driving circuitry, but limited output power [1.110]. By comparison, in traveling wave motors, the stator and rotor contact continuously and the contact area moves along with the wave crest of the stator (sometimes the traveling-wave type is referred to as a surfing type). This type of motor requires two piezoelectric vibration sources to generate one travel wave, leading to low efficiency (usually less than 50%) and complex structure, but it can generate a relatively larger power output.

A typical standing wave, single-mode USM was developed by Physik Instrumente [1.111, 1.112]. The stator of this USM consists of a piezoelectric hollow cylinder with pushers attached to the cylinder face for contacting with the rotor. Segmented electrodes are arranged on the outer surface of the cylinder, evenly divided into 10 sections. Combining every other electrode into a group, thus it forms two groups of electrodes, market as the first and second groups, respectively. At any time, only one electrode group is active. Each pusher is located at the junction of a pair of electrodes, separated by a pair of electrodes. In such a way, the electrodes allow for excitation of the stator in a special Eigen mode (coupled with tangential axial mode), in which the pushers move back and forth at an angle of 135° or 45° with respect to the cylinder face, as shown in Figure 1.45. This configuration allows the direction of motor rotation to be readily changed by changing the activation of a corresponding electrode group.

By applying a single-phase sine wave voltage on one of the electrode groups and keeping another electrode group inactive, an oblique motion on friction tips (*i.e.*, pushers) is obtained. The rotor is pressed against the pushers by means of a preloading force. Then due to the oscillations of the stator when excited at the resonance frequency, the pushers impart micro-impulses to the rotor, making the rotor rotate [1.113, 1.114].

More recently, Ryndzionek *et al.* [1.115] proposed a multicell piezoelectric motor (MPM) using three rotating-mode actuators. The motor consists of three individual cells that are integrated into the stator, two rotors, a shaft, and a preload system. Two rotors locate at the two ends of the stator, forming a symmetrical sandwich structure. The stator consists of two pairs of piezoceramics and counter-masses with three rotating-mode actuators. A proper mix of the performance of the three rotating-mode actuators will generate three traveling waves. The rotational motion of the rotor is transmitted to the shaft by Smalley springs and two end plates (Figure 1.46).

A rotary piezoelectric motor using three bending transducers was proposed by Liu *et al.* [1.116]. These piezoelectric transducers are used to drive a disk-shaped rotor together by the elliptical movements of their driving tips; these motions are produced by the hybrid of two first bending vibration

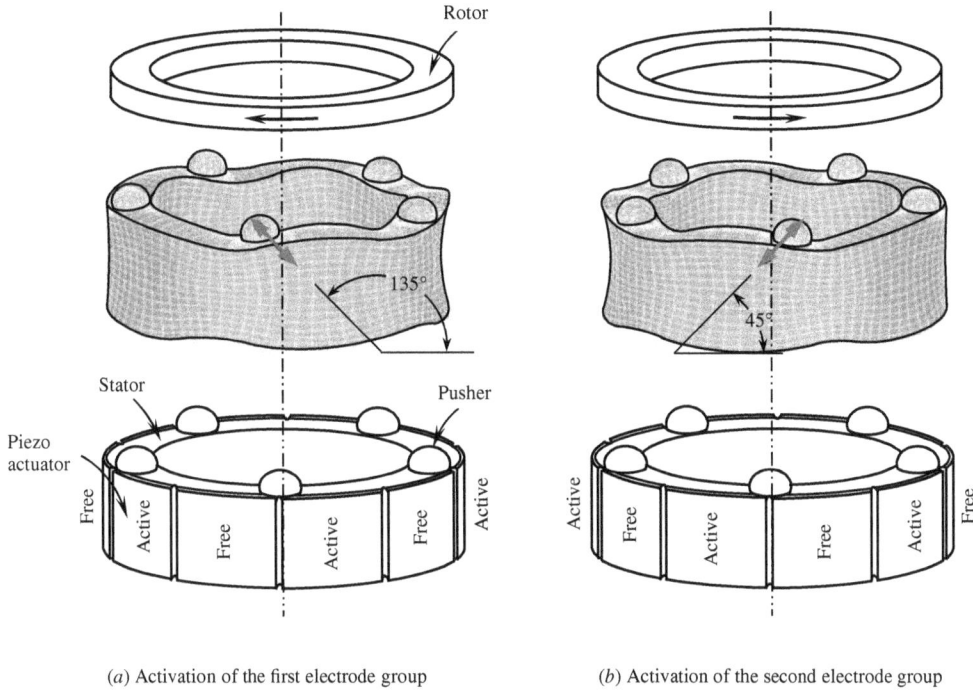

(*a*) Activation of the first electrode group   (*b*) Activation of the second electrode group

FIGURE 1.45 Operating principle of a single-mode excitation type USM with a cylindrical piezoelectric element using tangential-axial standing waves [1.110], showing the arrangement of the pushers, electrode groups, and the resulting deformation of the cylinder for excitation of different electrode groups. The motor direction, as well as the direction of the oblique motion, is controlled by changing the driven part.

modes (Figure 1.47). Each piezoelectric transducer has a simple structure as it only contains an aluminum alloy beam and four pieces of lead zirconate titanate plates. Under a working frequency of 53.2 kHz, the maximum no-load speed and the maximum torque of the prototype are obtained as 53.3 rpm and 27×10^{-3} Nm, respectively.

Similarly, a rotary USM based on the use of traveling waves has been designed, fabricated, and tested [1.117]. In this motor, four sandwich-type transducers are connected by thin beams as the stator, and two orthogonal longitudinal vibration modes are actuated to generate elliptical trajectories of teeth and to drive the rotor. The tested results from a prototype motor have shown the maximal no-load speed of 157.9 rpm and the maximum output torque of 11.76×10^{-3} Nm, at an exciting voltage with an amplitude of 300 p-p and a frequency of 50.93 kHz.

1.3.15 MOTOR CLASSIFICATION ACCORDING TO POWER RATING

According to the motor nominal power, electric motors can be briefly divided into five categories:

- Micromotors are electric motors with a rated output power of 0.05 hp (<37 W) or less. Micromotors have been used across a wide range of commercial and industrial applications for light duty, especially in microelectronics, computer, and precision instrument industries.
- Small motors are larger than 0.05 hp but less than 1 hp (37 W–746 W). Their applications focus primarily on power hand tools, appliances, medical devices, small fans, optical devices, electrical cars, precise motion control systems, and other small machinery.

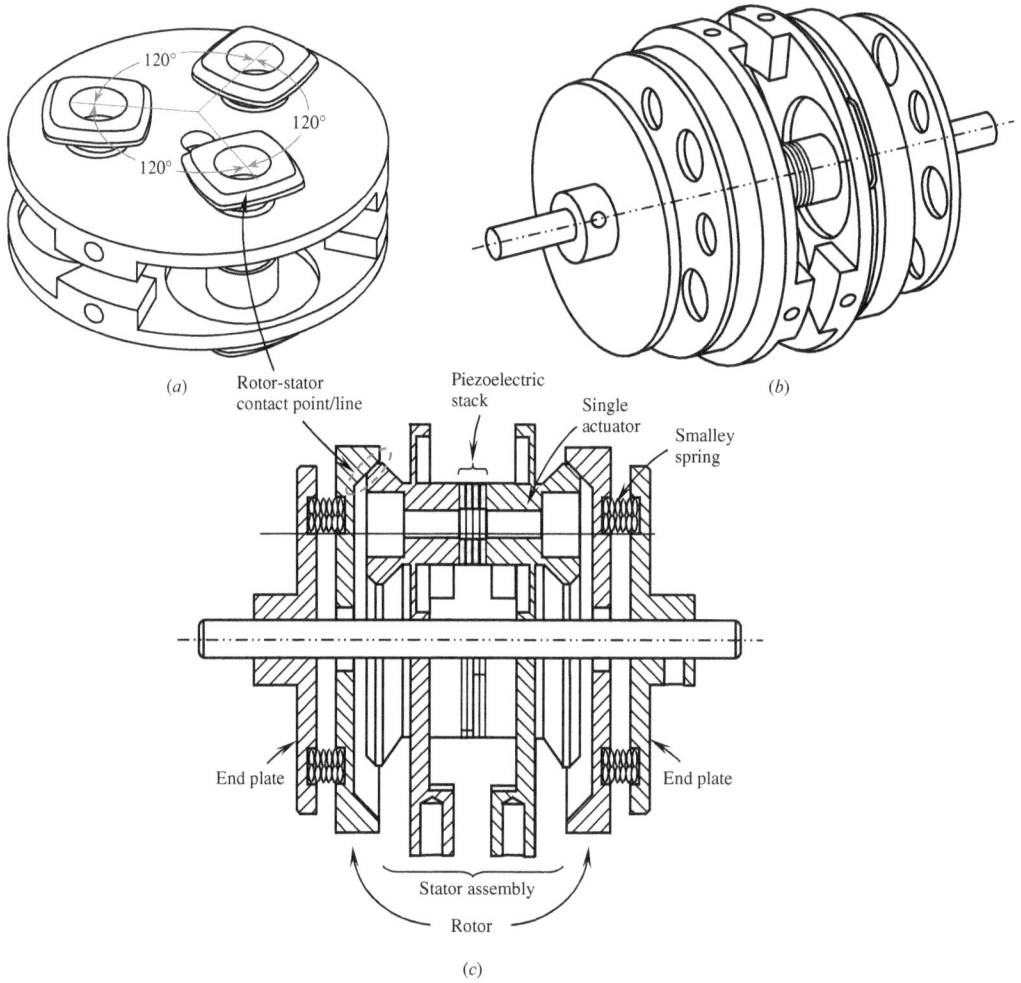

FIGURE 1.46 Configuration of MPM: (*a*) stator structure, (*b*) motor configuration, and (*c*) motor assembly.

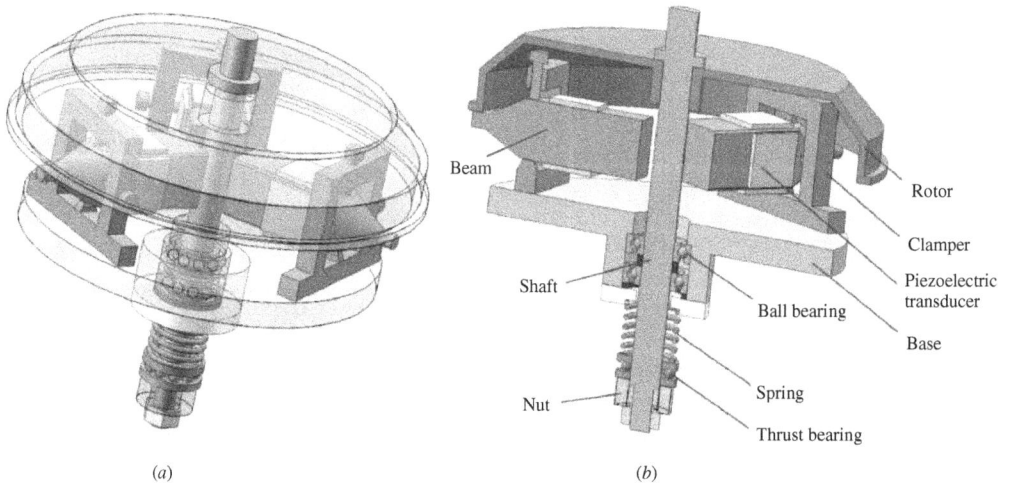

FIGURE 1.47 Structure of the piezoelectric motor: (*a*) three-dimensional model and (*b*) cut-away view [1.116].

- Medium motors are considered to be in the range of 1–100 hp (746 W–74.6 kW). The majority of medium motors are used in various industrial applications such as industrial fans, pumps, motion-control systems, machine tools, and vehicles.
- Large motors occupy the power range of 100–1,000 hp (74.6 kW–746 kW). They are normally designed for use in medium-duty applications as in elevators, large vehicles, industrial blowers/fans, printing machines, package machines, air compressors, and other industrial large machines.
- Extra-large motors range from 1,000 to 10,000 hp (746 kW–7.46 MW), which are usually used in heavy-duty applications, such as large rolling mills, large machine tools, high-speed trains, skyway elevators, ship propulsion, and mining machinery.
- Ultra-large motors are considered to be larger than 10,000 hp (>7.46 MW). NASA used a motor that is rated 135,000 hp for a wind tunnel. In addition, the industry's largest players such as GE, Siemens, TECO-Westinghouse, ABB, and Toshiba have developed their own ultra-large motors.

It is worth to note that the range for each motor category shown earlier has not been clearly defined by international, national, or professional standards yet.

1.4 MOTOR DESIGN AND OPERATION PARAMETERS

In designing, selecting, and repairing electric motors, it often involves a number of constants that reveal the relationships of torque, current, speed, voltage, power loss, and other operating characteristics. Among these constants, the back EMF constant and torque constant are the two most important constants for evaluating motor performance.

1.4.1 BACK EMF CONSTANT, K_E

The induced voltage in motor conductors under a rotating magnetic field is defined as the back EMF V_e and is directly proportional to the angular velocity of the motor. Thus, the proportionality constant, referred to back EMF constant, is defined as the ratio of back EMF to motor rotational speed, in the unit of V/(rad/s) or V/rpm (in some applications, V/krpm):

$$K_e = \frac{V_e}{\omega} \tag{1.60}$$

Thus, K_e is a measure of how many volts per rpm the motor would produce if driven as a generator. It can also be used to determine how fast a motor will run with a given voltage applied to it.

1.4.2 TORQUE CONSTANT, K_T

K_t is called torque constant (Nm/A) or torque sensitivity, which is defined as the torque T (in Nm) generated by a motor to the motor input current I (in A), that is,

$$K_t = \frac{T}{I} \tag{1.61}$$

For a sinusoidal commutated motor, the root-mean-square (*rms*) current i_{rms} should be used to replace I in the aforementioned equation, where i_{rms} is related to the peak AC current i_p as

$$i_{rms} = \frac{i_p}{\sqrt{2}} \tag{1.62}$$

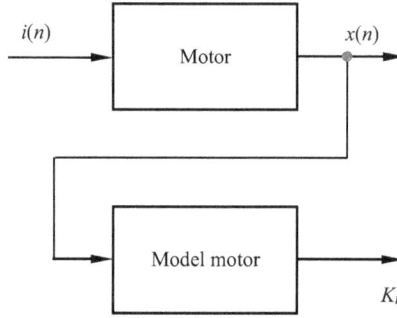

FIGURE 1.48 The block diagram of a conventional circuit for estimating motor torque constant.

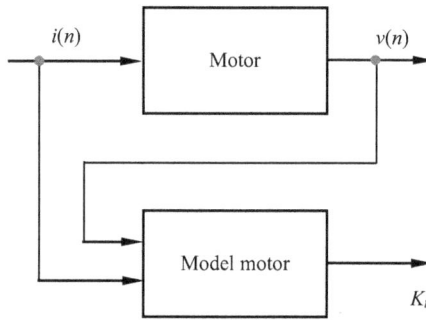

FIGURE 1.49 The block diagram of the proposed new circuit for estimating the motor torque constant.

Figure 1.48 presents the block diagram of conventional circuit for estimating the motor torque constant. This conventional engineering method assumes that when the motor is accelerated, the motor torque constant is proportional to a position change as the output of the motor, regardless of the input current of the motor. However, the motor torque constant actually depends on the input current during the acceleration of the motor. Hence, the conventional method cannot provide an accurate result for the motor torque constant. Based on a new block diagram shown in Figure 1.49, a new method was proposed to calculate motor torque constant by a multiplicity of measured current and speed values [1.118]:

$$K_t = \frac{\sum_1^n i(k-1)[v(k)-v(k-1)]}{\sum_1^n i(k-1)^2} \qquad (1.63)$$

where
  $i(k)$ and $v(k)$, respectively, indicate the current and the speed of the motor at a sampling time k
  n indicates a natural number greater than 1

1.4.3 VELOCITY CONSTANT, K_v

K_v is the motor velocity constant, measured in rpm or rad/s per volt. In fact, K_v is the reciprocal of K_e. For brushless motors, K_v is the ratio of the motor's unloaded rpm or rad/s to the peak voltage. This parameter is used for selecting the winding in the motor.

It can be shown that if the torque constant K_t is in Nm/A and the back EMF constant K_e in V/(rad/s), then

$$K_t = K_e = \frac{1}{K_v} \tag{1.64}$$

1.4.4 MOTOR CONSTANT, K_M

K_m is the motor constant, defined as the torque constant K_t divided by the square root of the resistive power loss P_r:

$$K_m = \frac{K_t}{\sqrt{P_r}} \tag{1.65}$$

- For DC motors,

$$P_r = I^2 R_{DC} \tag{1.66a}$$

- For three-phase AC motors,

$$P_r = i_{rms}^2 R_{AC} = \frac{3 r_p^2 R_{AC}}{2} \tag{1.66b}$$

where R_{DC} and R_{AC} are the winding resistances for DC and AC motors, respectively.
Substituting (1.66a) into (1.65) yields

$$K_m = \frac{K_t}{I\sqrt{R_{DC}}} = \frac{T}{I^2 \sqrt{R_{DC}}} \tag{1.67}$$

Since $P_{in} - P_{out} = \sum P_{loss} > I^2 R_{DC}$, it follows that

$$\sqrt{P_{in} - P_{out}} > I\sqrt{R_{DC}} \tag{1.68}$$

Therefore, the minimum motor constant is obtained as

$$K_{m,\min} = \frac{T}{I\sqrt{P_{in} - P_{out}}} = \frac{K_t}{\sqrt{P_{in} - P_{out}}} \tag{1.69}$$

Similarly, substituting P_r for the corrected 3-phase AC motor and thus obtaining another K_m equation as

$$K_m = \frac{K_t}{1.225 i_p \sqrt{R_{AC}}} \tag{1.70}$$

In fact, K_m represents the ability of a motor to convert electrical power into mechanical power. A motor with a higher value of K_m can generate torque more efficiently. K_m is winding independent and is used as a rating comparison factor in selecting the motor size in various motor applications. In some references, K_m is also called motor size constant.

1.4.5 MECHANICAL TIME CONSTANT, τ_M

In design and analysis of servomotors, it often deals with two kinds of time constant: mechanical time constant τ_m and electrical time constant τ_e. A servomotor's dynamic motion response is controlled by these two time constants. Usually, τ_m and τ_e are listed in the servomotor specifications. However, it should be cautioned that these two time constants in the specifications are for the motor alone with no load inertia connected to the motor shaft. It is important to understand the impact of actual load conditions on the time constants [1.119].

The mechanical time constant τ_m (in unit of seconds) is usually defined as the ratio of motor moment of inertia to the damping factor with a zero-impedance power source:

$$\tau_m = \frac{RJ_m}{K_t K_e} \tag{1.71}$$

where
  R is the motor winding resistance
  J_m is the motor moment of inertia

To take into account the impact of load on τ_m, the previously mentioned definition may be modified as

$$\tau_m = \frac{R_t J_t}{K_t K_e} \tag{1.72}$$

where
  R_t is the total motor winding resistance of all phases plus the external circuit resistance and the total inertia
  J_t is the sum of the motor inertia and the reflected inertia from the load to the motor shaft

According to NEMA [1.75], the mechanical time constant refers to the time taken by an unloaded motor to reach $(1 - 1/e)$ (where $1 - 1/e \approx 63.2\%$) of its maximum rated speed in a no-load condition. τ_m can be measured by applying a constant voltage to the motor, then measuring the velocity, and determining the time it takes to reach 63.2% of the maximum rated speed.

One important factor to affect the mechanical time constant is temperature. During a normal operation process, the motor temperature rises from the ambient time (usually $40°C$) to its normal operation temperature that is below the allowable maximum value. The electric resistance of the winding changes accordingly as

$$R(T) = R(T_o)\left[1 + \alpha(T - T_o)\right] \tag{1.73}$$

where α is the temperature coefficient of resistance ($°C^{-1}$ or K^{-1}).

Similarly, for PMMs, the back EMF constant K_e and torque constant K_t are also affected by temperature. Both K_e and K_t have the same functional dependence on the motor's airgap magnetic flux density that is produced by the motor's magnets. All PMMs are subject to both reversible and irreversible demagnetization. It has been reported [1.120] that within the temperature range of $-60°C$ to $200°C$, all four types of PMs (alnico, Sm-Co, Nd-Fe-B, and ferrite/ceramic magnets) exhibit linear, reversible thermal reduction in field strength such that the amount of magnetic flux density produced by each magnet decreases linearly with increasing magnet temperature. Hence, the expression for the reversible decrease in both $K_e(T)$ and $K_t(T)$ with increasing magnet temperature is given as

$$K_e(T) = K_e(T_o)\left[1 + \alpha(T - T_o)\right] \tag{1.74a}$$

$$K_t(T) = K_t(T_o)\left[1 + \alpha(T - T_o)\right] \tag{1.74b}$$

where the temperature coefficient of resistance α takes different values according to the different types of PM:

$$\alpha(\text{Alnico}) = 0.0001 / {}^\circ C$$

$$\alpha(\text{Sm-Co}) = 0.00035 / {}^\circ C$$

$$\alpha(\text{Nd-Fe-B}) = 0.001 / {}^\circ C$$

$$\alpha(\text{Ferrite}) = 0.002 / {}^\circ C$$

Thus, a $100^\circ C$ rise in magnet temperature causes a reversible decrease in K_e and K_t as -1% for alnico, -3.5% for SmCo, -10% for NdFeB, and -20% for ferrite or ceramic magnets. This indicates that the motor operation temperature has the strongest impact on ferrite magnets.

By taking into account both load and temperature, the mechanical time constant becomes

$$\tau_m(T) = \frac{R_t(T)J_t}{K_t(T)K_e(T)} \tag{1.75}$$

Taking the ambient temperature of $40^\circ C$ ($104^\circ F$), the normalized mechanical time constant ratio, $\tau_m(T)/\tau_m(40^\circ C)$, for each type of PMs is plotted in Figure 1.50 as a function of temperature. The increase in the mechanical time constant ratio has resulted from the combined effects of increasing electrical resistance and thermal demagnetization. It can be seen from the figure that the largest increase occurs in ferrite magnets with the mechanical time constant ratio increasing by a factor of 2.45 at $155^\circ C$. At the same temperature, it increases by a factor of 1.85, 1.57, and 1.49 for Nd-Fe-B, Sm-Co, and alnico magnets, respectively. These results have confirmed that the operation temperature has a strong impact on servomotor's dynamic motion response. Therefore, ignoring the temperature effect can lead to erroneous predictions of the servomotor's dynamic motion response to the applied voltage command. Rare earth and ferrite/ceramic magnets have increased in magnetic flux at cold temperatures. Rare earth magnets have improved resistance to demagnetization at cold temperatures, whereas ferrite/ceramic magnets have an increase in demagnetization risk at low temperatures.

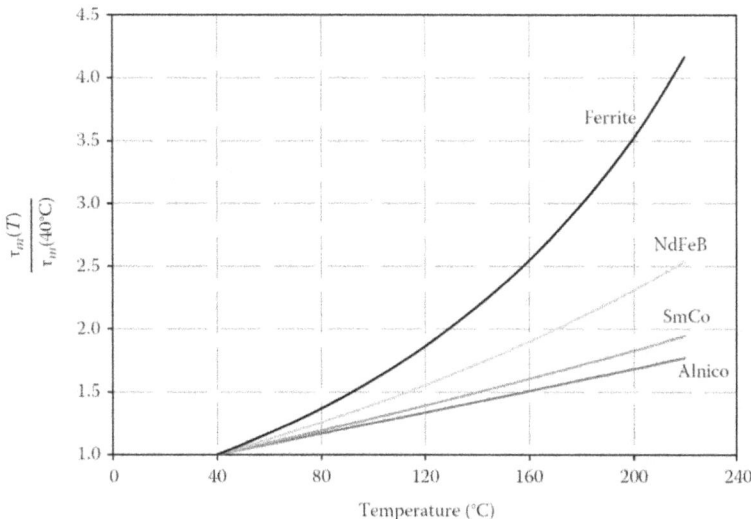

FIGURE 1.50 The temperature influence on mechanical time constant ratio for four types of PMs, with the ambient temperature of $40^\circ C$.

1.4.6 ELECTRICAL TIME CONSTANT, τ_E

According to NEMA [1.75], the electrical time constant of a servomotor is the time required for the current to reach $(1 - 1/e)$ (*i.e.*, 63.2%) of its final value after a zero source impedance, and the stepped input voltage is applied to the motor that maintained in the locked rotor or stalled condition (*i.e.*, $\omega = 0$). A small electrical time constant indicates the high dynamic response of a servomotor.

Mathematically, the electrical time constant τ_e is defined as the ratio of armature inductance L to its winding electric resistance R:

$$\tau_e = \frac{L}{R} \tag{1.76}$$

Similarly, by taking into account the load and temperature effect, the equation becomes

$$\tau_e(T) = \frac{L}{R_t(T)} \tag{1.77}$$

where $R_t(T)$ is the total resistance, including the resistance of all phase windings and the external circuit resistance, at the temperature T. This equation indicates that with the increasing temperature, electrical time constant τ_e decreases due to the increase in $R_t(T)$.

1.4.7 THERMAL TIME CONSTANT, τ_{TH}

The thermal time constant τ_{th} is an indicator of the heat capacity of a motor, showing how fast or how slow the generated heat can be built up in the motor and can be effectively dissipated to the environment. In other words, it is a measure of how long it takes a motor to reach thermal equilibrium.

The thermal time constant τ_{th} can be expressed as

$$\tau_{th} = \frac{\rho c_p V_e}{h A_s} \tag{1.78}$$

where
ρ is the density
c_p is the specific heat
V_e is the effective motor volume (this is because a motor is not a solid)
h is the convective heat transfer coefficient
A_s is the motor outer surface area

This equation indicates that larger mass m ($m = pV_e$) and specific heat c_p lead to larger τ_{th}, that is, larger heat storage capability and slower changes in temperature. Higher heat transfer coefficient h and larger motor outer surface area A_s help dissipate the generated heat quickly from the motor to the ambient, leading to smaller τ_{th}, that is, faster changes in temperature. Obviously, a lower value of the thermal time constant is highly desired.

The motor thermal time constant may be used to estimate the temperature rise of a motor. The motor temperature rise AT can be predicted as [1.121]

$$\Delta T = T - T_a = P_{loss} R_{th} + \left(T_a + T_i - P_{loss} R_{th} \right) e^{-t/\tau_{th}} \tag{1.79}$$

where
T_a is the ambient temperature
T_i is the motor initial temperature at the start of operation
P_{loss} is the motor total power losses that must be dissipated from the motor
R_{th} is the overall motor-to-ambient thermal resistance
t is the motor operation time

If ΔT is specified, the aforementioned equation can be used to determine how long the motor could reach the specified value of temperature rise.

Mathematically, as t approaches infinity (*i.e.*, $e^{-t/\tau_{th}}$ approaches zero), the system attains thermal stability and the motor temperature reaches its final temperature T_f. Hence, the temperature rise ΔT becomes

$$\Delta T = T - T_a = P_{loss} R_{th} \tag{1.80}$$

The variation of the temperature ratio T/T_f with respect to motor operation time is plotted in Figure 1.51. At $t = 0$, the power is supplied to the motor and the motor temperature starts to rise exponentially until it reaches its final temperature at the thermal equilibrium. According to the engineering definition, the thermal time constant is the time required to reach 63.2% of the final temperature.

It is worth to note that in practice, it has been found that the thermal time constant of the motor is not constant. It varies correspondingly to the change in heat load rates. For example, motors running in intermittent operation will have a shorter time constant than a motor running up to temperature in a continuous load operation. This is due to the definition of τ_{th} that assumes no internal heat generation and uniform heat distribution. Both are not purely true in motor operation.

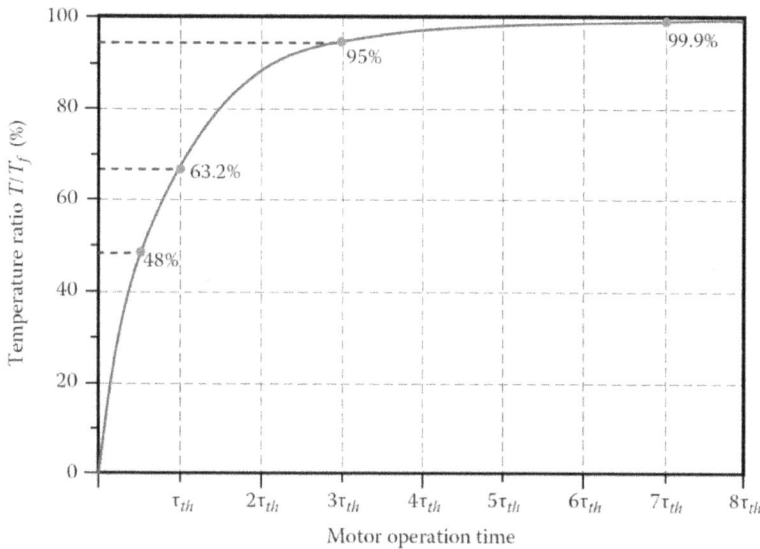

FIGURE 1.51 Variation of temperature ratio T/T_f as a function of motor operation time for constant applied power.

1.4.8 VISCOUS DAMPING, K_{VD}

In general, damping can be divided into three types: viscous damping, coulomb or dry-friction damping, and hysteretic or structural damping. From the standpoint of physics, viscous damping is the dissipation of energy as occurred in liquid or air between moving parts. An example of viscous damping is ball-bearing lubrication. It results in lower torque delivered at the output shaft to the torque developed at the rotor. Viscous damping in a single-degree-of-freedom torsional system is referred to as torsional viscous damping $K_{vd,t}$, which is directly proportional to the damping torque T_d and inversely proportional to the angular velocity ω and is always opposite to the direction of motion, that is,

$$K_{vd,t} = \frac{T_d}{\omega} \tag{1.81}$$

The unit of viscous damping is Nm/rpm or Nm/(rad/s).

For reciprocating or linear motion systems, viscous damping $K_{vd,r}$ is defined as the ratio of the damping force F_d over the velocity $\dot{x}(t)$ with the unit of N-s/m:

$$K_{vd,r} = \frac{F_d}{\dot{x}(t)} \tag{1.82}$$

Another example of viscous damping is iron losses in a motor that are functions of frequency, circulating currents, and iron mass.

1.5   SIZING EQUATIONS

The determination of appropriate motor size is an essential task in motor design. An oversized motor can provide itself a higher safety factor in stress and structural firmness but waste energy and can potentially create performance problems with the driven equipment, especially in turbomachinery such as fans or pumps. In some circumstances, an oversized motor may compromise the reliability of both the components and the entire system.

Empirical sizing equations are very useful for motor engineers to preliminarily determine the motor design at the early stage of motor design. Various sizing equations have been developed by different researchers and designers in the history of motor development. A set of sizing equations was proposed by Honsinger [1.122] for induction machines. This method focuses on the optimal electrical loading and the machine internal geometry for a given power level:

$$\frac{P}{n_s} = \xi_s D_{s,o}^3 L_e \tag{1.83}$$

$$\frac{P}{n_s} = \xi_r D_{r,o}^2 L_e \tag{1.84}$$

Based on these two sizing equations, the $D^{2.5}L_e$ sizing equations can be derived as

$$\frac{P_o}{n_s} = \xi_s' D_{s,o}^{2.5} L_e \tag{1.85}$$

$$\frac{P_o}{n_s} = \xi_r' D_{r,o}^{2.5} L_e \tag{1.86}$$

where
  $D_{s,o}$ and $D_{r,o}$ are the OD of stator and rotor, respectively
  P_o is the output power
  n_s is the rotor rotating speed

L_e is the effective stack length and ξ_s, ξ_s', and ξ_r, ξ_r' are coefficients for stator and rotor, respectively. These coefficients contain the information regarding motor structure.

However, traditional sizing equations are based on the premise that the excitation of the electrical machine is provided by a sinusoidal voltage source to produce a sinusoidal EMF. In order to eliminate the deficiencies of traditional sizing equations, scientists at the University of Wisconsin [1.123–1.125] have developed a general-purpose sizing equation that could take into account different waveforms of back EMF and machine characteristics. A particular effort has been made to express the sizing equation that characterizes the output power P_o as a function of overall volume of the machine. This sizing equation is easily adjustable for different motor topologies, such as RF, AF, and transverse-flux motors. The general-purpose sizing equation takes the form of

$$P_o = \frac{1}{1+K_\phi}\frac{m}{m_1}\frac{\pi}{2}K_e K_i K_p \eta B_g A \frac{f}{p}\lambda_o^2 d_{s,o}^2 L_e \tag{1.87}$$

where

$K_\phi = A_r/A_s$ is the ratio of electrical loading on rotor and stator

$A = A_r + A_s$ is the total electrical loading

m is the number of phases of the motor

m_1 is the number of phases of each stator (if there is more than one stator, each stator has the same)

K_e is the EMF factor

K_i is the current waveform factor, defined as the peak phase current I_p to the *rms* value of the phase current I_{rms}

K_p is termed the electrical power waveform factor

η is the motor efficiency

f is the frequency

p is the motor pole pairs

λ is the ratio of the airgap surface diameter to the stator OD ($\lambda = d_g/d_{s,o}$)

The application of this general-purpose sizing equation can provide motor engineers a useful tool in designing new high power density motors.

1.6 MOTOR DESIGN PROCESS AND CONSIDERATIONS

The design of an electric motor is a complex task, involving multiple disciplines such as electromagnetics, mechanics, thermal science, material science, sensing technology, rotordynamics, vibrations, acoustics, electronics, tribology, control theory, and mathematics. The design process of electric motors involves continuous iterations between electromagnetic, thermal, structural, rotordynamic, and systematic designs based on a variety of theoretical analyses, numerical simulations, and lab tests. To achieve an optimum design, all design parameters and criteria must be considered comprehensively.

There are two basic approaches in motor design: a subsystem/component approach and a system approach. Traditionally, a common engineering approach is to break down the system into subsystems or components (*e.g.*, stator, rotor, feedback, cooling, and coupling), design and optimize these subsystems, and then assemble them as a whole system. This is more likely the conventional bottom-up design strategy in engineering design. In the subsystem/component approach, the design of each subsystem/component is essentially independent of each other and all the work carried out in parallel. Thus, one of the benefits of this approach is its short design time with the simplified problem. However, this approach ignores the interactions among different subsystems/components. This leads to a possible result that even if each subsystem/component performs well, the system as a whole may not perform well. This is because the sum of the functioning of the individual subsystem/component is quite often not equal to the functioning of the whole.

A system approach focuses on overall system performance and creates a technical solution that satisfies the functional requirements for the system. This is the result of the synthetic mode of thinking applied to physical problems. This approach takes account of the intrinsic connections and interactions among different subsystems/components. In this approach, the design engineers evaluate the entire system to determine how end-use requirements can be provided most effectively and efficiently. The system approach is actually based on a top-down design strategy, in which the requirements are always satisfied at every step of the design process.

In modern motor designs, engineers often involve some degree of compromise and trade-offs among various competing requirements and design features such as motor efficiency, operation reliability, torque density, IP rating, cooling, noise, and cost-effectiveness, to name a few. Decisions on trade-offs involve systematic comparisons of all benefits and costs for achieving a satisfactory overall design.

It is worth noting that the two design approaches are not always absolutely opposite. Under some circumstances, engineering designs may gainfully employ both methodologies [1.126].

1.6.1 Design Process

In recent years, the sixth wave of technological innovation has emerged interest in the role of technology in the modern society to grow up continuously. The concept of innovation is directly related to the exploration of successful idea that can generate profitable products (*e.g.*, smartphones, commercial jets, automobiles, drones, and robots), processes, services, or profitable business practices [1.127].

Good design of electric motor involves many engineering aspects upon which the success of the design work depends, including customer's specifications, motor operation condition, material selection, system integration, technical analysis, and manufacturing process, as demonstrated in Figure 1.52. In finding the best design solution, engineers must make trade-offs among many factors that determine the final design and cost of the motor. In addition, identifying and understanding the design constraints and limitations are critical to the success of the motor design.

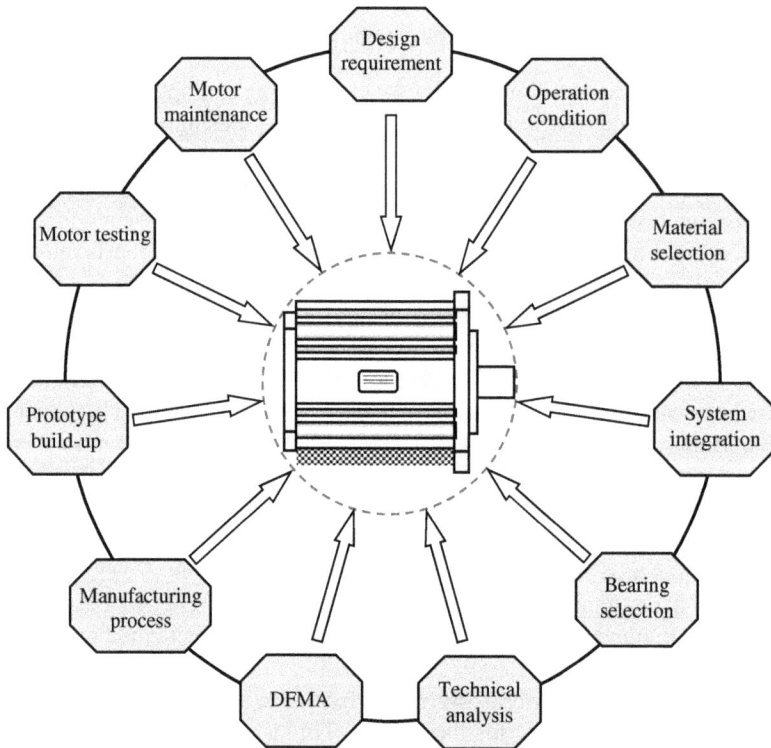

FIGURE 1.52 Motor design chart, indicating that in the motor design process many different design elements must be taken into account.

The general design flow chart of an electric motor is presented in Figure 1.53. The design process starts with the careful reviewing of customer's specifications and requirements, especially the requirements of motor operation and performance. Based on the information, the motor type and topology can be determined, such as PMM or IM, synchronous or asynchronous motor, and RF or AF motor. Then, the main motor design parameters are primarily set by using empirical equations.

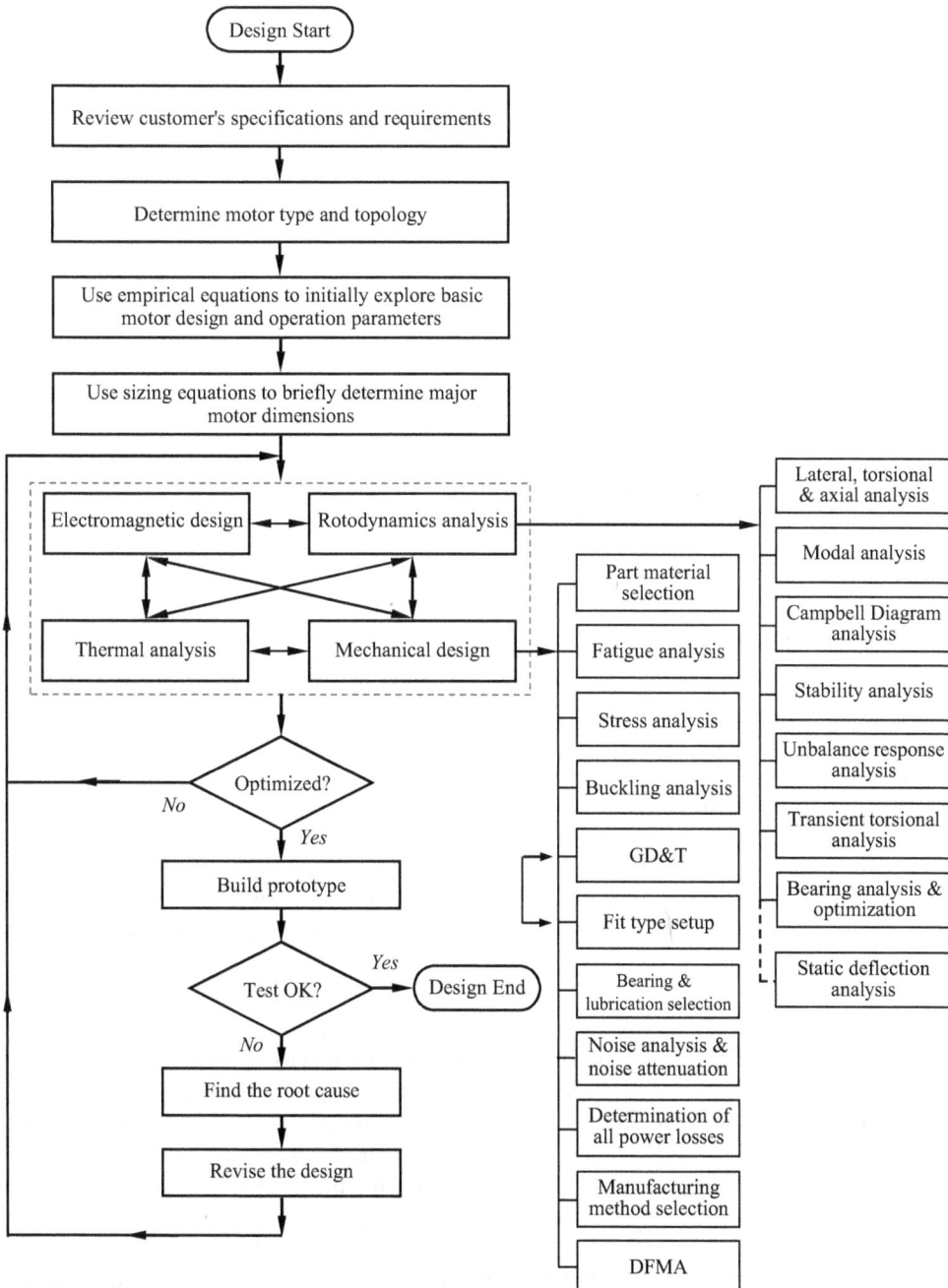

FIGURE 1.53   The flowchart of motor design.

The overall sizing of the machine is of great importance in terms of performance, structural integrity, cost, and weight. The major motor dimensions are determined via sizing equations available in the literature.

In the design of modern electric motors, none of the technical design is isolated from the other designs. In fact, all designs in different fields have strongly influenced each other. Hence, these design processes are neither in series nor in parallel, rather they form a crossover network. For instance, thermal simulation and analysis are required to ensure that the design meets all thermal requirements. The thermal analysis must be carried out in the early design stage, keep pace with electromagnetic and mechanical design. Thus, the overall design process can be shortened and result in a more robust product design. During this design stage, engineers/designers must communicate frequently to exchange their design concepts, innovative ideas, experience, testing results, and other useful information, especially the modifications that could impact other designs. The design iterations between different design processes continue to carry out until the initial design is fully implemented.

Conventionally, the design practice of an electric motor has been based on trial and error approaches, where designers iteratively change the values of the design variables and reanalyze the system until acceptable design criteria are achieved. This is not only tedious, but also time-consuming and hence expensive. The current trend in design is to follow a parallel approach as opposed to a sequential approach. The modern technique dictates the involvement of all the disciplines of design throughout the whole design process, from the conceptual phase to the production stage. Due to advances in computational techniques and simulation software, the focus of motor design has switched to multidisciplinary design synthesis using optimization approaches for obtaining the best possible design at the system level. Today, multidisciplinary design optimization (MDO) becomes a mature methodology involving modeling the interactions of complex systems across a variety of disciplines for realizing more effective and optimized solutions of designed systems. Most of today's commercial FEA software has built-in capabilities of engineering optimization, allowing for performing the analysis and optimization simultaneously. According to specific applications, a multidisciplinary FEA is performed often involving mechanical structures, electromagnetics, rotordynamics, motor performance, heat transfer analysis, *etc.* to study the behavior of the system and to serve as a basis for an MDO procedure. For this purpose, a set of objective functions for the optimization, ranges of design variables (so-called *design space*), and various constraints must be defined prior to the analysis.

Electromagnetic design is the core part of the motor design for determining motor key operating and performing characteristics. The electromagnetic design is highly dependent on the motor structure and rotordynamics analyses performed on the rotor. Recently, with the increasing demands for high-efficiency and high-reliability motors, one of the trends in novel motor development is to integrate the electromagnetic structure and mechanical transmission structure into a functional system for not only simplifying the motor structure but also increasing torque density, motor efficiency, and operation reliability.

The calculated power losses from electromagnetic design are passed into thermal analysis. With the fast development of computational technologies in the last several decades, it becomes more popular today for thermal analysis to be executed by means of advanced computational fluid dynamics (CFD) software package. In the current CFD market, many powerful commercial software packages are available, either for general-purpose or for specific applications, to solve heat transfer and fluid dynamics problems. CFD methods are used to provide detailed information of motor cooling such as temperature distribution, velocity field, pressure drop, and thermal interaction between solid surfaces and fluid flows in electric motors and to confirm and refine the preliminary designs. However, performing a CFD analysis remains a challenging task and requires highly skilled engineers due to the complexities involved, such as the complexity of highly nonlinear partial differential equations, CFD model meshing, large amount demanding of computational resources, determination of models and parameter of turbulent flows, convergence of simulation, and complexity of inputs and related tasks. Extensive design iterations exist between the thermal, mechanical, and other design processes for achieving the optimized motor design.

The mechanical design of an electric motor consists of a number of design elements:

- Selection of materials of motor components—Material selection in the motor industry is an artful balance among material performance, motor reliability, manufacturing process, and total cost. Apparently, material selection can heavily influence the motor performance, reliability, and lifetime. None of the materials is perfect to satisfy all requirements. It is the responsibility of engineers to make trade-offs among various design factors, for example, material performance, cost, strength, and durability.

- Fatigue analysis—As a motor component is subject to cyclic loads, a fatigue crack nucleus may be initiated on a microscopically small scale, and then the crack grows to a macroscopic size, finally causing the component failure suddenly at a stress level below its ultimate tensile strength or even yield tensile strength. During normal operation, some motor components are subject to alternative loads. One example is the motor shaft that is subjected to gravitational force from the rotor core and drag force from a load machine (*e.g.*, using belt or gear driving). When the rotor rotates, the shaft experiences cyclic loads creating a compression-expansion cycle for every revolution.

- Mechanical stress/strain analysis—The performance of mechanical stress/strain analysis is to determine whether the motor structure or components can withstand various loads without failure. The analysis allows for design optimization to achieve the best performance of electric motors. Due to the increasing complexity of modern motors, the stress/strain analysis today is mostly carried out using 3D finite element models. Recently, 3D finite element models have used to perform a fully coupled thermal-mechanical-fluid analysis for obtaining more complete and accurate simulation results.

- Buckling analysis—Bucking is a phenomenon that a structural member subjected to compressive stress suddenly fails due to the loss of its stability. The actual compressive stress of the structural member at the point of failure is much lower than the ultimate strength that the material can withstand. For instance, some motors have a number of ventilation ducts within the lamination core (either stator or rotor or both). These ventilation ducts are constructed by I-beams or π-beams that function like spacers. It is highly desired to perform bucking analysis to avoid the instability failure of beams.

- Determinations of geometric dimension and tolerance (GD&T)—The objective of GD&T is to define the requirements of geometry and tolerance for parts and assemblies. Proper applications of GD&T can ensure allowable part and assembly geometry defined on the drawings leads to right size, shape, form, location, orientation, and location, define the proper assembly of mating parts, and enable assembled motors to function as intended. Mechanical engineers are responsible to provide the information for drafters and manufacturing engineers for the fabrication of motor components and the layout of motor assembly. Based on the information, 3D solid models can be generated to describe explicitly nominal geometries and their allowable variations at both the component level and the system level. There are several GD&T standards available worldwide, including those that are defined by American Society of Mechanical Engineering and International Organization for Standardization (ISO).

- Setting up appropriate fits among mechanically mating components—Improper fits among mating parts may greatly reduce the motor lifetime, deteriorate motor performance, or even lead to motor failure. For instance, a loose fit between a shaft and a bearing can cause the relative movement between the two components, resulting in a fast wear of the shaft and high heat generation due to the sliding friction. In addition, an improper fit may cause motor vibration.

- Selection of motor bearing and lubrication—It has been reported that more than a half of motor failures can be attributed to the failure of bearing. Among them, improper selection of bearing is one of the major causes. During motor operation, bearings must be properly lubricated to prevent metal-to-metal contact among the rolling elements, raceways, and retainers. Inadequate lubrication (*e.g.*, insufficient or over lubrication) or adoption of wrong lubricant grade can directly lead to bearing damage or premature failure.

- To meet the government regulations and national standards, acoustic noise generated from electric motors should be controlled below a certain level. Acoustic noise in electric motors is predominantly related to electromagnetic forces in the airgap, forced cooling flows, and vibration-induced acoustic fields. There are existing noise prediction methods that can be applied to electric machines, including analytical, computational, or empirical methods. For many motor manufacturers, perhaps the easiest way is to measure motor noise in acoustic test laboratories.

- Determination of mechanical, electrical, and electromagnetic power losses—The mechanical losses in a motor usually consist of friction losses (bearing, seal, brake, *etc.*) and windage losses. The electric losses (also referred to as copper losses) represent all the resistive losses in motor windings and cables and are typically the main power loss component in electric motors. The electromagnetic losses (*i.e.*, iron losses), including eddy current and hysteresis losses in motor cores, PMs, and other components, are often obtained through electromagnetic analysis conducted by electrical and electromagnetic engineers. All the information is requested by thermal engineers for performing CFD or numerical heat transfer analysis.

- Determination of manufacturing process for motor components—Each motor part can be fabricated by many different manufacturing processes. The selection of the manufacturing process is based on many factors, including productivity, cost-effectiveness, raw material utilization, energy consumption, process cycle time, availability, part geometry, requirements for part strengths, vibration and dampness, surface finishing, material properties, porosity/void, carbon emission, and environment friendliness. As an example, for a large volume of motor end bell housings, the most suitable manufacturing method is diecasting for its high productivity and relatively low cost. In recent years, additive manufacturing processes, such as 3D printing, have been widely used in a variety of industries because of their capabilities of manufacturing components with complex structures and geometries, elimination of raw material waste, cost-effectiveness, and versatility.

- DFMA is the combination of two distinct design methodologies: design for manufacturing (DFM) and design for assembly (DFA), each methodology has its own desired object, scope of application, and performing process. DFM focuses on selecting cost-effective raw materials and attempting to minimize the complexity of manufacturing operations during the product design phase. Generally, DFM is to reduce the overall manufacturing time and costs for product components. As an example, the performance of DFM can avoid over-constraint and tolerance staking issues [1.128]. By contrast, DFA is a tool used to assist engineering teams, which contains engineers and designers from various groups such as design, manufacturing, quality, testing, service, and tooling, in the design of products, focusing on minimizing costs, assembly operations, and complexities by means of part count reduction, assembly process simplification, and assembly variability management. The part minimization and assembly process optimization can directly lead to the reduced bill of material costs, less labor required to build products, shorter manufacturing cycle time, simpler assembly instructions, higher quality, and higher profit margin. Recently, more and more engineers tend to combine DFM and DFA as a single methodology for effectively applying them to gain the greatest benefit.

Operation of rotating machinery at critical speeds can frequently cause a high level of mechanical vibration, noise, and excessive wear. In almost all cases, the major objective regarding rotating machinery is either to reduce or eliminate actual resonances or to avoid operating the equipment at the critical speeds. All rotating shaft-bearing systems have a number of frequencies at which they tend to vibrate. These frequencies are referred to as natural frequencies, or eigenfrequencies, which are determined by the mass and stiffness distribution of shafts and the location and stiffness/damping of the supporting bearings. Sometimes, the compliance of the housing also has a significant influence [1.129].

Rotordynamics is an engineering branch that deals with the vibrational behaviors of rotating systems or components in rotating machinery, such as rotors, shafts, bearings, and couplings. With the advanced computing techniques and powerful computing resources today, a variety of rotordynamics analyses can be effectively performed.

- Performing these analyses can help avoid resonance and vibration issues to improve the lifetime and performance of modern electric motors. In the design of a single rotor system, the torsional, lateral, and axial vibrations are typically decoupled and considered separately. However, for a geared rotor system with multiple rotors, the torsional, lateral, and axial vibrations may be fully coupled through the gear mesh and/or rider ring [1.130] and then these coupled vibration analyses are performed together to obtain the complete dynamic characteristics of the system.

  When a motor is in normal operation, it is often subject to unexpected vibration due to various causes. Lateral vibration occurs perpendicular to the axis of rotation, either side-to-side, or up-and-down, or both. Lateral analysis is typically used for rotating machinery rather than reciprocating units. Lateral analysis is performed to assess the potential occurrence of high vibration, associated degradation, and eventual motor failure. The analysis is used to gain damped and undamped critical speeds, mode shapes, and unbalance responses. The data of undamped critical speed can help check the bearing stiffness, as well as the behavior and sensitivity of rotor near its critical speed.

  Torsional vibration is an angular vibration of the shaft. Torsional analysis is often performed for both reciprocating actuators and rotating motors, as well as any associated rotating components such as gearbox shafts and couplings. For instance, it has been commonly used to evaluate the twisting interaction between rotors and couplings. In analytically determining the torsional response, it requires to calculate the torsional resonance frequencies of the motor. To perform the analysis, it requires the torsional stiffness and mass inertia of both the motor and load systems. Furthermore, a transient torsional analysis is recommended when synchronous motors or variable frequency drives are involved, in order to determine if the machinery can tolerate the stress and torque levels developed during transients such as motor start-up or short-down events [1.131].

  Axial vibration occurs along the axis of rotation. However, except for special circumstances, axial vibrations are often given less attention in electric machines, compared to vibrations in torsional and lateral directions.
- Modal analysis is the study of dynamic response of a system or structure under vibrational excitation. The object of modal analysis for a rotary machine or operating system is to determine the natural mode shapes and frequencies to avoid resonant conditions at operation speeds. Modal analysis is understood as the ensemble of analytical and experimental techniques intended for the modeling of the dynamic behavior of vibrating systems that derive from the fact that, under certain conditions, the dynamic response can be represented as a superposition of the dynamic responses of elementary mechanical systems, in terms of the modal characteristics [1.132]. As the computer technology advances and computing capability improves in recent decades, it becomes more popular for engineers to use advanced FEM techniques to perform modal analysis. Using FEM permits to quickly

obtain the solutions of large complex structural dynamics problems. Today, with efficient and comprehensive commercial FEM codes, the dynamics response calculations of complex linear and nonlinear structural systems under a variety of dynamic excitation conditions, and environmental conditions including temperature and fluid effects, can be performed.

- However, it has been pointed out that the dynamic response of the FEM model may differ substantially from that of the actual structure. This can occur because of errors in setting model parameters, and when the finite elements do not approximate the real world situation well enough [1.133]. Therefore, dynamic modal testing, also referred to as experimental modal analysis, is necessary to confirm the validity of FEM results. In practice, modal testing is a formalized method for identifying natural frequencies, mode shapes, impedance data of structures, *etc.* It utilizes dedicated modal test equipment and a variety of sensors and requires a formalized procedure for disturbing the structure into motion, and then recording the distribution of the resulting motion throughout the structure. Moreover, by conducting dynamic modal testing, the modal mass, stiffness, and damping matrices of the structure can be obtained. These matrices can be directly used in a finite element model for subsequent problem solving (*e.g.,* structural buckling) or redesigning the structure for a more optimum dynamic response. This method has been widely used for evaluating the designs and modification of structural problems. However, the dynamic modal testing is subjected to some limitations, such as induced nonlinear errors and the lack of capability for addressing forced response, transient response, and random response problems. In addition, this method is unable by itself to predict structural stability problems, such as buckling. Consequently, the common practice in industries is to use both FEM analysis and modal testing in a combined manner to achieve fully rotordynamic characteristics of structures or systems [1.134, 1.135].

- The Campbell diagram was introduced by Wilfred Campbell in 1924 when he worked in General Electric [1.136], with which the frequency of propagated bending waves in the circumferential direction of an elastic body is plotted against the rotational speed. The Campbell diagram is an overall regional vibration excitation that can occur in an operating system. A transient Campbell diagram based on either experimental tests or numerical simulations of a complicated rotor system allows analyzing the vibration characteristics of the system in real time. It is frequently used to determine the effect of multiple excitation frequencies in high-speed electric motors.

- Under normal operating conditions, high-speed servo systems may become unbalanced if there is even a mass unbalance in the rotating rotor system. In a model of an unbalanced rotor-bearing system, the exciting force is the centrifugal force produced by the unbalance mass, the unbalance response is the vibration of the rotor system, and the unbalance is reflected by the dynamic vibration characteristics of the system [1.137]. In this way, the vibration characteristics and unbalance vibration response law of the rotor system is the theoretical basis of active balancing research [1.138]. The vibration response obtained from unbalance response analysis can be used to calculate the mass of unbalance and provide a theoretical basis for active balancing and vibration control of high-speed rotating systems.

- Vibration characteristics of unbalance response are the research basis for active balancing control. Unbalance response analysis, as an indispensable and critical part in rotordynamic analysis, is also used to verify that there is sufficient margin between operating speeds and critical speeds. Continuous operation at or near critical speeds can produce amplified vibrations that can potentially result in component failure.

- Along with the rotor, bearings are the most influential components in rotating equipment vibrations. Therefore, the accuracy of rotordynamic analysis relies on the analysis of the bearing. In rotating machinery, many different types of bearings are adopted, including rolling element bearings, tilting pad bearings, fluid film bearings, and active magnetic bearings. Although rotor-bearing systems can be complicated to the point where their response appears random, but rotordynamic analysis can simplify the complexity and deepen understanding. The stiffness and damping properties of the bearings are important factors to affect the dynamic characteristics of the whole rotating system. Today's sophisticated rotordynamic analysis can take into account the stiffness and damping of shaft, bearings, and housing. This enables motor engineers to design rotor-bearing systems with critical speeds located away from the operational speed by a safe margin. Moreover, the results of the bearing analysis can greatly benefit the design and optimization of bearing for reducing the system vibration and ensuring optimal performance of electric machines.

Design optimization is the top goal of motor design. When the preliminary design is complete, the engineers need to examine whether the predetermined optimum objectives of the design have been reached. For different applications, these optimum objectives can be different, namely, the maximum peak torque, the highest torque density ratio, the most efficient cooling, and the total weight. If the design does not satisfy certain requirements, the design engineers need to review their designs to identify the design gap and do the design iteration again until all optimum objectives are gained.

As an important milestone within a product development process, a design review is a comprehensive assessment of the overall product design with performance, efficiency, reliability, functionality, and optimization as the driving factors. In a design review, the engineering team evaluates the product design against its requirements and outcomes, identifies any existing issues, and prioritizes the next steps. There are different levels of design review. Design reviews at the initial stage of product development, also known as initial or preliminary design reviews, help clarify the design goals, customer's specifications, application scope, failure modes, and potential design risks. Design reviews in the middle design stages are to correct product design flaws for avoiding any need for redesign, retooling or rework late, review detailed design work and analysis results, solicit suggestions and feedback for any possible design change, and ensure the product performing desired functions. The final design review, referred to as critical design review, is typically held as the majority of the design work has been completed. It offers every engineering team member a last chance for optimization and adjustment of the product prior to the fabrication of prototypes or pre-production models. A chief engineer or a team leader usually leads design reviews, and the participants include all engineering team members.

The next step is to build a prototype motor for validating the motor design as a whole unit. This is the critical step for the success of products. No matter how perfect the results achieved from engineering analyses are, prototype motor testing is an essential and necessary step to validate the conceptual design and ensure normal motor operation. Only after a series of successful tests, the motor can start mass production.

1.6.2 DESIGN INTEGRATION

An electric motor usually consists of dozens, hundreds, or even thousands of individual components. In order to increase motor operation reliability, reduce the manufacturing cost, and simplify the assembly process, it is desired to integrate some of the components together in one self-contained package, including (a) motor-transmission device (e.g., gearbox, belt, and chain) integration, (b) mechanical-electromagnetic system integration, and (c) motor control (e.g., feedback and drive).

As one of the most important constituents in an automated electric driving system, an electronic drive is used to deliver a usable form of power to control the performance of a servomotor for an end user. As presented in Figure 1.54, the integrated motor drive assembly system includes a motor, a fan, and a drive unit. One remarkable advantage of such integrated assembly systems is their compactness in size and ease of installation into a small industrial or other application. Generally, the drive is disposed on the motor or arranged in an integral housing with the motor. Both the motor and the drive are shared with the same cooling fan [1.139].

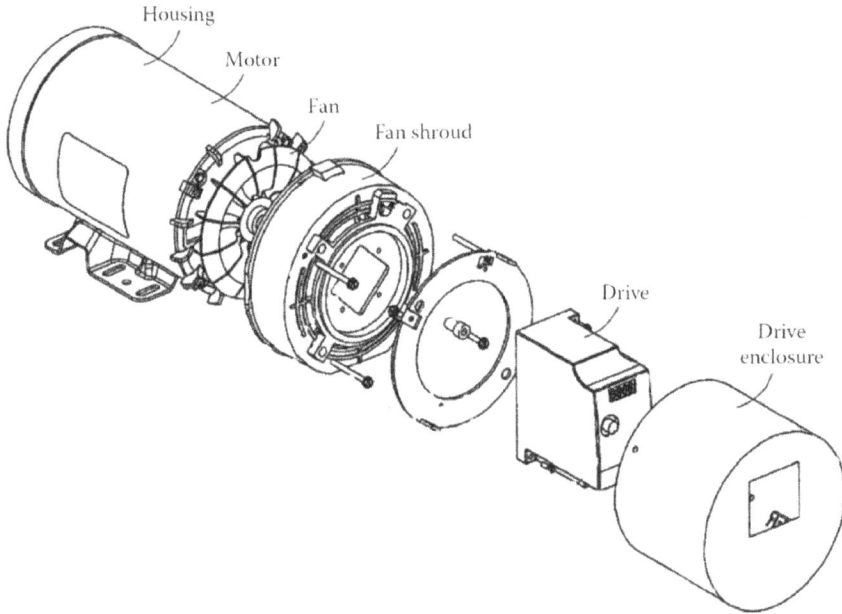

FIGURE 1.54 Integration of electric motor, drive, and fan as one unit (US patent 7,362,017) [1.139]. (Courtesy of the U.S. Patent and Trademark Office, Alexandria, VA.)

A growing trend in motor control is the integrated system that combines motor, feedback, and controller in one self-contained package. By integrating the controller with the motor, the possibility of mismatch is eliminated. Based on the integrated modular motor drive concept, an integrated traction drive has been developed recently. The integration of motor and drive offers a number of attractive features such as reduced drive volume and the elimination of power transmission cables. Correspondingly, it reduces radiated electromagnetic interference and voltage transient due to power transmission over long cable distances [1.140]. Furthermore, the integration of motor and drive can offer fault-tolerant features not possible with conventional drives [1.141, 1.142]. On the caution side, significant thermal management and analysis are required to protect the electronic drive and motor combination.

1.6.3 MECHATRONICS

Mechatronics is the confluence of classical engineering disciplines such as mechanical engineering, electrical and electronic engineering, electromagnetism, sensor technology, drive and actuator technology, control theory, and computer science. In recent years, one of the design trends in the electric machine industry is to design intelligent mechatronics products. In fact, mechatronics is not only a modern design strategy but also a new way of doing business to gain a competitive advantage in the global market. As a result, it has been extensively used to design improved products and processes [1.143].

In fact, the mechatronic design approach that involved multidisciplinary is expected to become a key technology to gain a competitive advantage in the era of modern manufacturing. The development of mechatronics will therefore be crucial to the continued competitiveness of national economies.

The magnetic hard disk drive (HDD) is believed to be one of the most successful examples of modern mechatronics. In a state-of-the-art HDD servo system, a high-speed servomotor, a read/write head, a data storage disk, and other components are designed as one unit. The precision

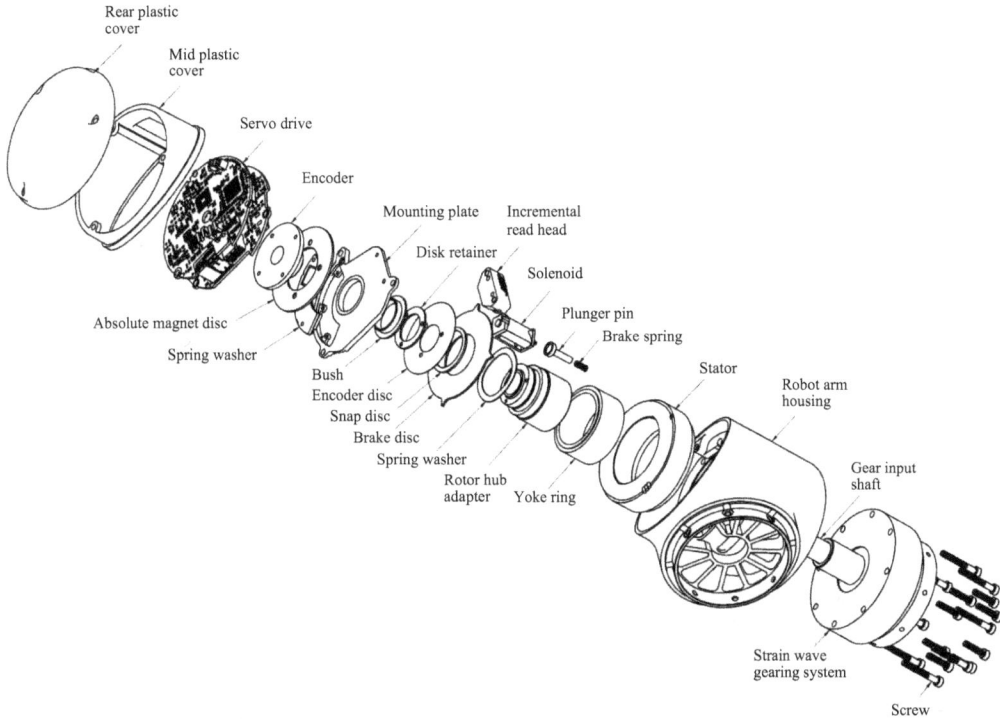

FIGURE 1.55 Exploded view for mechatronic design of robot joint arm (US patent 11,161,258) [1.145]. (Courtesy of the U.S. Patent and Trademark Office, Alexandria, VA.)

position control of HDD enables the tolerance to be less than one micrometer while operating at high speeds [1.144].

Over the past decade, industrial robots have experienced rapid development in many countries. Industrial robots have a key role to play in manufacturing, product assembly, packaging, warehouse, and logistics management operations. Today, almost all robots incorporate PM servomotors due to their high efficiency, precise control and positioning capability, force density, and compact size. As a perfect example of mechatronics, all components of a robot joint, including servomotor, gearing system, brake, feedback device, servo drive, and other auxiliary parts, are integrated as a package contained in a casted robot arm housing (Figure 1.55) [1.145].

1.6.4 Temperature Effect on Motor Performance

The influence of temperature on motor performance is multifaceted. When a motor operates at extreme temperatures, it is subjected to a variety of risks and uncertainties to affect motor performance, including:

- At high temperatures, both remanent flux density and coercivity of PM decrease, deteriorating their ability to produce sufficient magnetic flux and increasing demagnetization risk. This is because when temperature increases atomic vibrations in PMs cause once-aligned magnetic moments to randomize, resulting in a decrease in magnetic flux density. When the highest allowable temperature is exceeded, the magnets will partially demagnetized. If the temperature reaches the Curie point, all magnetism will be lost. Therefore, to ensure stable operation of PMMs, the PM temperature must be kept as low as possible.

- Electric motors are usually assembled at room temperature. When a motor experiences dramatic changes in temperature, the fit condition among mating parts (*e.g.*, shaft and rotor core, stator core and motor housing, and bearing and bearing bore) changes significantly due to the differential thermal expansion/contraction among the mating parts. For example, because the coefficient of thermal expansion of aluminum is much higher than that of steel at high temperatures (*e.g.*, 21.9 μm/m-°C vs 13 μm/m-°C at 150°C), the larger expansion of the aluminum housing may greatly reduce the fit strength between the stator core and housing. Therefore, the effect of differential dimension changes must be carefully considered at the early motor design stage.

- Motor winding resistance is a critical factor in determining motor resistive power losses. From a microcosmic perspective, electrical resistance in a conductor reflects essentially the intensity of collisions between valence electrons and vibrating ions. An increase in temperature causes more collisions and thus an increase in resistance. The winding resistance R at an arbitrary temperature T can be calculated as

$$R(T) = R_o(T_o)\left[1 + \alpha(T - T_o)\right] \tag{1.88}$$

  where R_o is the resistance at the reference temperature T_o and α is the temperature coefficient of resistance (as a reference, $\alpha_{Cu} = 0.0040$, $\alpha_{Al} = 0.0043$, and $\alpha_{Ag} = 0.0038$).

- It can be seen from this equation, a higher temperature causes a higher winding resistance and, in turn, a higher power loss in the motor. Actually, the continuous torque is inversely proportional to the square root of the product of motor winding resistance and thermal resistance.

- Motor sensors are typically specified for normal operation within certain temperature ranges. This is because that sensor measuring errors can change significantly with large temperature variations. In fact, temperature has different effects on different types of sensors. For optical encoders, the increase in temperature causes a decrease in the light output of the LED. Moreover, thermal expansion is another important factor that affects the performance of an optical encoder. It has been reported that thermal expansion can narrow the airgap between the disk and the source/detector by 0.020 in (0.51 mm). Under an extreme temperature, this thermal expansion can bring the components in contact with one another, causing encoder damage or even catastrophic failure [1.146]. For magnetic encoders, thermal expansion and contraction of the magnetic wheel can alter the pitch of the magnetic poles and thus alter the output. For resolvers, temperature-induced resistive changes in resolver copper windings have an influence on phase shift, which results in changes in the transformation ratio, impedances, input current, and input power. In addition, minute mechanical changes with temperature may cause a shift in the electrical zero, referred to as EZ shift.

- Many motor failures often originate with insulation damages due to thermal, electrical, mechanical, and environmental deterioration mechanism. A well-accepted rule of thumb claims that as a motor temperature exceeds its rated insulation temperature, every increase of 10°C will reduce the insulation life by half. It is worth to note that thermal deterioration resulted from overheating is irreversible. Once it occurs, reducing the temperature merely stops further thermal deterioration.

- Motor bearings are usually lubricated by mineral oil-base greases. It has been widely recognized that bearing friction is affected by several factors such as temperature, velocity, load, lubricant properties, and operating conditions. A change in temperature can cause a significant change in viscosity of the grease and in turn its molecular friction. For instance, low temperatures can significantly increase the viscosity of the grease, leading to insufficient lubrication and to wear and potential system failure. As a comparison, high temperatures can sheer/crack the oil molecules into smaller molecules, which causes a decrease in viscosity. This may lead to oil leakage or loss from the bearing. In fact, high temperatures can trigger many different grease failure mechanisms, directly affecting the effective useful grease performance life.

- As one of the major sources of electric noise, thermal noise arises from the thermal fluctuations in electronic circuits. Thermal noise highly depends upon the temperature, *i.e.*, the higher the temperature, the higher the thermal noise level. The effective way to reduce thermal noise is to enhance heat dissipation capability in electronic devices.
- The mechanical properties (such as yield strength and ultimate strength) of steel and steel alloys are generally functions of temperature. Changes in material mechanical properties due to the change in temperature can alter the stress distribution and factor of safety of motor components.

All these problems must be addressed properly in the early stage of design to avoid any issues that may arise later in sustainability performance of motor.

1.7 MOTOR FAILURE MODES

There are a number of different approaches to assess motor failure modes. Generally, there are three primary failure modes identified from motor root cause analyses. The most important failure mode is electrical-related failure, caused by the breakdown of motor winding insulations, overloading condition, winding short circuit, and other issues. In a typical case, the breakdown of insulation material is attributed to exceeding the temperature limit of windings. An overloading condition can cause the winding current exceeding its limit to damage coils. The second failure mode is magnetic failure, which usually resulted from the thermal demagnetization of PM (for PMMs). The thermal demagnetization of PM may be attributed to high motor operation temperature, shock, vibration, or strong magnetic field generated by a stator. The third failure mode is mechanical failure of motor structure and components. In this failure mode, bearing failures are encountered frequently. Another common failure mode is loose/broken rotor bars. Usually, most mechanical failures result from slow processes such as mechanical wear and gradual degradation of material properties. During the process, the defective parts continue to display characteristic signs until they completely fail. This is different from most electronic component failures that happen suddenly and without warning.

Statistics have shown that despite the reliability and simplicity of construction of IMs, the annual motor failure rate is conservatively estimated at 3%–5% per year, and in extreme cases, up to 12%, as in the pulp and paper industry [1.147]. The downtime in a factory due to motor failure may lead to a high cost, even exceeding the cost of motor replacement. In some instances, a motor failure may cause the entire production line to stop and thus interrupt the production process.

Based on the data collected from 1,141 motors, those larger than 200 hp and less than 15 years in service, *IEEE* published a long report in three parts [1.148–1.150] to address motor failure modes according to the root causes of failure. In this report, Part I presented general results based on categories of motor types. Part 2 combined various categories and addressed some questions resulting from Part 1. Part 3 of the survey results was to address new questions and comments and to add more specific analyses of areas not yet explored previously.

A similar investigation of motor failure modes has been conducted by *Electric Power Research Institute (EPRI)* [1.151]. Though the approach of *EPRI* is different from *IEEE* (*EPRI* focused on failed motor components), their results are consistent. The integration of these two surveys was made by Venkataraman *et al.* [1.147], as presented in Table 1.7.

From the table, it can be seen that the most common reasons of the motor failures are bearing failures and failures related to the stator insulation. High vibration can lead to mechanical failures. Furthermore, many failures are related directly or indirectly to high operation temperature.

TABLE 1.7

Comparison of Failure Modes for Large-Size Motor

IEEE Survey		EPRI Survey		
Failure Contribution	**%**	**Failure Contribution**	**%**	**Average (%)**
Persistent overload	4.2	Stator ground insulation	23	Electrical-related failure
Normal deterioration	26.4	Turn insulation	4	
		Bracing	3	
		Core	1	
		Cage	5	
Electrical-related total	30.6	*Electrical-related total*	36	33.3
High vibration	15.5	Sleeve bearings	16	Mechanical-related failure
Poor lubrication	15.2	Antifriction bearings	8	
		Thrust bearings	5	
		Rotor shaft	2	
		Rotor core	1	
Mechanical-related total	30.7	*Mechanical-related total*	32	31.35
High ambient temperature	3.0	Bearing seals	6	Failures related to environmental maintenance and other reasons
Abnormal moisture	5.8	Oil leakage	3	
Abnormal voltage	1.5	Frame	1	
Abnormal frequency	0.6	Wedges	1	
Abrasive chemicals	4.2			
Poor ventilation cooling	3.9			
Other reasons	19.7	Other components	21	
Environmental reasons and other reasons total	38.7	*Maintenance-related and other parts total*	32	35.35

Source: Data from *IEEE* and *EPRI* motor reliability surveys.

1.8   IP CODE

Electric motors are designed to withstand various harsh environments such as flammable and explosive environments, high ambient pressure or vacuum, extreme temperatures, and high shock loads and vibrations. In some cases, they may operate in space or under deep water.

The IP code is an international standard for electric devices. It applies to the classification of degrees of protection provided by enclosures for electrical equipment against the insertion or intrusion of solid and liquid objects, external influences, or conditions such as dust, moisture, water, icing, corrosive solvents, and mechanical impacts. The IP code consists of two digits and optional letters. The indications of digits and optional letters are shown in Tables 1.8 and 1.9, respectively [1.152].

The IP code has a strong impact on the motor design. With the IP defined, a number of design factors, such as motor construction, insulation, sealing method, coating, cooling technique, and component material, must be specifically determined to satisfy the requirements and regulations of the IP code.

As a reference, common IP codes used in electric motors in practice are summarized in Table 1.10. It is noted that IP69K is the IP code for a very high level of protection as indicated in ISO 20653 [1.157].

TABLE 1.8
Indications of IP Digits

	First Digit: Protection against Ingress of Particles		Second Digit: Protection against Ingress of Water	
IP	**Protection of Human/Tool Contact**	**Protection of Equipment**	**IP**	**Protection of Equipment**
0	No protection	No protection	0	No protection
1	Protected against contact with large areas of the human body (back of hand)	Protected against objects over 50 mm in diameter	1	Protected against vertically falling drops of water, *e.g.*, condensation
2	Protected against contact with fingers	Protected against solid objects over 12 mm in diameter	2	Protected against direct sprays of water up to 15° from vertical direction
3	Protected against tools and wires over 2.5 mm in diameter	Protected against solid objects over 2.5 mm in diameter	3	Protected against sprays to 60° from vertical direction
4	Protected against tools and wires over 1 mm in diameter	Protected against solid objects over 1 mm in diameter	4	Protected against water sprayed from all directions (limited ingress permitted)
5	Protection against entry of dust in sufficient quantity to interfere with satisfactory operation of equipment	Protected against dust (limited ingress, no harmful deposit)	5	Protected against low-pressure jets of water from all directions (limited ingress permitted)
6	Complete protection against entry of dust	Totally protected against dust	6	Protected against strong jets of water
			7	Protected against the effects of immersion between 15 cm and 1 m
			8	Protected against long periods of immersion under pressure
			9K	Protected against ingress of high-temperature (steam) and/or high-pressure water

TABLE 1.9
Indications of Optional Letters in IP

First Letter (Optional)		Second Letter (Optional)	
Letter	**Protected against Access to Hazardous Parts with**	**Letter**	**Indication**
A	Back of hand	H	High-voltage device
B	Finger	M	Device moving during water test
C	Tool	S	Device standing still during water test
D	Wire	W	Weather condition

TABLE 1.10

Typical IP Codes Used for Different Motor Applications

IP Classification	Application
IP00	Open motor
IP12	Open drip-proof motor [1.153]
	Drip-proof motor [1.153]
IP13	Splash-proof motor [1.153]
IP21	Elevator motor with open enclosure
IP23, IP24	Weather protected motor [1.153]
IP42	Commercial refrigeration application
IP44	Totally enclosed fan-cooled motor [1.153]
	Totally enclosed pipe-ventilated motor [1.154]
IP54	Regular for commercial applications (fans, blowers)
	Totally enclosed force-ventilated motor [1.153]
	Totally enclosed air-to-air-cooled motor [1.154]
	Totally enclosed air-over motor [1.153]
	Totally enclosed air-to-water-cooled motor [1.153]
	Totally enclosed water-cooled motor [1.154]
IP55	Waterproof motor [1.154]
	Built-in invert motor [1.155]
	Motor used for automatic industrial doors [1.155]
IP65	Motor used in hybrid vehicles
	Elevator motor
	High-speed spindle servomotor
	Some direct drive motor
IP67	Motor used under severe operation conditions (*e.g.*, pitch and yaw control motors in wind turbines, driving motors in chemical reactors and in machine tools)
	Motor used outdoor under harsh environments
	Motor used in aerospace and defense applications (military vehicle, submarine, satellite, *etc.*)
	Motor used in high-pressure, high-temperature washdown applications (*e.g.*, food processing, beverage, and pharmaceutical processing) or applications in the presence of highly aggressive corrosive agents
IP69/IP69K [1.156]	Motor operating under ultra-harsh environments (*e.g.*, autoclaving, high-pressure spray, and frequent washdown with caustic chemicals)
	Motor used in deep water such as in propelling systems
	Liquid immersion pumps

REFERENCES

1.1.   IEA. 2016. International Energy Agency: World energy outlook 2016. Paris, France.

1.2.   Waide, P. and Brunner, C. U. 2011. *Energy-Efficiency Policy Opportunities for Electric Motor-Driven Systems*. International Energy Agency, Paris, France.

1.3.   Malcolm, D. R. 2006. Turn energy waste into profit. Plant Services. http://www.plantservices.com/articles/2006/313/?show=all.

1.4.   Scheihing, P. E. 2007. A national strategy for energy efficient industrial motor-driven systems. In *Energy Efficiency Improvement in Electric Motors and Drives. Part VIII* (Eds.: A. de Almeida, P. Bertoldi, and W. Leonhard). Springer, Berlin Heidelberg, pp. 377–389.

1.5.   Zhang, S. and Ding, X. L. 2014. Development of high-efficiency remanufacturing industry of electric motors (电机高效再制造业发展). *Development Strategy* (发展战略) **2014(15)**: 1.

1.6.   Hu, B., Zheng, T., Zhao, F. Y., Tieben, R., Brunner, C. U., and Wang, K. 2015. Topmotors China: Improving motor system efficiency with motor-systems-check in Zhenjiang. *Proceedings of 9th International Conference on Energy Efficiency in Motor Driven Systems*. Helsinki, Finland.

1.7. Moreels, D. and Leijnen, P. 2019. Turning the electric motor inside out. *IEEE Spectrum*, September issue, pp. 40–45.

1.8. Motor Intelligence. 2019. Industrial motor market – growth, trends, and forecast (2020–2025). https://www.motorintelligence.com/industry-reports/industrial-motor-market.

1.9. World Trade Organization. 2021. World trade primed for strong but uneven recovery after COVID-19 pandemic shock. Press/876. https://www.wto.org/english/news_e/pres21_e/pr876_e.htm

1.10. Gordon, A. 1745. *Versuch einer Erlärung der Electricität*. *Nonne*, Enfurt, Germany.

1.11. McInally, T. 2012. *The Sixth Scottish University: The Scots Colleges Abroad: 1575 to 1799*. Koninklijke Brill, Leiden, Netherland.

1.12. Gee, W. 2004. Sturgeon, William (1783–1850). *Oxford Dictionary of National Biography*. Oxford University Press, Oxford, U.K.

1.13. Doppelbauer, M. 2018. The invention of the electric motor 1800–1854. Karlsruhe Institute of Technology. https://commons.princeton.edu/josephhenry/wp-content/uploads/sites/71/2019/08/electric-motor-history.pdf.

1.14. Davenport, T. 1837. Improvements in propelling machinery by magnetism and electromagnetism. U.S. Patent 132.

1.15. Tesla, N. 1888. Electromagnetic motor. U.S. Patent 381,968.

1.16. Hubbell, M. W. 2011. *The Fundamentals of Nuclear Power Generation: Questions and Answers*. AuthorHouse, Bloomington, IN.

1.17. Friedel, R. D. and Israel, P. 2010. *Edison's Electric Light: The Art of Invention*. The Johns Hopkins University Press, Baltimore, MD.

1.18. Armstrong, A. W. Ir. 1998. Load to motor inertia mismatch: Unveiling the truth. *Drive and Controls Conference*, Telford, U.K. http://www.diequa.com/download/articles/inertia.pdf.

1.19. Voss, W. 2007. *A Comprehensible Guide to Servo Motor Sizing*. Copperhill Technologies Corporation, Greenfield, MA.

1.20. IEEE Power Engineering Society. 2004. *IEEE 112-2004 Standard Test Procedure for Polyphase Induction Motors and Generators*. IEEE, New York.

1.21. Hanselman, D. C. 1997. Effect of skew, pole count and slot count on brushless motor radial force, cogging torque and back EMF. *IEE Proceedings of Electric Power Applications* **144**(**5**): 325–330.

1.22. Hanselman, D. C. 2006. *Brushless Permanent Magnet Motor Design*, 2nd edn. Magna Physics Publishing, Lebanon, OH.

1.23. Lu, K. Y., Rasmussen, P. O., and Ewen Ritchie, E. 2006. An analytical equation for cogging torque calculation in permanent magnet motors. *17th International Conference on Electrical Machines*, Chania, Crete, Greece.

1.24. Hitachi Metals, Ltd. 2013. Permanent magnets. http://www.hitachi-metals.co.jp/products/auto/el/pdf/hg-a27.pdf.

1.25. Hsu, I. S., Scoggins, B. P., Scudiere, M. B., Marlino, L. D., Adams, D. J., and Pillay, P. 1995. Nature and assessments of torque ripple of permanent-magnet adjustable-speed motors. *Proceedings of IEEE Industry Application Conference* **3**: 2696–2702.

1.26. Bianchi, N. and Bolognani, S. 2002. Design techniques for reducing the cogging torque in surface-mounted PM motors. *IEEE Transaction Industry Applications* **38**(**2**): 1259–1265.

1.27. Zhu, Z. Q., Ruangsinchaiwanich, D., Ishak, D., and Howe, D. 2005. Analysis of cogging torque in brushless machines having non-uniformly distributed stator slots and stepped rotor magnets. *IEEE Transaction on Magnetics* **41**(**10**): 3910–3912.

1.28. Aydin, M., Zhu, Z. Q., Lipo, T. A., and Howe, D. 2007. Minimization of cogging torque in axial flux permanent magnet machines—Design concepts. *IEEE Transaction on Magnetics* **43**(**9**): 3614–3622.

1.29. Salminen, P. 2004. Fractional slot permanent magnet synchronous motor for low speed applications. PhD thesis, Lappeenranta University of Technology, Lappeenranta, Finland.

1.30. Martin, J. P., Meibody-Tabar, F., and Davat, B. 2000. Multiple-phase permanent magnet synchronous machine supplied by VSIs working under fault condition. *IEEE Industry Application Conference. 35th IAS Annual Meeting*, Rome, Italy.

1.31. Li, T. and Slemon, G. 1988. Reduction of cogging torque in permanent magnet motors. *IEEE Transactions on Magnetics* **24**(**6**): 2901–2903.

1.32. Dosiek, L. and Pillay, P. 2007. Cogging torque reduction in permanent magnet machines. *IEEE Transactions on Industry Applications* **43**(**6**): 1565–1571.

1.33. Ishikawa, T. and Slemon, G. R. 1993. A method of reducing ripple torque in permanent magnet motors without skewing. *IEEE Transaction on Magnetics* **29**(**2**): 2028–2031.

1.34. Lateb, R., Takorabet, N., Meibody-Tabar, F., Enon, J., and Sarribouette, A. 2006. Design technique for reducing the cogging torque in large surface-mounted magnet motors. In *Recent Developments of Electrical Drives* (Eds.: S. Wiak, M. Dems, and K. Komeza). Springer, Dordrecht, the Netherlands, pp. 59–72.

1.35. Muljadi, E. and Green, J. 2002. Cogging torque reduction in a permanent magnet wind turbine generator. *ASME 2002 Wind Energy Symposium*. Paper no. WIND2002–56, Reno, NV, pp. 340–342.

1.36. Lao, Y. D., Huang, D. R., Wang, J. C., Liou, S. H., Wang, S. J., Ying, T. F., and Chiang, D. Y. 1997. Simulation study of the reduction of cogging torque in permanent magnet motors. *IEEE Transactions on Magnetics* **33**(**5**): 4095–4097.

1.37. Soleimani, J., Vahedi, A., and Mirimani, S. M. 2011. Inner permanent magnet synchronous machine optimization for HEV traction drive application in order to achieve maximum torque per ampere. *Iranian Journal of Electrical & Electronic Engineering* **7**(**4**): 241–248.

1.38. Holtz, J. and Springob, L. 1996. Identification and compensation of torque ripple in high-precision permanent magnet motor drives. *IEEE Transactions on Industrial Electronics* **43**(**2**): 309–320.

1.39. Parker Hannifin Electromechanical Division. 2011. Factors affecting velocity and current ripple. http://www.parkermotion.com/dmxreadyv2/blogmanager/blogmanager.asp?category=drives.

1.40. Kim, K.-C. 2014. A novel method for minimization of cogging torque and torque ripple for interior permanent magnet synchronous motor. *IEEE Transactions on Magnetics* **50**(**2**): 793–796.

1.41. Fei, W. Z. and Luk, P. C.-K. 2012. Torque ripple reduction of a direct-drive permanent-magnet synchronous machine by material-efficient axial pole pairing. *IEEE Transactions on Industrial Electronics* **59**(**6**): 2601–2611.

1.42. Fei, Q., Deng, Y. T., Li, H. W., Liu, J., and Shao, M. 2019. Speed ripple minimization of permanent magnet synchronous motor based on model predictive and iterative learning controls. *IEEE Access* **7**: 31792–31800.

1.43. Hunt, G. How to select a DC motor: The different characteristics of each group of DC motors, http://insidepenton.com/machinedesign/nl/MicroMO-DCmotor-select.PDF.

1.44. Feese, T. and Hill, C. 2009. Prevention of torsional vibration problems in reciprocating machinery. In *Proceedings of the Thirty-Eight Turbomachinery Symposium*, Houston, TX, pp. 213–238.

1.45. Eshleman, R. L. 2009. Chapter 37: Torsional vibration in reciprocating and rotating machines. In *Harris' Shock and Vibration Handbook* (Eds.: A. G. Piersol and T. L. Paez), 6th edn. McGraw-Hill, New York.

1.46. Stephens, L. 2011. Servo controls deliver performance benefits for medical imaging systems. Kollmorgen Paper 2-09-11.

1.47. Stephens, L. 2010. The significance of load to motor inertia mismatch. Kollmorgen Paper 08-12-10.

1.48. Shu, H.-C. and Chen, C. H. 2011. Servo motor with large rotor inertia. U.S. Patent 7,911,095.

1.49. International Electrotechnical Commission (IEC). 2004. *International Standard IEC 60034-1. Rotating Electrical Machines—Part 1: Rating and Performance*, 11th edn. International Electrotechnical Commission (IEC), Geneva, Switzerland.

1.50. Mirchevski, S. 2012. Energy efficiency in electric drives. *Electronics* **16**(**1**): 46–49.

1.51. International Electrotechnical Commission (IEC). 2008. IEC 60034-30. Rotating Electrical Machines—Part 30: Efficiency Classes of Singlespeed, Three-Phase, Cage-Induction Motors. International Electrotechnical Commission (IEC), Geneva, Switzerland.

1.52. International Electrotechnical Commission (IEC). 2007. IEC 60034-2-1. Rotating electrical machines—Part 2-1: Standard methods for determining losses and efficiency from tests. International Electrotechnical Commission (IEC), Geneva, Switzerland.

1.53. International Electrotechnical Commission (IEC). 2010. IEC 60034-31. Rotating electrical machines—Part 31: Guide for the selection and application of energy-efficient motors including variable-speed applications. International Electrotechnical Commission (IEC), Geneva, Switzerland.

1.54. Berger, L. I. 2006. Chapter 15: Dielectric strength of insulating materials. In *CRC Handbook of Chemistry and Physics* (internet version 2006) (Eds.: D. R. Lide). Taylor & Francis, Boca Raton, FL, pp. 42–46.

1.55. Skanavi, G. I. 1958. Fizika dielektrikov: Oblast' sil'nykh poleĭ (Physics of dielectrics: Strong fields domain). *Moscow: Fizmatgiz* **26**: 609–612.

1.56. Vedensky, B. A. and Vul, B. M. 1965. *Encyclopedic Dictionary in Physics*, Vol. 4. Soviet Encyclopedia Publishing House, Moscow, Russia.

1.57. Suzuki, H., Mukai, S., Ohki, Y., Nakamichi, Y., and Ajiki, K. 1997. Dielectric breakdown of low-density polyethylene under simulated inverter voltages. *IEEE Transactions on Dielectrics and Electrical Insulation* **4**(**2**): 238–240.

1.58. Mori, T., Matsuoka, T., and Muzitani, T. 1994. The breakdown mechanism of poly-p-xylylene film. Prestress effects on the breakdown strength. *IEEE Transactions on Dielectrics and Electrical Insulation* **1**(**1**): 71–76.

1.59. Bjellheim, P. and Helgee, B. 1994. AC breakdown strength of aromatic polymers under partial discharge reducing conditions. *IEEE Transactions on Dielectrics and Electrical Insulation* **1**(**1**): 89–96.

1.60. Zheng, J. P. Cygan, P. J., and Jow, T. R. 1996. Investigation of dielectric properties of polymer laminates with polyvinylidene fluoride. *IEEE Transactions on Dielectrics and Electrical Insulation* **3**(**1**): 144–147.

1.61. Shugg, W. T. 1995. *Handbook of Electrical and Electronic Insulating Materials*, 2nd edn. IEEE Press, New York.

1.62. National Electric Manufacturers Association (NEMA). 2011. NEMA MG 1–2011 motors and generators. NEMA, Rosslyn, VA.

1.63. International Electrotechnical Commission (IEC). 2007. *IEC 60085 Electrical Insulation – Thermal Evaluation and Designation*, 4th edn. International Electrotechnical Commission (IEC), Geneva.

1.64. Bonnett, A. H. and Soukup, G. C. 1992. Cause and analysis of stator and rotor failures in three-phase squirrel-cage induction motors. *IEEE Transactions on Industry Applications* **28**(**4**): 921–937.

1.65. Wilson, D. S. and Smith, R. 1977. Electric motor reliability model. Report number RADC-TR-77-408. Rome Air Development Center, Griffiss Air Force Base, New York.

1.66. U.S. Department of Energy. 2008. *Improving Motor and Drive System Performance: A Sourcebook for Industry*. National Renewable Energy Laboratory. https://www1.eere.energy.gov/manufacturing/tech_assistance/pdfs/motor.pdf.

1.67. Chapman, S. J. 2005. *Electric Machinery Fundamentals*, 4th edn. McGraw-Hill, New York.

1.68. NMB Technologies Corporation. The emergence of brushless DC motors within medical applications. http://www.nmbtc.com/pdf/engineering/motors_emergence_of_brushless_dc.pdf.

1.69. Smith, J. S. and Watson, A. P. 2006. Design, manufacturing, and testing of a high speed 10 MW permanent magnet motor and discussion of potential application. *Proceedings of 35th Turbomachinery Symposium*, Houston, TX, pp. 19–24.

1.70. Parviainen, A. 2005. Design of axial flux permanent-magnet low-speed machines and performance comparison between radial flux and axial flux machines. PhD dissertation, Lappeenranta University of Technology, Lappeenranta, Finland.

1.71. Chan, C. C. 2002. The state of the art of electric and hybrid vehicles. *Proceedings of the IEEE* **90**(**2**): 247–275.

1.72. Schultz, J. W. and Huard, S. 2013. Comparing AC induction with permanent magnet motors in hybrid vehicles and the impact on the value proposition. Parker Hannifin Corporation. http://www.parkermotion.com/whitepages/Comparing_AC_and_PM_motors.pdf.

1.73. Jian, L. N., Chau, K. T., Yu Gong, Y., Yu, C., and Li, W. L. 2009. Analytical calculation of magnetic field in surface-insert permanent magnet motors. *IEEE Transactions on Magnetics* **45**(**10**): 4688–4691.

1.74. Lewotsky, K. 2012. Interior permanent magnet motors power traction motor applications. Motion Control Association. http://www.motioncontrolonline.org/i4a/pages/index.cfm?pageID=4379.

1.75. National Electric Manufacturers Association (NEMA). 2004. Industrial control and systems: Motion/position control motor, controls and feedback devices. ICS 16–2001. Section I.

1.76. Schneider Electric Motion. 2012. Electric motors: General information for all motors. http://motion.schneider-electric.com/downloads/whitepapers/Electric_Motors_whitepaper.pdf.

1.77. McClelland, W. 1927. The application of the electricity in warships. *Journal of the Institution of Electrical Engineers* **65**(**369**): 829–859.

1.78. Feiertag, K. M. and Donahoo, J. T. 1952. Dynamoelectric machine, U.S. Patent 2,589,999.

1.79. Coughlin, P. 2014. Tracing the steps of the first industrial stepping motor. Kollmorgen blog-in-motion article. https://www.kollmorgen.com/en-us/blogs/_blog-in-motion/articles/paul-coughlin/tracing-the-steps-of-the-first-industrial-stepping-motor/.

1.80. Condit, R. and Jones, D. W. 2004. Stepping motors fundamentals. Microchip Technology Inc. Publication no. AN907. http://www.bristolwatch.com/pdf/stepper.pdf.

1.81. Tsuchiya, E. and Shamoto, E. 2017. Pulse drive: A new power-transmission principle for a compact, high-efficiency, infinitely variable transmission. *Mechanism and Machine Theory* **118**(**2017**): 265–282.

1.82. Kollmorgen Corporation. 2016. The history of Kollmorgen 1916–2016: 100 years of innovation. https://www.kollmorgen.com/uploadedFiles/kollmorgencom/Company/HistoryofKollmorgenAnniversaryBook-mobile.pdf.

1.83. Wavre, N., Vaucher, J.-M., and Piaget, D. 1998. *Drive Systems for Demanding Applications—Linear and Torque Motors in the Industrial Environment*. Carl Hanser Verlag, Munich, Germany.

1.84. Electropaedia. 2005. Electric drives—Brushless DC/AC and reluctance motors (description and applications). http://www.mpoweruk.com/motorsbrushless.htm.

1.85. Ahn, J.-W. 2011. Switched reluctance motor. In *Torque Control* (Ed.: M. T. Lamchich). InTech, Rijeka, pp. 201–252.

1.86. Vaithilingam, C. A., Misron, N., Aris, I., Marhaban, M. H., and Nirei, M. 2013. Electromagnetic design and FEM analysis of a novel dual-airgap reluctance machine. *Progress in Electromagnetics Research* **140**: 523–544.

1.87. McMahon, J. 2010. Piezo motors and actuators: Streamlining medical device performance. *European Medical Device Technology*. http://www.emdt.co.uk/article/piezo-motors-and-actuators.

1.88. Kollmorgen. 2011. Servodisc catalog. http://www.electromate.com/db_support/downloads/KollmorgenGearmotorSeries.pdf.

1.89. Underwriters Laboratories. 2013. *UL 1203: Standard for Explosion-Proof and Dust-Ignition-Proof Electrical Equipment for Use in Hazardous (Classified) Locations*, 5th edn. Underwriters Laboratories, Northbrook, IL.

1.90. Appleton. 2014. NEC® 2014 code review: A guide for use of electrical products in hazardous locations.

1.91. von Amerom, U. 2011. Choose the right electric motors for hazardous locations. *Chemical Engineering Progress (CEP) Magazine* **107**(11): 18–23.

1.92. International Electrotechnical Commission (IEC). 2014. 60079-1: 2014 Explosive atmospheres – Part 1: Equipment protection by flameproof enclosures "d". International Electrotechnical Commission (IEC), Geneva, Switzerland.

1.93. Figliulo, B. 2018. Matching seals to dynamic sealing applications. *Hydraulics & Pneumatics*. https://www.hydraulicspneumatics.com/technologies/seals/article/21887824/matching-seals-to-dynamic-sealing-applications

1.94. Zwyssig, C., Kolar, J. W., and S. D. Round, S. D, 2009. Megaspeed drive systems: Pushing beyond 1 million r/min. *IEEE/ASME Transaction on Mechatronics* **14**(**5**): 564–574.

1.95. Zwyssig, C., Duerr, M., Hassler, D., and Kolar, J. W. 2007. An ultra-high-speed, 500,000 rpm, 1 kW electrical drive system. *Proceedings of 2007 Power Conversion Conference*, Nagoya, Japan, pp. 1577–1583.

1.96. Borisavljević, A. 2011. Limits, modeling, and design of high-speed permanent magnet machines. Ph.D. Dissertation, Technische Universiteit Delft. Delft, the Netherlands.

1.97. Peirs, J., Reynaerts, D., and Verplaetsen, F. 2003. Development of an axial microturbine for a portable gas turbine generator. *Journal of Micromechanics and Microengineering* **13**(**4**): S190–S195.

1.98. Shen, J. X., Qin, X. F., and Wang, Y. C. 2018. High-speed permanent magnet electrical machines – applications, key issues and challenges. *CES Transactions on Electrical Machines and Systems* **2**(**1**): 23–32.

1.99. Zhao, L., Ham, C., Zheng, L., Wu, T., Sundaram, K., Kapat, J., Chow, L., and Siemens, C. 2007. A highly efficient 200,000 rpm permanent magnet motor system. *IEEE Transactions on Magnetics* **43**(**6**): 2528–2530.

1.100. Mellor, P., Al-Taee, M., and Binns, K. 1991. Open loop stability characteristics of synchronous drive incorporating high field permanent magnet motor. Electric Power Applications, *IEE Proceedings B* **138**(**4**): 175–184.

1.101. Kähkönen, O.-P., Makinen, S., Talvitie, M., and Manninen, M. 1992. Radiation damage in Nd-Fe-B magnets: Temperature and shape effects. *Journal of Physics: Condensed Matter* **4**(**4**): 1007–1014.

1.102. Li, S. M., Wang, H. F., Zheng, Y. F., and Cao, L. 2017. Radiation effect on the performance of robot manipulator. *International Journal of Mechatronics and Automation* **6**(**1**): 10–19.

1.103. Hageman, J. P. 2015. Answer to question #11162 submitted to "Ask the Experts". Health Physics Society (HPS) - Specialists in Radiation Protection. https://hps.org/publicinformation/ate/q11162.html#.

1.104. Kushpil, V., Mikhaylov, V., Kugler, A., Kushpil, S., Ladygin, V. P., Svoboda, O., and Tlusty, P. 2016. Radiation hardness of semiconductor avalanche detectors for calorimeters in future HEP experiments. *Journal of Physics: Conference Series* **675**(**2016**): 012039.

1.105. Curie, J. and Curie, P. 1880. Développement par compression de l'électricité polaire dans les cristaux hémièdres à faces inclinées. *Bulletin de la Société Minérologique de France* **3**(**4**): 90–93.

1.106. Lippmann, G. 1881. Principe de la conservation de l'électricité. *Annales de Chimie Et de Physique* **24**: 145.

1.107. Tsukahara, S., Torres, J., Neal, D., and Asada, H. H. 2012. Design method for buckling amplified piezoelectric actuator using flexure joint and its application to an energy efficient brake system. *Proceedings of ASME 2012 5th Annual Dynamic Systems and Control Conference (Joint with JSME 2012 11th Motion and Vibration Conference)*. DSCC2012-MOVIC2012-8711. Fort Lauderdale, Florida, USA.

1.108. Uchino, K. 1998. Piezoelectric ultrasonic motors: Overview. *Smart Materials and Structures* **7**(**3**): 273–285.

1.109. Li, X., Kan, C. H., Cheng, Y., Chen, Z. W., and Ren, T. 2019. Performance evaluation of a bimodal standing-wave ultrasonic motor considering nonlinear electroelasticity: Modeling and experimental validation. *Mechanical Systems and Signal Processing* **141**: 106475.

1.110. Spanne, K. and Koc, B. 2016. Piezoelectric motors, an overview. *Actuators* **2016(5–6)**: 1–18. doi:10.3390/act5010006.

1.111. Vyshnevskyy, O., Kovalev, S., and Wishnewskiy, W. 2004. New type of piezoelectric standing wave ultrasonic motors with cylindrical actuators. *Proceedings of the 9th International Conference on New Actuators*, Bremen, Germany, pp. 451–455.

1.112. Spanner, K. 2006. Survey of the various operating principles of ultrasonic piezomotors. *Proceedings of the 10th International Conference on New Actuators*, Bremen, Germany, pp. 414–421.

1.113. Flüeckiger, M., Fernandez, J. M., and Perriard, Y. 2008. Computer-aided design and optimization of piezoelectric ultrasonic motors. *Electrimacs 2008*, Quebec, Canada.

1.114. Fernandez, J. M., Flüeckiger, M., and Perriard, Y. 2008. Study of a hollow ultrasonic rotary motor. *The Proceedings of 2008 IEEE Ultrasonics Symposium*, Beijing, China, pp. 1449–1452.

1.115. Ryndzionek, R., Sienkiewicz, Ł., Michna, M., and Kutt, F. 2019. Design and experiments of a piezoelectric motor using three rotating mode actuators. *Sensor* **2019(19)**: 5184. doi:10.3390/s19235184.

1.116. Liu, Y. X., Xu, D. M., Yu, Z. Y., Yan, J. P., Yang, X. H., and Chen, W. S. 2015. A novel rotary piezoelectric motor using first bending hybrid transducers. *Applied Sciences* **2015(5)**: 472–484.

1.117. Zhou, X. Y., Chen, W. S., and Liu, J. K. 2015. A new rotary ultrasonic motor using longitudinal vibration transducers. *Advances in Mechanical Engineering* 7(**5**): 1–8.

1.118. Kang, C.-I. 2001. Method of estimating motor torque constant. U.S. Patent 6,320,338.

1.119. Younkin, G. W. Electric servo motor equations and time constants. http://www.ctc-control.com/customer/eleaming/younkin/driveMotorEquations.pdf.

1.120. Welch, R. H. Jr. and Younkin, G. W. 2002. How temperature affects a servomotor's electrical and mechanical time constants. *IEEE Industry Applications Conference. 37th IAS Annual Meeting*, Pittsburgh, PA, Vol. 2, pp. 1041–1046.

1.121. Mazurkiewicz, J. Check temperature when specifying motors. http://www.motioncontrolonline.org/files/public/Check_Temperature_when_Specifying_Motors.pdf.

1.122. Honsinger, V. B. 1987. Sizing equations for electrical machinery. *IEEE Transaction on Energy Conversion* **EC-2(1)**: 116–121.

1.123. Huang, S., Luo, J., Leonardi, F., and Lipo, T. A. 1998. A general approach to sizing and power density equations for comparison of electrical machines. *IEEE Transactions on Industry Applications* 34(**1**): 92–97.

1.124. Huang, S., Aydin, M., and Lipo, T. A. 2001. Torque quality assessment and sizing optimization for surface mounted permanent magnet machines. *IEEE Industry Applications Conference*, Chicago, IL, Vol. 3, pp. 1603–1610.

1.125. Huang, S., Aydin, M., and Lipo, T. A. 2002. A direct approach to electrical machine performance evaluation: Torque density assessment and sizing optimization. *Proceedings of International Conference on Electrical Machines, ICEM'02*, Brugge, Belgium.

1.126. Misra, K. B. 2008. Engineering design: A system approach. In *Handbook of Performability Engineering* (Eds.: K. B. Misra). Springer, London, U.K., pp. 13–24.

1.127. Silva, G. and Serio, L. C. D. 2016. The sixth wave of innovation: Are we ready? *RAI Revista de Administração e Inovação* 13(**2**): 128–134.

1.128. Anderson, D. M. 2020. *Design for Manufacturability: How to Use Concurrent Engineering to Rapidly Develop Low-Cost, High-Quality Products for Lean Production*, 2nd edn. CRC Press, Boca Raton, Louisiana.

1.129. Zeillinger, R. and Köttritsch, H. 1996. Damping in a rolling bearing arrangement. https://evolution.skf.com/us/damping-in-a-rooling-bearing/.

1.130. Chen, W. J. 2015. *Practical Rotordynamics and Fluid Film Bearing Design*. Trafford Publishing, Bloomington, IN.

1.131. Simons, S., Hinchliff, M., White, B., Talabisco, G., Kurz, R., and Ji, M. 2019. Chapter 11: Compressor system design and analysis. In *Compression Machinery for Oil and Gas* (Eds.: K Brun and R. Kurz). Gulf Professional Publishing, Cambridge, MA, pp. 427–447.

1.132. Rade, D. A. and Steffen, V. Jr. 2008. Structural dynamics and modal analysis. In *Experimental Mechanics, Encyclopedia of Life Support Systems*. UNESCO chapter.

1.133. Ramsey, K. A. 1983. Experimental model analysis, structural modifications and FEM analysis on a desktop computer. *Sound and Vibration*, February issue, pp. 1–12.

1.134. Chandravanshi, M. L. and Mukhopadhyay, A. K. 2015. Experimental modal analysis of the vibratory feeder and its structural elements. *International Journal of Applied Engineering Research* 10(**13**): 33303–33310.

1.135. Assis, M., Neto, R. M., Sicchieri, L. C., Kazeoka, T. F., Sanches, L., Guimarães, T. A. M., and Cavalini, A. 2015. Modal and flutter analyses using finite element model for an aerodesign airplane wing. *Proceedings of 23rd ABCM International Congress of Mechanical Engineering*, Rio de Janeiro, Brazil.

1.136. Campbell, W. 1924. *The Protection of Steam-Turbine Disk Wheels from Axial Vibration.* General Electric Company, Schenectady, New York.

1.137. Didier, J., Simon, J. J., and Faverjon, B. 2012. Study of the non-linear dynamic response of a rotor system with faults and uncertainties. *Journal of Sound and Vibration* **331**(3): 671–703.

1.138. Xu, J., Zheng, X. H., Zhang, J. J., and Li, Xuan. 2017. Vibration characteristic unbalance response for motorized spindle system. *Procedia Engineering* **174**(2017): 331–340.

1.139. Piper, J. A., Dudas, M. J., and Sudhoff, D. H. 2008. Motor with integrated drive unit and shared cooling fan. U.S. Patent 7,362,017.

1.140. Choi, G., Xu, Z., Li, M., Gupta, S., Jahns, T. M., Wang, F., Duffie, N. A., and Marlino, L. 2011. Development of integrated modular motor drive for traction applications. *SAE International Journal of Engines* **4**(1): 286–300.

1.141. Brown, N. R., Jahns, T. M., and Lorenz, R. D. 2007. Power converter design for an integrated modular drive. *IEEE Industrial Applications Society Annual Meeting*, New Orleans, FL, pp. 1322–1328.

1.142. Sylora, B. J., Jahns, T. M., and Lorenz, R. D. 2008. Development of a demonstrator model of an integrated modular motor drive. *2008 NSF-CPES Annual Conference*, Blacksburg, VA.

1.143. Ashley, S. 1997. Getting a hold on mechatronics. *ASME Mechanical Engineering Magazine*, May issue, pp. 60–63.

1.144. Oonsivilai, A. and Meeboon, N. 2008. Verification skip writes head-positioning error mechanism using skip writes problem detection. *Eighth World Scientific and Engineering Academy and Society (WSEAS) International Conference on Robotics, Control and Manufacturing Technology*, Hangzhou, China.

1.145. Boyland, J., Dawson, B. S., Dunkleman, S., Tong, W., Trago, B. A., and Winesett, B. L. 2021. Robot arm joint. U.S. Patent 11,161,258.

1.146. Dynapar. 2018. How temperature and humidity affect encoder performance. https://www.dynapar.com/knowledge/encoder-temperature-humidity.

1.147. Venkataraman, B., Godsey, B., Premerlani, W., Shulman, E., Thakur, M., and Midence, R. 2005. Fundamentals of a motor thermal model and its applications in motor protection. *Proceeding of 58th Annual Conference for Protective Relay Engineers*, College Station, TX, pp. 127–144.

1.148. Motor Reliability Working Group. 1985. Report of large motor reliability survey of industrial and commercial installations, Part I. *IEEE Transactions of Industry Applications* **IA-21**(4): 853–864.

1.149. Motor Reliability Working Group. 1985. Report of large motor reliability survey of industrial and commercial installations, Part II. *IEEE Transactions of Industry Applications* **IA-21**(4): 865–872.

1.150. Motor Reliability Working Group. 1987. Report of large motor reliability survey of industrial and commercial installations, Part III. *IEEE Transactions of Industry Applications* **IA-23**(1): 153–158.

1.151. Cornell, E. P., Owen, E. L., Appiarius, J. C., McCoy, R. M., Albrecht, P. F., and Houghtaling, D. W. 1982. Improved motors for utility applications: Final report. The Institute, Palo Alto, CA.

1.152. International Electrotechnical Commission (IEC). 2013. International Standard IEC 60529. Degree of protection provided by enclosures (IP Code), 2.2 Edition (Consolidated version). International Electrotechnical Commission (IEC), Geneva, Switzerland.

1.153. Teco-Westinghouse. Wound rotor motor technology. http://www.tecowestinghouse.com/PDF/woundrotor.pdf.

1.154. NEMA. 2002. NEMA standards publication MG 1-1998 (revision 3, 2002) interfiled: Motors and Generators. http://www.homewoodsales.com/PDFs/tech-library/Motors/NEMA/CompleteMG1-1998Rev3.pdf.

1.155. MGM Electric Motors. 2020. Brake motor general catalogue. http://www.mgmrestop.com/wp-content/uploads/2020/06/MGM-cat-EN-2020-web-JUNE.pdf.

1.156. Jolles, P. 2018. IP69 vs. IP69K. F2 Tech Notes. https://f2labs.com/technotes/2018/01/05/ip69-vs-ip69k/.

1.157. ISO. 2013. International Standard ISO 20653 Road vehicles — Degrees of protection (IP code) — Protection of electrical equipment against foreign objects, water and access. International Organization for Standardization (ISO), Geneva, Switzerland.

2

Rotor Design

Rotor design involves a wide range of knowledge, skills, and experience across various disciplines including material science, rotordynamics, electromagnetics, structural mechanics, aerodynamics, manufacturing technology, thermal control and management, tribology, and fluid dynamics. Furthermore, during the rotor design, a large number of design parameters must be carefully considered and determined.

As discussed previously, an AC motor can be primarily classified as either an induction motor (IM) or a permanent magnet motor (PMM), depending on how the rotor is constructed. Because the structure and the working principle of these two types of motors are fundamentally different, IM and PMM are addressed separately in the following sections.

2.1 ROTOR IN INDUCTION MOTOR

A typical rotor in an IM comprises a cylindrical core that is made up of silicon steel laminates and pressed onto the cylindrical motor shaft. There are two types of IMs: wound rotor IMs and squirrel cage rotor IMs. The primary difference between these two types of IMs is the design of the motor rotor. While the design of stators for all IMs is almost the same, the rotor construction is quite different for different types of IMs.

2.1.1 WOUND ROTOR

In a wound rotor, a set of insulated rotor windings, which are similar to the stator windings, are utilized to accept external impedances. A set of rotor windings are correspondingly connected to a set of insulated slip rings mounted on the rotor shaft. The carbon-composite brushes riding on the slip rings connect to a set of external rheostats (Figure 2.1). This rotor construction design allows the external rheostats to be placed in series with the rotor windings. These rheostats, varying from almost short-circuit condition to an open-circuit condition with infinite external resistance, can remarkably improve motor starting characteristics, of which low inrush current is being the most significant.

Wound rotor IMs have some distinct advantages over other types of IMs. A special feature of the wound rotor motor is its capability to control the motor velocity at different load levels by controlling the rotor currents. By varying the rheostat resistance, it is possible to change the slip and thus to change the rotor rotating velocity. This feature has been used to improve the motor performance. For instance, a wound rotor has the ability to gradually bring up to speed for high-inertia and large load equipment. Long motor life is ensured with the use of external resistor banks during motor startup. For these reasons, the wound rotor IMs are conventionally used for high-power industrial applications.

Using this feature can optimize the torque–speed characteristics for different applications [2.1]. The impact of the rotor resistance on the torque–speed characteristics is shown in Figure 2.2. It can be seen that increasing rotor resistance can result in the shift of the breakdown torque toward zero speed. This can eventually alter motor performance characteristics: (*a*) The increase in the rotor resistance causes a significant increase in the starting torque and the decrease in the corresponding starting current. (*b*) A higher rotor resistance leads to the reduction of the rotating speed of the rotor, that is, the increase in slip. (*c*) For a given full-load torque, the motor full-load speed decreases as the motor rotor

DOI: 10.1201/9781003097716-2

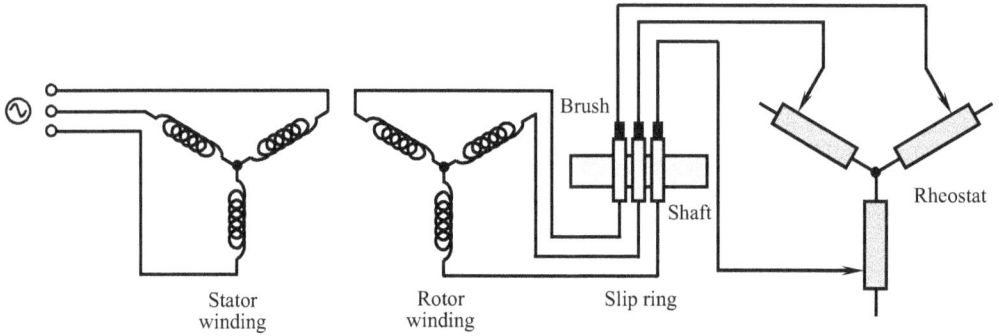

FIGURE 2.1 Wound rotor IM using slip rings and brushes to contact rotor winding and external resistances for improving motor start characteristics.

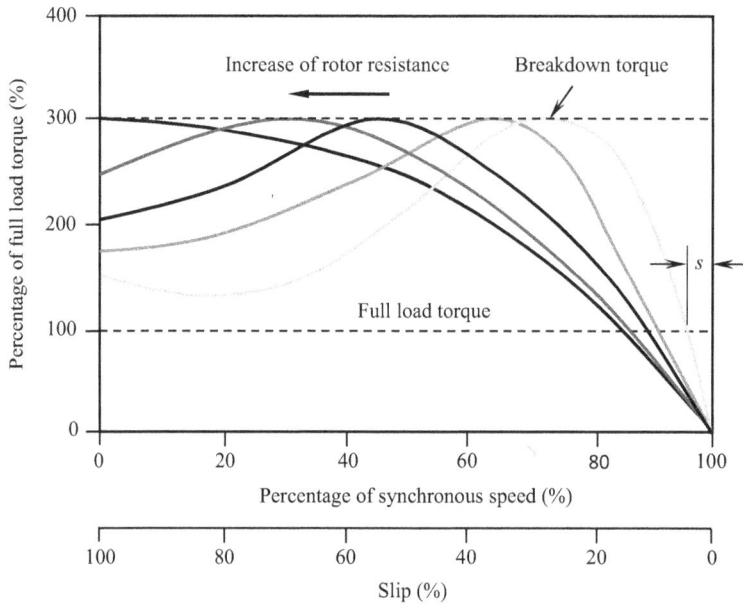

FIGURE 2.2 Variations of torque–speed characteristics of wound IMs as a function of rotor resistance. Increasing rotor resistance results in the shift of the breakdown torque toward zero speed, causing the increase in slip and starting torque and the decrease in full-load speed.

resistance increases. However, the reduction of the full-load speed in such a way is highly undesirable since it lowers the motor efficiency and power output as the increase in the resistive losses (*i.e.*, I^2R losses). (*d*) Under high rotor resistance conditions, the pull-up torque may be eliminated from the torque–speed curves. In fact, a high rotor resistance causes the motor torque to decrease more rapidly than those with lower rotor resistances. (*e*) The higher rotor resistance is always related to the larger variation in the rotation speed as the torque varies. This indicates that wound rotor IMs having high rotor resistances are more susceptible to speed variations as the load torque changes [2.2].

The rotor windings are made up of copper coils inserted into rotor slots. Essentially, the rotor windings exhibit inductance and resistance, and these characteristics can effectively depend on the frequency of the current flowing in the rotor. Today, the wound rotor IMs are less used in industrial applications due to frequent maintenances for slip rings and brushes, high rotor inertia, and requirement of bulky resistor banks.

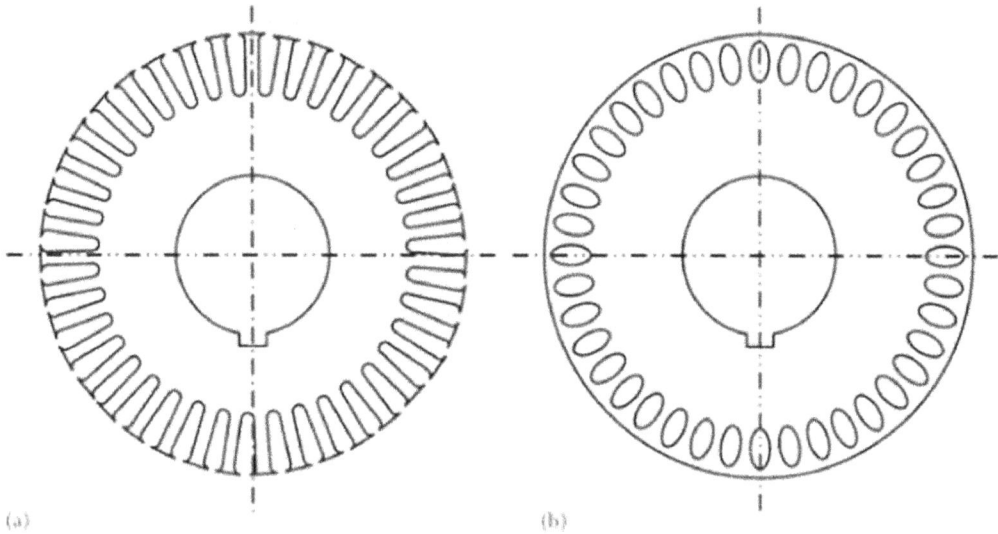

FIGURE 2.3 Rotor slots of a squirrel cage motor: (*a*) open slots and (*b*) closed slots.

2.1.2 SQUIRREL CAGE ROTOR

Most IMs used today are of squirrel cage type. A squirrel cage rotor consists of a rotor core made by stacked laminations and a shaft fitted in the rotor core. A number of equally spaced open or closed slots are punched at the outer circumference of the laminations (Figure 2.3). In fact, the selection of open or closed rotor slot design involves the trade-off because each design has its own advantages and disadvantages, depending on different applications. Open rotor slots allow a substantial reduction of the rotor leakage inductance with the possibility to increase the constant power range speed regulation. In contrast, closed slots tend to produce lower surface loss and have higher rotor leakage inductance for a better filtering in the presence of inverter supply [2.3, 2.4].

A transient 2D finite element analysis (FEA) has shown that the closed-slot rotor geometry gives less spatial rotor harmonics and less stator current harmonics, leading to a significantly lower level of stator vibrations for frequencies above 1 kHz of induction machines [2.5]. In addition, a closed slot rotor helps improve the die-casting process.

Rotor conducting bars are made by casting a conducting material (usually aluminum or copper) directly into rotor slots. These casted conducting bars are usually slightly skewed to the axis of the rotor for reducing cogging torque and connected together with an end ring at each of the rotor ends to form a closed electric circuit. Thus, the stator windings set up a rotating magnetic field around the rotor to induce electric current in the conductive bars and end rings. In turn, the induced currents in the conducting bars/end rings react with the magnetic field of the stator to produce forces acting at a tangent orthogonal to the rotor, resulting in continuous torque on the motor shaft to drive external loads. Figure 2.4a shows an assembled casted copper rotor. By removing the steel laminations, the conducting bars and end rings form a squirrel cage (Figure 2.4b). Thicker end rings can not only reduce current densities but also increase the heat dissipation capability of the rotor.

2.1.2.1 Factors Affecting Resistance of Squirrel Cage Rotor

There are a variety of factors impacting the characteristics and performance of squirrel cage rotors, including the conducting bar material, the number of conducting bars, cross-sectional area, bar configuration and shape, skew angle, and bar position relative to the rotor surface.

A conducting bar with a large cross-sectional area exhibits a low resistance. Moreover, the electric conductivity of copper is nearly 60% higher than that of aluminum. Thus, it is commonly

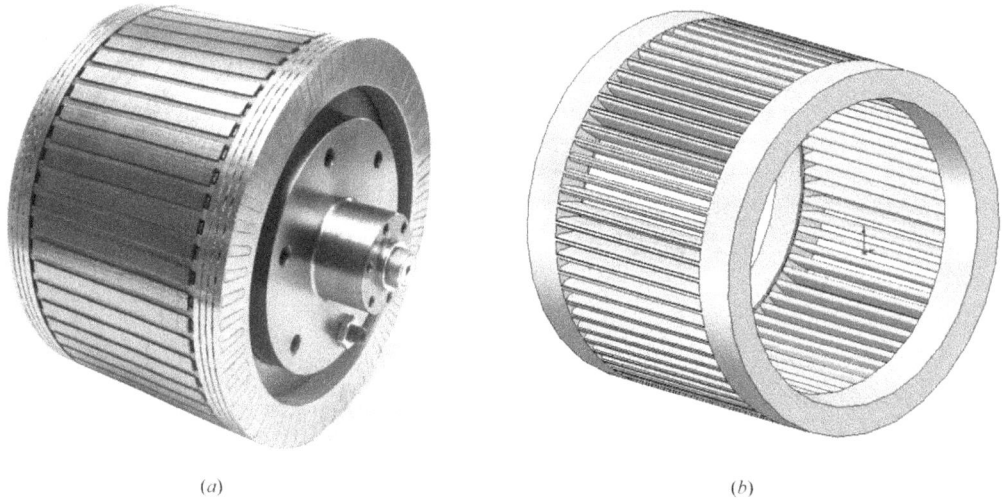

FIGURE 2.4 Casted copper rotor: (a) the rotor is assembled by integrating a shaft with the casted copper rotor core; (b) conducting bars and end rings form a closed electrical circuit for providing the inducted current path, where the electric steel rotor laminations are removed from the rotor core.

accepted that copper rotor bars are superior to aluminum bars because of lower power losses, better performance, and higher motor efficiency. It has been reported that the incorporation of casting copper for conducting bars and end rings in place of aluminum would result in attractive improvements in motor energy efficiency through reductions in motor losses ranging from 15% to 20% [2.6]. With an FEA software, Daut *et al.* [2.7] have analyzed and investigated the performance of the three-phase 0.5 hp IM using either copper or aluminum rotor bars. The FEA results fall into four groups: torque versus speed, torque versus slip, power loss versus speed, and power loss versus slip. Comparisons between copper and aluminum rotor bars have been performed in these four groups and have shown that the copper rotor bars excel in all investigated performance issues.

The position of conducting bars in a squirrel cage IM is an important design parameter to affect the motor performance. The impedance of a conducting bar is comprised of both resistance and inductance. According to the position of the conducting bar relative to the rotor surface, the conducting bar exhibits different magnetic leakage flux and inductance. Positioning the conducting bars near the rotor surface (*i.e.*, near the stator) will produce a relatively small leakage flux and a small inductance. In contrast, positioning the conducting bar deeper radially into the rotor slots (*i.e.*, away from the stator) will result in more leakage and a higher inductance and consequently a higher impedance. Due to the skin effect, at high frequencies, the AC impedance of the outer portion of the conducting bar is lower than that of the inner portion.

2.1.2.2 Double-Cage Rotor

The invention of double-cage rotor was motivated to develop a type of motor that is characterized by a variable rotor resistance during motor operation. This type of motor offers preferably high rotor resistance at the motor start-up, high locked torque, low starting current, and high pull-up torque, while maintaining higher efficiencies when compared to the same power ratings with single-cage designs. For the specific rotor designs and operating conditions, the comparison of torque–speed curves of double-cage and single-cage rotors is presented in Figure 2.5.

As presented in Figure 2.6, a double-cage consists of a large cage buried deeply in the rotor and a small cage located near the rotor surface. The conducting bars in the large cage are loosely linked with the stator and have low resistance and high leakage inductance. On the contrary, the conductors in the small cage are closely linked with the stator and have high resistance and low leakage

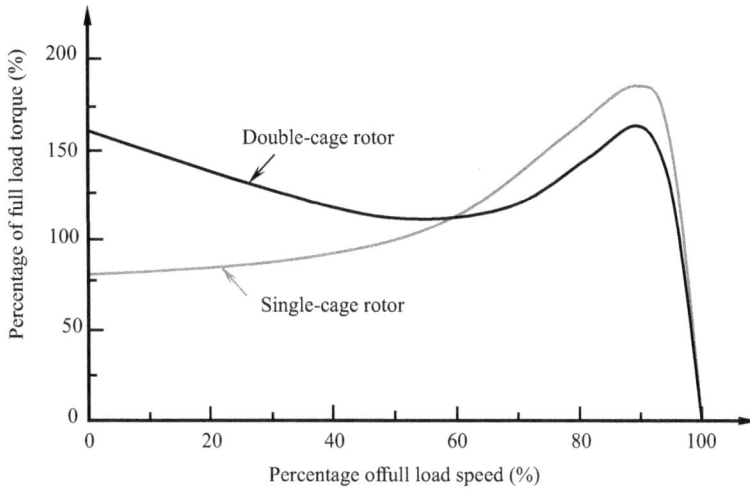

FIGURE 2.5 Comparison of torque–speed curves between double-cage and single-cage rotors.

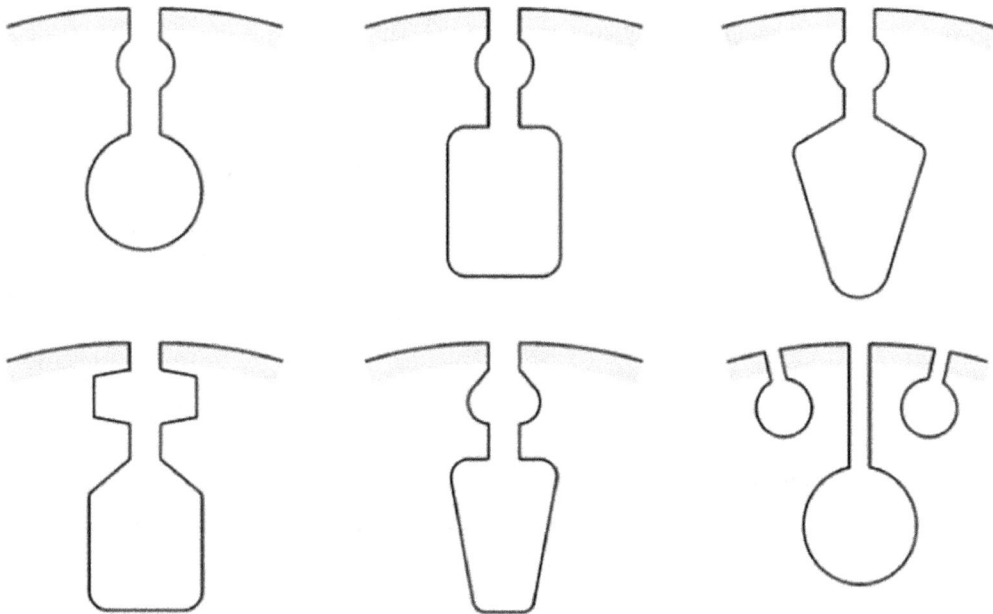

FIGURE 2.6 Various double-cage rotor designs for improving motor start-up and normal operation.

inductance. At the motor start-up stage, only the conductor bars in the small cages that have high resistance are effective, providing high starting torque. In the motor normal operation stage, the conducting bars in both large and small cages become effective. Because the two resistances are connected in parallel, the rotor resistance is

$$R_r = \frac{R_i R_o}{R_i + R_o} \tag{2.1}$$

where R_i and R_o are the resistances of the conductors in the inner and outer cages, respectively. In a case that $R_o \gg R_i$, $R_r \approx R_i$. Therefore, under normal operation conditions with low rotor frequencies, the rotor current mainly goes through the conducting bars in the inner cage with lower I^2R losses. For this reason, some motor manufacturers use a high-resistance material such as brass for the outer cage and a low-resistance material such as copper for the inner cage.

Some large-size motors with double-cage rotors may employ separated end rings to connect the inner and outer cages, respectively. One of the reasons for this is to reduce thermal stresses induced in the end rings due to the different coefficients of thermal expansion of the inner and outer cages.

2.1.2.3 Casting Squirrel Cage Rotor

Several casting processes can be used for making squirrel cage rotors, including centrifugal casting and high- and low-pressure die casting. Low-pressure die casting is usually used for large rotors. In a casting process, the stack of laminations is placed in a die-casting die or a permanent mold containing a space at the top and bottom for the simultaneous casting of end rings. The molten material is poured or forced into the mold.

Two conducting materials are commonly used in casting squirrel cage rotors: (*a*) aluminum, which is the popular casting material of choice by most motor manufacturers due to its low melting temperature (660°C), low cost, and low density which means that an aluminum rotor is subject to less starting inertia, less stress from centrifugal forces, and less vibration, and (*b*) copper, which is the promising material to significantly reduce rotor power losses as the result of high electric conductivity. The melting temperature of copper is 1,083°C. The comparison of properties of copper and aluminum is presented in Table 2.1.

It is well known that incorporation of copper for the rotor bars and end rings in place of aluminum would result in attractive improvements in motor energy efficiency. Manoharan *et al.* [2.10] have reported that for the same output power, by merely replacing an aluminum rotor with a copper rotor and keeping all other parameters unchanged, the motor efficiency is increased by nearly 2.6%. It is observed that this increment in efficiency in the copper rotor is attributed to the higher electrical conductivity of the copper rotor. For instance, under a temperature of 20°C, the electrical resistance of copper is 44% lower than that of aluminum. Hence, for the same current, the substitution of aluminum by copper can result in 44% reduction in the resistance loss.

TABLE 2.1
Comparison of Properties of Copper and Aluminum

Property	Copper UNS C10100[a] [2.8]	Aluminum 383 [2.9]
Density (kg/m³)	8,890–8,940	2,700
Hardness, Vickers	75–90	85
Yield tensile strength (MPa)	69–365	152
Ultimate tensile strength (MPa)	221–455	310
Elongation at break (%)	55	3
Modulus of elasticity (GPa)	115	71
Poisson ratio	0.31	0.33
Shear modulus (GPa)	44	27
Machinability (%)	20	70
Melting temperature (°C)	1083	660
Thermal conductivity (W/m·K)	391	96.2
CTE (μm/μm/K)	1.692×10^{-5} (20°C–100°C)	2.11×10^{-5} at 20°C
	1.728×10^{-5} (20°C–200°C)	2.25×10^{-5} at 250°C
	1.692×10^{-5} (20°C–300°C)	

[a] Due to casting, C10100 is assumed the annealed properties.

FIGURE 2.7   Rotor with casted copper conducting bars and end rings. To reduce cogging torque and magnetic noise, the rotor conducting bars are skewed slightly along the length of the rotor.

Figure 2.7 presents a copper rotor used in vehicles. It is benefited with low resistance losses and low starting torque and increased torque density. This leads to a small motor size with the same power output or more power output for the same motor size. Furthermore, the much simpler design of a copper rotor compared to an aluminum rotor can significantly lower the manufacturing costs [2.11].

It was reported [2.12] that the advantage of using diecasting copper motors is not only the reduction in power losses but also the increase in power density. The diecasting copper rotor motor would be 18%–20% lighter and 14%–18% less expensive than the aluminum rotor machine at the same efficiency, with a possible reduced frame size.

However, two major issues must be addressed before cast copper rotors are fully commercialized: (a) Casting copper into a rotor remains a challenge due to high casting temperatures (>1,000°C) that could be detrimental to the rotor laminations. (b) Copper has a higher coefficient of thermal expansion (CTE) than aluminum. As a motor passes through temperature cycles during its lifetime, the difference in CTE between copper conductors and steel laminations may develop fatigue stresses at the interfaces between the rotor conducting bars and stator core slots. The fatigue stresses may initiate cracks on the conducting bars and eventually lead to fatigue failure of the conductor bars. In some cases, it may result in the separation of the conducting bars/end rings from the rotor core. For the same reason, sometimes breaks may occur at the joints of the conducting bars to the end rings.

The advantages of squirrel cage IMs include their low cost, ruggedness, and maintenance-free operation. However, since the rotor conducting bars and end rings are permanently short-circuited, it is unable to add any external resistance in series with the rotor circuit for starting purposes, leading to high starting current and low starting torque. In fact, this is the major disadvantage of the squirrel cage motors. Because squirrel cage IMs have good constant speed characteristics, they are preferred for driving machinery with constant operating speeds such as fans, blowers, water pumps, grinders, lathes, and printing and drilling machines.

As a rotor is subjected to a very high rotating speed, a retention mechanism may be required to keep the end rings connected to the rotor against high centrifugal forces. A number of solutions are effective in reducing the stress concentration in certain areas and are also effective in keeping the rotor together at a high speed. These include the following: (a) Each end ring can be secured using a band of a high-strength material. (b) Rivets can be cast into the rotor to secure the end rings to the body. (c) Specially designed laminations can be used for securing the end rings in rotors used today with the rotor body. (d) By breaking the end ring into sections, limiting the thermal growth of each section, thus reducing the thermal growth of the entire the ring. The specially designed geometry of the end ring in rotor used today helps contain the end ring at high speed by acting like a band as used in rotors today [2.13].

2.1.2.4   Skin Effect

In power systems, the skin effect refers to the tendency of AC to concentrate near the outer surface of a conductor and to decay exponentially toward the center. This indicates that the induced current in conducting bars is unevenly distributed over the cross-sectional area of the conducting bars. The current density increases from the center to the outer surface of the bar. Because of the skin effect, as well as the proximity effect between the conductor bars, the nonuniform current distribution causes more resistive loss. As a result, the AC resistance of a conducting bar is much higher than its DC resistance. Generally, the AC resistance increases with the increase in the frequency of rotor currents. The distribution of current density in a round conducting wire is given as

$$J = J_o e^{-d/\delta} \tag{2.2}$$

where

J_o is the current density at the conductor surface
e is the base of the natural logarithm ($e = 2.718$)
d is the distance measured from the surface toward the center of the conductor
δ is the skin depth, which is defined as the distance at which the current decreases to e^{-1} ($e^{-1} = 0.3679$) of its original value

The skin depth δ is inversely proportional to the square root of the frequency (in Hz).

2.1.3 INDUCTION MOTOR DESIGN TYPES AND THEIR PERFORMING CHARACTERISTICS

The National Electrical Manufacturers Association standard has specified four major design classifications of AC IMs: designs A, B, C, and D. These design classifications apply particularly to the rotor design and hence affect the performance characteristics of the motors. Figure 2.8 displays a series of torque–speed curves for these design classifications with identical power rating and rotating speed [2.14]. It illustrates that the design characteristics such as starting torque, speed, and breakdown torque can be combined in different ways:

- Design A motors are used for applications that may undergo frequent but short-time overloading and thus require high breakdown torque, such as injection molding machines. This type of motor has normal starting torque (typically 150%–170% of the rated torque) and a relatively high starting current.
- Design B motors are general-purpose motors, designed for applications that require normal starting and running torques, such as machine tools, fans, and centrifugal pumps. This

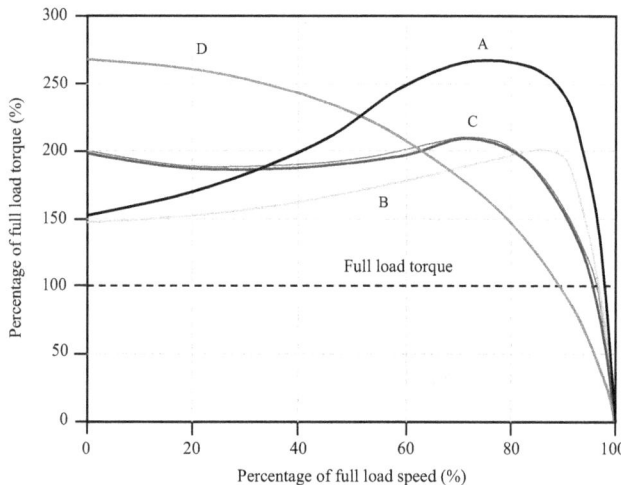

FIGURE 2.8   Torque–speed characteristics for design A, B, C, and D motors defined by NEMA.

type of motor usually has a comparable starting torque with Design *A* but offers the low starting current. The locked rotor torque is good enough to start many loads encountered in industrial applications. The slip is about 5%.

- Design *C* motors are selected for applications that require high starting torque, such as inclined conveyors and gyratory crushers.
- Design *D* motors, called *high slip* motors, are sometimes used to power hoists and cycling loads, such as oil well pump jacks and low-speed punch presses.

Note that the starting torque ranges broadly among the design classifications, from about 150% of full-load torque for Design *A* motors to about 260% of rated torque for Design *D* motors. In addition, the speed at full-load for Design *A* motor is several percent less than the Design *D* motor.

2.2   PERMANENT MAGNET ROTOR

A PMM adopts permanent motors (PMs) attached to the rotor in place of rotor windings or conductors/end rings to generate magnetic fields. The first motor invented by Faraday in 1821 was practically a primitive PM DC machine [2.15]. The first attempt to introduce rare earth PM motors into the market of electric machines was in the 1980s using samarium–cobalt (Sm–Co) magnets that were invented in the 1960s. Since then, many manufacturers have successfully launched rare earth PM motors into the market. Neodymium–iron–boron (Nd–Fe–B) magnets were added into the rare earth PM family in the mid-1980s. Like Sm–Co, Nd–Fe–B has demonstrated many design advantages and superior performance over common PMs. Because Nd–Fe–B is more abundant than Sm–Co, it is extensively used in energy-conversion devices today. In recent years, modern PM motors have presented a momentum of rapid development. However, PM motor applications have been generally limited to relatively low-power motors, usually less than 500 kW, in most cases, less than 100 kW.

2.2.1   Discovery of Phenomenon of Magnetism

Ancient Chinese discovered the phenomenon of magnetism. Utilizing the magnetism of natural lodestone, the first compass was invented around the fourth century BC and was called *south pointer* or *Si Nan* (司南). In the book of *Guiguzi* (鬼谷子), which was written in the early third century BC, it stated that "the lodestone makes iron come or it attracts it." The author Wang Yu (a thinker and strategist at approximately 300 BC) noted that in addition to its main purpose of serving as designators of direction, the Chinese primarily used it to order and harmonize their environments and lives (it refers to as *Feng Sui*—风水). The compass could be carried with jade hunters to prevent them from getting lost during their journeys.

The earliest Chinese records show the spoon-shaped compass *south pointer* could be dated back to sometime during the Han Dynasty (from the second century BC to the second century AD). As shown in Figure 2.9, a spoon-shaped lodestone is placed on a cast bronze plate that had the 8 trigrams (*Ba Gua*—八卦), as well as the 24 directions (based on the constellations) and the 28 lunar mansions (based on the constellations dividing the equator). Because the shape of the spoon bottom is spherical, the spoon can rotate freely on the bronze plate. Toggling the spoon to rotate, when it stops, its handle will point to the south.

2.2.2   Permanent Magnet Characteristics

The use of PMs in electric motors is highly desirable since they have great potential for reducing motor weight and increasing efficiency. In a PMM, the magnetic field of rotor is produced by PMs. Because there is no need of rotor windings or conducting bars, the motor structure is greatly simplified. However, due to the limited magnetic strength of PMs, IMs are still dominant in large-size electric motors.

FIGURE 2.9 Invented compass in ancient China: (*a*) Appeared on a Chinese stamp of 1953. (*b*) A replica.

PMs are key parts in PM motors to generate magnetic fields along with stator windings. The advantages of PM motors are as follows:

- Having higher efficiency than IMs
- Simplifying the rotor structure
- Requiring less maintenance
- Lowering the rotor inertia under the same power rating
- Eliminating the need for field exciter and slip rings
- Reducing cooling requirements due to less rotor losses
- Offering more precise speed control
- Lowering bearing currents
- Achieving higher power/torque density

With the fast developments of advanced magnetic materials and continuous improvements in operation performance and state-of-the-art manufacturing methods, PM motors have been extensively used in various military and industrial applications. However, in comparison with conventional IMs, PM motors are generally more expensive due to high costs of magnets, especially for the rare earth magnets.

A thorough understanding of the basics of magnetism and magnetic field is a key foundation in designing optimal modern PM motors. It is the interchange and interaction between electric field and magnetic field that construct the theoretical basis for electric machine design. The magnetic flux density vector **B**, magnetization vector **M**, and applied magnetic field strength vector **H** can be related according to

$$\mathbf{B} = \mu_o(\mathbf{H} + \mathbf{M}) \tag{2.3}$$

where μ_o is the magnetic permeability in a vacuum. As a universal constant, it has a value of $\mu_o = 4\pi \times 10^{-7}$ Wb/(A·m), or H/m. The relative magnetic permeability μ_r is defined as the ratio of the permeability in a material to the permeability in a vacuum,

$$\mu_r = \mu/\mu_o \tag{2.4}$$

The permeability or relative permeability is a measure of the resistance of a material against the formation of a magnetic field, or the ease with which a **B** field can be induced in the presence of an external field **H**.

For linear materials that do not contain ordered paramagnetic ions, the magnetization **M** is related linearly to magnetic field strength **H**,

$$\mathbf{M} = \chi_m \mathbf{H} \tag{2.5}$$

where χ_m is the magnetic susceptibility ($\chi_m = \mu_r - 1$). Combining the above equations, it yields

$$\mathbf{B} = \mu_o(1 + \chi_m)\mathbf{H} = \mu_o\mu_r\mathbf{H} = \mu\mathbf{H} \tag{2.6}$$

The most important relationship of a magnetic material is the nonlinear behavior between the magnetic flux density **B** and the magnetic field strength **H,** where **B** and **H** are measured in gauss and oersteds and are analogous to current and voltage in an electric circuit, respectively. A typical **B–H** hysteresis loop of a magnetic material is shown in Figure 2.10. This hysteresis loop is useful for evaluating the response of a magnetic material to the applied magnetic field. It is readily noted that the **B–H** hysteresis loop is skew-symmetric with respect to the *H*-axis. An unmagnetized PM is composed of a large number of small magnetic domains that randomly oriented and cancel one another, resulting in a net magnetic moment of zero on a large scale. This indicates that for the material in the unmagnetized state, both **B** and **H** are at the origin in Figure 2.10. During the magnetizing process, these randomly oriented magnetic domains are forced to align with the applied external magnetic field, and the magnet is thus magnetized following the initial magnetization curve (marked as curve #1). As it reaches the point of magnetic saturation, an additional increase in **H** will produce little increase in **B.** When driving magnetic field strength **H** to zero (curve #2), the magnet still retains a considerable degree of magnetization because of hysteresis of the magnet. This phenomenon is referred to as residual magnetism and the corresponding magnetic flux density is denoted as **B**$_r$. To move the magnetic flux density **B** to zero, **H** must be reversed. This is referred to as coercivity (curve #3 – also refers to demagnetization curve). The force required to remove the residual magnetism from the magnet is called the coercivity force, expressed as −**H**$_c$ in Figure 2.10.

As the magnetic field strength continues to decrease in the negative direction, the magnet is magnetized again but in the opposite direction (curve #4). Similarly, as it reaches the point of magnetic saturation (but in opposite direction), any changes in −**H** in the negative direction will result in little variation of −**B**. From this point, increasing **H** back in the positive direction will lead to an increase in **B** (curve #5). As **H** becomes zero again, the material remains negative flux density −**B**$_r$ Further increasing **H** will bring **B** to zero (curve #6) and reach the point of magnetic saturation (curve #7). It is noted that this curve did not pass through the origin because the positive magnetic field strength is required to remove the residual magnetism.

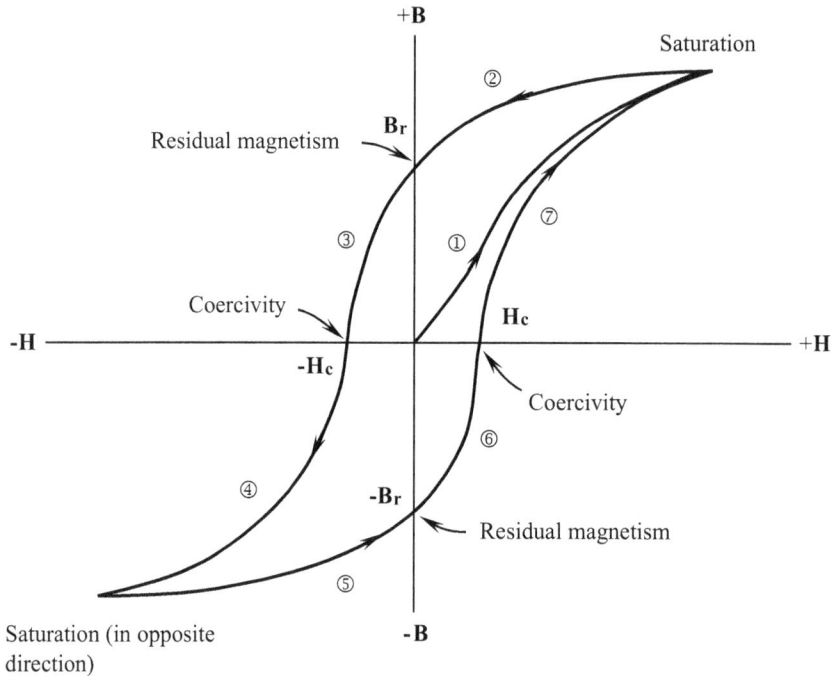

FIGURE 2.10   The magnetization hysteresis loop of a PM.

A magnetized PM may lose partly or completely its magnetism under certain circumstances. The demagnetization of the PMs in a motor can cause performance degradation or complete failure of the motor.

The electromagnetic properties of PMs are strongly influenced by temperature [2.16]. An increase in temperature can increase the kinetic energy of the molecules in a magnetic material, causing them to wiggle and jiggle around and the domains to lose alignment. Studies have shown that temperature variations normally produce changes in the hysteresis loop width, induction amplitude, maximum differential permeability, and other magnet properties [2.17], impacting the performance of PM motors. An FEA was conducted to investigate the influence of PM materials and temperature on the performance of interior PM motors [2.18]. The results show that along with the increase in temperature, remanent flux density $\mathbf{B}_r$ decreases, resulting in the decrease in the average torque and cogging torque but the increase in torque ripple.

2.2.3   PERMANENT MAGNET MATERIALS

PMs have been used in electric machines for more than a century. However, the modern history of PMs started in the 1930s with the introduction of alnico. As the strongest type of magnets at that time, alnico alloys dominated the market of electric machines due to their excellent remanence and high thermal and chemical stability until the development of rare earth magnets in the 1970s.

In the last several decades, dramatic improvements have been made in magnetic properties and availability of PMs. Applications of PMs in electric machines are increasingly growing. There are several main types of PM materials used in motor manufacturing, including ferrite PMs, alnico PMs, and rare earth PMs. The comparison of magnetic characteristics among various magnetic materials is presented in Table 2.2.

TABLE 2.2

Magnetic Characteristics for Various Magnetic Materials

Magnetic Material	Grade	B_r (Gauss)	H_c (Oe)	H_{ci} (Oe)	BH_{max}[a] (MGOe)	T_{max} (°C)	T_c (°C)	Reference
Ferrite magnet	8	3,900	3,200	3,250	3.5	300	450	[2.19]
Nd–Fe–B	39H	12,800	12,300	21,000	40	150	310	[2.19]
Sm–Co	26	10,500	9,200	10,000	26	300	750	[2.19]
Alnico	5	12,500	640	640	5.5	540	860	[2.19]
Ceramic	8	3,900	3,200	3,250	3.5	300	460	[2.19]

[a] BH_{max} represents energy product, given in MGOe. Nd–Fe–B magnets have the highest energy product of any magnetic materials available below 200°C.

2.2.3.1 Ferrite Permanent Magnets

Ferrite PMs are made from some oxides (SrO, BaO, or Fe_2O_3). Due to low costs of raw materials and simple manufacturing techniques, the magnets are considerably cheap. The corrosion resistance of the oxides is considered excellent, and no surface treatments are required. Ferrite PMs can work under severe environmental conditions such as high temperature, dust, salt fog, and moisture. Therefore, they are extensively used in automobile motors and power tools.

Alnico magnets are alloys made of aluminum, nickel, cobalt, copper, iron, and other materials by either sintering or casting. The remarkable advantage of alnico magnets is that they have the widest range of temperature stability and the lowest temperature coefficient of any commercial magnet materials (0.02% per degree centigrade). Their maximum working temperatures can be up to 450°C–550°C. Due to good corrosion resistance, there are no surface treatments required. This type of magnet is usually used in temperature-sensitive devices such as instruments and sensors.

2.2.3.2 Rare Earth Permanent Magnets

Rare earth elements refer to the 15 lanthanide series elements of atomic numbers from 57 through 71, plus scandium (atomic number 21) and yttrium (atomic number 39) in the periodic table of chemical elements. Rare earth PMs are made from alloys of rare earth elements such as neodymium (Nd), dysprosium (Dy), and samarium (Sm). In comparison to regular ferromagnets, rare earth magnets have higher residual magnetic flux density and much higher coercivity but lower Curie temperature. Today, rare earth magnets are extensively used in high technologies, clean energies, and especially high-end defense platforms. They play a vital role in various PM motors for producing high efficiency, large torque density, and high-performance motors with reduced motor vibration and noise. Because Sm–Co magnets can work at higher temperatures (up to 250°C–350°C) than Nd–Fe–B magnets (usually 80°C–220°C), they are more suitable for manufacturing various high-performance PM motors. In addition, due to the strong anticorrosion ability, electroplating of Sm–Co magnets is generally unnecessary. However, because of the low reserve of samarium on the earth, Sm–Co magnets are very expensive.

Among all magnets, Nd–Fe–B magnets by far are the strongest magnets in the world, offering the highest energy product of any magnetic material available below 170°C (338°F). Figure 2.11 describes the temperature dependence of $(BH)_{max}$ for most commercial permanent magnets [2.20]. Except for a few cases, the $(BH)_{max}$ decreases with the increase in temperature. Among them, Nd–Fe–B is strongly temperature-dependent, its $(BH)_{max}$ falls sharply once the temperature exceeds 100°C. Nevertheless, compared to Sm–Co and alnico magnets, Nd–Fe–B magnets have the $(BH)_{max}$ of 45 MGOe at 25°C, which is 1.4 and 9.0 times higher than those of Sm_2-Co_{17} (32 MGOe) and alnico (5 MGOe) magnets, respectively. Permanent magnets based on Nd–Fe–B alloys emerged in the middle of 1980s. Production expanded rapidly and was concurrent with the growth of the personal computer market. After nearly 40 years of development, today Nd–Fe–B magnets have not

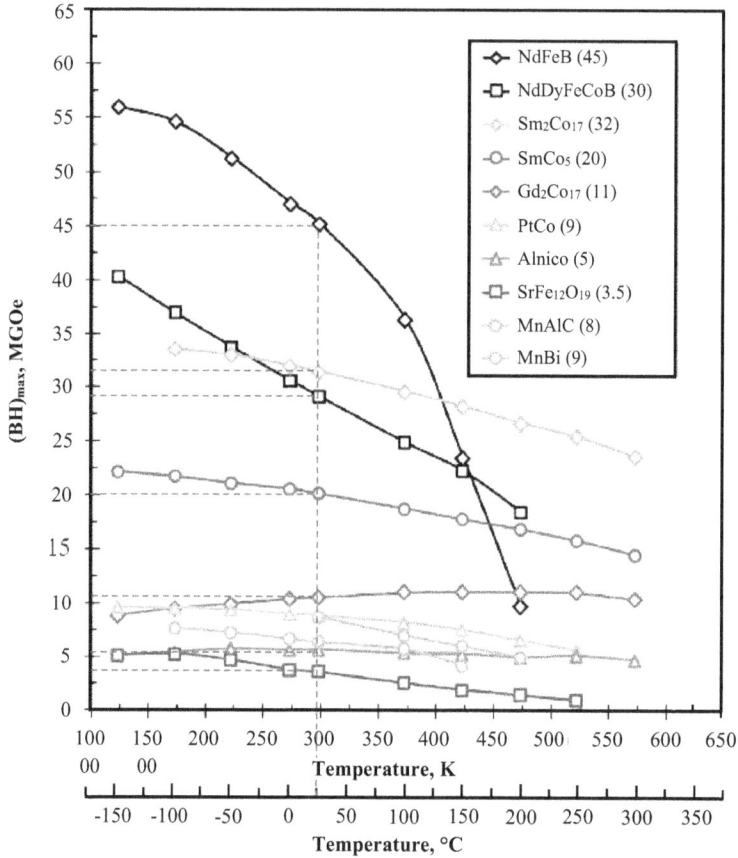

FIGURE 2.11 Comparison of temperature dependence of $(BH)_{max}$ among various commercial permanent magnets. Among them, Nd–Fe–B decreases the most for temperature $T > 100°C$ ($373K$). The $(BH)_{max}$ values in parentheses are measured at $25°C$ ($298K$) [2.20].

only the excellent qualities of high remanent magnetization, high coercive force, and high magnetic energy product, but also the better mechanical properties for processing of complex shapes.

According to the current trend, the Nd–Fe–B magnetic material may become the dominant magnetic material in the future in electric machines, energy transportation, medical devices, computers, home appliances, and other fields. However, because the material contains large amounts of neodymium and iron, the Nd–Fe–B magnets are easily corroded and thus must be coated with nickel (Ni), zinc (Zn), gold (Au), chromium (Cr) epoxy resin, *etc.* One of the disadvantages of Nd–Fe–B magnets is their low performance temperature. The magnetic loss at high temperature is relatively large.

According to the manufacturing process, Nd-Fe-B magnets can be categorized into two groups: sintered magnets and bonded magnets. While sintered Nd-Fe-B magnets are produced by using the powder metallurgy techniques to obtain anisotropic and fully dense sintered magnets with high $(BH)_{max}$, bounded Nd-Fe-B magnets are typically made from rapidly-quenched Nd-Fe-B magnetic powder mixed with a polymer binder (*e.g.,* thermo-elastomer and thermo-plastic resin) at an appropriate mass ratio by *compression mold* (for simple geometries) or *injection mold* (for complex geometries).

Generally, sintered Nd-Fe-B magnets are heated at high temperatures through complex processes, forming a crystal microstructure, they have better magnetic properties and higher magnetic performance than bonded Nd-Fe-B. Consequently, they are widely employed in various fields such as electric machines, robots, energy-saving appliances, electronic devices, hybrid electric vehicles, and medical equipment. However, because sintered Nd-Fe-B magnets have poor dimensional accuracy, they require secondary processing such as grinding, machining, wire cutting, *etc.* to achieve desired shapes and dimensions, leading to a higher processing cost. In contrast, bonded Nd-Fe-B magnets can be made into various shapes with very tight tolerances. As a result, it eliminates the need for secondary processing and greatly reduce the processing cost. In addition, bonded magnets have high resistive, resulting in suppressing eddy current. However, the density of bonded magnets is lower than that of sintered magnets because it contains a resin matrix (which is usually lighter than rare earth magnet materials), and the density varies according to the content of the resin matrix, yielding lower and resin content-dependent magnetic properties. It is worth noting that the characteristics of bonded magnets are a function of both the magnet power and the binder utilized. Because of their low cost, high dimensional accuracy, high mechanical strength, and large degree of freedom in shape, bonded Nd-Fe-B magnets are suitable for small motors, office automation equipment, instrumentation, audiovisual equipment, mobile phones, automated instruments and other products.

2.2.3.3 New Developments of Alternative PMs and Reductions of Reliance on Rare Earths

In the last two decades, new materials have played a major role in the technological innovations needed to propel the rapid development of high-performance, high-powered electric machines that are ever lighter and smaller.

Rare earth magnets as presently the strongest permanent magnets in the world have been extensively employed in a variety of industrial, commercial, and military applications, for instance, electric machines, computer hard discs, wind power generators, smartphones, electrified vehicles, magnetic resonance imaging scans, sensors, drones, and military equipment. Among all applications, about a quarter of rare earth magnets are used for electric motors and about 11% for wind power generators. Although the rare earth elements may not be as scarce as their name indicates, they are unevenly distributed on the Earth. Accounting for 71%–77% of global production of rare earths (2018–2019) and 36.7% of world reserves [2.21], China plays a pivotal role in the manufacturing and sales of rare earth materials.

The disruption to supply chains caused by tariffs, lack of sustainability, and uncertainty over the future for the stable supply of critical rare earths have prompted industries to be less reliant on rare earth magnets. Fully replacing rare earth magnets or even just lowering the demand for them will gain both economic and environmental benefits. Over the past few decades, scientists and researchers in universities, national labs, and R&D centers worldwide have been dedicated to developing new novel PM materials for replacing existing rare earth-based magnets. Various approaches have been proposed, including using rare earth-free magnets, making magnets with less-expensive, more-plentiful rare earth materials (*e.g.*, cerium), or making magnets with reduced critical rare earths (*e.g.*, neodymium, europium, terbium, dysprosium, yttrium, and praseodymium). Among them, the direct substitution of critical rare earths in PMs has received considerable attention.

Aiming at developing a new generation of novel materials for high-performance rare earth-free PMs and reducing the demand for rare earths, several different but relevant strategies are often adopted by researchers and scientists:

- It has been widely acknowledged that one of the major factors governing magnetic properties is the nanostructures (or microstructures) of magnetic materials. The key structural characteristics that explain rare earth magnets' exceptional strength is their microscopic anisotropy. In order to obtain a complete understanding of it, attention has been given to the study of nanostructures of magnets [2.22]. Well-designed and controlled nanostructures of alternative materials, particularly high magnetic moment materials (magnetic moment mm is induced by a current *i* flowing about a loop of area A) with high anisotropy, can remarkably increase the coercivities of the materials. High anisotropy materials are attractive candidates for high-coercivity application and shape anisotropy is predicted to produce the large coercivity forces [2.23]. In fact, anisotropy in nanosystems helps reduce the magnitude of thermal fluctuations that can randomize the magnetic moment. The most common types of anisotropy are as follows: (*a*) magnetocrystalline anisotropy; (*b*) surface anisotropy; (*c*) shape anisotropy; (*d*) exchange anisotropy; and (*e*) induced anisotropy. All these anisotropies have an influence on the magnetic properties to a certain extent. In nanoparticles, shape anisotropy and magnetocrystalline anisotropy are the most important [2.24]. In nanostructured materials, anisotropy depends not only on the band structure of the parent material, but also on the shape of the nanoparticles. Researches in this direction are ongoing using nanoparticles, nanotubes, and nanorods.

- To explore new materials that exhibit good magnetic characteristics comparable to the rare earth magnets, it generally uses a bottom-up approach to discover alternative materials based on state-of-the-art theoretical modeling and simulation methods. Experimental studies of material's magnetic properties are considered not only expensive but also extremely lengthy. Hence, using the theoretical analysis and numerical simulation as the powerful tools allows to identify new phases and compounds that are suitable candidates for rare earth-free PMs [2.25]. Evaluations of a wide range of material combinations while simultaneously analyzing their magnetic properties have been carried out by means of a systematic computational high-throughput screening technique. With this approach, a very large number of promising materials and compounds of interest can be effectively and rapidly predicted and identified.

- Besides developing new rare earth-free magnets, there are a variety of other approaches for indirectly reducing the demand for rare earths, including but not limited to: (*a*) Improving the magnetic properties of the existing types of magnets. An example concerns replacing the bonded Nd–Fe–B magnets using modified ferrites with lanthanum and cerium to increase the remnant flux density. Although this is also a rare earth element, lanthanum-based ferrite costs remain much lower than Nd–Fe–B products. However, the coercivity of ferrite magnets H_c is about 1/5–1/3 of Nd–Fe–B magnets. It becomes a critical challenge for researchers to improve H_c of modified ferrites. (*b*) Optimizing the production process of rare earth magnets to reduce the manufacturing cost and the use of raw materials. Fraunhofer researches have targeted the manufacturing process using injection molding, in which the magnetic material is brought directly into a desired shape together with a plastic binder and then sintered [2.26]. This also eliminates time elaborate reworking. (*c*) Designing more efficient electric motors and wind turbine generators with powerful cooling systems could indirectly reduce the use of rare earths in the machines. (*d*) The race to repatriate rare earth supply chains in the wake of the COVID-19 pandemic have given fresh impetus to industry's efforts to reconsider the use of conventional induction and reluctance motors in some applications. As an example, Tesla has used this type of motor in its electric vehicles with its new generation electric motors intends to abandon PM motors. Optimizing the design of such types of motor with newly emerged technologies such as 3D printing may reduce the gap between PM and IM motors.

- Developing advanced rare earth recycling techniques has opened up another way to reduce industry's reliance on rare earths. Cost-effective methods for separating rare earth elements that have been processed into products could provide a steady domestic source of rare earths while reducing waste. However, the bottleneck of recycling rare earth elements is the cost required to purify the mixtures obtained from end users' devices. Moreover, recycling in some cases could create greater environmental harm than mining for the virgin metals. Typically, only around 1% of the rare earths in US are recycled from the end-products [2.27]. In the EU, only 6% of the heavy rare earths and 7% of the light rare earths are recovered. A group of researchers from the University of Pennsylvania [2.28] has proposed a simple, solubility-based separation process of rare earth elements. This method makes purifying recycled rare earth elements much less expensive.

Many investigations focus on improving the magnetic properties, such as energy product, coercivity, remanence, and Curie temperature in a variety of ways. Besides nanostructures, heat treatment and magnetic alignment are found to be extremely important factors in controlling magnetic properties. The rapid developments of nanotechnology in recent years shed a new light on research and development of nanoalloy magnets that are especially interesting to the motion control industry because they can further reduce the power density of electric machines and increase the diversity and availability of PMs. Previous studies showed that arrays of nanomagnets displayed various magnetization patterns that are closely related to the nature of the magnetic interactions between neighboring nanomagnets [2.29, 2.30].

It is well known that the strength of a magnet is measured in terms of coercivity (the ability of a magnet to resist demagnetization) and saturation magnetization (the maximum magnetization at full saturation condition) values. These values increase with a decrease in the grain size and an increase in the surface to volume ratio. Therefore, the magnets made of nanocrystalline yttrium-samarium-cobalt grains may possess very unusual magnetic properties due to their extremely large surface area [2.31].

Bi-magnetic core/shell nanoparticles have gained increasing interest due to their foreseen applications. Inverse antiferromagnetic (AFM)/ferrimagnetic (FiM) core/shell nanoparticles are particularly appealing because they may overcome some of the limitations of conventional FiM/AFM systems. These inverse structures, where the core is AFM and the shell is FiM, have exhibited very large coercivities and loop shifts. The coercivity H_c and exchange bias shift H_e demonstrate a non-monotonic dependence with the core diameter and the shell thickness. Both H_c and H_e exhibit maximum values for rather small core diameters (in a range of 3–6 lattice spacings) [2.32]. Lottini et al. [2.33] have devised a new way to synthesize AFM/FiM core/shell nanoparticles as a feasible alternative for use in rare earth-free magnets. The research team has used a mixed iron-cobalt oleate complex ($Co_{0.3}Fe_{0.7}O/Co_{0.6}Fe_{0.4}O_4$) in a one-step synthetic approach to produce magnetic core-shell nanoparticles, with mean diameter ranging from 6 to 18 nm. The resulting material demonstrates strong magnetic properties that induce a big improvement of the energy-storing capabilities and therefore could replace rare earths in magnets. Their studies suggest that the combination of highly anisotropic AFM/FiM materials can be an efficient strategy toward the realization of novel rare earth-free PMs.

A research team led by the University of Leeds [2.34] has made a breakthrough in an advanced material that may potentially replace rare earth-based magnets. The scientists have developed a hybrid film from a bilayer of cobalt and nanocarbon molecule C_{60} (Co-C_{60}). Because the presence of C_{60} dramatically boosts cobalt's magnetic energy product, this material exhibits significantly enhanced coercivity with minimal reduction in magnetization. The measured magnetic strength increases by a factor of five at low temperatures. Arising from asymmetric magnetoelectric coupling in the metal–molecule interface, the form of anisotropy is outlined, called π-anisotropy. The properties of Co/C_{60} bilayer have been measured and showed an extremely strong anisotropy enhancing effect arising from the C_{60} film. Bilayers of this type may represent a means to create thin films and multi-layers with extremely high marked $\mu_o MH$ energy products. However, of particular note is that this type of hybrid magnet may take a long time to practically apply to electric motors and wind turbine generators.

Some R&D efforts are focusing on finding new magnet compositions instead of the direct substitution of rare earth magnets. Manganese (Mn) has emerged to be a potential candidate as a key element in rare earth-free magnets. Its five unpaired valence electrons give it large magnetocrystalline energy and the ability to form several intermetallic compounds [2.35]. Mn-based magnets, such as MnBi, MnAl, and MnGa, have been developed and fabricated to achieve the predicated magnetic properties. More recently, by adopting a combination of surfactant-assisted milling and resin-bonding techniques, multi-walled carbon nanotubes (MCNTs) in processing rare earth-free MnAl nanocomposite magnets were exploited [2.36]. The surfactant-coated MnAl nanopowders are characterized with respect to their structural and magnetic properties. The MnAl(MCNT) with 5 wt.% Fe bonded nanocomposite magnets processed with the surfactant-coated powders exhibit a significant improvement in the magnetic parameters such as saturation magnetization M_s and coercivity H_c.

Strong permanent magnets based on cobalt nanorods are most suited to small-scale applications in microelectronics and will not replace rare earth magnets in large engineering projects like wind turbines and maglev systems. Researchers have investigated magnets from various combinations of cobalt, iron, copper, and cerium (a less-expensive, more-plentiful rare earth material than neodymium). The research group at Ames Laboratory has identified promising candidates with the rehabilitative approach. The cerium and cobalt-based alloys such as $CeCo_3$ or $CeCo_5$

may potentially replace rare earth magnets. By adding magnesium, paramagnetic $CeCo_3$ could be transformed into ferrimagnets. Polycrystalline copper, iron, and refractory metal-doped $CeCo_5$ systems can achieve $(BH)_{max}$ of 13.0–15.5 MGOe. Ce-based magnets become comparable to Sm-Co magnets at the curie temperature of $600°C–650°C$ [2.37]. In addition, $CoFe_2C$ in the form of nanoparticles is a highly magnetic material in a specific direction, leading to nanomagnets that work at room temperature [2.38].

In the effort of reducing critical rare earths in industrial applications, Toyota has successfully developed electric motors that use up to 50% less rare earth magnets. The new magnet uses lanthanum (La) and cerium (Ce) to replace a portion of neodymium, which costs 20 times less than neodymium. In fact, merely reducing the amount of neodymium and replacing it with lanthanum and cerium can result in a decline in motor performance. To address this issue, Toyota engineers have adopted new technologies that suppress the deterioration of coercivity and heat resistance [2.39].

Pathak *et al.* [2.40] have proposed a new way to fully replace dysprosium and partly replace neodymium (by about 20%) in the Nd–Fe–B magnet using cobalt and cerium, showing excellent high-temperature magnetic properties, with intrinsic coercivity among the highest known for $T \geq 180°C$.

Other researches are looking at optimization of the lower-performance magnets, *e.g.*, Alnico, and their iron homologs or combining well-known magnetic materials such as iron nitrides with hard ferrites at the nanoscale level [2.20]. Compared with rare earth magnets, Alnico magnets have higher magnetic remanence B_r but lower coercivity H_c and thus are subjected to the risk of demagnetization. Alnico magnets possess excellent corrosion resistance and high-temperature stability. Researchers at Florida International University have studied an extended Alnico family of magnetic alloys with eight alloying elements (*e.g.*, Fe, Co, Ni, Al, Ti, Hf, Cu, and Nb) and without rare earths [2.41]. With advanced semi-stochastic algorithms for constrained multi-objective optimization in combination with experimental testing, magnetic properties of 180 alloys are captured and compared. Among them, 80 optimized magnetic alloys are verified experimentally. Based on the testing results, the research is highly recommended for determining which of the alloying elements could be replaced with small amounts of readily available, affordable rare earths so that $(BH)_{max}$, H_c, and B_r could be significantly increased and maintained at higher temperatures. The most promising direction on developing new high-performance magnets is the optimization of parameters defining thermal treatment and magnetic treatment protocols. With the new procedure, H_c increases from 1,100 to 1,350 Oe and $(BH)_{max}$ increases from 5,600 to 6,900 J/m³.

Over the past few decades, despite significant progress have been made in replacing rare earth magnets, the direct substitution of rare earths in PMs or the development of new materials with compatible properties to the rare earth magnets is still at the research stage. Many experts believe that unless there is a major R&D breakthrough, it is unlikely that the sintered Nd–Fe–B magnets will be replaced in wind power generators in the short term because other materials have yet to achieve the required energy density [2.42]. To date, no promising results are announced regarding the substitution of rare earth magnets in commercial applications or the discovery of new magnets with a similar high-energy product [2.43]. This implies that directly replacing rare earths still has a long way to go before reaching the stage of large-scale industrial production.

2.2.4 Magnetization

Magnetization refers to the process through which the crystal domains of a non-magnetized magnetic material are forced to align with one another under an applied external magnetic field. This makes all of the domains point in the same direction, creating its magnetic north and south poles.

Permanent magnets can be divided into two categories based on their magnetic property patterns: (*a*) Isotropic magnets have equal magnetic properties in all directions. Because this type of magnet has no preferred magnetic direction, it is possible to magnetize them in any direction. The remanent magnetization of magnets has the same direction as the applied external magnetic field. (*b*) Anisotropic magnets that exhibit optimum magnetic properties in a certain direction of orientation when magnetized along this preferred direction of orientation to saturate magnets. In terms of energy product, anisotropic magnets are much stronger than the isotropic magnets. Hence, in practice, the most useful commercial magnets are anisotropic magnets.

To reach the high magnetic energy output and achieve magnetic stability, magnets must be fully magnetized, called magnetic saturation. In fact, saturation of a material can be thought of the limit at which the material can carry magnetic flux. The magnetizing energy required to saturate a magnet depends on the intrinsic coercivity of a magnetic material. As a general rule of thumb, a peak field of 2–2.5 times the intrinsic coercivity is required for fully saturating magnets.

There are two magnetization methods commonly used in industries: *static field magnetization* (SFM) and *pulse field magnetization* (PFM). The SFM uses a created static electromagnetic field to magnetize magnets. The PFM is used when stronger magnetic fields are desired, or for specific magnetization patterns (*e.g.*, multipole magnetization). Each method is optimized for specific materials, magnet shapes, sizes, and certain magnetizing patterns.

In the SFM method, magnetization is achieved by exposing magnets to a high-intensity magnetic field, in which the strength depends on the type of magnetic materials. Among the magnetic properties, coercive force is the key parameter that decides the required magnetic field strength for a certain material.

The most common approach in the SFM method is to use DC coils to produce external magnetic fields. Usually, for small and simple shaped magnets, DC power sources (*e.g.*, a set of batteries) may be directly applied to the coils without using iron cores. For large-size magnets or magnets that require strong magnetizing fields, coils with iron cores are needed and energized with rectified AC power sources. Due to the high heat generation during the magnetization process, air- or water-cooling is often applied to the magnetizer. The major disadvantages of this method are the difficulty in obtaining proper magnetizing field direction in complicated magnet designs and the relatively slow magnetization rate due to coil inductance.

Most magnetizers need magnetizing fixtures in order to complete the process. While axial and diametrical magnetization can be made by standard coils, multipole, radial, and other complex magnetic patterns require specially designed coils that are suitable for different shapes and grades.

As invented by Pyotr Kapiza in 1924 [2.44], the PFM technique is the most efficient and effective way for magnetizing magnets. An advantage of PFM is that it enables to produce a high magnetic field with less energy. Due to the rapid development of digital technology, electrical current pulses (especially pulse width) can be precisely produced and controlled to meet various requirements in magnetizing processes. Generally, a material with high resistivity can be magnetized with a pulse of a few microseconds (*i.e.*, a narrow pulse), while a more conductive material may need a pulse lasting a few hundredths of a second (*i.e.*, a wide pulse). A wide pulse ensures that all domains are exposed to proper magnetizing field strength. However, a pulse wider than necessary results in power losses due to generated heat. In practice, the PFM can be typically done by using commercially available pulse magnetizers such as capacitor discharge magnetizers and waveform control pulse magnetizers.

In a capacitor discharge magnetizer, electrical charge is held by a capacitor bank. When a silicon controller rectifier (SCR) is closed, a discharging current flows through the magnetizing coil from the capacitor bank. The discharging current then gently decreases due to the effect of electromagnetic induction as the SCR opens. Thus, the pulsed current can be controlled by chopping as repeating the switching operation over a short time. Rare earth magnets are commonly magnetized with capacitor discharge magnetizers.

The full performance of PMs can only be guaranteed if it is magnetized properly to saturation. The magnetization of PMs can be implemented either prior to or posterior to rotor assembly. Premagnetized magnets can easily achieve 100% of saturation. This approach may be especially suitable for some cases. For example, since a large PM motor is difficult to place in a magnetizer, the premagnetization of magnets is the only solution. However, there are some apparent disadvantages for premagnetized magnets: (*a*) The magnetized magnets are difficult to handle during rotor assemblies because of their attraction to steel parts and/or attraction/repulsion with other

FIGURE 2.12   Postmagnetizing of a PM motor by means of a magnetizer.

magnets. (*b*) Premagnetized magnets may pick up magnetically permeable dirt in their assembly and transportation processes. (*c*) Magnets are usually brittle and easily damaged in shipment if they are not packaged properly. (*d*) Strong PMs in shipment may affect the performance of electronic devices such as navigation instruments in aircraft or damage some items in nearby packages such as data storage devices and smartphones. (*e*) PMs may be demagnetized during transportations due to temperature changes, mechanical impacts, and external magnetic fields. Premagnetized magnets cannot guarantee to maintain their original magnetizing strengths during assemblies. Therefore, it is often preferred to magnetize assembled systems at the assembly locations. However, in this way, the magnetized saturation level may be slightly lower (about 99%) due to the fact that the magnets are too close to one another.

Usually, postmagnetization works efficiently for surface-mounted PMs and single-layer interior PMs those are located near the rotor surface. In the motor industry, because of the high production efficiency and ease of assembly, a majority of small PM motors are postmagnetized by means of magnetizers, as depicted in Figure 2.12.

For multilayer buried magnet rotors, postmagnetization would result in a large amount of magnetic material buried deep within the rotor to be only partially magnetized or even could not be magnetized at all. To solve this problem, an innovative method is presented in Figure 2.13. It removes magnetic material from the regions of the rotor that cannot be effectively magnetized during the postmagnetization process and inserts magnetizing coils. Thus, the inserted magnetizing coils enhance the magnetizing field produced by the stator or other magnetizing fixture for improving the rotor magnetization. When it is complete, the magnetizing coils are removed from the rotor [2.45].

2.2.5   Factors Causing Demagnetization

During the assembly process, magnetic properties of PMs may change due to the improper installation processes and changes in environmental conditions. As a result, the variation in magnetic properties can directly affect the reliability and performance of electric motors. A primary cause of demagnetization of PMs is high temperature. When a PM is heated, the internal energy increases to make the atoms in the PM vibrate actively. The vibration of the atoms may shift the alignments of some atoms from an ordered pattern to a nonaligned disordered pattern. As a result, the PMs may be partially demagnetized. In a worse case, excessive temperatures may lead the PMs to completely lose their magnetic properties, causing electromechanical failure, power loss, or erratic power fluctuation. This is especially true for rare earth PMs.

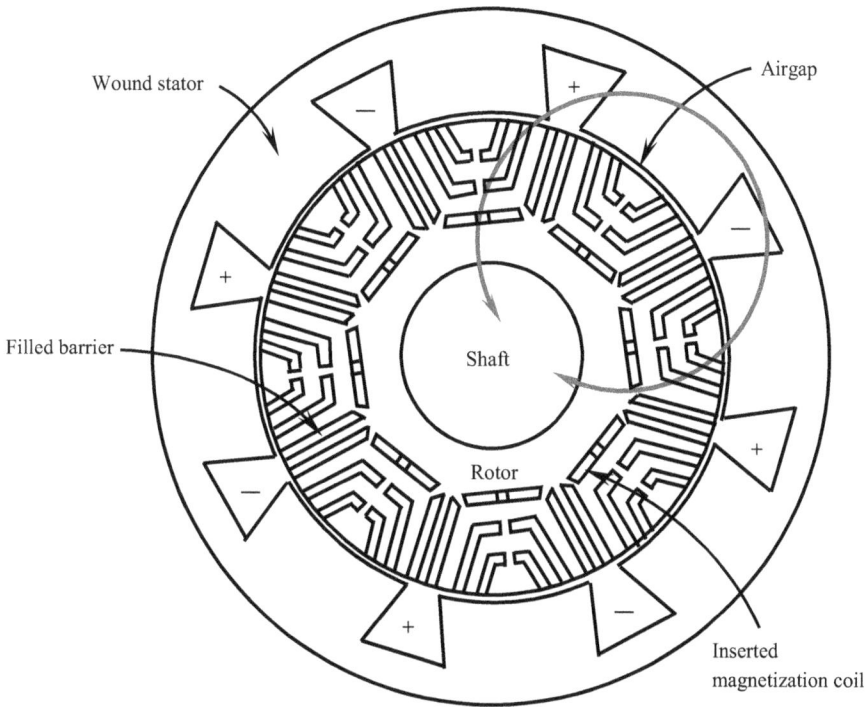

FIGURE 2.13 Cross-section of a multilayer interior magnet motor with a magnetizing auxiliary winding for magnetizing interior magnets (U.S. Patent 6,674,205) [2.45]. (Courtesy of the U.S. Patent and Trademark Office, Alexandria, VA.)

The temperature effect on magnet demagnetization usually falls into three categories:

- Reversible losses: At the early stage, the rise in temperature will gradually decrease the magnetism of a magnetic material. For example, the reduction in magnetism of Nd–Fe–B magnets is 0.13% per degree Celsius. At this stage, the losses in magnetism can be recovered automatically as the magnetic material returns to its original temperature.
- Irreversible but recoverable losses: Continuously increasing the temperature will result in the irreversible drop in the magnetism due to the reduction in the coercive force. The losses in magnetism can be only recovered by remagnetizing. This is also applicable for the partial demagnetization due to the interference of external magnetic fields.
- Irreversible and unrecoverable losses: Each magnetic material has a specific temperature, the Curie temperature T_c, at which all magnetic properties are completely lost. Because the crystal microstructure of magnetic materials changes at the Curie temperatures, remagnetization cannot bring its magnetism back again. When the temperature is much higher than the Curie temperature, the lattice of the magnetic material will be melted and recrystallized, resulting in the permanent loss of the characteristics as the permanent magnetic material.

The impact of temperature on the demagnetization characteristics of PMs has been extensively investigated by many researchers and engineers. This is normally studied by looking at the **B-H** hysteresis loop at the second quadrant. A demagnetization curve represents the relationship of the magnetic flux density generated by a magnet and the demagnetization force imposed on the magnet. As examples, the demagnetization curves for a typical ferrite PM under various temperatures are shown in Figure 2.14 [2.46]. It can be seen from this figure that the shapes of all demagnetization curves are similar. These curves decrease slowly initially with the increase in $-\mathbf{H}$ and then drop off abruptly when the *knees* are passed. At the knee points, the irreversible demagnetization starts to

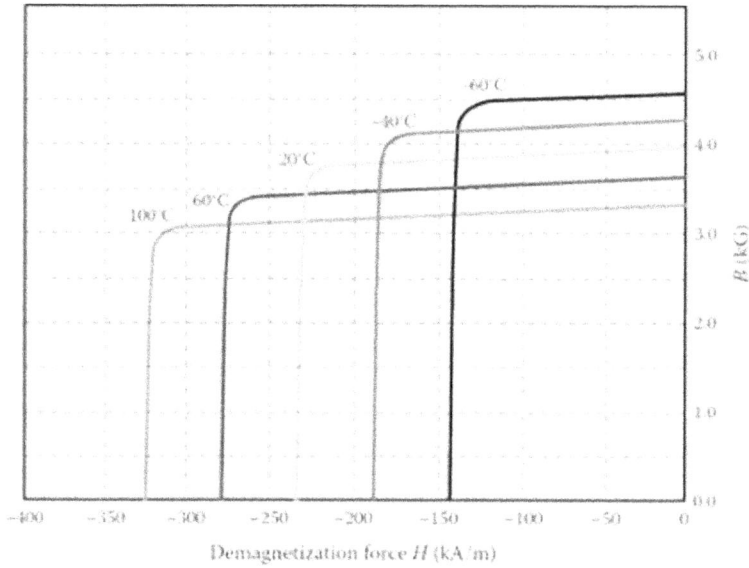

FIGURE 2.14 Impact of temperature on demagnetization characteristics of a typical ferrite magnet.

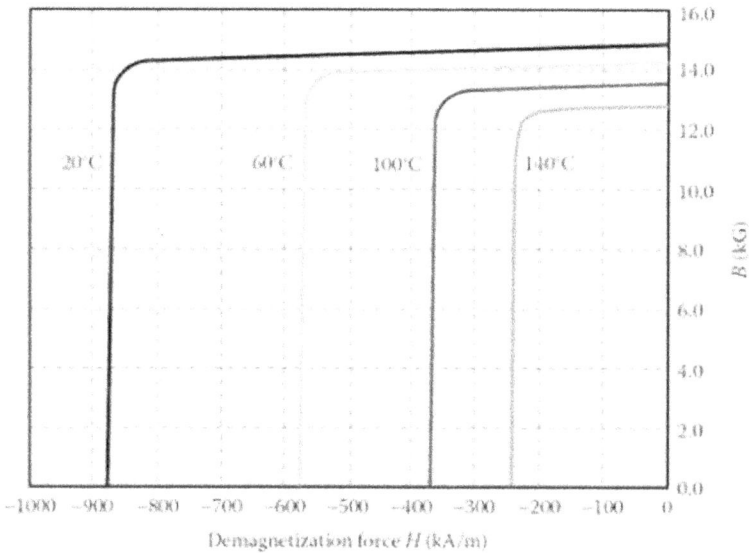

FIGURE 2.15 Impact of temperature on demagnetization characteristics of a typical Nd–Fe–B magnet.

take place. The knee occurs at approximately −130 kA/m with the temperature of −60°C but occurs −320 kA/m as the temperature is raised to 100°C. Increasing operation temperature from −60°C to +100°C, the demagnetization curves in sequence reduce in both their magnetic field strength **H** (*i.e.*, move toward −**H**) and magnetic flux density **B** (*i.e.*, move toward −**B**).

For a typical Nd–Fe–B PM, the demagnetization curves are shown in Figure 2.15 [2.47]. Unlike in ferrite magnets, the temperature impact on the magnetic characteristics of Nd–Fe–B magnets is somewhat different. When the operation temperature increases from 20°C to 140°C, the demagnetization curves increase in their magnetic field strength **H** (*i.e.*, move toward +**H**), while they decrease in their magnetic flux density **B** (*i.e.*, move toward −**B**). This indicates that for a PM motor subject to a large range of operating temperature, its maximum torque capacity and efficiency of the motor can vary over a wide range.

Furthermore, it has been observed that the magnetic properties change with temperature due to the change in the magnetic domain structure. The effect of the operational temperature on Nd–Fe–B magnets has been investigated on a 2.5 kW, 24-pole surface PM prototype motor by Beniakar *et al.* [2.48]. Their results show that the rise of the operational temperature from 20°C to 140°C causes a significant decrease in the airgap flux density thus deteriorating the machine's overall performance.

In addition to temperature, other factors can also cause demagnetization of magnetic materials. A strong external magnetic field can distort the magnetic field of a PM, causing some atoms to orient their electron spins to conform to the external magnetic field. This will turn some magnetic domains in the PM from an aligned orientation to a random orientation. In fact, it is possible to demagnetize a PM by applying an external magnetic field in a direction opposite to the direction of magnetization of the PM.

Mechanical disturbances also tend to randomize the magnetic domains and thus demagnetize PMs. Thus, it is important to avoid strong mechanical shocks and violent impacts such as hammering or jarring on PMs. It is usually preferable to ship PMs in an unmagnetized state to prevent damages to themselves and things close to them. In order to safely and reliably ship magnetized PMs, special care must be taken to ensure that the PMs are packaged well with damping materials.

Radiation-induced demagnetization of PMs has been studied for various insertion devices [2.49, 2.50]. It has been reported that Sm–Co magnets exhibit significant demagnetization when irradiated with a proton beam of 10^9–10^{10} rad. Nd–Fe–B test samples were shown to lose all of their magnetization at a dose of 7×10^7 rad and 50% at a dose of 4×10^6 rad [2.51]. When a PM motor is subjected to electron beam or γ-ray irradiation, this type of loss must be carefully considered.

Studies have demonstrated that PMs are affected by time, known as magnetic creep. PMs undergo changes immediately after magnetization. At this stage, the PMs are less stable, showing significant variations in magnetic properties by fluctuations in thermal and magnetic energy. After a period of time, the PMs become more stable as unstable magnetic domains decrease.

The time effect on modern PMs is minimal. It has been reported that over 100,000 h, the losses are in the range of essentially zero for samarium-cobalt materials to less than 3% for alnico five materials at low permeance coefficients [2.52]. Due to the extremely high coercivities, rare earth magnets experience a much less time effect than other types of magnets. Furthermore, most losses of rare earth magnets are reversible or recoverable.

2.2.6 Maximum Operating Temperature

Maximum operating temperature is defined as the temperature below which magnets can operate normally with a slight reduction in demagnetization. For PM motors, the maximum operation temperature provides an important reference temperature for the design of the motor cooling system. It is important to maintain the maximum operating T_{max} well below the Curie temperature T_c for preventing losses in magnetism, as well as the structural or mechanical damage of magnets. In fact, for a particular magnetic material, T_{max} is determined by not only the magnetic material but also the material fabrication process.

2.2.7 Permanent Magnet Mounting and Retention Methods

A significant disadvantage of using PMs in a motor is the thermal and mechanical vulnerability of PM materials. Because of the inability of sintered Ne-Fe-B and Sm-Co magnets to withstand rotationally induced tensile stresses, they are often contained in retaining devices such as retaining sleeves and wrapped Kevla® fiber bands for converting the stresses in magnets from tensile to compressive.

In practice, PMs may be either mounted on the rotor outer surface or inserted into retention slots passing axially through the rotor core. A motor with surface-mounted PM is simple in construction and easy to manufacture. However, the magnets in this design are always subjected to large centrifugal forces and spatial harmonics of the stator winding magnetomotive force.

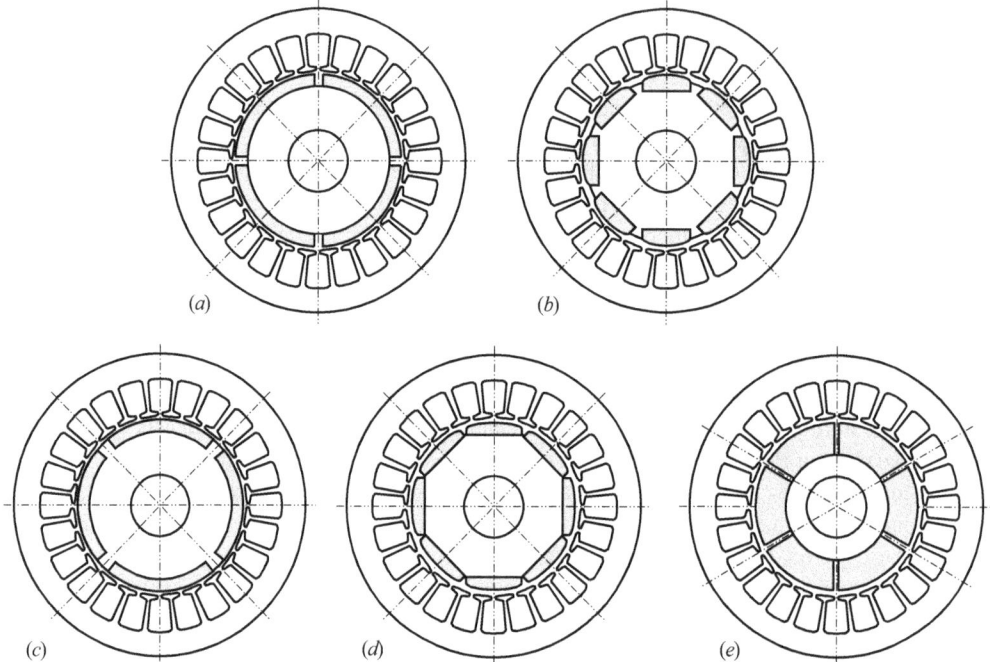

FIGURE 2.16 Various mounting methods on rotor surface: (*a*) tile-shaped magnets mounted on the rotor surface, (*b*) bread loaf magnets partially positioned in the sunken portions of the rotor, (*c*) tile-shaped magnets fully positioned in the sunken portions of the rotor, (*d*) bread loaf magnets mounted on the surface of an octagon rotor, and (*e*) thick magnets mounted on the thin rotor core, generating a strong magnetic field.

Surface-mounted PM rotors, which offer higher performing characteristics than interior magnet rotors, are the most common rotor configuration for PM machines. To minimize eddy-current losses in PMs, it is desirable to use multiple small magnets. However, PMs on the rotor surface are prone to centrifugal force, which may break their mechanical integrity and result in dynamic imbalance of the rotor, particularly at high-speed operation. For this type of machine, one of the most important issues is to hold the magnets in place firmly to prevent them from flying off during motor operation [2.53]. Some surface mounting methods are presented in Figure 2.16, where PMs are positioned either directly on the rotor surface or on the sunken portions made on the rotor surface.

PMs can be mounted on rotor surfaces by mechanical retaining methods (*e.g.*, using clamps, wedges, and wrapped fibers), chemical retaining methods (*e.g.*, using adhesives), physical retaining methods (*e.g.*, using soldering), or combinations of these methods. In practice, the common magnet retaining methods include (*a*) to glue arc magnets on the rotor core surface with strong reliable adhesives (Figure 2.17), (*b*) to secure the magnets with reinforcing fine fiber (*e.g.*, Kevlar® aromatic polyamide fiber) or fiberglass band/string wrapped around the magnets (Figures 2.18 and 2.19), (*c*) use high-strength nonmagnetic metal sleeves or composite (made of non-metallic materials) sleeves such as carbon-fiber/epoxy sleeves, and (*d*) utilize thin stainless steel or other metal sleeves on the outside of PMs. Though each method has its own pros and cons, all these techniques offer simple and effective solutions for retaining magnets.

Among all techniques, the stainless steel and other metal sleeves play multiple roles. In addition to retain PMs, they also effectively shield PMs from stator's harmonic currents. Although eddy current is induced in the metal sleeve to raise its temperature, the sleeve is readily cooled by the airflow through the airgap. On the contrary, carbon-fiber sleeves do not generate eddy current. With the low density, some carbon-fiber sleeves are far stronger than their metal counterparts. However, using this type of sleeve has some drawbacks: (*a*) This type of sleeve does not provide any harmonic

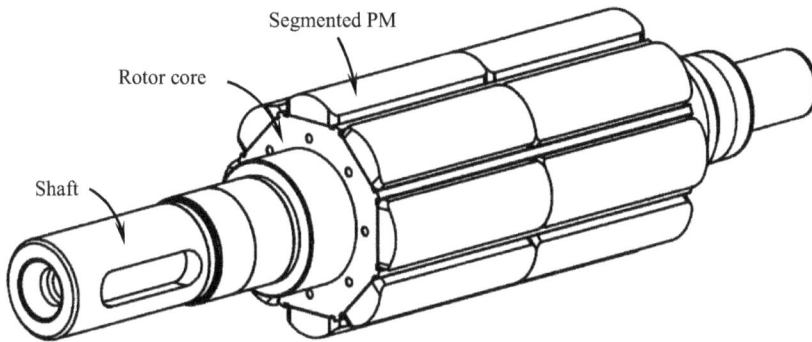

FIGURE 2.17   Segmented permanent magnets are adhesively bonded onto the surface of a rotor core.

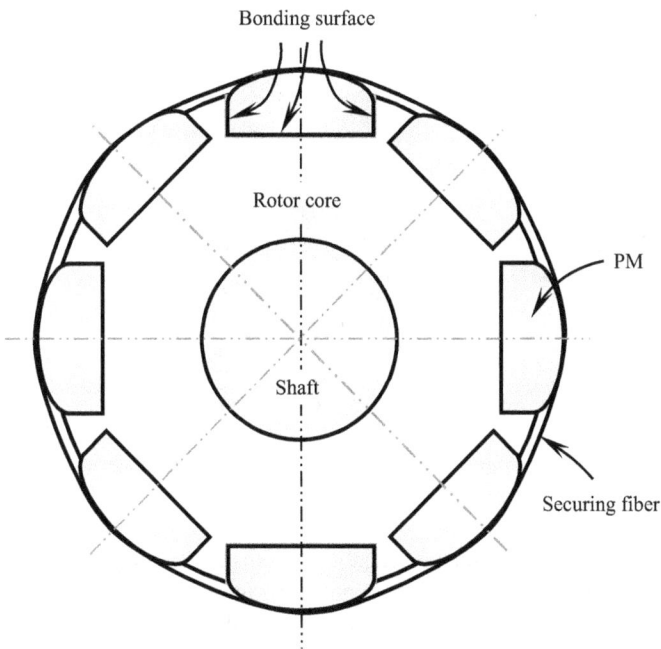

FIGURE 2.18   Gluing PMs on the rotor surface and securing with wound special fiber strings outside the magnets.

filtering. (*b*) Due to their low thermal conductivity, they act as thermal barriers to heat generated in the magnets [2.54]. It is noted that regardless of its material, a sleeve reduces the effective airgap and thus increases the magnetic leakage flux at the airgap.

An adhesive provides strong bonds between magnets and a rotor core. However, under high rotating speeds, both PMs and their bounding layers are subjected to large centrifugal forces that cause high tensile stress in them. If the induced stresses in these bodies exceed their strengths, the PMs could fall off from the rotor surfaces or break into pieces, causing the motor failure. Furthermore, the bonded magnets become permanently attached to the rotor, making it difficult or otherwise infeasible to replace them, wholly or individually. Moreover, adhesive is also susceptible to failure due to extreme temperatures, aging, and chemical exposure [2.55].

FIGURE 2.19 PM rotors are commonly wound with reinforcing fine fibers to provide large mechanical strength for retaining PMs on rotors.

In operation, a retaining device must be capable of withstanding not only high centrifugal forces but also axial forces that may arise during manufacturing, assembly, or installation processes. In addition, properly designed retaining devices may remarkably reduce the power losses of the rotor and PMs, especially for high-speed PM synchronous machines. Thus, the risks of overheating due to eddy current in PMs and the PM demagnetization resulted from high temperature are mitigated.

Shen *et al.* [2.56] have investigated various methods to reduce the eddy current losses and proposed a simple method by grooving metal retention sleeves. In such a way, the flowing path of the eddy currents can be obscured by both circumferential and axial grooves on the metal retaining sleeve, and consequently the rotor and PM eddy current losses can be reduced. Compared among the circumferential, axial, and comprehensive grooves, it is found that the circumferential grooving is the most effective for overall performance improvement and the simplest for practical manufacture with little extra cost.

The selection of the material of retaining device can influence the rotor and PM losses of PMMs. It was reported that the PM loss decreases when a copper layer is put between a carbon fiber and a PM ring [2.57]. Chen *et al.* [2.58] have analytically investigated the performance of carbon fiber and nonmagnetic alloy sleeves. They found that at the static state ($\omega=0$) and 24,000 rpm both the radial stress and hoop stress of PMs with a carbon fiber sleeve always remain negative, indicating that these stresses are compressive stresses. Conversely, for the protective measure of nonmagnetic

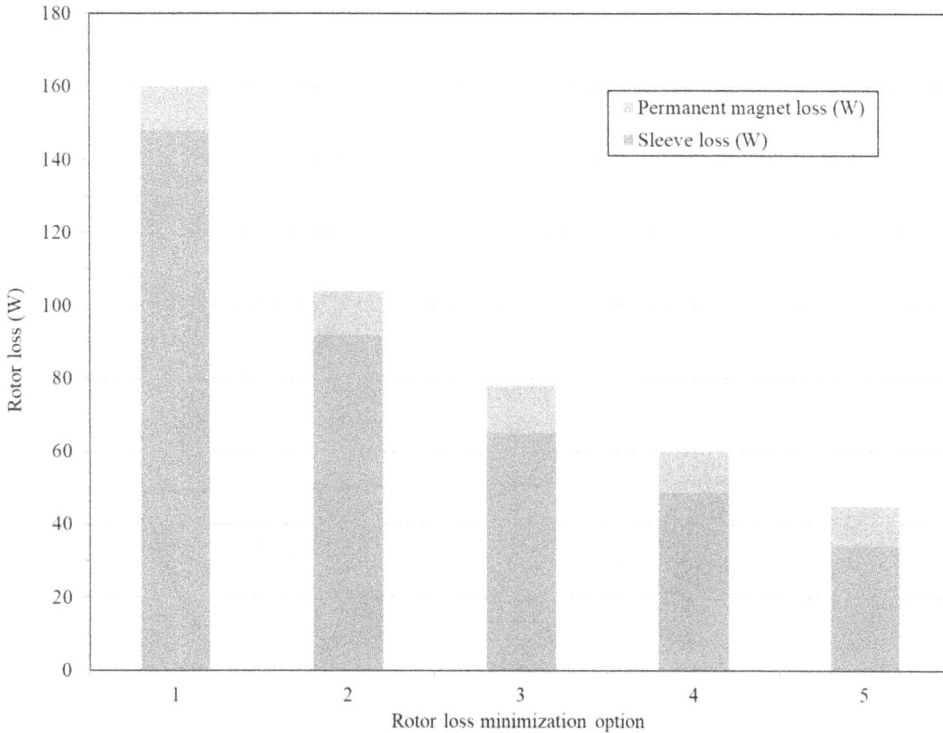

FIGURE 2.20 The rotor loss with varying machine structure (the airgap δ_a and sleeve thickness t_s): (1) $\delta_a=2$ mm and $t_s=3$ mm; (2) $\delta_a=2$ mm and $t_s=3$ mm, with wedges relative permeability $\mu_r=3$; (3) $\delta_a=2$ mm and $t_s=2.5$ mm, with wedges relative permeability $\mu_r=3$; (4) $\delta_a=2.9$ mm and $t_s=3$ mm, with wedges relative permeability $\mu_r=3$; (5) $\delta_a=3.4$ mm and $t_s=2.5$ mm, with wedges relative permeability $\mu_r=3$.

alloy sleeve at the same geometry size under the same condition, while the radial stress keeps compressive, the hook stress is turned to tensile stress at high rotating speed (24,000 rpm) and high temperature rise (80°C). It is concluded that the strength of PM rotors with the carbon-fiber sleeve is more reliable than that with the nonmagnetic alloy sleeve. To effectively shield PMs and rotor core and reduce their power losses, Merdzan et al. [2.59] studied the influence of three retention sleeve materials (Inconel 718, stainless steel, and titanium) on the performance of PM machines. By considering the penetration depth in a metallic material exposed to time changing magnetic field, they recommended to use the sleeve material with high electrical conductivity and permeability. In this way, eddy current is mostly generated in the sleeve itself, rather than in PMs and the rotor core. Due to the direct contact with the airgap, cooling of the sleeve is much easier than the magnets and rotor core. This implies that titanium (with an electrical conductivity of 2.38×10^6 S/m) is better than stainless steel (1.45×10^6 S/m) and Inconel 718 (0.83×10^6 S/m).

The thickness of the retention sleeve is another important parameter for influencing the motor performance and sleeve performance. On one side, to reduce the potential for leakage flux at the airgap and eddy current induced in the sleeve, the sleeve is desired to be as thin as possible. On another side, the sleeve with a very small thickness may loss its mechanical strength to hold the PMs. For a metal sleeve, a very thin sleeve may significantly reduce its shield function to the PMs. Aiming at minimizing rotor losses, Uzhegov et al. [2.60] calculated magnet and sleeve losses of a PM synchronous machine based on five rotor loss minimization options. With the optimization of main geometric parameters, i.e., the airgap thickness δ_a and the sleeve thickness t_s, a significant reduction in rotor losses was achieved, as shown in Figure 2.20. It can be seen from the figure that due to the shielding effect of the sleeve (made of AISI 304 stainless steel) to the rotor, the power loss of the sleeve is much larger than that of PMs.

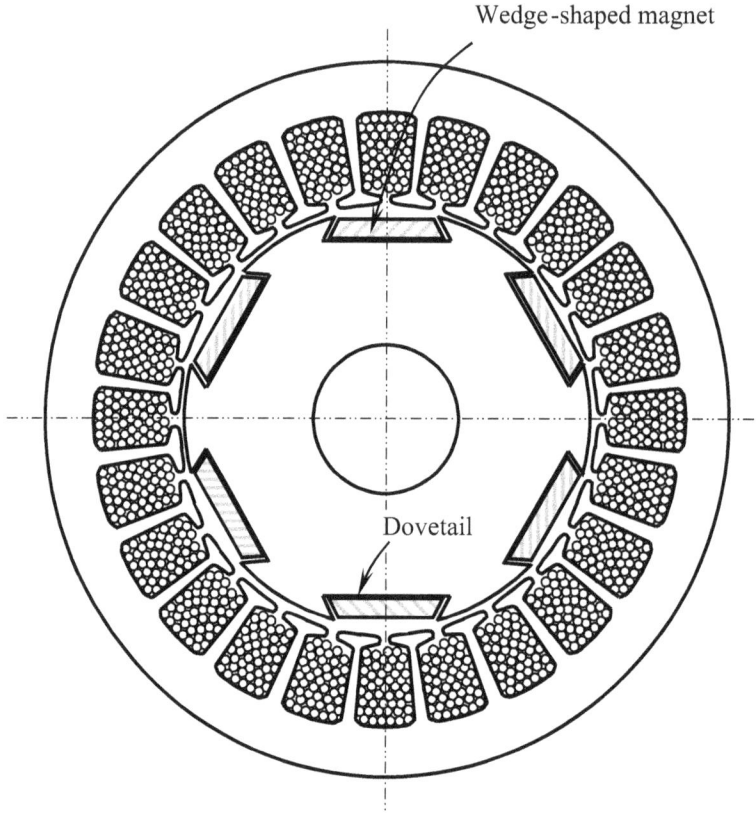

FIGURE 2.21 An innovative method for magnet retention by inserting wedge-shaped magnets into the dovetails on the rotor surface.

For all cases, while the PM losses remain almost constant (within 12±1 W), optimized main geometric parameters (δ_a and t_s), as well as the relative permeability of the wedge (μ_r), can greatly reduce the sleeve power losses.

In order to retain firmly retention sleeves on surface-mounted PMs and generate the compressive force among the sleeve and magnets, it is required to adopt interference fit joints in which sleeves are interference fitted onto PMs. To avoid the deformation of thin sleeve during the press fit process, the radial interference must be carefully analyzed and calculated. In designing the retaining sleeve for a 1.12 MW, 18,000 rpm PM machine, both carbon fiber sleeve and hybrid sleeve (which consists of both titanium alloy and carbon) are analyzed with FEM. By trading off the stresses in PMs and sleeve, the interference of approximately 0.1–0.12 mm may be suitable [2.61]. This interference value is also selected in the mechanical design and analysis of high-speed PM rotors [2.62].

An innovative retention method [2.63] combines the two magnet mounting methods, that is, surface-mounted and slot insertion methods, as shown in Figure 2.21. This method utilizes a number of dovetail slots that are made circumferentially on the rotor core surface to hold wedge-shaped magnets robustly. Correspondingly, the magnet shape must match with that of the dovetail slots. Because magnets are usually fragile and unable to withstand large forces, the interference fit is inapplicable in assembling the magnets into the dovetail slots. Instead, the magnets are fixed in the dovetail slots by an adhesive.

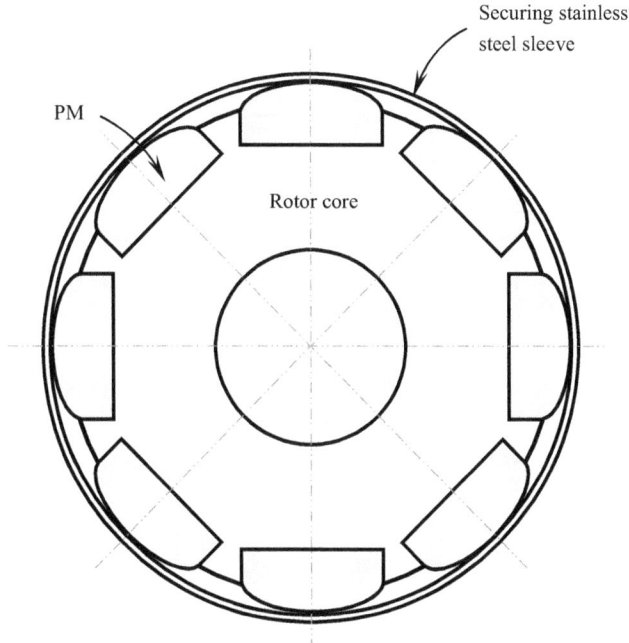

FIGURE 2.22 Using securing stainless steel sleeve for holding magnets on the rotor core.

Compared to mechanical retention methods, soldering takes less space that can be critical for high-power-density motors. It has a lesser outgassing rate than gluing.

However, the primary disadvantage of soldering in comparison to mechanical attachment and gluing is the necessity to raise the temperature of the PM to above the solder melting point during the soldering process, which can easily demagnetize the magnets. Therefore, special measures in magnet cooling must be carefully considered and taken.

Another alternative method for the magnet retention is to use high-strength metallic sleeves over the magnets [2.64], as shown in Figure 2.22. The metallic sleeve is advantageously resistant to high temperature and easy to process with a high productivity. The thin-wall metallic sleeve has a capability of restraining the magnets against centrifugal forces. By extending the sleeve beyond the ends of the rotor core and bending the sleeve toward the magnets, the metallic sleeve thus prevents the magnets from shifting in the axial direction (Figure 2.23). However, because of electrically conductive properties of the material, the metallic sleeve produces an eddy-current loss, which in turn reduces the efficiency of the motor. To minimize the influence of the metallic sleeve on motor performance, the sleeve is preferably made of the materials with low conductivity such as titanium or chromium alloys. However, these materials are expensive and hence impractical. In practice, stainless steel and corrosion-resistant Inconel™ alloys that contain primarily nickel, chromium, and iron are commonly used for metallic sleeves. For small motors, the thickness of the sleeve preferably ranges from about 0.1 mm (0.004 in.) to about 0.6 mm (0.024 in.). For large motors, the thickness can go up to 3 mm (0.12 in.). With such a thin wall, the sleeve can be mounted over magnets by various operations such as heat shrinking or pressing fit [2.65]. It is worth to note that the use of sleeves increases the thermal resistance and thus lowers the rotor cooling efficiency.

In addition to metal sleeves, an alternative option is to use carbon fiber composite sleeves for lowering the cost and eliminating eddy current in the sleeves.

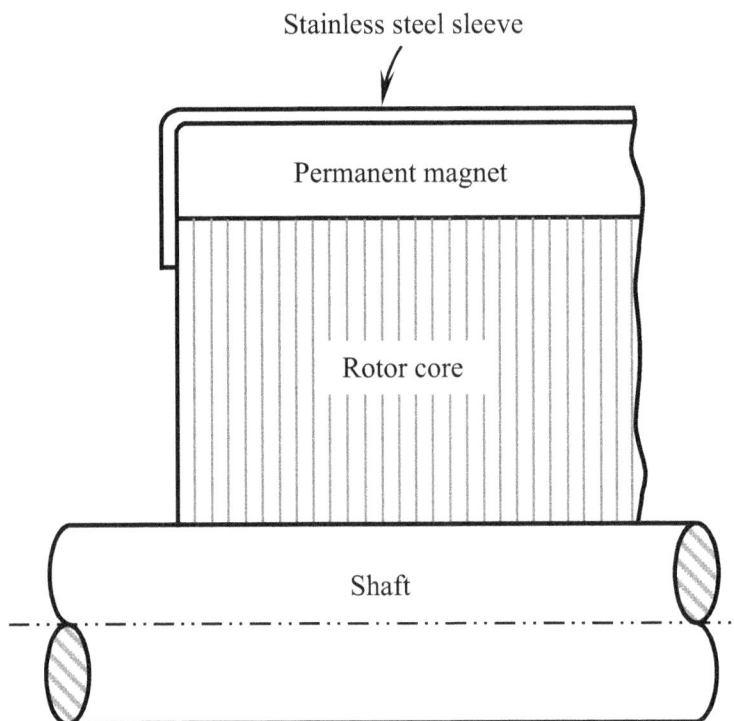

FIGURE 2.23 Bending of stainless steel sleeve at the end of rotor core for preventing magnets from shifting in the axial direction.

This type of sleeve has a high strength to weight ratio, high stiffness to weight ratio, and low cost, offering a cost-effective technical solution for magnet retention on high-speed rotors. However, this type of sleeve cannot shield the field harmonics and reduce eddy currents in rotors.

Magnets can be also retained by a molded plastic cylindrical sleeve. As illustrated in Figure 2.24, the magnets are encapsulated entirely by the plastic sleeve. This method provides a perfect solution for the magnet corrosion problem and increases the magnet retention strength against the centrifugal forces [2.66].

The schematic diagram for PMs embedded in the interior of a lamination rotor core is shown in Figure 2.25. In contrast to surface-mounted PM machines, a buried magnet design can offer certain electromagnetic and mechanical advantages: (*a*) With buried magnets, the magnetic flux concentration can be achieved for inducing higher airgap flux density and consequently higher motor torque. (*b*) In an interior-buried PM rotor, the magnetic flux can move tangentially above the magnets in the rotor core to provide a significant rotor-to-stator phase advance. (*c*) Because magnets are placed inside the rotor slots that pass axially through the rotor core, the magnet retention is thus enhanced, and the manufacturing process is greatly simplified. Furthermore, this arrangement of PM significantly increases the structural stability, especially for high-speed motors. (*d*) With this technique, rectangular magnets, which are less expensive than arc magnets, can be used. (*e*) During operation, the centrifugal forces acting on the magnets are borne by the rotor core. (*f*) When a magnet is broken due to corrosion, heating, shock load, or other causes, its fragments do not fall directly into the airgap to damage the motor. However, for the same power and same machine size, the surface-mounted magnet machine requires less magnet material than the corresponding embedded magnet machine.

For a motor with embedded magnets, the stator synchronous inductance in the *d*-axis is lower than the synchronous inductance in the *q*-axis.

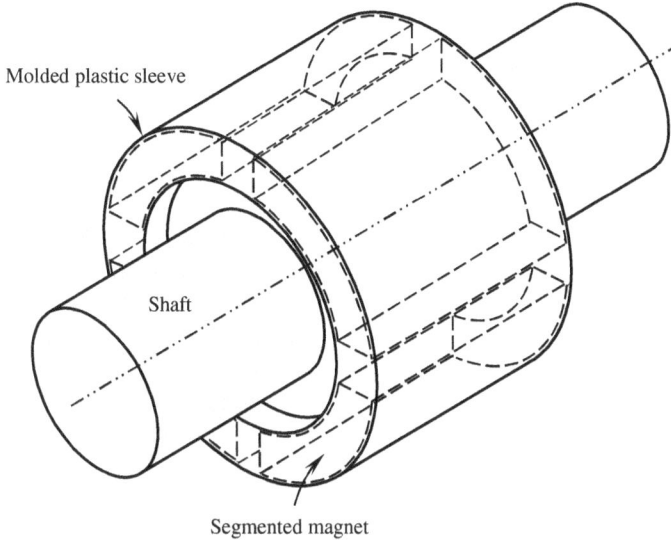

FIGURE 2.24 Retention of magnets with a molded plastic sleeve (U.S. Patent 4,973,872) [2.66]. (Courtesy of the U.S. Patent and Trademark Office, Alexandria, VA.)

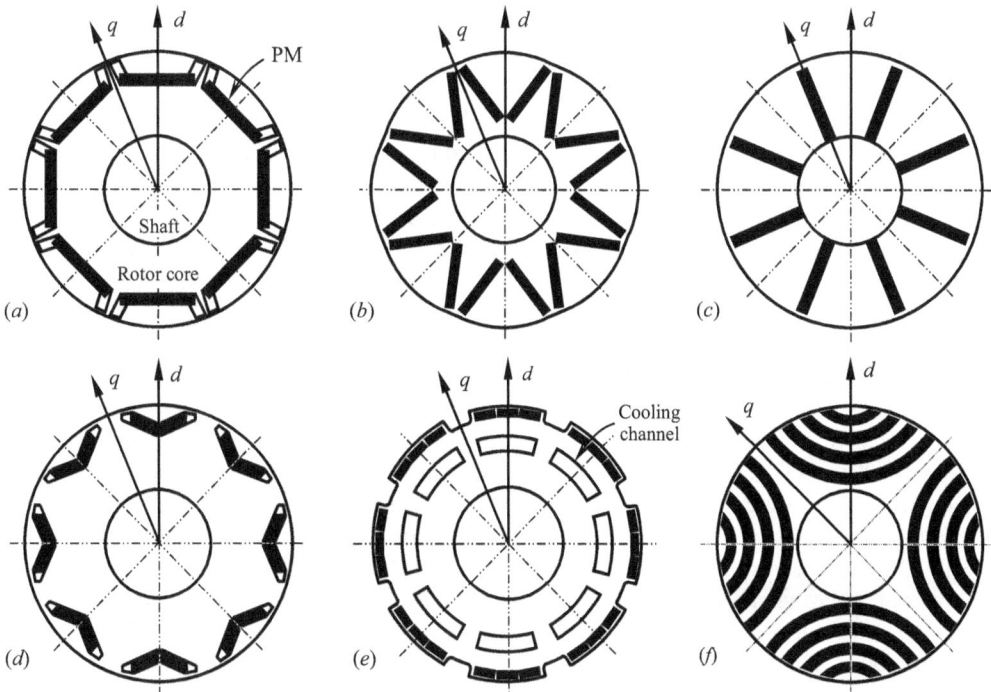

FIGURE 2.25 PMs placed in the interior of the rotor core, allowing for high-speed operation than surface-mounted magnet design: (*a*) buried tangential magnets near the rotor surface; (*b*) buried splitted V-magnets with 1/cosine-shaped airgap outline; (*c*) buried radial magnets so that their magnetization is azimuthal; (*d*) embedded splitted magnets in *V*-shaped slots near the rotor surface; (*e*) embedded splitted magnets in slots near the rotor outer surface (EP 1420501), similar to surface mount method but more secure magnet retention; (*f*) buried multilayer splitted arc magnets in saliency slots, exhibiting excellent performing characteristics of a motor.

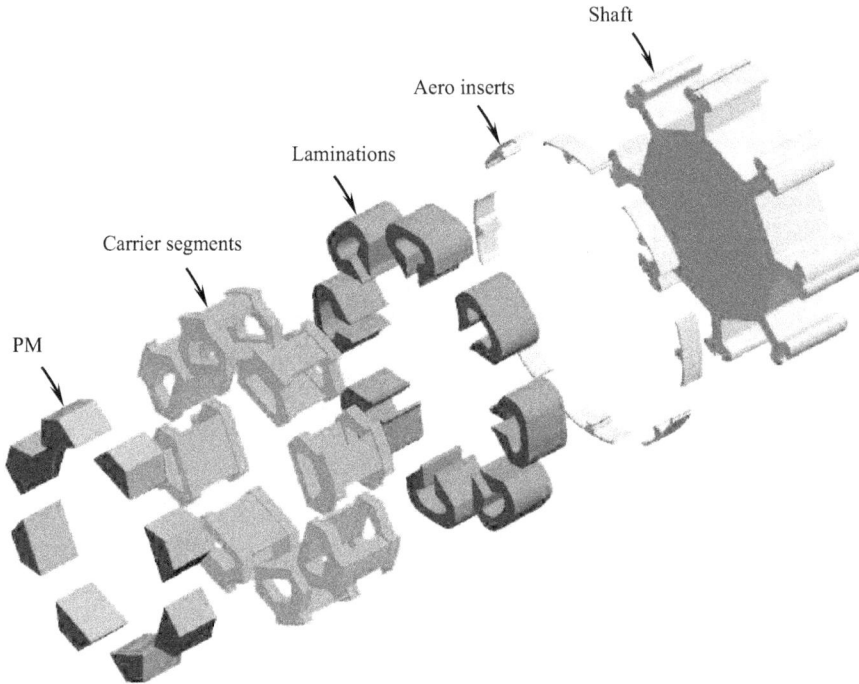

FIGURE 2.26 Explored view of large PM retention method used in a 10 MW, 6,200 rpm machine. Each piece of PM is held with an individual carrier to transfer the magnet's load to the rotor/shaft [2.67].

Motors with high rotational speeds, and thus high circumferential speeds, usually require specially designed retaining devices to secure PMs against centrifugal forces. For a high-speed 10 MW synchronous PM motor, a magnet retention technique has been developed to firmly hold the PMs at the rotor peripheral speed up to 250 m/s [2.67]. For such an ultra-high power motor, the size and volume of the PMs far exceed those in regular PMMs. In operation, these large-size PMs are subjected to very high centrifugal forces. Thus, one of the major mechanical challenges is to develop a retention mechanism to hold the PMs robustly. Because of the strong centrifugal forces, surface mounted PMs in conventional PMMs, which use an adhesive to attach the magnets to the surface of the rotor and wrap a high-strength composite material around the outside of the magnets to keep them in place, are no longer applicable. Another commonly used method to attach magnets to a rotor is called an embedded design, where PMs are held to the rotor core by some mechanical arrangement. This design typically relies on the pole pieces, which limits the speed and can restrict magnet circuit so that the full benefit of magnets is not realized. In response to this challenge and address the limitations in the prior arts, a new retention method has been developed by using the specially designed individual PM carriers to hold each piece of PM firmly, as shown in Figure 2.26. The material of the carrier not only needs to be nonmagnetic but also has high electrical resistance to minimize eddy current heating of the parts. The PM carriers also need to have a high strength-to-weight ratio to support magnets while minimizing adding additional load to the shaft. To meet these requirements, a titanium alloy may be a good choice. In such a design, the load acting on PMs can be transferred to the rotor/shaft.

2.2.8 RING MAGNETS

Ring magnets have been used in a variety of applications such as sensors, actuators, and servomotors. In sensor applications, because lower flux densities are sufficient for signal and sensing functions, low residual flux density materials, such as bonded Neo or ferrite materials, are commonly applied. For high-performance PM servomotors, ring magnets are typically made from sintered rare earth materials. Ring magnets have several major advantages over conventional segmented magnets from mechanical and electromagnetic design perspectives, including (*a*) Due to the high dimension accuracy of ring magnet, a smaller rotor-stator airgap can be achieved in a motor, resulting in reduced power losses. (*b*) A ring magnet eliminates the existing gaps between poles in conventional PM motors for improving motor performance. (*c*) Because of the cylindrical shape, ring magnets allow to be easily assembled with rotor cores without the need for retentions, leading to substantial cost savings on assembling the rotor and shorter assembling time. (*d*) With different magnetization patterns, several magnetic flux waveforms of ring magnets, *e.g.*, sinusoidal, square, or trapezoidal, can be controlled and achieved to meet specific customer requirements. (*e*) The number of magnetic poles, the magnetic flux distribution within the ring magnet, and the pole skew angle can be accurately controlled by properly designed magnetization fixtures. In such a way, it can flexibly provide customers with more design options. (*f*) Ring magnets are radially oriented with uniform magnetic performs along the circumference. Therefore, it allows motors to run more quietly with smoother torque output, compared to segmented arc magnets.

One of the main issues that needs to be addressed is the fit between a ring magnet and a rotor core due to the difference in coefficients of thermal expansion. In severe cases, the thermal stresses induced by temperature rises can cause ring magnets to crack or even break. To solve the problem, three countermeasures are often adopted: (*a*) The range of operating temperature is properly set for enabling safe operation. (*b*) In selecting the materials, the difference in the coefficients of thermal expansion between mating parts may be restricted to a certain percentage (*e.g.*, <15%). (*c*) To effectively dissipate heat from a motor, a cooling system may be adopted. For example, it may use cooled air to pass through a hollow shaft to cool the rotor core and ring magnet.

Early ring-shaped magnets were commonly made by bounding arc-shaped magnets, with alternating north and south poles around the circumference. The main drawbacks of this fabrication method are dimensional control (especially ring diameter and concentricity) and size limitation. In recent years, a new technique has been developed for fabricating single piece ring magnets. This manufacturing method offers advantages in terms of reduction of parts count and labor, geometric precision, assembly simplification, improvement of magnetic field, and miniaturization. However, there are some potential problems in using ring magnets: (*a*) Due to the relatively brittle nature of some magnet materials, press fit is not recommended for ring magnets. Assemblies can be accomplished by adhering ring magnets to rotor with an appropriate adhesive. Thus, the differences in CTE of the mating components (*i.e.*, ring magnet, rotor core, and adhesive) must be taken into account to avoid tiny cracks produced in ring magnets due to the pressure caused by temperature changes. (*b*) Under an identical electromagnetic field, higher eddy current is induced in ring magnets than in segmented magnets, resulting in higher eddy current losses. A study was carried out to investigate a technique to minimize eddy current losses in PM brushless machines by means of magnet segmentation in both circumferential and axial directions [2.68]. The results show that the eddy current loss is extremely high when the PM is solid (*i.e.*, no PM segmentation). When the magnet is subdivided into 5 pieces in the circumferential direction and 16 pieces along the axial length, the eddy current can decrease by 84%. (*c*) It is well known that convective heat transfer strongly depends on the shape of object, surface roughness, surface area for heat dissipation, and fluid properties. Segmented magnets, function like segmented fins, have better heat transfer characteristics over ring magnets. (*d*) Compared to segmented magnets that improve cogging torque by optimizing the magnet shape and geometry, ring magnets can effectively reduce cogging torque as changing the magnetization patterns of the PMs in the same motor structure.

Generally, segmented permanent magnets are supplied with a simple two-pole magnetization pattern, *i.e.*, a north pole on one face and a south pole on the opposite face. By appropriately arranging the configuration of magnetization coils, it enables to magnetize a ring magnet with

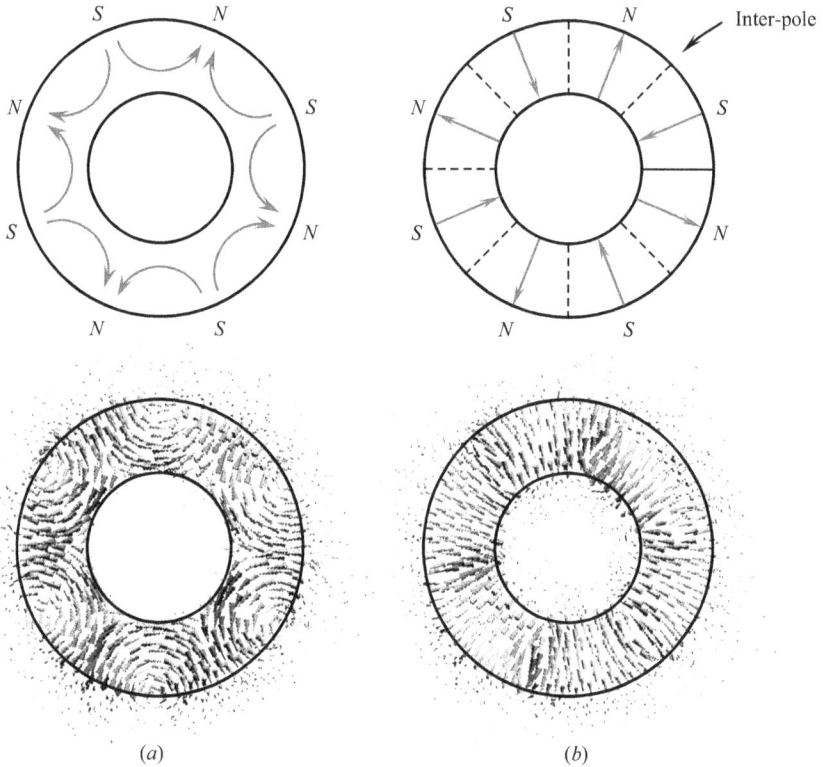

FIGURE 2.27 Magnetic fields in anisotropic ring magnets as the results of different magnetizing patterns: (a) Multipole polar anisotropic magnetization, with sinusoidal surface magnetic flux waveform that minimizes the cogging effect of motor. It shows that the intense magnetic field is within the ring magnet, with zero internal magnetic field. (b) Radial anisotropic magnetization, with trapezoidal surface magnetic flux waveform. The primary magnetic field is within the ring magnet. Generally, magnetic properties of polar anisotropic magnets are much higher than those of radial anisotropic magnets.

a multipole magnetization pattern. In such a way, the magnetic flux is oriented tangentially, and poles are formed at the center points where flux comes from the opposite directions converge. Unlike segmented PMs that are attached to rotor, ring magnets require no backiron. The entire magnetic flux path is contained within the ring magnet, so it could be bonded to a non-ferrous material such as aluminum.

The comparison of magnetic fields in different anisotropic ring magnets is shown in Figure 2.27. As can be seen in Figure 2.27a for multipole polar anisotropic magnets, the magnetic fluxes travel approximately in half circular paths from south to north poles inside the ring magnet with little leakage flux. Figure 2.27b shows a radial anisotropic magnetizing pattern, where it is achieved by injecting a high-intensity magnetic field into a continuous surface, resulting in alternating south and north poles. The magnetic fluxes travel in and out of the ring magnet nearly straight paths. A very narrow transition region occurs near the inter-pole lines between adjacent poles, at which the magnetic fluxes with opposite directions tend to cancel each other. It is to be noted that in these figures the darker the color, the higher the magnetic flux density. While magnetizing either multipole or radial magnetic patterns in the cross-sectional area, skew magnetizing in the axial direction of the magnets is allowed.

Similar to the step-skew structure in segmented magnets, ring magnets can be magnetized to gain skew poles for reducing cogging torque, as shown in Figure 2.28.

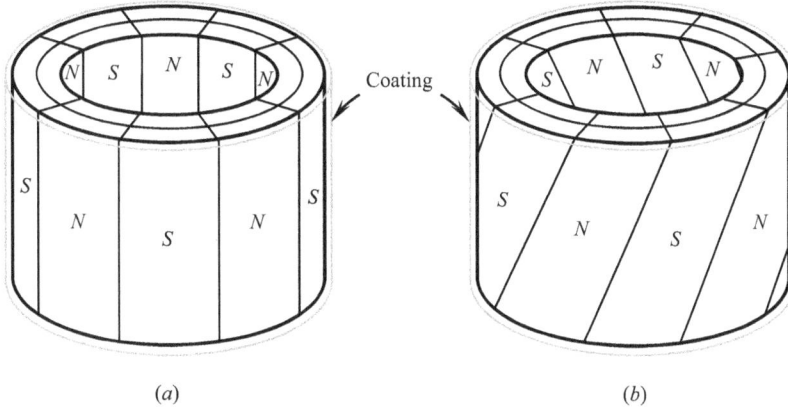

FIGURE 2.28 Magnetization patterns of ring magnets: (*a*) multipole magnetization with straight magnetic poles and (*b*) multipole magnetization with skewed magnetic poles.

2.2.9 CORROSION PROTECTION OF PERMANENT MAGNETS

Electric motors may work under severe environmental conditions. Therefore, the protection of PMs from corrosion could be critical for motor normal operation and lifetime in some applications. For instance, rare earth magnets exposed to acid/alkaline solution, salt, deleterious gas, or some chemical substances can result in corrosion. Even under a high-humidity condition, some magnets are easy to be eroded. It is noted that though the corrosion itself does not actually cause the part failure, the corrosion is the catalyst that starts a long chain of events over a long time of motor operation that eventually leads to the motor failure.

The detrimental effects of magnet corrosion include the following: (*a*) Excessive corrosion on magnet surfaces can cause premature wear of magnets. This in turn will cause magnet degradation and even failure of motor. (*b*) Debris and powders detached from corrosive magnets may block the airgap between the stator and the rotor. (*c*) Microscale cracks may develop on magnet surfaces as a result of corrosion. This will eventually lower the strength of magnets.

The corrosion process of a magnet begins with the diffusion of oxygen, water vapor, or hydrogen along grain boundaries inside the magnet. The most common method to protect magnets against corrosion is to use coatings on magnet surfaces. Coatings can isolate magnets from all harmful chemical substances and thus enhance the corrosion resistance. A number of materials can be used for coating, such as polyurethane, phtaluretane and epoxy lacquer, polymer epoxide, polyester, polyester–epoxy, zinc, nickel, copper, and chromium. Experimental results [2.69] show that polymer coatings demonstrate the best resistance. The lacquer coatings are resistant to the water environment. Metal coatings have a lower corrosion resistance compared to the polymer and lacquer in corrosive environments.

Another method is to improve the intrinsic properties of the magnetic material by controlling the chemistry of the grain boundary in the magnet. For example, the addition of certain transition elements (*e.g.*, Co, Ga, Nb, Mo, and V) can reduce the corrosion rate of rare earth magnets [2.70].

In order to increase the corrosion resistance of ring magnets, there are several surface treatment methods available: (*a*) plating or coating a film of metal, such as nickel, tin, aluminum, and Zn, on the entire surfaces of a ring magnet for effectively separating ring magnets from a corrosive environment; (*b*) coating a thin layer of either organic material such as epoxy and Teflon or inorganic materials such as silicate; and (*c*) applying passivation treatment. According to Hitachi Metals [2.71], a coated epoxy film (10–30 μm) is excellent for salt spray and a plated nickel film (5–25 μm) is excellent for thermal humidity. The comparison of these surface treatment methods is given in Table 2.3.

TABLE 2.3

Comparison of Surface Treatment Methods for Ring Magnets [2.71]

Item		Aluminum Coating	Nickel Plating	Epoxy Coating	Silicate Coating	Passivation
Standard coating thickness (μm)		2–20	5–25	10–30	-	-
Corrosion resistance	Moisture resistance	Good	Excellent	Good		
	Salt water resistance		Good	Excellent		
Adhesive durability		Excellent				Good
Insulation				Excellent		
Dimensional accuracy		Good			Excellent	Excellent
Application		Sensor, speaker, actuator/motor	Sensor, VCM, optical pickup	HEV, EPS, motor	Compressor, motor, HEV	Compressor, motor, HEV

2.3   ROTOR MANUFACTURING PROCESS

The manufacturing process for both induction and PM rotors starts from the material section. This is a critical step because it determines rotor core losses, rotor operating performance, and motor efficiency. Usually, motor manufacturers directly purchase thin lamination sheets from steel suppliers. The lamination sheets are punched by stamping machines or cut by laser devices to make desired lamination patterns. The stamped laminations are usually required to pass a heat treatment process to stabilize their electromagnetic, thermal, and mechanical properties and to release the internal stresses for dimension stabilization. Finally, these laminations are stacked together to form rotor cores.

The rotor assembly is achieved when a shaft is interference fitted into a rotor core. The design of the fit method is very important for reliable motor operation and securely transmitting its torque to an external loaded machine. An improper fit may lead to the separation of the shaft and the rotor core during motor operation. The final step in the rotor manufacturing is to carefully perform the rotor balance under a dynamic condition.

2.3.1   LAMINATION MATERIALS

It has been commonly accepted that high motor torque can be achieved with the magnetic core material having high saturation magnetic flux density, low coercivity, and low-core-loss characteristics. The selection of rotor core material is based on the electromagnetic, thermal, and mechanical properties of materials, as well as the material cost. It has long been recognized that no material is perfect in all aspects and optimum for all applications. During design processes, trade-offs are often made between material properties (*e.g.*, permeability saturation flux density and tensile strength), core losses, cost, fabrication processes, formation of lamination insulations, and other factors.

Of all the soft magnetic core materials, the most widely used materials are known as electrical steels, which are divided into several general classes. Among them, a major class is silicon steel, in which silicon is the principal alloying element. Alloying the steel with silicon can considerably increase the volume resistivity of the steel and thereby reduce the eddy-current loss. In addition, silicon can affect the grain structure of the steel and thus gives somewhat improved core loss by the reduction of the hysteresis loss in nonoriented electrical steels [2.72]. Hence, the electric and magnetic characteristics of silicon steel make it well suited for making rotor and stator lamination cores.

Silicon steel consists of body-centered cubic crystals. During a rolling process, these crystals are stretched and flattened. If they are left in that state, the magnetic properties are maximized in the

TABLE 2.4
Compositions of Some Electrical Steels (after Final Anneal)

Steel	Description of Material	Composition (%)				
		C	Mn	P	S	Si
M45	Low silicon steel	0.003	0.15	0.03	0.001	1.6
M27	Medium silicon steel	0.003	0.15	0.01	0.001	2.0
M15	High silicon steel	0.003	0.15	0.01	0.001	2.7
M4	Grain-oriented silicon steel	0.003	0.07	0.01	0.001	3.1

Source: Data from AK Steel, Selection of electrical steel for magnetic cores, http://www.aksteel.com/pdf/markets_products/
    electrical/Mag_Cores_Data_Bulletin.pdf.

rolling direction (*i.e.*, along the length of the strip). Hence, this type of steel is defined as oriented steel. Taking advantage of this characteristic, the maximum performance can be achieved by applying the material in the same direction that magnetic flux is expected. When an annealing heat treat (usually $850°C–1,100°C$) is performed, it may eliminate the grain structure in the material needed for achieving approximately isotropic properties within the material. This material is defined as nonoriented steel.

Silicon is the most important element in electrical steel. It is added to increase the electrical resistivity of steel and thereby reduces the eddy-current power loss. Silicon content in electrical steels is usually between 1.5% and 3.5%. According to the silicon content, silicon steel can be defined as low, medium, and high silicon steel (Table 2.4). Oriented silicon steel contains more silicon than nonoriented steel, approximately 3.0%–3.5%. However, it must be noted that very high silicon content (*e.g.*, >4%) may result in lowering of induction permeability saturation density. To minimize hysteresis losses, impurities such as oxides, nitrides, and sulfides in silicon steel must be tightly controlled.

Carbon content in all silicon steels is usually considerably low, approximately 0.003%. This helps minimize the hysteresis loss in the silicon steels, make them easier in rolling and other fabrication processes, and increase the lifetime of tooling.

Silicon steel is available in an array of grades and thicknesses, suited for applications in different types of electric motors. According to the American Iron and Steel Institute (AISI), the silicon steel is graded by core loss, represented by a series of M numbers. The lower the M number, the lower the core loss and thus the higher the cost.

As a reference, some flat-rolled silicon steels are listed in Table 2.5. Among all grades, M19 is probably the most common grade for motion control products, as it offers nearly the lowest core loss in this class of material.

An alternative material commonly used for rotating electric machines is nickel alloys due to their high permeability and low core losses. These characteristics make them especially ideal for motors. Nickel alloys typically contain 49%–80% pure nickel, resulting in corrosion resistance and heat resistance. Laminations made of nickel alloys are frequently used in the automotive, electrical, and electronics industries. However, its cost is significantly higher than silicon steel for the high content of nickel, together with the unavoidable annealing treatments. To achieve desired properties, nickel alloys require special measures in the annealing process. During the process, insulation films are formed on the surfaces of the nickel alloy laminations.

In some demanding applications that require very high flux density without saturation, cobalt alloys may be used to make motor laminations. The cobalt saturation point, which mainly depends on its annealing temperature and process, is significantly higher than that of silicon alloy. This type of alloy is good for weight-sensitive applications such as space shuttles and satellites. The commonly used cobalt alloys contain 48%–50% cobalt and 2% vanadium, making them high tensile

TABLE 2.5
Silicon Steel Grades

Silicon Steel	AISI Grade	Thickness, mm (in.)	Note
Nonoriented silicon steel	M15	0.36 (0.014)	The magnetic properties are
		0.47 (0.0185)	practically the same in any
	M19	0.36 (0.014)	direction of magnetization
		0.47 (0.0185)	in the plane of the material.
		0.64 (0.025)	
	M22	0.36 (0.014)	
		0.47 (0.0185)	
		0.64 (0.025)	
	M27	0.36 (0.014)	
		0.47 (0.0185)	
		0.64 (0.025)	
	M36	0.36 (0.014)	
		0.47 (0.0185)	
		0.64 (0.025)	
	M43	0.36 (0.014)	
		0.47 (0.0185)	
		0.64 (0.025)	
	M45	0.47 (0.0185)	
		0.64 (0.025)	
	M47	0.47 (0.0185)	
		0.64 (0.025)	
Orientated silicon steel	M2	0.18 (0.007)	The magnetic properties are
	M3	0.23 (0.009)	strongly oriented with
	M4	0.27 (0.011)	respect to the direction of
	M6	0.36 (0.014)	rolling.
High permeability orientated	—	0.23 (0.009)	Low core losses with very
		0.27 (0.011)	thin laminations.

Source: Data from AK Steel, Selection of electrical steel for magnetic cores, http://www.aksteel.com/pdf/markets_products/electrical/Mag_Cores_Data_Bulletin.pdf.

strengths, corrosion resistance, and thermal-fatigue resistance. Hence, for some high-speed, large-power motors, silicon steel is no longer applicable due to its low mechanical strengths, and cobalt alloys become an excellent choice for the lamination material due to the high saturation magnetization. Like nickel alloys, after stamping cobalt alloy laminations, it requires careful annealing in a protective atmosphere or vacuum environment at a preset temperature to provide an optimum combination of magnetic and mechanical properties. However, cobalt alloys usually use a separate process to add oxide coatings on lamination surfaces.

Since the density of cobalt alloy with 50% cobalt (about 8,120 kg/m^3) is about 4.1% higher than that of silicon alloy (about 7,800 kg/m^3), in designing motors that require a high power-to-weight ratio, the cobalt alloy must provide the same magnetic flux but with lower material in the yokes and teeth [2.73].

Other elements such as manganese and aluminum can also help reduce core losses due to the different mechanisms from silicon. The addition of these elements into steel will alter the metallurgical grain structure to contribute to lowering of the core loss.

Cold rolled lamination steel is the most cost-effective and most common material for core laminations of low-cost motors. Similar to carbon steel, this material has relatively high core losses. It usually requires annealing after stamping to develop optimum properties and to add oxide coatings on the lamination surfaces.

The rotor core is assembled onto the rotor shaft with an interference fit. The radial interference value is optimized to provide adequate radial contact pressure to transmit torque under full-load and full-rotating speed.

One of the most critical issues for a high-speed motor is the extremely high centrifugal force acting on the rotor core, resulting in a high level of spin stress. In addition, interference fit preload stress also contributes to the loading of the rotor core. Thus, it becomes critical to carefully select lamination materials based on not only their magnetic properties but also mechanical properties, especially for high-speed motors.

2.3.2 LAMINATION CUTTING

There are primarily two ways used by lamination manufacturers in lamination cutting: laser beam cutting and stamping machine cutting, depending on applications. For large-volume production manufacturing, stamping machine cutting is more appropriate for its high efficiency and low cost. For achieving high productivity, most lamination manufacturers adopted high-speed stamping machines in production lines. By supplying silicon steel sheets automatically, stamping machines operate continuously with press capabilities of over 250 strokes per minute and up to 1,250 tons of force. To fully utilize the material, some motor manufacturers stamp slotted laminations of the stator and rotor from the same sheet of core steel simultaneously with the outer doughnut-shaped punching as the stator lamination and the inner as the rotor lamination.

One problem often encountered in stamping cutting is that burrs are left at the lamination edges after the stamping process. Without passing through special treatments, these burrs may form electric paths for interlaminar eddy current under assembly pressures. Consequently, control of stamping operation must be carefully designed to keep the stamping burrs as small as possible. If necessary, all stamped edges are debarred when effective control is difficult or impossible.

As the non-contact cutting method, laser cutting is typically applied to large-size motor laminations in which stamping machines may not be suitable. This technique is also suitable for the fast fabrication of motor prototypes for its flexibility dimensional accuracy and short lead time. By concentrating on a very high temperature into a tiny spot during a short time, laser cutting provides superior cutting quality with small burrs, low deformation, and tight control of dimensions. During the cutting process, an oxide film is naturally generated on the burrs, thereby reducing the conductivity of burr contact and reducing interlaminar losses. The drawbacks of laser cutting are its low productivity, induced thermal residual stress near cutting lines, and high cost. As an alternative method, waterjet machines may also be used for lamination cutting.

2.3.3 LAMINATION SURFACE INSULATION

When a laminated core is subjected to an alternating electric field, eddy current is induced in each lamination and between the laminations (Figure 2.29). The resistance to the eddy current within a lamination depends mainly on the lamination thickness and material properties. In most circumstances, the eddy-current losses vary approximately as the square of the thickness of flat-rolled magnetic materials, that is, $P_e \propto t^2$, where t is the lamination thickness. This indicates that reducing the lamination thickness by half will lead to the reduction in the eddy-current losses by one-fourth. From a technical perspective, with today's advanced nanotechnology, magnetic core laminations might be possibly made extremely thin so that eddy-current losses within the laminations become negligible. However, in practice thin laminations are usually more expensive due to the increased manufacturing cost and the technical feasibility in cutting, coating, and stacking processes. Moreover, a stator or rotor core for a large electric machine may contain a huge number of laminations (e.g., a large-size turbine generator may have over 200,000 laminations), adopting thinner laminations will greatly increase the number of lamination, resulting in a much longer manufacturing lead time. Therefore, as a compromise between power loss reduction and cost, thin laminations are often used for stators and relatively thicker common laminations (with a possible higher saturation flux density) are used for rotors. This may improve the efficiency by 2%–2.5% [2.74, 2.75]. With the continuous breakthroughs of science and technology, today's users and designers of electric machines more consider the machines' life-cycle costs rather than their purchase prices, resulting in a trend toward thinner and higher quality steel laminations in next-generation machines.

Referring to Figure 2.29, in addition to the eddy current within each lamination, eddy current may occur between laminations, called interlaminar eddy current. In order to minimize eddy-current losses between laminations, silicon steel laminations must be insulated from one another to maximize the lamination surface resistance by coating a thin layer on two sides of each lamination with an organic, inorganic, or magnetite material.

The simplest way for making lamination insulation is to oxide laminations through annealing. This will create an oxide film on all lamination surfaces, as well as burrs. Thus, it greatly reduces the conductivity of laminations and minimizes eddy-current losses.

Lamination recoating restores interlaminate resistance to bring magnet core laminations back to optimal operating specifications. Recoating on stamped or laser cut laminations eliminates the possibility of interlaminate shorts.

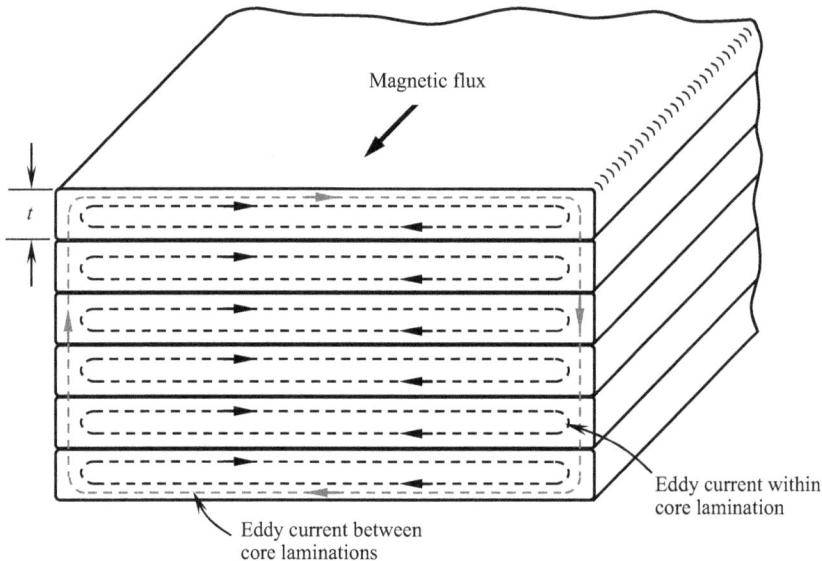

FIGURE 2.29 Eddy current within each core lamination and between core laminations.

2.3.4 LAMINATION ANNEALING

Lamination annealing is used to restore the electromagnetic, thermal, and mechanical properties of newly stamped or laser cut laminations. During the stamping process, plastic deformation, stress, and strain are produced at the cut zone, causing nonfavorable effects on magnetic properties of laminations. Though laser cutting produces little plastic deformation at cut edges, it induces a thermal shock wave and high temperatures to the cut laminations, resulting in thermal stresses and/ or permanent damage to magnetic properties, such as reduced remanence and permeability. It has been reported that annealing processes can increase permeabilities and reduce power losses for both stamping and laser cutting [2.76]. Usually, the annealing temperature for the recrystallization ranges 830°C–890°C. A higher annealing temperature up to 1,100°C is also applicable for some lamination materials.

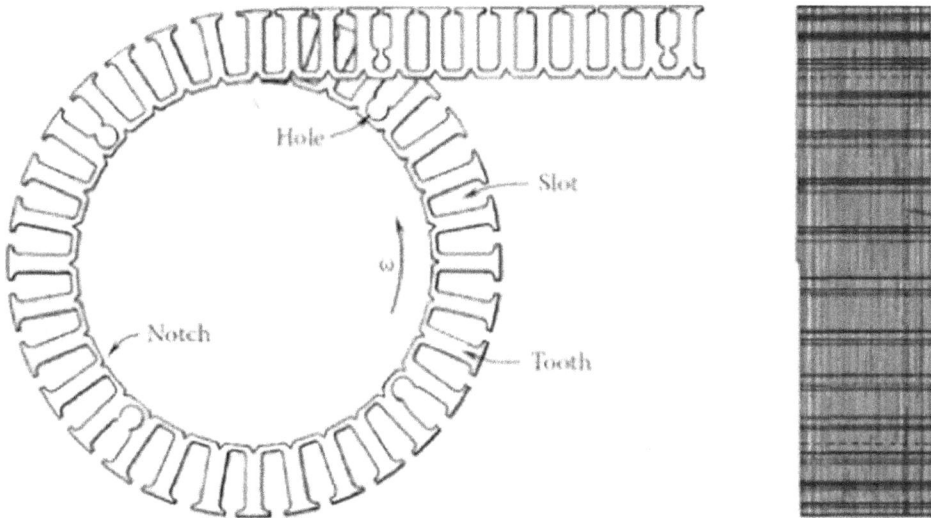

FIGURE 2.30 Applying the slinky method for fabricating motor core for using as either rotor or stator (U.S. Patent 3,188,505) [2.77]. (Courtesy of the U.S. Patent and Trademark Office, Alexandria, VA.)

2.3.5 LAMINATION STACKING

A stack of laminations is assembled for a rotor core. Laminations are often stacked on a mandrel and then compressed under high pressure for obtaining the rigidity of the rotor core in the axial direction. Traditionally laminations have been bonded using either adhesives or pins. Obviously, these stacking methods require additional operations and thus increase manufacturing costs and production lead time. In the last several decades, a number of manufacturing processes have been developed to simplify the lamination stacking operation without using adhesives or pins. A simple method is to weld the rotor core when it is compressed. Another process involves interlocking the laminations at their outer tips with a die-punching machine so that corresponding laminations can interlock one another during assembly. Interlocking dimples on internal regions is another method. For large motors, the finished rotor cores are kept tightly in the axial direction through bolts with a plate at each end of the stacked rotor core. A coated thin layer of insulation on each surface of each lamination reduces eddy current to minimize energy loss and boost performance.

An alternative method for fabricating rotor core is referred to as the slinky method. As demonstrated in Figure 2.30 [2.77], a rotor core is made by coiling up a straight continuous strip of sheet metal with respect to its centerline. To help bend the strip into a generally spiral and arcuate configuration, a V-shaped notch is formed between adjacent teeth at the ID of the core. The predetermined holes are used to receive suitable fastening means such as rivets to hold the rotor core firmly. All the slots, notches, and holes are punched together into a continuous longitudinally extending metal band of a relative ductile material such as silicon steel.

2.3.6 ROTOR CASTING FOR SQUIRREL CAGE MOTOR

A squirrel cage rotor has conducting bars and end rings made by casting either aluminum or copper into the rotor core slots and at the end surfaces of the rotor. The conducting bars are parallel to the rotor axis with a small skew angle, and the end rings serve to connect electrically all the rotor bars, forming closed-loop paths for electric current.

Cast rotors can be made by either permanent casting or die-casting process. In a permanent casting process, a rotor core is placed in a permanent mold with a desired space at the core top and bottom for the casting of end rings. Then, the mold is clamped together and a molten aluminum alloy is poured or forced into the mold. In a die-casting process, a rotor core can be cast either vertically or horizontally. One of the advantages of using a die-casting process is its high pressure acting on the molten aluminum to force it to fill the rotor slots in a very short period of time for minimizing porosities in cast rotors. To achieve high casting quality, the key casting parameters including the aluminum temperature, injection speed, casting pressure, and casting time must be controlled strictly. In addition, to achieve high casting efficiency and rotor strength, all conducting bars, end rings, and fans are cast as one integral unit, resulting in a robust and rigid assembly to withstand high loads.

In a rotor casting process, a shaft may or may not be inserted into a rotor core, depending on the succeeding finishing steps required and the particular manufacturing process being used. With the shaft insertion, high casting temperatures may result in the deformation of the shaft and the reduction of the hardness on the shaft surface.

2.3.7 Heat Treatment of Casted Rotor

Heat treatment of casted rotors is important for optimizing the motor performance. By heating casted rotors at $300°C–450°C$ ($572°F–842°F$) for about 1–2 h, it helps break metallurgical bonds between steel laminations and the conductor bars, as the result of a large differential in thermal expansion of the two metallic materials. This greatly increases the electrical resistance between the conducting bars and laminations. A theoretical analysis has shown that even a partial improvement in the insulation between the conducting bars and laminations can immediately result in noticeably reduced eddy-current losses. In addition, heating of the casted rotor creates a thin protective insulation of naturally formed oxide film at the cutting edges of the laminations to further reduce interlaminar eddy-current losses.

2.3.8 Rotor Assembly

The most economic and reliable rotor assembly method is to use interference fits to firmly integrate shafts and rotor cores together, such as press fit and shrink fit. Generally, press fit is used for relatively small rotors without exceeding the capacity of axial hydraulic pressing forces. However, the axial force required to assemble large rotors is considerably high in some instances and thus may bend or damage shafts or cause overstressed rotor cores. As a result, heat shrink processes are often used for larger rotors. The rotor assembling process is shown in Figure 2.31. In this case, shrink fit is made by heating rotor cores with induction heating (Figure 2.31a) while maintaining shafts at room temperature. When a rotor core reaches a preset temperature, its ID expands to become a little larger than the OD of the shaft, allowing the shaft to be easily inserted into the bore of the rotor core with little or no axial force (Figure 2.31b). The preset heating temperature is determined based on the rotor-shaft interference and ID of the rotor core. The rotor core is then shrunk as cold compressed air is applied to the rotor assembly. Once the shaft seizes to the core, shrinkage causes the core to pull away from the shaft shoulder. The end of the rotor core is hit with a rubber hammer to ensure the shaft shoulder is banked on the core face (Figure 2.31c). Finally, the rotor assembly is covered with a cooling hood for complete cooling of the rotor assembly.

2.3.9 Rotor Machining and Runout Measurement

A rotor assembly process involving heat shrinking or high mechanical pressuring may cause a change in shaft runout. Therefore, after rotor assembly is complete, it is preferred to check the runout of shafts to ensure the changes of runout are in the range of acceptance (Figure 2.32).

Kollmorgen engineers have investigated the effect of heat shrinking on shaft runout [2.78]. Thirty shafts were measured before and after the heat-shrinking assembly process. The runout of the shafts was inspected in seven locations with a 0.00254 mm (0.0001 in.) dial indicator using the bearing

FIGURE 2.31 Rotor assembly process with the shrink fit technique: (*a*) heating the rotor core, (*b*) inserting the shaft, and (*c*) quenching the assembly and hammering the shaft to reach its desired position.

journals as the datum. The mean change in runout for shafts was 0.00414 mm (0.000163 in.) with a standard deviation of 0.00676 mm (0.000266 in.). The average runout from the rotor OD (measured at two locations) to the bearing journals was 0.01707 mm (0.000672 in.) with a standard deviation of 0.01034 mm (0.000407 in.). Data collected from the runout testing have shown that the heat shrink process for rotor assembly has minimal to no effect on shaft runout.

FIGURE 2.32   Measurement of shaft runout after rotor assembly but before rotor machining.

FIGURE 2.33   Machining the outer surface of a casted rotor with a CNC.

Then, the rotor is placed on a lathe, where the rotor outer surface is machined to a smooth finish for ensuring a uniform airgap between the rotor and stator and for minimizing the runout of the rotor outer surface to the axis of rotation (Figure 2.33). However, machining the rotor outer surface may smear the casted conducting bars and steel laminations, resulting in an increase in the core losses, creating some hot spots and a decrease in motor efficiency. To reduce the smearing effect on motor performance, some motor manufacturers grind the rotor outside surfaces.

After the rotor is machined, it requires inspecting the runout on both the rotor and shaft for ensuring the centricity of the rotor to the rotating axis and uniform airgap, as demonstrated in Figure 2.34. The measured shaft runout data should be compared with those measured prior to the rotor assembly.

2.3.10   ROTOR BALANCING

Motor manufacturing involves a variety of mechanical processing techniques, such as machining, milling, forging, stamping, welding, grinding, drilling, and casting. When a rotor has been fabricated and assembled, unbalance of the rotor is likely experienced due to a number of factors, including improper manufacturing tolerances, material defects (*e.g.*, porosity, voids, and blowholes), unsymmetrical structures (*e.g.*, keys and keyseats/keyways), poor assembly, and improper fabrications

FIGURE 2.34 Runout measurement made on the outer surface of a cast rotor after it was machined.

(*e.g.*, misshapen casting and eccentric machining). Rotor unbalance can result in vibration and variable stress in rotors and their related supporting structures. Rotor balancing is extremely important to motors that have very high rotating speeds for ensuring safe and normal operation.

Rotor balancing involves the entire rotor structure, including a shaft, a rotor core, rotor conducting bars/end rings, fans, and other auxiliary components.

There are a number of international standards available for rotor balance. One commonly used standard in the motor industry is ISO Standard 1940/1 [2.79], which provides a method for applying unbalance tolerances based upon the static and coupled components. This standard defines common levels of acceptable unbalance for various types of machines and applications.

2.3.10.1 Type of Unbalance

There are three types of unbalance: (*a*) static unbalance is when the principal inertia axis of a rotor is offset from and parallel to the axis of rotation and is corrected only in one axial plane (Figure 2.35a); (*b*) coupled unbalance is when the principal inertia axis intersects the axis of rotation at the center of gravity (Figure 2.35b); and (*c*) dynamic unbalance is the vectorial summation of static unbalance and coupled unbalance (Figure 2.35c). Thus, it is equivalent to two unbalance vectors in two specified planes that completely represent the total unbalance of the rotor. Because dynamic unbalance is a multiplane unbalance, the correction of dynamic unbalancing requires at least two correction masses. In all cases, the center of gravity C_g is located on the principal inertia axis rather than on the axis of rotation.

It has been widely accepted that dynamic unbalance is one of the most common sources of motor vibration and noise. As reported previously, dynamic unbalance is the main source in about 40% of the excessive vibration situations [2.80]. In severe situations, this type of unbalance can cause failures of rotor and bearings.

As the simplest form of unbalance, static rotor unbalance U_s can be determined as

$$U_s = M_r e = m_u r_u \tag{2.7}$$

where
   M_r and m_u are the rotor and unbalance masses, respectively
   e is the mass eccentricity measured as the distance between the principal inertia axis and the axis of rotation
   r_u is the radius from the center of correction mass to rotation axis

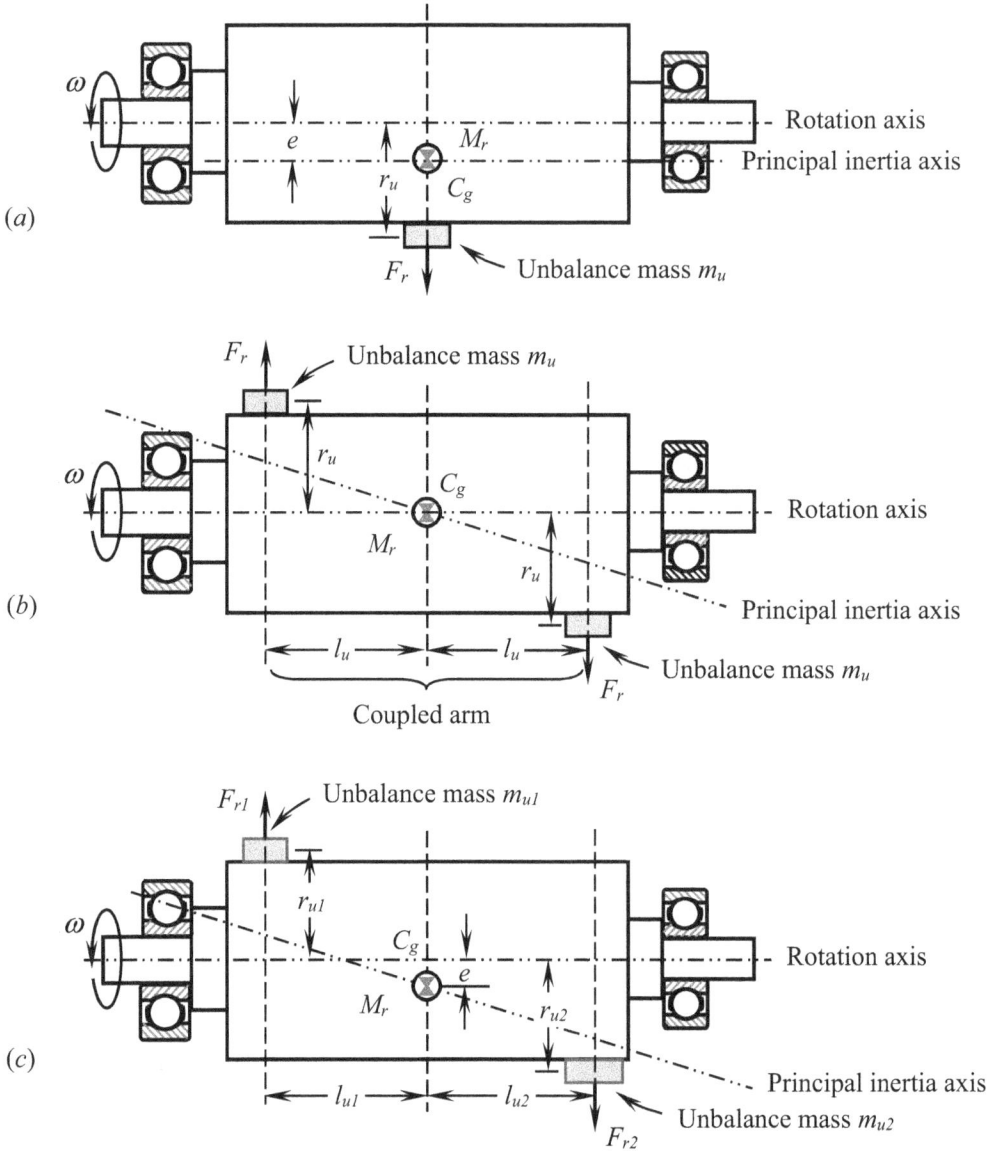

FIGURE 2.35 Three types of unbalance: (*a*) static unbalance, (*b*) coupled unbalance, and (*c*) dynamic unbalance. In all cases, the center of gravity is located on the principal inertial axis.

Thus, the mass eccentricity e can be obtained:

$$e = \frac{m_u r_u}{M_r} \tag{2.8}$$

Referring to Figure 2.35a, the magnitude of unbalance force or centrifugal force is expressed as

$$F = \left(m_u r_u + M_r e \right) \omega^2 \tag{2.9}$$

This indicates that centrifugal force is exponentially sensitive to the rotor rotating speed.

Coupled unbalance is expressed as

$$U_c = m_u r_u \left(2 l_u \right) \tag{2.10}$$

where $(2l_u)$ is defined as the coupled arm, which is the distance between the two unbalance masses. Thus, coupled unbalance is described as a mass times a length squared.

2.3.10.2 Rotor Balancing Machine

To prevent possible damages to motor due to vibration caused by rotor unbalance, balancing operation is a necessary corrective action before assembling rotors to motors. Many motor manufacturers use balancing machines to detect the amount and location of unbalanced masses on rotors. Some precise balancing machines can sensitively and accurately identify any mass axis 0.001 mm off the axis of rotation.

The simplest type of balancing machine is used for static balancing only. This type of machine is suitable for balancing disk-shaped rotors. For a majority of rotors, the type of balancing machine required must be capable of identifying dynamic unbalance in two axial planes. This type of machine is suitable for balancing rotors with long span length and small shaft diameter (*i.e.*, $L/D \gg 1$).

The principle of rotor balancing is very similar to that of car wheel balancing. With an advanced balancing machine, masses have to be either removed from or added to a rotor to minimize the uneven mass distribution so that rotor vibration can be minimized. It must be noted that the added mass must be firmly installed to the rotor and thus will not fall off when the rotor rotates at a high speed. For this reason, the material removal method is ideal.

An advanced balancing machine is shown in Figure 2.36. This machine can implement the whole balancing process with full automation, from measuring the location and amount of unbalance, displaying unbalancing results in polar graphs, drilling on the testing rotor, and repeating the process until unbalance is lower than an allowable level.

2.3.10.3 Balancing Operation

Not all unbalanced rotors need to be corrected. In common practices, a permissible unbalance is set by motor manufacturers as the balancing acceptance limit. The permissible unbalance can be determined based on experiments from similar machines, permissible bearing forces, or standards. Balancing corrections would be taken only when the existing balance is larger than the permissible unbalance.

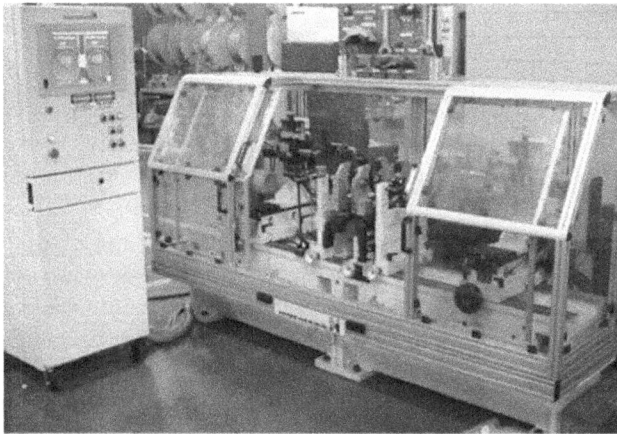

FIGURE 2.36 Rotor balancing machine.

The principle of rotor balancing is that all forces and moments acting on the rotor must be balanced under both static and dynamic conditions so that the resultant force and moment on the rotor must be zero, that is,

$$\sum_{i=1}^{n} F_i = 0 \qquad (2.11a)$$

and

$$\sum_{i=1}^{n} M_i = 0 \tag{2.11b}$$

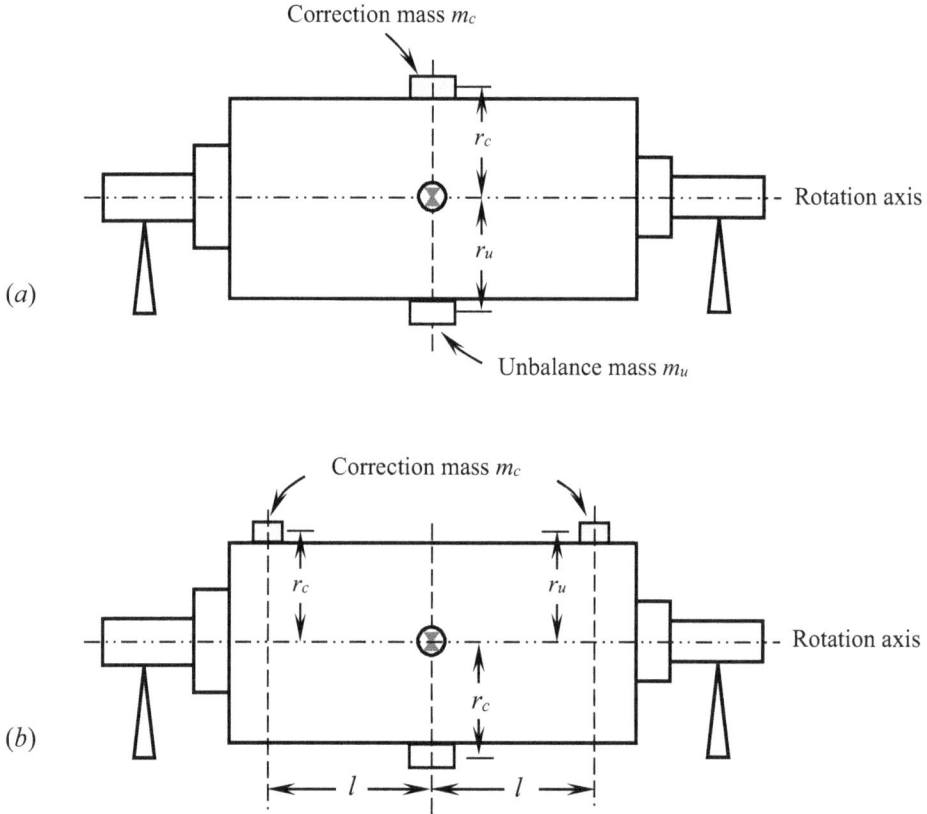

FIGURE 2.37 Two alternative balancing operations for static unbalance: (*a*) adding one piece of correction mass in the plane containing the center of gravity and (*b*) adding two pieces of correction mass in two planes with equal distance from the plane containing the center of gravity and equal radius from the rotation axis.

Static unbalance can be detected without spinning the rotor and can be most easily corrected. As shown in Figure 2.37, a static unbalance can be corrected in two ways. One method is to add directly a correct mass m_c at the opposite location of the unbalance mass m_u in the plane that contains the center of gravity (Figure 2.37a). As long as the product of the correct mass m_c and the radius from the center of the correct mass to the axis of rotation r_c is equal to the product of the unbalance mass m_u and the radius from the center of the unbalance mass to the axis of rotation r_u ($m_c r_c = m_u r_u$), the static balance is achieved. In a case that $r_c = r_u$, it leads to $m_c = m_u$.

An alternative method is to use two pieces of correct mass on the rotor in planes equidistant from the plane that contains the center of gravity at the radius r_c (Figure 2.37b). The static balance is achieved when $2m_c r_r = m_u r_u$. If $r_c = r_u$, then $m_c = m_u/2$.

An advantage of using these methods for correcting static unbalance is their simplicity and flexibility. Rotor balancing can be done without using a complex balance machine. However, these methods are only suitable for the disk-shaped rotors that are dominated by static unbalance.

Unlike static unbalance that is measured under a nonrotating condition, coupled unbalance can only be measured by spinning the motor rotor. There are two methods to correct coupled

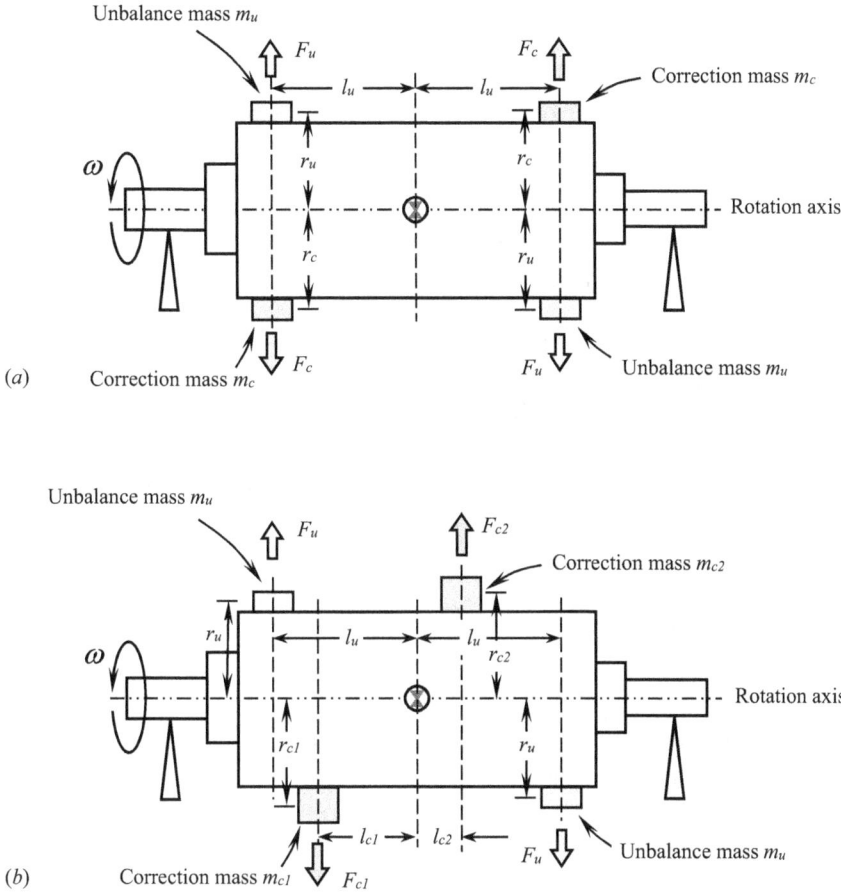

FIGURE 2.38 Balancing operation for coupled unbalance.

unbalance. The first method is to use two equal masses, in which each is to be added on the rotor at an angle of 180° apart from the unbalance mass, as shown in Figure 2.38a. In the second method, coupled unbalance is corrected in any two planes. However, each correcting mass and the distance to the plane that contains the center of gravity must be carefully determined, as shown in Figure 2.38b.

Because dynamic unbalance is the vectorial summation of static unbalance and coupled unbalance, the correction of dynamic unbalance is thus the combination of the corrections of static unbalance and coupled unbalance.

A real balancing operation is shown in Figure 2.39. An assembled rotor is placed in a rotor balancing machine. The bearing journal at each end of the shaft is put on a pair of bearing rollers. The rotor is driven by a driving belt that is placed over the rotor surface near the rotor center.

2.4 INTERFERENCE FIT

A rotor primarily consists of a rotor core and a shaft. In order to integrate these two parts together as the rotor to transmit the motor torque, the fit between the rotor core and shaft must be carefully designed. For example, too loose a fit could result in a corroded or scored rotor core and shaft, while too tight a fit could result in unnecessarily large mounting and dismounting forces, even damages of the mating surfaces or deformation/cracks of the fit parts.

FIGURE 2.39 Automatically controlled rotor balancing process.

The term *interference* refers to the fact that one part slightly interferes with the space that the other is taking up. Interference fit is often used to join two mating parts together either semipermanently or permanently. Interference fit can be generally achieved by shaping the two mating parts so that one or the other (or both) slightly deviates in size from the nominal dimension. In the motor industry, interference fit is extensively used to join rotor cores and shafts, as well as stator cores and housings, for its high joining strength and concentricity between the mating parts. Interference fit can be divided into several types as discussed in the following.

CTE is the material property of primary influence on dimensional stability as it represents the material response to changes in temperature. This material property can be exploited in an interference fit such as shrink fit. It is to be noted that in the interference fit design, it is critical to carefully consider the temperature effect on the fit strength and component stresses in the whole temperature range including both motor operation and storage. In some improper designs, interference fits work fine near room temperature. However, as the temperature becomes extremely high or extremely low, due to the large difference in thermal expansion between the assembly parts, the two assembled parts may separate from each other or pressed against each other, causing one of them to be broken.

Consequently to avoid failure of the interference-fitted assembly and minimize thermal stresses arising from the difference in CTE during temperature variations, it is highly desired to select materials of the mating parts that have close CTE values.

2.4.1 PRESS FIT

Press fit is generally chosen for its high operation reliability in rotor assemblies. Press fit can be obtained by selecting the proper interference between a shaft and a rotor core, that is, the OD of the shaft is slightly larger than the ID of the rotor core. Thus, at room temperature, the shaft must be pressed into the rotor core under an axial pressing force. During this insertion process, the shaft reduces its original OD due to the compressive force, and the rotor core increases its original ID due to the expansion force, producing a radial contact stress at the interface. This radial stress is referred to in the literature as contact pressure or interference pressure. In general, the contact pressure changes in response to the interference, ambient temperature, shaft dimensions, and material properties of both the shaft and rotor core.

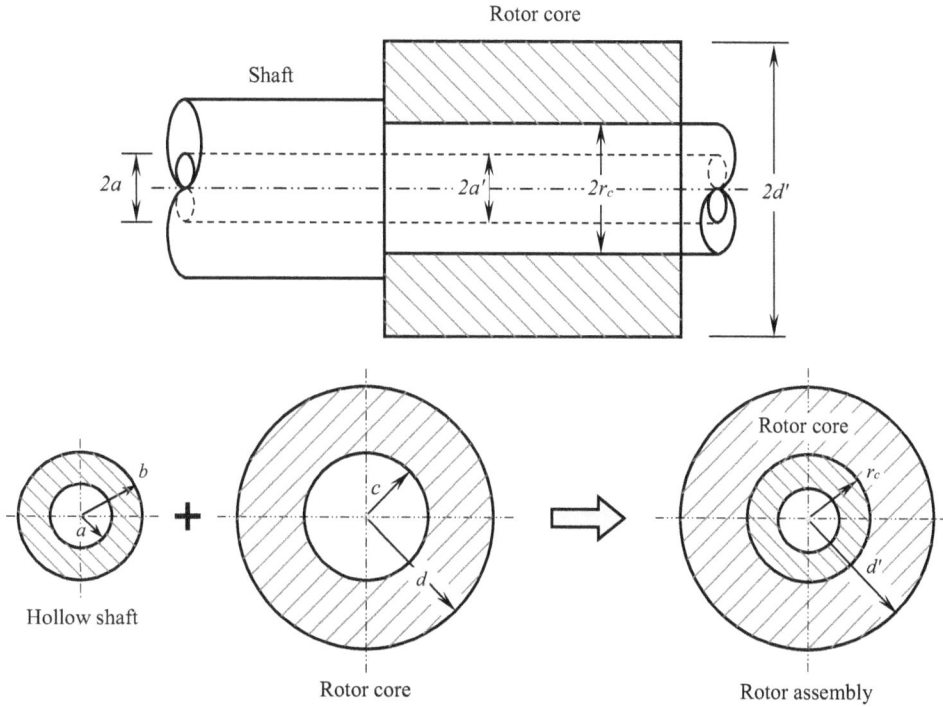

FIGURE 2.40 Dimensions of the hollow shaft and rotor core before and after assembly.

As shown in Figure 2.40, the original inner and outer radii of a hollow shaft and the rotor core are a and b and c and d, respectively. For the case that $b > c$, the interference between the shaft and the rotor core is defined as

$$\delta = b - c \tag{2.12}$$

when the shaft is pressed into the rotor core, the shaft radius becomes b' and

$$b' = b - \delta_s \tag{2.13}$$

where δ_s is the decrease in the shaft radius. As the same matter, the internal radius of the rotor core becomes c'

$$c' = c + \delta_{rc} \tag{2.14}$$

where δ_{rc} is the increase in the internal radius of the rotor core. Because that $b' = c' = r_c$, it follows that

$$\delta = b - c = \delta_s + \delta_{rc} \tag{2.15}$$

A contact pressure p is thus created at the contacting surface at $r = r_c$ (Figure 2.41). It can be derived that the radial stresses $\sigma_r = -p$ in each member at the contacting surfaces. The tangential stresses of the rotor core bore and the shaft surface are [2.81]

$$\sigma_{t,rc} = p \frac{d^2 + r_c^2}{d^2 - r_c^2} \tag{2.16}$$

$$\sigma_{t,s} = -p\frac{r_c^2 + a^2}{r_c^2 - a^2} \tag{2.17}$$

respectively.

The tangential strain of the rotor core is determined as the ratio of the change in circumference to the original circumference, that is,

$$\varepsilon_{rc} = \frac{2\pi(r_c - c)}{2\pi c} = \frac{\delta_{rc}}{c} \tag{2.18}$$

This gives that

$$\delta_{rc} = c\,\varepsilon_{rc} \tag{2.19}$$

Since

$$\varepsilon_{rc} = \frac{\sigma_{t,rc}}{E_{rc}} - \nu_{rc}\frac{\sigma_r}{E_{rc}} \tag{2.20}$$

where

ν is Poisson's ratio
E is Young's modulus (modulus of elasticity)

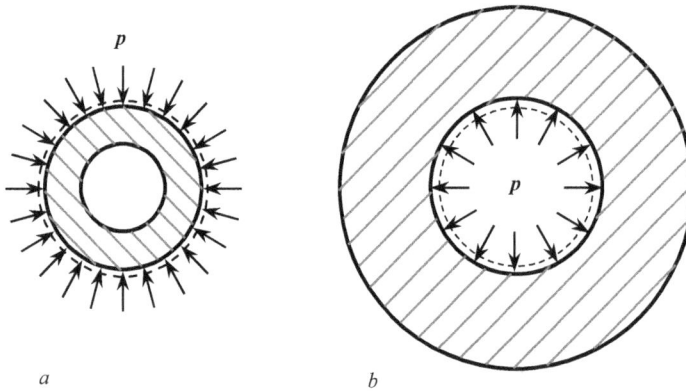

FIGURE 2.41   Contact pressure acting on (*a*) the hollow shaft and (*b*) the rotor core.

Combining Equations 2.18 and 2.20, it yields

$$\delta_{rc} = \frac{c}{E_{rc}}\left(\sigma_{t,rc} - \nu_{rc}\sigma_r\right) \tag{2.21}$$

Noting that $\sigma_r = -p$ and substituting Equation 2.16 into 2.21 yields

$$\delta_{rc} = \frac{cp}{E_{rc}}\left(\frac{d^2 + r_c^2}{d^2 - r_c^2}\nu_{rc}\right) \tag{2.22}$$

Similarly to Equations 2.18 and 2.20 for the tangential strain of the rotor core, the governing equations for the tangential strain of the shaft are as follows:

$$\varepsilon_s = \frac{2\pi(r_c - b)}{2\pi b} = -\frac{\delta_s}{b} \tag{2.23}$$

$$\varepsilon_s = \frac{\sigma_{t,s}}{E_s} - \nu_s \frac{\sigma_r}{E_s} \tag{2.24}$$

Therefore, from the previous two equations, the reduction in the shaft radius δ_s can be expressed as

$$\delta_s = \frac{bp}{E_s}\left(\frac{r_c^2 + a^2}{r_c^2 - a^2} - \nu_s\right) \tag{2.25}$$

Thus, Equation 2.15 can be rewritten as

$$\delta = \delta_s + \delta_{rc} = \frac{cp}{E_{rc}}\left(\frac{d^2 + r_v^2}{d^2 - r_c^2} + \nu_{rc}\right) + \frac{bp}{E_s}\left(\frac{r_c^2 + a^2}{r_c^2 - a^2} - \nu_s\right) \tag{2.26}$$

It must note that the interference δ is associated with radius, rather than diameter. This equation can be solved for the contact pressure p when the interference δ is known:

$$p = \frac{\delta}{\dfrac{c}{E_{rc}}\left(\dfrac{d^2 + r_c^2}{d^2 - r_c^2} + \nu_{rc}\right) + \dfrac{b}{E_s}\left(\dfrac{r_c^2 + a^2}{r_c^2 - a^2} - \nu_s\right)} \tag{2.27}$$

For a solid shaft (*i.e.*, $a=0$), the previous equation is simplified as

$$p = \frac{\delta}{\dfrac{c}{E_{rc}}\left(\dfrac{d^2 + r_c^2}{d^2 - r_c^2} + \nu_{rc}\right) + \dfrac{b}{E_s}(1 - \nu_s)} \tag{2.28}$$

The contact area between the shaft and the rotor core is

$$A_c = 2\pi r_c L \tag{2.29}$$

where L is the contact length between the rotor core and the shaft. Thus, for a given interference δ, the maximum torque that can be carried by the shaft is

$$T_{\max} = \mu_s F_c r_c = \mu_s p A_c r_c = 2\pi r_c^2 \mu_s p L \tag{2.30}$$

where μ_s is the coefficient of static friction between the shaft and the rotor core.

The axial force required for the press in assembly is (see Figure 2.42)

$$F_a = \mu_k p A_c = 2\pi r_c \mu_k p L \tag{2.31}$$

where μ_k is the kinetic coefficient of friction between the shaft and the rotor core.

It is noted that $b = r_c + \delta_s$ and $c = r_c - \delta_{rc}$. Because $(b+c) \gg (\delta_{rc} - \delta_s)$, r_c can be expressed as

$$r_c = \frac{b+c}{2} + \frac{\delta_{rc} - \delta_s}{2} \approx \frac{b+c}{2} \tag{2.32}$$

When a shaft is pressed into a rotor core, the force driving the shaft should be applied uniformly to the end of the shaft to avoid galling, peening, or damaging the rotor core. The mating surfaces of both the shaft and rotor core should be thoroughly cleaned and free of imperfections.

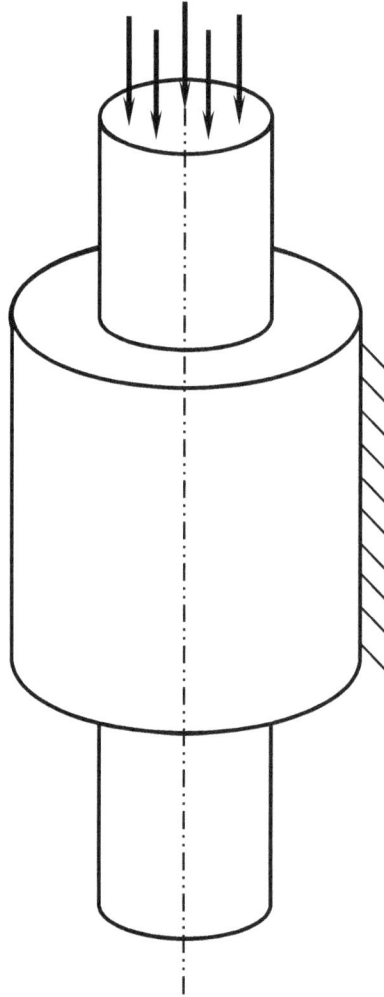

FIGURE 2.42   Axial pressing force applied in press fit rotor assembly.

2.4.2   SHRINK FIT

Shrink fit is a semipermanent assembly system that can transmit large torques through the creation of high contact pressure at the interface of its mating components. Being a tight joining method, shrink fit has been extensively used in various industries due to its high joining strength and reliability. There are a number of benefits to using shrink fit over press fit: (*a*) shrink fit can minimize mechanical stresses associated with the pressing operation and thus reduce the deformation of the rotor core; (*b*) it allows a rotor core to shrink onto a shaft symmetrically; (*c*) the requirement of surface finishing for shrink fit is relatively lower than that for press fit; and (*d*) the process of shrink fit is completely controllable.

Most materials are subjected to thermal expansion as the temperature goes up and thermal contraction as the temperature goes down. Shrink fit techniques utilize such material properties in machine assembly processes.

For a solid material, the coefficient of linear thermal expansion is typically a function of temperature, for measuring how much the material expands for a change in temperature. It is defined as the linear dimension change with respect to the change in temperature per unit linear dimension (*i.e.*, length) under a constant pressure p_o:

$$\alpha_L = \frac{1}{L}\left(\frac{\partial L}{\partial T}\right)_{p_o} \tag{2.33}$$

If the desired expansion in length ΔL is provided, the temperature rise ΔT can be determined as

$$\Delta T = \frac{1}{\alpha_L}\frac{\Delta L}{L} \tag{2.34}$$

The coefficients of linear thermal expansion for some common materials are listed in Table 2.6. Similarly, the area and volume thermal expansion coefficients are defined as follows:

$$\alpha_A = \frac{1}{A}\left(\frac{\partial A}{\partial T}\right)_{p_o} \tag{2.35}$$

$$\alpha_V = \frac{1}{V}\left(\frac{\partial V}{\partial T}\right)_{p_o} \tag{2.36}$$

In shrink fit applications, the ID of a rotor core is slightly smaller than the OD of a shaft. In an assembling process, to achieve a shrink fit joint between the two components, either the rotor core or the shaft (or both) must be treated. This can be done by either heating the core to increase the internal diameter or cooling the shaft to reduce the external diameter. Thus, according to the fit conditions, shrink fit techniques can be categorized into three groups: (*a*) heating technique, (*b*) cooling technique, and (*c*) mixing technique, which is the combination of (*a*) and (*b*).

Most motor manufacturers utilize heating techniques in rotor shrink fits, that is, heating rotor cores to a certain temperature and maintaining shafts at room temperature. With the desired expansion and shrinkage, the rise in temperature can be determined. For example, if the shaft OD is 31.75 mm and the rotor core ID is 31.70 mm. This gives the diametrical interference fit of 0.05 mm. Thus, by adding the minimum desired slip fit clearance of 0.07 mm ($\approx$ 0.003 in.), the total differential expansion is 0.12 mm. The temperature rise ΔT required on the rotor core to give 0.12 mm expansion in the diameter is calculated as

$$\Delta T = \frac{0.12/31.70}{10.8\times10^{-6}} = 350.5°C$$

Therefore, the total temperature would be 350.5°C plus the ambient temperature.

TABLE 2.6
Coefficient of Linear Thermal Expansion of Some Common Materials

Material	Coefficient of Linear Thermal Expansion $\times 10^{-6}$ (at 20°C) (mm/mm·°C)
Aluminum	23
Aluminum nitride	5.3
Brass	19
Carbon steel	10.8
Copper	17
Gold	14
Iron	11.8
Nickel	13
Stainless steel	17.3
Steel	11–13

(a)

(b)

FIGURE 2.43 Inducting heating process for rotor core prior to shaft insertion: (*a*) loading and inducting heating coil on the rotor core and (*b*) heating the rotor core.

Induction heating is a noncontact process of providing fast, consistent, and controllable heating for bonding, hardening, or softening metals or other electrically conductive materials. The process relies on induced electric currents within metallic components for heating themselves. By using this technique, shrink fit can be achieved by heating rotor cores to preset temperatures while maintaining shafts at room temperature. As demonstrated in Figure 2.43, a rotor core is placed in an induction heating equipment. As the induction heating coil is placed around the rotor core, it requires only a few seconds to achieve the desired temperature of the rotor core. The power supply to the heating coil is controlled and monitored with the front panel LCD and sealed touchpad.

As a rotor core is heated up, its internal diameter increases, allowing a shaft to insert into position. Then, as the rotor core is cool down, it shrinks back to its original size and holds the shaft tightly. By contrast, in a cooling technique, shrink fit is achieved by chilling shafts with a suitable median, such as liquid nitrogen or dry ice. An alternative way is to place shafts into a freezer.

During this time, the rotor core remains at room temperature. In such a way the shaft shrinks to allow it to pass through the rotor core bore. When the shaft restores to room temperature, it expands to its original size to gain a tight fit with the rotor core. One of the advantages of the cooling technique is that it causes little distortions on affected parts.

In some applications, it is preferred to cool down shafts rather than to heat up rotor cores. This is because high temperature may cause damage to the rotor winding insulation (as in IMs) or irreversible demagnetization of magnets (as in PMMs). In addition, since a shaft has a relatively low weight compared with that of a rotor core, it is easier to deal with the shaft than the rotor core. During the cooling shrink fit process, the shaft is usually cooled via exposure to a cryogen, typically solid carbon dioxide ($-78.5°C$ at normal atmospheric pressure), liquid carbon dioxide ($-56.6°C$ at 518 kPa), or liquid nitrogen ($-195.8°C$ at normal atmospheric pressure) in order to reduce its size through the contraction.

In comparison with heating, the cooling process cannot achieve as much change in diameter. Thus, for some special fit applications, both the heating and cooling techniques are used to provide the desired fit strength.

The mechanical design of a shrink fit set is based on either the classical Lamé elastic solution of a thick-walled cylinder or the elastoplastic solution that is based on the yield criterion of von Mises. Various analytical, experimental, and numerical studies have been performed by many investigators. Horger and Nelson [2.82, 2.83] discussed the detailed design of shrink fit assemblies based on linear elasticity solutions. An analysis of the shrink fit in the context of nonlinear elasticity was presented by Antman and Shvartsman [2.84]. Using the von Mises yield criterion and the Hencky deformation theory, Lundberg [2.85] presented the first elastoplastic solution for the shrink fit problem. Later, based on the work of Lundberg, Gao and Atluri [2.86] proposed an analytical solution for the axisymmetric shrink fit problem with a thin strain-hardening hub and an elastic solid shaft.

The difference between shrink fit and press fit is the method of assembly: press fit applies an axial force during the assembling process at room temperature and shrink fit takes advantage of thermal expansion or contraction of the materials without using a pressing force. In some cases, shrink fit is the only way to join parts that have low mechanical strengths.

2.4.3 SERRATION FIT

Serration fit is referred to as special press fit technique under partial interference conditions. In serration fit, a number of serrations are made symmetrically on the outer circumference of a shaft at the room temperature to form local deformations on the shaft (Figure 2.44). Due to the hardening effect of cold forging, the local strength near the serrations can increase 15%–30% of the material strength. When the shaft is pressed into a rotor core, the interference occurs at the vicinity of each serrated groove on the shaft. During the fit process, the serrations are deformed to generate a contact pressure at the small areas near deformed serrated grooves. Unlike in press fit and shrink fit, the requirements for surface finish and dimension tolerance of mating components can be significantly reduced because of the limited contact area in serration fit.

In the motor industry, the serration fit technique is adopted in rotor assemblies for relatively low torque applications. Figure 2.45 demonstrates the V-shaped serrated grooves disposed in the axial direction on the outer surfaces of the shafts.

2.4.4 FIT WITH KNURLING

Knurling is the process by which repeated patterns are rolled onto material surfaces, commonly accomplished on a lathe or a rolling machine with one or more hard rollers that contain a certain pattern. Knurling patterns comprise straight, diagonal (left-hand or right-hand), and diamond.

To increase the joint strength, it is often to use knurled surfaces in fit between two mating components. Compared to other kinds of fit methods, this fit technique offers several advantages, such as high transmission capacity, easy machinability and high resistance against combined axial and torsional loadings.

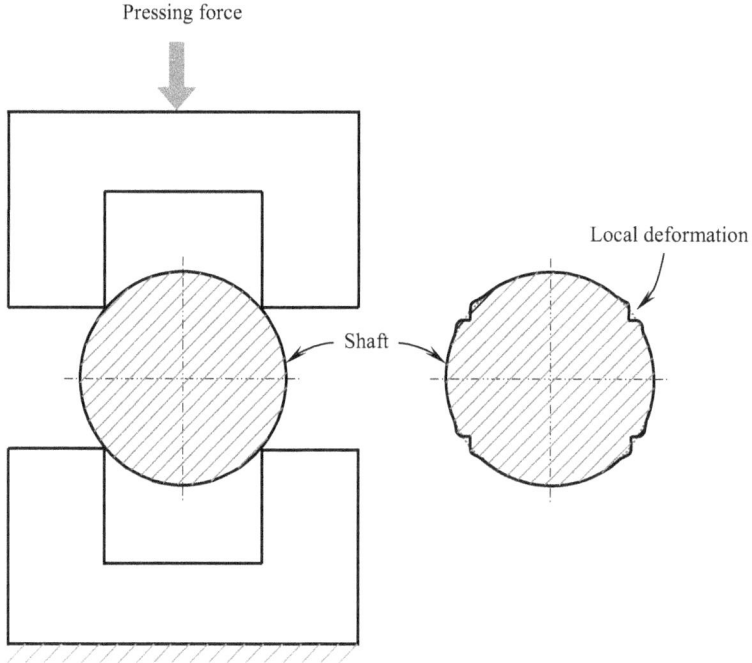

FIGURE 2.44   Fabrication of serrations on a shaft.

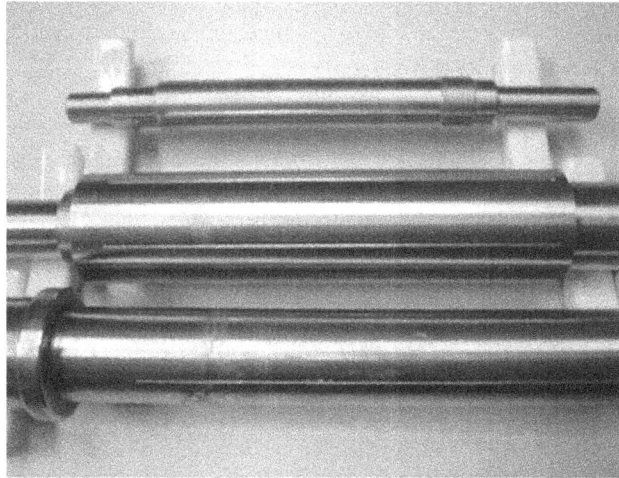

FIGURE 2.45   Serrations on motor shafts.

In the *knurled interference fit*, the connection of the components is established through the axial joining of a harder knurled component into a softer slightly undersized component (Figure 2.46). While joining, a counter profile is incised by performing a plastic deformation into a softer undersized component. In Figure 2.46*a*, the knurled shaft is axially pressed into a hub, which is relative softer than the shaft material. In fact, this joining technique has been successfully used in the power tool industry for decades. In contrast, Figure 2.46*b* shows the softer shaft is inserted into a harder hub with the inner-knurled hub bore.

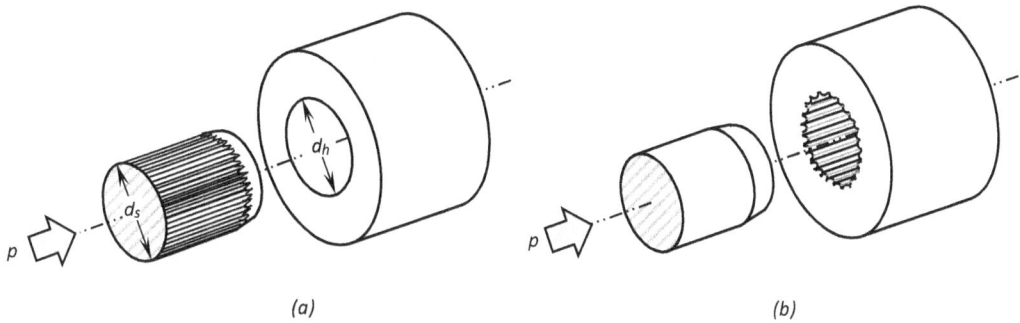

FIGURE 2.46 Knurled interference fit: (*a*) knurling the shaft surface and pressing the shaft into the smooth hub bore, where the hardness of the shaft is larger than that of the hub; and (*b*) knurling the hub surface to receive a smooth shaft, where the hardness of the hub d_h is larger than that of the shaft d_s. In both cases, the hub diameter is slightly smaller than that of the shaft.

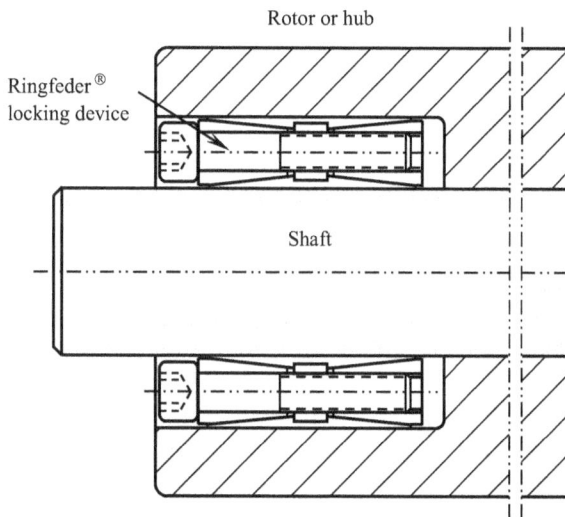

FIGURE 2.47 Keyless connection between shaft and rotor/hub using the Ringfeder® locking device.

In some applications, the mating components are made of the same material with the same hardness. In such cases, it may knurl the surface on either the shaft or the hub bore surface and use a light interference fit.

2.4.5 Fit with Adjustable Ringfeder® Locking Devices

Ringfeder locking devices generate easily adjustable and releasable mechanical fits. As shown in Figure 2.47, these locking devices feature either single- or double-tapped thrust rings with self-releasing tapers. Thus, they can bridge relatively large fit clearances between shafts and rotor cores. Because of their advantages over other interference fit methods, they have been successfully used in various industries for more than a half century.

FIGURE 2.48   Using tolerance rings between mating parts such as shaft and rotor core.

2.4.6   Fit with Tolerance Rings

Tolerance ring is a precision-engineered device usually made from a thin strip of spring steel or stainless steel. The application of tolerance rings can provide robust fit joints in rotor assemblies. Wave pitches pressed on tolerance rings are used to provide the radial contact pressure between mating components. When the tolerance ring is assembled between mating parts, the wave pitches are compressed and elastically deflected, resulting in a large contact pressure between the mating parts for tightly holding them together. In motor manufacturing, a tolerance ring presented in Figure 2.48 can be used for assemblies of shafts and rotor cores and bearings. The wave pitch of the tolerance ring is carefully designed so that the desired spring rate can be achieved.

Tolerance rings work on the two physics principles of spring and friction. The corrugations on a tolerance ring act like stiff radial spring. Like regular springs, the relationship between the spring force F_s applied on the tolerance ring and the deflection of the ring is given as

$$F_s = -kx \tag{2.37}$$

This minus sign means that the spring force F_s is always in the opposite direction of the displacement x. In fact, the compressive force acting on the tolerance ring always flattens the corrugations.

The friction force of tolerance springs is determined as

$$F_f = \mu N \tag{2.38}$$

where
   μ is the coefficient of static friction
   N is the normal force on the contacting surfaces

This force is important for holding the mating parts together firmly without sliding on each other. While motor torque capacities are related to the amount of interference and coefficient of static friction, motor radial load capacities rely on the yield limit of the material, the preload, and the cumulative compression of the corrugations caused by interference fit.

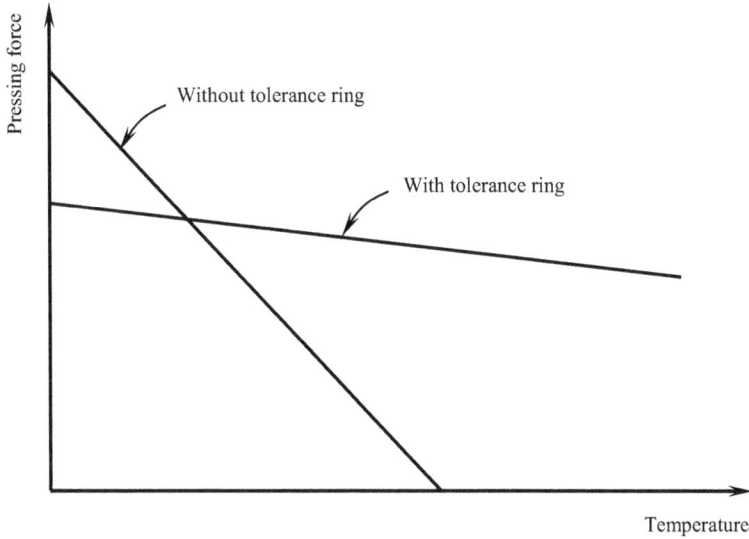

FIGURE 2.49 Comparison pressing forces at different temperatures with and without tolerance rings.

Tolerance rings provide many design advantages over other conventional fit methods. They are used with great success in compensating for different thermal expansion of mating parts. As shown in Figure 2.49, without using tolerance rings, the pressing force between mating parts reduces sharply as the temperature goes higher due to the thermal expansion of the mating parts. In contrast, with the tolerance rings, the pressing force only reduces slightly with the temperature increase. For instance, for a 50.8 mm (2 in.) diameter ball bearing on an aluminum housing, the installation pressing force is 2,224 N (500 lb$_f$) at room temperature. The bearing becomes loose at a temperature of 77°C (170°F) as the result of the higher thermal expansion of the aluminum housing than that of the steel bearing. With the tolerance ring, an initial installation pressing force is only (320 lb$_f$), and retention still remains high at 1,023 N (230 lb$_f$) even at the temperature of 132°C (270°F).

Because tolerance rings allow for a broad range of tolerance, this permits the use of loosen tolerances for the mating parts (*e.g.*, shaft and rotor core). As a result, the requirements for the surface finish on the mating parts are obviously reduced. Another advantage is that the use of tolerance rings can reduce the vibration of the system and lower the noise emission from the system. Finally, the use of tolerance ring can significantly simplify the assembly and disassembly processes.

However, the transmitted torque through the tolerance ring fit is not as high as those through press fit and shrink fit, indicating that the application of tolerance rings is limited to small-rating motors.

2.5 STRESS ANALYSIS OF ROTOR

One of the most important parameters in the rotor design is the length-to-diameter aspect ratio (*i.e.*, *L/D* ratio). A rotor with a low aspect ratio has high rotor stiffness and thus has a high critical speed. The determination of the *L/D* ratio is based on factors such as motor torque, spin stresses, rotor dynamic characteristics, and heat loads. Small *L/D* ratios mean high tangential speed and high centrifugal forces at the rotor surface. However, a rotor with a large diameter and a small *L/D* ratio may have large windage losses, especially for high-speed motors. The normal *L/D* ratio ranges 0.5–1.0 for wound rotor motors and 1–3 for PM motors [2.87]. The tangential speed on the rotor surface u_t is calculated as

$$u_t = r\omega = \frac{d\omega}{2} \tag{2.39}$$

To restrict the centrifugal force of a rotor, the upper limit of the rotor tangential speed is set in the range of 100–250 m/s, depending on motor applications.

During rotor rotation, centrifugal forces are produced in the rotor and in turn generate stresses in the circumferential and radial directions. In general, hoop tensile stresses are dominant and play a decisive role in the selection of rotor material. The stress resulted from centrifugal loading is governed by the radial equilibrium equation in the cylindrical coordination system [2.88]:

$$\frac{d\sigma_r}{dr} + \frac{1}{r}(\sigma_r - \sigma_\theta) + \rho\omega^2 r = 0 \tag{2.40}$$

where

σ_r and σ_θ are the radial and circumferential stresses, respectively
ρ is the density
ω is the rotational speed

Thus, a linear relationship between stress vector $\boldsymbol{\sigma}$ and strain vector $\boldsymbol{\varepsilon}$ can be written as

$$\boldsymbol{\sigma} = \mathbf{k}(\boldsymbol{\varepsilon} - \boldsymbol{\alpha}\Delta T) \tag{2.41}$$

where

$\mathbf{k}$ is the stiffness matrix
$\boldsymbol{\alpha}$ is the vector of thermal expansion coefficient
ΔT is the temperature difference

The previous equation can be expressed in the matrix form as

$$\begin{Bmatrix} \sigma_\theta \\ \sigma_z \\ \sigma_r \end{Bmatrix} = \begin{bmatrix} k_{11} & k_{12} & k_{13} \\ k_{21} & k_{22} & k_{23} \\ k_{31} & k_{32} & k_{33} \end{bmatrix} \begin{Bmatrix} \varepsilon_\theta \\ \varepsilon_z \\ \varepsilon_r \end{Bmatrix} - \begin{Bmatrix} \alpha_\theta \\ \alpha_z \\ \alpha_r \end{Bmatrix} \Delta T \tag{2.42}$$

By ignoring the quadratic terms of the deformation, the strains in circumferential and radial directions can be linearly related to the radial displacement u_r; the strain in the axial direction (z) is assumed to linearly vary along the radial direction:

$$\begin{cases} \varepsilon_\theta = \dfrac{u_r}{r} \\[2mm] \varepsilon_r = \dfrac{\partial u_r}{\partial r} \\[2mm] \varepsilon_z = \varepsilon_o + \varepsilon_1 r \end{cases} \tag{2.43}$$

Substituting Equations 2.42 and 2.43 into 2.40, the governing equation for radial displacement u_r is obtained. A closed-form solution is derived in detail by Ha et al. [2.89].

For a high-speed motor, the rotor is subjected to a high centrifugal force during normal operation. Generally, as long as the von Mises stress of the rotor is lower than the material yield strength, the rotor can operate safely. Thus, without considering the factor of safety, the mechanical failure criteria can be mathematically expressed as

$$\sqrt{\frac{\left(\sigma_x - \sigma_y\right)^2 + \left(\sigma_y - \sigma_z\right)^2 + \left(\sigma_z - \sigma_x\right)^2 + 6\left(\tau_{xy}^2 + \tau_{yz}^2 + \tau_{zx}^2\right)}{2}} \leq S_y \tag{2.44}$$

where

σ_i is the stress in the direction indicated by the subscript

τ_{ij} is the shear stress on the plane perpendicular to the first subscript and in the direction indicated by the second subscript

The term at the left-hand side is known as the von Mises stress.

By introducing the factor of safety n_s, the above expression can be rewritten as

$$\sqrt{\frac{\left(\sigma_x - \sigma_y\right)^2 + \left(\sigma_y - \sigma_z\right)^2 + \left(\sigma_z - \sigma_x\right)^2 + 6\left(\tau_{xy}^2 + \tau_{yz}^2 + \tau_{zx}^2\right)}{2}} \leq \frac{S_y}{n_s} \tag{2.45}$$

Although in practice some motor engineers use the ultimate tensile strength to determine the rotor operational safety, it is not recommended because the rotor plastic deformation can lead to the rotor imbalance. In a worse case, the deformed rotor may cause the rotor to touch the internal surface of the stator, resulting in severe motor failure.

It is worth to note that theoretical solutions are available only for rotors with simplified geometries (*e.g.*, cylinders). For more complicated rotor geometries, for instance, interior PM rotors, rotor stresses can be calculated using FEA. In this type of rotor, in order to reduce the magnetic flux leakage, it is desirable to reduce the thickness of iron bridges between permanent magnets. However, for high-speed applications, the rotor structure with narrow iron bridges may cause the stress concentration on these bridges. Thus, the design of buried magnet rotor is greatly affected by the tradeoff between the electromagnetic performance and mechanical robustness. One of the conventional practices to reduce bridge stresses is to use V-shaped poles where magnets are partially used to support the pole structure. The distribution of von Mises stress of a buried PM rotor is displayed in Figure 2.50.

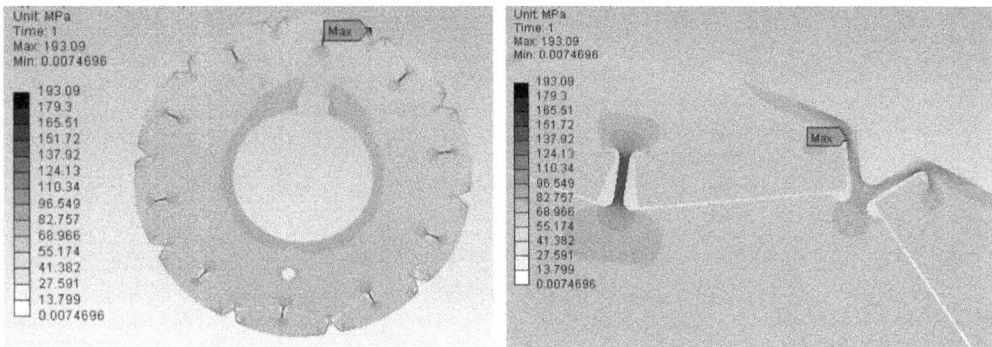

FIGURE 2.50   Von Mises stress distribution on an interior PM rotor.

2.6   ROTORDYNAMIC ANALYSIS

Rotordynamics is a specialized branch of applied mechanics concerned with the behavior and stability of rotating machinery. It plays an important role in improving the safety and reliability of electric motors. To design a robust motor, it is necessary to perform both steady-state and transient lateral and torsional calculations.

With the fast development of advanced computing techniques in recent decades, there are many software packages that are capable of solving the rotordynamic system of equations. Each software package has its specific capabilities and applications. Today, because of the complexity of the modern rotating machinery, rotordynamic software packages have been extensively used to analyze the behavior of structures ranging from gas and steam turbines to auto engines.

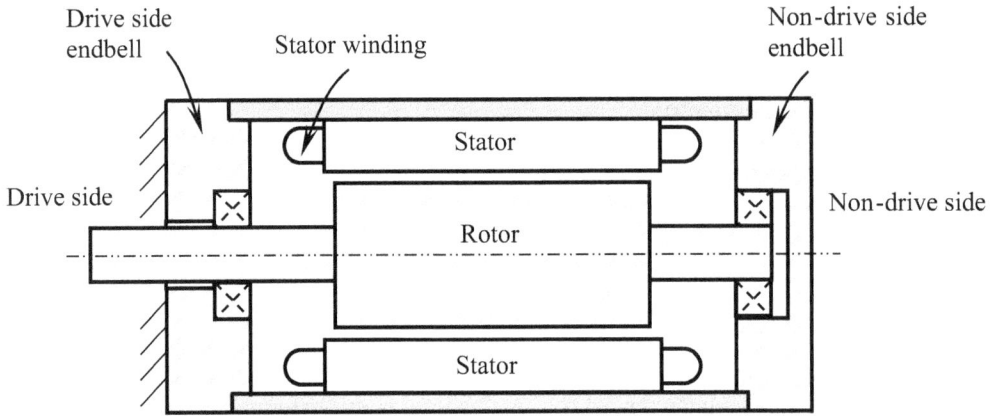

FIGURE 2.51 A motor is mounted to the driven machine via drive-side endbells.

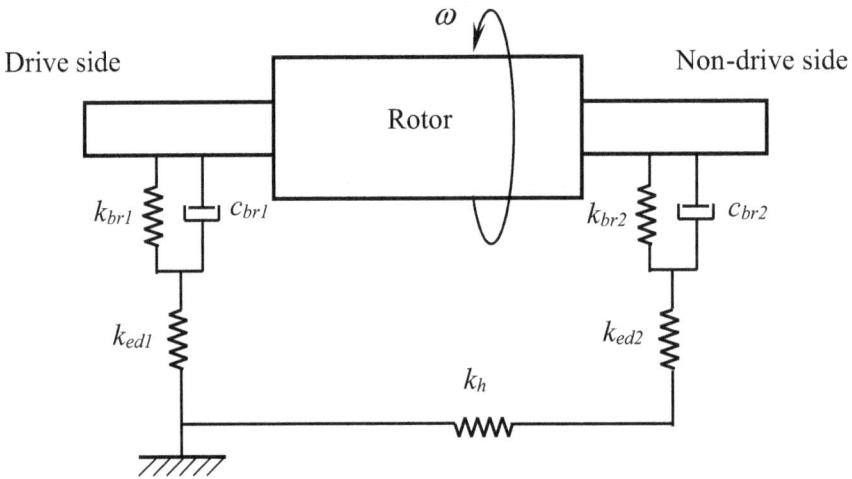

FIGURE 2.52 Rotordynamics model for a motor mounted at one end.

Rotordynamic results strongly rely on many factors such as motor structure, bearing type, rotor inertia, rotating speed, mass distribution in rotating system, motor component materials, and motor mounting pattern. As illustrated in Figure 2.51, a motor is fixed on the frame of a machine via the drive-side endbell of the motor. In order to perform rotordynamic analysis of the rotor, the stiffness and damping of the bearings, endbells, and the motor housing must be determined prior to the analysis. The rotordynamic model is shown in Figure 2.52, where S_{br1} and S_{br2} and c_{br1} and c_{br2} are the stiffness and damping of the bearings 1 and 2; S_{ed1} and S_{ed2} are the stiffness of the endbells 1 and 2, respectively; and S_h is the housing stiffness. The bearing stiffness and damping are functions of the rotor rotating speed ω. The support properties of the endbells and housing may also be speed dependent.

2.6.1 Rotor Inertia

In high-speed or large-scale motors, the rotor inertias can be considerably large. The total inertia of the motor rotating system can be obtained by summing the component inertias together, such as the rotor core inertia, shaft inertia, and bearing rotating component inertia:

$$J_r = J_{core} + J_{shaft} + J_{bearing} \tag{2.46}$$

The formulas for determining the inertia and spring constant of rotating bodies with various shapes are given in Table 2.7. The bearing inertia consists of two components: the inertias of the inner raceway and rolling elements

$$J_{bearing} = J_{ir} + NJ_{rolling} \tag{2.47}$$

where N is the number of rolling elements. The inertia of the inner ring is

$$J_{ir} = \frac{\pi l \rho \left(d_{ir,o}^4 - d_{ir,i}^4 \right)}{32} \tag{2.48}$$

where $d_{ir,o}$ and $d_{ir,i}$ are the outer and inner diameters of the bearing inner ring, respectively (Figure 2.53). For a rolling bearing, the inertia of each rolling element with respect to the bearing centerline is

---

TABLE 2.7

Polar Moment of Inertia/and Torsional Stiffness S_t for Various Types of Shafts

Shaft Type	Shaft Shape	Polar Moment of Inertia J_p (kgm²)	Torsional Stiffness S_t (Nm/rad)
Solid circular shaft		$J_p = \dfrac{\pi l \rho d^4}{32}$	$S_t = \dfrac{\pi G d^4}{32l}$
Hollow circular shaft		$J_p = \dfrac{\pi l \rho \left(d_o^4 - d_i^4 \right)}{32}$	$S_t = \dfrac{\pi G \left(d_o^4 - d_i^4 \right)}{32l}$
Tapped circular shaft		$J_p = \dfrac{3\pi}{32} \dfrac{l \rho d^4}{\left(n + n^2 + n^3 \right)}$ $n = \dfrac{d}{D}$	$S_t = \dfrac{3\pi}{32l} \dfrac{G d^4}{\left(n + n^2 + n^3 \right)}$ $n = \dfrac{d}{D}$
Stepped shaft		$J_p = \sum_i J_{p,i} = \sum_i \dfrac{\pi l_i \rho d_i^4}{32}$	$S_t = \dfrac{1}{\sum_i \dfrac{1}{S_i}}$

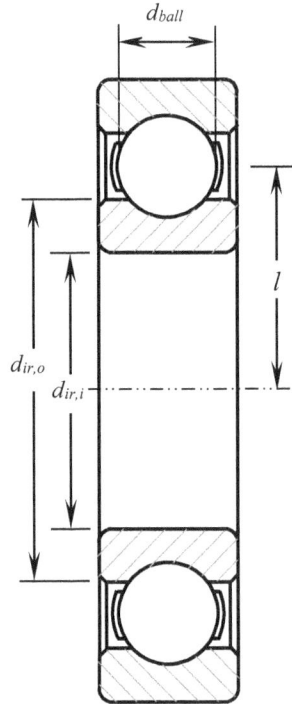

FIGURE 2.53 Calculation of inertia of rolling bearing with respect to the bearing centerline.

$$J_{roll} = m\left[K\left(\frac{d_{roll}}{2}\right)^2 + l^2 \right]$$ (2.49)

where
  m and d_{roll} are the mass and diameter of the rolling element, respectively
  l is the distance from the center of the rolling element to the bearing centerline
  K is the geometric constant of the rolling element, where $K=2/5$ for sphere bolls and $K=1/2$
    for rollers

2.6.2 MOTOR CRITICAL SPEED AND RESONANCE

A rotor assembly consists of a rotor core and a shaft, fitted together tightly. The rotor critical speed refers to the speed at which a system resonance is excited. In such a case, the centrifugal force associated with even a small mass unbalance can cause a vibration of high amplitude. In worse cases, it may lead to a disintegration of a motor within a few seconds.

   For some small lightweight motors with low rotating speeds, the operation speed is far below the critical speed. Therefore, the verification of critical speed for this type of motor may be unnecessary. Some motors are designed to operate at a rotational speed above the critical speed. It can work well if the motors accelerate quickly through the critical speed before the vibration buildup to an excessive amplitude.

Every rotating machine has its own natural frequency at which all attached components and the machine structure self-vibrate, known as resonance. Theoretically, natural frequency is directly proportional to the stiffness and inversely proportional to the mass of the machine. The accurate calculation of critical speed and natural vibration frequency is quite complex if all factors are taken into account. In practice, there are two methods used to predict critical speed and natural frequency: Rayleigh–Ritz and Dunkerley methods. Both Rayleigh–Ritz and Dunkerley methods are an approximation to the first natural frequency of vibration. Generally, the Rayleigh–Ritz equation overestimates and the Dunkerley equation underestimates the natural frequency. Good practice suggests that the maximum operation speed should not exceed 75% of the critical speed [2.90].

Both resonance and critical speeds are frequencies that are governed by natural frequencies, damping, and vibration forces. A resonance is a condition in a structure in which the frequency of the vibrating force is equal to the natural frequency of the system. For the rotation-excited vibration, the resonance is defined as the critical speed.

During operation, a motor is subject to a variety of forces, including mechanical and electromagnetic forces. Though the calculation of all motor forces is considerably complex, some factors can be identified as follows:

- When a rotor is spinning, any unbalance on the rotor causes a large centrifugal force deflection.
- Because of the finite number of stator and rotor slots (or magnets in a PM motor), it generates an unbalanced magnetic force.
- Force due to the torque is transmitted to the load.

Regularly, the torsional critical speed is less of a problem in regular motors but can be significant for motors with large L/D ratios.

Considering a rotor assembly in a motor as presented in Figure 2.54a, the rotor is positioned at the center of the system, supported by two roller bearings at each end of the rotor shaft. This physical model can be converted into a rotordynamic model, as shown in Figure 2.54b.

The system can be divided into three sections: the middle section, which contains the rotor core and one-half of the shaft, and two end sections, each containing one-quarter of the shaft. In order to perform the rotordynamic analysis, this system can be further simplified to a two-degree-of-freedom model with two masses, m_1 and m_2 [2.91]:

$$m_1 = \frac{1}{2}m_s \tag{2.50}$$

$$m_2 = m_r + \frac{1}{2}m_s \tag{2.51}$$

where m_s and m_r are the shaft and rotor mass, respectively. The exciting force F_u is resulted from the eccentric mass m_2 in the rotor system and is given by

$$F_u = m_2 e\omega^2 \sin(\omega t - \varphi) \tag{2.52}$$

The above system can be simplified as a two-degree-of-freedom rotordynamic model, as shown in Figure 2.55. There are two equations of motion for this system, one for each degree of freedom. These two equations are generally in the form of coupled differential equations:

(a)

(b)

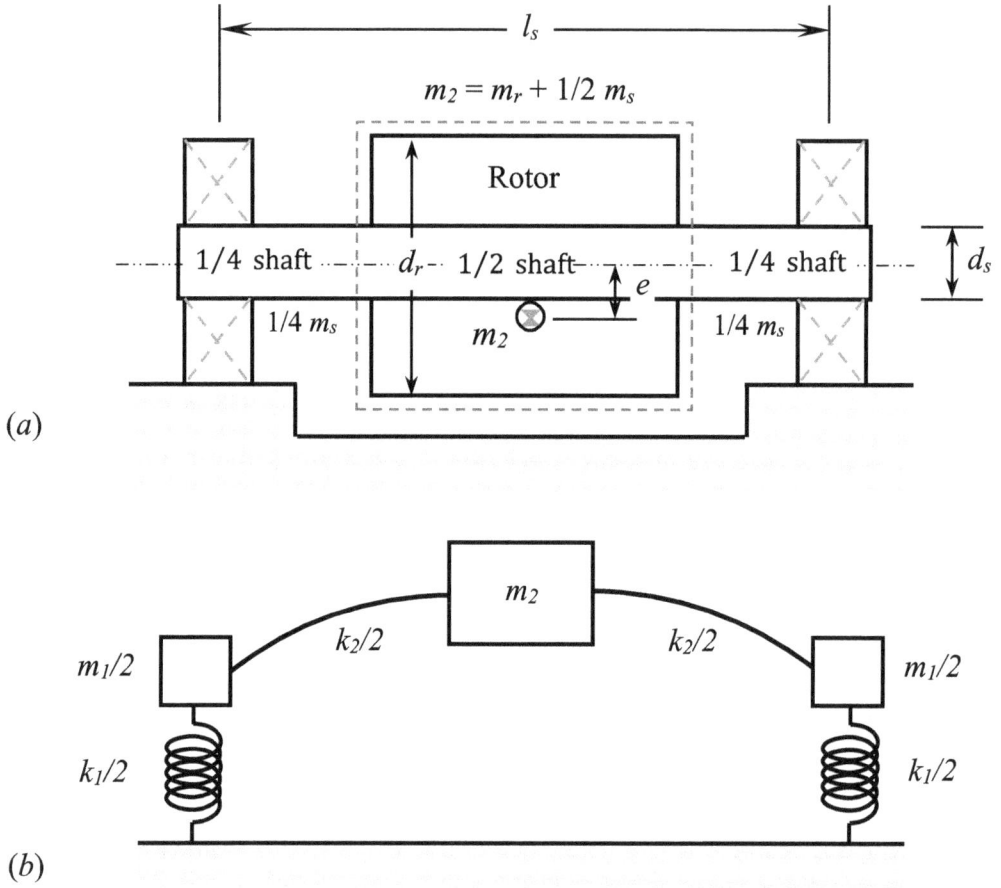

FIGURE 2.54 Rotor models: (a) physical model and (b) rotordynamic analytical model.

$$m_1 \frac{d^2 x_1(t)}{dt^2} + (k_1 + k_2) x_1 - k_2 x_2 = 0 \qquad (2.53)$$

$$m_2 \frac{d^2 x_2(t)}{dt^2} + k_2 x_2 - k_2 x_1 = \left(m_2 e \omega^2 \right) \sin(\omega t - \varphi) \qquad (2.54)$$

The solutions of the motion equations may take the form

$$x_1(t) = X_1 \sin(\omega t - \varphi) \qquad (2.55)$$
$$x_2(t) = X_2 \sin(\omega t - \varphi) \qquad (2.56)$$

where
   X_1 and X_2 are constants denoting the maximum amplitudes of $x_1(t)$ and $x_2(t)$
   φ is the phase angle

Thus, the motion equations become

$$\left[-m_1 \omega^2 + (k_1 + k_2) \right] X_1 - k_2 X_2 = 0 \qquad (2.57)$$

$$-k_2 X_1 + \left(-m_2 \omega^2 + k_2 \right) X_2 = m_2 e \, \omega^2 \qquad (2.58)$$

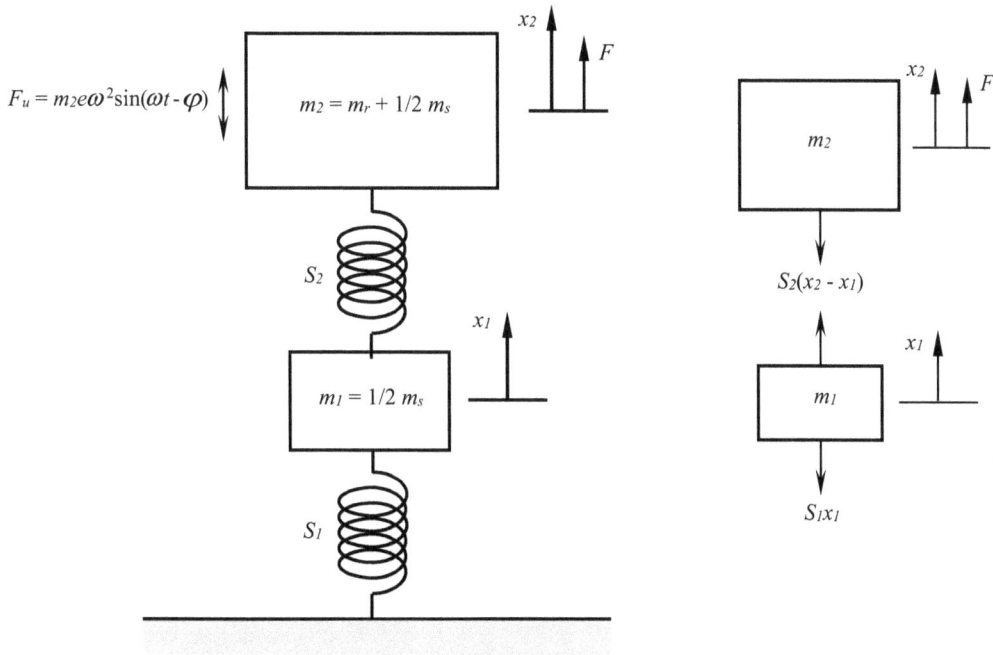

FIGURE 2.55 Reduced two-degree-of-freedom rotordynamic model for rotor assembly, where the exciting force is produced due to the rotor unbalance.

These two equations can be expressed in a matrix form

$$\begin{bmatrix} (k_1 + k_2) & -k_2 \\ -k_2 & k_2 - m_2\omega^2 \end{bmatrix} \begin{Bmatrix} X_1 \\ X_2 \end{Bmatrix} = \begin{bmatrix} 0 \\ m_2e\omega^2 \end{bmatrix} \tag{2.59}$$

This can be solved to obtain X_1 and X_2:

$$X_1 = \frac{k_2 m_2 e\omega^2}{AB - k_2^2} \tag{2.60}$$

$$X_2 = \frac{m_2 e\omega^2}{B - \dfrac{k_2^2}{A}} \tag{2.61}$$

where

$$A = (k_1 + k_2) - m_1\omega^2$$

$$B = k_2 - m_2\omega^2$$

By setting the determinant of the dynamic stiffness matrix in Equation 2.53 to be equal to zero, the fourth-order frequency equation can be obtained as

$$\omega^4 + \left(-\frac{k_1 + k_2}{m_1} + \frac{k_2}{m_2} \right)\omega^2 + \frac{k_2(k_1 + k_2)}{m_1 m_2} = 0 \tag{2.62}$$

This equation can be solved using the quadratic formula,

$$\omega^2 = \frac{-B \pm \sqrt{B^2 - 4C}}{2} \tag{2.63}$$

where

$$B = -\frac{k_1 + k_2}{m_1} + \frac{k_2}{m_2}$$

$$C = \frac{k_2(k_1 + k_2)}{m_1 m_2}$$

This leads to the two natural frequencies

$$\omega_1 = \left[\frac{-B + \sqrt{B^2 - 4C}}{2}\right]^{1/2} \tag{2.64a}$$

$$\omega_2 = \left[\frac{-B - \sqrt{B^2 - 4C}}{2}\right]^{1/2} \tag{2.64b}$$

2.7 ROTOR BURST CONTAINMENT ANALYSIS

For high-speed motors, a key technical issue is the operation safety of electric motor. This is especially true for some applications such as elevators, roller coasters, and cableways. According to rotordynamics, rotor burst refers to the phenomenon that a rotor breaks suddenly at high rotating speeds. When this failure occurs, the motor stator and housing must function as a containment to retain all debris to avoid severe personal injuries or disasters. For this reason, rotor burst containment is a critical design requirement regarding motor safety.

2.7.1 ROTOR BURST SPEED

The rotor burst speed can be predicted using different approaches. A logical approach is that rotor burst occurs when the hoop stress of a rotor is equal to the ultimate tensile strength of the rotor material. Consider a rotor core assembled with a shaft to form a rotor, with an inner and outer radii r_i and r_o, respectively, and a length l. The rotor is contained inside a stator and a housing (Figure 2.56). A differential element in the rotor has the radial thickness dr and the circumferential length $rd\theta$. The centrifugal force acting on the rotor element due to the rotor rotating is given as

$$dF = \left(r\omega^2\right)dm \tag{2.65}$$

where the mass of the element dm is found as the product of the material density ρ and the element volume dV:

$$dm = \rho dV = \rho l(rd\theta)dr \tag{2.66}$$

At an equilibrium condition, all forces acting on the element must be balanced, that is,

$$\sum_i F_i = 0 \tag{2.67}$$

It follows that

$$dF = 2l(dr)\sigma_h \sin\left(\frac{d\theta}{2}\right) \cong l(dr)\sigma_h\theta \tag{2.68}$$

where σ_h is the hoop stress on the element and $\sin(d\theta/2) \cong d\theta/2$ for small angle of $d\theta$. Thus, the hoop stress σ_h at any radius r becomes

$$\sigma_h = \rho(r\omega)^2 \tag{2.69}$$

This equation indicates that the hoop stress of the rotor is independent of the rotor length l. The maximum hoop stress occurs at the rotor outer surface where $r = r_o$:

$$\sigma_{h,\max} = \rho\left(r_o\omega\right)^2 \tag{2.70}$$

The rotor burst speed ω_b is thus determined as $\sigma_{h,\max}$ is equal to or larger than the ultimate tensional strength of the rotor material S_{ut}:

$$\omega_b = \frac{1}{r_o}\sqrt{\frac{S_{ut}}{\rho}} \tag{2.71}$$

This indicates that the rotor burst speed ω_b is both design-dependent (the rotor outer radius r_o) and material-dependent (square root of the material's ultimate strength-to-density ratio).

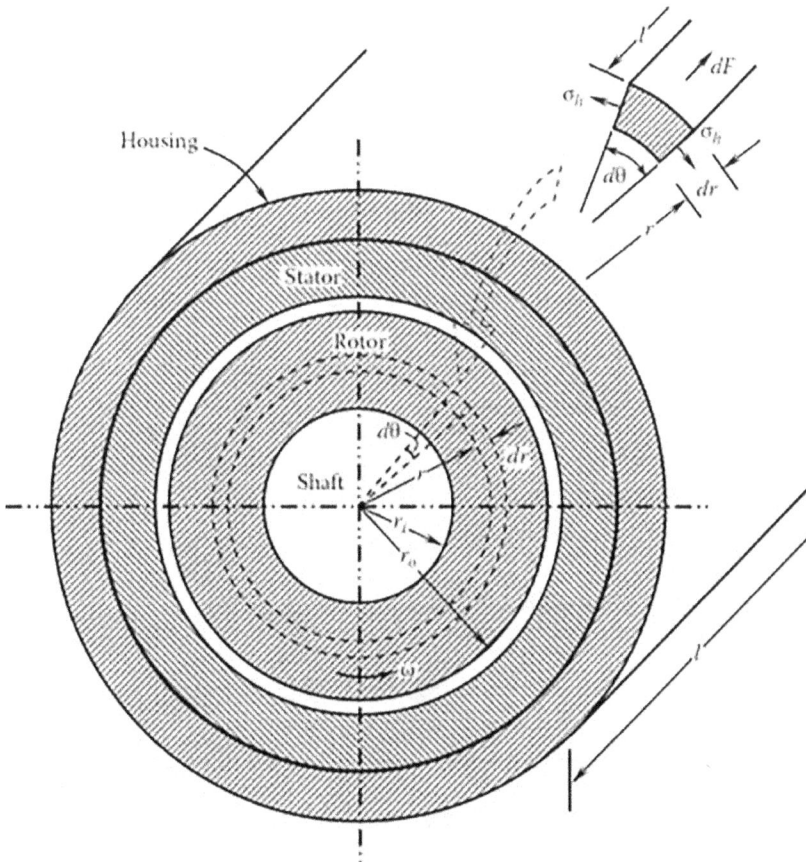

FIGURE 2.56 Rotor is contained inside a stator and a housing that functions as the containment of the rotor.

2.7.2 Energy in Rotating Rotor

For a rotating rotor, the dominant energy stored in the rotor is the kinetic energy about the center of the rotor mass. In addition, the elastic potential energy is also stored in the rotating rotor because of circumferential tensile stress and elongation during operation.

2.7.2.1 Kinetic Energy in Rotor

The kinetic energy of a rotating rotor is proportional to the product of the polar moment of inertia of the rotor J_p and the square of the rotor rotating speed ω, that is,

$$KE = \frac{1}{2}J_p\omega^2 = \frac{1}{2}J_p\left(\frac{\sigma_h}{\rho r^2}\right) \tag{2.72}$$

From this equation, it can be deduced that the kinetic energy of the rotor increases quadratically with the angular velocity ω. Furthermore, Equation 2.72 indicates that the rotor diameter also has a greater influence on kinetic energy. In the previous equation, J_p can be expressed as

$$J_p = \int r^2\,dm = \int r^2\rho\,dV \tag{2.73}$$

Referring to Table 2.7, the differential volume of the rotor dV is

$$dV = 2\pi rldr \tag{2.74}$$

Hence, for a hollow cylinder, the polar moment of inertia becomes

$$J_p = 2\pi l\rho\int_{r_i}^{r_o} r^3\,dr = \frac{\pi l\rho}{2}\left(r_o^4 - r_i^4\right) \tag{2.75}$$

The kinetic energy of a rotor at any rotating speed of ω is given as

$$KE = \frac{1}{2}J_p\omega^2 = \frac{\pi l\sigma_h}{4r_o^2}\left(r_o^4 - r_i^4\right) \tag{2.76}$$

At the burst rotating speed $\omega=\omega_b$, the hoop stress is equal to the ultimate tensile strength of the rotor material. The kinetic energy of the rotor is

$$KE_b = \frac{1}{2}J_p\omega_b^2 = \frac{\pi l S_{ut}}{4r_o^2}\left(r_o^4 - r_i^4\right) \tag{2.77}$$

2.7.2.2 Elastic Potential Energy in Rotor

The elastic potential energy in a rotating rotor is given by [2.92]

$$PE = \frac{1}{2}\int_V \sigma_h\varepsilon_h\,dV \tag{2.78}$$

where ε_h is the hoop strain (circumferential strain) in the rotor and

$$\varepsilon_h = \frac{\sigma_h}{E} = \frac{\rho r^2\omega^2}{E} \tag{2.79}$$

where E is Young's modulus. Thus, the elastic potential energy is expressed as

$$PE = \frac{\pi l\rho^2\omega^4}{6E}\left(r_o^6 - r_i^6\right) \tag{2.80}$$

At the rotor burst speed, the elastic potential energy reaches its maximum value

$$PE_b = \frac{\pi l \rho^2 \omega_b^4}{6E}\left(r_o^6 - r_i^6\right) = \frac{\pi l S_{ut}^2}{6E r_o^4}\left(r_o^6 - r_i^6\right)$$ (2.81)

This indicates that the elastic potential energy at the burst speed is dependent on the rotor geometry and material properties but independent of the rotor mass.

2.7.2.3   Ratio of Potential Energy to Kinetic Energy of Rotor

The ratio of the potential energy to the kinetic energy becomes

$$\frac{PE}{KE} = \frac{2}{3}\frac{\rho r_o^2 \omega^2}{E}\left[\frac{1-\left(r_i/r_o\right)^6}{1-\left(r_i/r_o\right)^4}\right]$$ (2.82)

When the rotor fails at the rotor burst speed, this ratio becomes

$$\frac{PE_b}{KE_b} = \frac{2}{3}\frac{S_{ut}}{E}\left[\frac{1-\left(r_i/r_o\right)^6}{1-\left(r_i/r_o\right)^4}\right]$$ (2.83)

As reported by Genta [2.93], the kinetic energy is at least one or two orders of magnitude greater than the energy needed to deform the rotor until failure occurs.

2.7.3   Rotor Burst Containment Design

At the rotor burst speed, a rotor splits into fragments that attempt to follow tangential trajectories to strike the internal wall of the stator. When the fragments impact the stator, it is the kinetic energy of the fragments that get converted into kinetic and strain energy of the stator. Also, part of the energy gets dissipated in the form of heat energy due to the fragments rubbed with the stator inner surface. The released fragments can be contained within the stator if the kinetic energy of the fragments is less than the sum of the shear and strain energy of the stator material. In many cases, stators have to withstand a very high-energy impact resulting from broken rotors.

However, the accurate prediction of the containment response is considerably complex due to the following factors: (a) the prediction of the shape and size of rotor debris would be very difficult, if not impossible; (b) there are uncertainties associated with the kinetics of the rotor debris; and (c) the determination of the impact parameters such as the impact area, points, and angle of each fragment relative to the containment wall is extremely difficult. Therefore, in practical containment design, rotor burst tests and FEMs are often used to determine the key design parameters of the containment structure.

Pichot et al. [2.94] have developed the loading models to analyze the containment response following rotor burst. With these models, the radial, axial, and torsional loadings on the containment wall are determined using fundamental energy and momentum principles. In order to overcome the difficulties due to the uncertainties associated with the kinetic energy of the debris, two separate models were developed. The first model is called the debris deflection model, in which the debris is assumed to impact the inside wall of the containment and to be immediately deflected axially without any accumulation on the wall. The second model is called the debris accumulation model, which assumes that all of the fragments pack into a debris bed against the wall without any axial deflection. In fact, these two models represent two extreme cases and thus provide the bounds for the real case. In analyzing the containment response, the debris deflection model tends to produce an impulsive loading that maximizes the magnitude of the axial loading. On the other hand, the debris accumulation model tends to produce longer duration loading and maximizes the torque on the containment.

Hagg and Sankey [2.95] have derived an analytical solution to predict the thickness of the containment shell in the event of turbine disk burst by comparing strain and shear energy with the total energy of disk fragmented missile before impact. Their solution is based on the assumption that the kinetic energy lost by the disk fragments during impact is converted into kinetic energy in the containment shell and energy loss to plastic strain and shear failure in the shell. The minimum perforation energy occurs with the normal force imposed by the translational motion of an impacting fragment. In their studies, the containment process consists of two stage events. The first stage accounts for localized perforation failure where the fragments perforate the containment shell upon initial impact. The equations assume that perforation failure occurs when the energy transfer during impact as measured by the energy loss of the burst fragment exceeds the maximum compression and shear strain energy that the contact zone of the containment shell is capable of absorption. The maximum compression and shear strain energies in the shell are based on the compressive flow stress, the failure strain, and the shear strengths of the shell material. Assuming that the impact is entirely inelastic and there are no losses to friction or heat, the energy lost during the impact process E_{loss} equals the kinematic energy of the fragment before the impact KE_f minus the residual energy E_{res} [2.96]:

$$E_{loss} = KE_f - E_{res} = \frac{1}{2} m_f v_f^2 \left(1 - \frac{m_f}{m_f + m_c} \right) \tag{2.84}$$

where the kinematic energy of the fragment KE_f is based on the normal velocity of the fragment to the wall of the containment shell v_f and m_f, and m_c are the mass of the fragment and the effective mass of the containment shell that responds to the initial contact and momentum transfer, respectively.

The energy required for straining the fragments E_{strain} consists of compressive strain energy in compression E_{comp} and shear strain energy in shearing E_{shear} [2.97]:

$$E_{strain} = E_{comp} + E_{shear} = At\varepsilon\sigma_d + C\tau_d pt^2 \tag{2.85}$$

where
   A is the striking area
   t is the thickness of the containment shell
   ε is the per unit plastic strain
   σ_d is the dynamic stress of the containment shell in compression
   τ_d is the dynamic shear stress of the containment shell in shearing
   p is the perimeter of the sheared area
   C is the experimental constant with a value in the range of 0.3–0.5

Hence, the necessary and sufficient condition for straining released fragments is

$$E_{loss} < E_{strain} \tag{2.86}$$

That is,

$$\frac{1}{2} m_f v_f^2 \left(1 - \frac{m_f}{m_f + m_c} \right) < At\varepsilon\sigma_d + C\tau_d pt^2 \tag{2.87}$$

Therefore, the minimum thickness of the containment shell is determined as

$$t_{min} = \frac{1}{2C} \left(\frac{A}{p} \right) \left(\frac{\varepsilon\sigma_d}{\tau_d} \right) \left[-1 + \sqrt{1 + \frac{2C\tau_d p}{(A\varepsilon\sigma_d)^2} \left(m_f v_f^2 \right) \left(1 - \frac{m_f}{m_f + m_c} \right)} \right] \tag{2.88}$$

This calculated value of t_{min} is to be compared to the total thickness of the stator and housing. In a case that t_{min} is larger than the total thickness, either the stator or housing must be redesigned.

If localized perforation failure does not occur, the process moves to the second stage. In this stage, the residual energy is dissipated in the form of tensile strain throughout an extended volume of the shell material. As discussed previously, the residual energy is

$$E_{res} = \frac{1}{2} m_f v_f^2 \left(\frac{m_f}{m_f + m_c} \right) \tag{2.89}$$

Failure occurs when the remaining energy exceeds the allowable strain energy in this extended volume.

Alternatively, several turbine manufacturers have presented their models to estimate the minimum thickness of the containment shell. Each of the models may involve different operation conditions and applications. These empirical formulas are listed here for the purpose of reference:

General Electric [2.98]

$$t_{min} = C_1 \sqrt{KE_f} \tag{2.90}$$

Pratt and Whitney [2.99]

$$t_{min} = C_2 \sqrt{\frac{B(KE_f)}{\tau_d p}} \tag{2.91}$$

Snecma [2.100]

$$t_{min} = \sin(\alpha) \sqrt{\frac{KE_f}{\tau_d p}} \tag{2.92}$$

where

C_1 and C_2 are the empirical constants
KE_f is the translational kinetic energy of the fragment
B is the blade buckling factor
α is the fragment impact angle

It is worth to note that no matter how advanced computational techniques have been developed for predicting the rotor burst, the development of a reliable and effective containment system for electric motor eventually requires extensive experimental testing to verify the analytical predictions. This usually requires a large amount of investment to build up a specific testing lab.

REFERENCES

2.1.   TECO-Westinghouse. 2009. Wound rotor motor technology. http://www.tecowestinghouse.com/PDF/woundrotor.pdf.
2.2.   Lab-Volt Ltd. 2011. Three-phase wound-rotor induction machines: Courseware sample. http://www.labvolt.com/downloads/86367_F0.pdf.
2.3.   Boglietti, A., Ferraris, P., Lazzari, M., and Profumo, F. 1997. About the design of very high frequency induction motors for spindle applications. *Electric Machines & Power Systems* 25(4): 387–409.
2.4.   Boglietti, A., Bojoi, R., Cavagnino, A., and Guglielmi, P. 2011. Analysis and modeling of rotor slot enclosure effects in high speed induction motors. *Proceeding of IEEE Energy Conversion Congress and Exposition*, Phoenix, AZ, pp. 154–161.
2.5.   Delaere, K., Belmans, R., and Hameyer, K. 2003. Influence of rotor slot wedges on stator currents and stator vibration spectrum of induction machines: A transient finite-element analysis. *IEEE Transactions on Magnetics* 39(3): 1492–1494.

2.6.   Peters, D. T., Cowie, J. G., and Brush, E. F., Jr. 1999. Die casting copper motor rotors: Mold materials and processing for cost-effective manufacturing. *Proceedings of EEMODS Second International Conference*, London, U.K. http://www.copper.org/applications/electrical/motor-rotor/pdf/eemods_paper.pdf.

2.7.   Daut, I., Gomesh, N., Ezanni, M., Yanawati, Y., Nor Shafiqin, S., Irwan, Y. M., and Irwanto, M. 2011. Modeling of 0.5 HP induction motor using AC analysis solver for rotor copper bar material. *Proceeding of the International Conference on Advanced Science, Engineering and Information Technology 2011*, Bangi, Malaysia, pp. 456–459.

2.8.   Azom.com. 2012. Oxygen free copper—UNSC10100. http://www.azom.com/article.aspx?ArticleID=6314.

2.9.   MatWeb. Aluminum 383.0-F die casting alloy. http://www.matweb.com/search/Datasheet.aspx?MatGU ID=2d5590682b514946b8f77c47f2908356&ckck=l.

2.10.   Manoharan, S., Devarajan, N., Deivasahayam, M., and Ranganathan, G. 2010. A comparative analysis of performance characteristics of 2.2 kW 3 phase induction motor using DAR and DCR technology. *International Journal of Computer and Electrical Engineering* **2**(**3**): 1793–8163.

2.11.   Kimmich, R., Doppelbauer, M., Kirtley, J. L., Peters, D. T., Cowie, J. G., and Brush, E. F., Jr. 2005. Performance characteristics of drive motors optimized for die-cast copper cages. *Conference Proceedings of the 4th International Conference on Energy Efficiency in Motor Driven Systems*, Heidelberg, Germany, Vol. I, pp. 110–117.

2.12.   Stark, C., Cowie, J. G., Peters, D. T., and Brush, E. F., Jr. 2005. Copper in the rotor for lighter, longer lasting motor. *Fleet Maintenance Symposium 2005*. San Diego, CA.

2.13.   Czebiniak, D. J. 2013. Die cast rotor with steel end rings to contain aluminum. U.S. Patent 8,368,277.

2.14.   NEMA Standards Publication MG 1-2011. Motors and generators.

2.15.   Atherton, W. A. 1984. *From Compass to Computer: A History of Electrical and Electronics Engineering*. Palgrave Macmillan, London, U.K.

2.16.   Robert, W. H. 1958. Performance of permanent magnets at elevated temperatures. *Journal of Applied Physics* **29**(**3**): 405–407.

2.17.   Kadochnikov, A. I., Ivanov, V. P., and Malyuk, V. P. 1979. Effect of temperature on the characteristics of reference materials for magnetic properties of iron-nickel alloys. *Measurement Techniques* **22**(**5**): 584–586.

2.18.   Wang, A.-M. and Li, H.-M. 2008. Influence of permanent magnet material and temperature on interior permanent magnet machine performance and torque ripple (in Chinese). *Journal of North China Electric Power University* **35**(3): 24–32.

2.19.   Magnet Sales & Manufacturing Inc. 1995. High performance permanent magnets. Catalog 7. http://www.magnetsales.com/Info_R2.htm.

2.20.   Cui, J., Kramer, M., Zhou, L., Gabay, A., Hajipanayis, G., Balasubramanian, B., and Sellmyer, D. 2018. Current progress and future challenges in rare earth-free permanent magnets. *Acta Materialia* **158**: 118–137.

2.21.   Shen, Y. Z., Moomy, R., and Eggert, R. E. 2020. China's public policies toward rare earths, 1975–2018. *Mineral Economics*. doi:10.1007/s13563-019-00214-2.

2.22.   Sabbe Mosa, I. S. 2014. History and development of permanent magnets. *International Journal for Research & Development in Technology* **2**(1): 18–26.

2.23.   Leslie-Pelecky, D. L. and Rieke, R. D. 1996. Magnetic properties of nanostructured materials. *Chemistry of Materials* **8**(8): 1770–1783.

2.24.   Obaidat, I. M., Issa, B., and Haik, Y. 2015. Magnetic properties of magnetic nanoparticles for efficient hyperthermia. *Nanomaterials* **5**(1): 63–89.

2.25.   Vishina, A., Vekilova, O. Y., Björkman, T., Bergman, A., Herper, H. C., and Eriksson, O. 2020. High-throughput and data-mining approach to predict new rare earth free permanent magnets. *Physical Review B* **101**(9): 094407-1–094407-9.

2.26.   Tangemann, C. 2020. Developing electric motors less dependable on rare earth magnets. https://www.automotive-iq.com/electrics-electronics/articles/developing-electric-motors-less-dependable-on-rare-earth-magnets

2.27.   Jowitt, S., Werner, T., Weng, Z. H., and Mudd, G. 2018. Recycling of the rare earth elements. *Current Opinion in Gree and Sustainable Chemistry* **13**: 1–7.

2.28.   Bogart, J. A., Cole, B. E., Boreen, M. A., Lippincott, C. A., Manor, B. C., Carroll, P. J., and Schelter, E. J. 2016. Accomplishing simple, solubility-based separations of rare earth elements with complexes bearing size-sensitive molecular aperture. *Proceedings of the National Academy of Sciences of the United States of America*. doi:10.1073/pnas.1612628113.

2.29.   Gatteschi, D., Sessoli, R., and Villain, J. 2006. *Molecular Nanomagnets*. Oxford University Press, Oxford, UK.

2.30. García, S. M. 2008. The nanofabrication and magnetic properties of nanomagnets patterned with thin films. PhD thesis, University of the Basque Country, Basque Country, Spain.

2.31. Cowburn, R. P., Koltsov, D. K., Adeyeye, A. O., and Welland, M. E. 1998. Probing submicron Nanomagnets by magneto-optics. *Applied Physics Letters* **73**(**26**): 3947–3949.

2.32. Vasilakaki, M., Trohidou, K., and Nogués, J. 2015. Enhanced magnetic properties in antiferromagnetic-core/ferrimagnetic-shell nanoparticles. *Scientific Reports* **5**: Article No. 9609. doi:10.1038/srep09609.

2.33. Lottini, E., López-Ortega, A., Bertoni, G., Turner, S., Meledina, M., Van Tendeloo, G., de Fernández, C. J., and Sangregorio, C. 2016. Strongly exchange coupled core\shell nanoparticles with high magnetic anisotropy: A strategy toward rare earth-free permanent magnets. *Chemistry of Materials* **28**(12): 4214–4222.

2.34. Moorsom, T., Alghamdi, S., Stansill, S., Poli, E., Teobaldi, G., Beg, M., Fangohr, H., Rogers, M., Aslam, Z., Ali, M., Hickey, B. J., and Cespedes, O. 2020. π-anisotropy: A nanocarbon route to hard magnetism. *Physical Review B* **101**: 060408.

2.35. Patel, K., Zhang, J. M., and Ren, S. Q. 2018. Rare earth-free high energy product manganese-based magnetic materials. *Nanoscale* **10**: 11701–11718.

2.36. Saravanan, P., Saju, S., Vinod, V. T. P., and Černík, M. 2020. Structural and magnetic properties of rare earth-free MnAl(MCNT)/Fe nanocomposite magnets processed by resin-bonding technique. *Journal of Materials Science: Materials in Electronics*. doi:10.1007/s10854-020-03532-2.

2.37. Ames Laboratory. 2019. The 15 MGOe gap magnet without critical rare earths. https://www.ameslab.gov/cmi/research-highlights/the-15-more-gap-magnet-without-critical-rare-earths.

2.38. El-Gendy, A. A., Bertino, M., Clifford, D., Qian, M. C., Khanna, S. N., and Carpenter, E. E. 2015. Experimental evidence for the formation of CoFe2C phase with colossal magnetocrystalline-anistopy. *Applied Physics Letters* **106**(**21**): 213109.

2.39. Toyota Motor Corporation. 2018. Toyota develops new magnet for electric motors aiming to reduce use of critical rare earth element by up to 50%. https://global.toyota/en/newsroom/corporate/21139684.html

2.40. Pathak, A. K., Khan, M., Gschneidner Jr., K. A., McCallum, R. W., Zhou, L., Sun, K. W., Dennis, K. W., Zhou, C., Pinkerton, F. E., Kramer, M. J., and Pecharsky, V. K. 2015. Cerium: An unlikely replacement of dysprosium in high performance Nd-Fe-B permanent magnets. *Advanced Material* **27**(**16**): 2663–2667.

2.41. Dulikravich, G. S. 2016. Direct and inverse design optimization of magnetic alloys with minimized use of rare earth elements. Final performance report submitted to Air Force Research Laboratory, AFRL-AFOSR-VA-TR-2016-0091.

2.42. Pavel, C. C., Lacal-Arántegui, R., Marmier, A., Schüler, D., Tzimas, E., Buchert, M., Jenseit, W., and Blagoeva, D. 2017. Substitution strategies for reducing the use of rare earths in wind turbines. *Resources Policy* **52**: 349–357.

2.43. Pavel, C. C., Thiel, C., Degreif, S., Blagoeva, D., Buchert, M., Schüler, D., and Tzimas, E. 2017. Role of substitution in mitigating the supply pressure rare earths in electric road transport applications. *Sustainable Materials and Technologies* **12**(**2017**): 62–72.

2.44. Kapiza, P. L. 1924. A method of producing strong magnetic field. *Proceedings of the Royal Society A: Mathematical, Physical and Engineering Sciences* **105**: 691–710.

2.45. Biais, F. J. and Rahman, K. M. 2004. Auxiliary magnetizing winding for interior permanent magnet rotor magnetization. U.S. Patent 6,674,205.

2.46. Hitachi Metals Corporation. High energy ferrite magnets NMF series—Demagnetization curve (NMF-3C). http://www.hitachi-metals.co.jp/e/products/auto/el/pdf/nmf_a.pdf.

2.47. Hitachi Metals Corporation. 2014. Neodymium-Iron-Boron magnets – NEOMAX Series: Demagnetization curve (NMX-S52). http://www.hitachi-metals.co.jp/e/products/auto/el/pdf/nmx_a.pdf.

2.48. Beniakar, M., Kefalas T., and Kadas, A. 2011. Investigation of the impact of the operational temperature on the performance of a surface permanent magnet motor. *Materials Science Forum* **670**: 259–264.

2.49. Simos, N. and Mokhov, N. 2008. An experimental study of radiation-induced demagnetization of insertion device permanent magnets. *Proceedings of 11th European Particle Accelerator Conference*, Genoa, Italy, pp. 2112–2114.

2.50. Alderman, J., Job, P. K., Martin, R. C., Simmons, C. M., Owen, G. D., and Puhl, J. 2000. Radiation-induced demagnetization of Nd-Fe-B permanent magnets. U.S. Department of Energy Report LS-290, Washington, DC.

2.51. Magma Magnetic Technologies. 2010. Permanent magnet stability. http://www.magmamagnets.com/permanent-magnet-stability.

2.52. Integrated Magnetics. 2013. Magnetics 101. http://www.intemag.com/magnetics_101.html.

2.53. Degner, M. W., Van Maaren, R., Fahim, A., Novotny, D. W., Lorenz, R. D., and Syverson, C. D. 1996. A rotor lamination design for surface permanent magnet retention at high speeds. *IEEE Transactions on Industry Applications* **32**(**2**): 380–385.

2.54. Andonian, A. T. and Huynh, C. 2020. Rotor retention and loss reduction for high-speed permanent magnet motor generators. Calnetix Technologies, LLC. https://www.calnetix.com/.

2.55. UPM Raflatac. 2012. *The Adhesive Book*. UPM Raflatac, Tampere, Finland. http://www.upmraflatac. com/europe/eng/images/51_783.pdf.

2.56. Shen, J.-X., Hao, H., Jin, M.-J., and Yuan, C. 2013. Reduction of rotor eddy current loss in high speed PM brushless machines by grooving retaining sleeve. *IEEE Transactions on Magnetics* **49**(7): 3973–3976.

2.57. Etemadrezaei, M., Wolmarans, J. J., Polinder, H., and Ferreira, J. A. 2012. Precise calculation and optimization of rotor eddy-current losses in high speed permanent magnet machine. *Proceedings of the 2012 20th International Conference on Electrical Machines*, Marseille, France, pp. 1399–1404.

2.58. Chen, L.-L., Zhu, C.-S., Zhong, Z. X., Liu, B., and Wan, A. P. 2018. Rotor strength analysis for high-speed segmented surface-mounted permanent magnet synchronous machine. *IET Electric Power Applications* **12**(7): 979–990.

2.59. Merdzan, M., Paulides, J. J. H., and Lomonova, E. A. 2015. Comparative analysis of rotor losses in high-speed permanent magnet machines with different winding configurations considering the influence of the inverter PWM. *Proceedings of Tenth International Conference on Ecological Vehicles and Renewable Energies (EVER)*, Monte Carlo, Monaco, pp. 1–8.

2.60. Uzhegov, N., Pyrhönen, J., and Shirinskii, S. 2013. Loss minimization in high-speed PM Synchronous Machines with tooth-coil windings. *Proceedings of IECON 2013 – 39th Annual Conference of the IEEE Industrial Electronics Society*, Vienna, Austria, pp. 2960–2965.

2.61. Zhang, F. G., Du, G. H., Wang, T. Y., Liu, G. W., and Cao, W. P. 2015. Rotor retaining sleeve design for a 1.12-MW high-speed PM machine. *IEEE Transactions on Industry Applications* **51**(5): 3675–3685.

2.62. Vărăticeanu, B. D., Minciunescu, P., and Fodorean, D. 2014. Mechanical design and analysis of a permanent magnet rotors used in high-speed synchronous motor. *Electrotehnică, Electronică, Automatică* **62**(1): 9–17.

2.63. Gan, J., Chau, K. T., Chan, C. C., and Jiang, J. Z. 2000. A new surface-inset permanent brushless DC motor drive for electric vehicles. *IEEE Transactions on Magnetics* **36**(5): 3810–3818.

2.64. Johnson, R. N., Kliman, G. B., Liao, Y. F., and Soong, W. L. 1998. Rotors with retaining cylinders and reduced harmonic field effect losses. U.S. Patent 5,801,470.

2.65. Schaefer, E. J. and Antrim, T. K. 1988. Permanent magnet rotor for electric motor. U.S. Patent 4,742,259.

2.66. Dohogne, L. R. 1990. Dynamoelectric machine rotor assembly with improved magnet retention structure. U.S. Patent 4,973,872.

2.67. Smith, J. S. and Watson, A. P. 2006. Design, manufacturing, and testing of a high speed 10 MW permanent magnet motor and discussion of potential applications. *Proceedings of the 35th Turbomachinery Symposium*, Houston, Texas, pp. 19–24.

2.68. Mlot, A., Lukaniszyn, M., and Korkosz, M. 2016. Magnet loss analysis for a high-speed PM machine with segmented PM and modified tooth-tips shape. *Archives of Electrical Engineering* **65**(4): 671–683.

2.69. Dobrzanski, L. A., Drak, M., and Trzaska, J. 2005. Corrosion resistance of the polymer matrix hard magnetic composite materials Nd–Fe–B. *Proceedings of 13th International Scientific Conference on Achievements in Mechanical and Materials Engineering*, Gliwice-Wista, Poland.

2.70. Trout, S. 2004. Optimum corrosion protection of Nd–Fe–B magnets. *Proceedings of Advanced in Magnetic Application, Technology and Materials*, Dayton, OH.

2.71. Hitachi Metals (2015). Permanent magnets. https://www.hitachi-metals.co.jp/products/auto/el/pdf/hg-a27-f.pdf.

2.72. AK Steel Corporation. 2007. Selection of electrical steel for magnetic cores. Product data bulletin 7180-0139. http://www.aksteel.com/pdf/markets_products/electrical/Mag_Cores_Data_Bulletin.pdf.

2.73. Henke, M., Narjes, G., Hoffmann, J., Wohlers, C., Urbanek, S., Heister, C., Steinbrink, J., Canders, W.-R., and Ponick, B. 2018. Challenges and opportunities of very light high-performance electric drives for aviation. *Energies* **11**(2): 344.

2.74. Huynh, T. A. and Hsieh, M.-F. 2018. Performance analysis of permanent magnet motors for electric vehicles (EV) traction considering driving cycles. *Energies* **11**(6): 1385.

2.75. Huynh, T. A. and Hsieh, M.-F., 2017. Performance evaluation of interior permanent magnet motors using thin electrical steels. *IEEJ Journal of Industry Application* **6**(6): 422–428.

2.76. Emura, M., Landgraf, F. J. G., Ross, W., and Barreta, J. R. 2003. The influence of cutting technique on the magnetic properties of electrical steels. *Journal of Magnetism and Magnetic Materials*, **254–255**: 385–360.

2.77. Wiley, J. B. 1965. Dynamoelectric machine means. U.S. Patent 3,188,505.

2.78. Rosenquist, S. 2005. Test of the effects of the heat shrink process on shaft run-out and torsional strength of EVS rotor assemblies. Kollmorgen report.

2.79. International Organization for Standardization. 2003. ISO 1940-1:2003 Mechanical vibration. Balance quality requirements for rotors in a constant (rigid) state.

2.80. Mobley, R. K. 2004. *Maintenance Fundamentals*, 2nd edn. Elsevier Butterworth-Heinemann, Amsterdam, the Netherlands.

2.81. Budynas, R. G. and Nisbett, K. J. 2010. *Shigley's Mechanical Engineering Design*, 9th edn. McGraw-Hill, New York.

2.82. Horger, O. J. and Nelson, C. W. 1937. Design of press- and shrink-fitted assemblies: Part I. *Journal of Applied Mechanics* **4**: A183–A187.

2.83. Horger, O. J. and Nelson, C. W. 1938. Design of press- and shrink-fitted assemblies: Part II. *Journal of Applied Mechanics* **5**: A32–A32.

2.84. Antman, S. S. and Shvartsman, M. M. 1995. The shrink-fit problem for aeolotropic nonlinear elastic bodies. *Journal of Elasticity* **37**: 157–166.

2.85. Lundberg, G. 1944. Die festigkeit von pressitzen. *Das Kugellager* **19**: 1–11.

2.86. Gao, X.-L. and Atluri, S. N. 1995. An elasto-plastic analytical solution for the shrink-fit problem with a thin strain-hardening hub and an elastic solid shaft. *Mathematics and Mechanics of Solids* **2**: 335–349.

2.87. Bianchi, N. and Lorenzoni, A. 1996. Permanent magnet generators for wind power industry: An overall comparison with traditional generators. *International Conference on Opportunities and Advances in International Power Generation*, Durham, United Kingdom, Conference Publication No. 419.

2.88. Lekhniskii, S. G. 1968. *Anisotropic Plates*. Gordon and Breach Science Publishers, New York.

2.89. Ha, S., Kim, D., and Sung, T. 2001. Optimum design of multi-ring composite flywheel rotor using a modified generalized plane strain assumption. *International Journal of Mechanical Sciences* **43**(4): 993–1007.

2.90. Ameridrives Power Transmission. 2009. Americardan universal joints: 5000 series high torque density. P-1751-APT.   http://www.gmbassociates.co.uk/downloads/Americardan   UniversalJoints5000Series Catalogue.pdf.

2.91. Kirk, R. G. and Keesee, J. 1989. Influence of active magnetic bearing sensor location on the calculated critical speeds of turbomachinery. *Proceeding of ASME Design Technical Conferences, Montreal*, Quebec, Canada, DE-Vol. 18–1, pp. 309–316.

2.92. Kass, M. D., McKeever, J. W., Akeman, M. A., Goranson, R L., Litherland, R S., and O'Kain, D. U. 1996. Evaluation of demo of 1C component flywheel rotor burst test and containment design. ORNL/TM-13159. Oak Ridge National Laboratory, Oak Ridge, TN.

2.93. Genta, G. 2005. *Dynamics of Rotating Systems*. Springer, New York.

2.94. Pichot, M. A., Kramer, J. M., Thompson, R. C., Hayes, R. J., and Beno, J. H. 1997. *The Flywheel Battery containment Problem*. SAE Technical Paper Series, Warrendale, PA, Book Number: SP-1243, Document Number: 970242.

2.95. Hagg, A. C. and Sankey, G. O. 1974. The containment of disk burst fragments by cylindrical shells. *ASME Journal for Engineering for Power* **96**(2): 114–123.

2.96. Stamper, E. and Hale, S. 2008. The use of LS-DYNA® models to predict containment disk burst fragments. *10th International LS-DYNA® Users Conference*, Dearborn, MI.

2.97. Jain, R. 2010. Prediction of transient loads and perforation of energy casing during blade-off event of fan rotor assembly. *Proceedings of the IMPLAST 2010 Conference*, Providence, RI.

2.98. Stotler, C. L. 1981. Development of advanced lightweight systems containment: Final report. NASA report no. CR-165212. http://www.dtic.mil/cgi-bin/GetTRDoc?AD=ADA305415.

2.99. Gunderson, C. O. 1977. Study to improve airframe turbine engine rotor blade containment. U.S. Department of Transportation, Federal Aviation Administration. Report No. FAA-RD-77-44. http://www.tc.faa.gov/its/worldpac/techrpt/rd77-44.pdf.

2.100. Payen, J. M. 1983. Containment of turbine engine fan blades. *6th International Symposium on Air Breathing Engines*. Symposium papers A83-35801 16-07, Paris, France, pp. 611–616.

3
Shaft Design

A shaft in an electric motor is used to transmit torque and power from the motor to an external loading machine. Because motor shafts are subjected to various combined effects of tension, compression, bending, and torsion during operation, they are typically designed for maximum stiffness and rigidity and minimum deflection to maintain shaft stress/strain well below allowable limits under various loading and operating conditions. The achievement of such design objectives relies on the selection of shaft material, determination of suitable shaft dimensions (especially the shaft diameter and span between bearings) and structures, mitigation/elimination of stress concentrations, and other design activities. Obviously, each of these can affect the long-term reliability of motor.

As a rotating component, the motor shaft is subject to a completely reversed bending load that results in an alternating bending stress in the shaft. Furthermore, some motors require to frequently change the direction of rotation or experience frequent starts and stops during operation, causing cyclic torsional stress in motor shafts. Therefore, the prevention of fatigue failure is an important consideration in the shaft design.

A typical shaft of an electric motor is illustrated in Figure 3.1. Typically, the motor shaft is supported by two bearings at its ends and a rotor core mounted on the shaft between the two bearings. Motor shafts usually carry keyed or splined power-transmitting components (*e.g.*, sheaves, pulleys, couplers, and sprockets) on the overhanging end to transmit torque to an externally driven machine. The shaft span is generally designed as short as possible to increase the shaft stiffness and reduce the shaft stress and lateral deflection. In addition, minimizing the shaft overhang is highly desired for lowering shaft bending stress.

The importance of engineering analysis and design of shafts has been widely recognized by engineers and designers. However, most machine design textbooks have mainly focused on shafts with uniform diameters. Such shafts are easy to produce but rare in practice. To help design motor shafts, this chapter emphasizes the design for stepped motor shafts.

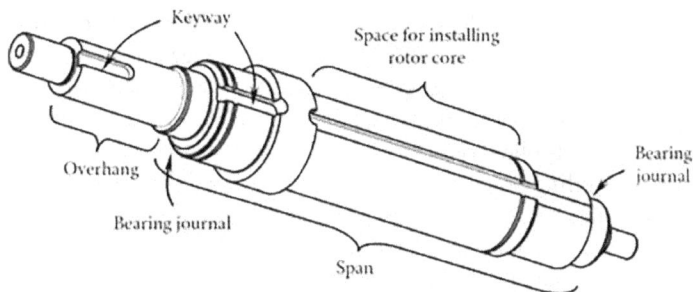

FIGURE 3.1 Structure of a motor shaft.

3.1 SHAFT MATERIALS

The selection of proper shaft material is critical for ensuring motor normal and safe operation. Depending on different applications, motor shafts can be made of low- to medium-carbon steel, cast

DOI: 10.1201/9781003097716-3

iron, stainless steel, aluminum alloys, brass, and bronze. The criteria of the shaft material selection include material mechanical and thermal properties, rigidity, hardness, wear resistance, machinability, noise absorption, manufacturing process, and cost.

By far the most widely used shaft material is carbon steel. The mechanical and thermal properties of carbon steel primarily depend on the amount of carbon it contains. Increasing carbon contents can result in an increase in hardness, yield tensile strength, and ultimate tensile strength. However, higher carbon contents also increased material brittleness and reduced weldability, machinability, and elongation.

According to the content of carbon in steel, carbon steel can be categorized into three types: low-carbon, medium-carbon, and high-carbon steel. Low-carbon steel contains carbon up to 0.25%. Medium-carbon steel has carbon content ranging from 0.25% to 0.70%. High-carbon steel contains carbon in the range of 0.70%–1.50%. Most motor shafts are made of steel containing 0.2%–0.5% carbon. With a single-digit elongation, high-carbon steel is too brittle to be used as the shaft material.

To improve the mechanical properties of shafts, such as hardness, yield and ultimate tensile strength, fatigue strength, and other material properties, motor shafts are often heat-treated. Heat treatment processes include spheroidizing, annealing, normalizing, carburizing, quenching, martempering, and austempering. Each process has its own purpose and is applied to certain types of materials.

Spheroidizing is a form of heat treatment for carbon steel, especially for high-carbon steel with a carbon content of more than 0.6%. The spheroidite structure, which contains sphere-like cementitie particles, forms when carbon steel is heated to approximately $700°C$ for more than 30 h. The purpose of this process is to soften high-carbon steels for improving machinability and increasing ductility of high-carbon steel. An annealing process is often used to relieve residual stresses in the cold-worked steel with a carbon content of 0.3% or above. Fully annealed steel has no residual stresses and becomes ductile. Normalizing helps steel with a fine and more uniform pearlitic structure. Normalized steel still maintains high strength and durability. For low-carbon steel, carburizing can increase the hardness and improve other mechanical properties. During the process, carbon molecules penetrate into the material surface. The affected area and depth of carbon content are dependent on the heating time and temperature. Generally, higher temperatures and longer heating time lead to greater carbon diffusion into steel. The carbon content of quenchable steel must be 0.3% or above. A quenching process can increase yield and ultimate tensile strength, fatigue strength, and hardness but decrease material elongation. Usually, quenched steel is about three or four times harder than the normalized steel. However, residual stresses can be introduced during the quenching process in the bulk of the shaft that may cause cracks on the shaft surface. In all heat treatment processes, the heating and cooling rate and temperature holding time are critical parameters to impact the metallographic microstructure of steel.

Either cold-rolled or hot-rolled carbon steel can be used for making motor shafts. Hot-rolled steel is produced at temperatures above the recrystallization temperature of steel. In contrast, cold-rolled steel is produced under the recrystallization temperature of steel (more regularly, at room temperature). The benefits of cold-rolled steel over hot-rolled steel are as follows: (a) The strength of cold-rolled steel can increase by 15%–30% due to the strain-hardening mechanism. (b) Cold-rolled steel has a smooth surface finish compared to hot-rolled steel. (c) Cold-rolled steel products have tight dimensional tolerances. However, the cold rolling process can generate residual stresses remaining in the rolled steel. Furthermore, the cost of the cold rolling process is usually higher than that of hot rolling.

For some applications that require high material strengths or operate under severe environmental conditions (e.g., extremely low or high temperatures, high humidity, acid and alkali corrosion), motor shafts may also be made of special steel alloys.

Stainless steel can be used as a shaft material when motors work under corrosive environments or for some special applications such as food industry and medical equipment. With a minimum chromium (Cr) content of 10.5%, stainless steel is continuously protected by a passive layer of chromium oxide that forms naturally on the surface through the combination of chromium and moisture in the air. This enhances the corrosion resistance of stainless steel. The addition of nickel (Ni) can also raise the corrosion resistance of stainless steel. However, machining stainless steel is more difficult than carbon steel because of heat buildup during the machining process and difficulties in chip breaking. Stainless steel alloy 304 is one of the most widely used stainless steel that contains 18% of chromium, 8% of nickel, and lower carbon (0.08%) to minimize carbide precipitation. This type of stainless steel possesses an excellent combination of strength, corrosion resistance, and fabricability [3.1]. However, because the coefficient of thermal expansion (CTE) of stainless steel is approximately 25%–50% higher than that of carbon steel, high thermal stresses on stainless steel shafts may be developed as the result of temperature rise in motors.

Cast iron is a large family of ferrous alloys in which four types can be categorized: gray cast iron, ductile iron, white iron, and metallurgic iron. Cast iron has been widely used to make crankshafts in internal combustion engines. Crankshafts convert the reciprocating motion into rotation by crank-pins. One of the main reasons to use cast iron is that it has excellent vibration damping properties to reduce the pulsation influence of the four-stroke cycle. In addition, cast iron exhibits low notch sensitivity, low modulus of elasticity, high thermal conductivity, moderate resistance of thermal shock, and outstanding castability. It is interesting to note that unlike most ferrous materials, shear strength of gray cast iron is much higher than ultimate tensile strength, indicating that gray cast iron can withstand higher shear forces than tensile forces. However, until now, it is still rare to use cast iron as the shaft material in electric motors.

Aluminum is a versatile and corrosion-resistant material. Advanced high-tensile aluminum alloys such as scandium- and titanium-enhanced aluminum alloys have been used to replace steel today in making shafts, vehicle body, and other important components in the automotive industry. Their advantages are ease of fabrication, high strength-to-density ratio, and resistance to the corrosive atmospheres. High-performance race cars utilize extensively lightweight aluminum parts as engineers strive to improve speeds and reduce corrosion under high-temperature and high-stress conditions. For example, high-grade 7075-T6 aluminum alloy may be acceptable as a shaft material in some applications due to its high strength, low weight, and good fatigue strength properties [3.2]. In the motor industry, the use of aluminum shafts is limited to small-size and low-torque motors.

Carbon fiber is one of modern materials that has been developed and employed in applications where a large strength-to-weight ratio, high rigidity, and strong corrosion resistance are required, such as in aviation and automotive industries. Carbon fiber has several exceptional advantages over conventional metallic materials, including high tensile strength (five times higher than steel), high stiffness (two times higher than steel), lightweight, corrosion-free, high-temperature tolerance, high thermal conductivity (equivalent to copper), and low thermal expansion. Moreover, as a non-metallic material, the application of carbon fiber as the shaft material can help mitigate shaft voltage and other electrical problems. In practice, carbon fibers are usually composited with other materials (*e.g.*, thermosetting resin such as epoxy, polyester, or vinyl ester) to form composite materials such as carbon fiber reinforced polymer (CFRP) and glass fiber reinforced polymer (GFRP). Among all fiber-reinforced materials, CFRP and GFRP are increasingly replacing conventional materials in recent years. Compared to aluminum and steel, CFRP composite structure of equal strength would likely weight 1/5 that of steel structure and 1/2 of aluminum structure. Therefore, they have been extensively used in a wide range of contemporary applications particularly in space and aviation, automotive, civil construction, renewable energy (*e.g.*, wind turbine blades), and sports equipment.

However, there are several concerns in taking CFRP as an alternative of steel. One is about its long-term durability due to the brittle nature of CFRP. Another issue is that CFRP composites are most costly materials than their counterparts presently. In addition, CFRP has a poor machinability. Delamination damages commonly occur during the machining of CFRP. These issues must be properly resolved before CFRP can be extensively utilized in engineering and commercial applications.

In recent years, CFRP composite is used to fabricate hybrid shafts of motor for replacing traditional metal shafts [3.3]. The greatly reduced weight and inertia of the shaft lead to benefits in reduced overall motor mass and improved motor dynamic performance. This material is likely to be used to make motor housings in the near future. Furthermore, CFRP can be used to construct composite wind towers for decreasing the total weight of the wind towers, leading to substantial savings in transportation and erection costs [3.4]. It was reported that the cost of composite towers, based on a 25-unit wind farm, is 28% less than the cost of steel towers.

The properties of some typical shaft materials are listed in Table 3.1. The machinability rating is an indicator of how easy or how difficult a material can be mechanically processed based on the reference material AISI 1212 steel as 100% machinability.

3.2 SHAFT LOADS

Motor shafts are generally subject to combined, variable loads during operation. The motor loads can be categorized into various types:

- Because a shaft transmits torque from a motor to an externally driven machine, the shaft is subject to the torsional load. Hence, the shaft must have adequate torsional strength to withstand the load.
- Transverse loads are most significant in various motor applications. As a shaft is subject to transverse loads, it undergoes bending or flexural deformations and possibly shearing at the joints of the shaft and the rotor core. There are several types of transverse forces acting on the shaft: (a) The gravitational force acting on the mass of the rotor assembly (the rotor core, shaft, and other components) is perpendicular to the shaft axis for a horizontally installed motor and (b) the shaft usually carries power-transmitting components such as sheaves, gears, pulleys, and chain sprockets, exerting forces at the end of the shaft in the transverse direction. (c) For a nonuniform airgap between a stator and a rotor, the unbalanced magnetic pull may develop and act on the rotor and shaft. As the shaft rotates, all these transverse forces cause reverse bending moments on the shaft. Thus, it is required to perform fatigue analysis to avoid fatigue failures.
- A motor shaft is usually subject to low axial loading compared to its transverse loads. However, for some applications such as high-speed PCB drilling machines that experience high acceleration and deceleration in a very short time period in the motor axial direction, the axial load on the shaft can become considerably high. Furthermore, buckling analysis may be required for assessing rotor stability.
- The coupling between a motor and a driven machine can be achieved by using flanges, splines, belts, chains, and other flexible or rigid coupling devices. For using belts or chairs, a bending moment is applied on the overhanging portion of the shaft.
- One of the major concerns during motor design is the motor's resonant behavior. As a motor starts up, the motor shaft is subject to a shearing force. Mechanical resonance occurs as an external source amplifies the vibration level of a motor at its natural frequency. The increase in the amplitude of vibration level leads to high shear stress on motor shaft and other components.
- The preload acts on a shaft due to interference fit of rotor core, bearing, and other mechanical components.
- A motor may encounter shock loads at normal operation. Shock loads can be created when motors experience interrupted motions in rapid sequences, such as repeated sudden start and stop, high acceleration and deceleration, forward-reverse, and indexing motions. In addition, with high rotating speeds, electric motors act like fly wheels. Typically, they represent approximately 80%–90% of the total kinetic energy in the drive systems. This flywheel effect can produce inertia shock loads far above the rated torque when the driven equipment is abruptly stopped [3.9].

If shock loads cannot be completely avoided, the motor components must be designed by considering an additional factor of safety (FS) (shock factor).

Earthquakes can always cause shock loads on motors. For the safe operation of the motor, this scenario must be carefully considered at the motor design stage. This is extremely important for some types of motors, such as elevator motors.

TABLE 3.1
Properties of Some Shaft Materials

Material	Density (kg/m³)	Yield Tensional Strength (MPa)	Ultimate Tensional Strength (MPa)	Modulus Elasticity (GPa)	Shear Modulus (GPa)	Shear Strength (MPa)	Fatigue Strength (MPa)	Elongation (%)	Hardness Brinell (HB)	Machinability Rating[a] (%)	Vibration Damping
AISI 1008 cold rolled	7,870	285	340	190–210	80	196	170	20	95	55	Acceptable
AISI 1020 cold rolled	7,870	350	420	205	80	242	193[b]	15	121	65	Acceptable
AISI 1045 cold rolled, annealed	7,870	505	585	206	80	338	268[b]	12	170	65	Acceptable
304 SS	8,000	241	586	193–200	86	334	241[c]	55	149	45	A little better than steel
Ductile iron 65-45-12	7,100	310	448	336	179	336	179	12	131–220	160	Good
Gray cast iron 40		—	293	110–138	44–54	393	128[c]	<1	235	70	Best
A7075-T6	2,810	503	572	71.7	26.9	331	159[d]	11	150	120	Poor
CFRP	1,580[e]	—	1,440 – 4,100[f]	142[e, g] – 231[f]	7.2[e]	4.05– 22.43	443[h]	<1.3	112–122	Low[i]	Good

a The American Iron and Steel Institute (AISI) has determined AISI 1212 carbon steel a machinability rating of 100%.

b Based on 10^7 cycles.

c Based on 10^8 cycles.

d Based on 5×10^8 cycles.

e The data are cited from reference [3.4], where ultimate tension strength and elastic modulus are in the fiber direction.

f The data are cited from reference [3.5].

g The data are cited from reference [3.6], where the test temperature is at 20°C.

h This value is obtained from a correlated S-N curve at $N = 10^7$ cycles [3.7].

i CFRP is considered a low machinability material because of the abrasive character of the carbon fibers causing strong tool wear. In fact, delamination damage is commonly observed during machining of CFRP. This defect affects the structural integrity of the laminate for long-term performance and results in poor assembly tolerance [3.8].

3.3   SOLID AND HOLLOW SHAFTS

According to the shaft structure, motor shafts can be briefly classified as solid or hollow shafts. Solid shafts have long been used in the motor industry for less production costs. Motors with hollow shafts, known as hollow bore motors, have useful features for some applications. The hollow space through the center portion of the motor provides a path for cooling flows. This can greatly improve heat dissipation from the motor.

In some motion control systems (*e.g.*, industrial robots), motor power and feedback cables must be integrated inside the systems for not only design optimizations but also safety considerations. The use of hollow shafts makes it possible to enable these cables passing through the bores. Therefore, this design could significantly reduce the cable length and increase system operating reliability and safety. To minimize torsional stress, the cables are typically made in a helical or double helical shape [3.10].

A hollow shaft has a higher strength-to-weight ratio and stiffness than a solid shaft with equal weight and length. This is because the outer diameter (OD) of the hollow shaft is larger than that of the solid shaft. If both shafts have the same OD, the solid shaft has more torsional resistance and less deformation than the hollow shaft under the combined torsional and cantilevered loading.

3.4   SHAFT DESIGN METHODS

There are several methods available in the design of motor shafts; each of them has its specific scope of application and advantages/disadvantages when compared with others.

3.4.1   MACAULAY'S METHOD

This method is relatively simple to deal with the stiffness, radius of curvature, deflection, and bending moment in a beam. The method enables discontinuous bending moment functions to be represented by a continuous function. It allows the contributions, from individual loads to the bending moment at any cross section to be expressed as a single function, which takes zero value at those sections where particular loads do not contribute to the bending moment [3.11].

As shown in Figure 3.2, when a beam is subject to a uniform load along its length, the bending moment M (in units of Nm) is given by

$$M = \frac{EI_a}{R} \tag{3.1}$$

where
   E is Young's modulus of elasticity, Pa, MPa, or GPa
   I_a is the second moment of area, m^4
   R is the radius of curvature, m

In some textbooks, EI_a is defined as the beam flexural rigidity.

The curvature $1/R$ equals the reciprocal of the radius and is expressed as the second derivative with respect to the beam deflection y, that is,

$$\frac{1}{R} = \frac{d^2 y}{dz^2} \tag{3.2}$$

It gives that

$$EI_a \frac{d^2 y}{dz^2} = M \tag{3.3}$$

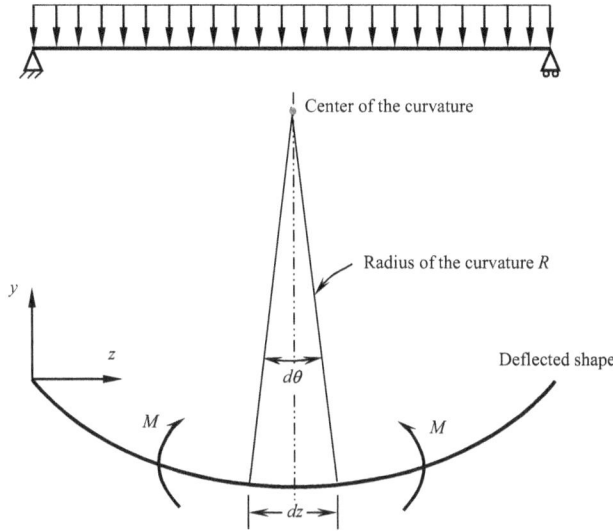

FIGURE 3.2 Beam deflection under uniformly distributed load.

Hence, the differential equation of an elastic curve may be given as

$$\frac{d^2y}{dz^2} = \frac{M}{EI_a} \tag{3.4}$$

Macaulay's method enables discontinuous bending moment functions to be represented by a continuous function, thus avoiding the need to deal with the beam section by section between discontinuities in the bending moment function. This is very desirable since it avoids the need to evaluate, and therefore eliminate, a large number of constants of integration.

3.4.2 Area–Moment Method

The area–moment method is commonly used to determine the deflection and slope of a shaft based on two theorems: the first is used to find rotations and the second is used to find displacements. It is effective for calculating the deflections of shafts with various cross sections.

In Figure 3.3, consider a section of an elastic curve between points A and B. Both the slope and M/EI_a change along the elastic curve from A to B. Thus, integrating both sides of Macaulay's equation with respect to dz yields

$$\int_A^B \frac{M}{EI_a}\,dz = \int_A^B \frac{d^2y}{dz^2}\,dz = \left[\frac{dy}{dz}\right]_A^B = \theta_B - \theta_A \tag{3.5}$$

This indicates that the difference in slope between points A and B on a beam is equal to the area of the M/EI_a diagram between these two points.

Multiplying both sides of Macaulay's equation by zdz and integrating from A to B, it gives that

$$\int_A^B \frac{M}{EI_a} z\,dz = \int_A^B \frac{d^2y}{dz^2} z\,dz = \int_A^B \left[\frac{d}{dz}\left(z\frac{dy}{dz}\right) - \frac{dy}{dz}\right]dz = \left[z\frac{dy}{dz}\right]_A^B - \left[y\right]_A^B$$

$$= z_B\left(\frac{dy}{dz}\right)_B - z_A\left(\frac{dy}{dz}\right)_A - y_B + y_A \tag{3.6}$$

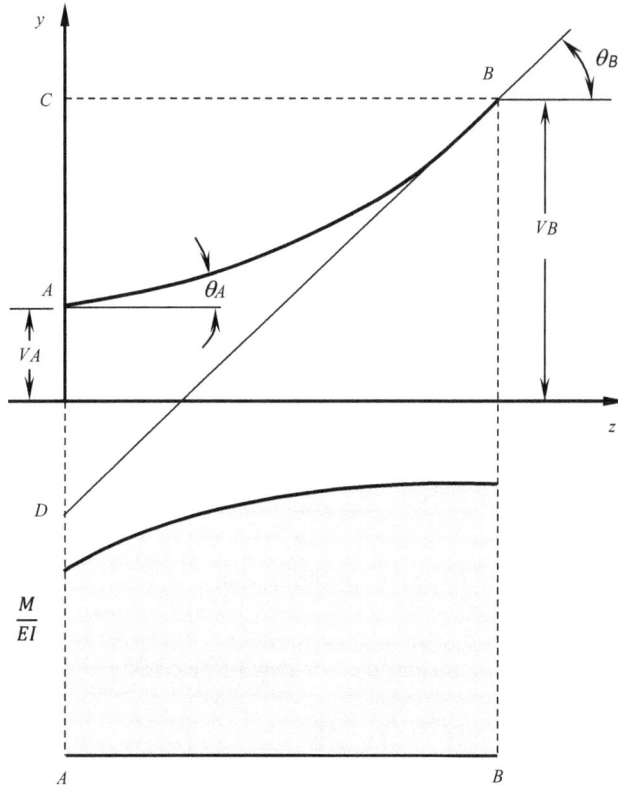

FIGURE 3.3   The changes in the slope and M/EI along the elastic curve between points A and B.

If the origin is shifted below point A (*i.e.*, $z_A = 0$), the second term in the previous equation vanishes, giving that

$$\int_A^B \frac{M}{EI_a} z\, dz = z_B \theta_B - y_B + y_A \tag{3.7}$$

where $z_B \theta_B$ is represented by the distance CD in Figure 3.3. This indicates that the moment about a point A of the M/EI_a diagram between points A and B provides the deflection of point A relative to the tangent at point B.

3.4.3   CASTIGLIANO'S METHOD

Castigliano's method is widely used to determine displacements of linear-elastic systems based on the strain energy of the systems. When such a system is subject to external loads, the deflection δ_i at any loading point in a direction of any external load Q_i can be determined by taking the partial derivative of the total strain energy U stored in the system with respect to the applied load Q_i in that direction, that is,

$$\delta_i = \frac{\partial U}{\partial Q_i} \tag{3.8}$$

If the displacement is in the same direction of the force, the displacement is positive.

TABLE 3.2

General Deflection Equations Using Castigliano's Method

Load Type	General Energy Equation	General Deflection Equation
Axial force	$U = \int_0^l \dfrac{F^2}{2EA} dx$	$\delta = \int_0^l \dfrac{F}{EA}\left(\dfrac{\partial F}{\partial Q}\right) dx$
Bending moment	$U = \int_0^l \dfrac{M^2}{2EI_a} dx$	$\delta = \int_0^l \dfrac{M}{EI_a}\left(\dfrac{\partial M}{\partial Q}\right) dx$
Torsional moment	$U = \int_0^l \dfrac{T^2}{2GJ} dx$	$\delta = \int_0^l \dfrac{T}{GJ}\left(\dfrac{\partial T}{\partial Q}\right) dx$
Transverse shear	$U = \int_0^l \dfrac{f_s V^2}{2GA} dx$	$\delta = \int_0^l \dfrac{f_s V}{GA}\left(\dfrac{\partial V}{\partial Q}\right) dx$

The general energy and deflection equations corresponding to different load types are listed in Table 3.2, where F, M, and T are the force, bending moment, and torque acting on the shaft, respectively, E is the modulus of elasticity of the shaft material, A is the cross-sectional area of the shaft, I_a is the second moment of area, G is the shear modulus of elasticity, J is the second polar moment of area (sometimes improperly referred to as the *polar moment of inertia*), and f_s is the form factor for shear:

$$f_s(x) = \frac{A(x)}{I_a^2(x)} \int_A \frac{Q^2(x,y)}{t^2(y)} dA \qquad (3.9)$$

The form factor is a dimensionless quantity that depends only on the shape of the cross section of shaft. In most cases, f_s is constant. For a circular shaft, $f_s = 10/9$.

3.4.4 Graphical Method

The graphical method has been primarily used in an earlier time when computational techniques have not been fully developed [3.12]. With this method, design engineers graphically analyze shaft stresses and deflections. Today, with the fast development of sophisticated computational techniques, design engineers concentrate more on the analytical approach. However, the graphical approach to the shaft design and analysis has not lost its utility, especially in some applications where the graphical technique can provide the most efficient and quickest solution and physical insight to visualize the response of a shaft to applied loads.

3.5 ENGINEERING CALCULATIONS

Engineering calculation is a key part in motor shaft design, which is conducted to ensure proper operation of shafts, including stress and strain, shaft deflection, critical speed, and stiffness.

For most loading types (forces, bending moments, *etc.*), the stress developed in a shaft is linearly related to the loading. As long as the stress in the material remains within the linear-elastic region, the stress is also linearly related to the deflection.

3.5.1 Normal Stress for Shaft Subjected to Axial Force

When a shaft of uniform diameter is subjected to an axial force F (Figure 3.4), the normal stress σ of the shaft is obtained as

$$\sigma = \frac{F}{A} = \frac{4F}{\pi d^2} \qquad (3.10)$$

where A is the cross-sectional area of the shaft.

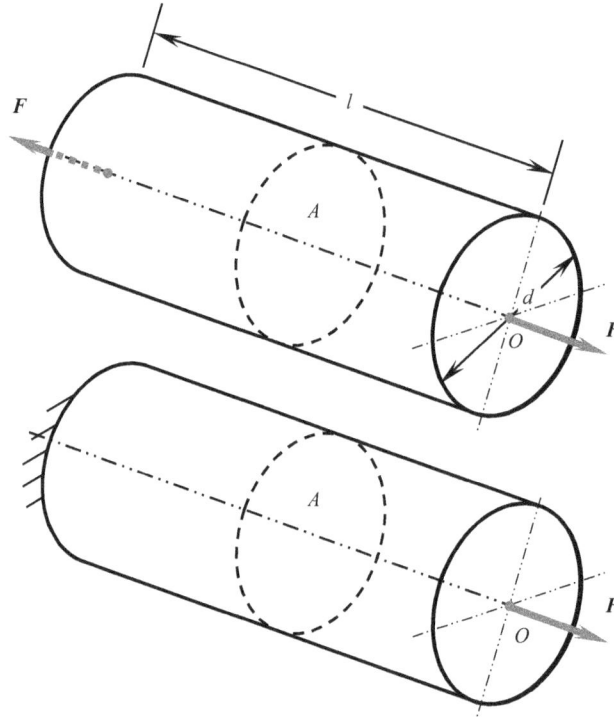

FIGURE 3.4 A single-section shaft subjected to axial force.

More generally, when the cross-sectional area of a shaft varies along the z direction, that is, $A = A(z)$, the normal stress becomes a function of z:

$$\sigma(z) = \frac{F}{A(z)} = \frac{4F}{\pi[d(z)]^2} \tag{3.11}$$

A common example is a tapered shaft that can be characterized by a small (or large) diameter d_1 (or d_2) and a cone angle α, as shown in Figure 3.5. Thus,

$$d(z) = d_1 + 2z\tan\alpha \tag{3.12}$$

Substituting Equation 3.12 into 3.11, it yields

$$\sigma(z) = \frac{4F}{\pi[d_1 + 2z\tan\alpha]^2} \tag{3.13}$$

For a stepped shaft that has an individual diameter at each segment, the normal stress at each segment i is given as (see Figure 3.6)

$$\sigma_i = \frac{F}{A_i} = \frac{4F}{\pi d_i^2} \tag{3.14}$$

This indicates that the maximum normal stress occurs at the segment having the smallest diameter.

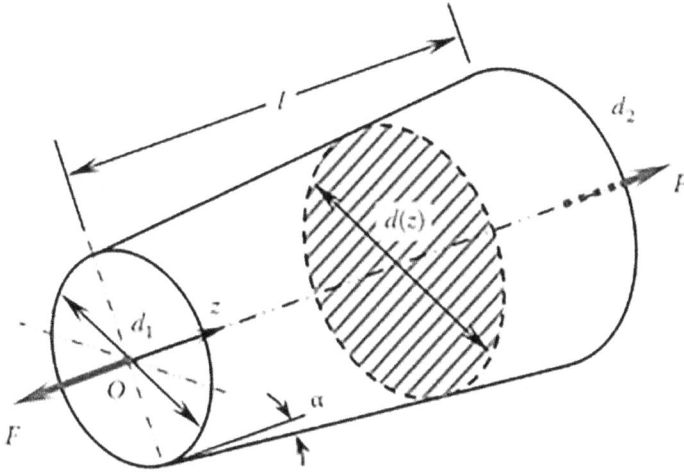

FIGURE 3.5 A tapered shaft with a cone angle α is subject to an axial force F.

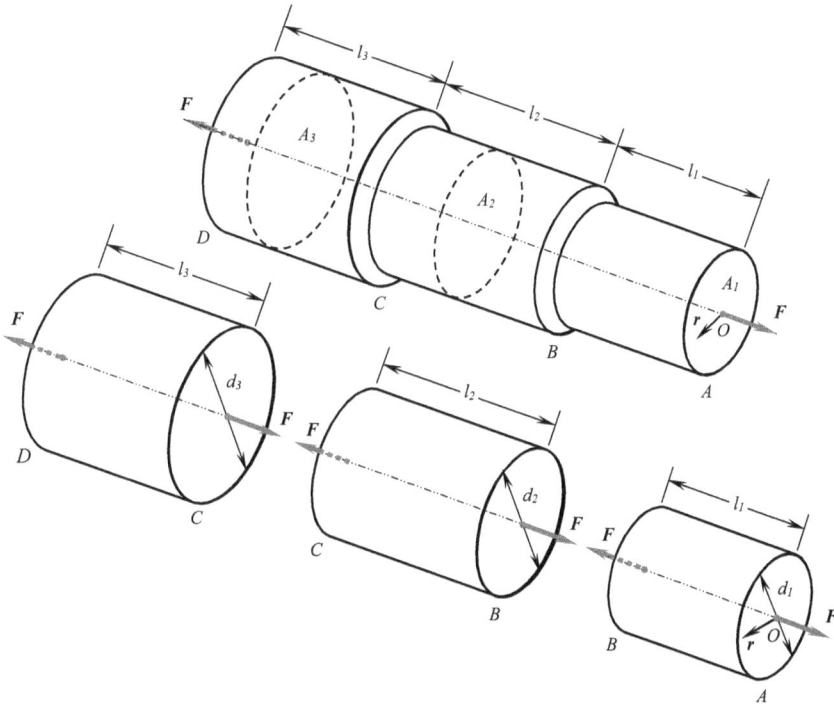

FIGURE 3.6 A stepped shaft subjected to an axial force.

3.5.2 BENDING STRESS FOR SHAFT SUBJECTED TO BENDING MOMENT

When a shaft is subjected to pure bending moment, there is bending stress generated in the shaft. As shown in Figure 3.7, compressive stresses are developed for $y > 0$ and tensional stresses for $y < 0$. At $y = 0$, there is neither compressive nor tensile stress. Thus, $y = 0$ is defined as the neutral axis. The classic formula for determining the bending stress is

$$\sigma(y) = \frac{My}{I_a} \tag{3.15}$$

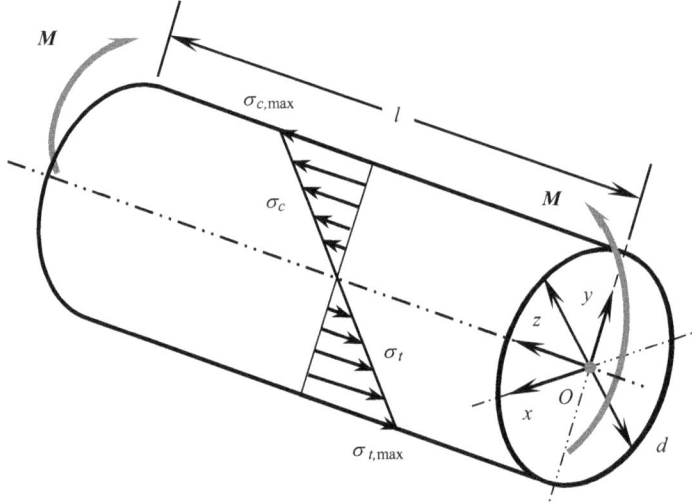

FIGURE 3.7   A single-section shaft subjected to pure bending moment.

- For a solid circular shaft,

$$I_a = \frac{\pi d^4}{64} \tag{3.16a}$$

- For a hollow circular shaft,

$$I_a = \frac{\pi \left(d_o^4 - d_i^4 \right)}{64} \tag{3.16b}$$

The maximum compressive and tensional stresses occur at $y = d/2$ and $-y = d/2$, respectively:
- For a solid circular shaft,

$$\sigma_{c,\max} = \sigma_{t,\max} = \frac{32M}{\pi d^3} \tag{3.17a}$$

- For a hollow circular shaft,

$$\sigma_{c,\max} = \sigma_{t,\max} = \frac{32 M d_o}{\pi (d_o^4 - d_i^4)} \tag{3.17b}$$

For a stepped shaft subjected to multiple bending moments (Figure 3.8), the bending stress at each segment is given as

$$\sigma_c(y) = \frac{My}{I_{a,i}} \tag{3.18}$$

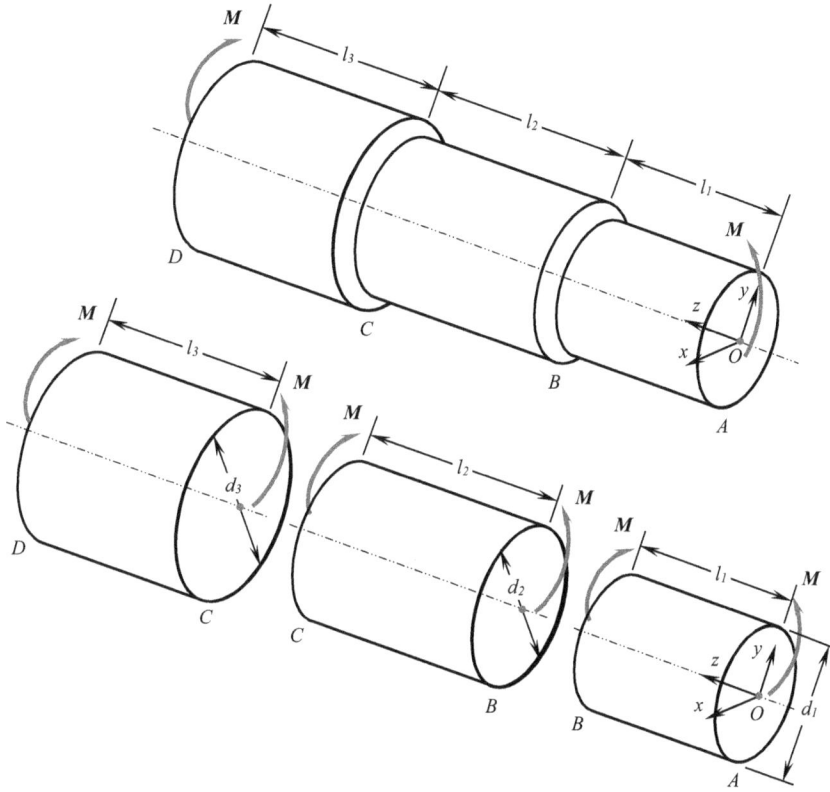

FIGURE 3.8   A stepped shaft subjected to pure bending moment.

- For a solid circular shaft,

$$I_{a,i} = \frac{\pi d_i^4}{64} \qquad (3.19a)$$

- For a hollow circular shaft,

$$I_{a,i} = \frac{\pi \left(d_{o,i}^4 - d_{i,i}^4 \right)}{64} \qquad (3.19b)$$

3.5.3   TORSIONAL SHEAR STRESS AND TORSIONAL DEFLECTION

Figure 3.9 shows that a pair of twist forces F is applied on a single-section shaft with a uniform diameter. This is equivalent to the case that the shaft is fixed at one end and a torque T is applied at the other end. In such a case, the radial line at the shaft end surface is rotated through an angle ϕ, defined as the shaft torsional deflection (or angular deflection in some textbooks). The length of the arc produced in the case is $R\phi$. Correspondingly, the straight line parallel to the shaft axis is rotated through an angle γ, defined as the shaft shear strain.

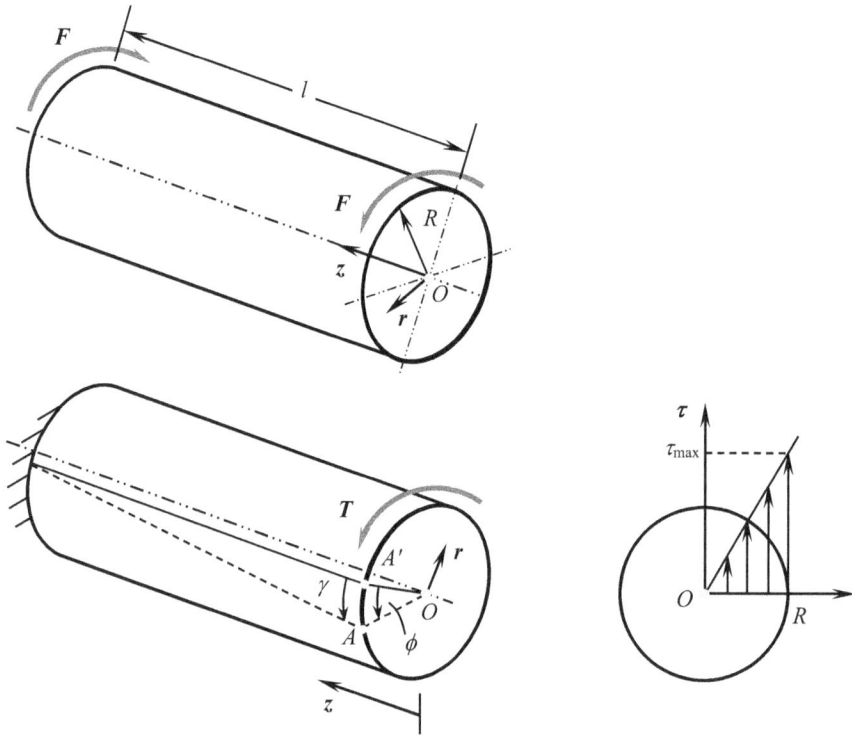

FIGURE 3.9 Torsional shear stress and torsional deflection of a single-section shaft subjected to torsional bending moment $M = F \times R$ or torque T.

The rate of twist $d\phi/dz$ can be derived as

$$\frac{d\phi}{dz} = \frac{T}{GJ} \tag{3.20}$$

where
  G is the shear modulus of elasticity, Pa, MPa, or GPa
  J is the second polar moment of area (in units of m⁴)

It can be seen from this equation that the higher the value of GJ, the lower the rate of twist $d\phi/dz$ for a given torque T. This means that the rigidity of the shaft increases with the increase in GJ. The quantity GJ is thus defined as torsional rigidity.

For a solid circular shaft,

$$J = \frac{\pi d^4}{32} \tag{3.21}$$

For a hollow shaft with an OD and ID, d_o and d_i, respectively, the second polar moment of area J is

$$J = \frac{\pi \left(d_o^4 - d_i^4 \right)}{32} \tag{3.22}$$

Integrating Equation 3.20 for obtaining torsional deflection of the shaft ϕ (in radians) yields

$$\phi = \frac{Tl}{GJ} \tag{3.23}$$

This equation shows that torsional deflection in a shaft is inversely proportional to the diameter of the fourth power. Taking a solid shaft as an example, a 10% increase in shaft diameter can result in a 31.7% reduction in torsional deflection (where $1/1.1^4 - 1 = -31.7\%$). However, for practical and economic reasons, shaft size is often minimized to just satisfy the basic requirements.

Torsional stiffness of a shaft is a measure of the resistance offered by the elastic shaft to torsional deformation. This design parameter is very important in positional systems. In practice, shaft torsional stiffness is specified as a torque per unit torsional deflection, that is,

$$S_t = \frac{T}{\phi} = \frac{GJ}{l} \tag{3.24}$$

Torsional stiffness may be expressed in several different units, but the most common and easiest to work with is Nm/rad.

Another design parameter closely related to torsional stiffness is torsional flexibility, which is defined as the reciprocal of the torsional stiffness.

The torsional shear stress τ at a given radius r of shaft is expressed as

$$\tau = \frac{Tr}{J} \tag{3.25}$$

This reveals that the torsional shear stress is independent of the choice of shaft material.

It can be seen from Figure 3.9 that the torsional shear stress is a linear function of shaft radius r; it vanishes at the shaft center (where $r=0$) and reaches the maximum at the outer surface of the shaft (where $r=R=d/2$):

$$\tau_{max} = \frac{TR}{J} = \frac{Td}{2J} \tag{3.26}$$

- For a solid shaft,

$$\tau_{max} = \frac{16T}{\pi d^3} \tag{3.27a}$$

- For a hollow shaft,

$$\tau_{max} = \frac{16Td_o}{\pi\left(d_o^4 - d_i^4\right)} \tag{3.27b}$$

This indicates that the shaft core has little contribution to the shaft torsional strength, and thus, the shaft can be designed as a hollow shaft minimal influence on its strength while significantly reducing weight.

Comparing Equations 3.25 and 3.26 gives

$$\tau = \frac{r}{R}\tau_{max} \tag{3.28}$$

The shaft shear strain γ, measured as the twist angle, can be expressed as the product of the radial coordinate r and the rate of twist $d\phi/dz$:

$$\gamma = r\frac{d\phi}{dz} \tag{3.29}$$

Since $d\phi/dz$ is a constant, the shaft shear strain γ is a linear function of the shaft radius r and reaches its maximum value γ_{max} at the outer surface (where $r=R$):

$$\gamma = \frac{r}{R}\gamma_{max} \tag{3.30}$$

From Hook's law, the shear stress τ can be expressed as

$$\tau = G\gamma = Gr\frac{d\phi}{dz} \tag{3.31}$$

Substituting Equation 3.20 into 3.31, it becomes

$$\tau = \frac{Tr}{J} \tag{3.32}$$

This is identical to Equation 3.25.

For a stepped shaft having a different diameter and length in each segment, when the shaft is subjected to a single torque (Figure 3.10), the torsional deflection at each shaft segment becomes

$$\phi_i = \frac{Tl_i}{GJ_i} \tag{3.33}$$

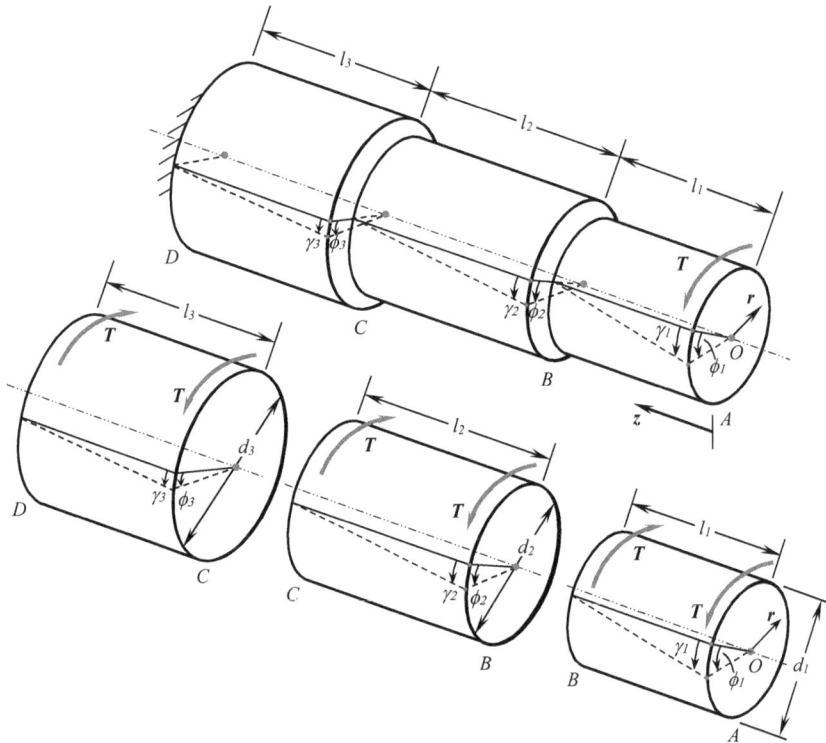

FIGURE 3.10 Torsional deflection of a stepped shaft subjected to a single torque and made from of single material.

It is noted that all torsional deflections ϕ have the same sign. The total torsional deflection is

$$\phi = \sum_i \phi_i = \frac{T}{G} \sum_i \frac{l_i}{J_i} \tag{3.34}$$

Similarly, the torsional shear stress on each shaft segment is

$$\tau_i = \frac{T r_i}{J_i} \tag{3.35}$$

For a stepped shaft with individual diameter d_i, length l_i, torque T_i, and material shear modulus G_i at each segment (Figure 3.11), the total torsional deflection becomes [3.13]

$$\phi = \sum_i \phi_i = \sum_i \frac{T_i l_i}{G_i J_i} \tag{3.36}$$

For this case, the sign of each torsional deflection ϕ depends on the torque applied on the segment. Correspondingly, shaft torsional stiffness at each segment is expressed as

$$S_{t,i} = \frac{T_i}{\phi_i} = \frac{G_i J_i}{l_i} \tag{3.37}$$

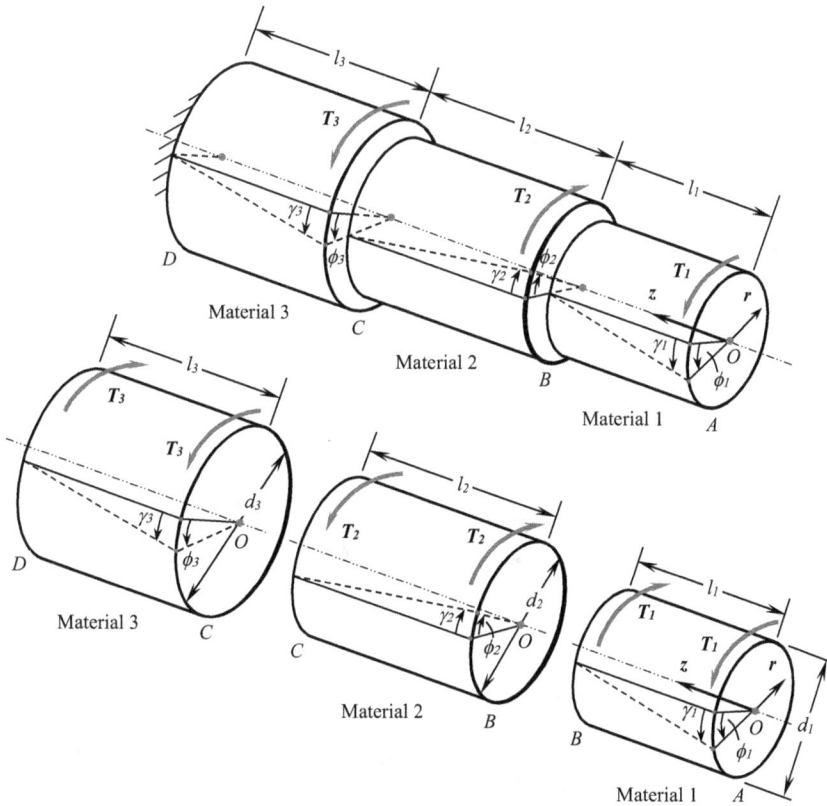

FIGURE 3.11 Torsional deflection of a stepped shaft having an individual diameter, length, torque, and material at each segment.

The total shaft torsional stiffness is given as

$$S_t = \left(\sum_i \frac{1}{S_{t,i}} \right)^{-1} \tag{3.38}$$

3.5.4 LATERAL DEFLECTION OF SHAFT

One of the critical design parameters of motor shaft is the maximum allowable deflection. If a motor shaft experiences a large lateral deflection, the rotating rotor may possibly contact the stator bore, causing a severe damage to the motor. Thus, the shafts must be rigid enough to avoid excessive deflection.

3.5.4.1 Lateral Deflection due to Bending Moment

When a shaft is subjected to a bending moment, the longitudinal centroidal axis of the shaft becomes a curve defined as an elastic curve, as shown in Figure 3.12 with an exaggerated shape. When the bending moment is a function of the position along the shaft, that is, $M = M(z)$, the lateral deflection y of the shaft that has a uniform diameter can be derived from the following equation [3.14]:

$$\frac{d^2 y}{dz^2} = \frac{M(z)}{EI_a} \tag{3.39}$$

Integrating the previous equation with respect to z, the slope of the shaft (in radians) is obtained as

$$\theta = \frac{dy}{dz} = \frac{1}{EI_a} \int M(z)\, dz \tag{3.40}$$

Continuously integrating the previous equation with respect to z, the lateral deflection of the shaft is found as

$$y = \int \theta\, dz = \frac{1}{EI_a} \int \int M(z)\, dz\, dz \tag{3.41}$$

In an electric motor, the shaft diameter is often nonuniform. For a stepped shaft having an individual diameter at each segment, $I_{a,i} = I_{a,i}(z)$, the lateral deflection y and the slope of the shaft θ can be determined, respectively:

$$y_t = \sum_i y_i = \sum_i \int \frac{d\theta}{dz} = \sum_i \frac{1}{E} \int \int \frac{M_i(z)}{I_{a,i}(z)}\, dz\, dz \tag{3.42}$$

$$\theta_t = \sum_i \theta_i = \sum_i \frac{dy_i}{dz} = \sum_i \frac{1}{E} \int \frac{M_i(z)}{I_{a,i}(z)}\, dz \tag{3.43}$$

The previous equations can be used in a stepwise fashion to solve the deflection problem for stepped shafts.

For the simple case, as shown in Figure 3.12, the deflection is given as

$$y(z) = \frac{Mz}{6EI_a I} \left(z^2 + 3a^2 - 6al + 2l^2 \right) \tag{3.44}$$

$$\theta_A = \frac{M}{6EI_a l} \left(3a^2 - 6al + 2l^2 \right) \tag{3.45}$$

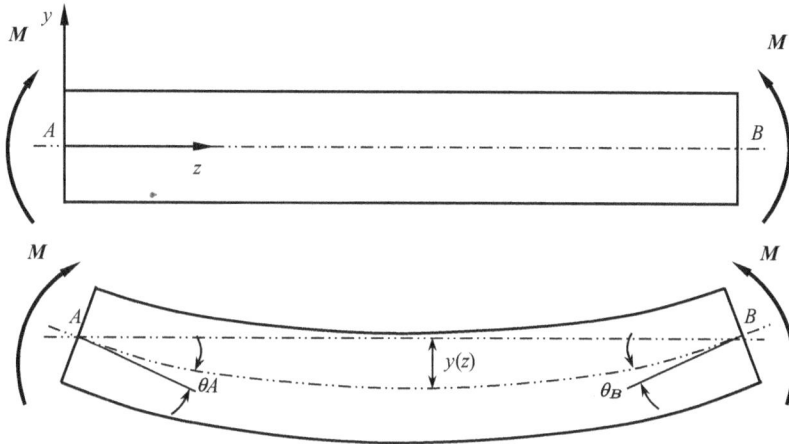

FIGURE 3.12   Deflection of a single-section shaft subjected to bending moment.

3.5.4.2   Lateral Deflection due to Transverse Force

As shown in Figure 3.13, a shaft is simply supported at the two ends and loaded by a transverse force F. At the equilibrium condition, all forces and bending moments acting on the shaft must be balanced:

$$\sum F = 0 \Rightarrow R_A + R_B - F = 0 \tag{3.46}$$

$$\sum M = 0 \Rightarrow R_B l - F_a = 0 \tag{3.47}$$

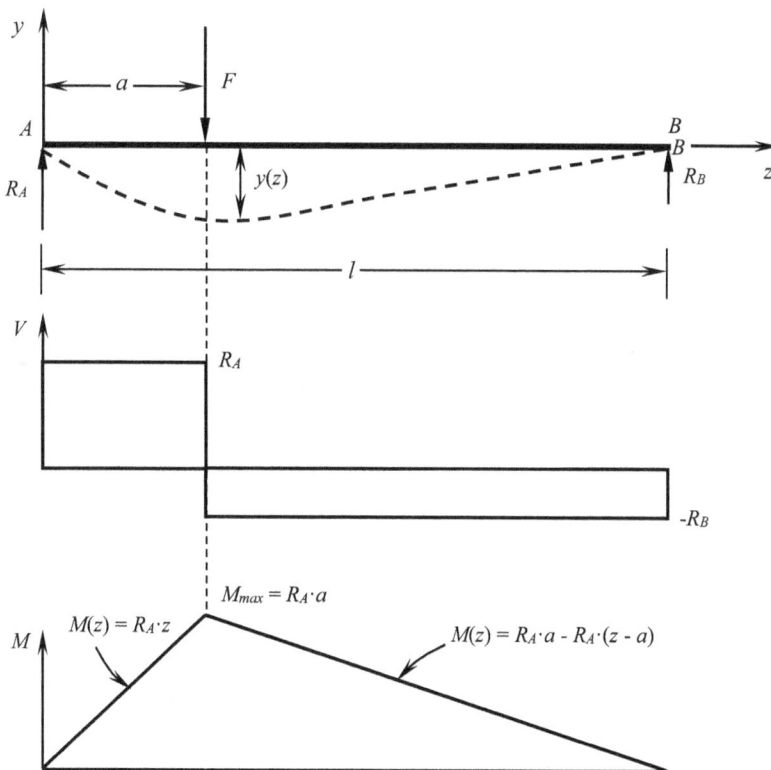

FIGURE 3.13   Deflection of a shaft subjected to a transverse force.

This gives

$$
\begin{cases}
R_A = \dfrac{F}{l}(l-a) \\[3mm]
R_B = \dfrac{Fa}{l}
\end{cases}
\tag{3.48}
$$

and

$$
\begin{cases}
M(z) = R_A z = \dfrac{Fz}{l}(l-a) & 0 \le z \le a \\[3mm]
M(z) = R_A a - R_B(z-a) = \dfrac{Fa}{l}(1-z) & a \le z \le l
\end{cases}
\tag{3.49}
$$

The deflection of the shaft is

$$
\begin{cases}
y(z) = \dfrac{F(l-a)z}{6EI_a l}\left[z^2 + (l-a)^2 - l^2\right] & 0 \le z \le a \\[3mm]
y(z) = \dfrac{Fa(l-z)}{6EI_a l}\left(z^2 + a^2 - 2lz\right) & a \le z \le l
\end{cases}
\tag{3.50}
$$

$$
\begin{cases}
\theta_A = \left(\dfrac{dy}{dz}\right)_A = -\dfrac{Fa}{6EI_a l}\left(2l^2 - 3al + a^2\right) \\[3mm]
\theta_B = \left(\dfrac{dy}{dz}\right)_B = -\dfrac{Fa}{6EI_a l}(l^2 - a^2)
\end{cases}
\tag{3.51}
$$

For this case, the maximum deflection occurs at $z = a$:

$$
y_{\max} = -\dfrac{Fa^2(l-a)^2}{3EI_a l}
\tag{3.52}
$$

In Figure 3.14, a shaft with a uniform diameter is subjected to multitransverse forces. In such a case, each deflection $y_i(z)$, which resulted from each of the transverse force F_i, is obtained as

$$
\begin{cases}
y_i(z) = \dfrac{F_i(l-a_i)z}{6EI_a l}\left[z^2 + (l-a_i)^2 - l^2\right] & 0 \le z \le a_i \\[3mm]
y_i(z) = \dfrac{Fa_i(l-z)}{6EI_a l}\left(z^2 + a_i^2 - 2lz\right) & a_i \le z \le l
\end{cases}
\tag{3.53}
$$

$$
y_t(z) = \sum y_i(z)
\tag{3.54}
$$

Therefore, the total deflection is obtained by superimposing each of the deflection $y_i(z)$:

$$
\begin{cases}
\theta_A = -\sum \dfrac{F_i a_i}{6EI_a l}[2l^2 - 3a_i l + a_i^2] \\[3mm]
\theta_B = -\sum \dfrac{F_i a_i}{6EI_a l}(l^2 - a_i^2)
\end{cases}
\tag{3.55}
$$

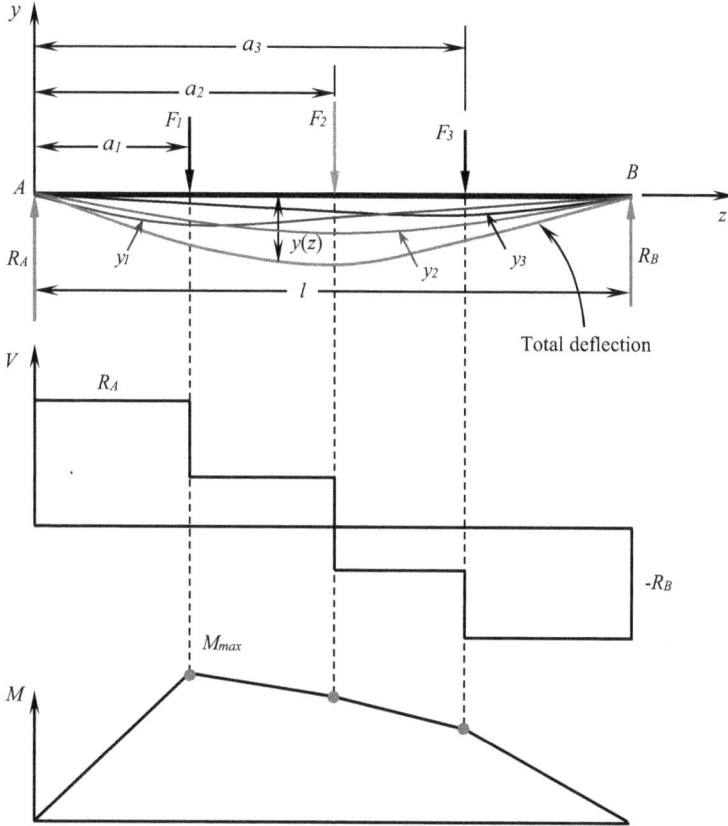

FIGURE 3.14 Deflection of a shaft subjected to multitransverse forces. The total deflection $y(z)$ is the superposition of each deflection $y_i(z)$, corresponding to each transverse force.

3.5.4.3 Lateral Deflection due to Shear Force

A shear condition is considered when a shaft is subject to a pair of equal, opposite, and parallel forces that are so close so that the bending moment is negligible. Typically, deflections due to direct shear are very small (<5%) compared to deflections due to bending moments. However, for short and heavily loaded shafts, this deflection may become significant.

As shown in Figure 3.15, two equal and opposite forces act on a shaft apart by Δz. Because Δz is small enough, the bending effect on the shaft can be ignored. The deflection due to each force is easily obtained. Thus, the total shaft deflection $y\,(F_x + F_2)$ can be determined using the superposition method.

3.6 SHAFT DESIGN ISSUES

In shaft designing, specific attention is often given to the arrangement of motor components and features on shaft, material selection and treatment, shaft deflection (*e.g.*, beading, torsional, and shear deflection) and rigidity, static/dynamic stress, fatigue life, FS, critical speed, connection of shaft to external load or gearing system, shaft dimensions, frequency response, and so on [3.15].

3.6.1 SHAFT DESIGN CONSIDERATIONS

The design of highly reliable shafts involves many aspects of mechanical design; some of them are listed in the following:

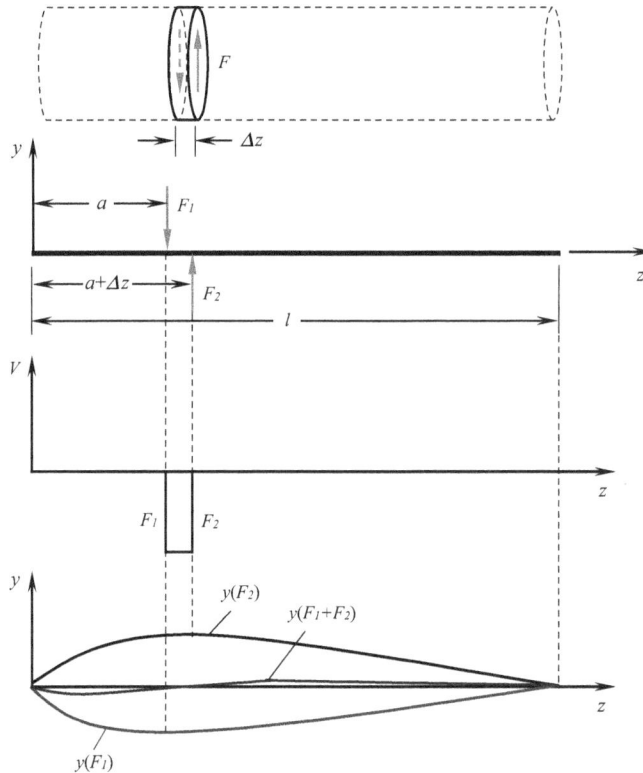

FIGURE 3.15 Deflection of a shaft subjected to shear load.

- The determination of shaft diameter is greatly affected by many design factors such as shaft loads, transmitted torque, support span, bearings, mating methods, stress concentrations, and motor cooling. Many of these factors have been discussed previously and integrated into the derivation of the equations for calculating the shaft diameter.
- In order to reduce stress concentration, the radius of the fillet between two adjacent steps must be larger than $0.2d$. Sharp corners are absolutely not allowed.
- To lower bending stress and shaft deflection, the shaft length must be kept as short as possible. So does the shaft overhang length.
- FS in shaft design can vary in a large range, mostly depending on applications. For some specific applications such as elevate motors, FSs are often mandated by applicable international, national, or industrial standards, policies, or law.
- The shaft shoulder in contact with the face of a bearing must keep perpendicular to the shaft centerline. To avoid interference, the shaft fillet radius must be smaller than the bearing radius. The shoulder height is an important parameter to provide enough support to bearing loads. This is especially true for tapered roller bearings subject to high axial loads. Sometimes, the shoulder height is roughly assumed to be 50%–80% of the thickness of the inner race of the bearing that directly contacts with the shaft. For more accurate design, shaft shoulders are determined using bearing catalogs for the specific bearing type and different load conditions. When the shoulder at a bearing becomes critical, it is necessary to check the stress and stress concentration on the shaft shoulder.
- Although there are many methods of fixing bearings in position on shafts, the most common way is to use retaining (snap) rings, which fit into grooves of shafts and take axial loads. However, the grooves can cause stress concentrations in the shafts.

3.6.2 SHAFT RIGIDITY

Shaft design involves two types of shaft rigidity:

- Torsional rigidity: Torsional rigidity is a measure of a shaft's ability to resist twisting loads. It is expressed as the product of the shear modulus of elasticity G (in units of Pa, MPa, or GPa) and the second polar moment of area (in units of m^4). Thus, the torsion angle (in rad) of the shaft that is subject to a torque T is determined:

$$\theta = \frac{Tl}{GJ} \tag{3.56}$$

where
T is the torque or torsional moment
l is the shaft length subjected to twist moment

This equation indicates that the torsion angle θ decreases with the increased shaft torsional rigidity GJ. In the unit of degree, the torsion angle can be expressed as

$$\theta = \frac{180}{\pi} \frac{Tl}{GJ} \tag{3.57}$$

The permissible angle of twist changes case to case. For instance, for machine tool applications, it is 0.25° per meter length of the shaft.
- Lateral rigidity: Lateral rigidity is a measure of a shaft's ability to resist forces or bending moments, defined as the product of modulus of elasticity E and second moment of area I. Generally, the higher the lateral rigidity EI, the lower the shaft lateral deflection y. In certain applications, the shaft is designed on the basis of lateral rigidity rather than on strength. Usually, the maximum permissible deflection for the transmission shaft is taken as +0.001/−0.003l, where l is the span between two bearings. For electric motors, the gap depth must be taken into account in the determination of the maximum shaft deflection. Generally, the shaft lateral deflection depends on shaft dimensions, load condition, material properties, bearing span, and supporting patterns.

3.6.3 CRITICAL SHAFT SPEED

A rotating shaft deflects during rotation due to its own weight, external loads, imperfect geometry, and uneven mass distribution. This deflection can significantly impact the motor performance. It has been found that there are a number of specific rotational speeds, at which the vibration amplitude increases dramatically. As the rotational speed passes beyond these speeds, the vibration amplitude decreases significantly. These speeds are defined as *critical speeds*.

The critical speed of a shaft generally depends on the magnitudes and locations of mass/load carried by the shaft, shaft support span, stiffness of the shaft, moment of inertia of the shaft, properties of shaft material (*e.g.*, density and modulus of elasticity), shaft varying diameter, type of end supports, bearing type, and system configuration. Typical shaft design requires the critical speeds to be much higher than the anticipated operating speeds of motors, usually 20%–25% of the operating speed.

As discussed in Chapter 2, two methods are commonly used to calculate the shaft critical speed. The Rayleigh–Ritz method is an approximate method of finding displacements that is based on the theorem of minimum potential energy. However, the Rayleigh–Ritz method overestimates the critical speed by a few percent [3.16]. Another method is called the Dunkerley method, which underestimates the critical speed.

3.6.3.1 Shaft with Uniform Diameter

For a shaft of uniform diameter, the calculation of critical speed is relatively straightforward.

- For a shaft itself, the critical speed is

$$n_c = \left(\frac{\pi}{l}\right)^2 \sqrt{\frac{gEI_a}{A\gamma}} \tag{3.58}$$

- For the shaft with one element (*e.g.*, rotor core) attached to it, the critical speed is

$$n_c = \frac{30}{\pi} \sqrt{\frac{g}{\delta_{st}}} \tag{3.59}$$

where
n_c is the critical speed in rpm
g is the acceleration of gravity
δ_{st} is the static deflection

3.6.3.2 Stepped Shaft

For a stepped shaft that has multiple elements, the critical speed can be calculated by the Rayleigh method after calculating the static deflection at each change of cross section [3.17]:

$$n_c = \sqrt{\frac{g \sum_{i}^{m} F_i y_i}{\sum_{i}^{m} F_i y_i^2}} \tag{3.60}$$

3.6.4 DIMENSIONAL TOLERANCE

The key tolerances of a shaft can strongly influence motor normal operation and reliability. Motor design engineers should pay special attention to the shaft dimensional tolerances at these locations:

- Bearing journals. It usually requires the press fits for assembling shafts with bearings.
- Rotor core. To ensure a rotor to withstand various loads, the interference fit (*e.g.*, press or shrink fit) adopted in the rotor–shaft assembly.
- Coupling attached to the shaft. Improperly installed couplings may initiate additional forces acting on shafts.
- Shaft seal. Because a seal may contact with the shaft, it generates a frictional force on the shaft. To help prevent scratches and nicks, the shaft must maintain certain levels of hardness (usually HRc>45, more desirable hardness is at the range of 60–65 HRc), smoothness, and roundness.
- Keyseats, retaining ring grooves. These may also use the press-fit method.

3.6.5 SHAFT RUNOUT

Two different kinds of shaft runouts are frequently observed in rotating machinery: (*a*) Radial runout, which is defined as a measurement of the shaft displacement that is perpendicular to the axis of rotation. (*b*) Axial runout, which is a measurement of the shaft displacement in the axial direction.

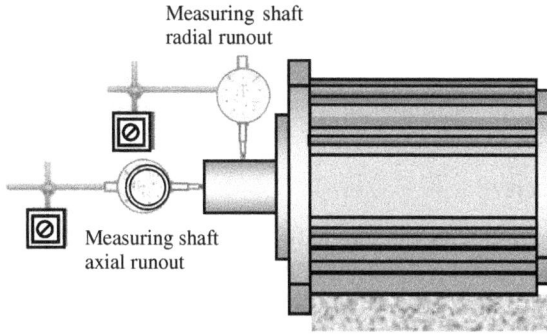

FIGURE 3.16   Measurements of radial and axial runout of motor shaft.

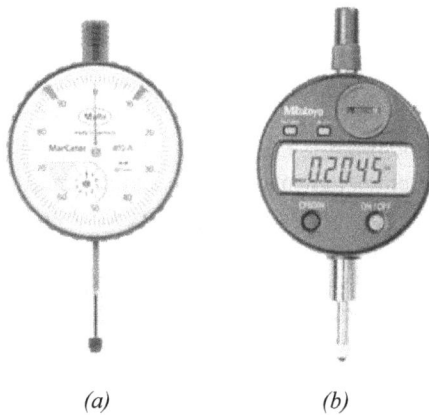

(a)   *(b)*

FIGURE 3.17   Runout measuring devices: (*a*) Dial indicator, which converts and amplifies the linear movement of a contact pointer into a dial needle; and (*b*) Electronic digital indicator, which converts a linear displacement into an electrical signal that is then amplified and transformed into a suitable data format.

This measurement is taken at the center of rotation to prevent shaft end flatness/squareness errors from affecting the measurement. Off-center measurements are often referred to as face runout in which the flatness and squareness become contributing factors to the measurement. The measurements of two kinds of shaft runout are shown in Figure 3.16.

According to the measuring condition, runout measurements can be classified as either static or dynamic measurements. As its name implies, a static runout measurement is carried out as a motor shaft is turned slowly by hands. This method is especially suitable for measuring runout on shafts with discontinuous surfaces, such as keyed shafts and spline shafts.

Dynamic shaft runout refers to the phenomenon that under normal operation a shaft deviates from the true circular rotation with respect to its central axis. A number of factors may cause shaft runoff such as shaft deflection, vibration, bearing clearance and misalignment, shaft imbalance, machining tolerance, and other inaccuracies. Shaft runouts can deteriorate motor operation, produce additional stresses inside the shaft, and enlarge sealing gap, which leads to a certain level of leakage.

Static or dynamic shaft runout measuring data can be captured using a dial indicator or an electronic digital indicator (Figure 3.17). The indicator with a magnetic base is attached to the motor (or a metallic platform) and positioned perpendicularly to the surface of interest. The readings are often expressed as total indicated runout (TIR), which is the difference between the two extreme measurements of the indicator in a full rotation of the shaft. For a runout measurement on a discontinuous surface, the measurement is usually taken at a number of positions on the outer shaft surface. Alternatively, runout measurements can be performed more efficiently by using a coordinate measuring machine (CMM) or non-contact sensors such as capacitive and eddy-current sensors, depending on runout measurement specifics and environmental conditions.

3.6.6 Shaft Eccentricity

Shaft eccentricity can be resulted from the shaft deformation due to unbalance forces on the shaft, the shaft-to-bore misalignment due to machining and assembly inaccuracies, as well as the bearing radial clearance, and dynamic runout from the shaft itself. Among them, the shaft-to-bore misalignment is involved by both the shaft and the bore and almost always exists to a certain extent in practice. Apparently, shaft eccentricity must be minimized to prevent vibration, noise, leakage, and possible motor failure, especially for high-speed motors.

Eccentricity is a measure of shaft bow that is caused by any or a combination of mechanical, thermal, and gravity bow. This bow is measured while a motor runs at a slow speed for detecting the peak-to-peak motion of the shaft. The measured data are critical for identifying the primary root cause of shaft eccentricity. There are a number of ways to reduce shaft eccentricity, including the following: (*a*) Both the shaft and its mating bore must be machined within the tolerance. (*b*) The shaft runout is controlled within the tolerance. (*c*) To reduce the gravity-induced shaft deformation, shafts that have been completely machined should be placed vertically. (*d*) All fits need to be carefully inspected, including bearing to bearing bore, shaft to bearing, and shaft to rotor core. (*e*) Rotor unbalance must be minimized. In addition, unbalance inspection includes all mechanical components mounted on the shaft, such as couplings and pulleys. (*f*) Bearings need to be inspected to ensure their performance. (*g*) Since one of the root causes of shaft eccentricity is overloading, sometimes reducing load can mitigate the extent of shaft eccentricity. (*h*) The selection of appropriate couplings is an important part impacting motor's overall performance. The adoption of flexible couplings helps reduce shaft misalignment and eccentricity.

Because shaft eccentricity is so critical for motor performance, it is highly desired to proactively solve it in the early stage of the design process.

3.6.7 Heat Treatment and Shaft Hardness

The purpose of heat treatment on motor shafts is to improve shaft mechanical properties as a result of the change in material microstructures. Medium- and high-carbon steel is heat-treated to increase strengths, hardness, wear resistance, and fatigue life. For low-carbon steel, the improvement of material properties can be achieved by means of carburization technology.

Hardness is a measure of a material's ability to resist mechanical abrasion and indentation. In fact, there is a positive correlation between the material hardness and the yield tensile and ultimate tensile strengths. Shaft surface hardness has a direct influence on its load capacity, performance, lifetime, and sealing.

It has been reported that a shaft of 40 HRc hardness has only 20% of the rated life of a system with a shaft at 60 HRc. To get the most rated life, heat treating shafts to a surface hardness of 60–65 HRc is essential [3.18]. For instance, by changing a shaft hardness of 50–55 HRc to 60–65 HRc, the shaft life can increase eight times.

The ability of the material to absorb a significant deformation before fracture is referred to as ductility. An indicator of the material ductility is its elongation under a tensile load. Toughness is the material's ability to sustain impact and shock loading. For a motor shaft, it is desired to maintain high hardness at the shaft surface and good ductility and toughness at its core.

There are many types of heat treatment processes available to motor shafts. The combination of induction heating and water quenching is the most effective heat treatment process for carbon steel shafts. As illustrated in Figure 3.18, during an induction heating process while a single induction

FIGURE 3.18 Induction heating and water quenching of a motor shaft for increasing the hardness and durability of the shaft. Induction is a no-contact process that quickly produces intense, localized, and controllable heat. With induction heating, only the part to be hardened is heated.

coil moves along the shaft axis to heat a certain length of the shaft, the shaft rotates slowly around its axis to ensure uniform heating all over the shaft surface and through a certain depth under the shaft surface. The surface temperature of the shaft during induction heating changes sharply from room temperature to above 1,000°C. The optimal heating time for required shaft properties is found out to be a few seconds, followed by a quenching period with oil-water emulsion.

The thickness of the surface hardened layer of steel parts through the induction hardening process is determined by many factors such as frequency of the AC field, heating and quenching time, current, moving speed of the heating coil, shaft material, and shaft shape, to name a few. Microscopic photos are commonly used to demonstrate the variation of the microstructure underneath the shaft surface. It has shown that the shaft surface is completely martensitic. At a certain depth, the microstructure is composed of martensite and ferrite. The ferrite in the microstructure indicates that the transformation to austenite was not completed during the induction heating period because the time for transformation was not long enough. Beneath the hardening depth, the microstructure is composed of pearlite and ferrite [3.19].

One of the most important factors in a shaft heat treatment is to minimize the distortions in shaft shape and size. The larger the mass of the metal heated, the greater the metal's thermal expansion and, thus, distortion.

3.6.8 Shaft Surface Finishing

Shaft surface finishing is important on the shaft fatigue life and seal performance, as well as the reduction of windage losses. Shaft surface finishing contains two features: surface texture and shaft lead. Surface texture is characterized by three parameters: average roughness R_a, average peak-to-valley height R_z, and average peak-to-mean height R_{pm}, as shown in Figure 3.19.

R_a is the average value of the deviations from the mean line over the evaluation length l and can be determined as

$$R_a = \frac{1}{l}\int_0^l |y(x)|\, dx \tag{3.61}$$

In general, R_a less than 0.5 μm and a shaft lead angle less than 0.05° can achieve the optimum seal performance.

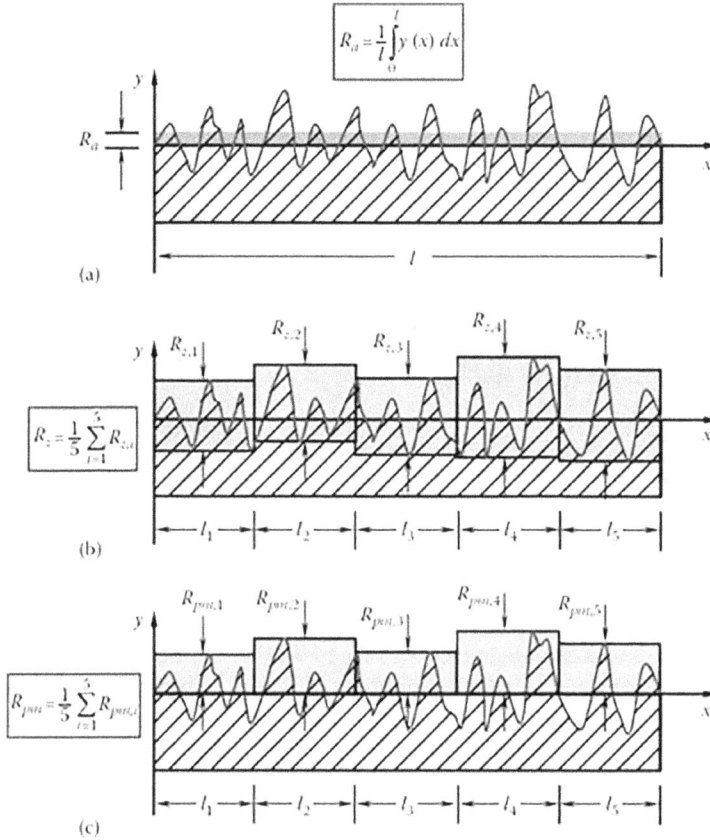

FIGURE 3.19 Three surface texture characteristic parameters: (*a*) average roughness R_a, (*b*) average peak-to-valley height R_z, and (*c*) average peak-to-mean R_{pm}.

The steel manufacturing process has a strong impact on surface finish. For example, cold-rolled stainless steel has a roughness of $R_a = 0.2$–$0.5\,\mu m$ and therefore usually does not need to be polished to meet surface roughness requirements. The surface treatment of stainless steel and the resultant surface topography are listed in Table 3.3 [3.20].

Similarly, the rms roughness R is defined as follows:

$$R_q = \sqrt{\frac{1}{l} \int_0^l y^2(x)\, dx} \tag{3.62}$$

R_z is the average peak-to-valley height and can be obtained by averaging $R_{z,i}$ values from a number of consecutive lengths along the contact surface, where

$$R_z = \frac{1}{N} \sum_{i=1}^N R_{z,i} \tag{3.63}$$

where N is the number of sections.

Besides R_a and R_z, another important parameter is the average peak to mean, defined as R_{pm}:

$$R_{pm} = \frac{1}{N} \sum_{i=1}^N R_{pm,i} \tag{3.64}$$

TABLE 3.3

Surface Treatments of Stainless Steel and Resultant Surface Topography

Surface Treatment	Approximate R_a Values (µm)	Typical Features of the Technique
Hot rolling	>4	Unbroken surface
Cold rolling	0.2–0.5	Smooth unbroken surface
Glass bead blasting	<1.2	Surface rupturing
Ceramic blasting	<1.2	Surface rupturing
Micropeening	<1	Deformed (peened) surface irregularities
Descaling	0.6–1.3	Crevices depending on initial surface
Pickling	0.5–1.0	High peaks, deep valleys
Electropolishing		Roundoff peaks without necessarily improving R_a
Mechanical polishing with aluminum oxide or silicon carbide		
Abrasive grit number		Surface topography highly dependent on process
500	0.1–0.25	parameters, such as belt speed and pressure
320	0.15–0.4	
240	0.2–0.5	
180	0.6	
120	1.1	
60	3.5	

By cutting surface roughness with the depth of C ($C=R_z/2$), the area can be distinguished as the contact areas and noncontact areas. The surface contact area ratio R_c is defined as the summation of all surface contact areas to the total surface area (Figure 3.20), that is,

$$R_c = \frac{1}{l}\sum_{i=1}^{N} x_i \tag{3.65}$$

3.6.9 SHAFT LEAD

Fluid leakage in electric motors may be encountered when defective shafts that were not properly manufactured are used. A number of factors can affect the shaft sealing performance such as shaft surface finish, dimensional accuracy, and operating conditions. During a shaft finishing process, spiral grooves may be generated on the shaft surface. As shown in Figure 3.21, while the shaft rotates counterclockwise at an angular velocity ω, a finishing tool (a lathe cutter or a grinding wheel) moves axially along the shaft toward the lathe chuck at a feeding speed u, forming the shaft lead angle α

$$\alpha = \tan^{-1}\left(\frac{u}{r\omega}\right) \tag{3.66}$$

and the magnitude of the resultant speed vector $\mathbf{U}$ is

$$\mathbf{U} = \sqrt{u^2 + (r\omega)^2} \tag{3.67}$$

Figure 3.21 indicates that the lead angle α is proportional to the feeding speed u but inversely proportional to the angular velocity ω and the shaft radius r. Regardless of shaft finishing conditions, the shaft lead angle α is generally recommended to be controlled less than 0.05°.

$$R_c = \frac{1}{l}\sum_{i=1}^{N} x_i$$

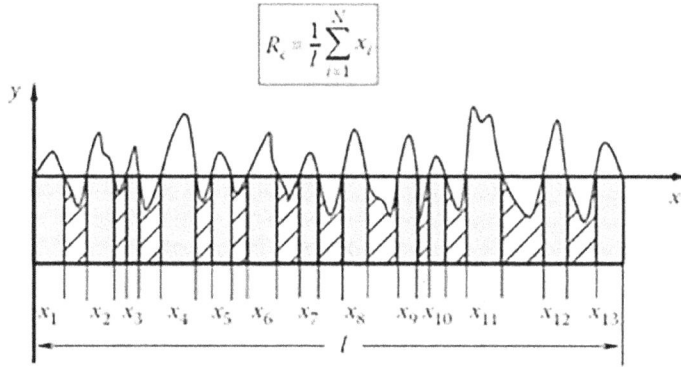

FIGURE 3.20 Surface contact area A_c, which is defined as the ratio of the summation of each contact area to the total area.

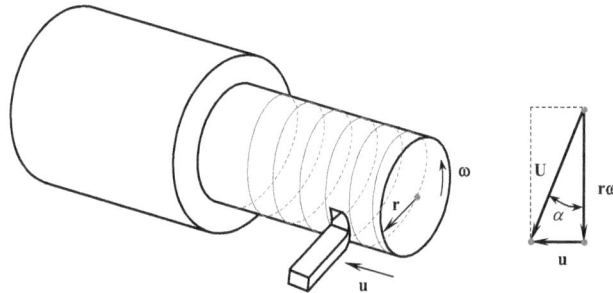

FIGURE 3.21 Formation of spiral grooves on shaft surface during shaft finishing machining.

With a large feed speed or a slow shaft rotating speed, or both, spiral grooves are formed on the shaft surface. This phenomenon, known as shaft lead, can result in shaft seal failure during motor normal operation.

A quick and simple test can be adopted by using the suspended weight method for establishing the presence of shaft lead. As illustrated in Figure 3.22, a shaft is mounted on a lathe chuck and a small weight (*e.g.*, 1 oz) is suspended on a testing shaft. As the lathe rotates slowly either counterclockwise or clockwise, observe whether the weight moves along the shaft and which direction it goes. The ideal case is that the weight remains in the same location with the rotating shaft, indicating there is no leakage problem.

3.6.10 SHAFT SEAL

Sealing is one of the main concerns in the design of rotating machinery such as motors, engines, gas turbines, generators, pumps, compressors, and other varieties of rotational machines. A sealing system in a rotating machine is used to control (*a*) fluid (liquid or gas) leakage, (*b*) dynamic performance, (*c*) tolerance to boundaries, (*d*) coolant or lubricant flow, and (*e*) windage losses. All these functions can directly affect the performance of rotating machines through parasitic losses, life, or limit cycles [3.21]. In a worse case, a failure of the sealing system may lead to the whole system failure or disaster (*e.g.*, exploration in a hydrogen-cooled machine).

In some motor applications, shaft seals can be critical for motor performance. For instance, in the food, pharmaceutical, and semiconductor industries, any leakages from motors that are associated with the production processes are not allowed. If using individual sealing elements, shaft seals may involve both static seals, where both the mating surfaces are stationary, and dynamic seals, where there is a relative motion between the mating surfaces. For a motor using an individual

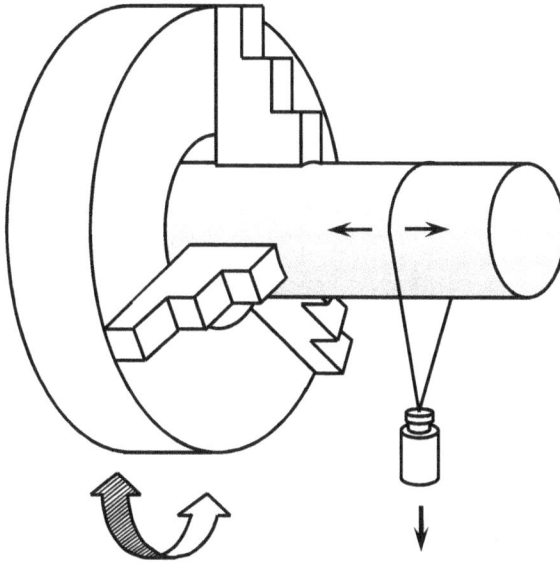

FIGURE 3.22 Motor shaft string testing.

sealing element, the static seal occurs between the stationary sealing element and the motor endbell and the dynamic seal occurs between the shaft and the sealing element. In most applications, seal elements are directly integrated into the motor components (*e.g.*, motor endbell). For such a circumstance, there exists only dynamic seal between the seal element and rotating shaft. Because of the dimensional variations in dynamic seal surfaces, the design of a dynamic seal system is much more difficult than a static seal system.

Seal performance and lifetime depend upon many parameters, including shaft hardness and surface finish, shaft spiral lead, runout, rotating speed, operation temperature, pressure, seal material, and lubrication conditions. To achieve a longer seal life, shaft hardness is expected to be higher than HRc 45. Most effective sealing is obtained with optimum shaft surface finishes. However, highly polished shafts may lead to high seal temperatures.

Many contacting and noncontacting shaft seals have been developed and used to rotating machinery, reflecting the need for specific solutions for individual applications. As always, each type of seal has its pros and cons. Some common seal types will be addressed in the following sections.

3.6.10.1 O-Ring Seal

O-ring seals are indispensable to a wide variety of applications, from vehicles, machine tools, pumps, and air conditioners to washing machines and vacuum cleaners. Advantages of O-rings include the following [3.22]:

- Having sealing ability over a wide range of pressure, temperature, and tolerance
- Normally requiring very little room and are light in weight
- Reusable in many cases
- Ease of service and installation
- Duration of life depending upon the normal aging period of its material
- Providing cost-effective sealing solution

The dimensions of O-ring are defined by the OD (or ID) and its cross section diameter d. O-rings are typically made of elastomers, including acrylonitrile–butadiene, carboxylated nitrile, ethylene–propylene rubber, fluorocarbon, and silicone rubber (high-temperature applications up to and above $300°C$), to name a few.

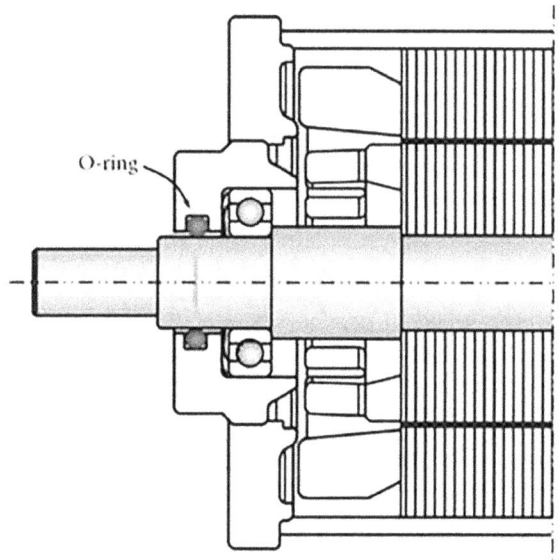

FIGURE 3.23 O-ring seal between a shaft and a motor endbell. Note that groove size prevents rotation of O-ring.

As demonstrated in Figure 3.23, an O-ring is radially squeezed in an O-ring cavity groove that is machined on the endbell, providing dynamic sealing ability for the bearing. To ensure effective sealing and prevent O-ring rotation, the O-ring's OD must be slightly larger than the groove's diameter. Thus, the O-ring is slightly squeezed in the groove. Furthermore, it is important to keep the shaft diameter no greater than the O-ring's ID. When the O-ring operating temperature increases, it tends to contract. This contraction can result in a tendency for the O-ring to seize the rotating shaft and cause more friction.

There are several parameters for quantitatively describing the squeezing degree of O-ring in rotary O-ring sealing applications:

- Relative interference, which is defined as the ratio of the differential diameter of the O-ring OD and the groove diameter to the O-ring OD, that is, $(OD_{O\text{-ring}} - d_{groove})/OD_{O\text{-ring}}$. The maximum relative interference is usually 2%–5%, depending on different applications.
- Compression squeeze, which is defined as the difference between the uncompressed O-ring cross-section diameter and its radial height after it is compressed. Although the recommended minimum value of compression squeeze by many O-ring manufacturers is about 0.005 in. (0.127 mm), reduced compression squeeze, as small as 0.002 in. (0.051 mm), may be used in some rotary sealing systems to minimize friction.
- Compression ratio, which is defined as the ratio between the compression squeeze and the uncompressed O-ring cross-section diameter. For rod- or piston-type seal, the recommended compression ratio ranges between 5% and 30%, with the target value of 20%.

In a continuously rotary application such as an electric motor, the shaft continuously rotates in the ID of the O-ring. Heat due to friction is generated at the same location of the contact and builds up inside the O-ring. Because elastomers are poor thermal conductors, high temperatures can cause O-ring swelling and finally lead to seal failure. This is especially true as the shaft surface speed exceeds 180 ft/min (0.91 m/s) or an O-ring is excessively squeezed.

However, O-ring seals are not recommended for rotary applications under the following conditions [3.23]:

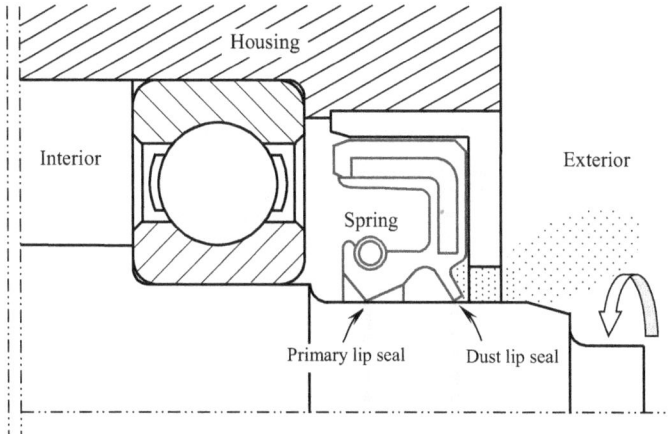

FIGURE 3.24   Universal lip seal with dual lip: a primary lip for sealing fluid and an auxiliary lip for sealing dusts.

- Pressure exceeding 5.52 MPa (800 psi)
- Temperature lower than $-40°C$ $(-40°F)$ or higher than $107°C$ $(225°F)$
- Surface speeds exceeding 3 m/s (9.8 ft/s)

Because O-ring seals are directly in contact with rotating shafts, O-ring wear and sealing life are some of the main concerns in O-ring seal design. Furthermore, rotary seals do not efficiently dissipate heat themselves compared to reciprocating seals. Thus, special care must be taken on the seal operating temperature to avoid seal failures during the service life.

3.6.10.2   Universal Lip Seal

A universal lip seal is used to seal a rotating shaft against the stationary motor housing. As an example presented in Figure 3.24, the seal system has a dual-lip structure: an auxiliary lip seal rides in the face of the motor endbell to establish the first barrier for solid particles such as dusts, and a primary lip seal is closer to the bearing to set the second barrier for fluid leakage. The dimensions of the dual lip are set based on material properties, sealing size, rotating speed, and lip geometry. Optimization of these ensures larger sealing capability, more reliable operation, and longer seal life. The use of a lubricant at the seal tip can promote seal durability and increase the seal life. Contaminations are always harmful for seal performance and should be avoided. With the well-designed seal and proper selections of seal material and lubricant, seals can often last over 10,000 h under clean environmental conditions [3.24]. This type of seal is primarily designed for low-pressure and low-speed applications.

3.6.10.3   V-Shaped Spring Seal

This type of seal differs from the previous one in some respects. First, the seal is stretched and mounted onto a rotating shaft rather than on a stationary motor endbell. Typically, the main function of the seal is to act as a face seal to prevent the ingress of dirt, dust, oil, and water. Second, the sealing face is the flat end surface of an endbell rather than the cylindrical shaft surface (Figure 3.25). Because the shaft radial runout is generally larger than its axial displacement, this sealing configuration can help improve the sealing performance. In fact, sealing on a flat face is easier than on a curved face. Third, the sealing pressure can be changed by adjusting the relative position of the seal to the sealing face. Finally, a high external pressure always pushes the seal lip closer to the sealing face, leading to the enhanced sealing performance but increased friction on the sealing face.

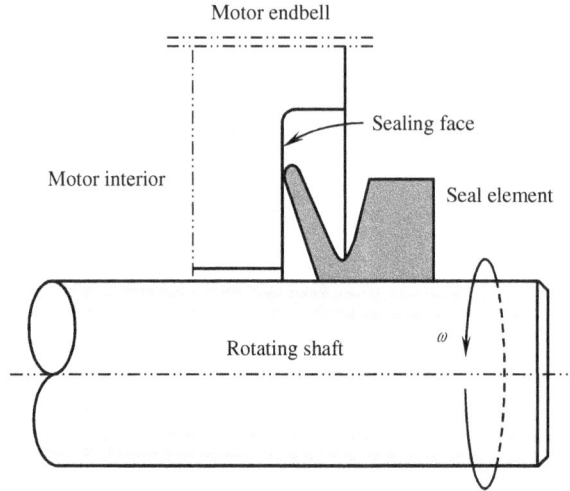

FIGURE 3.25 A V-shaped seal attached to a rotating shaft for sealing a motor at the flat end surface of the motor endbell.

3.6.10.4 Brush Seal

The initial concept of brush seal was proposed by General Electric in 1955. Since then, the brush seal technique has been greatly developed and found more applications in a variety of industries. In the last three decades, brush seals have attracted more industrial attention in replacing conventional labyrinth seals for their improved resistance to leakage. Today, the brush seal technology represents a promising advance in constructing seal systems with high sealing effectiveness and possible small size. The brush seal uses a biased pattern of wire Kevlar® fibers in contact with the surface of a shaft. This type of seal is remarkably effective in reducing leakage.

Brush seals typically comprise a plurality of elongated Kevlar fibers or wire bristles in contact with a rotating surface. The Kevlar fibers provide a tight rub-tolerant seal that experiences only slight degradation over time. The Kevlar fibers of the seal are compliant in use, and this minimizes damage due to transient impact between the sealing components. A typical brush seal is formed by sandwiching Kevlar fibers between a pair of supporting metal plates that are welded together on their top surfaces (Figure 3.26).

The Kevlar fibers or bristles of the seal are compliant in use, and this minimizes damage due to transient impact between the sealing components. Brush seal construction consists of three main components: Kevlar pack and back and front steel rings. The brush fibers lie at a cant angle affording the capability of accommodating radial excursions of the rotating shaft. For bidirectional rotating motors, the cant angle is set at zero.

Figure 3.27 presents an example that uses the combination of a flexible silicone rubber brush seal and a labyrinth seal to seal the shaft of rotating machinery for minimizing leakage flows. The brush seal comprises a plurality of elongated wire bristles in contact with the rotating shaft. The bristles provide a rub-tolerant seal that experiences only slight degradation over time. In order to use the brush seals under high-temperature conditions, they are fabricated with the flexible heat-resistant magnetic silicone rubber material. Therefore, this type of brush seal can fit a wide range of sealing dimensions and complex sealing geometry. By changing the brush bending pattern, the brush seal can be used for inner sealing where the bristles point radially inward, outer sealing where the bristles point radially outward, or axial sealing where the bristles point parallel to the rotor-shaft axis. Additionally, by forming the brush seal body from the magnetic silicone rubber, the seal may be readily adhered to the metallic sealing components. In fact, this type of brush seal is particularly suitable for applications with large sealing dimensions [3.25].

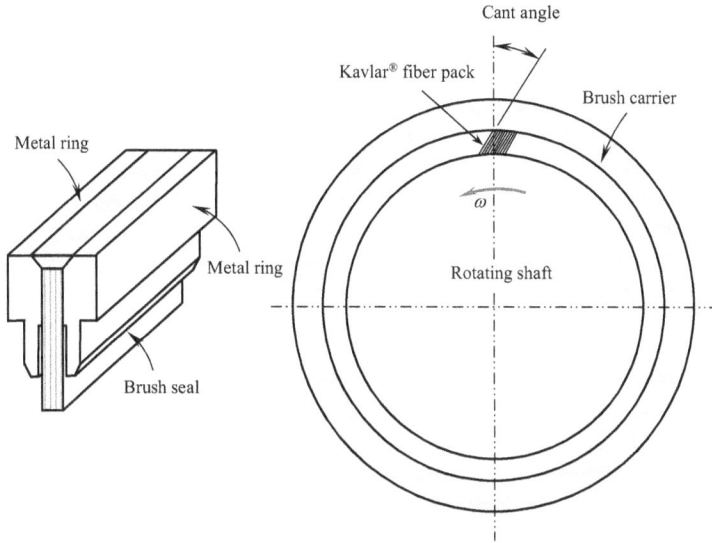

FIGURE 3.26 Brush seal employed Kevlar® fiber pack for sealing rotating shaft.

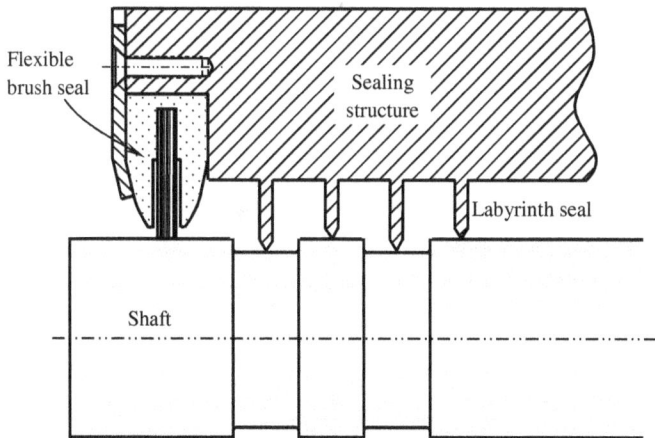

FIGURE 3.27 A flexible silicone rubber brush seal combined with a labyrinth seal to seal the shaft of rotating machinery. The brush seal body is made of heat-resistant magnetic silicone rubber material for high-temperature applications (U.S. Patent 6,390,476) [3.25]. (Courtesy of the U.S. Patent and Trademark Office, Alexandria, VA.)

3.6.10.5 PTFE Seal

PTFE, popularly known as its trade name Teflon, is the shortened name of the chemical polytetra-fluoroethylene. It was discovered by accident in 1938 by a DuPont scientist. This material has some unique and remarkable properties, including [3.26] the following:

- PTFE has thermal stability across a wide temperature range. It can be used in cryogenic temperatures to seal liquid nitrogen ($-196°C$), liquid hydrogen ($-253°C$), and liquid helium ($-269°C$). It can also keep normal service at a temperature up to $260°C$ for virgin PTFE.
- PTFE has the lowest friction coefficient of any known solid. The material has self-lubricating capability, making it an excellent seal material in dynamic seal applications.

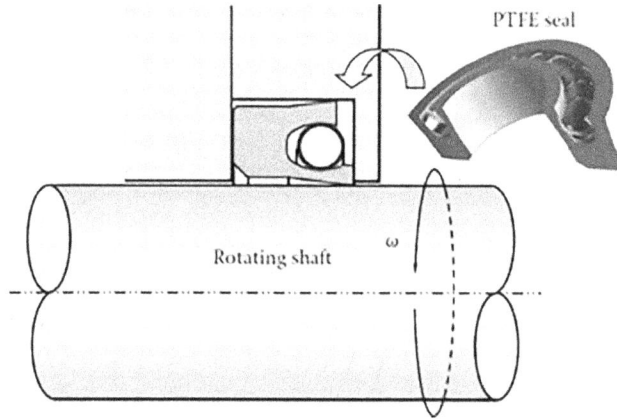

FIGURE 3.28   High-pressure, high-speed dynamic PTFE seal.

- Compared with other main polymeric materials (*e.g.*, UHMW-PE, PEEK, PCTFE, and PI), PTFE has the highest elongation, up to 300%.
- PTFE has very strong chemical resistance and chemical stability. It is inert to virtually all industrial chemicals and solvents even at elevated temperatures and pressures.
- The exceptional insulating properties of PTFE make the material ideally suited for use in electric machines.

High-pressure, high-speed shaft seals are often encountered in rotating machinery relevant to chemical processing under severe operating conditions. PTFE seals are especially suitable for these applications due to their superior material properties. Some specially designed PTFE seals can withstand water washdown pressures up to 55 MPa (7,977 psi) and surface speeds up to 15 m/s (49.2 ft/s) [3.26]. A high-pressure, high-speed PTFE shaft seal is presented in Figure 3.28.

Sealing in the food industry can be extremely challenging due to sanitary requirements. One of the most important commodities in food production is water, which is used throughout all stages of preparation, production, and cleaning. It is vitally important that all the seals that come into contact with water are manufactured from materials that will not contaminate the water in any way. To mitigate the contamination risk from a seal and lubricant, a PTFE seal is selected for sealing the motor shaft for its lowest friction coefficient and strong chemical stability (Figure 3.29).

3.6.10.6   Spring-Energized Seal

The sealing performance of a motor seal system is affected by a variety of factors, such as shaft surface roughness, shaft runout, seal structure (*e.g.*, sealing-lip design), contact pressure, seal material, operation temperature, lubrication condition, dynamic operating condition (*e.g.*, rotating speed and vibration), and environmental condition. To ensure leak-free performance, these influencing factors need to be taken into consideration when designing and choosing the perfect seal for applications.

The spring-energized sealing technique has been developed to provide superior sealing performance in a wide range of temperature, pressure, velocity, and medium. The canted coil spring seals employ a specialty wound wire spring installed in a polymer jacket (Figure 3.30) [3.27]. As versatile and energizing components, the canted coil springs exert near-constant forces throughout their normal working deflection zone, allowing them to operate in friction-sensitive applications (Figure 3.31). The use of canted coil spring is to maximize the seal lip contact area and compensate for misalignment, tolerance variations, and mating surface irregularities. By combining with a precision polymer seal (such as a PTFE seal), this type of seal can greatly enhance the sealing

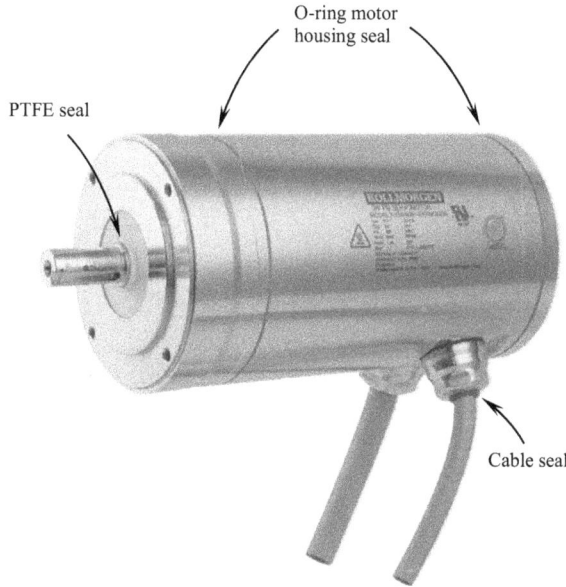

FIGURE 3.29 A stainless steel motor is designed to meet the toughest hygienic requirements in the food processing industry. Its sealing system is designed to meet the standards of IP69K with FDA-approved food-grade materials. The housing and cables can endure daily washdown with high pressure, high temperature, and caustic chemicals. (Courtesy of Kollmorgen Corporation, Radford, VA.)

FIGURE 3.30 Spring-energized seals: (*a*) Canted coil energized seal and (*b*) V-spring energized seal.

performance and minimize seal friction. In rotary sealing applications, canted coil springs can be designed to achieve the preferable loading characteristics (*e.g.*, friction, spring load, wear, rotary speed, pressure, vacuum/gas, temperature, and tolerance). On the other end of the spectrum, canted coil spring seals are often used with heavy load springs in highly viscous media such as biofluids.

To meet specific performance demands and enhance corrosion resistance in severe service applications, the canted coil springs are typically made from stainless steel, inconel, platinum iridium, and titanium alloys.

This type of seal has been used in many industrial, medical, and military applications. It is especially suitable for sealing submersible motors that are subjected to ultra-high pressures.

Over the past few decades, in addition to sealing applications, canted coil springs have found applications in connecting mechanical components, referred to as latching and holding. In regard to electromagnetic interference (EMI) and radio frequency interference shielding applications, canted coil springs can reduce radiated and conducted interference to prevent premature failure in electrical devices.

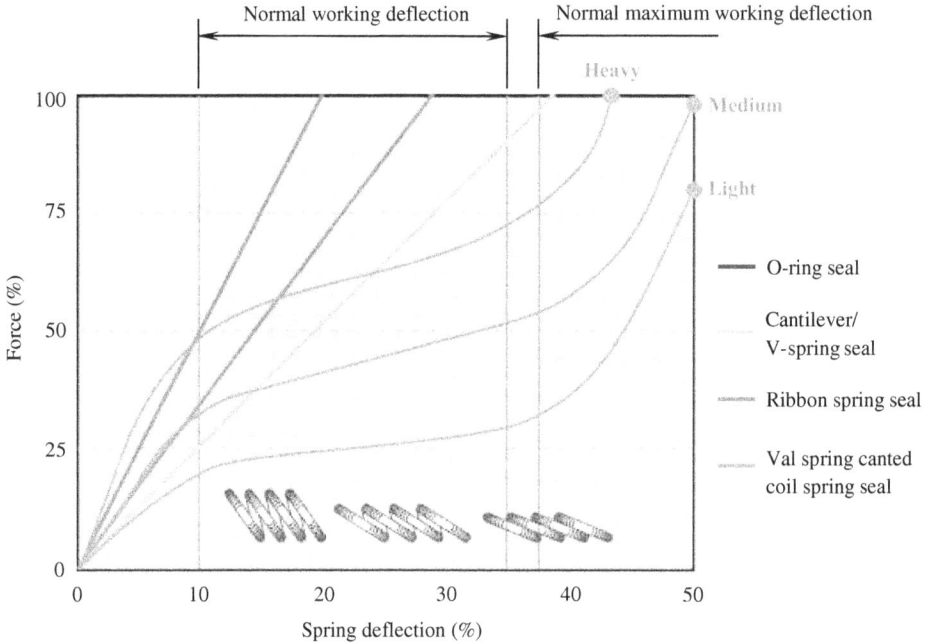

FIGURE 3.31 Comparison of spring compressive forces (in percentage) of canted coil spring seals to those of other types of seals. While the canted coil springs display near-constant forces throughout the normal working deflection range, the compressive forces for other seals are linear functions of the deflection with large slops [3.27].

3.6.10.7 Noncontact Seal

As one of the noncontact seal types, labyrinth seals are the most commonly used sealing configuration in rotating machinery to control the internal leakage of fluid. A typical labyrinth seal is generally characterized by a series of grooves formed along the adjacent surfaces of a rotating shaft. The circumferential grooves make a tortuous axial path that restricts fluid flow across it, as shown in Figure 3.32a [3.28].

In motor operation, the labyrinth seal on a rotating shaft provides noncontact sealing action by controlling the passage of fluid. With the labyrinth design, the leakage fluid has to pass through a long and tortuous path, producing high friction forces against the movement of the fluid. In addition, at high rotating speeds, the leakage fluid is subjected to a high angular momentum and hence is flung radially outward. In such a way, the leakage fluid along the tortuous path is disturbed and pushed away from the passage. Consequently, the combination of these two effects makes this type of seal have the good sealing capability.

Another type of labyrinth seal is presented in Figure 3.32b, where an elastomeric seal body has multiple radially outward lips in different directions, creating an even more complicated flow path for improving sealing performance [3.29].

This type of labyrinth seal is known for its low manufacturing cost, small fiction (only fluid friction involved), no wear, tight radial clearance between the seal and shaft, and acceptable leakage control effectiveness. One important issue in using labyrinth seals is that they are prone to developing cross-coupled forces, which can induce rotor dynamic instability. The effects of pressure differential, rotor speed, entry flow conditions, and seal geometry on the rotordynamic stability were addressed by Benckert and Wachter [3.30] in detail.

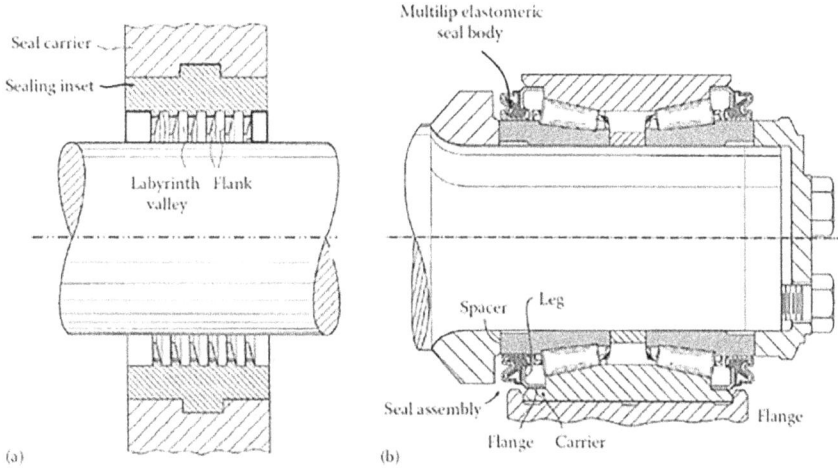

FIGURE 3.32 Noncontact labyrinth seals used in rotating machinery: (*a*) tooth-groove labyrinth seal (U.S. Patent 9,746,085) [3.28] and (*b*) multiple labyrinth seal (U.S. Patent 8,342,535) [3.29]. (Courtesy of the U.S. Patent and Trademark Office, Alexandria, VA.)

3.6.11 DIAMETRICAL FIT TYPES

There are two diametrical fit types: hole-based fit and shaft-based fit. In the hole-based fit, the dimension of the hole is taken as the basis and the clearance or interference is applied to the shaft dimension. In contrast, in the shaft-based fit, the dimension of the shaft is taken as the basis. Generally, the hole-based fit is considered better than the shaft-based fit. This is because holes are often machined by standard tools (drills or reamers) with fixed dimensions and shafts are machined by computerized numerical control (CNC) lathes to any given dimensions. In addition, some standard mechanical parts such as bearings, pulleys, and gears often have standard internal diameters with tight tolerances. However, the shaft-based fit may also be used in some special applications.

3.7 STRESS CONCENTRATION

It has long been recognized that any structural and discontinuity in machinery components can alter the stress or strain distributions in the vicinity of the discontinuity and significantly raise the local stress or strain levels. The structural discontinuities can be geometric discontinuity such as grooves, holes, notches, shoulders, keyseats, threads, splines, and sudden changes in geometry and material discontinuities such as voids, cracks, and porosities. As the sources of stress/strain intensification, these discontinuities affect only small regions of actual machinery components or structural members but have little impact on the overall stress/strain pattern. This is extremely dangerous because without any apparent signs, the machinery components or structural members may suddenly fail due to the cracks developed at the stress concentration locations.

In order to quantitatively describe the phenomenon of stress concentration, a dimensionless factor, known as stress concentration factor K, is introduced as the ratio of the actual maximum stress with the stress riser to the nominal stress without the stress riser, that is,

$$K = \frac{\text{Actual maximum stress}}{\text{Nominal stress}} \tag{3.68}$$

More specifically, the normal and shear stress concentration factors, K_t and K_{is}, are defined, respectively, as

$$K_t = \frac{\sigma_{max}}{\sigma_0} \quad \text{for normal stress} \left(\text{underbending, tension, or compression}\right) \qquad (3.69\text{a})$$

$$K_{ts} = \frac{\tau_{max}}{\tau_o} \quad \text{for shear stress} \left(\text{undertorsion}\right) \qquad (3.69\text{b})$$

where the stresses σ_{max} and τ_{max} are the real maximum normal and shear stresses, and σ_o and τ_o are the nominal normal and shear stresses without stress raisers, respectively.

Stress concentration factors for different discontinuity types, loadings, and geometries can be found in some typical mechanical engineering design textbooks [3.14, 3.31, 3.32].

3.8 TORQUE TRANSMISSION THROUGH MECHANICAL JOINTS

For a shaft to transmit torque or power to a driven machine, some mechanical components are needed to be mounted on the shaft, such as coupling, pulley, gear, hub, sheaves, sprocket, and/or cam, to name a few. To ensure that these mechanical components are firmly connected with the shaft, it must satisfy the following design requirements: (*a*) no relative motion between the shaft and the attached components, (*b*) easy and rapid assembly/disassembly, (*c*) easy to manufacture, (*d*) minimized stress concentration, and (*e*) induced less vibration and acoustic noise.

Motor torque is normally transmitted from the motor shaft to the mounted components using keys, splines, pins, or other torque-transmitting elements.

3.8.1 KEYED SHAFTS

Most motor shafts use keys for securing mating elements rotating together to transmit motor torque to external equipment. The advantages of using keys include cost-effectiveness, ease of installation, and high reliability. In addition, keys used in the joint systems may function as medianical *fuses* to prevent further damage to joints or other parts of the machine.

3.8.1.1 Selection of Key Material

In the design of keyed connections, it is important to select proper key material. An improper key material may cause damage to the key itself, shaft, hub, or even motor failures in severe cases. In transmitting power/torque, there are two basic failure modes: shear across the shaft–hub interface and compression due to the bearing forces between the key (lower portion) and shaft and key (upper portion) and hub. Since shafts are subjected to combined loads, the key material must have high strengths, high hardness, proper elongation, and tolerance to shock loads.

The required key strengths are closely associated with the key length. When the key length is 1.6 times the shaft diameter, the key stress is equal to the shaft stress. When the key length is comparable to the shaft diameter, the key stress becomes much higher than the shaft stress. However, because there is a high contact pressure between the shaft and the hub bore due to the interference fit to transmit a considerably large portion of power/torque, it is generally accepted that the key material can be chosen with the same strengths and hardness as the shaft. In practice, carbon steel is the most common material for keys.

3.8.1.2 Stress Analysis of Key and Keyseat

Numerous types of keys are available, including flat keys, square keys, round keys, gib-head keys, feather keys, and Woodruff keys. Among these, flat keys are widely used in the motor industry. To accept a key, longitudinal grooves are made into a shaft and its mating element (*e.g.*, sheave, hub), known as key ways or keyseats. According to the milling pattern, a shaft keyseat can be classified as either sled runner keyseat or profile keyseat. The sled runner keyseats are milled with a disk milling cutter that has the same width of the keyseat. The profile keyseat is produced by a circular endmill that has a diameter equal to the keyseat width (Figure 3.33). Though the type of profile keyseat is

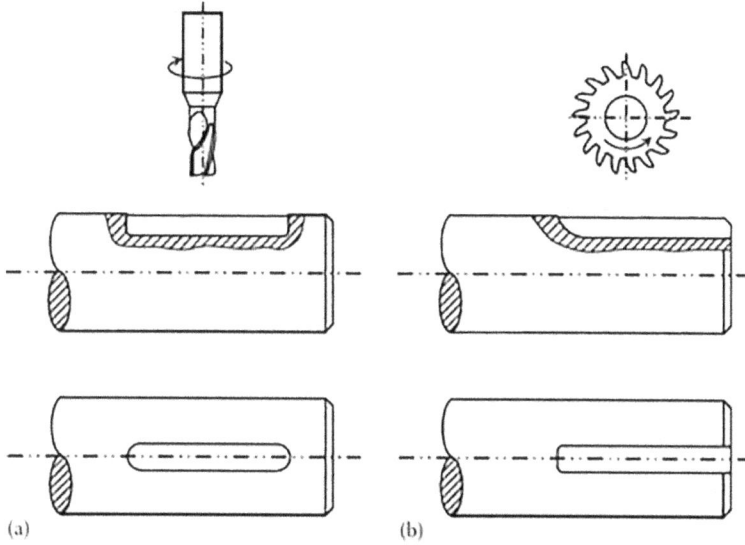

FIGURE 3.33 Keyseat types: (*a*) profile keyseat and (*b*) sled runner keyseat.

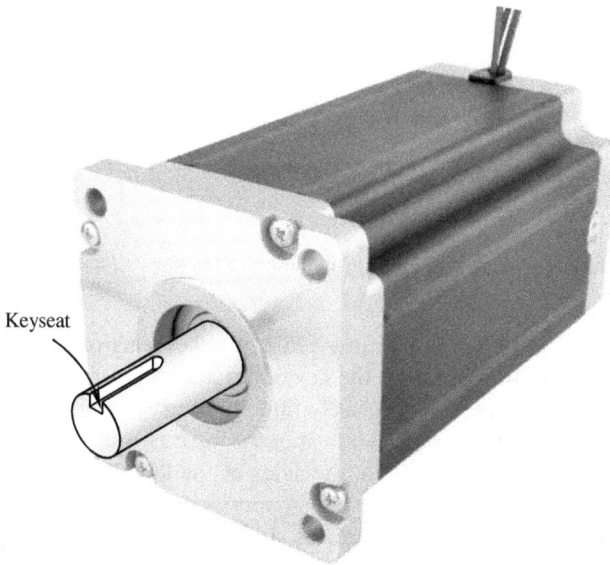

FIGURE 3.34 A keyseat is machined at the drive side of a PM motor shaft for transmitting the motor power.

widely used in industries, the type of sled runner keyseat is desirable for its lower stress concentration factor. Figure 3.34 presents a PM rotor with a profile keyseat.

A keyed shaft is primarily characterized by five variables: shaft diameter d, keyseat width w, keyseat depth h, effective keyseat length l, and the fillet radius at the keyseat bottom edge r (Figure 3.35). Usually, the fillet radius is about 2% of the shaft diameter. This gives the stress concentration factor of the keyseat approximately 3. If a keyseat has sharp corners at the bottom, the stress concentration factor can be significantly high. As reported by Peterson, typical stress concentration factors are approximately 1.1–2.0 for stepped shafts and 2.5–4.0 for keyseats [3.33]. It is noted that the rounded key ends avoid high local stresses.

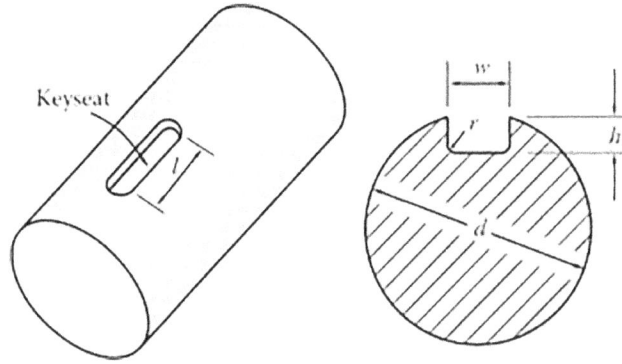

FIGURE 3.35 Dimensions of the profile keyseat made on the motor shaft. Small fillets are made at the keyseat corners to decrease stress concentration.

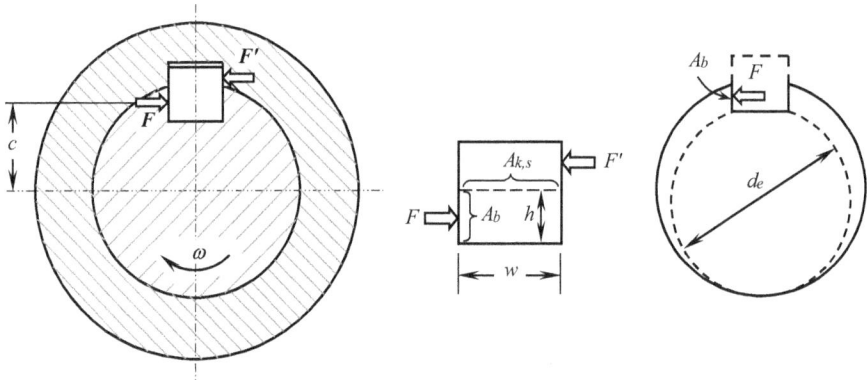

FIGURE 3.36 Shear failure and crushing failure applied on a flat key and a keyed shaft.

As illustrated in Figure 3.36, a key is inserted in the keyseats of the shaft and sheave/hub. To avoid interference with the keyseats at the key edges, the radii of the key must be larger than those of the keyseats. The key must be precisely fitted to the keyseat in the shaft and sheave/hub with a small radial clearance on the top. Thus, motor torque is transferred from the shaft to the sheave/hub through two contact surfaces at the sides of the key. While the contact force of the shaft F exerts on the lower section of the key, the opposite reaction force F' from the sheave/hub exerts on the upper section of the key (where F is slightly larger than F'). These resulting forces cause shear and compressive stresses. Thus, for this locking system, there are two major failure mechanisms: shear failure and crushing failure, which are applicable to all components (shaft, key, and sheave/hub).

In the shear failure mechanism, a key withstands opposite forces on its upper and lower sections, causing a shear stress on the key shear plane $A_{k,s}$,

$$A_{k,s} = lw \tag{3.70}$$

The maximum shear stress τ_{max} is reached when the force F approaches its maximum value F_{max}:

$$\tau_{max} = \frac{F_{max}}{A_{k,s}} \tag{3.71}$$

The FS for the shear failure is defined as

$$n_s = \frac{S_{sy}}{\tau_{max}} = \frac{0.577 S_y}{\tau_{max}} \tag{3.72}$$

where S_{sy} and S_y are shear yield strength and tensile yield strength (according to the distortion-energy theory, $S_{sy} = 0.577 S_y$), respectively. Thus, the maximum torque that can be transmitted by the key is

$$T_{k,max} = F_{max} c \approx F_{max} \left[\frac{1}{2}(d-h) \right] = \frac{0.289 l w (d-h) S_y}{n_s} \tag{3.73}$$

The stress analysis of a keyed shaft is different from the analysis of the key. As a good practice to estimate the torque that can be transmitted by the keyed shaft with a reduced shaft diameter, the concept of an effective diameter d_e is introduced into the analysis:

$$d_e \cong \sqrt{(d-h)^2 + W^2} \tag{3.74}$$

Correspondingly, shear stress resulting from the applied torque T is

$$\tau = \frac{Tr}{J} \tag{3.75}$$

where r is the shaft radius and

$$J = \frac{\pi d_e^4}{32} \tag{3.76}$$

The maximum shear stress is reached at $r = d_e/2$:

$$\tau_{max} = \frac{T_{max}(d_e/2)}{\pi d_e^4/32} = \frac{16 T_{max}}{\pi d_e^3} \tag{3.77}$$

Combining the previous equation, it follows that

$$T_{max} = \frac{0.1133 d_e^3 S_y}{n_s} \tag{3.78}$$

In the crushing failure mechanism, the compressive force F acts on the bearing area A_h of the shaft and the key, respectively. The maximum bearing stress is given as

$$\sigma_{max} = \frac{F_{max}}{A_b} = \frac{F_{max}}{hl} \tag{3.79}$$

The FS for the crushing failure is defined as

$$n_s = \frac{S_{cy}}{\sigma_{max}} \tag{3.80}$$

where S_{cy} is the compressive yield strength. Hence, the corresponding maximum torque becomes

$$T_{max} = F_{max} c \approx F_{max} \left[\frac{1}{2}(d-h) \right] = \frac{hl(d-h) S_{cy}}{2 n_c} \tag{3.81}$$

For steady loads, $n_s = n_c = 2$ is frequently adopted. As for shock loads, the FS can be chosen between 2.5 and 4.5 [3.34].

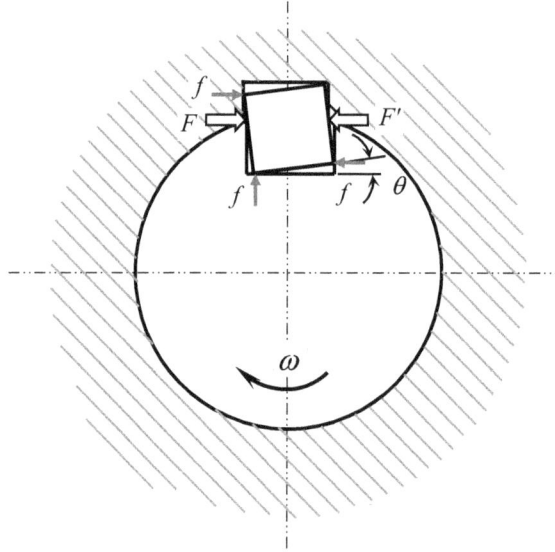

FIGURE 3.37 A key inserted loosely in a keyseat, causing high edge loadings between the key and keyseat.

3.8.1.3 Key Fit

Usually, higher motor torques require tight fits (*e.g.*, interference fits) between keys and keyseats. When a key is loosely fitted into the mating keyseat as a clearance fit. Thus, the key is subject to the transverse forces from both the shaft and hub. Thus, the key roles in the keyseat are to pit the edges and surfaces of the keyseat (Figure 3.37). This condition can cause high loadings between the key and keyseats, leading to the failure of either the key or shaft, or both. It becomes even worse if the motor frequently changes its rotating direction. On the other hand, too tight fitted keys and keyways may also result in mechanical failures. To ensure appropriate fit, the width and height dimensions of keys and keyways must be held to recommended tolerances, as defined in industrial or national standards.

Because the stress analysis of key fit is very complicated, it is usually performed using finite element analysis software to achieve accurate results.

3.8.1.4 Stress Concentration Factors of Keyed Shafts

A machined keyseat in a shaft does not only reduce the shaft strength and stiffness but also cause stress concentration in the shaft. The impacts of keyseats on the torsional stiffness of shafts were addressed first by Filon in 1900 [3.35]. Subsequent to that, a large number of theoretical, experimental, and numerical investigations have been conducted for determining stress concentration factors in keyed shafts [3.14, 3.36–3.44, as some examples]. Pilkey [3.31] proposed a correlation of stress concentration factor for a specific case of $w/d = 1/4$ and $h/d = 1/8$:

$$K_t = 1.9753 + 0.1434 \left(\frac{0.1}{r/d} \right) - 0.0021 \left(\frac{0.1}{r/d} \right)^2 \quad r/d \in [0.005 : 0.07] \tag{3.82}$$

This correlation is found overestimating the K_t values. More recently, Pedersen [3.43] reported a better correlation for the same case:

$$K_t = 1.8755 + 0.1397 \left(\frac{0.1}{r/d} \right) - 0.0018 \left(\frac{0.1}{r/d} \right)^2 \quad r/d \in [0.003 : 0.07] \tag{3.83}$$

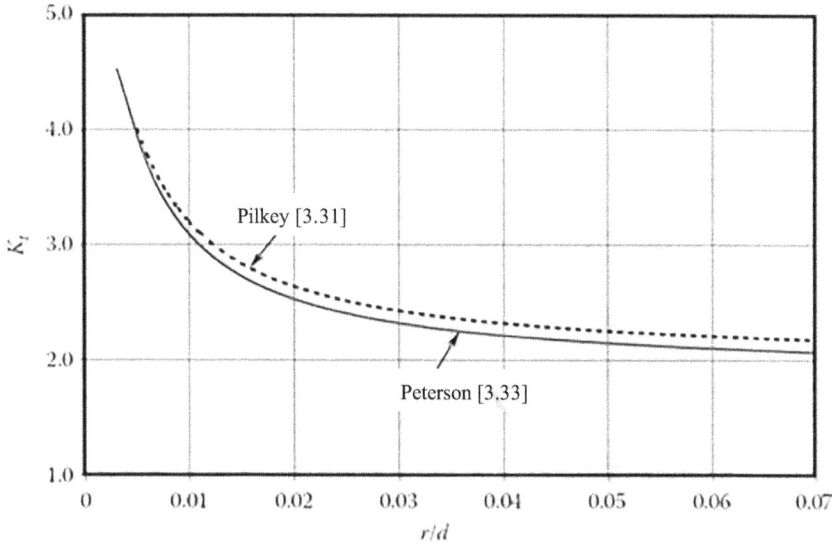

FIGURE 3.38 Comparison between two correlations of stress correlation of keyed shaft for the case of $w/d=1/4$ and $h/d=1/8$.

The comparison of the two previous correlations is given in Figure 3.38. It can be seen for $r/d<0.005$, the predicted K_t values from two correlations are almost identical. For large r/d ratios, the correlation of Pilkey is about 4%–5% higher than that of Pedersen.

More general, stress concentration factors of keyed shaft depend on three ratios: fillet radius to shaft diameter ratio r/d, keyseat width to shaft diameter ratio w/d, and keyseat height to shaft diameter ratio h/d. As discussed by Pedersen [3.43], the width ratio and height ratio can be linked together and fall naturally into two groups depending on the shaft diameter:

$$\frac{w}{d} = 1.2662\frac{h}{d} + 0.0866 \quad d \in [6:38]\text{mm} \tag{3.84a}$$

$$\frac{w}{d} = 1.6683\frac{h}{d} + 0.1055 \quad d \in [38:500]\text{mm} \tag{3.84b}$$

Corresponding to these two diameter ranges, stress concentration factors can be correlated as

$$K_t = \left(1.4786\frac{h}{d} + 0.6326\right)\left(\frac{r}{d}\right)^{\left[0.868(h/d)^2 - 0.4392(h/d) - 0.2369\right]} \quad d \in [6:38]\text{mm} \tag{3.85a}$$

$$K_t = \left(1.0428\frac{h}{d} + 0.5355\right)\left(\frac{r}{d}\right)^{\left[2.8074(h/d)^2 - 0.8091(h/d) - 0.2476\right]} \quad d \in [38:500]\text{mm} \tag{3.85b}$$

respectively.

Many researchers and investigators have contributed to determine shear stress concentration factor K_{ts} for keyed shaft. Leven [3.38] did the early work for the stress concentration arising from torsion in a shaft containing a keyseat and obtained K_{ts} values at the straight part of a keyseat as a function of r/d. Using the electroplating method, Okudo et al. [3.41] experimentally examined stresses in keyseats under the most practical conditions. To obtain the limit value for a shaft twisted only through a key, the fit is devised so as to minimize the surface friction at the fitted portion. From Peterson's charts [3.33], for a ratio of $r/d=0.02$, $K_{ts}=2.60$ without the key in place and $K_{ts}=3.0$ with the key in place.

FIGURE 3.39 Spline shafts used in induction motors.

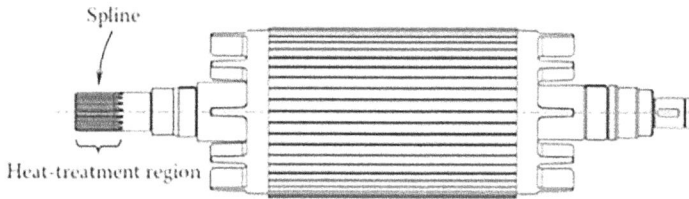

FIGURE 3.40 The spline is heat treated for increasing surface hardness, wear resistance, and fatigue life.

Thus, the allowable maximum torques transmitted by a keyed shaft in the shear failure and crushing failure become

$$T_{sh,\max} = \frac{0.289 d S_y}{n_s K_{ts}} A_{sh,s}$$ (3.86)

$$T_{\max} = \frac{hl(d-h)S_{cy}}{2n_c K_t}$$ (3.87)

3.8.2 SPLINE SHAFTS

Spline shafts are widely used in various industries such as automotive, aerospace, heavy machinery machine tool, textile machinery, and mining. Splines perform the same function as keys in transmitting torque from a motor to a driven machine. In fact, a spline can be viewed as a series of axial keys machined into the shaft (Figure 3.39), with corresponding grooves machined into the bore of the mating element. As demonstrated in Figure 3.40, the spline is machined at the drive side of the shaft. To increase the surface hardness, wear resistance, and fatigue life, shaft splines must go through heat treatment processes. It should be noted that to prevent shaft deformation, heat treatment is applied to the limited region that contains only spline. However, improper heat treatment of steel and alloys may lead to temper embrittlement, tempered martensite embrittlement, or hydrogen embrittlement, depending on different heat treatment processes [3.45]. Consequently, all of these embrittlement types could cause spline shaft failure, as shown in Figure 3.41.

FIGURE 3.41 Failure of spline shafts due to improper heat treatment.

3.8.2.1 Advantages of Spline Shafts

There are numerous advantages of spline connections over key connections: (*a*) Because the spline is machined directly on the shaft, no relative motion can occur between the spline teeth and the shaft. (*b*) With multiple spline teeth, the torsion and bending loads are uniformly distributed to the teeth. (*c*) Each spline tooth transfers a much lower torsion load compared with a keyed shaft. Thus, spline shafts can transmit large torque or power. (*d*) A spline shaft has an ability to accommodate large axial movements between the shaft and the mating part while simultaneously transmitting torque. (*e*) An involute spline shaft has an autocentering dynamics that allows a semi- or full-floating coupling. (*f*) The mating part can be indexed to various positions with respect to the shaft spline. (*g*) A helical spline shaft can drive the axial and rotary motion at the same time. (*h*) Using the hob to cut spline shafts can offer notable advantages in terms of both manufacturing efficiency and spline profile accuracy over the manufacturing method for keyed shafts. (*i*) Splines are usually heat-treated to increase the surface hardness and wear resistance. By contrast, it is very rare to do heat treatment to shaft keyseats. (*j*) Uniformly distributed spline teeth have little impact on shaft unbalance. For a keyed shaft, the keyseat can bring an unbalance problem. (*k*) While a shaft with an external key is subject to a high stress concentration at the mouth of keyseat, a spline shaft has much lower stress concentration due to the distribution of loads to all teeth.

3.8.2.2 Type of Spline

Although a number of spline types exist, the most commonly used splines across a wide variety of industries are involute splines and straight-sided splines (sometimes also referred to as parallel key splines), as shown in Figure 3.42. Involute splines typically have pressure angles of 30, 37.5, or 45. This type of spline tends to be self-centering that equalizes bearing and stresses among all teeth. This characteristic feature also helps achieve better concentricity between the shafts of the motor and its mating machine. In contrast with involute splines, straight-sided splines have ridges with a rectangular profile. This type of spline has been standardized by Society of Automotive Engineers (SAE) to have 4, 6, 10, or 16 splines, which is equally distributed around the circumference of a shaft. Today, straight-sided splines still remain popular in many applications, especially in the machine tool and automotive industries.

Involute splines are favored over straight-sided splines because of their self-centering capability, higher torque-transmitting capacity, and greater strength, as well as some manufacturing advantages (*e.g.*, for any given pitch, the spline cutting tools can cut any number of teeth).

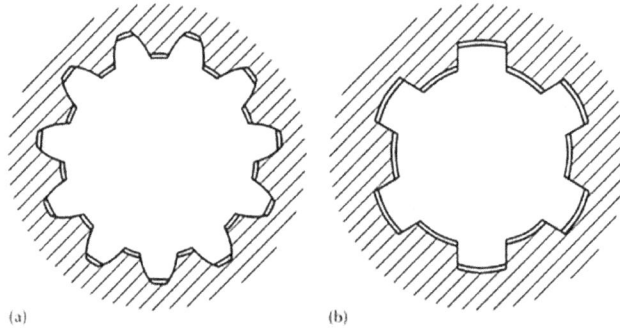

FIGURE 3.42 Spline types: (*a*) involute spline and (*b*) straight-sided spline.

3.8.2.3 Stress Concentration Factors of Spline Shafts

When spline motor shafts transmit torque or power to a driven machine, they are subject to torsion loads. A mathematical analysis was conducted by Okubo [3.46] for a shaft with a number of longitudinal semicircular grooves. It was found that as r/d approaches zero (where r is the fillet radius at the tooth root and d is the shaft diameter), the stress concentration factor $K_{ts} = 2$. By using the 3D photoelastic technique, Yoshitake [3.47] performed an experimental investigation on eight-tooth splines. In his work, the torsion stress concentration factors were measured by the wedge method with varied tooth fillet radius. The test results indicate that the involute splines are more advantageous than parallel side splines from the standpoint of torsion stress concentration factor. It was reported that the torsion stress concentration factor is 2.8 for an involute spline and is 3.1 ($r = 1.5$ mm), 3.4 ($r = 1.0$ mm), and 4.4 ($r = 0.5$ mm) for parallel splines.

3.8.3 Tapered Shafts

Since keys and splines usually produce stress concentrations on motor shafts, tapered shafts may provide a promising solution with keyless designs. As shown in Figure 3.43, a taper is made at the drive end of a motor shaft with a small cone angle α and inserted in the bore of the traction sheave/hub. A number of fasteners (screws or bolts) are engaged with the shaft to ensure the close contact of the tapered shaft with the sheave/hub. As the fasteners are tightened, the tapered shaft is further pushed into the bore of the sheave/hub, creating a contact pressure at the contact surface.

By comparing with conventional shafts, tapered shafts are more expensive and may require special tooling. In addition, this method may not be suitable for very high torque conditions.

Assume the tapered shaft is perfectly contacted with the sheave/hub, the contact diameter d is a function of the contact distance x, as

$$d(x) = d_1 + \frac{d_2 - d_1}{L} x \tag{3.88}$$

The area of the contact surface between two components is thus

$$A = \pi \int_0^L d(x) dx = \pi \int_0^L \left(d_1 + \frac{d_2 - d_1}{L} \right) x \, dx = \frac{\pi L}{2} (d_1 + d_2) \tag{3.89}$$

With the normal contact pressure p_n applying on the interface of the tapered shaft and the sheave/hub, the force acting on the contact surface can be decomposed as the normal contact force F_n, which is perpendicular to the contact surface, and the friction force F_{fi} which is parallel to the

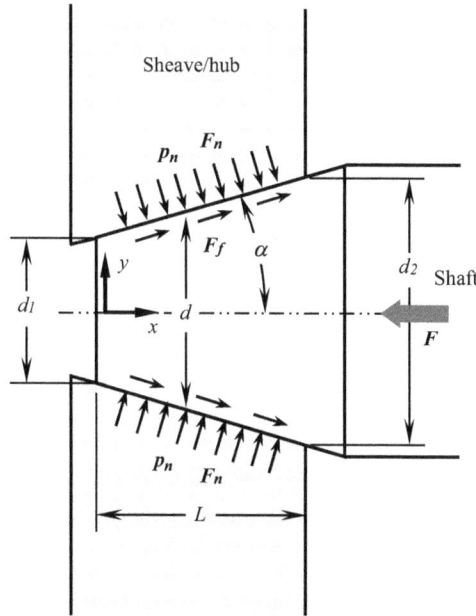

FIGURE 3.43 A tapered motor shaft against a sheave/hub.

contact surface. Since the two components are perfectly contacted each other, the contact pressure p_n is uniform on the interface, following that

$$F_n = \pi p_n \int_0^L d(x)\,dx = \pi p_n \int_0^L \left(d_1 + \frac{d_2 - d_1}{L}x\right) dx = \frac{\pi L p_n}{2}(d_1 + d_2) \tag{3.90a}$$

$$F_f = \mu F_n = \frac{\pi \mu L p_n}{2}(d_1 + d_2) \tag{3.90b}$$

where μ is the friction coefficient between two contact components.

At the equilibrium condition, all forces are balanced with each other, $\Sigma F = 0$. The axial pressing force F can be expressed as

$$F = p_n A(\sin\alpha + \mu\cos\alpha) \tag{3.91}$$

The torque transmitted by the shaft can be determined as

$$T = \int \frac{d}{2}\mu p_n\,dA = \frac{\mu}{2}\int_0^L \frac{\pi d^2 p_n}{\cos\alpha}\,dx = \frac{\pi \mu p_n L}{6\cos\alpha}\left(d_1^2 + d_1 d_2 + d_2^2\right) \tag{3.92}$$

In a similar manner, a governing equation for calculating the brake torque of a cone clutch was derived in Shigley's textbook [3.13]. By assuming uniform contact pressure, the brake torque can be expressed as

$$T = \frac{F_n \mu}{3\sin\alpha}\frac{d_2^3 - d_1^3}{d_2^2 - d_1^2} \tag{3.93}$$

where

$$F_n = \frac{\pi p_n}{4}\left(d_2^2 - d_1^2\right) \tag{3.94}$$

Combining Equations 3.93 and 3.94 yields

$$T = \frac{\pi \mu p_n}{12\sin\alpha}\left(d_2^3 - d_1^3\right) \tag{3.95}$$

Note that in Equation 3.92, the length of the tapered shaft L can be expressed as

$$L = \frac{d_2 - d_1}{2\tan\alpha} \tag{3.96}$$

Substituting Equation 3.96 into 3.92, Equation 3.92 is identical to Equation 3.95.

However, there is a difference in force (or pressure) equation between two systems. By replacing L in Equation 3.90a with 3.96, it gives

$$F_n = \frac{\pi p_n}{4\tan\alpha}\left(d_2^2 - d_1^2\right) \tag{3.97}$$

Comparing Equations 3.97 and 3.94, it can be clearly seen that a factor of $1/\tan\alpha$ is missing in Equation 3.94 (Shirgley's equation). This indicates that Equation 3.94 is a special case of Equation 3.97 with $\tan\alpha = 1$ (*i.e.*, $\alpha = 45°$). This has been confirmed by Keesee [3.48] using an FEA analysis to calculate the contact pressure and contact force on the interface. The numerical results have shown good agreement with the calculated data obtained from Equation 3.97.

3.9   FATIGUE FAILURE UNDER ALTERNATIVE LOADING

Rotating shafts are frequently subject to partial or complete reversing loads, resulting in fatigue failure.

The fatigue stress concentration factor is defined as

$$K_f = \frac{\text{Maximum stress with discontinuity}}{\text{Nominal stress without discontinuity}} \tag{3.98}$$

The relationship between K_j and K_t is found as

$$K_f = 1 + q\left(K_t - 1\right) \tag{3.99}$$

where q is the notch sensitivity of material and $0 \leq q \leq 1$. Thus, it gives that $1 \leq K_f \leq K_t$. If notch sensitivity data are not available, it is conservative to use K_t in fatigue calculations.

3.10   SHAFT MANUFACTURING METHODS

There are a variety of shaft manufacturing methods in manufacturing industries. None of the manufacturing methods is perfect. Each has its own set of advantages and disadvantages. The determination of the optimal manufacturing process depends on many factors, such as production efficiency, dimension accuracy, surface finishing, influence on part's strength, processing cycle, shaft material, shaft size, equipment availability, and cost.

3.10.1 MACHINED SHAFT

A majority of motor shafts are manufactured by machining for high productivity, low cost, and easy tolerance control.

A machined shaft usually passes through a series of machining processes. As the shaft is machined from a raw material, in order to increase the productivity, it is typical to use a common lath first to quickly machine it to the dimensions that are close to its final dimensions. Then, use a precise machine such as a CNC to machine the shaft to the final dimensions. For some important sections, such as bearing journals, the shaft may go through a heat treatment and finally grinding it to the final dimensions.

During these processes, the shaft has to be machined on different machines. Therefore, holding the shaft's geometric tolerances such as roundness, cylindricity, runout, concentricity, and straightness is critical in obtaining satisfactory shaft balance and lowing motor vibration. As a matter of fact, this depends entirely on the success of the initial centering phase of the machining operation. The key is providing a good reference point for machining between centers.

In motor shaft manufacturing, endworking of shafts is often incorporated as a primary operation of a series of machining processes. Endworking can eliminate some secondary machining operations by combining them with initial end finishing, thereby increasing productivity and helping to decrease the part process scrap rate for subsequent operations [3.49].

Major benefits of the endworking process include the improved concentricity, roundness, and squareness of the machined surfaces. This greatly reduces the shaft runout. Practically, the more accurate shaft centers allow for more accurate dimension and position controlling, in favor of shortening the production cycle time and costs.

Endworking is conventionally used for centering and facing a shaft that is loaded on a lathe between centers in high-volume production of shafts. Presently, the endworking process is often performed by a CNC double-end machine to work on both shaft ends at the same time for achieving cost savings and high productivity.

3.10.2 FORGED SHAFT

Forging is a manufacturing process that has been extensively used in industries for producing high-strength and high-quality machine parts. In a forging process, under high pressure, a metal workpiece undergoes large plastic deformation, resulting in an appreciable change in shape or cross section. Unlike in a conventional machining process that the material grain is cut off, forging increases the strengths and toughness of metal by aligning the grain along the line of potential stress. Moreover, because of the enormous pressure involved during the forging process, a forged part is denser in the bulk due to the reduction or elimination of porosities and voids. Thus, the forged shafts have higher strength-to-weight ratio and higher corrosion resistance. However, because of the high cost of forging tooling and required multiple steps, forging is generally more expensive than most manufacturing processes. Motor manufacturers often use medium-carbon steel to make forged motor shafts.

Several forging processes are available, including cold forging and hot forging. Cold forging is performed at room temperature and requires a very high forging pressure to overcome the deformation resistance of the material. This indicates that cold forging is limited to relatively small shafts with simple geometries. Cold forging can achieve product features such as smooth surfaces, tight tolerances, concentric diameters, and beneficial grain flow. It offers significant cost-saving advantages as a result of the material saving. In contrast, hot forging is performed above the material recrystallization temperature and thus generates oxide scales on the forged shaft surface.

A method for forging a stepped shaft through a combination of forging operations has been reported recently [3.50]. The method includes providing a billet of a predetermined mass; heating the billet; cross-wedge rolling the billet to form an intermediate workpiece having a first cylindrical

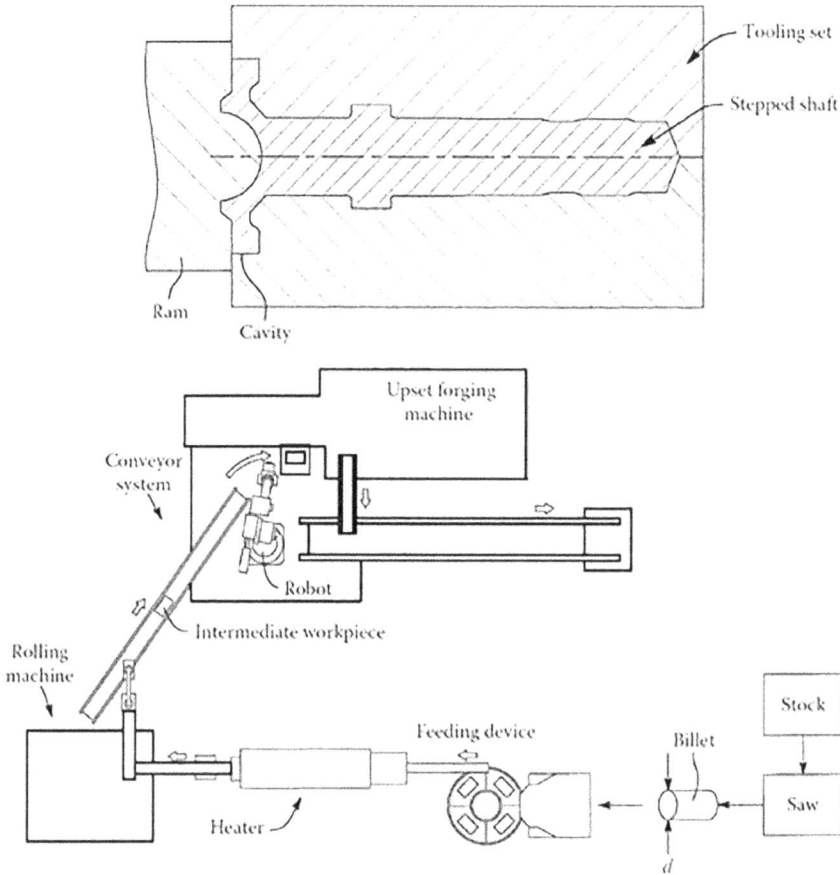

FIGURE 3.44 Fabrication system for producing forged stepped shafts (U.S. Patent 7,866,198) [3.50]. (Courtesy of the U.S. Patent and Trademark Office, Alexandria, VA.)

portion and a second cylindrical portion that are axially spaced apart by a neck that has a smaller diameter than the two portions; and performing at least one upset forging operation on the end of the intermediate workpiece to enlarge the first cylindrical portion such that in at least one location, its diameter is larger than a diameter of the billet and a diameter of any other shaft portion (Figure 3.44).

3.10.3 WELDED HOLLOW SHAFT

Motor weight can be very sensitive in some applications such as robots and aircrafts. Motor designers always take tremendous efforts to reduce the machine weight and increase its torque density. One of the effective methods is to use hollow shafts to lower shaft weight without affecting their function and performance. In fact, hollow motor shafts can not only reduce their weights and inertias but also provide the passage for cooling fluid (air or liquid) passing through the shaft interior to cool the rotor. In some applications, motor power cables and other connecting wires (*e.g.*, brake, encoder) can go through the central part of hollow shafts, mitigating EMI and improving operation safety of the system, like industrial robots.

Hollow shafts are often used in direct drive motors and vehicle motors that have large shaft diameters. Because of the larger diameter, a hollow shaft has higher stiffness and rigidity, higher resonant frequency, larger strength, more polar moment of inertia, and lower deflection, compared

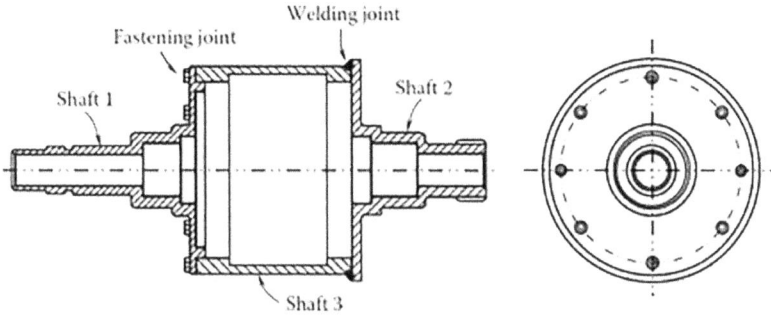

FIGURE 3.45 Three-piece hollow shaft design using fastening and welding joint methods.

to a solid shaft with the same mass. However, hollow shafts are often more expensive than solid shafts due to the complex manufacturing process.

The fabrication of hollow shafts may involve several different manufacturing processes such as welding, machining, and casting. Among them, welding is the most cost-effective way for making hollow shafts. Therefore, it requires that the shaft material has a good weldability.

The weldability of steel strongly depends on the chemical composition of steel, especially carbon. Higher quantities of carbon and other alloying elements result in a lower weldability. To quantitatively evaluate the influence of the contributions of the various alloying elements on the difficulties encountered in alloy welding, a concept of carbon equivalent (CE) is introduced in an attempt to measure weldability with a single number. CE is an empirical value expressed in weight percent. Several empirical formulas are available in the literature. One popular expression, which is recommended by the International Institute of Welding (IIW) and most often used in ASME applications [3.51, 3.52], is given in the following (all elements in mass percentage):

$$CE = C + \frac{Cr + Mo + V}{5} + \frac{Mn + Si}{6} + \frac{Ni + Cu}{15} \qquad (3.100)$$

Another popular formula for predicting CE is known as Dearden and O'Neill formula, which is similar to (3.100) but without the Si term. Dearden and O'Neill [3.53] first proposed the CE to assess the weldability of alloy steels. Later, their formula was revised slightly by the IIW.

A CE value of 0.30 or less is considered to be good for welding and no special precautions are required. For $0.30 < CE < 0.4$, modest preheat is necessary. Steel with a CE value in excess of about 0.5 cannot be easily welded because of their increased tendency to develop hydrogen-induced cracking in the heat-affected zone. Hence, welding of steel with a high CE value requires a special care for the preheat process. Figure 3.45 is an example of hollow shaft used in hybrid vehicles.

An alternative design of hollow shaft is presented in Figure 3.46. It can be seen that the hollow shaft is formed by integrating two shafts together with the rotor laminated core through the interference fit. In order to ensure the firmness of the rotor assembly, welding is used between shaft 1 and the rotor core.

A hollow shaft can be also fabricated by utilizing the friction welding technology. Friction welding is a melt-free process that generates heat through mechanical friction between a rotating and a stationary component. Because there is no melt involved, this processing method is not a truly welding process in the traditional sense but more like a forging process. One of the advantages of this technology is to allow different materials to be joined together, such as aluminum and steel and stainless steel and carbon steel. A hollow shaft generally consists of three segments: two solid segments and one pipe segment. Two solid segments are friction welded to both ends of the center pipe segment (Figure 3.47). The formed shaft is then machined to its final dimensions. With this manufacturing technique, the hollow shafts can achieve up to 50% or more weight reduction compared to the similar solid shafts.

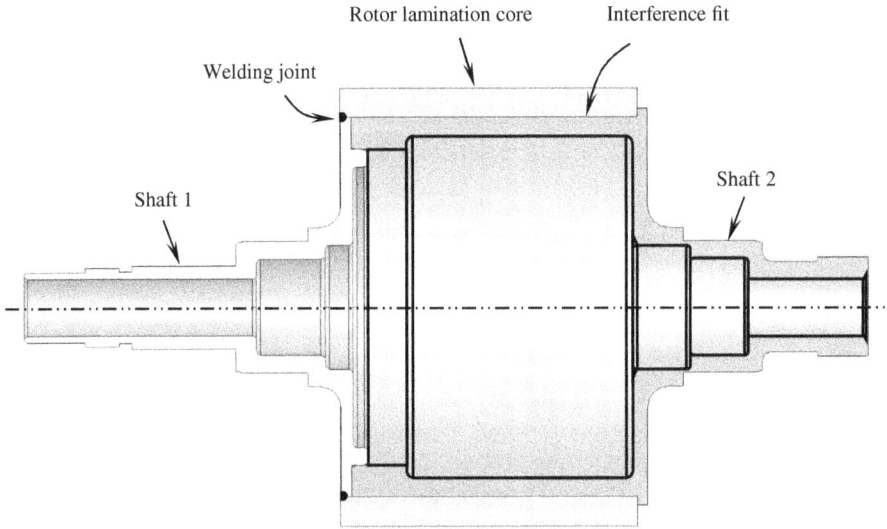

FIGURE 3.46   Alternative welded hollow shaft design by integrating two shafts together with the rotor lamination core.

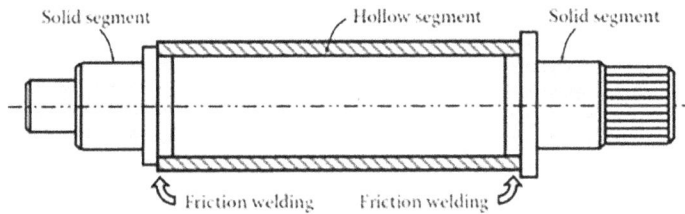

FIGURE 3.47   Hollow shaft fabricated by friction welding technology.

3.10.4   SHAFT MEASUREMENT

Confirming that manufactured shafts meet all specifications requires validating shaft hardness, geometric dimensions and tolerances, total/circular runout, concentricity, circularity, surface roughness, and other items. Figure 3.48 shows a CMM that is used to measure the geometric dimensions and other characteristics of shafts. The CMM is composed of three axes, that is, X, Y, and Z, forming a regular 3D coordinate system. In measurement, the shaft is put on a pair of V-shaped stands on the table, which is defined as the X-Y plane. Measurements are executed using a probe attached to the movable third axis, that is, Z-axis. As the probe touches the shaft surface, the CMM takes readings in six degrees of freedom. In such a way, a measuring point is generated with its three coordinates (X, Y, and Z). All collected points are analyzed via regression algorithms for the shaft constructive features. Usually, dimensional measurement capabilities of a CMM include position, perpendicularity, parallelism, angularity, profile of a surface, profile of a line, straightness, flatness, circularity, cylindricity, symmetry, concentricity, and total/circular runout.

3.11   SHAFT MISALIGNMENT BETWEEN MOTOR AND DRIVEN MACHINE

Shaft alignment has a strong impact on motor and driven machine operation and performance. Shaft misalignment, occurring as the centerlines of the motor and the driven machine shafts are not in line with each other, can result in excessive vibration, noise, and additional loads in motor and

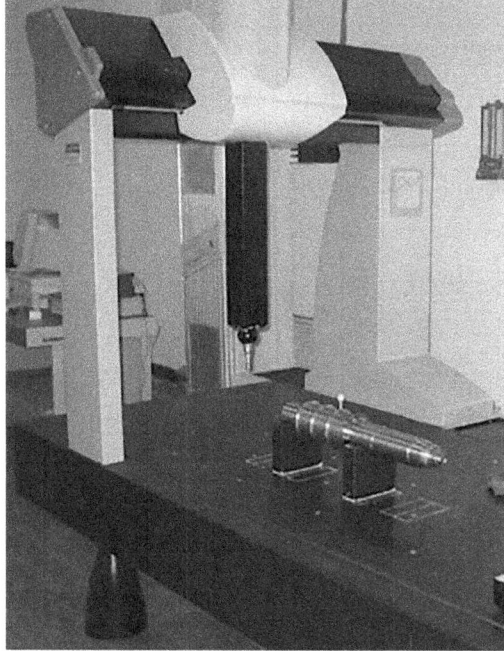

FIGURE 3.48 A coordinate measurement machine for measuring shaft geometric dimensions.

driven machine components, such as coupling, bearings, seals, shafts, and other machine components. It can lead to premature wear or even catastrophic failure of these components.

3.11.1 Type of Misalignment

Because the motor and driven machine are in 3D space, the misalignments between their shafts can exist in any direction. In actual applications, it is most convenient to break up this 3D space into the horizontal and vertical planes. As demonstrated in Figure 3.49, there are basically two types of motor misalignment:

- Angular misalignment can be further divided into the horizontal and vertical angular misalignments. Both of the misalignments occur when a motor is set at an angle to the driven machine in the horizontal or vertical direction, or both. It can be seen that from Figure 3.49a, as the centerlines of the shafts of the motor and driven machine are extended, the two centerlines cross each other at an angle. Angular misalignment can cause severe damage to motors and driven machines.
- Axial misalignment occurs when a motor has an axial distance with its mating machine (Figure 3.49b). This type of misalignment, sometimes referred to as *shaft end float*, often results from the dimensional variations due to the thermal expansion/contraction of the shafts. In practice, axial misalignment is easily measured and corrected compared to angular misalignment.
- Similar to angular misalignment, parallel misalignment can be further divided into horizontal and vertical parallel misalignments. These two misalignments occur when the two shaft centerlines are parallel do not line up either horizontally or vertically or both (Figure 3.49c). Parallel misalignment may be resulted from manufacturing processes such as improper tolerances and shaft runouts, as well as poor installation practices.
- In practice, angular, axial, and parallel misalignments may occur simultaneously, as shown in Figure 3.49d.

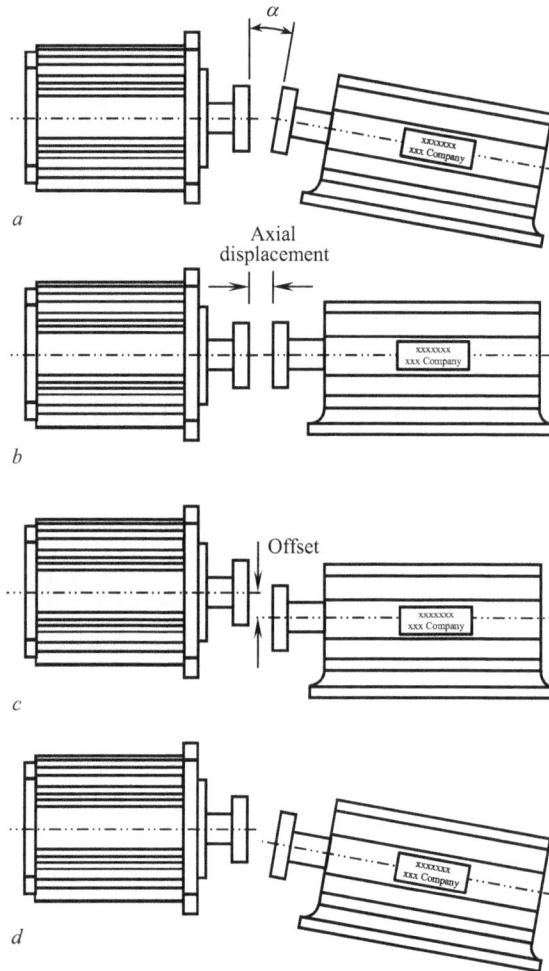

FIGURE 3.49 Misalignment of the motor shaft with the shaft of driven machine: (*a*) angular misalignment, (*b*) axial misalignment, (*c*) parallel misalignment, and (*d*) combination of angular, axial, and parallel misalignments. Only vertical misalignments are presented in this figure. Horizontal misalignments are in a similar manner.

3.11.2 CORRECTION OF SHAFT MISALIGNMENT

Misalignments have long been recognized as one of the leading causes of electric machine damage and have been responsible for huge economic losses. In order to achieve best results in correcting shaft misalignment, it must take into account the machine sizes, shaft geometries, coupling type and condition, rotating speed, and operating temperature of both the motor and driven machine.

During motor normal operation, the alignment condition may change as the result of variations in a number of operating characteristics, such as operating temperature, transmitted torque, load, coupling, bearing play, and mounting conditions of both the motor and driven machine. Therefore, it is highly desired to measure motor-driving machine alignments under actual operating conditions (sometimes referred to as dynamic alignments).

There are good practices for measuring and correcting shaft misalignments [3.54]. The first step is to check alignments using appropriate tools. Modern laser alignment systems have been developed since the 1980s. Today, laser alignment becomes an essential component of a viable maintenance strategy for rotating machines. One of the notable features of laser alignment systems is their capability of performing real-time measurements to monitor the changes in machine's alignments.

As the misalignments have been detected, the next step is to follow suitable procedures to correct them. Soft foot is one of the most prevalent conditions found in shaft misalignments. Some laser alignment tools have a soft foot operation capability that guides engineers through the correction procedure. In practice, a set of precision shims is often used to bring the machines back into both vertical and horizontal alignment [3.55].

3.12 SHAFT COUPLING

The term *coupling* refers to a device that is used to connect two shafts together at their ends. The primary function of a motor coupling is to transmit torque or power from a motor shaft to the shaft of the driven machine. However, in the real world, whatever precautions are taken to make alignments as precise as possible, the alignment between two separate shafts is never perfect. Some amount of residual misalignment will inevitably remain. Misalignments are resulted not only from the improper installation, but also from the variations of temperature and other factors. Consequently, it forces rigid components such as shafts to defect, causing vibration of the system and bring uneven loads on bearings and system frame. Therefore, it is vital to design couplings properly to compensate for misalignment to ensure normal operation of motor and driven machine.

According to the connection pattern and relative motion, motor couplings can be briefly classified as rigid and flexible couplings.

3.12.1 RIGID AND SEMIRIGID COUPLINGS

Rigid couplings are designed to draw two shafts together tightly so that no relative motion occurs between them. Rigid couplings are suitable when the alignment of the two shafts can be maintained very accurately. This type of coupling does not compensate for any misalignment between a motor and a driven machine. If there is any significant angular, radial, or axial misalignment between them, additional loads, such as bending moment and tensional or axial stress, will be applied on the shafts of the motor and the driven equipment, as well as the coupling itself. These loads may result in bearing, seal, or/and coupling premature failures, shaft cracking and bending, and excessive radial and axial vibrations.

One of the most commonly used rigid couplings is a flange coupling, as shown in Figure 3.50. The motor shaft and the driven machine shaft are connected tightly through two flanges using bolt fasteners. Thus, in addition to tensile forces, the bolts are also subject to shear forces. This type of coupling uses a sleeve to align the two shafts.

A semirigid coupling is shown in Figure 3.51. This coupling is similar to the previous one, but there is no common sleeve shared between two shafts. As shown in the figure, the motor shaft and driven machine shaft are firmly assembled together with their hubs/flanges, respectively. Each flange is bolted with a ring-shaped annular element, which is securely fixed on the outer surface of the split tube. To avoid distortions of the split tubes under loads, the split tubes are supported internally with two reinforcing half-rings. This type of coupling can provide a certain degree of freedom to accommodate a small misalignment between two shafts.

In order to further increase the coupling flexibility, the two ring-shaped annular elements may be made of flexible elastomeric materials such as polyurethane elastomer, as suggested by Zilberman *et al.* [3.56]. However, one of the unfortunate consequences of increasing the coupling flexibility is a corresponding reduction in torque transmission reliability. This indicates that such couplings may be restricted to use in relatively low-torque transmission applications.

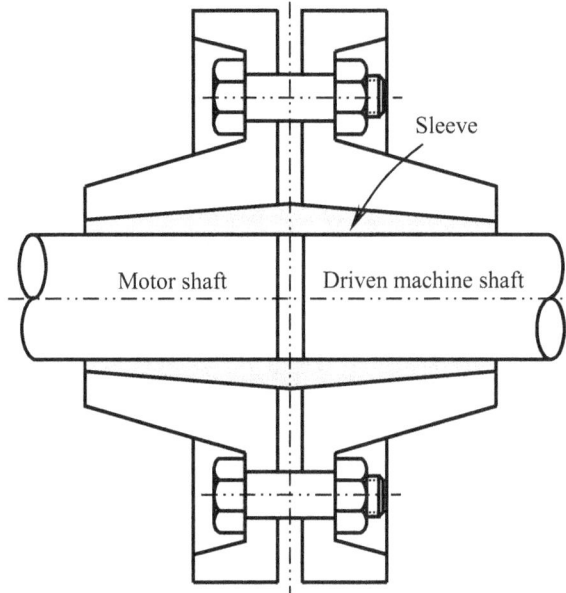

FIGURE 3.50 Flange rigid coupling. The sleeve is for aligning two shafts.

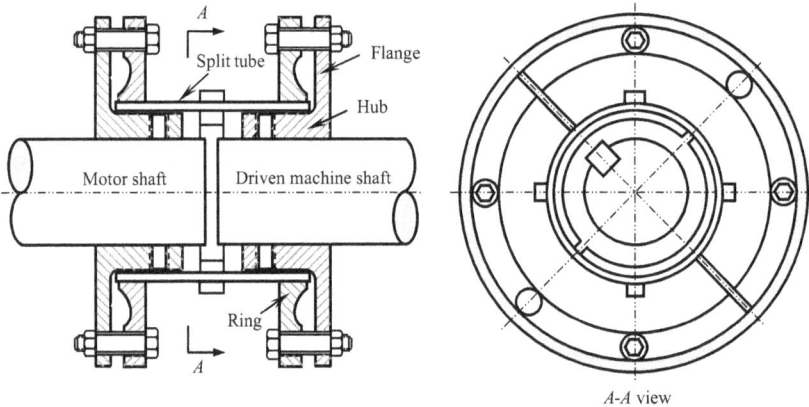

FIGURE 3.51 Semirigid coupling used in torque/power transmission. The motor shaft and driven machine shaft are indirectly connected together through two ring-shaped annular elements and split tubes for gaining a certain degree of flexibility.

3.12.2 FLEXIBLE COUPLINGS

Flexible couplings are designed to transmit motor torque to drive machine smoothly while accommodating minor misalignment in the axial, radial, or angular directions. The flexibility is such that when misalignments do occur, the coupling components move with little resistance to absorb the residual misalignment. It is noted that with today's technology, flexible couplings are designed to tolerate more misalignments.

A large variety of flexible coupling types are available, as displayed in Figure 3.52. Each of them has its special features and is suitable for particular applications. For example, depending upon the specific design, Thompson couplings, also known as universal joints, are used for maximum allowable misalignment up to 20°–30°. They are extensively applied to the drive shafts of vehicles to

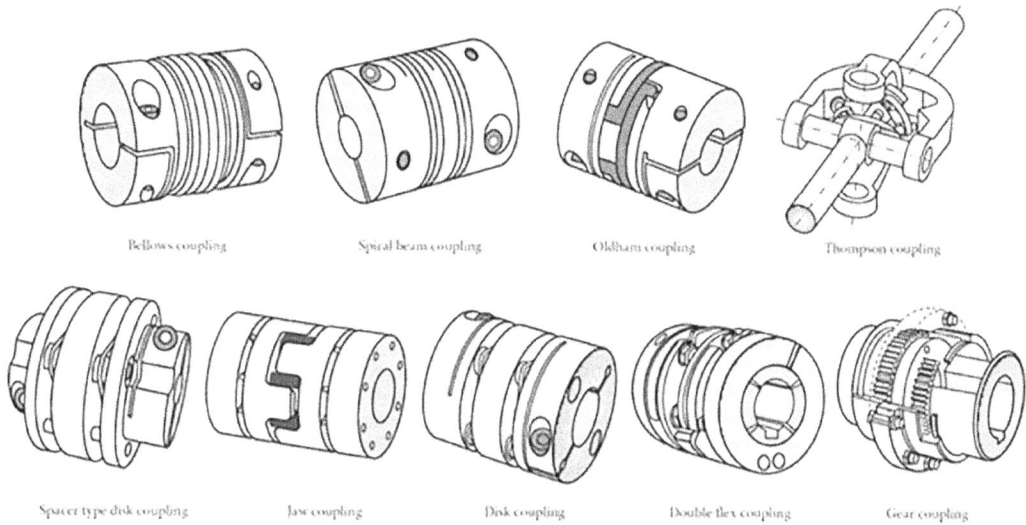

FIGURE 3.52   A variety of flexible couplings used in motor applications

allow the wheels to move with the suspension system. Motor engineers need to trade-off couplings' pros and cons to select the coupling that best fits the specific application.

Other types of flexible couplings are also available [3.57], such as geared, roller chain, and grid couplings (Figure 3.53). Gear couplings have two alternative designs: (*a*) mounted external gear teeth on one shaft mates with internal gear teeth that mount on another shaft; (*b*) mounted external gear teeth on both shafts mate with internal gear teeth on a housing (Figure 3.53a). In each design, the meshed teeth transmit torque between the coupled shafts while allowing some misalignment to exist between them. Misalignment results in sliding relative motion across mating teeth as they pass through each revolution. Thus, they have misalignment capabilities generally about 0.25–0.50 mm (0.01–0.02 in.) in parallel and 2° in angular [3.58]. However, this type of couple requires special care of maintenance, especially regular lubrication.

A roller chain coupling consists of a pair of sprockets mounting on each shaft and a pair of strand roller chains (Figure 3.53b). The teeth of sprocket are hardened to increase wear resistance. The sprockets are coupled when the chains are wound around the sprockets. A slip-fitted joint pin and a spring clip are used to connect the two roller chains as a single double-roller chain, allowing for easy coupling or decoupling. The clearance between the chains and sprockets as well as the clearance between chain components (*e.g.*, roller and bushing, bushing and pin, roller link plate and teeth, *etc.*) absorb the great positional misalignment of both shafts. Angular and parallel misalignment allowances are typically 2° and 0.38 mm (0.015 in.), respectively. For this type of coupling, torsional rigidity is required.

A grid coupling functions as a resilient coupling by damping torsional vibration and cushioning shock loads. The external grid coupling (Figure 3.53c) utilizes grooved discs attached to the ends of each shaft. A corrugated steel grid with a high tensile strength is bended and inserted into the grooves of the discs to transmit torque. When the coupling is subjected to a torsional load, the grid deforms accordingly, widening in some locations and narrowing in others over each revolution. The deformation of the grid in the disc grooves accommodates the slight variations in alignment between connected shafts, while dampening vibration and shock loads. This type of coupling is capable of handling parallel misalignment up to 7.6 mm (0.30 in.) and angular misalignment of about 0.25° [3.58]. The grid-groove design permits a rocking and sliding action. This occurs through the resilient grid without loss of power. Because the grid coupling relies on metal-to-metal contact, it is considered stiffer than jaw couplings but somewhat softer than gear couplings. An advantage it

(a)   (b)   (c)

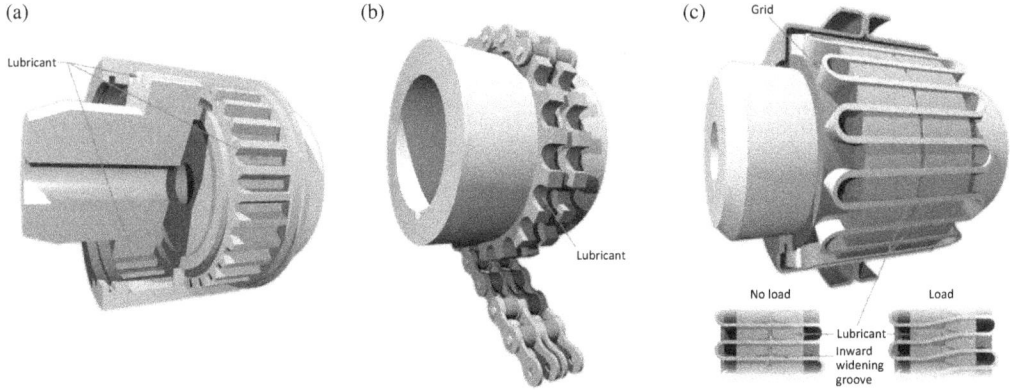

FIGURE 3.53 Other types of flexible couplings: (*a*) gear coupling, (*b*) roller chain coupling, and (*c*) grid coupling. All allow angular and axial misalignment. (Courtesy of Noria Corporation, Tulsa, OK.)

has over gear and disc couplings is its ability to reduce vibration between machines by as much as 30%. All these types of couplings require lubrication to ensure proper performance and longevity.

3.12.3 NON-CONTACT COUPLINGS

A non-contact coupling can transmit torque between a motor and a driven machine without any physical contact. Compared to conventional mechanical couplings, this type of coupling offers some substantial benefits such as wear free, vibration isolation, high tolerance to shaft misalignment, overload protection, flexible configuration, reduced maintenance, and allowance of speed variation and regulation between the driver (*e.g.*, motor) and load. This type of coupling is often used in applications requiring low or medium torque transmissions, such as pumps, conveyors, and blowers.

According to the operating mechanism, non-contact power transmission couplings can be generally categorized into the following types:

- As emerging flexible transmission devices, PM eddy-current couplings can realize torque transmission without physical contact between driving and driven machines. Based on the configuration of magnetic flux, PM eddy-current couplings can be further classified into an axial-flux type (*i.e.*, a disc type) and a radial-flux type (*i.e.*, a cylinder type), as shown in Figures 3.54 and 3.55, respectively.

  It can be seen in Figure 3.54, an axial-flux PM eddy-current coupling consists of two separated rotors: a PM-rotor where a set of PMs mount on the primary back iron disc and a conductor-rotor where a copper plate as the conductor mounts on the secondary back iron disc. The two rotors are put side by side so that a small airgap is formed between the PMs and the copper plate. The PMs with different polarities are alternatively distributed along the θ-axis on the primary back iron disc and magnetized in the z-direction. The combination of high conductivity of the copper conductor and high permeability of the back iron enables the eddy-current coupling to produce efficient transmission torque. In fact, the PMs, copper conductor, and back iron discs constitute a closed effective magnetic circuit. At the airgap, there is a magnetic field produced by the PMs and the eddy-current magnetic field. The superposition of the two magnetic fields constitutes a dynamic airgap magnetic field. When the PM rotor rotates at the annular speed ω_1, the magnetic field moves across the copper conductor and thus induces eddy-current loops in the conductor. The magnetic field resulted from induced eddy current opposes the magnetic field produced by

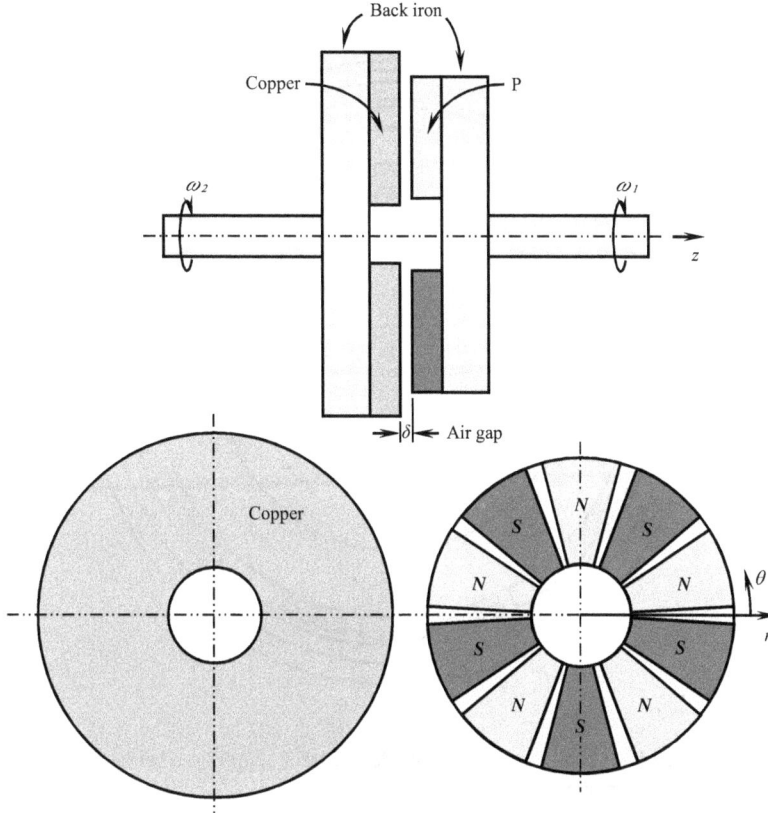

FIGURE 3.54 Structure of a typical axial-flux PM eddy-current coupling, where ω_1 and ω_2 are the primary and secondary rotating speed, respectively.

PMs. The interaction between the two magnetic fields thus couples the two components for producing and transmitting the magnetic power [3.59], enabling the conductor-rotor rotating at ω_2 in the same direction of ω_1.

The torque–speed curve of such a device is related to the induced eddy current in the copper plate. The value and the current distribution depend on the slip speed ω between the two discs [3.60]

$$\omega = \omega_1 - \omega_2 \tag{3.101}$$

The efficiency of the coupling is given by

$$\eta = 1 - s = 1 - \frac{\omega}{\omega_1} \tag{3.102}$$

where s is the slip, $s = \omega/\omega_1$. In normal operation, s ranges of 2%–5%.

It is noted that the induced eddy current in a coupling always leads to the eddy-current loss and consequently a temperature rise in the device [3.61]. In a coupling design stage, the thermal issue must be taken into account. It is highly desired to perform the coupled electromagnetic-thermal analysis to ensure normal operation of the coupling. Nevertheless, this type of asynchronous coupling has been widely used in mining, steel, energy, and military industries for small or medium torque transmission applications.

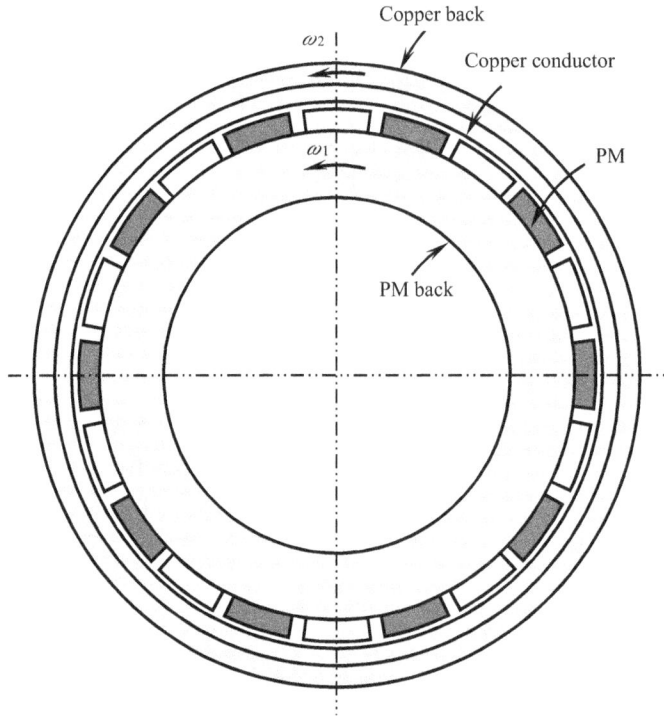

FIGURE 3.55   Structure of a typical radial-flux PM eddy-current coupling, where ω_1 and ω_2 are the primary and secondary rotating speed, respectively.

- As the name implies, magnetic couplings transmit torque and power through magnetic forces. Although the performance of both PM eddy-current and magnetic couplings relies on PMs, there are some fundamental differences between the two. The main difference is that a magnetic coupling uses two sets of PMs rather than one in a PM eddy-current coupling. That means that the copper conductor in the PM eddy-current coupling is replaced by an array of PMs. Therefore, the magnetic coupling is a synchronous system with zero slip ($s=0$), and consequently both the internal and external rotors rotate at the same rotating speed and direction. Synchronous couplings exploit the attractive and repulsive forces between PMs to produce motion.

  As shown in Figure 3.56, the magnetic coupling consists of two concentric rotors: an inner rotor and an outer rotor, fitted with magnets of alternating poles on the outer and inner surfaces, respectively. Each rotor contains the same number of magnets for achieving fully coupling between them. A stationary contaminate shroud is placed between the two rotors to separate two different mediums (*e.g.*, liquid and air). In an idle state, the magnets on the rotors are opposite, *i.e.*, each North and South pole on the driving rotor is linked to each respective South and North pole on the driven rotor, producing a completely symmetrical magnetic field. As the driving rotor rotates, the magnet poles start to overlap one another, leading to a push-pull effect and a consequent motion of the driven rotor. In such a way, torque is transmitted through the airgap without any magnetic flux leakage. The magnitude of the resultant coupling force depends on a number of parameters, including the rotating speed (*i.e.*, the amount of overlap in a unit time), characteristics of the magnetic material, and airgap between the PMs on the inner and outer rotors.

  The most significant advantage of magnetic coupling is that it allows the transmission of torque through the contaminate shroud, precluding the use of shaft seals and eliminating

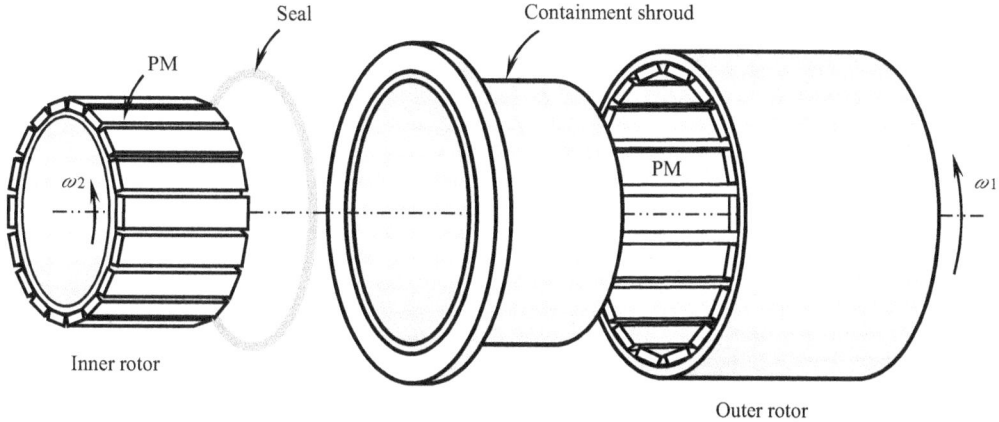

FIGURE 3.56 Exploded view of a magnetic coupling. PMs are attached to two concentric cylinders on either side of the stationary containment shroud. The outer rotor is attached to the motor shaft and the inner rotor is attached to the shaft of a driven machine (*e.g.*, a pump). The two rotors are isolated by a stationary containment shroud, avoiding the risk of leakage and allowing operation under pressure differentials.

the risk of fluid leakage. Therefore, magnetic couplings are especially suitable for applications where they are necessary to separate hermetically two medium zones to prevent liquid or gas leakage from one zone to another. For instance, in a pump system the pumped fluid is contained within a hermetically sealed containment shroud and isolated from motor components, achieving zero leakage. Torque is transferred through the containment shroud because of the coupled magnets of the outer and inner rotors. Other advantages of magnetic coupling include the removal of dynamic shaft seals, which have a finite lifetime and are prone to leakage, and the ease of maintenance.

The rotating magnets in a magnetic coupling can induce eddy current in the cylinders and the containment shroud if it is conductive, resulting in power loss and heating. To maximize coupling efficiency, it is highly desired to make the containment shroud with non-conductive materials, which may have a composite construction for improved chemical resistance and mechanical strength.

- Ferromagnetic materials exhibit a tendency to remain magnetized to some extent even after the external magnetic field is removed. This memory of magnetic history is referred to as magnetic hysteresis. To eliminate the residual magnetic field in a ferromagnetic material, an opposite magnetic field with a proper strength must be applied to the material. This property is useful in designing hysteresis couplings that engage and disengage during torque transmission operation. In fact, hysteresis materials are non-magnetized PM materials with high residual induction and low coercive field strength. They are easily magnetized and demagnetized, making them highly suitable for applications involving change magnetic fluxes.

As a hybrid of the magnetic and PM eddy-current couplings, the type of hysteresis coupling is typically used in an asynchronous operating state, but also can be utilized in a synchronous state in low torque transmission applications. When it exceeds its synchronous torque threshold, the coupling decouples from synchronous operation and smoothly switches to operate in an asynchronous state. In practice, hysteresis couplings are useful when an application needs to transmit a constant torque over a wide range of speed.

An axial-flux hysteresis coupling comprises of two discs: the hysteresis disc that is made of a non-magnetized hysteresis material; and the magnetic disc on which an array of magnetized magnets with alternating poles is attached (Figure 3.57). There is a small airgap

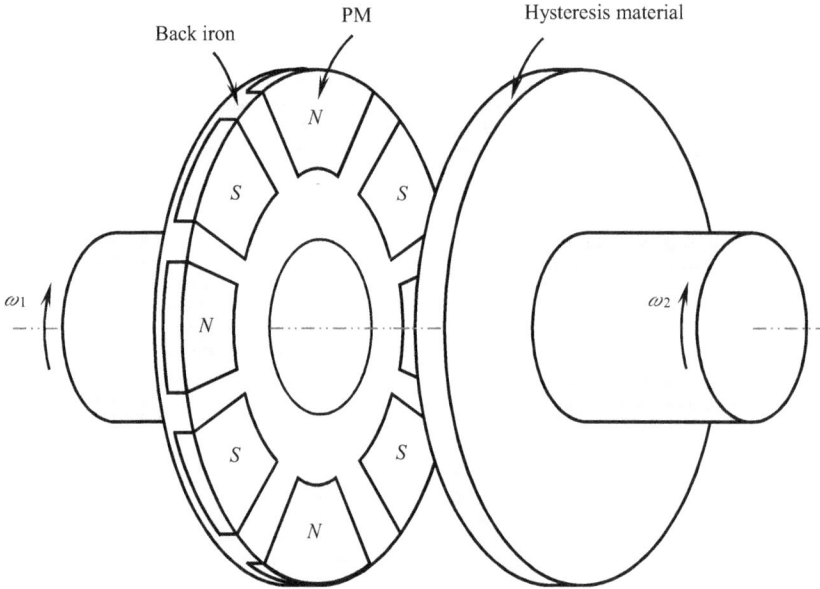

FIGURE 3.57 Configuration of an axial-flux hysteresis coupling. The hysteresis disc is opposed by the magnet disc, on which an array of alternating pole PMs is attached. The PMs are magnetized in the axial direction.

between the two discs. The magnitude of the transmitted torque can be easily adjusted by changing the airgap. When the coupling is at rest, the PMs magnetize the hysteresis disc, resulting in a synchronously coupled magnetic circuit. As the magnetic disc rotates, it causes the hysteresis material to go through its hysteresis (magnetization-demagnetization) loop. This cyclical progression around the hysteresis loop of ferromagnetic material repeatedly alters magnetic poles in the material. As a result, some of the magnetic energy is lost by continuously changing the orientation of material's domains. Due to the magnetic characteristics of ferromagnetic material (*e.g.*, ease of magnetization and demagnetization), hysteresis couplings are much less prone to Joule heating [3.62].

- The first wireless, non-radiative power transmission device was successfully tested in 2007 at MIT [3.63]. Using two self-resonant coils in a strongly coupled regime, the device could transfer 60 W power with approximately 40% efficiency over a distance in excess of 2 m. The two coils, the source coil and device coil, are aligned coaxially to achieve full coupling between them. This concept may be used to develop resonance inductive couplings for mechanical torque transmission. The coupling comprises a primary drive coil and a secondary receive coil. When the two coils are paired at the resonant frequency (usually ranged from 100 kHz to 10 MHz), the receive coil can pick up most of the energy stored in the drive coil. Thus, the phases of the magnetic fields of the drive and receive coils are synchronized [3.64]. It is noted that this type of coupling is still in the concept stage and there is a long way to get the coupling into practice.

3.12.4 OIL SHEAR COUPLINGS

Oil shear technology was developed by Force Control Industries in the 1960s. It has been successfully used in oil shear brakes, which transmit torque or power through shearing of oil molecules between friction discs. The viscous shearing of the oil dissipates the heat generated from operation of the brake and is carried away by the continuous circulation of the oil. This oil shear principle can also be used to design oil shear couplings.

As a semi-contact type of coupling, oil shear couplings offer advanced features, including the following:

- To maximize the life of components such as bearings and shafts, flexibility must be built in to absorb the residual misalignment. Increased flexibility of coupling in torsional direction improves fatigue life.
- Because of the lack of contact surfaces, it offers high tolerance of misalignment between the driving and driven shafts. Misalignment is accommodated by clearances between the moving components.
- It is possible to effectively compensate for shock loads, as there is no solid connection between input and output shafts.
- The difference in speed between the input and output shafts is known as slippage. Slippage occurs during an extreme overload condition. This coupling provides the overload protection to avoid damage of motor and connected equipment.

REFERENCES

3.1. AK Steel Corporation. 2013. 304/304L stainless steel. Product data bulletin. http://www.aksteel.com/pdf/markets_products/stainless/austenitic/304_304L_Data_Bulletin.pdf.

3.2. CRP Technology. Aluminum 7075-T6; 7075-T651. http://www.crptechnology.com/sito/images/PDF/Aluminum7075-T6-7075-T651.pdf.

3.3. Koch, S.-F., Peter, M., and Fleischer, J. 2017. Lightweight design and manufacturing of composites for high-performance electric motor. *Procedia CIRP* **66(2017)**: 283–288.

3.4. Polyzois, D. and Ungkurapinan, N. 2007. Design system for composite wind towers. World Intellectual Property Organization (WIPO) patent application WO 2007/012200 Al.

3.5. Alver, N., Tanarslan, H. M., and Tayfur, S. 2017. Monitoring fracture processes of CFRP-strengthened RC beam by acoustic emission. *Journal of Infrastructure Systems* **23(1)**: B4016002-1–B4016002-9.

3.6. Kablov, E. N., Kondrashov, S. V., and Yurkov, G. Y. 2013. Prospects of using carbonaceous nanoparticles in binders for polymer composites. *Nanotechnologies in Russia* **8(3–4)**: 163–185.

3.7. Im, Y. C., Kim, D. Y., Lim, S. W., Yoon, S. J., Choi, C. H., and Kim, M. H. 2021. Fatigue life prediction for Carbon-SMC and Carbon-FRP by considering elastic modulus degradation. *Journal of Composites Science* **5(2)**: 1–17. doi:10.3390/jcs5020054.

3.8. Feito, N., Muñoz-Sánchez, A., Díaz-Álvarez, A., and Loya, J. A. 2019. Analysis of the machinability of carbon fiber composite materials in function of tool wear and cutting parameters using the artificial neural network approach. *Materials* **12(17)**: Article number 2747.

3.9. PT Tech. Specialized clutches and brakes for heavy industry. http://www.pttech.com/images/company_assets/AA63FEEA-0BF3-4414-88El-FD86F132AE9C/PTTechSpecializedClutchesBrakesb320_4a3a.PDF.

3.10. Schuler, S., Kaufmann, V., Houghton, P., and Székely, G. S. (2006). Design and development of a joint for the Dextrous robot arm. *Proceedings of the 9th ESA Workshop on Advanced Space Technologies for Robotics and Automation*, Noordwijk, Netherlands.

3.11. Rostom, F. R. S. 2008. *Computer Analysis and Reinforced Concrete Design of Beams*. Fadzter Media. http://www.fadzter.com/engineering/downloads/Fadz_BeamDoc.pdf.

3.12. Richmond, E. L. and Feng, R. H. 1953. Graphical methods for determining beam deflections. *Machine Design*, September Issue, pp. 177–183.

3.13. Budynas, R. G. and Nisbett, J. K. 2008. *Shigley's Mechanical Engineering Design*, 8th edn. McGraw-Hill, Boston, MA, pp. 834–835.

3.14. Ugural, A. C. 1991. *Mechanics of Materials*. McGraw-Hill, New York.

3.15. Childs, P. R. N. 2019. Chapter 7: Shafts. In *Mechanical Design Engineering Handbook*, 2nd edn. Butterworth-Heinemann, Oxford, UK, pp. 295–375.

3.16. Norton, R. L. 2000. *Machine Design: An Integrated Approach*, 4th edn. Prentice-Hall, Boston, MA.

3.17. Central Machine Tool Institute. 1985. *Machine Tool Design Handbook*. Tata McGraw-Hill, New Delhi, India.

3.18. Ng, A. 2000. Better shafts for better linear motion performance. *Motion System Design*. http://machinedesign.com/linear-motion/better-shafts-better-linear-motion-performance.

3.19. Kosec, B., Karpe, B., Budak, I., Ličen, M., Đorđević, M., Nagode, A., and Kosec, G. 2012. Efficiency and quality of inductive heating and quenching planetary shafts. *Metalurgija* **51**(**2012**): 71–74.

3.20. Hauser, G., Curiel, G. J., Beilin, H.-W., Cnossen, H. J., Hofmann, J., Kastelein, J., Partington, E., Peltier, Y., and Timperley, A. W. 2004. *Hygienic Equipment Design Criteria*. Doc 8, 2nd edn. European Hygienic Engineering and Design Group (EHEDG), Frankfurt, Germany. http://www.ehedg.org/uploads/DOC_08_E_2004.pdf.

3.21. Hendricks, R. C., Braun, M. J., Canacci, V., and Mullen, R. L. 1991. Brush seals in vehicle tribology (Section IX). In Tribology Series *18: Vehicle Tribology* (Eds.: D. Dowson, C. M. Taylor, and M. Godet). Elsevier, Amsterdam, pp. 231–242.

3.22. Parker Hannifin Corporation. 2007. *Parker O-ring Handbook*. ORD 5700. http://www.parker.com/literature/ORD 5700Parker_0-Ring_Handbook.pdf.

3.23. R. L. Hudson & Company. 2008. *O-Ring Design & Material Guide*. http://www.rlhudson.com/O-RingBook/opening.html.

3.24. Simrit. 2011. *Radial Shaft Seal Technical Manual*. Publication no. 4100. http://www.simritna.com/news/brochures/RSSTechManual.pdf.

3.25. Tong, W. and Zhuo, Z. Q. 2002. Heat-resistant magnetic silicone rubber brush seal in turbomachinery and methods of application. U.S. Patent 6,390,476.

3.26. Parker Hannifin Corporation. 2011. *PTFE Seal Design Guide*. http://www.parker.com/literature/Packing/Packing-Literature/Catalog_PTFE-Seals_PDE3354-GB_1103.pdf.

3.27. Bal Seal Engineering. 2020. EMI shielding springs. https://www.balseal.com/spring/shhielding/.

3.28. Bode, R. 2012. Labyrinth seal and method for producing a labyrinth seal. U.S. Patent 9,746,085 B2.

3.29. Lattime, S. B., Dillon, K. M., Borowski, R., Tmtman, R., Brister, S. E., Meadows, C. F., Hupp, B. M., and Toth, D. G. 2013. Non-contact labyrinth seal assembly and method of construction thereof. U.S. Patent 8,342,535.

3.30. Benckert, H. and Wachter, J. 1980. Flow induced spring coefficient of labyrinth seals for applications in turbomachinery. NASA CP2133. Texas A&M University, College Station, TX, pp. 189–212.

3.31. Pilkey, W. D. and Pilkey, D. F. 2008. *Peterson's Stress Concentration Factors*, 3rd edn. John Wiley & Sons, Hoboken, NJ.

3.32. Ugural, A. C. 2004. *Mechanical Design: An Integrated Approach*. McGraw-Hill, Boston, MA.

3.33. Peterson, R. E. 1974. *Stress Concentration Factors*. John Wiley & Sons, New York.

3.34. Calistrat, M. M. and Calistrat, A. B. 1994. *Flexible Couplings: Their Design, Selection and Use*. Caroline Publishing, Houston, TX.

3.35. Filon, L. N. G. 1900. On the resistance to torsion of certain forms of shafting, with special reference to the effect of keyways. *Philosophical Transactions of the Royal Society of London, Series A* **193**: 309–352.

3.36. Peterson, R. E. 1933. Stress concentration phenomena in fatigue of metals. *Journal of Applied Mechanics* **1**(**4**): 157–171.

3.37. Hetényi, M. 1939. The application of hardening resins in three-dimensional photoelastic studies. *Journal of Applied Physics* **10**(**5**): 295–300.

3.38. Leven, M. M. 1949. Stresses in keyways by photoelastic methods and comparison with numerical solution. *Proceedings of the Society for Experimental Stress Analysis* **7**(**2**): 141–154.

3.39. Peterson, R. E. 1953. *Stress Concentration Design Factors*. John Wiley & Sons, New York.

3.40. Nisida, M. 1963. New photoelastic methods for torsion problems. *Proceedings of International Symposium on Photoelasticity* (Ed.: M. M. Fracht). Pergamon, New York, pp. 109–121.

3.41. Okubo, H., Hosono, K., and Sakaki, K. 1968. The stress concentration in keyways when torque is transmitted through keys. *Experimental Mechanics* **8**(**8**): 375–380.

3.42. Fessier, H., Rogers, C. C., and Staley, P. 1969. Stresses at end-milled keyways in plain shafts subjected to tension, bending and torsion. *The Journal of Strain Analysis for Engineering Design* **4**(**3**): 180–189.

3.43. Orthwein, W. C. 1975. Keyway stress when torsional loading is applied by the keys. *Experimental Mechanics* **15**(**6**): 245–248.

3.44. Pedersen, N. L. 2010. Stress concentration in keyways and optimization of keyway design. *The Journal of Strain Analysis for Engineering Design* **45**(**8**): 593–604.

3.45. Eliaz, N., Shachar, A., Tal, B., and Eliezer, D. 2002. Characteristics of hydrogen embrittlement, stress corrosion cracking and tempered martensite embrittlement in high-strength steels. *Engineering Failure Analysis* **9**(**2002**): 167–184.

3.46. Okubo, H. 1950. Torsion on a circular shaft with a number of longitudinal notches. *Transactions of ASME, Applied Mechanics Section* **72**: 359.

3.47. Yoshitake, H. 1962. Photoelastic stress analysis of the spline shaft. *Bulletin of Japanese Society of Mechanical Engineering* **5(17)**: 195–201.

3.48. Keesee, J. 2010. Shaft taper analysis—Peer review. Kollmorgen Corporation.

3.49. Felix, C. 2008. Endworking enhances shaft manufacturing process. *Production Machining Magazine*, September Issue. http://www.productionmachining.com/articles/endworking-enhances-shaft-manufacturing-process.

3.50. Chilson, T. R., Khetawat, M. P., and Pale, J. A. 2011. Method of producing a stepped shaft. U.S. Patent 7,866,198.

3.51. ISO/TR 17671-2:2002 (EN 1011-2). 2002. Welding—Recommendations for welding of metallic materials—Part 2: Welding of ferritic steels.

3.52. Yurioka, N. 2004. Comparison of preheat predictive methods (Report III). IIW Doc. IX-2135-04.

3.53. Dearden, J. and O'Neill, H. 1940. A guide to the selection and welding of low-alloy structural steels. *Transactions of the Institute of Welding (UK)* **3**: 203–214.

3.54. Piotrowski, J. 1995. *Shaft Alignment Handbook*, 2nd edn. CRC Press, Boca Raton, FL.

3.55. Michalicka, P. 2010. Correcting shaft misalignment: Three steps to smooth production. *Engineering and Mining Journal* **211(7)**: 82–84.

3.56. Zilberman, J., Munyon, R. E. and Meier, W. R. 2003. Split spool type flexible coupling. U.S. Patent 6,508,714.

3.57. Noria Corporation. 2002. Coupling lubrication and maintenance requirements. *Machinery Lubrication*, November issue.

3.58. Thomas Industry. Types of shaft couplings – A Thomas buying guide. https://www.thomasnet.com/articles/hardware/coupling-types/.

3.59. Yang, X. W., Liu, Y. G., and Wang, L. 2019 Nonlinear modeling of transmission performance for permanent magnet eddy current coupler. *Mathematical Problems in Engineering* **2019**: Article ID 2098725.

3.60. Michael, P. C., Hensley, S. L., Galea, C. A., Chen, E., Karmaker, H., Bromberg, L. 2016. Noncontact high-torque magnetic coupler for superconducting rotating machines. *IEEE Transactions on Applied Superconductivity* **26(4)**: 1487–1491.

3.61. Zheng, D. and Guo, X. F. 2020. Analytical prediction and analysis of electromagnetic-thermal fields in PM eddy current couplings with injected harmonics into magnet shape for torque improvement. *IEEE Access* **8**: 60052–60061.

3.62. Dexter Magnetic Technologies. 2020. Couplings. https://dextermag.com/products/magnetic-assemblies/couplings/.

3.63. Kurs, A., Karalis, A. Moffatt, R., Joannopoulos, J. D., Fisher, P., and Soljačić, M. 2007. Wireless power transfer via strongly coupled magnetic resonances. *Science* **317(6)**: 83–86.

3.64. Transistor Technology Special Editorial Department. 2017. How to Make a Cordless Power Supply that is Resistant to Water and Dust (in Japanese). *Green Electronics No. 19*. CQ Press, Thousand Oaks, CA.

4
Stator Design

The stator is usually the outer body (or the inner body in some applications) of an electric motor to house the stator windings on a laminated steel core for creating a rotating magnetic field. The stator core is made up of a stack of prepunched insulated steel laminations assembled into a motor housing that is made of aluminum or cast iron. The thickness of the laminations and the type of electrical steel, as two of the key factors in the stator design, are chosen to minimize the eddy current and hysteresis losses. To receive stator windings, the inner surface of the stator is made up of a number of deep slots or grooves, distributed either uniformly or nonuniformly on the circumference of the stator. The arrangement of the windings within the stator slots depends on the number of motor poles, the number of phases, and the number of slots. In practice, there are several types of windings commonly adapted by motor manufacturers, such as concentric winding, lap winding, spit winding, and wave winding. The two common winding configurations used for three-phase motors are Δ and Y.

4.1 STATOR LAMINATION

The considerations in lamination design include the selection of the lamination material, optimization of the lamination stamping/cutting process, and determination of the lamination profile and pattern, dimension tolerances, core losses, grain orientation, unique shape requirements, and, of course, lamination manufacturing costs.

4.1.1 STATOR LAMINATION MATERIAL

In practice, the base lamination material used for stators is fundamentally the same for rotors when it is desirable to punch the stator and rotor laminations simultaneously with the same die. However, some differences may exist between these two applications: (a) Because the main function of a stator is to generate a rotating magnetic field, the lamination material of the stator has higher requirements for electromagnetic properties. (b) Unlike a rotor that is subject to large centrifugal forces as it rotates, the forces acting on the stator are much lower, leading to lower requirements for mechanical strength properties. Based on these reasons, segmented cores are extensively used in stators for all types of motors but are restricted from use in high-speed rotors due to high centrifugal forces.

4.1.2 STATOR LAMINATION PATTERNS

One of the factors that affect the efficiency of electric motors is the ability of steel laminations used in the stator to carry magnetic flux. There are several lamination patterns available for either enhancing the magnetic flux or simplifying the manufacturing processes, or both, depending on different applications.

4.1.2.1 One-Piece Lamination

Stators and rotors of electric machines are usually manufactured using a stack of one-piece laminations, made by punching the desired pattern from large, insulated sheets of steel. One-piece lamination is fabricated from an undivided piece of steel sheet and continuous in the 360° circumference, as shown in Figure 4.1. Obviously, the one-piece lamination method offers the advantage of

DOI: 10.1201/9781003097716-4

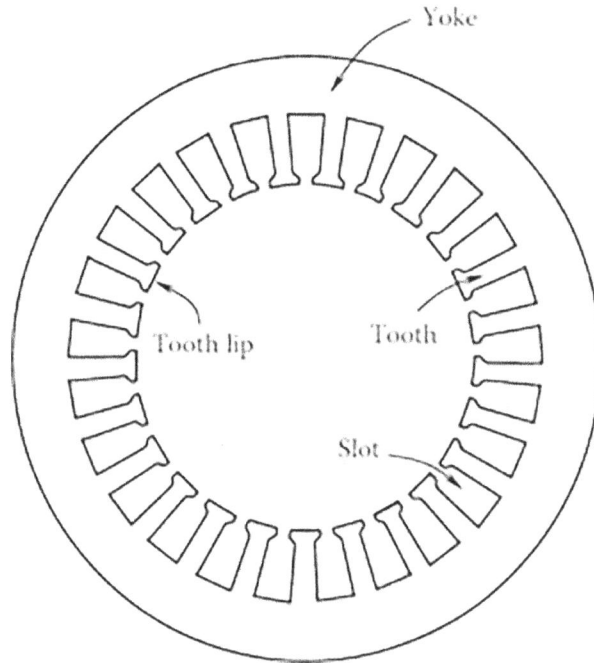

FIGURE 4.1 One-piece lamination is made by an undivided piece of steel sheet in full revolution. Stator slots are for receiving stator windings.

fabrication and assembly simplicity. Multiple pieces are always used over the stator circumference and will be discussed in the following sections.

As illustrated in the figure, a one-piece lamination contains a number of teeth facing inward and connecting with the yoke. Slots are formed between adjacent teeth for receiving stator windings. The tooth tips, as well as the insulation sleeves or wedges, help keep the stator windings inside the slots. Tooth tips also have many electromagnetic functions.

4.1.2.2 T-Shaped Segmented Lamination

Segmented stator laminations are frequently encountered in large-size motors (Figure 4.2) wherein stamping one-piece laminations is no longer practical, either due to the limitation of the capability of stamping machines or due to the incredibly high costs. For such cases, segmented laminations are the best choice for the stator core fabrication.

The recent trend toward using segmented laminations in small motors is a means to not only increase the slot-filling rate and facilitate automated fabrication of electric motors but also increase the continuous torque (well-designed segmented lamination stators may increase motor torque by up to 50% over conventional motors). The segmented lamination technology is probably best suited to electric machines where the size of the effective airgap is insensitive to small variations, as in permanent magnet (PM) machines. It has been developed for high power density, cost-effective modern motors with better heat transfer within small overall packages, offering some benefits over traditional one-piece lamination stator cores [4.1, 4.2]:

- The segmented design lowers the requirements for stamping machines and thus reduces both the machine and tooling costs. However, handling and assembly costs of the larger number of pieces to assemble increase or require automation tooling.
- The adoption of the segmented lamination technology can make full use of the lamination material for significantly reducing the material consumption and material cost.

FIGURE 4.2 A segmented stator lamination containing a number of teeth for a large-size motor. The wide radial slots are used for receiving stator windings and the narrow radial slots for reducing eddy-current losses.

- Because each lamination section can be machine wound it allows placing more copper wires in stator slots. Generally, segmented lamination stators allow for 20%–30% higher slot-filling rate. This makes it able to use a larger diameter copper wire to fabricate the stator windings, thereby reducing the *PR* power loss.
- The segmented type of stator may significantly increase in continuous motor torque over conventional integral type stator.
- The segmented lamination technology and single-tooth windings enable a motor to have significantly shortened end turns. This feature allows the motor to be much shorter, compared with the long end turns found in conventionally random-wound servomotors.
- In manufacturing segmented stator coils, there are more choices in coil insertion methods. Many innovative ideas have been awarded patents over the past decade.
- Eddy-current loss is lowered due to the restriction of eddy-current paths. Thus, it could increase the motor efficiency. However, there are increases in electromagnetic ampere-turn drops due to the increase in mechanical gaps that can also reduce the total performance gain.
- T-shaped or I-shaped segmented laminations allow forming stator cores with different diameters to greatly save the tooling costs on very large motors. Again, one has to factor in the increased part handling and possible automation costs if employed on small motors.
- By taking advantages of the permeability of a grain-oriented material, the motor performance is correspondingly enhanced.

However, the adoption of the segmented core structure in electric machines comes with a penalty of increased core losses at the segment joints due to several factors, such as the degraded material conditions, the increased number of punched edges with induced residual stress, and the effect of compressive stress applied on segmented stator cores. Furthermore, a potential cause of increased core losses may result from the increase in eddy-current losses at the edge-to-edge butt joints [4.1]. Punching electrical steel drastically alters its magnetic properties near the cutting edges, resulting in the increase in core losses in two different ways [4.3]:

- In a direct way, the cutting process creates residual stresses in the laminations at the vicinities of the cut edges, which can significantly increase the hysteresis losses and hence the total core losses at these locations.
- In an indirect way, cutting can alter the magnetization profile inside the lamination. Due to the permeability drop, the polarization obtained with a given excitation is significantly reduced at the cut edge. Hence, in order to have the same flux across the sample (*i.e.*, the same average polarization), a higher polarization in the bulk of the lamination is required. This higher polarization induces higher total losses.

Moses *et al.* [4.4] found that for a T-joint transformer core, the inner edges of the yoke and limb laminations are the highest power loss areas, up to 30% higher than the mean core loss. The third circulating harmonics causes additional losses calculated to be 20% of the nominal core loss [4.5].

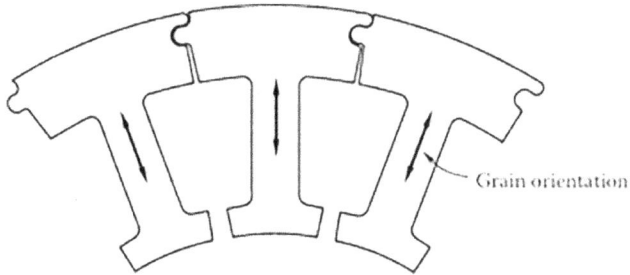

FIGURE 4.3 Segmented stator laminations, jointed with interlocking convex and concave portions.

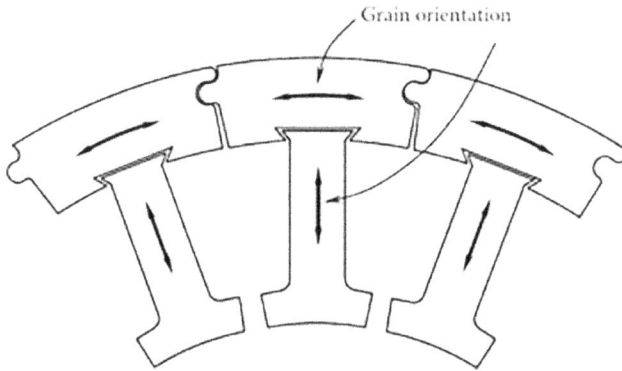

FIGURE 4.4 Grain-oriented segmented laminations.

Segmented laminations are made by punching one or several lamination teeth as an individually segmented piece. To save the tooling cost, all segmented laminations are usually designed to have an identical shape. As illustrated in Figure 4.3, T-shaped segmented laminations are jointed together by interlocking convex and concave portions to form a complete piece of a lamination. As all segmented lamination pieces are put into positions, the geometric pattern and magnetic circuit are approximately the same as for one-piece laminations. To enhance the motor performance, the grain orientation is arranged in the radial direction on all T-shaped segmented laminations.

A grain-oriented material has a superior performance in respect of less saturation and lower losses in the direction of the grain orientation. For this material, both core loss and permeability vary markedly, depending on the direction of the magnetic flux relative to the material rolling direction. To further take the advantage of the permeability and core loss of the grain-oriented material, it is to split the T-shaped segmented lamination into two pieces, as shown in Figure 4.4. Thus, by arranging the grain orientation in the radial direction on the tooth piece and in the circumstantial direction on the yoke piece, the stator performance is maximized.

A large number of patents are available on the segmented lamination design. As the design in Figure 4.3, an innovative design that was awarded as a US patent [4.6] is presented in Figure 4.5. The lamination pieces are secured by means of the curved arc mating surfaces. A patented design is illustrated in Figure 4.6 [4.7], where both teeth and yoke are made by segmented pieces. In Figure 4.7, the stator core consists of segmented large and small components, arranged alternately. Each stator slot is formed between a large and a small component, forming nonuniform slots in the circumference of the stator. It has been reported [4.8] that the periodicity of the cogging torque waveform for the motor with nonuniformly distributed slots is half that for the motor with uniformly distributed slots. However, the amplitude of the cogging torque is increased significantly because the tooth tips of the wider teeth are approximately equal to the pole pitch in order to maximize the flux linkage per coil.

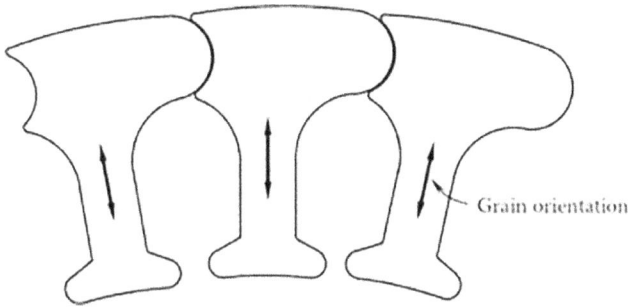

FIGURE 4.5 Patented segmented stator laminations with curved arc mating surfaces (U.S. Patent 5,212,419) [4.6]. (Courtesy of the U.S. Patent and Trademark Office, Alexandria, VA.)

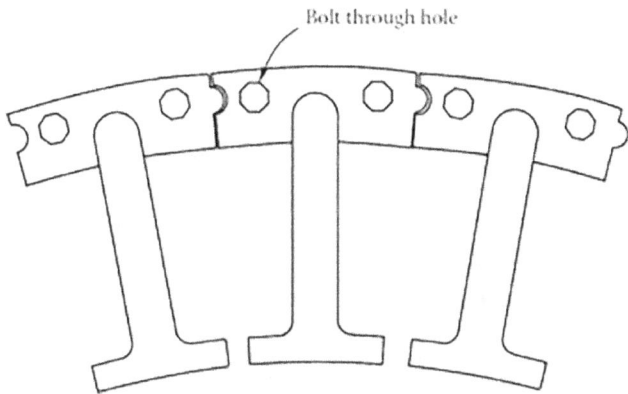

FIGURE 4.6 Patented segmented stator laminations in which both the tooth and the yoke are segmented (U.S. Patent 7,816,830) [4.7]. (Courtesy of the U.S. Patent and Trademark Office, Alexandria, VA.)

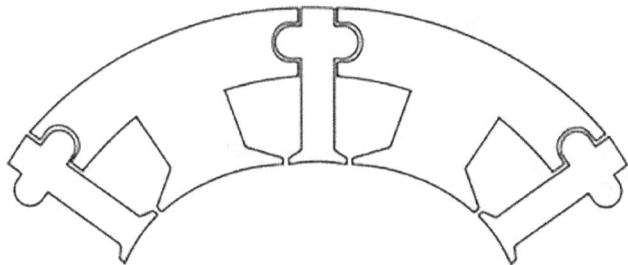

FIGURE 4.7 Alternative segmented stator laminations, showing each stator slot is formed between a large and a small segmented piece.

4.1.2.3 Connected Segmented Lamination

In order to increase the productivity of segmented laminations, the manufacturing process shown in Figure 4.3 can be improved. In this new process, instead of stamping individual segmented laminations, a long string of lamination pieces may be continuously stamped. These lamination pieces are connected at the lamination outer tips (Figure 4.8). Then, these pieces are easy to roll together to form a complete lamination.

An alternative design of punched segmented laminations that connected at the outer tips of the laminations is shown in Figure 4.9. The laminations are fabricated by stamping steel sheets in a

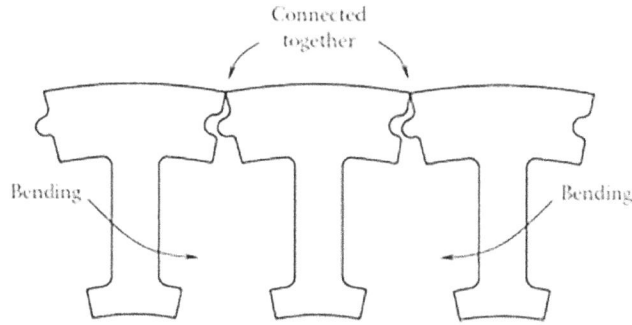

FIGURE 4.8   Punched segmented laminations are connected at the outer tips of the lamination.

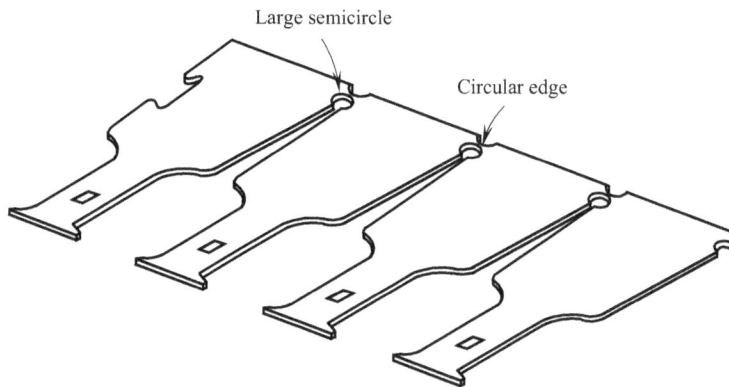

FIGURE 4.9   An alternative design of punched segmented laminations that connected at the outer tips of the lamination (U.S. Patent 7,062,841) [4.9]. (Courtesy of the U.S. Patent and Trademark Office, Alexandria, VA.)

progressive die. To reduce the bending force, circular edges and large semicircles are arranged at the tips of each lamination [4.9].

4.1.2.4   Two-Section Stator Lamination

In this design, a stator lamination is integrated by two lamination pieces: a section of teeth and a section of yoke (Figure 4.10). The stator windings are wound on the section of teeth. When all windings get in their positions, the section of yoke covers at the outer surface of the section of teeth, for housing the stator windings [4.10]. The section of teeth is positioned in the concave slots that are made at the yoke inner surface.

For this design, the use of non-oriented materials, in which the magnetic properties are primarily the same in all directions, is more appropriate.

4.1.2.5   Stator Lamination Integrated by Individual Teeth and a Yoke Section

An alternative design of a segmented laminated core is presented in Figure 4.11. It can be seen that all teeth are no longer connected together. Each tooth is individually positioned in the concave slot made at the inner surface of the yoke. Thus, this design can use the grain-oriented material to make the teeth, arranging the grain orientation in the radial direction. It should be noted that the yoke in Figure 4.11 can be made either with a whole piece or segmented pieces. The benefit of a segmented yoke is that it can arrange the grain orientation in the circumferential direction to optimize motor performance, as shown in Figure 4.12.

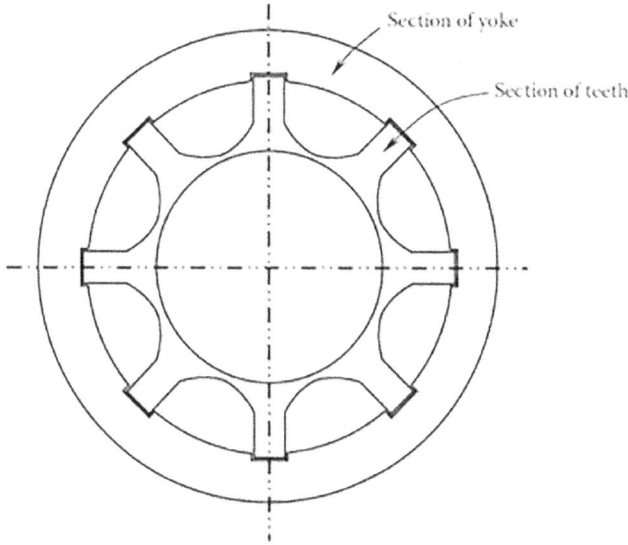

FIGURE 4.10 Integrated stator core with a section of teeth and a section of yoke; the section of teeth is positioned in the concave slots at the yoke inner surface.

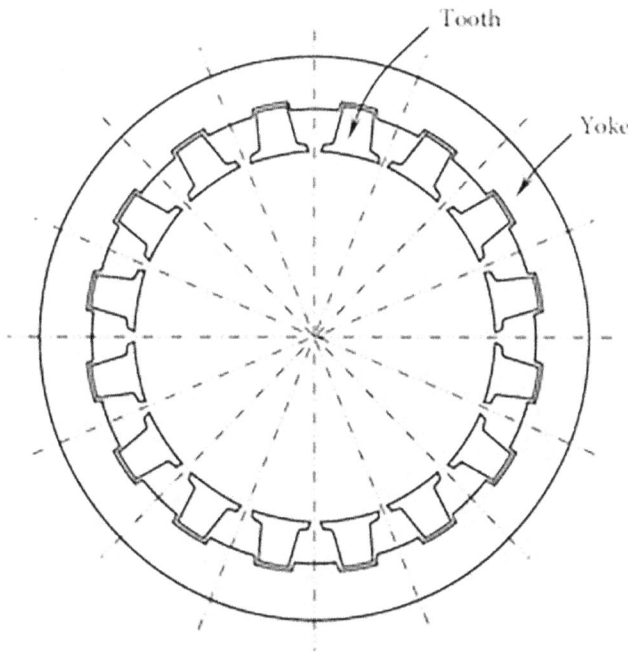

FIGURE 4.11 Integrated stator core with individual teeth and yoke; the teeth are positioned in the concave slots on the inner surface of the yoke.

The segmented lamination design in Figure 4.13 is similar to that in Figure 4.11 but in the opposite orientation of the yoke and the teeth. As a novel PM motor, the teeth are put outer forward and the yoke is situated in the inner of the stator core [4.11]. In the assembly process, the sheets for a tooth are glued together to form a component, which are then bonded onto the stator yoke (Figure 4.14). Rectangular-shaped concentrated fractional-pitch windings are inserted in the stator slots.

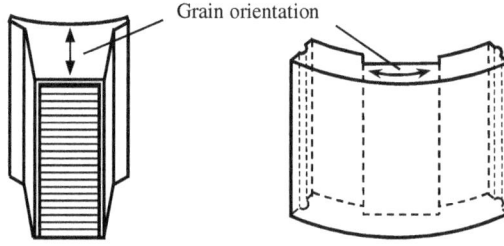

FIGURE 4.12 Utilization of grain-oriented electrical steel laminations to fabricate segmented stator core (both the yoke and teeth) for optimizing motor performance.

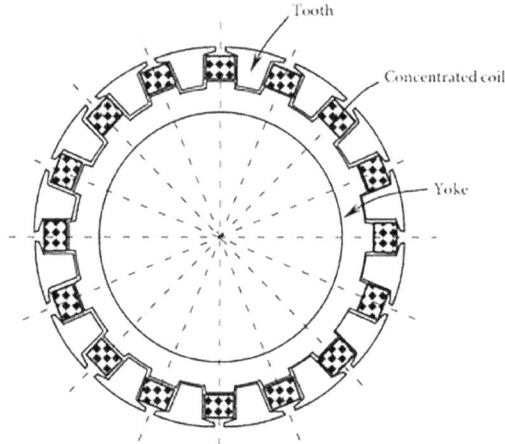

FIGURE 4.13 Integrated stator core with teeth and an internal yoke section, where the teeth are positioned in the concave slots on the outer surface of the yoke. As the coils are placed around the yoke and connected in a certain pattern to form the concentrated winding, the stacked teeth are glued onto the yoke to complete the concentrated stator winding.

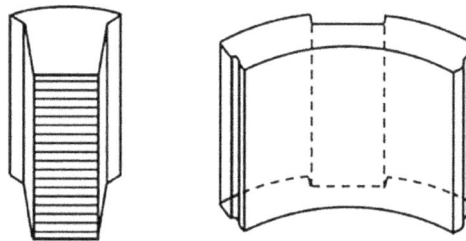

FIGURE 4.14 The sheets of teeth are held together as a subsection and then integrated with the yoke laminations.

4.1.2.6 Slotless Stator Core

In brushless PM machines, cogging torque is generated from the interaction between the PMs on the rotor and the stator slots. Cogging torque can result in torque/speed ripples, rough running, vibration, and acoustic noise. Motor engineers always struggle with reducing cogging torque. As a result,

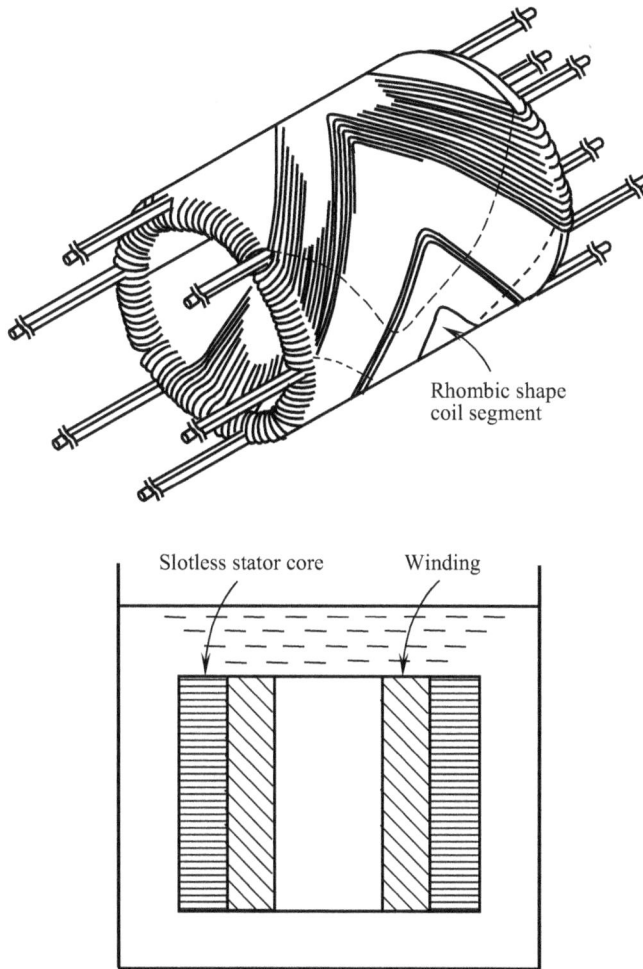

FIGURE 4.15   Fabrication of slotless stator (U.S. Patent 6,525,437) [4.12]. (Courtesy of the U.S. Patent and Trademark Office, Alexandria, VA.)

slotless stator designs have emerged as a solution to zero cogging in conventional PM motors. The slotless stators eliminate the cogging torque, simplify the lamination process, and smooth the motor performance. In this type of motor, stator wires are wound into a cylindrical shape and assembled inside the stator core in a radial inward compressed condition [4.12]. Since the stator winding effectively functions as a spring pressure at the internal wall of the stator core, it is mechanically secured to the inner periphery of the stator core and a special fixation mechanism is not required. Then, the stator assembly is immersed in high-temperature epoxy resins and finally cured in an oven. Thus, the stator winding maintains its orientation with respect to the stator laminations and housing assembly (Figure 4.15).

However, the slotless winding requires increasing the airgap and, in turn, reduces the PM excitation field. To maintain the same airgap flux density, the height of the PM must be increased. Therefore, slotless PM motors use more PM materials than slotted motors [4.13]. The cross-sectional view of the slotless stator is shown in Figure 4.16. Like all technologies, there are pros and cons to the differences.

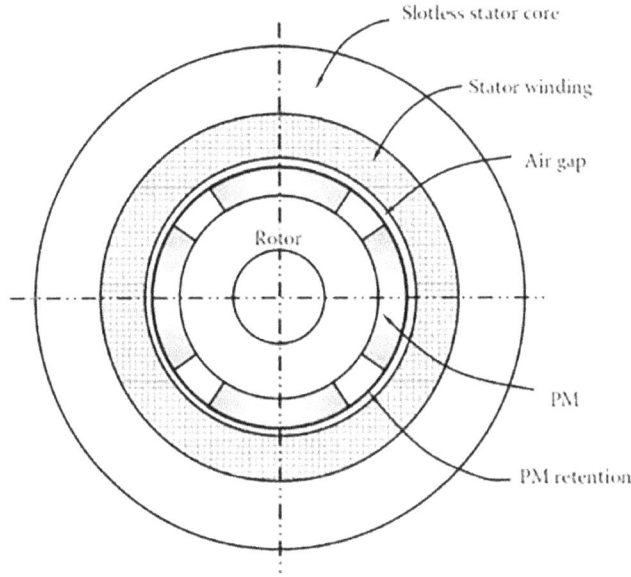

FIGURE 4.16   Structure of a four-pole slotless PM motor.

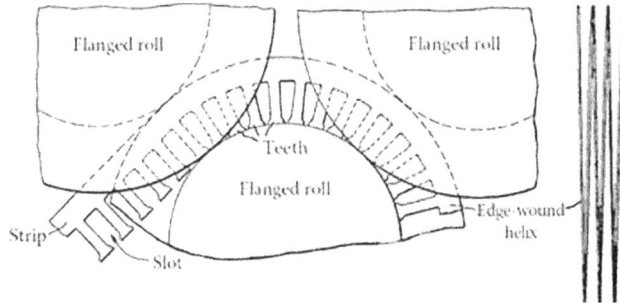

FIGURE 4.17   Fabrication of helical stator core with three flanged rolls (U.S. Patent 1,920,354) [4.11]. (Courtesy of the U.S. Patent and Trademark Office, Alexandria, VA.)

4.1.2.7   Slinky Lamination Stator Core

Using new methods for fabricating stator cores is highly desired for both cost reduction and manufacturing efficiency. One such design is known as the slinky method [4.14, 4.15]. With this method, a stator core is built up from a continuous slotted strip of silicon steel rather than cross-sectional laminations in a conventional manufacturing process (Figure 4.17). The strip is wound edgewise in a helical configuration by a coiling machine that consists of three flanged rolls. By making the stacked strip to the desired thickness, the rolled core is compressed longitudinally and welded at the outer circumferential surface of the core, avoiding noise caused by vibration. The advantages of this manufacturing method include the following: (*a*) Using the slinky design provides the induced magnetic field with a preferable grain orientation in the rotor rotation, which would subsequently boost the overall efficiency of the motor. (*b*) It reduces the iron losses, noise generation, and power draw as compared to the conventional process. (*c*) This method can lead to great savings in material and winding labor.

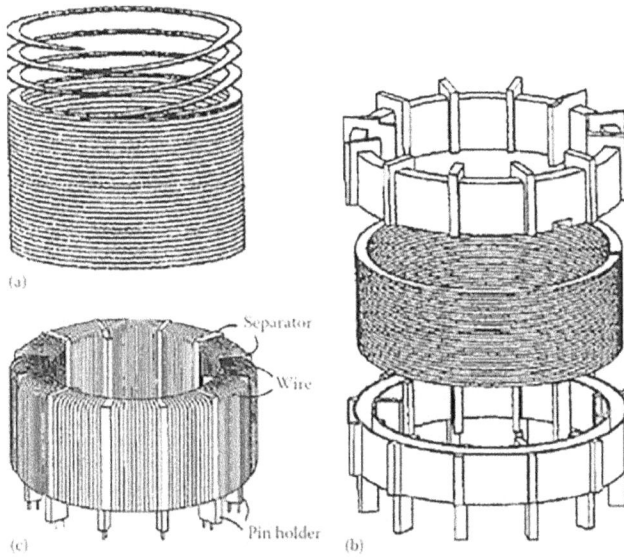

FIGURE 4.18 Using the slinky method in designing slotless stator: (*a*) fabricating a helical stator core, (*b*) adding a molded plastic annular insulator at each end of the stator core, and (*c*) making the stator winding over the insulators (U.S. Patent Application 2005/0073210) [4.15]. (Courtesy of the U.S. Patent and Trademark Office, Alexandria, VA.)

This technique can be used in a slotless stator design. With the additional elimination of the teeth on the steel back iron, cogging of the motor is eliminated. It also simplifies the manufacturing process, as shown in Figure 4.18 [4.15].

To gain full advantages of core loss reductions from a laminated core, an adequate surface insulation must be maintained between the laminations. Stator laminations are usually coated with inorganic nonconductive materials as the insulation. The selection of coating materials is based on material properties especially electrical resistance, operation temperature, cost, and processing time. Furthermore, the application of a suitable coating can promote a better stamping performance such as low burr and long tool life.

4.2 MAGNET WIRE

The most common materials used for stator windings are copper and aluminum for their excellent electrical properties, especially high electrical conductivity. By comparing with aluminum, copper has better electrical properties but a higher melting temperature. At the temperature of 20°C, the conductivity of copper is 70% higher than that of aluminum. Though silver has the highest electrical conductivity (its conductivity is 5.7% higher than that of copper at 20°C) and the highest thermal conductivity of any metal, its high cost and tarnishability have prevented it from being used for motor windings. Magnet wires are widely used in small and medium electric motors for fabricating stator windings. For heavy-duty large-size electric motors, electric conducting cables are more commonly used to make stator windings.

Magnet wire is coated with insulation materials (such as enamels) to allow winding wires to contact each other without forming electricity short circuits between them. The selection of the magnet wire is based on the motor operation conditions (working frequency, current, voltage, temperature, *etc.*), wire material properties, wire diameter for allowing maximum current density, thermal class, coating material, thickness of coating layer, and others. Magnet wire received its name undoubtedly from its use to make electromagnets.

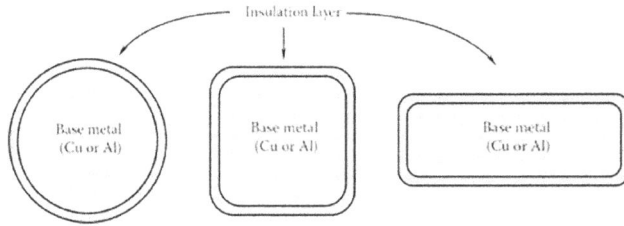

FIGURE 4.19   Modern magnet wires with different shapes.

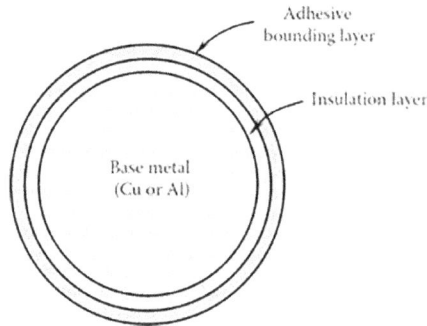

FIGURE 4.20   Self-adhesive magnet wire: an adhesive bounding layer is added to the magnet wire.

4.2.1   REGULAR MAGNET WIRE

Regular magnet wire consists of a base metal, commonly copper or aluminum, and coated one- or multilayer of insulation materials, such as enamel, fibrous polyester, fiberglass yarn, and polyamide. For multilayer coatings, it is often necessary to use different compositions to achieve the most optimal insulation result. Several cross-sectional shapes of magnetic wires are available in stator windings, including round, square, and rectangular (Figure 4.19). The use of square wires can significantly increase the slot-filling rate. This is especially useful where space constraints are concerned.

4.2.2   SELF-ADHESIVE MAGNET WIRE

As shown in Figure 4.20, a self-adhesive magnet wire has a thermoplastic adhesive film superimposed over standard film insulation. When activated by heat or solvent, the bond coating cements the winding turn-to-turn to create a self-supporting coil. This type of wire has opened up new avenues for some special applications, especially where regular magnetic wires are not suitable. For instance, bobbin-free deflection yoke coils in television sets can be conveniently made by self-adhesive wires. With a coated thermoplastic cement on the outer surface of the wire, a strong turn-to-turn bond throughout a winding can be achieved, and thus the need for varnish impregnation can be eliminated.

The adhesive can be activated by different ways. An easy way is to allow the wire to pass through a solvent while winding. As the solvent is evaporated, all turns are bonded together as the whole coil. Another effective way is called resistance heating, which is accomplished by passing the electric current through the coil. Due to the resistance of the coil itself, heat is generated in the coil, and

TABLE 4.1
Dimensions of Self-Adhesive Magnet Wires (Bondeze®M)

AWG Size	Nominal Bare Wire Diameter, mm (in.)	Minimum Insulation, mm (in.)	Minimum Cement, mm (in.)	Maximum OD, mm (in.)
15	1.4503 (0.0571)	0.0406 (0.0016)	0.0203 (0.0008)	1.5392 (0.0606)
16	1.2903 (0.0508)	0.0381 (0.0015)	0.0203 (0.0008)	1.3741 (0.0541)
17	1.1506 (0.0453)	0.0381 (0.0015)	0.0203 (0.0008)	1.2344 (0.0486)
18	1.0236 (0.0403)	0.0356 (0.0014)	0.0203 (0.0008)	1.1506 (0.0453)
19	0.9119 (0.0359)	0.0356 (0.0014)	0.0203 (0.0008)	0.9931 (0.0391)
20	0.8128 (0.0320)	0.0330 (0.0013)	0.0178 (0.0007)	0.8865 (0.0349)
21	0.7239 (0.0285)	0.0330 (0.0013)	0.0178 (0.0007)	0.7976 (0.0314)
22	0.6426 (0.0253)	0.0305 (0.0012)	0.0178 (0.0007)	0.7137 (0.0281)
23	0.5740 (0.0226)	0.0279 (0.0011)	0.0152 (0.0006)	0.6401 (0.0252)
24	0.5105 (0.0201)	0.0279 (0.0011)	0.0152 (0.0006)	0.5740 (0.0226)
25	0.4547 (0.0179)	0.0254 (0.0010)	0.0152 (0.0006)	0.5156 (0.0203)
26	0.4039 (0.0159)	0.0254 (0.0010)	0.0127 (0.0005)	0.4623 (0.0182)
27	0.3607 (0.0142)	0.0229 (0.0009)	0.0127 (0.0005)	0.4140 (0.0163)
28	0.3200 (0.0126)	0.0203 (0.0008)	0.0127 (0.0005)	0.3708 (0.0146)
29	0.2870 (0.0113)	0.0203 (0.0008)	0.0102 (0.0004)	0.3353 (0.0132)
30	0.2540 (0.0100)	0.0178 (0.0007)	0.0102 (0.0004)	0.2997 (0.0118)

the desired temperature can be reached by controlling the heating time. When the desired temperature of the coil is reached, the coil is compressed so that the wires contact closely and thus bounded together. In fact, high resoftening temperature of cement allows the self-adhesive wires to compare with varnish-impregnated heavy-grade magnet wires.

As an example, the dimensions of a self-adhesive wire are presented in Table 4.1. It can be seen from the table that the thickness of both the insulation layer and the cement layer is proportional to the bare wire diameter. The thickness of the cement layer is about half of that of the insulation layer. This type of wire can withstand 180°C, which is well suited for electric motors.

The optimum bonding of the wires is reached when the coil temperature is raised to the range of 220°C–240°C, measured by the change in the coil electrical resistance.

4.2.3 LITZ WIRE

Litz wire is basically used for high-frequency applications. It contains many thin wire strands that are individually insulated and twisted together. Litz wire utilizes the full cross-sectional area of the wire to carry current. Because each individual strand is thinner than the skin depth, the use of Litz wire can greatly minimize the skin effect and proximity effect losses in high-frequency windings.

The size of a Litz wire is often expressed in the abbreviated format of N/XX, where N is the number of strands and XX is the AWG size of each strand. The twist per inch or meter is also an option to specify.

4.3 STATOR INSULATION

As magnet wires are wound into stator slots to form stator windings, electrical insulation is required wherever there is a difference in electric potential between two electric conductors to prevent the conductors from short circuiting to one another, including (a) one turn of a coil to adjacent turns,

FIGURE 4.21   The damage of stator winding due to the failure of electric insulation.

(*b*) coils to adjacent coils, (*c*) stator windings to the ground (*i.e.*, the stator core), and (*d*) one phase winding to adjacent phase windings.

The insulation of stators can be done by various methods, for example, using aramid or mylar layered paper, fluidized-bed powder coating, or overmolding with thermoplastic materials. For low- or moderate-voltage stators, the magnet wire insulating varnish can usually satisfy the requirements of turn-to-turn insulation. However, as the stator voltage increases, an additional insulation becomes necessary. For instance, wrapping an insulating tape around the conducting wire can effectively enhance the dielectric strength. Similarly, for low-voltage stators, coil-to-coil insulation is provided by the wire insulation to avoid the risk of short circuiting in various series or parallel connected coils. The practical method for winding-to-ground insulation is usually to employ slot liners and slot wedges at the bottom and at the top of stator slots, respectively, to separate the stator windings to the adjacent lamination surfaces for preventing any part of the stator windings from shorting to the stator core. These are available in many shapes, sizes, and materials. Phase-to-phase insulation can be achieved by completely insulating the winding of one phase from those of adjacent phases. Failure of electric insulation can lead to the damage of the stator winding (Figure 4.21), causing permanent failure of the electric motor.

4.3.1   INJECTION MOLDED PLASTIC INSULATION

Injection molded plastic insulation provides highly reliable insulating properties to laminated stator stacks. The use of molded plastic insulation prior to winding assures consistent and durable insulation of wound winding to the stator cores. Figure 4.22 is an example of the molded plastic insulation, which consists of a main body and an end cap. Using the molded plastic insulation can offer proven engineering advantages in the stator assembly, including consistent insulation thickness, reliable and long-lasting insulation, and reduced labor time. The use of such insulation in small motors is presented in Figure 4.23.

A specially designed molded plastic insulator is applied at the motor end, providing the insulation for lead wires, as shown in Figure 4.24.

4.3.2   SLOT LINER

As shown in Figure 4.25, heat-resistant and mechanically stable insulation papers or thermoplastic materials are inserted in stator slots for preventing coils from shorting to the stator core. The typical

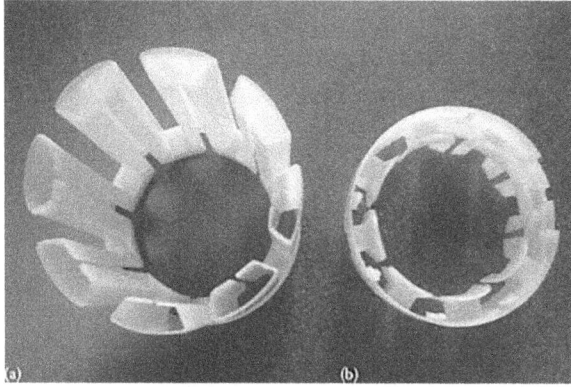

FIGURE 4.22 Molded plastic insulator: (*a*) main body; (*b*) end cap.

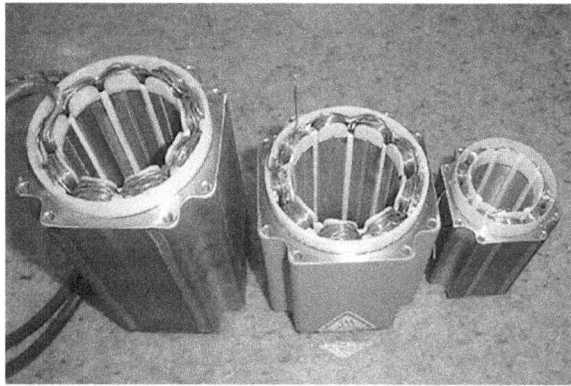

FIGURE 4.23 A variety of molded plastic insulators applied to stator windings.

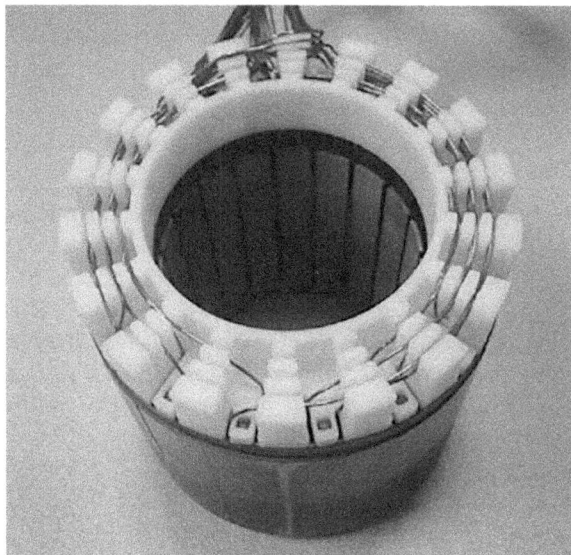

FIGURE 4.24 Specially designed molded plastic insulation part used at the motor ends for insulating the lead wires.

FIGURE 4.25 Insulation papers inserted in the stator slots for receiving winding coils.

thickness of insulation materials may have a range of 0.1–0.65 mm, depending on liner materials and stator operation conditions. Although the material is inexpensive, it takes a longer labor time for fabricating stator windings.

4.3.3 Glass Fiber Reinforced Mica Tape

For large-size K_V motors (where K_V is a parameter, used to measure the motor's rpm when the voltage goes up by one volt without load), stator windings are often made of conducting bars that are continuously wrapped with mica tape. The taped winding bars are placed in a vacuum impregnation tank and flooded with a low-viscosity epoxy resin. After being completely impregnated, the insulation is cured at high temperatures in large chamber ovens.

4.3.4 Powder Coating on Stator Core

Powder coating technology (fluidized-bed coating) has been successfully applied in motor manufacturing since the early 1950s and is recognized as a superior and powerful method of applying a protective finish on electrical components. As an example, motor stators are often shaped cylindrically with inwardly facing slots configured to receive stator windings. It is required to insulate the copper windings from the stator metal surfaces. This can be achieved by applying power coating techniques to provide uniform insulating coating layers on the stator slot surfaces, as well as partially on the stator end surfaces (Figure 4.26). There are a number of insulation materials suitable for powder coatings such as epoxy, Glyptal®, and Loctite®. Epoxy powder coating has been effectively used as a high dielectric insulator on copper/aluminum conductors by many motor manufacturers because of its high durability and superior dielectric strength. Numerous ways of distributing powder throughout motor parts are available. One approach that offers significant advantages over a wide range of other approaches is to use a specially designed apparatus that is equipped with a power fluidized bed to get uniform coating layers [4.16]. The thickness of the coating layer typically has a range of 0.1–0.5 mm.

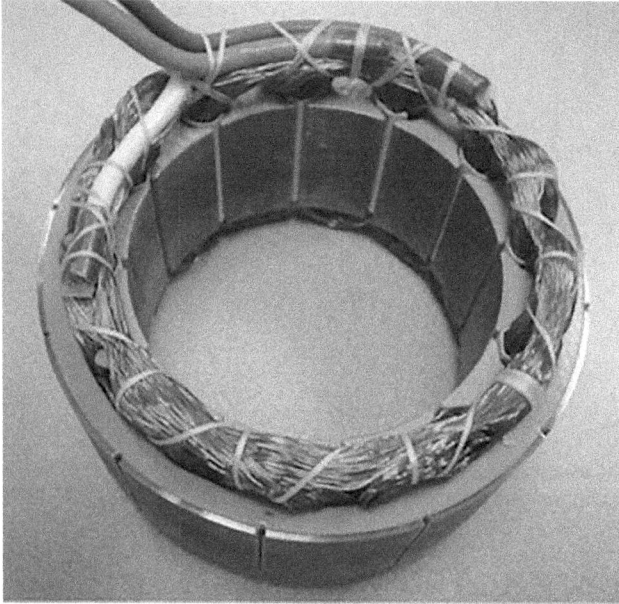

FIGURE 4.26 A stator core having a coated insulation layer on the slot surfaces and partially on the end surfaces with powder coating technique.

4.4 MANUFACTURING PROCESS OF A STATOR CORE

The manufacturing process of a stator core includes lamination stamping, annealing, stacking, winding, and impregnating/encapsulating. The selection of stator core materials is similar to that of rotor cores. Since there are no centrifugal forces acting on stator cores, the requirement for material strengths can be lower than that of rotor cores.

To effectively reduce eddy current and hysteresis losses, a stator core is manufactured by stacking a large number of thin laminations made of silicon steel. As a comparison, a solid steel core would function as an electrical heater. The addition of silicon to steel can significantly increase the permeability and volume resistivity to eddy current and thus enhance motor efficiency. It was shown that the iron loss of stator core can be reduced considerably by employing 6.5% silicon steel laminations instead of 3% silicon steel [4.17].

Silicon steel is available in an array of grades and thicknesses so that the material can be tailored for various applications. Generally, the thickness of motor laminations is in the range of 0.36 mm (0.014 in. for 29 gages) to 0.64 mm (0.025 in. for 24 gages). Thinner laminations are also available at 0.051 mm (0.002 in.), 0.102 mm (0.004 in.), and 0.178 mm (0.007 in.), considered for use in high-performance and high-frequency applications. It has been reported that at high flux densities and high frequencies, the specific iron loss is proportional to the square of the lamination thickness [4.18]. For demonstration purposes, the specific iron loss in a PM motor versus the lamination thickness for high saturated cobalt iron alloy at $B = 2T$ and $f = 1{,}200\,Hz$ is presented in Figure 4.27. This indicates for a lower iron loss; the lamination thickness should be less than 0.1 mm. However, the determination of the lamination thickness is a trade-off between core loss and fabrication cost. Thinner laminations are always more expensive and have lower stacking efficiency.

Two major distinctions of laminated materials are the fully processed electrical steel and annealed electrical steel. The annealed electrical steel is most common in North America and has higher magnetic and electrical properties than the fully processed steel. Usually, when a lamination supplier is set up with high output flow through an oven, the annealing process can be down

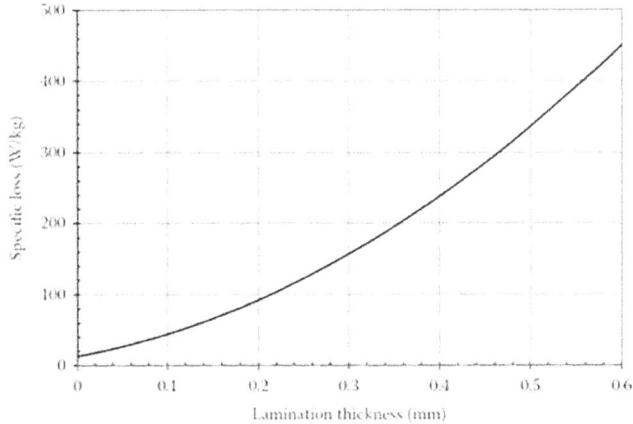

FIGURE 4.27   Specific iron loss in a PM motor versus lamination thickness at $B = 2T$ and $f = 1,200\,Hz$, showing that the reduction in lamination thickness can significantly reduce stator power loss.

very economically. Fully processed steel is usually delivered with an insulating coating, full heat treatment, and defined magnetic properties. To reduce the manufacturing time and cost, this type of steel does not need to be annealed after punching. Therefore, it is well suited for applications where the magnetic properties of steel laminations are not significantly degraded after the punching process.

4.4.1   Stator Lamination Cutting

Stamping is the most popular method used for lamination fabrication by most lamination manufacturers. The advantages of stamping include high productivity, low cost, and simplicity of the stamping process. The accuracy of the lamination cutting is typically controlled within ±0.0254 mm (±0.001 in.). However, burrs and warps near the lamination cutting edges due to stamping can lead to difficulties in stacking and tolerance control. As discussed by Brown [4.19] for a high-speed stamping machine, these inaccuracies sometimes can be 50% or more of the design gap between laminations. In addition, burrs can result in short circuits during the steel packaging and thus increase eddy-current losses. Moreover, the lamination insulation layer may be damaged as the result of residual and thermal stress produced from the stamping process. Using a magnetomechanical setup, Daem *et al.* [4.20] investigated the influence of the punching process on the magnetic properties of silicon steel. It was found that the magnetic properties dramatically degrade with increasing deformation levels under mechanical load. Releasing the mechanical load increased the hysteresis loss up to 270% at 10.4% pre-release strain. At this level of strain, the relative magnetic permeability decreased up to 45% after mechanical load release.

Laser cutting is often used in special cases, such as prototype buildup, stators with extra-large dimensions, or very complex geometries and shapes. Since the design information of lamination is computerized and loaded in a laser cutting machine, there is no need for tooling as in the stamping process. In laser cutting, a highly-energized laser beam is focused in a very tiny spot so that the local temperature rises extremely high to melt lamination sheets. Hence, this process generates much smaller residual stresses, lower distortion, better surface finishing, and high precise dimension control than the stamping process. However, laser cutting has lower productivity, higher power consumption, and higher costs. During the cutting process, some deterioration to the magnetic characteristics may occur, but it is too small to require annealing. This fabrication process is suitable for small quantities of laminations. The required tolerances for laser cutting are typically set at ±0.0254 mm (±0.001 in.).

Electrical discharge machining (EDM) is a manufacturing cutting process. An unwanted material is removed from the workpiece by a series of rapidly recurring current discharges between two electrodes—one is called the tool electrode and another is called the workpiece electrode. In a normal operation, the two electrodes are separated by a dielectric liquid. Like laser cutting, EDM can do a good job in cutting laminations but suffers from its low productivity, excessive tool wear, and high-power consumption.

Waterjet cutting is a fabrication process for cutting through virtually any material. In a cutting process, an ultra-high pressure steam of water (60,000 psi or higher) is applied to erode a narrow line in the stock material. To increase the cutting power, abrasive particles are often added into waterjets for cutting hard materials such as steel and glass [4.21]. Today, modern waterjet cutting machines extensively use computer numerical control systems to precisely control the jet nozzle in desired locations. For well-controlled cutting machines with microjet cutting heads, the cutting tolerances can be controlled within ±0.15 mm. One significant advantage of waterjet cutting is that during the cutting process, heat generated from cutting pieces is carried away by water, and thus there is no thermal stress produced in the cut parts. Compared to the stamping cutting method, this technique has the ability to cut parts with complex shapes and smooth edges (*i.e.*, burr-free). The disadvantages of waterjet cutting include: (*a*) long cutting time, especially for steel alloys such as electrical steel laminations with many layers, (*b*) prone of failing in cutting heads, and (*c*) the bonding effect on lamination materials being cut.

Chemical etching is a very low-cost procedure that requires only a few chemicals (such as sodium persulfate and hydrochloric acid) and very little equipment. In practice, this method has been extensively used in manufacturing printed circuit boards. The chemical etching process can generate burr-free and stress-free laminations that do not require annealing. The process starts to create a photographic image (*mask*) with the desired stator pattern. The mask is then illuminated with UV light onto lamination sheets. The lamination sheets need to be immediately placed into a developer solution such that the desired stator design is chemically etched onto the laminations.

The effects of different cutting methods (wire EDM, punching, laser, and abrasive waterjet) for electrical steel laminations on the performance of induction motor (IMs) have been experimentally studied by Bayraktar and Turgut [4.22]. In terms of motor performance, the best results are obtained with the wire EDM cutting. However, the results show that the power losses in electrical steel cutting with abrasive waterjet are higher than the other cutting methods.

4.4.2 Lamination Fabrication Process

An effective way to produce stator and rotor laminations together was presented by Isayama [4.23]. This method uses a sequence of stamping operation on strip materials to generate first the rotor sheet piece and then the stator sheet piece. As demonstrated in Figure 4.28, the stamping operation starts to punch pilot holes (Stage A) and then a shaft hole (Stage B), rotor ventilating holes (Stage C), and rotor teeth to complete the rotor lamination fabrication (Stage D). The following operation is for the stator lamination. The slots are made in Stage E. In Stage F, an annular punch is used to generate a concentric thin portion. In Stage G, round calking portions are formed. Following this, stator teeth are punched (Stage H). In the last stage (Stage I), the stator lamination is blanked away from the metal sheet.

This method has numerous advantages over conventional methods, including (*a*) reducing the residual stress in stator and rotor laminations, (*b*) reducing the lamination deformation, (*c*) lowering production cost because all operations complete on a single strip material with the use of a single press machine and a single die machine, (*d*) enabling to deal with complex shapes of stator and rotor, and (*e*) obtaining laminations that are excellent in such high shape precision as squareness and parallelism and highly suitable for use in accurate electric motors.

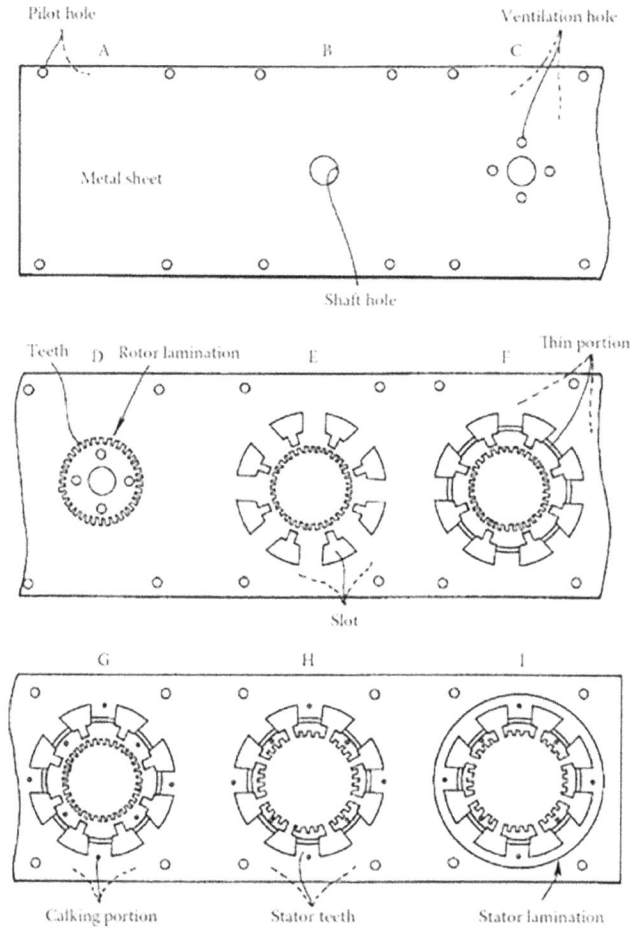

FIGURE 4.28 Rotor and stator laminations are made by a sequence of stamping operations (U.S. Patent 5,539,974) [4.23]. (Courtesy of the U.S. Patent and Trademark Office, Alexandria, VA.)

4.4.3 LAMINATION ANNEALING

During the lamination cutting process, residual stresses are introduced in processed laminations, leading to the degradation of material magnetic properties near the edges of the laminations and possible large cogging torque in motor operation. The smaller the stator size, the larger the cut-affected zone relative to the whole lamination area. Therefore, annealing is a necessary step for stress relief and for optimum properties of laminations, especially for stamped small laminations.

The annealing cycle requires to maintain processed laminations for several hours at the temperature range of $730°C–790°C$ under nonoxidizing and noncarburizing conditions (such as a vacuum or nitrogen-filled chamber) and then to naturally cool them to room temperature. Fully processed steels skip this step and perform very satisfactorily, but the highest performance is achieved with the annealing cycle.

4.4.4 LAMINATION STACKING

With the development of industrial automation, automatic lamination stacking becomes more popular today for modern motor manufacturers. This increases production efficiency and reduces the requirement for the storage capacity, particularly for a rotary stacking.

FIGURE 4.29   Stator laminations are integrated by welding.

A stator/rotor stack is formed by stacking laminations into a pack. Today, there are several joining processes to assemble stack laminations into stator/rotor cores [4.24]:

- Welding—This method robustly joins laminations together to withstand large forces without lamination separation (Figure 4.29) and thus is widely used in the motor industry. Commonly, the welding types used by lamination manufacturers include tungsten inert gas and metal inert gas welding. The laser welding technique has been developed to join stacked stator and rotor thin laminations with a fiber optic laser. The spot size of the laser is about 0.05 mm in diameter with a penetration depth of 0.5 mm, making it an excellent choice for small motor core assembly. However, the effects of residual stress, microstructure change, and inter-conduction among laminations introduced during the welding process result in a great deterioration in the magnetic properties of stator/rotor cores [4.25–4.27]. Furthermore, welding across laminations increases local core losses to impact motor performance, particularly for small motors. Sundaria *et al.* [4.28] investigated the effect of stator core welding on IMs. They observed that stator core losses increased about 10% due to the welding effect. Moreover, if improperly designed or welded, an increase in cogging torque can occur. Therefore, in order to avoid this undesired result, it must keep welds out of critical flux return path locations.
- Riveting—This is an economical and reliable method for securing stator cores. It requires punching small holes or slots on laminations. However, the heads of the rivets may obstruct the stator end windings.
- Bonding with adhesive agents—This method is to deposit a very thin layer of adhesive (a few μm) on lamination edges or surfaces to be bonded and assemble the laminations into a stator core. The core and winding are reheated to cure the adhesive material. Thus, it has bounded together the laminations of the core to provide adequate core strength (Figure 4.30). By comparing with other methods such as welding and riveting, the main advantage of this non-damaging joining process is that it has little impact on the key material properties of electrical steel. In contrast, welding and interlocking can substantially damage the microstructure and thus impair the magnetic properties of the material. Other benefits of this joining process include enhanced insulation, reduced corrosion, improved thermal conductivity, and reduced vibration due to its strong damping effect, leading to reduced acoustic noise. In practice, bonding often helps reduce production costs.

FIGURE 4.30 Stator core is fabricated using a spot bonding technique with epoxy adhesive.

FIGURE 4.31 Laminated electrical steel sheets are joined together with 10 dowels to form a stator core.

- Lamination interlocking—This is a widely used method in the automated fabrication of motor cores. In this method, interlocking dowels are used to join laminated steel sheets together. As can be seen in Figure 4.31, 10 dowels are placed at the back iron on each stator lamination. When the laminations are stacked together, they nest inside each other. However, the main disadvantage of this process is that interlocking can deteriorate core magnetic properties due to the induced elastic strains during the dowel formation and the dowel joint. The formation of short circuits could also contribute to the core power loss [4.29]. A study of the influence of interlocking on magnetic properties of ring cores reveals that the inverse of permeability and iron loss increase linearly with the number of interlocks [4.30]. Moreover, the interlocking process requires a complex die that is usually very expensive.

FIGURE 4.32 Stator core is secured with a thin sleeve at the outside of the core.

- Fastened by pins—This method is similar to riveting but without the problem of the obstruction of the rivet heads to the stator end windings.
- Self-cleating—This method rolls V-shaped strips and then flattens them down into dovetail slots at the OD of the stator laminations.
- Using slot liners—This method is commonly used for stator cores less than 150 mm. Cuffed slot liners hold the stator core together, as described in [4.31].
- Using thin sleeves—As demonstrated in Figure 4.32, this method for the stator core retention is to use thin metallic sleeves outside of the core. Due to the high strength of the sleeve, the stator core can withstand large forces with only tiny deformation. The installation of the sleeve usually relies on the shrink fit technique.
- Bolting—This method is similar to riveting for providing reliable lamination integration. A plurality of bolts is extended through boltholes in the lamination stack and engaged with nuts. Upon tightening of the bolds, the laminations are forced together to establish a generally axial compression on the lamination stack. However, bold heads have even larger heads than rivets to obstruct the stator end windings.

4.4.5 Stator Winding

One of the important parameters in stator winding is the slot fill ratio, which is defined as the percentage of the space occupied by magnet wires to the total available space of the slot. In order to lower the wire resistive loss and increase the power density, it is highly preferable to have the maximum copper fill, that is, the maximum slot fill ratio.

A winding end turn refers to the amount of the winding extending beyond each end of the stator's magnetic core structure. Though the end turns are necessary to complete the electrical path within the winding, they contribute little to the motor torque output. Motor torque is only generated by the winding that lies within the stator's magnetic core structure. Consequently, it is highly desired to minimize the length of the winding end turns. This can not only save the wiring material and lower the material cost but also reduce the copper loss and increase the motor efficiency. The shorter the winding stack length, the greater the impact of the end turn length on motor efficiency.

4.4.5.1 Random Winding by Hand

The advantages of random winding by hand include the flexibility and suitability to some manufacturing processes, such as prototypes, low-volume production, and/or motor rewinding. Slot fill can

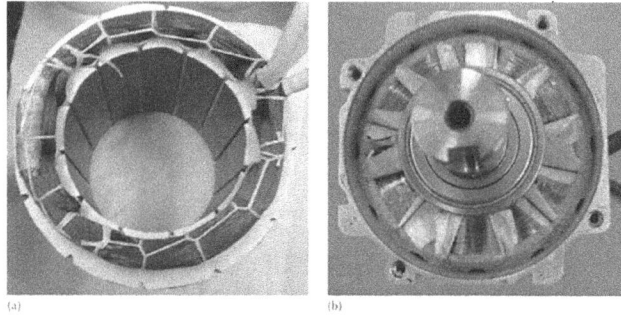

FIGURE 4.33 Coil formation: (*a*) distributed winding; (*b*) concentrated winding.

also be higher than single slot designs wound with typical needle winding machines. However, this winding method may have more winding variabilities and require highly skilled operators. Hand wiring can have different configurations and is best to clarify if (*a*) coils are hand wound and hand inserted, (*b*) coils are wound around stator teeth by hand, and (*c*) coils are machine wound and hand inserted.

4.4.5.2　Coil Formation—Distributed Winding

This type of winding is made by arranging winding turns in several full-pitch or fractional-pitch coils (Figure 4.33*a*). Then, this coil is housed in the slots spread around the airgap periphery. Because a distributed overlapping winding generally results in a more sinusoidal magnetomotive force (MMF) distribution and electromotive force (EMF) waveform, it is used extensively in PM brushless AC machines [4.32]. High slot fills can be achieved depending upon the motor size. Generally, the larger the motor, the higher the slot fill.

4.4.5.3　Coil Formation—Concentrated Winding

The interest in motors with concentrated windings is growing due to their short end windings and simple winding structures that are highly desired for high-volume automated manufacturing. For this type of winding, all the winding turns are wound together in series to form one multiturn coil. The concentrated winding motors have potentially more compact designs compared to conventional motors with distributed windings.

It has been reported that based on FEM calculations, the performance of an electric machine with concentrated winding is superior because the minimization of the copper volume greatly reduces power losses and manufacturing costs. Induced EMF in concentrated windings is greater than distributed windings. In addition, harmonic or noise in distributed windings is lesser than concentrated windings. Furthermore, distributed windings have less armature reaction. The concentrated windings are given in Figure 4.33*b*.

4.4.5.4　Coil Formation—Conductor Bar

In building large-size motors and generators, it is often necessary to bundle several individual conductors that are insulated against one another to form a so-called conductor bar (Figure 4.34). The insulation between the conductor bar and the stator core is defined as the main insulation. To fabricate stator windings, several conductor bars are connected to each other at the machine ends, as shown in Figure 4.35.

Conductor bars are prefabricated with either round or rectangular shape copper wires, coated with mica tape, and impregnated with epoxy resin. They are then heated up and cured resulting in high winding mechanical strength. Generally, a conductor bar with a rectangular cross section is preferred for its high filling rate in stator slots. In producing conductor bars, the individual conductor can be either twisted around each other or maintained parallel to each other. During the

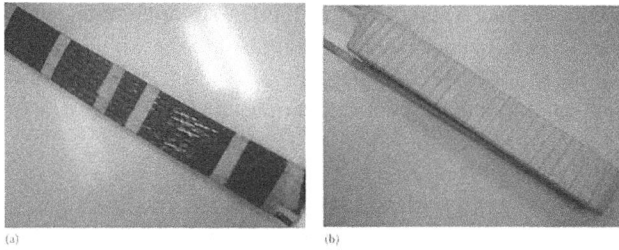

FIGURE 4.34 Bundled conducting bars used in electric motors: (*a*) rectangular magnetic wires are twisted to reduce eddy-current loss and (*b*) stator winding is wrapped by an insulating strip.

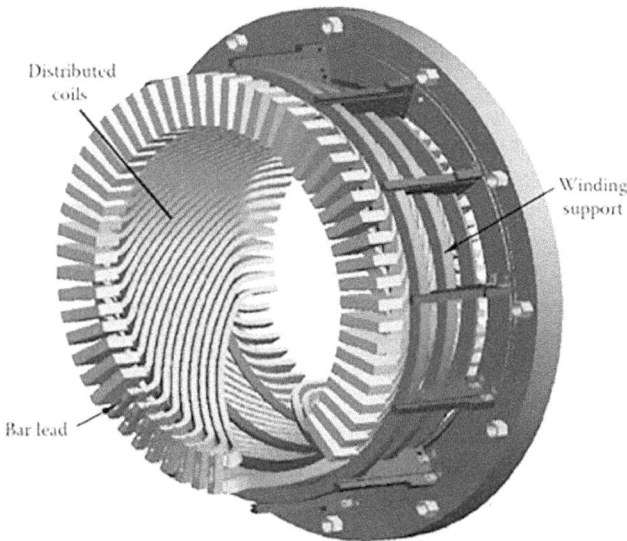

FIGURE 4.35 Conducting bars are commonly used in large electric machines. It shows the joints of conducting bars at the end turn region.

manufacturing process, sometimes, conductor bars are required to be compressed to make them into the desired shape and compact size. It has been found that an improper compressing method may potentially damage the conductor's insulation, causing stator winding failure. The connection efforts are more labor intensive with individual coils.

Before and after the winding impregnation, stators must undergo a series of tests such as hipot (dielectric voltage withstand test), surge (impulse test), resistance, and contact resistance to verify the stator condition.

4.5   STATOR ENCAPSULATION AND IMPREGNATION

Encapsulating and impregnating stators can strengthen stator winding electrical insulation, provide reliable protection to chemicals and harsh environments, enhance thermal dissipation, promote stator structure integrity, and stabilize motor operation.

4.5.1   ENCAPSULATION

Encapsulation has been widely used in electric machines. According to different applications, electric motors can be either partially or entirely encapsulated. Partial encapsulation is basically applied to the stator end windings for integrating them with other stator components against vibration and for enhancing heat transfer. Entire encapsulation is applied to the whole stator assembly for achieving better protection of the stator from moisture, dirt, debris, and erosions caused by chemicals such as acids. This treatment can provide very robust, durable, and long-life stators.

4.5.1.1 Encapsulation Materials

A wide variety of encapsulation materials are available for electric machines, as given in Table 4.2. These materials usually exhibit high dielectric strengths and relatively high thermal conductivity and are developed for different applications.

Thermoset plastics (also known as thermosets), such as epoxies, phenolics, and thermoset polyesters, have long been applied to electric machines as encapsulation materials. They can form irreversible chemical bonds during the curing process and have good mechanical and thermal properties. In recent years, with the recognition by IEC-85 and UL-1446, thermoplastics as encapsulation materials are used for some encapsulated motors, solenoids, and transformers. Thermoplastics offer some performance benefits such as high strength and shrink resistance.

Though these two types of encapsulation materials sound similar, they have very different properties and applications. One of the remarkable differences is that the volatile organic solvents used in thermoset plastic encapsulation processes are eliminated in thermoplastic processes. As a result, the harmful solvent emissions to the environment are eliminated. Another important characteristic feature is that the curing process of thermoplastics is completely reversible as no chemical bonding takes place. This indicates that thermoplastics can be remelted at high temperatures. This special characteristic feature allows thermoplastics to be remolded and recycled but restricts their use for relatively low-temperature applications.

4.5.1.2 Encapsulation Process

In practical operation, prior to the encapsulation process, the stator is preheated in an oven to drive off any moisture or volatile components.

Thermally conductive thermosets are ideally suited for the full stator encapsulation due to their outstanding electrical and thermal properties and mold flow characteristics [4.38]. Among them, epoxy resins are the most popular in the motor industry. They are generally hard and tough, exhibit low shrinkage on cure, and offer superior chemical and environmental resistance. The key features of encapsulating materials include high heat conductivity, high dielectric strength, and low shrinkage, thermal endurance, and chemical resistance. It is worth to note that thermal conductivity in an encapsulation system is determined by all its components, such as epoxy resin, hardener, and filler. Generally, thermal conductivity increases as more fillers are added. However, a high filler volume may significantly increase the viscosity of the mix, making it much more difficult to fill tight spaces. One way to get high filler volume in an encapsulation system is to use fillers with a spherical shape; angular fillers require much resin between them to maintain the necessary liquidity of the encapsulating materials.

Epoxy resins usually contain diluents or reducers to help their fluidity during the process. The hardener used with the epoxy resin plays a very important role in the final properties and has an impact on the speed of cure. Many epoxy resin systems use solid fillers as the constituent for cost reduction and increased hardness and stiffness of the cured product. Moreover, by carefully choosing the solid filler materials, it may lead to an increase in thermal conductivity in the cured encapsulation. It was reported that the types of filler that can help increase thermal conductivity include ceramic material, glass fiber, Kevlar® fiber, carbon fiber, silicon nitride fiber, and other materials [4.39]. The filler materials can be in the form of particles, granular powder, whiskers, fibers, or any other suitable forms. The thermally conductive polymer composition used to form the encapsulating polymer layer for the stator assembly comprises a base polymer and a thermally conductive filler material and has a thermal conductivity of at least 0.6 W/(m·K) and up to 5 W/(m·K) or higher.

TABLE 4.2
Encapsulation Materials

Encapsulation Material	Thermal Conductivity (W/m·K)	Dielectric Strength (V/mil)	CTE ($10^{-6}/K$) (%)	Note	Reference
E88 epoxy	1.049 (at 23°C)	780	3.49	A two-component flame retardant resin system is used for the impregnation and encapsulation of high-voltage electrical components.	[4.33]
C89 hardener	1.069 (at 50°C) 1.083 (at 75°C) 1.074 (at 100°C) 1.043 (at 125°C) 1.004 (at 140°C)				
Araldite CW 229-3 Aradur HW 229-1	0.75	508	27	It is a prefilled resin system that provides high crack and thermal shock resistance.	[4.34]
Altherm XB 2710 Aradur XB 2711	1.5	508	24	It is a good choice for stator encapsulation for its high thermal conductivity.	
Araldite XB 2252 Aradur XB 2253	0.7	508	60	The system is mineral filled with excellent flowability and impregnation capability.	
Araldite CW 1312 Aradur HY 1300	1.1	381	100	It has higher thermal conductivity and adequate impregnation capability. High filler loading results in higher viscosity.	
Arathane CW 5631 Arathane HY 5610	0.6	508	70	The curing reaction is fast and exothermal at room temperature. It is easy to process and has good impregnation capability.	
Aratherm CW 2731	3.0		20	One-component product. Developed for end turn encapsulation of motors and generators.	
Catalyst 11	1.28	380	97.9[a]	It is a two-component, thermally conductive epoxy encapsulant. It features a low coefficient of thermal expansion (CTE) and excellent electrical insulative properties.	[4.35]
EP234	3.77	415		It is room temperature cured and designed for medium-voltage devices.	
EP1121	0.14	440		Its low viscosity allows for good wicking and penetration into components and circuitry and also gives good air release.	[4.36]
EP1282	0.14	440		EP 1282 is a two-part unfilled electronic grade epoxy encapsulant. It cures at room temperature to a tough, semirigid polymer.	
EP 1285	1–1.27	365–400		Designed for applications requiring a high degree of thermal conductivity and a low CTE combined with a moderate free-flowing viscosity.	
Thermoset® SC-320 silicone	3.2			A two-component silicone elastomer that provides excellent thermal conductivity, high thermal shock resistance, and improved coefficient of expansion.	[4.37]

Note:
[a] Samples tested were cured for 24 h at 25°C.

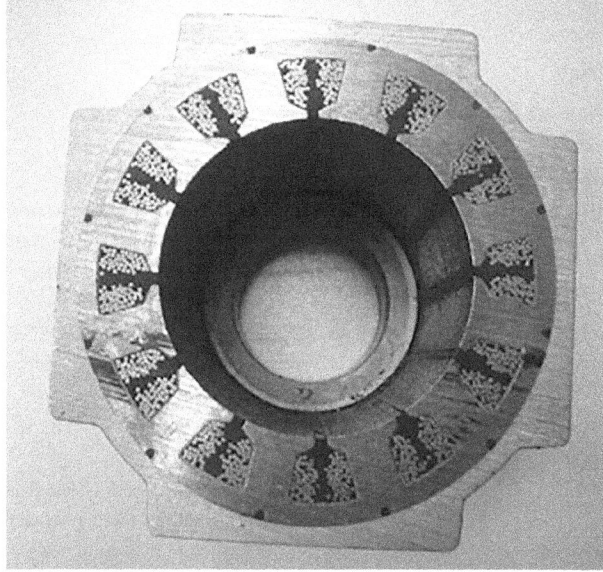

FIGURE 4.36   A cutaway view of an encapsulated stator shows that epoxy penetrates into stator coils and forms void-free encapsulation.

Different encapsulation processes are available such as vacuum casting, gravity casting, and pressure gelation to allow highly automated production lines with a short cycle time. Prior to casting, a sealing core is placed in the middle of the stator for preventing an encapsulant contamination of stator laminations.

A stator encapsulation process starts with the heating of an encapsulant to a certain temperature for the purposes of degassing and increasing the fluidity of the encapsulant. The higher the fluidity of the encapsulant, the shorter the filling time. With a gravity casting technique, liquid encapsulant is induced at the bottom of the stator to avoid air bubbles trapped in the encapsulation. Then, the stator is put in the oven for curing. The curing temperature and time can be optimized to minimize volume shrinkage and maximize mechanical strengths.

A standard casting process is typically controlled and executed by an encapsulating equipment, which consists of two independent systems for dealing with the encapsulant (*e.g.*, epoxy resin) and the hardener individually before they are mixed. The epoxy resin is stored separately with the hardener in its packaging drum and goes to the vacuum premixers for degassing and the metering mixer for further mixing. This is the same process for the hardener. After this step, both materials get in the static mixer to get them to be mixed together. The well-mixed epoxy resin is pumped to fill the mold.

The cutaway view of an encapsulated stator shows that the epoxy completely penetrated into the stator coils (Figure 4.36).

The entire stator assemblies are immersed into a liquid epoxy resin in a vacuum tank. The vacuum helps to not only completely fill voids in the windings but also exhaust moisture and other vapors from the windings.

4.5.2 Varnish Dipping

Varnishing dipping is an effective way of securing stator windings. In this method, a stator is immersed into an open varnish tank. After a certain time, the stator is removed from the varnish tank to allow excess varnish dipping. Then, place the stator in an oven to dry off the solvent.

However, during varnish dipping and baking processes, the solvent is continuously evaporated and released into the environment air, causing air pollution problems.

4.5.3 TRICKLE IMPREGNATION

Trickle impregnation is extensively used in many motor manufacturers. In practice, a trickle varnish machine is usually used to impregnate the stator winding with a varnish that rigidly secures the wires. Prior to the impregnation process, stators need to be preheated to the desired temperature. A thin stream of impregnating resin is poured/dripped onto the winding. In such a way, the impregnating resin fills the voids of the winding and gets polymerized in a short time. This method offers several advantages, including short processing time, high retention, no need of post oven curing, and consistent quality of impregnation.

The varnish improves heat transfer within the winding and between its surrounding magnetic core structures. This improves motor cooling and, in turn, increases the motor's continuous torque and power density. In addition, trickle impregnation helps the stator winding to withstand vibrations, avoiding motor failure due to the damage of wire insulation.

The utilization of a thermally conductive epoxy instead of a regular varnish to impregnate the stator winding can enhance motor cooling remarkably, for example, using a potting epoxy that is a recognized component in a UL-1446 insulation system that has a Class H ($180°C$) temperature rating. Actual measurement shows that potting the stator winding using this thermally conductive epoxy lowers its winding-to-ambient thermal resistance, R_{th} (in units of $°C$/W), by 50% compared to impregnating the winding with a typical varnish [4.40].

4.5.4 VACUUM PRESSURE IMPREGNATION

The vacuum pressure impregnation (VPI) technology has been extensively used in a wide range of applications in the electric machine manufacturing industry from insulating coil windings to sealing porous metal castings. The characteristic of this technology is to use a VPI tank (Figure 4.37) that is vacuumed first and then pressurized to achieve the best insulation effect on stators. This method can provide the highest industrial standards for electric machines. By driving out voids from the electric winding through the VPI process, the thermal conductivity of the winding is remarkably enhanced so that the hot spots are eventually eliminated. This also reduces the risk of partial discharge in the winding. In fact, the VPI can make the high stator mechanical integration to reduce the vibration of the motor. Since the resins contain no solvents and only a small amount of resin vaporizes during curing, the VPI process has a negligible environmental impact. Presently, this state-of-the-art method is primarily applied on heavy-duty motors.

The primary VPI process involves (Figure 4.38) the following steps:

- Place a stator in a VPI tank.
- Pull the vacuum of the container for a period of time to remove air and moisture from the stator winding.
- Introduce a preheated solventless epoxy resin into the vacuum container until it fully covers the stator and holds the vacuum of the container for a period of time.
- Release the vacuum of the container and pressurize the container to the desired level to achieve maximum impregnation for deep penetration of epoxy resin into voids.
- Draw the epoxy resin from the container.
- Place the impregnated stator into a temperature-controlled oven (Figure 4.39) and bake it to cure the epoxy resin and fully develop the properties of the insulation system for interlock at the desired temperature. In today's motor workshops, the baking process is fully controlled by a built-in computer in the oven to follow up the preset temperature chart. To prevent *thermal shock* to motor insulation systems, the oven temperature is brought up and down slowly Ovens are typically equipped with digital panel meters to monitor baking temperatures.

FIGURE 4.37   VPI tank used for large electric motors.

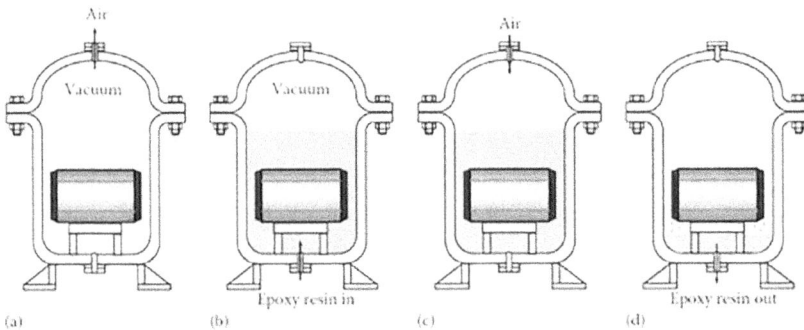

FIGURE 4.38   VPI process: (*a*) evacuating the container into deep vacuum, (*b*) introducing epoxy resin into the container until the stator is completely submerged, (*c*) releasing the vacuum of the container and applying pressure over the resin-covered stator to force resin completely into the stator, (d) draining epoxy resin out of the container, and (e) placing the stator into an oven and baking it for curing the epoxy resin (not shown).

This VPI process can achieve complete penetration of resin throughout turns, coils, slots, and insulation and thus is primarily applied on heavy-duty applications. The VPI technology is superior to solvent varnishes in many aspects. As discussed previously, solvent varnishes lose 50%–70% of volume during baking, leaving voids and air pockets in the windings. In contrast, VPI provides

FIGURE 4.39 The temperature-controlled oven for curing and baking epoxy resins of impregnated motors.

a 100% solid mass structure that increases the mechanical strength, minimizes the hot spots, and reaches the highest level of environmental protection. A motor winding that undergoes a VPI process is virtually impervious to oil, moisture, and chemical contaminants [4.41].

4.6 STATOR DESIGN CONSIDERATIONS

Recently, stator as one of the most critical motor components has drawn more attention from motor engineers, researchers, and end users. In order to optimize motor performance, one important step is to optimize the stator design, which is not only limited to the stator and its components (laminations, windings, *etc.*), but also includes the interactions between the stator and other motor components (*e.g.*, rotor, PMs).

4.6.1 COGGING TORQUE

Cogging torque is one of the inherent characteristics of PM motors, resulting from the interaction of the PM MMF harmonics and the airgap permeance harmonics due to slotting. Since cogging torque can cause speed ripples, induce motor vibration, and deteriorate motor performance, it is one of the major design goals for motor engineers to reduce cogging torque.

Electromagnetic design primarily determines the level of cogging torque. As addressed in Chapter 1, there are many methods for reducing the amount of cogging torque, for instance, skewed stator stack, optimized slot and pole combination [4.42], uneven distribution of stator slots, stator tooth notching, and pole shifting.

In addition to EM design, a cogging condition may be inadvertently set up from improper stator fabrication processes, including the following: (*a*) The lack of a proper annealing after lamination stamping will directly lead to high residual stresses in the stamped laminations, as well as the degradation of magnetic properties, and thus results in cogging torque. (*b*) The lack of deburring will increase eddy-current losses. (*c*) Machining defects (*e.g.*, concentricity and roundness) on the stator ID may result in unequal airgap and thus greatly induce a large amount of cogging torque. (*d*) Exceeding roundness or concentricity tolerance of the stator ID can significantly change cogging torque. A decrease in the airgap or airgap variations can considerably increase cogging torque. (*e*) Exceeding tolerance of slot opening is very sensitive to change cogging torque. (*f*) Welding a stack of stator laminations may increase the core loss at the welded area, cause the deformation of the stator core, and alter magnetic flux distribution. All these factors could change cogging torque. A more important issue is the welding location. Welding in the back iron between the teeth could increase cogging torque.

4.6.2 Airgap

The radial distance between the rotor and the stator in a motor is defined as the airgap. Normally, a smaller airgap provides a more efficient and powerful motor. Hence, it is highly desired to maintain the airgap dimension as small as possible and within a small variation in operation. The control of the airgap dimension involves the design of several components such as the stator, rotor, motor housing, and endbells. An important factor that affects the airgap dimension is the accuracy of the coincidence of the stator and rotor axes. Thus, to provide a motor with a small airgap dimension within only a small tolerance, preciseness in the manufacturing of these parts is required.

In attempting to obtain concentricity, it has been proposed [4.43] to mechanically fit the metal rotor bearing support members to the stator stack in a matter that forms a cavity. A hardenable plastic material is ejected to fill the cavity and bond the parts together. The assembly is then machined to the desired dimensions.

4.6.3 Stator Cooling

An important objective in motor design is to control the motor temperature below its allowable value. Increased motor temperature often reduces motor efficiency and affects bearing life. The Arrhenius equation predicts that the failure rate of an electric device is exponentially related to its operation temperature.

In designing a motor cooling system, thermal engineers always focus on stator cooling for several reasons: (*a*) The stator winding is usually the main heat source in a motor. Test data show that in most applications heat generated in a motor is primarily attributed to the stator. (*b*) Cooling in the stator end-winding region is particularly difficult and remains a challenge due to various factors [4.44]. (*c*) As a stationary component, the stator is much easier to be cooled compared with the rotor. In fact, the stator often serves as a heatsink for the motor. (*d*) For some electric motors, the pumping effect, which is resulted from the rotor rotation, is strong enough to generate turbulent circulating flows for making the motor self-cooling [4.45].

4.6.4 Robust Design of Stator

The root causes of motor failure are often related to mechanical deterioration such as vibration, static and cyclic loads (*e.g.*, centrifugal and differential expansion forces), insulation fracture, and bearing lubricant contamination and leakage, to name a few. Robust design techniques can optimize the stator design and ensure adequate motor performance over a wide range of operating conditions.

Because the rotor is supported on bearings located at the endbells of the machine, the stator design is significantly impacted by the dynamic behavior of the rotor. As a matter of fact, the loads carried by the rotor are essentially transmitted to the stator and finally to the motor base. In this load configuration, the structural vibratory loads caused by the rotor, and loading caused by stator vibration that drives rotor behavior, are intertwined. This requires a full understanding of the structural interactions for all motor components, as well as the interaction between the motor and its base.

Motor vibration is greatly influenced by its base. A weak motor base usually results in high vibration. Thus, it is essential to design a motor base strong enough to withstand all motor loads for preventing resonance and vibration.

4.6.5 Power Density Improvement

Power density is a measure of power output per unit volume, given in units of W/m^3. When designing electric machines, a key challenge is how to respond to the continuous market demand for more output power with less size. Historically, increasing power density of electric machines has been driven by emerging materials and technologies such as rare earth PM, graphene, miniaturization and MEMS technology, additive manufacturing process, and nanotechnology. In fact, over decades the trend toward high power density has sustained in the automation and motion control industry. High power density is always desirable for military, medical, industrial, and commercial applications, where the space and the weight are strictly restricted. These include aircrafts, missiles, robots, electrical vehicles, and medical equipment, to name a few.

It should be noted that in the literature there is another relevant but different parameter: power-to-mass ratio, or specific power, which is defined as the amount of power per unit of mass (in units of W/kg). The power-to-mass ratio is commonly applied to electric machines, engines, and power sources to enable the comparison of one unit to another. However, it is important to note that a high power density does not necessarily mean a high power-to-mass ratio, or vice versa. Unfortunately, the two are often confused by some investigators in previous studies.

There are a number of technical approaches to achieve high power density. First, power density is highly dependent on the motor type. Axial-flux (AF) motors have a significant higher power density than radial-flux (RF) motors due to several factors [4.46]: (*a*) For AF motors, the magnets are located further away from the central axis. This results in a larger lever on the central axis. (*b*) With dual PM rotors, AF motors have an inherently more efficient topology in the electromagnetic sense. Consequently, the magnetic flux path is much shorter than that in RF motors. (*c*) AF motors have the concentrated windings that nearly 100% of them are fully active for establishing the electromagnetic field, indicating more ability to increase the number of turns and less heat caused by the end effect. In contrast, a large portion of the winding (as much as 50%) in RF motors makes no contribution to produce the electromagnetic field. (*d*) The primary factor limiting engineers' ability to improve power density is the heat-dissipating capability of motors. As the motor size has continuously shrunk, effective motor cooling becomes more difficult. This indicates that the ability to get the heat out of a motor directly impacts power density. In AF motors, because the stator windings are directly in contact with the aluminum housing, the windings of AF motors stay cool while the resistance of the copper remains low. Compared to RF motors, heat must be dissipated through the stator to the environment of the machine. However, the cooling of winding at the end regions becomes very difficult because they have no contact with any solid parts and the heat transfer mainly relies on convection. Furthermore, in the case of RF machines, due to the superior magnetic properties of rare earth magnets such as Nd-Fe-B magnet, synchronous permanent magnet motors are considered to have higher power density and efficiency than asynchronous IMs.

An effective way to increase the motor power density is to choose an appropriate wire shape. To achieve a higher fill factor, windings may be made with rectangular or flat wires [4.47]. The higher fill factor can result in a more compact motor with the same power output. As a result, the winding heating can be greatly reduced and the utilization rate of winding copper materials can be increased by 15%–20%, which improves torque density, power density, and efficiency.

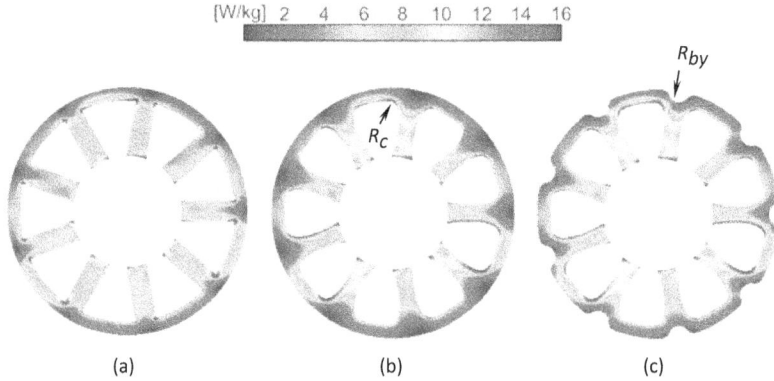

FIGURE 4.40 Comparison of power-to-mass ratio distribution among different models: (*a*) base model, (*b*) Bezier model, where R_c is the corner radius at the stator slot root, and (*c*) modified model by cutting the radii at the stator back yoke with the radius $R_{by} = 1.25$ mm.

An important parameter to increase the power density of electric motors is the thermal conductivity of the winding. For this purpose, Stöck *et al.* [4.48] suggest utilizing twisted windings to replace conventional parallel windings. The results show that the twisted winding has a 20%–30% higher thermal conductivity compared to the parallel winding at the same fill factor. Moreover, the increase in the thermal conductivity with the fill factor is more sensitive for the twisted winding than for the parallel winding. The significant increase in the thermal conductivity coupled with the lower losses of a twisted winding is a promising solution for the future motor development.

To improve the power-to-mass ratio, a special consideration is given to the optimization of the stator core shape [4.49]. The comparison of power-to-mass distribution among three models (the base model, the Bezier model with R_c, and the modified model with $R_{by} = 1.25$ mm) has been presented in Figure 4.40. The results show that the Bezier curve not only decreases the core loss of a motor, but it also has an effect that extends the low magnetic flux density region in the back yoke part. By cutting the radii at the back yoke, the increase in the power-to-mass ratio in the modified model with $R_{by} = 1.25$ mm and $R_{by} = 1.5$ mm from the base model is about 9% and 12%, respectively.

The final piece of the puzzle toward power density is the selection of the right manufacturing process. Additive manufacturing is a transformative approach to industrial production that enables the creation of lighter, stronger parts and systems. In recent decades, additive manufacturing such as 3D printing has experienced the transition from analog to digital processes. Using the computer-aided-design (CAD) software to direct hardware, 3D printing can deposit material, layer upon layer, in precise geometric shapes. In this way, 3D printing delivers a perfect trifecta of improved performance, complex geometries, and simplified fabrication. For example, with 3D printing, motor housing and endbells can be designed with more-complex architectures, such as porous media structure (either open cell or closed cell) for reducing the power-to-mass ratio and enhancing heat transfer of the motor. In addition, one of the advantages of 3D printing for electric machinery is the ability to innovate the design of components. An example of this is part consolidation, where several parts are designed as a single component, such as motor endbell, airflow deflector, and housing with embedded air ducts. Moreover, since each part is built independently, it can easily be modified to suit unique needs to accommodate customers' specific requirements.

The trend toward higher power density is expected to move forward. Overcoming thermal performance and power loss challenges require innovations in optimization of motor design, winding topology, cooling method, and manufacturing process. Each of these puzzle pieces by itself provides significant improvement opportunities in power density.

4.7 MECHANICAL STRESS OF STATOR

Electric motors are exposed to many kinds of disturbances and stress. Some disturbances are due to imposed external conditions such as mechanical or electric unbalanced system; harmonics; supply interruptions; variations in voltage, current, and frequency; unstable cooling system; and large variations in ambient temperature and humidity. Stress factors include frequent successive start-ups, abnormal use of the motor, stall, and overload situations. These stress and disturbances deteriorate the winding insulation mechanically and increase the thermal aging rate and thus may eventually lead to an insulation failure [4.50].

Stresses and strains in stator cores can not only cause mechanical problems but also change the magnetic properties of electrical steels (*e.g.*, silicon steel). A stator is subjected to a variety of forces during operation. When the forces are relatively small, only elastic deformations are generated on the stator core. Removal of the forces will permit the stator core to return to its original stress-free state. However, under large load conditions, the stator core must withstand high stresses that produce plastic deformations of the stator core. These plastic deformations can distort the metal crystals or atomic structure and, in turn, affect the magnetization characteristics of the material. As a result, a set of the maximum stress on magnetic core materials must consider both the mechanical and electromagnetic requirements.

There are three main sources of mechanical stress for stators. The first common mechanical stress is caused by power frequency current. When a motor starts, the current through the stator winding reaches its maximum value that gives rise to a magnetic force oscillating at twice the power frequency [4.51]:

$$F = \frac{k i_{rms}^2}{w} \tag{4.1}$$

where
i_{rms} is the rms current
w is the stator slot width
$k = 0.96$

If the current in a stator winding is

$$i_{rms} = A \sin \omega t \tag{4.2}$$

Equation 4.1 can be rewritten as

$$F = \frac{k A^2 (1 - \cos 2\omega t)}{2w} \tag{4.3}$$

Noting that $\omega = 2\pi f$, this equation describes the oscillation force in the radial direction at twice the power frequency. If the stator winding resonates at this frequency, the insulation of the winding will be damaged. More likely, the damage occurs at the stator end-winding regions. In fact, resonances cause serious vibration levels of end windings and develop high stress in these windings. As discussed previously, stator encapsulation can provide a good solution to avoid it.

In addition, there is also a force at twice the power frequency in the circumferential direction caused by the rotor's magnetic field interacting with the current in the stator winding. However, this circumferential force is only about 10% of the radial force.

The second mechanical stress is caused from the transient effect. For synchronous machines, switching on of motors or out-of-phase synchronization can give rise to a large transient power frequency current that may be five times (or even more) greater than the normal operating current in the stator. This results in that the magnetically induced mechanical force is 25 or more times stronger than the normal service. This alternative force tends to bend the stator winding, especially in the end-winding regions. If the force is too large, the stator winding may crack.

The third source of mechanical stress has resulted from rotors. In an electric motor, the rotor rotates at a high speed with a coupled gearing system or an externally driven machine for a DDR system. While the rotor is subjected to larger centrifugal force, it transmits its radial and axial loads to the stator via bearings.

REFERENCES

4.1. Klontz, K. W. and Li, H. D. 2008. Reducing core loss of segmented laminations. *Power Transmission Engineering*, June Issue, pp. 26–32.

4.2. Welch, R. 2009. Think thermal to increase motor efficiency. *Motion System Design— Motors & Drives* **51**(**8**): 32. http://motionsystemdesign.com/motors-drives/think_thermal_increase_0809/.

4.3. Vandenbosschel, L., Jacobs, S., Henrotte, F., and Hameyer, K. 2010. Impact of cut edges on magnetization curves and iron losses in e-machines for automotive traction. *The 25th World Battery, Hybrid and Fuel Cell Electric Vehicle Symposium & Exhibition*, Beijing, China.

4.4. Moses, A. J., Thomas, B., and Thompson, J. E. 1972. Power loss and flux density distributions in the T-joint of a three-phase transformer core. *IEEE Transactions on Magnetics* **8**(**4**): 785–790.

4.5. Jones, M. A. and Moses, A. J. 1974. Problems in the design of power transformers. *IEEE Transactions on Magnetics* **10**(**2**): 148–150.

4.6. Fisher, G. A. and Jacobs, J. T. 1993. Lightweight high power electromotive device. U.S. Patent 5,212, 419.

4.7. Dicks, G. 2010. Permanent magnet alternator with segmented construction. U.S. Patent 7,816, 830.

4.8. Zhu, Z. Q., Ruangsinchaiwanich, S., Ishak, D., and Howe, D. 2005. Analysis of cogging torque in brushless machines having nonuniformly distributed stator slots and stepped rotor magnets. *IEEE Transactions of Magnetics* **41**(**10**): 3910–3912.

4.9. Neuenschwander, T. R. 2006. Method of manufacturing a formable laminated stack in a progressive die assembly having a choke. U.S. Patent 7,062,841.

4.10. Magnussen, F., Svechkarenko, D., Thelin, P., and Sadarangani, C. 2004. Analysis of a PM machine with concentrated fractional pitch windings. *Proceedings of Nordic Workshop on Power and Industrial Electronics (NORPIE)*, Trondheim, Norway.

4.11. Carpenter, D. E. 1933. Edge-wound core. U.S. Patent 1,920,354.

4.12. Ozawa, M. and Fukuda, T. 2003. Rotating electrical machine stator. U.S. Patent 6,525,437.

4.13. Gieras, J. F. 2002. *Permanent Magnet Motor Technology: Design and Applications*, 2nd edn. Marcel Dekker, New York.

4.14. Aoki, K. and Yamaguchi, T. 1969. Method of making the stator core for rotary electric machinery. U.S. Patent 3,436,812.

4.15. Rocky, D. M., O'Connor, Jr., J. F., Marvin, R. H., Charwick, E. R., Tribodeau, P., and Won, B. 2005. Permanent magnet motor. U.S. Patent Application 2005/0073210.

4.16. Bellemare, D. J., Donahue, J., Fagan, M., and Vollono, R. 2002. Electrostatic fluidized bed coating method and apparatus. U.S. Patent 6,458,210.

4.17. Paulides, J. J. H., Jewell, G. W., and Howe, D. 2004. An evaluation of alternative stator lamination materials for a high-speed, 1.5 MW, permanent magnet generator. *IEEE Transactions of Magnetics* **40**(**4**): 2041–2043.

4.18. Nagomy, A. S., Dravid, N. V., Jansen, R. H., and Kenny, B. H. 2005. Design aspects of a high speed permanent magnet synchronous motor/generator design for flywheel applications. *IEEE International Electric Machines and Drives Conference*, pp. 635–641, San Antonio, TX.

4.19. Brown, C. P. 1996. Design for manufacturability of a high-performance induction motor rotor. MS thesis, Massachusetts Institute of Technology, Cambridge, MA.

4.20. Daem, A., Sergeant, P., Dupré, L., Chaudhuri, S., Bliznuk, V., and Kestens, L. 2020. Magnetic properties of silicon steel after plastic deformation. *Materials* (Basel, Switzerland) **13**(**19**): 4361.

4.21. Hashish, M. A., Kirby, M. J., and Pan, Y.-H. 1987. Method and apparatus for forming a high velocity liquid abrasive jet. U.S. Patent 4,648,215.

4.22. Bayraktar, Ş. and Turgut, Y. 2016. Effects of different cutting methods for electrical steel sheets on performance of induction motors. *Proceedings of the Institution of Mechanical Engineers, Part B: Journal of Engineering Manufacture* **232**(**7**): 1287–1294.

4.23. Isayama, M. 1996. Method for producing laminated iron cores. U.S. Patent 5,539,974.

4.24. Hendershot, Jr. J. R. and Miller, T. J. E. 1994. *Design of Brushless Permanent-Magnet Motors*. Magna Physics Publishing, Hillsboro, OH.

4.25. Zhang, Y., Wang, H., Chen, K., and Li, S. 2017. Comparison of laser and TIG welding of laminated electrical steels. *Journal of Materials Processing Technology* **247**(**2017**): 55–63.

4.26. Krings, A., Nategh, S., Wallmark, O., and Soulard, J. 2014. Influence of the welding process on the performance of slotless PM motors with SiFe and NiFe stator laminations. *IEEE Transactions on Industry Applications* **50**(1): 296–306.

4.27. Cui, R. G. and Li, S. H. 2020. Pulsed laser welding of laminated electrical steels. *Journal of Materials Processing Technology* **285**(**2020**): 116778.

4.28. Sundaria, R., Daem, A., Osemwinyen, O., Lehikoinen, A., Sergeant, P., Arkkio, A., and Belahcen, A. 2020. Effects of stator core welding on an induction machine -Measurements and Modeling. *Journal of Magnetism and Magnetic Materials* **499**(1): 166280. doi:10.1016/j.jmmm.2019.166280.

4.29. Senda, K., Toda, H., and Kawano, M. 2015. Influence of interlocking on core magnetic properties. *IEEJ Journal of Industry Applications* **4**(4): 496–502.

4.30. Imamori, S., Steentjes, S., and Hameyer, K. 2017. Influence of interlocking on magnetic properties of electrical steel laminations. *IEEE Transactions on Magnetics* **53**(11): 8108704.

4.31. Crawford, D. E. and Fields, H. T. 1990. Core and slot liner. U.S. Patent 4,922,165.

4.32. EL-Refaie, A. M. 2010. Fractional-slot concentrated-windings synchronous permanent magnet machines: Opportunity and challenges. *IEEE Transactions on Industrial Electronics* **57**(1): 107–120.

4.33. The RD. George Company. 2001. Technical data sheet: Pedigree No. E-88 epoxy and No. C-89 hardener. http://www.elantas.com/.

4.34. Hollstein, W. 2012. Thermosetting resin systems for stator encapsulation in electro-mobility and industrial motors. http://www.lindberg-lund.fi/files/PDFkataloger_FI/WHollstein_Inductica_Encapsulationof Statorsformobility_2012.pdf.

4.35. Emerson & Cuming. 2001. Technical data sheet—STYCAST® 2850 FT-thermally conductive epoxy encapsulant. http://lartpc-docdb.fnal.gov/0000/000059/001/stycas2850.pdf.

4.36. Resinlab. Thermally conductive encapsulants. http://www.resinlab.com/encapsulant/thermal.html.

4.37. Lord Corporation. 2005. SC-320—Highly thermally conductive encapsulating system. http://www.lord.com/Products-and-Solutions/Electronic-Materials/Product.xml/561.

4.38. Du Pont. 1997. Electrical/electronic thermoplastic encapsulation. H-58633. http://plastics.dupont.com/plastics/pdflit/europe/markets/TEncap_e.pdf.

4.39. Neal, G. D. 2002. Stator assembly. U.S. Patent 6,362,554.

4.40. Welch, R. Jr. 2012. A more efficient servomotor. http://www.exlar.com/pages/380-A-More-Efficient-Servomotor.

4.41. Inman Electric, Inc., 2012. Vacuum pressure impregnation (VPI). http://www.inmanelectric.com/vacuum.html.

4.42. Zhu, Z. Q. 2000. Influence of design parameters on cogging torque in permanent magnet machines. *IEEE Transactions on Energy Conversion* **15**(4): 407–412.

4.43. Kaufman, Jr., G. A. 1977. Electric motor with plastic encapsulated stator. U.S. Patent 4,048,530.

4.44. Tong, W. 2008. Numerical analysis of flow field in generator end-winding region. *International Journal of Rotating Machinery* **2008**: Article ID 692748.

4.45. Tong, W. 2001. Numerical analysis of rotating pumping flows in inter-coil rotor cavities and short cooling grooves of a generator. *International Journal of Rotating Machinery* **7**(2): 131–141.

4.46. Moreels, D. and Leijnen, P. 2018. High efficiency axial flux machines: Why axial flux motor and generator technology will drive the next generation of electric machines. White Paper V1.9. Magnax Axial Flux Machines.

4.47. Cai, W., Fulton, D., and Congdon, C., Multi-set rectangular copper hairpin windings for electric machines. U.S. Patent 6,894,417 B2.

4.48. Stöck, M., Lohmeyer, Q., and Meboldt, M. 2015. Increasing the power density of e-motors by innovative winding design. *CIRP 25th Design Conference Innovative Product Creation* (Edited by Moshe Shpitalni, Anath Fischer and Gila Molcho) **36**(**2015**): 236–241.

4.49. Soda, N. and Enokizono, M. 2020. Stator core shape design for low core loss and high power density of a small surface-mounted permanent motor. *Sensor* **20**(**2020**): Paper # 1418.

4.50. ABB. 2010. Power system protection: Motor protection (Section 8.11). In *Distribution Automation Handbook*. ABB Oy, Vaasa, Finland. http://www.abb.com/product/ap/db0003db004281/a40c9d1981558a3c-c1257a060018bfb1.aspx.

4.51. Stone, G. C., Boulter, E. A., Culbert, I., and Dhirani, H. 2004. *Electrical Insulation for Rotating Machines*. Wiley-Interscience, Hoboken, NJ.

5
Motor Frame Design

Electric motors are manufactured in a variety of types and configurations. Typically, an electric motor assembly includes a hollow, substantially cylindrical-shaped housing with a stator and a rotor disposed within the housing interior and an endbell at each end of the housing for providing protections to the motor components against the insertion or intrusion of solid/liquid objects under various environmental conditions. In addition to rotors and stators, some motors may have electronic components that are used to improve operating characteristics for particular applications. These electronic components also need to be well protected.

For radial airgap motors, the stator coils are wound axially through the stator slots, and the end turns of the stator winding are positioned adjacent to the stator end surfaces. The endbells and the motor housing form an enclosure to protect the stator windings and end turns from inadvertent contact and grounding while providing a mounting surface for rotor bearings and shaft bushings. The endbells also prevent debris from entering the hollow interior of the stator and interfering with operation of the motor.

5.1 TYPES OF MOTOR HOUSING BASED ON MANUFACTURING METHOD

Motor housings, also known as motor enclosures, are designed to support the stator, rotor, and other components. While totally enclosed housings protect motor components from foreign objects such as water, dust, sand, and moisture (see Figure 5.1), open housings allow ventilation flows passing through them to cool motors. When designing or selecting a motor housing, several factors must be taken into account:

- The housing is capable of securing the position of all motor components relative to each other under various operating conditions.
- To withstand and absorb the applied forces, impacts, and moments, the housing must have a rigid structure with necessary strengths.
- For better thermal management, the housing must possess a high heat dissipation capability for effectively dissipating heat generated by the motor.
- They are capable of damping and isolating vibration and noise for smooth motor operation.
- Material damage (usually metals) caused by corrosion can be seen everywhere. Corrosion degrades material properties, weakens base materials, and reduces component/system longevities. To reduce the effects of corrosion and decrease life-cycle costs, motor housings should have high corrosion resistance, which is particularly important in food, beverage, and medical processing applications. In addition to metals, corrosion can occur in materials such as ceramics or polymers.
- On the premise of ensuring the necessary component strength, reducing the housing weight is highly desired for some aerospace and medical applications.

Motor housing can be made by a variety of manufacturing processes, such as wrapping, casting, extruding, machining, and stamping. For rareearth magnet motors, either magnetic or nonmagnetic materials can be used to fabricate motor housings. However, for Alnico magnet motors, the

DOI: 10.1201/9781003097716-5

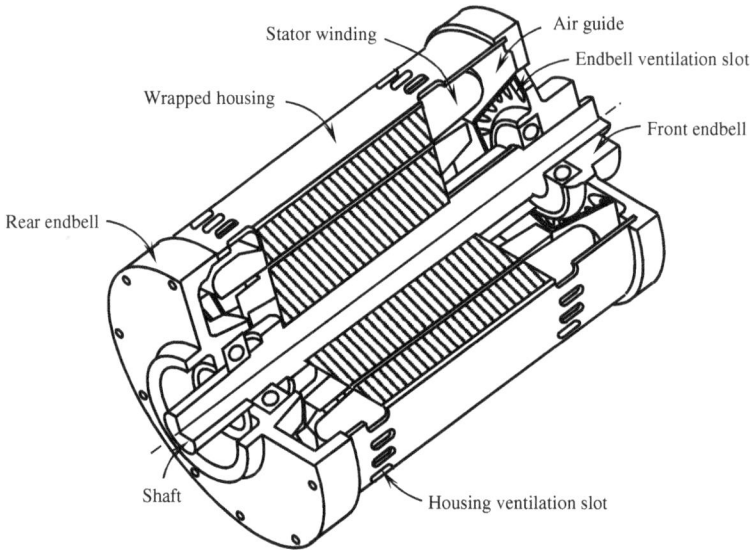

FIGURE 5.1   Cutaway view of an electric motor with wrapped open housing. There are two rows of ventilation slots that are evenly distributed on the circumference of the wrapped housing, allowing air to pass through the windings for heat dissipation.

housings must be made from only nonmagnetic materials such as aluminum, brass, and stainless steel to preserve specified performance characteristics of the motors.

Motor housings are often made of aluminum alloys due to their low density, high thermal conductivity, and good corrosion resistance. Today, making hybrid metal-composite housings for modern motor systems becomes a viable option, because the most useful features of each material can be meshed and placed at specific locations.

5.1.1   WRAPPED HOUSING

A wrapped motor housing is made from a flat sheet metal that passes through a 3-roll bending machine to form a cylindrical-shaped form. Figure 5.2 illustrates the manufacturing process of wrapped housing. The first step is to prepare a metal sheet with prepunched apertures for installing motor fasteners and notches for receiving electric connection cables (Figure 5.2a). At each edge of the sheet metal, a small notch is left. When the two edges come together nicely, these notches form a narrow clearance for receiving welding operation. Subsequently, the sheet metal is rolled by a bending machine with three adjustable rolls to control the radius of the rolled sheet metal (Figure 5.2b). When the stator core is inserted, the wrapped housing is pressed tightly with a special tool for welding along the welding seam (Figure 5.2c). Lastly, the two ends of the housing are machined for receiving the endbells that block off the axial ends.

During the welding process of the wrapped housing, a large amount of heat is produced and conducted into the motor housing and stator core. Excessive heating can cause the deformation and distortion of the housing and stator core, resulting in permanent damage to these parts. High heat input can also adversely affect the mechanical properties of the welded parts. Therefore, it is critical to control the temperature on the welded parts and enable efficient heat dissipation from them to the environment. For example, using a higher welding speed, a lower welding current, and/or an active/passive cooling technique (*e.g.*, cooling via cold plate) can essentially solve most welding overheating problems. While the control of heat input is highly desired in the welding process, it must be noted that the rapid cooling, on the other hand, may lead to the change of metallographic microstructure in welded materials, resulting in the formation of large grains and the reduction in resistance to cracking.

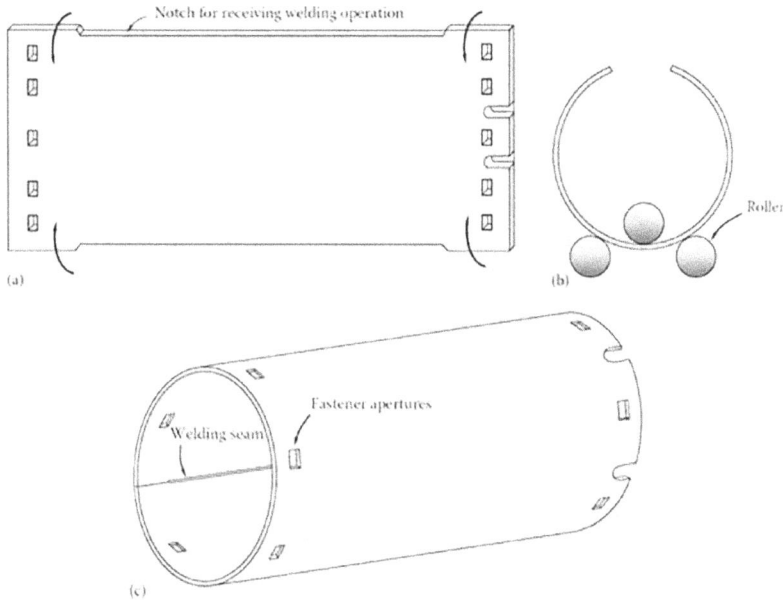

FIGURE 5.2 Manufacturing procedure of wrapped motor housing: (*a*) punching apertures for installing fasteners and notches for receiving motor cables on a sheet metal using a high-speed punching machine and cutting the sheet metal, (*b*) rolling the sheet metal to form an open hollow tube with a 3-roll bending machine, and (*c*) welding at the notched edges to form the motor housing.

5.1.2 CASTED HOUSING

Casting refers to a manufacturing process that involves a phase change of casting materials between liquid and solid. Metal casting processes can be traced back to approximately 3,600 BC as ancient Egyptians utilized casting to make bronzes. In China, the discovery of *Three-Star Piles* (三星堆) civilization in the last century has revealed that about 4,000 years ago, ancient Chinese casted out exquisite bronzes. The cast iron technique was invented in China in the seventh century BC. The earliest cast iron pieces in the world were discovered in Shanxi, China, by archaeologists in the 1980s.

A casting process begins by heating up of casting material to reach the material melting temperature or above, at which the casting material is in the liquid state. The molten casting material is maintained at the temperature for a period of time for degassing, then poured into a mold with a desired shape, and allowed to solidify. Therefore, casting technologies are especially suitable for producing motor housings, endbells, and other components with a wide variety of shapes and geometries. In fact, with the ability to effectively produce high-quality and low-cost mechanical components with a wide range of sizes and materials, various casting processes are extensively used in almost all industries, especially in automobile, aerospace, and power industries. Figure 5.3 is a typical servomotor with a casted housing. In this design, the housing is integrated with the front endbell as a single-casted part.

5.1.2.1 Casting Material

Common casting materials include aluminum, iron, copper, magnesium, zinc, and steel alloys. Among them, aluminum is the most popular material extensively used by motor manufacturers. Benefited with a low melting temperature, high liquidity, lightweight, high thermal conductivity, and other good casting characteristics, aluminum alloys have been commonly used for manufacturing motor housings, endbells, and other components. Furthermore, reusability is one of the principal advantages of aluminum. With the improvement of aluminum alloys and casting processes

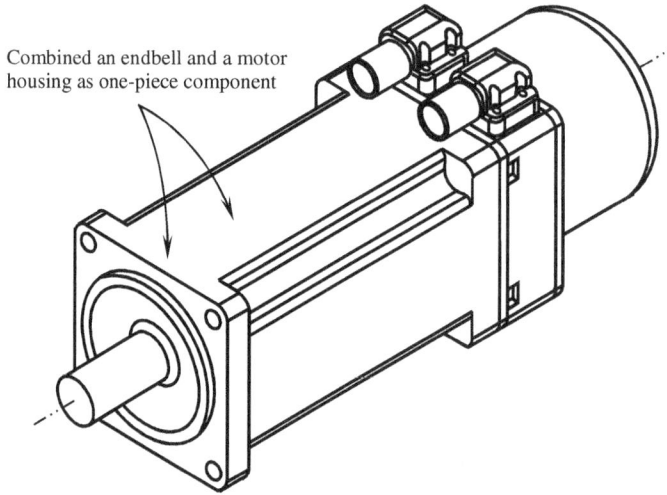

FIGURE 5.3   Casted servomotor housing, showing that the front endbell is integrated into the motor housing.

and good balance between mechanical properties and lightness, it is expected that more and more aluminum-casting parts will be demanded by the motor industry.

Today, cast iron housings are commonly used in heavy-duty and large-size electric motors, as cast aluminum is unable to offer the required strength and hardness. Cast irons are ternary alloys of Fe-C-Si. The microstructure and mechanical properties of cast irons vary considerably depending on the form of the carbon element presented in the irons. There are four main types of cast irons with different carbon forms: gray, malleable, ductile, and white cast iron. Carbon is mainly present as graphite in gray, malleable, and ductile cast iron, while exists as cementite (Fe_3C) in white cast iron. Gray cast iron contains 2.5%–4% carbon and 1%–3% silicon [5.1]. Silicon is essentially needed as a graphite stabilizing agent that makes cementite (Fe_3C) more difficult to form in the cast irons. Graphite is a relatively soft material, and its presence in gray iron gives it some unique and useful properties:

- Excellent vibration damping capacity, being 25–100 times better than that of a 1080 steel [5.2]. Damping is the ability of a material to absorb vibration energy by converting it into other forms of energy such as heat. The quelling of vibration is very important for structures and devices with moving parts. Damping capacity of gray iron strongly depends on its graphite structure and size. For instance, the damping capacity for a gray iron with coarse flake graphite can be 5–10 times higher than that for a gray iron with fine flake graphite. The strength and hardness of the gray iron is provided by the metallic matrix in which the graphite occurs. Generally, damping capacity decreases with increasing strength since a smaller amount of graphite present in the higher strength irons decreases the energy absorbed. The primary damping mechanism is localized microplastic behavior. Under cyclic loading conditions, this microplastic behavior produces a hysteresis loop whose area is proportional to the energy absorbed during each cycle [5.3]. This property is especially desirable as it reduces the acoustic noise and vibration emitted by motors.
- Superior wear resistance as the metallic matrix microstructure being primarily pearlite. Among the casting materials in Table 5.1, gray cast iron has the highest hardness.
- High compressive strength. Gray iron's compressive strength is typically three to four times higher than its tensile strength [5.4]. Thus, the parts made by gray cast iron are mainly designed to withstand compressive forces rather than tensile forces.
- Internal lubricating qualities and machinability. Graphite has little strength or hardness. The presence of the graphite provides several valuable characteristics to gray iron,

TABLE 5.1

Properties of Some Common Casting Materials for Electric Motors

Material Property	Aluminum 383	Aluminum 356-T6	Aluminum 6061-T6	Ductile Iron 65-45-12	Gray Cast Iron
Melting point ($°C$)	516–543	557–613	582–652	1,148	1,180
Density (kg/m³)	2,700	2,670	2,700	7,096	7,150
Yield tensile strength (MPa)	152	>165	276	310	—
Ultimate tensile strength (MPa)	310	>234	310	448	276
Elongation (%)	3.5	3.5	17	12	1
Poisson ratio	0.33	0.33	0.33	0.29	0.29
Hardness (Brinell)	75	70–105	95	170–207	183–234
Modulus of elasticity (GPa)	71.0	72.4	68.9	163	90–113
Shear modulus (GPa)	27	27.2	26.0	65	36–45
Fatigue strength (MPa)	145	90	96.5	210	110
Shear strength (MPa)	187	143	207	400	334
Machinability (%)	70	50	50	0	0
Thermal conductivity (W/m·K)	96.2	151	167	32.3	47
Coefficient of thermal expansion @ $20°C$ (mm/m·K)	21.1	21.4	23.6	14.6	10.8

including the ability to produce sound castings, dimensional stability, borderline lubrication, and excellent machinability. In fact, cast iron is the easiest ferrous material for dry machining.

- Improved thermal properties. Gray iron has a better thermal conductivity and a lower coefficient of thermal expansion compared with ductile iron.

Due to these superior properties, gray cast iron has been a staple in automotive manufacturing for many years. As a desired material in vibration damping for achieving low-noise performance, there is a steady increase in the use of cast iron in certain types of motors, such as those in medical equipment.

As a member of the cast iron family, ductile iron contains 3%–4% carbon and 1.8%–2.8% silicon. This type of cast iron has high strength. Unlike gray cast iron in which graphite is presented as flakes, ductile iron contains tiny graphite nodules, which make cracking more difficult. Therefore, ductile cast iron is much stronger and has much higher ductility than gray cast iron. In addition to high strength, the toughness and shock resistance also increase for ductile cast iron [5.5]. Furthermore, the damping capacity of ductile iron is considerably greater than most other ferrite materials (except for gray iron). These superior properties of ductile iron make it ideally suited for various industrial applications. In the motor industry, ductile iron is often used to make motor housings and endbells.

For ductile iron with high silicon contents, a silicon-rich surface layer is formed to resist oxidation, corrosion, and heat. Stabilization of the ferrite phase reduces the influence of high temperature in two ways: (*a*) silicon raises the critical temperature at which ferrite transforms to austenite; (*b*) the strong ferritizing tendency of silicon stabilizes the matrix against the formation of carbides and pearlite, thus reducing the growth associated with the decomposition of these phases at high temperatures [5.6].

Steel casting is a specialized form of casting, adapted for parts that must withstand heavy loads and shocks. There are generally two types of steel castings to meet different needs of various applications: carbon steel casting and alloy steel casting. Carbon steel casting is the most popular type, which uses low-, medium-, and high-carbon steel in casting to obtain different ranges of strength. For alloy steel castings, special alloy elements (*e.g.*, manganese and chromium) are added to casting steels to enhance the desirable material properties. Among alloy steel castings, stainless steel casting is designed to resist corrosion and thus widely used in the chemical, shipbuilding, and renewable energy industries. However, steel casting is seldom used for making motor components unless casting iron cannot deliver enough strength or motors have strict anticorrosion requirements.

Compacted graphite iron has been developed since the mid-1970s. Because the graphite in this type of iron takes a form between that in gray iron and ductile iron, it has the benefit of combining the high thermal conductivity of gray iron with the high strength and high modulus of elasticity of ductile iron [5.7]. Since it has high thermal-shock resistance, this type of iron is especially suitable in applications involving thermal shocks.

In the selection of materials for casting housings, a number of mechanical and thermal properties must be taken into account, including mechanical strength, elongation, hardness, modulus of elasticity, dimensional stability, shrinkage, machinability, thermal conductivity, among other indices. The properties of the most common casting materials are given in Table 5.1.

5.1.2.2 Casting Process

Casting processes can be divided into two broad groups, gravity-fed and pressure-fed processes, where the gravity fed involves pouring molten metal into mold cavities under gravity and the pressure fed involves injecting molten metal under pressure. The choice of the casting process is greatly influenced by product geometric complexity, size, material, dimensional accuracy, surface finish, tooling and other costs, and production volume.

As one of the oldest known manufacturing techniques, casting still remains a highly efficient and cost-effective manufacturing process today. Modern advances in casting technology have led to a wide array of specialized casting methods. In manufacturing industries, four main casting methods are commonly used to make a variety of cast products—die casting, sand casting, permanent-mold casting, and evaporative pattern casting (EPC)—each having its own unique fabrication advantages and limitations.

Die casting is a manufacturing casting process that can produce geometrically complex metal parts using reusable molds (i.e., dies). There is a traditional high-pressure die casting process but also using low pressures. The casting materials typically are nonferrous alloys such as aluminum, zinc, magnesium, and copper. In a die casting process, molten metal is injected under high pressure over a very short period of time into a hardened steel or cast iron die (Figure 5.4). Then, the ejection die is removed, and the die casting part is ejected by means of ejection pins. The total cycle time is very short, typically between a few seconds to less than a minute. High casting pressures help reduce porosities in casting parts, increase casting uniformity, and improve the surface finish and dimensional accuracy. Because die casting can provide good surface finishing and dimensional accuracy, only light machining is required to enforce the accuracy of a few critical dimensions. In

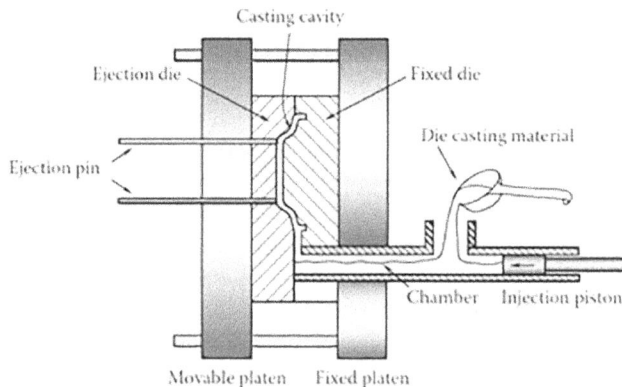

FIGURE 5.4   Illustration of the die casting process.

some cases, post machining can be completely eliminated. A key feature in die design is the positioning of the runners and gates. Well-designed gates permit rapid flow into the die cavity, minimizing turbulence and long flow paths for the molten metal.

Diecasted housing is most popular in the motor industry due to its inherent advantages of lightweight, low cost, high thermal conductivity, and high productivity. In the design of die casting housing, the most important thing is to determine the departing line as it has a direct impact on the success of the casting process. Second, it is better to make the wall thickness as uniform as possible to reduce casting deflects caused by nonuniform cooling rates and residual stresses in casting housings. Reinforced ribs are often employed in motor housings to reduce the housing weight and casting material. All corners of casting parts should be radiused generously to avoid stress concentrations and cracks. The set of the draft angle allows casting parts to be easily removed from dies. The minimum draft angle per side in the parting direction depends on the casting materials: usually 0.25° for zinc alloys, 0.5° for aluminum alloys, and 0.75° for brass. In practice, a draft angle of 1°–2° on all walls in the parting direction is suitable for most applications. The arrangement of ejection pins is critical for allowing casting parts to be effectively ejected from dies.

Sand casting applies reusable sand to make sand molds. The advantages of sand casting include low cost of mold, large-size casting part, and a wide variety of metals and alloys. However, sand-casted parts also suffer rough surface and poor dimensional accuracy. In this technique, the preparation and bonding of the sand mold are the critical steps. A typical structure of a sand-casting mold is shown in Figure 5.5. For casting parts with a simple geometry, the mold may consist of two components, that is, the upper mold and the lower mold. Depending on the geometry requirement, cores may be placed in the mold. The paths of molten metal are usually arranged on the department surface. Gas vents are left on the upper mold to allow air to escape from the mold during the casting process. In sand casting, molten metal is poured into the sand mold with the aid of the gravitational force.

In the metal processing industry, the surface finish is characterized by either the average root mean square of a surface measured microscopic peaks and valleys or the arithmetic average (Ra) of a surface measured microscopic peaks and valleys. The surface finish of sand casting parts largely depends upon the type and quality of foundry sand. Among many parameters, the sand grain size and shape have a marked influence on many properties of foundry sand as a whole body, such as permeability, cohesiveness, flowability, surface fineness, compression/shear strength. Generally, finely grained sand produces a good surface finish but possesses low permeability [5.8], as well as low flowability.

Permanent-mold casting refers to the casting process that employs reusable metal molds to produce casting products repeatedly. This casting process usually requires the preheat of mold. Therefore, when the molten metal is poured into the mold under the gravitational force, the expansion of the mold is smaller than that in sand casting without preheating. The molds are usually

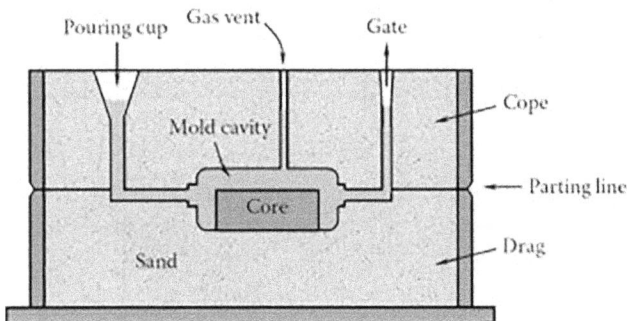

FIGURE 5.5 Typical sand-casting mold and casting process.

made of high-strength steel or iron alloys. To achieve the desired lifetime of the mold, this casting process is mainly used for lower-melting-temperature materials such as aluminum alloys. When the process is used to cast steel or iron alloys that have much higher melting temperatures, the mold life is extremely short [5.9].

For the ease of manufacturing and casting, the mold usually consists of two or more components. To reduce the temperature effect on the mold dimensions, a thin layer of heat-resistant material (such as clay or sodium sulfate) is coated on the inner surfaces of the mold.

Permanent-mold casting produces castings with excellent structural characteristics. There are some inherent advantages that the permanent-mold-casting process has over other casting processes: (*a*) because of the reusable molds, the unit tooling cost is lower than other casting processes under high-volume production conditions; (*b*) some mechanical properties of casted parts via permanent-mold casting are more favorable; (*c*) the dimensional accuracy is increased; (*d*) the surface finish is improved; and (*e*) the directional solidification rate can be controlled by varying the mold wall thickness or selectively heating or cooling certain positions of the mold. The fast-cooling rates created by using a metal mold result in a finer grain structure than sand casting [5.9]. However, because of the high tooling cost, this casting process is obviously uneconomical for low-volume productions. The permanent-mold-casting process is especially suitable for making motor housings with thick walls. Figure 5.6 illustrates the permanent mold for elevator motor housings.

The EPC technique has been developed for providing a cost-effective and low-pollution solution in the foundry industry. EPC takes advantage of certain materials that are easy to evaporate under high temperatures (*e.g.*, foam plastic) to fabricate casting patterns. The pattern is coated with refractory materials and buried into dry sand.

There are two main cast processes in EPC. The lost-foam-casting process is demonstrated in Figure 5.7 [5.10]. In this casting process, a pattern is positioned in a suitable container. Dry sand is placed in a container to surround and cover the pattern. Sand is vibrated to ensure that the pattern is completely surrounded without leaving undesirable voids. As the molten alloy is introduced into the mold through a sprue, the pattern material evaporates immediately upon contacting the molten alloy. As a consequence, the molten alloy replaces the space of the pattern and forms a seamless casting piece, which has exactly the same shape as the pattern. Therefore, there is no need for removing the pattern from the mold during the whole casting process.

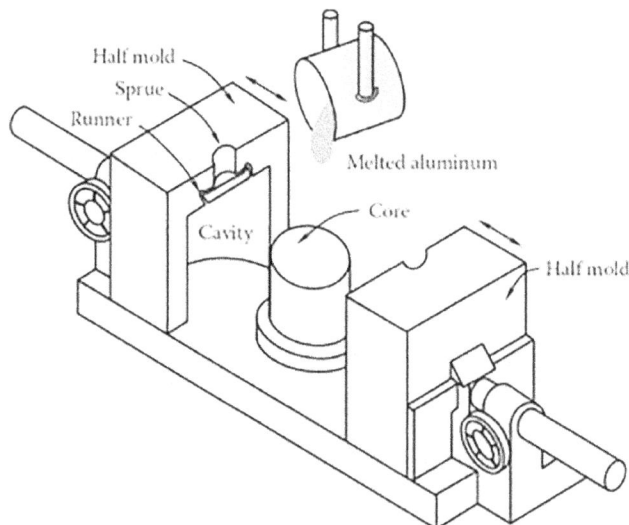

FIGURE 5.6   Permanent-mold casting for motor housing.

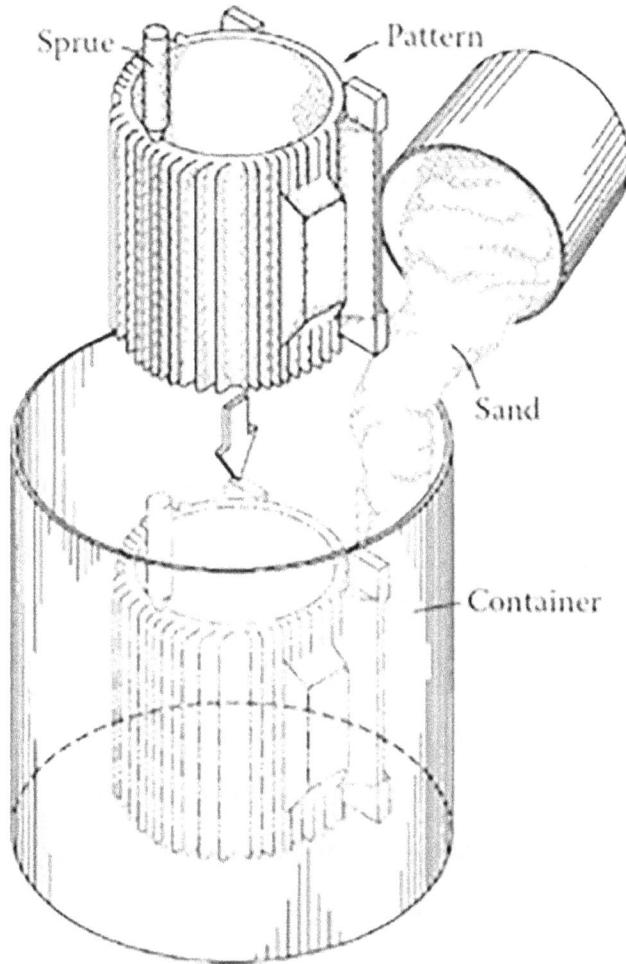

FIGURE 5.7 Making motor housing with the lost-foam-casting process (U.S. Patent 6,109,333) [5.10]. (Courtesy of the U.S. Patent and Trademark Office, Alexandria, VA.)

Full-mold casting combines sand casting and lost-foam casting. It employs an expandable poly-styrene (or foamed polystyrene) pattern, which is supported by sand in a single-piece sand mold. The molten metal is then poured through the sprue into the mold. With the progressive evaporation of the pattern material, the molten metal fills the available space. This casting process can make complex-shaped parts (*e.g.*, automobile engines) without using cores and drafts.

5.1.2.3 Pressure Casting

In order to reduce porosities in casting parts, pressure casting is widely adopted by many motor manufacturers, especially for large-size, high-power-density motors. In this casting technique, the molten metal is forced into a metal mold under pressure. The higher the applied casting pressure, the shorter the casting time and the less the porosities in casting parts.

Figure 5.8 presents a low pressure-casting process. While the molten metal is poured into the crucible, a pressurized gap forms. The pressure acts on the surface of the molten metal to force the molten metal into the mold through the refractory tube. The air pressure is maintained until the casting part has hardened within the mold. Once the casting part has solidified, the mold is opened and the casting part is removed. In this casting process, since the molten metal is drawn from the lower part of the crucible, it has less exposure to the environment and thus contains less trapped gas

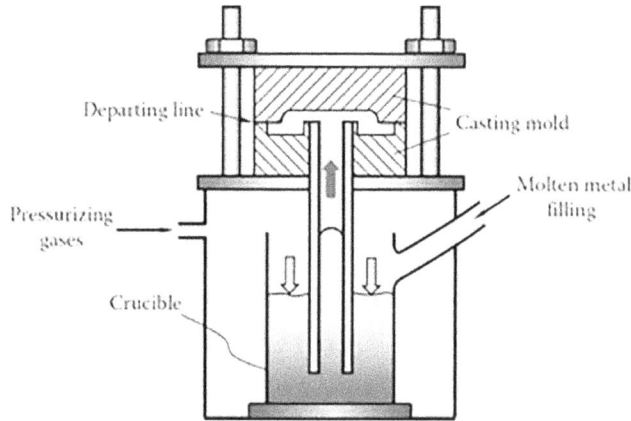

FIGURE 5.8   Low pressure-casting process.

and other undesired contaminants, compared with the molten metal at the upper part of the crucible. Therefore, casting quality is remarkably improved.

5.1.2.4   Heat Treatment

During a casting process, the solidification rate of the casting part is far from uniform. This is especially true for permanent-mold casting in which the liquid aluminum contacted with the metal mold surfaces solidifies much faster than that in the bulk of the part. Consequently, large residual stresses are created inside the casted part with nonuniform mechanical properties.

There are several heat treatment methods, each having a unique combination of temperature and time depending on the casting process, casting material, and cast requirements. Standard T6 heat treatment can enhance material properties such as improved ductility and fracture toughness through spheroidization of the eutectic silicon particles in the microstructure and increase yield strength through the formation of a large number of fine precipitates [5.11].

As part of T6 heat treatment, artificial aging at a relatively low-temperature range of $120°C–210°C$ can effectively reduce residual stresses. During this stage, the precipitation of dissolved elements occurs. These precipitates are responsible for the strengthening of the material [5.12]. It is interesting to note that a casting process normally takes less than 10 min, while a typical T6 heat treatment cycle may take more than 10 h. Figure 5.9 shows a casted aluminum housing of an elevator motor that is placed in the pit furnace, waiting for heat treatment.

5.1.3   Machined Housing

For fast fabrication of motor prototypes, motor housings are often machined to reduce the lead time and avoid high tooling costs in casting processes. In such a way, motor housings can be machined to exact dimensions from raw materials by means of computer numerical control (CNC) machining centers. However, this method is not suitable for housings with very large sizes and/or complicated structures.

5.1.4   Stamped Housing

In stamping processes, metal sheets are deformed to desired shapes and dimensions under high stamping forces at room temperature. Because of its high productivity and acceptable accuracy of shape and size, the stamping process is extensively used in various industries. In motor manufacturing, stamped housings and endbells are typically used for small-size, large-volume, and ordinary motors such as vacuum cleaner motors (Figure 5.10).

FIGURE 5.9   A pit furnace for heat treatment of large size sand casting elevator motor housing.

Stamped housing

Non-drive side   Drive side

FIGURE 5.10   Stamped motor housing of a low-voltage DC motor for vehicle applications, noting that the non-drive side endbell is integrated with the housing as one-piece component.

5.1.5   EXTRUDED MOTOR HOUSING

In various industries, aluminum extrusions provide engineering solutions for a diverse range of machinery and equipment needs due to their complex shapes, lightweight, high thermal conductivity, dimension precision, high strength, and minimal maintenance. For instance, properly designed and fabricated aluminum extrusions are an excellent choice for electric motor housings. An extruded motor housing with extruded fins and cooling channels on the housing wall can literally be obtained as a finished component by cross-cutting from extruded materials. Even with small or moderate production runs, the manufacturing costs of extruded motor housings are well below those of cast aluminum housings [5.13]. In addition, since cooling fins are designed on the extruded materials on all sides, as shown in Figure 5.11, the extruded housing design can give electric motor an outstanding cooling.

Like regular extrusion processes, the impact extrusion process is a type of specialty cold forming process. In an extrusion process, a metal blank is placed inside a die. A punch impacts the metal blank with an extremely high force to make it to extrude in one of the three matters (Figure 5.12): (*a*) in a forward impact extrusion, the punch pushes the metal through the orifice of the die, in the same

FIGURE 5.11 Motor housings are made by cutting from extruded aluminum. The fins are extruded in the axial direction on all sides of the extruded aluminum for enhancing heat dissipation.

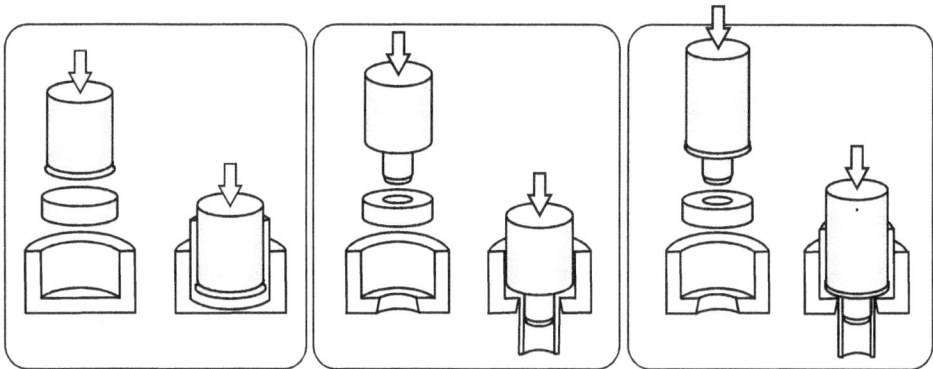

FIGURE 5.12 Schematic diagram of the formation of motor housing using impact extrusion.

direction of the applied force; (*b*) in a backward impact extrusion, the metal is forced to flow backward around the punch through an opening between the die and the punch, in the opposite direction of the applied force; and (*c*) in a combined impact extrusion, the metal flows in both forward and backward directions. The impact extrusion process is best suited for the fabrication of small motor housing with materials such as aluminum.

Compared to other manufacturing processes, the primary benefits of impact extrusion include the following:

- Minimized wastage of raw material
- Greatly lowed manufacturing cycle time
- Improved material properties (especially strengths) due to cold working
- Formed homogenous and linear grain structure without voids or porosity
- Altered wall thickness in specific areas of the blank
- Enhanced corrosion resistance
- Reduced machining operation
- Significant reduced part costs

5.1.6 Motor Housing with Composite Materials

The use of composite materials in motor housing facilitates the fabrication of motors with reduced weight, attenuated vibration, improved mechanical damping, increased corrosion resistance, enhanced heat dissipation, and thus improved motor performance. Variation of the thickness and fiber laying geometry in the composite housing may be used to control motor housing strength, stiffness, and damping characteristics. The motor housing contains both metal and composite resin material that may take more than 50%, making it lighter and easier to fabricate [5.14].

As discussed in Chapter 3, CFRP has shown excellent mechanical properties, including high rigidity, high strength, extreme corrosion resistance, and low density. This makes it well suited for motor housings. In recent years, the application of CFRP on airframes has spread rapidly [5.15].

5.1.7 Motor Housing Fabricated by 3D Printing and Other Additive Manufacturing Processes

The emergence of 3D printing technology is probably the most important technological advancement in the field of advanced manufacturing process over the past decade. 3D printing, as one of the additive manufacturing (AM) processes, is a newly emerged technology that fabricates parts by adding materials, layer upon layer, in precise geometries and shapes. It was originally appeared in some form in the 1980s [5.16] and developed by researchers at MIT in the early 1990s [5.17, 5.18] to produce relatively inexpensive prototype parts for industrial and commercial design work. Unlike traditional manufacturing processes such as machining, milling, and carving, this new manufacturing process allows computer-generated 3D models to be directly transformed into 3D physical objects through a layered printing process. Because of its cost-effectiveness and versatility, the advent of AM has ultimately the potential to become as important and game changing as the introduction of the assembly line a hundred years ago [5.19].

The emergence of 3D printing technology has posed challenges to the traditional manufacturing technologies such as subtractive manufacturing. The greatest benefit of this technology is the capability of producing incredibly complex products that cannot be manufactured by any traditional processing method. This is particularly valuable in the manufacture of human organs and some specially designed machine components. For instance, 3D printing can be used to make porous motor housings for enhancing heat dissipation and attenuating acoustic noise.

With dramatically decreased costs of 3D printers in recent years, 3D printing today becomes more popular across a variety of industries for not only building prototypes but also directly fabricating complex parts even whole systems in production lines. The survey data have shown that in the near future the market can expect growth spurts in tooling, production trial, and end-use production applications, with an emphasis on metals [5.20].

Like 3D printing, powder bed fusion technology is widely used in a variety of AM processes in recent years, including direct metal laser sintering, selective laser sintering (SLS), selective heat sintering, electron beam melting (EBM), and direct metal laser melting. These processes employ laser, electron beam, or thermal print heads to melt or partially melt ultrafine layers of materials (typically metallic alloys, ceramics, and plastics) in a 3D space for rapid prototyping in industries. For instance, the SLS process uses a laser as the power source to sinter powdered material into a solid structure based on a 3D model. In the process, the high-power laser selectively scans the thin layer of powder, sintering together powder particles in the shape of the cross-section of the first layer. As one layer is built up, the platform descends one layer depth, and another layer is built on the top of the existing layer. This process is continuously repeated until the part is complete. One of the major benefits of the SLS technology is that it requires no support structures to prevent the design from collapsing during production. This greatly saves the manufacturing cost in both material and production time. In addition, the SLS technology enables the creation of consolidated parts with complex geometries to simplify assembly processes.

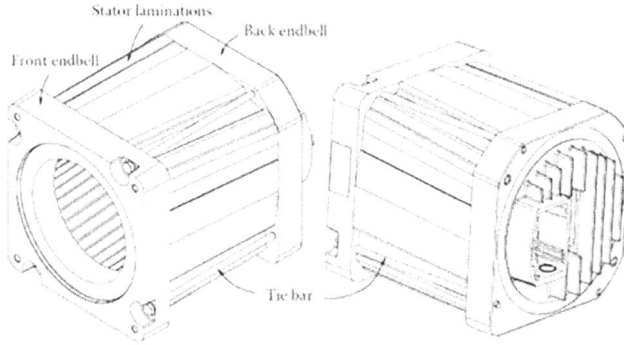

FIGURE 5.13 Frameless motor, assembled with tie bars on endbells.

In recent years, this technique has been developed to become a manufacturing process for producing solid parts. This technique is an extremely versatile and rapid process accommodating geometry of varying complexity in a wide range of industrial applications. Because this technique can support many types of materials, it is especially suitable for the fabrication of prototypes.

With newly developed 3D printing techniques such as laser sintering and EBM, this manufacturing process has been extended to produce metal parts. Although it has not been used for fabricating motor parts today, the 3D printing techniques, with the continuous improvement and cost reduction, may be applied to the fabrication of motor parts and casting toolings in the near future.

5.1.8 FRAMELESS MOTOR

Frameless motors (see Figure 5.13) are especially suitable for certain applications such as airplanes and satellites for their reduced volumes and weights. The advantage of frameless motors, as compared with conventional housed motors, resides in their high reliability and efficiency, as well as their capability to form a compact mechatronic system without using redundant parts or components.

A typical type of frameless motors is the frameless direct drive rotary (DDR) motor (Figure 5.14). A DDR motor generally consists of two separate components, that is, rotor (field) and stator (armature). Frameless motors are integrated directly with the load where the same bearings that support the load also support the motor. This configuration eliminates shaft, bearings, endbells, and couplings, which offer reduced volume, weight, and complexity, and results in improved servo stiffness and quicker response [5.21]. In addition, the frameless DDR motor eliminates mechanical transmissions, such as gearboxes, belts, and pulleys, resulting in zero backlash, quiet operation, and zero-maintenance servo solution.

Frameless motors are attractive to those customers whose applications demand a minimized motor size and weight along with a maximized dynamic performance. Since the load is often supported on its own bearing structure, the frameless motor can be integrated directly into the system/load shaft and be suspended on the same bearings with the load.

The advantages of the frameless motor include the following:

- By eliminating the motor housing, the thermal resistance between the heat sources (*e.g.*, stator and rotor windings) and the environment is greatly reduced.
- Compared with the conventional framed motor, a frameless motor has a smaller volume for a given power rating. Thus, frameless motors offer the most compact and lightweight solution for the motion control industry.
- Frameless motor has a high overall stiffness. Shaft stiffness is a function of the cube of the shaft diameter. For a frameless DDR motor, the shaft diameter can be approximately three times larger than that of a conventional motor.

FIGURE 5.14   Stators of frameless DDR motor.

- Hollow shafts with large diameters can be used in frameless motors for not only reducing rotor inertia but also improving motor cooling.
- Frameless motors can reduce not only the part cost but also assembly time. Therefore, the total motor cost is reduced.

For frameless motors, the motor sealing system must be carefully designed to prevent any potential damages to the motor parts.

5.2   TESTING METHODS OF CASTED MOTOR HOUSING

Testing machines are commonly used by motor manufacturers to determine mechanical properties of materials, such as yield strength, ultimate strength, strain, and elongation. As shown in Figure 5.15, a typical tension testing machine possesses a movable upper head and a stationary lower head. The grips, which are attached to the heads of the testing machine, hold the specimen firmly at each end. A tensile load is then applied to the test specimen by controlling the separating rate of the two heads until the specimen fractures. This process produces a stress–strain curve showing how the testing specimen behaves throughout the testing process. The test data are recorded by a monitoring/measuring system during the test and are used to determine the mechanical properties of the material.

In order to understand the strength distribution over the whole motor housing, the test specimens are specifically sampled at different locations of the housing. The dimension and size of the testing specimen are described in the standard [5.22] and are shown in Figure 5.16. The specimen has a reduced gage section in the middle and enlarged shoulders at the ends for being held by serrated grips. To avoid the effect from the shoulders, the length of the transition region x should be at least as big as the specimen width w, and the total length of the reduced section L should be at least four times the width w.

Figure 5.17 illustrates the tested specimens that were cut from the different locations of the same casted aluminum housing. It shows clearly that all fractures occurred in the test area between the gage marks.

The stress σ is determined from the measured load F and the original specimen cross-sectional area A_o:

$$\sigma = \frac{F}{A_o} \tag{5.1}$$

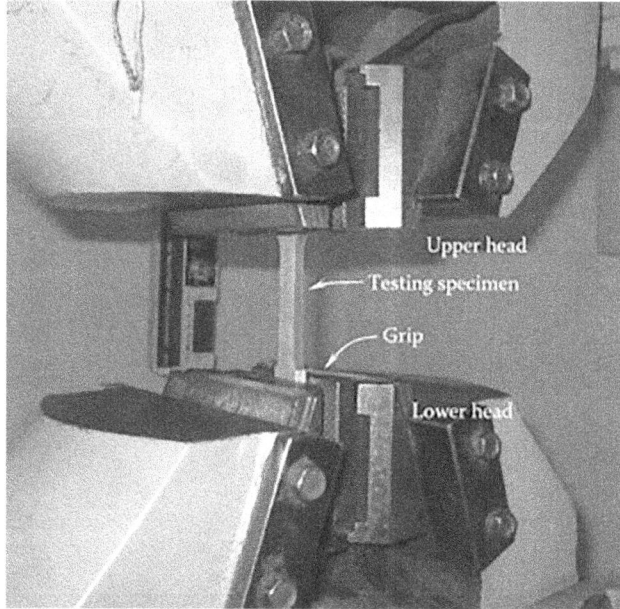

FIGURE 5.15 Standard configuration for tensile testing strip material cutting from casting housing and endbell.

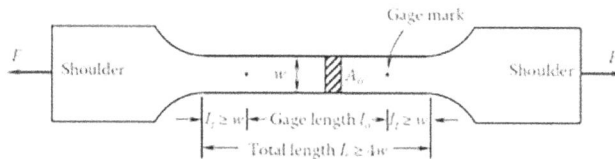

FIGURE 5.16 Typical tensile specimen, showing a reduced gage section at the middle and enlarged shoulders at the ends of the specimen.

The strain ε is calculated as the ratio of the length change Δl during the test to the original specimen length l_o, which is measured between the gage marks before the test:

$$\varepsilon = \frac{l - l_o}{l_o} = \frac{\Delta l}{l_o} \tag{5.2}$$

After fracture, the fractured specimen needs to be put together to measure its gage length l_1. Thus, the elongation of the material is calculated as

$$\varepsilon_e = \frac{l_1 - l_o}{l_o} = \frac{\Delta l_e}{l_o} \tag{5.3}$$

Stress–strain curves are an extremely important graphical measure of a material's mechanical properties. As an example shown in Figure 5.18, for low strains, the material obeys Hooke's law to a reasonable approximation so that stress is proportional to strain with a constant of proportionality (the line AB). Young's modulus E is computed as

$$E = \frac{\sigma}{\varepsilon} \tag{5.4}$$

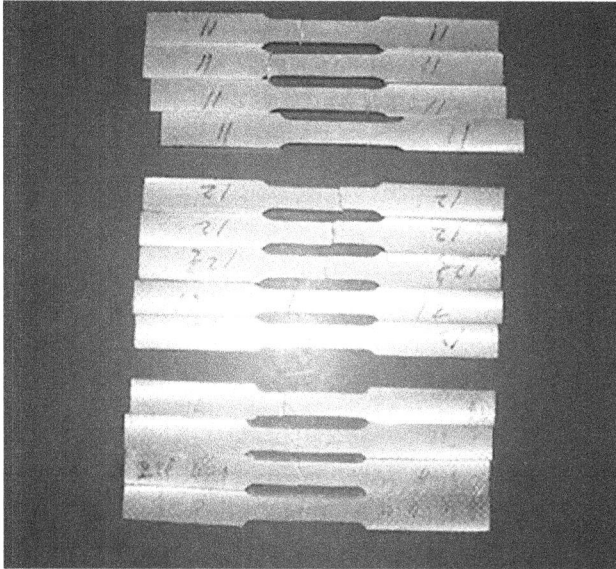

FIGURE 5.17 Testing specimens cutting from different locations of a casting housing.

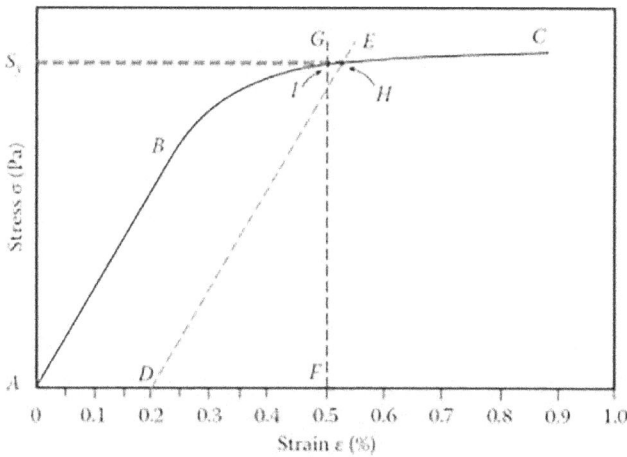

FIGURE 5.18 Determination of yield strength from the offset method and the extension under load method.

As strain is increased, the material eventually deviates from this linear proportionality and exhibits nonlinear behavior between stress and strain. The highest point attained before the stress–strain line begins to bend is defined as the proportional limit, which is the maximum point at which the material obeys Hooke's law.

Material yield strength refers to the point on the stress–strain curve beyond which the material experiences deformations plastically, as the molecular or microscopic structure undergoes a rearrangement and atom lattice planes slide over each other. This indicates that the yield point and the proportional limit refer to the same point. The only difference is that one is defined from the microscopic perspective and another from the macro perspective.

In practice, it is often difficult to pinpoint the exact yield strength at which plastic deformation begins to curve. There are primarily two common methods to determine yield strength from the stress–strain curve: the offset method and the extension under load method. The offset is the

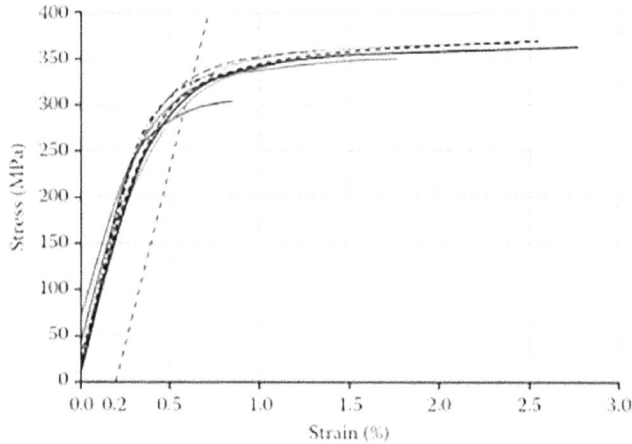

FIGURE 5.19 Stress–strain curves obtained from tested specimens cutting from different locations of the casted aluminum housing. Each curve represents the stress level at that location.

horizontal distance between the modulus line (the line *AB*) and any line running parallel to it. In industries, tensile yield strengths are often determined by this method at an offset of 0.2% of strain. The intersection of the offset line and the stress–strain curve (the point *H*) thus gives the yield strength.

The extension under load method involves drawing an ordinate line *FG* from the point on the *x*-axis where the strain equals the specified extension, typically 0.5%. The intersection of the line *FG* and the stress–strain curve (the point *I*) defines the yield strength. From Figure 5.17, it can be seen that the values determined by two methods are very close.

The measured stress–strain curves from the samples being cut from different locations of the housing are displayed in Figure 5.19. It can be seen that the ultimate tensile strength and strain data distribute in the range of 300–370 MPa and 0.7%–2.7%, respectively. These results are very important for optimizing the motor housing design and improving casting processes.

5.3   ENDBELL MANUFACTURING

As an integral part of the motor structure, motor endbells are used to cover the ends of motor housing, to house bearings at the ends of a motor for supporting the rotor, to provide paths for cooling airflows, and to protect internal electrical and mechanical parts from moisture and dirt. Motor endbells can be made by many manufacturing processes, such as casting, machining, stamping, forging, extruding, and powder metallurgy sintering. In considering the production efficiency, cost, appearance, and other factors, casting processes have been widely adapted to produce motor endbells.

5.3.1   CASTED ENDBELL

Similar to casting motor housings, motor endbells are often produced through a casting process. As shown in Figure 5.20, an endbell is made of die casting with an aluminum alloy. In many cases, endbells must be designed to provide a rigid and stable support for the stator and rotor assembly. For the consideration of safety, a minimum clearance gap of 1.5 mm must be provided between the coil end turns and the endbell wall.

Motor endbells can be made by either die casting or sand casting. Compared with sand casting, die casting produces parts with thinner walls, higher-dimensional accuracy, and smoother surfaces. Production is faster and labor cost per casting is lower.

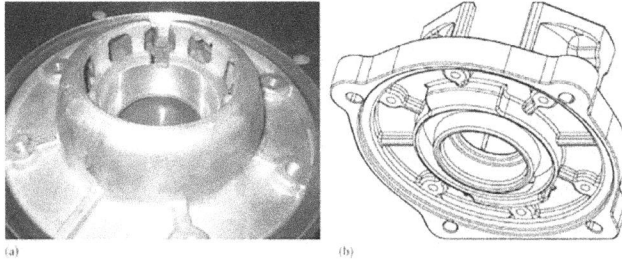

FIGURE 5.20 Aluminum endbells made by the die casting process: (*a*) with ventilation passages and (*b*) without ventilation passages.

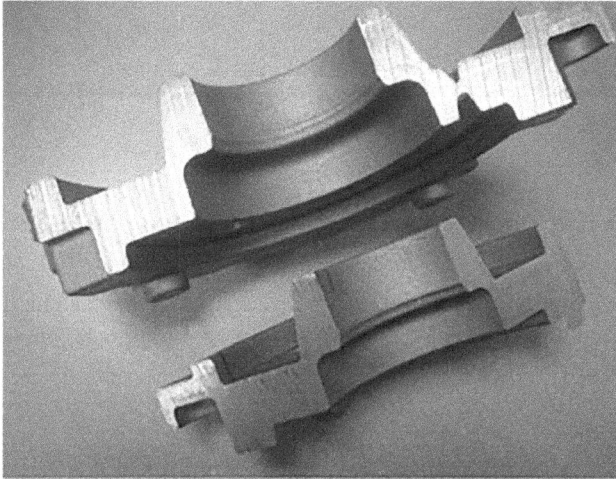

FIGURE 5.21 A casted endbell is cut to inspect the casting defects, particularly for porosities and voids.

For some motors with low IP requirements, endbells may be designed to have cooling openings serving as the passage of air over and around the windings. For totally enclosed motors, the endbells must have reliable and tight sealing capabilities. Therefore, no openings are allowed on the endbells.

In the design of casting endbells, it is important to consider the effect of material impurity and porosity on the strength of casting parts. In order to ensure good quality of casting, it is common to cut the casted parts to inspect casting defects, especially porosities and voids, as demonstrated in Figure 5.21. However, due to the multiparameter control in casting processes and the complex geometry of motor components, it is difficult to completely avoid casting defects. As an example, a broken endbell is presented in Figure 5.22.

5.3.2 Stamped Endbell

To provide an economic solution for endbell production, stamped endbells are often used in low- or mid-end motors. As a matter of fact, stamped parts are widely used in many advanced industries such as automotive, aerospace, and energy due to their good mechanical properties (*e.g.*, light-weight), low cost, and quick production. In practice, stamped parts have replaced many expensive casted, forged, and machined products in small motors.

5.3.3 Iron Casting versus Aluminum Casting

Several metal alloys are available for die casting such as aluminum, zinc, magnesium, iron, and copper. In most modern small motors, the endbells are made from aluminum alloys for their good

FIGURE 5.22 Broken aluminum die casting endbell due to the porosity and impurities in the part.

corrosion resistance, lightweight, high-dimensional accuracy, stability, and high thermal conductivity. Endbells made from high-grade cast iron are often used for large motors due to their strength, rigidity, and ability to attenuate motor vibration and noise.

5.3.4 MACHINED ENDBELL

Similar to machined housings, machined endbells are often used for motor prototypes or small quantities of products for a short lead time and no need of tooling. It is apparent that machined endbells are more expensive than those made by other manufacturing processes such as die casting for a mass production. With the rapid development of 3D printing technology, it is possible to use the 3D printing technology to replace the machining technology for making motor endbell prototypes in the future.

5.3.5 FORGED ENDBELL

Forging is a manufacturing process involving the shaping of metal by pressing or hammering. Metals can be either cold-forged or hot-forged depending on applications. Forging refines the grain structure in the direction of the deformation and develops the optimum grain flow. The modified structure gives forged parts better mechanical properties than machined parts in which the grain flow is broken by machining. Generally, forging technology is suitable for making parts that have simple geometries. For endbells with complicated structures, casting is more convenient than forging as the manufacturing process.

5.4 MOTOR ASSEMBLY METHODS

There are a number of different ways to secure motor assembly, each of which is applicable to certain applications depending on the motor size, motor type, IP, and mounting requirements.

5.4.1 TIE BAR

This method is to use nuts, in coordination with long bolts (or in the alternative key bars), which extend through a distal side of one endbell, the housing and the opposite endbell, to secure the

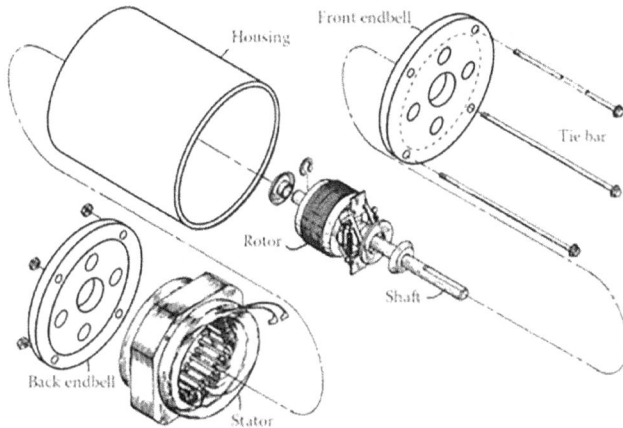

FIGURE 5.23 Motor assembly is secured by tie bars (U.S. Patent 5,412,270) [5.23]. (Courtesy of the U.S. Patent and Trademark Office, Alexandria, VA.)

assembly together, as shown in Figure 5.23 [5.23]. The high compressive forces produced by bolts act to maintain the end shields in a static position with respect to the housing. This assembly method is especially suitable for a motor with a low length-to-diameter ratio. For motors with large length-to-diameter ratios; however, this assembly method is prohibitive due to the inherent structural instability caused by the increased bolt length to match the long motor body.

5.4.2 Tapping at Housing End Surface

A frame assembly method incorporates tapping directly at the housing end surfaces, creating attachment sites (*i.e.*, threaded spaces) integrally as part of the housing itself (Figure 5.24). Specifically, short screws are threaded into the tapped housing through the endbells. However, this necessitates either a significant increase in the housing wall thickness or a housing design with complex structures formed with additional material that integrally provide a portion of the housing configured to receive the screws.

5.4.3 Forged Z-Shaped Fastener

Forging refers to the shaping of metal by plastic deformation. Forging is one of the most economical metal-forming methods for its advantages of significant material saving, simplified production process, high production rate, superior surface quality (especially for cold forging), high shear strength, and high fatigue resistance. During a forging process, internal grain deforms to follow the general shape of the part. As a result, the grain is continuous throughout the part, giving rise to a piece with improved strength characteristics [5.24].

Cold forging is done at room temperature while conventional hot forging is done at temperatures higher than material recrystallization temperature (for instance, up to $1,150°C$ for steel, $360°C–520°C$ for aluminum alloys, and $700°C–800°C$ for copper alloys). Motor manufacturers may choose cold forging over hot forging for a number of reasons, including little or no finishing work, higher-dimensional accuracy, better surface finishing, and improved reproducibility and interchangeability.

One of the most common types of cold forging is a process called impression-die forging, where the metal is placed into a die. Then, the metal is forced into the die by a pressing machine or a descending hammer. In cold-forging operations, pressures as high as $2,500\,MPa$ are developed at the forming tool-workpiece interface [5.25]. In order to minimize the friction between the tool and the processed workpieces, phosphate coatings together with lubricants are applied prior to cold forging.

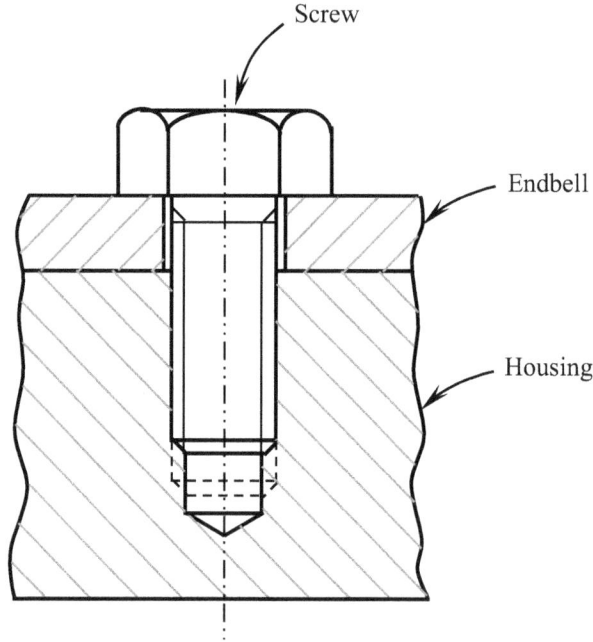

FIGURE 5.24 The motor housing is tapped for receiving short screws, which go through the endbell to complete the construct of the motor frame.

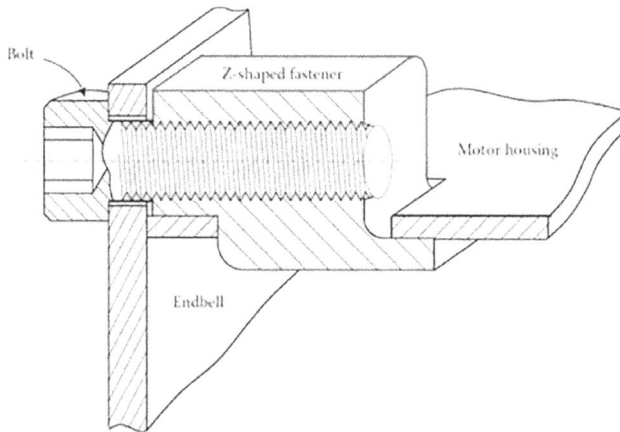

FIGURE 5.25 Jointing motor housing and endbell with Z-shaped fastener.

It is widely recognized that surface finish has a significant effect on fatigue behavior. During a forging process, oxide and scaling layers are formed on the surfaces of the forged workpieces. Thus, it requires using either chemical methods such as acid pickling or mechanical methods such as sand blasting and shot peening to remove such oxide and scaling layers. In addition, the residual stress, produced in forgings as a result of inhomogeneous deformation in the forged workpieces, must be relieved through the annealing process following the forging process.

To provide robust motor frame assemblies, Boyland *et al.* [5.26] invented Z-shaped fasteners for reinforcing the joints between endbells and motor housings (Figure 5.25). Fastening apertures are formed in the lateral ends of the housing. The fastener includes elements that engage the housing through a plurality of contact areas. Specifically, the fastener includes a central portion that

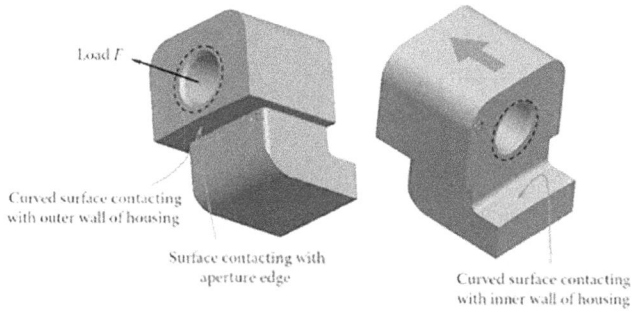

FIGURE 5.26 Contacting surfaces of Z-shaped fastener to housing.

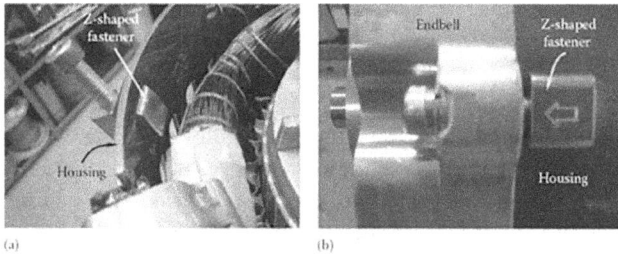

FIGURE 5.27 Configuration of Z-shaped fastener to housing (a) and housing endbell (b).

establishes a contact area in the plane of the aperture, as well as a base extension tab that establishes at least one contact area with the interior wall of the housing (Figure 5.26). The fastener also includes an extension block that establishes a contact area with the exterior of the housing and engages a securing bolt. The securing bolt threads through an aperture in the endbell and engages the fastener, thereby securing the endbell to the housing. The configuration of a Z-shaped fastener to a housing and housing endbell is given in Figure 5.27.

Advantageously, this type of fastener reinforces the motor housing so as to provide additional support to the assembly by reinforcing the joints between the endbell and the motor assembly main body. Furthermore, the invention secures the endbell to the housing by simply engaging the housing and the securing bolt. Accordingly, the fastener acts to simplify housing/endbell manufacturing and the assembly process.

5.4.4 Rotary Fasteners

Rotary fasteners refer to the fasteners that insert through the apertures of joint members and rotate certain angles to secure the joint members together. This type of fastener is capable of quickly and easily securing and releasing joint members.

5.4.4.1 Triangle-Base Rotating Fastener

Figure 5.28 illustrates a triangle-base rotating fastener with a triangle-shaped base extension tab. The fastener utilizes both horizontal and vertical forces in relation to the housing while engaging the housing's fastening aperture. The extension block and base extension tab may exert forces on the exterior and interior housing walls in a vertical direction. These vertical forces work in coordination with horizontal forces exerted by the central portion of the fastener and the force created by engaging a securing bolt with the extension block.

With regard to engaging the securing bolt with the extension block, the securing bolt used during the fastening process applies a load to the fastener. The load works, in coordination with the

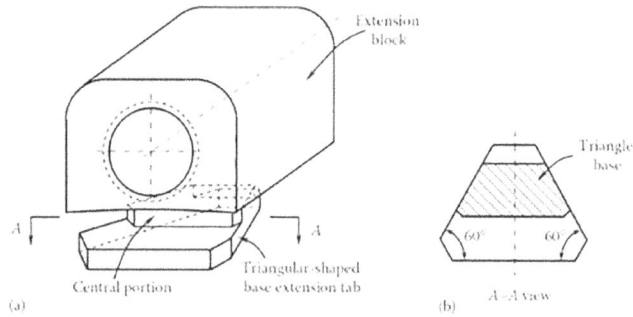

FIGURE 5.28   Triangle-base rotating fastener.

fastener contact areas, to maintain the vertical and horizontal fastener position with respect to the housing, the endbell, and the securing bolt. The contact areas apply a retaining force or pressure to the interior and exterior walls of the housing as well as apply a force in the plane of the fastening aperture portion of the housing.

The underside of the extension block joins the central portion to create a second section of the fastener. The central portion lies in the plane of the fastening aperture as the fastener engages the housing. This central portion also separates the base extension tab and the extension block in the fastener. The central portion is formed such that after the fastener is rotated, the central portion provides additional structural support along with an additional contact area between the fastener and the housing. Generally, both the extension block and the base extension tab extend beyond the central portion to engage the surface of the housing. This advantageously allows for the fastener to firmly engage both the interior and exterior sides of the housing wall, as well as a securing bolt.

The cross-sectional view of triangular base extension tab is shown at the right-hand side in Figure 5.28, taken along lines *A-A*. It is to be understood that for a given embodiment, both fastening aperture and the base extension tab are designed to match and work together. Generally, the base extension tab is formed with a similar contour as the fastening aperture. The shape of the fastening aperture on the housing also contributes to determining the shape of the central portion (shown shaded). The wedge shape of the central portion allows it to engage the triangular fastening aperture. It can be seen from the figure that making the fastening aperture a triangular shape leads to a complementary contoured triangular base extension tab. Varying the shape of the fastening aperture leads to subtle implementation variances in the placement of the retention forces securing the elements.

Figure 5.29 illustrates the steps of engaging the fastener having a triangular base extension tab with a motor housing. Initially, as shown in Figure 5.29a, the fastener is positioned above a fastening aperture, thereby aligning the complementary contours of the triangular base extension tab with the fastening aperture. After the elements are aligned, the fastener is inserted into the fastening aperture. In Figure 5.29b–d, the central portion and the triangular base extension tab are represented with dashed lines as they are hidden beneath the fastening aperture and in the interior side of the housing. To ensure a secure connection between the fastener and the housing, the thickness of the central portion is equal to the housing thickness. Figure 5.29b shows the fastener with the central portion, and the base extension tab is inserted into fastening aperture. The extension block prevents the fastener from fully passing through the aperture and advantageously provides an additional contact area between the fastener and the housing. As shown in Figure 5.29c, upon insertion, the fastener is rotated by 180° with respect to the axial edge of the housing. The degree of rotation is one of the subtle implementation-specific aspects of the fastening processes referenced earlier.

Figure 5.30 illustrates a fastener viewed from the top before and after its rotation. The 180° rotation illustrated creates staggered contact areas for the apex and the base corners of the base extension tab and the triangular aperture. In Figure 5.30, the base extension tab and the complementary

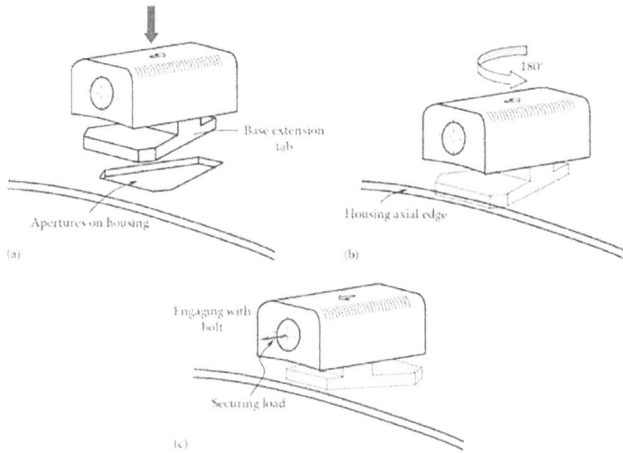

FIGURE 5.29 Fastening mechanism of triangle-base rotating fastener: (*a*) inserting a fastener into an aperture on the housing, (*b*) rotating 180° of the fastener, and (c) applying the securing load to the fastener.

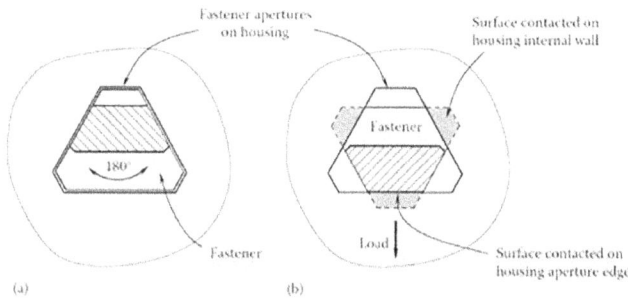

FIGURE 5.30 Installation of triangle-base rotating fastener: (*a*) inserting the triangle-base fastener and (*b*) rotating the fastener by 180° so that the three corners of the fastener base come in contact with the interior of the housing (the dark areas).

aperture are aligned to facilitate insertion of the fastener into the aperture. As shown, the central portion is set back from the leading edge of the base extension tab.

Rotating the fastener provides three contact areas on the interior surface of the assembly housing, which, in turn, provide additional structural support for the fastening assembly. Specifically, as the hook nut is inserted into the slot and turned by 180°, the three corners of the base extension tab engage the housing's internal wall, thereby providing radial support for balancing the bending moment acting on the fastener as the securing bolt is engaged.

Once the fastener is engaged with the housing, the other elements of the assembly may be secured to the housing. The fastener also includes an extension block that establishes a contact area with the exterior of the housing and engages a securing bolt (not shown). The securing bolt threads through an aperture in the endbell and engages the hollow portion of the extension block. The engaged securing bolt and fastener rigidly secure the endbell between the securing bolt head and the fastener's extension block. It is to be understood that the fastener may work alone or in coordination with other fasteners located at areas along the circumference of the axial end to secure the endbell to the housing.

5.4.4.2 Square-Base Rotating Fastener

A square-base rotating fastener is presented in Figure 5.31. This type of fastener is similar to the triangle-base rotating fastener, except that the central portion and the base extension tab of the

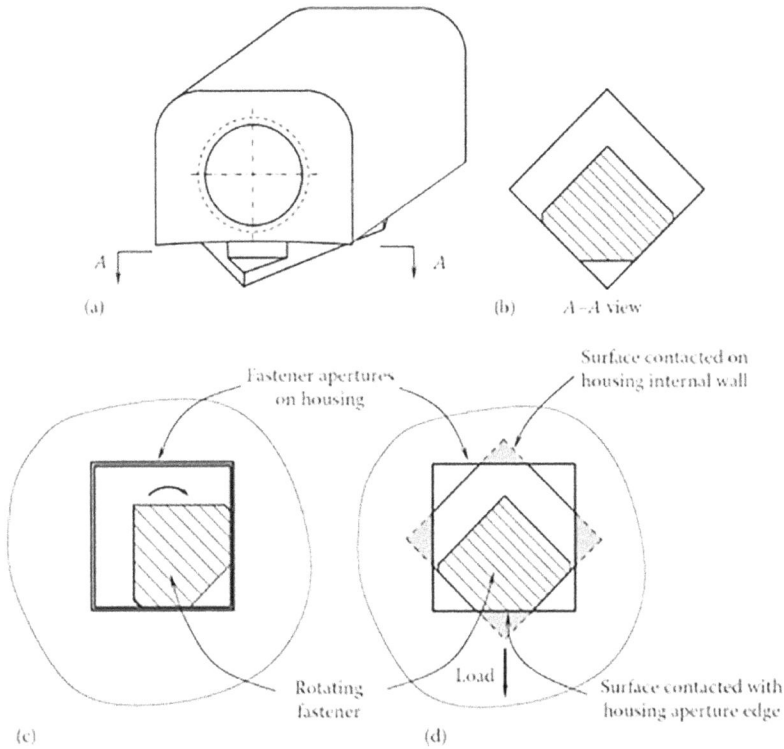

FIGURE 5.31   Installation of square-base rotating fastener: (*a*) a perspective view of square-base rotating fastener, (*b*) a cross-sectional view of the square-base extension tab, (*c*) inserting a fastener into an aperture on the housing, and (d) rotating 45° of the fastener and applying the load on the fastener.

fastener are substantially square shaped. As can be seen in Figure 5.31a and b, the central portion of the fastener is set back from the edge of the base extension tab, allowing for an additional contact area when the fastener engages the housing. Figure 5.31c and d illustrates the process of applying the square-base rotating fastener. As discussed previously, varying the geometry of the fastening aperture, the base extension tab, and the central portion leads to subtly different placement of the retention forces. Specifically, as the fastener engages the securing bolt (not shown) and the housing, there are different retaining forces exerted on the interior wall of housing by the square-base extension tab. The square-shaped base extension tab has four independent areas of contact with the interior wall of the housing. Further, due to the geometry associated with this embodiment, the fastener is rotated 45° to engage the housing. The other aspects of engaging the fastener, the housing, the endbell, and the securing bolt are similar to those discussed previously with regard to triangle-base fastener.

5.4.4.3   Butterfly-Base Rotating Fastener

A butterfly-base rotating fastener is demonstrated in Figure 5.32. It is similar to the triangle-base and square-base fasteners, except the base extension tab is substantially butterfly shaped. The contact surface to the housing is shown at the bottom horizontal wedge of the central portion.

Figure 5.32a illustrates a perspective view of the butterfly-base rotating fastener, and Figure 5.32b presents the cross-sectional view of butterfly-base extension tab taken along lines A-A in Figure 5.32a. Figure 5.32c and d presents the top views before and after the rotation of the fastener, respectively. The base extension tab and the complementary aperture are aligned to facilitate insertion of the fastener into the aperture. The final orientation of the fastener after its 90° clockwise rotation is shown in Figure 5.32d. It shows the interior contact areas at the front and the back of the fastener aperture.

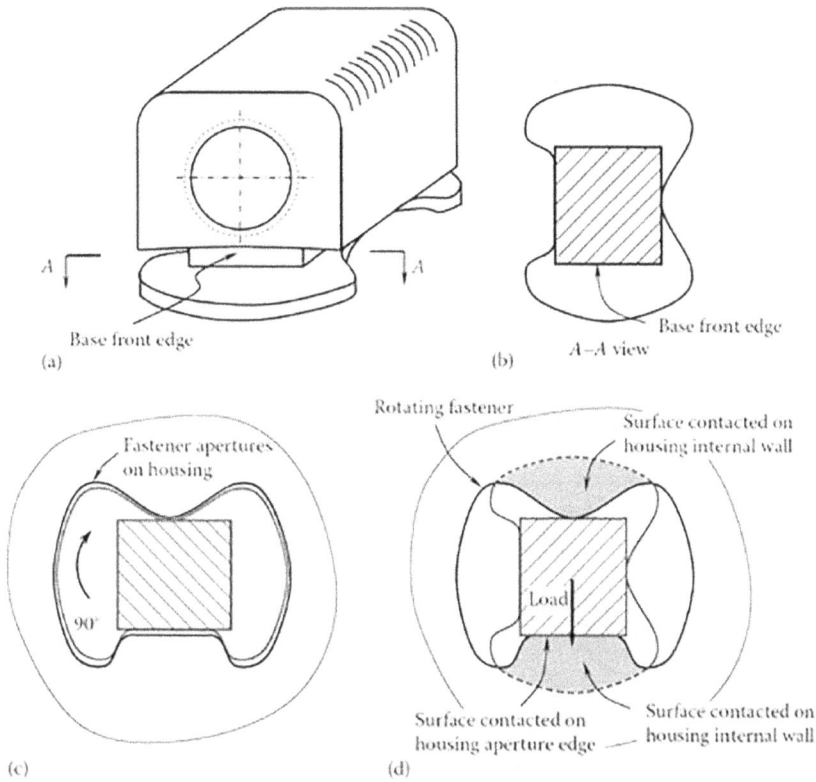

FIGURE 5.32 Installation of butterfly-base rotating fastener: (*a*) a perspective view of butterfly-base fastener, (*b*) a cross-sectional view of the butterfly-base extension tab, (*c*) inserting the fastener into an aperture on the housing, and (*d*) rotating 90° of the fastener and applying the load on the fastener.

5.4.5 OTHER TYPES OF FASTENERS

There are other types of fasteners that may be applied in motor assemblies, each type being appropriate for a certain type of motor.

5.4.5.1 Cylinder-Base Fastener Locked with Retaining Ring

Figure 5.33 demonstrates a cylinder-base fastener that is locked with a retaining ring. The fastener has two grooves at the cylindrical base to receive the retaining ring. To ensure normal performance of the fastener, the fastener base is designed to have a sliding fit with the hole on the motor housing, and the surface of the retaining ring must be in contact with the inner wall of the housing. This type of fastener may lead to less production costs associated with the simple elements, as well as a less complex manufacturing and assembly process.

An alternative locking mechanism may also be implemented as a securing pin engaged with a hole through the cylindrical base.

5.4.5.2 Cylinder-Base Fastener with Self-Locking Aperture

This design is essentially similar to the fastener design addressed previously. As shown in Figure 5.34, instead of using a retaining ring to lock the fastener, this type of fastener can be self-locked by engaging an aperture on the motor housing. For this purpose, the fastener consists of two cylinders having the diameter d_1 and d_2, respectively. The self-locking aperture consists of two circles that have diameter D_1 and D_2, respectively. The two circles are connected smoothly by two arcs to form a pear-shaped aperture, with the smaller end disposed toward the axial end of the

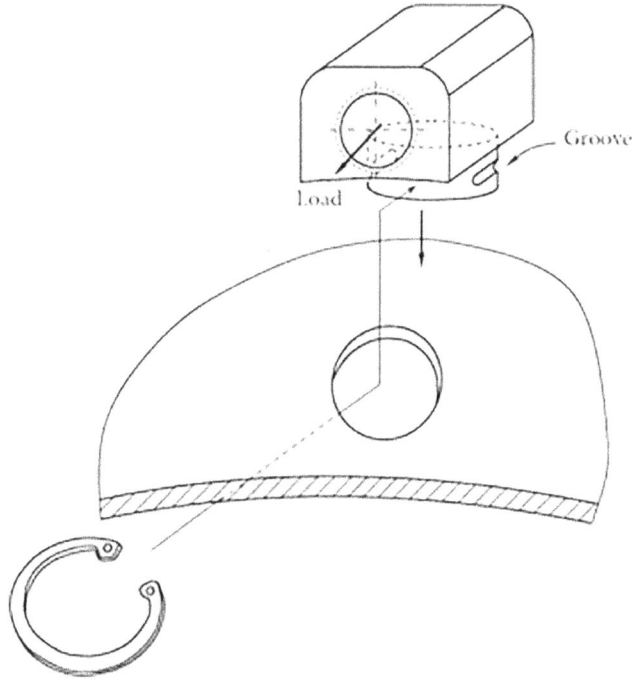

FIGURE 5.33   Cylinder-base faster, locked with a retaining ring at the grooves of the cylindrical base.

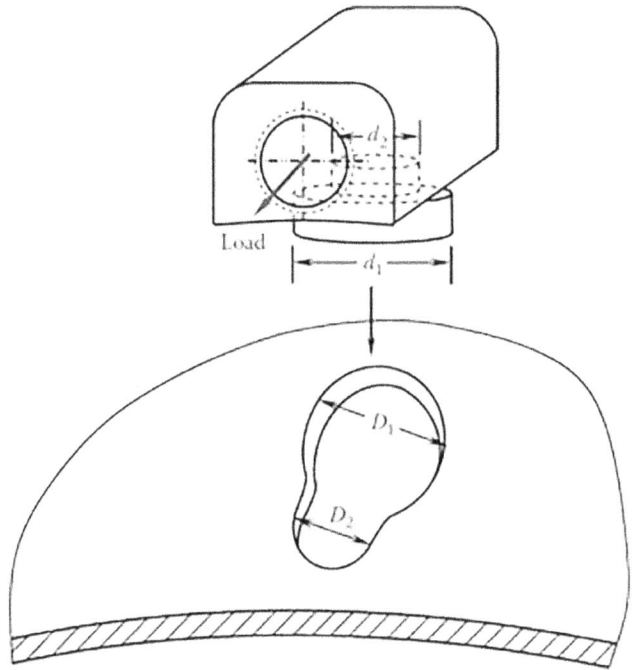

FIGURE 5.34   Cylinder-base fastener with a self-locking aperture on the motor housing.

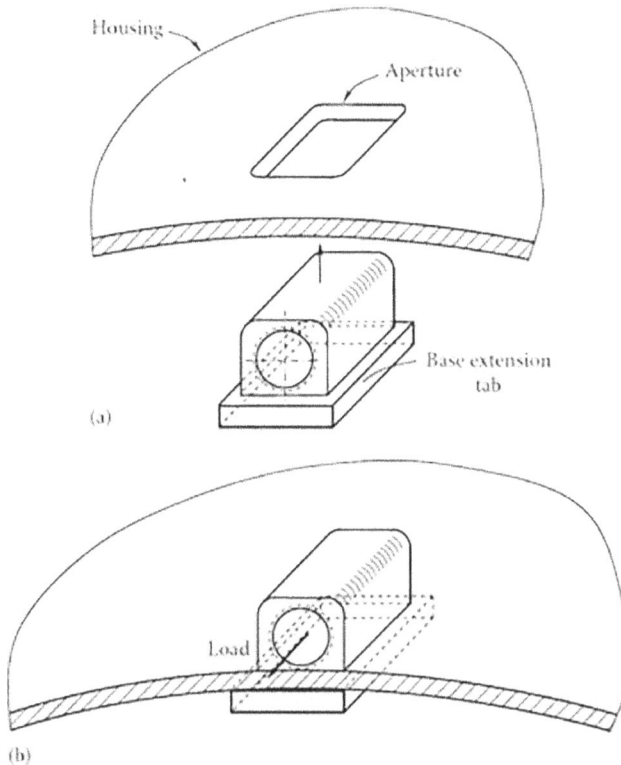

FIGURE 5.35 Installing process of the nut: (a) inserting the nut from the interior of the housing and (b) engaging the nut with the endbell using a bolt.

housing. The diameter d_1 is designed to be a little smaller than D_1, and d_2 is a little larger than D_2. This geometry facilitates the fastener's insertion through the larger circle of the aperture and the engagement to a securing bolt (not shown). With the tightening load applied on the fastener, the small cylinder achieves an effective interference fit with the small circle of the aperture.

5.4.5.3   Fastener Engaged with Housing from Housing Interior
Figure 5.35 illustrates a rectangular base fastener inserted into the fastening aperture from the interior side of the housing. As shown in the figure, the extension block of the fastener is inserted along the arrow into the fastening aperture. The continuous base contacts the entire circumference of the interior wall of housing along the fastening aperture. Once the fastener is inserted into the fastening aperture, a securing bolt (not shown) acts to apply a load along the arrow. The securing bolt acts in coordination with the continuous base extension tab to secure both the fastener and the endbell (not shown) to the housing.

5.4.5.4   Self-Clinching Fastener
Self-clinching fasteners, including self-clinching nuts and self-clinching bolts (Figure 5.36), offer solutions on multiple fronts. They take less space and require fewer assembly operations compared with conventional fasteners. In addition, self-clinching fasteners have greater reusability.

    As an example, a self-clinching nut is designed to become a permanent fixture of the component or device onto which it's installed. The nut incorporates a knurled platform and an angular recess. This platform, when pressed into ductile metal, causes the displaced host material around the mounting hole to cold flow into a specially designed annular recess in the shank of the nut (Figure 5.37). Thus, the serrated clinching ring, knurl, or hex head prevents the nut from rotating in

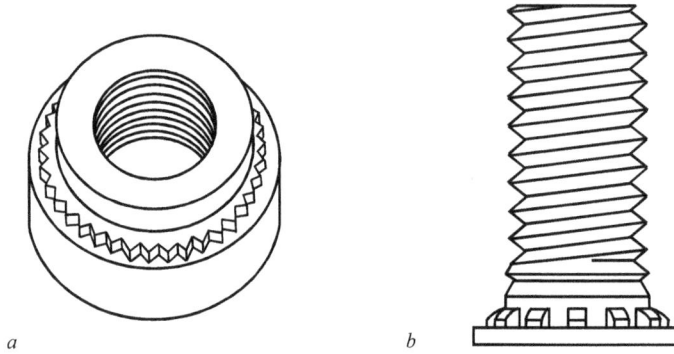

FIGURE 5.36 Self-clinching fastener: (*a*) nut and (*b*) bolt.

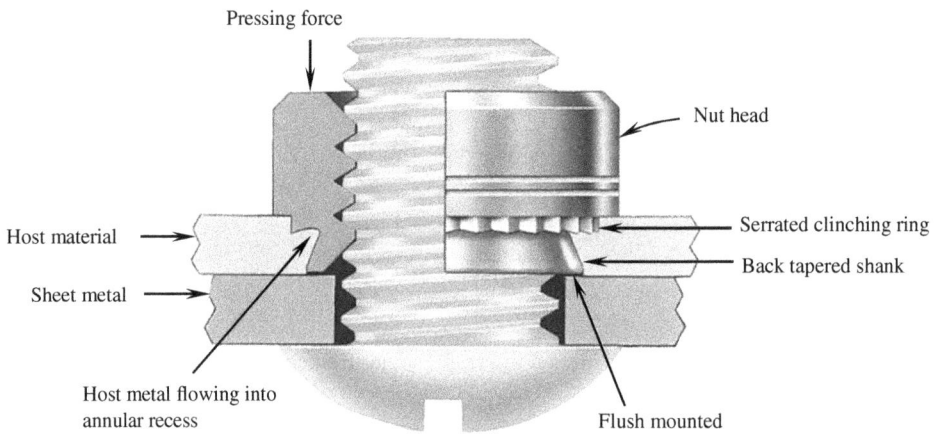

FIGURE 5.37 Self-clinching nut where the serrated clinching ring provides torque resistance (Courtesy of Penn Engineering & Manufacturing Corp., Danboro, Pennsylvania).

the host material once it has been properly inserted. Consequently, the nut allows for repeated screw removal and reattachment without compromising thread-locking performance [5.27].

5.5 FASTENING SYSTEM DESIGN

In the motor industry, it is very common to use fasteners (bolts, nuts, screws, *etc.*) for joining motor components together and for installing motors into places.

5.5.1 TYPES OF THREAD FASTENERS

Though a wide variety of threaded fasteners are used in engineering practice, two types of thread fasteners primarily used in motor assemblies and installations are as follows:

- Bolt-nut fasteners, in which a bolt passes through slightly larger holes in two joint members and is engaged with a hexagonal nut. The desired tightness of the fastener can be achieved by applying a wrench to either the bolt head or nut. In this way, the two joint members are clamped between bolt head and nut, as shown in Figure 5.38. The grip length l_g is the sum of the thicknesses of both joint members:

$$l_g = l_1 + l_2 \tag{5.5}$$

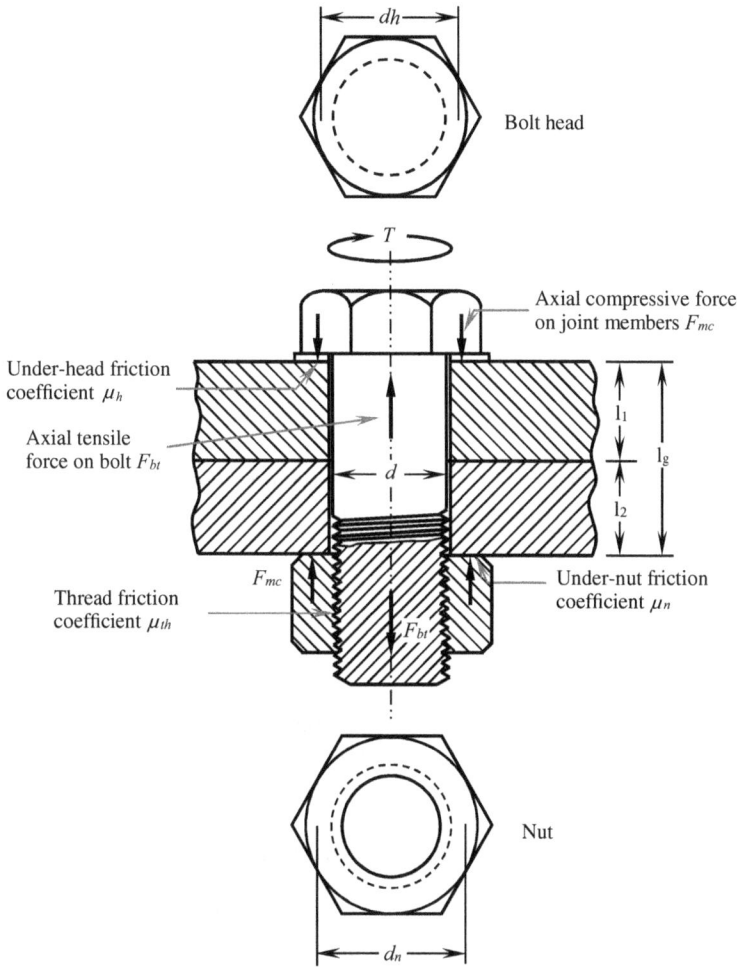

FIGURE 5.38 Bolt–nut fasteners with drilled holes through joint members.

- Screw-threaded member fasteners, in which the threads of the screw are threaded into an internally threaded hole of the lower joint member, as shown in Figure 5.39. The effective grip length is determined according to the bolt diameter relative to the thickness of the first joint member:

$$l_g = \begin{cases} l_1 + l_2/2 & \text{for} \quad l_2 < d \\ l_1 + d/2 & \text{for} \quad l_2 \geq d \end{cases} \tag{5.6}$$

In both cases, while bolts are subjected to tensile forces, joint members withstand compressive (clamping) forces.

5.5.2 THREAD FORMATION

Fastener threads can be produced by either rolling or cutting operations. Rolled threads are formed by rolling the reduced diameter portion of a shank between two reciprocating serrated dies. The advantages of rolled threads include more accurate and uniform thread dimensions, smoother thread surfaces, higher thread strength (usually 10%–20% in tensile strength), improved fatigue and wear

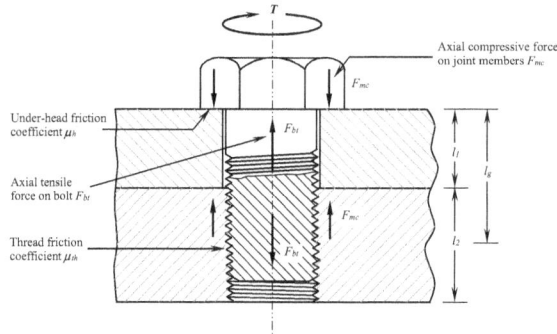

FIGURE 5.39 Screw-threaded member joint. Due to the elimination of a nut, the friction under nut-bearing surface no longer exists.

resistance, and material savings. In addition, the grain structure in rolled threads maintains continuous unbroken lines, as opposed to the broken grains in cut threads. Usually, cut threads are used for large diameter or nonstandard fasteners.

5.5.3 Fastener Preload

The preload P_p, or the clamp load, of a fastener is created as a tightening torque is applied on the fastener before receiving any external loads. The tightening process exerts an axial preload tension on the bolt and an axial compression on the joint members, where the generated bolt tension load exactly counterbalances the compression load applied on the assembled components. It has been reported that when an external load is applied on a preloaded joint, only a portion (usually 80%) of the external load acts on the jointed members [5.28].

In order to achieve high strength and long lifetime of a bolted joint, it is critical to set appropriate bolt tightening torque. Excessive tightening of bolts can cause thread damage or bolted joint failure, and insufficient tightening may lead to separation of jointed members or poor motor performance (such as large vibration). Many surveys have shown that among the possible causes of bolted assembly failure, the most frequent cause is poor assembly. Improper and irregular bolt tightening alone accounts for over 30% of all assembly failures and 45% of all fatigue incidents [5.29].

5.5.4 Fastener-Tightening Process

During a fastener-tightening process, a desired tightening torque is applied to a threaded fastener to stretch the bolt while compressing the clamped joint members elastically according to the effective spring rates of the bolt and the joint members. This process can be characterized by plotting the input torque against the angular displacement of the bolt. As illustrated in Figure 5.40, four distinct zones can be identified in a fastener-tightening process:

- Rundown or prevailing torque zone. In this zone, the fastener has not yet touched the bearing surfaces. There is no axial load acting on both the fastener and joint members. The torque needs only to overcome the friction between the fastener threads.
- Alignment zone. In this zone, the fastener and bearing surfaces are drawn into alignment, and the torque-angle curves are nonlinear. The compressive force acting on the joint members begins to increase but not at the same rate as the bolt tensile force.
- Elastic clamping zone. This is the working zone of the fastener. In this zone, the angular displacement of the fastener is proportional linearly to the applied torque on it. The slope of the torque–angle curve at this zone is a function of the stiffness of the assembly and the friction between the threads and on the bearing areas.

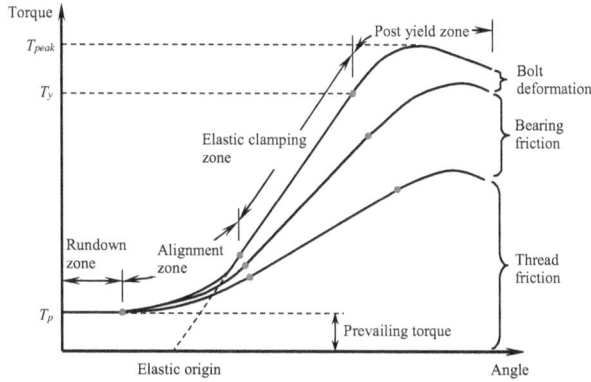

FIGURE 5.40 Four distinct zones in the fastener-tightening process.

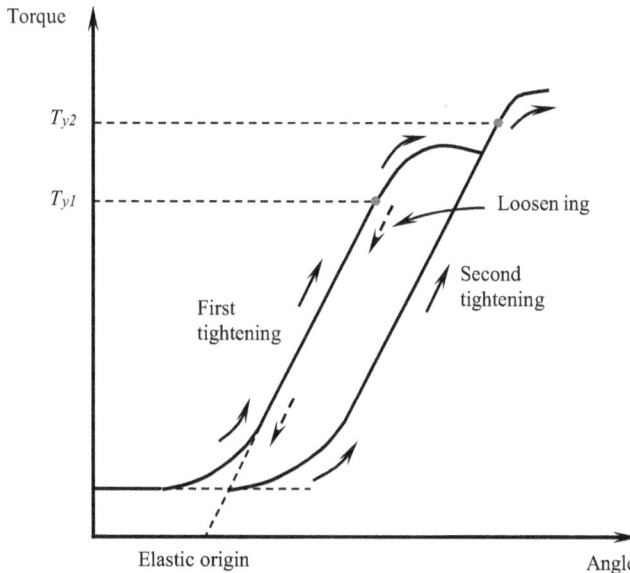

FIGURE 5.41 By loading the fastener beyond the yield point, then unloading and reloading again, yield strength, tensile strength, and hardness of the fastener increase, but its ductility decreases.

- Post-yield zone. This zone begins with the yield point. Yielding may occur in the bolt or in the joint members or both. Correspondingly, the slope of the torque–angle curve decreases. A continuous increase in the angular displacement will cause the failure of the joint.

The fastener yield torque improvement due to cold working is demonstrated in Figure 5.41. During a tightening process, both the torque applied to the fastener and the angular displacement of the fastener increase simultaneously along the torque–angle curve. Tightening the fastener beyond the yield point but below the ultimate tensile strength can achieve the maximum preload for a given fastener size [5.30]. Due to the plastic deformation of the fastener, releasing the load will not return the fastener to its original state. Then, retightening the fastener again, it is found that the torque and angular displacement increase along the offset curve that is approximately parallel to the first torque–angle curve and the yield torque appears at a higher level ($T_{y2} > T_{y1}$).

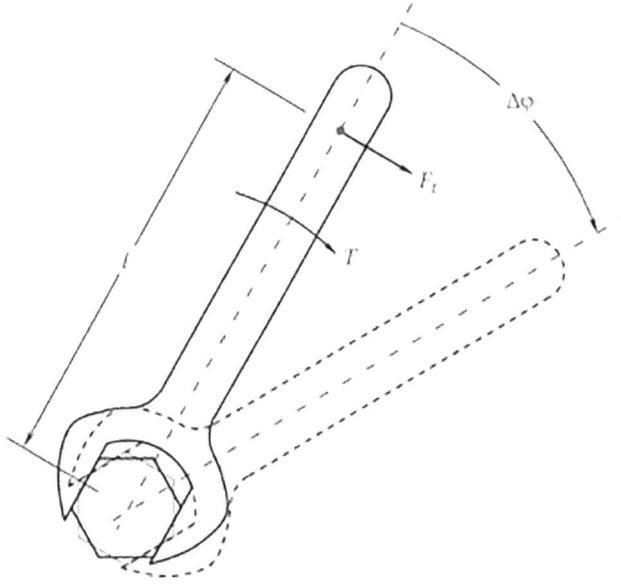

FIGURE 5.42 Work is done by tightening torque over a small angle of $\Delta\varphi$.

It should be noted that while the yield strength, tensile strength, and hardness of the fastener increase as a consequence of cold working, the ductility decreases along a path opposite to that of hardness. Since the yield strength increases faster than the ultimate tensile strength, the yield strength comes closer to the ultimate strength after each cold working cycle. After several repeated load/unload cycles, the fastener exhibits brittle behavior and may suddenly fail without warning signs. Therefore, for critical joints or applications, the reuse of fasteners (bolts, nuts, washers, and other mechanical fasteners) is not recommended.

5.5.5 TIGHTENING TORQUE

In the elastic clamping zone, when a bolt fastener turns an angle of $\Delta\varphi$ under a tightening torque T, the work done by the tightening torque is (Figure 5.42),

$$W_t = T\Delta\varphi = F_t l \Delta\varphi \tag{5.7}$$

In practice, the work W_t is balanced with three components: the work done by the axial tensile force F_{bt} toward the elastic elongation of the bolt and the clamping force F_{mc} ($F_{bt}=-F_{mc}$) toward the compression of the joint members, W_{def}; the work done by the thread friction between the threads of the fastener system (*e.g.*, bolt and nut), W_{th}; and the work done by the friction on the bearing surfaces under the bolt head and the nut, W_f, as referred to in Figure 5.38.

As shown in Figure 5.43, when a wrench tightens a fastener in a full circle (*i.e.*, 2π in radians) in the elastic clamping range, the corresponding change in the bolt length is the thread pitch p. Thus, as the wrench turns a small angle of $\Delta\varphi$, the change in the bolt length is δ:

$$\frac{\delta}{p} = \frac{\Delta\varphi}{2\pi} \tag{5.8}$$

As a result, the work done due to the bolt deformation becomes

$$W_{def} = F_{bt}\delta = F_{bt} p \frac{\Delta\varphi}{2\pi} \tag{5.9}$$

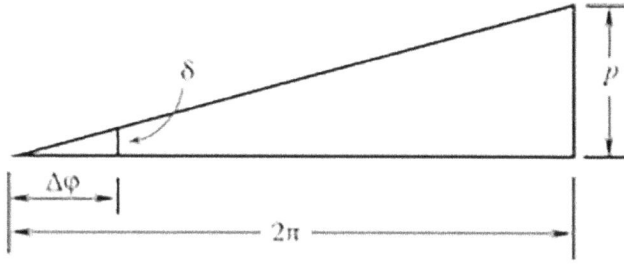

FIGURE 5.43 When a tightening wrench rotates a full circle in the elastic clamping zone, the change in the axial length of bolt is p (thread pitch). As it turns a small angle of $\Delta\varphi$, the change in length is δ.

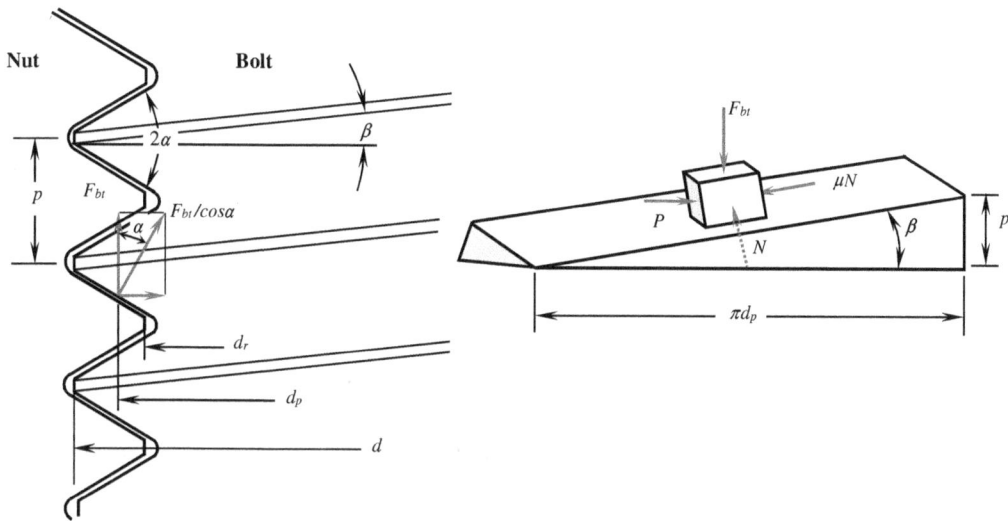

FIGURE 5.44 Work is done by thread friction. Noting that 2α is the thread angle; β is the thread lead angle; d, d_r, and d_p are the major diameter, minor (root) diameter, and pitch diameter, respectively.

During a fastener-tightening process, the thread surfaces of the bolt and nut remain in contact with each other where a relative movement exists. As demonstrated in Figure 5.44, the forces acting on the entire thread surface are the bolt axial compressive force F_{bt}, the tangential force P, the normal force N, and the thread friction force $\mu_{th}N$, where μ_{th} is the coefficient of sliding friction between the bolt and nut threads. It is to be noted that the friction force $\mu_{th}N$ acts on the inclined plane opposite to the direction of motion. In this figure, the inclined angle β denotes the lead angle of the thread, and the angle α is the half of the thread angle. As a function of angle α, the normal contact force N is expressed as $F_{bt}/\cos\alpha$, and hence, the friction force acting on the threads becomes $\mu_{th}F_{bt}/\cos\alpha$. This indicates that an increase in α results in an increase in the friction force.

When the system is in equilibrium, the sum of all forces acting on the system is equal to zero. Therefore, two force–balance equations can be derived as

$$\sum F_h = 0 \quad P - N\left(\mu_{th}\cos\beta + \cos\alpha\sin\beta\right) = 0 \tag{5.10a}$$

$$\sum F_v = 0 \quad F_{bt} + N\left(\mu_{th}\cos\beta - \cos\alpha\sin\beta\right) = 0 \tag{5.10b}$$

Combining these two equations yields

$$P = F_{bt} \frac{\tan\beta\cos\alpha + \mu_{th}}{\cos\alpha - \mu_{th}\tan\beta} \tag{5.11}$$

Thus, the torque required to overcome thread friction is

$$T_{th} = P\left(\frac{d_p}{2}\right) = \frac{F_{bt}d_p}{2} \frac{\tan\beta\cos\alpha + \mu_{th}}{\cos\alpha - \mu_{th}\tan\beta} \tag{5.12}$$

where d_p is the pitch diameter. Thus, the work done by the thread friction can be expressed as

$$W_{th} = T_{th}\Delta\varphi = \frac{F_{bt}d_p}{2}\left(\frac{\tan\beta\cos\alpha + \mu_{th}}{\cos\alpha - \mu_{th}\tan\beta}\right)\Delta\varphi \tag{5.13}$$

For a very small angle β, $\tan\beta$ approaches zero; thus,

$$W_{th} = \frac{F_{bt}d_p}{2}\frac{\mu_{th}}{\cos\alpha}\Delta\varphi \tag{5.14}$$

The necessary and sufficient condition for thread self-locking is that the coefficient of thread friction μ_{th} is greater than or equal to the product of the tangent of the thread lead angle β and cosine of the half thread angle α [5.31]:

$$\mu_{th} \geq \tan\beta\cos\alpha \tag{5.15}$$

This indicates that fine-thread fasteners are easier to reach self-locking than coarse-thread fasteners because of their smaller thread lead angles.

Other methods to prevent fastener slackening include using fastener retainers, bolts, nuts, and washers with serrated locking ribs, locked inserts, and adhesive retention materials (such as Loctite).

The work done due to the friction on the bearing surfaces under bolt head and nut becomes

$$W_h = \frac{F_{bt}}{2}\left(\mu_h d_h\right)\Delta\varphi \tag{5.16a}$$

$$W_n = \frac{F_{bt}}{2}\left(\mu_n d_n\right)\Delta\varphi \tag{5.16b}$$

respectively. In the above equations, μ_h and μ_n are friction coefficients on the bearing surface under the bolt head and nut; d_h and d_n are the effective diameters of the bolt head and nut, respectively:

$$d_h = d_n = \frac{d + 1.5d}{2} = 1.25d \tag{5.17}$$

Based on the preceding analysis, it gives that

$$W_t = W_{def} + W_{th} + W_h + W_n \tag{5.18}$$

That is,

$$T = \frac{F_{bt}p}{2\pi} + F_{bt}\frac{d_p}{2}\left(\frac{\tan\beta\cos\alpha + \mu_{th}}{\cos\alpha - \mu_{th}\tan\beta}\right) + 0.625F_{bt}d\left(\mu_h + \mu_n\right)$$

$$= \left[\frac{p}{2\pi d} + \frac{d_p}{2d}\left(\frac{\tan\beta\cos\alpha + \mu_{th}}{\cos\alpha - \mu_{th}\tan\beta}\right) + 0.625\left(\mu_h + \mu_n\right)\right]F_{bt}d \tag{5.19}$$

$$= KF_{bt}d$$

and

$$K = \sum_{i=1}^{3} K_i \tag{5.20}$$

where K_1, K_2, and K_3 are tightening factors due to the bolt tension, thread friction, and friction on bearing surfaces under the bolt head and nut:

$$K_1 = \frac{p}{2\pi d} \tag{5.21a}$$

$$K_2 = \frac{d_p}{2d} \left(\frac{\tan\beta\cos\alpha + \mu_{th}}{\cos\alpha - \mu_{th}\tan\beta} \right) \tag{5.21b}$$

$$K_3 = 0.625 \left(\mu_h + \mu_n \right) \tag{5.21c}$$

For metric threads,

$$d_p = d - 0.649519 p \tag{5.22}$$

It has been estimated that around 10% of the torque applied to the fastener is to produce the stretch of the bolt and compression of the joint, 40% of the torque is to overcome thread friction, and the remaining 50% is to overcome head- and nut-bearing friction (see Figure 5.40). This implies that approximately 90% of the total torque is to overcome friction, while only 10% does useful work [5.32].

5.5.6 THREAD ENGAGEMENT AND LOAD DISTRIBUTION

In bolted joints, thread engagement refers to the number or length of threads that are engaged between the bolt and nut threads. It has long been recognized that when a bolt is engaged with a nut, the axial load acting on the bolt and nut is distributed extremely nonuniformly between the threads of threaded joints. During a tightening process, the first engaged thread takes the majority of the applied load. This suggests that the first thread and subsequently some of the other threads may yield and proceed to plastic deformation. This plastic softness will redistribute the load over other engaged threads and, as a result, make the loading distribution more uniform.

A simple model has been proposed by Tseng [5.33] for determining the loading distribution on bolt threads. The results of his model are in good agreement with those of Timoshenko and Goodier [5.34] and Young et al. [5.35]. The data in Table 5.2 indicate that extra threads engaged do not increase joint strength.

TABLE 5.2
Loading Distribution of Bolt Threads

Thread Number	Loading Distribution (%)		
	Tseng [5.33]	Timoshenko [5.34]	Young [5.35]
1	58.9	54.5	55.0
2	24.2	24.6	24.8
3	10.0	10.7	11.2
4	4.1	4.7	5.1
5	1.8	2.2	2.5
6	1.0	1.3	1.5

Assuming both the bolt and nut are made with the same material, the minimum thread engagement length required is approximately 65% of the nominal diameter [5.36]. For instance, a $M8 \times 1.25$ bolt needs a minimum of 5.2 mm of thread engagement, that is, 4.16 threads. In addition, it is appropriate to apply a safety margin of approximately 30% to the calculated minimum length to account for high-order effects.

5.6 COMMON TYPES OF ELECTRIC MOTOR ENCLOSURES

The function of the motor enclosure is not only to support all motor components internally, but also to provide some kind of protection to these components from electrical shortage, chemical corrosion, and mechanical damage. The design of enclosure must meet specific requirements, including the following: (*a*) environmental requirements for restricting foreign objects (such as water, dust, moisture, *etc.*) to entering the motor, (*b*) thermal requirements for maintaining the motor temperature below its maximum allowable temperature, (*c*) safety requirements for personal protection, and (*d*) mechanical requirements for withstanding all external and internal loadings. There are a variety of motor enclosures in response to customer demanding. Each type has its own specific design requirements, cooling characteristics, different levels of protection, and certain applications. There are several specific types of electric motor enclosures available in the market, which are addressed in the following sections.

5.6.1 Open Drip Proof Enclosure

Open drip proof (ODP) enclosure is typically used for indoor application at clear and dry locations. This type of enclosure has ventilating openings to allow cooling air to be drawn from the surrounding environment, flowed over and around the motor windings and other internal components, and then expelled back through the openings into the surroundings. In some cases, to prevent liquid drops and solids from entering the motor within up to 15° angle from vertical, ventilation openings on the top section are shielded. A motor with an ODP enclosure corresponds to IP rating of IP22.

5.6.2 Totally Enclosed Non-Ventilated Enclosure

Totally enclosed non-ventilated enclosure is used wherever a motor requires a high degree of protection. Due to the lack of cooling fans, cooling of motor internal parts basically relies on heat conduction between contacted solid parts and then natural convection from the enclosure to the surroundings. This cooling method is often referred to as *closed circuit cooling*, since the primary coolant (*i.e.*, air or liquid) remains contained within the motor. The use of this type of enclosure is limited to the motor that is well designed with low enough temperature rise or has a low duty cycle to generate a small amount of heat during operation.

5.6.3 Totally Enclosed Fan Cooled Enclosure

As the most common type used in industries, totally enclosed fan cooled (TEFC) motor uses a fan that mounts on the rear shaft out of the enclosure to force cooling air passing across the motor frame's external cooling fins for dissipating heat from the motor to the environment. For safety reasons, the fan is protected by a fan cover. In such an approach, the volumetric flow rate of cooling air is proportional to the motor rotating speed. Since this design does not allow the free exchange of ventilating air between the interior and exterior of the enclosure, it can be used in applications that require high degrees of protection. They often feature IP ratings of IP54 or IP55, which permit a very limited amount of dust to intrude into the motor enclosure. However, TEFC motors may not enable to secure against high dynamic pressure water.

5.6.4 Totally Enclosed Air over Enclosure

The type of totally enclosed air over enclosure is similar to TEFC. The main difference is that the cooling flow in this enclosure comes from an external source, rather than produced by its own fan in a TEFC motor. This is especially suitable for some applications where a constant cooling airflow rate is necessary, regardless of the change in the motor's own speed.

5.6.5 Totally Enclosed Forced Ventilated Enclosure

Designed for motors that operate over an extended speed range, the totally enclosed forced ventilated enclosure utilizes an externally powered fan to blow external cooling air across the motor through a pipe connection. Heated air is exhausted through the pipe connection at the other side of the motor. In such a way, the airflow rate is determined by the external fan and has nothing to do with the motor speed. This cooling method is probably especially suitable for low-speed motors.

5.6.6 Totally Enclosed Washdown Enclosure

The type of totally enclosed washdown (TEWD) enclosure is extensively used in the food processing and beverage production industries. Typically, motors with TEWD enclosures are designed to withstand the high pressure of water or other liquids without the penetration of water/liquid into the motor. Depending on the water pressure they can withstand, their IP ratings can be ranged from IP55 to IP69K. The TEWD enclosure works well in environments in which regular cleaning or sanitizing is necessary.

5.6.7 Explosion Proof Enclosure

A major safety concern in industries is the occurrence of fires and explosions. In hazardous environments, flammable liquids/gases/vapors or combustible dusts/powders may exist in sufficient quantities to potentially produce a fire or explosion. Since the flammable substances are not always avoidable, it is of great importance to design and utilize explosion protected equipment, allowing to electrical and control devices to be used in environments where there is a danger of explosion.

An explosion proof motor is typically provided with an enclosure for: (*a*) preventing the ingress of an explosive substance and/or contact with sources of ignition arising from the functioning of the motor; (*b*) if an explosion is unavoidable, confining the explosion within the enclosure so that the explosion cannot spread to the surrounding environment, and (*c*) enhancing heat dissipation to maintain the surface temperature lower than the spontaneous ignition temperature and avoid hot spots on the housing surface.

5.7 ANTICORROSION OF ELECTRIC MOTOR AND COMPONENTS

Many metallic products produced by manufacturing industries may have bare metal surfaces, which require receiving some form of protection from corrosion. The classification of corrosion protection methods includes active, passive, permanent, and temporary protection. Active anticorrosion methods can alter or control the proceeding reaction during a corrosion process so that corrosion is avoided. Examples of such an approach are the development of corrosion-resistant alloys and the formation of an effective passivation layer on metal surfaces for impeding corrosion half-reactions. In passive protection, corrosion is prevented by mechanically isolating metal surfaces from aggressive, corrosive agents. A coating of this type of protection serves as a barrier layer precluding permeation of corrodent to the metal surface. Among various methods, protective layers, films, or other coatings have been usually chosen by motor manufacturers to provide overall protection against corrosion.

5.7.1 SURFACE TREATMENT METHODS

By coating a thin layer of a certain metal on a motor component (*e.g.*, a fastener) surface, the surface properties of the component can be changed to that of the metal applied. Thus, the coated component becomes a composite material, consisting of a durable and corrosion-resistant layer and a core having the load-bearing capability. A number of metals are suitable for protective coating, such as chromium, nickel, copper, zinc, and cadmium. However, coating processes may cause inherent pollution problems.

There are many different methods of metal surface treatment for preventing corrosion on motor components. Metal surfaces can be treated chemically, mechanically, or physically. Some of these methods are addressed in the following sections.

5.7.1.1 Electroplating

Electroplating is achieved by passing an electrical current through a solution containing dissolved metal ions and the metal object to be plated. The most widely used metallic coating method for corrosion protection is galvanizing, which involves the application of metallic zinc to iron and steel products requiring protection from corrosion. Its excellent corrosion resistance in most environments leads to its successful use as a protective coating on a variety of products and in many exposure conditions.

5.7.1.2 Electroless Plating

The electroless plating technology has been developed for many decades. Electroless plating is the process of plating a coating with the aid of a chemical reducing agent (*e.g.*, formaldehyde) in solution without the passage of external power. Compared with electroplating, electroless plating has some superior characteristics: (*a*) Without using electric current in the plating process, it is easier to obtain uniform coatings on parts. This feature is especially suitable for parts with irregular, complex-shaped geometries. (*b*) It is applicable to non-conductive substrates such as glass and plastic. (*c*) Electroless plating can deposit particles from different materials to obtain composite coatings readily for adapting to different application requirements. For instance, nickel–phosphorus coatings are used to enhance corrosion resistance. Alternatively, thin cobalt–phosphorus coatings offer superior sliding wear, enhanced lubricity and corrosion resistance, and improved fatigue properties. Because they exhibit some magnetic properties, they are also of interest to the magnetic recording community.

5.7.1.3 Physical Vapor Deposition

Physical vapor deposition (PVD) refers to the techniques used to deposit molecular thin films on solid surfaces in a vacuum environment. As its name implies, the PVD process is a physical process, without involving chemical reactions as in chemical vapor deposition. There are two common types of PVD processes: electron beam evaporation process and sputtering deposition process. During the electron beam evaporation process, a solid or liquid coating material evaporates into the vapor phase to travel in the vacuum space and then condensates on the solid surfaces, with the transition of its physical morphology to the solid phase. The sputtering deposition process, which is extensively used in the semiconductor industry, involves ejecting coating materials from the sputtering target to the substrates such as silicon wafers. One of its important characteristics is that it can deposit the materials with high melting points. The most outstanding feature of PVD is that the properties of one material can be imparted to the surface of the workpiece which is made from another material, thus creating a wholly different product. However, the PVD techniques require expensive equipment and instruments. Furthermore, because the PVD process is usually carried out in a vacuum chamber, the size of the coated members is subject to certain restrictions.

5.7.1.4 Inorganic Coating

Inorganic coating can be produced by chemical reaction, with or without electrical assistance. The treatments change the immediate surface layer of metal into a film of metallic oxide or compound that has better corrosion resistance than the natural oxide film and provides an effective base or key for supplementary protection such as paint.

Black oxide is a process that provides a conversion coating on surfaces of ferrous alloys. The coating is formed from chemical reactions between the iron of ferrous alloys and black oxide solutions such as the alkaline aqueous salt solution. When workpieces are immersed in a hot black oxide solution at the operation temperature of $132°C$–$143°C$ ($270°F$–$290°F$), a layer of magnetite (Fe_3O_4) is produced on the workpiece surfaces about 30 min.

Unlike other coating processes, black oxide has some specific properties and characteristics that make it ideal for certain applications. One of the remarkable characteristics is that for a properly controlled coating process, black oxide has little contribution to hydrogen embrittlement. This is because that the black oxide process does not require an acid activation nor is it an electro process. However, if the workpieces are scaled or rusty such that an acid pickle is required, it may lead to hydrogen embrittlement problems.

5.7.1.5 Phosphate Coating

Phosphate coating is a crystalline conversion coating consisting of an insoluble crystalline metal-phosphate salt formed in a chemical reaction between ferrous metals and a phosphoric acid (H_3PO_4) solution containing metallic ions (*e.g.*, zinc, iron, or magnesium). Phosphate coating can serve either as the foundation for subsequent coatings or the protective coating on ferrous parts for corrosion resistance. In addition, phosphate coating is often employed in the cold-forging process for lubricity.

The primary types of phosphate coatings include manganese, iron, and zinc coating. Among them, manganese phosphate coating has the highest hardness and superior corrosion and wear resistance. Therefore, manganese phosphate coating is extensively used to improve friction properties of sliding components such as engine pistons, gears, camshafts, and power transmission systems. Iron phosphate coating inhibits corrosion and improves the adhesion and durability of paint finishes. Due to the low cost and moderate corrosion resistance, this type of phosphate coating is basically suitable for indoor equipment. Zinc phosphate coating can provide the highest corrosion resistance. Hence, it is widely used in the automotive industry and in certain sectors of the appliance and electronics industries.

5.7.1.6 Electropolishing

As a reverse plating technique, electropolishing is an electrochemical process that is typically used for surface polishing and corrosion resistance enhancment. While electroplating deposits a thin metal layer on surfaces on either metallic or nonmetallic parts, electropolishing removes surface metal, beginning with the high points within the microscopic surface texture, by using a combination of chemicals rectified electrical current. As a result, surface imperfections (*e.g.*, surface crack, burr) and embedded contaminants (*e.g.*, rust, oxide scale) of the parts are eliminated. Electropolished parts are left in a homogenous and passive condition, which enhances surface resistance to corrosion tarnish or oxidation. These properties are especially evident on stainless steel but also found on carbon steel, brass, aluminum, copper, and inconel. Even though all these materials are more corrosion resistant, electropolishing is generally applied to stainless steel today

As a nondistorting process, electropolishing is especially suitable for applications in which mechanical deformation is not allowed. Other benefits of electropolishing include deburring, precise dimension control, ultracleaning, stress sensitivity reduction, and weldability improvement.

5.7.1.7 Nanocoating

The nanocoating technology has been developed for providing superior protective properties and multi-functionalities to various materials. Compared to other conventional coating methods that produce coating layers with the thickness of dozens or hundreds micrometers, a nanocoating is measured on the nanoscale (typically 1–100 nm). With this method, ultrathin layers or chemical structures can be built up on material surfaces for enhancing corrosion resistance of materials, as well as resistance to water, oil, friction, scratches, UV rays, and so on. Because nanocoatings are ultrathin, they are typically transparent, not visible to the human eyes. Because of this special feature, nanocoatings are particularly suitable for applications where opacity is a problem.

Nanocoatings can be extensively used in industrial, military, and commercial applications, such as electric motors, electronics, medical equipment, defense and aerospace, 3C (computer, communication, and consumer electronics) products, and home appliances. In addition, they can be applied to a wide variety of materials like metals, ceramics, plastics, and polymers.

In practice, many materials have been used for nanocoatings, for instance, zinc oxide (ZnO), titanium dioxide (TiO_2), silica dioxide (SiO_2), ceramic, grapheme, and carbon nanotube (CNT) [5.37].

5.7.2 Anticorrosion Treatment of Electric Motor

Almost all metals, with the exception of the common precious metals such as gold, silver, and platinum group metals (iridium, osmium, palladium, platinum, rhodium, and ruthenium), oxidize and corrode in corrosive environments, forming compounds such as oxides, hydroxides, and sulfides. In fact, corrosion has a great effect on motor performance and lifetime and may cause serious damage to motors or components in some severe cases. Therefore, corrosion control and treatment are of vital concern in motor design.

There are many anticorrosion methods in industrial technologies such as cathodic protection, chemical conversion, inhibition of surface reaction, rust preventive treatment, thermal spray, paint, electroplate, and anodic protection. Each of these methods can, to a certain extent, extend the life of electric motors. Among all the anticorrosion methods, the surface protective coatings are often adopted by motor manufacturers for their good rust inhibitive properties, durability, easy operation, and no requirement for special equipment.

The materials of the protective coatings can be generally categorized into two types: inorganic and organic materials. The inorganic materials include metal, glass, ceramics, and plastic. The organic materials include paint, resin, paraffin, ointment, rubber, and asphalt. Each of the protective coatings has its own properties and usage scopes. In selecting protective materials, the pros and cons of the material must be completely taken into account.

In practice, motor manufacturers commonly use thermally conductive paints on the external surfaces of motor frames at room temperature. The painting materials include plastic ferrite polyurethane, polyurethane acrylic varnish, expanded graphite, enamel, alkyd, and epoxy-based paints. The paint thickness is typically less than 0.5 mm. If a motor has been overpainted, the paint thickness may be greater than 1 mm. In such a case, low thermal conductivity paints could weaken the heat dissipation from the motor to the ambient. Black paint is commonly utilized on motor surfaces to achieve higher thermal emissivity (about 0.96) and thus higher thermal radiation from the motor. It should be noted that the primary thermal energy emitted by the motor is in the invisible infrared portion of the spectrum. All paint colors in the visible spectrum have small differences in thermal emissivity. All paint colors in the visible spectrum have small differences in thermal emissivity. For instance, white paint has an emissivity of 0.90–0.95, blue 0.94, green 0.92, and red 0.91.

The thermal spraying process involves the projection of small molten metal particles onto prepared motor surfaces to form a continuous coating. During the process, a certain amount of heat is transferred to the motor housing. It is thus important to access the distortion of the housing as a result of the heat transfer. This technique has been widely used in constructions, oil and gas plants, petrochemical equipment, and off shore wind turbines.

When a motor operates under harsh environmental conditions such as high temperature, high humidity, corrosive media, high G-shock, high contamination, and/or high pressure, adequate precautions shall be taken to prevent the motor from corrosion. Conventionally, the motor is well sealed with sealing devices on the shaft, housing, and endbells to avoid moisture, corrosive media, and other contamination particles to enter the motor.

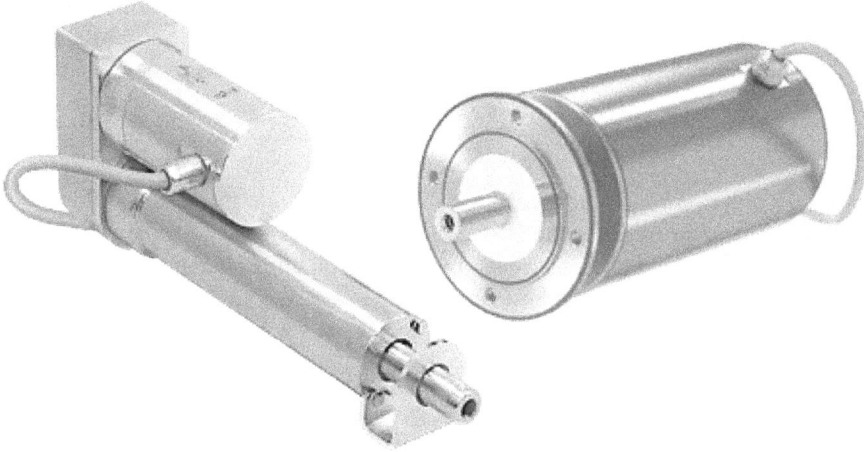

FIGURE 5.45 Stainless steel servomotors for food processing applications.

In fact, the use of seals can reduce but not completely eliminate moisture to penetrate into the motor. Usually, thermal cycling can develop tiny gaps in a motor. Thus, as the air inside the motor periodically experiences expansion and contraction, a small amount of moisture may be sucked inside the motor. Under this circumstance, the seals act as a barrier to the moisture removal, causing interior corrosion of the rotor, stator, PM, and other components. According to Lenzing and Chen [5.38], shaft seals typically cause a loss of 3%–15% of shaft torque due to friction.

A new technique for providing corrosion protection is to apply a protective coating to all the motor parts, including rotor and stator cores, PMs, windings, housing, and endbells, both internally and externally. Using the standard salt-spray test as defined in MIL-STD-202G [5.39], in which testing machines are subjected to a fine mist of salt solution, a conventional motor develops a locked rotor and is unable to operate after 48 h of the salt-spray test. By contrast, the motor treated with this new technique still runs normally after 140 h of the test.

Alternatively, motor manufacturers often use stainless steel to make motor housing and endbells for some special applications, such as food processing-related applications. Stainless steel has long been thought of as the best material for withstanding caustic chemicals and high pressure. It is thus ideal for applications requiring high corrosion and heat resistance, frequent washdown, and long life. The drawback of stainless steel is its high price.

Kollmorgen has recently launched its AKMH stainless steel servomotors (Figure 5.45) that fulfill the European Hygienic Engineering & Design Group (EHEDG), National Sanitation Foundation (NSF), and 3A standards [5.40]. This type of motor is specially designed for food processing manufacturers with superior performance, long-life operation, low maintenance, and great corrosion protection. These IP69K rated hygienic servomotors feature stainless steel housings and can withstand water pressure up to 100 bars.

5.7.3 Hydrogen Embrittlement Issues

During some surface finishing processes, such as electroplating, acid pickling, phosphating, electrocleaning, and heat treatment, atomic hydrogen can enter and diffuse through metallic materials (mainly high-strength steel and titanium/aluminum alloys) to degrade the fracture behavior or load-carrying ability of these materials. This phenomenon, known for more than 140 years, is often referred to as hydrogen embrittlement or hydrogen-induced brittle failure.

For instance, hydrogen can be produced from the reaction of the cathode with metal deposition in a zinc electroplating process [5.41],

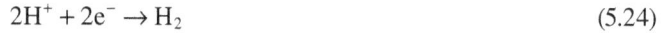

$$Zn^{2+} + 2e^- \rightarrow Zn^0 \tag{5.23}$$

$$2H^+ + 2e^- \rightarrow H_2 \tag{5.24}$$

An acid pickling operation prepares the surfaces of workpieces prior to some processes such as electroplating, black oxide, and phosphate coating. The acid pickling operation is usually completed with sulfuric acid, but hydrochloric acid is also used.

In the acid pickling process, oxide and scaling layers are dissolved in acid pickling solutions without generating hydrogen. Hydrogen is formed as acid attacking the base material [5.42]:

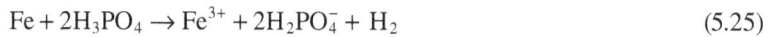

$$Fe + 2H_3PO_4 \rightarrow Fe^{3+} + 2H_2PO_4^- + H_2 \tag{5.25}$$

Unlike other deposit coating processes, black oxide is a chemical conversion coating to enhance corrosion resistance of many metals and their alloys. This surface treatment has positive outcomes in terms of corrosion protection, lubricity, and galling and smearing resistance [5.43, 5.44].

The major stage in the black oxide process is to immerse the parts in an aqueous solution of 60%–80% sodium hydroxide containing an oxidizing agent such as 15%–40% sodium or potassium nitrite at a temperature of about 130°C–150°C for approximately 30 min, leading to the following sequence of chemical reaction [5.45]:

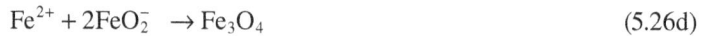

$$Fe^{2+} + 2H_2O \rightleftarrows Fe(OH)_2 + 2H^+ \tag{5.26a}$$

$$Fe(OH)_2 + OH^- \rightleftarrows Fe(OH)_3 \tag{5.26b}$$

$$Fe(OH)_3 + OH^- \rightleftarrows FeO_2^- + 2H_2O \tag{5.26c}$$

$$Fe^{2+} + 2FeO_2^- \rightarrow Fe_3O_4 \tag{5.26d}$$

Thus, a thin oxide film of magnetite (Fe_3O_4) is formed on part surfaces. Due to the alkaline nature of the solution, there is no hydrogen evolved in the chemical process [5.46] and thus, the black oxide process does not contribute to hydrogen embrittlement of ferrous materials [5.47]. Furthermore, since the thickness of the oxide film is typically 0.7–2.6 μm, this surface finishing method causes negligible dimensional changes.

The absorption and diffusion of hydrogen in steel are quite complicated [5.48]. When hydrogen atoms enter and diffuse through steel surfaces, they may present either as atomic or molecular hydrogen or in combined molecular form. Because these molecules are too large to diffuse through steel, pressure builds at crystallographic detects (dislocations and vacancies) or discontinuities (voids, inclusion/matrix interfaces) causing minute cracks to form [5.49]. The relief of hydrogen embrittlement is partially related to the nature of the coating applied. Simply speaking, if the coating is dense and nonporous, the escape of hydrogen may be difficult; if the coating is open and porous, the release of hydrogen is a matter of time and temperature [5.47].

Even today, hydrogen embrittlement still remains a major cause of fastener failure. In a root cause failure analysis, an essential step is to determine the failure mechanism. More specifically, it is important to distinguish hydrogen embrittlement and stress corrosion cracking from each other because both failures are associated with the interaction between hydrogen and metal workpieces (e.g., fasteners) and occur under tensile stresses well below the yield strength of the material. The major differences include [5.50–5.52] the following:

- While hydrogen responsible for embrittlement comes from manufacturing processes such as plating and cleaning, hydrogen in stress corrosion cracking is supplied from the environment.
- Stress corrosion cracking is associated with high-stress workpieces where the core hardness is larger than 32 HRC, more likely >36 HRC.
- Stress corrosion cracking is not only induced by hydrogen; it could also resulted from various corrosive substances that attack small cracks, leading to the final brittle fracture.
- Though both hydrogen embrittlement and stress corrosion cracking are delayed failures, the delay time for hydrogen embrittlement is likely from a few minutes to 24 h after installation, whereas the delay time for stress corrosion, as the primary indication of this type of failure, is typically longer than 24–48 h after installation.
- Stress corrosion cracking often exhibits more crack branching and less pronounced dimples than cracking produced by hydrogen embrittlement.

A number of methods can be taken to avoid hydrogen embrittlement failure, including the following:

- The hydrogen removal process must be carried out following the electroplating process immediately According to the ASTM standard [5.53], the hydrogen removal requires that the electroplated parts must be baked at 204°C–218°C (400°F–425°F) within 1 h after the plating process for at least 4 h. In ASTM B850-98 [5.54], the baking time varies from >8 to >22 h at 190°C–220°C (374°F–428°F), depending on the tensile strength and hardness of plated parts (see Table 5.3).
- It is worth to note that the baking temperature is limited for some metal coatings. For example, at 225°C, the zinc starts to oxidize in air.
- A stress relief process is necessary to reduce the residual stress in the parts. For low-carbon steels, it requires to anneal 3–4 h at about 400°C.
- A more direct and effective way for reducing the hydrogen embrittlement risk is to use low embrittling electroplating processes such as special solution compositions and operating conditions, which result in either a lower pickup of hydrogen or a deposit that allows easier removal of the absorbed hydrogen during the baking treatment.
- For high-stressed parts, it is recommended to use a pickling-free process such as dry sand blasting or other mechanical means.
- Some coating techniques can eliminate hydrogen embrittlement. One example is called PVD. This is because PVD processes are done in vacuum, and thus, the chance of embrittlement by hydrogen is precluded. Another example is the mechanical galvanizing process.

TABLE 5.3
Recommended Baking Process by ASTM B850-98

Tensile Strength		Hardness (HRC)	Temperature (°C)	Baking Time (h)
MPa	ksi			
1,000–1,100	145–160	31–33	190–220	>8
1,101–1,200	160–174	33–36	190–220	>10
1,201–1,300	174–189	36–39	190–220	>12
1,301–1,400	189–203	39–43	190–220	>14
1,401–1,500	203–218	43–45	190–220	>16
1,501–1,600	218–232	45–47	190–220	>18
1,601–1,700	232–247	47–49	190–220	>20
1,701–1,800	247–261	49–51	190–220	>22

In this process, zinc coatings are applied to workpieces at room temperature without hydrogen embrittlement, as described in ASTM B695-2000 [5.55]. The coating thickness can be controlled between 5 and 100 μm.

- It is generally agreed that in the plating of Cr, Zn, Cd, Ni, Sn, and Pb, hydrogen tends to remain in steel. By contrast, some elements such as Cu, Mo, Al, Ag, Au, and others lead to lower hydrogen diffusion and solubility and, consequently, lower hydrogen permeation through steel surfaces. Under the condition of honoring the technical requirements, it is preferred to use the coating materials that have low hydrogen permeation.
- Use low-strength steel to replace high-strength steel. This can effectively reduce the risk of hydrogen embrittlement.
- Because coating layers cover the surfaces of plated fasteners, they play the role as a hydrogen diffusion barrier. In fact, the coating thickness significantly affects the rate of hydrogen diffusion. Usually, the maximum coating thickness for fasteners with the hardness <32 HRC is 12 μm and 8 μm with the hardness >32 HRC.

While avoiding hydrogen embrittlement in some manufacturing processes, scientists and engineers have also taken full advantage of hydrogen embrittlement in other manufacturing processes. A typical example is the production of micro- and nanoparticles. This technique has been successfully adopted for making rareearth PMs.

REFERENCES

5.1. Groover, M. P. 2010. *Fundamentals of Modern Manufacturing: Materials, Processes, and Systems*, 4th edn. John Wiley & Sons, Hoboken, NJ.
5.2. Verhoeven, J. D. 2005. Metallurgy of steel for bladesmiths and others who heat treat and forge steel. Online eBook at http://www.feine-klingen.de/PDFs/verhoeven.pdf.
5.3. Rio Tinto Iron and Titanium, Inc. 2013. *Ductile Iron Data for Design Engineers*. http://www.ductile.org/ductile-iron-data/.
5.4. Davis, J. R. 1999. *Cast Irons*, 2nd edn. ASM International, Materials Park, OH.
5.5. Karwa, R. 2006. *A Textbook of Machine Design*, 2nd edn. Laxmi Publications Ltd., New Delhi, India.
5.6. ASM. 1990. *ASM Handbook (vol. 1): Properties and Selection: Irons, Steels, and High-Performance Alloy*. ASM International, Materials Park, OH.
5.7. Graham, D. 2006. Machining cast iron. *Manufacturing Engineering* **136**(2): 77–83.
5.8. Bawa, H. S. 2004. *Manufacturing Processes II*. Tata McGraw-Hill Publishing Company, New Delhi.
5.9. DeGarmo, E. P., Black, J. T., and Kohser, R. A. 2002. *Materials and Processes in Manufacturing*, 9th edn. John Wiley & Sons, New York.
5.10. Pontzer, J. H. 2000. Method of manufacturing electric motor housing frame and foam pattern therefore. U.S. Patent 6,109,333.
5.11. Zhang, D. L., Zheng, L. H., and St. John, D. H. 2002. Effect of a short solution treatment time on microstructure and mechanical properties of modified Al-7wt. %Si-0.3wt. %Mg alloy. *Journal of Light Metals* **2**(1): 27–36.
5.12. Manente, A. and Timelli, G. 2011. Chapter 9: Optimizing the heat treatment process of cast aluminum alloys. In *Recent Trends in Processing and Degradation of Aluminum Alloys* (Ed.: Z. Ahmad). InTech, Rijeka, Croatia, pp. 197–220.
5.13. Bauser, M., Sauer, G., and Siegert, K. 2006. *Extrusion*, 2nd edn. ASM International, Materials Park, OH.
5.14. Koyama, T., Abe, T., and Takagi, K. 2014. Development of CFRP (carbon fiber reinforced plastic) monolithic sandwich construction. *Mitsubishi Heavy Industries Technical Review* **51**(4): 16–19.
5.15. Van Dine, P., Odessky, V., Spencer, B. E., Smith, J. S., and Harring, W. R. 2000. Method for making a composite electric motor housing. U.S. Patent 6,125,528.
5.16. Masters, W. E. 1987. Computer automated manufacturing process and system. U.S. Patent 4,665,492.
5.17. Sachs, E. M., Haggerty, J. S., Cima, M. J., and Williams, P. A. 1993. Three-dimensional printing techniques. U.S. Patent 5,204,055.
5.18. Sachs, E. M., Haggerty, J. S., Cima, M. J., and Williams, P. A. 1994. Three-dimensional printing techniques. U.S. Patent 5,340,656.

5.19. Cima, M. J., Sachs, R. M., Fan, T. L., Bredt, J. F., Michaels, S. P., Khanuja, S., Lauder, A., Lee, S.-J. J., Brancazio, D., Curodeau, A., and Tuerck, H. 1995. Three-dimensional printing techniques. U.S. Patent 5,387,380.

5.20. Ghalotra, N. and Singh, A. 2016. Rapid growth and development of 3D printing. *International Journal of Engineering Science and Computing* **6**(6): 6305–6319.

5.21. Kollmorgen Corporation. RBE(H) motor series. http://www.clemson.edu/ces/crb/procedures/WAM_Data_Files/WAM_motors_datasheet_rbe_series.pdf.

5.22. ASTM Standard B557-06. Standard test methods for tension testing wrought and cast aluminum- and magnesium-alloy products.

5.23. Butcher, J. A. and Pitzer, M. A. 1995. Motor assembly with mounting arrangement. U.S. Patent 5,412,270.

5.24. Black, J. T. and Kohser, R. A. 2007. *DeGarmo's Materials and Processes in Manufacturing*, 10th edn. John Wiley & Sons, Hoboken, NJ.

5.25. Garietya, M., Ngaileb, G., and Altan, T. 2007. Evaluation of new cold forging lubricants without zinc phosphate precoat. *International Journal of Machine Tools and Manufacture* **47**(3–4): 673–681.

5.26. Boyland, J., Tong, W., Bartha, L., and Krogen, O. 2005. Hook nut connector assembly. U.S. Patent Application 2005/0123377.

5.27. Rossi, M. J. 2018. When should you use self-clinching locknuts? *Design World*, February issue, pp. 88–91.

5.28. Budynas, R. G. and Nisbett, J. K. 2008. *Shigley's Mechanical Engineering Design*, Chapter 8. McGraw-Hill, New York, p. 422.

5.29. SKF. 2001. *Bolt-Tightening Handbook*. SKF Group.

5.30. Shoberg, R. S. 1999. Tightening strategies for bolted joints: Methods for controlling and analyzing tightening. *11th Annual Technical Conference on Fastening Technology*, Cleveland, OH. www.pcbloadtorque.com/pdfs/Tightening%20Strategies.pdf.

5.31. Ugural, A. C. 2004. *Mechanical Design: An Integrated Approach*, Chapter 15. McGraw-Hill, Boston, MA, p. 608.

5.32. Fernando, S. 2001. An engineering insight to the fundamental behavior of tensile bolted joints. *Steel Construction* **35**(1): 2–13.

5.33. Tseng, S.-W. 2010. A simplified simulation on loading distribution at bolt threads. Bastion Technologies. http://www.bastiontechnologies.com/ogp/TechnicalPapers.html.

5.34. Timoshenko, S. P. and Goodier, J. N. 1970. *Theory of Elasticity*, 3rd edn. McGraw-Hill, New York.

5.35. Young, W. C., Budynas, R. G., and Sadegh, A. M. 2011. *Roark's Formulas for Stress and Strain*, 8th edn. McGraw-Hill, New York.

5.36. Fernando, S. 2001. Minimum thread engagement—what is the optimum engagement length? Ajax Fasteners Innovations. Technical note: AFI/01/002. http://www.ajaxfast.com.au/downloads/Technical%20notehowmanythreads.pdf.

5.37. Bao, W. W., Deng, Z. F., Zhang, S. D., Ji, Z. T., and Zhang, H. C. 2019. Next-generation composite coating system: Nanocoating. *Frontiers in Materials*. Article 10.3389. https://www.frontiersin.org/articles/10.3389/fmats.2019.00072/full.

5.38. Lenzing, R. and Chen, W. 2010. New stepper motors. *Design World*, February issue, pp. 62–64.

5.39. Department of Defense, USA. 1980. Test method standard: Electronic and electrical component parts. MIL-STD-202G.

5.40. Kollmorgen Corporation. 2013. HKM: New dimension stainless steel motors. http://www.heason.com/wp-content/uploads/2013/06/HKM-Servo-Motor-Flyerl.pdf.

5.41. Ferraz, M. T. and Oliveira, M. 2010. Steel fasteners failure by hydrogen embrittlement. *Ciência & Tecnologia dos Materiais* **20**(1/2): 128–133.

5.42. Bay, N. 1994. The state of the art in cold forging lubrication. *Journal of Materials Processing Technology* **46**(1–2): 19–40.

5.43. Brizmer, V., Stad;er. K., van Drogen, M., Matta, C., and Piras, E. 2017. The tribological performance of black oxide coating in rolling/sliding contact. *Journal of Tribology Transactions* **60**(3): 557–574.

5.44. Hager, C. H. Jr. and Ryan, E. 2015. Friction and wear properties of black oxide surfaces in rolling/sliding contacts. *Wear* **338–339**: 221–231.

5.45. Hurd, R. M. and Hackerman, N. 1957. Kinetic studies on formation of black-oxide coating on mild steel in alkaline nitrite solutions. *Journal of the Electrochemical Society* **104**(8): 482–485.

5.46. Bhadeshia, H. K. D. H. 2016. Prevention of hydrogen embrittlement in steels. *ISIJ International* **56**(1): 24–36.

5.47. Wolff, R. H. 1966. Hydrogen embrittlement of steel in metal finishing processes of black oxide and zinc phosphatize. Technical Report 66-2008. Rock Island Arsenal Laboratory, Rock Island, Illinois, USA.

5.48. Grabke, H. J. and Riecke, E. 2000. Absorption and diffusion of hydrogen in steels. *Materiali in Technologije* **34**(**6**): 331–342.

5.49. Herring, D. H. 2010. Hydrogen embrittlement. *Wire Forming Technology International* **13**(**4**): 24–17.

5.50. Eliaz, N., Shachar, A., Tal, B., and Eliezer, D. 2002. Characteristics of hydrogen embrittlement, stress corrosion cracking and tempered martensite embrittlement in high-strength steels. *Engineering Failure Analysis* **9**(**2**): 167–184.

5.51. Woodtli, J. and Kieselbach, R. 2000. Damage due to hydrogen embrittlement and stress corrosion cracking. *Engineering Failure Analysis* **7**(**6**): 427–450.

5.52. Bickford, J. H. 2008. *Introduction to the Design and Behavior of Bolted Joints*, 4th edn. CRC Press, Boca Raton, FL.

5.53. American Society for Testing and Materials. 1999. ASTM F1940-99 Standard test method for process control verification to prevent hydrogen embrittlement in plated or coated fasteners.

5.54. American Society for Testing and Materials. 2009. ASTM B850-98 Standard guide for postcoating treatments of steel for reducing the risk of hydrogen embrittlement.

5.55. American Society for Testing and Materials. 2000. ASTM B695-2000 Standard specification for coatings of zinc mechanically deposited on iron and steel.

6
Motor Bearing

The purposes of using bearings in an electric motor are to support and locate the rotating rotor in both radial and axial directions in the motor, maintain a uniform airgap between the rotor and stator over a wide range of operating speeds, reduce friction and wear between the rotating and stationary components, and transfer loads from the shaft to the motor frame and, in turn, to the motor base. It has been reported that two major root causes of motor failure are attributed to bearing and insulation failures. Statistical data have further indicated that more than half of all motor failures are due to bearing failure [6.1]. Therefore, it is important to understand bearing performance characteristics and bearing failure mechanisms.

The most important bearing rating factors are speed and load because they have a great impact on bearing life, especially load. Bearings should have good dynamic operating characteristics over a wide range of speeds, low noise and vibration levels, minimized friction, high fatigue strength, less maintenance, long service lifetime, and low cost. To ensure motor operation, it is critical to select the appropriate bearing type, lubricant, lubrication method, and suitable bearing arrangement, according to different operational conditions and motor applications.

In most cases, a motor shaft is supported by a pair of bearings at each end of the shaft. The use of more bearings to support the shaft is limited to very specific circumstances. Such situations occur for long shaft spans, small shaft diameters, or large external loads. However, the application of three or more bearings in a motor system can cause some problems. The biggest challenge is that the multi-bearing system requires precise and perfect alignment among all bearings. Otherwise, for an inadequately aligned bearing system, amounts of bending moments and loads are generated to act on the shaft. This situation, called *shaft overconstraint*, results in the shaft deformation and bearing failure. Moreover, it makes more difficult to replace bearings in regular maintenance. In fact, any additional shaft support could potentially cause shaft overconstraint. In practice, design engineers should avoid using multiple bearings to support a shaft unless there is good reason to do so.

6.1 BEARING CLASSIFICATION

There are generally three types of bearings used in industrial, military, and commercial applications: journal bearings (also known as sliding bearings), rolling-element bearings (*e.g.*, ball, roller, and needle bearings), and noncontact bearings. Each type has different performing characteristics and is intended for specific applications. Among them, noncontact bearings can be further categorized into different groups, such as fluid bearings (air and liquid bearings), maglev bearings (magnetic bearings), and superconducting bearings.

6.1.1 JOURNAL BEARING

A journal bearing commonly uses a grease or mineral oil to form a thin layer between the bearing surface and loaded shaft, withstanding the applied load on the shaft and separating the rotating shaft from the stationary bearing surface (Figure 6.1). The lubricant film thickness t_f is very small in comparison to the shaft radius R and thus has a high load-carrying capability. In most lubricated journal bearings with incompressible lubricants, the ratio (t_f/R) is typically on the order of 10^{-3}. Due to the eccentricity and the difference in radius between the shaft and bearing, a convergent

DOI: 10.1201/9781003097716-6

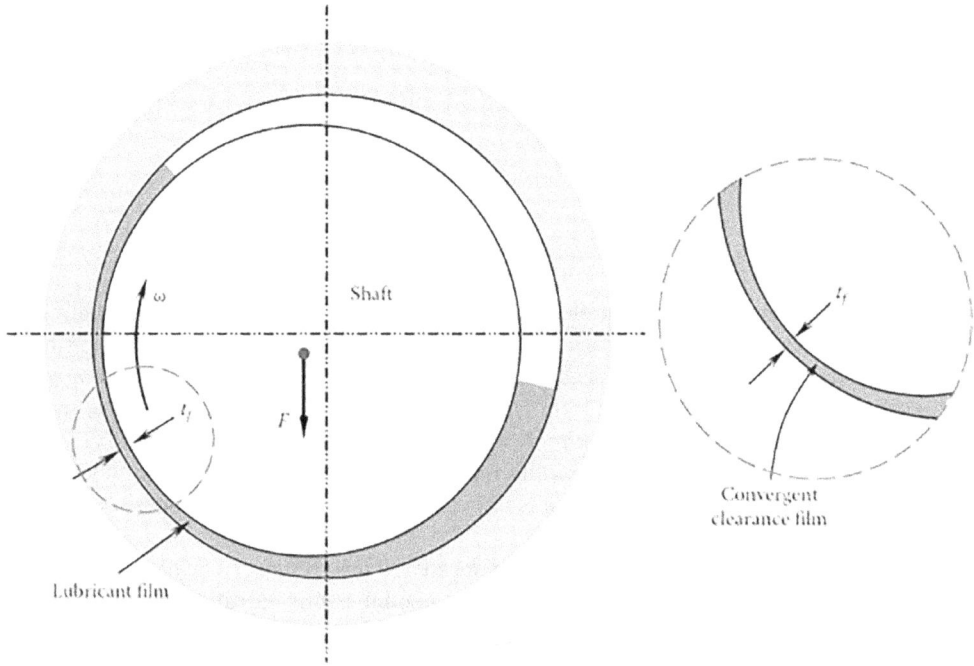

FIGURE 6.1 Lubricant films for supporting motor loads and separating the shaft and stationary bearing surface. Hydrodynamic pressure is developed in the convergent clearance film to lift the shaft.

clearance is formed between the two components. Hydrodynamic pressure is thus built up in this convergent film by the shaft rotation. It is the hydrodynamic pressure that allows journal bearings to tolerate dynamic loads or even momentary shock loads reasonably.

This type of bearing is often used in heavy-duty industrial machinery such as engines, steam/gas turbines, centrifugal compressors, pumps, and large-size motors. A journal bearing consists of a shaft that rotates freely in a supporting metal sleeve. There are no rolling elements in this type of bearing. Journal bearings have various designs: (*a*) The self-lubricating bearings, which are made of porous materials and soaked with lubricants. When the temperature rises due to bearing operation, the lubricant gets out of the pores for playing a role in lubrication. As the temperature decreases, the lubricant is sucked back into the pores due to the capillary effect. The most significant advantage of self-lubricating bearings is lubrication and maintenance-free. (*b*) The regular lubrication bearings, which require lubricants to form an oil film between the shaft and bearing to support the rotating shaft. (*c*) The hydrodynamic lubrication bearings, which require lubricants that elevates the rotating shaft by providing the pressurized lubricant from outside. Journal bearings are usually made from cast iron or Babbitt alloys and are widely used for engines and large-size, heavy-duty motors.

Journal bearings have some advantages over other types of bearings, including the following:

- Low cost for manufacturing
- Less sensitive to contamination
- Enabling operation under severe conditions
- Requiring small radial space
- Available to work in split halves
- Better shock load-sustaining capacity
- Low operation noise
- Less-precise mounting
- Easy for maintenance

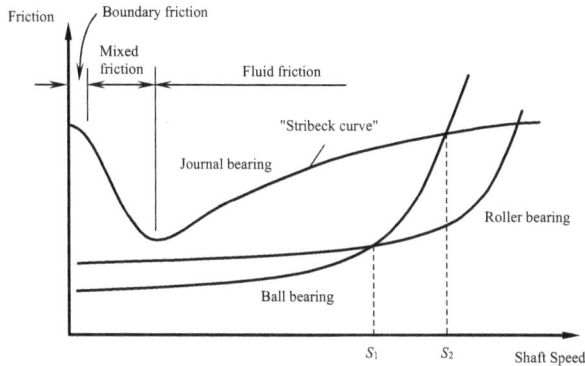

FIGURE 6.2 Comparison of friction between different types of bearings.

Tribology is a multidisciplinary science to study friction, lubrication, and wear in sliding lubrication systems. In fact, tribological behavior is a complex phenomenon in metallic materials. Many factors, such as loading, lubrication, surface roughness, and relative speed, can significantly affect the interaction between two sliding materials. Stribeck [6.2] had systematically studied the variation of friction between two liquid-lubricated surfaces. Based on the investigations, Stribeck claimed that the journal bearing friction is a function of load, lubricant viscosity, and shaft rotating speed. He showed that the friction in journal-type bearings started with high friction at very low speeds, decreased to a minimum friction when metal-to-metal contact was eliminated, and then increased again at higher speeds, which is well known as *Stribeck curve* today. As depicted in Figure 6.2, there are three friction regions: boundary friction, mixed friction, and fluid friction. In the boundary region, the hydrodynamic pressure between the mating surfaces (*i.e.*, the surfaces of the shaft and journal bearing) has not been developed adequately to support the load. In this region, the lubricant boundary film is partly discontinuous, and the thickness of the film is very thin, approximately $200\,\text{Å}$ [6.3]. There are solid contacts in some areas, causing high friction of the sliding system. With an increase in speed, the film thickness of the lubricant increases rapidly to reduce the direct contact of the shaft and the bearing and consequently reduce the friction until it reaches the minimum value. In the fluid friction region, the film thickness of the lubricant becomes much larger (at least four times) than the surface roughness so that the shaft is completely isolated from the bearing. As a result, the friction between the shaft and bearing is mainly determined by the dynamic viscosity of the lubricant and the surface speed. Since the lubricant viscosity is temperature dependent, it is important to select a lubricant that has a small change in viscosity over the full range of operating temperature. The friction factors of different regions are presented in Table 6.1 [6.4] to compare with the dry friction.

The data in Table 6.1 indicate that friction is strongly influenced by lubrication conditions. Dry friction occurs at the interface between two direct contact solids in relative motion without the presence of a lubricant. In any case, dry friction should be avoided due to its high friction factor and the resulted high wear rate.

Figure 6.2 also provides the comparison of friction between different types of bearings. It exhibits that ball bearings have the lowest friction over a wide range of rotating speeds, where the rotating speed is lower than a certain value of increasing the speed to S_1, the friction of ball bearings increases steeply and thus becomes larger than that of roller bearings. At very high rotating speeds ($> S_2$), the friction of journal bearings becomes lower than that of rolling bearings.

It should be noted that rotating machinery supported by journal bearings presents two kinds of self-excited vibrations, that is, oil whirl and oil whip; both can drive the rotating system to an unstable condition. Oil whirl is commonly masked by the rotor unbalance and occurs near half of the rotating speed, hence being rarely associated with instability problems. Oil whip is a severe vibration that occurs when the oil-whirl frequency coincides with the first flexural natural frequency

TABLE 6.1

Comparison of Friction Factors between Different Friction Regions

Friction Region	Friction Factor, f
Dry friction	0.15–0.8
Boundary friction	0.03–0.15
Mixed friction	0.005–0.03
Fluid friction	<0.005

Source: Mayer, E., *Mechanical Seals*, 3rd edn., Newnes-Butterworth, London, U.K., 1977.

of the shaft, hence near twice this natural frequency. When operating under this condition, the vibration amplitudes in the bearings are limited by the bearing's clearance. Based on the rotordynamic analysis, two different thresholds of fluid-induced instabilities have been reported by Mendes and Cavalca [6.5].

In designing journal bearings, the bearing static performance characteristics are related to a dimensionless parameter as Sommerfeld number S, which is defined as

$$S = \frac{\mu\omega}{P}\left(\frac{R}{\delta}\right)^2 \tag{6.1}$$

where
  μ is the viscosity of the lubricant, Pa-s
  ω is the shaft rotating speed, rad/s
  R is the shaft radius, m
  δ is the clearance between the shaft and bearing, m
  P is the specific load, Pa

$$P = \frac{F}{ld} \tag{6.2}$$

where
  F is the external force acting on the shaft, N
  l and d are the length and diameter of the bearing, indicating the ratio of applied load to bearing projected area, m

For an unloaded bearing condition, $F = 0$ and $P = 0$, so that Sommerfeld number S approaches infinity. Thus, a weightless shaft will run concentrically with the bearing. Either increasing the external load or decreasing the rotating speed will reduce Sommerfeld number S. Under such circumstances, the journal moves away from its concentric position to eccentric positions. As specific load becomes huge or speed approaches zero, S approaches zero. Under this condition, metal-to-metal contact occurs between the shaft and bearing.

It has been shown experimentally and theoretically that the lubricant film thickness and dynamic load-carrying capability are significantly affected by the lubricant temperature. At high temperature and high load pressure, the film thickness decreases with the dramatic increase in the load-carrying capability. By contrast, increasing the film thickness has proved to be beneficial at high shaft rotational speed but it is quite dangerous because it strongly increases the ripple amplitude of the film thickness, resulting in the decrease in the load-carrying capability [6.6]. In fact, the dynamic coefficient of friction increases with an increase in the ripple amplitude.

6.1.2 Rolling Bearing

Rolling bearings have high loading capacity and exhibit very low rolling friction torques. A rolling bearing usually consists of an inner and outer ring, a number of rolling elements, and a cage to hold the rolling elements. Raceways are made on the inner and outer rings, on which the rolling elements rotate freely. The performance of bearing strongly depends on the hardness, surface finish, and accuracy of raceways and other parameters such as internal radial and axial clearance. The ring surfaces must go through grinding without producing waviness. Typically, when the surface roughness is larger than $0.2\,\mu m$, the bearing cannot fully play to its carrying capacity. In today's motor industry, the great majority of bearings are rolling bearings.

There are two major types of rolling bearings, distinguished mainly by the rolling-element shape: ball bearings and roller bearings. According to the configuration of the bearing rings, ball bearings are further classified as deep-groove ball bearings and angular contact ball bearings. On the other hand, according to the shape of the rollers, roller bearings are classified as cylindrical, spherical, tapered, and needle roller bearings. To distinguish bearings from their functions, the bearing can be divided into radial bearings, which withstand primarily radial loads and axial loads to a certain extent, and thrust bearings, which withstand only axial loads.

6.1.2.1 Ball Bearing

As the most common type of bearing, deep-groove ball bearings are extensively used in electric motors. This type of bearing is not only capable of taking radial loads but also axial loads to some extent. As shown in Figure 6.3, a standard deep-groove ball bearing consists of a number of rolling balls and two concentric steel rings. Each of the concentric steel rings has a hardened ring on which the balls roll. The balls are separated from each other by a cage between the inner and outer rings. The cage is traditionally made of thin steel, but some bearing manufacturers now use plastic materials. To protect against contaminants, the bearing is sealed with plastic covers at the bearing-side surfaces between the inner and outer rings. To reduce the friction and wear in the rolling contacts, bearings are lubricated with greases or mineral oils. From Figure 6.3, it can be seen

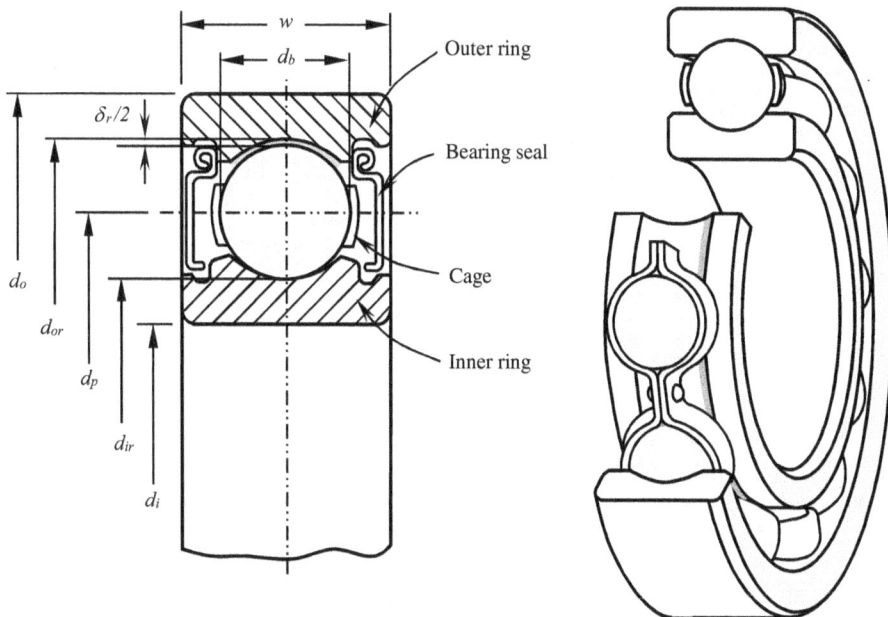

FIGURE 6.3 Deep-groove ball bearings (seals are removed from the right figure).

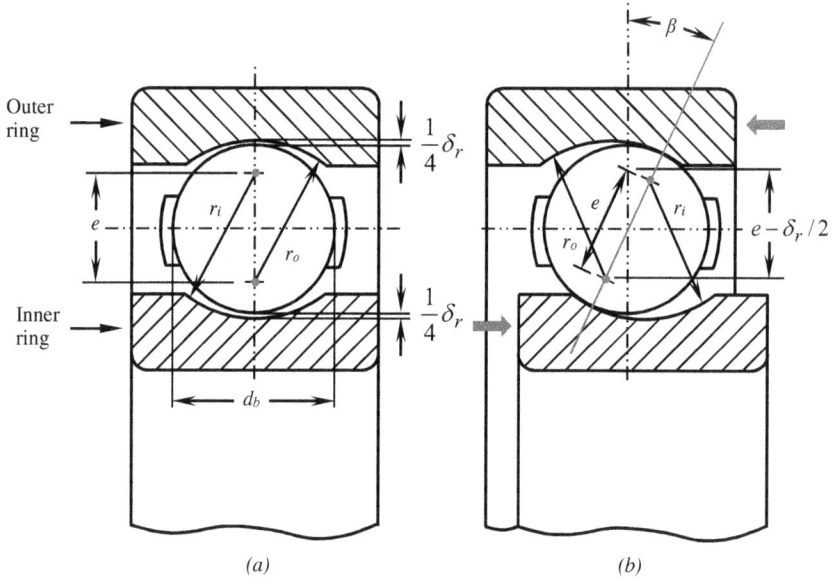

FIGURE 6.4 Cross section of a radial ball bearing, showing inner and outer ring radii (r_i and r_o) and the distance e between the centers of the radii r_i and r_o: (a) initial position and (b) shifted position under axial load.

that a deep-groove ball bearing is symmetric with respect to its centerline of cross section and is characterized by a number of primary dimensions, including the following:

- The bearing OD d_o is the fit dimension with the bearing bore on the motor endbell or housing, and the bearing ID d_i is the fit dimension with the rotor shaft.
- d_{ir} is the inner ring diameter at the lowest point of the raceway.
- d_{or} is the outer ring diameter at the highest point of the raceway.
- d_p is the bearing pitch diameter.
- d_b is the ball diameter.
- is the radial clearance of the bearing.
- w is the bearing width.

The bearing pitch diameter d_p is defined as the diameter going through the ball centers. From Figure 6.3, it can be easily derived that

$$d_p = d_{ir} + d_b \tag{6.3}$$

An alternative expression of the bearing pitch diameter d_p can be related to the bearing internal radial clearance δ_r as

$$d_p = \frac{d_{ir} + d_{or}}{2} - \frac{\delta_r}{2} \tag{6.4}$$

One of the key design features of the deep-groove ball bearing is the curvature of the outer and inner rings because they have a strong impact on the bearing stress, friction, and fatigue life. The curvatures of the outer and inner rings are r_o and r_i, as depicted in Figure 6.4. The ratio of the ring radius to the ball diameter is defined as race conformity, which is a measure of the geometrical conformity of the race to the ball in a plane passing through the bearing axis. Therefore, the outer and inner race conformities are expressed as

$$f_o = \frac{r_o}{d_b} \tag{6.5a}$$

$$f_i = \frac{r_i}{d_b} \tag{6.5b}$$

respectively. Usually, the outer and inner race conformities are around 52%.

Ball bearings are designed with a specific internal radial clearance because it provides free rotation of balls, compensation for thermal expansion, and optimum load distribution (Figure 6.4a). This parameter can significantly influence bearing operating characteristics. When the bearing is subject to axial load, the inner and outer rings shift relatively in the axial direction. As a result, the internal radial clearance no longer exists (Figure 6.4b).

The bearing contact angle β is defined as the angle between the bearing radial line and the contact line, which passes through the contact points between the rolling balls and the inner and outer raceways. From Figure 6.4b, the contact angle β can be expressed as

$$\cos \beta = 1 - \frac{\delta_r/2}{e} \tag{6.6}$$

where e is the distance between the radius centers of r_i and r_o and can be determined as

$$e = (r_o + r_i) - d_b \tag{6.7}$$

Combining Equations 6.5a and 6.5b with Equation 6.7, it follows that

$$e = (f_o + f_i - 1) d_b \tag{6.8}$$

where $(f_o + f_i - 1)$ is known as the *total conformity ratio*, which is the measure of the combined conformity of both the outer and inner rings to the balls [6.7].

By optimizing the contact geometry between the balls and rings, the misalignment between the outer and inner rings can be minimized, providing 35% less friction and 50% less noise [6.8]. These optimized bearings are ideal for applications that require precise rotation with a large range of torques, such as motors and machine tools. Other design features impacting bearing loss include the ring finish, cage material, and bearing lubrication.

Angular contact ball bearings are specially designed to accommodate combined radial and axial loads. Since this type of bearing uses axially asymmetric rings, the contact line that connects the contact points of the ball and the rings is no longer perpendicular to the axis of the bearing rotation. Rather, it forms an angle with the bearing radial line, called contact angle (Figure 6.5). The standard contact angles are 15°, 20°, and 25°, but the contact angles of 30° and 40° are also available for use in different loading conditions. Generally, the larger the contact angle, the higher the axial load that bearing can handle. It is very important to note that single-row angular contact ball bearings can take axial loads only in one direction. Under reverse loading conditions, the elliptical contact area on the outer ring is truncated by the low shoulder on that side of the outer ring, causing excessive stress and temperature rise at the contact area, and thus leading to high vibration and bearing failure. Therefore, at least a pair of bearings must be used together, as shown in Figure 6.6. As can be observed, there are two bearing arrangements: face-to-face (contact line inwards) and back-to-back (contact line outwards). In the face-to-face arrangement (Figure 6.6a), the cones formed by the contact lines point inwards. The axial loads converge toward the bearing axis and may cancel each other out. However, because the support base length H is shorter than the back-to-back arrangement (Figure 6.6b), this arrangement provides a higher tilting clearance (which is defined as the differential mutual ring displacement in the axial direction). As a result, the face-to-face arrangement is not as stiff as the back-to-back arrangement and is not appropriate for the tilting moment loading.

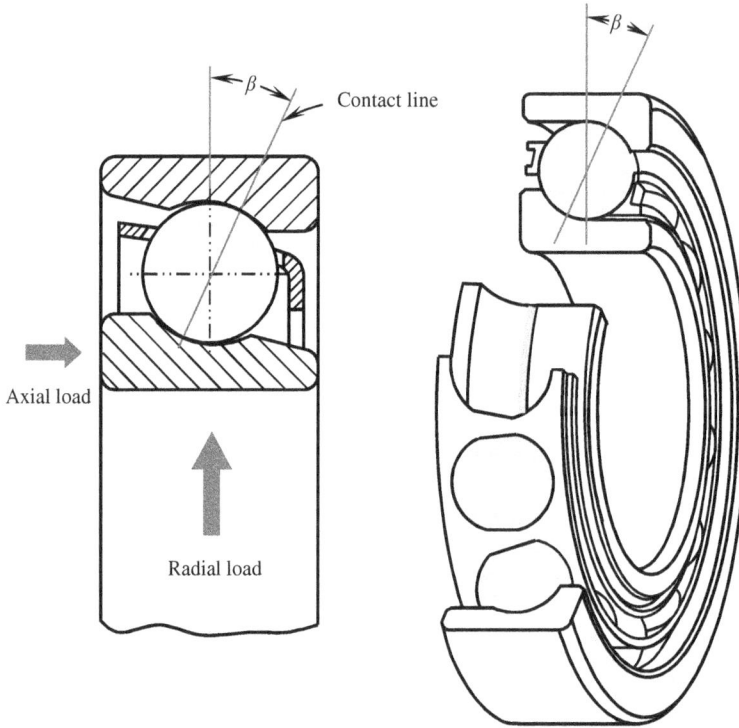

FIGURE 6.5   Angular contact ball bearing with the contact angle β. The bearing shown here has the outer ring cut away reveling the balls and ball cage. This type of bearing has the capability to take one-directional thrust or combined radial and axial loading. To support thrust loading in either axial direction, these bearings are generally mounted in opposing pairs. They are extensively used in electric motors, milling and drilling machines, machine tool spindles, large astronomical telescopes, radar aerials, worm gears, motorcycle engines, *etc.*

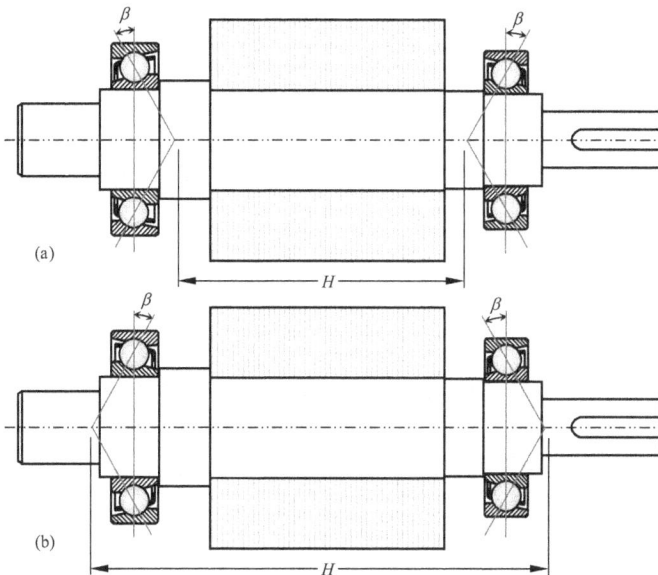

FIGURE 6.6   A pair of angular contact ball bearings used together to support rotor in a motor with two bearing arrangements: (*a*) contact line inwards (face-to-face) and (*b*) contact line outwards (back-to-back).

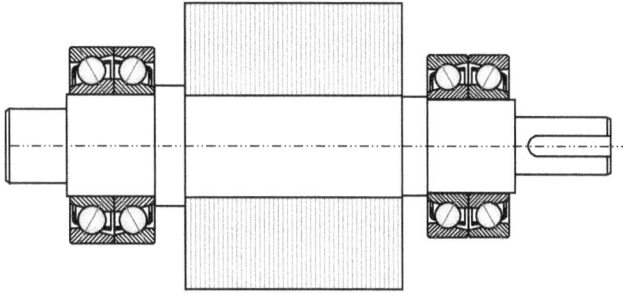

FIGURE 6.7 Matched angular contact ball bearings in a specific bearing arrangement for supporting heavy rotor.

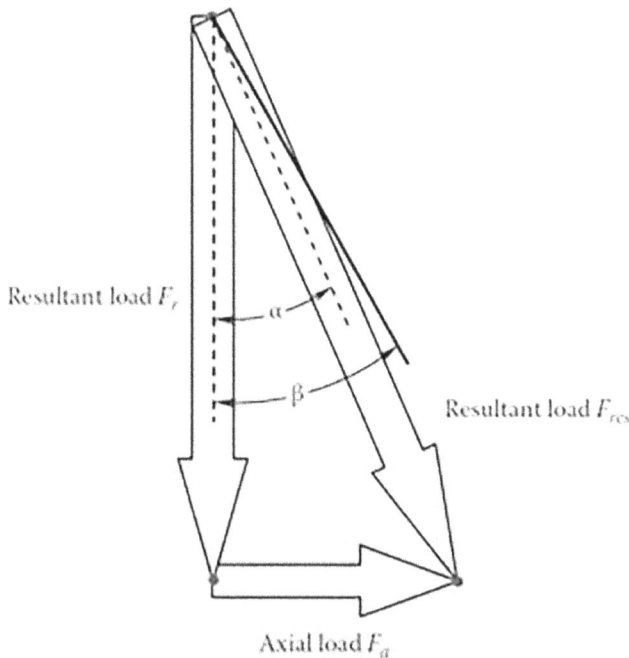

Resultant load F_r

α

β

Resultant load F_{res}

Axial load F_a

FIGURE 6.8 Loads acting on angular contact ball bearing. The load angle α is determined from the radial and axial loads.

The performance of the angular contact ball bearings is similar to that of the deep-groove ball bearings. However, this type of bearing is limited to lower rotating speeds and has slightly higher noise. For some applications, matched bearing sets that combine two, three, or four angular contact ball bearings together may be used to meet special requirements for load capacity and rigidity (Figure 6.7).

The load analysis of an angular contact ball bearing is shown in Figure 6.8. The radial load F_r acting on the bearing is perpendicular to the bearing axis of rotation, while the thrust load F_a is parallel to the axis of rotation. Thus, the resultant force is formed from these two bearing loads. The load angle α is defined as the angle between the resultant load and the radial load, determined as

$$\alpha = \arctan\left(\frac{F_a}{F_r}\right) \tag{6.9}$$

The radii of the outer and inner rings in an angular contact ball bearing are shown in Figure 6.9. It is noted that the centers of the radii are located at the bearing contact line.

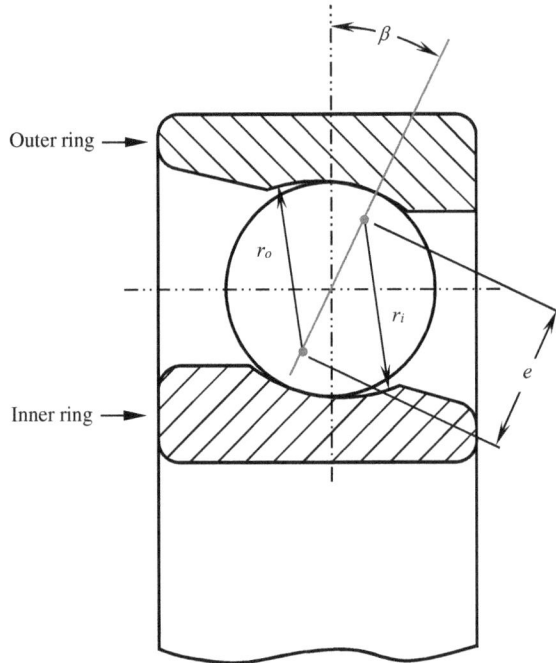

FIGURE 6.9 Cross section of an angular contact ball bearing, showing the inner and outer ring radii (r_i and r_o) and the distance e between the centers of the radii r_i and r_o.

6.1.2.2 Roller Bearing

In roller bearings, contact rollers may come in a variety of different shapes, including cylindrical, spherical, and conical (tapered roller). In general, the bearing running performance and noise generation depend to a large extent on the dimension precision and surface finish of rollers. Compared with ball bearings, roller bearings have much greater load-carrying capacity but also higher rolling resistance (the friction factor is typically 0.0015 for roller bearings and 0.0008 for ball bearings).

Cylindrical roller bearings improve load-carrying capacity and have been designed to support primarily high radial loads and, to a lesser extent, thrust loads. They have exceptionally low-friction torque characteristics that make them suitable for high-speed applications. The critical design features of this type of bearing include the surface finish, roundness, and waviness of the tracks and cylindrical rollers. As depicted in Figure 6.10, a single-row cylindrical roller bearing consists of an outer and an inner ring and a number of cylindrical rollers for enhancing the load-carrying capacity. The cylindrical rollers are parallel to the bearing axis of rotation and held in position by a bearing cage.

Tapered roller bearings can carry combinations of large thrust and radial loads. As shown in Figure 6.11, a taper angle is formed between the inner and outer rings. Nominal taper angles of 10° and 17° are commonly used in tapered roller bearings.

Thrust bearings are a particular type of rotary bearing, designed to support primarily thrust (axial) loads in a variety of applications (Figure 6.12). There are several thrust bearing types such as ball, crossed roller, cylindrical roller, tapered roller, needle, and spherical roller. Thrust bearings are often used in vertical motors. However, they are typically noisier than ball bearings.

6.1.3 Noncontact Bearing

Noncontact bearings, including air and magnetic bearings, are referred to as floating bearings that maintain a small gap with moving surfaces to completely avoid any kind of solid-to-solid contact

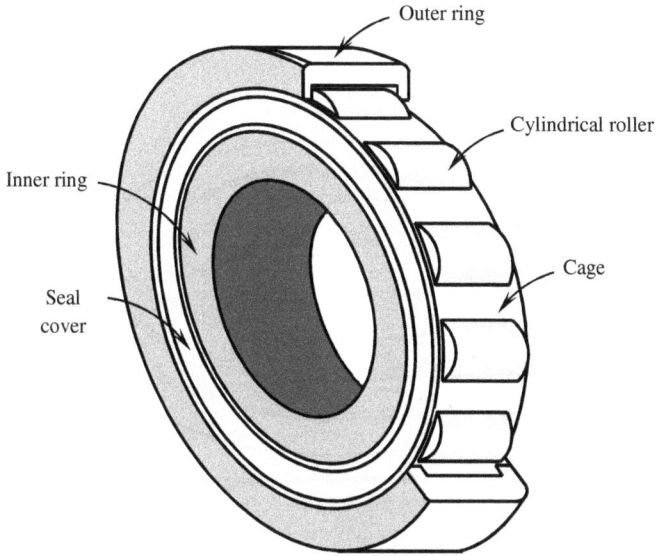

FIGURE 6.10 A single-row cylindrical roller bearing. The roller bearings have greater load-carrying capability than ball bearings of equivalent size, suitable for heavy and sudden loading, high speeds, and continuous service. Interestingly, the manufacturing cost of this type of bearing is lower than that of equivalent type of ball bearing.

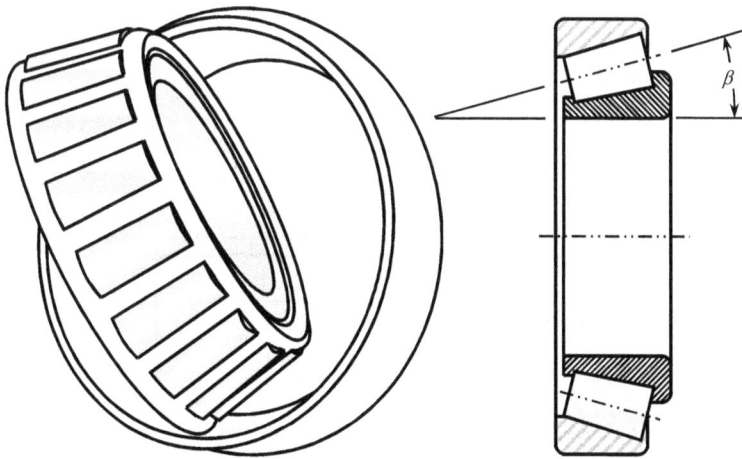

FIGURE 6.11 Tapered roller bearing, which can withstand both radial and axial loads. The cross-sectional view shows the specially designed tapered rollers and demonstrates their angular mounting, which enables their dual load-carrying capability. They are suitable for high-speed spindles of CNC machines and machining centers, vehicle wheels, heavy-duty gearing systems, bevel-gear transmissions, *etc.*

under typical operating conditions. Unlike conventional bearings, noncontact bearings prevent problems of friction. Air bearings use a thin film of pressurized air to support a load (Figure 6.13) and offer much higher stiffness than rolling bearings because the air film fully supports the rotating rotor load, as opposed to rolling elements (*i.e.*, balls or rollers) that have point or line contact. As air has a much lower viscosity than liquid, pressurized air bearings have lower load capacity and operate with essentially zero static and dynamic frictions as opposed to the liquid film bearings, which have much higher frictions and pumping losses within the bearings. These features, as well as

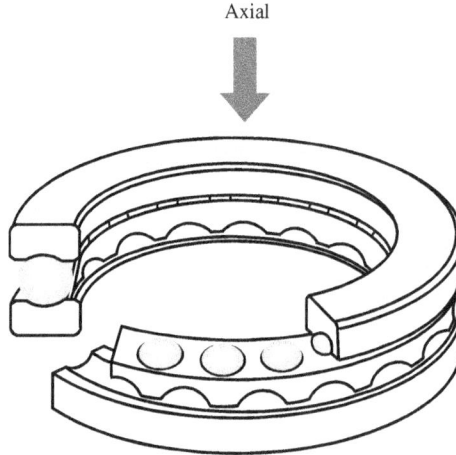

FIGURE 6.12 Single-row thrust ball bearing designed for receiving thrust (axial) loads. This type of bearing is used in large thrust load and low-speed applications, such as thrust spindles, vertical shafts, radio antenna masts, and gearboxes with helical gears. Double-row thrust ball bearings may be adopted in heavy thrust load applications, such as in-wheel motors.

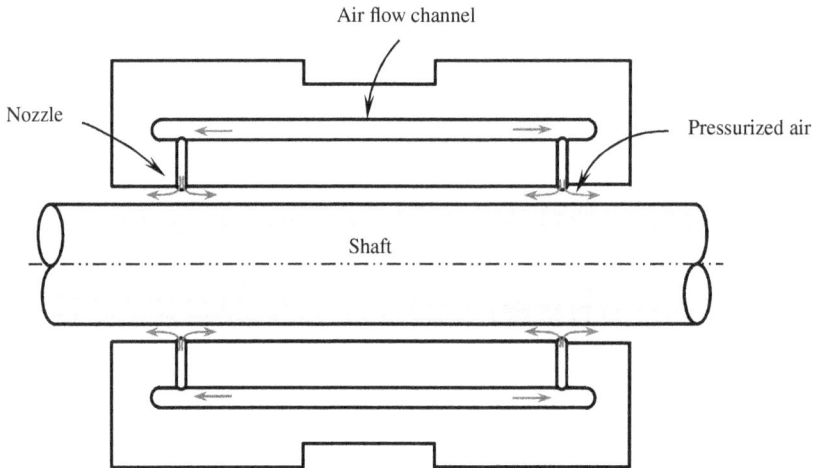

FIGURE 6.13 In an air film bearing, load is supported by pressurized air.

their high precision capabilities, high damping, and silent operation, make air bearings excellently suitable for high-speed, low-load, and low-noise applications such as dental drills and coordinate measuring machines (CMMs) [6.9]. However, air bearings require very tight bearing gaps (about 10 μm compared with 100 μm for liquid film bearings) and very high manufacturing accuracy. This directly leads to extremely tight tolerances, typically less than ±2.54 μm (±0.0001 in.) for the bearings with airgaps of 12.7 μm (0.0005 in.).

Magnetic bearings can provide superior performance over fluid film bearings and roller-element bearings. They generally have lower drag losses, higher damping properties, and moderate load capacity. The steady-state stiffness of magnetic bearings can be essentially infinite, depending on how the close-loop control system is designed. The dynamic stiffness depends on the frequency of applied load and the bandwidth of the control system [6.10]. Unlike other types of bearings, magnetic bearings do not require lubrication, thus eliminating lubricant, valves, pumps, filters, coolers, and other related components, which are typically responsible for adding system complexity and reducing machine's operating reliability.

Although magnetic bearings can use either attractive or repulsive magnetic forces to elevate the rotating shaft, magnetic bearings are efficient only in the attraction mode. In order to obtain high performance in systems with randomly oriented force components, magnetic bearings should be used in an opposed mode design. Air bearings use air pressure to support the rotating shaft. For this type of bearing, the ratio of δ_r/R is typically 0.0001. Noncontact bearings are often used for extremely high-speed applications such as gyroscopes and dental drills and for precision machinery such as CMM.

6.1.4 SENSOR BEARING

In some motor applications, it is required to monitor the operational status of rotor, including rotor position (number of revolutions), rotating speed and direction, as well as acceleration and deceleration. A sensor bearing is made by integrating shielded Hall effect sensors, a magnetic impulse ring, and a versatile ball bearing to form a sensor-bearing unit to detect the required motor operation information and convert it into electrical signals.

As shown in Figure 6.14, while a magnetic impulse ring divided into a sequence of north and south poles is attached to the inner ring of the bearing, small sensors are attached to the outer ring of the bearing. When the bearing inner ring rotates, the impulse ring moves past the stationary sensor ring, generating a magnetic field of alternating polarity. The sensor outputs a pulse whose frequency depends on the number of polarity changes per second. Via the sensor-bearing connection cable, the sensor output signal is transmitted to an electronic unit normally developed by the user. This unit analyzes the signal and provides application-specific information [6.11].

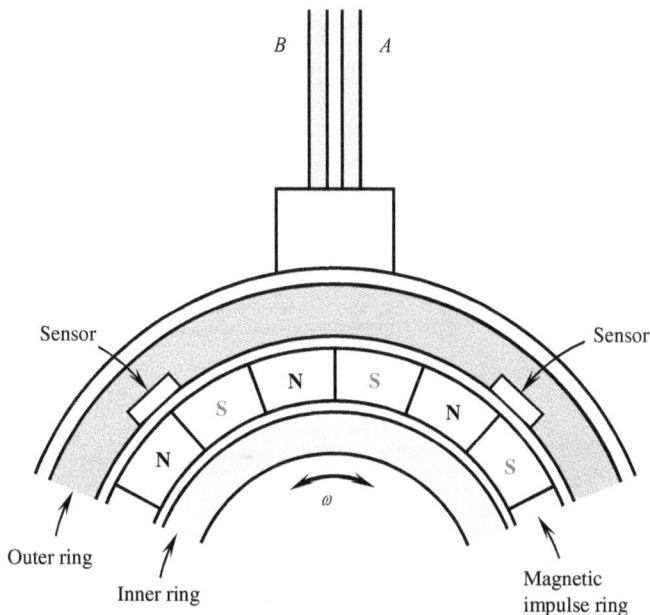

FIGURE 6.14 Operating principle of sensor bearing. As a mechatronic motor component that features a bearing and a sensor together, the sensor bearing provides a cost-effective and compact solution for reducing the number of parts, taking up the less space, lowering the unit weight, and cutting the manufacturing/assembly costs.

6.1.5 SLEWING RING BEARING

Specially designed four-point contact slewing ring bearings can simultaneously handle complex load spectrums that cover diverse combinations of axial (thrust), radial, and tilting moment loads. They are configured to perform both slewing (oscillating) and rotational movements with typically slow speeds. With large diameters, slewing bearings are widely used in heavy-duty applications, such as wind turbines, cranes, excavators, telescope antennas, medical equipment, machine tools, tunnel boring machines, and transportation systems. Because they handle all loads in one assembly, these bearings eliminate weight, space, and cost penalties of other rotational designs.

Slewing bearings can be produced in a variety of constructions and shapes, in terms of row number (*e.g.*, single-, double-, or multi-row), type of rolling elements (*e.g.*, balls, taped rollers, or cylindrical rollers), roller arrangement, and raceway geometry. Figure 6.15 shows a single-row ball slewing bearing with a simple structure. Like regular ball bearings, this type of bearing consists of an inner ring, an outer ring, a number of spacers and balls, and seals (now shown). The mounting holes are provided at both inner and outer rings.

A special type of slewing bearing is called the four-point contact wire-race bearing, where rolling elements run on the hardened surfaces of drawn and ground wire-like inserts (Figure 6.16). The wire-race slewing bearing was invented by Erich Franke in 1934 [6.12]. Because the wire-races are replaceable, this type of slewing bearing is ideal in applications where removal and replacement of the slewing bearing are difficult, where the weight of the slewing ring is sensitive to end users, and where ambient environment is corrosive to slewing bearing materials. This type of bearing is known for being very light and reliable and suitable for medical scanning equipment, radar antennas, wind turbines, and excavators.

The construction of wire-race slewing bearing provides several mechanical advantages over fixed raceway slewing rings [6.13]:

- The wire-races are free to twist in the grooves, providing a self-aligning capability in situations where the support rings are slightly distorted to conform to mounting structures.
- Because the wire races are not joined at the ends, the races can migrate in the grooves to accommodate differential expansion/contraction.
- Wire-race slewing bearings can be preloaded for accurate positioning without causing the development of high frictional torque. This ability provides a very stiff rotational system that can be easily driven.
- Wire raceways can be replaced without complete disassembly of the slewing bearing.
- The use of wire-race inserts allows supporting rings to be manufactured in materials (*e.g.*, aluminum alloys) that may be more appropriate to a specific application without sacrificing the integrity of the raceway.

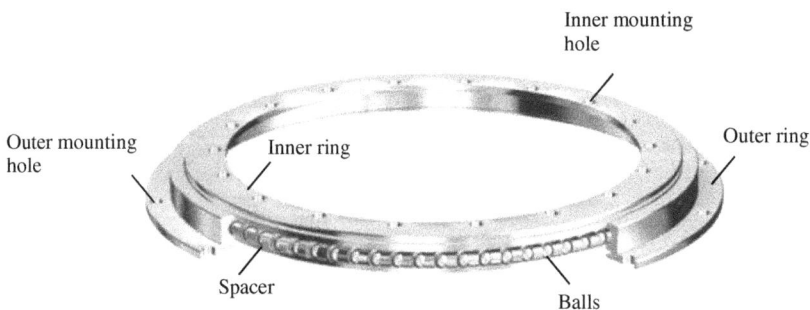

FIGURE 6.15 A single-row ball slewing bearing with a simple structure, similar to a regular ball bearing but with a large diameter-to-thickness ratio.

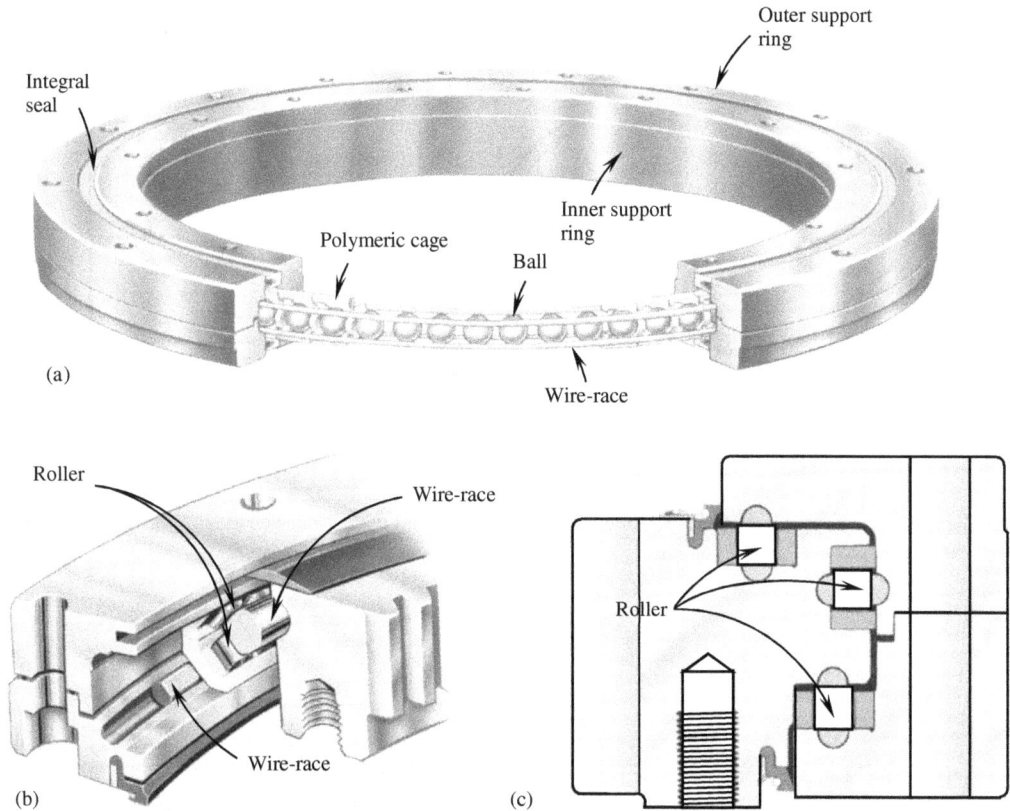

FIGURE 6.16 Wire-race slewing bearings with different configurations: (*a*) single-row four-point contact ball type, where balls roll within an X-shaped array of four wire-race inserts; (*b*) two-row roller type, where two rows of rollers that are angled at 45° to the cross section's vertical axis, and roll between raceways formed by a triangular array of wire-race inserts; (*c*) three-row roller type, where one row of rollers (vertically) carries radial load and other two rows of rollers (horizontally) carry thrust and overturning loads. (Courtesy of Rotek Incorporated, Aurora, Ohio.)

In the selection of slewing ring bearing, a number of factors must be thoroughly considered, such as rotational speed, load condition, accuracy, frictional resistance, operating temperature range, duty cycle, turning torque, protection from contamination, system inertia, mounting structure, and static/dynamic capacity for low/high duty cycle.

Slewing rings can be designed with or without gearing. In practice, many slewing rings are supplied with either internal or external spur gear teeth cut into one race ring. An integrated gear eliminates additional bolt-on gears, which helps to reduce design work and cost. Some slewing rings have induction-hardened gear teeth, providing a fine-grain martensitic layer on the gear surfaces and producing significant compressive residual stresses at the surfaces, as well in the subsurface region. Compressive stresses help inhibit crack development and resist tensile bending fatigue.

Over the years, manufacturers have continued to make substantial improvements in wear resistance, contact fatigue strength, endurance, and impact strength to help extend the lifespan and eliminate the risk of premature failure of slewing ring bearings.

6.1.6 Crossed Roller Bearing

Crossed roller bearings are designed to offer a high level of rotational accuracy and rigidity. As shown in Figure 6.17, a standard crossed roller bearing consists mainly of an outer ring, an inner

Inner ring

Outer ring

Split outer ring

Inner ring

Spacer retainer

Roller

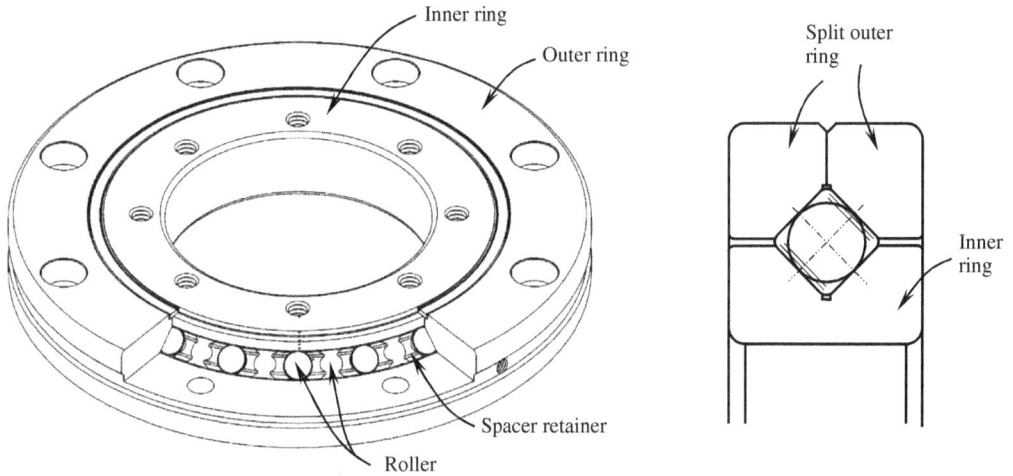

FIGURE 6.17 The structure of a cross-roller bearing with the split outer ring. The axes of the adjacent rollers lie at right angles to each other.

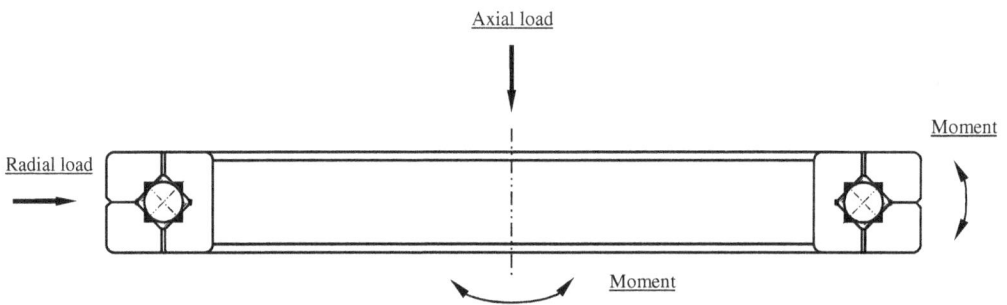

Axial load

Moment

Radial load

Moment

FIGURE 6.18 Cross-roller bearing can withstand loads in all directions simultaneously, most suitable for applications where high performance and high load-carrying capability are needed.

ring, and a plurality of rollers and spacer retainers. Cylindrical rollers are placed between the inner and outer rings and arranged orthogonally to each other in a 90° V-shaped groove, enabling the cross-roller bearing to handle axial, radial, and moment loads (Figure 6.18). Space retainers are placed between the rollers to prevent the mutual friction and decrease the torque resistance for rotation. According to the ring structure, cross-roller bearings can be classified as either the split outer ring type, which is suitable for applications where the rotational accuracy of the inner ring is required, or the split inner ring type, which is suitable for applications where the rotational accuracy of the outer ring is required.

Able to withstand high overturning moments with minimized dimensions, the cross-roller bearing is optimal for applications such as industrial robots, swiveling tables of machine tools, rotary units of manipulators, precision rotary tables, medical equipment, measuring instruments, and grinding machines. It is especially suited to many other pivot and pedestal applications where space is limited.

6.1.7 Ball Screw

A ball screw is a special motion-transfer actuator that combines a ball bearing and a screw together to convert rotational motion into linear motion with negligible rolling friction. During the motion-transfer process, loads are transferred from the screw shaft to the nut through a set of rolling balls.

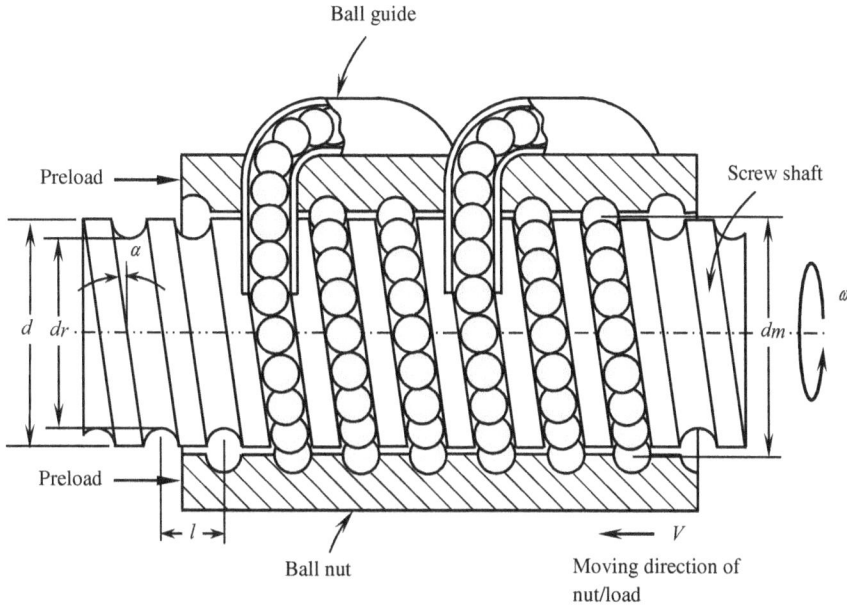

FIGURE 6.19 Structure of ball screw with design parameters. The balls are recirculated through the ball guides.

As shown in Figure 6.19, the ball screw assembly primarily consists of four components: a screw shaft, a ball nut, a ball recirculation system, and a plurality of bearing balls [6.14]. The screw and nut have matching helical grooves and balls roll between these grooves. To cope with the high axial forces as the screw shaft rotates in either direction, angular contact bearings or thrust bearings may be adopted at two ends of the screw shaft (not shown in the figure). Since the specially designed ball screw systems can take heavy loads and achieve high precision positioning, they are widely used in computer numerical control (CNC) machine tools and precision positioning tables.

Ball screw actuators have a long service life and high efficiency (approximately 80%). The ball screw can receive a preload, in the opposite moving direction of the axial load. Thus, the axial clearance can be minimized under an axial load and the high rigidity is achieved due to the preload.

The main design parameters of a ball screw system include lead angle α, screw lead l, ball diameter d_b, screw diameter d, pitch circle diameter d_m, screw root diameter d_r, thread length L, nut/load moving velocity V, and screw angular velocity ω. When designing a groove for a ball screw, the profile of the screw thread often adopts the so-called *gothic-arch profile*, which consists of at least two radii with offset centers. This enables each ball to touch both sides of the raceway simultaneously. This type of groove always allows for four-point contact and consequently, the balls can be loaded in any direction in that plane.

Like ball screws, roller screws use rollers to replace balls to offer higher thrust capabilities and dynamic load ratings. Because the rollers have more contact areas than the balls, the efficiency of roller screws is a little bit lower than that of ball screws, ranging from 75% to 80%. Thanks to more area of contact and improved force distribution, the roller screws' operation life can be several times longer than that of the ball screws [6.15].

6.2   BEARING DESIGN

Bearing design involves several fundamental elements: selecting correct bearing materials, designing proper bearing lubrication and cooling approach, understanding bearing service environment and operating conditions, and choosing suitable values of bearing design parameter.

6.2.1 Bearing Materials

Bearings materials are classified as through-hardened materials used largely for ball bearings and case-hardened materials used largely for roller bearings. The commonly accepted minimum hardness for steel bearing components is 58 Rockwell, and the hardness for ceramic bearings ranges from 78 to 81 Rockwell [6.16]. For normal applications, Society of Automobile Engineers (SAE) 52100 (100Cr6 in Germany and GCR15 in China) is an excellent general-purpose bearing steel, which has a carbon content of 0.95%–1.1% and a chromium content of 1.3%–1.6% [6.17]. This steel grade is generally suitable for applications in which high-strain strength and high resistance to wear under high alternate loads are required. Therefore, the steel is manufactured by the induction vacuum melting process for minimizing porosity that might be caused by gas released from the smelting process.

High-temperature and high-load applications often use hybrid bearings with rings made of SAE 52100 and balls made of ceramic (*e.g.*, silicon nitride Si_3N_4). Under corrosive environment conditions, stainless steel AISI 440C (SUS440C in Japan and 9Cr18 in China) hardened to 58 Rockwell and above is used as the standard rolling bearing material. Due to the lower hardness, the bearings made from this material have a load-carrying capacity 20% lower than those made from SAE 52100.

Nickel-chrome alloys are often used for making bearings. The addition of elements nickel and chrome can increase the hardenability of steel. In addition, chromium brings high-temperature strength and resistance to corrosion, oxidation, and abrasion. As a nickel-chrome alloy, SAE 8620 contains about 0.2% of carbon, 0.5% of chrome, and 0.55% of nickel.

Another alternative material is high-nitrogen steel (HNS), which is a new type of alloy with up to 0.9% nitrogen. This material has a good size stability, high strength, and enhanced resistance to pitting corrosion.

Because cylindrical roller bearings can take much higher loads than ball bearings, these bearings are widely used in large motors with heavy loads. In addition, the cylindrical roller bearing allows for floating of the inner race relative to the outer race. In some motor applications, the external load acting on a motor is far from uniform, that is, the motor drive end takes much higher load than the nondrive end. Thus, for such an application, it is common to use a cylindrical roller bearing on the drive end and a deep-groove ball bearing on the nondrive end.

For some special applications, spherical roller bearings may also be used for heavy radial load applications. This type of bearing has a good bearing misalignment capability and can carry even more radial loads. However, the performance of sphere bearings is very sensitive to their lubrication conditions, requiring three to five times relubrication than similarly sized cylindrical roller bearings. Statistic results have shown that lubrication failures with these types of bearings are common. Consequently, these bearings are rarely found in motors today [6.18].

6.2.2 Bearing Internal Clearances

Bearing clearance is defined as the total free distance that one bearing ring can be shifted relative to the other either in the radial direction, which is defined as the internal radial clearance δ_r, or in the axial direction, which is defined as the internal axial clearance δ_a, as demonstrated in Figure 6.20. Both internal radial and axial clearances are important parameters that significantly affect bearing operating characteristics such as vibration, noise, lubrication, heat generation, friction, wear, and bearing fatigue life. If the radial clearance of a bearing is too large, it may lead to bearing vibration and inaccurate motion. If the radial clearance is too small, the lubrication film may break down, leading to metal-to-metal contact between rolling balls and raceways.

The axial internal clearance is established by ball diameter, inner and outer raceway radius, and the radial internal clearance. Depending on the bearing geometry, the axial clearance is 8.5–10 times larger than the radial clearance [6.19].

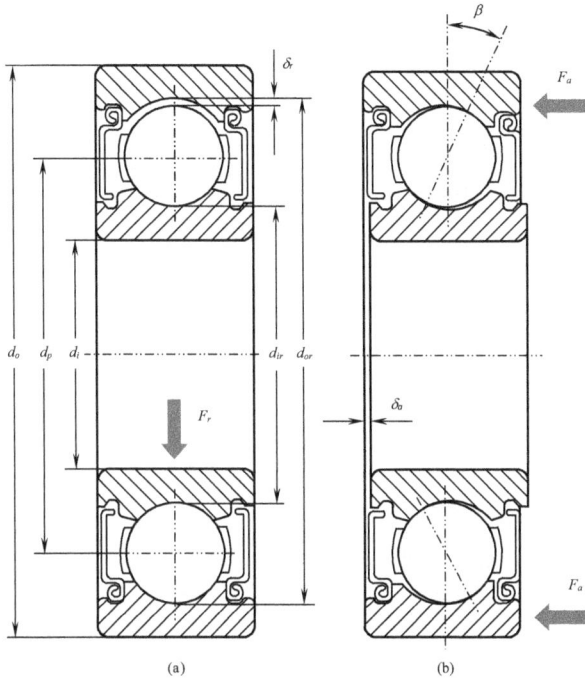

FIGURE 6.20 Bearing clearance: (*a*) when a bearing is subject to a radial load F_r, a radial clearance δ_r is formed between the inner and outer rings; (*b*) when a bearing is subject to axial load F_a, the axial clearance δ_a and contact angle β are formed, but the radial clearance disappears.

Bearing clearance is categorized as unmounted and mounted (effective) clearance. Unmounted clearance is referred to as the initial bearing clearance, which is provided by bearing manufacturers. As a reference value, the initial bearing clearance is chosen during the motor design and bearing selection process. Mounted clearance is referred to as the operational clearance as the bearing is fitted onto the shaft and into the housing bore and when the bearing reaches steady-state operating temperature. In practical applications, mounted clearance is difficult to measure in real time. Usually, the operational clearance is smaller than the bearing initial clearance due to several factors including bearing fit and the temperature difference between the inner and outer rings.

Because cooling the stator in a motor is typically easier than the rotor, the outer ring is often cooler than the inner ring by about 5°C–10°C. For water-cooled stator cases, the temperature difference between the two rings can even be greater to significantly reduce the bearing radial clearance. In some cases, the radial clearance can reach zero or even negative values. When the interference between the rolling elements and rings becomes quite large, it may damage bearing elements, accelerate bearing wear, or even lead to bearing failure. For each size of bearings, there is an optimum range of radial clearance depending on bearing applications.

The internal radial clearance δ_r is determined by the outer and inner ring diameters and ball diameter:

$$\delta_r = d_{or} - d_{ir} - 2d_b \tag{6.10}$$

where d_m and d_{ir} are the outer and inner ring diameters, respectively. From Equation 6.6, the internal radial clearance δ_r can also be expressed as

$$\delta_r = 2e(1 - \cos\beta) \tag{6.11}$$

The interference fits of an inner ring to a shaft and an outer ring to a bearing bore can reduce the bearing radial clearance. As described by Harris [6.20], the increase in the inner ring diameter Δd_{ir} due to the interference fit between a bearing inner ring and a hollow shaft (with the ID of d_2) is given as

$$\Delta d_{ir} = \frac{2\delta\, d_{ir}/d_i}{\left[\left(d_{ir}/d_i\right)^2 - 1\right]\left\{\dfrac{\left(d_{ir}/d_i\right)^2 + 1}{\left(d_{ir}/d_i\right)^2 - 1} + v_b + \dfrac{E_b}{E_s}\left[\dfrac{\left(d_i/d_2\right)^2 + 1}{\left(d_i/d_2\right)^2 - 1} - v_s\right]\right\}} \tag{6.12}$$

where
  δ is the interference between the inner ring and the hollow shaft
  v_b and v_s are Poisson's ratios of the bearing and shaft, respectively
  E_b and E_s are moduli of elasticity of the bearing and shaft, respectively

If both the bearing and shaft are made of the same material, v_b and E_b should be equal to v_s and E_s, respectively. The above equation can be simplified as

$$\Delta d_{ir} = \delta\,\frac{d_{ir}}{d_i}\left[\frac{\left(d_i/d_2\right)^2 - 1}{\left(d_{ir}/d_2\right)^2 - 1}\right] \tag{6.13}$$

Furthermore, for a solid shaft, $d_2 = 0$. By means of L'Hôpital's rule, it can be derived that

$$\lim_{d_2 \to 0}\left[\frac{\left(d_i / d_2\right)^2 - 1}{\left(d_{ir} / d_2\right)^2 - 1}\right] = 1 \tag{6.14}$$

Thus,

$$\Delta d_{ir} = \delta\,\frac{d_{ir}}{d_i} \tag{6.15}$$

Similarly, the decrease in the outer ring diameter $-\Delta d_{or}$ due to the interference fit between a bearing outer ring and a bearing bore (with the diameter of d_1) on the endbell or housing is given as

$$\Delta d_{or} = \frac{2\delta\, d_o/d_{or}}{\left[\left(d_o/d_{or}\right)^2 - 1\right]\left\{\dfrac{\left(d_o/d_{or}\right)^2 + 1}{\left(d_o/d_{or}\right)^2 - 1} + v_b + \dfrac{E_b}{E_h}\left[\dfrac{\left(d_1/d_o\right)^2 + 1}{\left(d_1/d_o\right)^2 - 1}\right] - v_h\right\}} \tag{6.16}$$

where v_h and E_h are Poisson's ratio and modulus of elasticity of the bearing bore, respectively.
    If both the bearing outer ring and the bearing bore are made of the same material, $v_b = v_h$ and $E_b = E_h$. The above equation can be simplified as

$$\Delta d_{or} = \delta\,\frac{d_o}{d_{or}}\left[\frac{\left(d_1/d_o\right)^2 - 1}{\left(d_1/d_{or}\right)^2 - 1}\right] \tag{6.17}$$

Taking into account the influence of interference fits, the bearing internal radial clearance becomes

$$\delta_r = \left(d_{or} - \Delta d_{or}\right) - \left(d_{ir} + \Delta d_{ir}\right) - 2d_b \tag{6.18}$$

When selecting motor bearings, thermal expansion must be taken into consideration because it can significantly change the internal clearances. A temperature differential between the shaft and the motor endbell/housing always leads to a change in the internal clearances. Ricci [6.7] has investigated the influence of the radial temperature gradient between the inner and outer rings on the internal radial clearance. Assuming the temperature rises in the outer and inner rings over the ambient temperature are $(T_o - T_a)$ and $(T_i - T_a)$, respectively, the increases in d_{or} and d_{ir} are Δd_{or}, and Δd_{ir}, respectively

$$\Delta d_{or} = \beta_{or} d_{or} (T_o - T_a) \tag{6.19a}$$

$$\Delta d_{ir} = \beta_{ir} d_{ir} (T_i - T_a) \tag{6.19b}$$

where β_{or} and β_{ir} are the coefficient of thermal expansion of outer and inner rings, respectively. Generally, $\beta_{or} = \beta_{ir} = \beta$.

Assuming the ball temperature takes the mean value of the outer and inner rings, the increase in the ball dimension is

$$\Delta d_b = \beta_b d_b \left(\frac{T_o + T_i}{2} - T_a \right) \tag{6.20}$$

Thus, the change in the bearing internal radial clearance due to the thermal expansion is

$$\Delta d_r = \beta d_{or} (T_o - T_a) - \beta d_{ir} (T_i - T_a) - \beta_b d_b \left(\frac{T_o + T_i}{2} - T_a \right) \tag{6.21}$$

The internal clearance is grouped into different classes, that is, C2, C3, C4, and C5, for which the lower number indicates the lower clearance. Some bearing manufacturers designate CN as the normal clearance which is between C2 and C3.

The clearance C3 is the most popular clearance used in most motor applications because it permits a sufficiently large tilting clearance even at high speeds. A bearing with a risk of increased thermal stressing for any reason needs to select the clearance of C4.

It is noted from Figure 6.20 that when a ball bearing is subject to an axial load, the inner and outer rings are dislocated axially, resulting in an axial clearance δ_a. Correspondingly, the contact line is no longer perpendicular to the axis of bearing rotation; instead, the contact angle β is formed between the bearing contact line and the bearing normal line. Under this circumstance, the radial clearance δ_r no longer exists. Generally, internal axial clearance δ_a can be established by internal radial diameter δ_r, ball diameter d_b, and outer and inner ring radii r_o and r_i, respectively

$$\delta_a = 2 \sqrt{\delta_r (r_o + r_i - d_b)} \tag{6.22}$$

6.2.3 ALLOWABLE BEARING SPEED

High bearing speed often leads to temperature rise in the bearing due to the friction heating effect. When the temperature becomes high enough, it may cause the decomposition and degradation of the lubricant, thereby deteriorating bearing performance. Therefore, it is a common practice for most bearing manufacturers to set a maximum bearing rotating speed to avoid such detrimental situations. The allowable speed of a bearing varies with the bearing type, size, material, lubrication method, manufacturing precision, sealing type, loading condition, noise level, operation configuration, both bearing and cage structures, cooling condition around the bearing, *etc.* As the bearing speed exceeds 70% of its allowable speed, high-speed lubricants (grease or oil) of high quality must be applied to the bearing. Generally, ball bearings have a higher allowable speed than roller bearings due to their lower friction.

As radial bearings are used in a vertically mounted motor, lubricant retention and cage guidance are not favorable compared to horizontal motor mounting. Thus, the allowable speed should be reduced to approximately 80% of the listed speed [6.21].

6.2.4 BEARING FIT

For the majority of bearings used in electric motors, the rotating inner rings of bearings are fitted tightly with the rotor shafts via the interference fit method for ensuring reliable torque transmissions. Meanwhile, the stationary outer rings are fitted less tightly (usually slip fit) with the stationary bearing bores at the motor frame. This ensures that at least one bearing in the motor is able to slide when the shaft expands or shrinks due to the change in operation temperature. Also, it facilitates the motor assembling and disassembling processes.

It is recommended by bearing manufacturers [6.22] that tight interference fit is used when (*a*) the operating conditions involve large vibration or shock loads, (*b*) hollow shafts or housings with thin walls are adopted, and (*c*) housings are made of light alloys or plastic materials. Similarly, loose interference fit is preferable for (*a*) the application requires high running accuracy and (*b*) it uses small-sized bearings or thin-walled bearings. While this fit is sufficient to hold mating parts together, it allows for disassembly.

The fit selection for bearings with shafts and housings is graphically presented in Figure 6.21. The bars designated by lower case letters and numbers (*e.g.*, g6, h6) represent shaft diameter and tolerance ranges to achieve various fitness conditions (*i.e.*, loosen and interference fits) required for various load and bearing rotation conditions. Similarly, the bars designated by upper case letters and numbers (*e.g.*, F7, G7) represent housing bore diameter and tolerance ranges. The recommendations for bearing fits are usually provided by bearing manufacturers in their catalogs for different operating conditions and a variety of bearing types and sizes.

The selection of proper bearing fits is crucial for some motor applications. The selection depends on both bearing and motor operating conditions (such as rotating speed, operation temperatures, and load conditions), shaft and housing materials, manufacturing accuracy of mating surfaces, among other factors. Too tight bearing fits can cause excessive stress, reduced radial clearance, and elastic deformation of rolling elements and raceways that may adversely influence the bearing performance. In a worse case, the bearing radial clearance may vanish, turning a loose interference into a tight interference. For such circumstances, the rolling elements may be excessively loaded, leading to rapid wear and temperature rise. On the other hand, too loose bearing fits can cause relative motion between mating parts, resulting in fretting corrosion on the matting surfaces.

6.2.5 PREVENTION OF BEARING AXIAL MOVEMENT

In normal motor operation, bearings must be held in their positions firmly. The axial movement of bearing may alter the load distribution on the bearings and thus possibly overload one of them, leading to the damage of the bearing, wave spring washer, endbells, or windings.

A motor shaft is generally supported at its ends by suitable bearings contained in motor endbells or motor frame. Typically, the inner ring of the bearing is pressed onto the shaft, and the outer ring is pressured into a bearing seat defined within an endbell/housing. A number of mechanical methods have been employed to prevent bearing axial movement. The most common way is to use shaft shoulders and retaining rings, which are inserted in machined grooves on the motor shaft to position the bearings precisely. In addition, proper fits of the bearing inner ring on the shaft and the outer ring in the bearing bore of the endbell/housing also help hold the bearings in their positions against axial movement.

Bischoff [6.23] proposed a method that uses a simplified end play control system to prevent excessive axial movement of the motor shaft in response to forces applied to the shaft. Figure 6.22 shows an endbell hub assembly, which consists of a rotor shaft, a self-aligning bearing carried on the shaft, and a bearing support structure surrounding the shaft and journaling the self-aligning

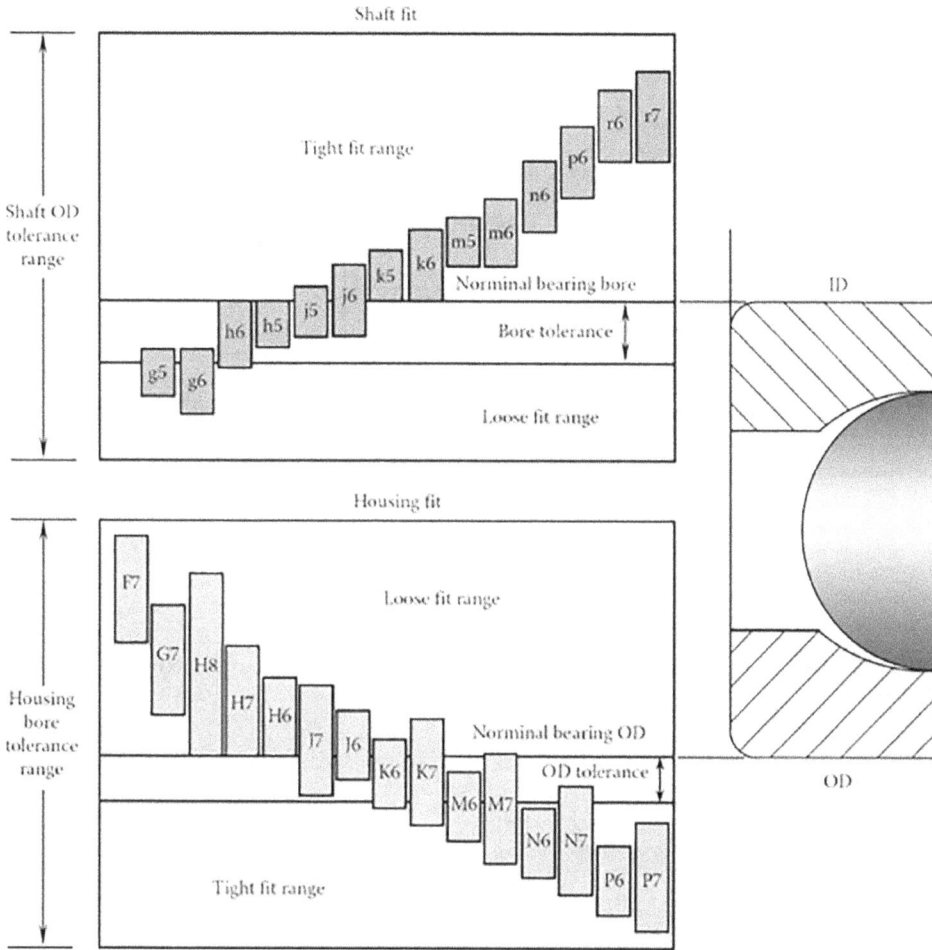

FIGURE 6.21 Fit section for bearings with shafts (top) and housings (bottom).

bearing member. A resilient retainer biases the self-aligning bearing into engagement with its journal. The end play control system includes a thrust collar, a retaining ring, and washers at the outboard (left) and inboard (right) side, respectively. While the thrust collar and retaining ring at the outboard side are for preventing inward movement (*i.e.*, movement to the right), the thrust collar and retaining ring at the inboard side are for preventing outward movement (*i.e.*, movement to the left) of the shaft. It is desirable to insert a seal washer and a pair of washers between the thrust collar and the bearing to position the bearing properly. The seal washer also serves to dampen axial vibration that may occur during motor operation.

An alternative approach is presented in Figure 6.23 [6.24]. This method uses a bearing holder to fix a bearing to a support structure such as a motor endbell or housing. As can be seen in the figure, the single platelike bearing holder is made from a single metal plate. A plurality of butt straps as a holding structure is formed on the side of the panel opposite to the bottom hooks. The butt straps are arranged to retain the bearing within the bearing holder and ideally to press the bearing axially against the bottom hooks.

Once the bearing has been inserted into the bearing holder with the bearing seated in contact with the bottom hooks, the butt straps are radially deformed to move the fingers radially inwardly so as to axially cover the bearing preventing its removal from the bearing holder. Centering lips extending inwardly from the hole edge may be provided to fix the lateral position of the bearing.

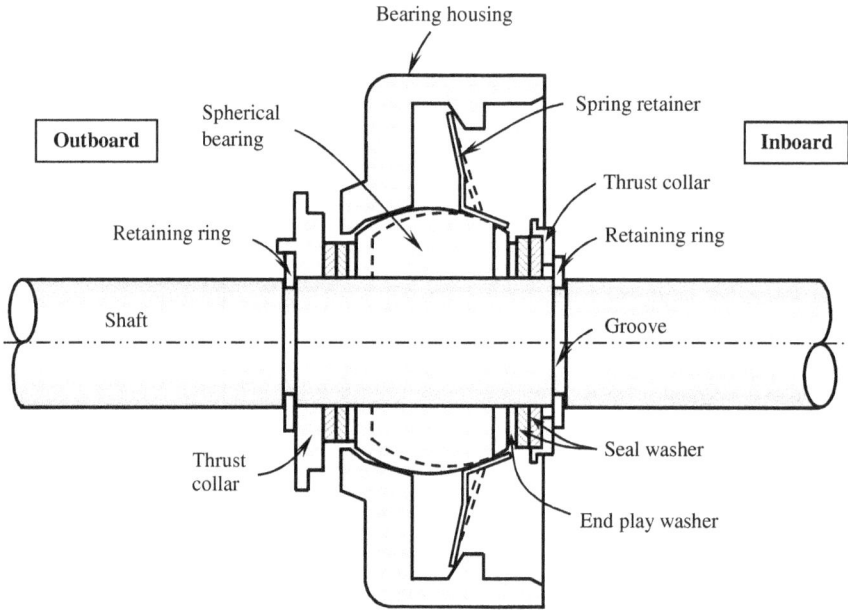

FIGURE 6.22　Using a simplified end play control system for preventing excessive axial movement of shaft and bearing.

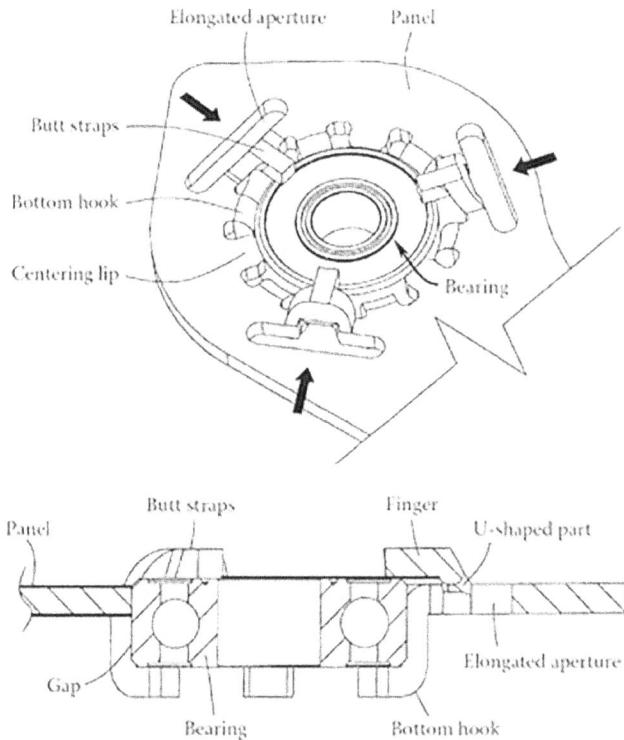

FIGURE 6.23　Bearing holder for fixing a bearing to a support structure using a single platelike body (U.S. Patent 8,894,291) [6.24]. (Courtesy of the U.S. Patent and Trademark Office, Alexandria, VA.)

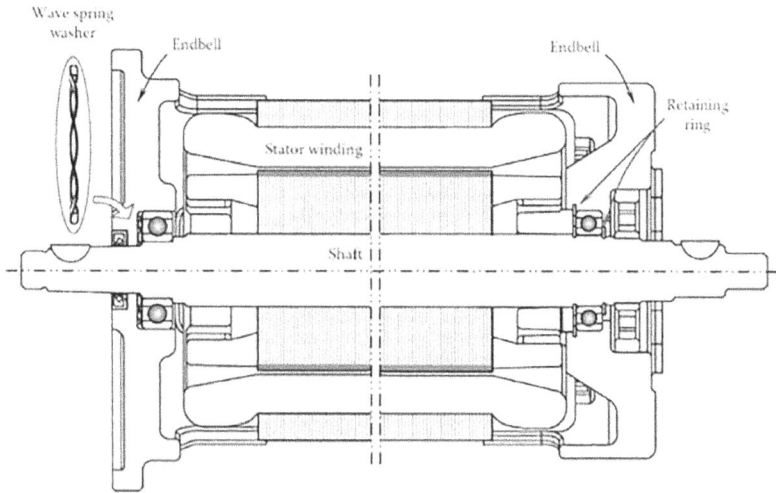

FIGURE 6.24 Using wave spring washer and retaining rings for providing preload to the shaft and positioning bearings axially.

6.2.6 Bearing Load

Bearing load and speed are two key factors in bearing selection and bearing design. The load acting on the bearing has a significant effect on the heat generation and temperature rise. This indicates that the higher the load, the higher the heat produced in the bearing, as well as the higher temperature. The maximum rotating speed is normally determined by the bearing operating temperature. Other factors influencing maximum bearing rotating speed include bearing type, bearing arrangement, lubrication condition, and mechanical limits of bearing components. To ensure normal operation, both bearing load and speed must be well below their allowable values.

6.2.6.1 Bearing Preload Arrangement

Wave spring washers are normally used to provide a preload on the outer ring of one of the bearings for compensating dimensional variations, absorbing impulse forces, and lowing motor operating vibration and noise (Figure 6.24). Many design variations have evolved to best serve these basic functions. The deflection rate for wave spring washers usually ranges from 20% to 80%, depending on the spring type, structure, material, and number of spring waves and turns (Figure 6.25).

The radial preload is often from an interference fit, usually between the motor shaft and the bearing inner ring and between the bearing bore and the outer ring. In some cases, improper fit can cause relative movements between the mating parts, leading to severe wear on either shaft or bearing born and reduced bearing life.

6.2.6.2 Radial and Axial Bearing Load

Even though most radial rolling bearings used in electric motors are designed to take primarily radial load, they can also withstand smaller axial load, as a certain portion of their rated radial load.

The most important parameters of bearing loads are the load magnitudes and their directions. The external load, which comes from the interaction of the rotor and stator, acts first on the inner ring and is subsequently transferred to the outer ring through the rolling elements and finally to the bearing bore on the motor endbell or housing.

6.2.6.3 Load Distribution

When a rolling bearing is subject to radial load, the rolling elements are not equally loaded. Usually, at any given time, only less than half of the rolling elements are loaded. The actual load on each

FIGURE 6.25 Various types of wave spring washers.

rolling element depends on its position in the bearing and other bearing characteristics such as motor internal clearance. Stribeck [6.25] conducted pioneering work on investigating the variation of loads on rolling elements of a radially loaded ball bearing. He found that the load zone for a radially loaded bearing with an internal clearance is less than $\pi/2$, and the load on the most heavily loaded ball is 4.37 times the average ball load for zero internal clearance. Later, Palmgren [6.26] determined that the theoretical value of Stribeck's constant for roller bearings at zero clearance is 4.08.

Oswald *et al.* [6.27] investigated the effect of motor internal clearance on the load distribution and the life of radially loaded ball and roller bearings. Their analysis was extended to negative clearance (interference) conditions for producing a curve of life factor versus internal clearance. It was found that for a small negative clearance, rolling-element loads can be optimized and bearing life maximized. To compare with the results from previous investigators, the concept of Stribeck number is introduced as

$$St = \frac{P_{\max}}{P_r/z} \tag{6.23}$$

where
  $P_{\max}$ is the maximum element load
  P_r is the applied radial load
  z is the number of rolling elements

Their analysis confirmed that $St=4.37$ for ball bearings at zero internal clearance and $St=4.08$ for roller bearings at zero clearance.

The load distribution of a rolling bearing is primarily determined by the load factor ε, the load angle, and the bearing internal clearance.

$$F(\phi) = F_{\max}\left[1 - \frac{1}{2\varepsilon}(1 - \cos\phi)\right]^n \tag{6.24}$$

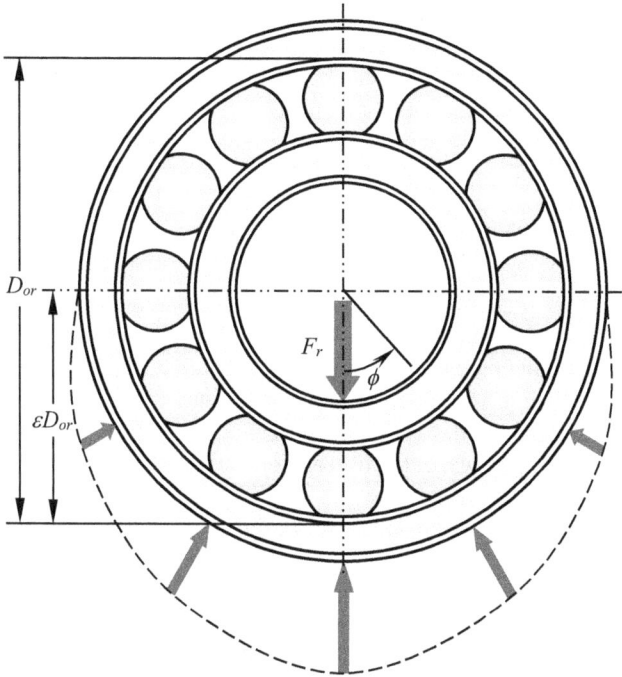

FIGURE 6.26 Load distribution of rolling bearing, $\varepsilon = 0.5$ in this case.

where

F_{max} is the maximum load

ε is the load distribution factor

ϕ is the angle from the bearing lower centerline (as shown in Figure 6.26)

$n = 10/9$ for ball bearings and $n = 3/2$ for roller bearings

For a bearing with a normal radial clearance, the maximum load F_{max} can be determined approximately as

$$F_{max} = \frac{5F_r}{Z \cos \alpha} \tag{6.25}$$

where

F_r is the applied radial load

Z is the number of rolling element

α is the mounted contact angle

6.3 BEARING FATIGUE LIFE

Each bearing has a limited life of service. The bearing life expectancy for industrial applications is typically quoted as 20,000–40,000 h, with theoretical mechanical life approaching 40,000–60,000 h, depending on the application. The factors that determine bearing fatigue life include (*a*) bearing rotating speed, (*b*) external radial and axial loads, (*c*) bearing radial and axial clearances, (*d*) lubrication condition (*e.g.*, formation of lubricant film, contamination), (*e*) lubricant type, (*f*) operation condition (*e.g.*, temperature, corrosive environment, machine vibration), (*g*) bearing type and size, (*h*) bearing material, (*i*) fit condition, (*g*) manufacturing process, (*k*) bearing installation condition, and (*l*) motor shaft misalignment. All these factors must be taken into account in the bearing section and application.

Bearing operating temperature has a strong impact on bearing life by affecting bearing radial and axial clearances, the viscosity of a lubricant, the rate of lubricant volatilization, and the bearing sealing performance. Motor vibration and shock loading can not only bring additional loading to motor bearings to deteriorate bearing operation but also alter bearing lubrication condition and thus are detrimental to bearing life.

6.3.1 CALCULATION OF BEARING FATIGUE LIFE

Bearing life is defined as the length of time, or the number of revolutions, until a fatigue spall of a specific size develops. This spall size is usually defined as an area of 0.01 in.^{2} (6.5 mm^2). This life depends on many factors as mentioned above. Due to all these factors, the life of an individual bearing is impossible to predict precisely. Also, bearings that may appear to be identical can exhibit considerable life scatter when tested under the same conditions.

Basic bearing rating fatigue life or L_{10} is defined as the life that 90% of a sufficiently large group of apparently identical bearings is expected to reach or exceed. By contrast, bearing average life is the life that 50% of the bearings exceed.

Either in theory or in fact, with sound and robust design, correct bearing material selection, accurate manufacture, proper lubrication, and diligent maintenance, bearing life can be extremely long. A calculation was made by Orlowski [6.28] for a medium-duty ball bearing with a dynamic load ration of 7,000 lb$_f$ at an operating speed of 1,800 rpm; the bearing life can be as high as 362 years!

Though there are several bearing fatigue life calculation methods available, the formula that is widely used by bearing manufacturers is given as

$$L_{10} = a\left(\frac{C}{P}\right)^b \frac{B}{N} \tag{6.26}$$

where
  C is the radial-rated bearing load rating
  P is the dynamic equivalent radial load applied on the bearing
  B is the factor that depends on the method and units used in the equation
  N is the bearing rotating speed
  a is the life adjustment factor
  $b = 3$ for ball bearings and $b = 10/3$ for roller bearings

In 2007, ISO [6.29] specifies methods of calculating the basic bearing rating life L_{10} of rolling bearings in its standard ISO 281: 2007, with commonly used high-quality material, good manufacturing quality, and conventional operating conditions. By introducing a correction factor a_{ISO}, which provides an estimation of the influence of lubrication and contamination on the bearing life, the adjusted rating life (in one million revolutions) for a different level of reliability is given as

$$L_{n,a} = a_1 a_{ISO} L_{10} \tag{6.27}$$

where a_1 is the life adjustment factor for reliability. The value of a_1 is entirely determined by L_n, where the subscript n represents the failure probability, that is, the difference between 100% and the requisite reliability. For instance, $a_1 = 1$ for L_{10} (*i.e.*, 90% reliability) and $a_1 = 0.21$ for L_1 (*i.e.*, 99% reliability). The introduction of a_{ISO} is based on the assumption that the fatigue load is directly linked with the bearing's static capacity and the contamination factor is constant whatever the lubrication conditions.

Another popular bearing life formula was also released by ISO in 1978, which is still used today in industrial applications. This model uses three adjustment factors by taking into account reliability, bearing properties, and operating conditions, respectively [6.30]:

$$L_{n,a} = a_1 a_2 a_3 L_{10} \tag{6.28}$$

where a_1 is the adjustment factor for reliability as in Equation 6.27. a_2 is the adjustment factor for bearing special properties such as material processing, forming method, heat treatment, and other manufacturing methods. For standard bearing material and manufacturing method, $a_2 = 1$. As special steel or advanced manufacturing method (*e.g.*, vacuum degassed or vacuum melting) is adopted for improving bearing quality, $a_2 > 1$. a_3 is the adjustment factor for operating condition, which takes into account a wide range of operating, mounting, and application conditions, as well as bearing features and design. For instance, it adjusts the impact of lubrication, contamination, mounting conditions, misalignment, *etc.*, on fatigue life. Under ideal lubrication where the rolling elements and raceways are completely isolated by lubricant films and normal operation conditions, $a_3 = 1$. For poor lubrication cases, such as incomplete lubrication or low lubricant viscosity, and unusual loading or mounting conditions, $a_3 < 1$. Under especially excellent lubrication condition with very smooth operation, a_3 becomes larger than 1. Logically, for a poor operating case where $a_3 < 1$, the life adjustment factor for bearing special properties a_2 cannot exceed unity.

In addition to these standards, some major bearing manufacturers such as Timken and SKF have developed their own bearing life models to best predict bearing life of their products. For example, SKF has introduced the SKF life modification factor a_{SKF} to adjust the bearing rated life for each type of bearings (*e.g.*, ball bearings, roller bearings, thrust ball bearings, and thrust roller bearings).

In recent years, with the advances in manufacturing, tribology, materials, bearing operation monitoring, and computational technology, the bearing life calculations continue to evolve and become increasingly accurate over time, reflecting the collective experience of the bearing industry.

6.3.2 BEARING FAILURE PROBABILITY DISTRIBUTION

Bearing reliability is the probability of bearing performing without failure over a certain period of time under the operation conditions encountered. Reliability is intimately linked with failure, failure rates, and component and system mortality Depending on the type of application machines, there are various models available for dealing with bearing failure probability distributions. These models can be briefly categorized as age-related failure models (*e.g.*, bathtub-shaped distribution, normal distribution, monotone distribution), and random failure models (*e.g.*, exponential distribution).

The Weibull probability distribution is one of the most widely used lifetime distributions in industries for modeling failure data. The main advantages of the Weibull distribution include the following: (*a*) it is used extensively in many different disciplines due to its ability to fit a variety of data and to represent a variety of distributions; (*b*) it can easily handle suspensions (nonfailure points); and (*c*) it can provide simple graphical solutions. The cumulative distribution function (CDF) of the Weibull distribution $F(t, k, \lambda)$ is

$$F(t,k,\lambda) = \begin{cases} 1 - e^{-(t/\lambda)^k} & t \geq 0 \\ 0 & t < 0 \end{cases} \tag{6.29}$$

where
 t is the bearing service time
 λ ($\lambda > 0$) is the scale parameter
 k ($k > 0$) is the shape parameter

By differentiating $F(t, k, \lambda)$ with respect to t, the probability density function (PDF) $f(t, k, \lambda)$ is obtained as

$$f(t,k,\lambda) = \begin{cases} k\left(\dfrac{t^{k-1}}{\lambda^k}\right)e^{-(t/\lambda)^k} & t \geq 0 \\ 0 & t < 0 \end{cases} \tag{6.30}$$

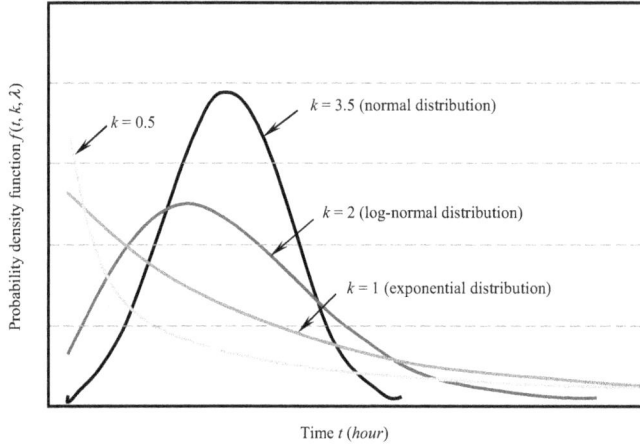

FIGURE 6.27   Weibull probability distributions with different values of the shape parameter k.

The Weibull probability distribution is a versatile distribution that becomes either the normal or exponential probability distribution, based on the value of the shape parameter k. Figure 6.27 illustrates the variations of the PDF versus time t. When $k=2$ and 3.5, the PDF $f(t, k, \lambda)$ becomes the lognormal and normal distributions, respectively. By setting the shape parameter k to 1, Equation 6.29 becomes the exponential probability distribution, which has a constant failure rate

$$F(t,\lambda) = \begin{cases} 1 - e^{-(t,\lambda)} & t \geq 0 \\ 0 & t < 0 \end{cases} \tag{6.31}$$

It can be observed in Figure 6.22 that the PDF curves exhibit obviously different behaviors for the $0 < k \leq 1$ and $k > 1$ cases. At $t=0$, $f(t, 0 < k \leq 1, \lambda)$ approaches infinity and $f(t, k > 1, \lambda) = 0$. The increase in t results in the decrease of $f(t, 0 < k \leq 1, \lambda)$ but the increase of $f(t, k > 1, \lambda)$ until it reaches the peak value. At the peak point, any increase of t causes a sharp decrease of $f(t, k > 1, \lambda)$. As the time t becomes very large, both $f(t, 0 < k \leq 1, \lambda)$ and $f(t, k > 1, \lambda)$ approach zero.

Many studies suggested that bearing failures due to the absence of lubrication or other negligence generally conform to a Weibull exponential distribution. As an extension of Weibull distribution, this model has wide applications in many disciplines, such as bearing fatigue and windshield break. As addressed by Brake [6.31], the exponential distribution is an excellent model for describing bearing fatigue failures. Based on this model, the bearing failure rate is constant. It indicates that the bearing failure occurs randomly, regardless of the age of bearings. According to Nowlan and Heap [6.32] and Moubray [6.33], among all failure patterns, random failure patterns account for 89%. The detailed information can be found in Table 6.2.

As a practical example, the bearing failure probability distribution versus the bearing operating time is presented in Figure 6.28, based on the real statistical data from a bearing manufacturer [6.34]. At the beginning of the bearing service, the bearing failure probability is rather low. Continuous bearing service results in an increase in bearing failure probability until it reaches the peak. Then, the failure probability decreases exponentially and asymptotically approaches a constant over a long time.

6.3.3   INFLUENCE OF UNBALANCE ON BEARING FATIGUE LIFE

Excessive vibration in rotating machinery such as motors can cause unacceptable levels of noise and, more importantly, substantially reduce the life of shaft bearings. Vibration essentially results from residual imbalance, which is caused by an effective displacement of the mass centerline from the true

TABLE 6.2

Failure Patterns of the Conditional Probability

Type	Failure Patterns	Share (%)	Note
Age-related failure (11%)	A	4	The bathtub curve consists of an infant mortality region, a constant or gradually increasing failure probability region, and a pronounced wear-out region. An age limit may be desirable.
	B	2	Constant or gradually increasing probability, followed by a pronounced wear-out region. An age limit may be desirable.
	C	5	Gradually increasing failure probability, but with no identifiable wear-out age. It is usually not desirable to impose an age limit in such cases.
Random failure (89%)	D	7	Low failure probability when the item is new or just out of the shop, followed by a quick increase to a constant level.
	E	14	Constant probability of failure at all ages (exponential distribution).
	F	68	Infant mortality, followed by a constant or very slowly increasing failure probability (particularly applicable to electronic equipment).

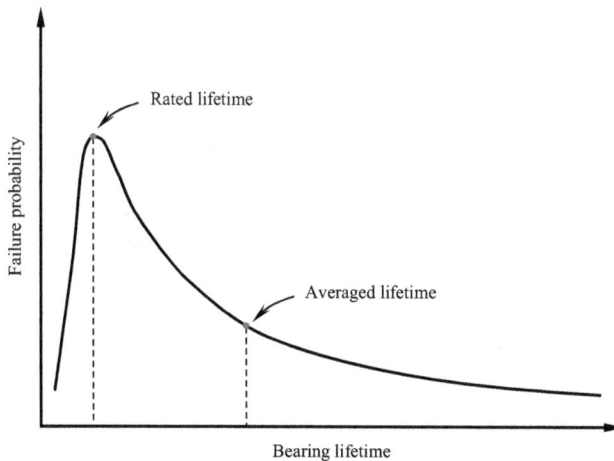

FIGURE 6.28 The bearing lifetime of motor as a function of bearing failure probability, suggested by NSK [6.34].

axis in a unit. Hence, it is desired to remove all inherent causes of vibration. Where vibrations are reduced, the size, mass, and/or complexity of the mounting structures can often be reliably reduced.

Mass unbalance in a motor commonly causes vibration of rotor-bearing systems. Usually unbalance is caused by an asymmetry in the rotor that results in an offset between the shaft centerline and the center of mass. Asymmetry can be attributed to an off-center weight distribution, or a thermal

TABLE 6.3

The Impact of Rotor Unbalance on the Bearing Life

Bearing Condition	Sensor Bearing at Rear Motor	Torque Bearing at Fount Motor	Sensor Bearing at Rear Motor	Torque Bearing at Fount Motor
	Hours		**Year**	
Bearing life at balance condition at 1 G	296,500,581	109,264,650	33,847.1	12,473.1
Bearing life at unbalance condition 1G	284,411,241 (−4.08%)	107,185,338 (−1.94%)	32,467.0 (−4.08%)	12,235.8 (−1.94%)
Bearing life at balance condition at 10 G	137,623	50,716	15.71	5.79
Bearing life at unbalance condition 10 G	132,012 (−4.08%)	49,751 (−1.90%)	15.07 (−4.08%)	5.68 (−1.90%)

mechanism that produces uneven heating and bowing of the rotor, or an electrical effect that produces uneven magnetic field.

One purpose for a rotor to be balanced is to reduce the force at the bearings. This is because that under an unbalanced condition, an additional centrifugal force will be added to the bearing load. The centrifugal force can be calculated as

$$F_c = 1.77 \left(\frac{n}{1,000} \right)^2 WR \tag{6.32}$$

where
n is the rotational speed in rpm
W is the unbalance weight in ounce (oz)
R is the radius of unbalance weight in inch

For example, for $n=600$ rpm, $W=0.42$ oz (11.9 g), and $R=1.26$ in. (32.0 mm), the calculated centrifugal force at 1 G is 0.58 lb$_f$ (2.58 N). With the increase in the centrifugal force to the radial load applied on the bearing (i.e., P) in Equation 6.32, the bearing fatigue life will be reduced for both 1 g and high-g environments. The calculated results are given in Table 6.3.

For the present example, the reduction of the bearing life is less than 5% because of the small magnitude and radius of the unbalance weight. Increasing either the magnitude or radius of the unbalance weight results in a more decrease in the bearing fatigue life. It shows that bearing fatigue life is very sensitive to the radial load acting on the bearing. For instance, for the sensor bearing, as the radial load increases from 46 lb$_f$ (205 N) to 460 lb$_f$ (2,046 N), the bearing life reduces from 33,847 to 15.7 years. The calculated results show that the reductions in bearing fatigue life make a little difference between 1 G and high-G environments.

6.3.4 INFLUENCE OF WEAR ON BEARING FATIGUE LIFE

During bearing operation, the frictional sliding and rolling motion of rolling elements along raceways usually cause rolling bearing to wear. The wear allows the rolling elements to move out of position, increases the internal radial clearance, and makes the load distribution on both the rolling elements and rings even more uneven. As a result, it significantly reduces the life of rolling bearings.

Mitrović and Lazović [6.35] investigated the influence of wear on bearing life for rolling bearings. Their results indicate that the bearing internal geometries are changed due to wear, that is, the diameters of the balls and inner raceway decrease while the diameter of the outer raceway increases.

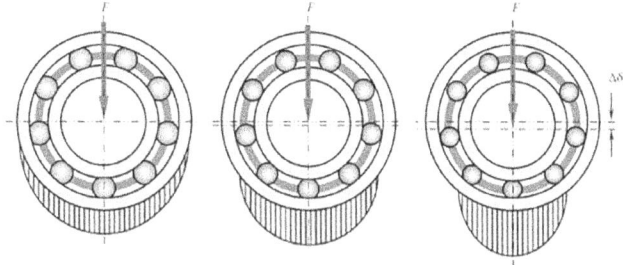

FIGURE 6.29 Influence of bearing wear on load distribution between rolling elements, where $\Delta\delta_w$ is the increased radial clearance due to bearing wear.

The change of these dimensions leads to the increase in the internal radial clearances of the bearing and the changes in bearing operational characteristics (friction torque, vibration, noise, static and dynamic load rating, *etc.*). With increasing internal radial clearance, load distribution between rolling elements becomes more unequal (Figure 6.29), causing the reduction of the bearing life. Base on the bearing life theories, a mathematical model was proposed to analyze the wear influence on bearing service time using a newly developed load distribution factor.

6.3.5 INFLUENCE OF INTERNAL RADIAL CLEARANCE ON BEARING FATIGUE LIFE

The internal radial clearance of radial loaded bearings can significantly alter the load distribution on rolling elements and thus impact the bearing operation reliability and fatigue life. The calculation of the bearing fatigue life is usually based on the assumption that the bearing internal clearance is zero. With a nonzero radial clearance δ_r, the function of load factor $f(\varepsilon)$ can be related to the bearing fatigue life with clearance L_ε [6.27]

- For deep-groove ball bearing

$$f(\varepsilon) = \frac{\delta_r d_b^{1/3}}{0.00044\left(F_r/Z\right)^{2/3}} \quad \text{in unit of Newton (N)} \tag{6.33}$$

- For cylindrical roller bearing

$$f(\varepsilon) = \frac{\delta_r l_{er}^{0.8}}{0.000077\left(F_r/Zi\right)^{0.9}} \quad \text{in unit of Newton (N)} \tag{6.34}$$

where
  δ_r is the radial clearance, mm or m
  F_r is the radial load, N
  Z is the number of rolling elements
  i is the number of rows of rolling elements
  d_h is the ball diameter, mm or m
  l_{er} is the effective roller length, mm or m

Under some circumstances, bearing radial internal clearance becomes zero or even negative (interference) at operating conditions. In fact, a slight interference helps reduce bearing vibration and noise, improve the load distribution on rolling elements, and achieve optimum bearing life and motor reliability. However, when the bearing radial interference becomes considerably large, high radial loads can act on rolling elements, causing damages to both rolling elements and raceways.

TABLE 6.4

Life Factors and Internal Clearances in Different ANSI/ABMA Clearance Groups

Group	Clearance Range (mm)	Corresponding *LF* Range
2	0.001–0.011	0.99–0.85
N	0.006–0.023	0.91–0.77
3	0.018–0.036	0.80–0.69
4	0.03–0.051	0.74–0.63

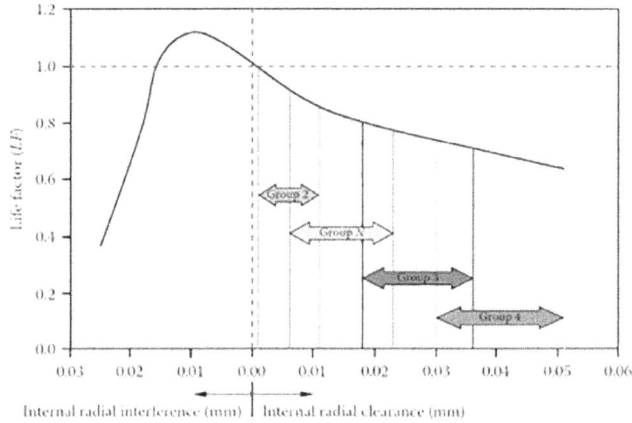

FIGURE 6.30   Effect of internal radical clearance on life factor for deep-groove bearings.

The bearing life factor is defined as the ratio of the bearing life L at any clearance condition to the life at zero clearance L_o

$$LF = \frac{L(\delta_r)}{L_o(\delta_r = 0)} \tag{6.35}$$

Hence, $LF = 1$ indicates zero clearance on any bearing.

ANSI/ABMA Standard 20-1996 [6.36] defines five ranges of radial clearance for various radial loaded bearings. Based on the investigation of 210-size, deep-groove ball bearings radially loaded to produce 1,720 MPa inner race maximum Hertz stress under the zero-clearance condition, Oswald *et al.* [6.27] provided radial clearance and the corresponding *LF* values for four ANSI/ABMA clearance groups, as shown in Table 6.4.

The influence of internal radial clearance and interference on bearing life is presented in Figure 6.30 (using the data from [6.27]) under the same loading condition mentioned earlier. It can be seen that bearing life increases with the decrease in radial clearance. At zero clearance, $LF = 1$. With a continuously increasing radial interference (*i.e.*, negative clearance), the bearing life increases accordingly until reaching its maximum at interference of about 0.009. Beyond this point, any increase in interference will cause a sharp decrease in bearing life.

6.4   BEARING FAILURE MODE

Bearings are considered to be the most critical components of electric motors for maintaining motor normal operation and ultimately reducing unnecessary motor downtime. Many factors can impact bearing life, including operating condition, loads, environment condition, bearing type, bearing

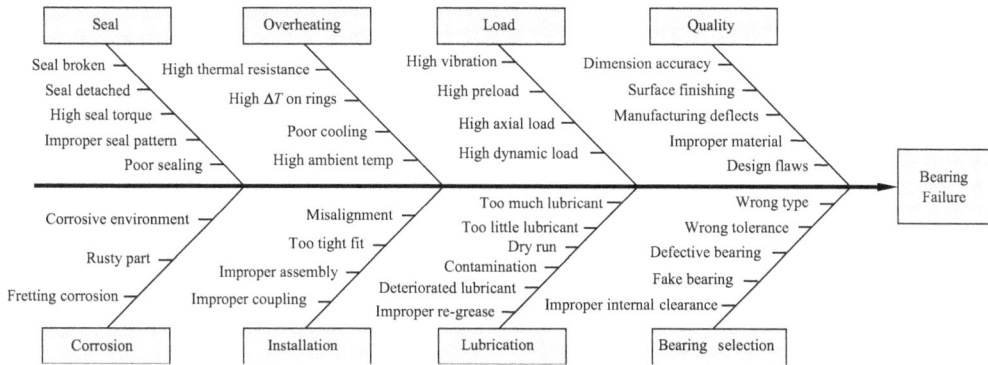

FIGURE 6.31 Fishbone diagram for root cause analysis of bearing failure.

structure and quality (seal, surface finishing, dimension accuracy, *etc.*), bearing materials, heat treatment, lubrication condition, installation, and maintenance. It is commonly accepted that bearing failures that resulted from normal rolling fatigue are less than five percent [6.37].

The failure of a bearing may result in a considerable amount of consequential damage to other motor components. There are tremendous expenses associated with bearing failure in industrial and military motor applications each year. Because the rotor is fully supported by the motor bearings, any bearing defect may directly lead to the variation of the rotor position in the radial direction. This mechanical movement causes the motor airgap to vary in a manner that can be described by a combination of rotating eccentricities moving in both directions [6.38].

Bearing failure can be caused by a variety of factors, including manufacturing defects, unsuitable lubrication, improper bearing selection, inefficient sealing, incorrect fits, defective installation, system misalignment, and careless shipment. Statistical data show that bearing failure due to material fatigue accounts for only a small portion among all bearing failures. In fact, the prediction of theoretical bearing life is based on material fatigue. Most bearing failures are caused by some avoidable factors.

In diagnosing root causes of bearing failure, a fishbone diagram is a useful tool for identifying the potential suspects. It can help understand systematically how system components/factors interact and how they lead to undesirable results such as bearing failure. A fishbone diagram allows engineers to identify, explore, and graphically display all of the possible causes related to a problem to reveal its real root causes. In fact, the fishbone diagrams work very well in many different areas.

As an example presented in Figure 6.31, the fishbone diagram has been successfully used in the root cause analysis for customers' bearing failure.

6.4.1 Major Causes of Premature Bearing Failure

Fatigue of rolling elements and raceways is an extremely complex and unpredictable process. Therefore, the analysis of bearing fatigue is based on statistical prediction depending on many different parameters, including steel type, steel processing, heat treatment, bearing precision, bearing type, lubricant, and operating conditions.

Bearing root cause analysis involves a structured investigative process that aims to identify the true cause of bearing failure and the actions necessary to mitigate and eliminate the failure. Among the factors that may lead to bearing failure are inadequate lubrication, improper bearing selection, excessive loading, overheating, contamination, high vibration, corrosion, misalignment, and loose or tight bearing fits. Among all failure modes for electric motor bearings, the lubrication-related bearing failure is the primary mode, as summarized in Table 6.5.

TABLE 6.5
Common Reasons for Bearing Failures

Ranking	Reason of Bearing Failure	Weight (%)
1	Incorrect or aged lubricant	22
2	Dirty bearing, coarse particles, or liquid	19
3	Too much or too little grease	14
4	Incorrect alignment	11
5	Bearing currents/bearing insulation	9
6	Vibration from motor or load machine	8
7	Installation/maintenance failure	7
8	Manufacturing/dimensioning	6
9	Consequential damages/others	4

Source: Data from Lawrie, R., *Electrical Construction & Maintenance (EC&M) Magazine*, February Issue, 2001.

TABLE 6.6
Bearing Failure Mode

Probable Cause for Bearing Failure	Weight (%)
Fatigue (surface and subsurface origin)	3
Cage	3
Bearing wear	6
Handling damage	7
Dimensional discrepancies	17
Debris denting/contamination	20
Corrosion pitting	27
Other	17

This table shows that the lubrication-related failure modes are the top three and account for 55% of all motor failures. The issues related to motor manufacturing and mechanical engineering account for about one-third.

Some more recent results of bearing failure modes have also been released, as shown in Table 6.6 [6.39]. Unlike the results in Table 6.5, the main bearing failure modes are corrosion, manufacturing defects, and bearing precision. The failure mode categorized under *other* includes bearing misalignment, true and false brinelling, excessive thrust/bearing overload, lubrication, heat and thermal preload, roller edge stress, cage fracture, element or ring fracture, skidding damage, and electric arc discharge. Some of these major causes will be addressed in the following sections.

For journal engine bearings, eight major causes of premature bearing failure are listed in Table 6.7 [6.40]. It is important to note that in many cases, a premature bearing failure is collectively caused by a combination of several of these causes.

6.4.2 Lubricant Selection

Suitable bearing lubricants are indispensable for ensuring normal operation of bearings. The function of bearing lubricants is multifaceted, including reducing the frictional resistance between various bearing elements, assisting in dissipating heat generated in the bearing, protecting the bearing surfaces from contaminants such as dust and moisture, and possibly blocking the path of

TABLE 6.7
Major Causes of Premature Bearing Failure

Cause of Premature Bearing Failure	Weight (%)
Dirt	45.4
Misassembly	12.8
Misalignment	12.6
Insufficient lubrication	11.4
Overloading	8.1
Corrosion	3.7
Improper journal finish	3.2
Other	2.8

high-frequency circulating bearing current. From reference [6.1], it is clearly shown that the vast majority of bearing failures in electric motors are associated with lubrication-related problems. All of the rolling bearing systems require appropriate lubrication to ensure normal operation. The metal-to-metal contact of the rolling elements against the raceways necessitates the presence of grease or oil at all times. If the external lubricant is not present, the balls or rollers will begin to make direct contact with the rail material, resulting in galling and brinelling damage. Many manufacturers attempt to overcome this weakness in the design by adding oil-impregnated seals to the ends of the bearing for extending the bearing life [6.41].

In the vast majority of cases, correct lubrication is the most important factor in obtaining good performance of bearings. As a matter of fact, the selection of lubricant and lubrication method is as important as the selection of the bearing itself.

In most cases, greases are considered the best lubricant for motor bearings. Today, grease manufacturers offer multitudes of greases suitable for a wide variety of bearing applications and operation conditions. The characteristic feature of greases is that they possess a high initial viscosity. The correct choice of grease depends on many factors, including grease viscosity, base oil type, additive requirements, thickener, and some other factors such as dropping point and operating temperature range.

For very high-speed applications such as spindle motors, the bearing temperature can be considerably higher due to friction between deformed rolling elements, raceways, and cages. For these circumstances, bearing cooling becomes more important than bearing lubrication. Obviously, the use of greases is no longer appropriate and oil lubrication should be provided. Oil, as an efficient coolant, can provide good solutions not only for bearing lubrication but also for bearing cooling. There are three types of oil lubrication:

- Oil mist lubrication. Oil mist is produced in an atomizer and conveyed to the bearings by an air current.
- Oil-air lubrication. Oil droplets are carried by compressed air to lubricate/cool bearings.
- Oil jet lubrication. Injection technique is often used in the cooling of high heat flux devices. Oil jet can carry away a large amount of heat from the bearings and is cooled by an oil-to-water or oil-to-air heat exchanger. This lubrication method is especially suitable for high-load, high-speed bearings.

6.4.3 Improper Bearing Lubrication

Among all causes of bearing failure, a primary cause is over or under bearing lubrication. Improper lubrication may result in high-temperature buildup in bearings, causing the grease to break down prematurely and thus leading to the excessive wear of the bearing until its being completely scrapped.

Lubricants are usually made from various oils and additives. There are many different types of motor lubricants, including mineral oils (pure and refined) and synthetic oils for higher temperatures. Animal and vegetable oils are not normally used for bearing lubrication due to the risk of acid formation after a short operating period. Solid lubricant nanoparticles like graphite and MoS_2 are frequently used as oil additives. Their role is to reduce friction and wear. Since the particles from which tribofilms form are so small, the film thickness is usually less than 100 nm. This mechanism of superlubricity caused by lubricant films is fairly studied and understood today. The effect of lubricant additives on friction reduction has been widely recognized. For example, under mixed lubrication conditions, using additive-free engine oil the coefficient of friction of steel-on-steel contact approaches 0.15, compared to that using the same engine oil with added solid lubricant nanoparticles is well below 0.05 [6.42].

Bearings with an overfilled lubricant can have adverse consequences: (a) Overfilling a rolling-element bearing with too much grease can cause excess churning of the grease during operation, resulting in power loss. Energy consumed by bearing rolling elements is then converted into heat that raises the bearing temperature. As a result, the viscosity of the grease reduces and, at elevated temperatures, the oxidation rate of the grease sharply increases. (b) The starting and running torques can become much higher, particularly at low temperatures. (c) Overgreasing bearing can damage bearing seals due to the excessive pressure generated from a grease gun. (d) Too much lubricant can effectively reduce the bearing internal clearance. On the other hand, too little lubricant cannot guarantee a sufficient lubrication of the bearing. With insufficient lubricant, an oil film with needed load-carrying capability cannot fully form. This may result in metal-to-metal contact between rolling elements and raceways, leading to adhesive wear and bearing failure. As a matter of fact, more motor bearing failures are due to overgreasing rather than undergreasing [6.43].

6.4.4 LUBRICANT CONTAMINATION

One of the leading causes of premature bearing failure can be attributed to lubricant contamination, which increases energy consumption by causing abrasive friction at bearing surfaces, raising lubricant viscosity, generating dents on the raceway and rolling-element surfaces, and consequently reducing the bearing operation life. Water from condensation of moisture-contained air can enter the lubricant via penetration through the bearing seal or some other ways. It has been reported [6.44] that as little as 0.002% water in the bearing lubricant will reduce the bearing life by 48%, while 6% water content in the lubricant will reduce the bearing life by as much as 83%.

Microscopic particles are the most harmful form of lubricant contamination as they initiate irreversible microscopic damages on the bearing contact surfaces. Over time, these microscopic damages may accumulate and progress into macroscopic damages and finally lead to premature bearing failure and breakdown of motors.

When bearings work under certain hazard conditions such as high temperature, high moisture, and other corrosive environments (e.g., steel or glass plants), corrosion on some bearing materials would be hardly avoidable. Figure 6.32 shows a rusty bearing after a 2-year service in an electric mold nonsinusoidal oscillation system for slab caster in a steel plant. Ferric oxide is an inorganic compound with the formula Fe_2O_3. During motor operation, Fe_2O_3 powder may enter critical areas of the bearing, where sliding or rolling contact takes place, causing fretting damage on the bearing.

6.4.5 GREASE LEAKAGE

In a field motor inspection, it is common to encounter grease leakage problems. Grease may leak from a sealed bearing between the inner seal groove and the seal, especially for vertically installed motors. While a small amount of grease leakage may exert little impact on motor performance, an aggravated level of grease leakage can significantly shorten bearing life, degrade bearing performance, or even cause bearing failure.

FIGURE 6.32 A rusty bearing with 2-year service in an electric mold nonsinusoidal oscillation system for slab caster in a steel plant.

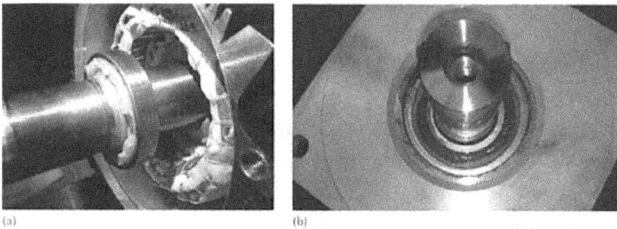

FIGURE 6.33 Grease leakage from the bearings in the vertically mounted motors: (*a*) with a 2-year service in an electric mold nonsinusoidal oscillation system for slab caster and (*b*) with a 3,600-h service in a CNC laser cutting machine.

When grease leakage occurs, it is important to check the questions listed in the following:

- Is the selected grease suitable? The properties of the grease (*e.g.*, viscosity, the allowable operation temperature) must be checked carefully.
- What is the operation temperature when the leakage occurs? As the operation temperature goes higher, the grease viscosity becomes significantly lower.
- What is the condition of the bearing seal? An improper or damaged bearing seal always results in grease leakage.
- What is the bearing load condition? An overloaded bearing tends to cause grease leakage.
- Is the bearing overfilled with grease? Grease overfilling can cause excess churning of the grease during operation and generate extra resistance in the bearing.
- What is the level of motor vibration? A high motor vibration can accelerate grease leakage from the motor bearings.

As demonstrated in Figure 6.33, lubricating grease leaks out of the bearings and accumulates on the bearing seals. It is noted that in both cases, the grease purge occurs from the vertically mounted motors with low-frequency vibration.

In order to improve the sealing capability and avoid grease purge from sealed deep-groove ball bearings, NTN engineers have designed new geometrical cages that differ in shape from that of the standard case [6.45]. The new type of cages has been confirmed experimentally to suppress the adherence of grease to the inner ring seal groove, improving the resistance to grease purge substantially.

6.4.6 BEARING SEALING AND BEARING SHIELDING

Bearing seals and shields are commonly used to prevent ingress of external contaminations and at the same retain the lubricant inside the bearing for ensuring bearing operation properly and expanding the bearing life. Both seals and shields are available in a variety of types, materials, and configurations. In practice, the selection of a seal or shield depends on bearing applications and environmental conditions. Briefly, seals are typically constructed from various elastomeric materials. The selection of seal material is based on the operation temperature and compatibility with lubricants and other adjacent substances. Seals are usually fixed into grooves on the bearing outer rings, but also make light contact with the rotating inner rings. Sealing grease is often filled in grooves for the purpose of further seal and lubrication. This sealing configuration provides a high degree of bearing sealing capability with a modest increase in friction and torque. Generally, when the sealing performance is enhanced, the torque required to turn the bearing increases with the increased friction caused by the enhanced seal. It is noted that some types of seals (*e.g.*, lip seals) may generate a high pressure at the lip, resulting in greatly increased torque and friction losses, especially in high-speed applications. Moreover, in some high-temperature and high-speed applications, specially designed noncontact seals are commonly employed, allowing operation without any friction. For the noncontact seal type, seals have no contact to the bearing inner rings, leading to the reduced rotational torque but less protection against contamination, as compared to the contact seal type. In fact, the design and development of bearing sealing is always a tradeoff process between the sealing performance and friction torque.

In contrast, a bearing shield is a noncontact closure piece, being staked to the bearing outer ring with a small gap between the shield core and the bearing inner ring. Thus, bearing shields have virtually little impact on bearing performance. As a result, in torque-sensitive applications, it is favorable to use shields rather than seals. Metallic shields are very common in industrial applications for withstanding high temperatures. To optimize bearing performance and meet special operation requirements, it is possible to get a combination of seal and shield on a single bearing to provide increased protection from contaminants compared to a double shielded bearing and reduced torque compared to a double sealed bearing. In this configuration, a seal should be installed at the side facing the contaminated environment.

Retention of bearing seals/shields has a critical impact on achieving the best bearing sealing performance. This is especially true for vertically installed motors subject to high levels of vibration. It is interesting to note that even for the same model and bearing brand, the retention methods could vary depending on in which countries they are manufactured. As shown in Figure 6.34, one bearing uses 19 knocks that are equally distributed along the circumference of the outer ring ID to retain the seal firmly. However, the other bearing has no such feature. Obviously, the former has a much better seal retaining capability.

Figure 6.35 shows an example of a bearing failure due to the bearing seal being pushed down in a vertically mounted motor as a consequence of improper sealing. The field inspection has revealed that as the seal was dropped off, the grease was completely lost from the bearing. Under a dry run condition, the bearing temperature was raised dramatically, leading to the disintegration of bearing cage into pieces.

6.4.7 EXCESSIVE LOAD

Bearing load and speed are the most common rating factors impacting bearing service life. Of the two, bearing load has a greater influence on its lifetime. Generally, by doubling the load of a bearing, the bearing life can be reduced by a factor of ten for the ball bearing and eight for the roller bearing. In contrast, doubling the speed reduces the bearing life by half.

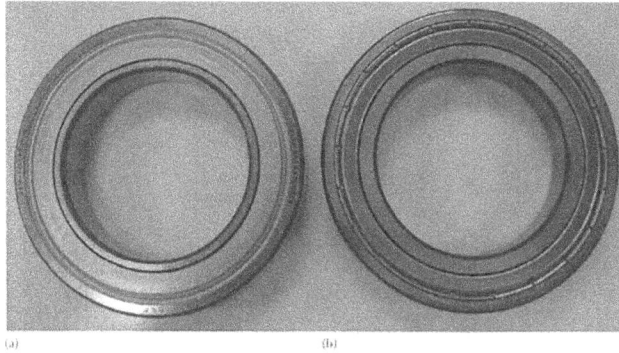

FIGURE 6.34 Different retention methods used for securing bearing seals for the same model and same brand bearings: (*a*) with a number of knocks equally distributed along the circumference of the outer ring ID and (*b*) a simple retention method without using knocks. It is obvious that the retention method used in (*a*) can provide better retaining capability than that in (*b*).

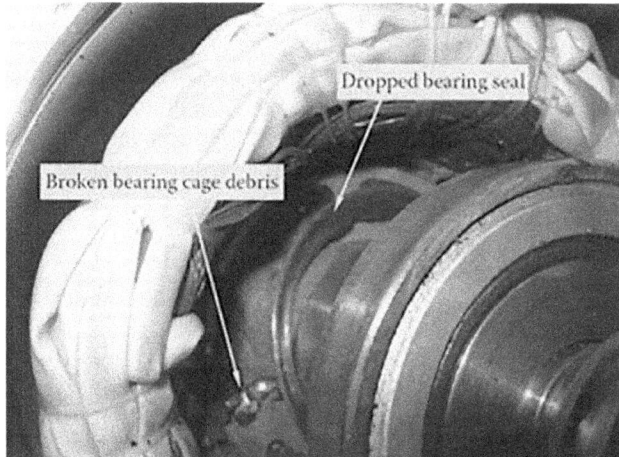

FIGURE 6.35 The bearing seal is dropped off from the bearing. The cage is severely destroyed, broken into pieces to scatter inside the motor.

Excessive loads usually lead to premature fatigue failure of the contact surfaces of the rolling elements. As excessively heavy loads repeatedly act on the contact rolling elements, microcracks start to develop on the surfaces of raceways or rolling elements (balls or rollers). After a period of time, these cracks grow and propagate on the fracture surfaces and finally cause flaking on the raceway or rolling-element surfaces. When bearings are subject to very high loads or shock loads, it may lead to large cracks or fractures of bearing components.

6.4.8 INTERNAL RADIAL INTERFERENCE CONDITION

Regularly, bearing manufacturers always design bearings with certain internal radial clearances. However, under some operating conditions, the internal radial clearance in a bearing may be reduced or even completely eliminated. In some cases, the radial clearance may be further turned into the interference. With the bearing internal radial interference, the rolling elements and raceways are tightly fitted together, applying large frictions and loads to the bearing components.

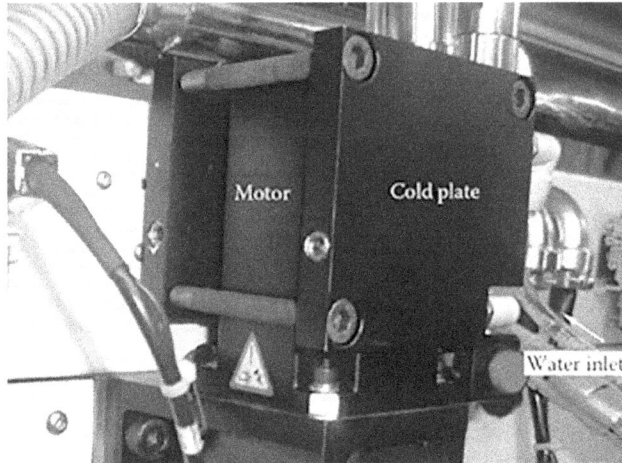

FIGURE 6.36 A motor is sandwiched by two water cold plates, causing a large temperature difference between the bearing outer ring and inner ring.

The bearing effective radial operating clearance δ_{eff} is defined as the amount of radial internal free movement during bearing normal operation. It can be determined by three clearance components: (a) the initial bearing radial internal clearance before mounting, δ_o; (b) the reduced amount of radial clearance due to interference fit, δ_f; and (c) the reduced amount of radial clearance due to the temperature difference between the outer and inner rings, δ_t [6.21], that is,

$$\delta_{eff} = \delta_o - \left(\delta_f + \delta_t\right) \tag{6.36}$$

When bearings are mounted on shafts and in bearing bores with interference fits, the inner ring will expand and the outer ring will contract, thus reducing the bearings' radial internal clearance to approximately 70%–90% of the effective interference.

$$\delta_f = (0.7 \sim 0.9)\Delta\varepsilon_{eff} \tag{6.37}$$

where $\Delta\varepsilon_{eff}$ is the effective interference.

During motor normal operation, the bearing outer ring is typically about $5°C$–$10°C$ cooler than the inner ring due to the smaller thermal resistance from the stator to the surrounding environment. However, as the motor housing is subjected to forced convective cooling, the temperature difference between the outer and inner bearing rings can be considerably larger. As shown in Figure 6.36, a motor is sandwiched by two water cold plates. A much lower temperature is measured at the bearing outer ring when the cooling systems start operating. Thus, the amount of radial internal clearance is further reduced due to the shrinkage of the outer ring and the expansion of the inner ring (Figure 6.37).

The decrease in radial internal clearance due to the temperature difference ΔT of the outer and inner rings δ_t can be calculated as

$$\delta_t = \alpha(\Delta T)d_{or} \tag{6.38}$$

where
   α is the material coefficient of thermal expansion of bearing, which, for stainless steel,
      $\alpha = 12.5 \times 10^{-6} \, °C^{-1}$
   ΔT is the temperature differential between the inner and outer rings, $°C$
   d_{or} is the outer ring raceway diameter, m. For ball bearings, $d_{or} = 0.2 \, (d_i + 4d_o)$

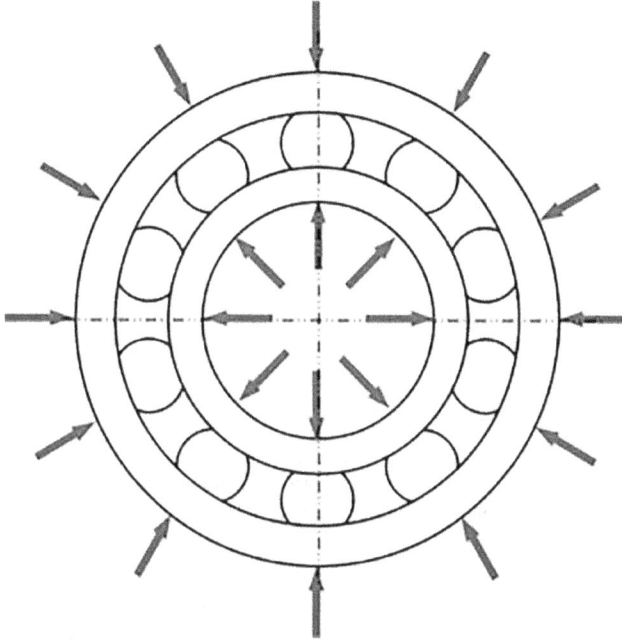

FIGURE 6.37 Due to the temperature difference in the bearing's outer and inner rings as a result of motor cooling, shrinkage occurs on the outer ring and expansion on the inner ring.

Using the above equations, the bearing effective radial operating clearance δ_{eff} can be displayed as a function of the temperature difference between the inner and outer rings. As an example, the results for a 6802 bearing with a normal clearance class CN (its radial clearance ranges of 3–18 μm) are shown in Figure 6.38.

For $\Delta T < 7.2°C$, the bearing has a positive radial clearance. At the point of $\Delta T = 7.2°C$, the radial clearance vanishes. As ΔT becomes larger than 7.2°C, it turns to the radial interference. Note that small interference is allowed and desirable for some applications for increasing bearing stiffness and reducing vibration and noise. However, excessive interferences result in extra rolling contact pressure, which can greatly reduce the fatigue life of the bearing. When the interference becomes high enough, the bearing balls may be completely held by the inner and outer raceways. As a result, these balls are unable to rotate freely but forcibly slide between the raceways, causing overheating, overwearing, and eventually the failure of bearing. It is to be noted that since large bearings usually have large radial clearances, they are unlikely experiencing overheating and overwearing situations. Even for the same bearing size, if the bearing has a higher radial clearance class, for instance, C3 or C4, it will significantly improve the bearing operation condition under such circumstances.

Most bearings are designed to withstand primarily radial loads, and only a certain extent of axial loads. As an example, for deep-groove ball bearings, the maximum axial load is typically about 50% of the maximum radial road. However, for small bearings, this value can be lowered to 25%. When a ball bearing is subjected to a large axial load, the internal radial clearance vanishes as the result of the shift of the inner ring axially relatively to the outer ring (see Figure 6.20). Correspondingly, the balls and raceways develop elastic deformation, resulting in an increase in the contact angle β (defined as the angle between the bearing contact line and bearing normal line) and contact width. This changes the bearing from the point contact to ellipse contact, which in turn causes asymmetric wear of the ring at their sides of the raceways, as demonstrated in Figure 6.39.

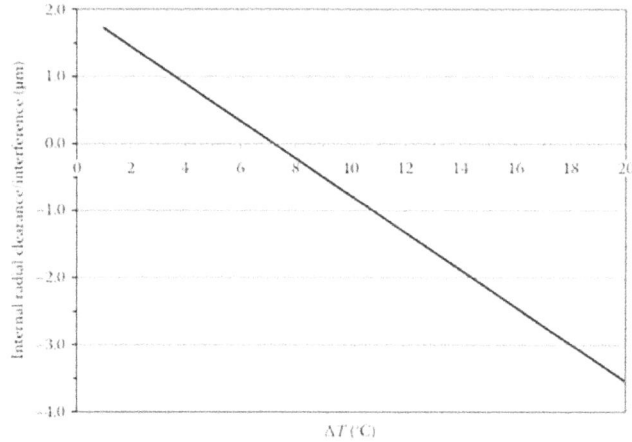

FIGURE 6.38 Effect of the temperature difference of the bearing inner and outer rings on the bearing radial clearance/interference for a specific bearing with the clearance class of *CN*.

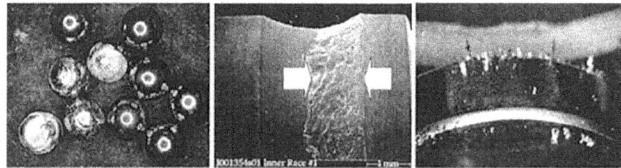

FIGURE 6.39 Damaged bearing balls, inner and outer raceways due to large axial load. It can be clearly seen that spalling occurs at one side of the inner raceway.

6.4.9 BEARING CURRENT

Motor bearing failures due to induced bearing currents and shaft voltages have been recognized for almost a century [6.46]. A shaft voltage arises from various sources in electric motor, such as winding faults, unbalanced supplies, electrostatic effects, magnetized shaft or other machine members, and asymmetries of the magnetic field [6.47]. The shaft voltage can cause the bearing current that flows through the bearings. Several surveys have indicated that 30% of all motor failures operated with 60 Hz sine wave voltage are due to bearing current damage [6.48]. In recent years, there is an increase in motor failures due to bearing current.

The most common underlying cause of bearing currents in three-phase IMs is alternating magnetic flux, which is a result of asymmetric magnetic properties of the stator or rotor core. The asymmetric flux through the steel results in time-varying flux lines that enclose the shaft. This can drive a current down to the bearings, endbell, housing, and back again to the bearings [6.49]. In addition, the use of variable frequency drives can induce the shaft voltage because of the extremely high-speed switching of the insulated gate bipolar transistors, which produce the pulse width modulation used to control AC motors. This shaft voltage can exceed the dielectric strength of the thin film of lubricant (typically 1–20 μm) in bearings, causing the bearings to discharge [6.50].

When electric current flows through IM bearings along internal closed-loop paths (Figure 6.40) either because the lubricant is momentarily bridged or electrical breakdown occurs, the electric current is concentrated through the contacting points and hence the local current density can be extremely high. The generated electric arc tends to melt and damage internal bearing surfaces, just like a series of small lightning strikes. This can cause bearing surfaces to flake away and spall out, leaving behind microcraters on the bearing surfaces. As rolling elements roll on these damaged

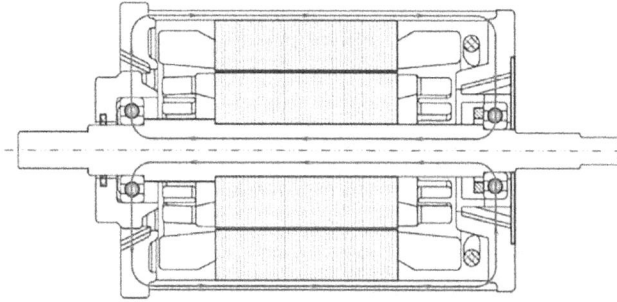

FIGURE 6.40 Bearing current resulted from time-varying magnetic flux in a motor.

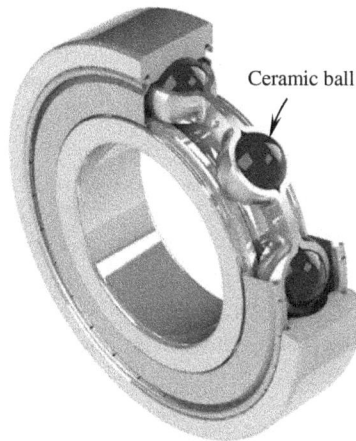

Ceramic ball

FIGURE 6.41 A ceramic hybrid bearing where the balls are made of silicon nitride (Si_3N_4) ceramics. While more expensive than traditional bearings, this type of bearing has some major benefits, including the elimination of electric arcing or fluting, up to 2–5 times the greater service life, running speed up to 50% higher, 20–30 times less coefficient of friction, lower adhesive wear and lower temperature rise at high rotating speeds.

surfaces with contact pressures, the microcraters will be increasingly enlarged, further leading to the deterioration of bearing contact surface, acceleration of wear, elevation of noise and vibration levels, and eventually bearing failure. In addition, wear debris may be introduced into the lubricant, causing lubricant contamination problems.

There are a number of strategies and methods available for reducing/eliminating bearing currents:

- An effective way to eliminate bearing currents is to break the current path by insulating either bearings or endbells from the motor frame. Insulating bearings may be accomplished by the following methods [6.51]: (*a*) Adding an insulation material to the bearing ring or bearing journal. An effective mitigation measure is to use the plasma spraying method to deposit a thin coating (50–300 μm) of high-performance insulating medium on the outer surface of the bearing. (*b*) Coating a ceramic material on bearing components. (*c*) Adopting bearings that are made entirely or partially from ceramic materials, as illustrated in Figure 6.41 [6.52]. In addition to high insulation, ceramic bearings are highly corrosion resistance and enable to operate at extremely high temperatures (up to 1,000°C or even higher). The effectiveness of the measure has been experimentally investigated and validated.
- Grounding shaft is another effective way to eliminate the shaft voltage. This can be done via the usage of metal brushes or grounding rings [6.53].

- Another solution to the bearing current problems is to improve the high-frequency grounding connection from the motor to the drive and from the motor to the driven equipment.
- Establishing an alternative current path directly from rotor to stator or from rotor to frame by means of bushes can effectively reduce bearing current.
- Theoretically, using a conductive Faraday shielded between the rotor and stator would prevent the current from being induced in the shaft. However, this method is difficult to implement in practice.

6.4.10 IMPACT OF HIGH TEMPERATURE ON BEARING FAILURE

The maximum permitted temperature in rolling-element bearings depends on many factors, including bearing material, bearing type, lubrication type, heat treatment, manufacturing and assembly accuracy, and bearing operation condition. Temperatures in excess of this limit can permanently cause bearing failure.

The standard stabilization temperatures are $120°C$ for deep-groove ball bearings and $150°C$ for angular contact ball bearings. Overheating bearings above their stabilization temperatures can anneal the ring and ball materials, leading to the rapid softening of the bearing steel, the reduction of load-carrying capacity, and possibly the subsequent failure.

High temperatures can lead to the decomposition, oxidation, or deterioration of some lubricates. In addition, bearing overheating also lowers the viscosity of bearing lubricates and thus makes it easier for lubricates to leak off from the bearings. It was reported [6.45] that grease purge from a greased and sealed bearing occurs between the seal and the inner seal groove that rotate against each other. An increase in bearing temperature causes a pressure on grease located on the inner ring seal groove, resulting in grease purge. In order to improve the prevention of grease leakage, the cage was redesigned with an optimized geometrical shape. The promising results were achieved experimentally using the improved cage.

6.4.11 BEARING FAILURE ASSOCIATED WITH MOTOR VIBRATION AND OVERLOADING

Vibration can occur any time when a motor is in normal or abnormal operation. Vibration can worsen motor performance and reduce the bearing life due to the alternating force producing both impact force and stress reversal. An excessive level of vibration may lead to premature failure of bearings.

Motor vibration can be activated by the combination of a variety of forces, including the following: (a) mechanical, electrical, and electromagnetic unbalancing forces, which can severely reduce the bearing life; (b) misalignment forces due to shaft misalignment; (c) belt/gear driving tension forces; (d) looseness forces due to looseness of motor components such as a large radial clearance in motor bearings; (e) gravitational force of rotor; and (f) reaction forces from other devices such as gearboxes. Increasing the dynamic forces caused by unbalance, misalignment, and looseness can remarkably result in higher vibration levels and thus significantly reduce the bearing life.

The impact of increased load/force on the bearing life is pronounced. As demonstrated in Figure 6.42, doubling the load/force on bearing can reduce the bearing life by 87% for ball bearings and 90% for other rolling-element bearings (e.g., cylindrical, spherical, tapered, and needle bearings) [6.54].

On the contrary, reducing the vibration level can greatly extend the bearing life. According to the data provided by Berry [6.54], by assuming the dynamic load as the major force component, reducing the vibration level by half will extend the bearing life by 700% for ball bearings and 908% for other rolling-element bearings, as shown in Figure 6.43.

As one of the leading causes of bearing failure, overloading often results in overstresses of bearing components. Overstress creates the possibility of motor component damage caused by a single extreme loading event. A common catalyst for this type of damage is improper belt tensioning.

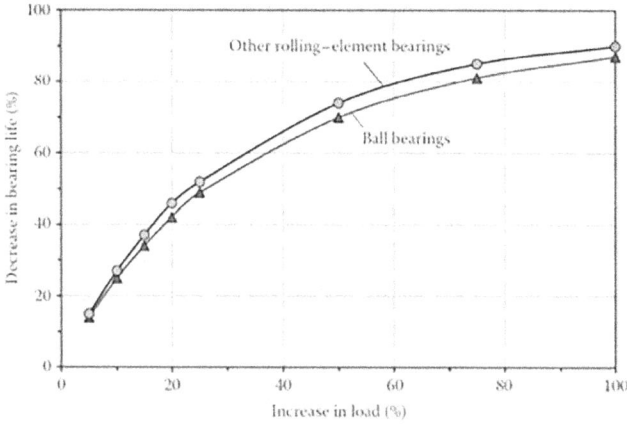

FIGURE 6.42   Impact of increased load on bearing life.

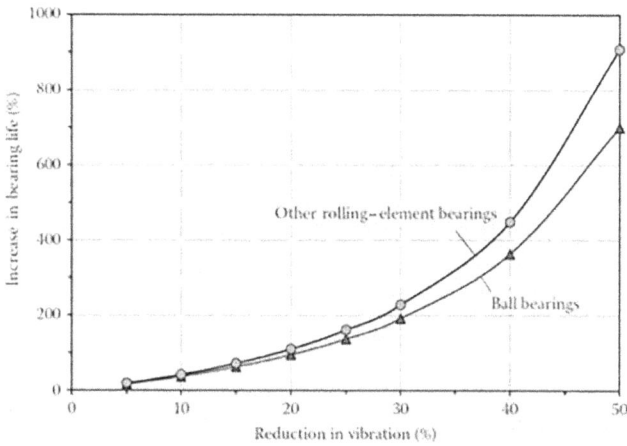

FIGURE 6.43   Impact of reduction in vibration level on bearing life.

Excessive load introduced by belt can drastically reduce the expected bearing life. It has been reported that a 10% increase in tension may reduce bearing life by up to 50%, depending on the system [6.55].

A tight fit between the bearing and the mating parts (*i.e.*, shaft and bearing bore) can eliminate the radial clearance of the bearing. Under this circumstance, the bearing balls and raceways become excessively loaded, leading to rapid wear on these components and resulting in a rapid temperature rise accompanied by high torque.

6.4.12   IMPROPER BEARING INSTALLATION AND BEARING MISALIGNMENT

Bearing failure may be also caused by improper bearing installation. In an installation process, bearings are forced onto shafts or in bearing bores of endbells/frames. This may produce physical damage in the form of brinelling or false brinelling of the raceways, which leads to premature failure. Bearing misalignment is another common consequence of improper bearing installation. As depicted in Figure 6.44, there are four types of bearing misalignment: out-off-line, shaft deflection, cocked or tilted inner raceway, and cocked or tilted outer raceway [6.56]. Each type of misalignment can cause abnormal forces acting on the bearing that may accelerate wear of the bearing raceways, cage, and rolling elements.

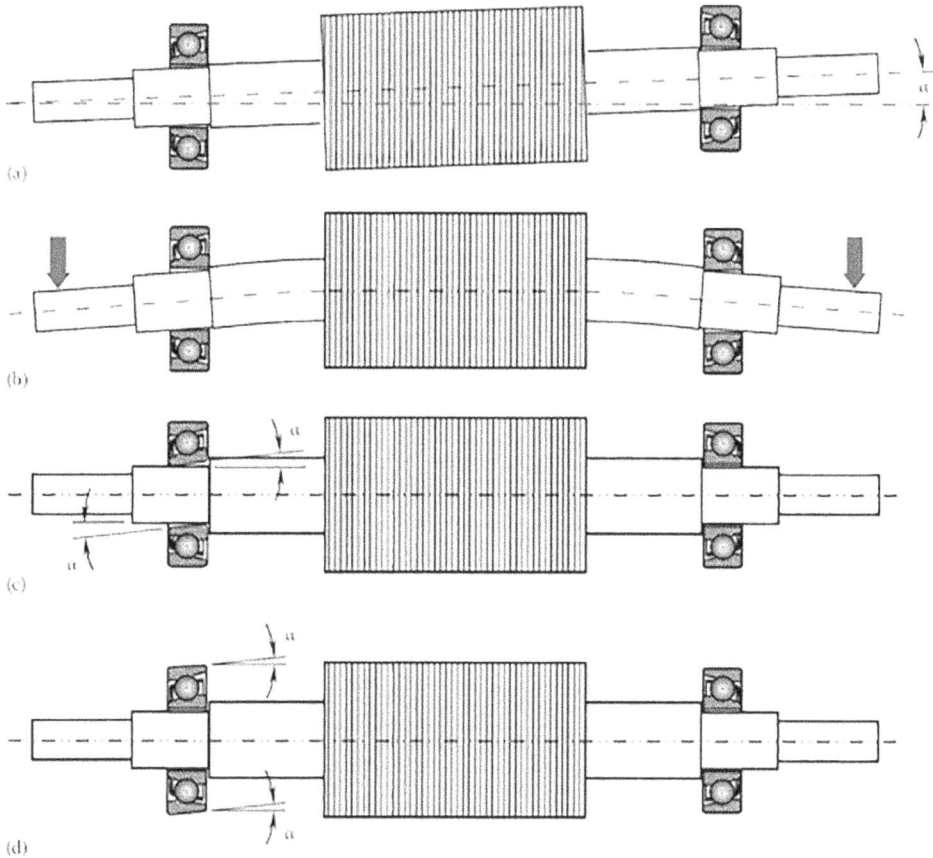

FIGURE 6.44 Four types of rolling-element bearing misalignment: (*a*) out-off-line; (*b*) shaft deflection; (*c*) cocked or tilted inner raceway; and (*d*) cocked or tilted outer raceway.

6.4.13 VERTICALLY MOUNTED MOTOR

As a motor is mounted vertically, the weight of the rotor itself as well as the possible external loads will act on the bearings in the motor axial direction. The increasing axial load can significantly change the bearing operation characteristics, for instance, resulting in larger pressure angle, pressure ellipse, and stresses on bearing rings and rolling elements.

Although many motor manufacturers claim in their motor manuals that their motors can be installed either horizontally or vertically it must be recognized that in most cases, electric motors are primarily designed for horizontal operation. Motor bearing surveys have revealed that a considerable number of motors use deep-groove ball bearings. This type of bearing can withstand large radial loads but relatively small axial loads. For a vertically mounted motor, it may consider using tapered roller bearings or dual bearing assemblies (which are similar to the front wheel bearing on a car) that have the ability to take loads in both radial and axial directions.

Lubricant retention is critical for a vertically installed motor for normal operation. Due to the gravity force, vibration, and temperature effects, the lubricant in motor bearings can gradually leak out from the bearings. Therefore, the sealing capability of bearing becomes one of the key parameters in the bearing selection for reducing the risk of premature bearing failure and enhancing bearing life. In addition, better bearing sealing can prevent external contaminants from entering the bearing interior.

For a vertically mounted motor, any movement of the shaft relative to the motor bearings, as well as to the motor endbells, is not allowed, especially under vibration and high-temperature operating conditions. To avoid this problem, it is critical to use interference fit between the shaft and bearings and tight fit between the bearings and bearing bores. In addition, retaining rings or snap rings must be added to the shaft to avoid the axial movement of the bearings and shaft in any direction with respect to the motor frame. Failing to do so will cause axial movement of either the shaft or bearing relative to the motor, resulting in severe bearing damages. For any cases, the axial load on motor bearings should not exceed the maximum allowed value.

6.5 BEARING NOISE

Bearing noise comes from many sources, including cage noise, race noise, seal noise, waviness noise, and lubricant noise. Some types of noise are unavoidable and acceptable, such as lubricant noise. However, some types of noise are associated with the manufacturing process and can be avoided, such as waviness and click noise. The various types of noise in rolling bearings are grouped into four categories [6.57]: structural (cage, squeal, race, and click noise), manufacturing (waviness noise), handling (flaw and contamination noise), and others (seal and lubricant noise).

Manufacturing precision has a strong impact on bearing noise. There are a large number of parameters affecting the bearing noise level, including roundness of bearing rings and balls, shape accuracy, surface finishing, dimensional tolerance, internal clearance control, and assembly accuracy. In fact, even a slight defect in shape may cause noise and vibration. Generally, noise due to manufacturing defects can be lowered with the improved manufacturing process. For instance, waviness noise can be reduced by decreasing the waviness in the circumferential direction on the finished surfaces of the bearing raceways.

Major bearing components are heat treated to render the hardness larger than 60 HRC. Lower quenching temperatures may give bearing components lower hardness or develop soft points on their surfaces. This reduces the wear resistance of the bearing components sharply. Other defects due to improper heat treatment include quenching microstructure, deformation, surface decarburization, and quenching crack. All these defects raise bearing noise levels.

Compared with ball bearings, rolling bearings are manufactured and processed according to higher precision standards and therefore generally produce a relatively low level of noise. For motors requiring low-noise operation, the selection of deep-groove ball bearings or cylindrical roller bearings can provide effective solution.

In some motor manufacturers, bearing noise can be simply detected with modified medical stethoscopes. This has been proven to be an effective and convenient method, particularly for operators on assembly lines. In order to find root causes of bearing noise, the suspected bearing is often sliced into halves to inspect manufacturing defects, abnormal wear, and other damages (Figure 6.45). It has been found that even very tiny manufacturing defects, which cannot be identified by naked eyes, can generate significant click or waviness noise. Thus, a microscope is a necessary apparatus for the bearing inspection.

6.6 BEARING SELECTION

The primary factors in the bearing selection are the total system reliability during the design life and the cost-effectiveness. The selection of appropriate bearings has become crucial to the overall performance of electronic motors. The selection process must carefully consider all factors that affect bearing performance and cost, including the following:

- Radial and axial load conditions. In fact, bearing fatigue life is sensitive to the magnitude of the load. Generally, if the load is doubled, the fatigue life is reduced by a factor of ten for the roller bearing and a factor of eight for the ball bearings.
- The change in the contact condition between the rolling elements and rings. When a bearing is running under loads, the forces are transmitted through the rolling elements from

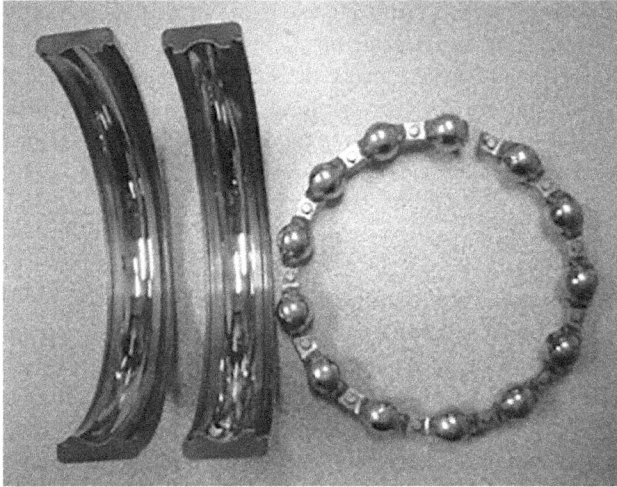

FIGURE 6.45   Bearing is cut for inspecting bearing damage and wear condition.

one ring to another. The magnitude of the load carried by an individual rolling element depends on the location of the element at any instant, the number of rolling elements, and the bearing type. Due to the plastic deformation of the bearing components, the contact condition of the bearing components can alter dramatically For instance, for a ball bearing, the contact between the balls and rings is no longer a point contact rather, it forms a contact area called contact ellipse. For a roller bearing, instead of a line contact, the contact area is a narrow rectangular.

- Rotor rotating speed range—The bearing speed has a strong impact on the selection of the bearing lubricant. Generally, the higher the bearing speed, the lower the lubricant viscosity.
- Expected bearing life and motor operation life—Usually, the life expectancy of a motor can exceed 20 years. This requires an extended bearing life. However, under some harsh operation conditions, the bearing life is much shorter than expected. Hence, it is important to adjust it according to the real situation. For instance, bearings used in oscillation systems for slab caster experience high temperature, moisture, and dust level. These bearings should be replaced after 2-year service, regardless of whether the bearings are damaged or not.
- Operating temperature—The variation in operating temperature may result in the expansion or contraction of rings and rolling elements, which alters the bearing internal clearance. In some cases, the radial clearance may be eliminated due to the temperature difference between the inner and outer rings. In the bearing selection, a wide operation temperature range is highly desired.
- Sealing requirement—Under harsh environmental operating conditions, good sealing is necessary for motor bearings to prevent dust, moisture, and other foreign matters to get in the bearings, as well as the leakage of lubricant from the bearings. This is especially true for outdoor motor applications.
- Lubrication method—Bearings typically use grease or oil as a lubricant. Grease lubrication is often used in applications where frequent replenishment of the lubricant is undesirable or impossible. Oil lubrication is suitable for applications that require low torque or a narrow range of torque variations. There are a number of oil lubrication methods for different applications, including oil bath, circulation oil, oil jet, oil drop, and oil mist. Because bearing lubrication is one of the major causes for bearing failure, it is highly

desired that bearings can be self-lubricated for extending the bearing life and reducing the maintenance cost.

- Bearing internal clearances in both radial and axial directions—For precisely controlled motion systems, bearings with small internal clearances are preferred for minimizing motor vibration and maintaining motor smooth operation. In the bearing section, it is important to take into account the variations of bearing internal clearances with the operating temperature.
- Bearing noise level—Bearing noise can be a critical factor for some applications such as medical devices. Specifically designed precision bearings generally produce very low levels of noise. However, low noise bearings are more expensive than regular bearings.
- Installation/disassembly requirements—In some harsh operating conditions, the bearing life is significantly shorter than the motor lifetime. This indicates that the motor bearings must be changed for a certain period of time during the motor lifetime. Easy bearing installation/disassembly is the basic requirement to ensure the changes of motor bearings.
- Maintenance requirements—Low-maintenance bearing is a prime advantage for some applications such as pitch control motors in offshore wind turbines. With continuous developments in bearing material, better designs, more advanced manufacturing process, lubrication technology, precise production methods, and improved surface finishing in the bearing industry, bearing life has been greatly increased with the reduced maintenance in the last several decades. A longer maintenance cycle and lower maintenance cost mean a reduced life-cycle cost of bearing.
- Shaft and endbell/housing designs—In an electric motor, the shaft and bearings, as well as the endbell/housing, constitute a rotating system, and thus, the selection of bearing can directly affect the designs of the shaft and endbell/housing. As a consequence, it is highly desired to take into account such influences in the earlier stage of the motor design.
- Allowable misalignment of inner/outer rings—Some types of bearing have a self-aligning capability, such as self-aligning ball bearings and spherical roller bearings. Using these types of bearing can help reduce loads that resulted from the misalignment between the motor and driven machine.

However, it is very difficult to satisfy all requirements listed above. Each application may have its specific requirements. It is critical for motor engineers to prioritize all these factors for the specific application, make trade-offs, and identify the relatively important factors as the design factors.

6.6.1 Bearing Type Selection Based on Load

Rolling bearings are available in a variety of types, configurations, and sizes. The correct selection of bearing type is mainly determined by the direction, magnitude, and characteristics of loads acting on the bearing. Generally, a thrust ball bearing can only withstand axial loads. By contrast, needle bearings are designed to take only radial loads. In practice, motor bearings have to take both radial and axial loads.

Bearings for practical applications often need to withstand both dynamic and static loads in a variety of magnitudes and directions. Usually, many bearing manufacturers use the equivalent load to determine the bearing load ratings. When both dynamic radial loads and dynamic axial loads act on a bearing simultaneously, the dynamic equivalent radial load P_{dr} (in N) can be determined as

$$P_{dr} = X_d F_{dr} + Y_d\ F_{da} \tag{6.39}$$

where
F_{dr} and F_{da} are the actual dynamic radial and dynamic axial load, N
X_d and Y_d are dynamic radial and dynamic axial load factor, respectively. X_d and Y_d can be found in bearing catalogs.

Similarly, the static equivalent radial load P_{sr} (in N) can be found from the following equation:

$$P_{sr} = X_s F_{sr} + Y_s \ F_{sa} \tag{6.40}$$

where

F_{sr} and F_{sr} are the static radial and static axial load, N

X_s and Y_s are static radial and static axial load factor, respectively

All values of X_d, X_s, Y_d, and Y_s are provided by manufacturers. The calculated values of P_{dr} and P_{sr} can be used to select the bearing for specific applications.

Couplings are used to connect the motor and driven machine together. The selection of flexible or rigid couplings can strongly influence bearing operation. Improper couplings may introduce additional loads to motor bearings. In addition, using belts or gears as the driving mechanism can produce the transverse force acting on the shaft, which is in turn transmitted to the motor bearings, especially on the drive end bearing. In this case, using roller bearings at the drive end is commonly recommended because of their higher load-carrying capability

In real bearing applications, there are always some small misalignments between the motor shaft and the driven machine shaft. Using flexible couplings can help accommodate the misalignments to some extent.

Different types of bearings have different load capacities. Deep-groove ball bearings can take more axial loads than ball bearings. Cylindrical roller bearings have a stable axial load-carrying capacity. This type of bearing can be safely used as long as the radial load exceeds the axial load by 2.5 times or more. In general, rolling bearings are capable of withstanding greater vibration and shock loads than ball bearings.

6.6.2 Bearing Type Selection Based on Speed

The ball bearing has relatively lower rolling resistances over a wide range of rotating speed than a roller bearing, making it perfectly suitable for medium-/high-speed and low-torque applications. Furthermore, compared with most other types of bearings, ball bearings have superior acoustic characteristics.

By contrast, a roller bearing is in line contact with the inner and outer raceways. Therefore, it can withstand much larger radial loads than the ball bearing. It is interesting to note that at very high speeds, the friction of roller bearing becomes lower than that of ball bearings.

For extremely high-speed motors, noncontact bearings such as magnetic bearings may be selected due to their extremely low friction and wear.

6.6.3 Selection of Bearing Size

Bearing size can be initially selected by the load ratings to determine the required load-carrying capacity with a given factor of safety n_s. Both dynamic and static loads must be evaluated prior to the selection. Dynamic loads are the loads related to the rotor rotation or system displacement (*e.g.*, vehicle). Shock loads are also dynamic loads, which are frequently encountered from heavy impact to earthquake. Usually, dynamic loads are higher than static loads for motors.

In order to ensure bearing normal operation, it has been recommended [6.58] by bearing manufacturers that bearing size should be selected on the basis of static load ratings for the following conditions: (*a*) the bearing is stationary and is subjected to continuous or intermittent loads, (*b*) the bearing makes slow oscillating or alignment movements under load, (c) the bearing rotates under load at a very slow speed (< 10 rpm) and only operates for short life, and (d) the bearing rotates and, in addition to the normal operating loads, has to sustain heavy shock loads.

6.7 BEARING PERFORMANCE IMPROVEMENT

The most recent significant advancement in bearing technology is the introduction of ceramic ball bearings. A ceramic bearing is usually made with ferrous inner and outer rings as well as ceramic balls made of silicon nitride steel, Si_3N_4. The advantages of ceramic bearings over steel bearings include lower friction, higher hardness, finer finish, lesser heat generation, and better acceleration capability. Ceramic bearings are usually used in high-speed and high-temperature (up to $500°C$–$800°C$) applications. High-speed electric motors that require voltage isolation often utilize ceramic bearings [6.59].

Other advantages of ceramic bearings over steel bearings include the following:

- They can withstand wear and tear better compared to steel bearings.
- The modulus of elasticity of silicon nitride is 1.5 times higher than steel, indicating that ceramic balls have a smaller contact area and thus lower rolling friction and heat generation.
- The smaller mass of the ceramic balls results in less inertia and gravitational force on the bearing bore.
- Ceramic balls are less vulnerable to lubricating film breakdowns and starved lubrication conditions than steel balls. They need less lubricant and exhibit less lubrication degradation, which contributes to an increased operation lifetime.
- Rotating bearing components may experience electric charging. This leads to superfluous noise and wear out the bearings. Because ceramic bearings usually have poor electrical conductance, electric charging hardly occurs in this type of bearing.

Bearings under contaminated lubrication condition suffer often from indentations that are generated on the bearing rolling surfaces by foreign particles. During bearing operation, high-stress concentrations occur in the vicinities of dents and eventually lead to cracking and surface-originating flaking. The stress concentration is expressed as

$$\frac{\sigma}{\sigma_o} \propto \left(\frac{r}{c}\right)^{-0.24} \tag{6.41}$$

where
  r is the radius at the shoulder of the dent
  c is the shoulder-to-shoulder half width
  σ_o is the stress far away from the dent

As shown in Figure 6.46, the greater the ratio of r/c, the smaller the stress concentration and the longer the bearing life.

In order to reduce the stress concentration around surface dents, NSK [6.60] has developed super-TF bearings using the TF technology. This technology is a heat treatment process for optimizing the level of retained austenite in bearing materials. It has been found that the increase in the retained austenite level can significantly raise the ratio of r/c and thus greatly improve bearing performance and extend the bearing life [6.61].

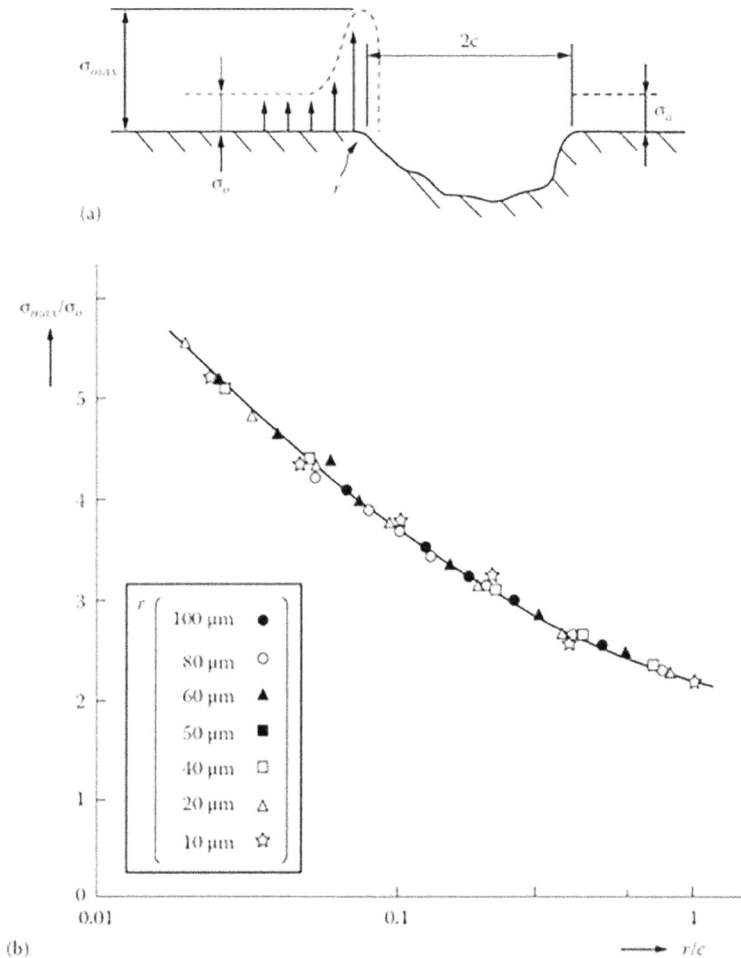

FIGURE 6.46 (a) Stress concentration around a surface dent and (b) the relationship between the stress ratio and the ration of r/c (U.S. Patent 4,904,094) [6.61]. (Courtesy of the U.S. Patent and Trademark Office, Alexandria, VA.)

REFERENCES

6.1. Lawrie, R. 2001. Bad bearings cause motor failure. *Electrical Construction & Maintenance (EC&M) Magazine*, February Issue. http://ecmweb.com/content/bad-bearings-cause-motor-failures.
6.2. Stribeck, R. 1902. Die wesentlischen eigenschaften der gleit- und rollenlager (The basic properties of sliding and rolling bearings). *Zeitschrift des Vereines deutscher Ingenieure* **46**(37): 1341–1348 (part I), **46**(38): 1432–1438 (part II), and **46**(39): 1463–1470 (part III).
6.3. Gu, Y. Y. 1990. *Fluid Sealing.* China Petrochemical Press, Beijing, China.
6.4. Mayer, E. 1977. *Mechanical Seals*, 3rd edn. Newnes-Butterworth, London, U.K.
6.5. Mendes, R. U. and Cavalca, K. L. 2014. On the instability threshold of journal bearing supported rotors. *International Journal of Rotating Machinery* **2014**: 1–17.
6.6. Tang, H. S., Yin, Y. B., Ren, T., Xiang, J. W., and Chen, J. 2018. Impact of the thermal effect on the load-carrying capacity of a slipper pair for an aviation axial-piston pump. *Chinese Journal of Aeronautics* **31**(2): 397–411.
6.7. Ricci, M. C. 2009. Internal loading distribution in statically loaded ball bearings, subjected to a combined radial and thrust load, including the effect of temperature and fit. *World Academy of Science, Engineering and Technology* **57**: 290–298.
6.8. FAG. 2011. Low noise, low friction, no dust, ever: FAG generation C deep groove ball bearings. http://www.hedan.pl/download/44/low-noise-low-friction-no-dust-ever.pdf.

6.9. New Way Precision. 2003. Air bearing application and design guide (rev D). http://www.olympic-controls.com/documents/Air Bearing Design Guide.pdf.

6.10. Xie, X. F. 2003. Comparison of bearings: For the bearing choosing of high-speed spindle design. Technical report, Department of Mechanical Engineering, University of Utah. http://home.utah.edu/~u0324774/pdf/Comparison_of_Bearings.pdf.

6.11. SFK. 2002. SFK sensor-bearing units—Concentrate intelligence in your motion control. Publication 5106 E. http://www.cementechnology.ir/Library/SKF.Sensor-Bearing.Units.pdf.

6.12. Franke, E. 1936. Kngellager (Ball Bearing). German Patent DE625461C.

6.13. Rotek Incorporated. 2010. Large-diameter anti-friction slewing rings. http://www.rotek-inc.com/files/Rotek_84pgCat_FINAL_ViewOnly.pdf.

6.14. Lange, D. and Saginaw, T. 2000. Ball screws that hang tough. *Machine Design*. Article number 21826814. http://machinedesign.com/technologies/ball-screws-hang-tough.

6.15. Mraz, S. J. 2021. What's the difference between roller and ball screws? *Machine Design*. Article number 21175536. https://www.machinedesign.com/mechanical-motion-systems/article/21175536/whats-the-difference-between-roller-and-ball-screws?

6.16. Burrell, M. 2008. Bearings: Don't let them drag you down. *National Kart News*, June Issue, pp. 56–61.

6.17. AST Bearings LLC. 2010. Technical information: Bearing materials. Document No. ENB-04-0553. http://www.astbearings.com/assets/files/Bearing-Materials-Technical-Information-Sheet_ENB-04-0553.pdf.

6.18. Finley, W. F. and Hodowanec, M. M. 2002. Sleeve versus antifriction bearings: Selection of the optimal bearing for induction motors. *IEEE Transactions on Industry Applications* **38**(4): 909–920.

6.19. GMN Bearing USA, Ltd. Deep groove radial ball bearings: Technical information. http://www.gmnbt.com/deep-groove-bearings-technical-info.htm.

6.20. Harris, T. 2001. *Rolling Bearing Analysis*, 4th edn. John Wiley & Sons, New York.

6.21. NIKO. Ball bearings catalogue. http://nipponkodobearings.com/English/catalogue/NIKO General Ball Bearing Catalogue.pdf.

6.22. NTN Corporation.1999. *Ball and Roller Bearings*. Cat. No. 2202-II/E.

6.23. Bischoff, R. F. 1982. Bearing movement preventing system. U.S. Patent 4,309,062.

6.24. Neuhaus, B. and Watzek, M. 2013. Bearing holder. U.S. Patent 8,894,291.

6.25. Stribeck, R. 1907. Ball bearings for various loads. Reports from the Central Laboratory for Scientific Technical Investigation. *ASME Transaction* **29**: 420–463.

6.26. Palmgren, A. 1959. *Ball and Roller Bearing Engineering*, 3rd edn. SKF Industries, Philadelphia, PA.

6.27. Oswald, F. B., Zaretsky, E. V., and Poplawski, J. V. 2012. Effect of internal clearance on load distribution and life of radially loaded ball and roller bearings. NASA/TM-2012-217115.

6.28. Orlowski, D. 2008. Extending motor bearing life. *Pumps and System Magazine*, June Issue. http://www.pump-zone.com/topics/motors/extending-motor-bearing-life.

6.29. The International Organization for Standardization (ISO). 2007. *ISO 281:2007: Rolling Bearings—Dynamic Load Ratings and Rating Life*. ISO, Geneva, Switzerland.

6.30. Harnoy, A. 2002. *Bearing Design in Machinery: Engineering Tribology and Lubrication*. CRC Press, Boca Raton, FL.

6.31. Brake, J. T. 2011. Finding the probability distribution for failures. *Maint World* **4**: 16–21.

6.32. Nowlan, F. S. and Heap, H. F. 1978. *Reliability-Centered Maintenance*. United Airlines Report Number AD-A066-579. Dolby Access Press, San Francisco, CA.

6.33. Moubray, J. 1997. *RCM II Reliability-Centered Maintenance*, 2nd edn. Industrial Press Inc., New York.

6.34. NSK. 2013. Rolling bearings. Catalog No. E1102m. http://www.jp.nsk.com/app01/en/ctrg/index.cgi?rm=pdfDown&pno=E1102.

6.35. Mitrović, R. and Lazović, T. 2002. Influence of wear on deep groove ball bearing service life. *Facta Universitatis. Series: Mechanical Engineering* **1**(9): 1117–1126.

6.36. ANSI/ABMA Standard 20-1996. 1996. *Radial Bearings of Ball, Cylindrical Roller, and Spherical Roller Types—Metric Design*. American Bearing Manufacturing Associate (ABMA), Washington, DC.

6.37. Woodard, M. and Wolka, M. 2011. Bearing maintenance practices to ensure maximum life. *Proceedings of the Twenty-Seventh International Pump Users Symposium*, Houston, TX.

6.38. Benbouzid, M. E. H. 2000. A review of induction motors signature analysis as a medium for faults detection. *IEEE Transaction on Industrial Electronics* **47**(5): 984–993.

6.39. Zaretsky, E. V. 2012. How to determine bearing system life. *Machinery Lubrication*, December Issue. http://www.machinerylubrication.com/Read/29228/bearing-system-life.

6.40. Dana Corporation. 2002. Engine bearing failure analysis guide. Form # CL77-3-402. http://www. studebaker-info.org/tech/Bearings/CL77-3-402.pdf.

6.41. PBC Linear™. 2010. The science of self-lubrication: Debunking the myth of "lubed for life". Technical Notes. http://www.pbclinear.com/Download/WhitePaper/The-Science-of-Self-Lubrication.pdf.

6.42. Österle, W. and Dmitriev, A. I. 2016. The role of solid lubricants for brake friction materials. *Lubricants* **2016(4–5)**. doi:10.3390/lubricants4010005.

6.43. Honeycutt, J. 2002. Reducing motor bearing failures. *Machinery Lubrication Magazine*, May Issue. http://www.machinerylubrication.com/Read/339/motor-bearing-failure.

6.44. McNally Institute. 2018. Ball bearing lubrication in centrifugal pumps. https://www.mcnallyinstitute. com/14-html/14-01.htm

6.45. Sato, N. and Sakaguchi, T. 2010. Improvement of grease leakage prevention for ball bearings due to geometrical change of ribbon cages. *NTN Technical Review* **78**: 98–105.

6.46. Alger, R L. and Samson, H. W. 1924. Shaft currents in electric machines. *Transactions of American Institute of Electrical Engineers* **43**: 235–245.

6.47. GAMBICA Association. 2016. Motor shaft voltages and bearing currents under PWM (pulse width modulated) inverter operation: A GAMBICA/BEAMA technical guide. http://www.brookcrompton. com/updoad/files/literature/MotorShaftVoltages_GAMBICA.pdf.

6.48. Prashad, H. 1991. Theoretical analysis of capacitive effect of roller bearings on repeated starts and stops of a machine under the influence of shaft voltages. *ASME Journal of Tribology* **114(4)**: 818–823.

6.49. Schiferl, R. and Melfi, M. 2005. Bearing current problems: Causes, symptoms, and solutions. *Electrical Construction & Maintenance (EC&M)*, September Issue. http://ecmweb.com/content/ bearing-current-problems-causes-symptoms-and-solutions.
6.50. Erdman, J., Kerkman, R. J., Schlegel, D., and Skibinski, G. 1995. Effect of PWM inverters on AC motor bearing currents and shaft voltages. *IEEE APEC Conference*, Dallas, TX.

6.51. Hoppler, R. and Errath, R. A. 2007. Motor bearings: The bearing necessities. *Global Cement Magazine*, October Issue, pp. 26–32. http://www05.abb.com/global/scot/scot393.nsf/veritydisplay/136d00b55a95 de57c1257b35003072fa/$file/GC_Oct07_ABB-motor-bearings_proof_v2b.pdf.

6.52. Design World Editor. 2021. Pandemic or not, IoT, predictive maintenace, and material remain the trend for 2021. *Design World*, March issue, pp. 65–68.

6.53. Oh, H. W. and Willwerth, A. 2008. Shaft grounding—A solution to motor bearing currents. *ASHRAE Transactions* **114(2)**: 246–251.

6.54. Berry, L. D. 1995. Vibration versus bearing life. *Reliability Magazine*, December Issue. http://www. atlan-tec.net/pdfs/Vibration.pdf.

6.55. Davis, J. and Golden, H. 2010. The impact on bearing life of overtensioned belts. http://reliabilityweb. com/index.php/articles/the_impact_on_bearing_life_of_overtensioned_belts/.

6.56. Schoen, R. R., Habetler, T. G., Kamran, F., and Bartfield, R. G. 1995. Motor bearing damage detection using stator current monitoring. *IEEE Transactions on Industry Applications* **31(6)**: 1274–1279.

6.57. Momono, T. and Noda, B. 1999. Sound and vibration in rolling bearings. *Motion & Control* **6**: 29–37.

6.58. SKF. Selecting bearing size using the static load carrying capacity. http://www.skf.com/group/prod-ucts/bearings-units-housings/ball-bearings/principles/selection-of-bearing-size/selecting-bearing-size-using-the-sta tic-load-carrying-capacity/index.html.

6.59. Bearing Catalogue. 2014. Learn the advantages of ceramic bearings over steel bearings. http://bearing-catalogue.com/ceramic-bearings.

6.60. NSK. Large super-TF bearings. http://www.nskamericas.com/cps/rde/xchg/na_en/hs.xsl/download-bearings.html.

6.61. Furumura, K., Matsumoto, Y., Murakami, Y., Nishida, S., Shiratani, T., and Takei, K. 1990. Ball-and-roller bearing system. U.S. Patent 4,904,094.

7

Motor Brake

Braking is a complex process of energy conversion during which kinetic energy of a moving object is primarily converted into thermal energy, which is eventually dissipated into the surrounding environment. The complexity may result from the fact that during the braking process numerous phenomena, such as thermal, mechanical, magnetic, and electrical effects, as well as the interactions among them, take place. For a motion control system, the braking system is one of the most important subsystems for precisely controlling the position and speed of the system.

Servomotors are extensively used in many motion control systems, such as CNC machine tools, industrial robots, material handling, packaging machinery, elevator machines, semiconductor equipment, wheel-driving systems, and printing machines. As the key components, motor brakes are designed to offer the capability of slowing down, stopping, or holding motors and meet the braking requirements of a wide range of driven machines. In some applications, motor brakes may play an extremely important role in preserving human life. For instance, in normal elevator operation, a built-in motor brake is to control the cage stopping at a desired position and holding it firmly. In an emergency, such as an interruption of electrical power, elevator malfunction, or any failure in the drivetrain or cables, the elevator brake instantly acts to slow down and stop/hold the elevator cage from preventing disaster. In real life, failure of elevator braking systems can invariably result in serious injuries or fatalities, as well as property damages. Consequently, in such an application a great deal of consideration has been given to reliability and safety improvements of brake systems. Practically, motor brakes are often customized to fit the needs of end users and offer optimal solutions for diverse applications.

In the motion control industry, the development of effective braking systems has been a major subject of mechanical engineers. The majority of brakes available today are spring-applied brakes that are designed and built to provide high braking torques to motor shaft in a compact size. For safety purposes, the brake coil is de-energized as a motor is held. The brake must be released (*i.e.*, the brake coil is energized) prior to motor rotation as determined by its drop-out time. The braking disc is usually mounted directly to the motor shaft. Thus, a deceleration of the motor shaft is achieved by applying friction or other means to the braking discs. In cases that require large braking torques, multiple discs may be used for delivering high braking torque within a small space without increasing the disc diameters. In certain applications such as elevator machines, instantaneous stopping during an emergency situation is not desired due to large deceleration G forces which can also hurt people so rapid slow down to a stop is preferred.

Brakes and clutches are mechanical devices used for transmitting mechanical power and rotary motion between the power source and driven machine. Because both operate on the same working principle, they are usually classified into the same group. Brakes and clutches are primarily made up of two components. If one component rotates and another is fixed to a non-rotating inertial frame of reference, the device functions as a brake. If both components rotate when engaged, then the device functions as a clutch [7.1]. Functionally brakes are designed to absorb mechanical energy and convert it into thermal energy during braking processes. Clutches are couplings that are used to engage and disengage the transmitted power between two connecting shafts on a common axis.

DOI: 10.1201/9781003097716-7

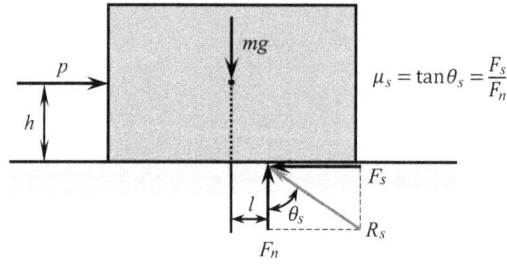

FIGURE 7.1   Coefficient of static friction μ_s is defined as the ratio of the maximum static friction force $F_{s,\max}$ over the normal force F_n, or the tangent of the static friction angle θ_s.

7.1   FUNDAMENTAL KNOWLEDGE OF MOTOR BRAKE

In designing high-performance brake systems or selecting proper brakes for specific applications, the most important step is to understand fundamental knowledge such as the nature of friction, wear phenomenon, kinetic energy of moving object, and brake friction material selection. Rigorous testing design assumptions and materials under specified environmental conditions are critical to verification of all production released designs.

7.1.1   STATIC AND DYNAMIC FRICTION

Friction causes wear and energy dissipation and is responsible for about one-fifth of global energy consumption [7.2]. In solid dynamics, friction is defined as the force that resists the relative sliding motion between two objects in contact. It is always in the direction opposite to the direction in which the object is moving or intending to move. Friction is a complex phenomenon depending on many physical parameters and working conditions. While friction is a highly unwanted phenomenon in most engineering applications, it is desirable for braking systems. As an example, the basic working principle of a fraction brake is to use friction between paired discs to stop or hold a motor and its associated load. Some of the main mechanisms of dry friction include adhesion, elastic and plastic deformation of asperities, ratcheting, fracture of asperities, and third body mechanism [7.3].

There are two main types of friction: static friction and dynamic (kinetic) friction. Static friction refers to the resistance between contacted stationary objects to prevent relative movement between them. As shown in Figure 7.1, when a stationary object placed on the ground is subjected to a small horizontal force p, a static friction force F_s is generated tangentially at the contact interface in the direction opposite to the tendency of motion. Static friction is typically considered as a self-adjusting force because the relationship $F_s=p$ is held until its maximum value $F_{s,\max}$ is reached.

The coefficient of static friction μ_s is defined as the ratio of the *maximum* static friction force (*i.e.*, shear force) between two contacted stationary objects before movement commences and the normal force pressing them together,

$$\mu_s = \tan\theta_s = \frac{F_s}{F_n} \tag{7.1}$$

where θ_s is the angle of static friction. The value of μ_s depends on the materials of two contacted objects.

The resultant force R_s can be determined by combining the friction force F_s and the normal force F_n, where F_n equals the gravitational force of the object ($f_g=mg$) plus possible external forces (f_e) that normally act on the object such as magnetic force and spring force, *i.e.*, $F_n=mg+f_e$. To simplify the problem, assume that $f_e=0$, it yields

$$R_s = \sqrt{F_n^2 + F_s^2} = mg\sqrt{1 + \left(\frac{l}{h}\right)^2} \tag{7.2}$$

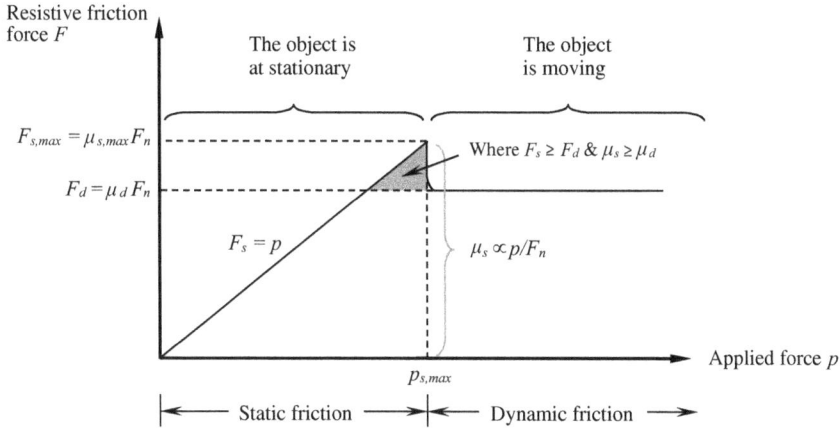

FIGURE 7.2 Transition from static friction to dynamic friction at the point $p = p_{s,\max}$. The coefficient of static friction μ_s is proportional to the externally applied force p, *i.e.*, it increases with p until reaches its maximum value $\mu_{s,\max}$ at the point that $p = p_{s,\max}$, where the corresponding μ_s value is defined as the nominal value of the coefficient of static friction between the two contacted objects.

It can be seen in Figure 7.1, as the applied force p is zero, the corresponding static friction force F_s is also zero and thus there is no friction between two contacted objects. Increasing the applied horizontal force p will lead to a linear increase in F_s until it reaches its maximum value $F_{s,\max}$ (this situation is called *impending motion*). This can be mathematically expressed as: at the point $p = p_{\max}$, $F_s = F_{s,\max}$ and $\mu_s = \mu_{s,\max}$, where $\mu_{s,\max}$ is normally referred to as the nominal value between two giving materials in engineering books. Once the object starts to move, the friction force drops momentarily to the dynamic friction force F_d (Figure 7.2),

$$F_d = \mu_d F_n \tag{7.3}$$

where μ_d is the coefficient of dynamic friction that depends on the nature of material, surface roughness, sliding speed, surface condition (dry or wet), and contact pressure. The coefficient of dynamic friction is defined as

$$\mu_d = \frac{F_d}{F_n} = \tan\theta_d \tag{7.4}$$

where θ_d is the angle of dynamic friction. It should be noted that $\mu_d < \mu_{s,\max}$, $\theta_d < \theta_s$, and $F_d < F_{s,\max}$ for the same contacted objects.

It is important to note that μ_d is independent of speed only at low sliding speeds. When the sliding speed becomes considerably high, μ_d can reduce remarkably. Experimental data show that for a steel-on-steel contact under the contact pressure of 40 MPa, the coefficient of dynamic friction decreases from 0.45 at the sliding speed of 0.01 m/s to 0.15 at the sliding speed of 76 m/s. For the same system under the contact pressure of 200 MPa, μ_d decreases from 0.28 at the speed of 0.01 m/s to 0.1 at the speed of 82 m/s. The combination of high sliding speed and high contact pressure appears to reduce the surface roughness of the rubbing surfaces due to the localized effect [7.4]. In general, a maximum sliding speed of 30 m/s is recommended for standard brake pad materials. Above this value, it is likely that the effective coefficient of dynamic friction is reduced, leading to a reduction in braking performance [7.5]. In cases where the sliding speed is high (up to 100 m/s), specifically fabricated pads using sintered friction materials are commonly employed.

7.1.2 Wear

Wear mechanisms are briefly classified by mechanical wear (*e.g.*, friction wear and fatigue wear), chemical wear (*e.g.*, corrosive wear), and thermal wear (*e.g.*, melt wear). Generally, for motor brakes under normal operation, sliding friction wear is the dominant mechanism. Sliding wear can be characterized as a relative motion between two surfaces in contact under loading. In sliding contact, wear can occur because of adhesion, surface fatigue, tribochemical reaction and/or abrasion, and contamination.

According to the Archard law of wear [7.6, 7.7], the wear volume of a material is proportional to the applied normal force F_n, sliding distance L_s, and inversely proportional to the hardness of the material σ_o

$$V_W = \frac{k_W F_n L_s}{\sigma_o} \tag{7.5}$$

where
V_w is the wear volume, m^3
k_w is the specific wear rate coefficient, m^3/Nm
F_n is the applied normal force, N
L_s is the sliding distance, m
σ_o is the material hardness

The equation in this form was postulated by Preston [7.8] and consequently it is known as the Preston equation. The sliding distance can be further expressed as

$$L_s = \bar{v}t_s \tag{7.6}$$

where
$\bar{v}$ is the main value of the sliding speed, m/s
t_s is the sliding time, s

Because k_w depends on many factors, it is to be determined experimentally.

Wear power loss P_w can be estimated by the following empirical formula [7.9]:

$$P_w = \alpha \, p^a \bar{v}^b t_s^c \tag{7.7}$$

where
α is the wear coefficient
p is the normal pressure, Pa

The exponents a, b, and c are the material pair-related values.

The sliding frictional energy E_s is calculated as

$$E_s = T_d \omega t_s \tag{7.8}$$

where
T_d is the load drag torque in the system, Nm
ω is the rotating speed, rad/s

In order to reduce wear and increase the brake lifetime, sometimes the surfaces of the braking disc are coated with anti-wear materials.

7.1.3 KINETIC ENERGY OF ROTATING OBJECT

Kinetic energy describes the energy of an object moving along certain paths. In some cases, higher kinetic energy is desirable, such as launching a satellite and playing golf. However, from a braking perspective, objects with lower kinetic energy are more likely to brake. For a rotary system, rotational kinetic energy is determined as

$$E_k = \frac{1}{2}J_t\omega^2 \tag{7.9}$$

where ω is the angular rotating speed in the unit of rad/s and J_t is the total polar moment of inertia of the motor-load system in the unit of kg·m². As discussed previously, J_t consists of the inertia of motor rotor J_r, bearings J_b, coupling J_c, gear system J_{gb}, driven load system J_l, and other rotating components,

$$J_t = J_r + J_b + J_c + J_{i,gb} + \frac{J_l + J_{o,gb}}{\gamma_g^2} \tag{7.10}$$

where $J_{i,gb}$ and $J_{o,gb}$ are the inertia of the gearing system at the input (driving side) and the output (driven side), respectively. It should be noted when calculating the total polar moment of inertia; all inertias must be referred to the motor shaft. Thus, the inertia of the driven machine and the inertia of the gearing system at the output side must be divided by the square of the gear ratio (γ_g^2), due to the changes in speed and torque of these parts relative to those of the motor shaft. The last term in the above equation is usually referred to as reflected inertia.

Because rotational kinetic energy is proportional to the square of the angular speed, increases in the angular speed have an exponentially greater effect on rotational kinetic energy. Doubling the angular rotating speed will quadruple kinetic energy.

7.1.4 BRAKE FRICTION MATERIALS

One of the most critical issues in brake design is to choose suitable brake friction materials to ensure reliable and effective actions of brake. In order to achieve the properties required of brakes, most brake friction materials are composites containing numerous ingredients. The selection criteria of braking material include the static coefficient (for a holding brake) or dynamic coefficient (for a stopping brake) of friction, wear rate, thermal expansion, heat-absorbing capability, mechanical strength, corrosion resistance, noise level, environmental factors, and ability to perform consistently and safely through their life. Almost all brakes are designed with a high coefficient of friction materials for the frictional mating surfaces. Further studies reveal that the coefficient of friction depends on many factors, like the shape and roughness of each contact surface, the nature of the molecular adhesion in the points of contact, and the surface deformation due to changes in temperature and pressure.

Surface roughness depends on the scale of measurement. At a brake friction interface, the surfaces may appear to be smooth; however, on a microscopic scale, the surfaces are pitted and jagged. When two surfaces come in contact with each other, the irregularities get interlocked that create friction on the interface. Generally, the higher the surface roughness, the greater the frictional force.

Other parameters that need to be considered in selecting braking materials include:

- High coefficient of friction—Braking materials must have high and stable coefficients of friction for yielding more braking forces under varying conditions of load, velocity, temperature, humidity, and moisture. Experimental results have indicated that Kevlar fibers in braking friction composite materials improve the friction, wear, and recovery performance but depress the fade performance, whereas natural fibers enhance the fade performance but depress the friction and wear performance [7.10].

- Wear resistance—Wear is one of the most common phenomena encountered in daily life. From an engineering perspective, wear is a progressive loss of material at solid surfaces due to the mechanism of friction. The wear rate of braking material is directly proportional to the sliding speed and reversely proportional to the wear resistance, where the wear resistance is greatly affected by material composition, microstructure (such as grain size), hardness, and coating method.
- Heat resistance—Due to the conversion of kinetic energy into thermal energy during a braking process, a large amount of heat is generated by the braking system. Therefore, the braking performance can be significantly affected by the temperature rise of the system. High temperatures may cause premature wear, bearing failure, thermal cracks, and thermally excited vibration. The ability of the braking friction material to resist heat can directly affect the brake performance and safe operation. In addition, a brake friction material with high thermal conductivity helps eliminate hot spots on the brake discs and pads.
- Maximum pressure—When a brake is engaged, the brake pads press against friction discs, producing contact pressure on the sliding interface between the paired pad and disc. The contact pressure is restricted by the ability of friction materials to withstand the highest contact pressure without cracks or fractures.
- Coefficient of thermal expansion—Because of the rapid temperature rise during a braking process, the brake components expand beyond their normal *volume,* affecting the contact condition and consequently the braking performance.
- Mechanical strength—When a brake is subjected to a series of heating-cooling cycles, thermal fatigue may occur on the surfaces of the brake discs and pads, resulting in surface cracking and plastic deformation, and leading to the premature failure of the brake. The factor of safety for a braking system is typically based on the ultimate tensile strength or compressive strength of the materials, depending upon the specific application. A higher factor of safety is often adopted when higher tensile or compressive stress is produced in the brake.
- Corrosion resistance—Corrosion of friction discs can result in not only the decrease in the braking friction but also the increase in wear, consequently leading to the reduction in the brake service life.

In response to the increased braking requirements, most brake materials used today are composite materials for their superior properties. Numerous types of brake materials have been designed and utilized to intentionally increase the braking friction force, thermal dissipation rate, corrosion resistance, and service lifespan; reduce the wear rate and operation noise; and lower the overall brake cost. There are four main categories of brake friction materials:

- Non-metallic materials—These materials are made from combinations of various synthetic substances bounded into composites. The composite materials are made of glass, rubber, and resins, as well as small particles of metal fibers. They are manufactured and cured to hold up under a high level of heat. However, due to the material softness and rapid wear, the brakes made of such materials cannot last as long as other types of brakes.
- Semi-metallic materials—These materials are mixed with flaked metals and organic materials and thus are harder than non-metallic materials. This type of material, with decent friction and high thermal conductivity, has excellent heat resistance, long service life, and low coefficient of thermal expansion. For these reasons, the majority of today's cars use semi-metallic brake pads.
- Fully metallic materials—These materials are composed of sintered steel without synthetic additives. They are characterized by high resistance and thermal conductivity. However, these materials may experience rapid corrosion and undesired braking noise.
- Ceramic materials—These materials are compounds of metallic elements and ceramic fiber, carbon, nitrogen, or sulfur. This design ensures high durability, stable friction at high

speeds and temperatures, low wear rate, high heat tolerance, lightweight, quiet operation, and less maintenance. Their crystal structures and chemical ingredients result in universally recognized ceramic-like properties of enduring utility. Ceramic composite materials, such as carbon-fiber-reinforced ceramic and siliconized carbon fiber, have been developed to fabricate brake discs. However, due to the complexity of the production process, this type of solution may be very expensive and used often for high-performance applications. The modern ceramic brakes wear developed by British engineers for the locomotive industry in the late 1980s. Today, they are widely used on exotic vehicles where the cost is not prohibitive.

The role of friction additives is to increase the coefficient of friction, provide friction stability, and reduce wear of brake materials at elevated temperatures. Additives can be categorized into two groups: lubricants and abrasives. Lubricants stabilize the coefficient of friction, especially at high temperatures. Abrasives usually increase the coefficients of friction. However, they may cause a rise in the wear rate. Graphite and metal sulfides are the most widely used solid lubricants for friction composites, while metal oxides and silicates are used as abrasives.

Graphite as the well-known solid lubricant additive is widely used in different kinds of applications including friction materials. Its impact on tribological properties can be manifold, depending on structure variants, contaminates, and environmental conditions [7.11]. Classical solid metal sulfides with a layer structure, such as MoS_2, WS_2, SnS_2, and TiS_2, and soft sulfides without layer structure, such as Sb_2S_3, SnS, Bi_2S_3, and CuS, are frequently used as friction modifiers for obtaining the desired brake performance properties [7.12].

7.1.5 BRAKE OPERATION MODE

Braking operation modes mainly depend on the applications. There are three operation modes: slowing down, stopping, and holding. Each mode has different performing characteristics.

The main function of the brake system in motion control systems is to decelerate motors and driven loads, known as dynamic braking. In a slowing down process, the brake produces a dynamic retardation torque that is applied to a spinning motor shaft, resulting in a deceleration on the motor. The stopping mode refers to the process in which the driving motor and driven load are forced to stop completely by a brake. In order to size a brake for dynamic stopping, it is crucial to determine the braking torque needed to stop the overall system inertia within a certain time. With the dynamic stopping mode, the brake must absorb kinetic energy and safely convert it into thermal energy that dissipates from the braking components into the surroundings [7.13].

A holding brake is primarily designed to hold the load stationary. When a brake simply holds the load, all rotating components come to rest. As a result, there is little wear and no heat buildup. Theoretically, the holding torque is larger than the average dynamic torque because the static friction is always larger than the dynamic friction, i.e., $\mu_s > \mu_d$ and $F_s > F_d$. It should be noted that the torque rating on most manufacturer catalogs is static torque. The brief relationship between the static torque T_s and the dynamic torque T_d is $T_s = 1.25 T_d$ [7.14]. Thus, wear factors for friction materials are more critical when brakes work in the dynamic stopping mode. Applications calling for repeated dynamic stopping may necessitate the use of different friction materials better suited to high heat and wear situations [7.15].

7.2 KEY DESIGN PARAMETERS AND CONSIDERATIONS IN BRAKE DESIGN

To design safe and reliable motor brakes, a variety of studies have been focused on optimizing brake parameters. Although great progress has been achieved in the existing studies, there are still many issues that need to be further refined.

7.2.1 Braking Torque

As one of the most important parameters in brake design, braking torque (or retarding torque) is defined as the required torque that a brake system can provide to effectively stop a rotating system and hold the load. Braking torque is equal in magnitude but opposite in direction to the torque exerted by the motor-load system. More particularly, braking torque is proportional to the angular acceleration/deceleration (in rad/s^2), polar moment of inertia of the rotating system (in kg·m^2), and load torque and friction torque (in Nm). The dynamic braking torque T_b (in Nm) needed to stop the load is determined as

$$T_b = J_t \frac{d\omega}{dt} \pm T_l - T_f = T_J \pm T_l - T_f \qquad (7.11)$$

where
  J_t is the total polar moment of inertia of all of the rotating components (*e.g.*, rotor, coupling, brake disc, and inner rings of bearings), kg·m^2
  ω is the angular velocity, rad/s
  T_J is the inertia torque where $T_J = J_t (d\omega/dt)$, Nm
  T_l is the load torque, Nm
  T_f is the friction torque, Nm

When the motor-load system has any out-of-balance load applied on it, the effective out-of-balance load torque T_l must be calculated. In the case of a suspended load, the load torque is given as

$$T_l = mgr \qquad (7.12)$$

where
  m is the mass of the suspended load, kg
  g is the gravitational acceleration, m/s^2
  r is the radius at which the load acts, m

It should be noted that the load torque T_l is negative when T_l works in the direction that assists braking action (*e.g.*, rising loads) or positive when T_l works in the direction that hinders braking action (*e.g.*, descent loads). This is easy to understand by considering the operation of an elevator. When it moves upward, its gravitational force acts as the drag force against the motion of the elevator, making it stop in a shorter time. By contrast, when the elevator moves downward, the gravitational force accelerates the motion of the elevator, making it stop in a longer time, as demonstrated in Figure 7.3.

The friction torque T_f is usually a constant for a specific system. Due to the nature of the torque, it always helps stop the system. Even if there is no braking torque, when the power is removed from a motor, the motor and its driven machine will eventually stop due to the friction torque after a relatively long time.

The term $T_J = J_t (d\omega/dt)$ is defined as the inertia torque. This is because it exists only in a transient process where $d\omega/dt \neq 0$. During motor acceleration $d\omega/dt > 0$, the inertia torque directly hinders the braking of the system. In contrast, during motor deceleration $d\omega/dt < 0$, it helps reduce the required braking torque. When $d\omega/dt = 0$, the motor rotates at a constant rotating speed ω_o.

Assuming that in a short period time of t_{ab}, defined as the actual braking time, the speed of the rotating system reduces from ω_o to zero, the inertia torque T_J can be expressed as

$$T_J = J_t \frac{d\omega}{dt} = J_t \frac{\omega_o}{t_{ab}} \qquad (7.13)$$

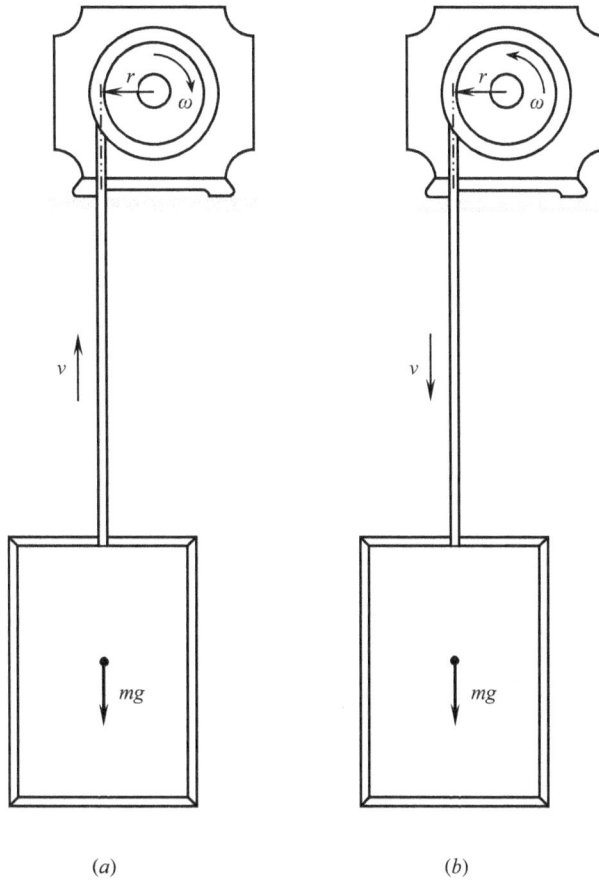

(a)   (b)

FIGURE 7.3 Load torque $T_l = mgr$: (a) a rising load assists braking acting; (b) a falling load hinders braking acting.

where ω_0 is the speed at the moment that the brake action takes place. Thus, Equation 7.11 can be rewritten as

$$T_b = \frac{J_t \omega_o}{t_{ab}} \pm T_l - T_f \tag{7.14}$$

From the relationship between the angular speed ω (rad/s) and n (rpm),

$$\omega_o = \frac{2\pi n_o}{60} \tag{7.15}$$

Substituting (7.15) into (7.14), it yields

$$T_b = \frac{\pi}{30} \frac{J_t n_o}{t_{ab}} \pm T_l - T_f \tag{7.16}$$

When information about the total system inertia is unavailable or cannot be reliably calculated, the required braking torque of a brake may be estimated using the rated power of the motor at full-load speed

$$T_b = \frac{P}{\omega} = \frac{9.55P}{n} \quad \text{in Nm} \tag{7.17a}$$

$$T_b = \frac{5,252 P_{hp}}{n} \quad \text{in lb}_f\text{-ft} \tag{7.17b}$$

where

P and P_{hp} are the rated motor power in units of W and hp (horsepower), respectively
ω and n are the rated motor rotating speed in units of rad/s and rpm, respectively

It should be noted that these equations might yield a braking torque that is larger than necessary.

To deal with the brake operation uncertainties resulted from a variety of factors such as variations of coefficient of friction due to the change in speed and temperature, manufacturing deviations, inhomogeneity of material properties, transient torque fluctuations, operation condition variations, *etc.*, the braking torque T_b must multiple the factor of safety n_s to gain the rated braking torque T_{rb},

$$T_{rb} = n_s T_b \tag{7.18}$$

where

T_{rb} is the rated braking torque listed in the catalog, Nm
n_s is the factor of safety

To provide stable operation and optimize brake performance, the substantial variation of the braking torque under the synergistic influence of brake actuation pressure, sliding speed, and brake interface temperature must be carefully controlled. The non-homogeneous temperature distribution on the contact surfaces of the friction discs, variations in the coefficient of friction, noise and induced vibrations are the main negative issues in the interaction of brake friction surfaces, particularly at a high sliding speed. These negative influences could be reduced through optimization of the brake actuation pressure during the braking cycle [7.16].

The major design objective of brake is to provide consistent and smooth braking torque under a variety of operation and environmental conditions. The optimal brake performance depends on the tribological behavior of the brake disc and pad. One of the approaches is to optimize the brake actuation pressure to achieve not only the optimal braking performance but also the stabilization of the braking torque. This is necessary when significant fluctuations in the brake performance occur during a braking cycle.

7.2.2 BRAKE OPERATION TIME

To fully understand the brake operation process, it is necessary to understand the brake operation time in detail. According to the action of the brake, the operation time can be further divided into different periods. Among them, the most important items are the actual braking time, total braking time, and total release time.

7.2.2.1 Definitions of Various Brake Action Time

Referring to Figure 7.4, each brake action time is defined as follows [7.17]:

- *Initial delay time t_{id}*
  The initial delay time t_{id} occurring at the brake engagement and the brake disengagement are called the delay time for brake engagement and the delay time for brake release, respectively. It is defined as the time from the start of the control command input to the start of the actuation input for the brake engagement event or the release input for the brake disengagement event. The initial delay time is the delay of operation of relays and the like. It is reasonable to adapt $t_d=0.025\,\text{s}$, the typical relay operation time.

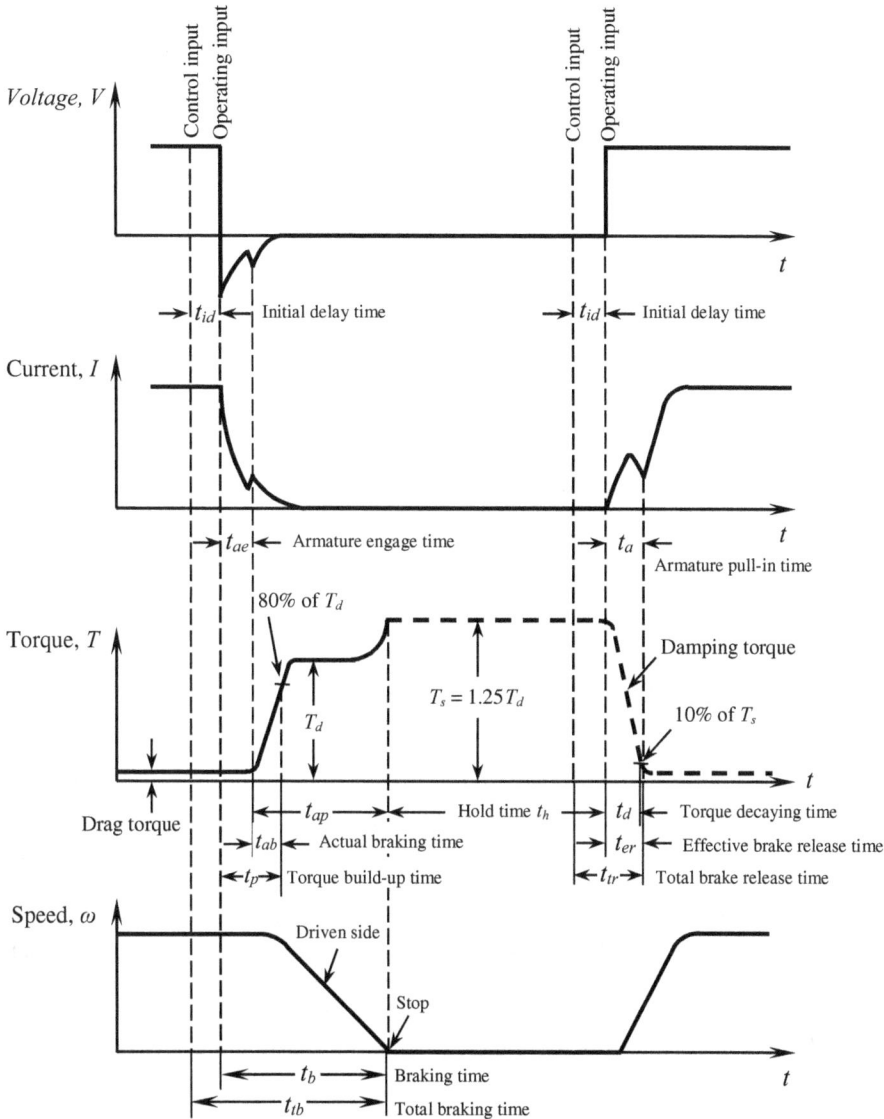

FIGURE 7.4 Brake operation process and related characteristics—brake operation time, braking torque, voltage, and current applied on the brake coil.

- *Armature engage time t_{ae}*

    It is the time from the shutdown of electrical current to the return of armature to its position prior to being pulled in and the braking torque being generated. The armature engage time mainly depends on the brake type, size, and structure, regularly ranging from 0.01 to 0.1 s even higher.

    Armature engage time depends on the magnetic field decay of the electromagnetic coil of an *energy release* brake. Closure force is usually created by springs or permanent magnets (PMs). That, in turn, depends on the type of arc suppression circuit used for preventing damage to the brake coil. Usually, the coil comes with an ordinary silicon diode or no protection at all. However, both of these circuit configurations have a relatively long time constant. When the circuit comes with a single silicon suppression diode,

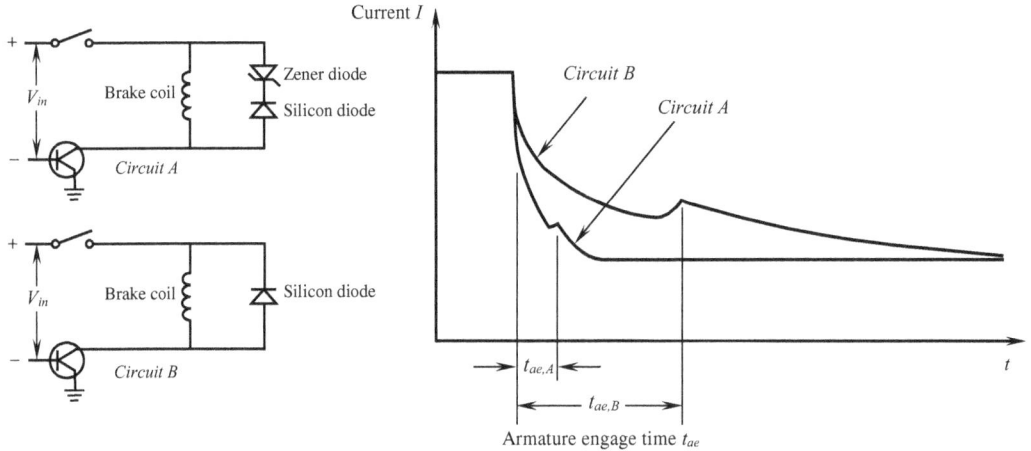

Note:
1. For the silicon diode with V_{in} = 24V or 90V, it may use 1N4004 diode, which has the RMS reverse voltage of 280V and the peak repetitive reverse voltage of 400V (all at 25°C).
2. The Zener diode with V_{in} = 24V may use 1N5368 - 1N5372 diodes, which have the voltage of 47V – 62V, respectively. The Zener diode with V_{in} = 90V may use 1N5378 – 1N5382 diodes, which have the voltage of 100V – 140V, respectively.

FIGURE 7.5   Comparison of armature engage time with different arc suppression circuits. It shows that the armature engage time t_{ae} with *Circuit A* is much shorter than that with *Circuit B*.

add a Zener diode in series with it, the armature engage time t_{ae} can be greatly reduced (Figure 7.5). A study shows that t_{ae} reduces from 28 ms with a single silicon diode (*Circuit B*) to 8 ms with the combination of a silicon diode and a Zener diode (*Circuit A*) [7.18]. Regularly, the voltage of the Zener diode is twice the coil voltage rating. Other testing results are shown in Table 7.1, which compares the armature engage time with different suppression circuits and input voltage V_{in} [7.19]. The results confirm that the armature engage times with the suppression circuit A are much smaller than those with suppression circuit B. However, the circuit input voltage V_{in} has shown little influence on t_{ae}. In addition, the comparison of brake armature release time with different input voltages V_{in} is presented in Table 7.2.

- *Actual torque build-up time t_{ap}*

   The t_{ap} value represents the actual torque build-up time of brake from the moment of first generating the braking torque to the brake delivering 80% of its rated dynamic torque value (*i.e.*, 64% of the rated static torque value, where $T_d = 0.8T_s$).

- *Torque build-up time t_p*

   The t_p value represents the effective engagement time of brake from the moment of interrupting the power supply to the brake delivering 80% of its rated dynamic torque value (*i.e.*, 64% of the rated static torque value).

- *Actual braking time t_{ab}*

   It is the time from the start of the braking torque build-up (where $\omega = \omega_o$) to the completion of braking (where $\omega = 0$).

- *Braking time t_b*

   It is the time from the shutdown of the electrical current to the completion of braking.

- *Total braking time t_t*

   Total braking time is calculated as the sum of the initial delay time t_{id}, armature engage time t_{ae}, and actual braking time t_{ab}.

TABLE 7.1

Comparison of Brake Armature Engage Time with Different Suppression Circuits and Voltages

| | Brake Armature Engage Time (s) | | | |
| | Suppression Circuit A | | Suppression Circuit B | |
Brake Model	$V_{in}=24V$	$V_{in}=90V$	$V_{in}=24V$	$V_{in}=90V$
#1	0.01	0.02	0.10	0.10
#2	0.02	0.02	0.15	0.15
#3	0.02	0.02	0.15	0.15
#4	0.03	0.03	0.20	0.20

TABLE 7.2

Comparison of Brake Armature Release Time with Different Voltages

| Brake Model | Brake Armature Release Time (s) | |
	$V_{in}=24V$	$V_{in}=90V$
#1	0.05	0.06
#2	0.07	0.08
#3	0.11	0.11
#4	0.16	0.20

- *Brake hold time t_h*

  Once the motor-load system is completely stopped ($\omega=0$), the braking torque is imminently converted from dynamic torque T_d to static torque T_s to hold the load, where $T_s=1.25T_d$. Correspondingly, the working mode of the brake is changed from the stopping mode to the hold mode. The brake hold time ranges from the first start of static torque until the introduction of electrical voltage/current to the brake coil for brake release.

- *Armature pull-in time t_a*

  It is the time from the first start of electrical current until the armature is pulled in and the braking torque disappears. During this time, energy is applied to the brake coil to release the braking force.

- *Torque decaying time t_d*

  It is defined as the time from the moment of applying the voltage on the brake coil to the braking torque falling to 10% of its rated static braking value. It is noted that the torque decaying time is a little shorter than the effective brake release time.

- *Effective brake release time t_{er}*

  Effective brake release time is the time from the moment of applying the voltage on the brake coil to the ceasing braking action. It is equal to the armature pull-in time.

- *Total brake release time t_{tr}*

  Total break release time is calculated as the sum of the initial delay time t_{id} and effective brake release time t_{er}.

- *Brake cycle time t_c*

  The brake cycle time t_c represents the time for completing the whole brake cycle. It consists of the total braking time t_{tb}, brake hold time t_h, and the total brake release time t_{tr}.

7.2.2.2 Brake Response Time

The brake response time t_r is the time from the control order being issued until the brake delivers its 80% of the rated dynamic torque. The value of t_r represents the engagement time of the brake and is calculated as the sum of the initial delay time t_{id}, armature engage time t_{ae}, and actual torque build-up time t_{ap},

$$t_r = t_{id} + t_{ae} + t_{ap} \tag{7.19}$$

Power-off brakes are typically used as safety brakes that engage when power is suddenly failed or interrupted. This makes the brake response time particularly important. In fact, the brake response time is highly dependent on the brake type. PM friction brakes have short response times, less than tens of milliseconds. Spring-engaged brakes tend to be slower. For instance, a three-inch spring-engaged brake has a response time ranging from 60 to 100 ms [7.20]. The brake response time is generally proportional to the brake size and the rated torque. The larger the brake size/rated torque, the longer the brake response time.

In selecting a proper type of brake, it is necessary to consider both their performances and costs. In the cases where a sudden drop of a robot arm could endanger personnel or damage product and equipment, a PM brake is the best choice due to the short brake response time. If an arm is not likely to encounter any object during a 100 ms move, then a spring-engaged brake provides a more cost-effective solution.

7.2.2.3 Actual Braking Time

When designing brakes, it is very important to distinguish among three different but interrelated braking-related times: braking time t_b, actual braking time t_{ab}, and total braking time t_{tb}. The actual brake stopping time t_{ab} is defined as the time from the start of the brake torque build-up (where $\omega = \omega_o$) to the completion of braking (where $\omega = 0$). From the formula of the braking torque, the actual brake stopping time can be determined as

$$t_{ab} = \frac{J_t \omega_o}{T_b \pm T_l - T_f} = \left(\frac{\pi}{30}\right) \frac{J_t n_o}{T_b \pm T_l - T_f} \tag{7.20}$$

where
   ω_o and n_o are the rotating speeds at the time when the brake actuates, in units of rad/s and rpm, respectively

7.2.2.4 Total Braking Time

The total braking time t_t is the amount of time from the brake receiving the command (e.g., for a power-off brake, when the current is switched off) to the motor-load system completely stopping. Generally, t_t can be viewed as the sum of several components of time, such as the brake response time, actual stopping time, and others. Taking a power-off brake as an example, the brake engages the load within a few milliseconds after the current passing through the brake coil is interrupted. The actual stopping time is the time from the brake actuation to the completion of load stopping. Generally, total braking time depends on brake operation and load conditions, including total system inertia, motor speed, braking disc/pad material, spring or PM actuating characteristics, and others.

The total braking time t_t is calculated as the sum of actual braking time t_{ab}, armature engage time t_{ae}, and initial delay time t_{id},

$$t_t = t_{ab} + t_{ae} + t_{id} \tag{7.21}$$

7.2.3 BRAKING ENERGY FOR SINGLE OPERATION AND OPERATION FREQUENCY PER MINUTE

Once the braking torque T_b and the total polar moment of inertia J_t are determined, the amount of braking energy required for a single operation E_b (in J) can be calculated as

$$E_b = \frac{J_t \omega^2}{2} \frac{T_b}{T_b \pm T_l - T_f} \tag{7.22}$$

Noting that $\omega = 2\pi n/60$ in units of rad/s, the above equation can be rewritten as

$$E_b = \frac{\pi^2 J_t n^2}{1{,}800} \frac{T_b}{T_b \pm T_l - T_f} \tag{7.23}$$

where n is the rotating speed in units of rpm. For the consideration of operation safety, the calculated value must be sufficiently smaller than the allowable braking energy $E_{b,a}$ that provided in the manufacturer's catalog, i.e., $E_b \ll E_{b,a}$.

The frequency of operation f_b represents the number of braking operation per minute,

$$f_b = \frac{60 E_{b,a}}{E_b} \tag{7.24}$$

It is worth to note the value of $E_{b,a}$ in a catalog is the value under ideal conditions, so the desired operation frequency needs to be sufficiently small. In cases that $E_b/E_{b,a} \geq 70\%$, it should allow the brake to cool down sufficiently after emergency braking before resuming use.

The service life of the brake can be estimated for calculating the total number of braking operation N_b,

$$N_b = \frac{E_t}{E_b} \tag{7.25}$$

where E_t is the total rated braking energy of the brake, provided by brake manufacturers in their catalogs.

7.2.4 MEAN HEAT POWER

During the actual braking time t_{ab}, the rotating system slows down until it is completely stopped. In the braking phase, the system kinetic energy is converted into thermal energy to dissipate into the surroundings. Thus, the mean heat power $\bar{q}$ (in W) is determined as

$$\bar{q} = \frac{E_k}{t_{ab}} \tag{7.26}$$

It is noted that the calculated heat power is an average value. The peak value could be doubled, occurring at the onset of braking,

For the evaluation of the brake pad performance, the heat flux q'' (in W/m²) through the pad surface A_p is calculated as

$$q'' = \frac{\bar{q}}{A_p} = \frac{4\bar{q}}{\pi \left(d_o^2 - d_i^2 \right)} \tag{7.27}$$

where d_o and d_i are the outer and inner diameters of the pad, respectively.

A heat flux value of 7 W/mm² has been shown to be acceptable for emergency stops of around 10-second duration. For a tension brake that provides a continuous torque on material passing through the machine, the value of the heat flux is more typically around 0.6 W/mm². Failure to observe these basic selection criteria may result in poor braking performance and limited pad life [7.21].

7.2.5 TEMPERATURE RISE AND THERMAL CAPACITY RATING

During a braking process, as kinetic energy is converted into thermal energy due to friction at the braking interface, a great amount of heat is generated and absorbed by the braking system, leading to a sharp temperature rise of the brake components. The rise in temperature is highly undesirable since it reduces the coefficient of friction, deteriorates the brake performance, and induces creep in disc material. In addition, excessive thermal loading, *e.g.*, thermal fatigue or thermal shock, can result in disc cracking, distort, thermal judder, and high wear of the rubbing surfaces. In fact, thermal fatigue is the main kind of failure of the brake disk apart from wear. Thermal fatigue is generated when the brake disk is heated and cooled repeatedly during braking, resulting in thermal fatigue fracture on the surfaces of brake disks during repeated braking [7.22]. Moreover, high temperatures can cause overheating of brake coil, seals, bearings, and other components.

The temperature rise ΔT of a brake system during brake operation depends on a number of factors including the mass of the motor system, the rate of retardation, and the duration of the braking event. Generally, it is proportional to the amount of converted thermal energy E_{th} and inversely proportional to the sum of the product of the mass of each brake component m_i, and its specific heat $c_{p,i}$,

$$\Delta T = \frac{E_{th}}{\sum_{i=1}^{n} c_{p,i} m_i} \tag{7.28}$$

This equation provides a brief estimation of the brake temperature rise. It should be noted that other parameters such as the rate of retardation, brake structure, and duration of braking event could also influence the temperature rise.

In transient heat transfer, the brake temperature can be estimated using *Newton's law of cooling*

$$\frac{T - T_a}{T_i - T_a} = \exp\left(-\frac{hAt}{\sum_{i=1}^{n} c_{p,i} m_i}\right) \tag{7.29}$$

where
   T is the temperature at any time τ, °C
   T_i is the initial temperature at time $\tau = 0$, °C
   T_a is the ambient temperature, °C
   h is the overall heat transfer coefficient, W/(m²·K)
   A is the area of heat dissipation, m²
   t is time, s

The ability of a given brake to absorb and dissipate heat without exceeding temperature limitations is known as thermal capacity. Brake thermal capacity is determined by a number of factors, such as ambient temperature, cooling mode, brake design, mounting configuration, disc material, and altitude of working location. A higher thermal capacity ensures cooler running of brake that prolongs brake life. It is noted that the brake thermal ratings listed in the manufacturer's catalog apply to brakes mounted in a horizontal position with an ambient temperature of 20°C–40°C. A brake operates at a higher ambient temperature requires derating of the thermal capacity rating. For instance, at 90°C, the thermal capacity is reduced by approximately 50%. Furthermore, for brakes mounted in a vertical position, or 15° or more from horizontal, the thermal capacity rating decreases due to friction disc drag. For vertical operation brakes, the thermal capacity ratings are derated 25% for two- and three-disc brakes and 33% for four-disc brakes. Generally, four- and five-disc brakes are not recommended for vertical use. Moreover, brakes with brass stationary discs are derated 25% [7.23].

The thermal capacity of a brake is limited by the maximum energy it can absorb in a single braking operation. This factor is important when stopping extremely high inertial loads at infrequent intervals. In any situation, calculated system energy should not exceed the maximum kinetic energy rating of the brake. System energy exceeding the brake's maximum rating may result in overheating of the brake to a point where the braking torque falls appreciably and the brake materials may glaze, carbonize, or fail.

Frictional heat generation in brakes may consist of repeated cycles. The characteristic feature of such a process is that the brake cannot cool down to the initial temperature after short periods of heat dissipation [7.24]. When analyzing the repetitive braking process, one approach is to solve it firstly as a single braking process. The calculated results are then used as the initial values for the subsequent calculation, allowing determining the temperature field of brake components. For very complicated braking processes, computational fluid dynamics simulations are performed to obtain simultaneously the temperature distribution, fluid field, and von Mises stresses in a friction brake.

The key part in the brake design is the effectiveness of heat dissipation. The heat generated during deceleration must be dissipated effectively and efficiently from the brake components to avoid problems of overheating. For this purpose, a maximum operation temperature must be set to prevent brake failure.

High temperatures and high repeated or sustained braking loads can cause brake fade and increased wear of rubbing surfaces, as well as local thermal-induced stresses. In order to ensure the normal performance of both brake and motor, the brake system must be carefully designed to ensure absorbing and dissipating the generated heat effectively to the surroundings without exceeding the temperature limitation of the braking system.

In industrial sectors, there are several effective ways to lower brake temperature and non-uniformity of disc temperature distribution. One way is to design brake discs with pin-fins that are distributed radially or circumferentially in the ventilated channel to minimize temperature gradients on brake discs and to enhance heat dissipation from brakes. Another is to use porous ventilated brake discs, which incorporate open cellular cores. The test results show that with wire-woven bulk diamond cores, the brake disc temperature reduces by up to 24% compared to the pin-finned brake discs. The results also reveal that in typical operation ranges (≤ 1,000 rpm), the wire-woven bulk diamond core provides up to 36% higher steady-state overall cooling capacity over that obtainable by the pin-finned core. In addition, the 3D morphology of the core gives rise to a tangentially and radially more uniform temperature distribution [7.25].

The modern automobile industry often uses aerodynamics to design ventilated brake rotors to replace the traditional solid discs. Ventilated rotors function as centrifugal fans to draw cooling airflow from the inboard side, pass through the rotor passages, and exhaust at the outer diameter. In practice, ventilated rotors are commonly designed with curved vanes or pin fins. It is recognized that finned rotors offer a high mass flow rate and create turbulent flows inside brakes but are subjected to high manufacturing costs. In comparison, for vented rotors, the vane number has a greater influence on heat transfer than other parameters. However, these cooling techniques have not been applied to the motor brakes so far.

7.2.6 Factor of Safety

Operation safety and risk avoidance/mitigation are major concerns in the design of key components and complex systems. In practice, the life cycle of brake is subjected to multiple uncertainties. For this reason, safety and design margins must be taken into account to ensure brake safe operation.

Factor of safety, also known as desired factor of service in industries, is determined by many factors for handling known risks and unspecific uncertainties in applications. For example, safety factor for motor brakes is chosen according to the motor-load overall inertia, load type (constant, variable, or impact), load magnitude and fluctuation, environmental condition, and duty cycle. However, it is also important to avoid overdesigning systems/products because it might waste work force and materials, increase the manufacturing difficulty, extend the production cycle time, and increase the manufacturing and assembly costs.

Many documentation and motor selection guides state that engineers should choose a safety factor of around 1.5–2.25. Generally, if the system is tightly controlled and highly tuned, then it is safer to go with a lower safety factor. As such, engineers can gain cost savings along with energy savings and reduce over-engineering. However, if the system is more uncertain, with more variable loads and operation in less than ideal conditions, selecting a higher safety factor is usually the safer choice [7.26]. In fact, the factor of safety compensates for any tolerance variation, data inaccuracy, unplanned transient torque, and potential variations of the friction disc. With non-overhauling loads, a factor of safety of 1.2–1.4 is typically used. Overhauling loads with unknown factors such as reductions may use a factor of safety of 1.4–1.8. Brake manufacturers suggest the factor of safety for electric machines and other industrial machines as [7.17]

- Low inertia/small load fluctuation, $n_s = 1.5$
- Ordinary use with normal inertia, $n_s = 2$
- High inertia/large load fluctuation, $n_s = 3$
- High inertia, light shock load and intermittent operation, $n_s = 3$
- Spring-set holding brake combined with variable frequency drives, $n_s = 2$
- Diesel engine drive, $n_s = 4$–5 [7.27]
- Compressor drive, $n_s = 5$–6 [7.27]
- Elevator motor brake, $n_s \geq 3.5$

Under various load conditions, the factor of safety for spring-set brakes is suggested as [7.27]

- Low masses, equal loading and non-intermittent operation, $n_s = 2.0$
- Low masses, light shock load and intermittent operation, $n_s = 2.5$
- Medium masses, light shock load and intermittent operation, $n_s = 3.0$
- Large masses, light shock load and intermittent operation, $n_s = 3.0$
- Non-overhauling loads, $n_s = 2$–3
- Overhauling loads, $n_s = 3$–4

Kernwein *et al.* [7.28] developed a computer-implemented method for determining safety factors of brake systems. Although this method focuses on the train brake systems, it can be easily modified to use in the motor brake systems.

7.2.7 BRAKE BACKLASH

Precision motion control applications require mechanically tight systems among their components. This is especially true for robotic surgery systems and tactile-controlled robots. In these applications, it is critical to adopt backlash-free brakes.

Brake backlash is defined as how many degrees (plus or minus) the shaft can rotate (known as the *lost motion*) while the brake is holding. For power-off brakes, this lost motion mainly depends upon the type of drive hub (also known as brake drive) that is used to connect brakes to shafts. In fact, brakes generally rely on a floating hub interface so that the brake's rotor assembly actually floats when a motor is at full rotating speed. Generally, a hub is mounted on a shaft and a brake is mounted on the hub. There are four types of hub designs (Figure 7.6) as follows: (*a*) Hex

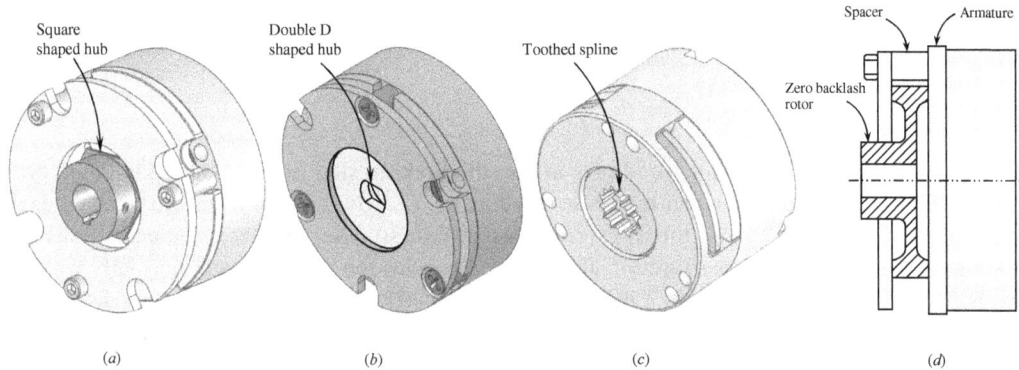

FIGURE 7.6 Brake backlash is strongly influenced by types of hub designs: (*a*) Hex or some polygon shape (square, octagon) design. (*b*) "D" or "double D" (a shaft with flats machined on opposing sides) design. (*c*) Toothed spline design. (*d*) Zero backlash drive design. (Courtesy of Sepac Inc., Elmira, NY.)

or some polygon shape (square, octagon) design. (*b*) D or *double D* design, where a D drive has a round shaft with a flat ground into it, and a *double D* simply put two D drives back to back. These *double D* drives, similar to square drives, have two points of contact that can overstress the rotor under high inertial load conditions. (*c*) Toothed spline design, where both the straight splines and involute splines are available. Involute splines provide a better sliding fit and less manufacturing tolerances. The spline hubs have the least backlash among all hub designs, typically ranging from 0.25° for small brakes (brake disc diameter $d \leq 38$ mm) to 1° for large brakes (d is up to 127 mm). (*d*) Zero backlash drive design, where the hub is fixed to a flexible diaphragm spring to transfer torque. Therefore, the rotor assembly has the least backlash because there is a rigidly mounted diaphragm spring mounted to a hub [7.29, 7.30].

It is worth to note that the hub must provide sufficient clearance to enable the armature to move axially along the shaft. As a result, a spring-engaged brake will always have some degree of backlash. Total unit backlash is typically less than one degree of rotation.

The level of acceptable brake backlash depends upon the application. For high-accuracy, high-precision applications, permanent-magnet brakes offer zero-backlash performance. Examples include semiconductor material fabrication and medical device manufacturing. For most other industrial applications, spring-engaged brakes perform effectively at a reasonable price.

7.2.8 BRAKE NOISE

Brake-related noise has been an issue of concern for brake manufacturers. In most cases, brake noise has little or no effect on the performance of the brake system. However, most customers perceive brake noise as a problem because it is an indication of unusual wear and users' comfort. Most brake noise is caused by frictionally induced dynamic instabilities in the brake, creating vibrations of brake components that radiate the sound. There are different kinds of brake noise classified according to their sound frequency. For instance, *squeal* noise is caused by self-excited oscillation or dynamic instability of the brake discs or pads or other brake components. *Judder* is a forced vibration, the forcing function being due to a cyclic non-uniformity of the friction force. The variety in the forms of judder lies in the various mechanisms by which the friction force non-uniformity is produced [7.31]. Clapping noise is the sound of brake discs engaging or releasing. This noise is reduced by soft landing of materials whether in disc friction or support structures. This is often a large concern to elevator occupants who do not like any noises outside the car.

In practice, a number of approaches can reduce or control brake noise. One of them is the robust design of brake through structure optimization for minimizing brake vibration and enhancing heat transfer. Another approach is to use layered thin metal plates riveted together on the back of the brake pads, acting as efficient dampers for brake squeal.

7.2.9 MAXIMUM SLIDING SPEED

With an increase in the rotating speed of a motor, the braking stop time is correspondingly prolonged. For a friction brake, when the brake is actuated, *i.e.*, the friction discs are touched each other, the sliding speed among the discs reaches the highest value and then decreases until the machine stops. The set of upper limit of sliding speed enables the brake to work normally without overheating. Above the speed limit, it is likely that the effective coefficient of dynamic friction will be reduced, leading to a reduction in braking performance. In addition, at very high sliding speeds, hot spots may occur on the friction surfaces due to generated high temperatures.

7.2.10 RELIABILITY AND DURABILITY

The reliability of a brake system is the ability to provide optimum performance within a range of operating conditions and over a specified time duration. It usually depends on a number of factors, including the reliability of components, environmental interference, integration of control algorithm, and actuation mechanism. The reliability of brake systems in some applications may directly affect the safety of humans, for example, passages in elevators and workers beside collaborative robots.

In general, brake durability problems are related to four elements: loading condition, brake structural design, material, and manufacturing. In brake operation, failures may occur at either the system level or component level. While the loading condition is usually viewed as a system-level parameter, the brake structure and material strength are component level factors to affect the brake performance and failure. For long-term sustainable brake operation, motor engineers need to select suitable brakes that can withstand the required brake loads. In addition, they need to assess the stability of the brake structure and the strength of brake materials.

7.2.11 BRAKE OPERATION CYCLE

Brake operation cycle describes how frequently a brake is used. A high operation cycle is always related to a high heat generation, increased heat dissipation demanding, and thus affects brake performance, operation reliability, and lifetime.

7.2.12 BRAKE MOUNTING ARRANGEMENT

If a motion control system involves a gearing system, the motor brake should be always on the high-speed shaft of a power transmission system. This is because by using a gearing system, the output speed ω_{out} decreases and the output torque T_{out} increases. Therefore, the brake should mount at the motor side, rather than the load side. This allows using a smaller brake and permits the brake with the lowest possible torque to be selected for the system.

For installed outdoor vertically mounted motors with the shaft downwards, to protect motors from water ingress and possibilities of ice formation on the motor brake, the assembly should be fitted with a protective roof.

7.2.13 BRAKE SIZE

The brake size is a crucial parameter in some applications such as industrial robots, missiles, and space shuttles. During the past few decades, many efforts have been focused on designing more compact and lightweight brakes without sacrificing braking torque.

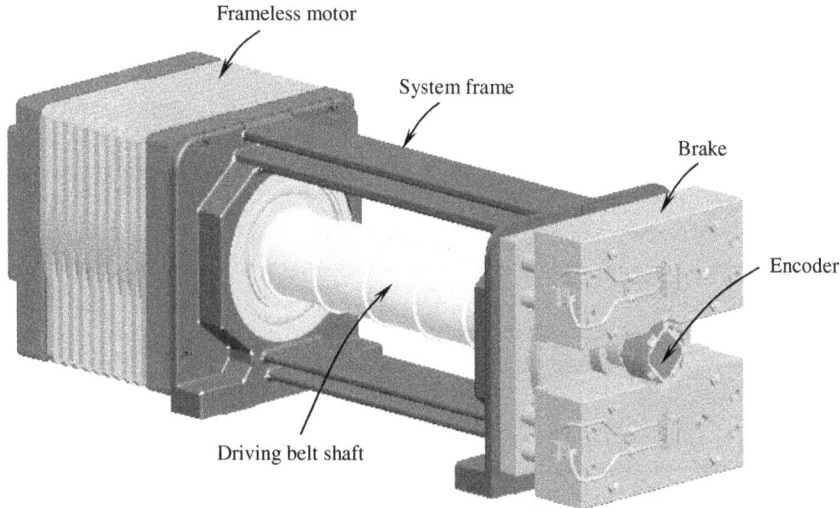

FIGURE 7.7 An elevator-driving unit, showing that the brake and the driving motor are placed at the two ends of the belt-driving shaft and the system frame.

Many factors can affect brake size, including the frequency of brake operation (the brake cycle rate), heat dissipation capability, braking torque required, and ambient temperature.

7.2.14 BRAKE INTEGRATION WITH ELECTRIC MOTOR

In most industrial and commercial applications, a brake is usually integrated into a motor within an enclosed motor housing. The benefits of such integration include: (*a*) This design feature greatly increases the power/torque density, especially suitable for the applications that have a tight requirement for space. (*b*) It provides the brake protection against foreign matter entering from outside. (*c*) It simplifies the motor installation process. However, for some applications, a brake may be separated with a motor. As shown in Figure 7.7, for an elevator system, the brake and the elevator motor mount on the opposite sides of the frame.

Motor brakes typically mount at the rear of motor, adjacent to the feedback device. However, this configuration may cause some negative impacts on the performance of the feedback device. For example: (*a*) During the brake engaging and disengaging processes, it may produce electromagnetic interference to the feedback device. (*b*) As a heat source, brakes may generate an amount of heat that can transfer to the feedback device to cause it heat up. (*c*) As a brake and feedback are placed side-by-side, the assembly may be subject to inaccuracies when their tolerances are stacked up incorrectly. To address these problems, some motor manufacturers mount brakes at the front of the motor, simply separating the two parts.

7.2.15 BRAKE INGRESS PROTECTION RATING

The brake ingress protection (IP) ratings can be widely ranged primarily depending on their applications, from open housing design with low IP ratings (*e.g.*, IP21) to IP68 for high protection. Brakes can be designed to operate from either AC or DC voltage source. Some brake manufacturers recommend that IP54 is a standard rating for AC brakes, with an optional choice of IP55 or IP65. Meanwhile, IP55 is a standard rating for DC brakes, with an optional choice of IP56, IP66, IP67, or IP68 [7.32]. In general, for use in standard industrial environments, IP54 is sufficient. For outdoor applications or for application that involves contact with water (not directly washdown), the protection rating IP55 or IP56 is advisable. However, it is recommended to adopt appropriate additional protections [7.33]. Here are some IP rating examples:

- IP21 is intended for general purpose, indoor applications with a ventilated enclosure.
- IP23 is intended for indoor applications with a non-ventilated enclosure.
- IP54 is applied to combine a clutch-brake system that consists of a spring-engaged brake and an air-engaged clutch.
- IP54–IP67 are commonly applied to DC brakes in pitch control systems of wind turbines.
- IP66 is used in areas where combustible dusts are presented during normal processing, handling, or cleaning.
- IP67 is used for submersible servomotor brakes.

7.2.16 ACCUMULATION OF BRAKE WEAR PARTICLES

For the type of friction brake, wear occurs between two contacted surfaces when they experience relative movement under load. During braking operation, wear particles/powders are produced and accumulated inside the motor housing. This may potentially cause dust contamination problems that not only decrease the coefficient of friction but also deteriorate the performance of other motor components such as optical encoders. If brakes are used for holding purposes only, this risk is eliminated or reduced significantly. Furthermore, in order to prevent the brake contact surfaces to be contaminated with oil or grease, many brake manufactures offer contamination shields on brakes.

In a dynamic braking system, kinetic energy is eventually transformed into thermal energy. While a majority of heat is dissipated to the surrounding structure and finally to the environment, a portion of energy will raise the temperature of the brake discs and pads. Because of thermal resistance constituted by the accumulation of wear particles at the contact surface between a pair of disc and pad, as well as the lack of necessary provisions for ventilation of the disc, there is a heat partition between two components, resulting in the significant temperature difference between the two. In order to remove wear particles from the contact zone of sliding components, several provisions are taken into account. One of these is to contrive slots on the surface of the pad. Another effective way is to design vane type of brake discs with ventilation holes on the surfaces for replacing solid discs.

7.3 CLASSIFICATION OF BRAKING SYSTEM

Motor brakes come in many different structures, configurations, and mechanisms. According to these characteristics, motor braking systems can be classified in different ways.

According to the braking mechanism, brakes can be briefly categorized as either frictional type or non-frictional (contactless) type. Friction brakes have long been used to regulate speeds and loads in most industrial applications. The non-frictional brakes, on the other hand, use electrical or electromagnetic methods to decelerate or hold electric machines/loads, such as eddy-current brakes, magnetic particle brakes, and hysteresis brakes, as well as regenerative brakes. Due to the contactless feature, these types of brakes are ideal for applications requiring precise repeatability, wear free, no friction heating, and zero dust emission.

Each braking cycle involves two opposite braking mechanisms: engagement and disengagement, these can be implemented electrically, magnetically, mechanically, pneumatically, or hydraulically. According to the mechanism of brake operation, there are two basic types of brakes: (*a*) Power-off brake. When a motor is energized during normal operation, a brake is also energized to move the friction faces apart (*i.e.*, brake disengaged), allowing the rotor to run freely. When an electric power is either accidentally lost or intentionally disconnected (*i.e.*, power off) from the brake coil, the brake is engaged to stop (with a dynamic torque) or hold (with a static torque) the motor and load by using springs, PMs, or other mechanisms. (*b*) Power-on brake. Opposite to the power-off brake, this type of brake is designed to hold the motor and load when a brake coil is energized. When power is removed from the brake coil, the brake is released to allow the motor to rotate freely. Generally, power-on brakes are used with clutches to stop or slow down at a precise position. Compared to power-on brakes, power-off brakes provide extra safety in applications where the load must remain in position in the event of power loss or equipment failure.

Spring-set types are less expensive than PM types. While spring-set types are well suited for static holding applications and low-cycle dynamic operation, PM types are generally better suited for applications that require high cycling rates, high torque capacity, or accurate positioning [7.34].

According to the working principle and the source of actuating force, electric machine brakes can be broadly categorized into four groups: electromagnetic, mechanical, pneumatic, and hydraulic brakes. Each group may contain several types of brakes, with different features and design specifications. While hydraulic brakes are extensively used in the automotive industry, electromagnetic and mechanical brakes are mainly adopted in electric motor applications.

7.3.1 ELECTROMAGNETIC BRAKE

Electromagnetic brakes are most popular in the automation industry. An electromagnetic brake is an actuator that uses the generated electromagnetic field to activate an electric machine brake. The brake coil functions like an electromagnet when energized, where the electromagnetic field releases the brake so that the motor can rotate freely.

Over the past few decades, while hydraulic and mechanical braking systems are still available in some applications, most of the motor manufacturers have adopted electromagnetic braking solutions.

7.3.1.1 Spring-Engaged, Electromagnetically Released Brake

This type of brake, also known as *fail-safe* or *power off* brake, is predominant in servo systems. They stop or hold a servo system and its driven load machine when electric power is either accidentally lost or intentionally disconnected. The braking torque of a spring-applied brake is produced by utilizing the spring mechanism to apply pressure on the friction disc against the motor shaft. As its name implies, this type of brake is disengaged by releasing the pressure of springs via electromagnetic actuation [7.35].

As shown in Figure 7.8, a spring-engaged, electromagnetically released brake consists of a field coil, a lining disc with attached friction pads on its two sides (Figure 7.9), an armature, and a set of springs. The lining disc is designed to permit relative axial movement. The friction pads are made from asbestos-free materials, which are highly resistant to wear and have high thermal conductivities.

The dimensions of the friction disc and pad are shown in Figure 7.10. The theoretical contacting friction area A_c and braking torque T_b are thus determined by

$$A_c = \int_{r_{p,i}}^{r_{p,o}} 2\pi r\, dr = \pi \left(r_{p,o}^2 - r_{p,i}^2 \right) \tag{7.30a}$$

$$T_b = \mu_d p A = \pi \mu_d p \left(r_{p,o}^2 - r_{p,i}^2 \right) \tag{7.30b}$$

When the electrical coil is energized by a low-voltage (usually 12–48 VDC) power supplier, the magnetic force generated by the coil overcomes the spring force, compressing the springs and pulling the armature away from the friction disc surface, allowing the motor shaft to rotate freely. When electrical power is removed, the electromagnetic field dissipates and thus the springs clamp the friction disc surfaces together to generate braking torque on the rotating rotor and motor shaft. In practice, the braking force can be accurately controlled by selecting suitable spring materials,

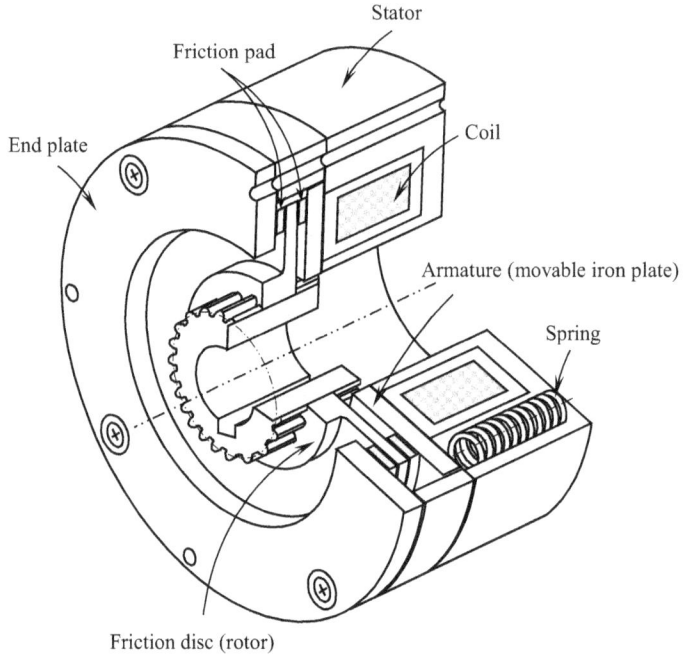

FIGURE 7.8 The structure of spring-engaged, electromagnetically released brake. When the coil is energized by a power supplier, the magnetic force overcomes the spring force, compressing springs to release the brake. As power is removed, the springs push the armature to clamp the friction disc for generating the braking torque.

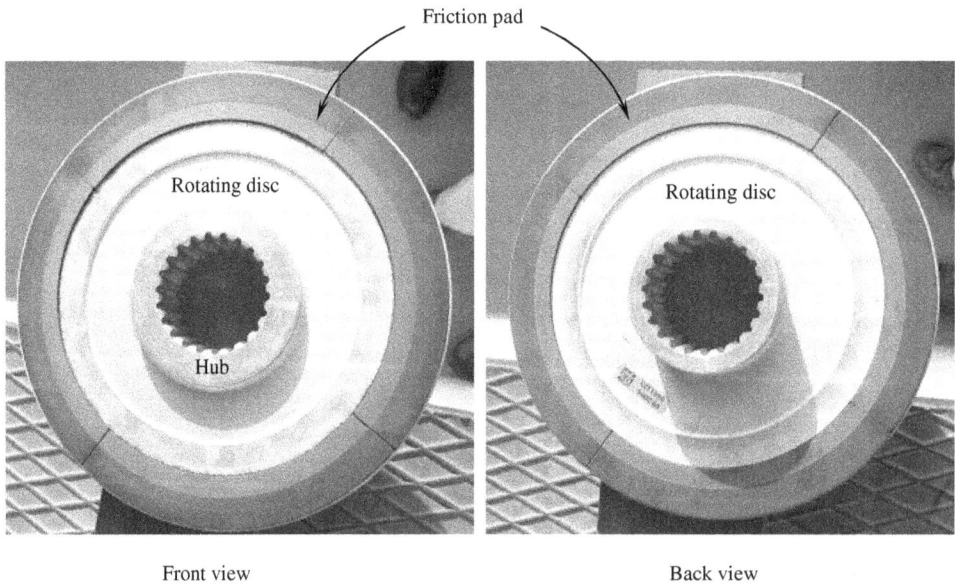

FIGURE 7.9 Segmented friction pads attached on two sides of a rotating disc (rotor).

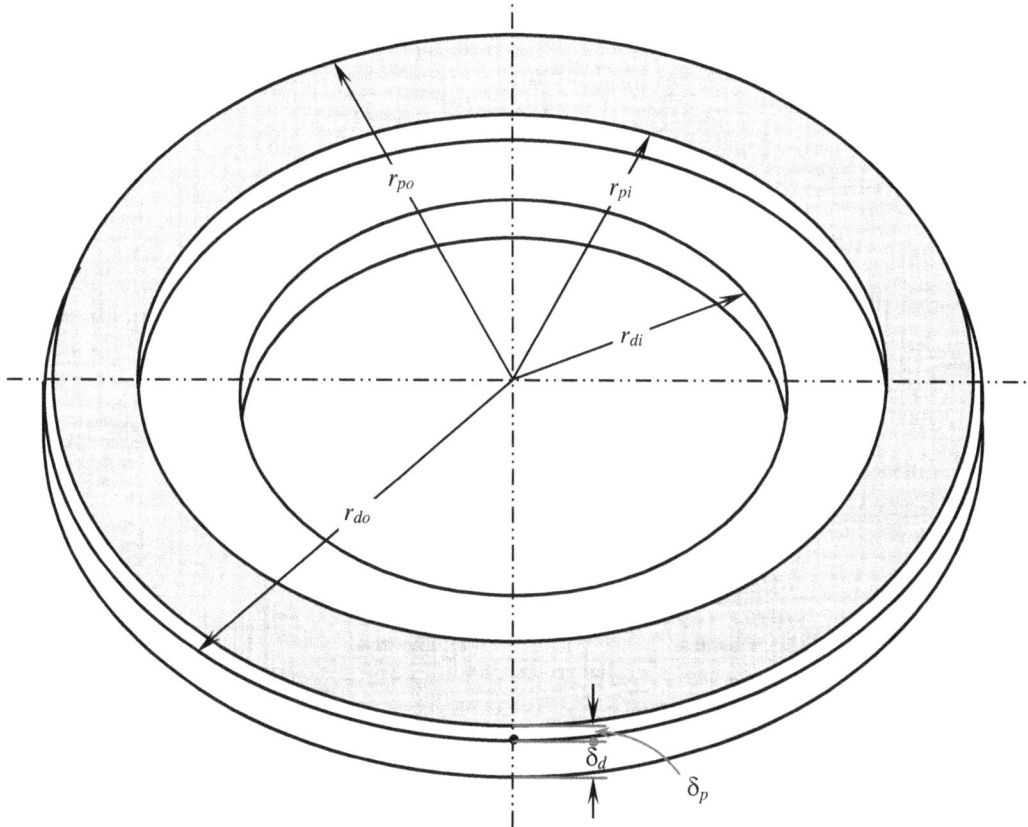

FIGURE 7.10 The dimensions of friction disc and pad.

geometries, and the number of springs. Generally, this type of brake has a compact size and a short response time. With high heat dissipation capabilities and adjustable braking torques, electromagnetic brakes are particularly suitable for dynamic braking applications.

The coil can be powered by either a DC power supplier or an AC power supplier using a bridge rectifier to convert AC to DC. The DC power supply usually provides higher speed and reliable brake operation.

7.3.1.2 Solenoid-Actuated Brake

Solenoids have been widely used as electrical actuators due to their advantages, such as fast response, simple structure, low cost, and easy control in the magnitude and direction of electromagnetic force. A solenoid typically consists of an electromagnetically inductive coil wound around a spring-loaded, movable plunger made of ferromagnetic material. When the coil is energized, a magnetic field is created inside the coil, providing an electromagnetic force on the plunger to move it along the solenoid axis against the spring force. When the coil is de-energized, the spring helps the plunger return to its original position. Because the moving direction of the plunger and the magnitude of the magneto-motive force applied to the plunger can be easily controlled by changing the input current applied to the solenoid coil, solenoids have been widely used among electromagnetic actuators [7.36].

The term solenoid-actuated brake refers to a group of braking mechanisms that rely on electrically controlled solenoids for their actuation. In many industrial applications, there are two types of solenoid-actuated brakes, namely, conventional solenoid shoe brakes and solenoid-actuated blocking brakes. These two types of brake have similar names but substantially different structures and braking mechanisms.

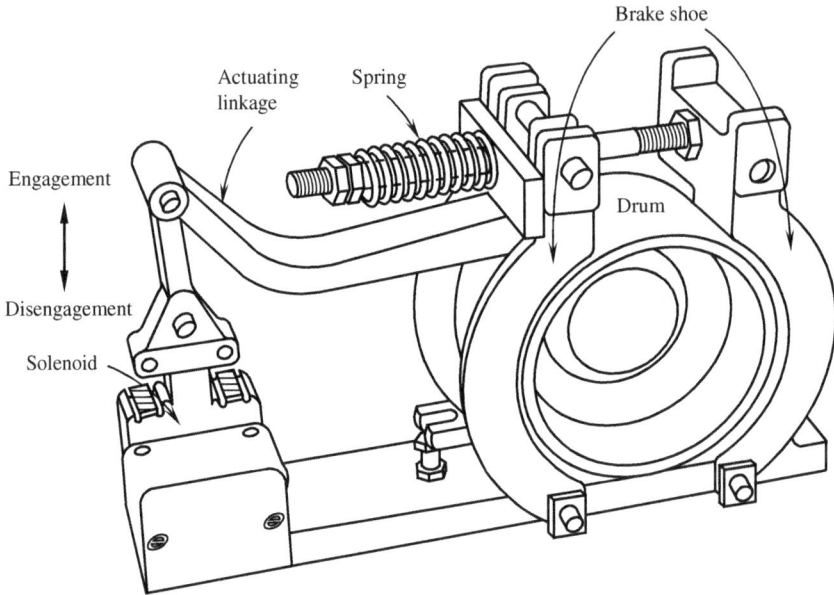

FIGURE 7.11 A conventional solenoid shoe brake is essentially a drum brake. The brake is engaged by a compressive spring when the solenoid is unpowered and disengaged by the solenoid as it is energized.

A solenoid shoe brake primarily consists of a drum, a pair of brake shoes, an electrically controlled solenoid, and an actuating linkage (Figure 7.11). The frictional pads are attached to the internal surfaces of the brake shoes. A brake is typically engaged by a compressed spring when the solenoid coil is unpowered and disengaged by the solenoid as it is energized. For a brake engagement, the compressed spring pushes the brake shoes against the drum that is connected to an electric motor or other equipment. Thus, the braking torque is generated through the friction between the shoe and the drum. When power is applied to the solenoid coil, the solenoid actuating linkage executes to compress the spring and release the shoes from the brake drum, disengaging the brake. This type of brake is usually designed and used in heavy-duty industry applications such as cranes, hoists, conveyors, mining machines, steel rolling mills, paper rollers, and many older elevator applications.

A solenoid-actuated blocking brake is based on the activation mechanism of solenoid to provide safe-fail operation. As shown in Figure 7.12, this type of brake has the simplest structure, consisting of two main components: a solenoid and a rotating brake wheel. The brake wheel, mounted on the motor shaft, has several teeth that are equally disposed on the outermost circumference. The solenoid has a plunger that can be controlled to move along the solenoid axis. When the brake is implemented using the solenoid, the plunger extends forward to the brake wheel for blocking the tooth. When the plunger retracts, the brake is disengaged. To reduce the impact of shock loads and avoid damages to brake components, a thin friction band is tightly fitted between the brake wheel and the shaft, allowing the brake wheel to slip relative to the shaft under overloaded conditions.

The major difference between two solenoid-actuated brakes is their braking mechanisms. As described above, the solenoid-actuated blocking brake uses the mechanical blocking mechanism for directly stopping driven units rather than through friction. Thus, it must operate in low rotating speed applications. In practice, it is usually mounted on the output shaft of a motor gearing system.

There are some considerations in designing a solenoid-actuated block brake. First, because the brake wheel can be stopped only at discrete teeth, the brake wheel allows a limited amount of backward motion, which equals to the space among the adjacent teeth (this refers to backlash). The backlash l_b (in the unit of mm) can be determined as

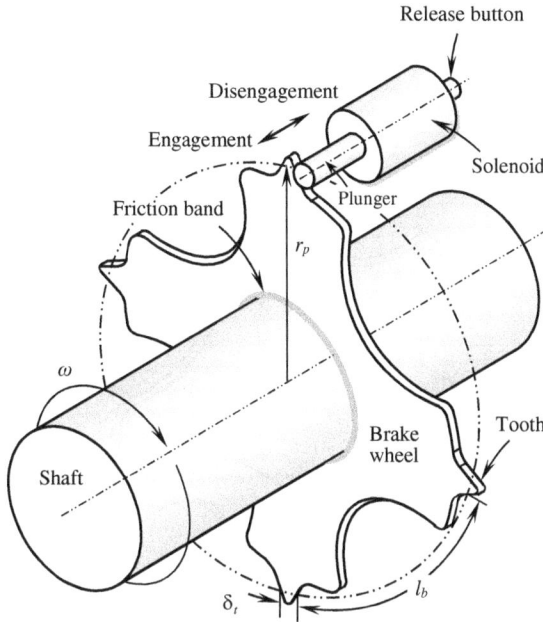

FIGURE 7.12 Solenoid-actuated blocking brake, relying on an electric solenoid for actuation in its axial direction. The backlash between adjacent teeth is denoted as l_b.

$$l_b = \frac{2\pi r_p}{N} - \delta_t \tag{7.31}$$

where
    r_p is the pitch radius of teeth, mm
    N is the number of teeth
    δ_t is the thickness of teeth, mm

In the term of radian, the backlash can be expressed as

$$\theta_b = \frac{2\pi}{N}\left[1 - \frac{N}{360}\tan\left(\frac{t_t}{r_p}\right)\right] \tag{7.32}$$

An effective way to reduce backlash is to increase appropriately the number of teeth on the brake wheel.

Second, during brake operation, the brake wheel teeth and solenoid plunger are subjected to repeated impact/shock loads. Therefore, these components must be designed strong enough to withstand dynamic impacts.

Third, during the process of brake disengagement, a mild vibration of the solenoid plunger may be needed to help it effectively disengage from the wheel tooth.

One of the alternative solenoid-actuated blocking brakes is shown in Figure 7.13. In this design, the movement of the plunger is in the radial direction of the brake wheel. Another alternative design is presented in Figure 7.14. The engagement of the brake is achieved when the plunger moves axially and touches to the bulge on the brake wheel. The brake disengages as the plunger moves back due to the removed power from the solenoid.

Solenoid actuated brakes are designed for long time, high reliability, minimum maintenance, and extremely low cost. They have been successfully used in robots [7.37] and electrical vehicles as parking brakes [7.38].

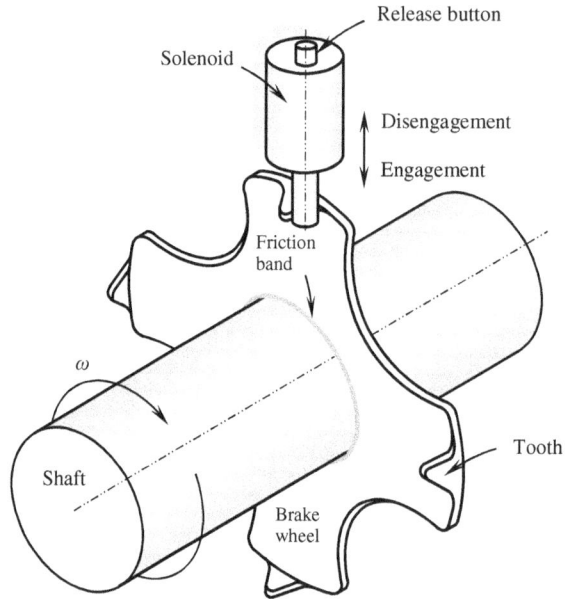

FIGURE 7.13   An alternative solenoid-actuated blocking brake, actuating in the radial direction.

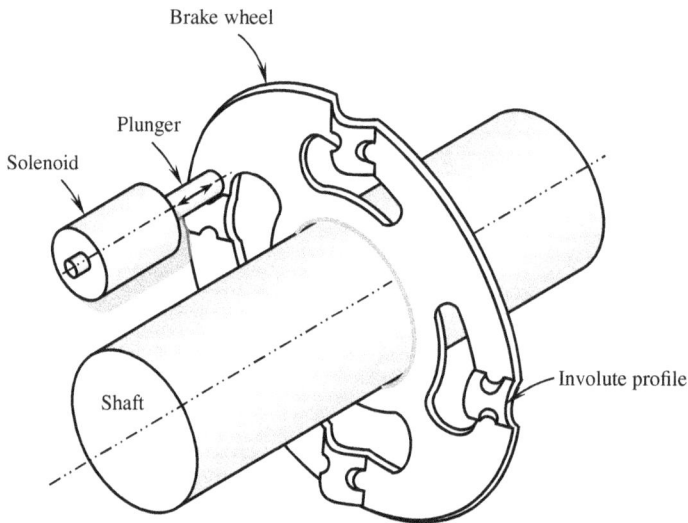

FIGURE 7.14   An alternative solenoid-actuated blocking brake, actuating in the axial direction.

One of the main drawbacks of solenoids actuated brake is heat generation in solenoid coils. For instance, as one of the heat sources in a robot, heat is generated when the solenoid coil is energized. If the generated heat can be effectively dissipated, it will be very impactful on the temperature rise of the robot, and thus provide longer life, better reliability, and more return on investment for the robot owner. One of the most effective ways to solve this problem is to use stacked piezoelements as the replacement for the electrical solenoid. Thanks to their capacitive principle, piezoelements require virtually no energy to maintain an active state. In addition, they do not generate heat if high-frequency control is not used. In fact, piezotechnology is especially ideal for use in the *very-low-power* range of battery-power devices.

7.3.1.3 Multiple Disc Friction-Blocking Brake

While a friction brake relies on the friction between the paired friction discs and brake discs for providing braking forces to halt the driving unit, an electromagnetically actuated blocking brake relies on the blocking action of the solenoid plunger to the brake wheel to stop the electric machine. As a comparison, a multiple disc friction-blocking brake is essentially a combination of a friction brake and a blocking brake. As described in Figure 7.15 [7.39], the multiple disc brake consists of a ring-shaped electromagnetic coil, a hub that is fixed on the motor shaft, a number of friction discs mounted on the hub so that they rotate together with the hub, a number of brake discs arranged freely rotatable relative to the hub, a brake ring that has a plurality of bumps formed among the machined incisions in the radial direction, a upper and lower housing, and a plurality of coil or wave springs that compress braking discs and friction discs tightly in a braking process.

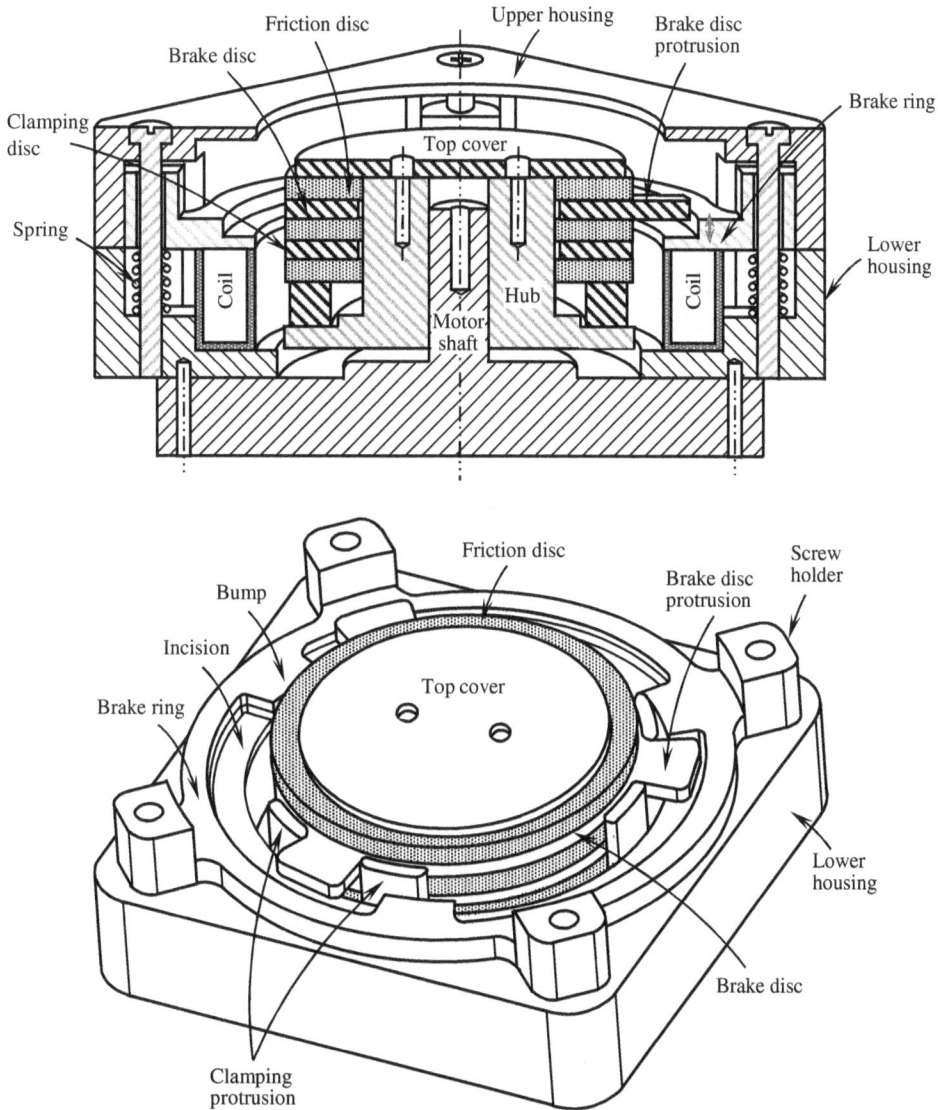

FIGURE 7.15 Schematic diagram of a multiple disc friction-blocking brake, which is essentially a combination of a friction brake and an electromagnetically actuated blocking brake. The brake is disengaged when the electromagnetic coil is powered and engaged when power is removed from the coil.

The movable braking ring includes four screw holders each with a through-hole in the axial direction of the hub. The four screw holders are equally spaced around the circumferential of the braking ring. While one of the braking discs has a plurality of brake protrusions protruding in a radial direction of the brake disc, which is uniformly distributed in the circumference of the disc, another braking disc (*i.e.*, the clamping disc) has a plurality of clamping protrusions protruding in the radial direction of the disc and then bending in an axial direction. Therefore, the clamping protrusions hold the clamping disc to the braking disc and strengthen the engaged brake protrusions in both rotational directions of the braking disc.

When the electromagnetic coil is energized, it attracts the brake ring and allows brake discs to rotate freely. Once the power is off due to the power failure or other reasons, the coil releases the brake ring thus the springs push on the brake ring in the axial direction of the hub, blocking the braking discs and preventing them from rotating.

Compared to other multiple disc brakes, this type of brake offers a cost effective and compact design solution for many industrial applications, such as robots.

7.3.1.4 Eddy-Current Brake

Most conventional motor brakes utilize friction forces to decelerate or stop rotating machines. However, due to friction, such brakes are subjected to wear on the contact surfaces and thus require regular inspection/maintenance to keep them in a good operating condition. Furthermore, brake wear debris represents the potential hazard to optical encoders, as well as to the environment. More seriously, as these debris stick to friction pads and discs, the braking friction will be remarkably reduced and consequently affect the safe and reliable performance of brakes.

As non-contact braking systems have no direct contact between braking components, eddy-current brakes use magnetic forces for inhibiting the motion of the rotating motor, and thus are preferable replacements for friction brakes. Actually, this type of brake tends to be much more robust than friction brakes. Advantages of this type of brake include fast control dynamics, short response time (in milliseconds), free of pollution and noise, precise braking torque control, and relatively inexpensive.

According to Faraday's law of induction, when a conductive material is subjected to a change in time or space of magnetic flux, eddy currents are induced in the bulk of the material with closed loops. By Lenz's law, these eddy currents produce an opposite magnetic field against the original magnetic field creating an opposing force. In practice, the time-varying magnetic field can be induced either by movement of the conductor in the field or by changing the strength or position of the source of the magnetic field [7.40]. If the non-ferrous material is in a moving state, the generated opposite magnetic field decelerates the motion of the conductor.

Although in most cases eddy current is undesired as an energy loss, it can be used to implement in developing a braking system. A characteristic of eddy-current brake is that the brake torque is proportional to the speed and is zero at zero speed. This indicates that eddy-current brakes cannot be used as holding braking systems.

According to the source to generate a magnetic field, eddy-current brakes can be classified as either a PM type or an induction type. The configuration of a typical PM eddy-current brake is presented in Figure 7.16, consisting of a rotating disc and a pair of PMs. The disc is made of a non-magnetic material (typically aluminum or copper), with the thickness δ and radius r. When the disc spins with respect to its axis in a constant angular velocity ω and is subjected to a constant magnetic field **B**, generated by an array of spaced apart PMs, eddy currents are induced in the disc, resulting in the deceleration of motion. If the PMs in Figure 7.15 are replaced by electromagnetic coils, the brake is called the induction eddy-current brake.

The experimental results have shown that aluminum is the best material for conductive discs, compared to copper and zinc. Aluminum has the highest speed reduction during the braking process over the other two materials [7.41].

The braking torque generated in an eddy-current brake is affected by a number of variables, including disc geometry, angular speed, material electrical properties, strength of applied magnetic

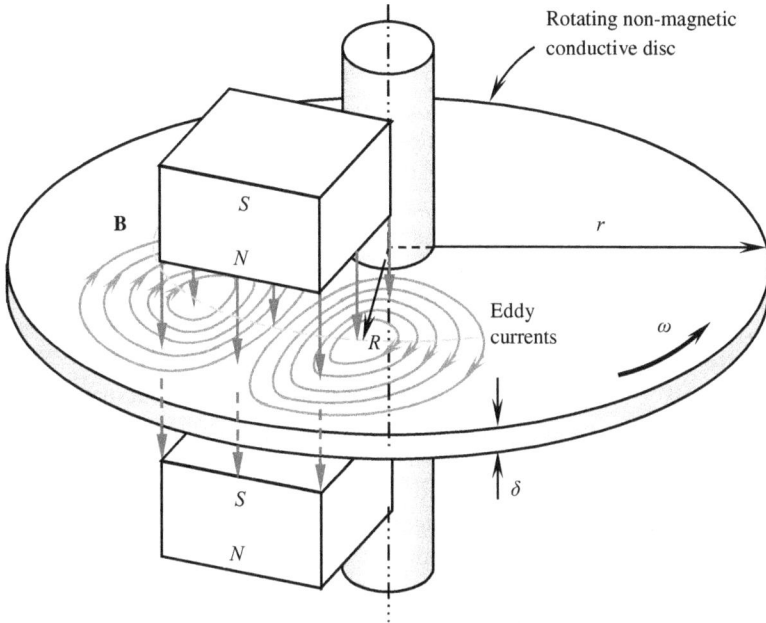

FIGURE 7.16 Schematic diagram of working principle of PM eddy-current brakes. The eddy current produced in the rotating wheel is induced by PM.

flux, and rotating speed of the disc. There are several braking torque models available in the literature. Smythe's model [7.42] and Schieber's model [7.43] are good at low rotating speeds. While Wouterse's model [7.44] offers a global solution for covering all speed regions, it has to use two different expressions for low- and high-speed regions. Later, Ming's model [7.45] was proposed to predict the braking torque, with five unknown parameters. This model agrees well with the experimental results. In the model of Gasline and Hayward [7.46], the braking torque T_b (Nm) generated in the rotating disc under the influence of the magnetic field is given as

$$T_b = \frac{\pi\sigma}{4} nD^2 \delta_d B^2 r_d^2 \omega \tag{7.33}$$

where

σ is the electric conductivity of the disc, $(\Omega \cdot m)^{-1}$
n is the number of pair of magnets
D is the diameter of the magnet array, m
δ_d is the thickness of the disc, m
B is the magnetic flux density, T
r_d is the radius of the disc, m
ω is the angular speed, rad/s

The influence of rotating speed ω on braking torque T_b is determined by two opposite factors: (a) The magnetic flux flowing in the disc changes substantially with the increase in the rotating speed, thereby increasing the eddy currents generated in the disc, and (b) according to the theory of skin effect, the skin depth is proportional to $1/\omega^{1/2}$. At high rotating speeds, eddy currents are concentrated on the surface, making higher electrical resistance and shrinking the amount of eddy currents. Consequently, at very low velocities with a minor skin effect, the break torque is proportional to the disc's rotating speed until it reaches the peak value. After this point, the further increase in the rotating speed will cause a decrease in braking torque due to the enhanced skin effect [7.47].

Generally, the greater the magnetic field produced by the PMs, the greater the amount of braking torque produced by the motor.

In addition to PMs, the magnetic field can be alternatively created by electromagnetic coils, and consequently the braking torque can be easily adjusted by varying the electric current through the brake coils. This gives flexibility to the braking system.

It is important to note that an eddy-current brake delivers a speed-dependent braking torque. Because the braking torque is directly proportional to the speed of the rotating rotor, the brake might not provide enough braking force at low speed. When a rotor is stationary, the brake generates no holding force. This indicates that eddy-current brakes cannot be used in low-speed applications, as well as in applications that frequently require holding loads such as elevators.

7.3.1.5 Magnetic Particle Brake

Magnetic particle brakes use the principle of induced losses in magnetic particles. The resistive torque in this type of brake is approximately proportional to the power loss induced by a magnetic field surrounding particles. Because the magnetic field is created by electrical current passing through the field coil, it enables to precisely control the braking torque. They are ideal for applications operating in the low to middle speed range.

As shown in Figure 7.17, the magnetic particle cavity between the rotor and stator is filled with very fine, dry ferrite particles. When the field coil is power-off, the particles are free flowing and thus the brake is disengaged. As DC passes through the field coil, the electromagnetic field is created and applied to the ferrite particles, forming chains along the lines of magnetic flux and linking the

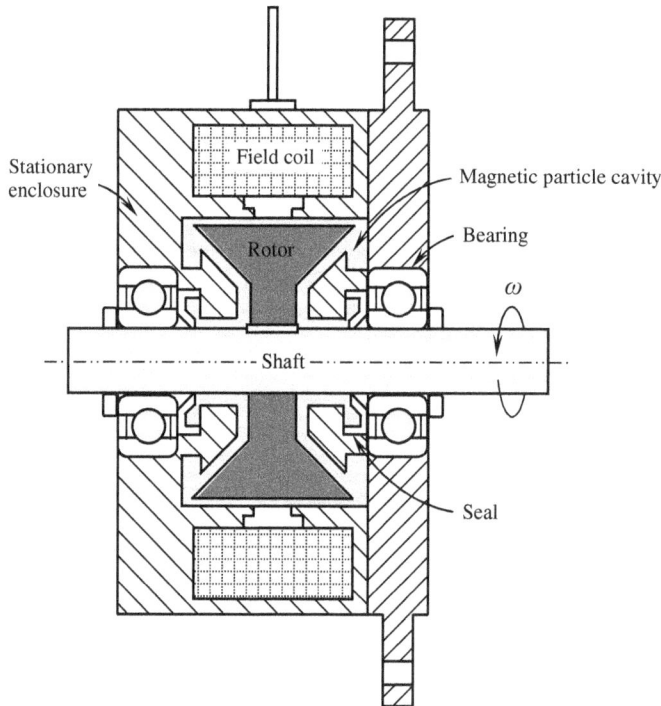

FIGURE 7.17 Magnetic particle brakes are unique in their design from other electromechanical brakes because of the wide operating torque range available. In a magnetic particle brake, torque can be controlled very accurately within the operating rotating speed range. This makes this type of brake ideally suited for tension control applications, such as wire winding, foil and film tension control, and tape tension control. Because of their fast response, they can also be used in high cycle applications, such as magnetic card readers, sorting machines, and labeling equipment.

rotor and stationary components nearby. The amount of current in the coil determines the strength of the magnetic field, which in turn determines the resistive torque imposed on the rotor.

Since the brakes do not involve unstable friction, they provide many quality advantages:

- With no sliding components, this type of brake has a virtually unlimited life expectancy.
- Unlike conventional friction brakes, noiseless magnetic particle brakes are suitable in applications that require quiet working environments, such as medical facilities, labs, and offices.
- Operation at a fixed current, the brake shows no significant brake torque variation even over temperature extremes.

7.3.1.6 Hysteresis Brake

Instead of using any friction components or magnetic particles, a hysteresis brake utilizes the hysteresis effect to produce a braking torque. As a contact-free device (Figure 7.18), the brake has two key components: a field coil embedded in the brake stator and a spinning disc (rotor) with the drag cap attached to it, without physical contact. When the power is disconnected from the field coil, the drag cap can spin freely on its shaft axis. As the field coil is energized, it produces an electromagnetic field across the airgap between the stator and drag cap to apply electromagnetic force on the rotating disc, resulting in a braking torque on the rotor. In fact, in this system, the braking torque is proportional to the input current. This type of brake has several advantages over friction devices including a wide range of braking torque, good heat dissipation, and wear-free.

However, hysteresis brakes have some drawbacks [7.48]: (a) An increase in electrical current by a certain amount leads to an increase in the braking torque. However, when the input current

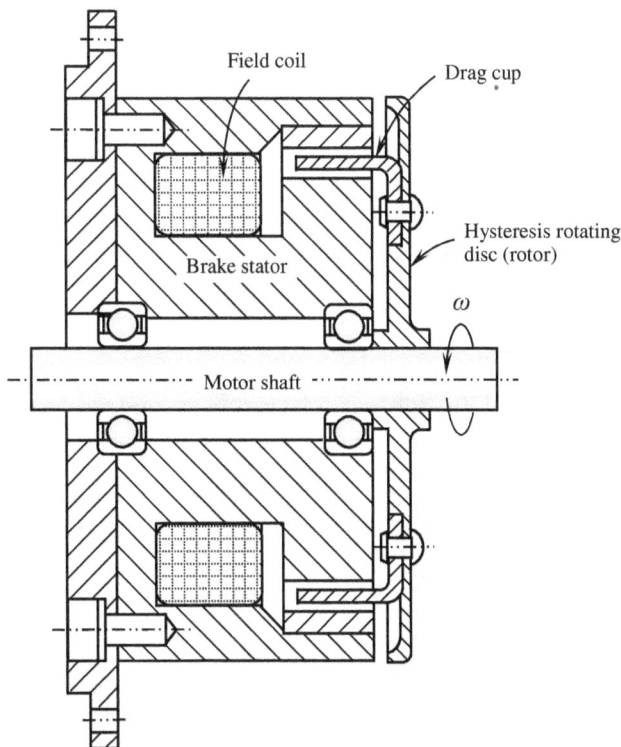

FIGURE 7.18 A hysteresis magnetic brake operates on a frictionless design principle. The braking torque is adjusted and controlled by adjusting the electrical current to the field coil.

decreases by the same amount, the torque decreases slower along a different path, showing the electric hysteresis. (*b*) The braking torque shows a nonlinearity behavior with the input current. (*c*) The cogging effect, or residual magnetism, encountered in hysteresis brakes is the main drawback and reason for not using one for a particular application. (*d*) The braking torque of hysteresis brake increases with rotor speed, rather than often incorrectly stated that it is independent of speed in some sources. Thus, to address these drawbacks, a new type of magnetic brake has been proposed and developed by combining a hysteresis brake and an eddy-current brake.

7.3.1.7 Permanent Magnet Brake

The engagement and disengagement of PM brake rely on the interaction between the magnetic field that is generated by PMs and the electromagnetic field that is generated by the brake coil.

As shown in Figure 7.19, in regular motor operation, to disengage the brake, power is applied to the brake coil that creates an electromagnetic field to cancel out the magnetic field produced by PMs. When the electric power is either accidentally lost or intentionally disconnected from the coils, the magnetic actuating forces generated by the PMs are utilized to produce the braking torque.

The advantages of PM brakes include their backlash-free properties, high reliability, high power density, and smooth operation with stepless braking torque change.

One of the challenges in brake design is to keep the two opposing magnetic fields in balance. This can be done by using an adjustable power supply for accurately adjusting the electromagnetic field against that by the PMs. In general, PM brakes are especially suited for applications that require frequent on-off cycling and consistent performance.

7.3.1.8 Magnetorheological Brake

Magnetorheological (MR) fluid is a smart fluid due to its outstanding properties, such as the reversible properties between the liquid phase and the solid phase. The discovery of MR fluid can be traced back to 1948 by Rainbow at the US National Bureau of Standards [7.49]. One of the first

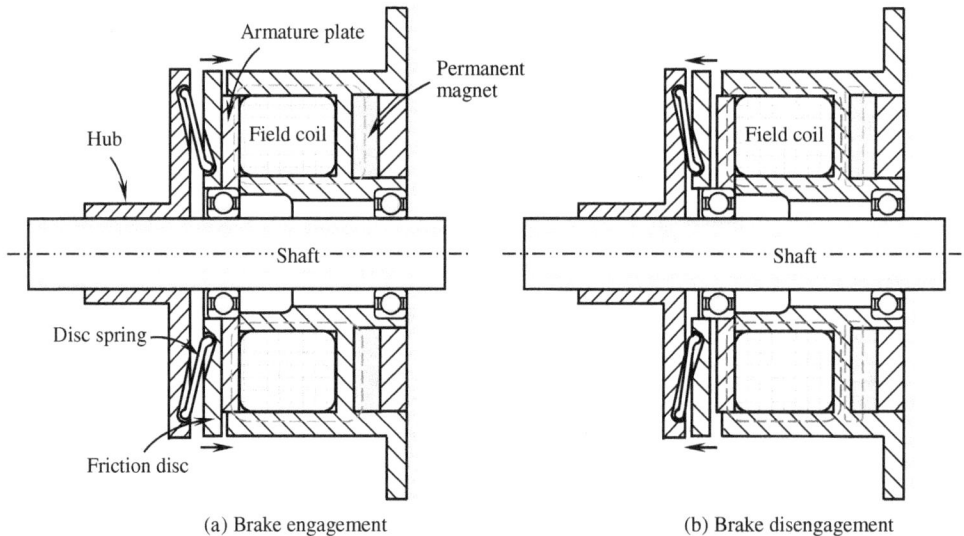

(a) Brake engagement   (b) Brake disengagement

FIGURE 7.19 The actuation of a PM brake relies on the interaction between the magnetic field generated by PMs and the electromagnetic field generated by the energized brake field coil: (*a*) The brake is engaged with the unpowered field coil. The magnetic force generated by the PMs attracts the armature plate to touch with the friction disc. The disc spring is stretched. (*b*) The disengaged brake with the energized field coil, the electromagnetic field cancels the magnetic field. The armature plate is pushed away from the friction disc by the disc spring.

MR fluid-based devices is a magnetic clutch invented in 1951 [7.50]. The MR fluid is composed of microscale or even nanoscale ferromagnetic particles suspended within the carrier oil and distributed randomly under normal circumstances. When subjected to an external magnetic field, the microscopic particles align themselves along the lines of magnetic flux, assuming properties comparable to a solid. The torque level of the brake is controlled by precisely controlling the magnetic field intensity. Therefore, very accurate torque transmission and stopping can be achieved by simple, but very effective MR brakes. Given these favorable properties, MR fluids are widely used in mechanical components, such as brakes, dampers, valves, clutches, and shock absorbers [7.51]. These MR fluid-based devices are characterized by low power consumption, fast response time, and design simplicity. In practice, MR fluids are used in high-end vehicle shock absorber systems for variably controlled damping.

A large number of studies about magnetorheological brakes (MRBs) have been extensively conducted for applications such as motorcycles and vehicles. While conventional hydraulic brakes have been successfully used in road vehicles, MRBs have been developed over the past few decades for possible substitution of hydraulic brakes because of their performance advantages over the conventional hydraulic brake systems.

This brake system consists of rotating discs immersed in an MR fluid that is enclosed in an enclosure (Figure 7.20). When a magnetic field is applied, the suspended particles in the MR fluid form iron powder clusters that solidify the MR fluid among the active magnetic poles. Thus, the controllable yield stress causes friction on the rotating disc surfaces, generating a retarding brake

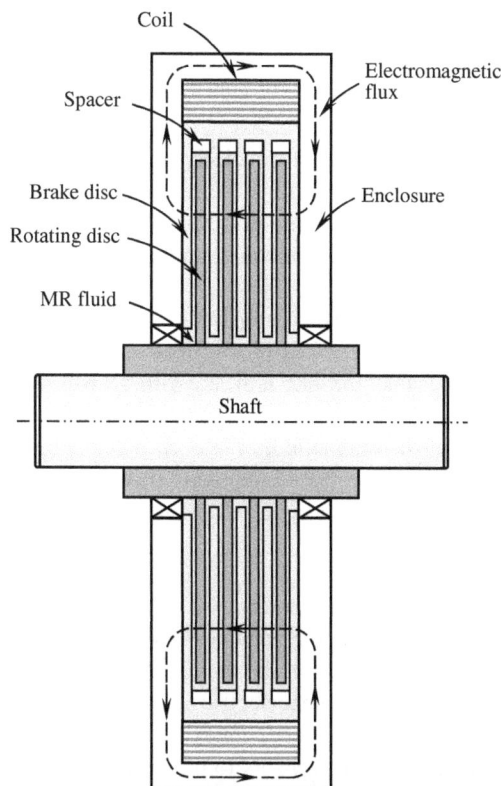

FIGURE 7.20 Structure of MR brake with multiple rotating discs. The rotating friction discs are immersed in an MR fluid that is encapsulated in an enclosure. When the electrical coil is energized, a magnetic field is produced to act on the MR fluid, yielding shear stresses on the rotating discs. The use of multiple discs is to increase the effective contact area among the MR fluid and rotating discs to increase the braking torque.

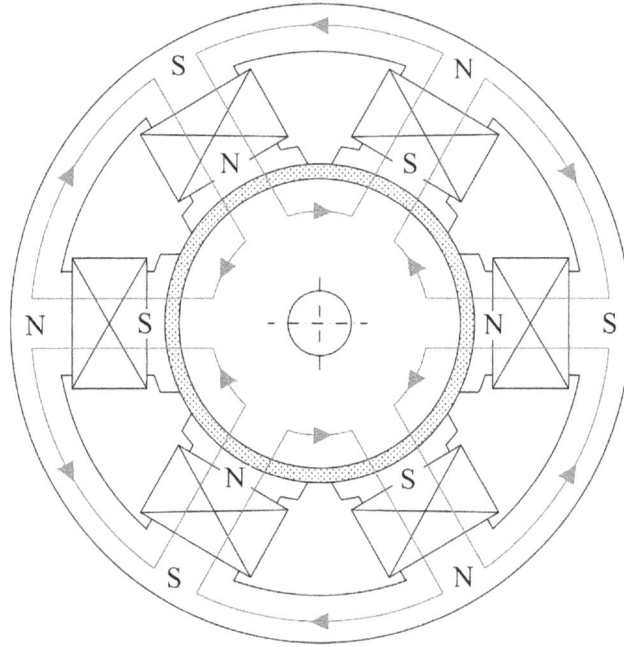

FIGURE 7.21 Operating principle of the multi-pole MR brake, where the gap between the stator and rotor is filled with an MR fluid. As the electromagnetic coils are energized, the magnetic flux travels in a closed loop between the adjacent poles, from one pole, through the MR fluid gap, into the rotor, back to the MR gap and into another pole, producing the shear force in the MR fluid to act on the connected machine.

torque. In such a way, the braking torque can be precisely controlled by changing the current applied to the electromagnet. The most notable feature is the response time: MRBs have a response time of 10–20 ms, compared with the response time of 200–300 ms for conventional hydraulic brakes [7.52]. Correspondingly, the brake stopping time is greatly reduced, providing significant improvements in the brake system performance.

An MR brake is similar to a motor, which consists of a magnetically permeable stator with a multi-pole and a rotor, as shown in Figure 7.21 [7.53]. Instead of the *airgap* in a typical motor, the MR brake uses the *fluid gap* because the gap is filled with an MR fluid. When the multi-coil is energized, the magnetic flux travels in a close loop between two adjacent poles, following the path of the red lines. The direction of the magnetic flux in each pole is opposite to those of its adjacent magnetic poles. In such a design, the produced magnetic flux penetrates the MR fluid in the fluid gap, yielding the resistance as the braking torque acting on the connected machine.

7.3.1.9 Piezoelectric Brake

Piezoelectric effect is the ability of certain materials (so-called *piezoelectric materials*) to generate an electric charge in response to applied mechanical stress. In fact, the piezoelectric effect is a reversible process. The same materials exhibit the ability to directly convert electrical energy into mechanical motion, known as the inverse piezoelectric effect. Today, there are a variety of applications that use piezoelectric technology, from cellphones, sonars, acoustic guitar pickups, diesel fuel injectors, ultrasonic transduces, vibrations sensors to piezoelectric motors, actuators, and medical equipment.

Piezoelectric brakes are designed based on the inverse piezoelectric effect for their fast response, compact size, less energy consumption, extreme precision, and self-locking at the rest position. Another key advantage is that they always work proportionally. The primary challenge in designing piezoelectric brakes is that the geometric displacement Δl generated by most piezoelectric materials

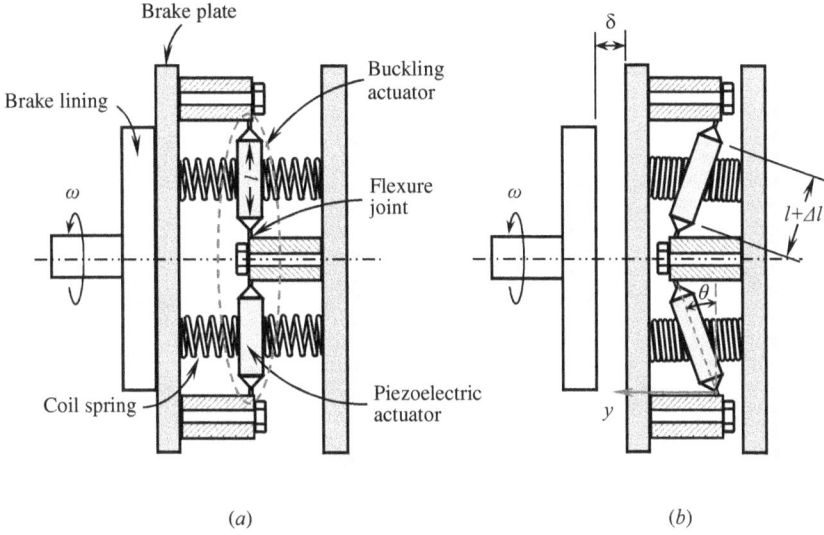

FIGURE 7.22 Basic architecture of a piezoelectric brake, which is actuated by buckling actuators using flexure joints for amplifying piezoelectric displacement: (*a*) brake engagement; (*b*) brake disengagement.

from their original length l is very small, usually $\Delta l/l$ less than 0.2%. To address this problem, piezoelectric brakes are usually made by stacking a number of piezoelectric elements in order to obtain a sufficient brake stroke. In addition, they often adopt some kind of leverage mechanism that amplifies the displacement of piezoelectric actuator geometrically.

Many types of piezoelectric brakes are designed and applied in various industrial and commercial applications. A novel piezoelectric brake was proposed by Tsukahara *et al.* at MIT [7.54]. The brake consists of three bucking actuators, arranged in an equilateral triangle, to provide the push-pull actuation for the brake. As shown in Figure 7.22, each bucking actuator comprises two piezoelectric actuators with the original length l. The flexure joints are employed at the ends of these piezoelectric actuators for maximizing the brake stroke δ. To obtain enough force for the braking function, coil springs are added at each side of the piezoelectric actuator, in parallel with the bucking actuators.

When the piezoelectric actuations are powered off, the brake plate contacts the brake lining closely under the compressive pressure generated by coil springs, producing the braking torque due to friction. When the piezoelectric actuators are energized, their lengths increase from l to $(l+\Delta l)$. Due to the bucking effect, the piezoelectric actuators rotate accordingly to compress the springs and thus separate the brake plate from the brake lining, disengaging the brake.

The displacement amplification ratio of the piezoelectric actuator is defined as the ratio between the output and input displacement. Among various piezoelectric actuators, the buckling type actuator can remarkably increase the displacement amplification ratio R,

$$R = \frac{dy}{d(\Delta l)} = \frac{dy/d\theta}{d(\Delta l)/d\theta} = \frac{l(1+\tan^2\theta)}{l(\sin\theta/\cos^2\theta)} = \csc\theta \tag{7.34}$$

where

$$\begin{cases} y = l\tan\theta \\ \Delta l = l(\sec\theta - 1) \end{cases} \tag{7.35}$$

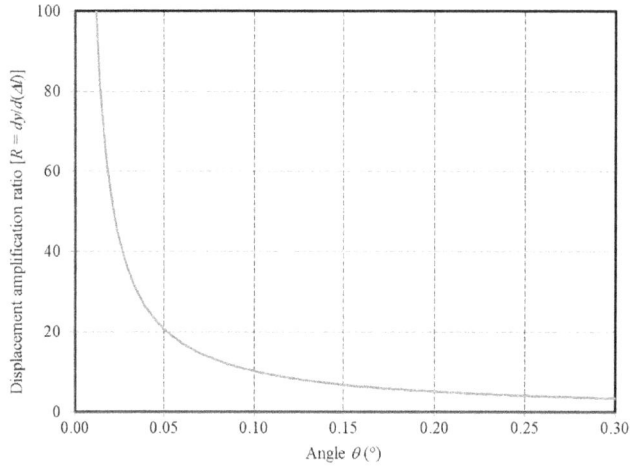

FIGURE 7.23 The displacement amplification ratio of a buckling actuator.

According to Equation 7.34, the displacement amplification ratio R with respect to the joint angle θ is shown in Figure 7.23. Due to the singularity, infinite amplification occurs when $\Delta l = 0$ and $\theta = 0$. In the proximity of the singularity, much higher amplification ratios can be obtained by this mechanism, compared to other one-step leverage mechanisms. Thus, the stroke of the piezoelectric brake becomes

$$\delta = R\Delta l = R\varepsilon l \tag{7.36}$$

where ε is the piezoelectric strain, $\varepsilon = \Delta l/l$.

The experimental results from a brake prototype show that the brake can produce more than 2.5 Nm in the displacement range of 0.5 mm. With the buckling-type displacement amplification mechanism, the amplification ratio R can be higher than 50.

An alternative piezoelectric brake is shown in Figure 7.24 [7.55]. This piezo-actuated brake includes a disc fixed on a motor shaft, a clamping structure that has two brake elements positioned in alignment on either side of the disc and a piezoelectric element. The piezoelectric element is made from numerous piezocrystals stacked together and connected to the clamping structure to urge the brake elements toward one another. The excitation of the piezoelectric element thus produces a braking force acting on the rotating disc. For a compact design, the disc may be alternatively positioned inside the motor.

A so-called *mechanical amplifier* is located at the opposite end of the brake elements to amplify the generally small expansion of the piezoelectric element into a movement sufficient to provide braking action to the braking system. On receiving electrical input, piezoelectric element undergoes mechanical expansion and that movement is translated and amplified by the mechanical amplifier to apply mechanical force to the clamping structure. Consequently, the clamping structure clamps down on the disc, impeding the motion of the motor.

7.3.2 MECHANICAL BRAKE

In a general sense, pure mechanical braking systems have been historically used in some applications such as vehicles and bicycles. However, they are uncommonly used in modern servomotors and motion control systems today.

One of the drawbacks of electromagnetic brakes is that they must energize coils to keep the brakes disengaged, which consume the amount of electrical power and generate heat in the coils.

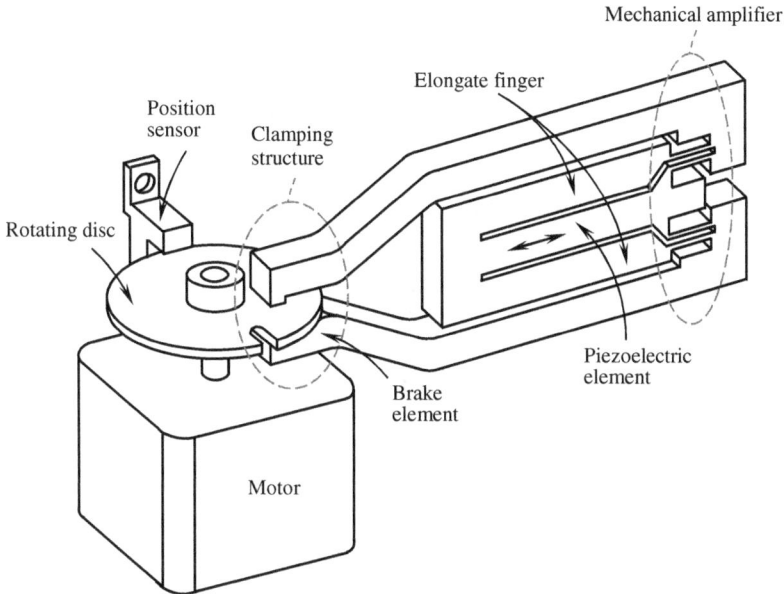

FIGURE 7.24 Piezo-actuated brake designed for an electric motor.

Consequently, the excess heat degrades the performance and ability of brake in stopping and holding the loads. To overcome this drawback, as well as to maximizing brake performance, mechanical brakes that consume less energy have been developed for some applications.

7.3.2.1 Compressed Air-Engaged, Spring-Released Brake

As a type of mechanical brake, a compressed air-engaged, spring-released brake is engaged with compressed air that acts on a piston to move it axially and collapses the springs, pushing the friction disc and drive disc to contact each other to generate the friction that transmits torque. To seal high-pressure air, the piston is fitted with three O-rings. The brake disengagement occurs when air pressure is removed, allowing the springs expand and pushing the contact discs apart.

Compressed air-engaged friction brakes have some special features that distinguish them from many other brake systems. This type of brake is the best choice for dynamic stopping and cycling applications because its design compensates for wear. Unlike conventional spring-engaged brakes, the output of braking torque of this type of brake is proportional to the applied air pressure, as an example shown in Figure 7.25. Even after a long period of operation, the friction discs become thin and the piston simply travels farther so that the torque output remains the same at a given air pressure. The brake failure occurs when the friction discs wear to the point that the piston O-ring seals actually travel beyond the cylinder walls, causing air to leak out. With all parts being enclosed, the brakes are suitable for working in unfavorable environments.

Compressed air-engaged friction brakes are the best choice for dynamic stopping and cycling applications, for wet or dry operations. In addition, because this type of brake is pneumatically powered by compressed air rather than electrical power, they are particularly suitable to operate in fire- and explosion-proof environments (*e.g.*, coal mines).

In practice, the opposite design, *i.e.*, spring engaged, compressed air released brake is also available. In such a design, a brake features a series of springs that push the drive discs against the friction discs, resulting in friction force on the contact surfaces that produce sufficient torque to stop or hold the load. The brake remains engaged until air pressure is applied to compress the springs and open a gap at the interfaces. Compared to the abovementioned brakes, this type of brake can be used to hold loads without continuously supply of high-pressure air.

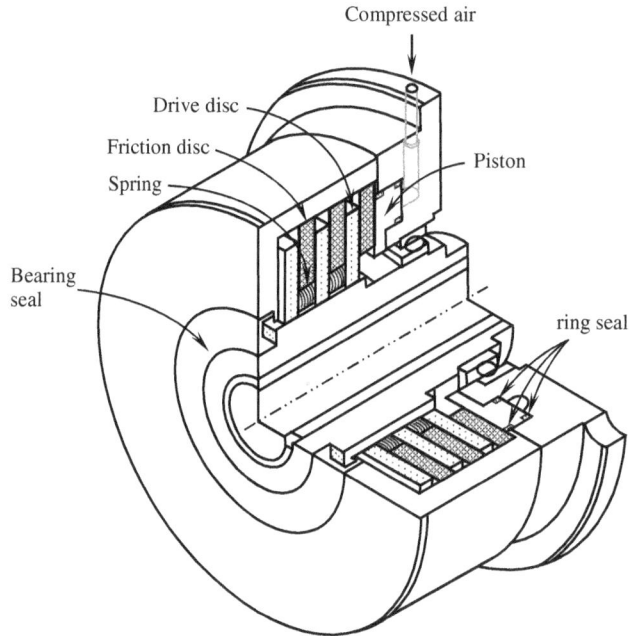

FIGURE 7.25 The cut-away view of the compressed air-engaged, spring-released brake. When the air pressure is applied to the piston, it forces the piston to move to compress the spring, pushing the friction disc and drive disc to contact together for the brake engagement. As the air pressure is removed, springs force the friction and drive discs to separate, releasing the brake.

7.3.2.2 Spring-Engaged, Hydraulic Pressure Released Brake

The structure of the spring-engaged, hydraulic pressure released multi-disc brake is shown in Figure 7.26. It consists of an annual piston, a hub, springs, friction discs, oil seals, and an enclosure. The brake operates with a hydraulic pressure. With filled high-pressure oil in the disc-pack chamber, the springs are compressed to release the friction discs, keeping the brake disengaged. As the hydraulic pressure drops, the springs push the annual piston moving axially toward the friction discs to produce the braking torque that is applied to the motor shaft. In practice, the hydraulic pressure is generated by a hydraulic pump with a pressure control system. The hub is fitted to the end of the motor shaft. The enclosed nature of the brake makes this type of brake suitable for external mounting, even in unfavorable environments.

Multi-disc brakes are superior in terms of reliability and performance over single-disc brakes. They can handle higher torque loads with compact constructions. This type of brake is the perfect choice for heavy-duty applications such as construction machinery, machine tools, agricultural and mining machines. However, they may generate more heat than single-disc brakes. To solve the issue, a through flow of oil is used for lubrication of internal components, increase in energy capacity, and dissipation of heat from friction discs during engagement. In addition, the hydraulic pressure applied to keep the brake disengaged also results in an axial force on the motor bearings. Therefore, when using this type of brake, it is suitable to select angular contact ball bearings as the motor bearings for withstanding both radial and axial loads.

7.3.2.3 Pneumatic Brake

The pure pneumatic brake system invented by George Westinghouse in 1868 revolutionized the railroad industry. This compressed air controlled brake is a great advancement in railroad safety and efficiency, making it possible to stop an entire train by pulling just one lever, and thus permitting

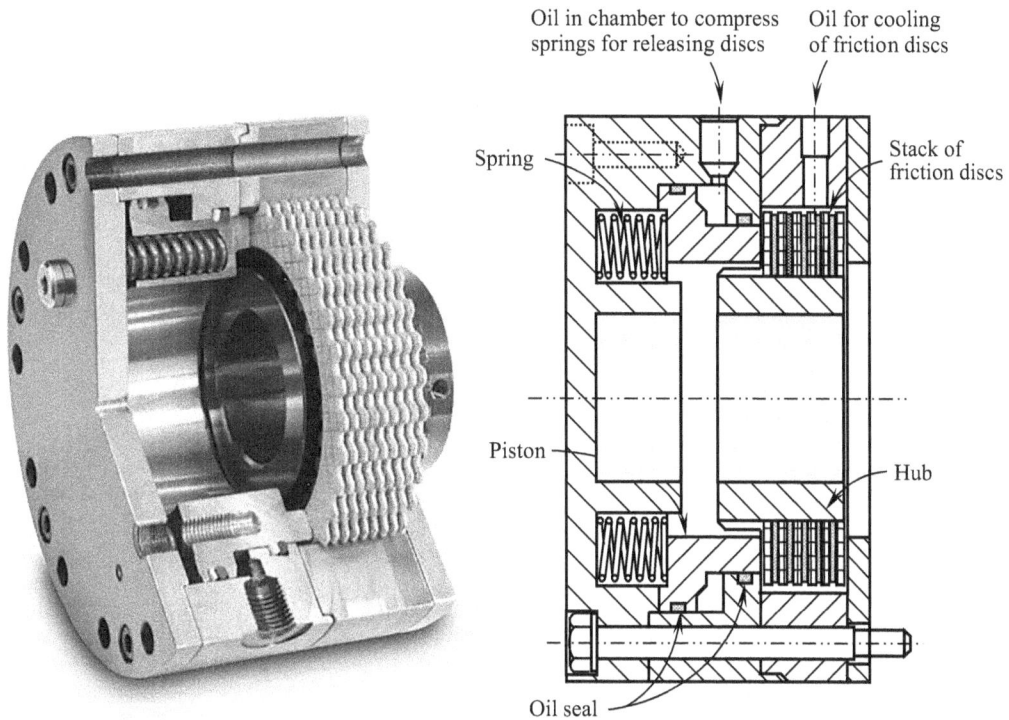

FIGURE 7.26 The structure of spring-engaged, oil pressure released multi-disc hydraulic brake. When high-pressure oil in the disc-pack chamber compresses the springs to release a stack of friction discs, the brake is disengaged. As the pressure is removed from the chamber, the springs clamp a stack of friction discs together to engage the brake. The hub is usually fitted to the end of the motor shaft. (Courtesy of Matrix International, South Beloit, IL.)

trains to travel at higher speeds. In the following years, Westinghouse made many alterations to improve his invention, leading to various forms of the automatic brake. His air brake was granted a U.S. patent in 1873 [7.56].

Today, many modern trains use brakes in various forms based on the design. Statistics show that more than 80% of urban transport services worldwide use air brakes. The same conceptual design of pneumatic brake is also found on heavy-duty vehicles, mining machinery, and other applications. Since air brakes use compressed air to actuate braking rather than electromagnetic coils or solenoids, this type of brakes generates less heat during the braking process. Therefore, they are more efficient and economical than similar-sized electromagnetic brakes, allowing for greater torque transmission. Moreover, the higher thermal capacity of the air brakes gives them a longer service life. However, this type of brake is rarely used for motor braking.

As an improvement of the compressed air brake, the electro-pneumatic brake was introduced in the early years of the twentieth century in an attempt to overcome the drawbacks of the air brake system on long freight trains. Its main advantage over the air brake is its short reaction time, permitting the instantaneous control to the whole train. In recent years, electronically controlled pneumatic brakes have been developed and tested in the US. With the modern electronic technology, the brake control of rolling stock can be more accurate, fast, and sensitive.

7.3.2.4 Hydraulic Brake

The advantages of hydraulic brakes include their quick response, high braking torque generation capacity, compact size, reduced weight, non-dependency on power unit, and greater feedback [7.57].

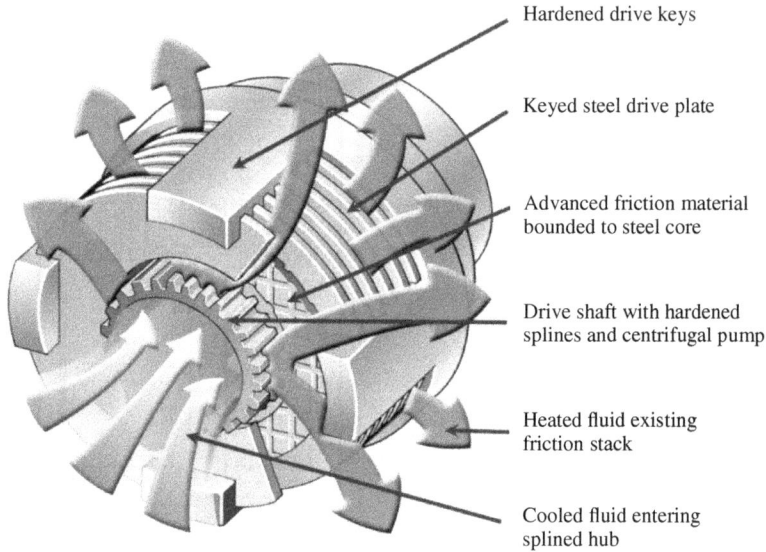

Hardened drive keys

Keyed steel drive plate

Advanced friction material
bounded to steel core

Drive shaft with hardened
splines and centrifugal pump

Heated fluid existing
friction stack

Cooled fluid entering
splined hub

FIGURE 7.27   The oil shear brake transmits torque through the molecules of specially formulated transmission fluid. The fluid acts as both the torque transmission medium and coolant for cooling the friction discs and drive plates during the dynamic braking phase. (Courtesy of Force Control Industries, Fairfield, Ohio.)

Hydraulic brakes are extensively used in the automobile industry. They can be implemented in both disc and drum forms. The disc-type brakes are more favorable due to their efficient heat dissipation.

7.3.3   Oil Shear Brake

Oil shear brake is a very special type of brake that is based on oil shear technology, which was invented by Force Control Industries in the middle of 1960s [7.58, 7.59], for solving three major problems regarding dry brake: friction material wear, actuating components wear, and heat transfer. With the technology, brakes transmit torque through the shearing of oil molecules among the brake friction discs and driving plates. Based on its working principle, this type of brake may be defined as *semi-contact* brakes.

Unlike traditional dry brakes, oil shear technology utilizes a thin layer of mineral oil between the brake disc and the drive plate to act as a medium for transmitting power. As the oil is compressed in a very narrow gap between the disc and plate, the fluid molecules shear each other to transmit torque from one disc to another. Meanwhile, heat generated from brake operation is carried away from the friction area by the continuous circulation of the oil (Figure 7.27). Then, the amount of heat carried by the internally circulated oil flow is further removed by using an external fan cooling or indirect water-cooling scheme. Since most of the work is done by the fluid particles in shear, wear is virtually minimized, leading to greatly reducing the need for brake adjustments. Consequently, the brake life is greatly extended in comparison with those of conventional friction brakes. In addition, this type of brake allows operating at high cycle rates with much less downtime.

These brakes accommodate severe duty applications, such as mining, steel, marine, power plants, lumber, agriculture, and more. They typically mount to the back of motor within totally enclosed housings, enabling them to work in harsh environments.

7.3.4   Regenerative Brake

Regenerative braking is based on the energy recovery mechanism to retard or stop an electric motor by converting kinetic energy into electric energy. When braking, the drive motor is turned to the power generation state to generate the braking torque to the driven load. During the process, the produced electricity can be fed back into the power supply for immediate use or stored for the future use. For this purpose, several energy storage technologies have been developed. For example, it can be stored chemically in batteries, electrically in capacitors, and mechanically in rotating flywheels, depending upon different applications.

As a type of electric brake, regenerative brakes can deliver a number of advantages over conventional friction brakes. The most remarkable advantage is that they can improve energy usage efficiency. It was reported that with regenerative brakes for electrical vehicles, it could improve the battery efficiency by 16%–25%, depending on the vehicle speed and the motor size [7.60]. However, the generated braking force from a regenerative brake decreases as the motor speed decreases. When the motor speed approaches zero, the braking force also approaches zero. This indicates that at low motor speeds a regenerative brake is unable to provide sufficient braking torque, *i.e.*, a regenerative brake can neither bring a motor to a complete stop nor hold a load at rest [7.61]. Therefore, for reliable braking, a friction brake or other types of brake are often needed to use along with a regenerative brake in applications. If the drive motor is also used as the regenerative brake and is connected to an inverter, the inverter can supply current into the windings to provide retarding torque at low or zero speed.

This type of brake is commonly found in hybrid vehicles and electric railways. This technique has also been successfully applied to elevators, where both a regular elevator brake and a regenerative brake work together to stop elevators safely and efficiently in regular operation.

7.4 BRAKE FAILURE

The development of reliable motor brake is one of the great achievements of the automation/motion control industry. However, due to the fundamental limitations of material lifetime, wear, deterioration, contamination, manufacturing defects, or damages in operations, a brake will eventually fail after a certain time of service. An unexpected brake failure might lead to a costly standstill in industry or in worst cases the loss of valuable human life. The brake failures are attributed to a broad variety of factors.

7.4.1 Overheating of Mating Friction Surfaces

When fraction discs and pads are subjected to overheating under exceeding load conditions, hot spots occur on the brake contacting surfaces that deteriorates the brake performance and possibly lead to brake failure. High temperature during braking may cause brake fade, premature wear, bearing failure, thermal cracks, and thermally excited vibration.

7.4.2 Excessive Wear on Friction Surfaces

After a long-time operation, fraction discs and pads may wear seriously. For brakes used in motor vehicles, discs are usually damaged in one of the four ways: scarring, cracking, warping, or excessive rusting.

7.4.3 Failure due to Corrosion

As brakes operate in an open environment, severe rust formation may occur on brake disc surfaces, leading to several possible consequences: (*a*) The coefficient of fraction reduces significantly, leading to poor brake performance. (*b*) It may result in overheating of brake discs and pads caused by rust spots. (*c*) It generates a higher level of noise. (*d*) Rust powders can contaminate optical encoders to degrade their performance. However, for small servomotors where brakes are integrated into motor housing, disc corrosion is no longer a problem.

Generally, mechanical components such as gears, shafts, or bearings must be made of materials that resist fatigue, wear, and corrosion. Precipitation hardened or austenitic stainless steels offer good protection against corrosion from saline or steam but may not have the required wear resistance for all components. Martensitic stainless steels, which have lower chromium content than the austenitic grades, have less corrosion resistance but increased strength and surface hardness, making them suitable for components in metal-to-metal contact such as gears and bearings [7.62]. Some newly designed alternative materials like polyetheretherketone or polyamide-imide can improve both corrosion and wear resistance, used for lightly loaded components. With excellent corrosion and wear resistance, cemented carbides or ceramics can also be considered. However, these materials are relatively brittle and sensitive to shock loads.

7.4.4 Runout of Friction Disc

When friction discs have run out with respect to their end surfaces due to improper installation, manufacturing errors, or other causes, brake pads are pressed unevenly against discs, resulting in reduced brake performance, unpleasant noise, and uneven disc/pad wear that may lead to premature failure.

7.4.5 Thermomechanical Fatigue

In material science, cyclic loading at moderate or high stresses can lead to the formation of fatigue crack and possibly ultimate failure of material. The fatigue cracking process can be made up of two stages: cracks initiation and crack propagation. In electric machines, brakes are exposed to repetitive thermomechanical load cycles due to simultaneous variations of temperature and mechanical strain during the operation of the machines. The stresses formed in the course of braking and subsequent cooling attain the plasticity limit of the brake materials in tension and compression. It is shown that, for locally heated and cooled discs, the cycles of thermomechanical loading and unloading are responsible for the growth of fatigue cracks [7.63]. Practically, thermomechanical fatigue is the main failure type of brake disc, apart from wear.

There are several common causes leading to thermomechanical fatigue failure of brake: (*a*) heavy braking loads that are required for high-speed, heavy-duty machines, resulting in high heating and cooling rates on brake components; (*b*) hot sports and high dynamic temperature gradients occurring on brake discs and pads, and (*c*) high-frequency alternating heating-cooling cycles acting on the brake system due to frequent braking. All can result in thermal fatigue cracks on the braking disc and pad surfaces.

Cycle rate describes how frequently a brake is used. In some heavy-duty load applications such as elevators, cranes, conveyors, grinders, and laundry washing machines, frequent stops and starts are commonly required, resulting in a high cycle rate of the brake.

7.5 BRAKE DESIGN AND SELECTION CONSIDERATIONS

The selection of brake is affected by a number of factors relating to a specific application. In practice, a selection of motor brakes can be a challenging task for motor engineers due to the wide range of options available. In order to choose the most appropriate unit, all features and performing characteristics of the brake system must be traded off and comprehensively analyzed.

When selecting brakes for an application, available options include mechanical, electrical, magnetic, electromagnetic, pneumatic, hydraulic, or self-actuated brakes. While electrical and electromagnetic brakes are easy to control and cycle, even to a couple of thousand cycles per minute, fluidic-actuated brakes (*e.g.*, pneumatic- and hydraulic-actuated brakes) have the advantage of running cooler with minimal power consumption. Friction brakes with discs, drums, and cone morphologies work well as emergency brakes because of their fail-safe holding in safety-critical systems [7.64].

7.5.1 Dynamic Stopping Brake or Holding Brake?

The main function of a dynamic stopping brake is to stop all of the rotating components of the motor and driven machine dynamically. After the completion of the stopping process, it is often required to hold the load firmly. Therefore, it is important to assess brake heat generation, power dissipation, and expected pad wear and service life. In comparison, holding brakes work under static conditions, with little wear and heat generation. The holding torque is usually 25% higher than the dynamic torque. The considerations are required to ensure pad/disc materials have enough coefficient of static friction, the brake design with an appropriate safety margin, and necessary measures to prevent contaminations on the brake surfaces.

7.5.2 AC or DC Brake?

Electrical brakes can operate on either AC or DC power. Generally, with DC brakes, the brake coils are energized via either an independent DC supply such as a battery or a rectifier that converts AC power to DC power. For AC brakes, the brake coils are fed with a three-phase or single-phase power supply. A general guidance of brake usage is summarized in Table 7.3 [7.32].

There are some important differences between AC and DC brakes. AC brakes are characterized by fast reaction time, low heat generation, but design complexity, in-rush current, more common coil failure, and chattering noise. By contrast, the advantages of DC brake include quiet operation, simple construction, no in-rush current, and either online or battery-powered operation. Furthermore, DC brakes are known for requiring less maintenance and fewer adjustments than AC brakes. However, they typically suffer from slow reaction time, high heat generation, and requiring the use of the rectifier for direct line voltage.

Therefore, for applications requiring a quicker engagement time or a lower operating temperature, AC brakes are highly preferred.

TABLE 7.3
Brake Usage Recommendation [7.32]

Application	DC Brake	AC Brake
Crone		
Hoist	√[a]	√
Long travel	√	
Cross travel	√	
Gear	√	
Hoist	√[a]	√
Elevator	√[a]	√
Guillotine	√[a]	√
Conveyor	√	
Machine tool	√	√
Large inertia	√	√
Woodworking	√	
Fast operation	√[a]	√
Accurate load positioning	√[a]	√
Deck watertight (IP56)	√	
Inverter operation	√	

Note:
[a] DC brake is switched on the DC circuit to power and control the brake.

7.5.3 Braking Torque

The braking torque is perhaps the most important factor in the selection of high-performance braking systems. As the first step, it is important to determine the braking torque required to stop a rotating system, where the torque is defined as the dynamic braking torque, or maintain the static position of the system, where the torque is defined as the static or holding braking torque. Generally, spring-applied brakes are dominant for the brake torque lower than 27,200 Nm due to the relatively low cost. Hydraulic and pneumatic brakes can easily achieve up to 67,800 Nm. These brakes can be either multiple-disc-style or caliper-disc-style brakes, tending to be better for stopping high inertia loads [7.35]. For elevator and hoist applications, electromagnetic brakes are often used for their operation reliability and safety. In fact, the operation stability and safety of an electric machine are determined by the braking torque that acts on the rotor of the machine.

7.5.4 Overall Inertia of System

As the measure of an object's resistance to change in motion, inertia is one of the governing parameters affecting the brake performance. Rotational inertia depends on both the mass of an object and how the mass is distributed relative to the axis of rotation. In order to determine the braking torque, it is necessary to know the total inertia of the servo system, component efficiency, and load torque acting on the system. The total system inertia is the sum of all the inertias of the individual rotating components and must be determined early in the sizing process. During the braking process, an additional braking torque is required to stop all the rotating parts of the servo system and the driven machine. This additional torque is defined as the inertia torque T_J, where $T_J = J_t \, (d\omega/dt)$. The total inertia of the system includes all moving parts, such as motor shaft, bearing (inner rings and balls), brake disc, gearbox, and rotating parts in the load machine. It is worth to note that each inertia component should be calculated and reflected back through the ration to the brake output shaft. Nevertheless, from the standpoint of brake performance, lower system inertia is always desirable for lower inertia torque, shorter stopping time, higher positioning accuracy, and longer brake lifetime.

7.5.5 Thermal Consideration in Brake Selection

Frictional heat generated during the braking process can have a detrimental effect on the performance of brake and other components of motors. Once the temperature of the brake disc and pad becomes very high, the brake torque can be significantly reduced, leading to brake fade, premature wear, thermal cracks, and disc thickness variation. Therefore, motor engineers must consider choosing the brakes that are made of heat-resistant materials. In addition, the brake materials need to absorb large heat rapidly and produce less heat during braking.

Thermal capacity of friction brake is referred to as the ability to absorb and transfer heat by conduction, convection, and radiation into the surrounding structure or environment during the braking process. In general, thermal capacity is proportional to the amount of heat generated and inversely proportional to the temperature change in the system, in the unit of J/°C. Heat generation due to friction in the sliding contact between the braking disc and the pad can strongly influence friction and wear characteristics of brake systems. Frequent cycling may cause heat buildup in some motor brakes. At the beginning of the braking action, the generated heat due to friction is very high. After reaching its peak value, the frictional heat generation reduces gradually due to the reduced relative sliding speed between the disc and the pad. At the end of the braking process, the heat generation is equal to zero.

7.5.6 Other Factors Affecting Brake Selection

Brake design involves a process of tradeoffs among design parameters. The selection of the right motor brake is governed by a number of major that must be balanced and analyzed based on industrial and national standards. In addition to the factors mentioned above, other key factors should be carefully considered in evaluating brake performance.

- The brake response time is defined as the time when the control signal is issued by a servo drive until the brake delivers its 80% of the rated dynamic torque. Power-off brakes are typically used as safety brakes that engage when power fails. In such a case, the braking response time is particularly important. PM brakes tend to have faster response times and thus are often the preferred choice in safety-critical applications.
- A brake with precise positioning capability is especially important for some applications. Typical examples include surgical robots and pick-and-place robots. Compared to spring-applied brake technology, PM brakes offer zero-backlash performance, thus suitable for precision motion control applications.
- Backlash in a friction brake is referred to as the amount of movement that takes place between the brake hub and friction disc. Zero backlash brakes can be required in high precision, high accuracy applications such as robots and machine tools. In a motor system, the brake backlash depends on the type of a brake hub and its mating shaft. Different hub/shaft designs exhibit different degrees of backlash. Generally, a D or *double-D* hub has the greatest amount of backlash. Furthermore, for the involute spline hub and shaft, while involute splines provide a better sliding fit and closer tolerances, backlash can normally be between 0.25° for small brakes (the brake disc diameter d is up to 38 mm) and 1° for large brakes (where 38 mm $< d \leq 127$ mm). There are several methods to design zero backlash brakes. A common way is to fix rigidly a hub to a flexible diaphragm spring. Thus, the friction disc assembly has the least backlash. Another way is to use a split-hub shaft collar to prevent backlash for smooth stopping and precision holding. Furthermore, multiple disc brakes have to move on a spline so that it is difficult to make them with zero backlashes.
- Brakes with a compact size and lightweight are often required for modern motion control systems where available space is quite limited. This is especially true for space and deep-diving applications such as spacecrafts, satellites, and deep unmanned submersibles.
- In industrial sectors, the cost is always a major factor influencing the decision-making process. Among all types of brakes, spring-engaged friction brakes are the cost-effective solution. They offer effective performance for a wide range of industrial applications, making them the go-to solution for many systems. PM friction brakes are next in terms of cost. That makes them less appealing from a budgetary standpoint, so they should be reserved for applications that demand the specific set of characteristics that they offer. Finally, spring-engaged tooth brakes are most expensive because of the precision manufacturing required for the tooth interface.
- Brake mounting configurations depend on the motor type and size. Generally, for a small motor, the brake disc (*i.e.*, brake rotor) is directly mounted on the motor shaft, eliminating misalignment problems between the brake and motor. For a large motor, the brake and motor are often separately foot mounted to a common base and their shaft are connected via flexible couplings.
- The brake lifetime mainly depends on the brake type. Non-contact types of brakes, such as eddy-current brakes, hysteresis brakes, magnetic particle brakes, and PM brakes, usually have a very long lifespan. For friction brakes, a wide set of factors can affect the brake lifetime, including motor duty-cycle, rotating speed, pad and disc materials, operating temperature, *etc.* Due to the nature of friction, wear occurs at the contact surfaces of pads and discs. In such a case, the brake life is limited by the wear level of the pads/discs. The solenoid shoe brakes have a limited lifetime because the moving parts (linkage and solenoid plunger) are prone to wear.
- For some applications that require high reliability and operating safety, double brake motors may be adopted. A double brake motor consists of a three-phase AC motor with two DC

brakes working independently of each other. Power is supplied to the two brakes through independent rectifiers. The rectifiers are equipped with an over-voltage protection device and EMC filter [7.65]. In an event of unintentional power failure, each brake secures the braking of the motor through a very quick and precise braking action. The braking system provides equal braking torque in both directions of rotation. In addition, the brake components and assembly are designed with special features to be noiseless during stops. This type of motor is particularly suitable for lifting applications such as hoists, elevators, and cranes.

- The maintenance of a motor brake is usually not as important as that of a vehicle brake system. Motor brakes, in general, need little or no maintenance throughout their working life. In fact, non-contact motor brakes are maintenance-free. For small servomotor systems, brakes and motors are often integrated into a motor package. In such cases, brake maintenance may be unnecessary. For a failed brake, the common way is to replace it directly with a new one. However, for large motors with separate brakes such as elevator motors regular inspections and routine maintenances are required. The preventive maintenance actions are aimed at reducing the probability of failure or damage to brakes under intensive use and elongating their service lifetime.

REFERENCES

7.1. Childs, P. R. N. 2021. *Mechanical Design*, 3rd edn. Elsevier Butterworth-Heinemann, Oxford.
7.2. Holmberg, K. and Erdemir, A. 2015. Global impact of friction on energy consumption, economy and environment. *FME Transactions* **43**(3): 181–185.
7.3. Nosonovsky, M. and Bhushan, B. 2007. Multiscale friction mechanisms and hierarchical surfaces in nano- and bio-tribology. *Materials Science and Engineering R: Reports* **58**(3–5): 162–193.
7.4. Arnous, J. J., Sutter, G., List, G., and Molinari, A. 2011. Friction experiments for dynamical coefficient measurement. *Advances in Tribology* **2011**: Article ID 613581.
7.5. Twiflex LLC. 2018. Industrial disc brakes. File number P-1648-TF. https://www.twiflex.com/-/media/ Files/Literature/Brand/twiflex-limited/Catalogs/P-1648-tf-ashx?la=en&hash=1A2A38BAF28709DF FE22D0D2B89FBB5570FFC75C.
7.6. Archard, J. F. 1953. Contact and rubbing of flat surfaces. *Journal of Applied Physics* **24**(8): 981–988.
7.7. Popov, V. 2019. Generalized Archard law of wear based on Rabinowicz criterion of wear particle formation. *Facta Universitatis - Series: Mechanical Engineering* **17**(1): 39–45.
7.8. Preston, F. W. 1927. The theory and design of plate glass polishing machines. *Journal of the Society Glass Technology* **11**: 214–256.
7.9. Phee, S. K. 1970. Wear equation for polymers sliding against metal surfaces. *Wear* **16**(6): 431–445.
7.10. Kuman, N., Singh, T., and Grewal, G. S. 2018. Tribo-performance evaluation of ecofriendly brake friction composite materials. *Proceedings of 2nd International Conference on Condensed Matter and Applied Physics* **1953**(1): 090083.
7.11. Blau, P. J. 2001. Compositions functions and testing brake materials and their additions. Technical Report 64. Oak Ridge National Laboratory, Oak Ridge, TN.
7.12. Österle, W. and Dmitriev, A. I. 2016. *Lubricants* 2016(4–5). doi:10.3390/lubricants4010005.
7.13. Talati, F. and Jalalifar, S. 2009. Analysis of heat conduction in a disk brake system. *Heat and Mass Transfer* **45**(8): 1047–1059.
7.14. Dragone, R. V. 2007. Holding loads with power-off brakes. *Machine Design*. September 27, 2007. http://machinedesign.com/archive/holding-loads-power-brakes.
7.15. Dragone, R. V. 2007. Making sure brakes don't break. *Machine Design*. October 25, 2007. http:// machinedesign.com/news/article/21829314/making-sure-brakes-dont-break.
7.16. Aleksendrić, D. and Carlone, P. 2015. Composite materials – modelling, prediction and optimization. In *Soft Computing in the Design and Manufacturing of Composite Materials: Applications to Brake Friction and Thermoset Matrix Composites* (Eds. D. Aleksendrić and P. Carlone, Woodhead Publishing, Cambridge, UK, pp. 61–289.
7.17. Miki Pulley. Clutches and brakes, pp. 250–369. http://www.mikipulley.co.jp/data/pdf/en/cb_of_ct.pdf.
7.18. Pieri, J. L. *How to Size and Apply Dynamic Spring-Set Brakes*. Thomson Industries, Inc. https://www. thomsonlinear.com/downloads/articles/How_to_size_and   _Apply_Dynamic_Spring_Set_Brakes_ taen.pdf.

7.19. Warner Electric. Electrically released spring-set brakes & unibrake AC motor brakes. https://www.warnerelectric.com/-/media/Files/Literature/Brand/warner-electric/catalogs/p-8589-we.ashx.

7.20. Dragone, R. V. 2018. Choosing the right brake for robotic applications. *Motion Control Magazine*, April Issue. https://www.techbriefs.com/component/content/article/tb/supplements/md/features/articles/28812.

7.21. Twiflex – Altra Industrial Motion. 2021. Industrial disc brakes. https://www.altraliterature.com/-/media/Files/Literature/Brand/twiflex-limited/Catalogs/p-1648-tf.ashx.

7.22. Gao, C. H., Huang, J. M., Lin, X. Z., and Tang, X. S. 2007. Stress analysis of thermal fatigue fracture of brake disks based on thermomechanical coupling. *ASME Journal of Tribology* **129**(3): 536–543.

7.23. Stearns Brake Catalog. Introduction to solenoid actuated brakes: Stearns brakes set the standard for excellence. https://www.stearnsbrakes.com/files/catalog-section-sab-brakes.

7.24. Grzes, P. 2019. Maximum temperature of the disc during repeated braking applications. *Advances in Mechanical Engineering* **11**(3): 1–13.

7.25. Yan, H. B., Mew, T., Lee, M.-G., Kang, K.-J., Liu, T. J., Kienhöfer, F. W., and Kim, T. 2015. Thermofluidic characteristics of a porous ventilated brake disk. *ASME Journal of Heat Transfer* **137**(2): 022601–022612.

7.26. Khan Z. 2016. FAQ: How to choose a safety factor so a motor design lasts? *Motion Control Tips*. https://www.motioncontroltips.com/faq-choose-safety-factor-motor-design-lasts/.

7.27. Torque Technologies. 2011. Industrial electromagnetic brake solution. http://www.emtorq.com/pdf/Trifold_EMTorq_Spring_Set_Brakes_2011.pdf.

7.28. Kernwein, J. D., Sutherland, D. W., Ruhland, K. M., and Oswald, J. A. 2016. Braking systems and methods of determining a safety factor for a braking model for a train. U.S. Patent 9,283,945 B1.

7.29. Dragone, R. V. Power off Brake backlash. Thomson Industries white paper. http://www.thomsonlinear.com.cn/downloads/articles/Power_Off_Brake_Backlash_taen.pdf.

7.30. Sepac Inc. Spring engaged brakes and backlash: Understanding safety/power-off/fail-safe brakes. https://sepac.com/wp-content/uploads/protected/SEPAC-Tech-Series_Brakes-Backlash.pdf.

7.31. Glišović, J., Demić, M., Miloradović, M., and Ćatić, D. 2011. System approach to solving brake NVH issues. *Proceedings of the 7th International Scientific Conference Research and Development of Mechanical Elements and Systems*, Zlatibor, Republic of Serbia, pp. 27–32.

7.32. Brook Crompton. 2010. Brake motors: Frames 63 to 355L. http://www.brookcrompton.com/upload/files/products/2010e_brakes_v1.pdf.

7.33. MGM Electric Motors. 2020. Brake motor general catalogue. http://www.mgmrestop.com/wp-content/uploads/2020/06/MGM-cat-EN-2020-web-JUNE.pdf.

7.34. Brown, J. 2000. A brake in the action. *Machine Design*. February 1, 2000. https://www.machinedesign.com/automation-iiot/article/21818476/a-brake-in-the-action.

7.35. Song, C.-W. and Lee, S.-Y. 2015. Design of a solenoid actuator with a magnetic plunger for miniaturized segment robots. *Applied Sciences* **2015**(5): 595–607.

7.36. Flemming, F. 2016. What to look for when choosing holding brakes? *Machine Design*. June 9, 2016. http://machinedesign.com/motion-control/what-look-when-choosing-holding-brakes.

7.37. Kossow, K. Østergaard, E. H. and Støy, K. 2013. Programmable robot and user interface. U.S. Patent 8,614,559.

7.38. Brennen, D. B., Mescher, P. A., Klode, H., and Bock, G. P. 2002. Solenoid based park brake method. U.S. Patent 6,435,320.

7.39. Ahmad, A., Novkovic, T., and Trangärd, A. 2017. A multiple disc brake for an industrial robot and an industrial robot including the multiple disc brake. World Intellectual Property Organization WO 2017/148499 A1.

7.40. Sodano, H. A. and Bae, J.-S. 2004. Eddy current damping in structures. *The Shock and Vibration Digest* **36**(6): 469–478.

7.41. Baharom, M. Z., Nuawi, M. Z., Priyandoko, G., Harris, S. M., and Siow, L. M. 2011. Eddy current braking study for brake disc of aluminum, copper and zinc. *Regional Engineering Postgraduate Conference (EPC)*, Banqi, Malaysia.

7.42. Smythe, W. R. 1942. On eddy currents in a rotating disk. *Transactions of the American Institute of Electrical Engineers* **61**(9): 681–684.

7.43. Schieber, D. 1974. Braking torque on rotating sheet in stationary magnetic field. *Proceedings of the Institution of Electrical Engineers* **121**(2): 117–122.

7.44. Wouterse, J. H. 1991. Critical torque and speed of eddy current brake with widely separated soft iron poles. *IEE Proceedings B: Electric Power Applications* **138**(4): 163–158.

7.45. Ming, Q. 1997. Sliding mode controller design for ABS system. Master thesis. Virginia Polytechnic Institute and State University.

7.46. Gasline, A. and Hayward, V. 2008. Eddy current brakes for haptic interfaces, design identification and control. *IEEE/ASME Transactions on Mechatronics* **13**(6): 669–677.

7.47. Zhuo, Q., Guo, X. X., Tan, G. F., Shen, X. M., Ye, Y. F., and Wang, Z. H. 2015. Parameter analysis on torque stabilization for the eddy current brake: A developed model, simulation, and sensitive analysis. *Mathematic Problems in Engineering* **2015**: Article ID 436721.

7.48. Bogdanowics, J. M., Michelson, M. G., and Michelson, M. G. 2012. The development of a new magnetic brake. *Design World*. https://www.designworldonline.com/the-development-of-a-new-magnetic-brake/.

7.49. Rabinow, J. 1948. The magnetic fluid clutch. *AIEE Transactions* **67**(12): 1308–1315.

7.50. Rabinow, J. 1951. Magnetic fluid torque and force transmitting device. U.S. Patent 2,575,360A.

7.51. Yu, L. Y., Ma, L. X., Song, J., and Liu, X. H. 2016. Magnetorheological and wedge mechanism-based brake-by-wire system with self-energizing and self-powered capability by brake energy harvesting. *IEEE/ASME Transactions on Mechatronics* **21**(5): 2568–2580.

7.52. Park, E. J., Stoikov, D., da Luz, L. D., and Suleman, A. 2006. A performance evaluation of an automotive magnetorheological brake design with a sliding mode controller. *Mechatronics* **16**(7): 405–416.

7.53. Shiao, Y. J. and Nguyen, Q.-A. 2013. Development of a multi-pole magnetorheological brake. *Smart Materials and Structures* **22**(2013): 065008.

7.54. Tsukahara, S., Torres, J., Neal, D., and Asada, H. H. 2012. Design method for buckling amplified piezoelectric actuator using flexure joint and its application to an energy efficient brake system. *Proceedings of ASME 2012 5th Annual Dynamic Systems and Control Conference (joint with JSME 2012 11th Motion and Vibration Conference)*. DSCC2012-MOVIC2012-8711, Fort Lauderdale, Florida, USA.

7.55. Burnett, D. H. 2013. Piezo-actuated braking system and method for a stepper motor. U.S. Patent 8,534,429 B2.

7.56. Westinghouse, G., Jr. 1873. Steam and air-brakes. U.S. Patent 144,006.

7.57. Milliken, W. F., Milliken, D. L., Kasprzak, E. M., and Metz, L. D. 2003. *Race Car Vehicle Dynamics - Problems, Answers and Experiments*. SAE International, Warrendale, PA.

7.58. Sommer, G. M. 1965. Clutch-brake unit. U.S. Patent 3,182,776.

7.59. Sommer, G. M. 1972. Clutch-brake unit. U.S. Patent 3,696,898.

7.60. Lakshmi, N. D., Kanwar, D. P., Sandhyaa, R. and Priya, S. L. 2017. Energy efficient electric vehicle using regenerative braking system. *International Journal of Advance Research, Ideas and Innovations in Technology* **3**(3): 55–58.

7.61. Zhang, W., Yang, J., Zhang, W. M., and Ma, F. 2019. Research on regenerative braking of pure electric mining dump truck. *World Electric Vehicle Journal* **10**(2): 1–17.

7.62. Culp. J. 2021. How to protect your powered surgical tools from moisture and corrosion. *Machine Design*. Article 21156302. https://www.machinedesign.com/medical-design/article/21156302/how-to-protect-your-powered-surgical-tools-from-moisture-and-corrosion.

7.63. Wu, S. C., Zhang, S. Q., and Xu, Z. W. 2016. Thermal crack growth-based fatigue life prediction due to braking for a high-speed railway brake disc. *International Journal of Fatigue* **87** (**2016**): 359–369.

7.64. Eitel, L. 2015. Motion systems application examples: Brakes, clutches and torque limiters. *Linear Motion Tips*. https://www.linearmotiontips.com/motion-systems-application-examples-brakes-clutches-and-torque-limiters-in-motion/.

7.65. MGM Electric Motors. 2016. BMBM series stage motors. http://www.mgmrestop.com/wp-content/uploads/2014/03/BMBM-depliant-inglese-2016.pdf.

8

Servo Feedback Devices and Motor Sensors

Evolution of motion control and automation systems has highlighted the need for modern sensors that provide higher precision, higher efficiency, higher sensitivity, lower cost, and faster response than the venerable solution. Sensor technology plays a variety of essential roles in a wide range of industrial automation, instrument, and motion control systems. Today, many kinds of sensors are used to determine various performance characteristics (*e.g.*, position, speed, force, torque, temperature, and vibration) of electric machines or provide information about products during the manufacturing process.

The primary objective of servo control systems is to not only precisely control the motor speed and position but also achieve the desired motor dynamic response, synchronization, and stability. In servo systems, feedback devices are commonly adopted with other control elements to automatically adjust the output to maintain operating stability, reduce variations, and minimize errors via motion controllers. Fundamentally, a servo system can perform no more accurately than the accuracy of the feedback device. In addition, errors in position or speed can be introduced into the system by the less-than-perfect mechanisms that transfer the motor power to the load [8.1]. In most high-performance servo systems, the position/speed errors are not acceptable.

To ensure optimum performance of servo systems, there are also other kinds of sensors used for continuously monitoring real-time operating parameters like temperature, pressure, force, and vibration. Real-time monitoring can discover potential risks over time and prevent system failure to avoid major personnel and economic losses.

Even though a large number of research articles have been published in the field of sensors, sensing technologies, and specialized applications, there are still ambiguities in the definition and classification of sensors. It is worth noting that *sensor* and *transducer* are two similar but easily confused terms. In order to clarify the confusion, sensors and transducers can be divided into *input* and *output* groups, depending upon the type of signal or process being *sensed* or *controlled*. Devices that perform an *input* function are commonly known as sensors since they *sense* physical quantities and convert them into digital or analog signals. As a comparison, devices that perform an *output* function are generally called actuators because they are used to actuate external devices.

Sensor is defined as a device that can detect or *sense* physical, chemical, or biological quantities (*e.g.*, temperature, force, pressure, voltage, pH, and odorant) and convert the measured data into electrical signals. Transducer is defined as a device that can convert energy from one form to another. According to this definition, typical examples of transducers include loudspeakers (converting electrical signals into acoustic energy), light-emitting diodes (LEDs) (converting electric energy into visible light), and heaters (converting electrical energy into thermal energy). Therefore, a transducer can be a *sensor* when it is used to detect or measure a certain physical quantity (*e.g.*, microphone) or an *actuator* when the electrical input is converted to, for example,

DOI: 10.1201/9781003097716-8

sound wave, force, and other physical forms, to control an external device (*e.g.*, loudspeaker). In such a sense, a sensor can be an integral part of a transducer but a transducer is not necessarily a sensor.

Sensors and transducers can be classified in a variety of ways, such as by sensing mechanisms, technologies, measurands, detection means, conversion phenomena, materials, or fields of application [8.2]. Different techniques can be employed to operate sensors, such as optical, electrical, piezoelectric, electromagnetic, and electrochemical. It is noted that different types of sensors can be used for the same parameter measurement; each type is based on its own measuring mechanism. For instance, the position of a rotary shaft can be measured by optical, capacitive, inductive, or electromagnetic sensors. In addition, one measuring mechanism enables multi-parameter measurements. As an example, strain gauges can measure stress, strain, displacement, force, load, torque, pressure, tension, vibration, *etc.*

Many types of feedback devices have been designed and developed to meet performance and environmental requirements for different motion control systems. The key step in selecting and designing optimal feedback devices is to understand their performing characteristics and operating conditions, for example, resolution, repeatability, accuracy, frequency, temperature stability, reliability, life expectancy, mounting constraints, cost, and operating environments. Today, most modern servo systems utilize encoders and resolvers as feedback devices that measure the rotary position of motor shaft by converting mechanical motion into electrical signals. A comparison of various types of motion feedback devices is presented in Table 8.1 [8.3–8.6]. Due to the significant differences between manufacturers' products, the data listed in this table are for reference only.

Encoders and resolvers are normally attached to motor shafts to convert the shaft angular position into electrical digital or analog signals (Figure 8.1). These signals are transmitted to the motion controller to control the movement of the motor. Recent advancements in motion control techniques have greatly improved the accuracy and reliability of feedback devices with reduced costs.

In a rotary system, three motion parameters, angular position θ (usually in units of rad), angular velocity ω (in units of rad/s), and angular acceleration a (in units of rad/s^2), can be mathematically related and expressed as

$$\omega = \frac{d\theta}{dt} \tag{8.1}$$

$$a = \frac{d\omega}{dt} = \frac{d^2\theta}{dt^2} \tag{8.2}$$

Resolvers were invented by MIT scientists in the 1940s. Originally, they were used in military applications to measure and control the angle of gun turrets on tanks and warships and were late adopted for industrial use because of their rugged design and ability to provide accurate measurements under severe conditions. Without solid-state electronics, resolvers can operate at much higher temperatures and electronic disturbance environments than encoders. Furthermore, compared to optical encoders, resolvers can withstand higher vibrations and shock loads.

The first optical encoder was developed in 1951 to address a need for higher resolutions [8.7]. In 1958, Baldwin Electronics released 18-bit optical encoders used in the Atlas missile guidance system [8.8]. Today, encoders have replaced resolvers in many applications for their high accuracy of measurement and easy integration into control systems as digital devices. Due to the lower rotational inertia than typical resolvers, encoders are suitable for applications with high acceleration and deceleration rates.

TABLE 8.1

Comparison of Various Types of Servomotor Feedback Devices

Feature	Feedback Device					
	Encoder					
	Optical Absolute	Optical Incremental	Magnetic Absolute	Magnetic Incremental	Resolver	Hall Effect Sensor
Output signal type	Digital	Digital	Digital	Digital	Analog	Digital (threshold device) or analog (linear device)
Wave	Square	Square	Square	Square	Sine/cosine	Square (threshold device) Trapezoidal (linear device)
Angular accuracy	Typical 36–360 arc-sec (0.01°–0.1°)	Up to 2.5 arc-min (0.04°)	20–60 arc-min (0.017°–1.0°)	20–60 arc-min (0.017°–1.0°)	2–50 arc-min (0.03°–0.83°)	20–60 arc-min (0.33°–1°)
Resolution	Up to 22 bit for single-turn[a]; Up to 12 bit for multi-turn	10,000–25,000 PPR	Up to 12 bit for single-turn; Up to 16 bit for multi-turn[b]	1,024–2,048 PPR[c]	10–16 bit[d] (1,024–4,096 PPR)	Typical 6–9 bit Up to 500 PPR
Frequency response	Up to 500 kHz but possible to 1,000 kHz	Up to 300 kHz	Up to 180 kHz	Up to 120 kHz	200 Hz–10 kHz[e]	Typically up to 100 kHz
Reliability	Good	Good	Good	Good	Excellent[f]	Acceptable
Electric noise	Cable length is limited by signal voltage drop and EMI disturbance	Cable length is limited by signal voltage drop and EMI disturbance	Cable length is limited by IR drop	Cable length is limited by IR drop	Cable length is limited by induced cable capacitance	Cable length is limited by induced signal noise and EMI disturbance

(Continued)

TABLE 8.1 (*Continued*)
Comparison of Various Types of Servomotor Feedback Devices

Feature	Feedback Device						
	Encoder				Resolver	Hall Effect Sensor
	Optical Absolute	Optical Incremental	Magnetic Absolute	Magnetic Incremental		
Operation temperature (°C)	−20°~+100°	−30°~+115°	−40°~+100°	−40°~+120°	−50°~+200°	−40°~+150°
Mechanical shock	≤100[g] G	≤50 G	≤200[h] G	≤200 G	≤200 G	>150 G
Vibration	≤20 G	≤20 G	≤20 G	≤18 G	≤40 G	>50 G

(a) PPR—Pulses per revolution.
(b) CPR—It may be explained as cycles per revolution or counts per revolution and thus, the confusion must be clarified before using it.
(c) LPR—Lines per revolution.
(d) Bandwidth can be limited by the rigidity of the coupling between the motor shaft and encoder shaft, as well as the natural frequency of the coupling.

Note:

a　Resolutions of absolute optical single-turn encoders can be made up to 22 bits [8.3]. The highest resolution may reach 32 bits.

b　Resolutions of magnetic encoders can be made up to 12 bits for single-turn encoders (absolute encoder) and up to 16 bits for multi-turn encoders (incremental encoder).

c　For typical magnetic incremental encoders, the resolution is in the region of 1,024–2,048 PPR. The maximum resolution can be as high as 25,000 PPR [8.4].

d　Theoretically, resolvers have infinite resolution, since every rotational position of the shaft is associated with a unique sine/cosine voltage. However, the device used to process the output signal (such as the resolver-to-digital converter) limits resolver accuracy (typically in the range of 3–15 arc-minutes) and rotational speed (typically ≤5,000 rpm) [8.5]. With the resolver-to-digital (RDC) device, the resolution for one mechanical revolution is the product of number of pole-pairs and the RDC resolution.

e　The frequency response of resolvers from input to output is similar to that of a transformer with a high leakage reactance. The low frequency is usually less than 100 Hz. The response is fairly flat from about 200 Hz to 10 kHz. A peak normally occurs between 10 and 100 kHz [8.6].

f　Resolvers have neither onboard electronics nor sensitive elements in their structure. As a result, they are much more reliable and can work optimally in extremely harsh environments.

g　Phased-array optical encoders, which use an array detector to capture the signal, can withstand up to 400 G shock and 20 G vibration.

h　Magnetic encoders (especially absolute magnetic encoders) are inherently shock and vibration resistant due to large gap, immunity to light impeding contaminant, and basic materials of construction that are rated to 135°C [8.4]. In fact, the effects of shock, stress, and vibration (below certain destructive limits) on most PM materials are minor.

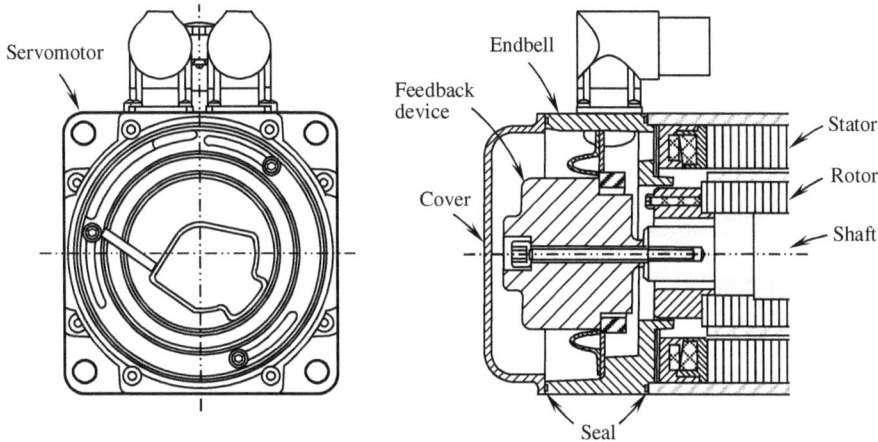

FIGURE 8.1 A feedback device is attached to the rotation shaft of a servomotor at the shaft end to measure the shaft position, rotating speed, and direction.

8.1 ENCODER

An encoder is a digital sensor that generates a coded reading of measurement to provide motion information like position, speed, and rotational direction. The main advantages of digital sensors over their analog counterparts include high accuracy, high resolution, ease of adaptation in digital control systems, effective signal transmission and data storage, capability to work under high noise environments, and free from observational errors.

According to their applications, encoders come in two mechanical configurations: rotary encoder that responds to rotational motion and linear encoder that responds to motion along a path. Some typical applications of rotary encoders include monitoring motor shaft position and velocity, tracking satellites with telescopes, and detecting robot arm movements. Linear encoders are used in linear motion systems, metrology instruments, precision jig borers, and high precision machine tools, ranging from digital calipers, CMM, CNC milling machines, to X-Y stages, gantry tables, and semiconductor steppers.

8.1.1 Type of Encoder

Encoders can be categorized in a number of ways, generally by their working principle, signal generation, method of signal interpretation, and so on. As an example, according to the working principle, encoders can be grouped into several categories such as optical, magnetic, capacitive, and inductive.

8.1.1.1 Optical Encoder

Optical encoders rely on optoelectronic components to detect rotary or linear motion. Optical encoders have long been recognized as indispensable displacement/position sensors due to their advantages of high resolution, high accuracy, fast dynamic response, lightweight, and long life (the lifetime is ultimately limited by the LED). Thus, they are particularly suited to high precision industrial automated systems, such as semiconductor fabrication. However, optical encoders often suffer from reduced reliability due to environmental contaminants. Dust, moisture, dirt, and other foreign debris can significantly degrade the accuracy of the measurements and lower the operation reliability.

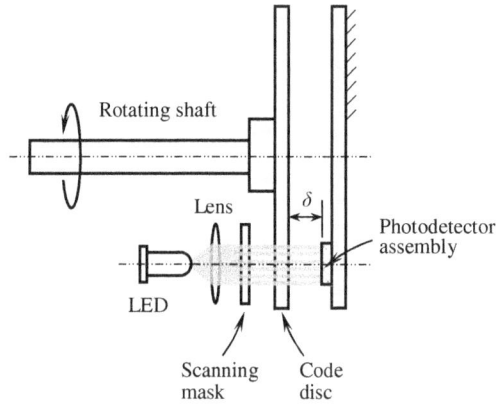

FIGURE 8.2 A transmissive optical encoder consists of an LED light source, lens, scanning mask, code disc, and photodetector assembly. The disc has alternating opaque and transparent segments, so it intermittently interrupts the LED's beam as it rotates. The photodetector responds to the series of light pulses and converts them into corresponding electric pulses. The gap between the code disc and photodetector can be as small as 0.5 mm.

An optical motor encoder is a motion detector that provides feedback signals to a closed-loop control system. Mounted on a motor shaft, the encoder converts the mechanical position of the shaft to digital or analog signals, used to determine the position and velocity of the shaft. A typical optical encoder design includes an emitter/detector module, which can be designed in the transmissive, reflective, or imaging configuration.

As shown in Figure 8.2, a transmissive optical encoder consists of a number of components: a light emitter such as LED, a rotating code disc, a stationary scanning mask, a set of photodetectors for sensing the presence or absence of light, a lens that focuses the light beam falling on the photodetectors, and signal processors. Among them, the most important component is the code disc that is made of an opaque material with a specially designed pattern. The code disc is mounted on the encoder shaft, rotating between the LED/mask and the photodetector assembly. The pattern of the code disc contains alternating opaque and transparent segments and spokes radially from its center (Figure 8.3). The code disc is illuminated by a light source (LED) positioned perpendicularly to the code disc. The photodetectors have the task to sense the variations of light resulted from the rotation of the code disc and convert them into corresponding digital or analog electrical signals.

The transmissive encoder locates the emitter and photodetectors at the opposite sides of the code disc. Light emitted from the light source is illuminated into parallel beams by means of a lens and passes through the transparent areas of the rotating code disc toward the photodetectors to produce electronic output signals from a photodiode array according to the pattern on the code disc. Then, the electronic signals are converted into several channels by the electronics board and fed into a motion controller to calculate the position and velocity of the motor.

An optical encoder employing reflective technology is also based on the photoelectric scanning of a code disc. Unlike the transmissive encoders, in a reflective encoder, the light emitting (e.g., emitter) and light receiving (e.g., photodetectors) elements are located at the same side of the code disc and often integrated into a single package. A beam of light is emitted from the emitter toward the code disc, reflected back from the opaque areas of the code, and received by the photodetectors (Figure 8.4). It is worth to note that only the reflective areas (i.e., the opaque areas) on the code disc can reflect light back and no light is reflected from the non-reflective areas (i.e., the transparent areas). The variations of light are thus detected by the photodetectors and transformed into corresponding electrical signals in the digital form. Due to its compact structure, it allows significantly

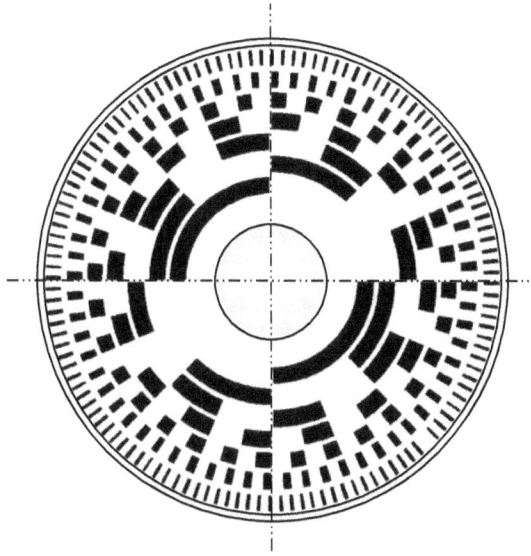

FIGURE 8.3 The pattern of a typical code disc in an absolute optical encoder with alternatively spaced opaque and transparent segments. The seven tracks are evenly distributed, radiating outward from the center of the disc. Each track has double the number of opaque-transparent segments than the inner track, giving the encoder resolution $360°/2^7 = 2.8125°$.

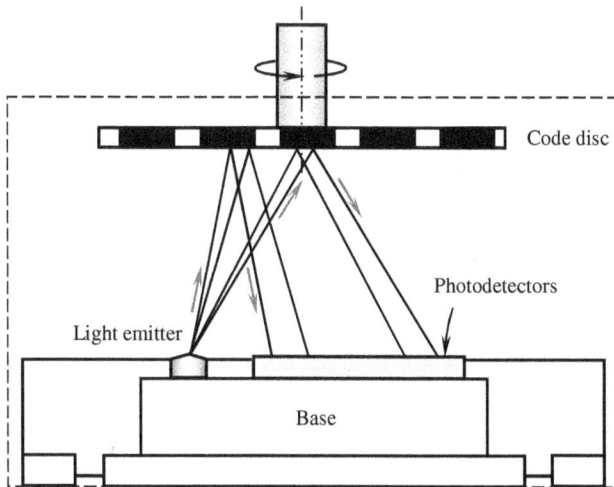

FIGURE 8.4 Working principle of the reflective optical encoder, where both the light emitter and photodetectors are placed at the same side of the code disc. A light beam is emitted from the emitter toward the code disc, reflected back from the opaque areas on the disc and received by photodetectors.

reducing the encoder size while maintaining its performance, making it especially suitable for some commercial and industrial applications where space is a primary concern (*e.g.*, robots, drones, portable medical devices, and spacecrafts). Moreover, this type of encoder features other advantages of relaxed alignment tolerances, wide airgap, low power consumption, and easy assembly as compared to transmissive and imaging encoders. However, reflective encoders may suffer from their low image contrast, low speed, and low resolution [8.9, 8.10].

Generally, optical encoders can better meet the requirements for temperature, humidity, shock, and vibration, but they cannot work under nuclear environmental conditions because of the sensitivity of semiconductor to radiation.

8.1.1.2 Magnetic Encoder

Beyond the optical encoders, there is an ever-expanding array of innovative techniques to provide reliable operation, small package size, and low cost. In addition to optical sensing techniques, magnetic sensing is increasingly popular for most applications.

As an attractive and viable alternative to optical encoders, magnetic encoders employ a signal detecting system to detect the variations of the magnetic field generated by permanent magnets (PMs). They are rugged and less susceptible to dust, oil, moisture, and condensation and more tolerant of shock and vibration loads. Compared to conventional optical encoders, magnetic encoders have a number of advantages, including greater durability, reliability, and compact package. This type of encoder can operate reliably under harsh environmental conditions. Therefore, they are highly suitable for applications that are subject to rugged environments where dirt, humidity, high temperature, high level of vibration, and shock load are present.

However, magnetic encoders are easily affected by external magnetic fields (*e.g.*, the unshielded magnetic interference created by motor) and unable to operate at very high temperatures due to the demagnetization of magnets. Although many improvements have been made to magnetic encoders, they are typically less accurate than optical and capacitive alternatives.

A magnetic encoder uses PMs placed on the edge of a wheel attached to the rotating shaft of an electric machine. It spins over an array of magneto-resistive sensors and causes predictable responses in the sensor. Based on the variations of the magnetic field, electrical signals are generated correspondingly. Generally, the more the magnetic poles and sensors are used, the higher the resolution of the encoder.

Magnetic encoders are robust, compact, and generally less expensive, making them particularly suitable for heavy-duty applications. Furthermore, because the magnetic technology is the absence of contact in the detection system, magnetic encoders require no maintenance and thus have a theoretically infinite durability.

A magnetic encoder consists of a rotor, a sensor, and a number of evenly spaced magnets around the circumference of the rotor (Figure 8.5). When the rotor rotates, the change in the magnetic

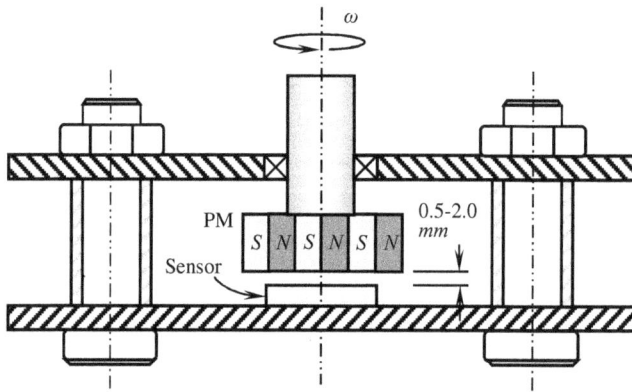

FIGURE 8.5 A magnetic encoder consists of a rotor and a sensor. A wheel, on which a number of PMs alternatively are attached on its edge, mounts at the end of a rotating shaft. When the rotor rotates, the sensor detects variations of the magnetic field generated by PMs.

field is detected by the sensor. In industrial applications, two primary types of sensors are commonly used: (*a*) Hall effect sensor, which detects the change in voltage by magnetic deflection of electrons; and (*b*) magnetoresistive sensor, which detects the change in resistance caused by the magnetic field.

Magnetic encoders, often used in rugged applications as steel, coal, textile, and paper mills, provide good resolution, high operating speed, and maximum resistance to dust, grease, moisture, and mechanical shock and vibration [8.11]. However, this type of encoder is subject to magnetic or radio interference.

8.1.1.3   Capacitive Encoder

Capacitive encoders adopt capacitive-sensing principles to detect changes in capacitance using high-frequency reference signals. A capacitive encoder is typically comprised of three main components: a stationary high-frequency transmitter, a rotating disc etched with a sinusoidal metal (*e.g.*, copper) pattern, and a stationary receiver board. As the rotor rotates, the sinusoidal pattern modulates the high-frequency electrical signal emitted by the transmitter in a predetermined way, causing the change in capacitance between the capacitively coupled transmitter and receiver. Then, the receiver detects the change in capacitance (*e.g.*, modulated signal) and the onboard electronics convert it into digital position information.

In industrial applications, a two-part configuration is also available, where the transmitter and receiver are combined as one part. As shown in Figure 8.6, a two-part capacitive encoder consists of a stationary transmitter/receiver and a rotor stamped with a sinusoidal pattern. The electronics is fabricated on the backside of the stator for producing high-frequency reference signals.

Capacitive encoders are expected to offer several advantages over optical and magnetic encoder designs. Due to the absence of an LED light source in the design, there is no concern regarding possible LED failure as with optical versions. This type of encoder is less susceptible

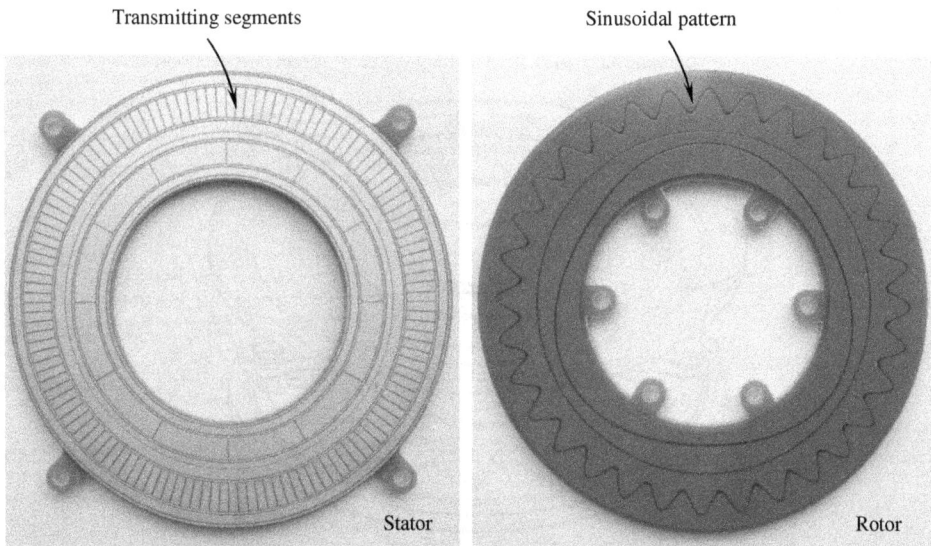

FIGURE 8.6   A two-plate capacitive encoder consists of a stationary transmitter (*a*) and a rotor (*b*). The PCB is fabricated on the backside of the stator (not shown). The high-frequency signal emitted by the transmitter is modulated by the sinusoidal pattern stamped on the rotating rotor. The concept of this type of encoder is to use planar electrodes excited by a high-frequency signal to measure capacitance.

to vibration and environmental contaminants such as dirt, dusts, and oil and consumes significantly less energy than comparable optical or magnetic encoders, which is especially beneficial in battery power supply applications. The digital nature of the design allows for increased flexibility through programmability of various features. In fact, capacitive encoders combine the accuracy of optical encoders with the durability of magnetic encoders. Moreover, the highly flexible encoder design allows for configuration into a wide range of communication protocols. Compared to electromagnetic resolvers, the capacitive encoders realize low cost, simple structure, and low power consumption.

Zheng *et al.* [8.12] have developed an innovative capacitive encoder for detecting the angular position and speed. As described in Figure 8.7, the capacitive encoder is composed of a disc-shaped stator and a rotor; both of them are made from standard printed circuit board (PCB) technology. The stator is fixed on housing with four screws and the rotor is connected to the motor shaft. The stator contains the transmitting segments, a pair of receiving electrodes and a pair of guarding electrodes on one side which faces the rotor, and the signal process circuit on another side. Opposite the stator, a pair of conductive reflecting electrodes, a pair of guarding electrodes, and a sinusoidal pattern are made on the rotor face. All electrodes are electrically isolated from each other. The test results show that the precision of the encoder is less than 0.006° and the resolution is 0.002°. The dynamic nonlinearity is evaluated at ±0.4° when the rotor is rotating at 1,000 rpm.

However, the primary concern when using capacitive encoders is that they are sensitive to changes in temperature and humidity. The effect of condensation or buildup of electrostatic charge can lead to a large deviation in position measurements. Moreover, capacitive encoders are prone to mounting misalignment. In order for them to perform as intended, mounting the encoder according to the tight mounting tolerances can ensure proper operation and allow the encoder to remain in service longer. In addition, since the airgap in capacitive encoder is very small, it requires a special care in encoder installation. Nevertheless, capacitance-based encoders today are used in a wide variety of industrial and automotive applications.

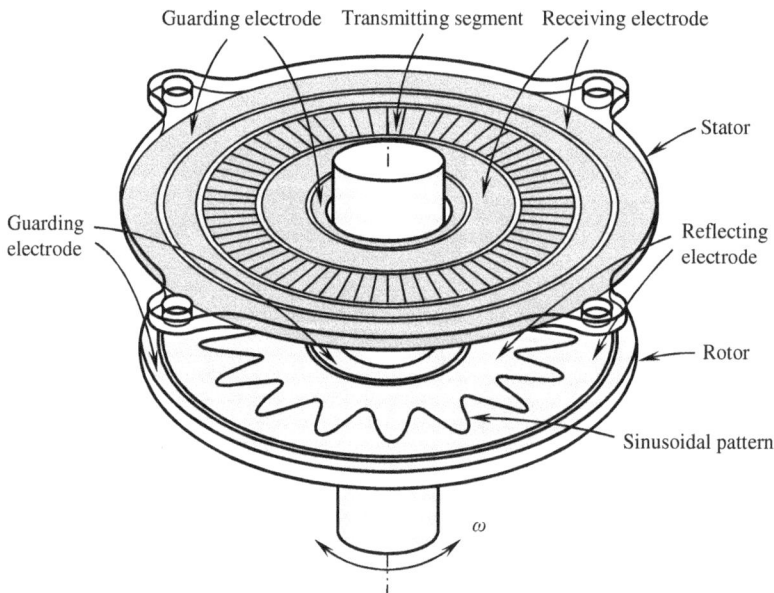

FIGURE 8.7 An innovative two-part capacitive encoder based on the quadrature modulation and demodulation.

8.1.1.4 Inductive Encoder

An inductive encoder, also known as *incoder*, can be viewed as a hybrid of an optical encoder and a resolver. Inductive encoders are based on the same electromagnetic induction principle as resolvers. Because inductive encoders employ PCB traces rather than coil windings in resolvers, they bring a number of benefits, such as superior accuracy along with extraordinary environmental ruggedness, reduced size, weight and cost, good operation stability, and elimination of errors from the winding process.

An inductive encoder typically consists of two main parts: a stator and a rotor; each shaped as a flat ring. All necessary electronics are integrated into the stator for receiving power and generating an output signal. Unlike capacitive encoders, inductive encoders are generally unaffected by the ingress of foreign matter and are less susceptible to the variation in temperature and humidity.

Inductive encoders offer considerable benefits over optical or magnetic encoders. They can operate reliably under harsh environmental conditions. These encoders are not affected by magnetic fields and a high level of vibration of the positioning element does not influence the output signal. It has been reported that the inductive encoder can run at up to 12,000 rpm, and take 100 G of shock and 10 G of vibration [8.13]. In practice, inductive rotary encoders are suitable for applications where accuracy requirements are relatively low, generally in the range of 10 arc minutes or higher.

The ultimate goal in selecting a suitable encoder for a specific application is to find the most cost-effective solution that meets the requirements of precision, resolution, accuracy, package size, and environmental ruggedness. A brief comparison of several types of encoders is given in Table 8.2 [8.14, 8.15].

8.1.2 ABSOLUTE AND INCREMENTAL ENCODERS

Depending on the signal interpretation method, encoders can be broken down into two main groups: absolute encoder and incremental encode. An absolute encoder provides the actual physical position within one revolution and maintains the record of the position within some absolute coordinate system, whereas an incremental provides position change information from a pre-defined home position. As a result, an incremental encoder requires additional electronics to count pulse and convert the data into speed or motion [8.16]. Absolution encoders are usually bigger and more expensive than incremental devices.

The advantage of absolute encoders is that they can provide absolute position values as soon as they are switched on by scanning the position of a coded element. Incremental encoders have the disadvantage that the position information is lost if a power supply is interrupted or removed. To regain machine location information with an incremental encoder after a power outage, it usually requires a homing routine.

8.1.2.1 Absolute Encoder

As the name implies, absolute rotary encoders have the capability to provide the actual shaft position, store the position data, and prevent loss of information in case of restart of the system or power loss. This makes them ideal for industrial motion control applications, such as robots, CNC machines, and precision automation equipment. An absolute encoder typically uses a unique binary pattern that does not repeat itself within the revolution, giving the encoder its absolute attributes. The absolute encoder has multiple concentric tracks of opaque-transparent segments on the code disc; each track has an independent light source. These tracks are evenly distributed, radiating outward from the center of the disc. As they go outward, each track has double the number of opaque-transparent segments than the inner one. In fact, the number of tracks determines the resolution of the encoder.

TABLE 8.2

Technology Comparison among Various Types of Encoder

Encoder Key Feature	Encoder Type			
	Optical	Magnetic[a]	Capacitive	Inductive[b]
Resolution[c]	Very High	Medium	High	High
Resistance to dirt, dust, oil, humidity, or condensation	Low	Very high	Low[d]	High
Resilience to electrostatic effect	Yes	Yes	No	Yes
Accuracy	High	Low	High	High
Temperature range	Medium	Narrow	Wide	Wide
Current consumption	High (50–100 mA)[e]	Very high (60–160 mA)[f]	Low (6–18 mA)	Low (≥ 6 mA)
Programmability	No[g]	No[g]	Yes	Yes
Package size	Medium	Very small	Small	Small[h]
Electromagnetic compatibility (EMC) immunity	High	High	High	High
Magnetic field immunity	High	Low	High	High
Vibration and shock resistance	Low	High	Medium	High[i]
Resolution range	Wide	Narrow	Wide	Wide
Cost	High	Medium	Medium	Low

Notes:

[a] Compared with optical encoders, magnetic encoders have a large airgap (up to 4 mm) without affecting signal accuracy.

[b] Inductive encoders lead with environmental ruggedness.

[c] Except for special cases, the resolutions of various motor feedback devices are briefly arranged in descending order: optical encoders, capacitive encoders/inductive encoders/resolvers, magnetic encoders, and Hall effect sensors. Because each type of feedback device has a wide range of resolution, there are overlaps among capacitive, inductive, magnetic encoders, and resolvers.

[d] Capacitive encoders are susceptible to condensation and electrostatic built up. In fact, capacitance varies with temperature, humidity, surrounding materials, and foreign matter, which makes engineering stable and high accuracy position encoder challenging.

[e] The current consumption of some optical encoders can go as high as 150 mA.

[f] While capacitive encoders draw much less current to comparable optical encoders, magnetic encoders can consume much more (in some cases almost twice) power than those of optical encoders.

[g] A new generation of programmable encoders has become more common recently, providing a lot more flexibility for users. The latest programmable encoders offer both incremental and absolute functionality on a common hardware platform and can be programmed by any Wi Fi-enabled electronic device.

[h] Some inductive encoders may require relatively large diameters.

[i] Inductive encoders may withstand vibration of 20–30 G and shock of 60–100 G.

The outputs of each track make up a unique binary signal for a particular shaft position. This type of encoding enables maintaining position information even when the power goes out.

The resolution of an absolute encoder can be determined as

$$\Delta\theta = \frac{360°}{2^n} \tag{8.3}$$

where $\Delta\theta$ is the resolution of the encoder in degrees (°) and n is the number of tracks (*i.e.*, n bit— where bit is a binary unit) on the code disc. The increase in the number of bits can lead to an increase in the encoder resolution. As a result, the resolution of an absolute encoder is often measured in

TABLE 8.3

Relationship among the Encoder Resolution in Bits, Measuring Discrete Position, and Smallest Increment in Angular Degree

Encoder Resolution (n bit)	Measuring Discrete Position (2^n)	Degree (°)	Encoder Resolution (n bits)	Measuring Discrete Position (2^n)	Degree (°)
8	256	1.406°	16	65,536	0.0055°
9	512	0.703°	17	131,072	0.0027°
10	1,024	0.352°	18	262,144	0.0014°
11	2,048	0.176°	19	524,288	0.00069°
12	4,096	0.088°	20	1,048,576	0.00034°
13	8,192	0.044°	21	2,097,152	0.00017°
14	16,384	0.022°	22	4,194,304	0.000086°
15	32,768	0.011°	23	8,388,608	0.000043°

terms of bits n, with the encoder measuring 2^n discrete positions per revolution (Table 8.3). A standard absolute encoder has 12 bits, which has an angular resolution of 0.08789° or 4,096 (2^{12}) pulses per revolution (*PPR*).

Resolution can also be expressed as arc-minute or arc-second, which is often used for rotational plane measurements. Arc-minute is defined as 1 degree=60 minutes. Arc-second is the smallest resolution unit with 1 degree=3,600 arc-second.

Unlike resolution, encoder accuracy depends on the whole system's interactions with the application and is traceable to the encoding disc and the deviation between actual and theoretical position [8.17].

Absolute rotary encoders can be further distinguished by whether they are single-turn or multi-turn. A single-turn encoder uses one code disc to provide a measurement range of one revolution. When the encoder shaft rotates more than one revolution, the measured position values are repeated for each revolution. This indicates that single-turn version is suitable for applications where the full range of positions does not exceed one revolution. However, in applications where the encoder makes more than one revolution, the single-turn encoder is unable to know how many revolutions it has completed. For these circumstances, a multi-turn encoder, which has multiple code discs that are appropriately geared together, is required to deliver the position within one revolution as well as the number of total revolutions of the encoder shaft. Multi-turn encoders are able to track the absolute position because the digital position value is not repeated until the maximum number (typically up to 4,096) of rotations is reached. The resolution of this type of encoder is the sum of the output of each disc. Taking the simplest version of multi-turn encoder as an example, it contains a primary disc for monitoring shaft rotation at 360° and a secondary disc for monitoring full rotations of the primary code disc. The two discs are connected by a complex gearing system. If the primary disc gives the standard 12-bit output and the secondary disc gives 8-bit output, the total encoder resolution becomes 20 bits, *i.e.*, $360°/2^{20}=0.00034°$, or 1,048,576 PPR.

Multi-turn encoders are suitable for servo systems, especially when the coordinated axes are offset. With a multi-turn encoder, the offset position can be programmed into the logic of the machine.

8.1.2.2 Incremental Encoder

The most common feedback devices used in servomotor control systems today are rotary incremental encoders because they can offer a standard interface on virtually any servo drive. This type of encoder produces electrical pulses (usually as a square wave) or increments with rotary motion for converting directly shaft position angles into digital signals. However, when power is interrupted, incremental encoders lose their position reference and must start over via a re-homing sequence to a reference point. Thus, incremental encoders are typically used for applications that do not require absolute positions.

It is noted that the code disc in an incremental optical encoder requires only one primary track with evenly spaced transparent and opaque sectors. Thus, as the light emitted from the light source passes through the code disc and detected by a photodetector, a train of equally spaced pulses is produced. The resolution of an incremental optical encoder is often measured in *PPR*. For a particular system and encoder, the encoder resolution *PPR* can be calculated by the formula [8.18]:

$$PPR = \frac{60 f_r}{n_{max}} \tag{8.4}$$

where

f_r is the maximum frequency response of encoder electronics (in Hz), which limits how rapidly it can generate output pulses. In another word, the frequency response places a practical upper bound on the resolution that can actually be achieved.

n_{max} is the maximum rotating speed in units of rpm

Incremental encoders may use one or two channels to generate square wave pulses. If there is only one channel, the encoder is limited to detect position only. Therefore, most incremental encoders have quadrature output channels, typically referred to as channel *A* and channel *B*. Each channel produces a specific number of equally spaced pulses per revolution. This allows the encoder to detect the direction of motion by assessing which channel leads another. For example, if channel *A* leads channel *B*, the encoder rotates clockwise. Otherwise, it rotates counter-clockwise. As shown in Figure 8.8, the pulses generated from the two channels are 90° (one-quarter of a cycle) out of phase and are known as quadrature signals. Often a third output channel, called *Index* (also known as *Z channel* in some manufacturers' manuals), yields one pulse per revolution, which is useful in counting full revolutions. Therefore, this type of encoder is suitable for bidirectional position sensing applications. However, they must be re-zeroed or reset every time when power is lost.

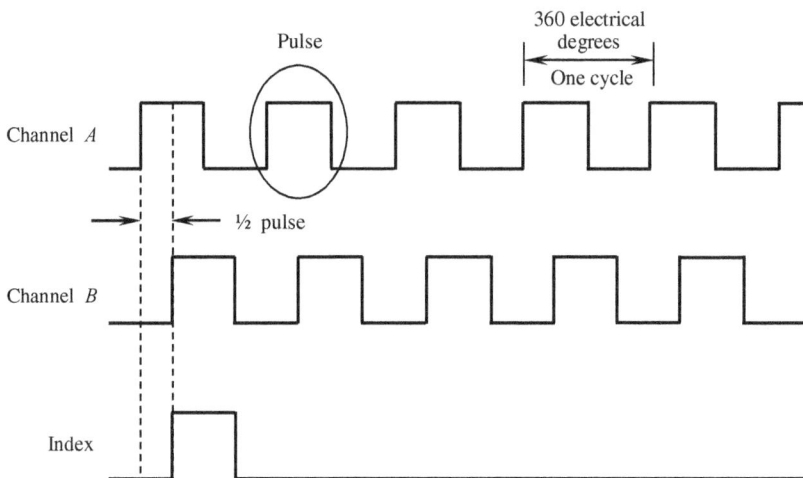

FIGURE 8.8 An incremental encoder uses quadrant output channels, channel *A* and channel *B*, producing pulses that are $1/2$ pulse out of phase. Because channel *A* leads channel *B*, the encoder rotates clockwise. A third output channel, called *Index* (or Channel *Z*), yields one pulse per revolution. It also shows that the resolution expressions *pulses per revolution* (PPR) and *cycles per revolution* (CPR) are identical.

It is noted that the resolution of the absolute encoder depends upon the number of tracks on the code disc. Once they are manufactured, the resolution is unchangeable. However, for incremental encoders, resolution can increase through signal decoding. By counting both leading and trailing edges of one signal (*e.g.*, signal *A*) or two signals (signals *A* and *B*), resolution can be multiplied by a factor of 2 or 4, respectively [8.19].

8.1.3 RESOLUTION OF ENCODER

Encoder resolution represents the number of mechanical degrees the encoder turns between each pole of square wave. It defines the smallest position increment that can be measured and is typically expressed in several ways, such as pulses per revolution (*PPR*), counts per revolution (*CPR*), or lines per revolution (*LPR*). While these terms are commonly used to specific resolution for encoders, some manufacturers use terms like cycles per revolution (*CPR*) or periods per revolution (*PPR*) and other phrases. Obviously, these can easily cause some confusion. Nevertheless, each expresses the encoder's granularity, although there are subtle differences between them [8.20]. As to which resolution expression is adopted, it usually depends on encoder manufacturers.

When using different expressions of resolution, it is important to be cautions on *CPR* acronym, because it can be explained as either *counts per revolution* or *cycles per revolution*, depending on different manufacturers. Actually, *cycles per revolution* (*CPR*) is equivalent to *pulses per revolution* (*PPR*) but *counts per revolution* (*CPR*) could be different from *pulses per revolution* (*PPR*). Taking an incremental encoder as an example, if the incremental encoder counts each signal rising and falling edge of both channels *A* and *B* to get 4 counts per cycle, *i.e.*, the resolution is multiplied by a factor of 4 (Figure 8.9). This is called *resolution multiplication*. Notice that in this case the resolution of the disk did not change. The resolution multiplication is achieved only by decoding the output waveforms in different ways.

The precision of an encoder is much higher than that of a resolver. Among all types of feedback devices, *sine* encoders generally offer the best performance in terms of resolution and accuracy (typically ±25 arc-second), particularly when compared with a resolver at ±2–10 arc-minute.

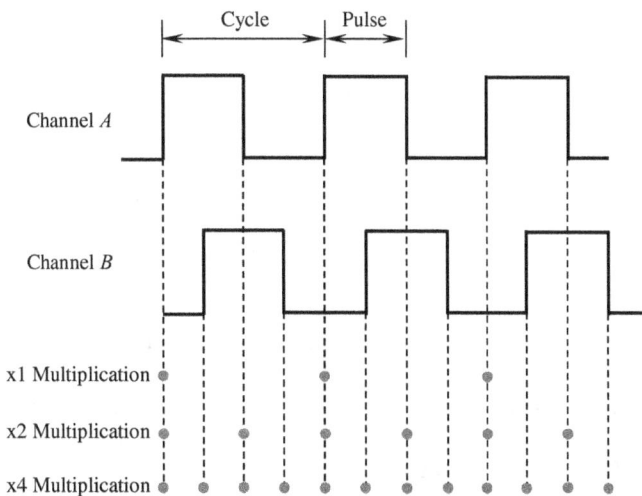

FIGURE 8.9 For incremental encoders, resolution multiplication can be achieved by decoding the output waveforms in different ways: (*a*) x1 – count each rising edge of pulse of Channel *A* as the disc rotates, getting 1 pulse per cycle. (*b*) x2 – count each rising and falling edge of Channel *A*, getting 2 pulses per cycle. (*c*) x4 – count each rising and falling edge of both Channel *A* and Channel *B*, getting 4 pulses per cycle.

8.1.4 ROTARY AND LINEAR ENCODER

While rotary encoders that respond to rotational motion currently dominate in the motion control and automation industries, linear encoders that respond to motion along linear or curved paths are found suitable in some applications. Though the operating principles of both encoders are very similar, there are some differences between them.

Most commonly, a rotary version of encoder is attached to a motor shaft that drives an external machine. Alternatively, a linear encoder is usually mounted on the moving component of the system for motion tracking. Instead of the code disc in a rotary encoder, a linear encoder employs a *codestrip* that has a series of black and white tracks on it.

As discussed previously, for a rotary application, resolution is the total number of steps representing one revolution of motion, often measured as the number of pulses per revolution (*PPR*) for rotary encoders or cycles per revolution (*CPR*) commonly for incremental encoders, where one cycle is equal to 360° electrical degrees. The final output resolution of linear encoder is determined by encoder count density (in terms of line per inch, *LPI*). Converting a rotary encoder resolution to a linear resolution using code disc size (which is often expressed as optical radius R_o) is expressed as

$$LPI = CPR\frac{\pi R_o}{2} \tag{8.5}$$

All that has been addressed for incremental rotary encoders can apply to incremental linear encoders. Linear encoders use a liner strip that is equivalent to a circular disc, which is cut along a radius and straightened out. Its resolution designates pulses per unit distance, pulses per millimeter (*PPM*), pulses per inch (*PPI*), and cycles per inch (*CPI*). *PPM* is calculated as

$$PPM = \frac{60f_r}{1,000\omega} = \frac{0.06f_r}{\omega_{max}} \tag{8.6}$$

where
  f_r is the maximum frequency response, Hz
  ω_{max} is the maximum linear speed, m/s

8.1.5 ENCODER MOUNTING

There are two standard rotary encoder mounting options: coupling mounting and direct mounting. Coupling mounting offers a reliable solution to servo systems. In addition to a flexible coupling, it requires interface components such as a bracket or adapter. Direct mounting eliminates the need for mounting bracket and flexible shaft coupling, does not require motor shaft alignment with respect to the encoder, and allows for mounting at various radii from the center of the motor shaft with slotted tethers. Encoder inserts and spring tethers help make direct mounting more robust than coupling mounting [8.21].

The optimal mounting approach depends upon the encoder type, application, and mechanical structure of electric machines. For a shafted encoder, its shaft is connected to a motor shaft through a flexible coupling. This approach provides some mechanical benefits: (*a*) The flexible coupling isolates the encoder from vibration and shock and movement in the motor shaft. (*b*) It offers compensation for the shaft misalignment. (*c*) It compensates for the radial and axial shaft movements, as well as installation tolerances without significantly affecting the accuracy of hollow shaft encoders. However, using the coupling mounting approach increases the overall length and cost of the motor system. In addition, connecting shafts with the coupling requires careful alignment between the encoder and motor shafts, leading to a long installation time and complexity. The direct mounting approach is suitable for hollow- or hub-shaft encoders. The hollow bores in hollow shaft encoders or the encoder caps in hub-shaft encoders allow motor shafts insert into the bores or caps and then

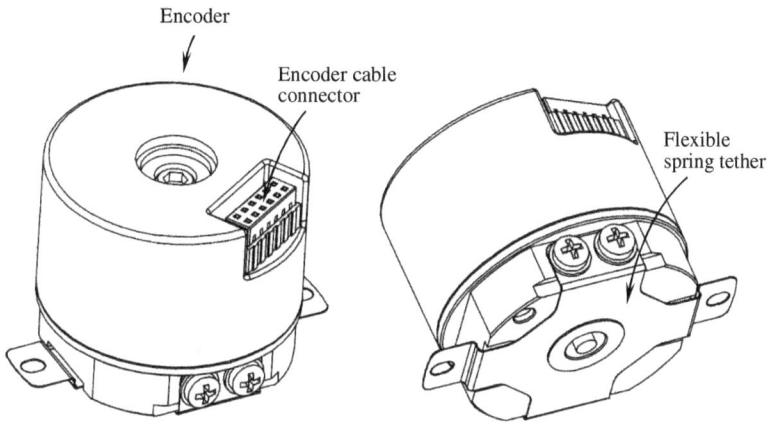

FIGURE 8.10 For a hollow-shaft encoder, direct mounting is adopted. A flexible spring tether is attached to the motor face to help minimize the influence of vibration and shock load on the encoder performance and prevent the encoder body from rotation with the motor shaft.

the shafts are fixed with clamps. In this mounting configuration, a flexible spring tether is attached to the motor face to help absorb vibration and shock and prevent the encoder body from rotating with the motor shaft (Figure 8.10). Consequently, the use of floating shaft mount and spring tether minimizes bearing loads for reducing wear and maintenance. Flex springs need to be analyzed for fatigue cycles of springs from radial and axial displacements.

8.2 RESOLVER

Resolvers are inductive feedback devices that work on the principle of a rotating transformer. Resolvers, like small induction motors, have windings on the rotor and stator. When a rotor shaft rotates relatively to an electrically excited stator winding, the output voltage across the resolver output winding varies sinusoidally, converting mechanical motion into analog signals. Unlike encoders that produce square waves, resolvers produce a set of *sine/cosine* waves (analog voltage). These analog signals can be modulated by a resolver-to-digital converter (RDC) to produce position information in digital form.

Resolvers come in a wide range of choices for features, fundamental mechanisms, types, winding topologies, sizes, and structural configurations (Figure 8.11). As an analog device, it outputs rotational angles as two-phase AC voltages. A single-speed resolver usually consists of a rotating winding as the primary windings and two stationary orthogonal windings as the secondary windings. By varying the degree of coupling between the primary and secondary windings, an analog signal can be generated, reflecting the annular displacement of the rotor. This coupling variation is accomplished by rotating the resolver rotating winding. With the simple and robust design, resolvers feature high environmental and shock resistance compared with other feedback devices such as optical encoders and are often preferred in military applications.

One of the technical specifications in resolver is its number of speed p, which refers to the number of electrical cycles per mechanical revolution. The number of resolver speed depends on the number of resolver pole-pairs (usually 1–3). A single-speed resolver features two magnetic poles (*i.e.*, one pole-pair). For this type of resolver, the electrical cycle is identical to its mechanical cycle. A multipole resolver is wound to produce multiple magnetic poles, which create multiple *sine/cosine* cycles in a mechanical revolution. For instance, while a single-speed resolver ($p=1$) produces one *sine/cosine* wave cycle in a mechanical revolution, a 3-speed resolver ($p=3$) produces one *sine/*

FIGURE 8.11 Resolvers, as a special type of rotary feedback device, come in a wide range of choices for performance characteristics and structural features and are suited to applications in harsh environments. Resolvers are designed as frameless or housed style to meet customers' requirements. (Courtesy of API Delevan, East Aurora, New York.)

cosine wave cycle for each 120° of mechanical revolution ($360°/p = 120°$). As a result, this resolver provides higher accuracy over that of a single-speed resolver.

While it is beneficial in accuracy by increasing the speed number, the p value is limited to the resolver size (*e.g.*, the resolve diameter). In addition, the increase in p can significantly increase the complexity of the device and result in an increase in cost, as well as possibly the decrease in operational reliability.

Resolver systems typically mount the resolver power supply and resolver interface board near the drive electronics, allowing the resolver to withstand higher temperature environments. Recent advances in technology have enabled the integration of a resolver and on-board electronics into a single housing.

Both brush and brushless resolvers are available in industrial applications. Like brush motors, electrical connections in a brush resolver rely on brushes and slip rings. Because of brush wear, it is very difficult to maintain its operational reliability, especially at high speeds. To solve the problem, brushless resolvers have been developed that employ the inductive technology to couple excitation energy to the rotor and essentially require no maintenance. In this design, energy is supplied to the primary winding in the rotor through a rotary transformer. Because there are no contact parts, brushless resolvers can withstand high vibration and shock loads and are insensitive to contaminations. Therefore, they have a much longer lifetime than the brush type.

Resolvers are well suited to harsh environments such as high temperature, high humidity, vibration, and shock load. Because resolvers are analog devices, their high voltage range makes them less susceptible to noise. However, as a type of magnetic field-based sensor, when they are used in applications of electromagnetic actuator control, the sensors can be affected by the magnetic field generated by the actuator. Hence, resolvers require magnetic shields in order to work correctly [8.22]. Furthermore, since some resolvers are little slow to respond to dynamic changes, one must be careful when using them to close a high-speed positioning control loop.

8.2.1 Type of Resolver

Three basic types of resolvers have been originally developed for various automation and servo systems: (*a*) Amplitude-modulated *resolver transmitter* (*RX*) converts mechanical position into an electrical signal. (*b*) Phase-modulated *resolver transformer* (*RT*) translates electrical signals into angular position in conjunction with a servo amplifier and electromechanical drive. (*c*) Amplitude-modulated *resolver differential* (*RD*) provides a system correction by acting as a variable electrical coupling between *RT* and *RX*. The three types of devices can be used as individual machines or can be integrated together as a system.

Today, resolver transmitters (*RX*) are extensively used in motion control systems for monitoring shaft annular position, speed, and rotational direction of electric machines. Like a small induction motor, an *RX* typically contains a wound rotor and stator, as shown in Figure 8.12. A primary winding and a reference winding are mounted on the resolver rotor, forming a rotary transformer. The reference winding is normally shorted internally to improve the accuracy. Two secondary windings (*sine* and *cosine* windings) are mounted on the stator and are mechanically displaced by 90° from each other. The primary winding is excited by an AC voltage, referred to as the excitation voltage V_e,

$$V_e = V_o \sin(\omega t) \tag{8.7}$$

where V_o is the amplitude of the excitation voltage and $\omega = 2\pi f$ is the angular frequency (*i.e.*, the resolver driving frequency). As the primary winding is energized, it generates a magnetic field that induces a voltage in two secondary windings based on the principle of Faraday's law of induction.

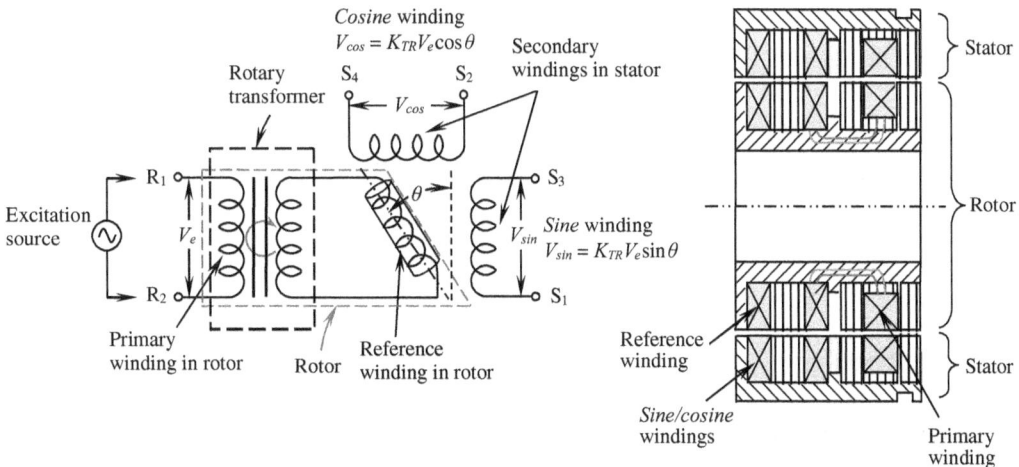

FIGURE 8.12 In a brushless resolver transmitter (*RX*), the primary winding is placed in the rotor, whereas two secondary orthogonal windings (*sine* and *cosine*) are put in the stator. The rotor contains a diagonal section of highly permeable material that varies the magnetic field across the stator as the rotor rotates. The shaft position angle θ is determined by two sinusoidal signals (*i.e.*, *sine* and *cosine* signals). The resolver has three pairs of lead wires: the primary winding (R_1 and R_2), *sine* winding (S_1 and S_3), and *cosine* winding (S_2 and S_4).

The induced output voltages (*i.e.*, V_{sin} and V_{cos}) are proportional to the transformation ratio K_{TR}, the excitation voltage V_e, and $\sin\theta$ and $\cos\theta$, respectively,

$$V_{sin} = K_{TR}V_e \sin\theta = K_{TR}V_o \sin(\omega t)\sin\theta \qquad (8.8a)$$

$$V_{cos} = K_{TR}V_e \cos\theta = K_{TR}V_o \sin(\omega t)\cos\theta \qquad (8.8b)$$

where θ is the absolute shaft position angle from a mechanical zero point. These indicate that when the induced voltage in the *sine* winding reaches its maximum value, the voltage in the *cosine* winding reaches its minimum value and vice versa (Figure 8.13). It can be seen from this figure that the frequency of the output voltages is identical to that of the excitation voltage.

Combining Equations 8.8a and 8.8b, the absolute shaft position angle θ can be solely determined as

$$\theta = \tan^{-1}\left(\frac{V_{sin}}{V_{cos}}\right) \qquad (8.9)$$

Theoretically, since θ is determined from the ratio of the *sine* and *cosine* output signals, this ratio is independent of the frequency and amplitude of the reference excitation. More specifically, any changes in operation parameters, such as the phase shift, transformer ratio, and operation temperature, have little impact on the measuring accuracy. Because the shaft angle is absolute, in a case that power is removed from the rotor when it is rotating, the resolver will report the new angular position value of the shaft when power is restored.

The working principle of *resolver transformer* (RT), also known as *resolver receiver*, is described in Figure 8.14. This type of resolver is used in the opposite way to *resolver transmitters* (RX). In this type of resolver, two input windings (*sine* and *cosine*) are energized with well-regulated reference voltages that vary as $\cos\phi$ and $\sin\phi$, respectively, where ϕ is an electrical input angle. The system turns the rotor to obtain a zero voltage in the rotor winding. At this position, the mechanical angle of the rotor equals the electrical angle applied to the stator. Thus, it produces a phase-shifted *sine* signal at the rotor output winding terminals (R_1 and R_2). The induced output voltage V_{out} varies with the mechanical angular position of the shaft θ in the same manner as those of the resolver transmitters, where θ is measured from the reference shaft position called zero point.

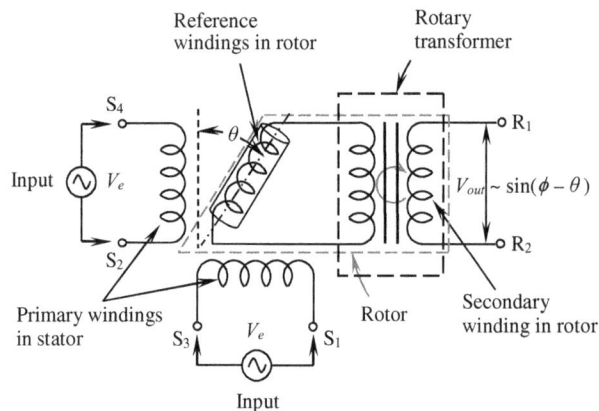

FIGURE 8.13 In a brushless resolver transformer (*RT*), two input stator windings (*sine* and *cosine*) are excited to produce a phase-shifted sine signal at the rotor winding terminals (R_1 and R_2). Thus, the output voltage is proportional to the sine of the angle difference between the electrical input angle ϕ and the mechanical angular position of the shaft θ, *i.e.*, $V_{out} \sim \sin(\phi - \theta)$.

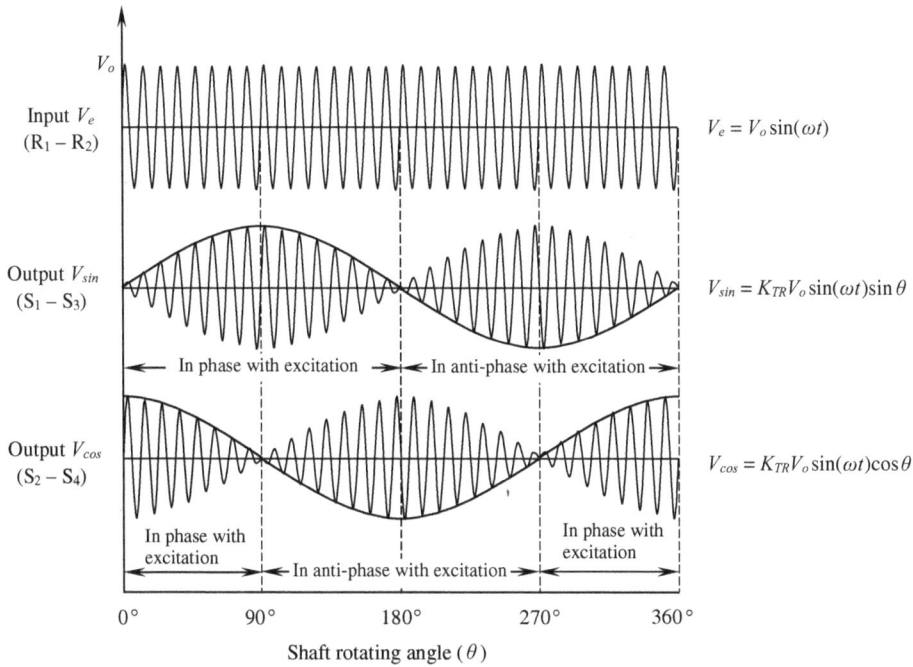

FIGURE 8.14 Resolver excitation and output signals versus shaft rotating angle for the *RX* type with single speed.

When the shaft rotates, the *sine* and *cosine* windings individually induce the components of the output voltage, V_{sin} and V_{cos} with different phases, respectively [8.23],

$$V_{sin} = K_{TR}V_o \sin(\omega t)\cos\phi\sin\theta \qquad (8.10a)$$

$$V_{cos} = K_{TR}V_o \sin(\omega t)\sin\phi\cos\theta \qquad (8.10b)$$

The overall output voltage is obtained by differencing these two voltage components V_{cos} and V_{sin}, and consequently,

$$V_{out} = K_{TR}V_o \sin(\omega t)(\sin\phi\cos\theta - \cos\phi\sin\theta) \qquad (8.11)$$

Using a trigonometric identity,

$$\sin\phi\cos\theta - \cos\phi\sin\theta = \sin(\phi - \theta) \qquad (8.12)$$

it yields

$$V_{out} = K_{TR}V_o \sin(\omega t)\sin(\phi - \theta) \qquad (8.13)$$

This means the induced output voltage is proportional to the sine of the angular difference between the electrical input angle ϕ and the mechanical angular position of the shaft θ. In this way, *RT* translates electrical signals into angular position in conjunction with a servo amplifier and electromechanical drive.

In a case that all three types of revolvers are integrated into a system, *RD* acts as a variable electrical coupling between a *RT* and a *RX* (Figure 8.15). The primary function of *RD* is to provide

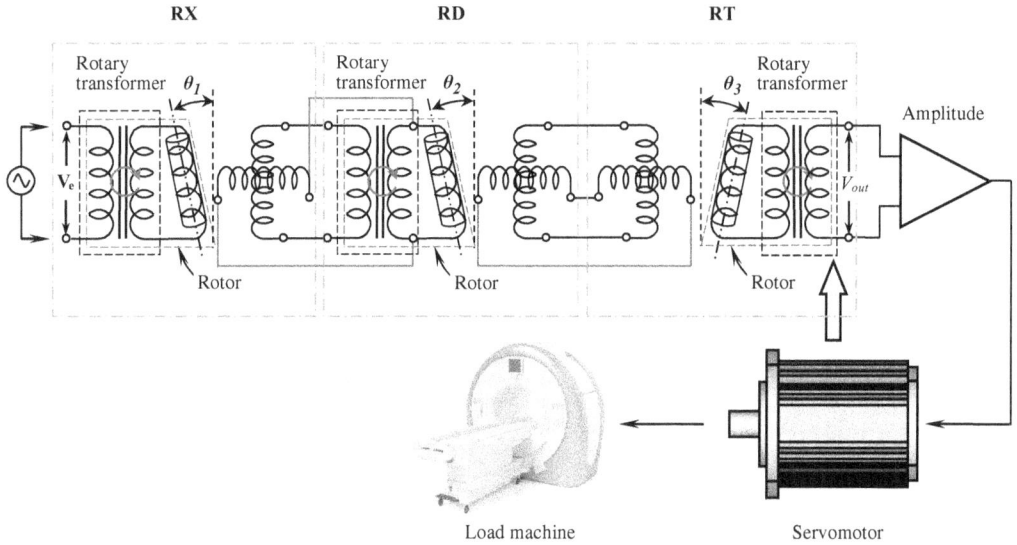

FIGURE 8.15   Three types of resolvers are integrated as a system, where the system output V_{sys} is a function of the input excitation voltage V_e and three angles θ_1, θ_2, θ_3 and three transformer ratios $K_{TR,RX}$, $K_{TR,RD}$, $K_{TR,RT}$ of RX, RD, and RT, respectively.

system correction. In addition, with a RD, a second angular vector can be added into the input signal [8.24]. In this system, the system output V_{out} is a function of the excitation voltage V_e, three rotating angles θ_1, θ_2, θ_3, and three transformer ratios $K_{TR,RX}$, $K_{TR,RD}$, $K_{TR,RT}$ of RX, RD, and RT, respectively.

8.2.2   Resolver Operating Parameters

There are a number of functional operating parameters defining resolver operation [8.25], including resolver accuracy, input excitation frequency/voltage/current, transformation ratio, phase shift, winding impendence, velocity ripple, and null voltage.

8.2.2.1   Resolver Accuracy

Resolver accuracy is based on the ability of the output voltages to define the shaft rotating angle. Since resolvers are analog devices, they are usually associated with a conversion process from analog to digital signals and a subsequent R/D circuit error. For this reason, resolvers are less accurate than encoders.

8.2.2.2   Input Excitation Frequency, Voltage, and Current

The input excitation frequency, voltage, and current for test resolvers depend on the type of servo drive. The excitation frequency for most resolvers usually ranges from 400 Hz to 20 kHz (up to a wide range of 60 Hz–100 kHz), the voltage from 2 to 40 V_{rms} (up to a wide range of 0.5–115 V_{rms}), and the current from 20 to 100 mA.

Generally, a high-performance resolver requires high input voltages, which in turn require high-power electronics to meet high output range and fast slew rate conditions [8.26].

8.2.2.3   Phase Shift

Phase shift $\Delta\phi$, expressed in electrical degree, is defined as the time-phase difference between the primary and secondary voltages at maximum magnetic coupling. Generally, the phase shift can be expressed as the arctangent of the ratio of the primary winding DC resistance R_{DC} to its reactive component X_o (see Figure 8.16),

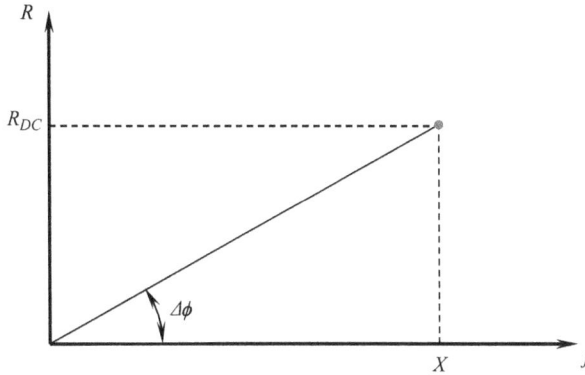

FIGURE 8.16 Determination of resolver phase shift $\Delta\phi$.

$$\Delta\phi = \tan^{-1}\left(\frac{R_{DC}}{X}\right) = \tan^{-1}\left(\frac{R_{DC}}{X_L - X_C}\right) \tag{8.14}$$

In most cases, X_L is much larger than X_C. However, in some special cases (*e.g.*, the length of a resolver feedback cable is larger than 100 m), the capacitive reactance in the feedback system cannot be ignored.

In practice, the phase shift can be determined from measured parameters,

$$\Delta\phi = f_e \Delta t \quad \text{(in cycle)}$$

$$= 360 f_e \Delta t \quad \text{(in degree)} \tag{8.15}$$

$$= 21,600 f_e \Delta t \quad \text{(in arc-minute)}$$

where f_e is the input excitation frequency and Δt represents the time difference between the excitation frequency and the output resolve *sine/cosine* frequencies (at their intersections).

8.2.2.4 Transformation Ratio

Transformation ratio K_{TR} is the ratio of resolver output voltage to its input voltage when the output is at maximum coupling. From Equations 8.8a and 8.8b, it can be derived that

$$K_{TR} = \frac{\sqrt{V_{sin}^2 + V_{cos}^2}}{V_e} \tag{8.16}$$

In practice, K_{TR} usually ranges between 0.2 and 1.0. For bipolar resolvers, common values of K_{TR} in industrial applications are 0.454, 0.5, and 1.0. In fact, K_{TR} is essentially determined by winding designs. Manufacturers usually customize resolver windings to meet customers' special needs for the transformation ratio.

The change in phase shift due to the temperature variation will lead to a change in transformation ratio correspondingly. The change is proportional to the ratio of the cosine of the phase shift at a temperature T, $\cos[\Delta\phi_2(T)]$, to the cosine of the phase shift at the reference temperature T_o, $\cos[\Delta\phi_1(T_o)]$, *i.e.*,

$$K_{TR1} = K_{TR2} \frac{\cos[\Delta\phi_2(T)]}{\cos[\Delta\phi_1(T_o)]} \tag{8.17}$$

where

K_{TR1} is the transformation ratio at the reference temperature T_o, °C or K

K_{TR2} is the transformation ratio at a temperature T, °C or K

$\Delta\phi_1(T_o)$ is the phase shift at the reference temperature T_o, degree

$\Delta\phi_2(T)$ is the phase shift at a temperature T, degree

8.2.2.5 Winding Impedance

The concept of impedance in AC circuits is introduced because two additional impeding mechanisms need to be taken into account besides the normal resistance of DC circuits—reactance and capacitance. Impedance is typically represented as a complex quantity Z and is given in the complex phase form,

$$Z = R + jX$$
$$= (R_{DC} + R_{AC}) + j(X_L - X_C) \tag{8.18}$$

where j is the imaginary unit. The real component R consists of the DC resistive component R_{DC} and AC resistive component R_{AC} as magnetic circuit losses (eddy current and hysteresis). The imaginary component X represents the winding reactance.

When an AC voltage is applied on a resolver winding, energy may be stored and released in the form of a magnetic field, in which case the reactance is inductive and $X=X_L$, where X_L is proportional to the product of the frequency f and the inductance L,

$$X_L = 2\pi f L = \omega L \tag{8.19}$$

where ω is the angular frequency ($\omega = 2\pi f$). If energy is stored and released in the form of an electric field, in which case the reactance is capacitive and $X = -X_C$, where X_C varies inversely with the product of the frequency f and the capacitance C,

$$X_C = \frac{1}{2\pi f C} = \frac{1}{\omega C} \tag{8.20}$$

Four primary impedances are often listed in resolver specifications

- Z_{po}—the impedance of the primary winding at rated frequency with the secondary windings open circuited.
- Z_{ps}—the impedance of the primary winding at rated frequency with secondary windings short circuited, with a voltage applied to produce the same current as for Z_{po}.
- Z_{so}—the impedance of the second windings at rated frequency with the primary winding open circuited.
- Z_{ss}—the impedance of the second windings at rated frequency with the primary winding short circuited, with a voltage applied to produce the same current as for Z_{so}.

To improve resolver performance and reduce its power loss, it is desired that the resolver impedances are as low as possible.

8.2.2.6 Speed Ripple

When monitoring the motion parameters of rotating machines, various errors could be introduced by resolvers to reduce the accuracy of the measurements. For different types of resolvers, speed ripple can be induced by various causes, such as deviations from a sinusoidal flux density around the airgap, deficiencies of feasible winding geometries, and the variable magnetic reluctance in the airgap due to the stator slot, which is defined as the *slot effect* that can introduce limit cycles in the control [8.27].

Measured speed ripples are typically expressed in terms of percentage (%) of deviations from the rated value, *i.e.*, $\Delta\omega/\omega \times 100\%$. Generally, when measured speed ripples are less than 1%, resolvers are very good. When speed ripples range between 1% and 2.3%, resolvers are usable. As speed ripples are larger than 2.3%, resolvers are definitely undesirable in precision motion control.

8.2.2.7 Null Voltage

Null voltage is the residual voltage at the minimum magnetic coupling between the primary and secondary resolver winding. Theoretically, when the primary and secondary windings are perpendicular, there is no voltage induced in the secondary windings. However, due to some factors, such as mechanical imperfections, winding errors, and distortions in the magnetic circuit, some degree of induced voltage may appear in the output windings.

The null voltage comprises three components: in-phase fundamental, quadrature fundamental, and harmonics. The in-phase fundamental component is an angular inaccuracy that can be canceled by re-nulling the rotor, thereby introducing an error (sometimes called null spacing error). Quadrature voltage is 90° out of phase with the in-phase component and cannot be nulled by rotor rotation. The harmonic voltages consist predominantly of the third harmonic, which is three times the excitation frequency [8.28].

8.2.3 RESOLVER TESTING

Resolver testing is often used to validate the resolver performance. The most important testing items include the resolver accuracy, input excitation voltage and frequency, transmission ratio (TR), velocity ripple, and phase shift.

8.2.3.1 Test Equipment and Instruments

In order to achieve accurate test data, certain test equipment and instruments are required, including the following:

- An AC signal generator capable of producing *sine*, *cosine*, *square*, *triangle* waves of varying amplitude and frequency up to 2 MHz.
- An oscilloscope with a bandwidth of 200 MHz.
- One or two servo drives, depending on the type of test item.
- A number of servomotors in which tested resolvers are mounted onto the shafts.
- A precise absolute encoder required for resolver position error measurements.
- A digital multimeter for measuring voltage, current, and resistance.
- A clamp-on ammeter for measuring current without cutting open the electrical circuit.
- An adjustable AC voltage power supply.

8.2.3.2 Determination of Resolver Position Error

The test setup used for measuring resolver position errors is described in Figure 8.17. The resolver being tested is directly mounted on the motor shaft. In order to obtain satisfactory measuring results, a precise absolute encoder (*e.g.*, optical or *sine* encoder) is used to be mounted on the motor shaft through a shaft extension (Figure 8.18) for providing accurate shaft angular positions. Two servo drivers are used to control the encoder and the test resolver, respectively. Before taking measurements, it is important to align both the encoder and resolver to the electrical zero, which refers to the angle at which the *sine* output voltage is at an in-phase null. Thus, the resolver position error can be determined by comparing the measured positions of the resolver and encoder at each measuring point.

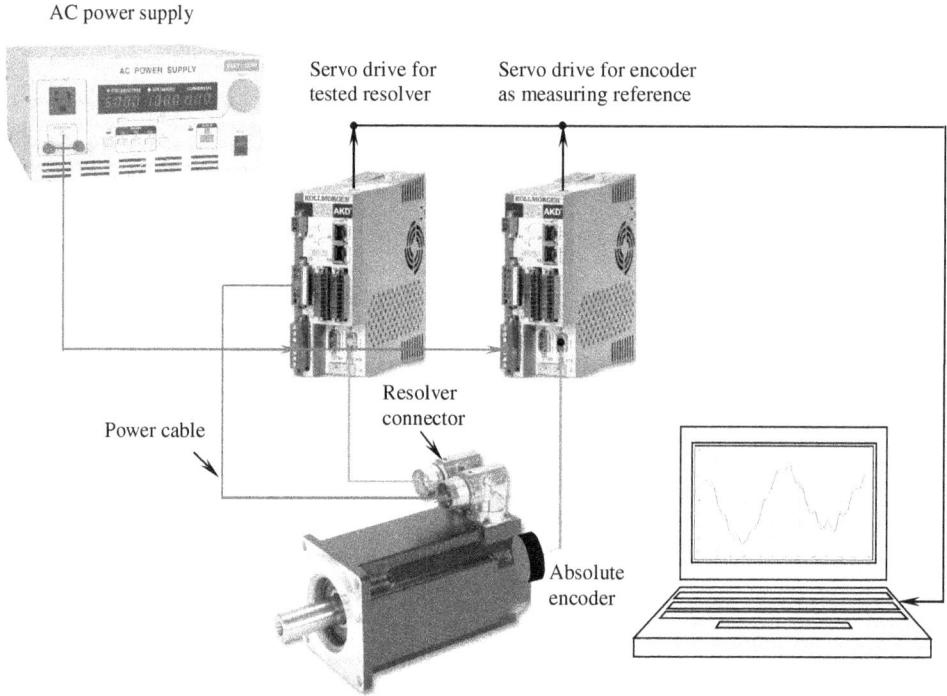

FIGURE 8.17   Test setup for measuring resolver errors from the electrical zero using a precise absolute encoder as the standard of measurement.

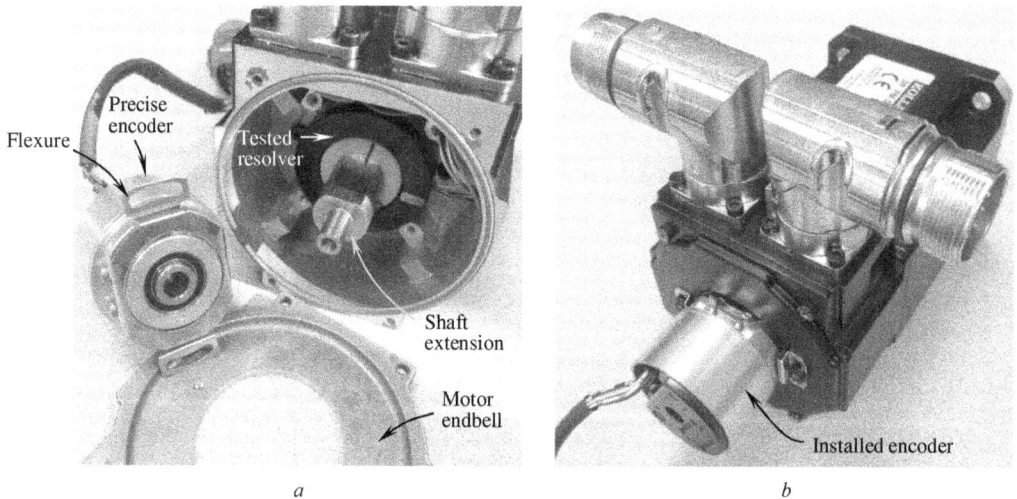

FIGURE 8.18   Resolver test setup for measuring resolver position errors: (*a*) the resolver being tested mounts on the motor shaft; (*b*) the rotor of the precise encoder mounts on the shaft extension that is fixed on the motor shaft and the stator of the encoder mounts on the motor endbell through a flexure.

The position measurement is performed in the stepping mode. Position errors have been measured in 120 points per revolution. Each measured position error is the averaged value of five readings. Then, the position errors are converted into minute of arc (also known as arc-minute, 1/60 of a degree) and plotted against the shaft angular position. For the purpose of demonstration, a set of

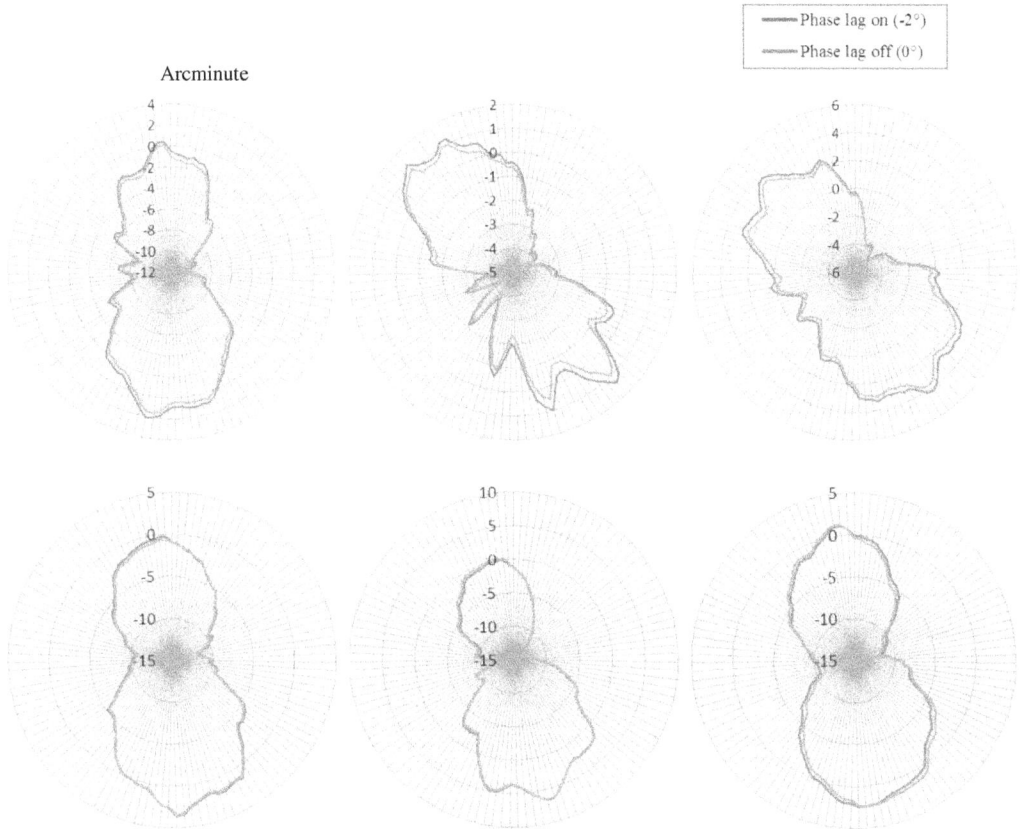

FIGURE 8.19 Resolver position error measurements from six resolvers with an identical model and size, showing the largest peak-to-peak position errors range of 20.5 arc-minute.

measurements of position error is shown in Figure 8.19, clearly showing the locations of the largest peak-to-peak position error.

8.2.3.3 Measurement of Transformation Ratio

The most important electrical parameters of resolvers are the position accuracy, transformation ratio, phase shift, and velocity ripple. Transformation ratio is the ratio of the output voltages (*e.g.*, V_{sin} and V_{cos}) to the input voltage (*e.g.*, V_e), as shown in Equation 8.16.

Alternatively, if the transformation ratio is measured through the *sine* winding, the voltage of the *cosine* winding should be adjusted to null voltage (by manually rotating the motor shaft) and thus the voltage of the *sine* winding reaches its maximum value. Before taking readings, it is important to make the phase of the *sine* wave the same as the excitation signal, especially for the case with a long resolver feedback cable. The setup for measuring transformation ratio is described in Figure 8.20.

Transformation ratios of resolvers vary with both the excitation frequency and supply voltage. With a fixed supply voltage, transformation ratios increase sharply at low excitation frequencies (<4,000 Hz) and remain almost constant for frequencies up to 8,500 Hz. At high frequencies, they decrease slightly (Figure 8.21).

Figure 8.22 displays the variation of transformation ratio with the supply voltage at a specific excitation frequency. For all test resolvers, their transformation ratios slightly increase (<12%) as the supply voltage changes from 1 to 10 V_{rms}.

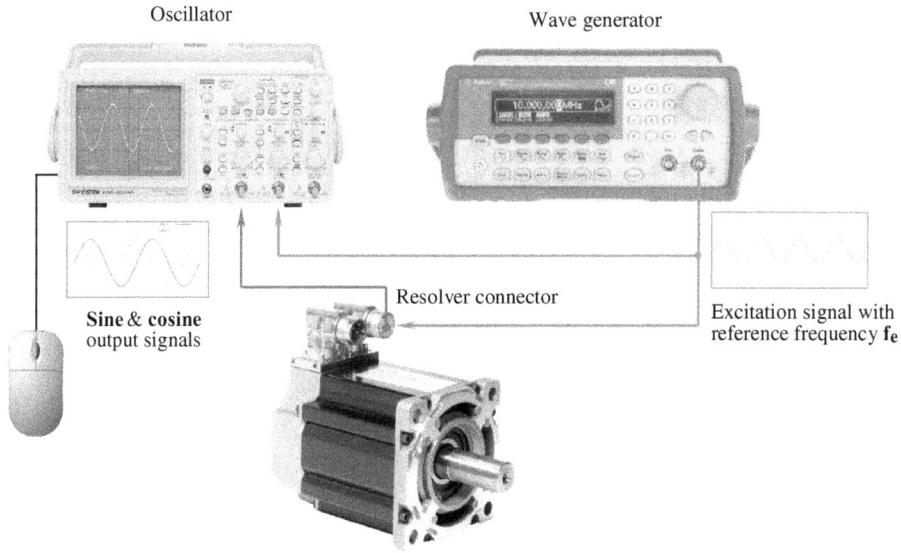

FIGURE 8.20 Test setup for measuring the transformation ratio and phase shift of resolver. During the test, the motor is de-energized.

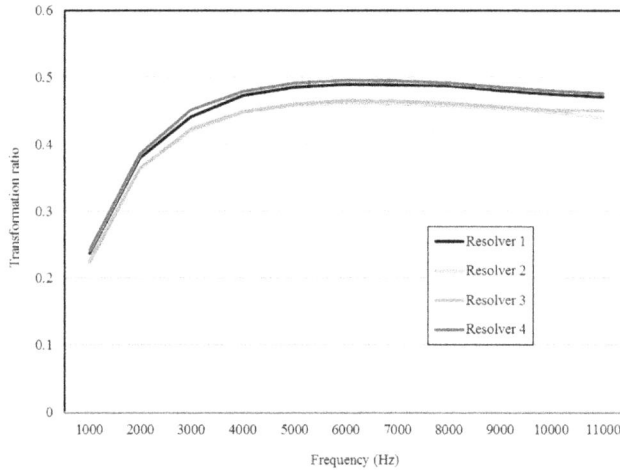

FIGURE 8.21 Transformation ratio K_{TR} versus excitation frequency f_e at 2.85 V_{rms}.

8.2.3.4 Measurement of Phase Shift

The setup of the phase shift test is the same as that used in the transformation ratio test. When a wave generator sends an excitation signal to the resolver being tested, it generates *sine* and *cosine* output signals.

Figure 8.23 presents the variations of phase shift with an increase in excitation frequency as the supply voltage is fixed. In all cases, the increase in frequency leads to a significant increase in phase shift, with little differences among all tested resolvers. As shown in Figure 8.24, for a fixed excitation frequency, the supply voltage has a weak influence on resolver phase. However, relatively large fluctuations occur at low (< 2 V_{rms}) and high (> 7 V_{rms}) voltages.

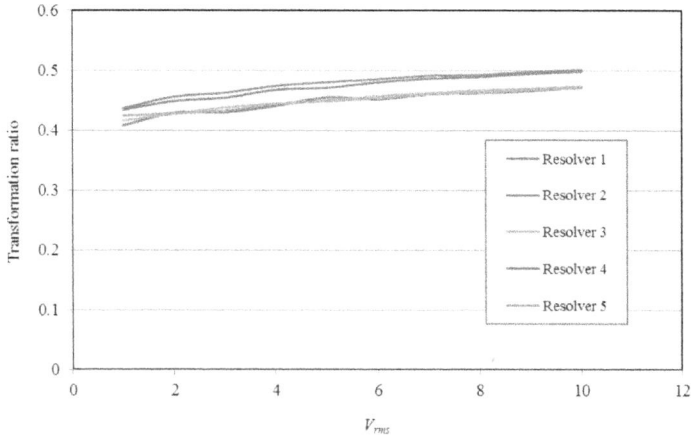

FIGURE 8.22 Transformation ratio K_{TR} versus supply voltage V_{rms} at 13,000 Hz.

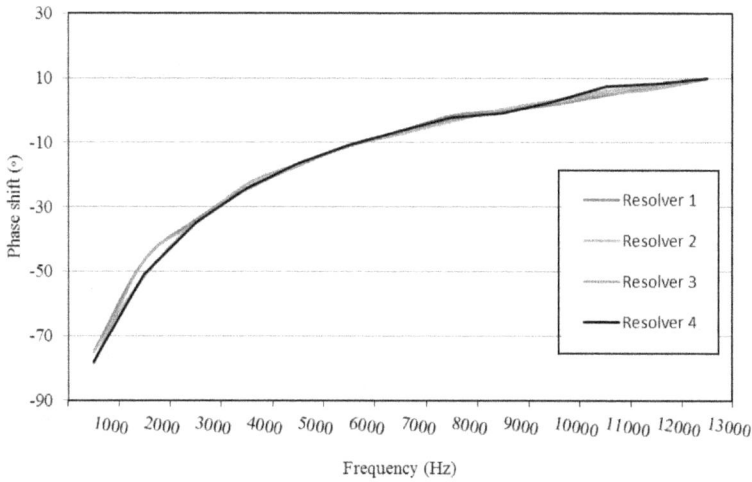

FIGURE 8.23 Phase shift $\Delta\phi$ versus excitation frequency f_e at 2.85 V_{rms}.

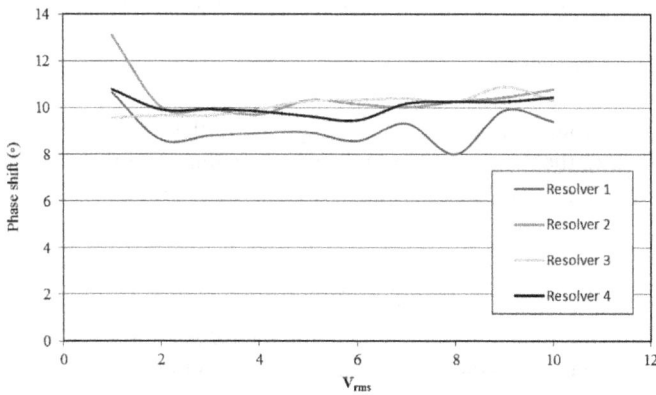

FIGURE 8.24 Phase shift $\Delta\phi$ versus supply voltage V_{rms} at 13,000 Hz.

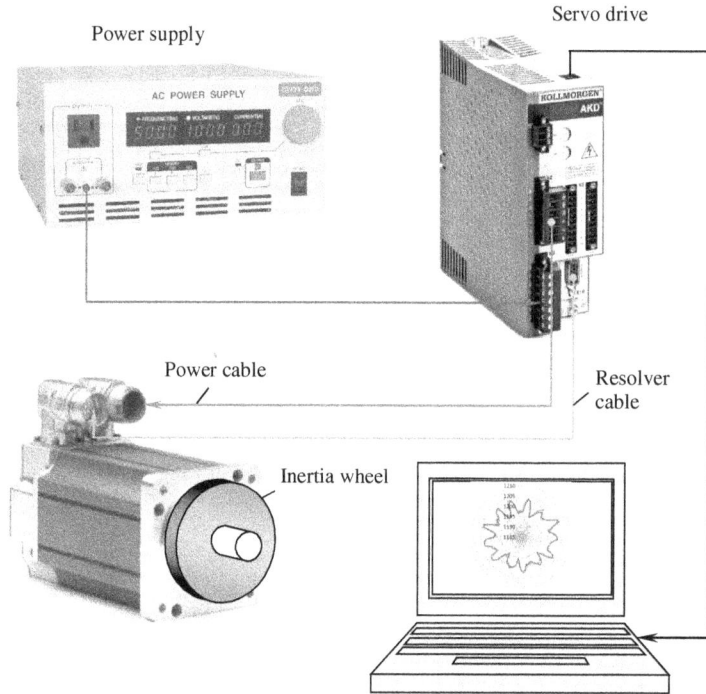

FIGURE 8.25   Test setup for measuring velocity ripple. The inertia wheel is installed on the motor shaft.

8.2.3.5   Measurement of Velocity Ripple

The test setup of velocity ripple measurement is shown in Figure 8.25. To enhance the smoothness of motor operation, an inertia wheel is mounted at the end of the motor shaft. The operation of the motor is controlled by a servo drive.

Some measuring results from six tested resolvers that have an identical model and size are presented in Figure 8.26. In this test, the rotating speed is set at 1,200 rpm. The largest peak-to-peak velocity ripple (in percentage) ranges 0.8%–1.6%.

8.2.3.6   Measurement of Resolver Impedance

To calculate the impedance of a certain resolver winding, the current through the winding and the voltage across it must be predetermined. The principle of measuring resolver impedances is illustrated in Figure 8.27. A shunt resistor or a decade resistance box is connected with the resolver winding in series to form a RL series circuit. An oscilloscope or AC voltmeter is connected across the winding to measure the voltage drops of the resistor and resolver winding and thus provides fast, accurate impedance measurements.

As an excitation AC voltage $V_{e,rms}$ (e.g., 8 V) is applied to the RL circuit with an excitation frequency f_e (e.g., 8,000 Hz), the AC generated in the circuit is thus determined from the voltage drop across the shunt resistor $V_{R,rms}$ and the resistance R,

$$I_{rms} = \frac{V_{R,rms}}{R} \tag{8.21}$$

According to the Kirchoffs second law, the excitation voltage is the sum of the voltage drop across the shunt resistor $V_{R,rms}$ and the voltage across the resolver winding $V_{w,rms}$,

$$V_{e,rms} = V_{R,rms} + V_{w,rms} \tag{8.22}$$

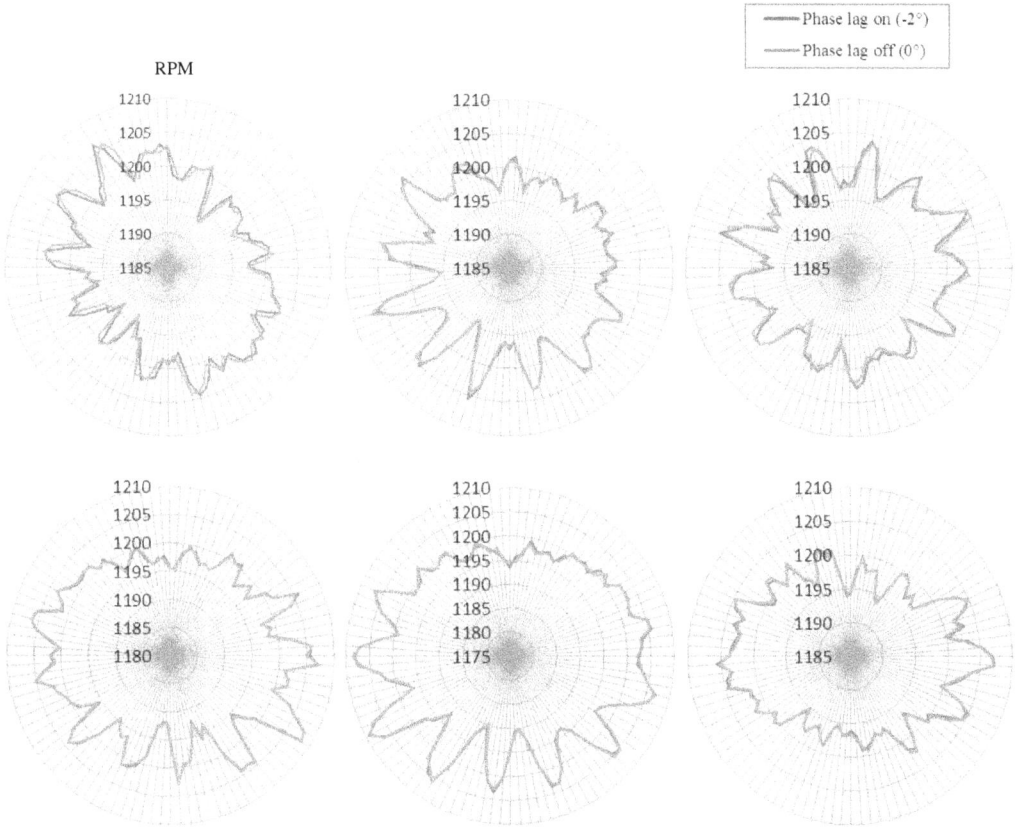

FIGURE 8.26 The measured velocity ripples of six resolvers with an identical model and size. The largest peak-to-peak velocity ripple (in percentage) ranges of 0.8%–1.6%

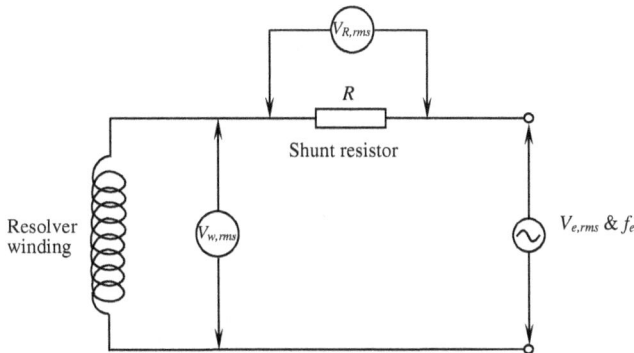

FIGURE 8.27 The measuring method for resolver impedances. A shunt resistor is connected with the resolver winding in series to form a *RL* series circuit.

Thus, the impedance of the resolver winding is given as

$$Z_w = \frac{V_{w,rms}}{I_{rms}} = \frac{V_{e,rms} - V_{R,rms}}{I_{rms}} \tag{8.23}$$

More detailed test setups for measuring primary winding impedances Z_{po} (with the second windings open) and Z_{ps} (with the second windings shorted) and secondary winding impedances Z_{so}

(with the primary winding open) and Z_{ss} (with the primary winding shorted) are described in Figure 8.28.

It is noted that since a resolver has two secondary windings (*sine* and *cosine*), Z_{so} and Z_{ss} must measure separately for these two windings.

8.2.3.7 Effect of Cable Length on Resolver Performance

Feedback device cables are used to transmit signals from feedback devices to servo drives or motion controllers. Under normal circumstances, the cable length is no more than a few meters. However, some applications may require longer cables, from several dozen meters to several hundred meters. Generally, a cable length of up to 10 m can be handled without any problems. If some proper measures are taken into account, even longer cables can be successfully adopted. Nevertheless, using very long cables for feedback devices may raise some problems. First, the increase in the cable length enhances the electromagnetic interference (EMI) disturbance as the signal noise, affecting the transmission of signal. Second, the primary cause of signal distortion is cable length, or more specifically, cable capacitance. The longer the cable, the higher the cable capacitance, and the greater the potential for signal distortion.

The most important characteristics of feedback cables are capacitance, impedance, attenuation, and shielding, all of these have dramatic effects on the feedback's overall performance. Cable capacitance, usually measured as picofarads per unit length (*pf/m* or *pf/ft*), is a type of stray or parasitic capacitance that is highly unwanted. It indicates how much charge the cable can store within itself. When a voltage signal is transmitted through a cable, a charge builds up across the insulation between the conductors as a result of the cable capacitance. It takes a certain amount of time for

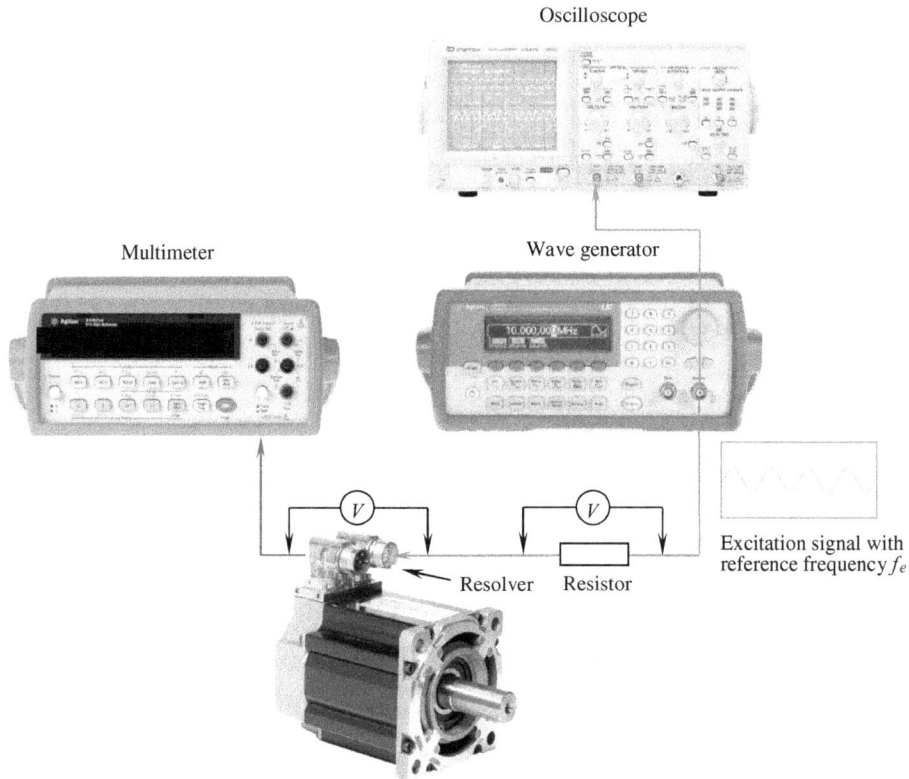

FIGURE 8.28 Test setup for measuring primary winding impedances Z_{po} (with the second windings open) and Z_{ps} (with the second windings shorted) and secondary winding impedances Z_{so} (with the primary winding open) and Z_{ss} (with the primary winding shorted).

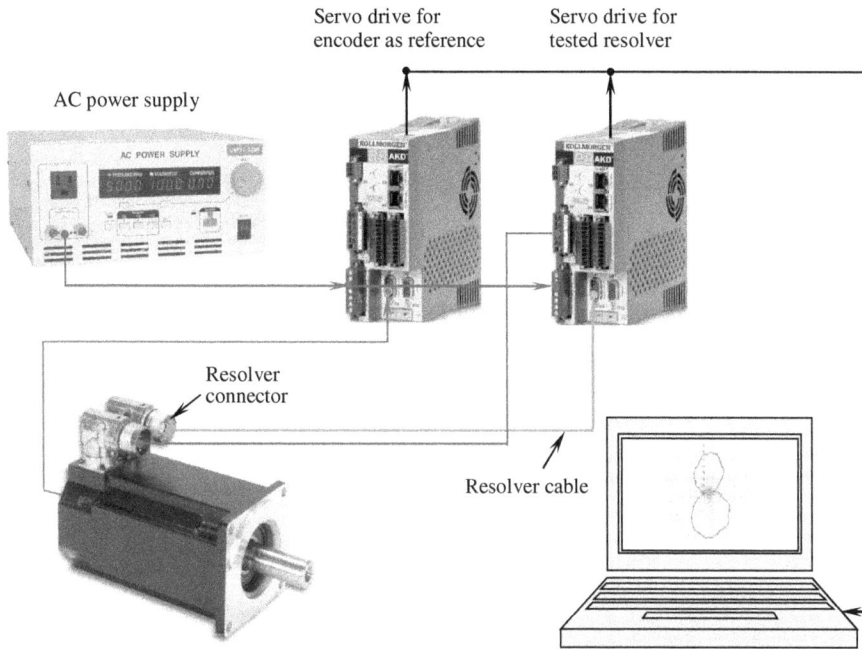

FIGURE 8.29   Test setup for determining the effect of cable length on resolver performance. In the test, both 1-m and 50-m resolver cables are used for the purpose of comparison.

the cable to reach its charged level. This lag in charging time slows down the signal transmission and even leads to the loss of the signal. Generally, the higher the signal frequency, the greater the reactance caused by the cable capacitance and the greater the signal loss. Modern high-speed servo-control systems operate with encoders that use data rates up to several megahertz (*MHz*). At such high rates, the encoder-signal cable must be properly terminated with a terminating resistor or network at the receiver end [8.29].

A long cable with a high *pf/m* value leads to the distortion of the output signal waveform, introduces electromagnetic noise, and affects measurement accuracy. To minimize distortion, low capacitance cable (typically less than 40 pf/ft) is desired, especially at high signal frequencies. The test setup is the same as used in the position error test (Figure 8.29). It uses the same resolver and resolver cables but with different cable lengths.

The cables used in the test have the capacitance per unit length of 492 pf/m (150 pf/ft). The comparison of measured resolver position errors between 1 and 50-m cables is presented in Figure 8.30, indicating that a long cable generates a much larger position error than that of a short cable.

8.2.3.8   Influence of Motor Brake on Resolver Performance

Motor brakes are designed to slow down, stop, or hold electric motors. In normal motor operation, the current is supplied to coils of a brake to release the brake from holding the motor shaft. In many servomotor systems, resolvers are mounted next to the brakes. When the brake is energized (releasing the brake) or de-energized (engaging the brake), the power is turned on or shut off accordingly. It will produce transient EMIs acting on the magnetic field generated by the resolver coils. Therefore, the operating parameters of the resolver fluctuate due to the transient variation in the magnetic field, influencing the resolver performance.

The test setup is shown in Figure 8.31. This test is to energize a motor brake while monitoring the resolver velocity and current flowing through the brake coil on the oscilloscope. It is acceptable if any glitch on the resolver velocity is less than 1 rpm.

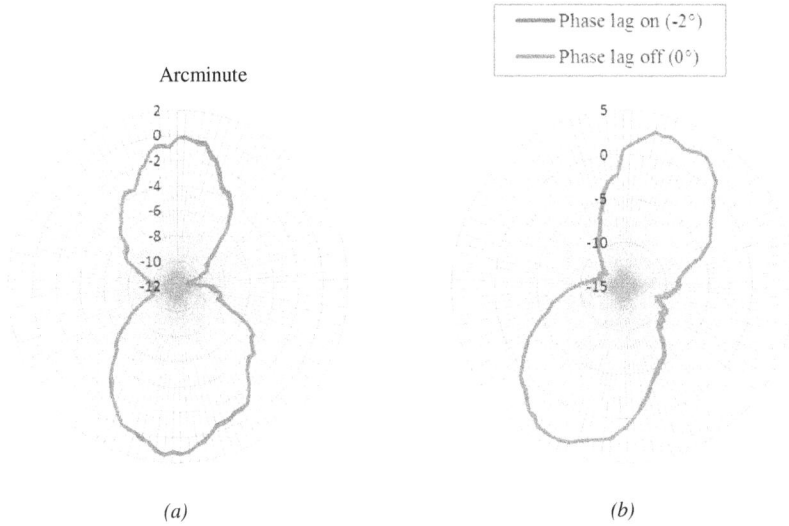

FIGURE 8.30   Effect of resolver feedback cable length on resolver position error: (*a*) 1-m cable and (*b*) 50-m cable.

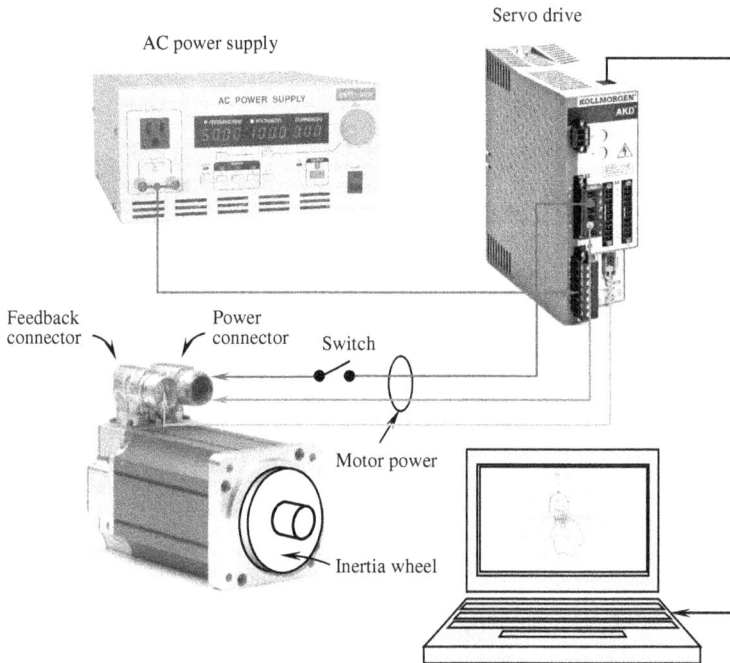

FIGURE 8.31   Test setup for determining the influence of EMI generated by brake coils during brake operation on resolver position error. An inertia wheel mounts on the motor shaft.

8.2.3.9   Influence of Mechanical Impacts on Resolver Performance

The purpose of the mechanical impact test is to identify the worst case where a resolver shaft repeatedly runs into a dead stop. In fact, this test is essential since it can verify not only the integration reliability of a resolver built-in motor, but also the robustness of other mating components (*e.g.*, rotor core and shaft, bearing, and shaft) in the motor.

8.3 HALL EFFECT SENSOR

When an electrical current-carrying semiconductor slab is subjected to an external magnetic field perpendicular to the direction of the current flow, the magnetic field exerts transverse forces on the moving charge carriers, *i.e.*, electrons and holes, to shift their moving directions toward the opposite sides of the slab, respectively. As a result, the build-up of charges between the two sides of the semiconductor slab produces a transverse potential difference, known as the Hall voltage V_H, where the voltage V_H is perpendicular to both the magnetic field and the flow of current (Figure 8.32). This phenomenon, called *Hall effect* today, was discovered by Edwin Hall in 1879 when he was a doctoral candidate at Johns Hopkins University [8.30].

The Hall effect provides information regarding the type of magnetic pole and magnitude of the magnetic field. Sensors based on the Hall effect principle are called Hall effect sensors. A Hall effect sensor employs a magnetic phased array sensor passing through a magnetic field to produce a signal that is then interpolated to resolution. Though Hall effect sensors essentially are magnetic field sensors, they can be applied in many other types of applications for measuring position, velocity, force, current, pressure, temperature, *etc.* Among the various sensing technologies, Hall effect sensors have long been used in numerous industrial segments for their small sizes, ruggedness, ease-of-use, and relatively low cost [8.31]. They can operate at high frequencies and are not affected by environmental contaminants. Compared with other motion feedback devices, however, Hall effect sensors feature low sensitivity, low breakdown voltage, and susceptibility to temperature; all limit sensors to use in high voltage power systems. In addition, Hall effect sensors are less precise. Therefore, Hall effect sensors provide the simplest and least expensive solution for applications that do not require precise speed or position controls.

Hall devices have been used for many years for brushless motor commutation of three-phase motors using three latching Hall devices. Velocity control and speed ripple are not precise and therefore an incremental encoder is commonly added to provide greater digital resolution per *rpm*

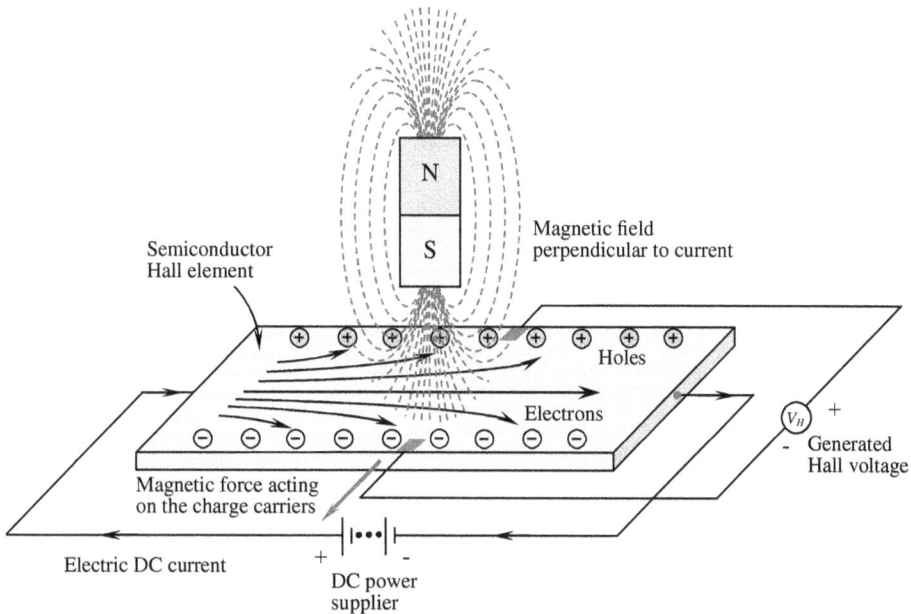

FIGURE 8.32 Schematic diagram of the Hall effect. When DC current passing through a semiconductor slab is subjected to a perpendicular magnetic field, magnetic force acts on the charge carriers, electrons, and holes, to shift their moving directions toward the opposite sides of the slab respectively, producing Hall voltage ($V_H \propto \mathbf{I} \times \mathbf{B}$) that is perpendicular to both the magnetic field and the current.

and improve speed ripple as well as position information. The Hall devices are used at motor initialization because three Hall sensors provide crude absolute position information in one revolution and then drive switches over to use the incremental encoder resolution for fine motion control and speed control. A commutation encoder includes both Hall and incremental channels on one disc at economical pricing.

Hall effect sensors come in two basic types: threshold (or digital) sensors and linear (or analog) sensors, each type has different output characteristics.

8.3.1 Linear Sensor

A linear sensor provides a continuous output voltage that is proportional to the electrical current and the intensity of the magnetic field, but inversely proportional to the sensor thickness, *i.e.*,

$$V_H = c_H \left(\frac{\mathbf{I}}{t} \times \mathbf{B} \right) \tag{8.24}$$

where
V_H is the Hall voltage in V (volt)
C_H is the Hall effect coefficient, depending on temperature and materials
$\mathbf{I}$ is the electrical current flow (vector) through the sensor in A (ampere)
$\mathbf{B}$ is the magnetic flux density (vector) in T (tesla)
t is the thickness of the sensor in mm

It can be seen from this equation that the Hall voltage is proportional to the vector cross product ($\mathbf{I} \times \mathbf{B}$). To ensure normal operation, in addition to the Hall element as a core component, a Hall effect sensor also consists of some auxiliary elements. A voltage regulator is commonly used to maintain the voltage/current constant so that the output of the sensor only reflects the variations of the magnetic field. In such an arrangement, the Hall voltage becomes a *linear* function of the magnetic field. Besides, the produced Hall voltage is regularly a low-level signal (in the order of a few to several dozen microvolts) and thus requires a differential amplifier to amplify the potential difference, *i.e.*, the Hall voltage. The key features of the differential amplifier include low noise, high input impedance, and moderate gain. With the standard bipolar transistor technology, the amplifier can be readily integrated with the sensor (Figure 8.33*a*). It is noted that errors involved in measurements are mostly due to temperature variations. Signal conditioning electronics are typically incorporated into sensors to compensate for the temperature effect.

Logically, the output of the amplifier cannot exceed the limits imposed by the power supply. In fact, the amplifier begins to saturate before the limits of the power supplier are reached. In linear output sensors, the output voltage increases with an increase in the magnetic field until it reaches the limits. After reaching the limits, any additional increase in the magnetic field will have no effect on the output once the amplifier reaches saturation [8.32].

8.3.2 Threshold Sensor

A threshold sensor, on the other hand, functions like an *On/Off* switch. For example, it may turn on when a positive field reaches a certain threshold (the operating point) and turn off when a negative field reaches a certain threshold (the release point). Because the digital output sensor has just two states, *i.e.*, *On* ("1") or *Off* ("0"), it generates square output waveforms. Compared with an analogy sensor, a Schmitt trigger circuit is added into the threshold sensor (Figure 8.33*b*). It compares the output of the differential amplifier with the preset reference. In a case that the amplifier output exceeds the reference value, the Schmitt trigger switches on, otherwise it switches off.

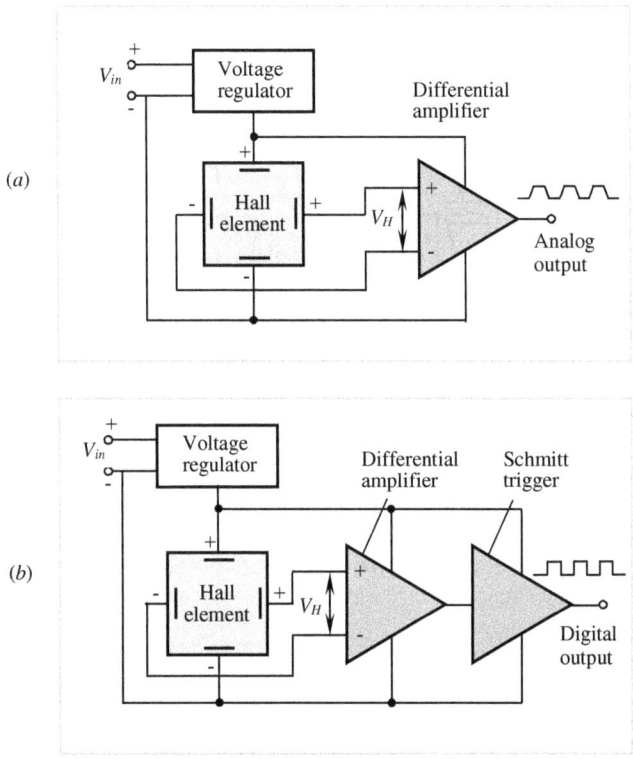

FIGURE 8.33 Two types of Hall Effect sensors: (a) Linear sensor with analog output in a trapezoidal waveform and (b) thresholder sensor with digital output in a square waveform.

As a digital on-off device, this type of Hall effect sensor is a sensor where output voltage varies in accordance with a magnetic field. It is usually made from semiconductor material and can operate at very high frequencies (up to thousands of motor rpm).

8.4 PROXIMITY SENSOR

A proximity sensor is a sensor that detects the movement or presence of nearby objects without any physical contact. Normally, a proximity sensor emits an electromagnetic or electrostatic field or a beam of electromagnetic radiation and detects any change in the field or return signal [8.33]. Based on the operation principle used in proximity sensors, there are a variety of proximity sensor types, each suited to specific applications and environments. Proximity sensors are widely used in industrial and automotive applications.

8.4.1 INDUCTIVE PROXIMITY SENSOR

Inductive proximity sensors use Faraday's law of electromagnetic induction to detect the nearness of conductive materials without actually coming into contact with them. Inductance is defined as the ratio of the induced voltage $v(t)$ to the rate of change of the current change di/dt, i.e., $L = - v(t)/(di/dt)$, where the negative sign indicates that voltage induced $v(t)$ opposes the change in current through a coil per unit time (di/dt). Inductance always has the tendency of an electrical conductor to oppose a change in the electric current flowing through it.

As shown in Figure 8.34, an inductive proximity sensor has four components: the inductive coil, oscillator, Schmitt trigger, and output amplifier. The oscillator generates a fluctuating

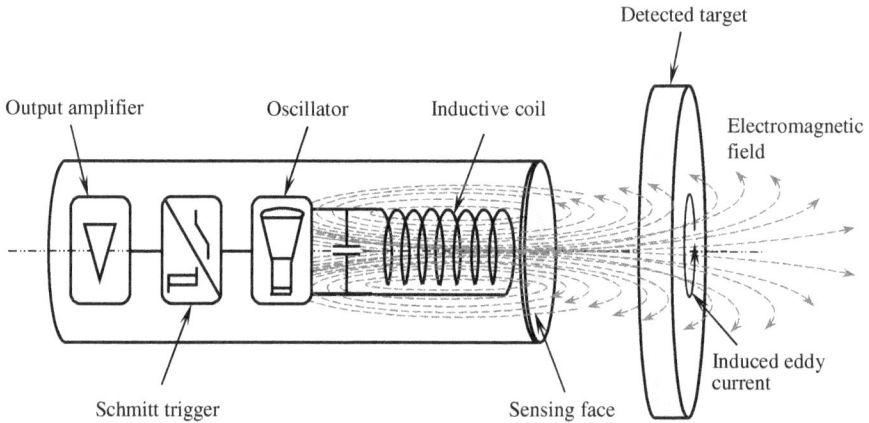

FIGURE 8.34 In an inductive proximity sensor, the oscillator excites the inductive coil to generate an electromagnetic field around the coil that locates in the device's sensing face. When a metal object moves into the detecting field of the sensor, eddy current is induced in the metallic object, creating a corresponding electrical field against the sensor's field. The detection circuit of the sensor monitors the strength of the oscillator's field and triggers an output from the output circuitry when its variation reaches the threshold.

magnetic field the shape of a doughnut around the winding of the coil that locates in the device's sensing face.

An inductive sensor system consists of four main components: an inductive coil for generating an electromagnetic field, an oscillator, a detection circuit for detecting the absence or presence of external objects, and an output circuit to transmitting detected signals to a control device. All the components are contained in sensor housing. When the sensor rotor with the teeth passes through the pole pin of the sensor, the magnetic field surrounding the coil is changed, resulting in the production of inductive voltage in the coil. The voltage signal produced by the sensor depends on the rotating speed of the sensor rotor, coil size, and the number of turns.

The advantages of an inductive proximity sensor include: (*a*) having high accuracy; (*b*) enabling to work in harsh environmental conditions; and (*c*) having a high switching rate. Their disadvantages are as follows: (*a*) the accuracy decreases when it comes to the measure of large displacement. This is because the linear relation between voltage and displacement is less at higher ranges; and (*b*) they can only detect metal conductors and different metal types can affect the detection range.

8.4.2 Capacitive Proximity Sensor

Capacitive proximity sensors use the variation of capacitance between the sensor and the object being detected. Unlike the inductive proximity sensor that senses only metallic objects, this type of proximity sensor can detect anything that is either conductive or has different dielectric properties. They are non-contact devices that detect the presence or absence of virtually any object regardless of metallic or non-metallic materials in the form of powder, granulate, liquid, or solid. This detection technique applies the electrical property of capacitance.

A capacitive proximity sensor contains a metal plate, which is placed at the end of sensor and electrically connected to an internal oscillator circuit. Corresponding to this, the object being sensed acts as the second plate to form a capacitor (Figure 8.35). When the target object approaches the sensor, the oscillator detects the change of the capacitance between the internal sensor plate and the target object. Essentially, this change reflects the change in the electrostatic field around the sensor. The closer the object to the sensor, the greater the change in capacitance. As the amplitude of the capacitance change reaches its threshold, the control circuit is activated to send the signal out. In this way, the intruding object is detected.

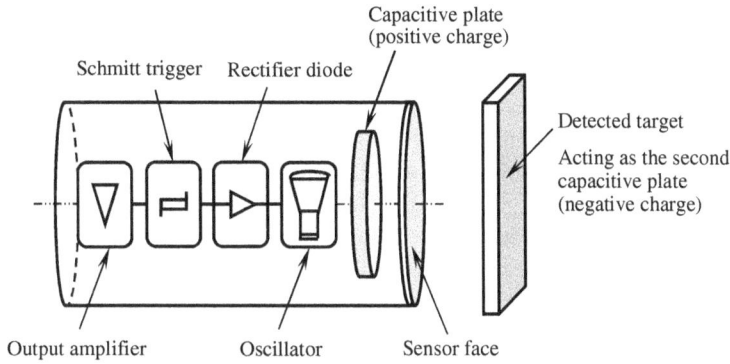

FIGURE 8.35 Structure of capacitive proximity sensor. While a capacitive plate with a positive charge is placed inside the sensor package, a detected target functions as the second capacitive plate with a negative charge.

There are also capacitive touch proximity sensors, which rely upon the proximity of a finger or stylus to initiate actions. In fact, touch is a very common mode of interface in today's electronic systems. This type of capacitive proximity sensor has found application in a wide variety of industrial and consumer products such as smartphones, iPads, appliances, and access controls.

Capacitive proximity sensors usually have high reliability and long functional life because of the absence of rotating parts and the lack of physical contact between the sensor and the sensed object. In addition, they possess functional stability and low cost. However, a large change in temperature or humidity may lower the accuracy of the device. When the leads become considerably longer in some applications, errors or distortions in signal may be introduced.

8.4.3 ULTRASONIC PROXIMITY SENSOR

Ultrasound can be used in detection and ranging applications using the time of flight principle to estimate the distance to an object. Ultrasonic sensors rely on the transmission and reception of sound waves with frequencies higher than that audible to the human ear. The ultrasonic proximity sensors are suitable for short- to medium-range applications at low speeds. Unlike their photoelectric counterparts, however, ultrasonic sensors have only two operating modes: reflection and through-beam. Because they typically operate at frequencies above 80 kHz, industrial ultrasonic sensors can offer robust performance for numerous short- and long-range applications.

Almost all materials reflect sound waves, so ultrasonic sensors are a fine choice for many tasks. Excellence in the detection and measurement of films, transparent objects, and liquids separate these sensors from their photoelectric counterparts. Target color or frequent color changes also have no effect on ultrasonic sensors.

The advantages of using ultrasonic proximity sensors are: (*a*) They are not affected by environmental conditions such as dust, snow, and rain. (*b*) The sensing distance is more than that of capacitive or inductive sensors. However, they have difficulty in sensing soft, curved, thin, or small objects.

8.4.4 PHOTOELECTRIC PROXIMITY SENSOR

Compared to capacitive sensors, photoelectric sensors detect the presence or absence of objects based on the principle of transmitting and receiving light. A typical photoelectric proximity sensor consists primarily of a light source (*e.g.*, LED or laser diode) as the emitter, a photodiode, or phototransistor receiver for detecting the emitted light, and associated electronics designed to evaluate and amplify the detected signal. Because photoelectric proximity sensors are based on the principle that an object interrupts or reflects light, they can be used to detect virtually any object made by various materials such as metal, wood, glass, ceramics, and plastic. In addition, due to the fast travel speed of light, photoelectric proximity sensors have a very short response time.

There are three major types of photoelectric sensors: through-beam, retroreflective, and diffuse-reflective (Figure 8.36). Each type has its own strengths and weaknesses, as well as an effective scope of application. Nevertheless, all offer reliable, precise, and wide sensing ranges.

In through-beam sensing, an emitter and a receiver are separately placed at two sides of the sensing object (Figure 8.36a). The unit is aligned in a way that the greatest possible amount of modulated light from the light source reaches the receiver. An object placed in the path of the light beam blocks the light to the receiver. Thus, the receiver detects the change in light intensity and converts it to output signal with electronic circuits. Modulated light, which is pulsed at a frequency range of 5–30 kHz, increases sensing range while reducing the effect of ambient light. The advantages of using a through-beam sensor are that it is the most accurate type of sensor and has the longest sensing range (up to 90 m) than other types. Furthermore, through-beam sensors are the best choice when using them in environments with airborne contaminants. They are suitable to detect opaque or reflective objects but cannot detect transparent objects.

Unlike the through-beam, a retroreflective sensor contains both the emitter and receiving elements in a single package. The light emitted from the light source is aimed at the retroreflector that is reflected to the light-receiving element (Figure 8.36b). The detection of an object relies on the change in reflected light intensity due to the light path interruption. This type of sensor is the best option when sensing transparent products. Compared with through-beam sensors, retroreflective sensors require wiring only on one side rather than two sides.

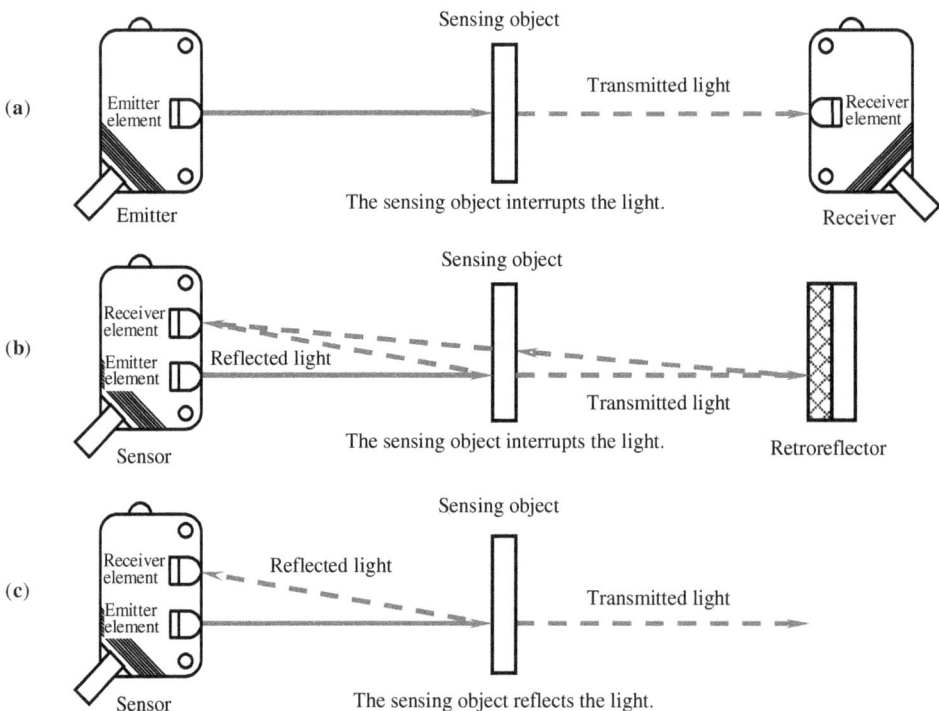

FIGURE 8.36 Photoelectric sensors can be classified into three types: (*a*) through-beam sensor, (*b*) retroreflective sensor, and (*c*) diffuse-reflective sensor.

In diffuse-reflective sensing, like the retroreflective sensor, the emitter and receiver elements are in one unit. Light from the light source strikes the target and the reflected light is diffused on the target surface at all angles. When the receiver gets enough reflected light, it generates output signals (Figure 8.36c). It makes the sensing system very compact since it requires installing only the sensor itself. This is ideal for situations where a sensor can be placed only on one side of the target. However, the response of the diffuse-reflective sensor is strongly affected by the surface reflectivity of the detected object. They are relatively less accurate compared to the through-beam and retroreflective sensors and are easily affected by color, texture, angle of incidents, and dirty environments. Moreover, they have the shortest detecting distance compared with other two types.

8.5 OTHER MOTOR SENSORS

In addition to the various sensors discussed above, other types of sensors are also used in electric motors.

8.5.1 Force/Torque Sensor

Force and torque are two interrelated mechanical parameters ($\mathbf{T} = \mathbf{r} \times \mathbf{F}$) in motor operation. If one is measured, the other can be determined accordingly. To precisely measure motor force and torque, many measuring methods have been developed, from traditional resistance strain gauge to non-contact electromagnetic and piezoelectric measurements.

Force and torque measurements can fall into two categories: static (also known as reaction) or rotary. A static or reaction force/torque measurement involves little (less than one revolution) or no rotation of the machine being measured. Its output signal varies proportionally to an applied torque. A rotary torque measurement is accomplished by using a free shaft, which functions as a coupling between the driving mechanism and the load. As the shaft is torsionally stressed, a proportional change in the output signal is converted into a force/torque signal. Thus, by monitoring the shaft torsional deflection, the shaft force/torque is determined.

Force/torque sensors are commonly embedded in electric motors for continuously monitoring the rotary shaft force/torque. It is worth noting that due to various losses (friction, drag, *etc.*) and rotor inertia, the shaft force/torque is always lower than the electromagnetic force/torque generated by the motor.

A number of techniques can be employed for measuring the shaft force/torque. The most common measurement is accomplished using a standard resistance strain gauge that is directly bonded to the torque-producing shaft. The strain gauge, either foil-type or wire-type, is made from a strain-sensitive material. When the material is stretched elastically, its length increases and its cross-section decreases by Poisson's effect, resulting in an increase in its electrical resistance (noting that $R = \rho L/A$). The change in resistance, which indicates the change in force/torque, is measured by a Wheatstone bridge circuit. The relationship between resistance change ΔR and strain ε can be expressed as

$$\varepsilon = \frac{\Delta L}{L} = \frac{1}{K} \frac{\Delta R}{R} \tag{8.25}$$

where K is defined as the gauge factor. For the three most commonly used strain gauge materials: Constantan, Iso-Elastic, and modified Karma, the gauge factors are 2.0, 3.5, and 2.1, respectively.

However, this type of force/torque sensor suffers from the disadvantages of relatively instability and limited reliability due to the direct contact with the rotating shaft. Additionally, they require frequent calibration.

In the last several decades, non-contact torque sensors have been developed for use with rotating shafts. Among them, magnetoelastic force/torque sensors became more popular in many applications than traditional strain gauges, due to their high operation reliability, compact size, easy installation, and low cost. One approach for this type of sensor is based on stress-induced anisotropy of the magnetic shaft or magnetic layer on nonmagnetic shaft [8.34]. Another approach is to use force/torque sensors that utilize the principle of inverse magnetostriction. In fact, the change in magnetic permeability of a material under stress (called inverse magnetostriction) offers the opportunity for fabricating high-performance and low-cost torque sensors. In an inverse-magnetostrictive sensor, a low-hysteresis magnetoelastic element is intimately attached to a rotating shaft. During machine operation, time-varying stresses or strains of the shaft are transmitted to the magnetoelastic element, causing measurable changes in the magnetic field. It operates on the premise that stresses and strains transmitted through the rotating shaft to the magnetoelastic element are proportional to the applied torque [8.35, 8.36].

Magnetorheological elastomers (MREs) are a relatively new class of smart materials that can undergo large deformations resulting from external magnetic excitation [8.37]. As composite materials, they consist of micron- or nano-sized magnetizable particles that are embedded into a non-magnetizable polymer matrix. MREs can change their mechanical behavior in response to an external magnetic field, making them to be promising candidates in developing actuators and sensing devices, such as force, temperature, and pressure sensors. A force sensor working with MRE was designed by Li *et al.* [8.38] for sensing normal loading. The MRE force sensor consists of three main parts: the mechanical assembly, electrical circuit, and LED display. The mechanical design is shown in Figure 8.37. The MRE is put near the bottom of the sensor. The metal spring is pressed to provide a recovery force to the sensor. As the upper plate is screwed, the O-ring is compressed, generating a preload acting on the MRE. When the reload is larger than the critical value, the sensor would generate a normal force to the MRE. Besides the preload, the external force results in the further deformation of the MRE, and the corresponding change in the resistance of the MRE can be precisely detected. The experimental testing shows that the sensor has a good linearity relationship and repeatability between the voltage output and loading force within some range of magnetic field. The MRE with the composition (55% carbonyl iron, 20% silicon rubber, and 25% graphite powder) has the best sensing capability.

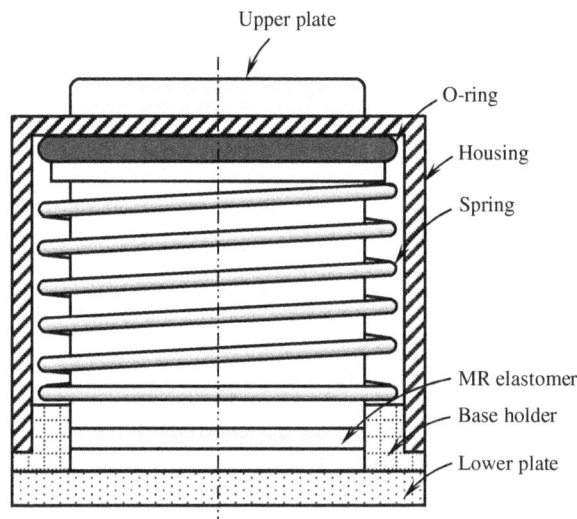

FIGURE 8.37 The mechanical design of the MRE force sensor, where the MR elastomer is placed near the sensor bottom. The metal spring is pressed to get a primary deformation, providing a recovery force to the sensor.

A typical example of a widespread use of force/torque sensors is collaborative robots that have developed rapidly in recent years. This new type of robot is specially designed to share a workspace with human workers. For consideration of safety, they have built-in force/torque sensors that provide instant feedback to the motion controller when any unusual impacts and abnormal forces are detected. When overloaded, the sensors immediately stop the robot or make it move in an opposite direction to avoid workplace accidents.

Another method is to measure the motor current through the inverter drive interface. A quick calibration of the application load and calculation can yield torque limiting information for the safety use for peak loading, incremental or continuous loads. The accuracy is probably ±5% but for safety purposes this information can be very possible.

8.5.2 TEMPERATURE SENSOR

Temperature sensors are used to measure the amount of heat energy. There are basically four types of temperature sensors: thermocouple, resistance temperature detector (RTD), thermistor, and monolithic integrated circuit (IC) sensor. All have different characteristics depending upon their actual applications. A comparison of these types of sensors is presented in Table 8.4.

Moreover, according to their operation mode, temperature sensors can be classified as contact type and non-contact type. Obviously, the contact type of sensor requires the sensor in physically contact with the object being sensed. They are widely used to detect solids, liquids, or gases over a wide range of temperatures. The non-contact type of sensor detects radiant energy being transmitted from an object in the form of infrared (IR) radiation.

8.5.2.1 Thermocouple

During 1821–1823, the German physicist Thomas Seebeck performed a series of experiments trying to directly convert heat into electricity [8.39]. Through his experiments, he discovered that a circuit made from two dissimilar metals (such as platinum, rhodium, iridium, chromel, alumel, iron, copper, constantan, and tungsten) with junctions subjected to a temperature gradient would deflect a compass magnet. After the discovery of the electron and its fundamental charge, it was quickly realized that the temperature difference produces an electrical potential that can drive an electric current in a closed circuit. The results were interpreted as a thermoelectric effect, which is now known as the Seebeck effect and is the basis of thermocouples. In essence, the Seebeck effect is the generation of a voltage across the conductor that counterbalances the applied temperature difference in open circuit conditions [8.40]. Thus, according to the Seebeck effect, thermocouples can only measure temperature differences, indicating that thermocouples are relative rather than absolute temperature sensors.

Among several types of temperature sensors, thermocouples are the most popular type of sensor used in electric motors. As passive sensors, they do not need any additional power source or excitation voltage. A thermocouple is fabricated by joining two different metallic alloy wires at one ends, referred to as the measurement junction or *hot* junction. The other end, where the wires are not joined, is connected to the signal conditioning circuitry traces. When the hot junction is placed on the object to measure its temperature, a voltage, known as the thermoelectric voltage or the Seebeck voltage V_s (in units of µV), is produced across the wire open ends. The voltage is proportional to the temperature difference between the measurement junction (*hot* junction) T_{mj} and non-heated wire ends T_{rj}, called the reference junction (*cold* junction), *i.e.*,

$$V_s = S\left(T_{mj} - T_{rj}\right) \tag{8.26}$$

where
   V_s is the differential voltage across the open terminals in µV or mV
   S is the Seebeck coefficient in µV/°C
   T_{mj} and T_{rj} are the temperature at the measurement and reference junction in °C, respectively

TABLE 8.4

Comparison of Various Types of Temperature Measurement Sensors

Sensor Type	Advantage	Disadvantage	Application	Temperature Range (°C)	Contact/ Noncontact	Note
Thermocouple	Self-powered, rugged capability, very fast response (less than 1 s), simple construction, inexpensive, widest temperature range, high long-term stability	Low output voltage in millivolt range, requiring reference junction compensation and signal amplifier, relatively low accuracy (typically ±1°C–2°C), least stable, least sensitive, susceptible to noise and corrosion, requiring frequent recalibration	Motors, generators steel-iron industry, engines, food processing equipment, semiconductor process, automotive, aerospace, industrial heat treating, diesel engines, gas turbines, 3D printers	Based on the specific type of thermocouples (−270°C–+2,320°C)	Contact	Base metal thermocouples for low-temperature measurement and noble metal/ refractory thermocouples for high-temperature measurement
RTD	Most accurate (up to ±0.01°C), most stable, good linearity (limited), high repeatability, high sensitivity (typically 0.4 Ω/°C), short response time (1–10s), wide temperature range	Usually expensive, external current source needed, signal conditioning required, susceptible to noise, affected by shock and vibration, low sensitivity, slow response time, self-heating problems, relative fragile	Generators, motors, aerospace, automotive, laboratory monitoring, HVAC systems, furnaces	−200°C to +850°C	Contact	
Thermistor	High output, very fast response, high accuracy (±0.1°C), high sensitivity (a 10 kΩ thermistor has the sensitivity of −1,699 Ω/°C), short response time (1–5 s), inexpensive, stable operation, not affected by shock or vibration, low cost	Highly nonlinear response, narrow temperature range, external voltage source needed, high-resolution ohms measurement required, self-heating problems, relative fragile, low long-term stability	Automotive, aerospace, medical, biological applications, HVAC systems, appliances, consumer devices, control systems	−50°C to +250°C (depending on the specific type, some can be used at temperature up to 300°C)	Contact	

(Continued)

TABLE 8.4 (Continued)
Comparison of Various Types of Temperature Measurement Sensors

Sensor Type	Advantage	Disadvantage	Application	Temperature Range (°C)	Contact/ Noncontact	Note
Monolithic IC	Excellent linearity (within ±1°C), no linearization required, better noise immunity through high level output signals, rugged as any IC housed in a plastic package, low cost	External power supply needed, lower accuracy than RTD and thermistor (±1°C), slow response (4–60s), narrow temperature range, limited configurations	Automotive and aerospace applications, telecommunications, notebook computers, HVAC systems, appliances, industrial immersion applications	−55°C to 150°C	Contact	
IR thermometer	Useful for measuring moving object, less material resources, high accuracy (±0.2°C), high resolution, flexible use, very fast response (in the ms range), good linearity, measurement on dot, line or area, measurement without process interruption	Unable to measure any gas or liquid, accuracy affected by intervening optical obstructions, reflective materials may skew reading, high cost for short term	Detecting temperature on operating machines, medical thermography, agricultural research	−20°C to 300°C (specially designed IR camera 300°C–1,500°C)	Noncontact	Enabling remote temperature measurement for avoiding risks in some applications, with wireless transmitter that transmit readings continuously to smartphone or laptop

The Seebeck coefficient, also known as thermoelectric sensitivity, of a material is a measurement of the magnitude of the induced electrical potential in response to the thermal gradient. It depends on the specific thermocouple type, more particularly, metal-lead materials. In practice, the Seebeck coefficient can be determined from the slope of a graph of the Seebeck voltage versus temperature difference, *i.e.*,

$$S = \frac{dV}{dt} \approx \frac{\Delta V}{\Delta t} \tag{8.27}$$

This indicates that the Seebeck coefficient can be calculated using available data sheets such as the National Institute of Standard and Technology (NIST) ITS-90 [8.41]. Alternatively, the Seebeck coefficient values for different thermocouple types can be easily calculated by means of a specially designed calculator developed by Fluke Corporation [8.42].

According to the constructive materials, thermocouples can be categorized into three main groups:

- Base metal thermocouples—This group of thermocouples, including types E, J, K, N, and T, is composed of common metals and metal alloys, such as iron, copper, aluminum, nickel, chromium, silicon, and manganese. As the most abundant thermocouples, they are often described as general-purpose thermocouples.
- Noble metal thermocouples—This group of thermocouples, including types B, R, and S, is manufactured using alloys of platinum and rhodium. They are specially designed for high-temperature measurements.
- Refractory metal thermocouples, sometimes known as tungsten thermocouples—They are manufactured from the exotic metals (*e.g.*, tungsten and rhenium), especially suitable for measuring extremely high temperatures (up to 2,300°C or higher). However, they are usually expensive, brittle, and have little or no oxidation resistance. Therefore, they are well suited for use in vacuum or no oxidizing environment.

As shown in Figure 8.38 [8.43, 8.44], while the *S-T* curves of the base metal thermocouples exhibit highly nonlinear behaviors, the curves of the noble metal and refractory thermocouples display

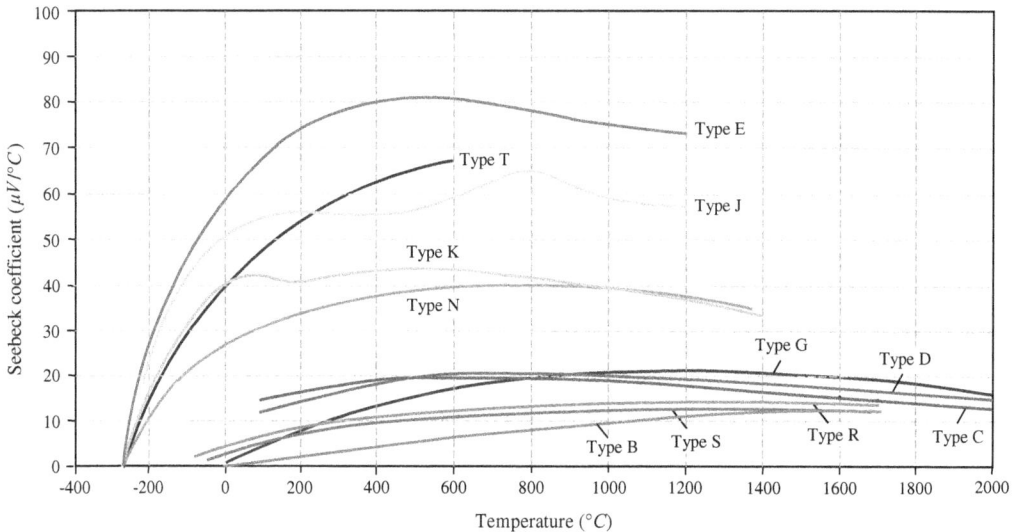

FIGURE 8.38 Variation of the Seebeck coefficient with temperature for a variety of thermocouple types. The *S-T* curves of thermocouple types C, D, and G are plotted based on the calculated data (where $S = \Delta V / \Delta T$) using thermocouple data sheets that contain the measured Seebeck voltages and their corresponding temperatures.

more linear behaviors. For the K type thermocouple, the Seebeck coefficient is approximately constant at 41 ± 2 $\mu V/°C$ from $0°C$ to $1,000°C$.

It is worth noting that the use of the terms *measurement junction* and *reference junction* rather than *hot junction* and cold junction in the traditional naming system is to avoid any possible confusion. This is because in many applications the temperature of the measurement junction can be lower than that of the reference junction.

As shown in Figure 8.39, due to the need to terminate the thermocouple wires, each thermocouple metal wire is connected to a copper wire at the reference junction. At the reference junction, these two connections are often mounted on an isothermal block made with a high thermal conductive material and are maintained at the same temperature. The Seebeck voltage produced at the reference junction depends on the voltages produced at both the measurement and reference junctions, where

$$V_s = V_{mj} - \left(V_{rj1} + V_{rj2}\right) \tag{8.28}$$

It follows that

$$V_{mj} = V_s + \left(V_{rj1} + V_{rj2}\right) \tag{8.29}$$

If the reference junction is truly at $0°C$, $(V_{rj1}+V_{rj2})$ is zero, and thus $V_{mj}=V_s$. The traditional method was to keep the reference junctions immersed in an ice bath ($0°C$), but that is clearly impractical for most modern applications. Today, reference junction compensation replaces the ice bath with an electronic circuit to compensate for the missing thermoelectric voltage because the thermocouple reference junction at the instrument is not at $0°C$. This allows electronics to use the established thermoelectric voltage table to determine the temperature at the measurement junction.

Based on the thermocouple materials, several thermocouple types are available. A comparison of operating characteristics of different thermocouple types is shown in Table 8.5.

Thermocouples are known for their versatility, lack of required excitation, simple structure, long-term stability, and low cost. Some disadvantages of thermocouples are: (*a*) Since the magnitude of the Seebeck voltage is of the order of a few μV, precise amplifiers are often required to

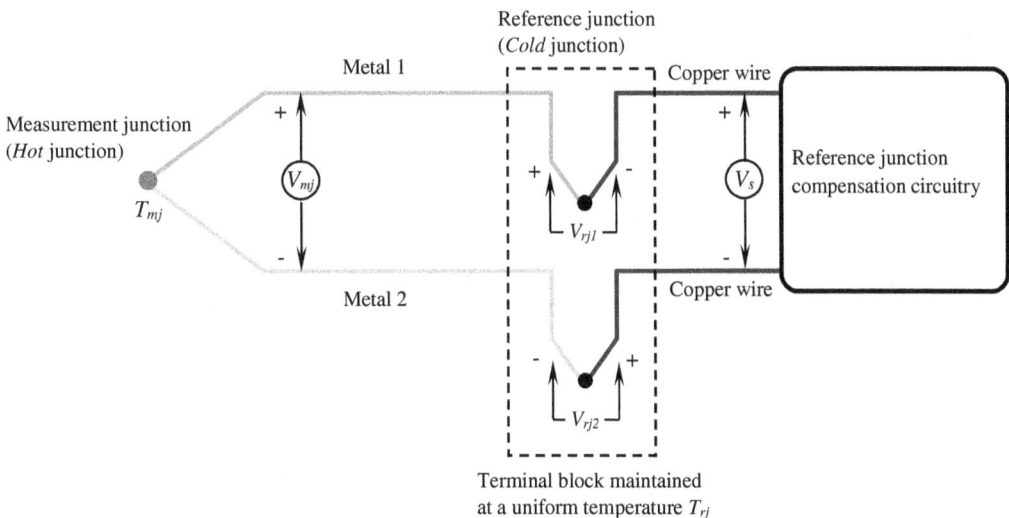

FIGURE 8.39 Typical thermocouple measuring circuit, where $V_s = V_{mj} - (V_{rj1} + V_{rj2})$.

TABLE 8.5
Thermocouple Types and Characteristics

Group	Type	Thermocouple Conductor	Seebeck Coefficient ($\mu V/°C$)	Temperature Range (°C)	Standard Accuracy (%)	Remark
Base metal Thermocouple	E	Nickel (90%)–Chromel (10%) (+) Constantan (−)	25 (at −200°C) 58.7 (0°C) 62 (20°C) 81 (500°C) 75 (1,000°C)	−270°C to 1,000°C	±0.50	Stronger signal and higher accuracy than type K or J, suitable for cryogenic use, producing highest EMF (mV)
	J	Iron-Constantan	22 (−200°C) 50.4 (0°C) 51 (20°C) 55 (150–500°C) 62 (700°C)	−210°C to 760°C	±0.75	Reliable, shorter lifespan, suitable for use in vacuum, oxidizing, reducing or inert atmospheres, oxidizing above 538°C (1,000°F)
	K	Nickel (90%)–Chromel (10%) (+) Nickel–Alumel (−)	22 (−200°C) 39.5 (0°C) 40 (20°C) 43 (600°C) 39 (1,000°C)	−270°C to 1,260°C	±0.75	Most common thermocouple in use, recommended for continuous oxidizing or neutral atmospheres, inexpensive, accurate, reliable, mostly used above 538°C (1,000°F)
	N	Nicrsil (+) Nisil (−)	18.3 (0°C) 27 (20°C) 39 (600°C) 37 (1,200°C)	−270°C to 1,300°C	±0.75	Very reliable and accurate at high temperatures, suitable for oxidizing or inert atmospheres, more stable than other base metal types E, J, K, and T
	T	Copper (+) Constantan (−)	16 (−200°C) 38.7 (0°C) 40 (20°C) 53 (200°C) 62 (400°C)	−270°C to 400°C	±0.75	Stable at low temperatures, suitable for cryogenic and vacuum measurements, useful in oxidizing, reducing or inert atmospheres, highly resistant to corrosion in moisture and condensation atmospheres

(Continued)

TABLE 8.5 (Continued)
Thermocouple Types and Characteristics

Group	Type	Thermocouple Conductor	Seebeck Coefficient (μV/°C)	Temperature Range (°C)	Standard Accuracy (%)	Remark
Noble metal Thermocouple	R	Platinum (87%)–Rhodium (13%) (+) Platinum 100% (−)	7 (20°C) 10 (300°C) 12 (700°C) 13 (1,000°C)	−50°C to 1,768°C	±0.25	High temperature, suitable for oxidizing or inert atmospheres, greater stability than types J, K, or N, industrial applications
	S	Platinum (90%)–Rhodium (10%) (+) Platinum 100% (−)	5.4 (0°C) 5.9 (20°C) 9.1 (300°C) 10.2 (600°C) 11.5 (1,000°C)	−50°C to 1,768°C	±0.25	High temperature, high resistance to oxidation and corrosion, good stability at high temperatures, non-magnetic conductor materials
	B	Platinum (70%)–Rhodium (30%) (+) Platinum (94%)–Rhodium (6%) (−)	1 (20°C) 6 (600°C) 11 (1,200°C) 13 (1,700°C)	0°C to 1,820°C	±0.50	High temperature, having lower sensitivity, suitable for continuous use in oxidizing or inert atmospheres, use above 50°C
Refractory thermocouple	C	Tungsten (95%)–Rhenium (5%) (+) Tungsten (74%)–Rhenium (26%) (−)	19.5 (600°C) 17.5 (1,200°C) 10.9 (2,200°C)	0°C to 2,320°C	±1.0	Extreme high temperature measurement, poor oxidation resistance, well suited for use in vacuum, highpurity hydrogen, or high purity inert atmospheres,
	D	Tungsten (97%)–Rhenium (3%) (+) Tungsten (75%)–Rhenium (25%) (−)	20.1 (600°C) 17.4 (1,200°C) 12.3 (2,200°C)	0°C to 2,320°C	±1.0	embrittlement due to oxidization, not for oxidizing atmosphere, not practice for temperature below 400°C
	G	Tungsten (100%) (+) Tungsten (74%)–Rhenium (26%) (−)	17.2 (600°C) 21.1 (1,200°C) 14.2 (2,200°C)	0°C to 2,320°C	±1.0	

Note:

(1) Constantan: Nickel (44%), Copper (55%), and Manganese (1%).

(2) Nicrsil: Nickel (84.4%), Chromium (14.2%), and Silicon (1.4%).

(3) Nisil: Nickel (95.5%), Silicon (4.4%), and Manganese (0.1%).

(4) Thermocouple types E, J, K, N, and T are defined as base metal thermocouples, with reasonably accuracy and low cost.

(5) Because thermocouple types R, S, and B are constructed of platinum and rhodium, they are referred to as noble metal thermocouples. They are more accurate and stable than base metal types but also more expensive.

(6) Thermocouple types C, D, and G are defined as the tungsten thermocouples.

(7) Tungsten–rhenium alloy thermocouples are well suited for measuring extremely high temperatures.

(8) Seebeck coefficient of K type thermocouple is approximately constant at 41±2 μV/°C from 0°C to 1,000°C.

amplify the magnitude of the output voltage. (*b*) Thermocouples are susceptible to external noise. When measuring microvolt-level signal changes, noise from stray electrical and magnetic fields can be a problem. (*c*) The outputs of thermocouples are not linearly proportional to any temperature scale. They require linearization or linearity calibration. (*d*) Because thermocouples are made from two dissimilar metals, in harsh environments corrosion may occur after long time services.

Modern thermocouples are made from semiconductors whose material properties and geometry have been tailored specifically to meet the intended application requirements. The voltage output of semiconductor thermocouples remains at hundreds of microvolts per degree, which is much higher than those of conventional thermocouples. In practice, a large number of semiconductor thermocouples are connected electrically in series and thermally in parallel by sandwiching them between two high thermal conductivity but low electrical conductivity ceramic plates to form a module [8.45].

8.5.2.2   Resistance Temperature Detector

RTDs are precise temperature sensors and are widely used in electric machines. They are structured in a coil or thin-film form made from high-purity conducting metals such as platinum, nickel, or copper. They measure temperature by correlating the electrical resistance of the RTD with temperature. In fact, RTDs are the most accurate temperature sensors with excellent stability and repeatability. They are relatively immune to electrical noise and therefore well suited for temperature measurements in industrial environments, especially around motors, generators, and other high-voltage equipment.

Carl Wilhelm Siemens made the first resistance thermometer in 1860 [8.46]. Ten years later, he introduced a platinum resistance thermometer for measuring high temperatures. The platinum RTD offers excellent accuracy over a wide temperature range (usually from −200°C to +850°C). Unlike thermocouples, it is not necessary to use special conducting wires to connect the sensor.

RTD output signals typically run in the millivolt range, making them susceptible to noise. Low pass filters are commonly available in RTD data acquisition systems and can effectively eliminate high-frequency noise in RTD measurements.

There are two ways to convert resistance to temperature. One way is to use the predefined thermocouple voltage-temperature table (*e.g.*, the NIST ITS-90 thermocouple database). Another is to use the Callendar-Van Dusen equation that describes the correlation between the resistance of the RTD and temperature for the temperature range $-200°C \leq T \leq 0°C$,

$$R(t) = R_o \left[1 + aT + bT^2 + cT^3(T - 100) \right] \tag{8.30}$$

where
  T is the temperature in °C
  R_o is the resistance at 0°C
  $R(T)$ is the resistance at temperature T
  a, b, c are known as the Callendar-Van Dusen constants

For the temperature range $0°C \leq T \leq 661°C$, the Callendar-Van Dusen equation can be simplified as

$$R(T) = R_o \left(1 + aT + bT^2 \right) \tag{8.31}$$

In fact, this equation can be used for subzero temperatures to −100°C. Below it, the error becomes very significant. For instance, at $T = -200°C$ the error can be 4.9%.

Most general-purpose RTDs are made of platinum wire. The resistances of platinum RTDs may range from tens to thousands of Ohms, but most have been standardized to a value of 100 Ω at 0°C (*i.e.*, $R_o = 100\ \Omega$).

8.5.2.3 Thermistor

Thermistors are passive semiconductor devices for measuring temperature. As a compound word, the name comes from the combination of the words *therm*ally sensitive and res*istor*. In fact, the electrical resistance of the thermistor changes with a change in temperature. Unlike RTDs, most types of thermistors have a negative temperature coefficient, *i.e.*, their electrical resistances decrease with an increase in temperature. This type of thermistor is referred to as negative temperature coefficient (NTC) thermistor. In addition, there is also another type of thermistor with a positive temperature coefficient, referred to as positive temperature coefficient (PTC) thermistor.

Thermistors are usually made from ceramic semiconductor material, which is formed into small pressed discs or balls and sealed to offer a fast response to the temperature PZT variation. Although the thermistors themselves are nonlinear elements, *i.e.*, their electrical resistances change with the temperature in a nonlinear matter, when they are used with a series such as in a Wheatstone bridge, the current obtained in response to a voltage applied to the bridge is linear with temperature. Thus, the output voltage across the resistor becomes linear with temperature. Thermistors are a popular device to add to a servomotor when temperature monitoring is critical.

One challenge in using thermistors is to calculate the temperature from the measured resistance value. To accomplish this, the three-term Steinhart–Hart equation is often adopted to convert a thermistor sensor's resistance to temperature. Compared with other methods, the Steinhart-Hart equation can provide more precise data across the sensors' temperature ranges,

$$\frac{1}{T} = a + b \ln R_T + c \left(\ln R_T \right)^3 \tag{8.32}$$

where

T is the absolute temperature, K
R_T is the resistance of the thermistor at temperature T, Ω
a, b, c are Steinhart–Hart coefficients for a given thermistor

8.5.2.4 Monolithic Integrated Circuit Temperature Sensor

In a monolithic IC sensor, while passive circuit elements such as resistors, inductors, and capacitors are fabricated in monolithic form on a silicon substrate, active amplifier stages are fabricated in the monolithic substrate that are coupled with the passive circuit elements as a temperature sensor. As a highly sensitive semiconductor type sensor, monolithic IC sensors can provide an output current proportional to absolute temperature and control complex thermal processes.

Monolithic IC sensors are normally employed in applications between −25°C and 105°C where conventional temperature sensors such as thermocouples, thermistors, and RTDs are currently being used. For some specially designed IC sensors, the operating temperature range can be expanded to −50°C to +150°C or even wider. The inherent low system cost, elimination of support circuitry (*e.g.*, linearization circuit), excellent linearity (usually ±0.15°C – ±0.8°C), high accuracy, repeatability and stability, wide operating temperature range, and minimal self-heating errors make them ideal for use in numerous applications, including motor temperature sensing, automotive HVAC temperature measurement, and industrial temperature control.

Researchers at Columbia University have invented a low power, wireless implantable temperature sensor that takes up less than 0.1 mm³. This sensor is made of a single chip and powered wirelessly by ultrasound. This sensor, equipped with a monolithically integrated PZT ultrasound transducer, is designed to replace larger implantable sensors that require several chips, packaging, wires, external transducers, and batteries [8.47]. Because of its extremely small size, it can be injected into the human body to monitor various biological parameters such as temperature.

Figure 8.40 illustrates a miniaturized, monolithically integrated, wireless temperature-sensing mote utilizing ultrasound for wireless power transfer and data transmission. The major components

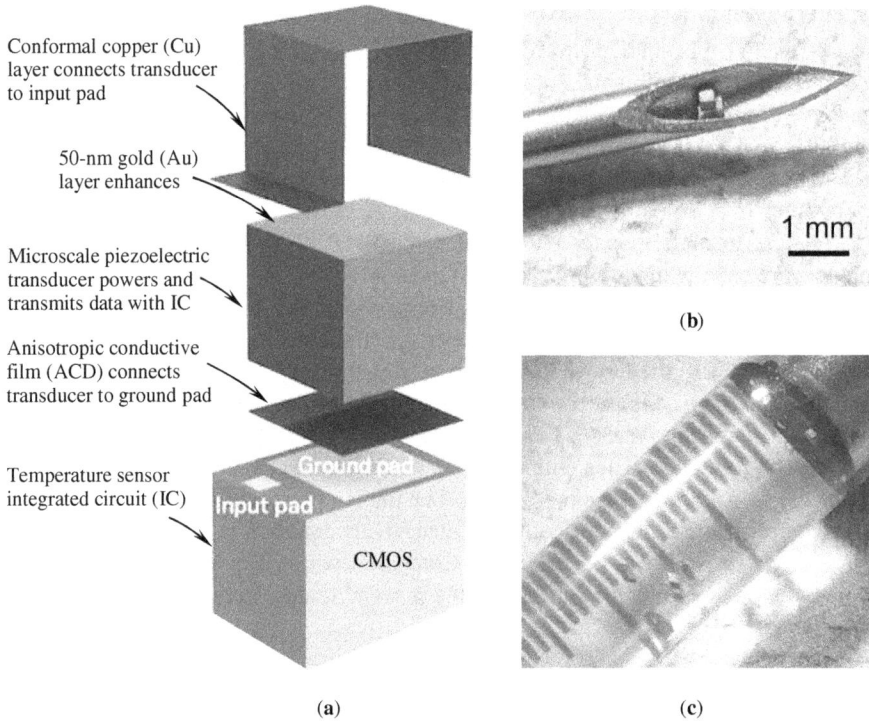

FIGURE 8.40 Miniaturized, monolithically integrated, fully wireless temperature-sensing motes with ultrasound powering and data communication [8.48]: (*a*) The illustration of the major components of a temperature-sensing mote. (*b*) A mote placed at the tip of an 18-G needle (inner diameter 0.84 mm). (*c*) Seven motes loaded in a 1-ml syringe filled with PBS solution.

of the mote include a complementary metal-oxide semiconductor (CMOS) temperature sensor chip with two exposed Al pads, a microscale piezoelectric transducer coated with a gold film on both sides to enhance the connections, an anisotropic conductive film that connects the transducer to the ground pad, and a conformal copper layer that connects the transducer to the input pad [8.48]. The small-displaced volume of these motes allows them to be implanted or injected using minimally invasive techniques with improved biocompatibility. Although the current sensor version is only used to measure body temperature, there will be many potential industrial applications in the near future.

8.5.2.5 Infrared Thermometer

Any object with a temperature above absolute zero ($> 0\ K$ or $>-273.15°C$) emits a continuous spectrum of electromagnetic radiation at all frequencies, referred to as thermal emission. The radiated energy is in the form of electromagnetic waves, which is directly dependent on the temperature and surface condition of the object. Generally, the shorter the radiation wavelength, the higher the radiation energy transferred from the object. The radiant energy, as well as the corresponding wavelength range, is listed in descending order: γ-rays (< 0.03 nm), X-rays (0.03–30 nm), ultraviolet rays (30 nm–0.40 μm), violet (0.38–0.45 μm), blue (0.45–0.50 μm), green (0.50–0.57 μm), orange (0.57–0.62 μm), red (0.62–0.75 μm), IR (0.75 μm–1.0 mm) [including near IR (0.75–1.4 μm), mid-infrared (1.4–3.0 μm), and far infrared (3.0 μm–1.0 mm)], microwave (1–300 mm), and radio wave (1 mm–100 km). The visible light has wavelengths ranging approximately from 0.38 to 0.75 μm.

A typical IR thermometer consists of a lens to focus the IR thermal radiation onto a detector (*i.e.*, thermopile), which converts the IR radiant energy into an electrical signal. As non-contact temperature sensors, the IR thermometers detect the IR energy emitted by an object and collected by the thermometer's detector filtered at a specified wavelength. The temperature reading is determined by comparing the measured IR energy to the theoretical amount of IR energy emitted by a blackbody at the same temperature. This working principle enables IR thermometers to sense temperature remotely from a safe distance away from the target to be measured, allowing for continuous temperature monitoring without the need to interrupt or stop system or component operation. Because of their operating flexibility, good accuracy, portability, and superior performing capability, they are useful in measuring temperature across a wide range of industrial, medical, military, astronomy, firefighting, border patrol crossing, and other applications. These non-contact temperature measurement sensors function well in circumstances where other types of temperature sensors are impractical.

IR thermometers can be made in different forms, such as IR thermal imaging cameras, IR scanning systems, spot IR thermometers (*e.g.*, IR forehead thermometers), and night-vision goggles. Today, modern thermal imaging cameras are IR cameras that are capable of measuring the heat radiating from objects. Instead of visible light, IR cameras operate with invisible light. This special feature makes them useful in total darkness, smog environment, and other low visibility, low-contrast situations.

Several key features should be considered when selecting an IR thermometer, including the following: (*a*) An important feature is the distance-to-spot ratio, which is defined as the detecting distance over the size of the area being evaluated by the IR thermometer. Depending on the thermometer construction, the most common distance-to-spot ratios are 4:1, 8:1, 30:1, and even up to 50:1. The larger the ratio value, the better the instrument's resolution and the smaller the spot size that can be measured. (*b*) Another consideration is the size of the object to be detected. Generally, the object must be optically visible and larger than the spot size of the IR thermometer. (*c*) As a non-contact sensor, IR thermometers receive all thermal radiant light (IR energy) within their field of view. Therefore, all thermal sources near the detected object can affect the measurement accuracy.

Another consideration in selecting an IR thermometer is self-calibration. Old models and low-cost models require calibration with a black body each time the meter is used; thus, large inaccuracies occur when the tune on calibration step is omitted. Meters with self-calibration provide ease of use and most user accuracy.

8.5.3 Vibration Sensor

For electric rotary machines, one of the parameters for judging their dynamic operating condition is vibration. Although in many cases vibration in industrial machinery is inevitable, excessive vibration can lead to serious consequences. Continuously monitoring and recording the vibration level of a machine allow prediction of problems before serious damage occurs.

Three parameters, that represent motion characteristics, are typically measured by vibration sensors: displacement s, velocity v, and acceleration a; all of them can be mathematically related to each other, $v=ds/dt$ and $a=dv/dt=d^2s/dt^2$. Generally, displacement sensors are used in low frequency (lower than 100 Hz) and low amplitude displacement for measuring shaft motion and internal clearances. Velocity sensors are for low- to medium-frequency measurements (up to 1,000 Hz). They are useful for vibration monitoring and balancing operations on rotating machinery. Acceleration sensors are the preferred motion sensors for high-frequency vibration applications [8.49].

For electric motors, vibration to be measured falls under one of the following types: (*a*) Axial (thrust) vibration, which is a longitudinal vibration in the direction of the shaft axis. This type of vibration can affect the shaft misalignment or vice versa. (*b*) Radial vibration, which occurs as a force applied outward from the rotor in the radial direction.

Accelerometers are devices that convert mechanical vibrations into electrical signals to measure or sense vibration or change in motion (*i.e.*, acceleration) of machinery, and thus widely used for monitoring and analyzing vibration in aerospace, automotive, and manufacturing applications. These devices are often utilized to diagnose problems with rotating equipment or assess the stability of structures that are subject to periodic stresses. They are designed to use in specific temperature and frequency ranges, as well as their orientation relative to the change in motion being measured. According to the power supply mode, accelerometers can be classified into either AC-response or DC-response accelerometers. While the AC coupled device is only suitable for measuring dynamic events, the DC coupled device is mainly used to measure static acceleration as well as dynamic acceleration with a very low frequency (typically < 1 Hz).

The most commonly used accelerometers can be categorized into six types, and they are each designed to efficiently function in their intended environments:

- The most common AC-response accelerometers utilize the piezoelectric effect to sense vibration or change in acceleration on machine surfaces. This can be done by converting mechanical force caused by vibration into electrical current. The piezoelectric sensors are reliable, versatile, unmatched for frequency, and amplitude range.

  There are two types of piezoelectric accelerometers: high impedance accelerometer that produces an electrical charge that is connected directly to the measuring instruments; and low impedance accelerometer that has a charge accelerometer as its front end as well as a built-in microcircuit and transistor to convert the charge into a low impedance voltage.

- Piezoresistive is the other commonly used DC-response sensing technology for handling static acceleration such as gravity and constant centrifugal acceleration and produces accurate velocity and displacement data. Piezoresistive accelerometers are versatile in terms of their frequency and dynamic range capabilities. A piezoresistive accelerometer produces resistance changes in the strain gauges that are part of the accelerometer's seismic system. The output of most piezoresistive designs is generally sensitive to temperature variation. It is therefore necessary to apply temperature compensation to its output internally or externally. Modern piezoresistive accelerometers incorporate the application specific integrated circuit (ASIC) for all forms of on board signal conditioning, as well as in-situ temperature compensation [8.50].

- Being DC-response devices, strain gauges are extensively used in many different measurements based on the same mechanism. As the name indicates, a strain gauge measures the strain on a machine surface. Because the measured strain value varies with applied force, it thus converts force, pressure, tension, *etc.* into an electric signal. This type of vibration sensor is versatile and accurate while is suitable for hazardous environments. However, it is hard to install correctly.

- As contactless AC-response devices, eddy-current sensors are used to measure the position and the change of position of a conductive object. The eddy-current sensor has a probe that creates a magnetic field at its tip. When the probe approaches the conductive object (*e.g.*, motor), it induces alternative current (called eddy-current) in the object. Thus, the sensor monitors the interaction of these two magnetic fields. Their high durability and contactless feature make them a good option for working under dirty environments.

- Like eddy-current sensors, capacitive accelerometers are also contactless sensors. Being a DC-response type of accelerometer, it senses changes in capacitance to determine an object's acceleration. The latest MEMS capacitive accelerometers are finding use in applications traditionally dominated by piezoelectric accelerometers and other sensors [8.51]. Because of their tiny size and affordability, they are embedded in a myriad of handheld electronic devices such as smartphones, digital cameras, and tablets. High-performing MEMS accelerometers offer low-cost solutions for a wide range of applications that incorporate vibration measurements.

  The basic structure of a sensing element in a capacitive MEMS accelerometer is shown in Figure 8.41 [8.52]. The sensing element consists of an upper plate of single crystal silicon

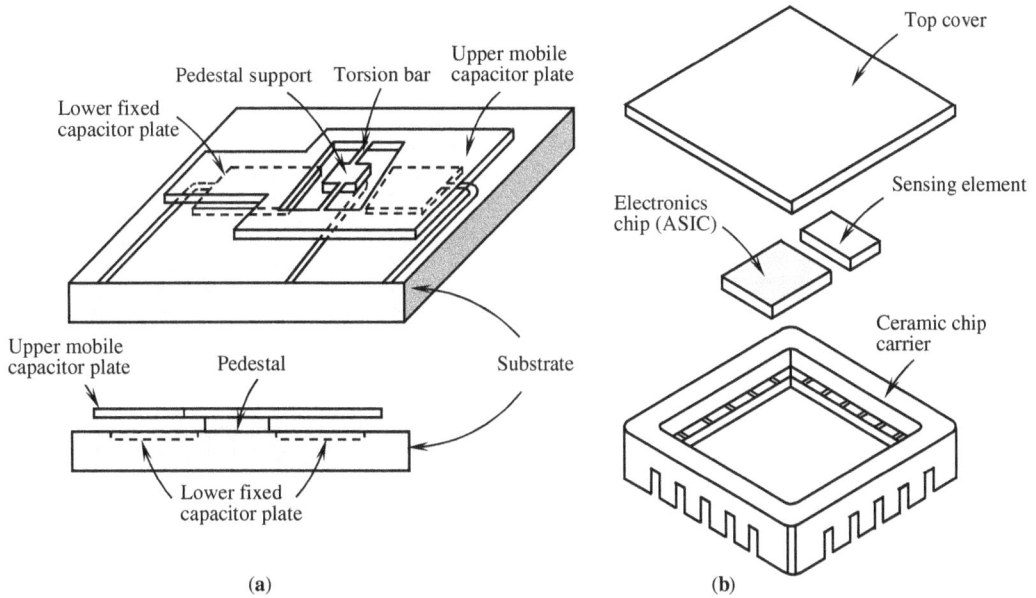

FIGURE 8.41 A capacitive MEMS accelerometer: (*a*) the mechanical structure of the sensing element and (*b*) an assembly of the capacitive accelerometer. The sensing element consists of an upper plate of single crystal silicon, which is free to rotate a small angle about a torsional axis, and two lower conductive plates, which are fixed on the substrate. The upper plate and two lower plates thus form two capacitors.

and two lower conductive plates. The torsion bar is made on the upper plate, enabling the upper plate to turn freely with respect to the torsion bar axis. The structure of the upper plate is asymmetrically shaped so that one end from the torsion bar is heavier than the other end, resulting in the center of mass that is offset from the torsion bar axis. Beneath the upper plate, two conductive plates are symmetrically located on each side of the torsion bar axis and attached on the substrate surface. Consequently, the upper plate and the two lower plates on the substrate form two capacitors with variable airgaps, creating a fully active capacitance bridge. When the torsion bar turns a small angle with respect to its axis, the average gap between the upper plate and one lower plate decreases, increasing the capacitance for the corresponding capacitor, while the average gap between the upper plate and the other lower plate increases, decreasing the capacitance for that capacitor. The sensing element and the electronics chip (ASIC) that is needed to convert the capacitance change of the sensing element into a useful electrical digital (or analog) signal are packaged in a ceramic chip carrier. The package is solder sealed to provide a strong and fully hermetic device.

Primarily, the sensitivity of the sense elements (the ratio of deflection to acceleration) is determined by the mass of the sensing element, the distance from the center of mass to the torsion bar axis, and the torsion bar stiffness.

- Ultrasensitive miniature optical accelerometers use IR laser light rather than electrical circuits to detect a change in the movement of equipment. The new IR laser-based accelerometers have been designed and developed by a NIST research team based on lasers and optomechanical principles [8.53]. As shown in Figure 8.42, this new optomechanical accelerometer consists of a pair of silicon chips and is only a millimeter thick: the bottom chip has a proof mass, with a mirrored coating on the surface, suspended by a set of silicon beams, allowing the proof mass to move up and down freely in response to acceleration in that axis. The top chip has an inset hemispherical mirror. Hence, the proof mass and the hemisphere mirror form an optical cavity. The IR laser is chosen to nearly match the cavity's resonant wavelength so that the laser beam builds in intensity as it bounces back and forth between the two-mirrored surfaces many times before exiting. IR laser light emitted from the laser emitter enters the bottom chip and exits from the top chip and is detected by the detector. When the accelerometer is subjected to acceleration, the cavity height changes correspondingly, shifting the resonant frequency. By continuously matching the laser to the resonant frequency of the cavity, engineers can determine the acceleration of the device.

The significant benefits of this new type of accelerometer include the following: (*a*) This type of accelerometer has no need to undergo the time-consuming process of periodic calibrations. (*b*) Because the instrument uses laser light of a known frequency to measure acceleration, it may ultimately serve as a portable reference standard to calibrate other accelerometers. (*c*) They are much more precise than the best commercial accelerometers in the market. (*d*) They can detect accelerations down to 32 billionths of a gravity. (*e*) Accelerometers have multiple applications in industry and science. While the optomechanical accelerometer is used for vibration measurement of rotating machinery, the device is equally well suited for inertial sensing, seismometry, and gravimetry [8.54]. It is especially useful when it is used in inertial navigation systems where GPS do not available, such as in submarines, aircrafts, satellites, and missiles. They also can be used in drones for flight stabilization.

Similar to the feedback devices, selecting a suitable sensor for a specific application depends on a number of sensor features and operation parameters, including the following: (*a*) environmental condition; (*b*) sensor frequency range; (*c*) sensor accuracy; (*d*) vibration and shock load level; (*e*) sensor temperature range; (*f*) signal level; (*g*) sensor transverse sensitivity; (*h*) detected shape of object; and (*i*) sensor mechanical features (*e.g.*, compact size, lightweight, and structure ruggedness). Using vibration data information often requires empirically understanding and documenting baseline data of new/normal use and abnormal operation for warning notifications.

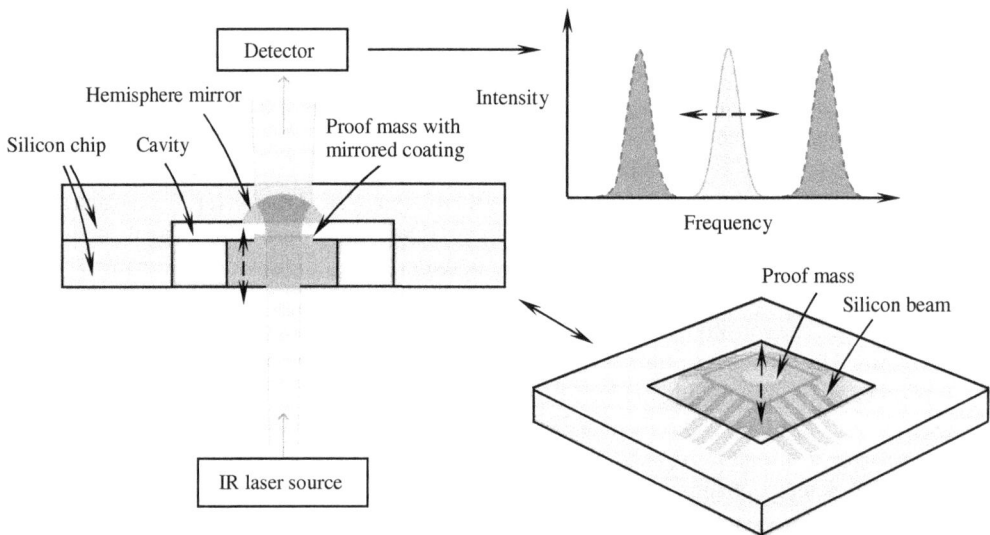

FIGURE 8.42 Operating fundamental of IR laser-based accelerometer. The proof mass is suspended by silicon beams on the silicon chip, allowing the proof mass to move up and down freely. A hemisphere mirror is inserted in the top silicon chip. Light transmitted through the top silicon chip is sensed by the detector.

8.5.4 Current Sensor

Current measurement is required for control, protection, monitoring, and power management purposes in electric machines. Because torque in a motor is directly proportional to current through its windings, monitoring the current in the motor via the drive inverter electronics and interfaces is one of the easiest methods and lowest cost to monitor motor torque. Current sensors are used to sense the amount of current in use by servomotors or particular equipment. Motor current is used as a control variable in motor drives. Vector control and direct torque control require current sensing for control purposes. In practice, motor current measurement is also used to reduce torque ripple in order to provide smooth torque.

There are several methods to sense current, each has its own advantages and disadvantages. The most commonly used current sensors in motor control systems are shunt resistors, Hall effect sensors, and current transformers. Among these sensors, shunt resistors, which use the resistive-based current sensing technique, are the most linear, the lowest cost, and are suitable for both *AC* and *DC* measurements. However, this method is unable to provide electrical isolation and incurs significant power losses at higher currents. As a comparison, Hall effect sensors and current transformers, both use electromagnetic-based techniques and provide inherent electrical isolation, allowing them to serve high current systems regardless of the level of pole voltage. However, they have a higher cost and result in a less accurate solution than shunt resistor usage. A comparison of the three types of current sensors is provided in Table 8.6 [8.55].

With today's advanced chip technology, it is more and more popular to use microelectronic chips, referred to as current IC sensors, to measure current and provide an integrated solution for end users. With an IC sensor, an output voltage, which is proportional to the current passing through it, is generated and transferred to a microcontroller. The current IC sensor is usually composed of chip circuits, a number of sense resistors and capacitors, RC low-pass filters, a non-inverting amplifier, and an output buffer.

8.5.5 Pressure Sensor

There are many methods to measure and monitor pressure. Regardless of the methods, all pressure sensors are designed to produce an output voltage or current that is dependent on the amount of pressure being sensed.

TABLE 8.6
Comparison of Three Types of Current Sensors [8.55]

	Sensor Type		
Feature	**Shunt Resistor**	**Hall Effect**	**Current Sensing Transformer**
Accuracy	Good	Good	Medium
Accuracy vs. temperature	Good	Poor	Good
Cost	Low	High	Medium
Electrical isolation	No	Yes	Yes
High current measuring capability	Poor	Good	Good
DC offset problem	Yes	No	No
Saturation/hysteresis problem	No	Yes	Yes
Power consumption/power loss	High	Low	Low
Intrusive measurement	Yes	No	No
AC/DC measurement	Both	Both	AC only

Pressure sensors find wide applications in a range of markets including medical equipment, motion control systems, automotive and aerospace industries, and HVAC systems, to name a few. In the motor industry, pressure sensors are commonly used to control operation of pump motors. In food processing applications, pressure sensors can sense the pressure inside a washable motor to detect sealing fault conditions.

There is a vast array of pressure sensors available on the market today, which function similarly but rely on different technologies. Conventional pressure gauges use diaphragms and strain gauges to provide a direct readable output reading of pressure in analog form. The strain gauge is attached to a diaphragm. As the diaphragm deflects under pressure, a change in electrical resistance of the strain gauge is measured by a Wheatstone bridge circuit, which converts a small change in resistance to a small change in voltage, in the order of a few millivolts. Thus, electronic circuitry is often required to amplify the voltage and produce a reading [8.56]. Due to the simple and robust construction, low cost, and durability, strain pressure gauges were the most widely used pressure measurement devices in the past.

As a comparison, pressure sensors produce an electrical signal that is proportional to the applied pressure. This output signal is then converted into a calibrated pressure reading. Along with the rapid development of digital technology, digital pressure sensors that help improve the system's safety, efficiency, and reliability in industrial and military applications become popular in recent years. In applications with battery-operated systems, battery-powered digital sensors with low-power consumption are preferred over analog pressure sensors [8.57].

Working on the principle of piezoelectric effect, piezoelectric pressure sensors are designed to measure dynamic rather than static pressure. Dynamic pressure measurements require sensors with special capabilities such as fast response, high stiffness, ruggedness, and extended ranges. The sensing elements are often made from crystals such as quartzes or specially formulated ceramics to ensure stable, accurate, and repeatable operation. These materials can tolerate very high temperatures (up to $1,000°C$), making them suitable for application such as measuring pressures in gas turbines and jet engines. When pressure is applied to a crystal, it is elastically deformed, resulting in a flow of electric charge. This electric signal is measured as an indicator of pressure.

Over the last few decades, microelectromechanical system (MEMS) and nanoelectromechanical system (NEMS) technologies have greatly facilitated the development of modern sensors. Many types of sensors can be miniaturized using silicon fabrication techniques and combined with electronics as MEMSs. Among them, MEMS piezoelectric pressure sensors are extensively used in a wide variety of applications. They are based on a micromachined silicon diaphragm as the piezoresistive sensing element. Some sensors include integrated electronics for signal conditioning and readout. Because of their small size and high degree of integration, MEMS piezoelectric pressure sensors have extremely low power consumption and fast response to change in pressure.

A resonant pressure sensor uses a pressure-sensing element to convert the applied pressure into the oscillation frequency change of the resonator. The resonator is fixed on a diaphragm or the edge of the cavity. When pressure is applied on a sensor, the silicon pressure-sensitive diaphragm is deformed, causing tensile and compressive stress acting on the resonator. A magnetic field is used to excite the resonator, and the inductive voltage is used to detect the frequency of the resonator. In such a way, the readout resonant frequency is related to the applied pressure [8.58].

Other types of pressure sensors are also available, such as inductive, capacitive, and reluctance sensors. In these sensors, a change in pressure results in a deformation or movement of the detecting element (*e.g.*, diaphragm), which in turn changes the corresponding inductance, capacitance, or reluctance. The variation of any physical parameter is detected and converted into output electrical signal.

8.5.6 MAGNETIC FIELD SENSOR

Magnetic field sensors, also known as magnetometers, are widely used to measure the direction and strength of magnetic field (**H**) and magnetic flux density (**B**) in modern industrial, medical, military, and consumer products, such as servo systems, nuclear magnetic resonance (NMR), magnetic resonance imaging, drones, smartphones, and cameras, to name a few. In addition to magnet technology, the measurement of the magnetic field is important in various technical areas. High sensitivity field measurements are used to detect ores underground, mines, submarines in the ocean; heart and brain activities; flux distribution inside superconductors; and data stored on magnetic support. In servo systems, magnetic field sensors are used to measure magnetic intensity around PMs, electromagnetic coils, and other electrical devices.

There are numerous types of magnetic field sensors based on different mechanisms. The most suitable sensor for an application can be selected by trading off among the sensor operating factors, including the field measurement range, reproducibility, accuracy, mapped volume, field geometry, and time bandwidth. The main approaches for magnetic sensing are briefly outlined below:

- The magnetic resonance devices, either NMR or electron paramagnetic resonance, provide the most reliable tools for measuring homogeneous magnetic fields with an accuracy of 0.1 ppm or better.
- The first fluxgate magnetometers were built in the 1930s and largely developed during World War II. One of the applications was submarine detection from low-flying aircraft [8.59]. The working principle of the fluxmeter sensor is based on the magnetic induction law in its integral form $\Phi = \int \mathbf{B} \cdot d\mathbf{A}$. A fluxmeter detects a loop voltage V, which is induced from the variation of the flux Φ with respect to time (*i.e.*, $V = -d\Phi/dt$), and uses it to measure the flux variation. Consequently, the magnetic field can be deduced from flux measurement. The fluxmeter is capable of providing a very accurate measurement of field direction, which is of great importance in some cases. The field measurement using a fluxmeter is generally performed either keeping the induction coil fixed, in the case of changing field, or moving the coil in a static field.
- Hall effect sensors are extensively used for proximity sensing, positioning, speed detection, current sensing, and magnetic field detection applications. Due to their high linearity between the output voltage and the magnetic field strength, the Hall effect sensors, as solid state-based magnetometers, are also designed to measure the magnitude of the magnetic field. In fact, among the various sensing technologies employed to detect magnetic fields, the Hall effect sensors have perhaps the most widespread usage [8.60]. However, Hall sensors are not ideal for sensitive magnetic field measurements in space and other environments because they have sensitivity proportional to magnetic field strength, making them ill-suited for near-zero field sensing. They are also prone to temperature drift and radiation [8.61].
- As non-contacting measuring systems, magnetoinductive (MI) sensors combine the advantages of both inductive and magnetic sensors. The MI sensing technology has some useful attributes that set them apart from other magnetic sensing technologies, specifically, (*a*) The technology provides a digital output so that it does not require additional hardware such as amplifiers and analog-to-digital converters. (*b*) A very high level of resolution can be obtained with electronic circuits. (*c*) The MI technology has very low power consumption. (*d*) The output is inherently stable over a range of temperatures, and (*e*) MI sensor design and the forward/reverse biasing nature of the MI circuit result in virtually no hysteresis [8.62]. Due to their flexibility, ruggedness, and miniature size, the MI sensors are ideal for many applications, such as compassing, inertial navigation, and security systems.
- Among a wide range of magnetic sensors, those based on magnetoresistive (MR) effects are particularly attractive for their low cost, small sense, and relatively high resolution at room temperature [8.63]. The MR effect is referred to as the change of resistivity of a current-carrying ferromagnetic material due to a varying magnetic field. Practically, the MR and Hall technologies are compatible with IC processing and are used to make

totally integrated single-ship sensors. Compared to a Hall effect sensor, an MR sensor is roughly 100 times more sensitive than a Hall effect sensor and its sensitivity is adjustable through selection of film thickness and line width. In addition, the MR sensor has small power consumption, but it has a narrow linear operating range. Furthermore, the MR sensors need to use a set/reset coil for their preset/reset operation. This will increase the complexity of their manufacturing process, while also increasing the sensor size and power consumption. The concept of the first MR sensor was proposed half a century ago by Hunt [8.64]. Today, there are three types of MR sensors: anisotropic magnetoresistive (AMR) sensor, giant magnetoresistive (GMR) sensor, and tunnel magnetoresistive (TMR) sensor. GMR effect involves small changes in magnetic fields creating major differences in electrical resistance. For the discovery of the GMR effect, the 2007 Nobel Prize in Physics was shared by the French scientist Albert Fert and German physicist Peter Grünberg.

A comparison of the technologies based on MR and Hall effects is given in Table 8.7 [8.65]. It can be seen from the table that among all magnetic sensing technologies, the newly developed TMR sensors have the lowest power consumption, most compact design, highest field sensitivity, finest resolution, and highest operating temperature.

- Fluxgate sensors are often used to measure a magnetic field with high precision. As a type of vector magnetometer, it can measure the strength of a magnetic field in a specific direction in 3D space. However, this kind of sensor is relatively large and always expensive. As depicted in Figure 8.43, a fluxgate sensor consists of a ferromagnetic material wound with two coils, a drive coil and a sense coil. It exploits magnetic induction together with the fact that all ferromagnetic material becomes saturated at high fields. This saturation can be seen in the hysteresis loops shown on the right side of Figure 8.43 [8.66].

- Magneto-optical (MO) sensors have been developed based on the Faraday effect that was discovered in 1845 by Michael Faraday. The effect describes that light passing through a transparent medium with an external-applied magnet field alters the light wave depending on the magnetic field. This discovery was the first indication of interaction between light and magnetism and later led to the establishment of Maxwell's equations [8.67]. This type of sensor offers a wide variety of application opportunities for research tasks regarding magnetic domain analysis and investigation.

- Superconducting quantum interference device (SQUID) is the most sensitive magnetometer in sensing magnetic flux Φ at low frequencies (where $\Phi = \int \mathbf{B} \cdot d\mathbf{A}$, *i.e.*, the product of magnetic flux density B and the area A of the SQUID loop). SQUID magnetometers usually detect the change in magnetic flux created by mechanically moving the sample through a superconducting pick-up coil that is converted to a voltage V_{SQUID}. However, special attention is needed to quantify and correct the residual fields of the superconducting magnet to derive useful information from integral magnetometry while pushing the limits

TABLE 8.7
Comparison of magnetic sensing technologies based on MR and Hall effects [8.65]

Technology	Hall Effect	AMR	GMR	TMR
Power consumption (mA)	5–20	1–10	1–10	0.001–0.01
Die size (mm^2)	1×1	1×1	1×2	0.5×0.5
Field sensitivity (mV/V/Oe)	~ 0.05	~ 1	~ 3	~ 100
Dynamic range (Oe)	~ 10,000	~ 10	~ 100	~ 1,000
Resolution (nT/Hz$^{1/2}$)	> 150	0.1–10	1–10	0.1–10
Operating temperature ($^\circ C$)	< 150	< 150	< 150	< 200

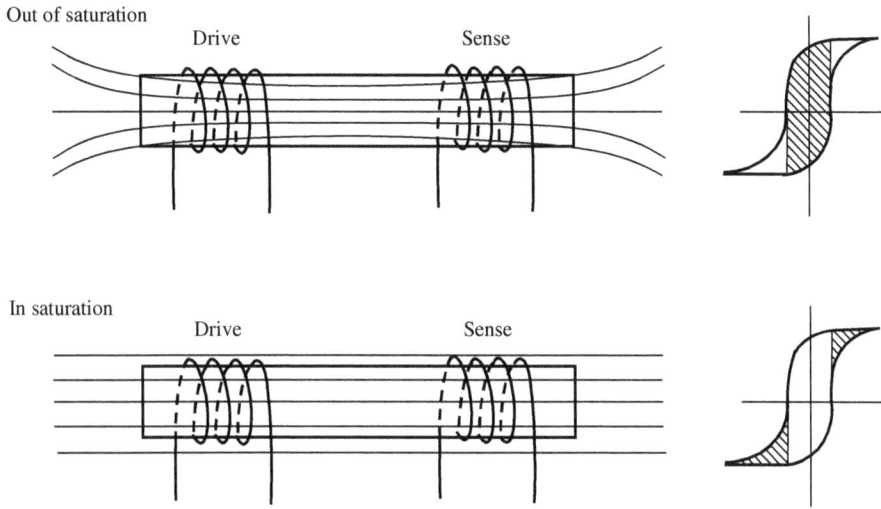

FIGURE 8.43 Illustration of the operating principles of fluxgate magnetometers. The output signal becomes modulated by driving the soft magnetic core into and out of saturation. The shaded regions indicate the regions of operation [8.66].

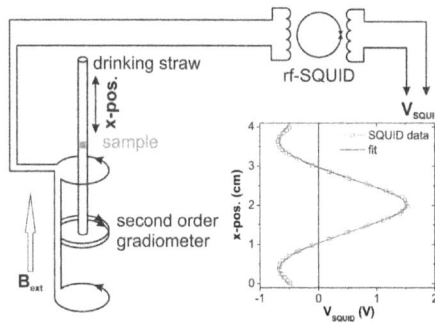

FIGURE 8.44 The schematic setup of a SQUID detection system with a second-order gradiometer. The inset exemplarily shows a single SQUID scan where the maximum of V_{SQUID} occurs at the x-position of 2 cm [8.68].

of detection and to avoid erroneous conclusions. The setup of a SQUID magnetometer is sketched in Figure 8.44 [8.68]. The insert in this figure shows the SQUID response V_{SQUID} versus sample position (x-position). It is worth to note that in order to maintain superconductivity, the entire SQUID needs to operate at cryogenic temperatures, for instance, less than 77 K (liquid nitrogen temperature) for high-temperature SQUIDs, or less than 4.2 K (liquid helium temperature) for low-temperature SQUIDs.

Figure 8.45 is the overview of typical accuracy attainable with different magnetic field measurement techniques depending on the magnetic field strength. It greatly helps control engineers assess the pros and cons of each sensing technology so that the most suitable sensors can be determined for their specific applications.

Although considerable progress has been made in the past few years, the exploitation of magnetic field sensors with high sensitivity and accuracy requires further development of the sensing technology.

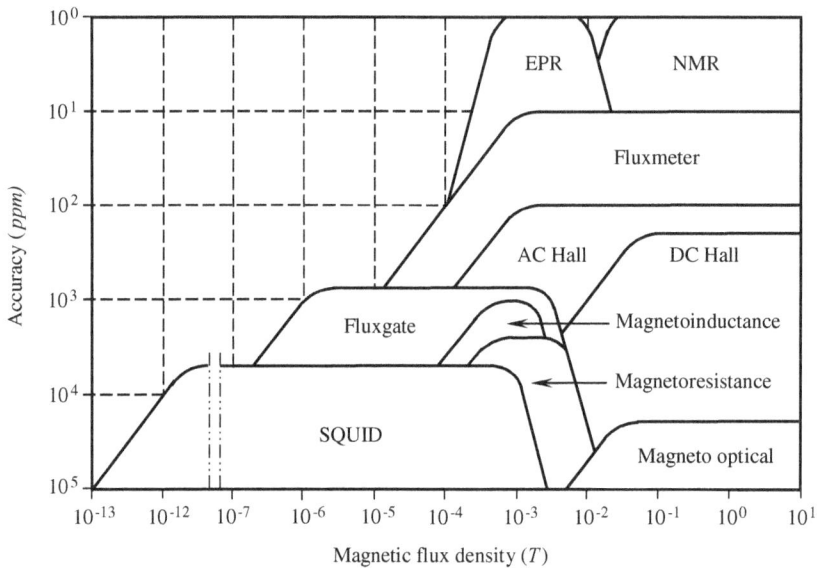

FIGURE 8.45 Overview of typical accuracy attainable with different magnetic measurement techniques depending on the measured flux density [8.59]. Note that 1 ppm = 0.0001%.

8.6 IMPROVING MOTOR SENSOR PERFORMANCE

During the control process of a servo system, errors in speed or position may be introduced into the system due to the less than perfect mechanisms that transfer the motor power to the load. Environmental factors may also introduce positioning errors, causing servo system to miss the target.

8.6.1 MITIGATION OF ELECTRICAL NOISE

Electrical noise in a sensor's output is the primary factor limiting its smallest possible measurement. It refers to a random fluctuation in an electrical signal, characterized by an undesired disturbance superimposed on useful information. In fact, all electronic components produce small random changes in voltage potentials that combine throughout the circuitry and appear as a band of noise.

There are two types of electrical noise according to the conduction mode: normal-mode and common-mode. Normal-mode noise (also known as transverse-mode noise) is the voltage noise that appears between two active conductors in an electrical system. On the other hand, common-mode noise in terms of AC power is the noise signal between the neutral and the ground conductor. In a three-phase connection, common-mode noise is noise-voltage appearing equally in-phase with each line-to-line or line-to-neutral voltage [8.69]. Typically, common-mode noise impulses tend to be higher in frequency than the associated normal mode noise signal. This is expected since the majority of the common-mode signals originate from capacitively coupled normal mode signals. The higher the frequency, the greater the coupling among the conductors, line, neutral, and ground [8.70].

Therefore, as discussed previously, long cable length by nature is more susceptible to noise. It is highly recommended to use shielded, twisted-pair cables with preferably low capacitance value.

EMI is an issue that affects signal transmission in servo feedback devices and drives. EMI sources can be power electronics switching, relays, power line surges, electrostatic discharges, lighting, RF equipment, *etc.*

8.6.2 SUPPRESSION OF TEMPERATURE RISE

The sensors must operate within specified temperature ranges for sustainable and reliable performance. This is because that change in temperature can significantly affect the performance of electric machines and their associated electronic devices and sensors. Without appropriate thermal
management, electronic devices degrade over time, causing problems ranging from minor errors to
critical failures.

There are several techniques for electronics cooling including various styles of heatsinks, cold
plates, forced air-cooling systems, impinging jets, immersion cooling techniques, heat pipes, and
others. Each of these options varies in cooling efficiency, complexity, cost, and heat dissipating
capacities. For servo drives, heatsinks are often used in combination with fans to keep the electronic
components of the drive in an acceptable temperature range and prevent them from overheating.
Thermoelectric devices are promising technology toward effective electronic cooling. There have
been many attempts to use thermoelectric devices to cool electronic devices with various types of
heatsinks. It has been reported that pulsed thermoelectric cooling for computer chips could achieve
a background heat flux of $40\,W/cm^2$ and a hotspot heat flux of $1,000\,W/cm^2$ [8.71].

In sensor cooling applications, the adopted cooling techniques are dependent on the sensor type,
amount of heat generation by the sensor, heat source nearby, sensor size, and specific application.
For instance, CMOS sensors, which are used in a wide range of digital imaging and other emerging
applications, usually use thermoelectric coolers for active spot cooling, enabling them to outper-
form at a lower cost point [8.72].

Thermoelectric coolers are solid-state heat pumps that utilize the Peltier effect to move heat
away from temperature-sensitive electronic devices. The Peltier effect, discovered by the French
physicist Jean Peltier in 1834 [8.73], is the reverse phenomenon of the Seebeck effect. It states
that when electrical current passes through an electrical junction connecting two conductors, heat
is released on one side and absorbed on the other side to balance the difference in the chemical
potential of the two conductors. This effect is the basis for many modern electronic devices, such
as thermoelectric coolers, electronic refrigerators, and IR detectors. As displayed in Figure 8.46,
a thermoelectric cooler consists of n-type (negative charge carriers with free electrons) and p-type
(positive charge carriers with holes) semiconductors that are set up alternately in an array and
connected through metallic electrical conductors, forming an electrical series connection, but a
thermal parallel connection (Figure 8.47). The system is sandwiched by two high thermal conductivity but low electrical conductivity ceramic plates acting as electrical insulation and offering rigidity to the assembly. When a current flowing through the semiconductors, it leads to a
temperature difference between two junctions, causing the hot side of the cooler to heat up and
the cold side to cool down. The cooling power of the thermoelectric cooler can be adjusted by
changing its operating current. By reversing the direction of the current, the cold and hot sides
are mutually exchanged. Therefore, a thermoelectric cooler can serve as either a cooler or heater.
Without the use of Freon™ and other refrigerants, thermoelectric coolers are environmentally
friendly devices. Moreover, thanks to the absence of any moving parts, they are especially useful
for electronics cooling.

Thermoelectric materials are commonly evaluated using a figure of merit Z, which characterizes
the thermal and electrical transport properties of the materials, defined as

$$Z = \frac{S^2}{k\rho} \tag{8.33}$$

where
    S stands for the Seebeck coefficient of thermoelectric material, $V/°C$
    ρ is the electrical resistivity of thermoelectric material, $\Omega \cdot m$
    k is the thermal conductivity of thermoelectric material, $W/(m \cdot °C)$

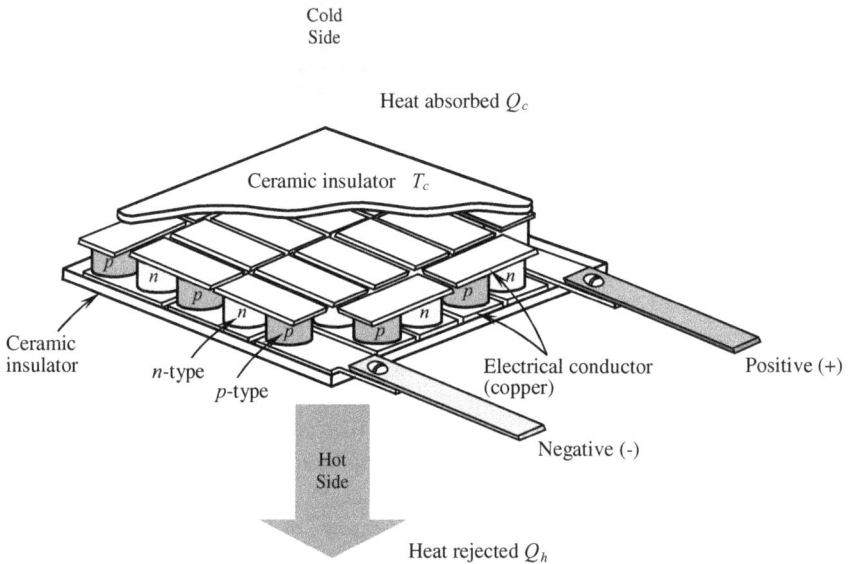

FIGURE 8.46 Cutaway view of a single-stage thermoelectric cooler. The thermoelements (n-type and p-type semiconductors) are connected electrically in series by copper conductors and the whole assembly is sandwiched by two ceramic plates. While the amount of heat absorbed from the cold side is given by Q_c, the amount of heat rejected from the hot side is given by Q_h.

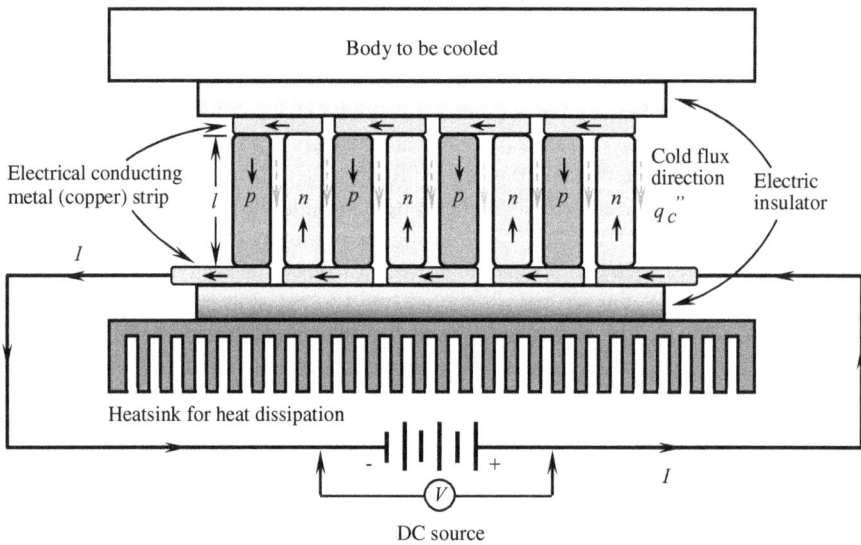

FIGURE 8.47 The n-type and p-type semiconductors, serving as thermoelements, are set up alternately in an array and connected electrically in series and thermally in parallel. For this case, the electrical current is from the n-type semiconductor to the p-type semiconductor. The direction of the cooling flux q''_c is always from the cold side to the hot side. The heatsink helps dissipate heat from the hot side to the surroundings.

Furthermore, a dimensionless parameter, the figure of merit ZT, can be defined as

$$ZT = \frac{S^2 T}{k\rho} \tag{8.34}$$

where T is the average temperature of the cold side and hot side (in K), given as $T=(T_h+T_c)/2$. It is perceived that increasing ZT value improves the performance of thermoelectric devices, $ZT>3$ is highly recommended for commercialization [8.74]. If the ZT could be raised to 2–3, the thermoelectric cooling device would be competitive with vapor compression cooling systems. If the ZT could be raised to 6, thermoelectric coolers are able to cool to cryogenic temperature ($77\ K$) from room temperature [8.75]. In recent years, the advances in semiconductor materials and processing technologies have raised the possibility of overcoming classical limitations and increasing ZT values significantly.

For conventional thermoelectric coolers, the maximum achievable temperature difference is estimated by

$$\Delta T_{max} = \frac{1}{2}ZT_c^2 = \frac{1}{2}\frac{S^2 T_c^2}{k\rho} \tag{8.35}$$

This indicates the highest cooling performance is attained at the highest value of Z and, in turn, at the highest value of S and lowest values of thermal conductivity k and electrical resistivity ρ. It can be seen that the cooling capability of a thermoelectric cooler improves with the square of the temperature at the cold side, yielding far better performance at elevated temperatures than at room temperature for thermoelectric materials whose properties are nearly invariant with the temperature [8.76].

By ignoring the temperature drop through the thermoelectric cooler, the heat absorbed at the cold side Q_c (in W) and the heat rejected at the hot side Q_h (in W) can be expressed as, respectively

$$Q_c = SIT_c - \frac{1}{2}I^2R - K\Delta T \tag{8.36a}$$

$$Q_h = SIT_h + \frac{1}{2}I^2R - K\Delta T \tag{8.36b}$$

where
$S=\Sigma(S_{p,i}-S_{n,i})$ is the Seebeck coefficient of the cooler, where S_p and S_n are the Seebeck coefficient of the p-type and n-type semiconductor, respectively, V/°C
I is the electrical current flowing through the thermoelements, A
$R=\Sigma(R_{p,i}+R_{n,i})$ is the total electrical resistance of the cooler in series, where R_p and R_n are the resistance of the p-type and n-type semiconductor, respectively, Ω
K is the combined thermal conductance of the cooler in parallel, W/°C
$\Delta T=(T_h-T_c)$ is the temperature difference between the cold and hot sides, °C

The first term in the above equations is the absorbed heat (for Q_c) or rejected heat (for Q_h). For the Joule heat term (I^2R), half of the overall Joule heat goes to the cold side and the other half goes to the hot side. The last term represents the conducted heat. The maximum heat pumping capacity is obtained when $\Delta T=0$,

$$Q_{max} = SIT_c - \frac{1}{2}I^2R \tag{8.37}$$

The coefficient of performance (COP) for operation in cooling mode is defined as the ratio of the cooling capacity Q_c over the input electrical power P_{in},

$$COP = \frac{Q_c}{P_{in}} = \frac{Q_c}{V_{in}I} \tag{8.38}$$

The maximum cooling flux q''_{max} of a thermoelectric cooler is defined as the maximum heat pumping out capacity per unit area that the thermoelectric cooler is capable of providing at a temperature difference across the cooler of zero, given by the equation below [8.77],

$$q''_{max} = \frac{\left(S_p - S_n\right)^2 T_c^2 f}{4\left(\rho_p + \rho_n\right)l} \qquad (8.39)$$

where
$\quad S_p$ and S_n are the Seebeck coefficients of the p- and n-type semiconductors, respectively, V/°C
$\quad \rho_p$ and ρ_n are the electrical resistivities of the p- and n-type semiconductors, respectively, Ω·m
$\quad T_c$ is the temperature of the cold side of the thermoelectric cooler, °C
$\quad f$ is the packing fraction
$\quad l$ is the thickness of the thermoelement (i.e., n-type and p-type semiconductors), m

This shows that q''_{max} of the thermoelectric cooler depends on the thermoelectric material properties, the thermoelement thickness, and the packing fraction.

Multistage thermoelectric coolers have been used when a higher temperature difference between the heat absorbing side (cold side) and the heat rejection side (hot side) is needed. The multistage design essentially places two or more single coolers with each other. With an increase in the number of stages, the COP increases for a given cold side temperature. It has been reported that at the subzero temperature range (−73°C to −23°C) for a given temperature difference Δt, using a multistage module with an appropriate number of stages may lead to a sharp increase in cooling efficiency [8.78].

In 2020, a new generation of high-performance thermoelectric coolers has been launched by Laird Thermal Systems, which offer a 10% boost in heat pumping capacity, greater temperature differential, and higher efficiency than standard thermoelectric coolers [8.79]. They can cool electronics well below ambient temperatures and can significantly reduce maintenance requirements and operating costs when compared to other cooling technologies. Consequently, sensors benefit from thermoelectric coolers in harsh, high-temperature environments to optimize performance and ensure long-term reliability.

It is anticipated the high-cooling-flux thermoelectric coolers will have reaching impacts in diverse applications, such as advanced computer processors, radio-frequency (RF) power devices, quantum cascade lasers, and DNA micro-arrays [8.80].

8.6.3 UTILIZATION OF DUAL-FEEDBACK SOLUTION FOR IMPROVING MOTION CONTROL ACCURACY AND RELIABILITY

Mechanical transmission elements such as lead screws and gearing systems, all contribute to the error between sensed and actual true position. Many of these problems can be adequately addressed by utilizing a dual-feedback solution. Dual-feedback is a technique where two feedback devices are used to control a single axis. One of the devices is attached to the load and the other to the motor. To provide good positional accuracy, the device attached to the load is used to measure the position and the encoder attached to the motor is used only to measure the motor's operation. This combination provides high positional accuracy and eliminates stability problems associated with mechanical compliance, backlash, and to some extent slip [8.81]. In practice, the use of two feedback devices provides some redundancy that could be used to detect device failure and comparison of their measuring results. For instance, robot manufacturers can significantly improve the absolute position accuracy by using a dual-encoder option at every robot axis, allowing the encoders to account for zero position error and backlash and thus resulting in 70%–80% improvement of the absolute position accuracy [8.82].

8.7 DEVELOPMENT OF INNOVATIVE SENSOR

The recent advances of sensor technologies have been powered by high-speed and low-cost electronic circuits, novel signal processing methods, electronics miniaturization, emerged technologies, and innovative advances in manufacturing technologies. The synergetic interaction of new developments in these fields allows the widespread use of innovative sensors in a variety of industrial applications. Significant advances in control technology have enabled the evolution of sophisticated sensors and servo feedback devices. To combat the complexity of today's industrial automation and motion control systems and the increasing demand for precise and durable sensors in many emerging applications, new technologies have been developed rapidly in recent years.

8.7.1 SENSOR MINIATURIZATION—MICROSENSOR AND NANOSENSOR

Miniaturization is a strong trend not only for sensing devices but also for motion control systems across almost all industrial sectors. Engineers and scientists always strive to pack more features into smaller packages. This creates a need for miniature feedback devices and sensors to meet the demand for reduced size.

The recent development of MEMS and NEMS technologies has become the main driving force for precise and reliable sensors. Today, in the global sensor market, about 25% sensors are made from MEMS-based devices. Among them, optical encoders are more flexible and best suitable for miniaturization. With advanced lithography technology, sensor manufacturers can significantly increase the resolution of code discs in transmissive and reflective optical encoders without increasing disc sizes (so-called scalability). However, LED as the light source in traditional optical sensors is no longer applicable to these miniaturized sensors. New light sources, such as lasers, may be considered as a replacement.

The rapid development of nanotechnology has led to the emergence of nanosensors that have various applications for non-destructive and in-situ measurements. Growing demand from healthcare, increased usage in military and defense, and raising R&D activities are the factors favoring the market growth. Furthermore, advanced technological developments in signal processing and microelectronics and cost efficiency of nanosensors are bolstering the market [8.83]. Newly developed nanostructured materials can offer extremely high surface-to-volume ratios, high surface activities, and high carrier mobility.

It must be kept in mind that as the size of the feedback devices and other motor sensors continues to shrink, maintaining high accuracy, resolution, and sensitivity at the micro- or nano-level becomes increasingly challenging. This indicates that motor and sensor engineers still have a long way to go to access novel technologies that can fit a broader spectrum of industrial applications.

More recently, scientists at the Oak Ridge National Laboratory have developed a method to print surface acoustic wave sensors on substrates of lithium niobate crystal using nanoparticle inks [8.84]. Using the printed RF technology, the sensors are fabricated via aerosol-jet printing. These sensors can offer a wide range of measurement opportunities such as temperature and pressure. The scientists have demonstrated that the sensor features can be printed at a resolution of about $10\,\mu m$, which increases their operating frequency and sensitivity. Ongoing research aims to reach $1\,\mu m$ resolution and to test the sensors in both a simulated nuclear plant application and on essential grid components such as transformers.

8.7.2 ADVANCED WIRELESS SENSOR TECHNOLOGY

In recent years, new wireless sensors have been developed to monitor and control industrial equipment and processes in efficient and effective ways. These sensors can help improve manufacturing processes, reduce lifecycle energy consumption of products, eliminate connector failure, allow greater physical mobility and freedom, and lower installation and maintenance costs; all of them ultimately boost a plant's productivity and profitability. In fact, wireless sensors not only eliminate the cost of the wiring and cabling needed for conventional sensor technologies, but also reduce cost by contributing to a plant's overall energy efficiency.

General Electric has developed battery-free, multi-detection wireless sensors that can be used for real-time monitoring and fault detecting of industrial motors. To operate without batteries, the power is obtained wirelessly from the sensor reader. They monitor motors' operating condition, performance, and efficiency. The central station analyzes data obtained from the wireless sensors and identifies which motor components need maintenance. The system also projects which motors will fail next, and when, so that repairs and replacements can be done without sacrificing productivity [8.85].

As an outcome of the convergence of MEMS technology, wireless communications, and digital electronics, wireless sensor networks (WSNs) represent a significant improvement over traditional sensors. Actually, the rapid evolution of WSN technology has accelerated the development and deployment of various novel types of wireless sensors [8.86].

Compared with conventional sensing methods using wires, constructing a monitoring system based on WSNs can provide several advantages such as low cost, convenient monitoring arrangements, collection of a variety of parameters, high detection accuracy, and high accountability of the monitoring network [8.87].

In the last decade, WSNs have been used in many application domains (industry, military, agriculture, transport, *etc.*), in order to monitor the environment and to prevent accidents. It can offer convenient and cost-effective solutions by providing data from sensors and controlling industrial applications.

8.7.3 SMART SENSORS

Conventional sensors are functionally simple devices that convert physical variables into electrical signals. In contrast with these sensors, smart sensors, also known as intelligent sensors, are typically equipped with one or more sensors, a processor with necessary signal conditioning circuitry, and a RF module for wireless communication. For some specific applications, memory and power supply can be embedded on board as well. All these features are expected to contribute to the smart sensor market growth. The capabilities to operate without onboard power supply and low power consumptions make them ideal for remote control systems.

Smart sensors have been developed in recent years for responding to the needs of modern society, especially for Internet of Things (IoT) applications. A smart sensor is a device that can convert input from the physical environment into digital data streams and use a built-in microprocessor unit (MPU) to perform predefined functions upon detection of specific input and then process data before passing them on. Moreover, the intelligence possessed by MPU can be used to execute many other functions. For instance, in industries, MPUs can spot any production parameters that start to drift beyond acceptable norms and automatically generate early warnings for avoiding major disasters. Today, sensor technology is becoming an integral part of modern energy-efficient electric machines.

The main differences between smart sensors and conventional sensors include that smart sensors have intelligent capabilities such as wireless communication, data processing, analog-to-digital conversion, and decision-making. Smart sensors perform self-diagnosis by monitoring internal signals for evidence of faults. Furthermore, smart sensors have other advantages over conventional sensors such as smaller size, minimal power consumption, and optimal performance. In fact, smart sensors have sensing, computational, data acquisition and communication, and diagnostic capabilities. They are often equipped with a built-in energy source. Because of their intelligent capabilities, smart sensors are especially suitable for monitoring and control mechanisms in a wide variety of environments, including traffic control, equipment fault diagnostics, automatic irrigation systems, medical diagnostic systems, smart grids, lighting systems, telecommunications, manufacturing process control, remote system monitoring, and many industrial, healthcare, and military applications.

MEMS encompass the process-based technologies used to fabricate tiny devices or systems that permit the integration of multiple sensor structures within a specially designed circuit board or a single miniaturized package for enabling multimode readings. The MEMS technology has demonstrated its potential to develop smart sensors fabricated by integrating sensing elements and microcontrollers together in a single package. The underlying idea is to integrate sensor technology at the silicon level. This is believed to improve power consumption while further reducing the sensor size and weight, as well as increasing sensor operating reliability.

8.7.4 COLOR-CHANGING DYE SENSOR FOR DETECTING MOTOR CONDITION

Motor windings account for about one-third of all AC electric motor repairs. Motor windings are made up of tightly wound magnet wires that are coated with a very thin layer of insulation material. Today, many insulation materials are available for magnet wire coating, such as polyvinyl formal, polyurethane, polyamide, polyester, polyester-polyimide, polyamide-polyimide, and polyimide, which are widely used on magnet wires in electric motors. Generally, polyimide resin is a standard insulation material, which is capable of operating up to $250°C$. Dual film insulation usually consists of a film of polyester resin and a film of polyamide-imide resin for delivering better electrical and mechanical properties.

Over time, insulation coatings on magnet wires may become brittle and inflexible due to heat and chemical processes. When the wires are subjected to strong vibrations, shocks, or other dynamic loadings, the wire insulation coating may be damaged or cracked, resulting in short circuits, and in the worst case, motor failures. Therefore, it is highly desirable to detect the degradation of insulation materials in the early time to prevent serious motor damage caused by winding insulation failure. However, because motor windings are situated inside the motor housing, the determination of the winding insulation condition remains a challenging task for motor engineers.

To address this problem, scientists at Martin Luther University Halle-Wittenberg (MLU) recently have developed an integrated dye process that helps detect wear and tear on electric motor wire insulation. The inline-monitoring system is induced for the thermal degradation of crosslinked poly(ester imide)s by embedding trifluoroacetyl functionalized stilbene molecules, serving as chemosensors to track the release of generated alcoholic byproducts [8.88].

In the process of research, the scientists have developed a test rig to analyze four different resin systems to determine which degradation products form at different temperatures. They found that the four resin systems consistently released a specific alcohol under different temperature conditions. According to this result, they choose a dye that normally glows reddish orange under ultraviolet light, but when alcohol binds to it, the color spectrum shifts to a light green. The different color spectra could then be analyzed using special devices that could be installed directly in the motor [8.89]. In this way, engineers can determine if the winding replacement is necessary without having to open up the motor.

8.7.5 SENSOR WITH NEWLY DEVELOPED MATERIAL

With the help of artificial intelligence and quantum computing, the discovery of new materials has accelerated exponentially in recent years. With these new discoveries, customized materials have grown dramatically to meet the needs of different industrial sectors. At the same time, the growing demand for modern sensors for use in static and dynamic magnetic measurement applications have pushed established sensor technologies to their limits.

Derived from graphite, graphene is a sheet of carbon just one atom thick. It is nearly weightless, but 200 times stronger than steel. Conducting electricity and dissipating heat faster than any other known substances. The high flexibility, impermeability, and strength of graphene membranes are key properties that can enable the next generation of nano-scale sensors.

On-chip temperature sensing on a micro- or nano-scale becomes more desirable as the complexity of nanodevices keeps increasing and their downscaling continues. The continuation of this trend makes thermal probing and management more and more challenging. In response to the challenges, a team of researchers from the University of Oxford, Delft University, and IBM Zurich has most recently demonstrated that graphene can be used to build sensitive, single-material, and self-powering temperature sensors. It is shown that U-shaped graphene stripes consisting of one wide and one narrow leg form a single material thermocouple that can function as a self-powering temperature sensor. By carefully tuning the geometry of the graphene legs and exploiting the effect of electron scattering at the edges of the graphene device, the team achieved a maximum sensitivity of $\Delta S \approx 39~\mu V/K$ [8.90]. The results could pave the way for the design of highly sensitive thermocouples with the possibility of integration in nanodevices and even living cells. Furthermore, this new type of temperature sensor is characterized by small footprint, high accuracy, a minimum amount of consumed power, and compatibility with established nanofabrication techniques, used in harsh or sensitive environments.

Scientists at the Lodz University of Technology in Poland have designed a transparent, flexible cryogenic temperature sensor with graphene structures as sensing elements. Such sensors could be useful for any field that requires operating in low temperatures, such as medical diagnostics, space exploration and aviation, processing and storage of food, and scientific research. By means of high thermal conductivity, the graphene temperature sensor is cooled from room temperature to cryogenic temperature. Graphene structures are characterized using Raman spectroscopy [8.91].

8.8 SELECTION OF MOTOR FEEDBACK DEVICES AND SENSORS

Among various engineering activities, choosing the right measuring instruments may have serious implications on the measurement results. Modern servomotor feedback devices and sensors are available with performance levels, communication interfaces, and mechanical features that can meet almost any requirement. Motor engineers have many choices when it comes to sensing and control technologies. The abundance of options can stall the decision-making of engineers and make the selection process even more challenging. Though there is no universal systematic procedure that can be followed, there are some basic principles that guide selection practices.

In practice, feedback device/sensor sizing for motion control systems is mainly driven by applications and corresponding working environments. When choosing a feedback device or a sensor, it is essential to consider how it will be used in a specified application. The intended outcome of the sensor selection process is to identify one or more sensors with properties that satisfy the functional requirements of product. In fact, the selection process is the process that permits tradeoffs among major factors. During the process, engineers must weigh the strengths and weaknesses of each approach and evaluate the technical capabilities and constraints of each technology with respect to the application.

First, one must look at the type of motion where a measuring device is applied, such as rotary motion, linear motion, reciprocating motion, and oscillating motion, with large, small, or micro displacements. This will point to a particular family of feedback devices or sensors.

The second consideration in selecting the most suitable feedback devices is about the environmental conditions where the devices are installed. In most applications, standard encoders are suited well enough to perform effectively. However, when some encoders are placed in operating environments subject to aggressive contaminants, strong shock and vibration, high temperature and humidity, intense radiation, and long-term submersion in liquids, they are more likely to fail. For instance, small particles, such as dust and soot, may enter and obstruct the light beam in optical encoders, leading to encoder failure. In a case where an encoder is submerged in liquid or exposed to pressurized water, its IC circuitry or other electronic components can be damaged [8.92]. In short, if an application is subjected to a higher temperature, humidity, dust, moisture, *etc.*, resolvers are preferred over encoders that may fail under the same conditions. Among encoders, magnetic encoders are typically utilized in harsh environmental conditions where optical encoders may suffer significant performance decrease. Under strong radiation conditions, resolvers have much higher resistivity to both nuclear radiation and electrical disturbances due to the lack of optics and solid-state electronics, as well as their very rugged structures. It has been reported that by exposing a robot to gamma radiation from a 20 terabecquerel (TBq) cobalt-60 source, optical encoder fails after a cumulative exposure of 164.55 *Gy* (*Gy* is the unit of absorbed dose of ionizing radiation, in units of J/kg) over a period of 16.8 h [8.93]. Practically, shielding of particularly vulnerable components such as encoders would increase the whole system's lifetime in a radioactive environment.

If the selection is limited to encoders, the first decision to be made is to select either incremental or absolute encoders. As mentioned previously, an absolute encoder provides a snapshot reading of the rotating angle and speed of the motor shaft and therefore determines the absolute position. By contrast, an incremental encoder indicates the relative position and direction of movement by adding incremental pulses to a known start position. It is ideal for speed control of motors since it provides a real-time reading of the shaft position and keeps track of absolute position by counting the number of pulses per revolution (*PPR*). However, this position count could be lost during a power failure or system shutdown. In such a case, it may be necessary to return the machinery to a known reference position and restart the position count before operations can resume [8.94]. For better understanding of the difference between the two types of encoders, consider the difference between a stopwatch and a clock. A stopwatch measures the relative time that elapses between its start and stop points. This is similar to an incremental encoder to measure the relative shaft position from the specified point. A clock, on the other hand, shows the exact current time within a day, similar to an absolute encoder to measure the absolute position of the shaft.

Operating environments (*e.g.*, ambient temperature, shock and vibration rating, contamination condition), performance requirements (*e.g.*, speed control, position control, or torque control), and communication protocols are all factors affecting the selection of feedback device for a rotary servomotor application. In fact, closed-loop control performance in a servo system depends on not only the dynamic response of controller, features of feedback device, measurement process, and so on but also the interfaces that facilitate the communication between the feedback device and the servo drive. To achieve the desired control objective, it is necessary to choose the most appropriate communication protocol. The protocols can be arranged based on functionality in groups. Among them, the transmission control protocol is often paired with the internet protocol (IP) together to prepare and forward data packets across a network such as Ethernet.

Ethernet was developed in the 1970s and emerged in products in the early 1980s. Nowadays, Ethernet is the dominant local area networking solution in the office environment and attracts a great deal of attention as a support for industrial communication. While real-time industrial Ethernet-based communication protocols (*e.g.*, EtherNet/IP, EtherCAT, and ProfiNet) are becoming popular for larger systems with many devices working together, fieldbus communications (*e.g.*, CANopen, DeviceNet, and ProfiBus) are well suited to standalone machine control systems. CANopen is a standardized, highly flexible, and highly configurable embedded network architecture used in industries such as servomotor, robot, medical, automotive, aerospace, and many others.

There are many other key factors affecting the choice of proper feedback devices, including the following:

- Accuracy—It is defined as the difference between a commanded position and an actual position. Among all types of feedback devices, optical encoders have the highest accuracy and thus are best suited to applications that require precise measurement of speed, position, and distance. By contrast, Hall effect devices can provide coarse speed and position feedback signals. When used alone, they are not suitable for applications that require precision or running at speeds below 500 rpm.

- Resolution—As one of the most important characteristics of feedback devices, resolution refers to the smallest increment of the position or distance that can be measured by a feedback device. Generally, absolute optical encoders have the highest resolution than their competitors. Incremental encoders commonly use quadrature. By decoding the output waveforms in different ways, their resolutions can multiples by a factor of up to four. In dealing with high-frequency signals, optical encoders with high resolutions are required. However, it is important to note that higher resolution does not mean higher accuracy. In addition, it may not be a good idea to always choose the feedback devices with the highest resolution for all applications, for a number of reasons: (*a*) High-resolution feedback devices are regularly more expensive. (*b*) Sensitive systems may over-respond to high-resolution information. (*c*) High-velocity applications need more time to read each cycle. (*d*) In some cases, higher resolutions may have size implications [8.95].
- Operation reliability and stability—Long-term operation reliability and stability of feedback devices are always desirable. To enhance the reliability and stability of servo systems, many motion control manufacturers today offer complete systems where the servomotor, feedback device, and drive are integrated into a single optimized package.
- Operating temperature range—Feedback devices operate effectively within a specified temperature range that depends on the device type, function, and application. Generally, the temperature range for military and aerospace applications is greater than that for industrial applications and the least strenuous temperature range is found in commercial products. In servo systems, the temperature range each feedback device is capable of operating is critical for reliable and accurate measurements. Due to the absence of imbedded electronic components, resolver devices can operate at the widest temperature range of $-50°C$ to $220°C$. Because high temperatures tend to shorten the life of electronics inside the encoder, regular encoders often operate at temperature lower than $120°C$ and higher than $-20°C$. The lowest operation temperature of magnetic encoders can be expanded to $-40°C$.
- Output signal—The output signal generated from a feedback device is often a time-varying voltage or current (either continuous or discrete) that is electromagnetic wave carrying information. There are two main types of signals: digital or analog. There is a key difference between digital and analog signals. A digital signal is a time-separated signal, denoted by square waves, whereas an analog signal is a continuous signal, denoted by *sine/cosine* waves. The most significant advantage of encoders is that they inherently produce digital signals, making them easy to interface to today's control systems and bringing them into the modern age of industrial IoT. Since the output signals generated from resolver are analog, analog-to-digital converters are often needed to translate analog measurements to digital form. Resolver feedback signals usually contain more position errors than encoder signals.
- Repeatability—It represents the maximum distribution from consecutive measurements taken under identical conditions. In practice, encoder repeatability reveals the encoder's likelihood to be consistent in its accuracies, as well as inaccuracies. Repeatability can be more important than absolute accuracy. The manufacturing process of the encoder, the equipment being positioned, mechanical errors (*e.g.*, gearing backlash), and even the sensing technology employed can contribute to errors in repeatability [8.96]. Optical and magnetic encoders are designed to prevent cumulative errors and thus have good repeatability.
- Response time—In control technology, fast response time in motion control is always desirable. The bandwidth, or response time, of the system is a measure of how fast it responds to the changing input command. In other words, the bandwidth of the control loop determines how quickly the servo system responds to changes in the parameter being controlled such as position, velocity, or torque [8.97]. Any components in the control loop, such as feedback device, control system, plant, servomotor, or servo drive, can affect servo loop bandwidth. Higher bandwidth generally improves transient response time, decreases

error, and provides stiffer motor performance. Actually, higher bandwidth means that the motor is commanded by the control system to respond at higher frequencies.

- Cost—Cost is always one of the important factors affecting the selection of feedback devices. Though high-resolution optical encoders provide excellent measurement accuracy, market competition has forced designers to look into less expensive solutions that have relatively large measurement errors, such as magnetic and inductive encoders. Resolvers are generally recognized as the preferred option for low-cost feedback devices and are extensively used in automation and servo systems. However, the electronics required for analog-to-digital signal conversion, which is indispensable for modern control and monitoring systems, may make the whole package expensive. Among several types of encoders, inductive encoders have the lowest cost. For optical encoders, absolute encoders are generally more expensive than incremental encoders. Hall effect sensors tend to be less costly than their magnetic resistive sensor counterparts.
- Service life—In general, the overall service life of an encoder is expressed as the total number of cycles. The service life is influenced by factors such as the environment, the installation location, the measuring range in use, and the travel speed and acceleration. If the value of one or more of these influencing factors is in the high range, the service life may be reduced proportionally [8.98].
- Tolerance to vibration and shock loading—While resolvers are available with shock loads to 200 G and vibration to 40 G, absolute and incremental optical encoders can withstand shock loads up to 100 G and 50 G, respectively. Both are rated for vibration of 20 G.
- Electrical noise immunity—Electrical noise can interfere with electrical feedback signals and distort data sent to the drive to varying degrees. One advantage of using optical transmission lines for feedback signals is immunity to electrical noise or EMI/radio frequency interference environments. Newly designed feedback devices utilize ICs to convert and interpolate signals into more robust waveforms that overcome noise and propagation loss in connecting cables [8.99].
- Magnetic field immunity—All but magnetic encoders are immune to extreme magnetic fields. Compared to other encoders, magnetic encoders are likely the best overall sensor for generating feedback signals. They are designed to respond to a wide range of magnetic fields in a variety of applications. Therefore, magnetic encoders are susceptible to magnetic or radio interference. Moreover, Hall effect sensors became popular for use in many different types of applications such as sensing position, velocity, or directional movement due to their non-contact operation mode, high reliability, high precision, and long lifetime. Similar to magnetic encoders, Hall effect sensors operate based on magnetic sensing to detect perturbations to a magnetic field and communicate the measured signal to a servo drive. Any interference from other magnetic fields can significantly affect the sensor performance and measurement accuracy.
- Energy consumption—The energy consumption of a feedback device is an important factor that not only helps achieve energy saving but also affects the thermal management of the device.
- Compact size—In some special applications, such as surgical robotics, drones, microsatellites, and space vehicles, there is a trend for miniaturization in servo and actuating systems. To respond to this trend, designing compact and lightweight feedback devices is highly desired.
- Structure ruggedness—A key measure of motor feedback systems is ruggedness. While some feedback devices are very rugged (*e.g.*, resolvers), targeting industrial and military applications, others are relatively fragile, intended more for precision laboratory equipment or semiconductor manufacturing equipment in a clean environment.
- Maintainability—Maintainability is the measure of the ability of the system or component to be retained in or restored to a normal condition through repairs and replacement. An increase in maintainability can lead to a reduction in operation and support costs.

When major factors have been identified and evaluated, there comes a time to trade off these factors to select the most suitable device for the application, even though several factors may not be qualified until a detailed design is completed.

8.9 CABLE TECHNOLOGY

A traditional servo system uses a multi-cable solution in which a servomotor is connected to a servo drive through separated power, feedback, and brake cables. This is because power and brake conductors near feedback conductors may degrade data transmission by inducing noise in the form of EMI and crosstalk. With the rapid development of cable technology in recent years, the single cable solution becomes increasingly popular in the motion control industry for its simplicity and efficiency [8.100]. This solution integrates all motor cables, including the power cable, feedback cable, and brake cable, in a single shielded cable (also called hybrid or composite cable), which is encapsulated in a moisture- and oil-resistant PVC or PUR jacket for flexible applications (Figure 8.48). This single cable solution can greatly simplify inventory management, cut down component and commissioning costs, boost equipment performance, and reduce installation and maintenance time. Today, the single cable solution has become one of the trends in the motion control industry.

These cables contain bundles of copper wire acting as conductors that transmit power, as well as conductor pairs that send power, temperature, and feedback signals to and from sensors.

The main challenge for the single cable solution is how to effectively prevent the feedback cables from EMI by the power and brake cables to impair data/signals transmission [8.101]. As it can be seen in Figure 8.49, braided shielding materials (*e.g.*, metal wire mesh and knitted gaskets [8.102]) are wrapped on the feedback and brake cable bundles and properly grounded throughout the end connectors to mitigate mutual EMI, as well as disturbances from the neighboring power cables. Meanwhile, the entire cable bundle is shielded by a tinned copper braided shield to minimize

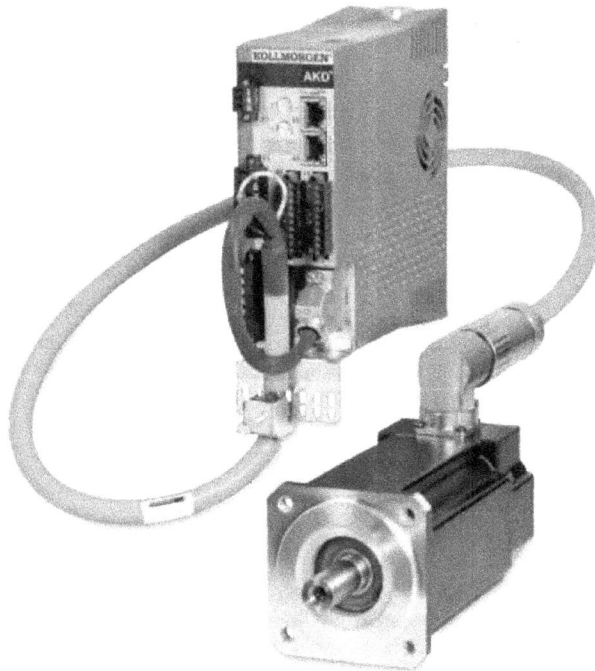

FIGURE 8.48 One cable solution for a servomotor system. It requires only one special connector on the motor.

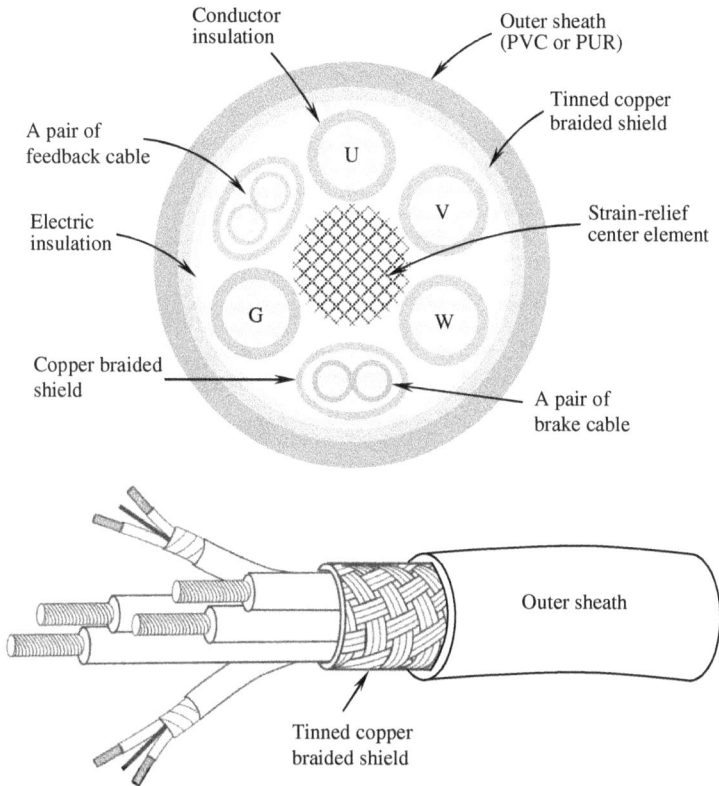

FIGURE 8.49 Cross-section of motor cable, containing three-phase power cables, a pair of feedback cable, and a pair of brake cable.

or eliminate external EMIs. To make bend-resistant cable, shielding is woven around conductor strands at an angle compatible with the cable's minimum bending radius (which is usually 10 times the cable diameter). High-quality shields are usually used to ensure the electromagnetic compatibility of the cable.

The majority of signal transmission problems involve electrical noise. An effective noise suppression method is to twist the two shielded lead wires. In this way, induced currents in the wires cancel each other. In practice, for very high-noise environments, differential encoders, which generate two outputs for each channel, with differential wiring provide an effective way to filter out noise.

Another challenge for the single cable solution is thermal management. Due to the joule heating effect, an amount of heat is always generated as electrical current flows through the power cables. Overheating can result in multiple hot spots that age localized areas of the conductor's insulation. To prevent cable overheating, it is critical to determine the size of power cables at the early stage of design to ensure their capability of carrying adequate electrical current. Furthermore, in order to enable the single cable to dissipate heat effectively, when choosing the right cable insulation/jacket and shielding materials, not only their electrical/electromagnetic and mechanical properties but also their thermal properties must be fully considered.

A well-designed single servo cable must have high strength and durability under excessive load conditions and industrial environments. While some cables are used in stationary machinery, many servo systems, such as automation systems, industrial robots, and handling equipment, operate under dynamic conditions. For these applications, the single servo cables need to offer continuous flex performance and compatibility with cable tracks. In fact, the flexible cables require more rigorous lifecycle testing under realistic conditions, with adequate bending cycles (up to 10 million alternate bending cycles).

As the power requirements of the motor increase, the cost-effectiveness and practicality of the single cable reduces because large current-carrying conductors splitting off in side motor or drive electronics become problematic with bend radii in tight enclosures and costs of special cable runs with suppliers.

REFERENCES

8.1. Matthews, G. 2013. Feedback sensors keep servomotors on target. Kollmorgen White Paper KM_WP_000200_RevA_EN. http://www.kollmorgen.com/uploadedFiles/kollmorgencom/Service_and_Support/Knowledge_Center/White_Papers/FeedbackSensors_02_14.pdf.

8.2. White, R. M. 1987. A sensor classification scheme. *IEEE Transactions on Ultrasonics, Ferroelectrics, and Frequency Control* **34**(**2**): 124–126.

8.3. Dynapar Engineering Team. Encoders and resolvers: How to choose the right feedback option. Danapar industrial white paper. https://kamanautomation.com/wp-content/uploads/dynapar-encoders-and-resolvers-how-to-choose-the-right-feedback-options.pdf.

8.4. Lacroix, M., Santos, J., and Stiffler, R. 2011. The advantages of magnetic encoder technology in harsh operating environments. Timken Company. https://www.timken.com/resources/the-advantages-of-magnetic-encoders-comparing-the-technologies.

8.5. Collins, D. 2019. What features make resolvers suitable for harsh environments? *Motion Control Tips* https://www.motioncontroltips.com/what-features-make-resolvers-suitable-for-harsh-environments/.

8.6. Moog Components Group. 2004. Synchro and resolver engineering handbook. https://www.moog.com/literature/MCG/synchrohbook.pdf.

8.7. Malcolmy, T. 1998. Chapter 14: Part 2: Optical encoders. In *Space Vehicle Mechanisms: Elements of Successful Design* (Ed.: P. L. Conley). John Wiley & Sons, New York, pp. 385–470.

8.8. US Digital. 2019. Who made the first optical encoder? https://www.usdigital.com/news/posts/encoders-006-who-made-the-first-encoder.

8.9. Rhodes, L. 2020. Best practices for encoder mounting. Manufacturing Tomorrow. https://www.manufacturingtomorrow.com/article/2020/04/best-practices-for-encoder-mounting/15149.

8.9. Wong, W. F., Cheang, C. K. 2012. Reflective optical encoder package and method. U.S. Patent 8,212,202.

8.10. Soo, Y.-F, Wong, W. F., and Yeap, W. J. 2014. Enhanced optical reflective encoder. U.S. patent 8,847,144.

8.11. Danaher control. 2003. *Encoder Application Handbook*. https://www.controlcomponentsinc.com/documents/dynapar-application-handbook.pdf.

8.12. Zheng, D. Z., Zhang, S. B., Wang, S., Hu, C., and Zhao, X. M. 2015. A capacitive rotary encoder based on quadrature modulation and demodulation. *IEEE Transactions on Instrumentation and measurement* **64**(**1**): 143–153.

8.13. Schroter, A. and Fuchs, P. 2002. As the coil turns. *Machine Design*. Paper number 12817257. https://www.mchinedesign.com/archive/article/12817257/as-the -coil-turns/.

8.14. Smoot, J. Capacitive, magnetic, and optical encoders – comparing the technologies. CUI Devices. https://www.machinedesign.com/archive/article/21817257/as-the-coil-turns

8.15. Celera Motion. A comparison of inductive and capacitive position sensors. Technical Paper TN-1515|191013. https://www.celeramotion.com/zettlex/support/technical-papers/inductive-and-capacitive-position-sensors/.

8.16. Jacobs, J. 2013. Incremental and absolute encoders: What's the best solution for your application? http://www.dynapar.com/uploadedFiles/_Site_Root/Technology/White_Papers/New_incremental%20absolute_7_29_13.pdf.

8.17. Eitel, E. 2014. Basics of rotary encoders: Overview and new technologies. *Machine Design*. http://machinedesign.com/sensors/basics-rotary-encoders-overview-and-new-technologies-0.

8.18. Ferbert, D. How to calculate encoder resolution. Dynapar Company. https://www.dynapar.com/knowledge/how-to-calculate-encoder-resolution.

8.19. Collins, D. 2017. Encoder resolution and accuracy: What's the difference? *Motion Control Tips*. https://www.motioncontroltips.com/encoder-resolution-and-accuracy-whats-the-difference/.

8.20. Kelly, J. 2019. Understanding the encoder resolution. https://www.electronicspecifier.com/sensors/understanding-the-encoder-resolution.

8.21. Rhodes, L. 2020. Best practices for encoder mounting. *Manufacturing Tomorrow*. https://www.manu-facturingtomorrow.com/article/2020/04/best-practices-for-encoder-mounting/15149

8.22. Kimura, F., Gondo, M. Yamamoto, A., and Higuchi, T. 2009. Resolver compatible capacitive rotary position sensor. *Proceedings of 35th Annual Conference of IEEE Industrial Electronics*, Porto, Portugal, pp. 1923–1928.

8.23. Crowder, R. 2019. *Electric Drives and Electromechanical Systems: Applications and Control*, 2nd edn. Butterworth-Heinemann, Oxford, UK.

8.24. General Dynamics Mission Systems. 2016. Pancake resolver handbook.

8.25. Hyatt, R. M. Jr. and Dayton, D. 2014. Synchro/resolver displacement sensors. In *Measurement, Instrumentation, and Sensors Handbook: Spatial, Mechanical, Thermal, and Radiation Measurement* (Eds.: J. G. Webster and H. Eren), 2nd edn. CRC Press, Boca Raton, FL, pp. 32-1–32-14.

8.26. Baker, B. 2019. How to precisely determine motor angular position and velocity with a resolver. Diki-Kay. https://www.digikey.com/en/articles/techzone/2019/may/how-to-precisely-determine-motor-angular-position-and-velocity-with-a-resolver.

8.27. Colombi, S. and Saghatchi, F. 1992. Digital position and velocity determination in controlled drive systems. In *Motion Control for Intelligent Automation* (Eds.: A. De Carli and E. Masada). Pergamon Press, Oxford, UK.

8.28. Spetzer, J. and Ekhaml, B. 2001. Things you need to know about sizing and applying: Resolvers. http://machinedesign.com/technologies/things-you-need-know-about-sizing-and-applying-resolvers.

8.29. Leyva, P. 2003. Design robust, fault-tolerant motion-control feedback systems. *Electronic Design*. https://www.electronicdesign.com/technologies/analog/article/21764671/design-robust-faulttolerant-motioncontrol-feedback-systems.

8.30. Hall, E. H. 1879. On a new action of the magnet on electric currents. *American Journal of Mathematics* **2(3)**: 287–292.

8.31. Ramsden, E. 2006. *Hall-Effects Sensors: Theory and Applications*, 2nd edn. Elsevier Inc., Burlington, MA.

8.32. Micro Switch Sensing and Control. *Hall effect sensing and application*. Honeywell Inc. https://sensing.honeywell.com/honeywell-sensing-sensors-magnetoresistive-hall-effect-applications-005717-2-en2.pdf.

8.33. Terzic, E., Terzic, J., Nagarajah, R., and Alamgir, M. 2012. *A Neural Network Approach to Fluid Quantity Measurement in Dynamic Environments*. Springer, London.

8.34. Buschow, K. H. J. 2009. *Handbook of Magnetic Materials*. Elsevier, Amsterdam.

8.35. Mohri, K. and Sugitani, N. 2000. Torque measuring device by integral shaft based upon inverse magnetostriction. U.S. 6,098,468.

8.36. Gandarillas, C. 2003. Magnetoelastic non-compliant torque sensor and method of producing same. U.S. Patent 6,516,508.

8.37. Mehnert, M., Hossain, M., and Steinmann, P. 2017. Towards a thermos-magneto-mechanical coupling framework for magneto-rheological elastomers. *International Journal of Solids and Structures* **128(2017)**: 117–132.

8.38. Li, W. H., Kostidis, K., Zhang, X. Z., and Zhou, Y. 2009. Development of a force sensor working with MR elastomers. *Proceedings of 2009 IEEE/ASME International Conference on Advanced Intelligent Mechatronics*, Singapore, pp. 233–238.

8.39. Seebeck, T. J. 1825. Magnetische polarization der metalle und erze durch temperature-differenz. (Magnetic polarization of metals and minerals by temperature differences). In Abhandlungen der Königlichen Akademie der Wissenschaften zu Berlin, pp. 265–373.

8.40. Rowe, D. M. 2006. *Thermoelectrics Handbook: Macro to Nano*. CRC Press, Boca Raton, FL.

8.41. NIST. 1993 (the page was modified on 09/07/2018). NIST ITS-90 thermocouple database (version 2.0). https://srdata.nist.gov/its90/main/its90_main_page.html.

8.42. Fluke Corporation. Thermocouple voltage to temperature calculator. https://us.flukecal.com/Thermocouple-Temperature-Calculator.

8.43. Laurila, H. 2017. Thermocouple cold (reference) junction compensation. https://blog.beamex.com/thermocouple-cold-junction-compensation#.

8.44. Duff, M. and Towey, J. 2010. Two ways to measure temperature using thermocouples feature simplicity, accuracy, and flexibility. *Analog Dialogue* **44-10**: 1–6. https://www.analog.com/media/en/analog-dialogue/volume-44/number-4/articles/measuring-temp-using-thermocouples.pdf.

8.45. Rowe, D. M. 2006. Chapter 1: General principles and basic consideration. In *Thermoelectrics Handbook: Macro to Nano* (Ed.: D. M. Rowe). CRC Press, Boca Raton, FL, pp. 1-1–1-14.

8.46. Siemens, C. W. 1861. On a new resistance thermometer. *Philosophical Magazine* 21: 73–74.

8.47. Mraz, S. J. 2021. Researchers develop a wireless implantable sensor small enough to be injected. *Machine Design*. Article number 21165153.

8.48. Chen Shi, C., Andino-Pavlovsky, V., Lee, S. A., Tiago Costa1, T., Elloian, J., Konofagou, E. E., and Shepard, K. L. 2021. Application of a sub–0.1-mm^3 implantable mote for in vivo real-time wireless temperature sensing. *Science Advances* 7(19): eabf6312. doi:10.1126/sciadv.abf6312.

8.49. Mathas, C. 2012. What you need to know about vibration sensors. Digi-Key. https://www.digikey.com/en/articles/techzone/2012/oct/what-you-need-to-know-about-vibration-sensors.

8.50. TE Connectivity Sensors. 2017. Choosing the right type of accelerometer. White paper. https://www.mouser.com/pdfdocs/choosing-the-right-accelerometer-white-paper.pdf.

8.51. Murphy, C. 2017. Choosing the most suitable MEMS accelerometer for your application – Part 1. *Analog Dialogue* **51**(**10**): 1–6.

8.52. SDI. 2020. SDI's capacitive accelerometer technology. https://www.silicondesigns.com/tech.

8.53. Mraz, S. J. 2021. NIST develops an IR laser-based accelerometer. *Machine Design*. Article 21157720. https://www.machinedesign.com/news/article/21157720/nist-develops-an-ir-laserbased-accelerometer.

8.54. Zhou, F., Bao, Y. L., Madugani, R., Long, D. A., Gorman, J. J., and LeBrun, T. W. 2021. Broadband thermomechanically limited sensing with an optomechanical accelerometer. *Optica* **8**(3): 350–356.

8.55. Lepkowski, J. 2003. Motor control sensor feedback circuits. White Paper AN894. Microchip Technology Inc. http://ww1.microchip.com/downloads/en/appnotes/00894a.pdf.

8.56. McAleese, S. 2000. Chapter 11: Pressure Gauges. In *Handbook of Petroleum Exploration and Production* (Ed: S. McAleese). Elsevier, Amsterdam, The Netherlands, **Vol. 1**, pp. 117–122.

8.57. Stopel, M. 2020. The advantages of digital pressure sensors in industrial applications. *Manufacturing Tomorrow*. https://www.manufacturingtomorrow.com/article/2020/11/the-advantages-of-digital-pressure-sensors-in-industrial-applications/16127/.

8.58. Song, P. S., Ma, Z., Ma, J., Yang, L. L., Wei, J. T., Zhao, Y. M., Zhang, M. L., Yang, F. H., and Wang, X. D. 2020. Recent progress of miniature MEMS pressure sensors. *Micromachines* **11**(1): 1–38.

8.59. Bottura, L. and Henrichsen, K. N. 2002. Field measurements. *Proceedings of the CERN Accelerator School on Superconductivity and Cryogenics for Accelerators and Detectors*, Erice, Italy, pp. 118–151.

8.60. Ramsden, E. 2011. *Hall-Effect Sensors: Theory and Applications*, 2nd edn. Elsevier, Oxford, UK.

8.61. Cochrane, C. J., Blacksburg, J., Anders, M. A., and Lenahan, P. M. 2016. Vectorized magnetometer for space applications using electrical readout of atomic scale defects in silicon carbide. *Scientific Reports* **6**: 37077.

8.62. Leuzinger, A. and Taylor A. 2010. Magneto-inductive technology overview. White Paper. PNI Sensor Corporation.

8.63. Grosz, A., Mor, V., Paperno, E., Smrusi, S., Faivinov, I. Schultz, M., and Klein, L. 2013. Planar Hall effect sensors with subnanotesla resolution. *IEEE Magnetics Letters* 4(**2013**): 6500104.

8.64. Hunt, R. P. 1971. A magnetoresistive readout transducer. *IEEE Transactions on Magnetics* 7(1): 150–154.

8.65. Multi Dimension Technology Co. 2015. Introduction to TMR magnetic sensors. http://www.dowaytech.com/en/1776.html.

8.66. Lenz, J. and Edelstein, A. S. 2006. Magnetic sensors and their applications. *IEEE Sensor Journal* **6**(3): 631–648.

8.67. Koschny, M. and Lindner, M. 2012. Magneto-optical sensors: Accurately analyze magnetic field distribution of magnetic materials. *Advanced Materials and Processes* **170**(2): 13–16.

8.68. Buchner, M., Höfler, K., Henne, B., Ney, V., and Ney, A. 2018. Tutorial: Basic principles, limits of detection, and pitfalls of highly sensitive SQUID magnetometry for nanomagnetism and spintronics. *Journal of Applied Physics* 124(**2018**): 161101-1–161101-13.

8.69. Violette, J. L. N., White, D. R. J., and Violette, M. F. 1987. *Electromagnetic Compatibility Handbook*. Springer Science, Berlin/Heidelberg.

8.70. Gaboian, J. 1993. A survey of common-mode noise. Texas Instruments Application Report SLLA057.

8.71. Alshehri, S. 2020. Optimizing the performance of thermoelectric for cooling computer chips using different types of electrical pulses. *International Journal of Computer and Information Engineering* **14**(8): 282–286.

8.72. Laird Thermal Systems. 2020. Thermoelectric cooling for CMOS sensors. https://www.lairdthermal.com/sites/default/files/ckfinder/files/resources/Application-Notes/Thermoelectric-Cooling-for-CMOS-Sensors/Thermoelectric-Cooling-for-CMOS-Sensors-Appnote-082520.pdf.

8.73. Peltier, J. C. A. 1834. Nouvelles expériences sur la caloricité des courants électrique (New experiments on the heat effects of electric currents). *Annales de Chimie et de Physique* **56**: 371–386.

8.74. Majumdar, A. 2004. Thermoelectricity in semiconductor nanostructures. *Science* **303(5659)**: 777–778.

8.75. Riffat, S. B. and Ma, X. L. 2004. Improving the coefficient of performance of thermoelectric cooling system: A review. *International Journal of Energy Research* **28**: 753–768.

8.76. Wang, P., Bar-Cohen, A., Yang, B., Solbrekken, G. L., and Shakouri, A. 2006. Analytical modeling of silicon thermoelectric microcooler. *Journal of Applied Physics* **100(1)**: 014501-1–014501-13.

8.77. Min, G. 2006. Chapter 11: Thermoelectric module design theories. In *Thermoelectrics Handbook: Macro to Nano* (Ed.: D. M. Rowe). CRC Press, Boca Raton, FL, pp. 11-1–11-15.

8.78. Anatychuk, L. I., Luste, O. J., and Vikhor, L. N., 1996. Optimal functions as an effective method for thermoelectric devices design. *Fifteenth International Conference on Thermoelectrics. Proceedings ICT '96*, Pasadena, California.

8.79. Bulman, G., Barletta, P., Lewis, J., Baldasaro, N., Manno, M., Bar-Cohen, A., and Yang, B. 2016. Superlattice-based thin-film thermoelectric modules with high cooling fluxes. *Nature Communications* 7: 10302. doi:10.1038/ncomms10302.

8.80. Laird Thermal Systems. 2020. UltraTEC™ UTX series, a new generation of high-performance thermoelectric coolers. https://www.allaboutcircuits.com/tech-days/summer-2020/laird-thermal-systems/new-products/ultratec-utx-series-a-new-generation-of-high-performance-thermoelectric-coolers/.

8.81. ABB. 2018. Application note: Dual encoder feedback control on motion drives. AN00262. https://library.e.abb.com/public/09e5178d4d0c442b9598ac3c92b1d44f/AN00262-Dual_encoder_feedback_control_on_motion_drives_Rev_A_EN.pdf.

8.82. Dougherty, J. 2018. Encoders for industrial robots: Secondary encoders improve robot accuracy in large assembly application. *Assembly Magazine*. January Issue. https://www.assemblymag.com/articles/94110-encoders-for-industrial-robots.

8.83. Stratistics Market Research Consulting. 2017. Nanosensors – Global Market Outlook (2016–2022). https://www.strategymrc.com/report/nanosensors-market/faqs.

8.84. Lariviere, B. A., Joshi, P. C., and Mcintyre, T. J. 2020. Surface acoustic wave devices printed at the aerosol-jet resolution limit. *IEEE Access* **8**: 211085–211090.

8.85. U.S. Department of Energy. DOE, industry create new generation of wireless sensors. https://www.reliableplant.com/Read/8799/doe-wireless-sensors.

8.86. Xia, F. 2009. Wireless sensor technologies and applications. *Sensors* **9(11)**: 8824–8830.

8.87. Jiang, P., Xia, H. B., He, Z. Y., and Wang, Z. M. 2009. Design of a water environment monitoring system based on wireless sensor networks. *Sensors* **9(11)**: 6411–6434.

8.88. Funtan, A., Michael, P., Rost, S., Omeis, J., Lienert, K., and Binder, W. H. 2011. Self-diagnostic polymers - inline detection of thermal degradation of unsaturated poly(ester imide)s. *Advanced Materials* **33(18)**: Article number 2100068.

8.89. Begg, R. 2021. Color-changing dye sensors detect electric motor condition. *Machine Design*. Article 21164675.

8.90. Harzheim, A., Könemann, F., Gotsmann, B., van der Zant, H., Gehring, P. 2020. Single-material graphene thermocouples. *Advanced Functional Materials* **30(22)**: 2000574.

8.91. Peleg, R. 2017. Polish team creates transparent cryogenic temperature sensor. *Graphene-infor*. https://www.graphene-info.com/polish-team-creates-transparent-cryogenic-temperature-sensor.

8.92. Kuffner, A. 2016. Choosing encoders for harsh environments. *Advanced Controls & Distribution*. https://blog.acdist.com/choosing-encoders-for-harsh-environments.

8.93. Zhang, K. Q., Hutson, C., Knighton, J., Hermann, G., and Scott, T. 2020. Radiation tolerance testing methodology of robotic manipulator prior to nuclear waste handling. *Frontiers in Robotics and AL* **7(6)**: 1–10.

8.94. Fell, C. 2018. Industrial rotary encoders: Selecting the right device for your application. *Machine Design*. April Issue, pp. 9–10.

8.95. Mathis, S. 2018. Resolution, accuracy and precision of encoders. *Assembly Magazine*. September Issue. https://www.assemblymag.com/articles/95187-resolution-accuracy-and-precision-of-encoders.

8.96. DickButturainson, K. 2020. How to understand encoder feedback. Optical and magnetic encoders are designed to prevent cumulative errors. *Control Design*. https://www.controldesign.com/articles/2020/how-to-understand-encoder-feedback/.

8.97. Collins, D. 2017. Why is the bandwidth of a servo control loop important? *Motion Control Tips*. https://www.motioncontroltips.com/why-is-the-bandwidth-of-a-servo-control-loop-important/.

8.98. SICK Inc. 2013. Selection guide for encoders. Minneapolis, Minnesota. http://www.rae.ca/wp-content/uploads/EncoderSelectionGuide.pdf.

8.99. Armstrong, R. 2005. Feedback for servos. *Machine Design*. Article 21816112. https://www.machinedesign.com/archive/article/21816112/feedback-for-servos.

8.100. Cannon, M. C. 2018. Turning two into one. *Design World*, March Issue, pp. 102–107.

8.101. Sidlyarevich, T. P. 2015. Single motor power and communication cable. U.S. patent 9,018,529 B2.

8.102. Tong, X. C. 2008. *Advanced Materials and Design for Electromagnetic Interference Shielding*. CRC Press, Boca Raton, FL.

9

Power Transmission and Gearing Systems

Mechanical power generated by an electric motor (where $P=T\omega$) can be transmitted to a driven machine using a certain type of transmission device between the motor shaft and the driven machine shaft. The primary function of power transmissions is to transmit torque and to adjust rotational velocity. Drive transmission devices are generally categorized into three groups: (*a*) flexible (non-rigid), indirect contact transmission such as belts, cables, and chains; (*b*) rigid contact transmission such as gears; and (*c*) non-contact transmission such as magnetic gears and hydraulic/hydrostatic drives. Each of these transmission mechanisms has its specific features and applications.

Belts and cables are flexible power transmission elements that are used to transmit power or motion (linear or rotary) over relatively long distances. Their elastic nature enables them to pass over pulleys typically with a high degree of efficiency. The belt and cable drives provide flexibility in the positioning of the motor relative to the load. By changing the diameters of the driving pulley, which is attached to a source of mechanical power (*e.g.*, a motor), and the driven pulley, which is attached to a driven load machine, the speed of the driven machine can be either decreased or increased. Because of the flexibility, the belt and cable drives have an inherent advantage to absorb amounts of shock and vibration. They can also accommodate some degree of misalignment between the transmitting shafts. However, since the power transmission with a belt or cable relies on friction between the belt or cable and pulleys, the belt or cable (excluding positive drives such as toothed synchronous belts running over toothed pulleys) is often subject to slip and creep. Thus, it can result in a decrease in the angular velocity of the driven machine and, in turn, a decrease in transmitted power. Furthermore, due to slippage of the belt, the belt drive helps protect the machinery from overload and jam. However, slippage may cause belts to run hot and age prematurely, leading to enormous wear of belts and pulleys.

Roller chain is the type of chain drive with a robust structure and strong overload capability. It is most commonly used for large power transmission on many kinds of industrial, military, medical, and agricultural machinery and equipment without slippage. This type of transmission system is used for rotary-to-rotary, rotary-to-linear, or linear-to-linear motion. In a chain drive system, a chain is engaged with a number of toothed sprockets, ensuring synchronization between the input and output shafts.

The disadvantages of the chain driving system are as follows: (*a*) Chains are much more sensitive to misalignment than belts and cables. Misalignment of both shafts and sprockets may potentially lead to uneven chain/sprocket wear and, in a worst-case scenario, premature failure of the transmission system. (*b*) It is not suitable for precision motion control due to wear, stretch, and slack. (*c*) It requires regular lubrication and proper maintenance. (*d*) A chain drive is usually noisier than a belt drive. (*e*) Contaminations by dirt, dust, rust, moisture, or exposure to corrosive environments prevent a lubricant from penetrating chain pins, rendering the well thought out lubrication system useless. (*f*) Excessive loads that exceed the chain's dynamic load-bearing capacity are a common cause of premature chain wear, especially under cyclic and shock overloading conditions. (*e*) It can only be used for power transmission between parallel shafts. Therefore, to reduce the premature damage and extend the service life of chain, it is required to regularly check the operating conditions regarding misalignment, lubrication, contamination, loading, *etc*.

DOI: 10.1201/9781003097716-9

As a rigid type of power transmission system, gearing systems are widely adopted when the drive and driven shafts are very close to each other, and the accurate velocity ratio is required. A typical example is a mechanical watch in which the gearing system must work with a very precise gear ratio. The use of gearing systems in motor-driven motion systems can provide numerous advantages including torque multiplication, speed reduction, inertia matching, and overall cost cut [9.1].

Gearing systems as essential and indispensable power transmission elements have been extensively applied to torque transmission and speed conversion across a broad spectrum of industries. The principal design goals of gearing systems include high efficiency and reliability, precise motion control, adequate level of safety, high load carryout capability, small or zero backlash, long service life, operating smoothness, moving angle change (for nonparallel gearing), compact size, lightweight, low noise, and manufacturing cost-effectiveness. However, these design goals are hardly achieved simultaneously. In fact, gearing systems are designed to meet some goals at the expense of others. Thus, it is important for engineers to cooperatively trade off among these goals to realize a result that maximizes their overall preference for the design.

Gearing systems come with many types, mechanisms, sizes, structures, and topologies. Over the last several decades, along with the development of advanced lubricants, more wear-resistant materials, and advanced manufacturing processes, modern gearing systems are designed to pack more power into less weight and size.

Many motor manufacturers have developed advanced solutions that integrate a motor and a gearing system together into a single package to respond to customers' needs for high-efficient motion control systems. This design configuration can help simplify the machine structure, reduce the number of components, increase torque density and operation reliability, and avoid any potential misalignment problems.

Due to the widespread use of gearing systems in motion control systems, this chapter will focus on the characteristics and operating features of gearing systems.

9.1　CHARACTERISTICS OF GEARING SYSTEMS

Except in certain cases, most gearing systems work in a similar fashion. To better design or select gearing systems, it is necessary to understand the basic terminologies and concepts associated with gearing systems.

9.1.1　GEARING SYSTEM EFFICIENCY

In a power transmission process using a gearing system, a portion of mechanical power is lost due to frictions of gear teeth, bearings and shaft seals, and lubrication churning in the gearing system. The efficiency of a gearing system η_g is thus determined as the ratio of the output power from the gearing system and the input power (*i.e.*, driving power) from a power source (*e.g.*, a motor or an engine), *i.e.*,

$$\eta_g = \frac{P_{out}}{P_{in}} = 1 - \frac{P_{loss}}{P_{in}} \tag{9.1}$$

where

$$P_{in} = T_{in}\omega_{in} \tag{9.2a}$$

$$P_{out} = T_{out}\omega_{out} \tag{9.2b}$$

The overall power losses P_{loss} in a gearing system consist of a number of components [9.2], including load-dependent and no load gearing losses, load-dependent and no load-bearing losses, shaft sealing losses, and lubricant-related losses. Among them, the main contributor is the friction losses among meshing teeth.

9.1.2   GEAR RATIO AND TORQUE RATIO

Gear ratio γ_g is defined as the input speed ω_{in} (the speed of the driving component) relative to the output speed ω_{out} (the speed of the last driven component) in a gearing system, *i.e.*,

$$\gamma_g = \frac{\omega_{in}}{\omega_{out}} \tag{9.3}$$

A gearing system with $\gamma_g > 1$ is defined as the speed reducer (where $\omega_{out} < \omega_{in}$). As a comparison, a gearing system with $0 < \gamma_g < 1$ is defined as the gear increaser (where $\omega_{out} > \omega_{in}$). When $\gamma_g = 1$ (*e.g.*, miter gears), it causes a change in the direction of rotation but without a change in the speed or torque.

Torque ratio ζ_T, which is defined as the ratio of the output torque (also known as resultant or driven torque) to the input torque (also known as applied or driving torque), can be derived from Equations 9.1 to 9.3,

$$\zeta_T = \frac{T_{out}}{T_{in}} = \frac{T_{out}\omega_{out}}{T_{in}\omega_{in}} \frac{\omega_{in}}{\omega_{out}} = \frac{P_{out}}{P_{in}} \frac{\omega_{in}}{\omega_{out}} = \eta_g \gamma_g \tag{9.4}$$

This indicates that the increase in either the gear ratio or efficiency leads to an increase in the output torque. For a gearing system with $\eta_g \gamma_g > 1$, the motor torque is amplified and correspondingly $T_{out} > T_{in}$. For a case that $0 < \eta_g \gamma_g < 1$, the output torque reduces so that $T_{out} < T_{in}$. Since the efficiency of the gearing system η_g is always less than 1, the magnitude of the torque ratio ζ_T mainly depends on the gear ratio γ_g.

The efficiency of the gearing system depends greatly upon the gear type, gear material, load torque, gear ratio, lubrication condition, and manufacturing process. Generally, higher gear ratios and light loading tend to produce poor gearing system efficiencies. With heavy loading and high gear ratios, gearing systems will approach their theoretical efficiencies.

For a servo system that consists of a servomotor and a gearing system, the overall system efficiency η_s is the product of the efficiency of the motor η_m and the efficiency of the gearing system η_g,

$$\eta_s = \eta_m \eta_g \tag{9.5}$$

This suggests that both the motor and gearing system efficiencies weight equally contribute to the overall system efficiency. Mathematically, in order to maximize the overall efficiency, the values of the two components should be as high as possible.

It is worth to note that the motor and gearing system have their own peak efficiencies, but the two may not occur at the same time. In low gear ratios, motors are more heavily loaded than gearing systems. As a result, while the motor reaches its peak efficiency, the gearing system remains at its low efficiency. At high gear ratios, the motor and gearing system follow similar curves because at that point the gearing system takes more load than the motor. This leads to peak efficiencies occurring in both the gearing system and motor. In practice, to use the least amount of power, it is critical to match the motor, gearing system, and load closely to get the best system efficiency [9.3].

9.1.3   INERTIA MATCHING

Given the motor inertia J_m and the load inertia J_l, the system is inertia matched if the gear ratio γ_g is chosen so that the load acceleration α_l is maximized for any given motor torque T_m. The load acceleration is expressed as [9.4]

$$\alpha_l = \frac{d\omega_l}{dt} = \frac{\gamma_g T_m}{J_l + \gamma_g^2 J_m} \tag{9.6}$$

where

ω_l is the load rotational speed, rad/s

The derivative of α_l with respect to γ_g gives that

$$\frac{d\alpha_l}{d\gamma_g} = \frac{\left(J_l - \gamma_g^2 J_m\right)T_m}{\left(J_l + \gamma_g^2 J_m\right)^2} \tag{9.7}$$

Solving $d\alpha_l/d\gamma_g = 0$ yields

$$\gamma_g = \sqrt{\frac{J_l}{J_m}} \tag{9.8}$$

This indicates that a motor and a gearing system is *inertia matched* with its load if the gear ratio γ_g is chosen to equal the square root of the inertia ratio (J_l/J_m). In other words, this *optimized gear ratio* provides a smooth power transmission between the gears. In practice, the inertia ratio plays a significant role in determining the motor's ability to efficiently control the load, especially during motor acceleration and deceleration processes. With this choice of gearing system, half of the torque goes to accelerating the motor's inertia and half goes to accelerating the load inertia. The motor inertia is a specification usually published by motors' manufacturers, including inertias of rotor and all other rotating components. The load inertia is calculated by adding together the inertia of all the moving parts (the external load, gearing system, and coupling).

It is important to note that Equation 9.8 is obtained based on the condition where the motor damping, load damping, and load torque are ignored. When friction torque, load torque, and system compliance are significant (Figure 9.1), the relationship among γ_g, J_l, and J_m becomes more complex, expressed implicitly as [9.5, 9.6],

$$\left[\gamma_g^2\left(J_m + J_{g1}\right) + \left(J_l + J_{g2}\right)\right]\frac{d\omega_l}{dt} + \left(\gamma_g^2 C_m + C_l\right)\omega_l = \gamma_g T_m - T_l \tag{9.9}$$

where

J_{g1} and J_{g2} are the inertia of driving gear and driven gear, respectively, kg·m²
C_m is the viscous damping between motor and ground, Ns·m/rad

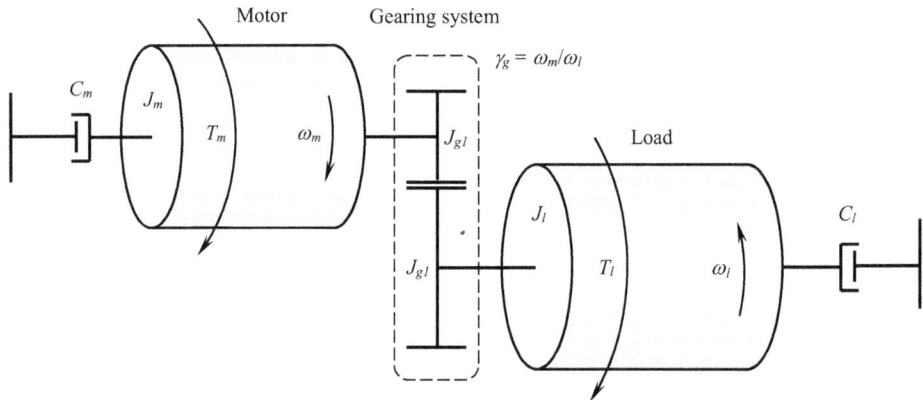

FIGURE 9.1 In a drive system where a geared motor drives a load to achieve the desired speed and torque, a number of parameters can affect the mechanical performance of the system, including motor inertia J_m, load inertia J_l, gearing system inertia (driving gear J_{g1} and driven gear J_{g2}), motor torque T_m, load torque T_l, viscous damping between motor and ground C_m, and viscous damping between load and ground C_l.

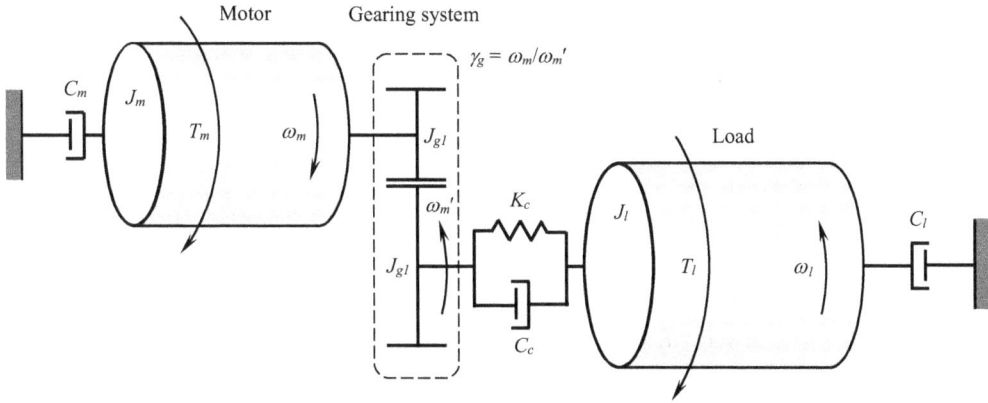

FIGURE 9.2 A compliantly coupled geared motor and load system, where coupling viscous damping C_c and elasticity K_c are taken into account for analyzing system performance characteristics.

C_l is the viscous damping between load and ground, Ns·m/rad
T_m and T_l are the motor and load torque, respectively, Nm
ω_l is the angular speed of load, rad/s

In the real world, all mechanical systems have compliance. Figure 9.2 shows a compliantly coupled geared motor and load system, where coupling elasticity is denoted by K_c and viscous damping by C_c. Thus, the governing equations that describe the relationship among γ_g, J_l, and J_m are derived as

$$\left[J_m + J_{g1} + J_{g2}\left(\frac{1}{\gamma_g^2}\right)\frac{d\omega_m}{dt} + C_m\omega_m + \frac{C_c}{\gamma_g}\left(\frac{\omega_m}{\gamma_g} - \omega_l\right) + \frac{K_c}{\gamma_g}\left(\frac{\theta_m}{\gamma_g} - \theta_l\right) \right] = T_m \tag{9.10}$$

$$J_l\frac{d\omega_l}{dt} + C_l\omega_l - C_c\left(\frac{\omega_m}{\gamma_g} - \omega_l\right) - K_c\left(\frac{\theta_m}{\gamma_g} - \theta_l\right) = -T_l \tag{9.11}$$

where
C_c is the viscous damping of coupling, Ns·m/rad
K_c is the coupling elasticity, Nm/rad
θ_m is the angular angle of motor, radian
θ_l is the angular angle of load, radian
ω_m is the angular speed of motor, rad/s

9.1.4 GEAR TOOTH PROFILE

There are two basic gear tooth profiles: the cycloid profile and the involute profile. If two intermeshed gears have teeth with the profile shape of either cycloids or involutes, they form either a cycloid or an involute gear system, respectively.

The cycloid refers to a curve generated by a point on a rim of a circular wheel as the wheel rolls along either a straight or circular path without slippage, as shown in Figure 9.3. Mathematically, a cycloid curve is described as

$$\begin{cases} x = R(\theta - \sin\theta) \\ y = R(1 - \cos\theta) \end{cases} \tag{9.12}$$

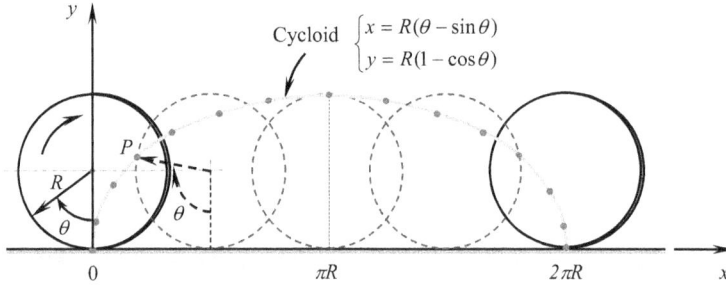

FIGURE 9.3 A cycloid curve is formed by a point P on a circular wheel as the wheel rolls along a straight path without slippage.

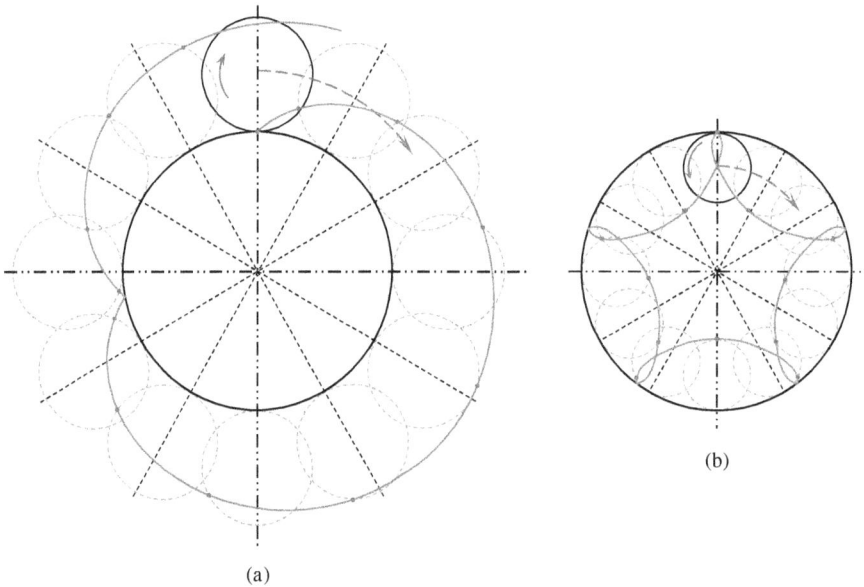

FIGURE 9.4 Formations of cycloid curves by rolling a circle along the circumference of a base circle without slippage: (a) epicycloid curve (rolling outside) and (b) hypocycloid curve (rolling inside).

If the circle is rolled along the outside and the inside of another circle without sliding, the curves traced by the point are known as epicycloid and hypocycloid, respectively (see Figures 9.4a and b). The epicycloid and hypocycloid curves are conjugated to each other to form the cycloid tooth profile.

An epicycloid curve can be mathematically expressed as

$$\begin{cases} x = r(k+1)\cos\theta - r\cos\left[(k+1)\theta\right] \\ y = r(k+1)\sin\theta - r\sin\left[(k+1)\theta\right] \end{cases} \tag{9.13a}$$

or

$$\begin{cases} x = R\left(1+\dfrac{1}{k}\right)\cos\theta - \dfrac{R}{k}\cos\left[(k+1)\theta\right] \\ y = R\left(1+\dfrac{1}{k}\right)\sin\theta - \dfrac{R}{k}\sin\left[(k+1)\theta\right] \end{cases} \tag{9.13b}$$

where k is defined as the radius ratio between the larger and small circles,

$$k = \frac{R}{r} \tag{9.14}$$

Thus, there exist three cases: (a) if k is an integer, the epicycloid curve is closed and has k cusps; (b) if k is a rational number and is expressed in the simplest terms (e.g., $k=p/q$), the curve eventually closes and has p cusps; (c) if k is an irrational number, the curve never closes on itself. Instead, it forms a dense subset in the space between the large circle and a circle with a radius of $R(1+2/k)$.

Similarly, a hypocycloid curve can be expressed as

$$\begin{cases} x = r(k-1)\cos\theta + r\cos\left[(k-1)\theta\right] \\ y = r(k-1)\sin\theta - r\sin\left[(k-1)\theta\right] \end{cases} \tag{9.15a}$$

or

$$\begin{cases} x = R\left(1-\dfrac{1}{k}\right)\cos\theta + \dfrac{R}{k}\cos\left[(k-1)\theta\right] \\ y = R\left(1-\dfrac{1}{k}\right)\sin\theta - \dfrac{R}{k}\sin\left[(k-1)\theta\right] \end{cases} \tag{9.15b}$$

The profile of a cycloid tooth consists of two separate curves: an epicycloid curve above the pitch circle and a hypocycloid curve below the pitch circle, as shown in Figure 9.5. In a cycloid gearing system, the lines of action between two gears vary in position at different points of contact during the course of action. That is, the path of contact formed by joining the different points of contact at different positions of the intermeshing teeth is curvilinear.

The involute profile is most commonly used in modern gearing systems, which is the spiraling curve traced by the endpoint P of an imaginary taut string unwinding itself from a stationary circle. The circle from which the involute curve is derived is called the base circle. Referring to Figure 9.5, an involute curve is mathematically described by

$$\begin{cases} x = R(\cos\theta + \theta\sin\theta) \\ y = R(\sin\theta - \theta\cos\theta) \end{cases} \tag{9.16}$$

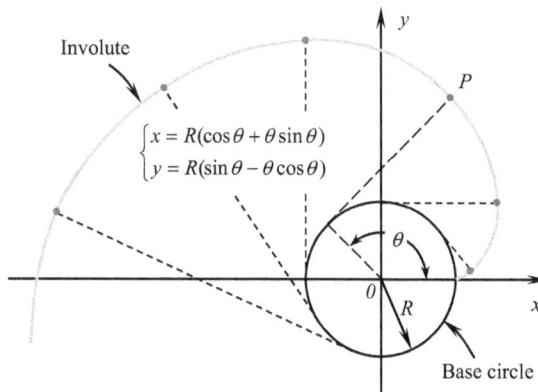

FIGURE 9.5 An involute curve is generated by the endpoint of a string that is unrolled from a basic circle with a radius R.

For a gearing system with involute profile, all points of contact take place on the same straight line [9.7]. The involute profile has an important property for maintaining a constant gear ratio between the meshed two gears.

Today, the involute curve is most commonly used for gear tooth profiles. The advantages include the following:

- For a pair of involute gears, the center distance can be varied within limits without changing the gear ratio.
- The pressure angle in mating involute gears remains constant throughout the engagement of teeth which results in smooth running. In comparison, the pressure angle in a cycloidal system varies from zero at pitch line to a maximum at the tips of the teeth.
- The face and flank of involute teeth are formed by a single curve, comparing with cycloidal gears that have double curves (*i.e.*, epicycloid and hypocycloid). This can greatly simplify the manufacturing process.
- Because the cycloidal teeth have wider flanks, they are stronger than the involute gears.
- There are no interferences occurring in cycloidal gears.

It is worth to note that when pinions have the minimum number of teeth, the gear tooth interference may occur in involute gears.

9.1.5 BACKLASH

Backlash is defined as the clearance between mating gear teeth, which compromises positioning accuracy and repeatability in precision systems (Figure 9.6). This implies that the output shaft in a gearing system can freely rotate bi-directional at a small angle without the input shaft moving. In fact, backlash is resulted from a number of factors, for instance, operation inability of machine tool, manufacturing imperfects, mounting tolerances, bearing play, and deflected workpiece under loads. For a gearing system that repeatedly changes the direction of rotation, backlash can lead to the loss in contact between the gear teeth and thus make the driven system difficult to achieve accurate positioning.

Although minimal or zero backlash in gearing systems is generally desirable for the purpose of precise motion control and smooth operation, a suitable amount of backlash may be allowed in some applications for absorbing manufacturing error, securing thickness of oil film on gear surfaces, compensating for thermal expansion of teeth, and preventing tooth interference due to the effects of machining and installation variations.

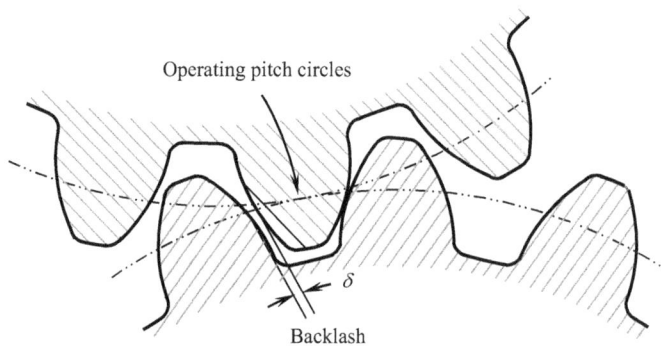

Operating pitch circles

δ

Backlash

FIGURE 9.6 Backlash is a clearance between mating gear teeth. While backlash generally reduces the transmission accuracy in motion control systems, adequate backlash can provide space for a film of lubricating oil between the teeth.

9.1.6 GEARING STAGE

Gearing systems are often categorized by the number of stages they utilized. The simplest gearing systems used in the motion control industry are those with a single stage, *i.e.*, the power is transmitted by a single set of gears. For applications where it is necessary to transmit a high torque in a small space, multi-stage gear trains must be adopted for converting the power and speed in multisteps. Compared to single-stage gearing systems, multi-stage gear trains have the advantages of higher gear ratio, larger load capacity, higher torque density, and more flexible configurations. A planetary gearing system with nth stages is demonstrated in Figure 9.7. For this case, at each stage while the ring gear is stationary, the sun gear and the planet carrier serve as the input and the output, respectively.

The compound gear ratio $\gamma_{g,c}$ and efficiency $\eta_{g,c}$ can be determined as the product of the individual stages, *i.e.*,

$$\gamma_{g,c} = \prod_{i=1}^{n} \gamma_{g,i} \tag{9.17}$$

$$\eta_{g,c} = \prod_{i=1}^{n} \eta_{g,i} \tag{9.18}$$

where $\gamma_{g,i}$ and $\eta_{g,i}$ are the gear ratio and efficiency at i^{th} stage, respectively. These equations imply that as the number of gear stages increases, the gear ratio of the system increases greatly, and the overall efficiency decreases remarkably. Since the power losses of the gearing system increase with the increased gearing stage, the tradeoff between the overall efficiency and the number of stages must be carefully considered when designing multi-stage gearing systems.

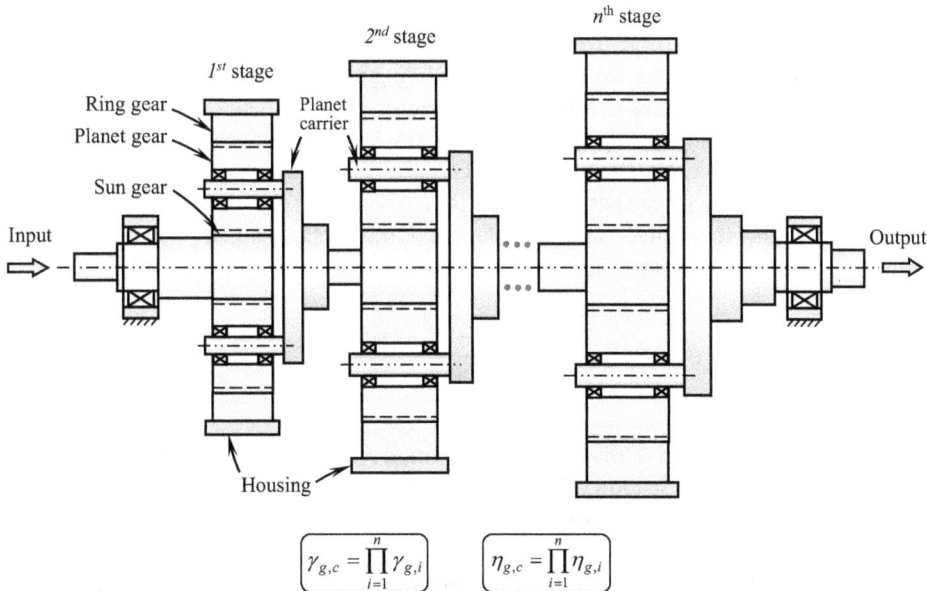

FIGURE 9.7 The structure of a planetary gearing system with n^{th} stages. For each stage, while the ring gear is stationary, the sun gear and planet carrier serve as the input and output, respectively. While the compound gear ratio is determined by multiplying each individual gear ratio from each gear stage, the overall efficiency of the gearing system is the product of the individual efficiency of each stage.

9.1.7   GEAR LUBRICATION

Like machine tools and engines, gearing systems rely on lubrication to prevent metal-to-metal contact among the moving surfaces, keep the moving parts from wearing out quickly, provide cooling and sealing for optimal system operation, help protect the metal parts from corrosion, and clean gear tooth surfaces (especially for oil lubrication). The importance of adequate lubrication to gearing systems cannot be overemphasized.

The selection of the gear lubrication method depends on the type of gearing system, the tangential speed at gear contact surfaces, gear dimension, and environmental conditions. Generally, there are three gear lubrication methods used in the gear industry:

- Grease lubrication

   This is the most popular lubrication method across a broad spectrum of industries. The technique is suitable for low-speed operation. However, due to its limited cooling efficiency, it is not recommended for continuous-duty or heavily loaded applications.
- Oil splash lubrication

   Splash lubrication, also referred to as oil bath, is used for light duty and low rotational speed applications. The effectiveness of oil splash lubrication is heavily dependent on the speed of the gears. A common rule of thumb is that a tangential speed of at least 3 m/s is required for effective lubrication. This lubrication technique has been extensively used in the automobile industry.
- Forced oil circulation lubrication

   This method, including oil mist, oil spray, and oil drop, is preferred for high-speed applications. This lubrication system requires an integrated pump, which supplies a lubricant from its reservoir and delivers it to the gears and bearings, and other auxiliary components such as a filter for particulates.

9.1.8   GEAR CONTACT RATIO

Gear contact ratio is defined as the average number of gear teeth in contact with one another as a gearset is in mesh. It accounts for the teeth that are sliding into the mesh, the teeth that sliding out of the mess, and the teeth that are in full contact when the gears are engaged. Theoretical analysis and experiments have shown that the contact ratio is a key factor affecting the carrying capacity and dynamic performance of gearset. High gear contact ratios can ensure smooth and quiet operation of gearset, improve gear carrying capacity, and increase gear durability (about 25%–50%) and strength ratings. Fundamentally, the contact ratio is the length of the contact path divided by the base pitch. The formula to calculate the gear contact ratio ε_{cr} for an external spur gearset is [9.8]

$$\varepsilon_{cr} = \frac{\sqrt{r_{ap}^2 - r_{bp}^2} + \sqrt{r_{ag}^2 - r_{bg}^2} - c\sin\alpha}{\pi m \cos\alpha} \tag{9.19}$$

where
     r_{ap} and r_{ag} are the addendum radii of the mating pinion and gear, respectively, mm
     r_{bp} and r_{abg} are the base circle radii of the mating pinion and gear, respectively, mm
     c is the central distance of the gearset
     α is the gear pressure angle, degree (°)
     m is the gear module ($m = d/z$, the ratio of gear reference diameter and number of teeth) mm

Similarly, for helical gears, the gear contact ratio is determined by

$$\varepsilon_{cr} = \frac{\sqrt{r_{ap}^2 - r_{bp}^2} + \sqrt{r_{ag}^2 - r_{bg}^2} - c\sin\alpha_t}{\pi m_t \cos\alpha_t} \tag{9.20}$$

where

α_r is the transverse pressure angle of helical gear, degree (°)

m_t is the transverse gear module, mm

The contact ratio for straight bevel gears is calculated using the following formula:

$$\varepsilon_{cr} = \frac{\sqrt{r_{ap,b}^2 - r_{bp,b}^2} + \sqrt{r_{ag,b}^2 - r_{bg,b}^2} - (r_{p,b} + r_{g,b})\sin\alpha}{\pi m \cos\alpha} \qquad (9.21)$$

where

$r_{ap,b}$ and $r_{ag,b}$ are the addendum radii at the back core of the mating straight bevel pinion and gear, respectively, mm

$r_{bg,b}$ and $r_{bg,b}$ are the base circle radii at the back core of the mating straight bevel pinion and gear, respectively, mm

$r_{p,b}$ and $r_{g,b}$ are the base circle radii at the back core of the mating pinion and gear, respectively, mm

Similar to straight bevel gears, the contact ratio for spiral bevel gears is calculated using the following formula:

$$\varepsilon_{cr} = \frac{\sqrt{r_{ap,b}^2 - r_{bp,b}^2} + \sqrt{r_{ag,b}^2 - r_{bg,b}^2} - (r_{p,b} + r_{g,b})\sin\alpha_t}{\pi m \cos\alpha_t} \qquad (9.22)$$

where α_r is the mean transverse pressure angle.

A common design goal of gear contact ratio is $\varepsilon_{cr} \geq 1.2$. Standard gears have a typical contact ratio of 1.2–1.6. High contact ratio is defined for $\varepsilon_{cr} > 2.0$. The gears with a high contact ratio perform especially well. Because there are more teeth in contact simultaneously during gear meshing, the load is spread across more teeth and thus the gear tooth stress is greatly reduced. As nonstandard gears, they have a larger number of teeth and a lower contact angle than standard gears for extending the length of the path of contact.

9.1.9 Temperature Rise and Thermal Effect on Gearing System Performance

In either continuous or intermittent operating conditions, heat is generated in gearing systems due to friction losses in gears, bearings, and lubricants. High temperatures can significantly affect the performance of the gearing system, resulting in a variety of issues:

- The main problem encountered in a hot-running gearing system is lubricant breakdown and its effect on various components such as gears and bearings. Hot operation can lead to the loss of lubricant film due to loss of lubricant viscosity and consequently lose lubricating efficiency (lubricity). According to the Arrhenius rate rule [9.9], once a lubricant exceeds its base activation temperature, it will degrade (oxidize) twice as fast for every $10°C$ increase in temperature, reducing the life expectance by half. In fact, even a partial loss of lubricant lubricity can cause metal-to-metal contact between meshing gears, resulting in gear abnormal wear. Be aware that grease contains fillers and mineral or synthetic oil. It is the oil that performs the actual lubrication. The fillers provide methods to suspend and meter out the oil in a variety of running and environmental conditions.
- A change in temperature causes a change in viscosity of oil and in turn its molecular friction. For instance, high temperatures can sheer/crack the oil molecules into smaller molecules, which causes a decrease in viscosity. A lubricant with very low viscosity cannot form a protective lubricant film to prevent metal-to-metal contact between the mating surfaces. In addition, low viscosity may lead to oil leakage. Furthermore, high temperatures

can trigger many different oil/grease failure mechanisms, directly affecting the effective lubrication performance life [9.10].

- Conventional gearing systems usually possess some level of backlash. While backlash lowers the accuracy of gearing system, it also provides spaces for lubricant and thermal expansion. However, for some gearing systems with very low or zero backlash, thermal expansion on gear elements may greatly reduce the space for lubricant and/or result in the interference between meshing gears, and thus considerably increase the tooth contact pressure, meshing friction, and reduced efficiency.

- Some types of gearing system are more sensitive to temperature due to their structural characteristics and working mechanisms. For instance, because a flexspline in a strain wave gearing (SWG) system is made of special alloy steel into a thin-walled cylinder with external teeth, the thermal deformation of the flexspline due to the temperature rise can significantly change the gear meshing condition, thus lowering the motion accurately, affecting the normal performance, and, in a worse case, even leading to gearing system failure.

- Scuffing or galling on gear tooth surfaces has found to be tightly related to interfacial temperature. As localized damage, scuffing or galling occurs when the interfacial temperature at a sliding surface reaches a critical level, leading to the thermal degradation of any existing boundary lubricant film and the eventual adhesive transfer of material [9.11]. Besides the thermal stress, other major influencing factors for the onset of scuffing damage include gear load, rotational speed, gearing geometry, quality of tooth flank, and properties of the lubricant [9.12].

- The change in temperature can result in the change in gear geometries due to thermal expansion or contraction of gears, and consequently the change in gear meshing conditions. At extreme temperatures, gear meshing deteriorates, leading to tooth surface damage and abnormal wear.

- It is well known that some rare earth permanent magnets (PMs), particularly Nd-Fe-B, are very sensitive to temperature. For example, the magnetic flux density of Nd-Fe-B magnet material decreases roughly by approximately $0.11\%/°C$ of temperature rise. This has a great influence on magnetic gearing (MG) systems to maintain normal operation.

- Increased temperatures at the sealing interface between the seal and shaft can reduce the service life of a seal through elastomer aging, swelling, and increased friction, leading to thermal and mechanical degradation of the seal, which can result in lubricant leakage, abnormal wear, or even premature failure of the seal.

9.1.10 COMPACT STRUCTURE

The optimal design of compact gearing systems without compromising their capabilities can offer a number of technical benefits, such as less occupied space, lighter weight, and lower moment of inertia. These features are especially desired for some applications where available physical space is rather limited (*e.g.*, robots). Many approaches for improving gearing system design have been posed in previous literature of references.

9.1.11 ACOUSTIC NOISE

During operation, acoustic noise is generated in a gearing system due to a number of factors, for instance, gear geometrical errors, dynamic meshing force, manufacturing and installation deflections, machining or placement of gear tooth centers in multipass gear train housing, and the interaction of gears and lubricant. Excessive noise may cause worker fatigue, strained communication, and possible hearing damage.

As in all cases, the machine designer must take into consideration the reactive forces caused by gear power transmission on the machine. For example, loads caused by radial reactive forces or axial forces need to be considered on shaft or bearing loading. This is especially true on the input pinion side, such as an electric motor bearing, as the pinion gear is typically cantilevered. The equations for calculating these radial or axial forces are also included below in several gear cases.

9.1.12 OPERATION RELIABILITY

One of the top objectives in designing gearing systems is to improve their operation reliability. An effective approach is to reduce the total part count in a gearing system. This has been done successfully by many manufacturers to continuously modify standard products with customized motor windings, electronics, and gearing systems to fit specific applications. In addition to improved reliability, the other benefits of the part count reduction include [9.13]: (*a*) lower tolerance stack-up, (*b*) required less inventory management and suppliers, (*c*) less rework at errors, (*d*) less inspections and associated fixtures, (*e*) simpler production control, (*f*) less assembly cost, material cost and tooling cost, and (*g*) higher quality production.

9.2 TYPES OF MODERN GEARING SYSTEMS

Gearing systems come in different types, shapes, sizes, configurations, and other variations. According to the gearing mechanism and mechanical structure, a number of gearing types being used in the motion control and automation industry can be identified. Each gearing type has its own characteristics such as speed-reduction ratio, permissible torque, backlash, operating temperature, and cost.

9.2.1 STRAIN WAVE GEARING SYSTEM

The concept of the SWG, also known as *harmonic drive*, was first introduced by Walton Musser in his US patent *Strain wave gearing* [9.14]. Unlike conventional gearing systems that use rigid elements (*e.g.*, gears) to transmit power throughout the gear mesh and follow the laws of *rigid-body mechanics*, the SWG uniquely utilizes elastic deformation characteristics of metal components for transmitting, converting, and/or changing mechanical power in the form of rotational motion. Because of unconventional gear-tooth meshing action, the SWG system can deliver a very high gear ratio in a very small package. Being a breakthrough in mechanical design with unique performing features, this kind of gearing system has captured the attention of engineers in many industrial applications such as robots and machine tools. However, while SWGs are beneficial for low or zero backlash, they have relatively low torsional stiffness compared to tooth–tooth contact of conventional gearing systems.

As shown in Figure 9.8, an SWG system typically consists of three main components: an elliptical wave generator, a flexspline, and a circular spline. The wave generator is an assembly of a thin-raced flexible ball bearing fitted onto the periphery of an elliptical cam. According to the structure, flexspline can be implemented into two groups: cup-shaped flexspline, used for cup-type SWG systems with solid shafts; and hat-shaped flexspline, used for hat-type SWG systems with hollow shafts, as shown in Figure 9.9. In industrial robot applications, hat-type SWG systems are favorable because they allow the power cables and data lines passing through the hollow shafts so that they can be fully protected. For either type of flexspline, the external small gear teeth are machined on the open end. The elliptical wave generator is inserted into the flexspline, distorting the flexspline into the elliptical form. The circular spline is a rigid ring with internal teeth that mesh with external teeth on the flexspline. The flexspline has fewer teeth (usually two) and consequently a smaller effective diameter than the circular spline. In practice, while the circular spline is fixed, the wave generator functions as the input and the flexspline as the output. A flexible ball bearing is placed between the flexspline and wave generator, allowing the two components to rotate at different but coupled speeds on the same axis. This bearing is specially designed and fabricated using thin metal rings for withstanding the deformation caused by the wave generator. Furthermore, due to the elliptical shape of the wave generation, the speed of the balls in the bearing varies according to their relative positions.

The design of flexspline is critical to the performance and service life of an SWG system. On the one hand, the flexspline must be thin enough to minimize the deflect stress caused by the strain wave. On the other hand, it must maintain adequate thickness to transmit the output torque tangentially. Because of its thin-wall structure, the flexspline often produces excessive noise and vibration at high rotational speeds.

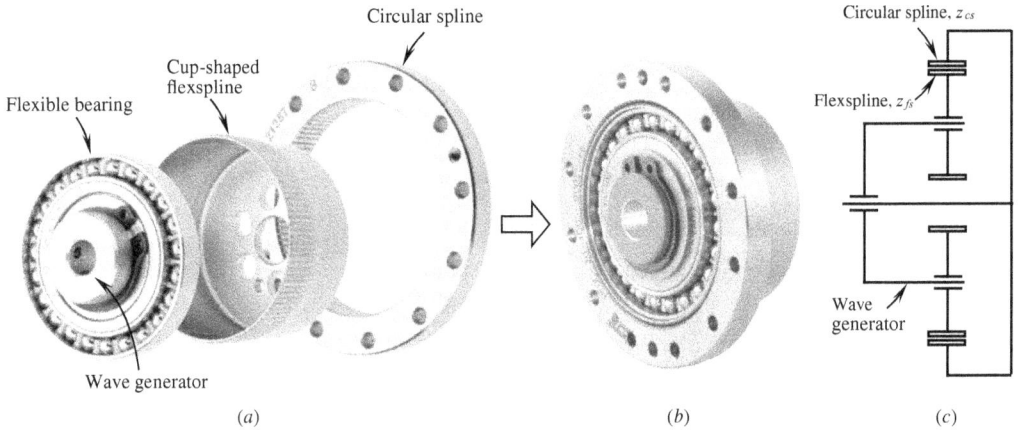

FIGURE 9.8 Structure of a cup-type SWG system: (*a*) major components: a wave generator, a cup-shaped flexspline, and a rigid circular spline; (*b*) the final assembly; and (*c*) the operating mechanism of the SWG system.

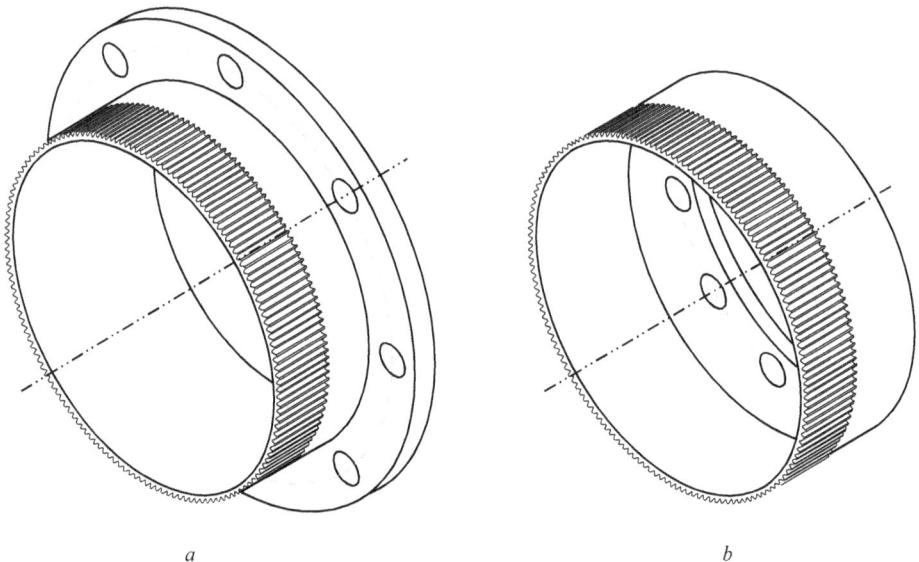

FIGURE 9.9 Two types of flexsplines: (*a*) hat-shaped flexspline, used with hollow shafts and (*b*) cup-shaped flexspline, used with solid shafts.

The teeth of cup-shaped flexspline can be fabricated progressively with tools such as laser cutter, electrical discharge machine, hobbing machine, and gear shaper. However, it is noted that some of these manufacturing methods may not be applicable for hat-shaped flexspline due to the interference at their roots.

During operation, as the flexspline is distorted into an elliptical shape by the wave generator, the external teeth of the flexspline will engage with the internal teeth of the circular spline at the major elliptical axis. Simultaneously, the teeth of the flexspline at the minor axis fully disengage from the internal teeth of the circular spline (Figure 9.10a). Upon rotating the wave generator clockwise by 90°, the flexspline is subjected to elastic deformation with the shifted major axis by 90°. Correspondingly, the tooth engagement and disengagement positions are shifted 90° (Figure 9.10b). When the wave generator further rotates clockwise to 180°, the tooth engagement occurs at 180° across the major axis of the ellipse (Figure 9.10c). As shown in Figure 9.10d, the wave generator completes one full revolution. For the configuration that the circular spline is fixed, each complete clockwise rotation of the wave generator causes the flexspline to rotate counterclockwise.

Crossed roller bearings are commonly used in SWG systems to withstanding loads from all directions (Figure 9.11). In the bearing, cylindrical rollers are arranged orthogonally to each other between the inner and outer rings. The spacers are placed among the rollers to prevent the mutual friction among them and thus decrease the torque resistance for rotation. To ensure the positioning accuracy, the bearing is often intentionally made with very tight or even negative clearance between rolling elements and bearing rings. This type of bearing has high rigidity, rotation accuracy, load-carrying capacity, and compact structure. The comparison of crossed roller bearings for the hat-shaped SWG and cup-shaped SWG is shown in Figure 9.12.

Fatigue testing is the most critical step for ensuring sustainable and reliable operation of SWG systems. Among all components, flexspline is the key component experiencing localized structure change and withstanding cyclic fatigue loading conditions during operation. To prevent fatigue failure, the fatigue strength check of the flexspline is the premise and basis of the design of SWG system. The fatigue test setup is presented in Figure 9.13. This device consists of a driving motor, a torque meter, a controllable torque load device, a flexible coupling, and an SWG system to be tested. It shows that a well-designed flexspline still maintains a good condition after 4,645 h operation. In addition, the supporting structures, bearings, and housings must be rigidly designed to react to small displacement *wobble* forces on both output and input sides due to the rotating wave action of gear components.

An SWG system has many advantages over conventional gearing systems, such as high gear ratio in single transmission, compactness, less noise, and perhaps the most notable, very low or zero backlashes [9.15]. The key features are its exceptionally high positioning accuracy and repeatability. Furthermore, because that power is transmitted through multiple teeth engagement, this gearing

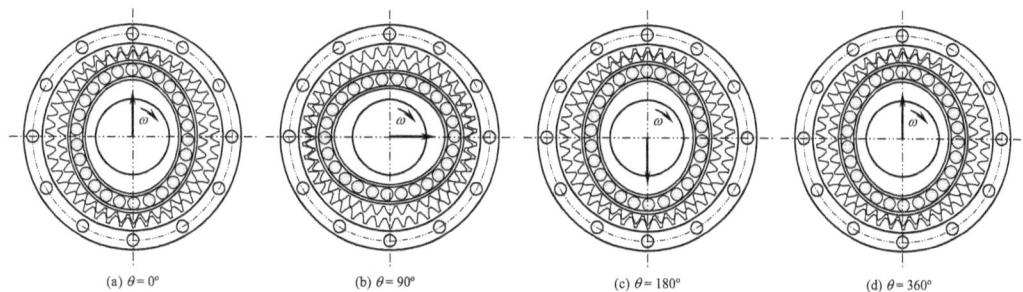

(a) $\theta = 0°$ (b) $\theta = 90°$ (c) $\theta = 180°$ (d) $\theta = 360°$

FIGURE 9.10 Engagement and disengagement among the flexspline teeth and circular spline teeth during operation of an SWG system: (*a*) At the beginning of the engagement, $\theta = 0°$; (*b*) the flexspline rotates clockwise by reaching $\theta = 90°$; (*c*) the flexspline continuously rotates up to $\theta = 180°$; (*d*) the flexspline completes a full revolution, where $\theta = 360°$.

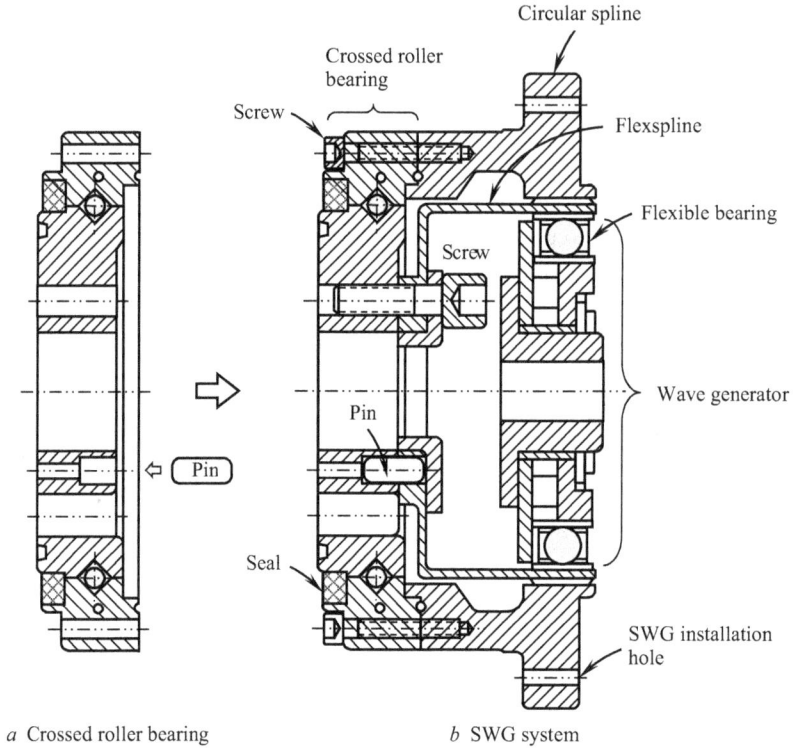

FIGURE 9.11 The built-in of an (a) crossed roller bearing in a (b) SWG system is to bear various kinds of loads (compression, tension, shear, bending moment, torsion, *etc.*) from any direction.

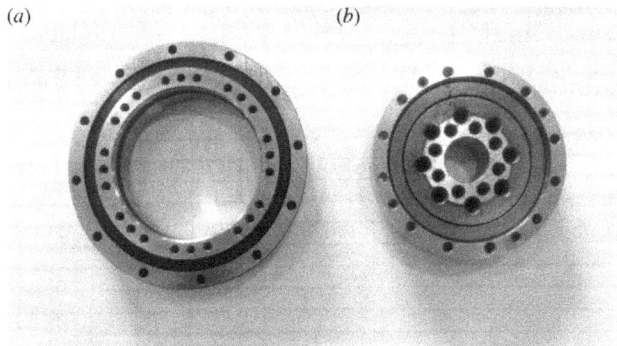

FIGURE 9.12 The comparison of crossed roller bearing subjected to identical torque for (a) hat-shaped strain wave gearing system that is associated with hollow shafts and (b) cup-shaped SWG system which is associated with solid shafts.

system can offer much higher torsional stiffness and output torque capacity than the conventional planetary gearing system.

The most commonly known disadvantages of SWG are the existence of structural flexibility [9.16], resonance vibration, and structural damping nonlinearities [9.17]. In addition, because the SWG systems require all kinds of precision machining, they come at a higher price point than conventional gearing. Although the preloaded teeth engagement prevents tooth-to-tooth backlash,

FIGURE 9.13 Fatigue test setup of the SWG system. The hat-shaped flexspline shown on the right-hand side has tested for 4,645 h and is still in a good condition.

there is a bidirectional hysteresis effect on positioning accuracy, so bidirectional motion needs to be treated by the position control system for *backlash* effects from the hysteresis.

In a typical gear driving mechanism, the wave generator is the input element, the flexspline is the output element, and the circular spline is fixed. Thus, the gear ratio is determined as

$$\gamma_g = \frac{\omega_w}{\omega_{fs}} = -\frac{z_{fs}}{z_{cs} - z_{fs}} \tag{9.23}$$

where

ω_w is the rotational speed of the wave generator
ω_{fs} is the rotational speed of the flexspline
z_{fs} is the number of teeth of the flexspline
z_{cs} is the number of teeth of the circular spline

The minus sign (–) in the above equation indicates that the wave generator and flexspline rotate in the opposite directions, similar to a single pass gear train.

It can be seen from Equation 9.23 the gear ratio is determined by the number of teeth of the flexspline z_{fs} and the difference between the numbers of teeth of the circular spline and the flexspline $(z_{cs} - z_{fs})$. Thus, the SWG system can be designed to achieve high gear ratios by choosing a large z_{fs} and a small value of $(z_{cs} - z_{fs})$. For other driving mechanisms of SWG systems, the gear ratios can be calculated, as shown in Table 9.1.

The efficiency of SWG is one of the main concerns for both SWG users and designers. Generally, the efficiency depends on a number of factors, including gear ratio, input speed, load torque, driving mechanisms from combinations of three basic elements (the wave generator, flexspline, and circular spline), the tooth number of the three elements, and operating conditions (*e.g.*, temperature and lubrication). Today, precise predictions of SWG efficiency remain a persistent challenge because it requires many complex calculations for various power losses in SWG. There are a number of models for calculating SWG efficiency. A model based on the FEM was proposed and validated from experimental results by Wu *et al.* [9.18]. The transmission efficiency of harmonic gear drive obtained by the numerical simulation is 87.1%, which is quite close to the type of XB1-120-100 harmonic drive efficiency (75%–90%). Zou *et al.* [9.19] presented a model considering the geometry, internal interactions, and assembly error. In this model, a single tooth pair is used to represent the transmission mechanism of harmonic drive. The meshing stiffness between the flexspline and the circular spline, the torsional stiffness of the flexspline cylinder, and the radial stiffness of the thin-walled ball bearing are included and formulated. The simulation results are in good agreement with the experimental results. More recently, a novel model has been developed for quickly estimating SWG efficiencies under various driving mechanisms [9.20]. The formulas, as functions of τ (where

$\tau = w_{fs}/w_{cs} = z_{cs}/z_{fs}$) and $\eta_{cs\text{-}fs}$ or $\eta_{fs\text{-}cs}$ (where $\eta_{cs\text{-}fs}$ and $\eta_{fs\text{-}cs}$ are reciprocals of each other), are listed in Table 9.1. Once the efficiency $\eta_{cs\text{-}fs}$ or $\eta_{fs\text{-}cs}$ is computed, the coefficients of other driving mechanisms can be determined using the corresponding formulas.

In designing a reliable SWG system, it is crucial to develop state-of-the-art flexspline tooth profiles for allowing continuous contact of teeth, enabling more tooth engagement, and improving the tooth torsional stiffness. A number of tooth profiles have been developed by major SWG manufacturers, including the "S" tooth profile [9.21, 9.22] and the 3D continuous tooth profile [9.23, 9.24]. These tooth profiles are optimized to make sure that up to 30% teeth are engaged simultaneously between the flexspline and circular spline. Furthermore, the increased tooth root radius enhances the tooth strength. These enable the use of fine-pith teeth and thin-walled flexspline for transmission of sizable torques [9.25]. More recently, a tooth profile design method for SWG without tip interference has been developed [9.26]. The tooth profile parameters are efficiently generated by the sample points determined from the conjugate points and tip points at the interference zone. The results demonstrate that the designed tooth profile has high precision and sooth transmission feature with good meshing qualities.

To ensure reliable operation, flexsplines are often made of special alloys such as 30CrMnSiA, 35CrMnSiA, 38CrMoAlA, 40CrNiMoA, 55Si2Mn, and 42CrMo4. In order to enhance the torque transmission capability and to reduce vibration and noise of flexspline, anisotropic composite materials are also considered [9.27]. For space applications, stainless steel may also be used for its low coefficient of thermal expansion and high corrosion resistance [9.28].

The factors affecting the efficiency of the SWG system are somewhat different from those in other gearing systems. Power losses have resulted from not only the friction due to gear teeth meshing but also the continuous stretching of the flexspline [9.29], *i.e.*:

- The friction loss due to gear meshing among the teeth of the flexspline and rigid circular spline in the three teeth meshing stages—engagement in, engagement, and engagement out. Unlike conventional gearing, teeth movement in a strain wave mechanism is primarily sliding due to a small phase shift among the corresponding teeth in engagement. The sliding effect among the teeth causes energy losses, heating up the unit to a temperature equilibrium of about $60°C–70°C$ [9.30].
- The molecular friction loss due to the viscous shear friction of a lubricant could become a significant part of the total loss, especially under high rotation speed, low temperature, and/or heavy load conditions.
- An SWG is usually equipped with several bearings, such as the cross-roller bearing, wave generator bearing, and others. The bearing friction loss is a major contributor to the overall mechanical loss of an SWG. In most cases, the prediction of the bearing loss is rather difficult because it highly depends on load conditions and certain tribological phenomena that occur in the lubricant film between the rolling elements, raceways, and cages. In the SKF model [9.31], the total friction loss is calculated as a sum of losses due to rolling, sliding, seal, and drag frictional moment.
- The power loss due to energy invested in repeatedly distorting the metallic flexspline every revolution. This is a special power loss occurring in SWGs but not applicable for other gearing systems.

More specifically, the transmission efficiency is affected by many factors such as gear ratio, input speed, load torque, temperature, quantity of lubricant, and type of lubricant [9.32]. Generally, the efficiency of SWG system is proportional to the percentile load (load torque/allowable average torque) and ambient temperature (typically $-10°C \le T_a \le 50°C$) and inversely proportional to the input speed and gear ratio. To ensure normal operation of SWG systems, the allowable housing surface temperature is usually set at 70°C by most major manufacturers.

TABLE 9.1

Gear Ratio and Efficiency for Different Driving Mechanisms of Strain Wave Gearing System

Operating Configuration				Speed Variation	Output and Input Directions of Rotation	Gear Ratio γ_g	Efficiency η_{i-j} $\left(\text{where } \tau = \dfrac{w'_{fs}}{w_{cs}} = \dfrac{z_{cs}}{z_{fs}}\right)$
Case	Input	Output	Fixed ($\omega = 0$)				
1	Wave generator	Flexspline	Circular spline	Decreasing	Opposite	$\dfrac{\omega_w}{\omega_{fs}} = -\dfrac{z_{fs}}{z_{cs}-z_{fs}}$	$\eta_{w-fs} = \dfrac{(\tau-1)\eta_{cs-fs}}{\tau-\eta_{cs-fs}}$
2	Wave generator	Circular spline	Flexspline	Decreasing	Same	$\dfrac{\omega_w}{\omega_{cs}} = \dfrac{z_{cs}}{z_{cs}-z_{fs}}$	$\eta_{w-cs} = \dfrac{\tau-1}{\tau-\eta_{cs-fs}}$
3	Flexspline	Circular spline	Wave generator	Decreasing	Same	$\dfrac{\omega_{fs}}{\omega_{cs}} = \dfrac{z_{cs}}{z_{fs}}$	η_{fs-cs}
4	Circular spline	Flexspline	Wave generator	Increasing	Same	$\dfrac{\omega_{cs}}{\omega_{fs}} = \dfrac{z_{fs}}{z_{cs}}$	η_{cs-fs}
5	Flexspline	Wave generator	Circular spline	Increasing	Opposite	$\dfrac{\omega_{fs}}{\omega_w} = -\dfrac{z_{cs}-z_{fs}}{z_{fs}}$	$\eta_{fs-w} = \dfrac{\tau\eta_{fs-cs}-1}{\tau-1}$
6	Circular spline	Wave generator	Flexspline	Increasing	Same	$\dfrac{\omega_{cs}}{\omega_w} = \dfrac{z_{cs}-z_{fs}}{z_{cs}}$	$\eta_{cs-w} = \dfrac{1-\tau\eta_{fs-cs}}{\eta_{fs-cs}(1-\tau)}$

Notes:

1. z_{fs} and z_{cs} are the tooth number of the flexspline and circular spline, respectively.
2. The minus sign in the above equations indicates that the output and the input members rotate in opposite directions.
3. The first two configurations make more sense than others do, from an engineering point of view.
4. Efficiency η_{i-j} assumes ith as driving element and jth as driven element.
5. The efficiency of a speed reducer is always greater than that of a speed increaser.

In recent years, the desire for designing more cost-effective and lightweight automation machines has led to the development of plastic SWG systems. As entirely made of plastic, this type of gearing system provides a number of advantages over its metal counterparts, such as lighter weight, less inertia, lower noise, and lower manufacturing cost. It is also resistant to certain corrosive environments. Some gearing systems are made of special materials for lubricant-free operation. However, this type of gearing system has some shortcomings relative to metal gears:

- Due to high wearing rates, the SWG system has low durability.
- Greater dimensional instability due to a larger coefficient of thermal expansion and moisture absorption.
- Since the flexspline must be flexible and deliberately deformed, it is typically made of special steel in a small thickness. The strain wave gearing system has lower rigidity with limited load-carrying capability.
- The flexspline is subjected to a cyclic elastic deformation during the rotation of the elliptical wave generator. This frequent deformation can fatigue the plastic material, causing the fatigue failure especially with long service time.

Therefore, the use of such a system is limited to non-demanding applications, such as light loads, low duty cycle, and less-precisely controlled automation systems.

9.2.2 Planetary Gearing System

Planetary gearing systems are classified into two categories: single and compound planetary systems. A single planetary gearing system, also known as *epicyclical gearing system* in some textbooks, is constructed with four main components: a sun gear which is at the center of the assembly, an outer ring gear which has internal teeth and wraps around the entire assembly, and a planet carrier on which a number of planet gears (usually 3–6) are attached (Figure 9.14). As planet gears rotate around the sun gear and mesh with both the sun and ring gear, the planet gears share the load that the system carries. In a planetary gearing system, any of these three main gears (sun gear, ring gear, and planet gears) can perform either the driving member (*i.e.*, the input), the driven member (*i.e.*, the output), or the stationary support (Figure 9.15) to meet the requirements of different applications.

If the angular velocities of sun gear, ring gear, and planet carrier are denoted by ω_s, ω_r, and ω_c, respectively, the following relationship is generally held,

$$z_s\omega_s + z_r\omega_r = (z_s + z_r)\omega_c \tag{9.24}$$

where
  z_s is the number of teeth of the sun gear
  z_r is the number of teeth of the ring gear

This indicates that the gear ratio of the planetary gear system depends not only on the tooth numbers of sun and ring gears but also on the operation condition. These six different operation cases are shown in Table 9.2 [9.33, 9.34].

As demonstrated in Table 9.2, the gear ratio, as well as the output torque, is determined by not only the teeth number of each gear but also the planetary component configuration/connection.

Compared to other gearing systems under same loads, planetary gearing systems feature high efficiency, compact size, low backlash, resistance to shock and vibration, coaxial arrangement, and high torque-to-weight ratio.

Ferguson's mechanism, invented by Scottish astronomer James Ferguson in 1764, refers to the mechanism that one gear engages with other gears ($n \geq 2$) simultaneously on the same tooth surface.

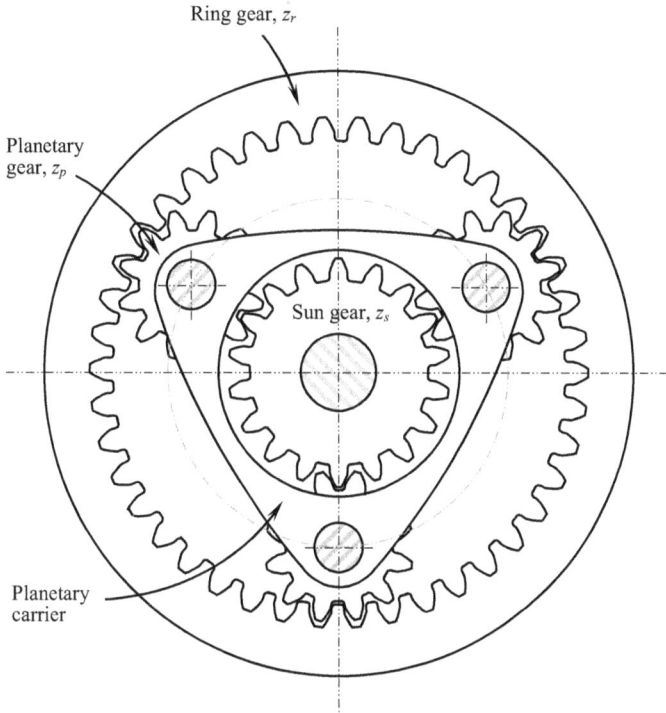

FIGURE 9.14 Structure of a planetary gearing system with three major components: a sun gear, a ring gear, and a planetary carrier on which several planetary gears are attached.

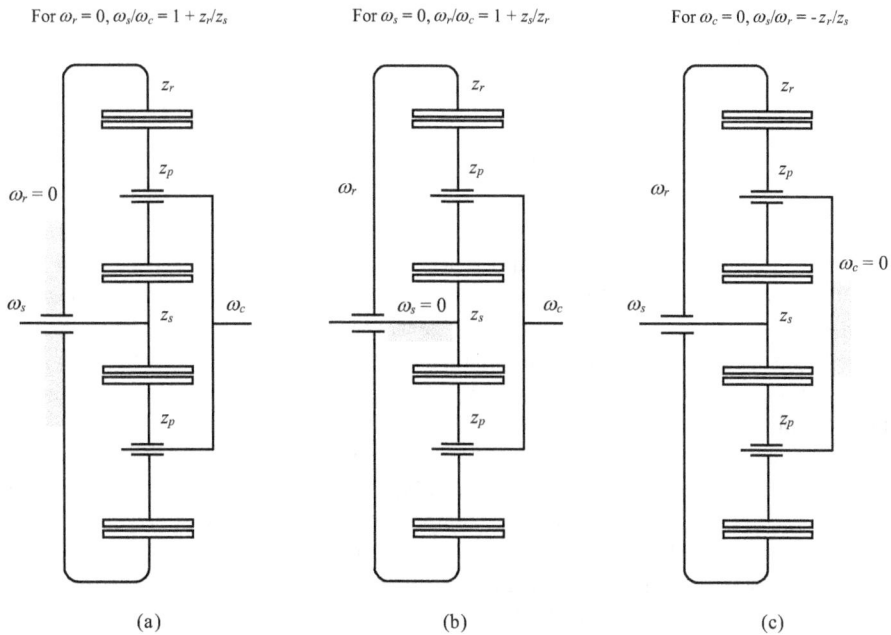

For $\omega_r = 0$, $\omega_s/\omega_c = 1 + z_r/z_s$

For $\omega_s = 0$, $\omega_r/\omega_c = 1 + z_s/z_r$

For $\omega_c = 0$, $\omega_s/\omega_r = -z_r/z_s$

(a) (b) (c)

FIGURE 9.15 Three planetary gearing system mechanisms: (a) planetary type (the ring gear is fixed); (b) solar type (the sun gear is fixed); and (c) star type (the planet carrier is fixed).

TABLE 9.2

Kinematic Analysis of Single Planetary Gearing System

Input	Output	Fixed<p> ($\omega = 0$)	Speed	Torque	Output and Input Directions of Rotation	Gear ratio γ_g
Sun gear	Planet carrier	Ring gear	Maximum reduction	Increase	Same	$\dfrac{\omega_s}{\omega_c} = 1 + \dfrac{z_r}{z_s}$
Ring gear	Planet carrier	Sun gear	Minimum reduction	Increase	Same	$\dfrac{\omega_r}{\omega_c} = 1 + \dfrac{z_s}{z_r}$
Planet carrier	Sun gear	Ring gear	Maximum increase	Reduction	Same	$\dfrac{\omega_c}{\omega_s} = \dfrac{1}{1 + z_r/z_s}$
Planet carrier	Ring gear	Sun gear	Minimum increase	Reduction	Same	$\dfrac{\omega_c}{\omega_r} = \dfrac{1}{1 + z_s/z_r}$
Sun gear	Ring gear	Planet carrier	Reduction	Increase	Opposite	$\dfrac{\omega_s}{\omega_r} = -\dfrac{z_r}{z_s}$
Ring gear	Sun gear	Planet carrier	Increase	Reduction	Opposite	$\dfrac{\omega_r}{\omega_s} = -\dfrac{z_s}{z_r}$

Notes:

1. z_r is the number of teeth of the ring gear.
2. z_s is the number of teeth of the sun gear.
3. The minus sign in the above equations indicates that the output and the input members rotate in opposite directions.

The planetary gearing system employing this mechanism is called the Ferguson-type planetary gearing system. As shown in Figure 9.16, z_s and z_p are the tooth numbers of the sun gear and planetary gear, respectively. Two internal ring gears, with the tooth number of z_{r1} and z_{r2}, respectively, are engaged simultaneously with the planetary gear, where the face width of the planetary gear is about twice that of the ring gear. The two ring gears are disposed on the same circumference but separated longitudinally along a common rotation axis. One ring gear is fixed and another is free to rotate. Thus, the gear ratio is determined as

$$\gamma_g = \frac{\omega_s}{\omega_{r2}} = \frac{1 + z_{r1}/z_s}{1 - z_{r1}/z_{r2}} \tag{9.25}$$

Unlike a regular planetary gearing system, the tooth numbers of z_{r1}, z_{r2}, z_p, and z_s in Ferguson's gearing system cannot be chosen freely as desired. To avoid tooth interference, the necessary and sufficient conditions for constructing this type of gearing system are (a) $z_{r2} > z_{r1}$ and (b) $z_{r2} - z_{r1} = z_p$ [9.35]. On comparing with the conventional planetary gearing system, the Ferguson type of planetary gearing system can offer much higher gear ratios. However, due to the above conditions, their gear ratios cannot be obtained free as desired.

Compound planetary gearing systems have been developed to increase the gear ratio and torque-to-weight (or torque-to-volume) ratio with more flexible configurations. They usually involve one or more of the three types of structures: (a) The Simpson gear set, which combines two single planetary gearing systems together with a common sun gear; (b) The Ravigneaux gear set, which uses two sun gears, two planet carriers with a common ring gear; and (c) The Lepelletier gear set, which connects a single planetary gear set with the Ravigneaux gear set.

In practice, compounded planetary trains have at least two planet gears attached in line to the same shaft, rotating and orbiting at the same speed while meshing with different gears. Compounded planets can have different tooth numbers, as can the gears they mesh with [9.36]. Though compound

$$\text{For } \omega_{rl} = 0, \frac{\omega_s}{\omega_{r2}} = \frac{1 + z_{rl}/z_s}{1 - z_{rl}/z_{r2}}$$

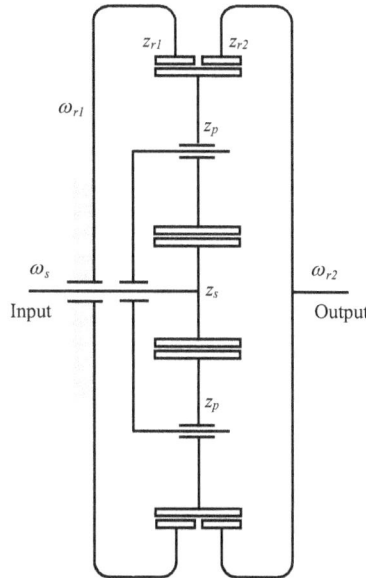

FIGURE 9.16 By utilizing the James Ferguson mechanism that was first presented in 1764, the Ferguson type of planetary gearing system can achieve a high gear ratio but suffer low load-carrying capacity. The necessary and sufficient conditions for constructing this type of gearing system are (a) $z_{r2} > z_{rl}$ and (b) $z_{r2} - z_{rl} = z_p$.

planetary gearing systems have some advantages over single planetary gearing systems, vibration and noise remain major concerns in applications of compound planetary gearing systems.

Depending on loads and applications, the planetary gearing systems may be constructed of carbon steel, stainless steel, aluminum, or brass. Generally, gears made of steel are likely noisy but have low wear. In slow and lightly loaded applications, non-metallic materials such as polymers may also be used for lowering meshing friction, reducing operation noise, and offering a long lubrication-free lifetime.

Planetary gearing systems are usually sealed to IP65 for complete protection against dust and washdown conditions, where the projection of water jet is at the water pressure of 30 kN/m² from a distance of 3 m.

9.2.3 CYCLOIDAL GEARING SYSTEM

The cycloidal gearing system was invented by the German engineer Lorenz Braren in 1927 [9.37]. As depicted in Figure 9.17, a cycloidal gearing system comprises several major components, including a high-speed input shaft to which an eccentric cam is attached and extended, a cycloidal disc in which the teeth (called lobes) have cycloidal profiles, a set of gear ring pin rollers that are mounted on a stationary ring gear housing (not shown), and a low-speed output shaft integrated with a set of output pin rollers. To assemble the gearing system, the eccentric cam is inserted into the central hole of the cycloid disc, gear pin rollers roll along the profile of the fixed gear housing, and the output pin rollers are inserted into the equally distributed holes of the cycloidal disc. When the input shaft rotates, it drives the eccentric cam that in turn drives the cycloid disc to rotate with respect to its axial centerline and revolve in a circle with respect to stationary ring gear pin rollers. It is the combination of these two movements that make the cycloidal disc progressively engage and disengage

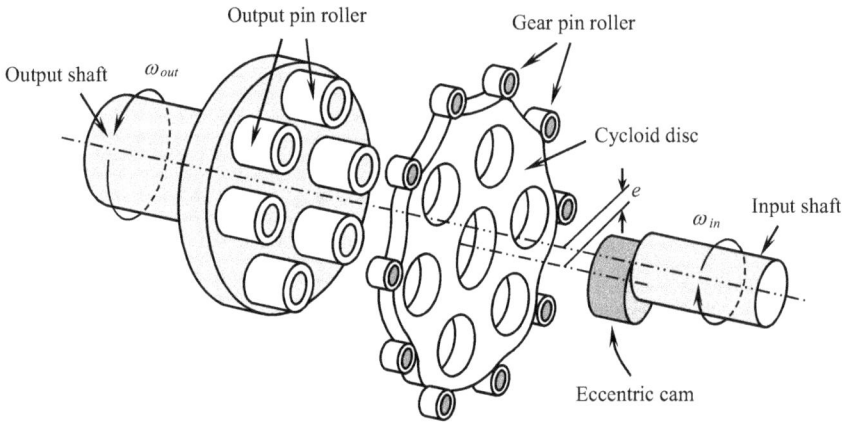

FIGURE 9.17   Structure of cycloidal gearing system (the ring gear housing is not shown here).

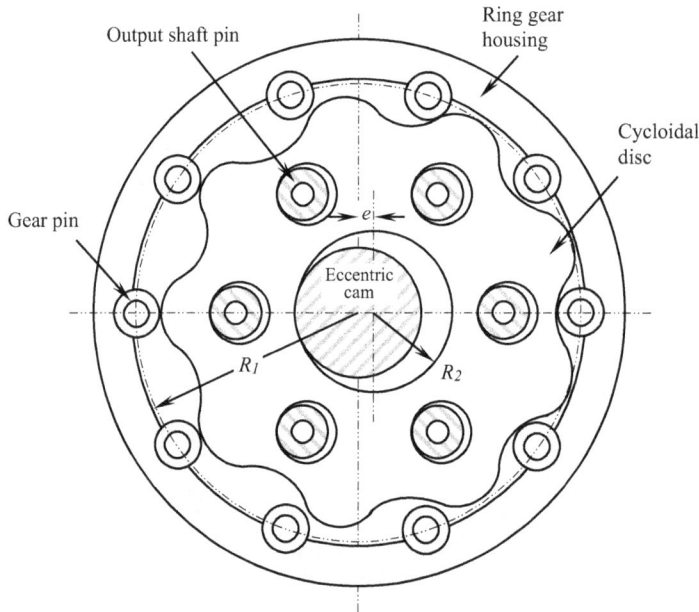

FIGURE 9.18   Working principle of the cycloidal gearing system, showing the eccentricity e between the eccentric cam and the cycloidal disc.

with the ring gear pin rollers, causing a reversed revolution of the output pin rollers at a reduced speed and transmitting to the output shaft (Figure 9.18).

Generally, the number of ring gear pin rollers z_p exceeds the number of lobes of the cycloid disc z_{cd}. The gear ratio of a cycloidal gearing system is determined as [9.38]

$$\gamma_g = \frac{\omega_{in}}{\omega_{out}} = \frac{z_{cd}}{z_p - z_{cd}} \tag{9.26}$$

It can be seen that when $z_p - z_{cd} = 1$, the gear ratio reaches the maximum value z_{cd}. If z_{cd} is made large enough, a very high gear ratio can be achieved.

In addition to the high gear ratio, the cycloidal gearing system is characterized by high torsional rigidity, large load-carrying and overloading capacities, high efficiency, and strong tooth root strength.

The maximum eccentricity is determined as [9.39]

$$e_{max} = \sqrt{\frac{27R^2 z_{cd} - R_p^2 (z_{cd} + 2)^3}{27 z_{cd} (z_{cd} + 1)^2}} \tag{9.27}$$

where

  R is the radius of the distributed gear pin roller circle, *i.e.*, the center distance between the pin roller and the cycloid disc

  R_p is the radius of pin roller

Compared with planetary gearing systems, cycloidal gearing systems offer a better choice in terms of backlash and positioning accuracy. As the gear ratio goes beyond 100:1, cycloidal gearing systems hold advantages because stacking stages are unnecessary, so the cycloidal gearing system can be shorter and less expensive [9.40].

Cycloidal gearing systems can work in high efficiencies at relatively high gear ration ($\gamma_g > 30$:1). Under normal working conditions, their efficiencies range from 75% to 85%.

A primary disadvantage of cycloidal gearing system is that it is not backdrivable, that is, the input and output shafts of the cycloidal gearing system cannot be reversed from the output shaft loadside. Another drawback is the eccentrically mounted cycloidal disc may cause vibration in the drive, leading to increased wear on the exterior teeth of the cycloidal disc. In addition, single-stage planetary gearing systems offer higher torque densities.

9.2.4 ROTATE VECTOR GEARING SYSTEM

Rotate vector (*RV*) gearing technology was introduced in the mid-1980s to address the customer needs for larger precise motion control systems, such as heavy-duty industrial robots, machine tools, semiconductor production equipment, and factory automation. This technology not only supports smooth and accurate movements of driven machines but also increases their resistance to shocks and vibration.

As shown in Figure 9.19, *RV* gearing system is characterized by integrating a planetary gearing system at the high-speed stage (first stage) and a cycloid gearing system at the low-speed stage (second stage). The planetary gearing system consists of a sun gear g_{sun} and several planet gears g_p. The sun gear typically connects the input shaft and drives the planet gears rotating in an opposite direction. The planet gears allocate evenly around the sun gear, allowing the transmission torque to be equally distributed among the planet gears. The crankshafts are directly connected to the planet gears and rotate at the same speed as the plant gears. The cycloid gearing system consists of cycloid gears, needle pins, needle wheel, and carrier. The rotation of the crankshafts results in the rotation of the cycloid gears. The carrier and output wheel are fixed by bolts as one component [9.41]. In order to achieve the equilibrium in the radial direction, it usually has two identical cycloid gears arranged symmetrically in the circumferential direction with an angle of 180°.

This type of gearing system possesses some advantages, including:

- As two-stage gearing systems, *RV* can offer a wide range of gear ratios.
- Because all main elements (*e.g.* crankshafts) are supported from both sides, the gearing system has high torsional stiffness and high shock load capability.
- The built-in angular ball bearing construction improves the ability of supporting external loads and increases moment rigidity and maximum allowable moment.

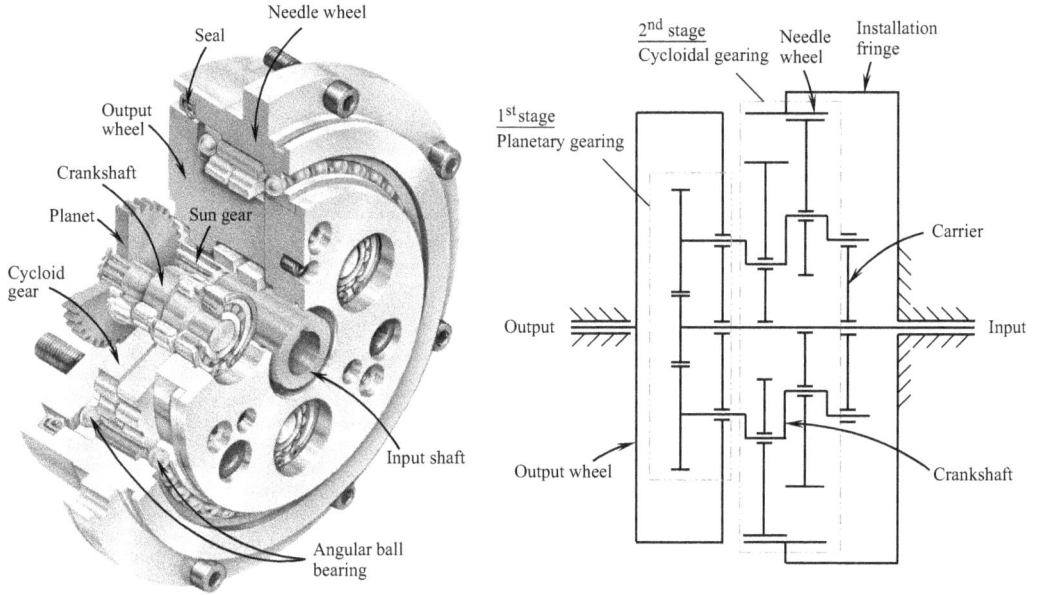

FIGURE 9.19   Structure of *RV* gearing system and its transmission mechanism.

- Since loads are transferred by means of rollers in the cycloidal gears, *RV* gearing systems feature high precision, low hysteresis loss, and long service life.
- *RV* systems use rolling contact rather than sliding contact for receiving low friction.
- With very low backlash and precise structure, the *RV* system has high positioning accuracy.
- Precisely fabricated *RV* systems generate only minimal vibration, achieving stable operation and performance.

During a transmission process, there are three types of energy loss for a *RV* gearing system: gear meshing friction loss, bearing loss, and hydraulic loss. The efficiency of the *RV* gearing system η_{RV} is given as

$$\eta_{RV} = \eta_{gc}\eta_b\eta_h \tag{9.28}$$

where
  η_{gc} is the efficiency of gear meshing, including gear contacting in the first stage and cycloid gear contacting in the second stage
  η_b is the efficiency of all bearings in the system
  η_h is the efficiency related to the hydraulic loss in the lubricant

9.2.5   MAGNETIC GEARING SYSTEM

All gearing systems mentioned above transmit power via gear meshing. As a result, these gearing systems suffer from some inherent problems, such as contact friction loss, generated noise and vibration, temperature rise, and lower operating reliability. In contrast, the MG system, using a contactless mechanism, has shed new light on these problems. This type of gearing system can offer significant advantages over conventional mechanical gearing systems: silent operation, lack of wear, minimized vibration, reduced maintenance, no need for lubrication and cooling, improved reliability, inherent overload protection, and physical isolation between the input and output shafts [9.42].

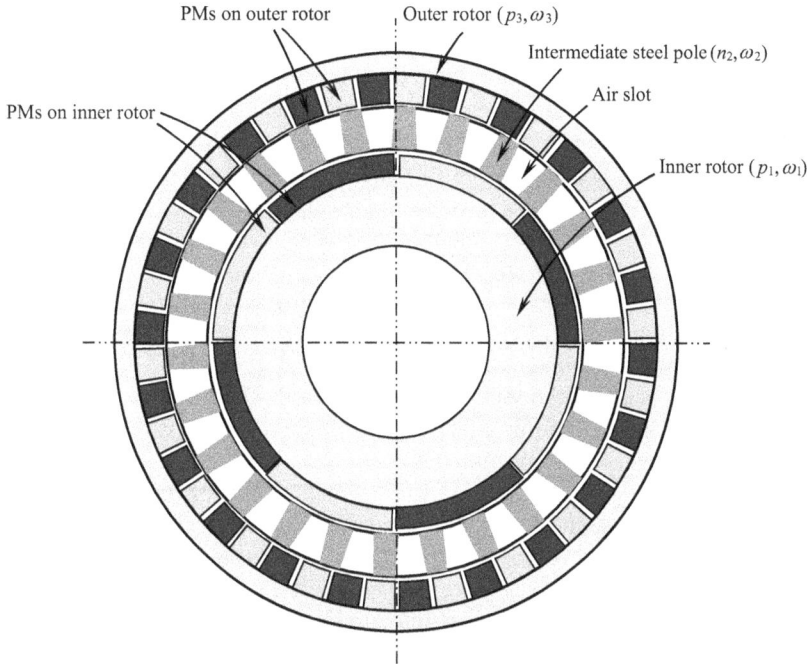

FIGURE 9.20 Structure of the coaxial magnetic gearing system.

The first concept of MG can be traced back to 1900s, which was initially presented in an U.S. patent by G. Armstrong [9.43]. It introduced a power-transmitting device by means of magnetic force, though the magnetism is produced by coils rather than by PMs. Although the MG technology continued to develop in the following decades, it really gained the interest of engineers and researchers when it began to use rare earth PMs (*e.g.*, Nd-Fe-B) at the beginning of this century [9.44].

The MG topologies can be generally divided into two groups. The first one is called conventional non-modulated MGs, in which PMs on different arrangements interact magnetically. The second group is called modulated MGs. This group of MG exhibits high efficiency and high torque density. The key merit of this type of MG is that all the PMs contribute to torque transmission, which helps achieve high torque density [9.45]. Among various MG topologies in the second group, the coaxial modulated configuration is preferred due to its better utilization of PM material and the capability of integration into PM machines to form the magnetic-geared PM machines.

An example of a coaxial magnetic gearing system is illustrated in Figure 9.20. It is composed of three main parts: an inner rotor with p_1 pole-pairs, an outer rotor with p_3 pole-pairs, and an intermediate part that contains n_2 steel pole pieces. In practical applications, each part can be held stationary while the others rotate. The inner and outer rotors carry PMs that are either mounted on the surface or embedded inside the rotors. The steel pole pieces are separated with the air slots and equally distributed in the circumferential direction. The magnetomotive forces of the inner and outer rotors are modulated by the permeance of intermediate pole pieces.

When the number of steel pole pieces is chosen to be [9.46, 9.47]

$$n_2 = p_1 + p_3 \tag{9.29}$$

The following relationship is held

$$p_1\omega_1 = n_2\omega_2 - p_3\omega_3 \tag{9.30}$$

where ω_1, ω_2, and ω_3 are the rotational speed of the inner rotor, intermediate part, and outer rotor, respectively. Therefore, the gear ratios for different operating configurations can be determined, as shown in Table 9.3. In practical applications, two cases are often used: (a) The inner rotor and outer rotor are set as the input and the output, respectively, while the intermediate part is kept stationary ($\omega_2=0$); and (b) the inner rotor and the intermediate part are set as the input and the output, respectively, while the outer rotor is kept stationary ($\omega_3=0$).

Typically, the gear ratio of a single-stage coaxial gearing system is rather low, usually less than 10:1. To achieve higher gear ratios, two or more MGs must be connected in series.

The MGs are typically used in low-speed applications. In a high-speed application, which generates a high centrifugal force on the rotor, the magnetic gearing system is no longer suitable because of the low mechanical strength of PM mounted on the rotor surfaces. Although there are a number of surface mount PM motors available under high-speed condition, they usually require complex and high-cost structures, such as the retaining sleeves [9.48]. To overcome this problem, a novel reluctance magnetic gearing system for attaching a high-speed motor is developed, based on the principle of the Vernier reluctance machine (Figure 9.21). The Vernier reluctance machine can operate by using the magnetic flux modulation without magnets on the rotor. Compared to the conventional MG system, the reluctance magnetic gearing system has the PMs on the low-speed rotor (outer rotor), and the high-speed rotor (inner rotor) has the salient structure constructed only by iron core. It was confirmed that the reluctance magnetic gearing system could operate at the maximum motor speed of 30,000 rpm and achieve the maximum efficiency of 92% in the high-speed region [9.49].

For the case that the inner rotor is the input, the outer rotor is the output and the intermediate part is fixed, the gear ratio is given as

$$\gamma_g = \frac{\omega_1}{\omega_3} = \pm\frac{p_3}{2p_1} \tag{9.31}$$

where p_1 and p_3 are the pole pair of the inner and outer rotor, respectively. The sign of the gear ratio depends on the magnetic coupling condition $2p_1 = n_2 + p_3$ or $2p_1 = n_2 - p_3$ (where $n_2 > p_3$).

TABLE 9.3

Gear Ratio for Different Operating Configurations of Magnetic Gearing System (where $n_2 = p_1 - p_3$)

Operating Configuration			Speed Variation	Output and Input Directions of Rotation	Gear Ratio γ_g
Input	Output	Fixed ($\omega = 0$)			
Inner rotor	Outer rotor	Intermediate part	Decreasing	Opposite	$\frac{\omega_1}{\omega_3} = -\frac{p_3}{p_1}$
Outer rotor	Inner rotor	Intermediate part	Increasing	Opposite	$\frac{\omega_3}{\omega_1} = -\frac{p_1}{p_3}$
Inner rotor	Intermediate part	Outer rotor	Decreasing	Same	$\frac{\omega_1}{\omega_2} = \frac{n_2}{p_1}$
Intermediate part	Inner rotor	Outer rotor	Increasing	Same	$\frac{\omega_2}{\omega_1} = \frac{p_1}{n_2}$
Outer rotor	Intermediate part	Inner rotor	Decreasing	Same	$\frac{\omega_3}{\omega_2} = \frac{n_2}{p_3}$
Intermediate part	Outer rotor	Inner rotor	Increasing	Same	$\frac{\omega_2}{\omega_3} = \frac{p_3}{n_2}$

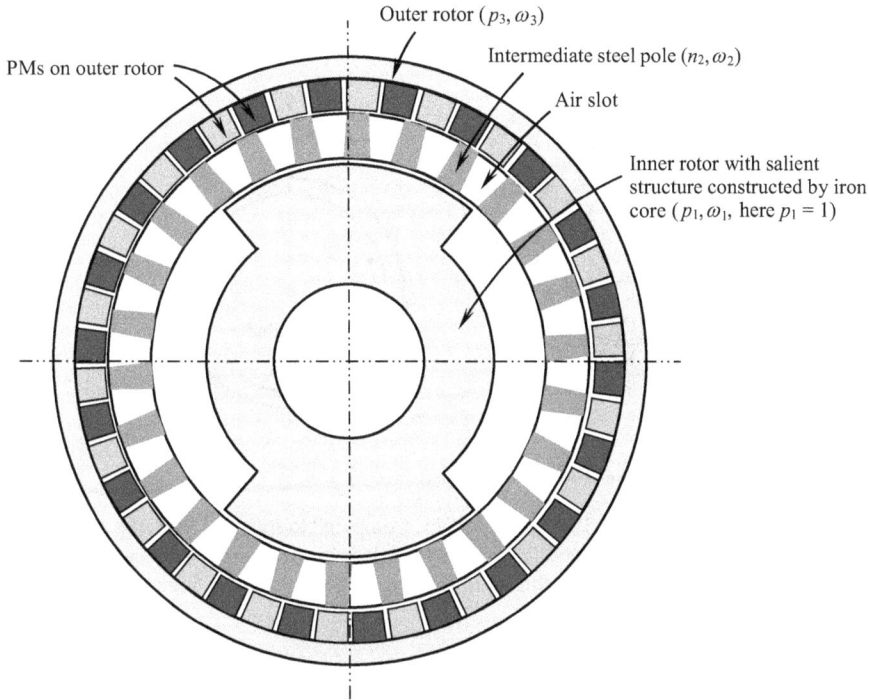

FIGURE 9.21 Structure of the reluctance magnetic gearing system, where the inner rotor has the salient structrure constructed by only iron core. It was confirmed by experiments that this type of gearing system can operate at a maximum rotational speed of 30,000 rpm.

PMs can corrode or demagnetize due to temperature, moisture, and shock load. Sintered rare earth magnets Nd-Fe-B must be protected via electroplating or an epoxy coating. Sm-Co magnets can also be used to avoid corrosion and demagnetization problems due to their higher corrosion resistance.

9.2.6 CONTINUOUSLY VARIABLE STRAIN WAVE TRANSMISSION

SWGs have the advantage of compact structure, lightweight, small size, and high gear ratio (up to 460:1 or higher) in a single stage. They also have features such as low- or zero-backlash, low friction loss, and high torque capacity. However, SWG has some limitations, particularly high cost and fixed gear ratio. While its high cost is attributed to the tight tolerances required in the geometries of the meshing teeth and the diameters of the rotating components, as well as rigorous manufacturing and assembly processes, the fixed gear ratio is a result of its reliance on intermeshing discrete teeth.

To resolve these problems, Naclerio *et al.* [9.50] have proposed a novel design of gearing system, namely, continuously variable strain wave transmission (CVSWT). This low-cost transmission maintains the high gear ratio of SWG and enables a continuously variable gear ratio. For this purpose, two major design changes are made compared to a standard SWG: (*a*) To address the challenge of continuous variation of the gear ratio, instead of discrete toothed contacts in the standard SWG, the CVSWT uses smooth frictional contact to allow continuous variation of gear ratio. A gecko-inspired dry adhesive is adopted to enhance friction at the torque transfer surface. (*b*) To create high friction, the CVSWT uses an inverted morphology to allow high torque transfer with low preload through capstan friction [9.51]. Instead of the flexible element inside the fixed component (*i.e.*, the circular spline in SWG), the fixed cylinder is wrapped with a flexible belt in a slightly larger diameter. The comparison between SWG and CVSWT is shown in Figure 9.22.

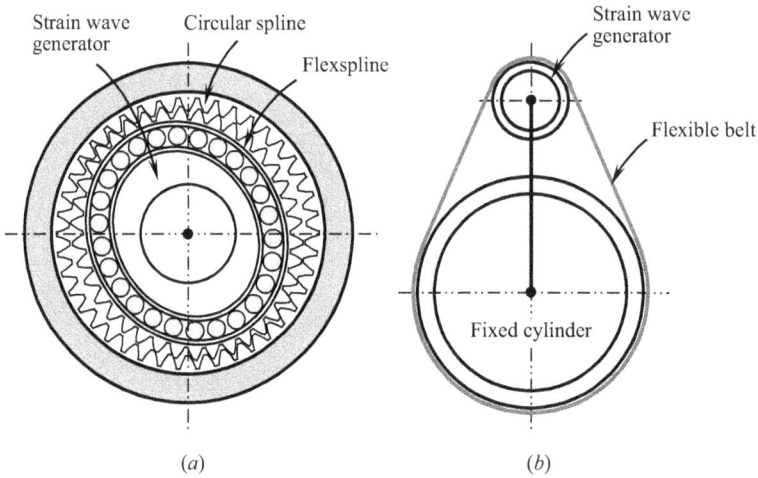

FIGURE 9.22 Comparison between SWG and CVSWT. Compared to standard SWG, two major changes are made in CVSWT: (*a*) using smooth contacting surfaces to replace discrete toothed contacts in SWG, allowing continuous variation of gear ratio, and (*b*) using an inverted morphology to allow high torque transfer with low preload through capstan friction.

As the key design concept in this study, the variation of continuous gear ratio is achieved by varying the diameter of the fixed cylinder. The researchers cleverly apply the mechanical actuator to control the cylinder diameter. Figure 9.23 illustrates the three major subsystems, *i.e.*, the motor drum, the orbiting roller and transmission arm, and the output cylinder, as well as the assembly of the CVSWT. The motor drum consists of four components: a tapered core that contains the drive motor, an outer shell with a matching taper that is cut radially into six segments and compressed by hoop springs, a base constraining the core and shell to each other, and an actuator to change the diameter. As the core moves toward the base, the outer shell segments are pushed radially outward by the taper, increasing the motor drum diameter. Conversely, as the tapered core moves upward, the spring around the shell segments returns them to their original position (Figure 9.23a). As can be seen in Figure 9.23b, the orbiting roller is used to move the belt slack around the drum. Its ends are attached to the transmission by two arms: one to the output shaft of the drive motor, and another to the base of the drum. The arms are hinged at their centers to allow the roller to maintain tension on the belt as the drum diameter changes. In Figure 9.23c, the rigid output cylinder is attached to the output belt by 24 connector belts (where 12 belts are wrapped clockwise and other 12 counter-clockwise to prevent backlash) for providing a well-constrained output. Each belt is attached at one

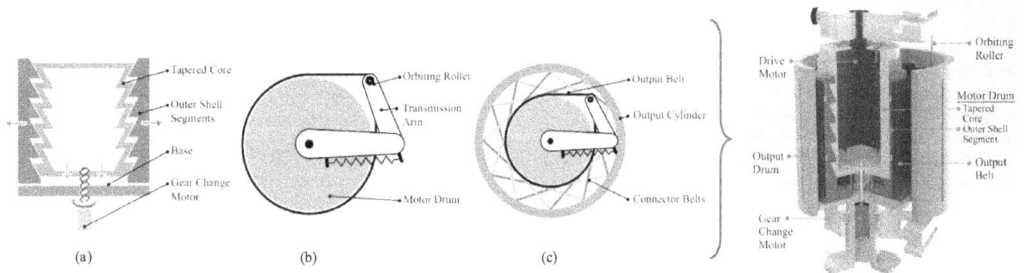

FIGURE 9.23 Major CVSWT components: (*a*) the variable circumference motor drum, (*b*) the orbiting roller and transmission arm, and (*c*) the output cylinder. In an assembly of CVSWT (right), the motor drives the orbiting roller that results in the output belt undergoing precession around the rotating axis [9.50].

end to the output drum and at another end tangentially to the output belt. This operating mechanism is completely different from SWG.

In order to derive a mathematical model for calculating the maximum torque T_{max} that CVSWT can deliver, it is necessary to consider important factors, such as the geometries of the orbiting roller and motor drum and the working conditions (Figure 9.24). Ignoring the derivation process, the equation is given as follows:

$$T_{max} = \left[\left(\frac{F_r}{2\sin\alpha} + \frac{wr\sigma_o}{\mu}\right)e^{\mu\phi} - \frac{wr\sigma_o}{\mu}\right]r \qquad (9.32)$$

where

F_r is the tension applied by the orbiting roller, N
α is a half of the core angle, degree (°)
ϕ is the wrap angle, degree (°)
σ_o is the adhesive-controlled friction stress, N/m^2
μ is the coefficient of friction between the flexible belt and the motor drum
r is the drum radius, m
w is the belt width, m

Experimental results have shown that the maximum torque between the motor drum and flexible belt (before the belt slip) is a function of wrap angle and the drum material (e.g., plastic, rubber, and dry adhesive).

It is important noting that because of the flexible elements, the CVSWT naturally has some imprecision in gear ratio and angular position. This makes it unsuitable for applications that require high positional accuracy. A feasible method for this issue is to use a dual encoder system to locate the encoders at both the transmission input and output, forming a closed loop for reliably controlling position and gear ratio.

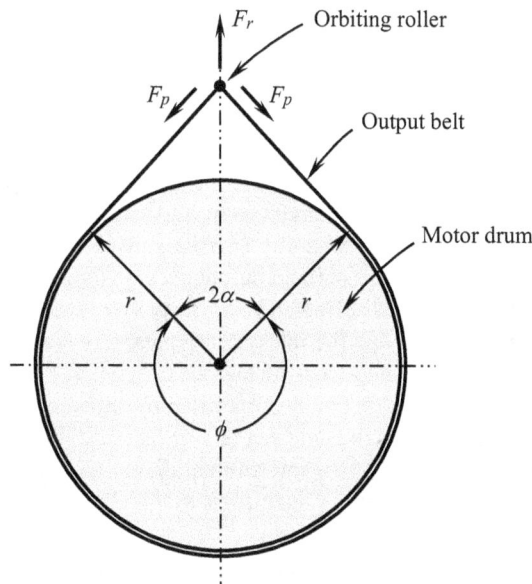

FIGURE 9.24 Geometries of motor drum and orbiting roller and the tensions applied on the belt and roller.

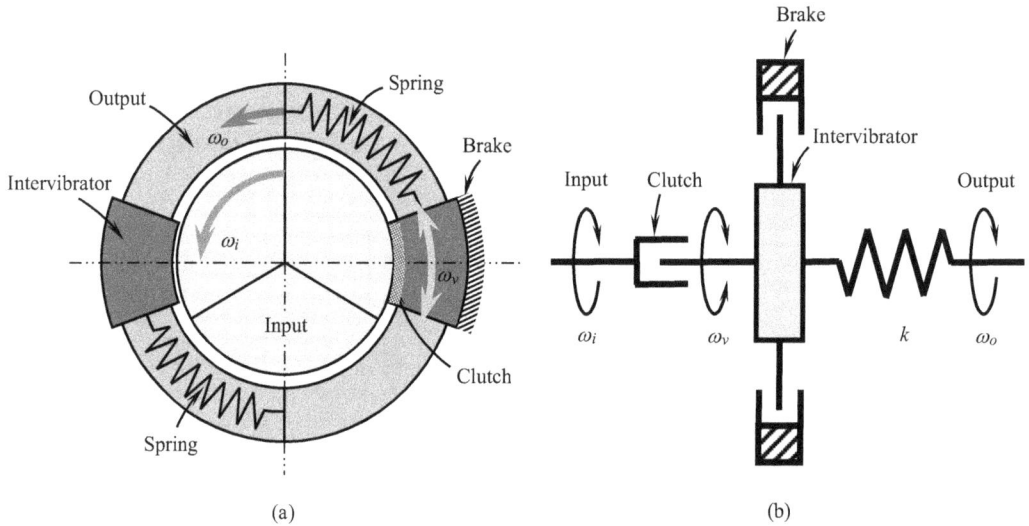

(a)   (b)

FIGURE 9.25 In a pulse drive, intermittent power transmission between the input and output axes that rotate at different speeds can be achieved by the intermittent interlocking engagement using the intervibrator, clutch, brake, and springs: (*a*) structure of pulse drive system and (*b*) operation principle of the transmission [9.52].

9.2.7 PULSE DRIVE

A very special power transmission mechanism named *pulse drive* was proposed by Toyota R&D engineers for realizing a highly compact, high-efficiency, and infinitely variable transmission in their electric vehicles. The driving power is intermittently transmitted through periodic vibration control [9.52]. Conventionally, rotational power can be transmitted between two systems with different rotational speeds through two mechanisms: the radius ratio mechanism (*e.g.*, gearing systems) and slip mechanism (*e.g.*, friction clutches). In a gearing system, the circumferential speeds of two rotating gears are identical at their meshing lines. The rotational speeds rely on the ratio of the radius at their pitch circles. In the slip mechanism, a friction clutch allows slip between two axes and transmits torque through friction. Unlike these two mechanisms, in the pulse drive transmission torque is transmitted between two axes rotating at different speeds through intermittent interlocking engagement. Theoretically, this system has high torque density because the entire circumference is engaged in torque transmission. The intermittent engagement-based mechanism has the following characteristics: the average torque can be continuously controlled using the duty ratio (*i.e.*, the relative clutch engagement period) and the high drive frequency smoothens the rotational speed of the output axis, which typically has a relatively large inertia and thus low-frequency response.

Figure 9.25 illustrates the method of engaging two axes using an intervibrator. A pulse drive transmission consists of two concentric input and output axes, two torsion springs, an intervibrator that is connected to the output axis via a spring and can freely vibrate along the output axis, a clutch that is located between the input axis and the intervibrator, and a brake that is located between the intervibrator and the non-rotational component such as a casing.

The magnitude of torque transmission can be controlled by adjusting the duration of clutch and brake engagements. Prolonging the clutch engagement will increase transmission torque due to the increase in the duty ratio, indicating that the intermittent drive can realize torque and speed controls. In fact, this is similar to duty control in electronic voltage converters, which utilize the ratio of engagement time rather than that of turns of coils.

Figure 9.26 demonstrates the speeds of the input, output, and intervibrator vary with time. The upper left of the figure shows absolute rotational speeds of the input, output, and intervibrator. While ω_i and ω_o are constants, ω_v varies with time sinusoidally. At the point A, $\omega_v = \omega_i$, and at the point of B, $\omega_v = 0$. The upper right of the figure shows the relative rotational displacement of the intervibrator to the output axis (*i.e.*, the spring displacement), to which the spring torque is proportional. The four operation states (*a*)–(*d*), corresponding to the velocity and displacement curves in the figure, are displayed in the lower part of the figure.

Theoretically, the pulse drive transmission can work at any reduction ratio, *i.e.*, from one to infinitive. Experiments on a prototype have successfully validated the principle at ratios 2 and 4. The transmission efficiency of 78% was attained in the experiment. This type of transmission seems to be promising since it exhibits high torque density, compact structure, and a wide range of speed reduction ratio. However, since the prototype transmitted very low power in the experiments, it remains a challenge for future work to expand the transmission power.

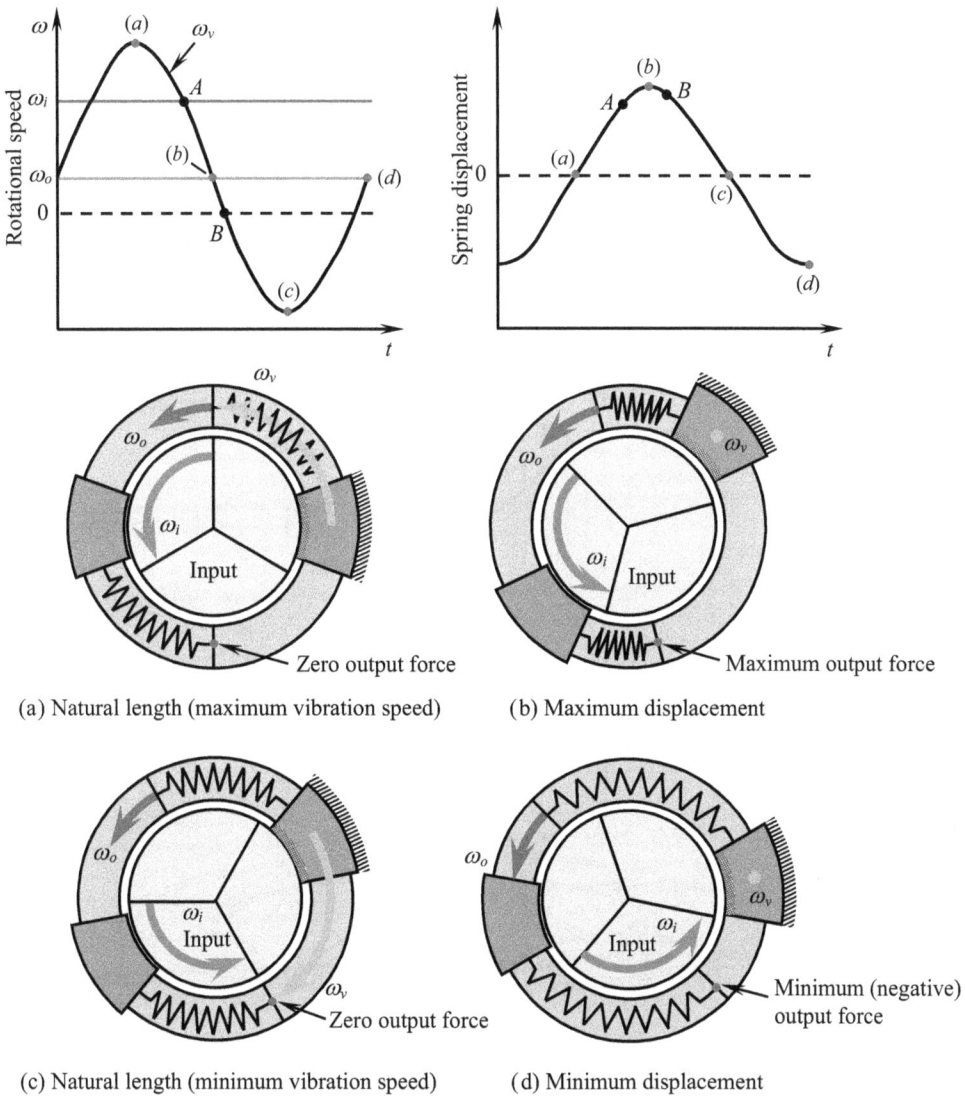

(a) Natural length (maximum vibration speed)   (b) Maximum displacement

(c) Natural length (minimum vibration speed)   (d) Minimum displacement

FIGURE 9.26   The upper left shows absolute rotational speeds of the input, output, and intervibrator; where ω_i and ω_o are constant but ω_v displays sinusoidally. The upper right shows the relative rotational displacement of the intervibrator to the output axis (*i.e.*, the spring displacement). The four states of intervibrator (*a*)–(*d*) corresponding to the upper figure are presented in the lower part of the figure. At the point A, $\omega_v = \omega_i$, and at the point of B, $\omega_v = 0$.

9.2.8 ABACUS DRIVE

A new type of rotary transmission system, called abacus drive, has been innovated and developed by *SRI International* [9.53, 9.54]. It was named from the shape of the core components that look like the beads in a Chinese abacus. While a cycloid transmission system has a significant amount of sliding contact under load, the abacus drive varies its geometry in a direction perpendicular to the traditional one, which allows for pure rolling motion. Thus, there are no parts in this type of transmission system that rub or slide against each other (Figure 9.27). Compared to some gearing systems, this type of transmission system generates less heat during operation.

During operation, the beads roll in and out of the groove with varying diameter, causing the beads to effectively change their diameters upon their positions in the groove and resulting in a cycloidal motion.

The gear ratio of this system is determined solely on the number of beads, z_b,

$$\gamma_g = \frac{z_b - 1}{2} \tag{9.33}$$

Due to the limitation of the system structure, the number of beads z_b in practice cannot be too large. Therefore, this type of gearing system is generally used in low gear ratio applications.

Other challenges in designing and manufacturing the abacus drive include achieving low/zero backlash and holding tight tolerances within the specifications.

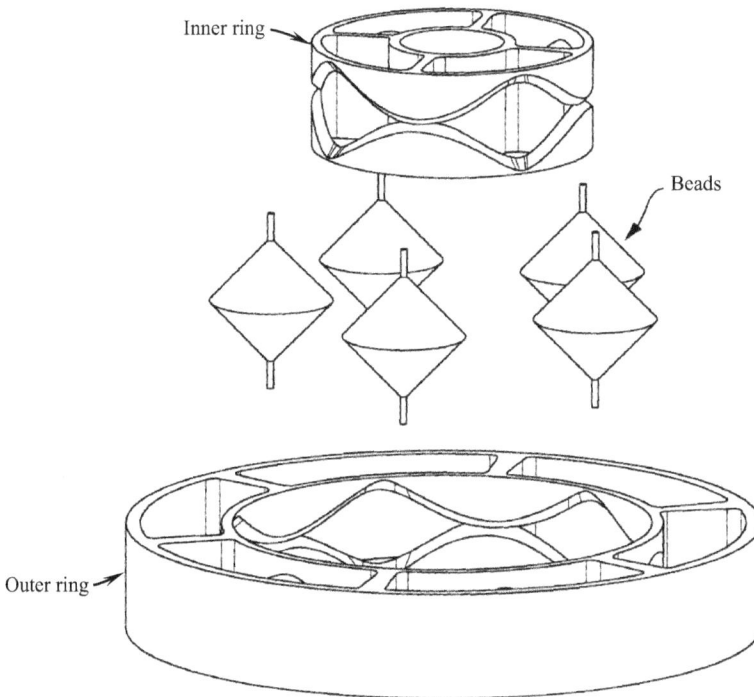

FIGURE 9.27 An abacus drive gearing system contains a number of beads that roll in and out of a groove with a variable diameter, enabling the beads to change effectively their diameters depending on where in the groove they are, resulting in a cycloidal motion. (Courtesy of WIPO, Geneva, Switzerland.)

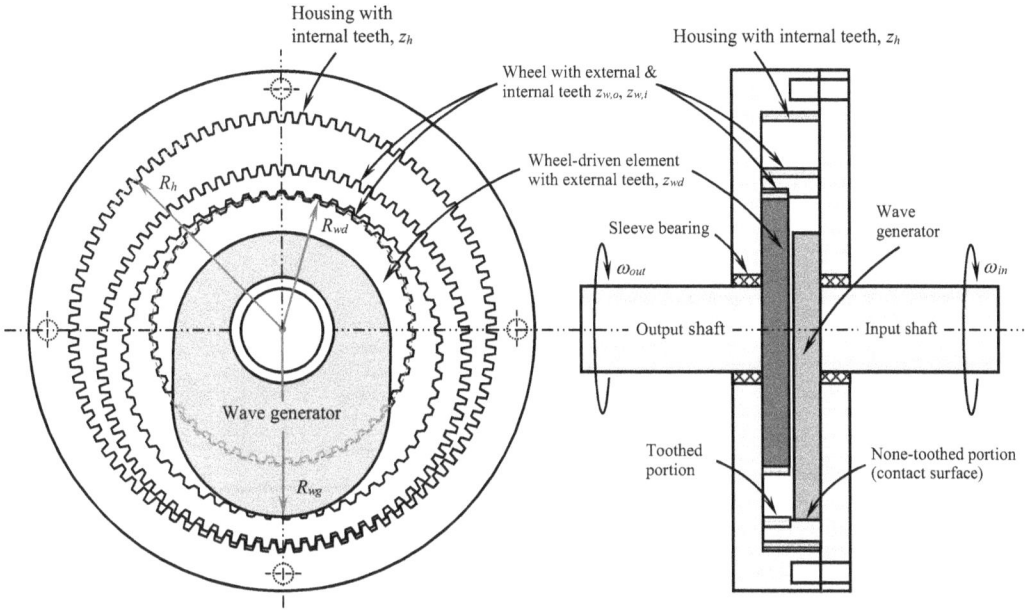

FIGURE 9.28 Working principle and structure of CWD. The input shaft drives the wave generator rotating at the speed ω_{in}. Due to the friction between the contact surfaces of the wave generator and the wheel, while the wheel performs an eccentric movement at the speed ω_{in}, it drives the wheel-driven element rotating with respect to the output shaft axis.

9.2.9 CIRCULAR WAVE DRIVE

The circular wave drive (CWD) concept has been proposed and developed in recent years, leading to awarded US patents [9.55, 9.56]. The fully developed design of the CWD allows for simple modular design, a large range of reduction ratios, very low and easily controllable backlash, and simple construction leading to streamlined design and increased durability [9.57].

As depicted in Figure 9.28, a CWD primarily consists of four key components: a wheel, a wave generator, a wheel-driven element, and a housing. The housing is disc-shaped and substantially hollow, with the internal teeth disposed along the peripheral wall. The wheel is ring-shaped with a hollow central portion and resides within the chamber of the housing. It has external teeth that mesh with the teeth of the wave generator. The internal wall of the wheel is circumferentially divided into a non-toothed portion, which is adapted to contact with the smooth surface of the wave generator, and a toothed portion, which is adapted to mesh with the external teeth of the wave-driven element. The input and output shafts are attached to the wave generator and the wheel-driven element, respectively.

The wave generator has an oval or truncated circle (*e.g.*, racetrack) cross-sectional shape. Since the centerline of the wave generator is offset from the main central axis (which goes through the input and output shafts) of the system, it drives the wheel to perform an eccentric motion along the internal teeth of the housing when it rotates and, in turn, the wheel drives the wheel-driven element rotating with respect to the main central axis.

An alternative design of CWD is shown in Figure 9.29 by adding a thin ball bearing between the wave generator and the wheel for reducing friction between them.

The gear ratio of the CWD is calculated via the following equation:

$$\gamma_g = \frac{R_h - R_{wd}}{R_{wd}} = \frac{z_h - z_{wd}}{z_{wd}} \tag{9.34}$$

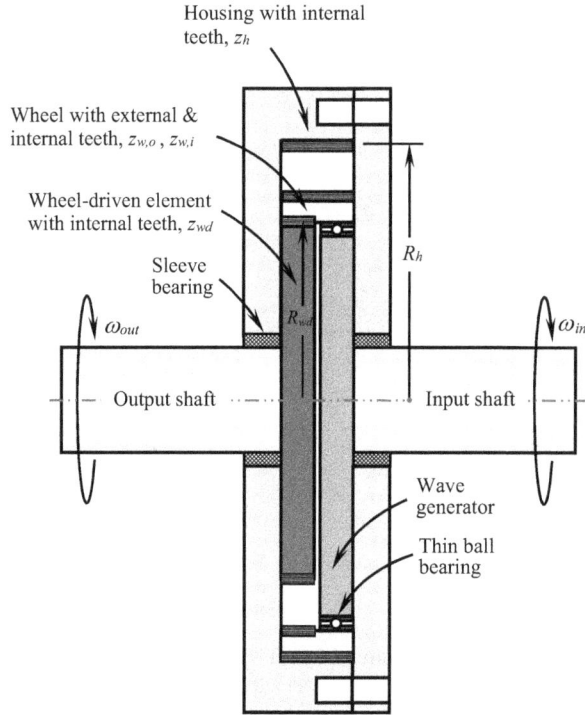

FIGURE 9.29 Alternative design of CWD. A thin ball bearing is added between the wave generator and wheel for reducing friction and eliminating slippage between them.

where

R_h is the radius of the housing, measured from the shaft axis to the pitch of the internal gear teeth.

R_{wd} is the radius of the wheel-driven element, measured from the shaft axis to the pitch of the external gear teeth.

z_h is the number of internal teeth of the housing

z_{wd} is the number of external teeth of the wheel-driven element

9.2.10 ARCHIMEDES DRIVE

Archimedes drive, also known *compound planetary friction drive* in the US patent 10,113,618 [9.58], is designed based on the same configuration as a planetary transmission. However, unlike a conventional planetary gearing system, it uses smooth frictional hollow rollers instead of gear teeth to transmit torque. The Archimedes drive combines the architecture of a compound drive with a friction drive to achieve gear ratios of up to 10,000:1 [9.59].

This type of system has broad application prospects. To meet various customer requirements, multiple drives are often connected in series; that is, the rotational output shaft of the first stage is linked to the input shaft of the next stage. For such a compound gearing system with multiple stages, the gear ratio for the overall system is found by multiplying all the individual gear ratios together. Because the drive harnesses traction rather than relying on cogs with fragile teeth, this compound drive exhibits very low or zero backlash, making it ideal for precision applications such as industrial and surgical robots. In addition, the lack of gear teeth also enables the system almost maintenance-free and there is no need for lubrication.

Figure 9.30 shows the assembled and exploded views of Archimedes drive, as well as the cross-sectional side views of the first and second stages of the drive. This compound planetary friction

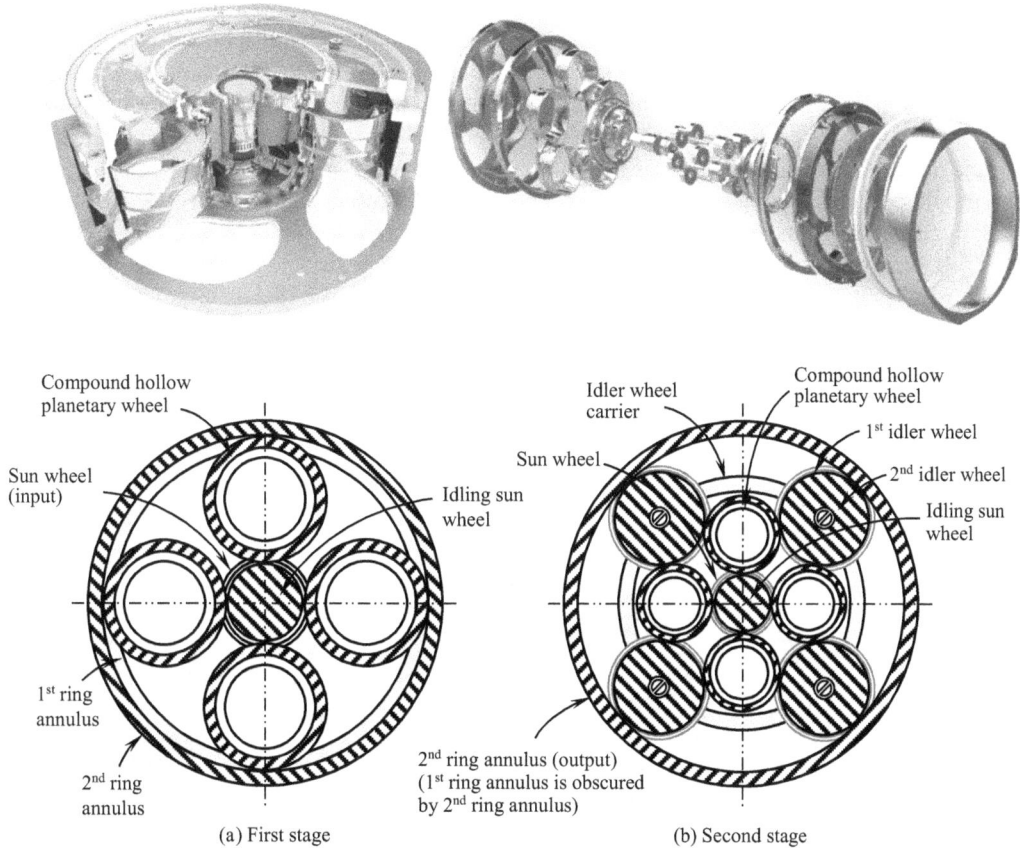

FIGURE 9.30 Top: The assembled and exploded views of Archimedes drive. (Courtesy of IMSystems, Delft, Netherlands.) Bottom: the cross-sectional side views of the first and second stages of Archimedes drive.

drive comprises two stages. In the first stage, an input sun wheel is engaged to four planetary wheels that are in frictional engagement with a ring annulus. The structure of the second stage is similar to that in the first stage. However, the planetary hollow wheels do not directly engage the ring annulus; instead, they engage second idler wheels that are attached to a carrier structure. Then, the second idler wheels engage the second ring annulus (as the output), forming a complete drive chain. The idling wheels are axially aligned via the carrier structure that allows the idling wheel to rotate freely. These two stages are connected in series.

When two wheels are in contact, power is transmitted between them as long as the tangential force exerted by the driving wheel is less than the maximum frictional resistance between them. If the tangential force exceeds the maximum frictional force, the wheel will slip, preventing the drive from permanent failure. Moreover, the device is backdrivable and easier to stop, contributing to safety-related objectives in robot designs.

9.2.11 SPIRAL CAM GEARING SYSTEM

Most recently, an innovative design of a new type of gearing system, known as spiral cam gearing system, has been developed on principles of kinesiology and kinematics by Motus Labs [9.60, 9.61]. The motivation of this innovation is to provide a more precise, compact, and lightweight alternative for automation applications such as industrial robots.

As shown in Figure 9.31, this design uses a plurality of cam-actuated (even numbered) gear block assemblies that transfer power from a motor shaft to a shaft of a driven machine. Each gear block assembly includes a gear block having a surface that periodically engages the output gear element.

The most critical component of the design is the cam, which is made by cutting two smooth spiral-shaped grooves on the outer surface of a cylinder, 180° apart from each other. In Figure 9.31a, r_1 is the radius of the cylinder and r_2 is the minimum radius of the spiral groove, where $r_1 > r_2$. This cam profile provides a unique, undulating motion of cam followers as the cam rotates. Each gear block includes three linkage assemblies, which are pivotally coupled to the gear block, and a cam roller follower, which maintains contact with the cam (Figures 9.31b and c). The two-dimensional circuit includes urging the gear block to engage the output gear a specified quantum distance prior to disengaging from the output gear and returning back the specified quantum distance to again reengage the output gear once again and repeat the process. The travel path of each gear block is

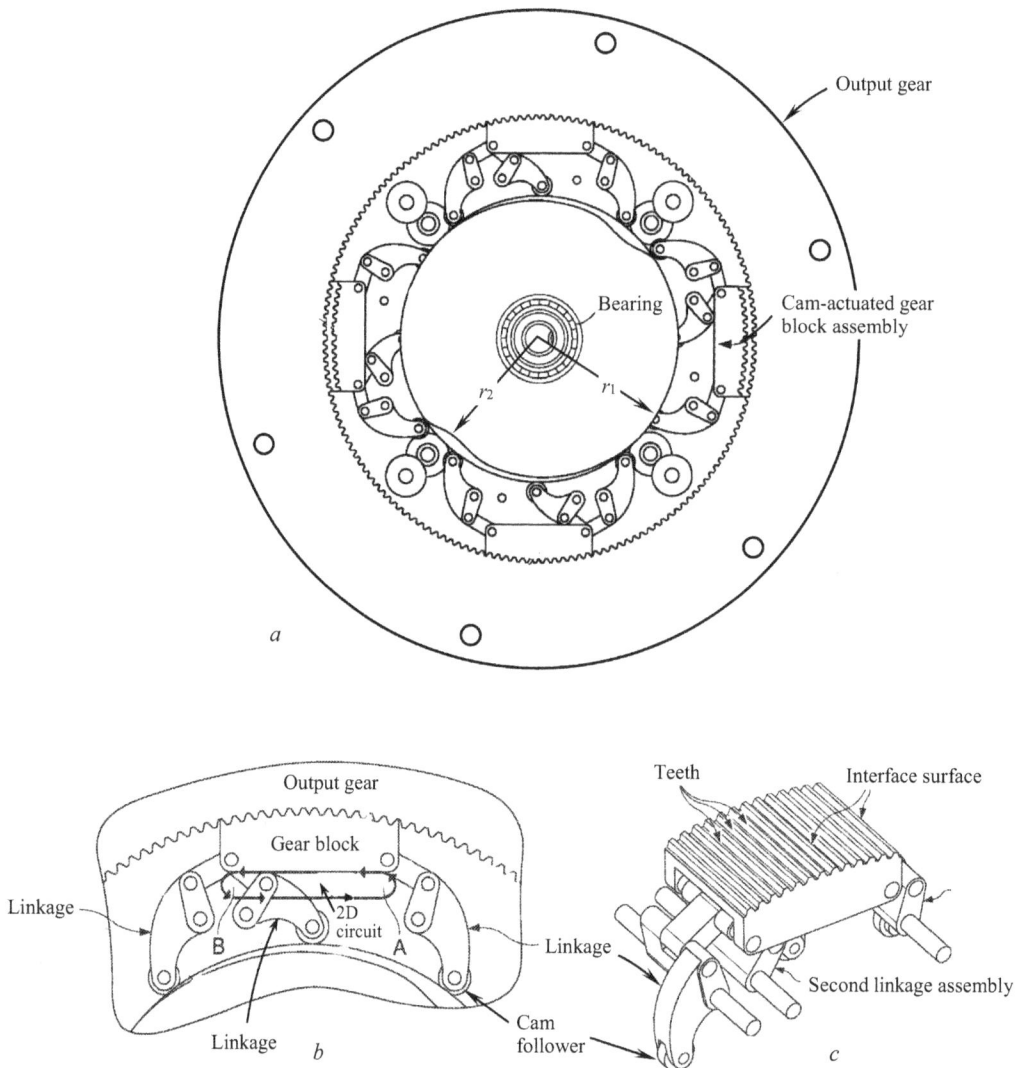

FIGURE 9.31 A spiral cam gearing system: (a) The structure of the spiral cam gearing system (as the outer stationary plate is removed). (b) A side elevation view of a gear block assembly. (c) A perspective view of a gear block assembly. (Courtesy of the U.S. Patent and Trademark Office, Alexandria, VA.)

controlled by adjusting the length and configuration of the various linkage assemblies. In such a way, the speed reduction may be regulated and controlled by adjusting the circuit of movement of each gear block. However, a formula for calculating the gear ratio of this gearing system has never been released.

In fact, cam-follower mechanisms are particularly useful when a simple motion of one part in a machine is converted to a more complicated prescribed motion of another part. For instance, in the automotive industry, this mechanism is used to drive fuel pumps.

9.2.12 CLUTCH-TYPE STEPLESS SPEED CHANGER

As the name implies, this clutch-type stepless speed changer can instantly go from zero to the maximum speed. As shown in Figure 9.32, the structure consists of a link mechanism and one-way clutch, which are arranged in four rows in the axial direction. An eccentric disc is attached to the input shaft, and the eccentric discs in the rows are arranged on the circumference at an angle of 90° to each other surrounding the input shaft [9.62]. Thus, the following mechanism is created:

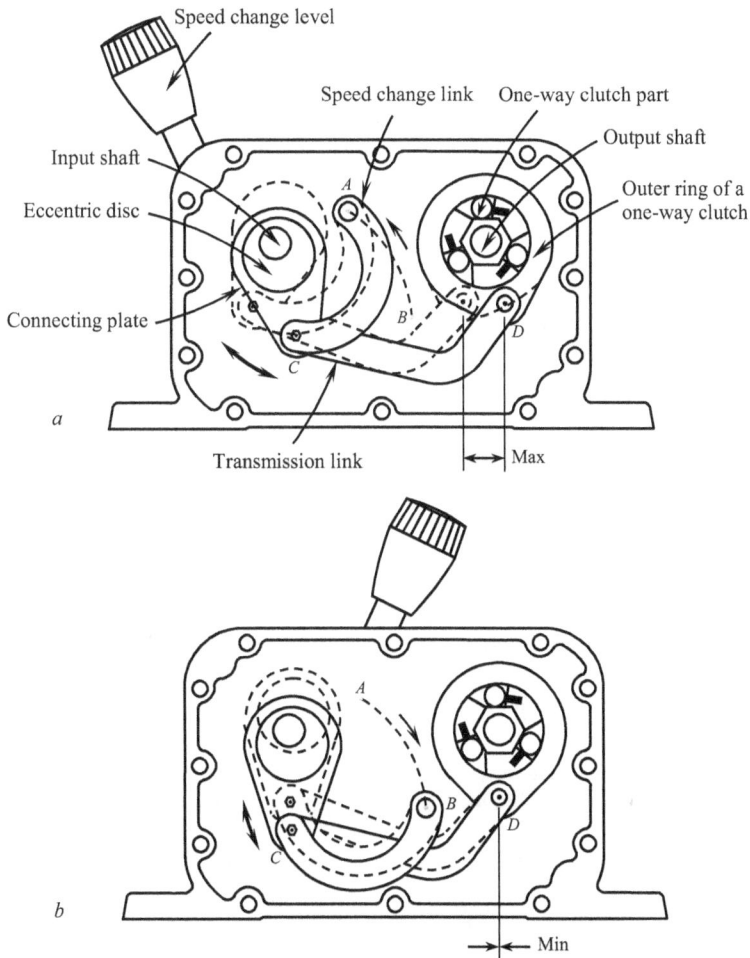

FIGURE 9.32 Clutch-type stepless speed changer combines a link mechanism and a one-way clutch mechanism to transmit motion and power from the input shaft to the output shaft. The output rotation speed can be changed steplessly by freely changing the amplitude of the reciprocating motion with a speed change level: (a) rotating at a maximum speed and (b) rotating at a minimum speed.

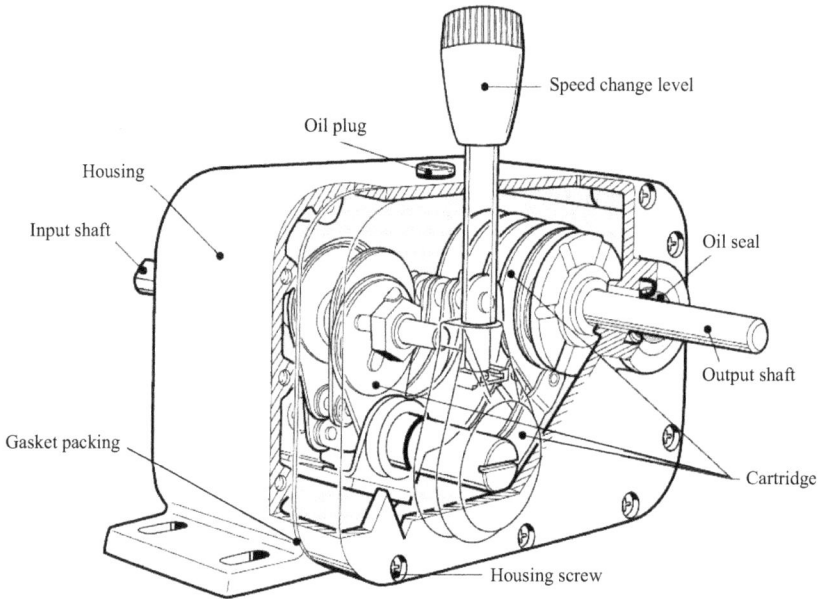

FIGURE 9.33   Section view of the clutch-type stepless speed changer. It is filled with oil for the purpose of lubrication and cooling. (Courtesy of Miki Pulley, Kanagawa, Japan.)

the rotational motion of the input shaft is once converted to the reciprocating motion, which then is converted to the rotational motion by each of the one-way clutches of the output shaft part. The output rotation speed can be changed steplessly by freely changing the amplitude of this reciprocating motion with a speed change lever. The speed change ratio is up to 12:1. The section view of the 3D speed changer is shown in Figure 9.33.

9.3   CONVENTIONAL GEARING SYSTEMS

Conventional gear transmission systems often appear in the form of gearbox to directly connect motors and driven machines. Until today, they are still the most widespread among the mechanisms intended for the transmission of rotational or linear motion. However, because these conventional gearing systems fall short in terms of repeatability and positioning accuracy, they are suitable for applications that do not require high-precision motion control.

The major design goals of conventional gearing transmissions include high efficiency, high strength, less transmission error, high durability, quiet operation, and low temperature in the gear mesh. These goals may be achieved by the introduction of optimal gear geometries especially tooth surfaces, precisely controlled manufacturing processes, improved lubrication conditions, and well-installed gearing units. All of these measures are taken to minimize transmission errors and mesh friction losses.

According to the relationship of shaft axes between meshing gears, conventional gearing systems can be classified as parallel, intersecting, and nonparallel nonintersecting types.

9.3.1   Spur Gear

As the simplest and most cost-effective type of gear, spur gears transmit rotational motion between two parallel shafts (Figure 9.34). They consist of a cylinder or disk with teeth projecting radially. The edge of each tooth is straight and aligned parallel to the gear's axis of rotation. As the primary

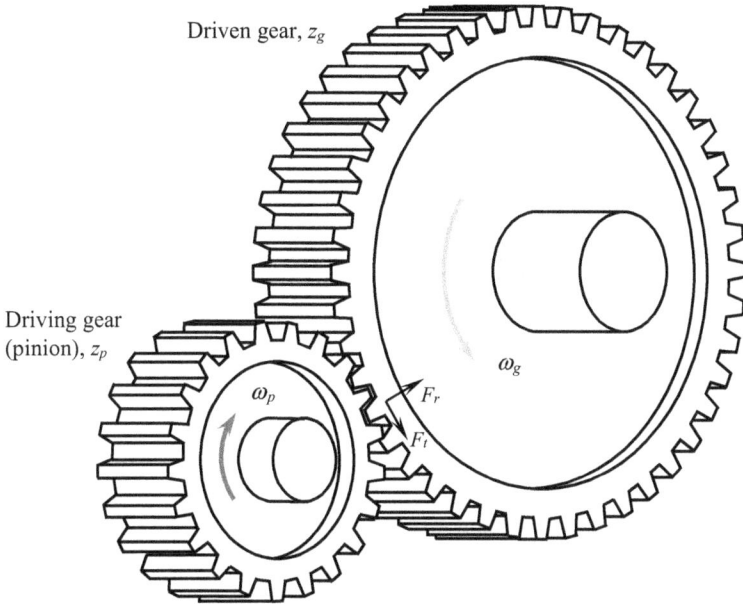

FIGURE 9.34 As the simplest and most common type of the gearing system, the spur gears have straight teeth that are oriented parallel to the shafts. The simplicity of the spur gear tooth design allows for both a high degree of precision and easy manufacturability.

form of mechanical power transmission, this type of gearing system is often used in multiple stages to provide desired gear ratios. Because of their high reliability, high precision, easy manufacturability and low cost, spur gearing systems have been widely used in a broad spectrum of industrial and commercial devices, from washing machines, fuel pumps, gearmotors, marine engines, automobile gearboxes, material handling equipment, to mechanical watches and DVD players.

The tooth profile of spur gear is based on the involute curve, which is generated during gear machining processes using gear cutters. During the gear meshing process, the contact among the teeth occurs mostly as rolling rather than sliding. Rolling is continuous throughout meshing, from root to tip on the driving tooth and from tip to root on the driven tooth. Sliding varies from a maximum velocity in one direction at the start of mesh through zero at the pitch line to a maximum velocity in the other direction at the end of mesh [9.63]. Therefore, spur gears can achieve very high efficiency, up to 99%.

The pressure angle of a standard gear is defined as the angle at the pitch point between the line normal to the tooth profile and the line tangent to the pitch circle. It plays an important role in the gear performance. A pressure angle of 14.5° was historically selected due to the consideration of reducing noise in the gear mesh and exhibiting lower rates of wear. Late, American Gear Manufacturers Association (AGMA) had recommended the pressure angle of 20°, because it is more suited for most applications. The benefits of selecting a 20° pressure angle include additional power transmission capacity, better lubrication in the gear mesh, and reduced numbers of teeth for the pinion without undercutting [9.64].

It has been broadly recognized that gears with higher pressure angles can yield better mechanical behavior regarding contact and bending stresses [9.65, 9.66]. Therefore, it is uncommon today to find a new design of cylindrical gear drives considering a pressure angle of 14.5°. Instead, new designs of cylindrical gears generally adopt pressure angles of either 20° or 25°. However, this widely accepted point of view has been challenged by some investigators [9.67]. The results gathered from finite element analysis (FEA) have shown that high pressure angle (30° and 35°) spur and helical gears do not show a better behavior regarding contact pressures, contact stresses, and bending stresses than spur and helical gears with a conventional pressure angle of 25°. It is important to note that in the published paper it claims, "The results gathered from the FEA are *not* in agreement with the improvement on the pitting and bending behavior of the gears that were anticipated by using analytical models as provided by international standards."

The forces acting on the meshed spur gears include tangential force F_t and radial force F_r,

$$F_t = \frac{2T}{d_p}$$ (9.35a)

$$F_r = F_t \tan\alpha$$ (9.35b)

where
α is the pressure angle of spur gear, for standard spur gears, $\alpha = 20°$
T is the transmitting torque (*i.e.*, input torque), Nm
d_p is the pitch diameter, which is the product of the number of teeth z and the module m, *i.e.*,
$d_p = zm$

The gear ratio of a given pair of spur gears is obtained as the number of teeth on the driven gear divided by the number of teeth on the driving gear, *i.e.*,

$$\gamma_g = \frac{\omega_p}{\omega_g} = \frac{z_g}{z_p}$$ (9.36)

The advantages claimed for spur gearing systems are as follows: (*a*) simplest gear design; (*b*) its tooth profile is parallel to the axis of rotation, transmitting power, and motion between parallel shafts; (*c*) straight cut tooth.

However, spur gears have some significant drawbacks. Due to the impact force generated at gear meshing, spur gears produce a high level of noise. The sound levels increase in proportion to an increase in speed. For this reason, spur gears are often used in low- or moderate-speed applications, as well as the situations where noise is not the main consideration. Furthermore, a normal gear ratio of a single-stage gearing system is relatively low, in most cases less than 10:1. For specially designed spur gear sets, the gear ratio may reach up to 20:1.

9.3.2 HELICAL GEAR

Helical gears are different from spur gears due to that their teeth are curved in the shape of helix. When a pair of teeth comes in contact with each other, the engagement begins at the point of leading edge of the curved teeth and maintains contact as the gears rotate into full engagement. During the process, the contact progresses across the tooth along the diagonal line [9.68]. This gradual engagement makes helical gears operate more smoothly and quietly than spur gears, where straight teeth suddenly meet at a contact line across their entire width. In addition, since helical gears have a high tooth contact ratio, the tooth load and surface stress are greatly reduced and consequently the load-carrying capability of the gearing is significantly enhanced (Figure 9.35).

A pair of helical gears can be arranged in parallel or crossed configurations. For parallel shafts, axes, the cross-axis angle between two shafts is zero. If the cross-axis angle is not equal to zero, in the standard case, both the meshing gears have the same kind of helix and helix angle (*i.e.*, $\theta_{h1} = \theta_{h2} = \theta_h$), which typically ranges from 15° to 30°. The cross-axis angle is the sum of the two meshing gears' helix angles, that is, $\Sigma = \theta_{h1} + \theta_{h2} = 2\theta_h$. While gears with parallel axes have opposite hands (*i.e.*, right-hand teeth vs. left-hand teeth), gears with nonparallel and nonintersecting axes have the same hand.

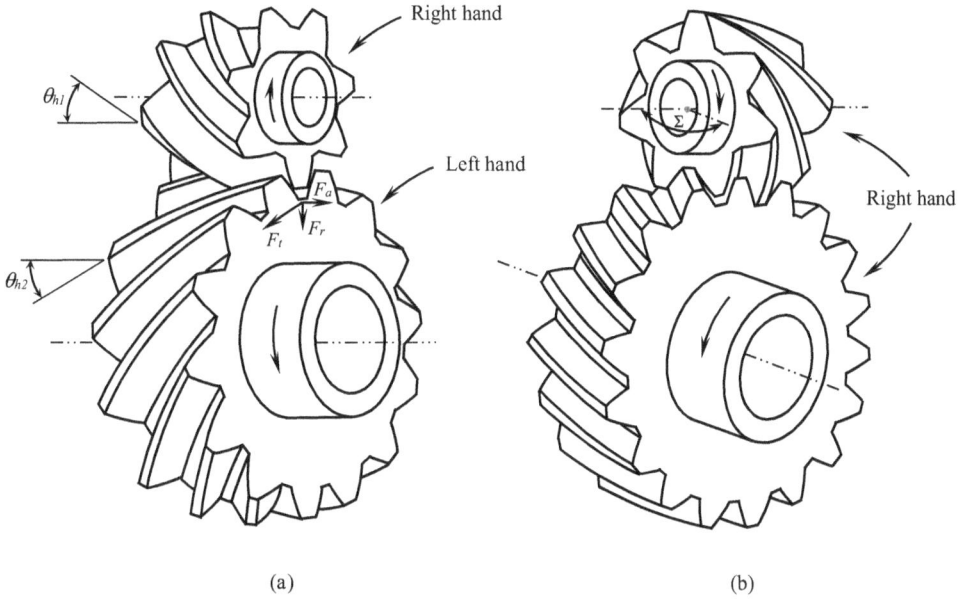

(a)   (b)

FIGURE 9.35 Configuration of the helical gear set: (*a*) a pair of helical gears with the opposite hands to drive parallel shafts and (*b*) a pair of helical gears with the same hand to drive nonparallel and nonintersecting shafts. Generally, the capacity of load transmission of cross-axis shafts is less than that of parallel shafts. This is because in the parallel configuration there is a line contact, and in the crossed configuration there is only a point contact between meshing gear teeth.

A pair of helical gears with the same hand and the helix angle of 45° on nonparallel, nonintersecting shafts are often referred to as *screw gears*. Clearly, their shaft angle Σ is 90°. Like other helical gears, the load-carrying capacity of screw gears is low due to the point contact.

Crossed helical gears have a virtual length crowning, which causes the point contact and reduces the sensitivity of shaft angle errors [9.69]. Because of point contact on the meshing teeth, their surfaces are subjected to high surface stress and the gear sets are used under light loads.

In comparison with a spur gear, a helical gear has a longer width of each tooth. Therefore, helical gears transfer loads more smoothly than spur gears. They are less wear and tear as the load is distributed among several teeth. Unlike spur gears, due to the angled teeth, helical gears are subjected to an additional thrust force F_a along the axes of the gears, which must be supported by appropriate bearings. The thrust (axial) force is calculated from the formula

$$F_a = F_t \tan\theta_h \tag{9.37}$$

where F_t is the tangential force, as the same as in Equation 9.35a. This suggests that the thrust force F_a varies directly with the magnitude of tangent of helix angle θ_h. The larger the helix angle, the higher the thrust force. Because of this thrust force and increased sliding friction among helical teeth, helical gears operate at slightly lower efficiencies than spur gears.

For helical gears, there are two pressure angles: the normal pressure angle α_n and the transverse pressure angle α_t, where two pressure angles can be related as

$$\alpha_t = \tan^{-1}\left(\frac{\tan\alpha_n}{\cos\theta_h}\right) \tag{9.38}$$

To eliminate the thrust forces, a special type of helical gears, known as herringbone gears, has been developed by combining two helical gears with the same helical angle but opposite hands side by side. The advantage of the herringbone gears over the helical gears is that each half of the gear produces the thrust in the opposite direction, which results in the net thrust of zero. Therefore, no special bearings are required for the herringbone gears. As a result, this type of gear is extensively adapted in gas and steam turbines, ship propulsion systems, heavy-duty vehicles, and heavy load machines.

Helical gearing systems have been used in a wide variety of applications that require high speeds or high loads, such as elevators, compressors, steel and rolling mills, textile machinery, and vehicles.

9.3.3 Bevel Gear

Bevel gears are conically shaped gears that transmit power between two intersecting orthogonal axles. Manufactured in pairs, they are widely used in the machine tools, mining machinery, aerospace equipment, robots, forklift trucks, boat actuators and propellers, high-speed offset printing, packaging machinery, polyethylene sheets, automobile differentials, and railroad transmissions. However, this type of gearing system provides limited gear ratios; the maximum gear ratio is about 10:1 per gear set. Moreover, bevel gears are not recommended at high operating speeds due to high noise.

Bevel gears have cones as their pitch surfaces. The teeth of a bevel gear are cut along the conical cones, with a straight, spiral, or hypoid tooth profile. Straight bevel gears have the teeth arranged on their pitch cones along straight lines. They are usually used to operate at relatively low speeds, typically less than 2 m/s circumferential speed. When the speed exceeds 5 m/s, it becomes very difficult to achieve quiet operation.

As a comparison, spiral bevel gears have curved spiral teeth. After heat treatment, the spiral teeth are often grinded to achieve a high level of geometric accuracy and surface finish. This finishing process not only allows removing heat treatment-induced distortions but also obtains accurate flank surfaces with predesigned and optimized *ease-off topographies* [9.70]. When the spiral gear teeth are engaged, the gear meshing line is changed so that the gear load is gradually and smoothly added and removed from one end of the tooth to the other, meshing with a rolling contact similar to helical gears. Therefore, compared to straight bevel gears, the power transmission of spiral bevel gears is relatively more stable and smooth, producing much less vibration and noise. In fact, quiet operation is particularly advantageous in environments such as hospitals, theatres, elevators, and airplanes. In addition, because of the increased tooth contact ratio, spiral bevel gears are stronger and more durable than straight bevel gears, allowing for higher load operation. Consequently, they are better suited for high-speed (for the peripheral velocity > 10 m/s) applications. To allow greater tooth contact, a pair of spiral bevel gears has equal but opposite spiral angles.

While the bevel gears are standardized with 20° pressure angle, other pressure angles, such as 14.5°, 16°, 17.5°, 22.5°, 25°, and 35°, are also used to meet the requirements in special applications. It is widely recognized that a lower pressure angle produces a lower noise level and an increased arc of action.

The pitch cone angles of a pair of bevel gears are

$$\phi_1 = \tan^{-1}\left(\frac{z_1}{z_2}\right) \tag{9.39a}$$

$$\phi_2 = \Sigma - \phi_1 \tag{9.39b}$$

where Σ is the shaft angle between two axes.

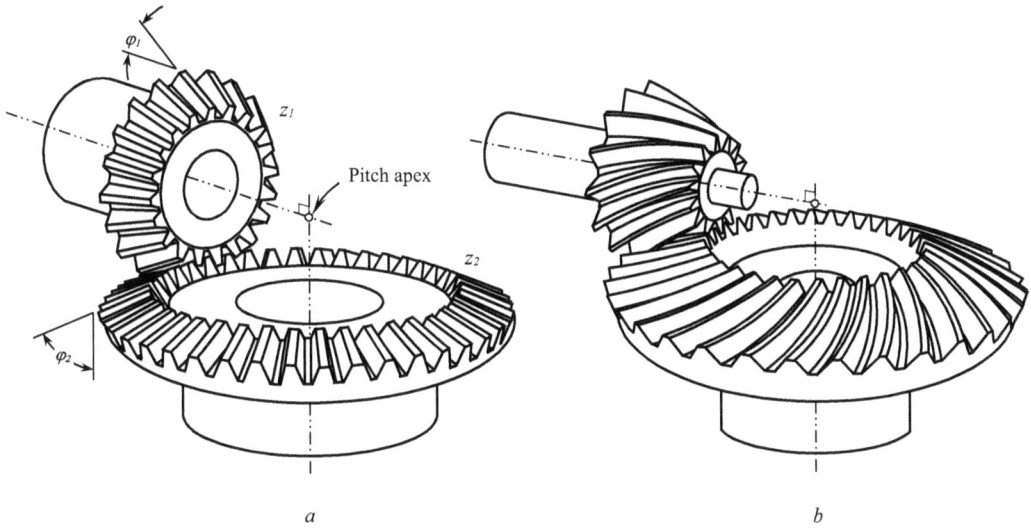

FIGURE 9.36 Two different configurations of bevel gears: (*a*) straight bevel gears with the straight tooth profiles and pitch cone angles φ_1 and φ_2, respectively; and (*b*) spiral bevel gears (or helical bevel gears) with the spiral tooth profile, which produces much less vibration and noise, especially at high speeds. Furthermore, spiral teeth can provide relatively stable and smooth power transmission, compared with straight teeth.

Though bevel gears can transmit power between nonparallel shafts at any angle desired, theoretically $0° < \theta < 180°$, they are most commonly used to transmit power at a crossed angle of 90° (Figure 9.36).

When the mating gears have the same number of teeth and the shafts are positioned at right angles, they are referred to as miter gears. Since the gear ratio is always 1:1, the miter gears are not used to change speed or torque, but to change the direction of rotational motion (Figure 9.37). Like bevel gears, there are two types of miter gears: straight miter and spiral miter gears. When adopting

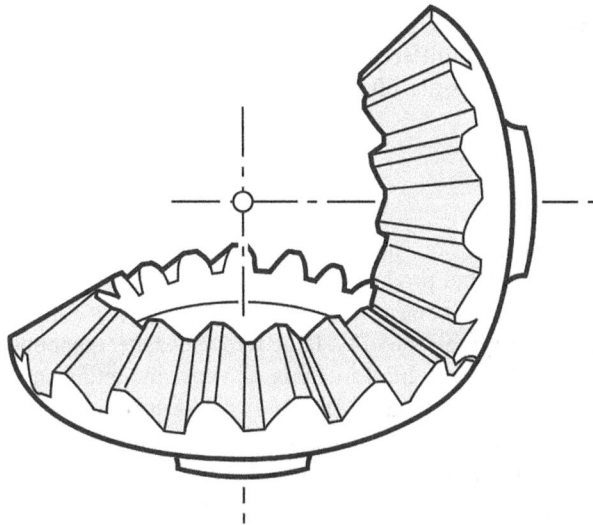

FIGURE 9.37 Straight bevel miller gears have the same number of teeth with a shaft angle of 90° and a gear ratio of 1:1. They are used to change the direction of the power transmission.

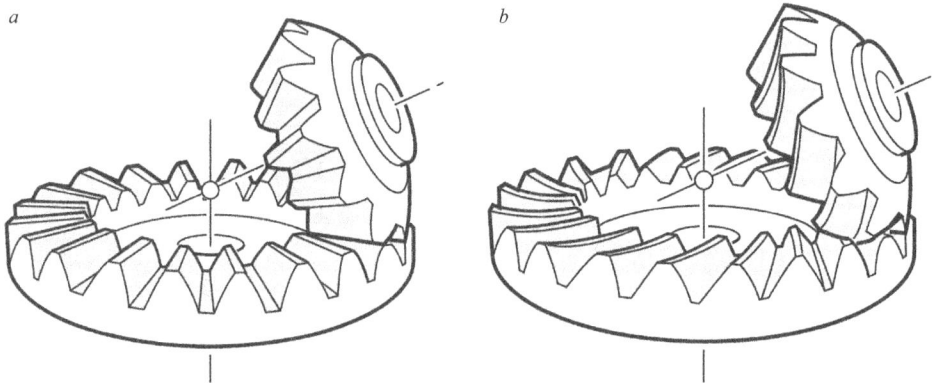

FIGURE 9.38   Bevel gears have a shaft angle of $\Sigma \neq 90°$ for (a) straight bevel gears and (b) spiral bevel gears. The arrangement of the gearset provides the flexibility in power transmission.

spiral miter gears, it is important to consider using thrust bearings due to the large thrust force produced by the miter gears in the axial direction. Miter gears with any other shaft angle (where $\Sigma \neq 90°$) are referred to as angular miter gears.

Figure 9.38 shows angular straight and spiral bevel gears that have a shaft angle of $\Sigma \neq 90°$. This offers more flexibility in power transmission in some special applications.

Hypoid gears are similar to spiral bevel gears except that the pinion is offset from the horizontal centerline of the face gear. Unlike bevel gears that the shafts are in the same plane, hypoid gears can engage with the axes in different planes. This design allows the pinion to be larger in diameter and provides a high contact ratio. Therefore, it permits the use of high gear ratios. A pair of hypoid gears always has opposite hand and the spiral angle of the pinion is usually larger than the angle of the gear [9.71]. Hypoid gears combine the rolling action and high tooth pressure of spiral bevels gears with the sliding action of worm gears. Therefore, their efficiencies range between spiral bevel gears and worm gears.

Because both the driving and driven gears are made of steel, the hypoid gear set often requires special lubrication. The gear shape is revolved hyperboloid instead of conical shape. This type of gearing system is often used in applications that require very smooth movement or automatic operation, such as rear-drive automobile differentials.

Spur and bevel gearing systems are often sealed to IP54 for protection against dust and other contaminations.

9.3.4   Spiroid Gear

Spiroid gears, often referred to as *skew axis gearing*, were invented by Oliver Saari in 1954 [9.72]. The spiroid gears are designed and produced to operate on nonintersecting and nonparallel axes for transmitting large power in less space. The key characteristics of this type of gearing system include high load/overload carrying capability, large torque density, quiet operation, positive backlash control, compact and lightweight, high stiffness, and easy mounting and assembly. Gear ratios are available ranging from 3:1 to more than 400:1 [9.73]. These superior features make them ideal in large power, high speed, and miniaturization applications, particularly in aerospace, robotic, medical, energy, packaging, and machine tool industries.

A spiroid gear set comprises a tapered pinion (or worm) and a complementary cooperating face gear, both having helical shape teeth, as shown in Figure 9.39. Thus, there is a concave side and a convex side of each tooth. The concave side, known as the driving side, has a higher pressure angle than the convex side. The high pressure angle α_h typically ranges from 20° to 40°, while the low-pressure angle α_l is in the range of 0°–20°.

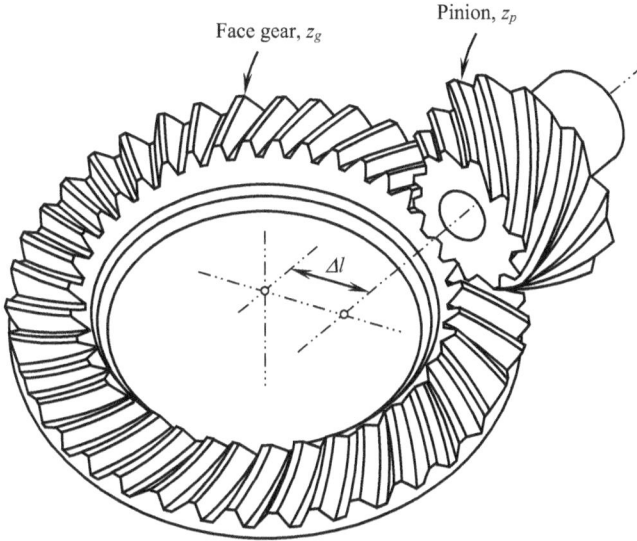

Face gear, z_g Pinion, z_p

Δl α

FIGURE 9.39 Spiroid gears are designed to transmit large power smoothly and quietly. Unlike the spiral bevel gears, the axis of the pinion is offset from the centerline of the face gear.

In fact, the spiroid gear set is very similar to a spiral bevel gear set except that the pinion axis is positioned a certain distance Δl from the horizontal centerline of the face gear. In such a way, it makes possible to hold the pinion shaft on both ends to increase stiffness and operating stability [9.74].

The spiroid pinion is conical in shape. Though the typical taper angle may vary in the range of 5°–10°, 5° is considered as the standard taper angle. The face angle is also conical in shape, where the typical cone angle is 8° and may vary by a few degrees [9.75].

Similar to other gear types with sliding mesh characteristics, the efficiency of the spiroid gearing system is determined by the gear normal pressure angle α_n, coefficient of dynamic friction μ_d, and spiral angles of the pinion and face gear θ_p and θ_g, respectively. As the pinion is the driving member, the efficiency η_p is given as

$$\eta_p = \frac{\cos\alpha_n + \mu_d \cot\theta_g}{\cos\alpha_n + \mu_d \cot\theta_p} \tag{9.40}$$

For the gearing system is in backdriving, *i.e.*, the face gear becomes the driving member, the efficiency η_g is

$$\eta_g = \frac{\cos\alpha_n - \mu_d \cot\theta_p}{\cos\alpha_n - \mu_d \cot\theta_g} \tag{9.41}$$

When $\eta_g = 0$, it yields that $(\cos\alpha_n - \mu_d \cot\theta_p) = 0$. Hence,

$$\theta_p = \cot^{-1}\left(\frac{\cos\alpha_n}{\mu_d}\right) \tag{9.42}$$

This leads to a criterion for determining the pinion spiral angle that ensures the self-locking of spiroid gears

$$\theta_p \geq \cot^{-1}\left(\frac{\cos\alpha_n}{\mu_d}\right) \tag{9.43}$$

Similar to many types of conventional gearing systems, the gear ratio of a spiroid gearing system is given as

$$\gamma_g = \frac{z_g}{z_p} \tag{9.44}$$

With today's digitized precision machining technology and advanced CAD/CAE techniques, it is no longer a challenge to manufacture spiroid gears with precise 3D tooth profiles. Using modern FEA software, desired 3D tooth profiles of spiroid gears could be analyzed and optimized. The generated gear tooth data are then entered into CAD software to generate a true 3D model of the gear, which allows for direct machining using multi-axis CNC machines. This novel manufacturing process greatly enhances gear quality, improves gear accuracy and performance, increases contact ratio, and reduces backlash.

9.3.5 HELICON GEAR

Another type of skew axis gearing, known as *helicon gear* today, was invented by Oliver Saari in 1960 [9.76]. The differences between helicon and spiroid gears are all attributable to geometry. While a spiroid pinion has a conical shape and engages a bevel-faced gear, a helicon pinion has a cylindrical shape and engages a flat-faced gear. This type of gearing system is particularly advantageous when it is desired to transmit relatively low power at fairly low gear ratios. It represents another powerful design option for engineers looking for transmitting higher torques within smaller spaces and broadens gearing design opportunities restricted by the physical limitations of conventional gearing systems.

Spiroid gear sets offer a finer backlash control over helicon gear sets. This is because a spiroid gear set allows for axial shimming of both the pinion and face gear to achieve backlash adjustability and a helicon gear set allows for axial shimming of only the face gear. However, spiroid gear sets have a higher overall cost than helicon gear sets due to the geometry complexity. Nevertheless, helicon is commonly preferred over spiroid except in some applications that need finite backlash control and higher torques.

9.3.6 WORM GEAR

Worm gears are specially designed to transmit power between nonintersecting nonparallel shafts that have axes of rotation offset from each other. With worm gearing systems, very high gear ratios can be realized in a single stage. This type of gearing system is extensively used in various industries for its ability to transmit high power with a high gear ratio, for instance, automotive transmissions, process machinery, steam turbines, conveyors, *etc.*

Worm gears must come in pairs: a worm (also known as worm screw) and a worm wheel (also known as worm gear), where the worm is the driving member with screw-like threads and the worm wheel is the driven member with helical teeth (Figure 9.40). In a worm gearing system, the worm and worm wheel must have the same pitch, hand, pressure angle, and lead angle. To enhance the engagement between the worm pair, the worm wheel is often made of a throated shape to wrap around the worm. This type of gearing system can offer a very large gear ratio in a single mesh with a less amount of space, compared with other conventional types of gearing systems.

While other conventional gears experience primarily rolling contact as teeth mesh, worm gear meshing is dominated by sliding between the worm and worm wheel teeth. This predominantly sliding contact results in significant friction and very high operation temperatures, leading to low efficiencies of worm gearing system. Furthermore, compared to other types of gearing systems, worm gearing systems face some unique lubrication challenges. The sliding motion tends to rub off the lubricant film on the meshing surfaces, causing worm gears operating under boundary lubrication conditions. To reduce sliding friction and lower operating temperatures, worm gears are typically lubricated with high viscosity lubricants such as compound oil or synthetic lubricants.

The most notable feature of the worm gearing system is that it prohibits dynamic reverse torque transmission as long as the static friction coefficient between the worm and worm wheel is larger than the tangent of the lead angle of the worm. This phenomenon is referred to as *self-locking*.

In order to make the self-locking feature of a worm gearing system, the following condition must be met,

$$\mu_s > \tan\theta_w \qquad (9.45)$$

This clearly shows that the static friction angle θ_s (where $\mu_s = \tan\theta_s$) must be larger than the worm's lead angle θ_w to prevent backdriving. Although it is not absolute, the self-locking can be expected for the worm lead angle θ_w less than 5° (where $\tan 5° = 0.0875$, less than the static coefficients of friction for most different material pairs under lubricated conditions). Thus, in normal operation only the worm can drive the worm wheel, but not inversely. Any torque reversal on the output shaft will cause the worm mesh to lock up instantly. As a result, this type of gearing system is suitable for applications where no backdriving is required.

It is important to note that the self-locking condition in the above formula is derived theoretically. In reality, when a worm gearing system is subjected to shock or vibration, the friction coefficient between worm gear and worm wheel may suddenly drop to its dynamic friction value (where $\mu_s > \mu_d$), possibly causing the gearing system to lose its self-locking ability and backdriving may occur. In addition, other factors such as temperature, surface condition (finish, oxidation, *etc.*), and lubrication may change the coefficient of friction.

Worm gearing systems feature offset drive capability, structure simplicity, low operation noise, good load capacity and stiffness, and high gear ratio. The main disadvantages of worm gearing system are as follows: (*a*) the positioning accuracy is limited due to the large backlash; (*b*) the transmission efficiency is relatively low; and (*c*) high contact stress and high heat generations on the sliding surfaces are two major concerns for reliable operation of this type of gearing system.

Characterized by the number of threads, the worm can have a single- or multiple-thread (usually up to 4). With a single-thread worm, for each revolution (*i.e.*, 360° turn) of the worm, the worm wheel advances one tooth. With a 3-thread worm, as the worm makes 360° turn, the worm wheel moves three teeth. As other parameters are fixed, a higher number of threads results in high efficiency. The increase in the gear ratio will lead to a drop in efficiency.

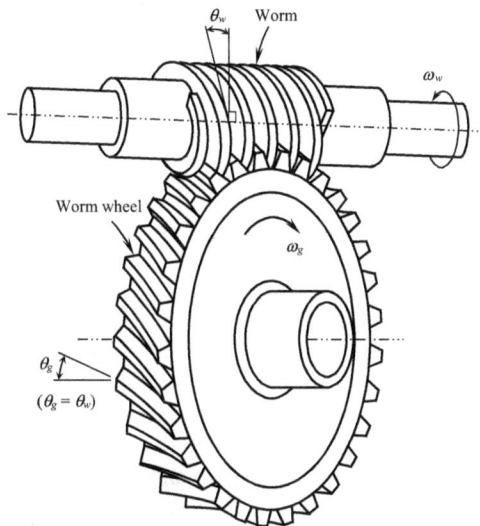

FIGURE 9.40 A cylindrical worm gearing system consists of a single thread worm and a worm wheel, with the axes at 90° to each other. Both the worm and worm wheel have the same hand. It is noted that the axes of the worm and worm wheel are nonintersecting. The lead angle θ_w (where $\theta_w = \theta_g$) is an important parameter for ensuring the system self-locking ability.

The worm gearing systems can provide a wide range of gear ratios, varying from 10:1 to 500:1. The gear ratio of the worm gearing system can be calculated as

$$\gamma_g = \frac{\omega_w}{\omega_g} = \frac{z_g}{z_p} \tag{9.46}$$

where

z_w is the number of threads (also known as the number of starts) in worm
z_g is the number of teeth in worm wheel

This suggests that the increase in the number of threads directly leads to a decrease in the gear ratio. The lead angle of worm θ_w (where $\theta_w = \theta_g$), which is measured on the pitch cylinder, is calculated as

$$\theta_w = \tan^{-1}\left(\frac{m_a z_w}{d_w}\right) = \sin^{-1}\left(\frac{p_n z_w}{\pi d_w}\right) \tag{9.47}$$

where
m_a is the axial module of worm
p_n is the normal circular pitch of worm
d_w is the pitch diameter of worm

Thus, with a fixed gear module m and pitch diameter d_w, the lead angle θ_w is proportional to the number of threads z_w. For this reason, the selection of the number of threads must be carefully considered to prevent the loss of self-locking ability.

Since worm gears have a lead angle, they do produce thrust loads that act on the shafts and bearings. The thrust loads vary upon the direction of rotation and the direction of the threads. For instance, a right-hand worm will pull the worm wheel toward itself if operated clockwise and push the worm wheel away from itself if operated counterclockwise [9.77].

The relative motion between a pair of worm gears is sliding rather than rolling. Due to the sliding contact nature, the coefficient of friction on the tooth surfaces and forces acting on the mating components have a great effect on its efficiency, which is determined by the worm's lead angle θ_w, and the tangential forces of worm and worm wheel, $F_{t,w}$ and $F_{t,g}$, respectively

$$\eta_g = \tan\theta_w\left(\frac{F_{t,g}}{F_{t,w}}\right) = \tan\theta_w\left(\frac{1 - \mu_d \tan\theta_w}{\mu_d + \tan\theta_w}\right) \tag{9.48}$$

where
μ_d is the coefficient of dynamic friction for worm gears

It should be pointed out that the motion of the worm and worm wheel is a mixture of sliding and rolling actions. The high rate of sliding friction tends to cause a temperature rise at sliding surfaces, wipe the lubricant along the convergent zone, and in a worse case, make the sliding surfaces stick together. This is especially true if the materials of the worm and worm wheel are the same. To reduce the sliding friction between the worm pair, the worm and worm wheel are always made of different materials. Practically, the worm is commonly made of steel and the worm gear is made of bronze, where the static coefficients of friction are 0.34 and 0.16 for dry and lubricated contacts, respectively. When gearing systems are used under high loads or high speeds, it is common to harden and grind worms. On the contrary, low loads or low life rotating components may use plastic or nylon materials for worm gears.

9.3.7 COMPARISON OF CONVENTIONAL GEARING SYSTEMS

Conventional gears can be classified into different types by their spatial relationship of the axis. The axes relationship can be parallel, intersecting, or skewed (nonparallel and nonintersecting). Of all gear types, the last one is probably the least understood.

There are four major types of skew axis gearing: spiral bevel, spiroid/helicon, hypoid, and worm gear. Their pinions and gears must have the same pressure angle but may have different helix angles.

In skew axis gear design, the pinion is the primary member because all calculations rely on it. Above all, the determination of *pitch point* location is the most important. Pitch point is the point where gear teeth actually make contact with each other as they rotate. With the chosen pitch point, all other contact parameters in a pair of gears can be determined. The pitch point angle σ is defined as the angle that is formed by connecting the pitch point and the center point of the face gear (*i.e.*, the pitch radius of the face gear) above the horizontal centerline of the face gear. For spiroid gears, this angle can range from 30° to 60°, but for regular design, 40° is found to work the best and thus is considered as the standard pitch point angle. For helicon gears, the pitch point angle is expanded from 10° to 70°. The offset Δl is directly related to the pitch point angle,

$$\sin\sigma = \frac{\Delta l}{r_g} \tag{9.49}$$

where r_g is the pitch radius of the face gear. This suggests that for a given pitch radius r_g, the center distance Δl is proportional to the pitch point angle σ.

Figure 9.41 shows the relations among the pinions and the face gear meshing for different skew axis gearing systems. The axis of a spiral bevel gear is perpendicular to the axis of the face gear, with $\Delta l=0$ (*i.e.*, $\sigma=0$). Hypoid gears resemble spiral bevel gears except the shaft axes do not intersect.

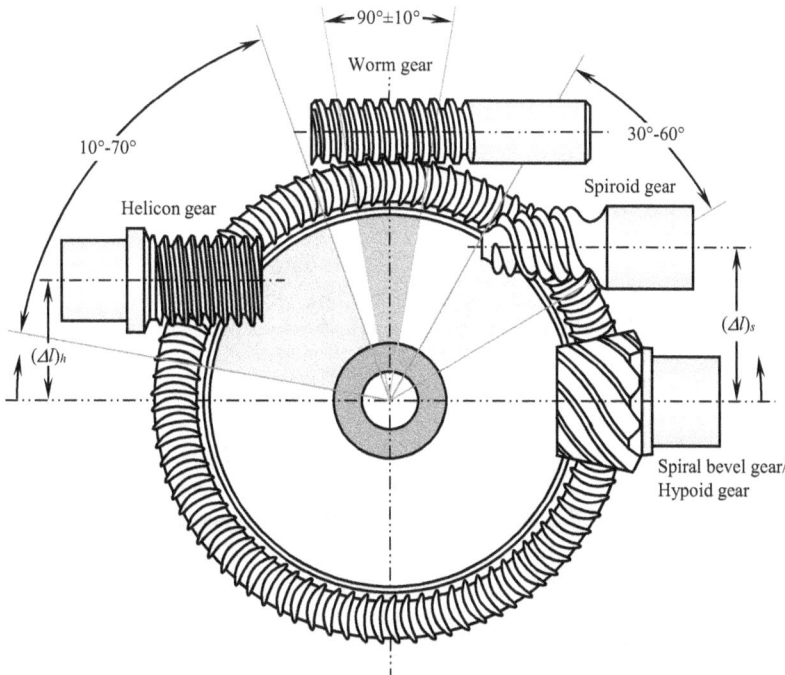

FIGURE 9.41 Comparison of various types of skew axis gears. According to the spatial relationship of the axis of the pinion to the axis of the face gear, skew axis gearing systems can be classified into spiral bevel/hypoid, spiroid/helicon, or worm gearing systems.

The hypoid pinion normally has an offset from the centerline of the mating gear. The increase in the offset results in an increase in the pinion diameter and spiral angle. Furthermore, the increase in the gear ratio requires a decrease in the number of teeth on the pinion, and finally, the pinion spiral angle approaches the thread-like appearance of a worm gear. Generally, the higher the gear ratio γ_g, the larger the center distance Δl, and the lower the efficiency η_g.

The main difference between a spiroid gear and a helicon gear is their geometry. While the spiroid pinion has a conical shape, the helicon pinion is cylindrical. Generally, spiroid gears are suitable for applications that require high torque transmission and need for finite backlash control. However, because of the complexity of geometry, spiroid gears face various challenges in design, manufacturing, and overall gear development. All of these have the potential to affect the overall cost of spiroid gears.

The shaft axes of a worm gearing system are perpendicular to each other in space but not intersecting. The standard pitch point angle is 90°, with a possible deviation of ±10°.

Although the efficiency of a gearing system is attributed to many sources of power losses, the mesh friction is considered one of the major factors to strongly affect the efficiency of gearing systems. In most cases, gear mesh action is neither pure rolling nor sliding, rather it involves both rolling and sliding action. An increase in sliding action causes an increase in friction losses and thus a decrease in efficiency. For instance, as the gear ratio increases, the sliding contact among the teeth in mesh increases, resulting in higher friction power losses, and eventually lower efficiency. For example, a gear set with crossed shafts (*e.g.*, worm gears) increases the sliding movement by 20%–90% than that with intersecting shaft (*e.g.*, bevel gears). Generally, for gearing systems with a similar size and gear ratio, bevel gears have the highest efficiency and worm gears the worst, with hypoid in the middle. This is because the relative sliding movement tends to increase along the path: bevel → hypoid → worm.

The comparison of the conventional gearing systems is presented in Table 9.4.

9.4   GEARHEAD AND GEARMOTOR

9.4.1   GEARHEAD

Many motion control systems employ separate motors and gearheads. This approach offers end users a selectivity and flexibility to choose the motor and gearhead most suitable for the application, even when they come from different manufacturers. In fact, many different gearheads can match with a specific motor. This configuration is more flexible than an integrated gearmotor. Since gearheads wear out more quickly than the motor itself, so when a gearhead fails, it only needs to replace the failed gearhead rather than the whole gearmotor [9.85].

9.4.2   GEARMOTOR

A gearmotor is an all-in-one combination of an electric motor, a gearing system, and associated components such as bearings, integrated into a single package. In such a way, it provides a simple and cost-effective solution for high-torque, variable-, or fixed-speed applications. The prominent advantage of this approach is that the overall volume of gearmotor is smaller than that of an assembly with a separate gearing system and a motor, particularly its overall length. This approach is especially suitable for some applications where space is the primary consideration. In addition, the gearmotor eliminates the risk of possible motor and gearhead misalignment, and the guesswork of sizing a motor and gearing system for users. Moreover, integrated gearmotors are designed to work well in harsh environments such as those found in the food processing industry, which usually requires IP 69K protection for resistance to the ingress of high-pressure water.

According to the relative position of the motor shaft (*i.e.*, driving shaft) to the gearing system shaft (*i.e.*, driven shaft), gearmotors can be categorized into two groups: parallel shaft gearmotor

TABLE 9.4

Comparison of Conventional Gearing Systems

Gearing System	Pinion	Gear	Pitch Point Angle σ	Pressure Angle α	Tooth Contact	Handedness in Gear Pair	Action between Meshing Gears[a]	Shaft Angle Σ	Gear Ratio γ_g	Efficiency η_g^b
Spur gears (parallel shafts)	Straight teeth	Straight teeth	–	Symmetric α 14.5°–25° Standard $\alpha=20°$	Line contact	–	Mostly rolling	Parallel shafts: 0°	≤ 20:1 for 1 stage[c] ≤ 45:1 for 2 stages ≤ 200:1 for 3 stages	94%–99% [9.71, 9.78]
Helical gears (parallel shafts)	Helix angle θ_1	Helix angle θ_2 $\theta_2=-\theta_1$	–	Normal α_n Transverse α_t Standard $\alpha_n=20°$	Line contact	Opposite[d]	Mostly rolling	Parallel shafts: 0°	3:2–10:1 [9.71]	93%–97%
Helical gears (crossed shafts)	Helix angle θ_1	Helix angle θ_2 $\theta_1=\theta_2=\theta$	–	Symmetric α Mainly $\alpha=20°$	Point contact	Same[e]	Sliding+rolling	Projected shaft angle $\Sigma=\theta_1+\theta_2$ Nonparallel Nonintersecting	≤6:1 [9.79]	70%–95%
Straight bevel gears (intersecting shafts)	Pitch angle φ_1	Pitch angle φ_2	0°	Symmetric α Standard $\alpha=20°$	Line contact	–	Mostly rolling	$\Sigma=\varphi_1+\varphi_2$ Typically $\Sigma=90°$	1:1–10:1 [9.80]	93%–97% [9.79]
Spiral bevel gears (intersecting shafts)	Pitch angle φ_1 Helix angle θ_1	Pitch angle φ_2 Helix angle θ_2	0°	Normal α_n Transverse α_t Standard $\alpha_n=20°$	Line contact	Opposite	Mostly rolling	$\Sigma=\varphi_1+\varphi_2$ Typically $\Sigma=90°$	1:1–10:1 [9.81]	94%–99% [9.71]
Spiroid gears (skewed shafts)	Taper angle 5°	Cone angle 8°	30°–60°	Asymmetric α α_n: 20°–40° α_i: 0°–20°	Line contact	Opposite	Sliding+rolling	Projected shaft angle $\Sigma=90°$ Nonparallel Nonintersecting	3:1–400:1	60%–94% [9.77]

(Continued)

TABLE 9.4 (Continued)
Comparison of Conventional Gearing Systems

Gearing System	Pinion	Gear	Pitch Point Angle σ	Pressure Angle α	Tooth Contact	Handedness in Gear Pair	Action between Meshing Gears[a]	Shaft Angle Σ	Gear Ratio γ_g	Efficiency η_g^b
Hypoid gears (skewed shafts)	Taper angle	Cone angle	10°–40°	Symmetric α	Line contact	Opposite	Sliding + rolling	Projected shaft angle Σ=90° Nonparallel Nonintersecting	10:1–200:1 [9.71]	60%–90%
Helicon gears (skewed shafts)	Taper angle 0°	Core angle 0°	10°–70°	Asymmetric α α_h: 20°–40° α_i: 0°–20°	Line contact	Opposite	Sliding + rolling	Projected shaft angle Σ=90° Nonparallel Nonintersecting	3:1–400:1	78%–95%
Worm gears (skewed shafts)	Lead angle θ_p	Lead angle θ_g $\theta_p=\theta_g$	90°±10°	Normal α_n Transverse α_t Standard α_n=20	Line contact	Same	Sliding	Projected shaft angle Σ=90° Nonparallel Nonintersecting	5:1[f]–500:1[g]	50%–90% [9.80]

Notes:

[a] While parallel axis gearing systems are characterized by predominantly rolling action in the gear mesh, skew axis gearing systems are characterized by mainly sliding action between meshing teeth.

[b] The gear ratio has a strong influence on efficiency, *i.e.*, $\eta_g \propto \gamma_g^{-1}$. This is especially true for gearing systems that have a wide range of gear ratios. An example demonstrates that for hypoid gears, when the number of threads of the pinion $z_p=1$, the efficiency $\eta_g=75\%$ for $\gamma_g=15:1$; and $\eta_g=59\%$ for $\gamma_g=70:1$. As $z_p=3$, the efficiency $\eta_g=85\%$ for $\gamma_g=15:1$ and $\eta_g=69\%$ for $\gamma_g=70:1$.

[c] Generally, the maximum gear ratio of spur gearing system is 10:1 for a single stage but possible to achieve 20:1 for special designs.

[d] For the parallel helical gears, the handedness for the meshed gears is opposite, for example, a right-hand pinion and a left-hand gear.

[e] For the crossed helical gears, the handedness for the meshed gears is the same.

[f] Worm gearing systems usually have a minimum gear ratio of 5:1 [9.71, 9.82]. Efficiency of a worm gearing system primarily depends on its gear ratio (*i.e.*, $\eta_g \propto \gamma_g^{-1}$). An example shows that the worm gear efficiency is 49% for $\gamma_g=300:1$ [9.83] and 90% for $\gamma_g=5:1$.

[g] Worm gearing systems accommodate a wide range of gear ratios. Generally, $\gamma_g=60:1$ and higher can be obtained from a single reduction and can go as high as 500:1 [9.84].

FIGURE 9.42   A cutaway of a right-angle gearmotor with the worm gearing system [9.86].

and right-angle shaft gearmotor. For a parallel shaft gearmotor, the shaft centerlines of motor and gearing system are parallel. In general, parallel shaft gearmotors have a higher output torque, efficiency, and lower backlash. By contrast, a right-angle gearmotor, where the shaft of the gearing system is at 90° from the motor shaft either using worm gearing (Figure 9.42) [9.86] or hypoid gearing, is commonly used when it is necessary to fit a servomotor into a tight space. Well-designed worm gearing systems have self-locking ability and high shock load capability. Right-angle gearmotors with hypoid gearings have several advantages when compared to those with standard worm gearing systems, including higher efficiency, less heat generation, higher torque, and more compact design. However, it is more expensive due to the higher manufacturing cost.

9.5   FAILURE OF GEARING SYSTEM

Failure analysis is the process of collecting and analyzing data to determine the root cause of failure of a system or components and prevent the failure from recurring. Failure causes of gearing systems can be generally attributed to eight classifications: (*a*) faulty design, (*b*) material defects, (*c*) processing and manufacturing deficiencies, (*d*) assembly or installation defects, (*e*) off-design or unintended service condition, (*f*) maintenance deficiencies, (*g*) abnormal wear as a result of lubricant deficiency, (*h*) improper operation, and (*i*) improper load and duty specifications [9.87]. In practice, these failure causes are usually determined by relating them to one or more specific failure modes.

For contact gearing systems, the gear failure mode can be briefly categorized by three types: (*a*) the damage to the tooth surface, (*b*) the breakage of the gear tooth, and (*c*) plastic flow. Among them, surface fatigue is a failure of gear material that has been under repeated surface stress beyond the endurance limit of the material. Surface fatigue forms include the following [9.88]:

- Scuffing is the surface damage as the lubricant film between meshing teeth is destroyed. This leads to metal-to-metal contact and results in surface damage, ranging from a lightly etched appearance to severe welding and tearing of engaging teeth. In gearing systems, this happens when the sliding velocity and load are increased and the local temperature reaches its upper limit.
- Pitting is surface damage from cyclic contact stress transmitted through a lubrication film that is in or near the elastohydrodynamic regime. Pitting is one of the most common causes of gear failure. It also affects other machine components (*e.g.*, bearings and cams) in which surfaces undergo rolling/sliding contact under heavy load [9.89]. Contact fatigue failures

such as micro pitting and spalling compete with surface wear as potential failure modes along the lifespan of gear components.

- Case crushing is typically associated with heavily loaded casehardened gears. Case crushing appears as long longitudinal cracks on the tooth surface, which can cause pieces of the tooth to subsequently break away.

Gear tooth fracture is one of the most common ultimate failures in industrial applications and occurs in various ways. Most bending failures happen when excessive acceleration/deceleration or shock loads are applied to the teeth, which result in root stresses higher than the endurance limit of the material. Fatigue failures of gear occur when the gears are subjected to high repeated load cycles. Bending fatigue failure is the result of cyclic bending stress at the tooth root. The best way to avoid fatigue breakage is to either design gear teeth strong enough so that the transmitted load will result in stresses well below the endurance limit of the material, or specify high strength materials for gears.

With high contact pressure applied on tooth surfaces under rolling or sliding action during the gear mesh process, plastic flow of tooth surfaces could take place. This plastic deformation is resulted from the yielding of the material based on the same mechanism in cold forging. After a long time operation or with heavy load, the pair of gears in ductile steel often exhibit ridges along the pitch line of wheel and groove in the pitch line of the pinion due to plastic flow of the material, attracted to low viscosity of the lubricant and lack of surface hardness. Moreover, rippling is a periodic wave-like formation that appeared on hardened gear surfaces. It is essentially a kind of wear or plastic deformation in microform with very thin oil films. This type of surface damage can lead to high noise and vibration level of gear.

Surface wear is one of the failure modes experienced at the contact surfaces of gear teeth under combined sliding and rolling motions [9.90]. Wear is a surface phenomenon in which layers of material are wore away with some degree of uniformity. Generally, wear rate is a function of the load intensity, material compatibility, tooth surface hardness, load-carrying characteristics of the lubricant, and operational duty cycle. Apart from the direct material loss that leads to functional failure, surface wear is a continuous abrasive process of material removal from contact gear teeth to cause the change in the tooth geometry and involute profile, thus affecting the patterns of gear contact and altering contact stresses and load distributions.

Corrosive wear, also known as chemical wear, is induced on gear tooth surfaces as a result of chemical or electrochemical action attacking the gear material. The active factors attacking the materials may come from sources external to the gearing system or from a chemical breakdown within the lubricant itself. In corrosive wear, when the corrosive layer is removed through sliding or abrasion, another layer begins to form, and the process of removal and corrosive layer formation is repeated [9.91]. Because the environment plays a great role in corrosive wear, material selection is crucial before designing a gearing system.

During gearing system operation, when the tip of a driving gear tooth contacts the root of the driven gear tooth, it is defined as tip-to-root interference. Tip-to-root interference is strictly not a failure mode; rather is abnormal mesh condition that could lead to other failure modes. The interference between meshing gears is often resulted from faulty gear design, manufacturing deficiency, and/or improper installation. In fact, properly designed gears can effectively avoid the interference between the meshing gears. In addition, by specifying more backlashes for the assembled gear pair, tight centers can be avoided and thus eliminate the interference problem.

Over the past few decades, a variety of advanced manufacturing technologies have been developed and introduced into production, and correspondingly gearing system failures have reduced dramatically. Many progresses have been made for enhancing the performance, operating reliability, and lifetime of gearing system, in the areas of heat treatment, surface finish, material selection, tolerance precision, quality control, *etc.*

9.6 SELECTION OF GEARING SYSTEM

The basic function of most gearing systems is to reduce the relatively high motor speed to the lower speed required by a driven machine and simultaneously increase the operating torque for the driven machine. Having a strong influence on the overall efficiency, performance, and cost of a motion control system, selecting a suitable gearing system that delivers reliable performance with the lowest total cost of ownership is a complex process because it requires paying attention to many factors simultaneously.

Gearing systems come in a variety of types, gear configurations, structures, sizes, and operating characteristics. Each of the gearing systems offers different behaviors and advantages, but the requirements and specifications demanded by a particular motion or power transmission application determine the type of gearing most suitable for use. In order to achieve this goal, it is critical for motor engineers to not only choose the best gearing system for a specific application but also comprehensively consider its technical advantages and challenges. It has been widely recognized that while a given machine may utilize different types of gearing systems, the same gearing type can be extensively used in different types of machines and devices. The selection process of gearing system often involves the balance among a number of factors, for instance, required transmitted power/torque, output speed, gear ratio, accuracy, structural complexity, installing configuration, inertia, size, weight, lifetime expectation, vibration and noise level, and cost. In fact, manufacturers' catalogs often list detailed selection criteria, formulas, and special conditions, which can serve as the practical guidance of gearing system selection.

Perhaps, the most important factor in the selection of the gearing system type is its load-carrying capacity (nominal power rating) that a gearing system can deliver to meet customer requirements. To ensure normal operation, the load-carrying capacity of gearing should be larger enough than the load of driven machines. Conversely, if the load in an application exceeds the allowed value, it may lead to the failure of the gearing system. In such a situation, it is required to increase the size of the gearing system to withstand the higher load.

The duty cycle is important because it determines if gearbox and motor size are based on average or peak torque and speed. Engineers should carefully and critically compare these values with the motor's nominal and peak torque and speed ratings. The duty cycle can be cyclical or continuous. Generally, when duty cycle is less than 60% and each move takes less than 20 minutes, it is considered cyclical. If average torque and speed exceed nominal ratings, the motor requires auxiliary cooling for dissipating generated heat [9.79].

Gearing systems not only change the rotation speed, but also change the torque on the axle. Gear ratio is one of the most critical concerns with gearing systems. Most gearing systems are used to decrease the angular speed ($\omega_{out}=\omega_{in}/\gamma_g$) and increase the operating torque ($T_{out}=\gamma_g T_{in}$) of motor. If the gearing system efficiency $\eta_g < 1$ is taken into account, the output torque is $T_{out}=\eta_g\gamma_g T_{in}$, indicating the continuous operating torque increases by a factor $\eta_g\gamma_g$. The range of gear ratio is mainly dependent on the type of gearing system. Many conventional gearing systems (*e.g.*, spur, helical, bevel gears, *etc.*) have low gear ratios; some modern gearings such as SWG can offer a high gear ratio. Generally, the efficiency of gearing system is inversely proportional to gear ratio, *i.e.*, $\eta_g \propto \gamma_g^{-1}$. However, since the overall efficiency of gearhead motor is the product of the motor and gearing efficiency, i.e., $\eta=\eta_m\eta_g$, the maximum efficiency is achieved when both η_m and η_g reach their peak values together. It has reported that when the gear ratio is in the range of 150:1–200:1, the motor and gearing efficacies reach approximately their peak values.

Gearing system efficiency, which depends on many factors such as gearing system type, loading, gear ratio, lubrication, and operating condition, is one of the determining factors. Generally, light loading and high ratios tend to produce poor gearing system efficiencies. Conversely, under heavy loading and with high ratios, a gearing system approaches its theoretical efficiency. In many cases, high efficiency cuts drive and operating costs. Among all factors, gear mesh and intersect are the main factors to affect the efficiency of the gearing system. Depending on the type of gearing system and the number of reduction stages, gearing efficiencies are ranged very scattered. Service factor takes into account other operational parameters, such as duty cycle, numbers of starts and stops, load characteristics, and power sources (*e.g.*, motor, engine, or turbine). Engineers should use the most-efficient gearing system that meets their application's needs.

In selecting gearing system, the gearing size is also considered to meet the space requirement. Sometimes, this can be a determining factor for the applications where the available space is limited. In addition, some gearing systems are available with special features such as self-locking capability that prevents reverse rotation. The manufacturer's selection guidance and procedure that applies to these features should be followed. If manufacturer's selection guidance is lacking detail, a critical review is needed before selecting.

The type of load can significantly affect the performance of a gearing system. Shock loads and high vibrations always deteriorate the operating condition of a gearing system and electric motor and shorten their service lifetime. This should not be underestimated.

Many gears such as spur, bevel, and helical gears, the maximum operating temperature is set about 65°C. For worm gears and hypoid gears, the operating temperature limits are 90°C and 95°C, respectively [9.92]. There are several possible methods to improve the thermal management, including the following: (*a*) Increasing the thermal heatsink mass, but it requires a trade-off of increased weight and reduced payload capacity. (*b*) Greater axial distance to bearings, electronic encoder, or gearing, which also requires a trade-off of increased weight and reduced payload capacity. (*c*) Reduce the maximum winding temperature limit of the frameless motor design, which also reduces the available torque from the motor. (*d*) Because the lubricant carries heat from gearing components to the gearing housing where heat is dissipated to the surrounding environment, the housing design can play a major role in enhancing thermal capacity. Features such as cooling fins, ventilation channels through the housing, air fans, and even the optimized shape of the housing itself can make a big difference in heat dissipation. Nevertheless, the thermal power limit must always be taken into account when designing and selecting gearing systems. It represents the maximum permissible power that can be transmitted by a gear unit at the ambient temperature in a continuous operation mode.

The high precision positioning requirement implies that a gearing system must have low or zero backlashes, especially in bidirectional and cyclic motion applications. As discussed previously, some gearing systems, such as SWG, RV, planetary, and cycloidal gearing systems, have very low or zero backlashes, but may have bidirectional hysteresis motion loss.

Most end users are very concerned about the lifespan of gearing systems. For example, SWGs are extensively used in spacecrafts. Reliability of SWG is of great importance to the functioning of spacecraft. Failure of SWG might cause malfunctions of spacecraft and significant economic losses [9.93]. The regular lifespan of SWG ranges 2,000–4,000 h, depending on manufacturers and SWG types and sizes. These values are obviously much less than the regular lifespan of servomotors (usually 10–30 years depending on power ratings and averaged lifespan of 13.3 years for all ratings [9.94]). This indicates that SWGs fail much earlier than servomotors. Therefore, in selecting a gearing system, engineers must carefully consider and balance the expected lifespans of the system and its components.

Environmental conditions can significantly affect gearing system performance. Under harsh operating conditions, proper sealing is necessary to maintain adequate lubrication, eliminate oil leaks, and avoid contaminates entering gearing systems for preventing damages to gears, bearings, shafts, and other components.

Once the gearing system is chosen and installed in the application, perform several test runs in sample environments that replicate typical operating scenarios. If the design exhibits unusually high heat, noise, or stress, repeat the gearing selection process until the desired outcome is obtained [9.81].

REFERENCES

9.1. Nazzaro, J. 2013. The benefits of gearboxes – and when to pick integrated gearmotors. *Machine Design.* Article 21833641. https://www.machinedesign.com/motors-drives/article/21833641/the-benefits-of-gearboxes-and-when-to-pick-integrated-gearmotors.

9.2. Deutsches Institut für Normung E. V. 2019. DIN 3996: 2019-09 Tragfähigkeitsberechnung von zylinder-schneckengetrieben mit sich rechtwinklig kreuzenden achsen (Calculation of, K.load capacity of cylindrical worm gear pairs with axes crossing at right angle).

9.3. Faulhaber, F. 2002. A second look at gearbox efficiencies. *Machine Design.* Article 21834659. https://www.machinedesign.com/archive/article/21834659/a-second-look-at-gearbox-efficiencies.

9.4. Lynch, K. M., Marchuk, N., and Elwin, M. L. 2016. Chapter 26: Gearing and motor sizing. In *Embedded Computing and Mechatronics with the PIC32 Microcontroller* (Eds.: K. M. Lynch, N. Marchuk, and M. L. Elwin). Newnes, Waltham, MA, pp. 427–437.

9.5. Kraig, K. 2013. Inertia mismatch: fact or fiction? *Design News.* https://www.designnews.com/inertia-mismatch-fact-or-fiction.

9.6. Lewotsky, K. 2015. Understanding the mysteries of inertia mismatch. *Motion Control & Motor Association.* https://www.motioncontrolonline.org/content-detail.cfm/Motion-Control-News/Understanding-the-Mysteries-of-Inertia-Mismatch/content_id/404.

9.7. Maitra, G. M. 1989. *The Handbook of Gear Design,* 2nd edn. Tata McGraw-Hill, New Delhi.

9.8. Dengel, B. 2021. The importance of contact ratio: How to calculate the contact ratios for various styles of gearing. *Gear Solutions,* March Issue, pp. 24–25.

9.9. Arrhenius, S. A. 1889. Über die dissociationswärme und den einfluss der temperatur auf den dissociationsgrad der elektrolyte. *Zeitschrift für Physikalische Chemie* **4U(1)**: 96–116.

9.10. Exxon Mobil Corporation. 2012. Lubricating grease basics. https://www.mobil.com/en/industrial/lubricant-expertise/resources/oil-grease-lubricant-basics.

9.11. Kannel, J. W. and Barber, S. A. 1990. Analysis of scuffing temperatures with consideration of wear. *ASME Journal of Tribology* **112(1)**: 123–121.

9.12. Miroslav, V., Adam, K., and Miroslava, N. 2014. Analysis of the HCR gearing from warm scuffing point of view. *FME Transaction* **42(3)**: 224–228.

9.13. Meuleman Electronics. 2017. 9 Benefits of part count reduction in your product. https://blog.meuleman.io/en/9-benefits-of-part-count-reduction-in-your-product.

9.14. Musser, C. W. 1959. Strain wave gearing. U.S. Patent 2,906,143.

9.15. Musser, C. W. 1960. Breakthrough in mechanical drive design: The harmonic drive. *Machine Design.* April 14, pp. 160–173.

9.16. Talebi, H. A., Abdollahi, F., Patel, R. V., and Khorasani, K. 2010. *Neural Network-Based State Estimation of Nonlinear Systems: Application to Fault Detection and Isolation.* Springer, New York.

9.17. Taghirad, H. D. and Belanger, P. R. 1996. An experimental study on modeling and identification of harmonic drive systems. *Proceeding of 35th IEEE Conference on Decision and Control* **4**: 4725–4730.

9.18. Wu, X. D., Bai, M. L., Li, Y., Cao, Q. H., and Du, X. C. 2014. Analysis of harmonic gear drive efficiency based on the 3D simplified model. In *Advanced Materials Research* (Ed.: X. Liu). Trans Tech Publications, Freienbach, **vol. 1006–1007**, pp. 249–252.

9.19. Zou, C., Tao, T., Jiang, G. D. Mei, X. S., and Wu, J. H. 2015. A harmonic drive model considering geometry and internal interaction. *Proceedings of the Institution of Mechanical Engineers, Part C: Journal of Mechanical Engineering Science* **231(4)**: 728–743.

9.20. Gravagno, F., Mucino, M., and Pennestri, E. 2021. The mechanical efficiency of harmonic drives: A simplified model. *ASME Journal of Mechanical Design* **143(6)**: 063302–063308.

9.21. Ishikawa, S. 1989. Tooth profile of spline of strain wave gearing. U.S. Patent 4,823,638.

9.22. Kiyosawa, Y., Sasahara, M., and Ishikawa, S. 1989. Performance of a strain waive gearing using a new tooth profile. *Proceeding of ASME 1989 International Power Transmission and Gearing Conference* **2**: 607–612.

9.23. Ishikawa, S. 2006. Wave gear drive with wide mesh three-dimensional tooth profile. U.S. Patent 7,117,759 B2.

9.24. Ishikawa, S. 2014. Wave gear device having three-dimensional continuous contact tooth profile. U.S. Patent 8,661,940 B2.

9.25. Klebanov, B. M. and Groper, M. 2016. *Power Mechanisms of Rotational and Cyclic Motion.* CRC Press, Boca Raton, FL.

9.26. Dong, H., Zhang, J., and Wang, D. 2019. A tooth profile design method for harmonic drive without tip interference. In *Mechanism Design of Robotics* (Eds.: A. Gasparetto and M. Ceccarelli). Springer, Cham, pp. 138–146.

9.27. Oh, H. S., Jeong, K. S., and Lee, D. G. 1994. Design and manufacture of the composite flexspline of a harmonic drive with adhesive joining. *Composite Structure* **28**(3): 307–314.

9.28. Ueura, K. and Slatter, R. 1999. Development of the harmonic drive gear for space applications. *Proceedings of the 8th European Symposium of Space Mechanisms and Tribology* **438**: 259–264.

9.29. Tong, W. 2019. Factors affecting robot performance and life. Kollmorgen White Paper. Also in *Design World* with the new title: Environmental effects on motion components in robotics, November Issue, pp. 10–15.

9.30. Kircanski, N. M. and Goldenberg, A. A. 1997. An experimental study of nonlinear stiffness, hysteresis, and friction effects in robot joints with harmonic drive and torque sensors. *International Journal of Robotics Research* **16**(2): 214–230.

9.31. SKF. 2017. The SKF model for calculating the frictional moment. https://www.skf.com/binaries/pub12/Images/0901d1968065e9e7-The-SKF-model-for-calculating-the-frictional-movement_tcm_12-299767.pdf.

9.32. Harmonic Drive LLC. 2018. Harmonic Drive® reducer catalog. http://www.harmonicdrive.net/_hd/content/documents/CSG-2UK_GearUnits.pdf.

9.33. Bennett, S. 2016. *Heavy Duty Truck Systems*, 6th edn. Cengage Learning, Boston, MA.

9.34. Ferguson, R. J. 1983. Short cuts for analyzing planetary gearing. *Machine Design*, May Issue, pp. 55–58.

9.35. Li, S. T. 2014. The latest design technologies for gear devices with great transmission ratios. *Power Transmission Engineering*, December Issue, pp. 70–76.

9.36. Kaim, C. S. 2000. The world of planetary gears. *Machine Design*. Article 21834331. https://www.machinedesign.com/mechanical-motion-systems/article/21834331/the-world-of-planetary-gears.

9.37. Braren, L. K. 1930. Gear transmission. U.S. Patent 1,773,568.

9.38. Sensinger, J. W. 2010. Unified approach to cycloid drive profile, stress, and efficiency optimization. *ASME Journal of Mechanical Design* **132**: 024503-1–024503-5.

9.39. Ye, Z., Zhang, W., Huang, Q., and Chen, C. 2006. Simple explicit formulae for calculating limit dimensions to avoid undercutting in the rotor of a cycloid rotor pump. *Mechanism and Machine Theory* **41**(4): 405–414.

9.40. D'Amico, J. 2011. Comparing cycloidal and planetary gearboxes. *Machine Design*. Article 21829580. https://www.machinedesign.com/news/article/21829580/comparing-cycloidal-and-planetary-gearboxes.

9.41. Chen, C. and Yang, Y. H. 2017. Structural characteristics of rotate vector reducer free vibration. *Shock and Vibration*, Article ID 4214370.

9.42. Li, X., Chau, K.-T., Cheng, M. and Hua, W. 2013. Comparison of magnetic-geared permanent magnet machines. *Progress in Electromagnetics Research* **133**: 177–198.

9.43. Armstrong, C. 1901. Power-transmission device. U.S. Patent 687,292.

9.44. Atallah, K. and Howe, D. 2001. A novel high-performance magnetic gear. *IEEE Transactions on Magnetics* **37**(4): 2844–2846.

9.45. Wang, Y. W., Filippini, M., Bianchi, N., and Alotto, P. 2019. A review on magnetic gears: topologies, computational models, and design aspects. *IEEE Transactions on Industry Applications* **55**(5): 4557–4566.

9.46. Jian, L., Chau, K. T., Gong, Y. Z., Yu, C., and Li, W. 2009. Comparison of coaxial magnetic gears with different topologies. *IEEE Transactions on Magnetics* **45**(10): 4526–4529.

9.47. Li, K., Modaresahmadi, S., Williams, W. B., Wright, J. D., Som, D., and Bird, J. Z. 2019. Designing and experimentally testing a magnetic gearbox for a wind turbine demonstrate. *IEEE Transactions on Industry Applications* **55**(4): 3522–3533.

9.48. Zhang, F., Du, G., Wang, T., Liu, G. G., and Cao, W. 2015. Rotor retaining sleeve design for a 1.12-MW high speed PM machine. *IEEE Transactions on Industry Applications* **51**(5): 3675–3685.

9.49. Aiso, K., Akatsu, K., and Aoyama, Y. 2019. A novel reluctance magnetic gear for high-speed motor. *IEEE Transactions on Industry Applications* **55**(3): 2690–2699.

9.50. Naclerio, N. D., Kerst, C. F., Haggerty, D. A., Suresh, S. A., Singh, S., Ogawa, K., Miyazaki, S., Cutkosky, M. R., and Hawkes, E. W. 2019. Low-cost, continuously variable, strain wave transmission using Gecko-inspired adhesives. *IEEE Robotics and Automation Letters* **4**(2): 894–901.

9.51. Jung, J. H., Pan, H., and Kang, T. J. 2008. Capstan equation including bending rigidity and non-linear frictional behavior. *Mechanism and Machine Theory* **43**(6): 661–675.

9.52. Tsuchiya, E. and Shamoto, E. 2017. Pulse drive: A new power-transmission principle for a compact, high-efficiency, infinitely variable transmission. *Mechanism and Machine Theory* **118**(2017): 265–282.

9.53. Kernbaum, A. S. 2017. Pure rolling cycloids with variable effective diameter rollers. World Intellectual Property Organization. Publication Number WO 2017/044171 A2.

9.54. Ackerman, E. 2016. SRI demonstrates abacus, the first new rotary transmission design in 50 years. *IEEE Spectrum.* https://spectrum.ieee.org/automaton/robotics/robotics-hardware/sri-demonstrates-abacus-rotary-transmission.

9.55. Zheng, Y. F. 2016. Circular wave drive. U.S. Patent 9,494,224 B2.

9.56. Zheng, Y. F. 2017. Circular wave drive. U.S. Patent 9,677,657 B2.

9.57. Center for Design and Manufacturing Excellence, The Ohio State University. 2017. Precision, low profile gear system. https://cdme.osu.edu/sites/cdme.osu.edu/files/uploads/0009a_gearsystem.pdf.

9.58. Schorsch, J. F. 2018. Compound planetary friction drive. U.S. Patent 10,113,618 B2.

9.59. Brown, A. 2019. New compound traction drive promises high ratios, no slipping. *American Society of Mechanical Engineering (ASME).* https://www.asme.org/topics-resources/content/new-compound-traction-drive-promises-high-ratios-no-slipping.

9.60. Hoefken, C. A. 2018. Motorized gearbox mechanism. U.S. Patent 10,151,375 B2.

9.61. Hoefken, C. A. 2021. Spiral cam gearbox mechanism. U.S. Patent 11,028,910 B2.

9.62. Miki Pulley. Speed changers and reducers, pp. 487–496. https://www.mikipulley.co.jp/data/pdf/en/md_zm_ct.pdf.

9.63. Whitby, R. D. 2013. Lubrication of worm gears. *Society of Tribologists and Lubrication Engineers.* https:www.stle.org/files/TLTArchives/2013/05_May/Worldwide.aspx.

9.64. Dengel, B. 2017. Under pressure: understand the choice of pressure angle in the design of spur or helical gearing. *Gear Solutions.* https://gearsolutions.com/media/uploads/uploads/assets/Digital_Editions/2017/201709/0917-TT.pdf.

9.65. Handschuh, R. F. and Zakrajsek, A. J. 2011. High-pressure angle gears: Comparison to typical gear design. *ASME Journal of Mechanical Design* **133**(11): 114501.

9.66. Miller, R. 2016. Designing very strong gear teeth by means of high pressure angles. *AGMA* Technical Paper 16FTM13.

9.67. Fuentes-Aznar, A. and Gonzalez-Perez, I. 2017. High pressure angle ears are not expected to show better mechanical behavior in high power transmissions that gears with conventional design. *AGMA Fall Technical Meeting*, Columbus, OH.

9.68. Ambekar, A. G. 2007. *Mechanism and Machine Theory.* Prentice-Hill of India Private Limited, New Delhi.

9.69. Stadtfeld, H. J. 2019. Relationship between misalignment and transmission error in cross-axes helical gear assembly. *Gear Technology*, April Issue, pp. 44–46.

9.70. Artoni, A., Gabiccini, M., and Guiggiani, M. 2014. Grinding face-hobbed hypoid gears through full exploitation of 6-axis hypoid generators. In *International Gear Conference 2014* (Ed.: P. Velex), Elsevier, pp. 107–117.

9.71. Gonzalez, C. 2016. What's the difference between spur, helical, bevel, and worm gears? *Machine Design.* Article 21832142. https://www.machinedesign.com/learning-resources/whats-the-difference-between/article/21832142/whats-the-difference-between-spur-helical-bevel-and-worm-gears.

9.72. Saari, O. E. 1954. Speed-reduction gearing. U.S. Patent 2,696,125.

9.73. Paul, D. 2013. Spiroid® and Helicon® gearing. In *Encyclopedia of Tribology* (Eds. Q. J. Wang and Y. W. Chung). Springer, Boston, MA.

9.74. Kazkaz, G. 2015. Mathematical modeling for the design of spiroid, helical, spiral bevel and worm gears. *Power Transmission Engineering*, April Issue. https://www.powertransmission.com/articles/0415/Mathermatical_Moderling_for_the_Design_of__Spiroid,_Helical,_Spiral_Bevel_and_Worm_Gears/.

9.75. Evertz, F., Gangireddy, M., Mork, B., Porter, T., and Quist, A. 2020. Spiroid high torque skew axis gearing: A technical primer. Spiroid Gearing White Paper. https://www.spiroidgearing.com/wp-05/Spiroid-Technical-Paper_Final.pdf.

9.76. Saari, O. E. 1960. Skew axis gearing. U.S. Patent 2,954,704.

9.77. Dengel, B. 2018. Worms and worm wheels – A primer. *Gear Solutions.* https://gearsolutions.com/departments/tooth-tips/worms-and-worm-wheels-a-primer/.

9.78. Summervill, R. 2016. Two-stage helical bevel gearing. *Power Transmission Engineering.* February Issue, pp. 24–27.

9.79. Landry, J. 2001. New rules for sizing servos: Throw out traditional guidelines when matching a servomotor and gearhead. *Machine Design.* Article 21814745. https://www.machinedesign.com/archive/article/21814745/new-rules-for-sizing-servos.

9.80. Regal Power Transmission Solutions. 2001. Gears and gear drives. *Motion System Design*, pp. A145–A160. https://www.regalpts.com/PowerTransmissionSolutions/ProductFundamentals/gears2.pdf.

9.81. Eitel, L. 2012. How to size and select a gearbox: A motion engineer's guide. *Motion Control Tips*. https://www.motioncontroltips.com/how-to-select-a-gearbox/.

9.82. Popp, C. 2001. Speed reducers. *Machine Design*. Article 21827591. https://www.machinedesign.com/automation-iiot/article/21827591/speed-reducers.

9.83. Stoeber, B. and Schumacher, J. 2000. Gear efficiency – Key to lower drive cost. *Machine Design*. Article 21834720. https://www.machinedesign.com/motors-drives/article/21834720/gear-efficiency-key-to-lower-drive-cost.

9.84. Mobley, R. K. 2001. Gears and gear drives. In *Plant Engineer's Handbook* (Ed.: R. K. Mobley). Elsevier, Amsterdam, pp. 1029–1042.

9.85. Multimedia, T. 2019. How to choose the right gearhead. *Sure Controls Inc*. https://www.surecontrols.com/how-to-choose-the-right-gearhead/.

9.86. Avalos-Vasquez, J. 2017. Six key elements of gearmotor optimization. *Power Transmission Engineering*, August Issue, pp.20–21.

9.87. Geitner, F. K. and Bloch, H. P. 2012. *Machinery Failure Analysis and Troubleshooting: Practical Machinery Management for Process Plants*, 4th edn. Butterworth-Heinemann, Oxford.

9.88. Ludwig Jr., L. and Beckman, M. 2021. Lubrication and its role in gear-failure analysis: It's important to employ best practices in order to identity gearset problems before they fail. *Gear Solutions*, March Issue, pp. 28–35.

9.89. Kren, L. 2007. Recognizing gear failures. *Machine Design*. Article ID 21816731. https://www.machinedesign.com/news/article/21816731/recognizing-gear-failures.

9.90. Kahraman, A. and Ding, H. L. 2013. Wear in gears. In *Encyclopedia of Tribology* (Eds.: Q. J. Wang and Y. W. Chung). Springer, Boston, MA.

9.91. Alojali, H. M. and Benyounis, K. Y. 2016. Advances in tool wear in turning process. In *Reference Module in Materials Science and Materials Engineering*. Elsevier, Amsterdam.

9.92. Whitby, R. D. 2022. *Lubricant Analysis and Condition Monitoring*. CRC Press, Boca Raton, FL.

9.93. Zhang, C., Wang, S. P., Wang, Z. M., and Wang, X. J. 2015. An accelerated life test model for harmonic drives under a segmental stress history and its parameter optimization. *Chinese Journal of Aeronautics* **28(6)**, 1758–1765.

9.94. U.S. Department of Energy. 1980. Classification and evaluation of electric motors and pumps. Report No. DOE/CS-0147.

10
Motor Power Losses

Motor power losses refer to the consumption of electrical energy not converted to useful mechanical energy. Motor power losses are important in motor design for a number of reasons: (*a*) Power losses in the electric motor can substantially increase the temperature rise in the windings and deteriorate the performance characteristics of the motor. The vast majority of power losses are converted into heat energy, which must be eventually dissipated from the motor to the surrounding environment. Consequently, it is highly desired to understand the mechanisms of various power losses for providing the best suitable solution for motor cooling. (*b*) The motor efficiency is defined as the ratio of the power output (which equals the power input minus the power losses) to the power input. Thus, the higher the power losses, the lower the motor efficiency. From an economic standpoint, higher power losses are always associated with increased motor costs. (*c*) The conversion of power losses into heat energy may cause an excessive internal temperature of motor. Thus, power losses have a strong impact on the selection of cooling modes that are required to keep the motor temperature below the maximum allowed value. (*d*) Temperature is a major cause of degradation of insulation materials. Excessive motor temperature can accelerate the aging of winding insulations and thus reduce their lifetime. (*e*) High temperature can significantly reduce magnetic properties of PMs to lower the performance of permanent magnet (PM) motors. As the temperature becomes higher than the Curie temperature of the magnet, all magnetic properties are eventually lost. (*f*) High temperature may cause bearing grease to deteriorate, leaving motor bearings without adequate lubrication. (*g*) A small portion of the total power loss may be converted to sound energy, leading to high acoustic noise emitted from the motor.

In recent years, with a continuous and ever-increasing demand for high-performance and high-efficiency electric motors, a great deal of effort has been directed toward determining power losses of rotating electric machines.

The motor power losses can be briefly defined as no-load and load losses. The major no-load losses, which do not require load currents, include (*a*) core losses, also called magnetic losses, because alternating magnetic flux produces both hysteresis losses and eddy-current losses in the stator and rotor cores, magnets, and other motor components and (*b*) mechanical losses that include friction and windage losses. The primary load losses are (*a*) resistive losses in the stator windings, (*b*) resistive losses in the rotor windings, and (*c*) stray load losses that consist of various kinds of losses that are not mentioned previously.

Benhaddadi *et al.* [10.1] have provided the estimation of the contribution of each type of loss to the total power loss. Their results show that the largest contribution comes from the stator winding (35%). The following are the rotor winding and magnetic losses; each accounts for 20%. The mechanical losses are the lowest, only accounting for 10%. GE [10.2] recently released the typical power loss distribution data for a four-pole motor. The largest contribution is found from stator windings, taking 33% of the total loss. This is consistent with the value of Benhaddadi as 35%. The mechanical (14%) and stray load losses (22%) are moderately higher than but the rotor windings (15%) and magnetic losses (16%) slightly lower than the data in [10.1]. The comparison of the contribution data from these two references is presented in Figure 10.1.

DOI: 10.1201/9781003097716-10

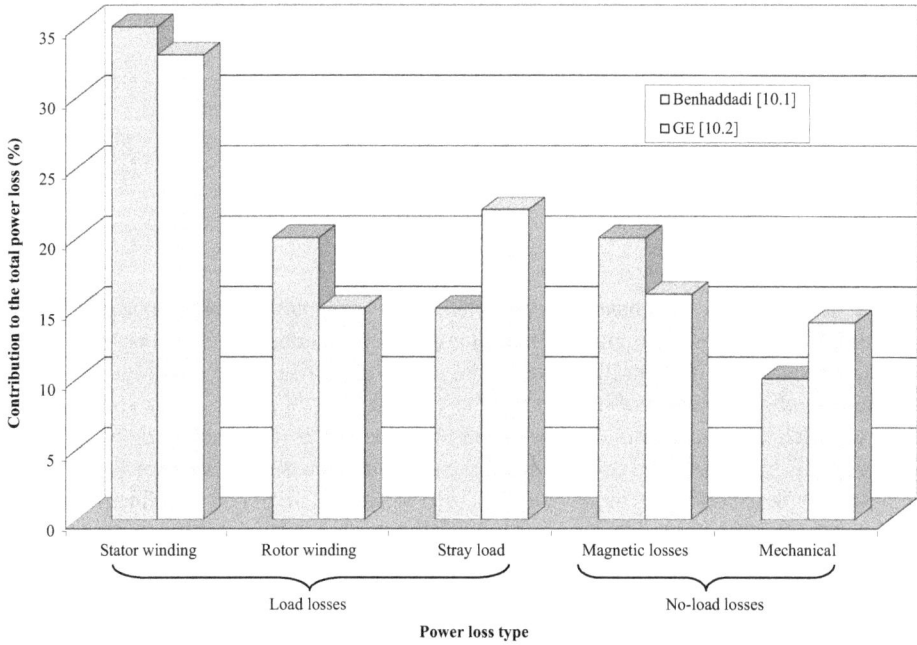

FIGURE 10.1 Comparison of power loss distributions in electric motors.

10.1 POWER LOSSES IN WINDINGS DUE TO ELECTRIC RESISTANCE IN COPPER WIRES

The most significant energy loss in a motor comes from electrical resistivity of the stator winding and either the rotor winding for a wound-rotor induction motor or the squirrel cage for a squirrel cage induction motor. For common conducting materials such as copper and aluminum, electric current is produced by the movement of electrons from one atom to another under the influence of an electric field. During this process, the migrations of free electrons cause many collisions with the captive electrons in the conducting materials. These collisions consume some energy of electrons as the basic cause of resistance.

In a DC motor, current flows uniformly across the entire cross section of the winding wires. The electrical resistance R_{DC} of a conductor is proportional to the resistivity of the material ρ (in $\Omega \cdot m$) and the conductor length l (in m) and inversely proportional to the cross-sectional area of the conductor A (in m^2), that is,

$$R_{DC} = \frac{\rho l}{A} \tag{10.1}$$

It must be noted that material resistivity is strongly influenced by temperature. A higher temperature typically results in more collisions between electrons, and hence, the higher resistance to electric current. For a common conducting material such as copper and aluminum, electric resistivity is approximately expressed as

$$\rho(T) = \rho_o(T_o)\left[1 + \alpha(T - T_o)\right] \tag{10.2}$$

where
  $\rho(T)$ is the electrical resistivity of material at temperature T, $\Omega \cdot m$
  $\rho_o(T_o)$ is the electrical resistivity of material at the reference temperature T_o, $\Omega \cdot m$

T_o is the reference temperature, $°C$

α is the temperature coefficient of resistance of material, $°C^{-1}$ or K^{-1}

Thus, the resistive winding loss can be simply calculated as

$$P_r = I^2 R_{DC} - \frac{V^2}{R_{DC}}$$ (10.3)

where

l is the electric current flowing through the winding

V is the voltage across the winding

If there is enough space in the winding slots, the resistive winding loss can be reduced by adopting a conductor with a larger diameter.

For an AC motor, Equation 10.3 can be expressed as

$$\overline{P_r} = i_{rms}^2 R_{AC} = \frac{v_{rms}^2}{R_{AC}}$$ (10.4)

where

P_r is the averaged power loss

R_{AC} is the AC electrical resistance

i_{rms} and v_{rms} are the RMS current and voltage, respectively

For AC with sine waves, i_{rms} and v_{rms} can be expressed as

$$i_{rms} = \frac{i_{peak}}{\sqrt{2}}; \quad v_{rms} = \frac{v_{peak}}{\sqrt{2}}$$ (10.5)

respectively In Equation 10.5, i_{peak} and v_{peak} are the peak value of AC and voltage, respectively.

The resistance loss in an AC motor increases with frequency due to the skin effect, which refers to the tendency of AC to distribute in the thin layer at the conductor's outer surface. The skin depth is defined as the distance below the conductor surface where the current density has decayed to e^{-1} ($e^{-1} = 0.368$) of its value at the surface (Figure 10.2). The skin depth δ (in m) varies as the square root of the electrical resistivity and the inverse square root of frequency and relative permeability:

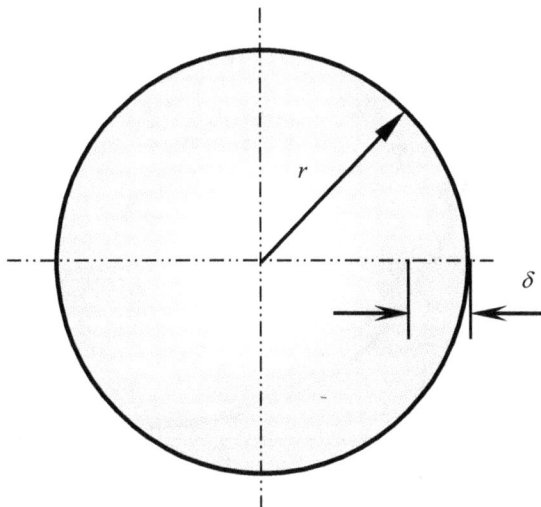

FIGURE 10.2 Skin depth of a conducting magnet wire, where $\delta < r$.

$$\delta = \sqrt{\frac{\rho}{\pi f \mu_r \mu_o}} \qquad (10.6)$$

where

f is the frequency, Hz

μ_r is the relative permeability

$\mu_o = 4\pi \times 10^{-7}$ is the permeability of free space, H/m

The variations of skin depth with frequency for different materials are displayed in Figure 10.3. It has shown that the skin depths for all materials reduce sharply for $f < 1\,\text{kHz}$. As f becomes larger than $100\,\text{kHz}$, the changes in the skin depth can be ignored. For good conducting materials, skin depth varies as the inverse square root of the material conductivity. This indicates that better conducting materials have a reduced skin depth.

The ratio of AC resistance to DC resistance can be simply determined as

$$\frac{R_{AC}}{R_{DC}} = \frac{1}{2\delta/r - (\delta/r)^2} \qquad (10.7)$$

In case that $\delta \ll r$, Equation 10.7 can be simplified as

$$\frac{R_{AC}}{R_{DC}} \approx \frac{r}{2\delta} \qquad (10.8)$$

This equation indicates that the smaller the skin depth, the higher the ratio of AC to DC resistance. The more advanced and complete expression of the resistance ratio was derived by Ramo *et al.* [10.3].

The skin effect appears as the skin depth δ is smaller than the wire radius r. In order to minimize the resistance loss for high-frequency applications, the copper windings are made with multiple strands, insulated from each other and in a weaving or twisting pattern (so-called *Litz wire*, see Figure 10.4). It was found that the slotless structure combined with a 75-strand *Litz wire* in a

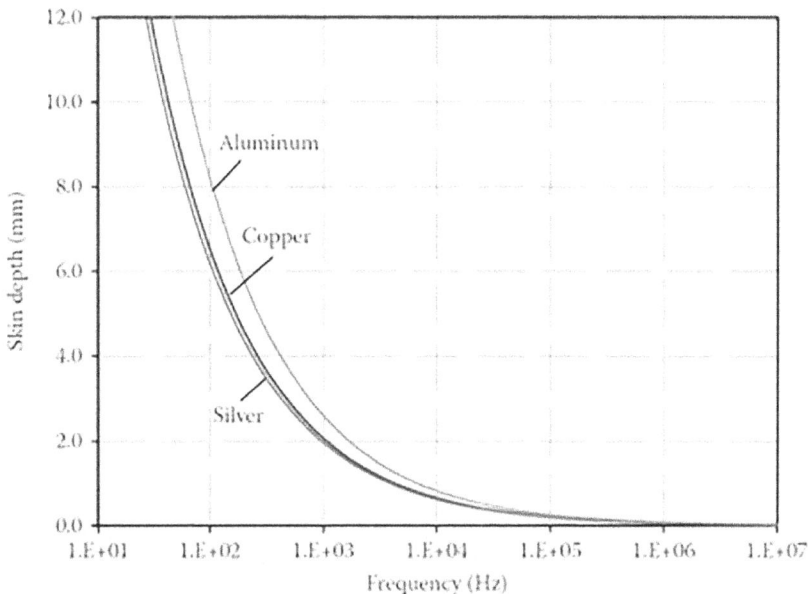

FIGURE 10.3 Variations of skin depth with frequency for different materials.

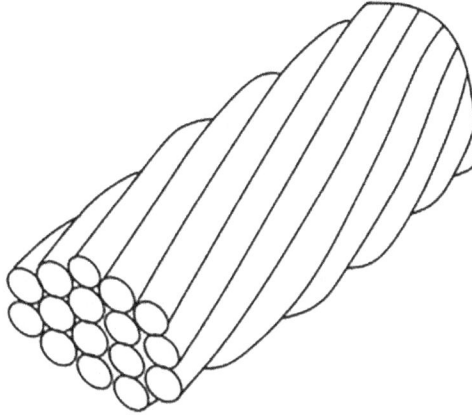

FIGURE 10.4 Litz wires used for motor windings for minimizing the skin effect.

superhigh-speed cryogenic PM motor can significantly reduce iron loss in the rotor without an increase in eddy-current loss in the winding [10.4].

The resistive winding loss is the loss that is proportional to the motor load. With the increase in the motor torque, the electric current through the motor windings increases so does the resistive winding loss. In most applications, the resistive winding loss is predominant among all losses, except motors having very high rotating speeds.

10.2 EDDY-CURRENT AND MAGNETIC HYSTERESIS LOSSES

10.2.1 Eddy-Current Loss

From Faraday's law of induction, when magnetic flux changes through a loop of area due to any reasons (the magnetic field change, the loop area change, or the orientation of the loop relative to the magnetic field change), an induced EMF (electromagnetic field or electromotive force) is developed along the loop. Thus, circulating currents, that is, eddy currents, are induced in bulk conductors moving through a magnetic field.

Eddy currents are often undesirable because they consume a considerable amount of energy to internal energy without making useful work. For electric motors, eddy-current loss is caused by local circulating currents induced in conductive core components and results in a rise in temperature.

Eddy currents are affected by the electrical resistance of ferromagnetic materials in which eddy currents flow. In order to reduce eddy-current loss, both stator and rotor cores are made of a stack of laminations, which are insulated by nonconducting materials such as lacquer or metal oxide. This layered structure prevents large current loops and effectively confines eddy currents to small loops in individual layers. The sum of the individual eddy current in each piece of the laminated core is much less than that in the solid iron core. Furthermore, the addition of the silicon element into the core steel increases electrical resistance, leading to a further decrease in eddy current.

10.2.2 Magnetic Hysteresis Loss

Magnetic hysteresis phenomena occur in ferromagnetic materials. As an external alternating magnetic field is applied to a ferromagnetic material, it forces the atomic dipoles in the material to align themselves with the magnetic field. When the magnetic field suddenly changes its orientation relatively with the material, the atomic dipoles must realign themselves to accommodate such change. Because the ferromagnetic material tends to retain some degree of magnetization, known as the magnetic hysteresis, it must take a certain amount of energy to overcome such a hysteresis to adjust the atomic dipoles according to the change of the external magnetic field. In fact, hysteresis losses depend on several factors including the power frequency, the peak flux density, the material of core steel, and the orientation of the magnetic flux with respect to the grain structure of the steel.

10.2.3 CALCULATIONS OF EDDY-CURRENT AND MAGNETIC HYSTERESIS LOSSES

Maxwell's equations, which perfectly describe classical electromagnetic phenomena, consist of four equations: Faraday's law of induction, Ampère–Maxwell's law, and Gauss's laws for describing electric and magnetic fields, that is,

Faraday's law of induction

$$\nabla \times \mathbf{E} = -\frac{\partial \mathbf{B}}{\partial t} \tag{10.9}$$

Ampère–Maxwell's law

$$\nabla \times \mathbf{H} = \frac{\partial \mathbf{D}}{\partial \mathbf{t}} + \mathbf{J} \tag{10.10}$$

Gauss's law for electric field

$$\nabla \cdot \mathbf{D} = \rho_e \tag{10.11}$$

Gauss's law for magnetic field

$$\nabla \cdot \mathbf{B} = 0 \tag{10.12}$$

In Maxwell's equations, $\mathbf{E}$ and $\mathbf{H}$ are the electric and magnetic field intensities, $\mathbf{D}$ and $\mathbf{B}$ are the electric and magnetic flux densities, respectively, $\mathbf{J}$ is the electric current density, and ρ_e is the electric charge density of any external charges.

The relationships between the current density $\mathbf{J}$ and the electric field intensity $\mathbf{E}$, electric flux density $\mathbf{D}$ and the electric field intensity $\mathbf{E}$, and the magnetic flux density $\mathbf{B}$ and magnetic field intensity $\mathbf{H}$ are given as

$$\mathbf{J} = \sigma \mathbf{E} \tag{10.13a}$$

$$\mathbf{D} = \varepsilon \mathbf{E} \tag{10.13b}$$

$$\mathbf{B} = \mu \mathbf{H} \tag{10.13c}$$

where
  σ is the electrical conductivity
  ε is the electric permittivity
  μ is the magnetic permeability

Using Equations 10.9 and 10.10, it can be derived that

$$\nabla \cdot (\mathbf{E} \times \mathbf{H}) = \mathbf{H} \cdot (\nabla \times \mathbf{E}) - \mathbf{E} \cdot (\nabla \times \mathbf{H}) = -\mathbf{H} \cdot \frac{\partial \mathbf{B}}{\partial t} - \mathbf{E} \cdot \left(\mathbf{J} + \frac{\partial \mathbf{D}}{\partial t} \right) \tag{10.14}$$

Integrating both sides over a closed volume V surrounded by the surface S, Equation 10.14 is expressed in the integrated form,

$$\int_V \nabla \cdot (\mathbf{E} \times \mathbf{H}) dV = - \int_V \left[\mathbf{H} \cdot \frac{\partial \mathbf{B}}{\partial t} + \mathbf{E} \cdot \left(\mathbf{J} + \frac{\partial \mathbf{D}}{\partial t} \right) \right] dV \tag{10.15}$$

Using the divergence theorem, it yields

$$\int_S (\mathbf{E} \times \mathbf{H}) d\mathbf{S} = - \int_V \left[\mathbf{H} \cdot \frac{\partial \mathbf{B}}{\partial t} + \mathbf{E} \cdot \left(\mathbf{J} + \frac{\partial \mathbf{D}}{\partial t} \right) \right] dV \tag{10.16}$$

The left side of Equation 10.16 represents the energy flow into the volume per unit time. The first and second terms at the right-hand side represent the energy stored and dissipated per unit time, respectively. Thus, the total energy loss in the volume over a complete cycle of time becomes

$$P = \int_V \int_t \left[\mathbf{H} \cdot \frac{\partial \mathbf{B}}{\partial t} + \mathbf{E} \cdot \left(\mathbf{J} + \frac{\partial \mathbf{D}}{\partial t} \right) \right] dt \, dV \tag{10.17}$$

Since the displacement current $\mathbf{D}$ is typically relevant only at radio frequencies (in the MHz regime), the term $\partial \mathbf{D}/\partial \mathbf{t}$ can be ignored in Equation 10.110.

Using Equation 10.13a, it gives that

$$\mathbf{E} \cdot \mathbf{J} = \frac{J^2}{\sigma} \tag{10.18}$$

Equation 10.17 can be rewritten as

$$P = \int_V \left(\int_B H dB \right) dV + \int_V \left[\int_t \left(\frac{J^2}{\sigma} \right) dt \right] dV = P_h + P_e \tag{10.19}$$

where P_h and P_e are the power losses due to hysteresis and eddy current, respectively.

Using a lumped circuit approach and assumed eddy-current paths, eddy-current loss of magnetic core P_e (in units of W) can be derived as [10.5]

$$P_e = \frac{\pi^2}{6} V_c B^2 f^2 a^2 \sigma = K_e V_c B^2 f^2 \tag{10.20}$$

$$K_e = \frac{\pi^2 a^2 \sigma}{6} \tag{10.21}$$

where
V_c is the volume of magnetic core, m³
a is the lamination thickness, m
B is the peak flux density, T

Equation 10.20 has shown that eddy-current loss is proportional to the square of lamination thickness a. This indicates that eddy-current loss can be greatly reduced if thinner laminations are adopted. For instance, by reducing the lamination thickness by 20%, eddy-current loss can be reduced by 36%.

Some textbooks use eddy-current loss per unit volume p_e, in units of W/m³.

$$p_e = K_e B^2 f^2 \tag{10.22}$$

Hysteresis refers to the phenomenon in which the magnetic induction of a ferromagnetic material lags behind the changing magnetic field. When a ferromagnetic material is subjected to a magnetic field, the magnetic particles in the material tend to line up with the magnetic field. As the magneticfield keeps changing its direction, the magnetic particles try to align themselves with the magnetic field. The continuous movement of the magnetic particles thus produces molecular friction, resulting in energy loss.

Hysteresis loss is strongly affected by material electrical and electromagnetic properties. The empirical formula expressing hysteresis loss P_h was developed by Steinmetz and other researchers [10.6–10.8] as

$$P_h = K_h V_c B^n f \tag{10.23}$$

where
K_h is the hysteresis coefficient
n is the Steinmetz coefficient, which has a value between 1.6 and 2.3 for most modern magnetic materials (most commonly around 2)
f is the frequency of magnetization

The hysteresis coefficients of some materials are listed in Table 10.1 [10.9].

From Table 10.1, it can be seen that permalloy has the lowest hysteresis coefficient, and, thus, it has been extensively used in transformer laminations and magnetic recording heads with medium or high frequencies. Silicon steel is the material having the second lowest hysteresis coefficient. As a result, hysteresis loss can be mitigated by making the motor cores with silicon steel. In practice, industrial applications of silicon steel vary in quantities from the few ounces used in small relays or pulse transformers to tons used in generators, motors, and transformers.

It is apparent that eddy-current loss P_e increases with the square of the frequency ($P_e \propto f^2$), and hysteresis loss P_h increases linearly with the frequency ($P_h \propto f$). This indicates that for high frequencies, eddy-current loss P_e becomes dominant among the total energy loss P.

Magnetic hysteresis and eddy-current losses are sometimes collectively known as iron losses or core losses. This type of loss is a significant fraction of total losses of electric motors. With advanced finite element analysis (FEA) tools, the contributions of the hysteresis and eddy-current losses in motor components can be determined fairly well.

TABLE 10.1
Hysteresis Coefficients of Some Metallic Materials

Material	Hysteresis Coefficient (J/m³) $\times 10^{-2}$
Cast iron	210.63–40.2
Sheet iron	10.05
Cast steel	10.54–30.14
Hard cast steel	63–70.34
Silicon steel (4.8% in Si)	1.91
Hard tungsten steel	145.7
Good dynamo sheet steel	5.02
Mild steel castings	10.54–22.61
Nickel	32.66–100.5
Permalloy	0.25

10.2.4 Losses in Stator and Rotor Iron Cores

The power losses in the stator and rotor cores consist of both hysteresis and eddy-current losses. In electric motors, the core or iron losses usually take about 20%–25% of the total power losses [10.10]. For PM motors, interior PM machines have significantly higher full-load iron losses than surface-mounted PM (SPM) machines.

By dividing eddy current into classical and excess eddy currents for more accurate analysis, power losses in stator and rotor iron cores can be determined as [10.11, 10.12].

$$P_{core} = k_h B^2 f + K_c \left(Bf\right)^2 + K_e \left(Bf\right)^{3/2} \tag{10.24}$$

where

K_h, K_c, and K_e are the coefficients of hysteresis loss, classical eddy-current loss, and excess eddy-current loss, respectively

F is the frequency of sinusoidal field excitation, Hz

B is the peak flux density, T

The coefficients K_h, K_c, and K_e can be obtained using the curve fitting of the iron loss data from manufacturers. For instance, K_h can be derived from a sinusoidal flux variation over time and is commonly reported in the lamination data sheet.

There are a number of ways to reduce the core losses: (a) Utilizing a *soft* magnetic material. This is because the hysteresis loss is proportional to the shaded area enclosed by the hysteresis loop on the *B–H* curve. A *soft* magnetic material has a smaller shaded area within the hysteresis loop than that of a *hard* magnetic material (see Figure 10.5). (b) Using silicon grain-oriented steel. (c) Using thinner steel laminations. (d) Designing longer cores to reduce magnetic flux density.

Figure 10.6 compares the hysteresis and eddy-current losses of the rotor in a small PM motor. Both the eddy-current and hysteresis losses are functions of the rotor rotating speed (or equivalently frequency). While the eddy-current loss shows a parabolic behavior, the hysteresis loss follows a power law behavior as expected. Therefore, as the rotor speed continues to increase, the growth rate of hysteresis loss far exceeds that of eddy-current loss. It shows that at 10,000 rpm, the eddy-current loss is about 50 W, and the hysteresis loss reaches 135 W, approximately 2.7 times higher than the

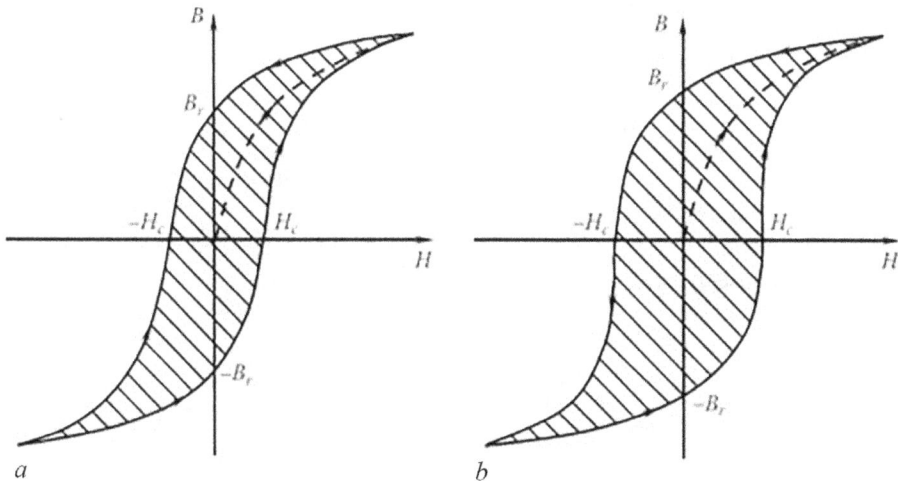

FIGURE 10.5 Magnetization hysteresis loops for different ferromagnetic materials: (a) *soft* magnetic materials with low hysteresis loss, which are desirable for motor cores to minimize power losses and (b) *hard* magnetic materials with high hysteresis loss.

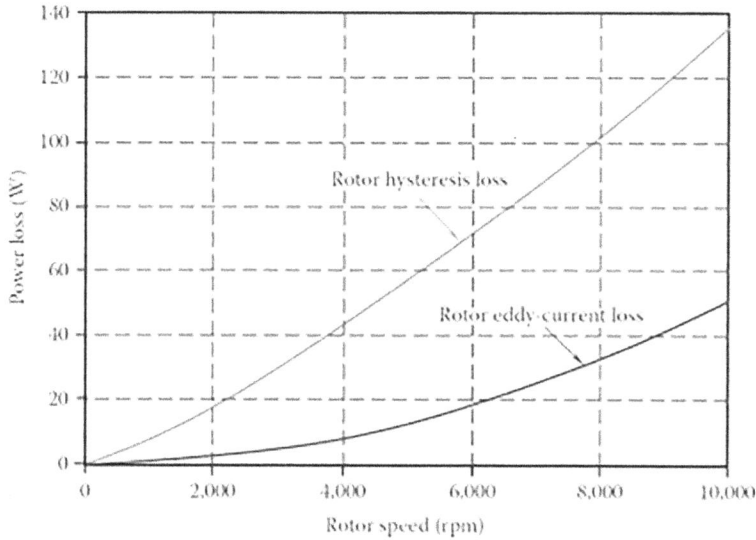

FIGURE 10.6 Rotor hysteresis and eddy-current losses as functions of the rotor rotating speed for a sampler brushless PM motor.

eddy-current loss. This indicates that for extra-high-speed motors, the hysteresis loss dominates the rotor loss. In such cases, the main design focus should be on how to reduce the hysteresis loss. Because hysteresis loss is caused by the magnetization and demagnetization of the lamination core as current flows in the forward and reverse directions, the most effective way for reducing core hysteresis loss is to select the electrical steel with a *narrow* magnetization hysteresis loop (*i.e.*, low coercive force), as shown in Figure 10.5a.

10.2.5 Losses in PMs

PM motors are extensively used in many applications for their excellent characteristics such as high power density, high efficiency, and good dynamic performance. However, eddy-current loss in PMs is often overlooked. For high-speed PM motors, eddy-current loss in PMs cannot be ignored. Eddy-current loss can be estimated as [10.13].

$$P_m \approx \frac{V_m b_m^2 B^2 f^2}{12 \rho_m} \tag{10.25}$$

where
    V_m is the magnet volume
    b_m is the magnet width
    ρ_m is the magnet resistivity ($\rho = 1/\sigma$)

Correspondingly, eddy-current loss becomes

$$P_m = K_m V_m B^2 f^2 \tag{10.26}$$

where

$$K_m = \frac{b_m^2}{12 \rho_m} \tag{10.27}$$

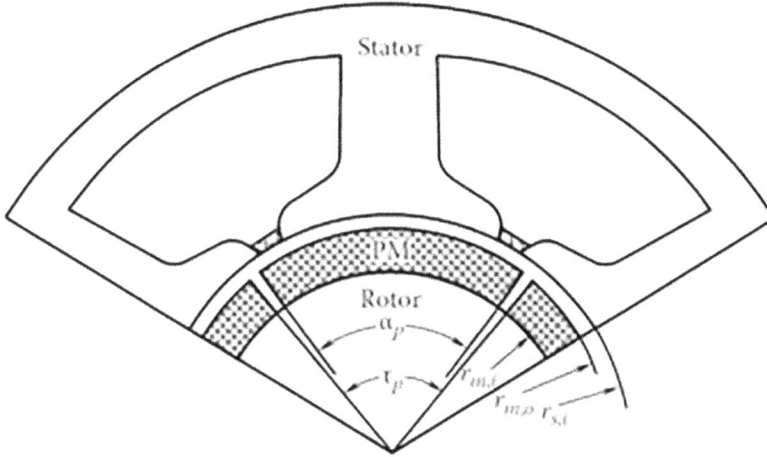

FIGURE 10.7 The model for predicting eddy-current loss in surface-mounted PMs. The motor has a fractional number of slots per pole.

An analytical model for predicting eddy-current losses in SPMs of PM motors was presented by Ishak *et al.* [10.14]. As shown in Figure 10.7, PMs are attached on the surface of the rotor that rotates at the speed of ω_r. The magnet has the inner radius $r_{m,i}$ and outer radius $r_{m,o}$. The inner stator bore radius is $r_{s,i}$. Thus, the airgap is given as the difference between $r_{s,i}$ and $r_{m,o}$. The magnet central angle corresponding to the magnet width denotes α_p.

For a PM motor with a slot number N_{slot} and pole number $2p_r$ (where p_r is the number of pole pairs), the number of slots per pole gives

$$n_{slot} = \frac{N_{slot}}{2p_r} \tag{10.28}$$

When n_{slot} is a noninteger number, the eddy-current loss per unit axial length in each magnet can be predicted as

$$P_m = 2p_r \frac{\omega_r}{2\pi} \int_0^{2\pi/\omega_r} \int_{r_{m,i}}^{r_{m,o}} \int_{-(\alpha_p/2)}^{\alpha_p/2} \rho_m J_m^2 r\, dr\, d\theta\, dt = \sum_u \sum_v (P_{cuv} + P_{auv}) \tag{10.29}$$

where
u is the time-harmonic order in the phase current waveform
V is the space-harmonic MMF order
ρ_m is the electrical resistivity of the magnets

The variables P_{cuv} and P_{auv} are given by

$$P_{cuv} = \frac{\mu_0^2 \alpha_p}{\rho_m} p_r \sum_u^\infty \sum_u^\infty \frac{J_{uv}^2 \left(up_r \pm up_s\right)^2 \omega_r^2}{\left[1 - \left(r_{m,i}/r_{s,i}\right)^{2vp_s}\right]^2 (vp_s)^2}$$

$$\times \left\{ \left(\frac{r_{m,o}}{r_{s,i}}\right)^{2vp_s} \frac{r_{s,i}^2 r_{m,o}^2}{2vp_s + 2} \left[1 - \left(\frac{r_{m,i}}{r_{m,o}}\right)^{2vp_s + 2}\right] + \left(\frac{r_{m,o}}{r_{s,i}}\right)^{2vp_s} r_{s,i}^2 \left(r_{m,o}^2 - r_{m,i}^2\right) + \left(\frac{r_{m,i}}{r_{s,i}}\right)^{2vp_s} r_{s,i}^2 r_{m,i}^2 F_v \right\}$$

$$\tag{10.30}$$

$$p_{auv} = \left[-\frac{8\mu_0^2}{\alpha_p \rho_m} p_r \sum_u^\infty \sum_u^\infty \frac{J_{uv}^2 H_v^2}{\left[1 - \left(r_{mi}/r_{s,i}\right)^{2vp_s}\right]^2 (2vp_s)^4} \frac{\left(vp_r \pm vp_s\right)^2 \omega_r^2}{\left(r_{m,o}^2 - r_{m,i}^2\right)} \sin^2\left(2vp_s \frac{\alpha_p}{2}\right) \right] \qquad (10.31)$$

where p_s is the number of pole pairs in the stator winding MMF. In the above equations,

$$F_v = \begin{cases} \dfrac{\left(\dfrac{r_{m,o}}{r_{m,i}}\right)^{-2vp_s+2} - 1}{-2vp_s + 2} & \text{for } vp_s \neq 1 \\[2em] \ln\left(\dfrac{r_{m,o}}{r_{m,i}}\right) & \text{for } vp_s = 1 \end{cases} \qquad (10.32)$$

$$H_v = \left(\frac{r_{m,o}}{r_{s,i}}\right)^{vp_s} \frac{r_{s,i} r_{m,o}^2}{vp_s + 2} \left[1 - \left(\frac{r_{m,i}}{r_{m,o}}\right)^{vp_s+2}\right] + \left(\frac{r_{m,i}}{r_{s,i}}\right)^{vp_s} r_{s,i} r_{m,i}^2 E_v \qquad (10.33)$$

where

$$E_v = \begin{cases} \dfrac{\left(\dfrac{r_{m,o}}{r_{m,i}}\right)^{-vp_s+2}}{-vp_s + 2} & \text{for } vp_s \neq 2 \\[2em] \ln\left(\dfrac{r_{m,o}}{r_{m,i}}\right) & \text{for } vp_s = 2 \end{cases} \qquad (10.34)$$

This model has been validated by FEA.

10.2.6 POWER LOSSES IN OTHER CORE COMPONENTS

Eddy-current and hysteresis losses may occur in other core components such as stator end plates. In most cases, the contributions of the losses from these components to the total system power loss are relatively small and thus can be ignored in design.

10.3 MECHANICAL FRICTION LOSSES

Mechanical losses in an electric motor refer to the losses due to mechanical friction between two or among more motor components as they contact each other and have relative movements.

Whenever there is relative motion of two surfaces in contact, there is frictional resistance between two surfaces. The friction force opposes the relative motion of two bodies. The friction losses are independent of load and dependent on the speed of the machine. For the electric rotating machinery, a minimum of friction is highly desired.

10.3.1 BEARING LOSSES

Bearings are key components for rotating machinery to support rotating components with as little friction as possible. In normal operation, bearing rolling elements (balls or rollers) spin and slide between the inner and outer raceways (Figure 10.8). Therefore, the frictional resistance is produced between the rolling elements and their contact components such as the raceways, the cage, the

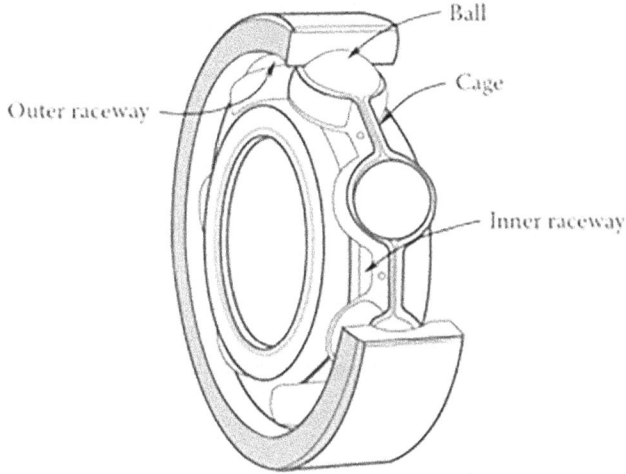

FIGURE 10.8 Ball bearing structure.

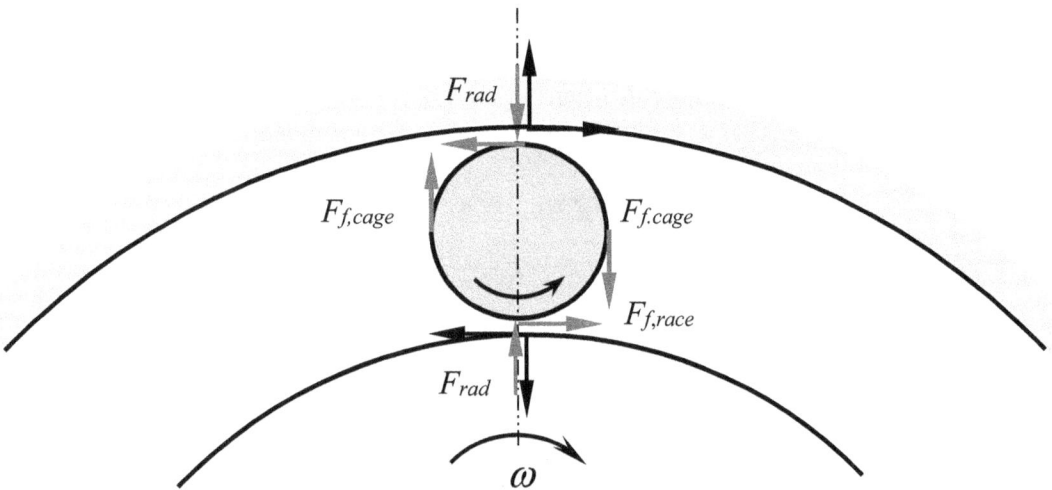

FIGURE 10.9 Bearing losses due to the friction between balls and raceways, balls and the cage, and balls and the lubricant.

rubbing seal (if present), and the lubricant (Figure 10.9). It has been estimated that among the total power losses of motor-driven equipment (e.g., pumps, fans, and compressors), about 20% of the loss is due to various mechanical losses, and of this, roughly 20% is attributed to bearings. This indicates that the bearing loss counts for about 4% of the total power losses [10.15].

There are a number of factors impacting the resistance to rotation of the bearing, for instance, the bearing load, the shaft rotating speed, the bearing dimensions, the roughness of the rolling elements and raceways, the lubrication condition, and the lubricant properties.

As an important performance index, the friction torque, defined as a resistance torque when a rolling bearing rotates, has a strong impact on the bearing's working condition and lifetime. The early approach to calculate the friction torque was proposed by Palmgren [10.16]. More detailed calculations of rolling bearing friction torque, including the torque due to applied load, lubricant viscous friction, and roller end-ring flange siding friction, are addressed by Harris and Kotzalas [10.17].

In this approach, the total friction torque is the sum of the load friction torque T_l and the viscous friction torque T_v, that is,

$$T = T_l + T_v \tag{10.35}$$

The load friction torque is determined empirically as

$$T_l = f_1 F_\beta \, D_m \tag{10.36}$$

where
    F_β depends on the magnitude and direction of the applied load
    D_m is the bearing mean diameter, $D_m = (D_i + D_o)/2$
    f_1 is a factor depending on the bearing type and bearing load expressed as

$$f_1 = z \left(\frac{F_s}{C_s} \right)^y \tag{10.37}$$

where
    F_s is the static equivalent load
    C_s is the basic static load rating
    z and y are constants, depending on the bearing type

For radial deep-groove ball bearings, $z = 0.0004$–0.0006 and $y = 0.55$.
    For bearings that operate at moderate speeds, the viscous friction torque can be estimated as

$$T_v = 10^{-7} f_o \left(v_o n \right)^{2/3} D_m^3 \qquad v_o n \geq 2,000 \tag{10.38}$$

$$T_v = 160 \times 10^7 f_o D_m^3 \qquad v_o n < 2,000 \tag{10.39}$$

where
    $v_o n$ is the kinematic viscosity of lubricant in centistokes
    n is the bearing rotating speed in rpm
    f_o is a factor depending on the bearing type and the lubrication method

In recent years, based on more advanced computational models, the new approach has been developed by SKF to calculate the total frictional moment (in units of N-mm) with the identified four sources of friction in every contact occurring in the bearing [10.18]:

$$M = \phi_{ish} \phi_{rs} M_{rr} + M_{sl} + M_{seal} + M_{drag} \tag{10.40}$$

where
    ϕ_{ish} is the inlet shear heating reduction factor
    ϕ_{rs} is the kinematic replenishment/starvation reduction factor
    M_{rr} is the rolling frictional moment
    M_{sl} is the sliding frictional moment
    M_{seal} is the frictional moment of seal
    M_{drag} is the frictional moment of drag losses, churning, splashing, *etc.*

The reduction factors ϕ_{ish} and ϕ_{rs} are introduced in the new friction model to account for the effects of inlet shear heating reduction and high-speed replenishment/starvation and of rolling friction, respectively.

The rolling frictional moment M_{rr} can be estimated as

$$M_{rr} = G_{rr} \, (\upsilon n)^{0.6} \tag{10.41}$$

where

n is the bearing rotating speed in rpm

υ is the kinematic viscosity of the lubricant at the operating temperature

G_{rr} depends on the bearing type, the bearing mean diameter D_m [where $D_m = (D_i + D_o)/2$], the radial and axial loads, F_r and F_a, respectively.

The sliding frictional moment M_{sl} is given as

$$M_{sl} = \mu_{sl} G_{sl} \tag{10.42}$$

where

μ_{sl} is the sliding friction coefficient (see Table 10.2)

G_{sl} depends on the bearing type, the bearing mean diameter D_m, and the radial and axial loads, F_r and F_a, respectively

The bearing seal frictional moment M_{seal} can be estimated as

$$M_{seal} = K_1 D_s^\alpha + K_2 \tag{10.43}$$

where

K_1 is a constant depending on the bearing type

K_2 is a constant depending on bearing and seal type

D_s is the seal counterface diameter

α is an exponent depending on bearing and seal type

The friction torque of drag losses T_{drag} are

$$M_{drag} = V_M K_{ball} D_m^5 n^2 \quad \text{for ball bearing} \tag{10.44a}$$

$$M_{drag} = V_M K_{ball} D_m^5 n^2 \quad \text{for roll bearing} \tag{10.44b}$$

where

V_M is a function of the oil level

w is the bearing width

K_{ball} and are the ball- and roller-related constants, respectively,

TABLE 10.2
Sliding Friction Coefficient μ_{sl}

Bearing Type	Lubrication	μ_{sl}
Ball	Mineral oils	0.05
	Synthetic oils	0.04
	Transmission fluids	0.1
Cylindrical roller	–	0.02
Tapped roller	–	0.002

$$K_{ball} = \frac{i_{rw}K_z(D_o + D_i)}{D_o - D_i} \times 10^{-12} \tag{10.45}$$

$$K_{roll} = \frac{K_L K_z(D_o + D_i)}{D_o - D_i} \times 10^{-12} \tag{10.46}$$

where

i_{rw} is the number of ball rows

K_z and K_L are the bearing-type and roller-bearing-type related geometry constants, respectively

The total power loss P_b (in W) of a bearing as a result of bearing friction can be obtained using the empirical formula [10.19]

$$P_b = (1.05 \times 10^{-4})Mn \tag{10.47}$$

where

n is the bearing rotating speed, rpm

M is the total frictional moment of the bearing, Nmm

10.3.2 Sealing Losses

Some electric motors work under harsh environmental conditions. In order to ensure motor's normal operation, mechanical seals are installed on motor shafts to prevent foreign contaminants such as dust, dirt, and moisture entering from the outside. When a shaft rotates, the seal tip of a contacting seal exerts a certain radial load on the shaft surface, resulting in friction losses at the sealing locations.

The friction force on the shaft is obtained by integrating the shear stress τ over the shaft surface:

$$F_f = R\int_0^L \int_0^{2\pi} \tau \, d\theta \, dz \tag{10.48}$$

Thus, the sealing loss becomes

$$P_{seal} = 2\pi R F_f \mu_d n = 2\pi R^2 \mu_d n \int_0^{W_c} \int_0^{2\pi} \tau \, d\theta \, dz \tag{10.49}$$

where

R is the shaft radius, mm or m

μ_d is the dynamic coefficient of friction between seal and shaft

n is the shaft rotating speed, rpm

W_c is the effective contact width on the shaft, mm or m

A simple equation for estimating the frictional losses due to shaft sealing is given as [10.20]:

$$P_{seal} = \frac{\pi^2 R^2}{15} n W_c \mu_d \bar{p}_c \tag{10.50}$$

where $\bar{p}_c$ is the average contact pressure acting on the shaft.

10.3.3 BRUSH LOSSES

In a brushed motor such as an induction DC motor, brushes are in contact with the commutator to carry current to coils. The commutator is the rotary electrical switch, which reverses the direction of the electric current periodically between the rotor and the external circuit. In general, brush losses consist of two parts. One is the electric resistance, consisting of the contact resistance occurring at the brush–commutator interface, and the electric resistance through the brushes. The interfacial electrical contact resistance R_i depends on the brush material, contact pressure, contact surface area, surface roughness, and contamination condition. It is the contact resistance R_i that causes a voltage drop ΔV at the brush–commutator interface. As reported by Hamdi [10.8], ΔV mainly depends on the brush material; for instance, $\Delta V = 0.4$–0.7 V for metal graphite, 0.7–0.8 V for electro-graphite, 0.7–1.2 V for natural graphite, and 0.7–1.8 V for hard carbon. The brush contact power loss $I^2 R_i$ can be thus alternatively determined as $\Delta V^2/R_i$. The $I^2 R$ power loss depends on the brush material and brush temperature. Another part of brush losses is the mechanical friction resistance due to the sliding contact of the brushes on the commutator, causing the friction loss. This type of power loss is proportional to the brush contact pressure p_c, kinetic friction coefficient between the brush and commutator μ_f, contact area A_c, and commutator tangential speed $r\omega$, i.e.,

$$P_b = \mu_f p_c A_c (r\omega) \tag{10.50}$$

where ω is the commutator rotating speed in units of rad/s.

The combination of these two effects results in heating at the brush contact interface, as well as the brush itself. Brush wear has resulted from both the mechanical friction and electrical erosion.

10.4 WINDAGE LOSSES

Windage in a rotating electric machine is defined as the resisting influence of fluid (air or liquid) against not only rotating but also stationary components, creating power losses. In an electric motor, the windage power loss P_w increases exponentially as a function of the rotor rotating speed ω and the rotor radius r_r. In practice, windage losses of electric motors are very important in the motor design and optimization, primarily consisting of the following components:

- The friction loss due to the viscous shear effect in the fully developed turbulent flow. It was reported that such a loss could be up to 30% of the total windage loss.
- The dynamic loss (or rotor body loss in some references) due to driving air in the airgap. The dynamic loss depends strongly upon the thermophysical properties of air such as density and viscosity and, in turn, the air temperature in the rotor–stator airgap.
- The loss associated with the stator slots with unfilled openings. The penetration of the airflow into the slot openings and the formation of the flow recirculations in such slot openings can have a complicated influence on energy loss.
- The loss associated with the roughness on rotor surfaces.
- The windage loss of some motor components such as fans.
- The ventilation path loss.

In practical engineering design, it is highly desired to minimize these losses to enhance the machine performance and efficiency.

Windage losses, also known as drag losses, can be very significant for high-speed motors due to the high windage friction. There are three components of velocity in the airgap of electric motors: (a) tangential flow due to the rotor rotation, (b) axial cooling flow passing through the airgap, and (c) Taylor vortices due to centrifugal forces. It is the reaction among these velocity components that determines the complex velocity field in the airgap. The importance of each velocity component depends on rotor rotating speed, cooling airflow rate, thermophysical properties of coolant (particularly density, specific heat, and thermal conductivity), rotor and stator geometries, and airgap dimension. The velocity and pressure fields at the end-winding regions are extremely complex because of a number of factors: a large number of components, windings with irregular geometries and existing voids inside the windings, complexity in cooling flow paths, splitting and mixing of coolant, and interaction between rotor-induced flows and nonrotating flows from stationary sections [10.21]. Modern motors are often designed to small-volume, high-speed machines to increase power design and reduce overall weight. These high-speed machines may lead to a small airgap between the rotor and stator. All these factors can result in high windage loss.

Windage losses vary on a case-by-case basis, depending largely on machine type, size, rated power, coolant type, cooling methods, and many others. For a large-size, high-power electric machine, the overall windage loss accounts for 15.1% to the total power loss of the machine (Figure 10.10). However, for some improperly designed machines, the contribution of windage losses to the total power loss can be very large. As shown in Figure 10.11, the overall windage loss accounts for more than one-third, up to 37.6%. The comparison of power losses of these two machines is presented in Table 10.3.

Because of the complicated nature of the flow field inside a motor, it becomes more popular today to use computational fluid dynamics (CFD) tools to predict and analyze the windage losses of the electric motor at the early stage of the design. In-depth understanding of windage loss helps design high-efficiency motors, leading to more optimized motor structure and configuration. Experimental testing is also needed to calibrate the CFD model and verify CFD results.

10.4.1 WINDAGE LOSS DUE TO ROTATING ROTOR

In a high-speed rotating system, windage loss is associated with pumping forces of the rotor in imparting energy to air at the rotor–stator annular gap. The magnitude of windage loss in electric motors varies widely from motor type, rotor geometry, rotor surface speed, airgap depth, fluid properties, and rotor and stator surface finishes. In practice, windage loss due to the rotating rotor accounts for a large portion of the overall windage loss in the machine, and thus, it must be addressed carefully.

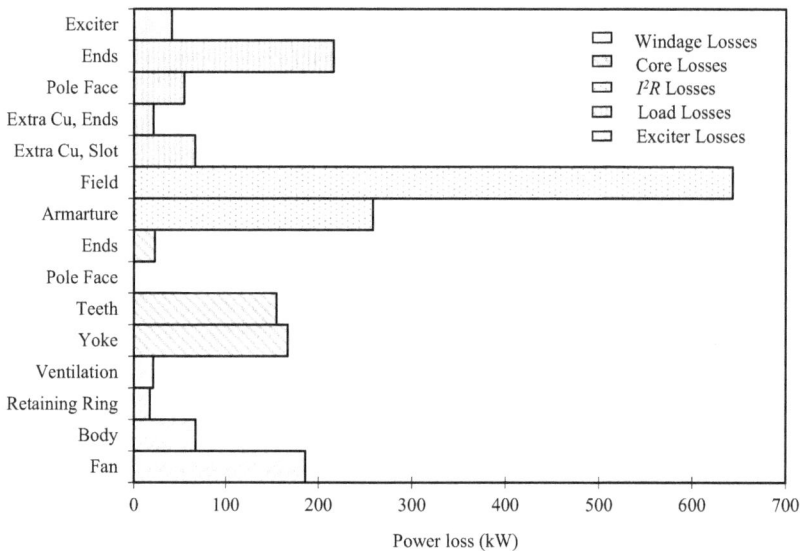

FIGURE 10.10   Various power losses for a high-power electric machine (machine *A*).

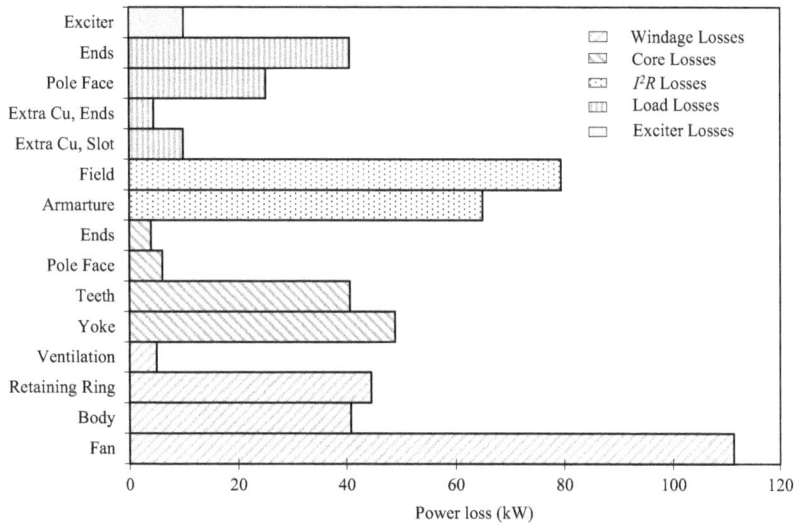

FIGURE 10.11 Various power losses for a high-power electric machine (machine *B*).

TABLE 10.3

Comparison of Power Losses in Two Large-Size, High-Power Electric Machines

Component Loss	Machine 1		Machine 1	
	Power Loss (kW)		Power Loss (kW)	
Fan	186 (9.6%)	Windage	111.3 (20.8%)	Windage
Rotor body	610.2 (3.5%)	291.7 (15.1%)	40.8 (10.6%)	201.5 (310.6%)
Retaining ring	110.5 (0.9%)		44.5 (8.3%)	
Ventilation	21 (1.1%)		4.9 (0.9%)	
Yoke	1610.5 (8.7%)	Iron (core)	48.9 (9.1%)	Iron (core)
Teeth	155.3 (8.0%)	345.4 (110.8%)	40.6 (10.6%)	99.6 (18.6%)
Pole face	0 (0%)		6.1 (1.1%)	
Ends	22.6 (1.2%)		4 (0.8%)	
Armature	258 (13.3%)	Copper (*PR*)	65 (12.1%)	144.4
Field	643 (33.2%)	901 (46.5%)	79.4 (14.8%)	(210.0%)
Extra Cu, slot	66.6 (3.4%)	Load	9.9 (1.9%)	Load
Extra Cu, ends	21 (1.1%)	358 (18.5%)	4.5 (0.8%)	80.1 (15.0%)
Pole face	54.7 (2.8%)		25.1 (4.7%)	
Ends	215.7 (11.1%)		40.6 (10.6%)	
Exciter	41 (2.1%)	Exciter (2.1%)	10 (1.9%)	Exciter (1.9%)
Total	19310.1		535.6	

10.4.1.1 Taylor Vortex

In fluid dynamics, Taylor–Couette flow refers to the flow between concentric cylinders with different rotating speeds. Couette [10.22], Mallock [10.23], and Taylor [10.24–10.26] were among the pioneers who studied this type of flow both theoretically and experimentally. At low rotating speeds, the flow is steady and purely azimuthal, which is known as circular Couette flow. In this situation, the centrifugal force due to the azimuthal velocity is balanced by the radial pressure gradient force. Taylor found that the angular velocity of the inner cylinder exceeds a critical value, a steady distinctive vortex structure is developed, as demonstrated in Figure 10.12. These vortices, named Taylor

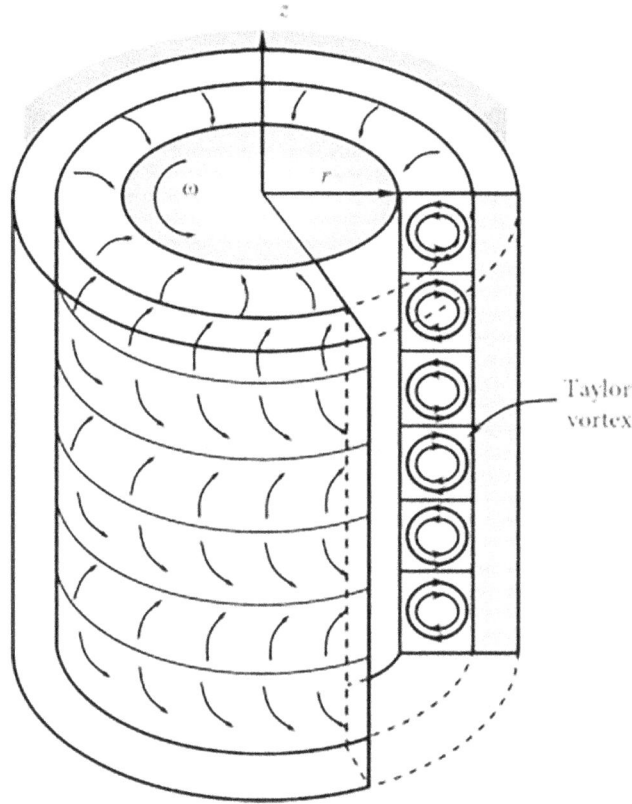

FIGURE 10.12 Flow configuration of Taylor vortices in the annular gap between two coaxial cylinders, with inner cylinder rotating and outer cylinder at rest.

vortices, are generated due to the centrifugal instability in the shear flow. Because of the vortices, high-speed fluid near the rotating inner cylinder is carried outward in the outflow regions between vortices, while low-speed fluid near the fixed outer cylinder is carried inward in the inflow regions between vortices, redistributing angular momentum of the fluid in the annulus [10.27]. The axial and radial velocities related to the Taylor vortices are relatively small, typically only a few percent of the surface speed of the inner cylinder [10.28].

Continuously increasing the rotational speed of the inner cylinder can result in wave vortex flow, which is characterized by azimuthal waviness of the vortices. The waves travel around the annulus at a speed that is 30%–50% of the surface speed of the inner cylinder, depending on the Taylor number and other conditions [10.29]. At higher Taylor numbers, the vortices become axisymmetric, but the flow is turbulent at small scales. When the Taylor number becomes high enough, the turbulent vortices disappear, and the flow is fully turbulent.

The flow instabilities that arise in Taylor–Couette flow and the related theoretical framework to describe these instabilities have provided valuable insight into the fundamental physical phenomena and analytic methods. Today, Taylor–Couette flow, as well as Plane Couette flow, has become the paradigm of hydrodynamic stability theory.

For a special case that the inner cylinder rotates and the outer cylinder is in rest, following Taylor's pioneering studies, considerable research on the Taylor vortex flow has been devoted to the investigations for a better understanding of the flow phenomena and energy losses between concentric rotating cylinders [10.30–10.36]. Ustimenlo and Zmeikov [10.34] experimentally investigated the hydrodynamics of a flow in an annular channel with an inner rotating cylinder. Based on their work, the comprehensive data of shear stress, velocity, total and static pressure were presented and

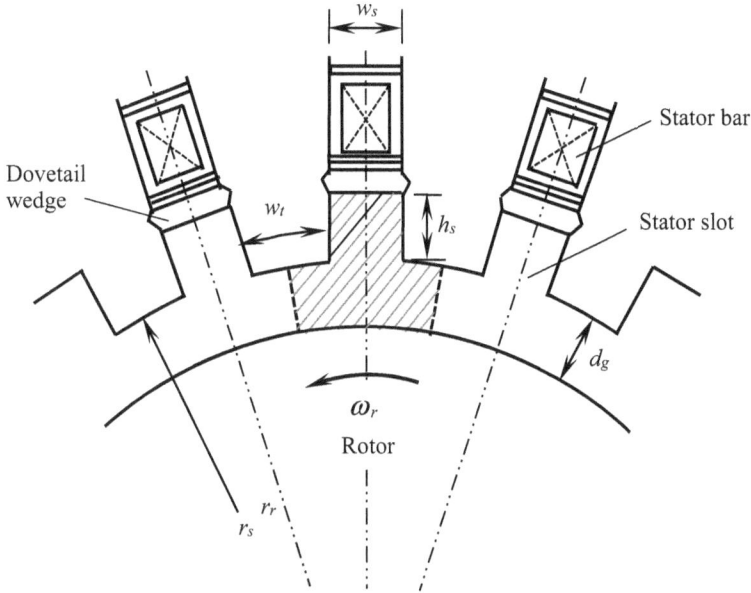

FIGURE 10.13 Geometries and configurations of airgap, state slots and teeth, dovetail wedges, and stator slot openings. The shaded area is the computational domain.

correlated. A theoretical analysis of the vortex flow and heat transfer was conducted by Leont'ev and Kirdyshkin [10.35]. The comparison of the theoretical formula with the experimental data on the friction coefficient has shown good agreement in the range of Taylor numbers between 4×10^3 and 10^{10}.

As shown in Figure 10.13, radially oriented stator slots are used to mount stator windings. The stator winding in each slot is secured by a dovetail wedge, which sustains forces from the stator winding. Conventionally, the wedge is short in height, leaving the slot partially unfilled near the slot mouth with the slot opening height h_s. The stator slots distribute uniformly on the stator circumference with the slot width w_s and tooth width w_t. The airgap depth in the radial direction is denoted by d_g. The rotor radius and stator inner radius are represented by r_r and r_s, respectively

In fluid dynamics, the Taylor number characterizes the relative importance of inertia forces to viscous force in rotating viscous fluids. As a dimensionless quantity, Taylor number is related to Couette Reynolds number Re_d as

$$T_a = \text{Re}_d^2 \left(\frac{d_g}{r_r} \right) = \frac{(\rho \omega)^2 r_r d_g^3}{\mu^2} \tag{10.51}$$

where Re_d is defined as

$$\text{Re}_d = \frac{\rho \bar{u}_\theta d_g}{\mu} \tag{10.52}$$

where
ρ is the density of air, kg/m^3
μ is the dynamic viscosity of air, Pa·s
d_g is the airgap depth in the radial direction, m

It is important to note that $\bar{u}_\theta$ is the mean swirl velocity across the stator–rotor gap, which can be obtained by integrating u_θ along the radial direction,

$$\bar{u}_\theta = \frac{1}{d_g} \int_{r_r}^{r_s} u_\theta(r)\,dr \tag{10.53}$$

The data and correlations in references [10.34] and [10.35] were combined and represented in a useful chart for friction factor for smooth rotating cylinders in a smooth coaxial enclosure without axial flow [10.36]. According to the Taylor number values, four regimes have been distinguished with a formula in each regime: laminar ($Ta \leq 41$), transition ($41 < Ta \leq 63$), vortex flow ($63 < Ta \leq Ta_{V-T}$, where Ta_{V-T} is the value at which the flow transits from the vortex-dominant flow to the turbulence-dominant flow), and turbulent ($Ta_{V-T} < Ta < \infty$).

Taylor–Couette flows between concentric rotating cylinders have been extensively studied as rotational power loss and flow instability problems. Due to the complexity of the flows in rotating machinery, a majority of the studies have assumed uniform rotor–stator gaps and smooth rotor and stator walls. In practice, windage losses and friction characteristics in electric machines are generally determined through experimental tests and numerical simulations. Analytical models that precisely predict flow fields are rather limited.

Bruckner [10.37] studied experimentally windage losses in the rotor–stator gap and gas foil bearings in electric machines. The results show an exponential rise in power loss as mean operating density is increased for both the machine windage and gas foil bearings. These losses can become increasingly significant for high-speed machines operating in high-pressure environments, especially in narrow rotor–stator gaps. Bruckner also reported that the machine windage can be non-dimensionalized in terms of Taylor number and moment coefficient. Rough stator surfaces have higher torque than the smooth-walled stator.

With the CFD technique, Wild *et al.* [10.38] have performed experimental and computational investigations of Taylor–Couette flow. The computed flow field shows significant variations in the axial distribution of the azimuthal shear stress due to the secondary flow associated with Taylor vortices. Three functional relations, developed by other researchers for windage torque of infinite cylinders, are found to be in good agreement with the experimental results when coupled with a correction for end wall effects. These relations are the most useful with respect to the design of rotating equipment.

It is noteworthy that rotors and stators of electric motors are not so smooth. They generally have grooves, slots, cavities, *etc.*, on their walls. For instance, the rotor of a switch reluctance motor is shaped with several poles on the rotating cylinder. This geometry can result in a more complex mechanism of windage loss. Thus, it is important to understand and predict the effect of such structural textures on windage losses and friction characteristics.

Recently, with an increase in demands for higher performance and higher efficiency of turbomachinery and other rotary systems, a great deal of effort has been directed toward the determination of friction factors on rotating and stationary surfaces.

10.4.1.2 Friction Factor

There are two types of friction factors used in the literature: Darcy–Weisbach friction factor and Fanning friction factor. Darcy–Weisbach friction factor is commonly used in the Moody diagram for pipe friction flows, and Fanning friction factor is primarily used for rotating or swirling flows [10.39]. Obviously, for electric rotating machines, the Fanning friction factor is more appropriate.

Fanning friction factor is defined as the ratio of the shear stress acting on the solid surface to the mean dynamic pressure at the same location, that is,

$$f = \frac{\tau_o}{\rho \bar{u}_\theta^2 / 2} \tag{10.54}$$

The shear stress τ_o can be evaluated from the gradient of the swirl velocity profile at the wall:

$$\tau_o = \mu \left| \frac{\partial u_\theta}{\partial r} \right|_{r=r_o} \tag{10.55}$$

The friction factor in an annular gap is a function of slot geometry, stator and rotor radii, rotational speed, and fluid properties, that is,

$$f = F\left(w_s, w_t, h_s, d_g, r_s, r_r, \omega, \rho, \mu\right) \tag{10.56}$$

To reduce the number of variables, two length ratios, the slot width-to-pitch ratio η and the slot aspect ratio A_s, are introduced as

$$\eta = \frac{w_s}{w_s + w_t} \tag{10.57}$$

$$A_s = \frac{h_s}{w_s} \tag{10.58}$$

Thus, applying Equations 10.57 and 10.58 in Equation 10.56 gives

$$f = F\left(\eta, A_s, \mathrm{Re}_d\right) \tag{10.59}$$

10.4.1.3 Windage Loss Due to Rotating Rotor

As a rotor rotates at high speeds, frictional windage losses associated with the airflow in the annular airgaps are generated on the surfaces of both the rotor and stator as they interact with the flows. In practice, this type of windage loss plays a significant role in motor performance.

For viscous flows of incompressible fluids, the governing equations are the continuity equation and Navier–Stokes equations,

$$\nabla \cdot \mathbf{u} = 0 \tag{10.61}$$

$$\frac{\partial \mathbf{u}}{\partial t} + (\mathbf{u} \cdot \nabla)\mathbf{u} = -\frac{1}{\rho}\nabla p + \nu\nabla^2 \mathbf{u} + \mathbf{f} \tag{10.62}$$

where $\mathbf{f}$ represents various body forces (*e.g.*, gravitational, centrifugal, and electromagnetic forces) per unit mass. It should be noted that these equations are usually solved using numerical methods. Exact solutions exist only for very special cases. The primary difficulty in obtaining exact solutions lies in the nonlinear terms in Navier–Stokes equations.

Consider a flow in an annulus of two coaxial cylinders with the inner cylinder rotation (Figure 10.14), where r_r and r_s are the radii of the rotor and stator, respectively, and ω_r is the rotor rotating speed. For this type of flow, the nonlinear term $(\mathbf{u} \cdot \nabla)\mathbf{u}$ is identically zero. In addition, $u_r = 0$ and $\partial()/\partial\theta = 0$. Therefore, the Navier–Stokes equations for steady-state flows can be simplified as

$$-\frac{1}{\rho}\frac{dp}{dr} + \frac{u_\theta^2}{r} = 0 \tag{10.63}$$

$$\frac{d}{dr}\left(\frac{du_\theta}{dr} + \frac{u_\theta}{r}\right) = 0 \tag{10.64}$$

The exact solutions, known as Couette solutions, are thus obtained [10.40]:

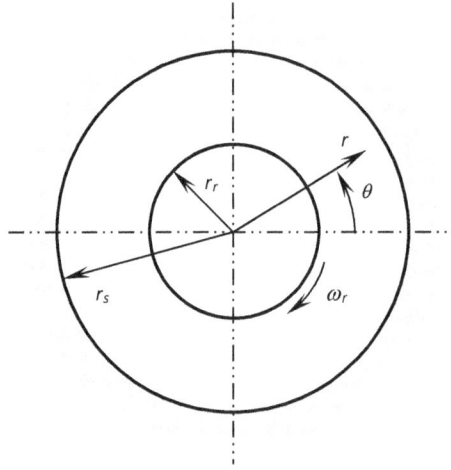

FIGURE 10.14   Flow in two coaxial cylinders with the inner cylinder rotating and outer cylinder at rest.

$$u_r = 0 \tag{10.65}$$

$$u_\theta(r) = Ar + \frac{B}{r} \tag{10.66}$$

where

$$A = -\frac{r_r^2 \omega_r}{r_s^2 - r_r^2}; \qquad B = \frac{r_r^2 r_s^2 \omega_r}{r_s^2 - r_r^2}$$

Thus, the velocity distribution in the fluid between the cylinders becomes

$$u_\theta(r) = \frac{\omega_r r_r^2}{r_s^2 - r_r^2} \left(-r + \frac{r_s^2}{r} \right) \tag{10.67}$$

Equation 10.67 indicates that on the rotor surface, $u_\theta(r_r) = r_r \omega_r$, and on the stator surface, $u_\theta(r_s) = 0$. The radial pressure gradient is thus obtained as

$$\frac{dp(r)}{dr} = \frac{\rho u_\theta^2}{r} = \rho \left(A^2 r + \frac{2AB}{r} + \frac{B^2}{r^2} \right) \tag{10.68}$$

Therefore, the radial pressure distribution is obtained as

$$p(r) = \int_{r_r}^{r} \frac{\rho u_\theta^2}{r} dr = \rho \left[\frac{A^2}{2} \left(r^2 - r_r^2 \right) + 2AB \ln \frac{r}{r_r} - B^2 \left(\frac{1}{r} - \frac{1}{r_r} \right) \right] \tag{10.69}$$

It is important to note that the Couette solutions are basically derived for laminar flows.
   As reported by Dou *et al.* [10.33], the shear stress of laminar flow is given as

$$\tau = \mu \left(\frac{du_\theta}{dr} - \frac{u_\theta}{r} \right) = -\mu \frac{2B}{r^2} \tag{10.70}$$

Thus, on the rotor and stator surfaces, the wall shear stresses τ_w become

$$\tau_{w,r} = -\mu \frac{2B}{r_r^2}; \qquad \tau_{w,s} = -\mu \frac{2B}{r_s^2} \tag{10.71}$$

respectively. Thus, the friction torques T_r and T_s that act on the rotor and stator surfaces with the area A_r and A_s, respectively, can be obtained as

$$T_r = \frac{\tau_{w,r} A_r D_r}{2} \tag{10.72a}$$

$$T_s = \frac{\tau_{w,s} A_s D_s}{2} \tag{10.72b}$$

where D_r and D_s are the rotor outer diameter and stator inner diameter, respectively.

Windage loss is generated by the shearing interaction between air and rotor/stator. For smooth rotor and stator surfaces, frictional windage loss P_w can be directly determined from the formula

$$P_w = (T_r + T_s)\omega_r = (T_r + T_s)\frac{n\pi}{30} \tag{10.73}$$

where

ω_r is the rotor rotating speed, rad/s
n is the rotor rotating speed, rpm

When the rotor and stator have irregular geometries (*e.g.*, cavities), T_r and T_s in equation (10.72) may be replaced by their average values, $\overline{T}_r$ and $\overline{T}_s$, respectively, which can be obtained through CFD analysis.

For fully turbulent Taylor–Couette flow, the viscosity terms in the governing equations must be replaced by the terms for turbulent frictional stresses. The normalized mean angular momentum (ru_θ/r_rU_o) profile becomes approximately skew-symmetric in the annulus between the two cylinders, as given in Figure 10.15. According to the pattern of the curve, three distinct flow regions can be

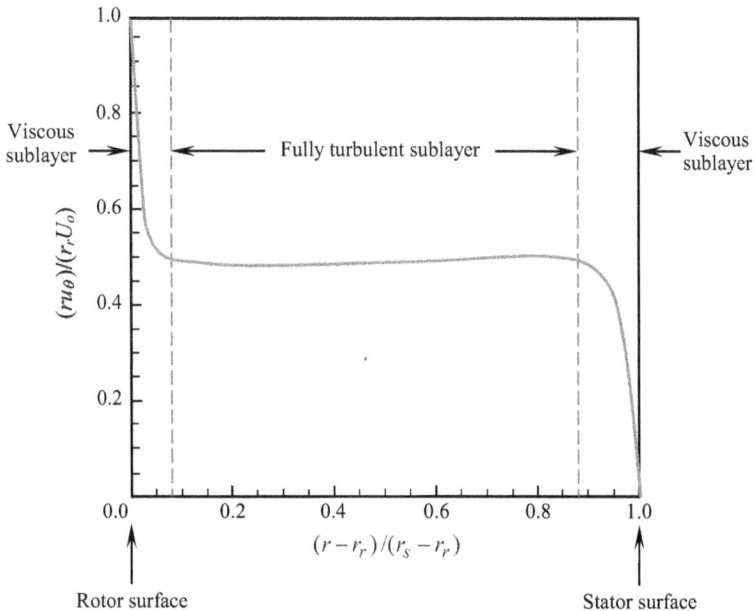

FIGURE 10.15 Normalized mean angular momentum profile of turbulent Taylor–Couette flow at $Re = 8,000$.

identified as follows: the viscous sublayer near the rotor wall, the fully turbulent sublayer in the bulk of flow, and the viscous sublayer near the stator wall. The result shows that the core of the flow has an essentially constant mean angular momentum $0.5r_r U_o$ (where $U_o = r_r \omega_r$), indicating that the mean velocity is close to that for potential flow [10.41, 10.42]. High gradients of angular momentum occur in the two viscous sublayers, leading to high turbulent shear stresses, as well as torques and windage losses, on both the rotor and stator walls.

The total shear stress τ in turbulent Taylor–Couette flow can be expressed as

$$\tau = \left(\mu + \rho \varepsilon_M \right) \frac{\partial \bar{u}_\theta}{\partial r} \tag{10.74}$$

where ε_M is the eddy diffusivity of momentum and $\rho \varepsilon_M$ is the eddy viscosity. In the fully turbulent region, the shear stress is set by the chaotic motion of fluid molecules and $\rho \varepsilon_M$ plays a dominant role over fluid viscosity μ in energy and momentum transfer. In the viscous sublayer close to wall, $\mu \gg \rho \varepsilon_M$.

Therefore, the shear stress on the wall becomes

$$\tau_o = \mu \frac{d \bar{u}_\theta}{dr} \bigg|_{r=r_r \, \& \, r=r_s} \tag{10.75}$$

Alternatively, as friction factor is determined, windage power loss P_w can be related to friction factor as

$$P_w = \pi \rho \bar{u}_\theta^2 r_r^2 f \omega l \tag{10.76}$$

where l is the length of the rotor.

As proposed by Saari [10.43], the windage loss of radial flux motor is associated with the resisting drag torque on the rotor,

$$P_w = k_r \pi f \rho \omega^3 r_r^4 l \tag{10.77}$$

where

    k_r is the surface roughness coefficient (for a perfect smooth surface, $k_r = 1$; for a lamination stacked surface, $k_r = 2–4$)
    f is the friction factor
    ρ is the density of fluid (air or liquid), kg/m³
    r_r and l are the rotor radius and axial length, respectively, m

Equation 10.77 is essentially derived from the definition of the friction factor. It can be seen from the equation that the windage loss on the rotor is proportional to ω^3. The roughness coefficient k_r is expected to increase with the Couette Reynolds number Re_d.

Assumed that the fluid velocity varies proportionally to the distance from the airgap midpoint with the power of 1/7, Yamada [10.44] developed an equation for the friction factor,

$$f = \frac{0.0152}{Re_d^{0.24}} \quad 800 < Re_d < 6 \times 10^4 \tag{10.78}$$

Based on their measurements, as well as the experimental data from other investigators. Bilgen and Boulos [10.45] presented equations for the friction coefficient on smooth surfaces,

$$f = 0.515 \frac{\left(d_g / r_r \right)^{0.3}}{Re_d^{0.5}} \quad 500 < Re_d < 10^4 \tag{10.79a}$$

$$f = 0.0325 \frac{\left(d_g/d_r\right)^{0.3}}{\mathrm{Re}_d^{0.2}} 10^4 < \mathrm{Re}_d \qquad (10.79b)$$

The experimental friction actors were found to deviate less than 8.35% from the results calculated by the above equations.

For the turbulent flow that $Ta \gg 400$, the friction factor f is [10.46]

$$f = \frac{0.0095}{(Ta)^{0.2}} \qquad (10.80)$$

The windage loss of axial flux motor is given by [10.47]

$$P_w = \frac{\pi \mu \omega^2 \left(r_o^4 - r_i^4\right)}{2l_g} \qquad (10.81)$$

where
  μ is the viscosity of fluid (air or liquid), Pa·s
  ω is the rotor rotating speed, rad/s
  l_g is the axial length of the airgap, m
  r_o and r_i are the shaft outer and inner radii, respectively, m

10.4.2 WINDAGE LOSS DUE TO ENTRANCE EFFECT OF AXIAL AIRGAP FLOW

Assuming that the air cooling flow does not have a tangential velocity component before entering the airgap, Polkowski [10.48] underlined that a rather large friction torque may be associated with the entrance effects of the axial airgap flow. He suggested that the torque needed to accelerate the axial cooling flow into a tangential movement is presented as

$$T_a = \frac{2}{3} \pi \rho \left(r_s^3 - r_r^3\right) \upsilon_m u_m \qquad (10.82)$$

where
  r_s and r_r are the radii of the stator and rotor, respectively
  υ_m and u_m are the mean axial and tangential air velocities, respectively

Thus, windage losses may be associated with the cooling flow through the airgap of a high-speed motor:

$$P_a = \frac{2}{3} \pi \rho \left(r_s^3 - r_r^3\right) \upsilon_m u_m \omega \qquad (10.83)$$

According to Saari [10.43], the power losses predicted by the above equations agree well with the measured results. The surface roughness caused by the stator slot openings does not significantly increase the friction losses in the airgap.

10.4.3 WINDAGE LOSS DUE TO STATOR SURFACE ROUGHNESS

The opposite observation to Saari's conclusion [10.43] that the stator roughness has little effect on the rotor body windage loss was presented by Tong and Gott [10.49]. They investigated the influence of stator slot openings on the rotor windage loss in large-sized electric machines. Their results suggest that the roughness of the stator surface could significantly alter the velocity field in the airgap and thereby change windage losses.

Referring to Figure 10.13, when the rotor rotates with angular velocity ω_r, the rotor rotation-driven flow may penetrate these unfilled slots so that the flow field in the stator–rotor airgap is distorted and leads to the variations in the friction factor on the rotor surface and correspondingly the rotor body windage loss.

Friction factor is defined as the ratio of the shear stress acting on the surface to the dynamic pressure applied on the same surface. For a rotating electric machine, friction factor is associated with the pumping action of a rotor in imparting energy to the cooling medium at the rotor–stator annular gap. An important characteristic of rotor rotation-driven flows is the tendency of fluid with high angular momentum to be flung radially outward.

With the defined unfilled aspect ratio A_s (where $A_s = h_s/w_s$) and friction factor ratio f_r/f_{ro}, where f_r is the friction factor for the rotor and f_{ro} is the friction factor for the smooth stator wall (*i.e.*, $h_s = 0$), the effect of the aspect ratio A_s on the friction factor ratio can be obtained by numerical simulations using a commercial CFD code FLUENT. As depicted in Figure 10.16, the friction factor ratio is zero as the wedge is flush with the adjacent stator surfaces. The friction factor ratio exhibits a rapid rise by increasing in A_s and then declines sharply with the further increase of A_s. When A_s approaches approximately 0.8, it reaches its local minimum. Thus, the preferred way to reduce windage loss is to redesign the wedge with either $A_s = 0$ or $A_s = 0.8$ for this specific machine. It has been found that the selection of correct A_s values can reduce rotor body windage loss by about 25% (Figure 10.16).

This may be similar to the golf ball design. As a simple, passive means of drag reduction, small dimples are made on the golf ball surface. In such a way, the dimpled surface manipulates the flow and creates vortices that can prevent or delay the boundary-layer separation on the ball surface [10.50].

When both the stator and rotor surfaces are smooth, the friction factors on the rotor and stator have shown a relative difference of about 12%, mainly resulted from the curvature effect and rotor pumping effect (Figures 10.17 and 10.18).

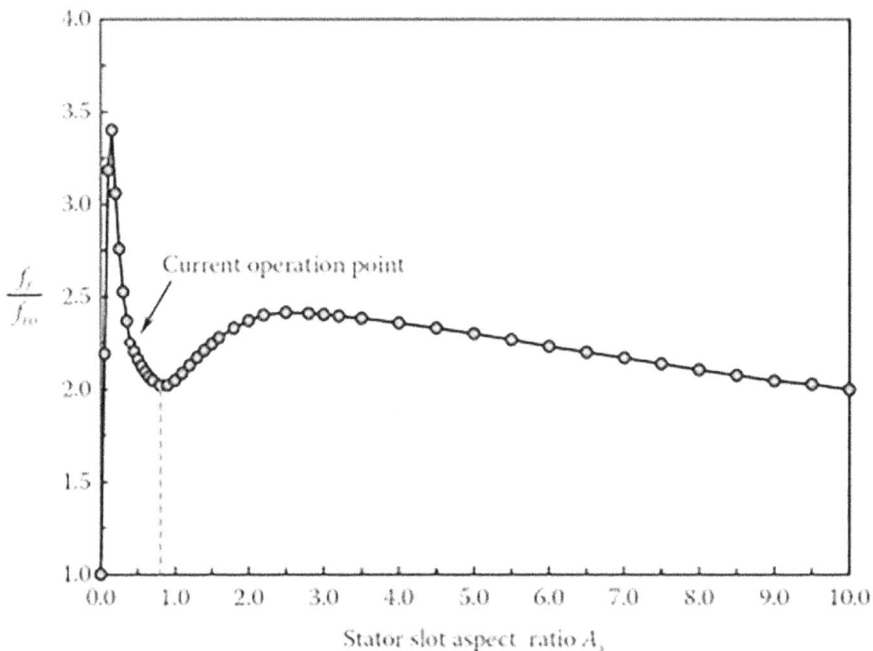

FIGURE 10.16 Effect of the stator slot aspect ratio on the rotor friction factor ratio.

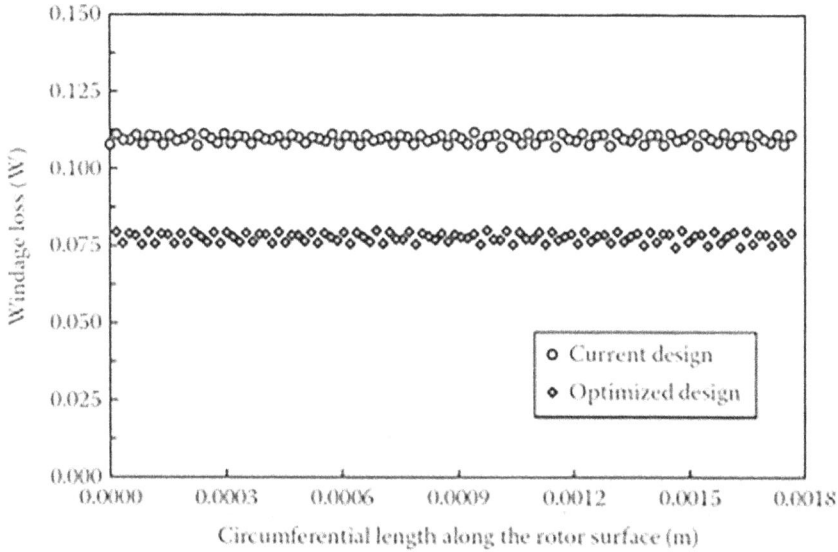

FIGURE 10.17 Comparison of rotor windage loss data between the current and optimized design.

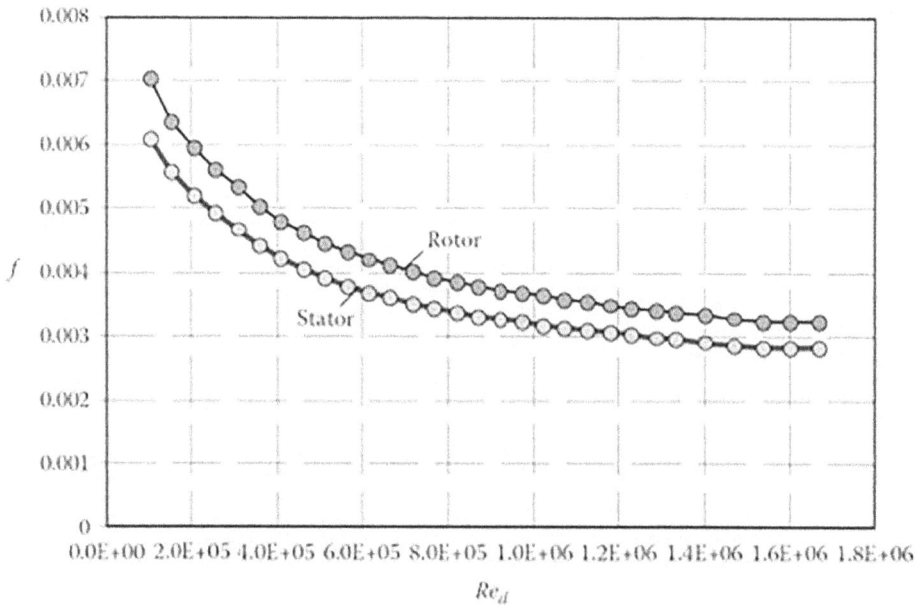

FIGURE 10.18 Comparison of friction factor on the smooth rotor and stator surfaces.

10.4.4 ENERGY LOSS DUE TO FLUID VISCOSITY

In a viscous fluid, viscosity is due to the friction between neighboring molecules in the fluid that are moving at different velocities. Therefore, any relative motion of molecules results in frictional or drag forces. The work done by these friction/drag forces in turn causes the loss of mechanical energy in the fluid. As proposed by Dou *et al.* [10.33], the energy loss per unit volumetric fluid in unit length along the flow streamlines in the fluid between the coaxial cylinders with the inner cylinder rotating is obtained from the formula

$$\frac{dH}{ds} = \frac{\tau}{u_\theta}\frac{du_\theta}{dr} - \frac{\tau}{r} = \mu\frac{4B^2}{r^4}\left(Ar + \frac{B}{r}\right)^{-1} \tag{10.84}$$

where

  H is the energy loss of unit fluid volume, J/m³
  s is the length in streamwise direction, m

It is worth noting that the energy loss increases with increasing radial r across the airgap. This indicates that the flow on the rotor surface has the lowest energy loss and thus the lowest damping mechanism in the response to any disturbance.

At the rotor surface where $r = r_r$,

$$\left.\frac{dH}{ds}\right|_{r=r_r} = \frac{4\mu\omega}{r_r\left(r_s - r_r\right)^2}\frac{r_s^4}{\left(r_s + r_r\right)^2} \tag{10.85}$$

10.4.5 FAN LOSSES

Fans are devices that utilize electric energy to drive fan blades rotating to achieve both the air movement and the increased air total pressure. Motor cooling fans are often attached to rotor shafts at one or two ends to cool motors. As shown in Figure 10.19, a centrifugal cooling fan is mounted on

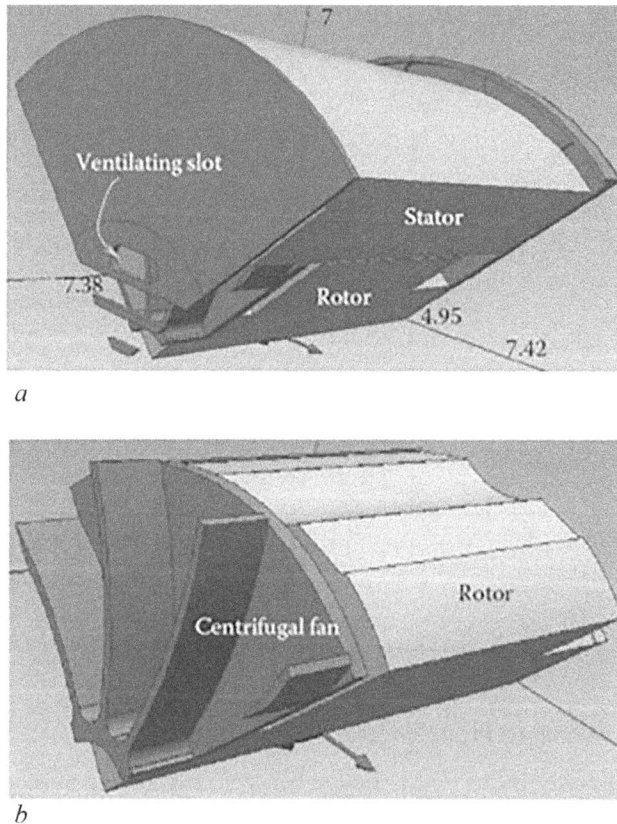

FIGURE 10.19 A centrifugal fan is mounted on one end of the rotor to suck the cooling flow into the motor: (*a*) the motor assembly and (*b*) the centrifugal fan and rotor assembly, where the rotor with scallop-shaped structure for providing ventilation paths of axial cooling flow.

FIGURE 10.20 Velocity field through the internal cooling paths.

one end of a rotor. When the rotor rotates, it drives the fan rotating to suck the cooling air into the machine through the ventilating slots on the endbell. Due to the centrifugal effect, the cooling air flows radially to cool the stator winding and then turns 180° to flow toward the rotor. Then, the cooling air turns 90° to get into the scallop-shaped cooling channels in the axial direction and finally exhausts at the other end of the machine (Figure 10.20). The exhausted air is guided by the endbell at the other end to flow over the outer surface for cooling the motor housing.

The pressure field in the machine has shown that the high static pressure zone occurs in the vicinity of the fan (Figure 10.21). A large static pressure differential is observed across the fan. In practice, the cooling fan represents a significant load on the machine, reducing available power. Generally, the fan power loss increases with an increase in the rotating speed of the fan. In addition, the fan power loss is a function of the duty cycle of the fan. As a duty cycle is larger than 50%, the fan power loss is about a constant regardless of its rotating speeds. As the duty cycle is less than 40% (*i.e.*, the *on time* is less than 40%), the fan power loss reduces sharply in the reduction of the duty cycle [10.51].

The loss in pressure between any two points in a system is always equal to the difference between the total pressures at these points. Therefore, the total pressure loss of fan ΔP_t is the difference between the total pressure at the fan inlet $P_{t,in}$ and outlet $P_{t,out}$:

$$\Delta P_t = P_{t,in} - P_{t,out} \tag{10.86}$$

The total pressure at any point consists of two components: static pressure P_s and dynamic pressure P_d. Consequently,

$$P_{t,in} = P_{s,in} + P_{d,in} = P_{s,in} + \frac{1}{2}\rho V_{in}^2 \tag{10.87a}$$

$$P_{t,out} = P_{s,out} + P_{d,out} = P_{s,out} + \frac{1}{2}\rho V_{out}^2 \tag{10.87b}$$

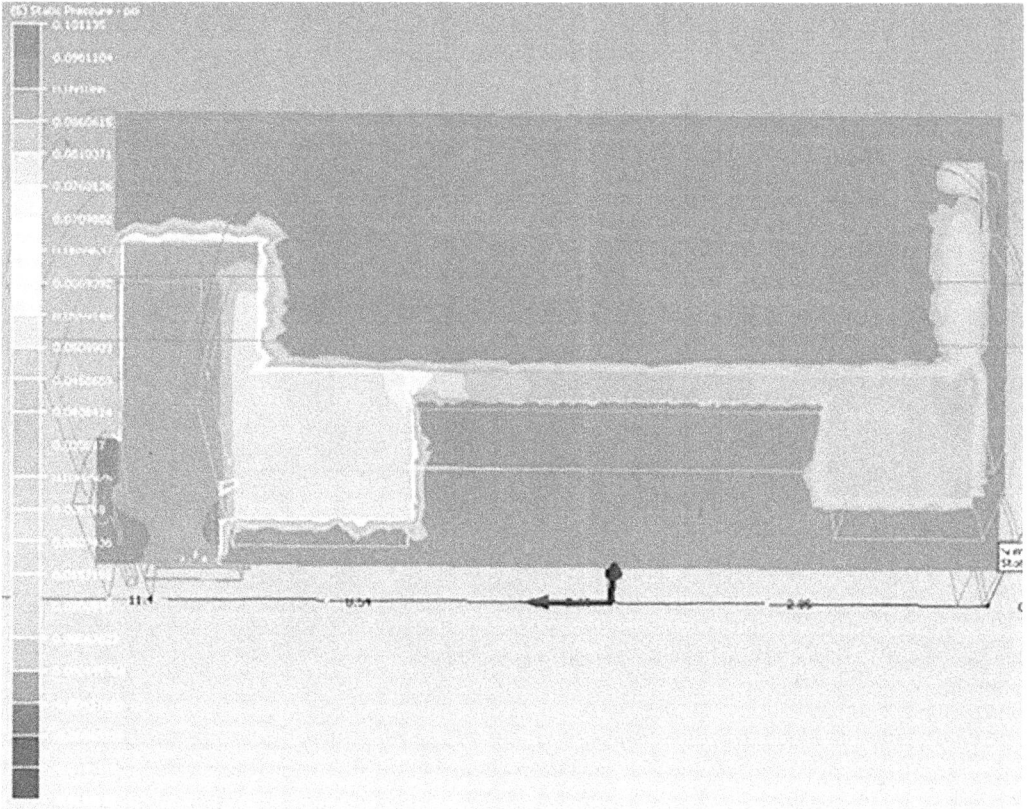

FIGURE 10.21 Pressure distribution inside the machine. The high-pressure zone is shown in the vicinity of the fan.

where

$P_{s,in}$ and $P_{s,out}$ are the static pressures

$P_{d,in}$ and $P_{d,out}$ are the dynamic pressures at the fan inlet and outlet

V_{in} and V_{out} are the velocities at the fan inlet and outlet, respectively

Substituting these two equations into (10.86) yields

$$\Delta P_t = \left(P_{s,in} - P_{s,out}\right) + \frac{1}{2}\rho\left(V_{in}^2 - V_{out}^2\right) = \Delta P_s + \Delta P_d \qquad (10.88)$$

All these pressures are defined in ANSI/AMCA Standard 320-07 [10.52]. The relations of these pressures and pressure differentials at the inlet and outlet of a fan are presented in Figure 10.22. With a ducted inlet and outlet, the airflow velocity along the ducted inlet is constant, and thus, the dynamic pressure remains unchanged. The static pressure decreases (relative to the pressure datum) as the airflow approaches the fan. Hence, the total pressure reaches its minimum (relative to the pressure datum) at the point of the fan inlet. Similarly, along the ducted outlet, both the airflow velocity and dynamic pressure remain constant. The static pressure decreases when the airflow leaves away from the fan, causing a drop in the total pressure along the ducted outlet. It is worth to note that the total pressure of the airflow experiences a big jump when air flows from the fan inlet to the outlet due to gained mechanical energy from the fan. If the total pressure is plotted against the flow path, it can be seen that the total pressure decreases everywhere except across the fan.

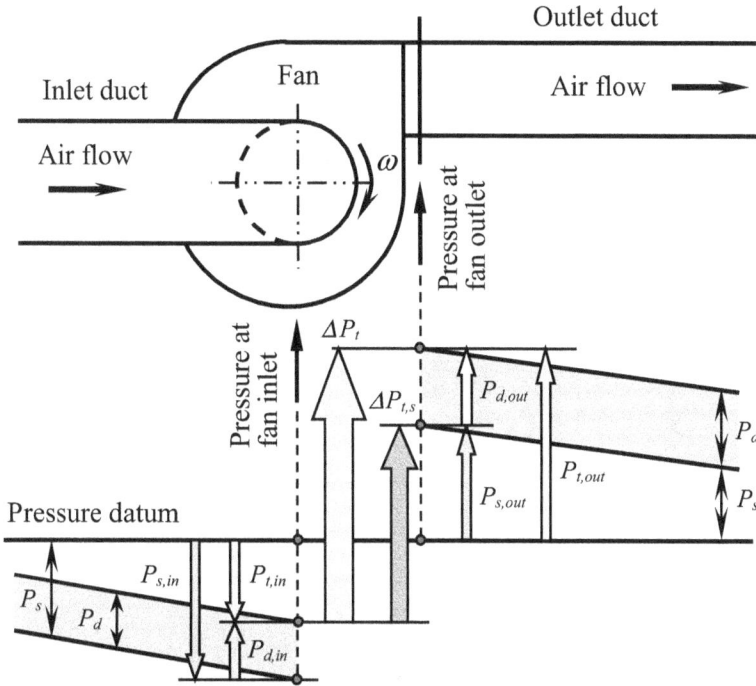

FIGURE 10.22 Relations of various pressures and pressure differentials at the inlet and outlet of a fan.

Fan losses may come from (*a*) head loss due to the changes in flow area and direction; (*b*) energy lost in some components such as filters, sound attenuators, and grills; (*c*) friction of the duct and fan walls; (*d*) airflow recirculation and reentrance; and (*e*) windage loss of the fan itself.

10.4.6 VENTILATING PATH LOSSES

Ventilating paths inside an electric rotating machine are considerably complex due to irregular shapes, rough surfaces, and internal obstructions. Pressure losses occur at sharp turns, sudden contractions and expansions, and any torturous paths in the end-winding regions and in the gap between the rotor and stator, resulting in large ventilation windage losses [10.53].

Because ventilating paths are so complicated and vary case to case, no theoretical solutions or empirical equations are available. In order to predict the windage loss resulted from ventilating paths, it is recommended to use CFD tools.

10.4.7 METHODS FOR REDUCING WINDAGE LOSSES

The increase in electric motor efficiency is extremely important for applications in modern society because any power losses ultimately imply the consumption of energy. As motor power densities decrease and output power levels increase to meet a high demand of high-performance, high-speed, and cost-effective motors, power loss due to windage in some motion control systems can become significantly large. The excessive windage loss can effectively lower the machine capacity and performance. In some cases, the derating due to the total windage loss may consume over half of the available shaft power in the machine.

One effective way to reduce windage loss in a motor/generator, which is flooded with fluid, is to apply the so-called *film divider* [10.54]. The film divider is positioned within the airgap and freely rotatable within the gap. As the embodiment shown in Figure 10.23, the film divider is essentially a

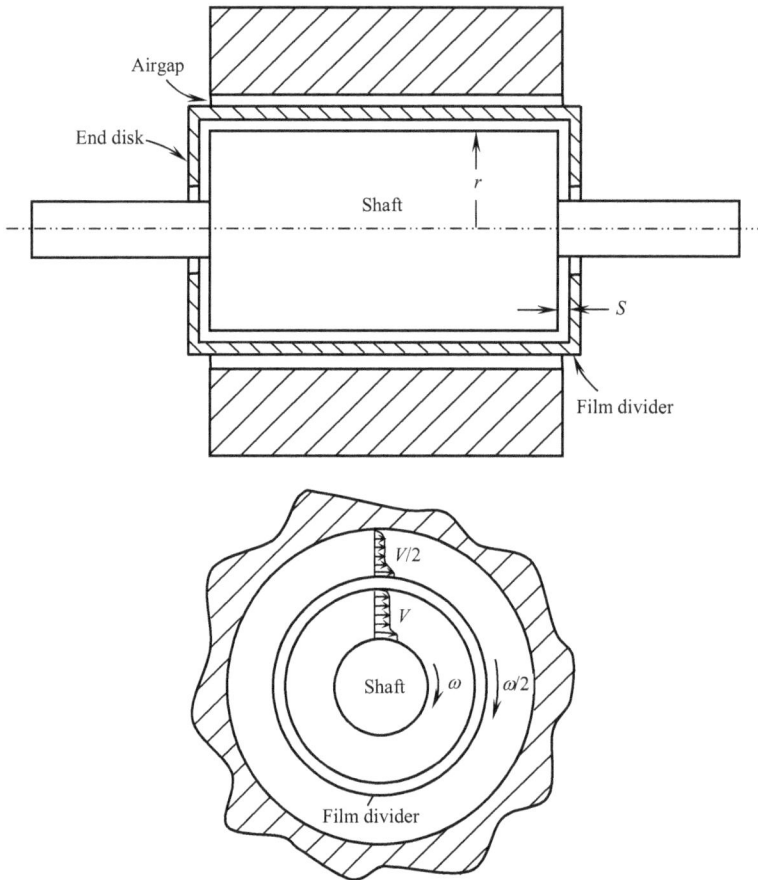

FIGURE 10.23 Using a film divider to lower windage loss in the airgap (U.S. Patent 5,828,148) [10.54]. (Courtesy of the U.S. Patent and Trademark Office, Alexandria, VA.)

thin sleeve, which may be made of any magnetically permeable material such as fiberglass or epoxy composite. It is concentric on both the rotor and stator. During machine operation, the rotor rotates with an angular velocity of ω. As the rotor spins, the fluid in the airgap causes the film divider to spin. Ideally, the angular velocity of the film divider is $\omega/2$. Thus, the differential vectors in each half of the gap are half of the vector flow without the film divider. Because windage loss is proportional to the 2.7 power of the velocity in the turbulent flow regime (*i.e.*, $P_w \propto \omega^{2.7}$), the overall windage loss is lower with the film divider in place.

The comparison of windage loss in high-speed motors with and without the film divider is presented in Figure 10.24. It can be seen that the motor with the film divider has much lower windage loss than that without the film divider.

In a high-power, high-speed electric machine, windage losses are created by a large rotor body The peripheral velocity on the rotor's surface causes significant windage losses. In order to reduce windage losses, it was proposed to optimize the rotor construction [10.55]. As shown in Figure 10.25, the rotor has a central cylindrical body provided with circumferentially spaced, axially extending surface recesses at the ends of the rotor, which define a desired number of salient poles at each end of the rotor. The inner and outer shrouds are provided on either side of salient poles to reduce windage losses. The outer shrouds block axial airflow into the recesses and thus reduce windage losses. Furthermore, with the reduced diameter of the central cylindrical body, the peripheral velocity of the cylindrical body is reduced, and, thereby, the windage loss is also reduced.

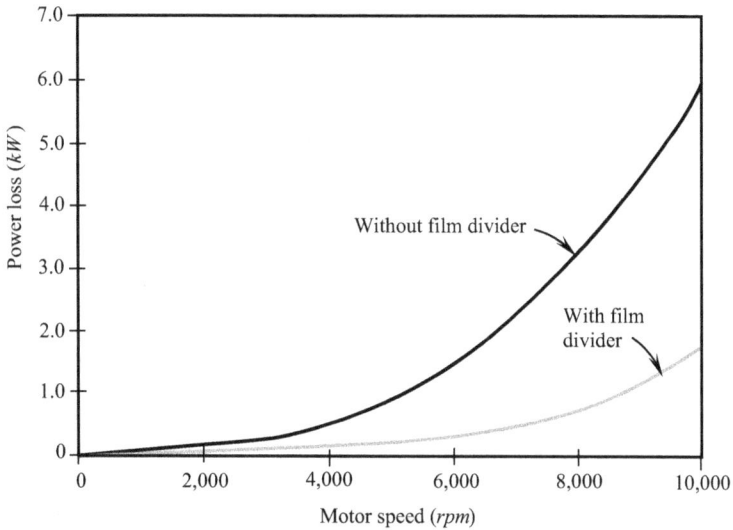

FIGURE 10.24 Comparison of windage power loss of motors with and without film divider in high-speed machines (U.S. Patent 5,828,148) [10.54]. (Courtesy of the U.S. Patent and Trademark Office, Alexandria, VA.)

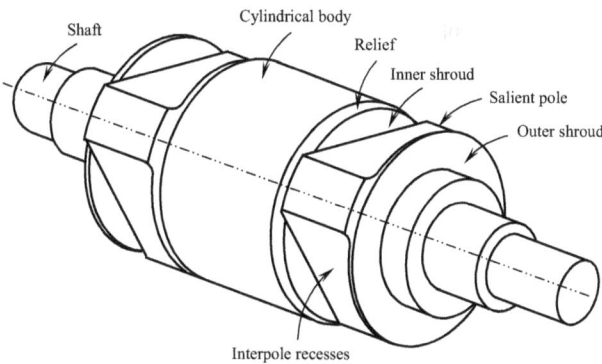

FIGURE 10.25 Optimized rotor construction for reducing windage loss (EP0489859) [10.55]. (Courtesy of European Patent Office, Munich, Germany.)

To further minimize windage losses, an outer cylindrical surface can be produced in the pole region of the rotor by filling interpole recesses with nonmagnetic material.

A similar idea is to use a rotor containment shell surrounding the rotor to lower windage losses [10.56]. The outer cylindrical surface of the shell is formed with a plurality of concavities in the form of annular turbulator grooves axially spaced from one another along the rotor length (Figure 10.26). These grooves will enhance the local air mixing, improve the axial heat transfer, reduce windage and friction losses, and reduce the heat penetration from the heat generated by windage into the rotor body. The optimal depth of the groove depends on a trade-off between increased resistance to tangential motion and decreased resistance to axial flow.

Smith *et al.* have proposed several methods for reducing power losses in motor [10.57]. One of them is to maintain constant pressure throughout the motor cavity. A pressure valve can be placed within the motor cavity to release higher-pressure air built up during operation. The maintenance of constant pressure in the cavity increases the motor efficiency due to the reduction of windage loss.

Outer surface of shell Annular groove Containment shell

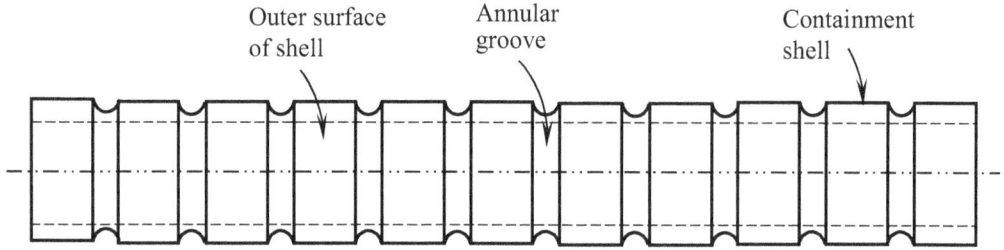

FIGURE 10.26 Rotor containment shell with a plurality of surface concavities (U.S. Patent 7,057,326) [10.56]. (Courtesy of the U.S. Patent and Trademark Office, Alexandria, VA.)

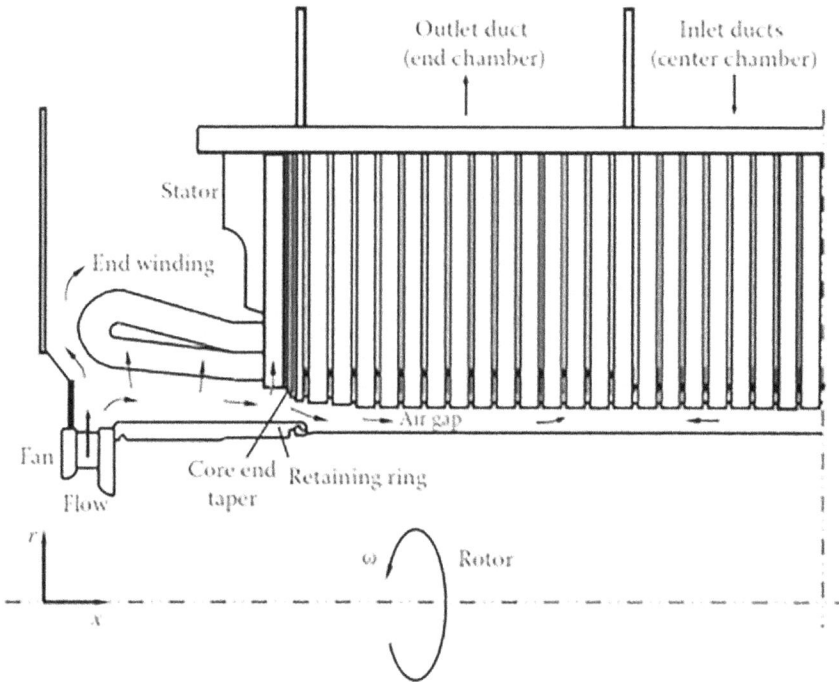

Outlet duct (end chamber) Inlet ducts (center chamber)

Stator

End winding

Air gap

Fan

Flow

Core end taper Retaining ring

r

x

ω Rotor

FIGURE 10.27 Schematic diagram of a ventilating cooling system in a large-size, high-power electric machine. Airflow is distributed to cool the end windings and the stator core.

One obvious way for windage loss reduction is to smooth cooling flow paths inside the motor [10.58]. All sharp edges, rapid expansions and contractions, and obstacles in flow paths should be avoided. A cooling air ventilation circuit in a large-size, high-power electric machine is presented in Figure 10.27. The inlet flow is accelerated in the radial direction by a ventilating fan, which is mounted on the end of the rotor. About one-third of the total volumetric airflow rate enters the airgap. The remainder proceeds through and around the armature end winding. This flow allocation is thus designed and controlled to ensure sufficient airflow for the heat removal. It can be seen that at the airgap entrance, a *bottle neck* is formed between the retaining ring inboard end and the end-core taper on the stator. As cooling gas passes through the gap entrance, the significant flow acceleration causes a large local pressure drop that is only partially recovered downstream.

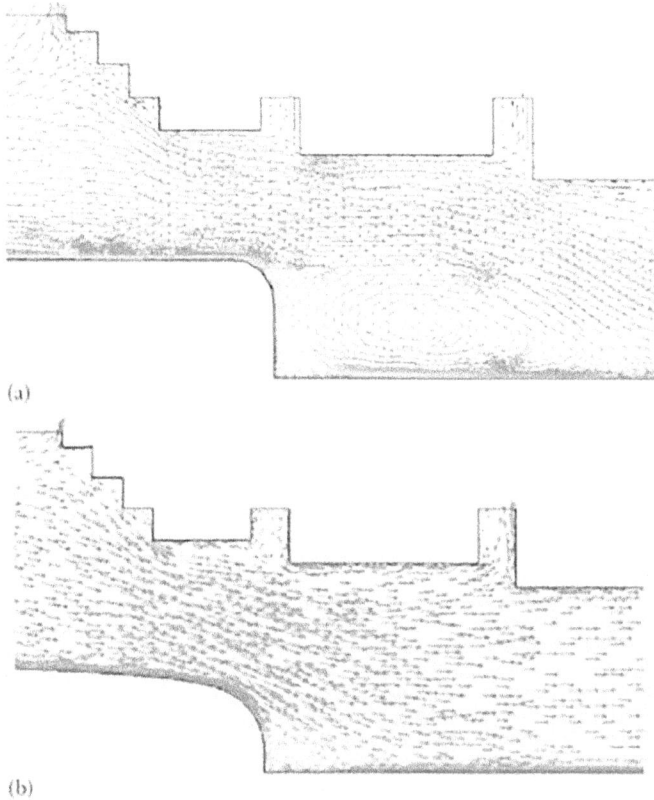

FIGURE 10.28 Reducing windage loss at the airgap entrance by smoothing the cooling flow path: (a) original design with a large flow recirculation and (b) improved design with an optimized flow path to eliminate the flow recirculation and thus reduce windage loss.

By optimizing the retaining ring with the preferable spline profile and other components, the flow recirculation is eliminated, and thus, local windage loss is remarkably reduced (Figure 10.28). In addition, the enlarged airgap entrance allows more flows to get into the airgap.

More recently, Pal [10.59] has proposed a new idea for reducing windage loss in liquid-cooled electric machines. This method uses injecting oil flow to cool the machine windings. The machine has two rotors and two stators. A number of baffles are coupled with the shaft at the center and each end of the shaft. The baffles extend radially outward from the shaft toward baffle cavities at the back iron. As the shaft rotates, oil is urged toward the baffle cavities along the baffles via centrifugal forces (Figure 10.29). The baffle directing oil to the baffle cavity substantially improves the scavenging performance. As a result, it reduces the oil level in a sump and lowers windage and friction losses.

An effective method to reduce windage losses is to use hydrogen gas to cool electric machines. The high thermal conductivity of hydrogen has proven to be a key advantage in its use as a cooling medium in electric power machines. Hydrogen has a thermal conductivity of nearly seven times higher than that of air, and its ability to transfer heat through forced convection is about 50% better than air [10.60]. At a given temperature and pressure, the density of hydrogen is only 1/14 the density of air. Due to its superior thermophysical properties, the use of hydrogen can not only reduce windage friction losses but also greatly enhance heat transfer superior properties.

Using hydrogen as a coolant can be traced back to the late 1930s. Since then, this technique has been developed and successfully used in generator cooling but still not applied to electric motors.

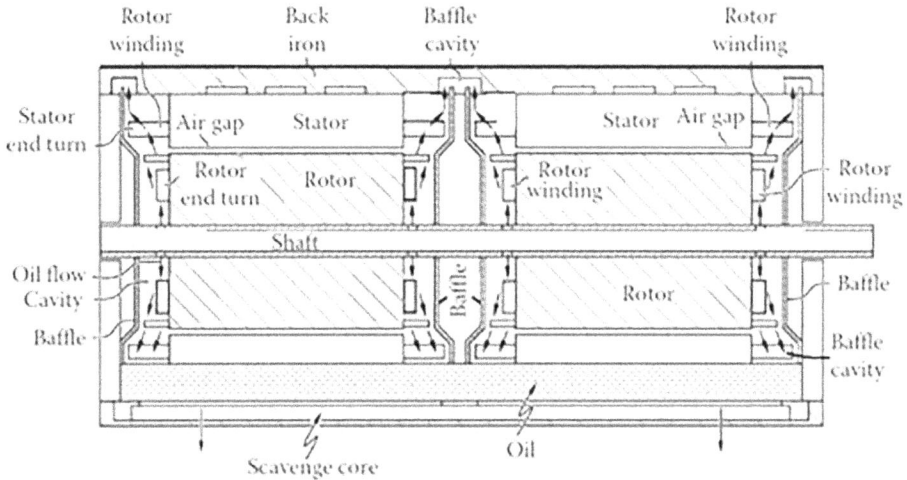

FIGURE 10.29 A liquid-cooled electric machine is cooled by injecting oil radially along a plurality of baffles via centrifugal force toward rotor and stator end windings [10.59]. (Courtesy of the U.S. Patent and Trademark Office, Alexandria, VA.)

10.5 STRAY LOAD LOSSES

Stray load losses consist of a variety of minor losses not accounted previously, for instance, the losses referring to the sources such as stray flux, the cage in the slots, interbar currents, and harmonic rotor bar currents from the nonsinusoidal distribution of the slotted stator winding [10.61]. Stray load losses appear only when motors operate under load conditions and are very difficult to predict analytically and to measure accurately. The largest contribution to stray losses is harmonic energies. These energies are dissipated as currents in the copper windings, harmonic flux components in the iron parts, and leakage fluxes in the laminate core. The detailed discussions and analyses of stray losses have been provided by Aoulkadi [10.62] and Englebretson [10.63].

It is commonly accepted that the investigation of stray load losses in electric rotating machines is far from complete. Most motor engineers rely on empirical equations for the estimation of stray load losses.

Stray load losses represent a small portion of the total motor power loss. The ratio of stray load loss to motor rated power decreases with the increased motor rated power. IEEE Standard 112-2004 [10.64], IEC Standard 60034-2-1 [10.65], and GB 18613-2002 [10.66] provide assumed values of stray load losses under different power ratings. Some of them are shown in Tables 10.4 and 10.5.

For other than rated load, stray losses are assumed to be proportional to the square of the rotor current [10.64].

TABLE 10.4
Assumed Values of Stray Losses

Machine Rating (kW)	Stray Load Loss (% of Rated Power)
1–90	1.8
91–375	1.5
376–1850	1.2
>1,850	0.9

Source: IEEE Power Engineering Society, IEEE 112-2004 Standard test procedure for polyphase induction motors and generators, 2004.

TABLE 10.5
Values of Stray Losses Proposed by GB 18613-2002

Machine Rating (kW)	Stray Load Loss (% of Rated Power)
0.55–1.1	2.5
1.5	2.4
3	2.3
4	2.2
5.5–10.5	2.1
11	2.0
15–18.5	1.9
22–30	1.8
37–45	1.7
55–75	1.6
90–110	1.5
132–160	1.4
200–315	1.3

Source: Standardization Administration of China (SAC), GB 18613-2002, Limited values of energy efficiency and evaluating values of energy conservation of small and medium three-phase asynchronous motors, 2002.

10.6 INFLUENCE OF POWER RATING ON MOTOR POWER LOSSES

One of the major factors affecting motor power losses is the motor power rating. As shown in Table 10.6 [10.67], the stator resistance or I2R losses decrease with the increase in power rating. On the contrary, friction and windage losses increase with the increased power rating. The stray load losses display a V-shape variation where it reaches its minimum value at the medium power rating.

It can be seen from the table that for all power ratings, the stator I^2R loss appears the highest compared to other power losses. For low-power ratings, the second largest loss is the rotor I^2R loss, followed by the core and stray load losses. For high-power ratings, the stray load loss becomes the second, which is about 33% larger than the rotor I^2R loss.

TABLE 10.6
Typical Distribution of Power Losses (%) among Motors with Different Power Ratings

	Motor Power Rating		
Type of power loss	25 hp (18.64 kW)	50 hp (23.79 kW)	100 *hp* (74.57 kW)
Stator resistance or I^2R losses	42	38	28
Rotor resistance or I^2R losses	21	22	18
Core losses	15	20	13
Friction and windage losses	7	8	14
Stray load losses	15	12	27

Note: The motor rotating speed is 1,800 rpm. All motors are equipped with the open drip proof enclosures.

REFERENCES

10.1. Benhaddadi, M., Olivier, G., Ibtiouen, R., Yelle, J., and Tremblay, J.-F. 2011. Chapter 1: Premium efficiency motors. In *Electric Machines and Drives* (Ed.: M. Chomat). InTech, Rijeka. http://www.intechopen.com/download/pdf/14083.

10.2. GE Industrial Systems. 2002. Evaluation and application of energy efficient motors. White paper publication number: e-GEA-M1019. http://www.geindustrial.com/publibrary/checkout/e-GEA-M1019?TNR=White Papers|e-GEA-M1019|generic.

10.3. Ramo, S., Whinnery, J. R., and Van Duzer, T. 1994. *Fields and Waves in Communication Electronics*, 3rd edn. John Wiley & Sons, New York.

10.4. Zheng, L. P., Xu, T. X., Acharya, D., Sundaram, K. B., Vaidya, J., Zhao, L. M., Zhou, L., Ham, C. H., Arakere, N., Kapat, J., and Chow, L. 2005. Design of a superhigh-speed cryogenic permanent magnet synchronous motor. *IEEE Transactions on Magnetics* **41**(**10**): 3823–3825.

10.5. Zhu, Z. Q., Ng, K., and Howe, D. 19910. Design and analysis of high-speed brushless permanent magnet motors. *Eighth International Conference on Electrical Machines and Drives*, Cambridge, UK, pp. 381–385.

10.6. Steinmetz, C. P. 1892. On the law of hysteresis. *AIEE Transactions* **9**: 3–64.

10.7. Mulder, S. 1995. Power ferrite loss formulas for transformer design. *Power Conversion & Intelligent Motion* **21**(**7**): 22–31.

10.8. Hamdi, E. C. 1994. *Design of Small Electrical Machines*. John Wiley & Sons, New York.

10.9. Rajput, R. K. 2008. *Basic Electrical Engineering*, 2nd edn. Laxmi Publications, New Delhi.

10.10. Mohan, N., Undeland, T. M., and Robbins, W. P. 1991. *Power Electronic Converters. Applications and Design*, 2nd edn. John Wiley & Sons, Hoboken, NJ.

10.11. Smith, A. C. and Edey, K. 1995. Influence of manufacturing processes on iron losses. *Proceedings of Seventh International Conference on Electrical Machines and Drives*, Durham, UK, pp. 77–81.

10.12. Binesti, D. and Ducreux, J. P. 1996. Core losses and efficiency of electrical motors using new magnetic materials. *IEEE Transactions on Magnetics* **32**(**5**): 4887–4889.

10.13. Deeb, R. 2012. Prediction of eddy current losses of surface mounted permanent magnet servo motor. *20th International Conference on Electrical Machines*, Marseille, France, pp. 1797–1802.

10.14. Ishak, D., Zhu, Z. Q., and Howe, D. 2005. Eddy-current loss in the rotor magnets of permanent-magnet brushless machines having a fractional number of slots per pole. *IEEE Transactions on Magnets* **41**(**9**): 2462–2469.

10.15. SKF Group. 2009. SKF energy efficient bearings. Publication 6860 EN. http://www.skf.com/binary/12-31950/6860_EN.pdf.

10.16. Palmgren, A. 1959. *Ball and Roller Bearing Engineering*, 3rd edn. SKF Industries Inc., Philadelphia, PA.

10.17. Harris, T. A. and Kotzalas, M. N. 2007. *Rolling Bearing Analysis: Essential Concepts of Bearing Technology*, 5th edn. CRC Press, Boca Raton, FL.

10.18. SKF Group. 2006. Using a friction model as an engineering tool. https://evolution.skf.com/us/using-a-friction-model-as-an-engineering-tool-2/#related-articles

10.19. SKF Group. 2017. Bearing friction, power loss and starting torque. https://www.skf.com/us/products/rolling-bearings/principles-of-rolling-bearing-selection/bearing-selection-process/operating-temperature-and-speed/bearing-friction-power-loss-and-starting-torque

10.20. Tietze, W. and Riedl, A. 2005. *Taschenbuch Dichtungstechnik*. Vulkan-Verlag, Essen.

10.21. Tong, W. 2009. Numerical analysis of flow field in generator end-winding region. *International Journal of Rotating Machinery* **2008**: Article ID 692748. doi:10.1155/2008/692748.

10.22. Couette, M. 1890. Études sur le frottement des liquids. *Annales de Chimie et de Physique* **21**(**6**): 433–510.

10.23. Mallock, A. 1896. Determination of viscosity of water. *Philosophical Transactions of Royal Society A* **187**: 41–56.

10.24. Taylor, G. I. 1923. Stability of viscous liquid contained between two rotating cylinders. *Philosophical Transactions of the Royal Society of London. Series A* **223**(**605–615**): 289–343.

10.25. Taylor, G. I. 1935. Distribution of velocity and temperature between concentric rotating cylinders. *Proceedings of the Royal Society of London. Series A: Mathematical and Physical Sciences Series* **151**(**874**): 494–512.

10.26. Taylor, G. I. 1936. Fluid friction between rotating surfaces: I and II. *Proceedings of the Royal Society of London. Series A: Mathematical and Physical Sciences Series* **157**(**892**): 546–578.

10.27. Lueptow, R. 2009. Taylor-Couette flow. *Scholarpedia* **4**(**11**): 6389.

10.28. Wereley, S. T. and Lueptow, R. M. 1998. Spatio-temporal character of non-wavy and wavy Taylor-Couette flow. *Journal of Fluid Mechanics* **364**: 59–80.

10.29. King, G. P., Li, Y., Lee, W., Swinney, H. L., and Marcus, P. S. 1984. Wave speeds in wavy Taylor-vortex flow. *Journal of Fluid Mechanics* **141**: 365–390.

10.30. Wendt, F. 1933. Turbulente Strömungen zwischen zwei eorierenden konaxialen zylindem (Turbulent flow between two rotating coaxial cylinders). *Ingenieur-Archiv* **4**: 577–595.

10.31. Lathrop, D. P., Fineberg, J., and Swinney, H. L. 1992. Turbulent flow between concentric rotating cylinders at large Reynolds number. *Physical Review Letters* **68**(10): 1515–1518.

10.32. Okaya, T. and Hasegawa, M. 1939–1940. On the turbulent boundary layers at the surface of two rotating co-axial cylinders. *Japanese Journal of Physics* **13**(2): 29–49.

10.33. Dou, H.-S., Khoo, B. C., and Yeo, K. S. 2010. Energy loss distribution in the plane Couette flow and the Taylor–Couette flow between concentric rotating cylinders. *International Journal of Thermal Science* **46**(3): 262–275.

10.34. Ustimenko, B. P. and Zmeikov, V. N. 1964. Hydrodynamics of a flow in an annular channel with an inner rotating cylinder. *Teplofizika Vysokikh Temperatur* **1**(2): 220–228.

10.35. Leont'ev, A. I. and Kirdyashkin, A. G. 1966. The theory of the convective heat transfer for the vertical flow of fluid. *Proceedings of Third International Heat Transfer Conference*, Chicago, IL, **vol. 1**, pp. 216–224.

10.36. General Electric. 1996. *Fluid Flow–Data Book.* Section 408.7: Drag-Rotating Cylinder Enclosed. Genium Publishing, New York.

10.37. Bruckner, R. J. 2009. Windage power loss in gas foil bearings and the rotor–stator clearance of high speed generators operating in high pressure environments. *ASME Turbo Expo 2009: Power for Land, Sea, and Air*, Orlando, FL. Paper No. GT2009-60118, pp. 263–270.

10.38. Wild, P. M., Djilali, N., and Vickers, G. W. 1996. Experimental and computational assessment of windage losses in rotating machinery. *Journal of Fluids Engineering* **118**(1): 116–122.

10.39. Black, H. 1969. Effects of hydraulic forces on annular pressure seals on the vibrations of centrifugal pump rotors. *Journal of Mechanical Engineering Science* **11**(2): 206–213.

10.40. Chossat, P. and Iooss, G. 1994. *The Couette-Taylor Problem*, Springer-Verlag, Berlin.

10.41. Dong, S. 2007. Direct numerical simulation of turbulent Taylor–Couette flow. *Journal of Fluid Mechanics* **587**: 373–393.

10.42. Li, J. D. 1998. Simulating the turbulent Taylor-Couette flows using DNS. *13th Australasia Fluid Mechanics Conference*, Melbourne, Australia.

10.43. Saari, J. 1998. Thermal analysis of high-speed induction machines. PhD thesis, Helsinki University of Technology, Espoo, Finland.

10.44. Yamada, Y. 1962. Torque resistance of a flow between rotating co-axial cylinders having axial flow. *Bulleting of JSME* **5**(20): 634–642.

10.45. Bilgen, E. and Boulos, R. 1973. Functional dependence of torque coefficient of coaxial cylinders on gas width and Reynolds numbers. *ASME Journal of Fluids Engineering* **95**(1): 122–126.

10.46. Awad, M. N. and Martin, W. J. 1997. Windage loss reduction study for TFTR pulse generator. *17th IEEE/NPSS Symposium on Fusion Engineering* **2**: 1125–1128.

10.47. Dorfman, L. A. 1963. *Hydrodynamic Resistance and the Heat Loss of Rotating Solids.* Oliver & Boyd, Edinburgh.

10.48. Polkowski, J. W. 1984. Turbulent flow between coaxial cylinders in the inner cylinder rotating. *Transactions of ASME—Journal of Engineering for Gas Turbines and Power* **106**(1): 128–135.

10.49. Tong, W. and Gott, B. E. B. 2002. Method of minimizing rotor body windage loss. U.S. Patent 6,438,820.

10.50. Johnson, T. J. 2009. Drag measurements across patterned surfaces in a low Reynolds number Couette flow facility. Master thesis, The University of Alabama, Tuscaloosa, AL.

10.51. Hawkins, J. S., Avery, Jr., R. M., and Super, L. 2005. Method of estimating engine cooling fan power losses. U.S. Patent 6,904,352.

10.52. American National Standards Institute (ANSI)/Air Movement and Control Association (AMCA). 2010. ANSI/AMCA Standard 320-07 Laboratory methods of testing fans for certified aerodynamic performance ration.

10.53. Tong, W. and Vandervort, C. L. 2002. Optimization of ventilating flow path at airgap exit in reverse flow generators. U.S. Patent 6,346,755.

10.54. Niggemann, R. E., Thomson, S. M., and Schneider, M. G. 1998. Method and apparatus for reducing windage losses in rotating equipment and electric motor/generator employing same. U.S. Patent 5,828,148.

10.55. Brook, R. W., Grant, J. J., and Miller, H. W. 1994. Rotor with reduced windage losses. European Patent Office EP0489859.

10.56. Ren, W.-M. and Carl, R. J. 2006. Rotor body containment shell with reduced windage losses. U.S. Patent 7,057,326.

10.57. Smith, S. H. and Stump, D. E. 2009. System and method for reducing windage losses in compressor motors. European Patent Application EP2024691.

10.58. Tong, W. and Vandervort, C. L. 2001. Spline retaining ring. U.S. Patent 6,285,110.

10.59. Pal, D. 2013. Electrical machine with reduced windage loss. U.S. Patent Application 2013/0076169.

10.60. Speranza, J. and Skoczylas, T. 2006. Increasing generator efficiency and capacity: A case study. *Electric Power 2006*, Atlanta, GA.

10.61. Ketteier, K. H. 1984. Über den einfluss der Wicklungsschaltung in induktionsmaschinen auf die zusatz-verluste und den einseitigen magnetischen. *Zug. ETZ-Archiv* **106**(**3**): 99–106.

10.62. Aoulkadi, M. 2011. Experimental determination of stray load losses in cage induction machines. PhD thesis, Technische Universität Darmstadt, Darmstadt, Germany.

10.63. Englebretson, S. C. 2009. Induction machine stray loss from inter-bar currents. PhD thesis, Massachusetts Institute of Technology, Cambridge, MA.

10.64. IEEE Power Engineering Society. 2004. IEEE 112-2004 Standard test procedure for polyphase induction motors and generators.

10.65. International Electrotechnical Commission (IEC). 20010. IEC 60034-2-1 Standard on efficiency measurement methods for low voltage AC motors.

10.66. Standardization Administration of China (SAC). 2002. GB 18613-2002. Limited values of energy efficiency and evaluating values of energy conservation of small and medium three-phase asynchronous motors.

10.67. Office of Energy Efficiency and Renewable Energy, U.S. Department of Energy. 2014. Premium efficiency motor selection and application guide: A handbook for industry. https://www.energy.gov/sites/prod/files/2014/04/f15/amo_motors_handbook_web.pdf.

11
Motor Cooling

The power output rating of an electric motor is often limited by the ability to provide additional electrical current through stator and rotor windings. Generally, the higher the current density, the larger the motor torque and output power. However, increasing current through the motor windings is constrained by temperature limitations imposed on the electric conductor insulation and permanent magnets (PMs) (for a PM motor). According to the Arrhenius equation [11.1, 11.2], the failure rate of an electric/electronic device is exponentially related to the reciprocal of the operating temperature. A fairly accurate approximation of the Arrhenius equation states that an operating temperature exceeds the thermal limit by $10°C$ and reduces the lifetime of the device by half. Therefore, effective cooling of stator and rotor windings contributes directly to the output capability, operation reliability, and lifetime of electric motors. This is especially true in motor end-winding regions where the configuration of cooling flow paths is extremely complicated and direct forced cooling is very difficult [11.3, 11.4]. In addition, the relatively poor heat dissipation from the rotor can cause an increase in rotor core losses, leading to an excessive temperature rise in the rotor and magnets. This may result in partially irreversible demagnetization of the magnets, particularly of sintered Nd–Fe–B magnets [11.5] and thus a decline in the performance of the motor.

In recent years, the development and production of high performance, high efficiency, and high power density electric motors have been accompanied by increasing heat fluxes at both the component and system levels. Over the years, significant advances have been made in the applications of air cooling and liquid cooling techniques to manage increased heat fluxes. As the prevailing market demands electric motors with better performance, lower power consumption, and smaller size, thermal management has emerged as an increasingly important aspect in motor design.

11.1 INTRODUCTION

Modern electric motors generate a large amount of heat during their normal operation as the result of various electrical, electromagnetic, mechanical, and windage losses. The great majority of motor losses are converted into heat, causing the temperature rise in windings, cores, magnets, bearings, and other components. Excessive temperatures are generally detrimental to motor performance and reliable operation. Hot spots in a motor due to nonuniform heat fluxes and improper cooling may result in damage to winding insulation or premature bearing failure and, in turn, greatly reduce the motor lifetime. For some special applications, thermal management is critical to motor performance. For instance, a telescope tracking motor requires an extremely tight control in the temperature rise relative to the ambient temperature because any sharp and excessive temperature changes may significantly influence the telescope's tracking accuracy Therefore, to ensure the safe and reliable operation of electric motors, all heat generated in the motor must be efficiently dissipated to the ambient air.

A variety of different cooling mechanisms may be used to cool electric motors to maintain the temperatures of stator and rotor windings, as well as PMs and bearings, below the maximum allowable limit. Today, cooling of an electric motor remains a major challenge for motor engineers and designers because of a number of factors: (*a*) a great number of motor parts with irregular geometries, (*b*) the nonuniform heat generation in different heat sources (stator and rotor windings, cores, bearings, *etc.*) due to various power losses, (*c*) the complexity in cooling

DOI: 10.1201/9781003097716-11

flow paths, (*d*) cooling flow splitting and mixing, (*e*) interactions between the rotor-induced rotating flows and the nonrotating flows at the stationary components, (*f*) a variety of thermal exchange modes (conduction, convection, radiation, *etc.*) that are involved simultaneously, and (*g*) the compact and lightweight motor design, as a result of growing global demand for high-power-density motors. Consequently, the thermal phenomena inside an electric motor are considerably complex.

Cooling of high-heat-flux heat sources can be generally divided into two steps [11.6]: (*a*) heat extraction, in which heat is removed from a high thermal density region (*e.g.*, motor windings) and reformatted to a larger area, and (*b*) heat rejection, in which the heat is dissipated convectively to the surrounding environment.

Remarkable progress in heat extraction and heat rejection have been made in the last several decades, including adoptions of advanced cooling techniques such as heat pipe, microchannel, heat pump, high-conductivity material, thermoelectric cooler, carbon nanotube, and recently proposed ionic wind generator in electric/electronic cooling. However, in cooling of electric motors, technologies have not yet made significant progress in the past several decades. As a result of the technology stagnation, the design and development of highly efficient cooling systems remain challenges today to motor engineers and manufacturers.

11.1.1 PASSIVE AND ACTIVE COOLING TECHNIQUES

Cooling techniques are often divided into two categories: passive cooling and active cooling, depending on what the heat transfer mode would be or how cooling flows are generated. Passive cooling refers to cooling with no active mechanism such as fan or pump. Passive cooling techniques have been developed for a long time for motors that have low power losses. In a passive cooling mode, motor cooling primarily relies on heat conduction to transport the generated heat from the heat sources (*e.g.*, stator winding) to the motor frame (*e.g.*, motor housing and endbells) and then to dissipate the heat to the environment by natural convection and radiation. By eliminating fans or pumps, motor operation reliability is significantly increased. Compared with active cooling techniques, passive cooling techniques have lower costs, lower noise emission, and no power consumption. However, passive cooling techniques are restricted to low thermal load conditions. For large-size, heavy-duty motors, active cooling is the only choice.

In contrast, active cooling techniques require the use of external devices such as fans, blowers, and pumps to force a coolant (air or liquid) flowing through cooling channels either inside or outside, or both, of the motor. With the development of modern electric motors, which have high power density and consequently high power losses, active cooling is required to remove the waste heat produced by motor components to keep these components within permissible operating temperature limits.

The magnitude of the heat transfer coefficient is determined by a number of parameters such as heat transfer mode, geometry, flow path dimensions, coolant type, flowrate, and flow condition (laminar or turbulent). The comparison of heat transfer coefficients for different heat transfer modes is presented in Table 11.1.

The selection of appropriate and feasible cooling techniques depends basically on the required cooling load, cooling efficiency, cooling system operation reliability, cost of cooling system, temperature difference between the heat source and the environment, coolant thermophysical properties, geometries of heated sources, noise, and vibration. All these determining factors must be taken into account. The trade-off between these factors is highly desired in order to choose the best cooling technique for each particular application. Three liquids, namely, air, fluorocarbon (FC), and water, are chosen to compare their heat transfer coefficients for natural and forced convection and forced jet.

The results show that all cooling modes that involve a phase change can make very high heat transfer coefficients. Water is always the best coolant. However, for motor cooling applications, water has to be restricted in indirect cooling application. In all cases, air cooling has the lowest heat transfer rate, compared with FC and water.

TABLE 11.1
Heat Transfer Coefficients of Different Heat Transfer Modes with Different Coolants

Heat Transfer Mode	Heat Transfer Coefficient (W/m²·K)	References
Conduction	5–25	
Natural convection (air)	5–30	[11.7, 11.8]
Natural convection (FC)	10–100	[11.7, 11.8]
Natural convection (water)	30–300	[11.7, 11.8]
Forced convection (air)	20–200	[11.7, 11.8]
Forced convection (FC)	50–300	[11.7, 11.8]
Forced convection (water)	300–8,000	[11.7, 11.8]
Forced jet (air)	200–800	[11.7, 11.8]
Forced jet (FC)	2,000–20,000	[11.7, 11.8]
Forced jet (water)	8,000–50,000	[11.7, 11.8]
Pool boiling (water)	2,500–3,500	[11.9]
Forced boiling (water)	5,000–100,000	[11.9]
Condensing (water)—vertical surface	4,000–11,300	[11.9]
Condensing (water)—Outside of horizontal tube	9,500–25,000	[11.9]
Radiation	5–30	

Note: Conductive heat transfer coefficient and radiative heat transfer coefficient are defined as $h_c = \dfrac{-\nabla(kT)}{(T_h - T_c)}$ and $h_r = \sigma\varepsilon\dfrac{T_h^4 - T_c^4}{T_h - T_c}$, respectively. They are estimated as the same order as natural convection.

11.1.2 HEAT TRANSFER ENHANCEMENT TECHNIQUES

A race for increasing power density of electric motors has provided the driving force behind the scientific research and technological advancement in the field of heat transfer enhancement. It is now widely accepted that the thermal management in high-power-density motors becomes critically important. Efficient cooling plays an increasingly key role that controls the performance and stability of electric motors.

It has been recognized that the primary limitation to performance of a convective cooling system is the boundary layer of motionless air that adheres to and envelops all surfaces of the cooling system. Within this boundary layer region, diffusive transport is the dominant mechanism for heat transfer. The resulting thermal bottleneck largely determines the thermal resistance and convective heat transfer coefficient and thus cooling efficiency of the cooling system. The heat transfer enhancement can be achieved by disturbing cooling flows to lower the thickness of the boundary layer and increasing the convective contact area.

A variety of heat transfer enhancement techniques have been developed for adapting to the rapid increase in thermal load in electric and electronic devices [11.10]. For passive techniques, heat transfer enhancements are achieved by employing special surface geometries or fluid additives. Common surface geometries include coated surfaces, rough surfaces, and extended surfaces to increase the film heat transfer coefficient and to reduce the thermal resistance. Additives for liquid include solid particles or gas bubbles in single-phase flows and liquid trace additives for boiling systems.

Active techniques employ turbulators such as displaced inserts, pin fins, dimpled surfaces, surfaces with arrays of protrusions, and swirl chambers. All of these turbulators act to increase rotating/secondary flows or vortices on heat surfaces for augmenting forced convective heat transfer. These rotating/secondary flows and vortices not only increase rotating/secondary advection of heat away from surfaces but also increase 3D turbulence production by increasing shear and creating gradients of velocity over significant flow volumes.

The selection and application of heat transfer enhancement techniques are core concerns for modern motor engineers and designers. Many researchers have contributed their efforts toward a better understanding of heat transfer techniques.

11.2 CONDUCTIVE HEAT TRANSFER TECHNIQUES

Heat conduction is a mechanism of heat transfer, taking place if there is a temperature gradient in a solid body or between contacted solid bodies. From the microscopic standpoint, conduction is relevant to the movement of particles (*e.g.*, atoms or molecules) in a material. The particles vibrate randomly about fixed positions. Since the rise in temperature will result in an increase in the particle kinetic energy, the higher the temperature, the greater the particle vibration. When particles collide with their neighbors, part of the kinetic energy will be transferred from the particles that have higher energy to the particles that have lower energy until thermal equilibrium is reached. At the state of thermal equilibrium, there are an equal number of collisions resulting in an energy gain as there are collisions resulting in an energy loss. On average, there is no net energy transfer resulting from the collisions of particles.

11.2.1 CONDUCTIVE HEAT FLUX AND ENERGY EQUATIONS

The conductive heat flux vector $\mathbf{q}''$ is proportional to the thermal conductivity k_n and the temperature gradient:

$$\mathbf{q}'' = -\nabla \left(k_n T \right) \tag{11.1}$$

where T is the temperature. Generally, thermal conductivity is a function of temperature. In anisotropic materials such as insulations and silicon steel laminations, thermal conductivity k_n may have different values in different directions of the material spatial orientation at a specific temperature T, that is,

$$k_n = k_n \left(n_1, n_2, n_3, T \right) \tag{11.2}$$

The negative sign in Equation 11.1 denotes that the heat flow is always from the region of high temperature to the region of low temperature. Equation 11.1 in 3D Cartesian, cylindrical, and spherical coordinate systems (see Figure 11.1) can be expressed as

$$q'' = -\left[\frac{\partial \left(k_x T \right)}{\partial x} + \frac{\partial \left(k_y T \right)}{\partial y} + \frac{\partial \left(k_z T \right)}{\partial z} \right] \tag{11.3a}$$

$$q'' = -\left[\frac{\partial \left(k_x T \right)}{\partial r} + \frac{1}{r} \frac{\partial \left(k_\phi T \right)}{\partial \phi} + \frac{\partial \left(k_z T \right)}{\partial z} \right] \tag{11.3b}$$

$$q'' = -\left[\frac{\partial \left(k_r T \right)}{\partial r} + \frac{1}{r \sin \theta} \frac{\partial \left(k_\phi T \right)}{\partial \phi} + \frac{1}{r} \frac{\partial \left(k_\theta T \right)}{\partial \theta} \right] \tag{11.3c}$$

respectively.

Energy equation governing conductive thermal energy transport in solids is based on energy conservation in a differential control volume through which energy transfer is exclusively by conduction and can be generally expressed as

$$\rho c_p = \nabla^2 \left(k_n T \right) + q''' \tag{11.4}$$

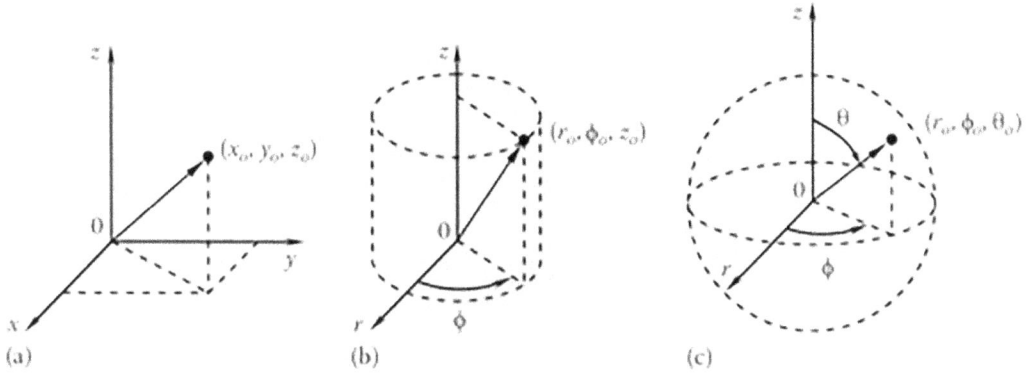

FIGURE 11.1 Coordinate systems: (*a*) Cartesian, (*b*) cylindrical, and (*c*) spherical.

where

ρ is the density, kg/m^3

c_p is the specific heat, J/(kg·K)

q''' is the heat generation in the control volume, W/m^3

Similarly, Equation 11.4 can be expressed in Cartesian, cylindrical, and spherical coordinate systems as

$$\rho c_p \frac{\partial T}{\partial t} = \frac{\partial^2 (k_x T)}{\partial x^2} + \frac{\partial^2 (k_y T)}{\partial y^2} + \frac{\partial^2 (k_z T)}{\partial z^2} + q''' \tag{11.5a}$$

$$\rho c_p \frac{\partial T}{\partial t} = \frac{1}{r} \frac{\partial}{\partial r}\left[r \frac{\partial (k_r T)}{\partial r} \right] + \frac{1}{r^2} \frac{\partial^2 (k_\phi T)}{\partial \phi^2} + \frac{\partial^2 (k_z T)}{\partial z^2} + q''' \tag{11.5b}$$

$$\rho c_p \frac{\partial T}{\partial t} = \frac{1}{r^2} \frac{\partial}{\partial r}\left[r^2 \frac{\partial (k_r T)}{\partial r} \right] + \frac{1}{r^2 \sin^2 \theta} \frac{\partial^2 (k_\phi T)}{\partial^2 \phi} + \frac{1}{r^2 \sin \theta} \frac{\partial}{\partial \theta}\left[\sin \theta \frac{\partial (k_\theta T)}{\partial \theta} \right] + q''' \tag{11.5c}$$

For isotropic materials, thermal conductivity is identical in all directions (*i.e.*, $k_x = k_y = k_z = k$, $k_r = k_\phi = k_z = k$, $k_r = k_\phi = k_\theta = k$). Taking this condition and assuming k is a constant, the aforementioned equations can be greatly simplified as

$$\frac{1}{\alpha} \frac{\partial T}{\partial t} = \left(\frac{\partial^2 T}{\partial x^2} + \frac{\partial^2 T}{\partial y^2} + \frac{\partial^2 T}{\partial z^2} \right) + \frac{q'''}{k} \tag{11.6a}$$

$$\frac{1}{\alpha} \frac{\partial T}{\partial t} = \left[\frac{1}{r} \frac{\partial}{\partial r}\left(r \frac{\partial T}{\partial r} \right) + \frac{1}{r^2} \frac{\partial^2 T}{\partial \phi^2} + \frac{\partial^2 T}{\partial z^2} \right] + \frac{q'''}{k} \tag{11.6b}$$

$$\frac{1}{\alpha} \frac{\partial T}{\partial t} = \left[\frac{1}{r^2} \frac{\partial}{\partial r}\left(r^2 \frac{\partial T}{\partial r} \right) + \frac{1}{r^2 \sin^2 \theta} \frac{\partial^2 T}{\partial \phi^2} + \frac{1}{r^2 \sin \theta} \frac{\partial}{\partial \theta}\left(\sin \theta \frac{\partial T}{\partial \theta} \right) \right] + \frac{q'''}{k} \tag{11.6c}$$

where α is the thermal diffusivity of material, $\alpha \equiv k/\rho c_p$.

FIGURE 11.2   Encapsulation of stator winding.

11.2.2   Encapsulation and Impregnation of Electric Motor

As discussed in Chapter 4, the objectives of encapsulation and/or impregnation of electric motors are as follows:

- The encapsulation/impregnation of an electric motor can remarkably enhance electrical insulation of motor windings from damage during normal operation and thus extend the motor lifetime.
- By filling voids in motor windings with epoxy resin and creating a rigid mass with optimum dielectric strength, an encapsulated and/or impregnated motor can significantly lower the thermal resistance from internal heat sources (*e.g.*, the stator winding) to the motor housing.
- The encapsulation/impregnation increases the overall protection to motors, especially chemical and environmental protections.
- By integrating stator components (*e.g.*, stator winding, laminations, and insulations) into a rigid mass, it increases the stator stabilization and thus reduces motor vibrations and noise levels.

Though many materials can be used in motor encapsulation/impregnation, the most popular encapsulation material is epoxy resin, used by many motor manufacturers in a wide range of applications. In the selection of encapsulation materials, thermal conductivity is one of the most important parameters. An epoxy-encapsulated stator is shown in Figure 11.2.

11.2.3   Enhanced Heat Transfer Using High-Thermal-Conductivity Material

One of the effective ways in motor winding cooling is to place high-thermal-conductivity materials between winding coils inside the winding slots and/or to interleave these materials into stacked motor laminations [11.11]. As shown in Figure 11.3a, the thermally conductive strips provide low thermal-resistant paths through which heat is transferred from internal high-temperature components (*i.e.*, stator winding coils) to the stator laminations and housing and then dissipated to the

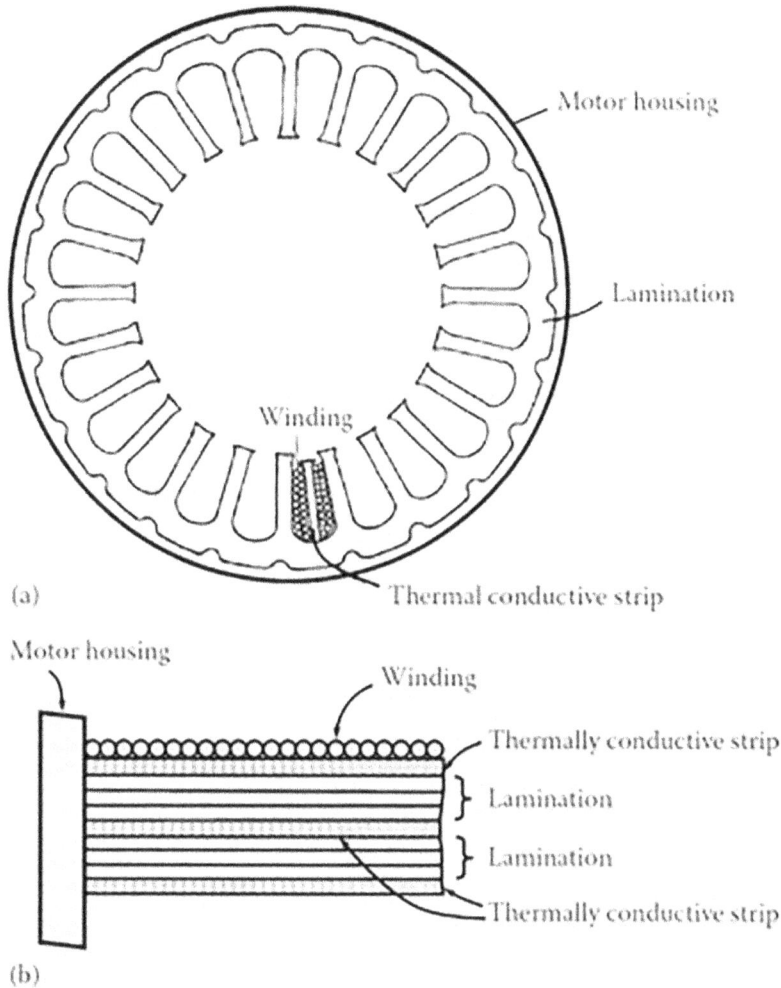

FIGURE 11.3 Using thermally conductive strips in motor cooling: (*a*) placed between winding coils and (*b*) interleaved into stacked motor laminations (U.S. Patent 6,777,835) [11.11]. (Courtesy of the U.S. Patent and Trademark Office, Alexandria, VA.)

environment. In Figure 11.3*b*, the thermally conductive strips are sandwiched between stacked stator laminations. In this way, heat generated inside the stator core due to the losses from eddy currents and hysteresis is conducted directly from the interior of the stator to the motor outer portions by means of these strips. Because these thermally conductive strips are directly contacted with heat sources, the cooling efficiency is considerably high. As a fact, this cooling technique can result in a significant reduction of weight and volume of motor, along with a substantial increase in the power density while operating at a moderate temperature above ambient.

There are a variety of advanced materials having exceptionally high thermal conductivities. Among them, the monolithic carbonaceous material is well known as the best thermally conductive materials, such as diamonds that are made from chemical vapor deposition (CVD) at low pressures and low temperatures. One of the remarkable properties of diamond is its unsurpassed thermal conductivity. With a value of $1800\,W/m{\cdot}K$ and higher at room temperature, it exceeds that of copper (approximately $400\,W/m{\cdot}K$ at room temperature) by a factor of 4.5 [11.12]. Furthermore, diamond had been the world's hardest material until recent years [11.13].

Coefficient of thermal expansion (CTE), in units of μm/(m·K) or 10^{-6} K^{-1}, describes how the dimensions of an object change with a change in temperature. In most cases, the phenomenon of thermal expansion or contraction is isotropic, *i.e.*, the material has the identical expansion or shrinkage rate in all directions. However, anisotropic materials such as crystals and steel laminates have different CTEs in different directions. Under these circumstances, it is necessary to treat the CTE as a tensor with up to six independent elements [11.14].

In addition to thermal conductivity, thermal stress arising from differences in CTEs between mating parts is a key issue in motor design. In general, materials with low CTE are highly desirable for their inherent dimensional stabilities. At a temperature of 300 K (26.85°C), the CTE of CVD diamond is 1×10^{-6} K^{-1}, which is much lower than that of copper (17×10^{-6} K^{-1} at 20°C) and aluminum (24×10^{-6} K^{-1} at 20°C).

The superior properties of carbon fibers such as high thermal conductivity, low thermal expansion, high strength, lightweight, and corrosion resistance have made these materials especially useful for applications where heat removal is important. Some commercial carbon fibers have nominal thermal conductivities as high as 1,100 W/m·K. Experimental discontinuous fibers reportedly have thermal conductivities of 2,000 W/m·K [11.15]. The development of carbon nanotubes with extremely high thermal conductivities is a major breakthrough and will be discussed in later sections.

More recently, a new advanced material called Pyriod® HT pyrolytic graphite has been developed for enhancing heat transfer in electronic cooling. With a single crystalline structure and a high purity (99.999%), the thermal conductivity of the material can be as high as 1,700 W/m·K [11.16], which is about four times higher than that of copper and seven times of aluminum (205–250 W/m·K at room temperature). Moreover, this material exhibits four times the ability to sustain tensile load than natural graphite material, nearly five times the flexural load and six times Young's modulus. The material can work at extremely high temperature, up to 3,300°C. Because of its superior thermal and mechanical properties, Pyriod HT can be used for thermal management applications in a variety of industries including aerospace, defense, electronic, automobile, medical device, and power generation.

11.2.4 Using Self-Adhesive Magnet Wire for Fabricating Stator Winding

The use of self-adhesive magnet wires to make stator windings has a long history in linear motor manufacturing. In a linear motor, each stator coil is produced by winding self-adhesive wire into a coil shape, subjecting the coil to rapid electric heating to activate a bounding layer on the wire and pressing the coil into its final shape. When the coil is cooled down, it forms the self-supported robust coil (Figure 11.4).

This winding technique can be easily adopted into electric motors. The use of self-adhesive wires may offer advantages over regular wires in some winding applications: (*a*) As a semisolid body with minimized voids inside the coil, conductive heat transfer in the self-adhesive winding

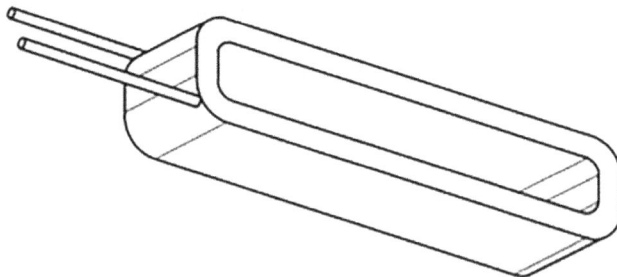

FIGURE 11.4 Stator coil made of self-adhesive magnet wire.

is greatly enhanced, and in turn, hot spots of the winding are mitigated or eliminated. (*b*) Using self-adhesive wires allows coils to be self-supported so that bobbins as well as coil taping are no longer necessary. (*c*) A stator winding made from self-adhesive wires has a much smaller volume compared to a conventional winding made from regular wires, leading to higher power density.

11.3 NATURAL CONVECTION COOLING WITH FINS

With very low heat removal capacity, natural convection cooling techniques are limited to applications with low thermal loads. Natural convection flows are induced by buoyancy forces due to density gradients caused by temperature variations in the air layer adjacent to the heated surfaces. As shown in Table 11.1, without applying heat transfer enhancement techniques, the heat transfer coefficient of natural air convection only ranges from 5 to 30 W/m²·*K*. As a result, natural convection cooling has a limited role in motor cooling.

11.3.1 COOLING FIN

The rapid increase in power density of advanced motors and the miniaturization of motion control devices have created a need for improved cooling technologies to achieve high heat dissipation rates. Under low or moderate heat flux conditions, the most common approach in passive cooling techniques is to utilize natural convective flows that are induced and developed along hot surfaces due to density gradients in fluids (air or liquid).

In order to achieve reliability and optimal performance of motion systems, appropriate thermal management is imperative. Thus, it is desirable to increase the overall convective surface area exposed to the cooling fluid for enhancing heat transfer rates between the heat-dissipating surface and the convective cooling fluid. One of the effective methods to accomplish this objective is to directly employ various cooling fins on the external surfaces of the motor housing and endbells. Such augmentation can result in as much as an order-of-magnitude increase in the heat transfer rate without altering other motor components.

There are many techniques available to enhance heat dissipation from electric machines. Heatsinks are widely employed in electronic systems where space is limited. The use of passive natural convection-cooled longitudinal straight plate fin heatsinks offers substantial advantages in cost and reliability but is often accompanied by relatively low heat transfer rates.

A heatsink performance is characterized by the *j*-Colburn factor,

$$j = \frac{\text{St}}{\text{Pr}^{2/3}} \tag{11.7}$$

where Stanton number St is a dimensionless number that measures the ratio of heat transfer into a fluid to the thermal capacity of the fluid, used in forced convection

$$\text{St} = \frac{\text{Nu}}{\text{Re}\,\text{Pr}} = \frac{h}{\rho u c_p} \tag{11.8}$$

Prandtl number Pr is defined as the ratio of momentum diffusivity (kinematic viscosity v) to thermal diffusivity α, used in natural and forced convection

$$\text{Pr} = \frac{v}{\alpha} = \frac{c_p \mu}{k} \tag{11.9}$$

Reynolds number Re is defined as the ratio of inertia to viscous forces

$$\text{Re} = \rho u l / \mu \tag{11.10}$$

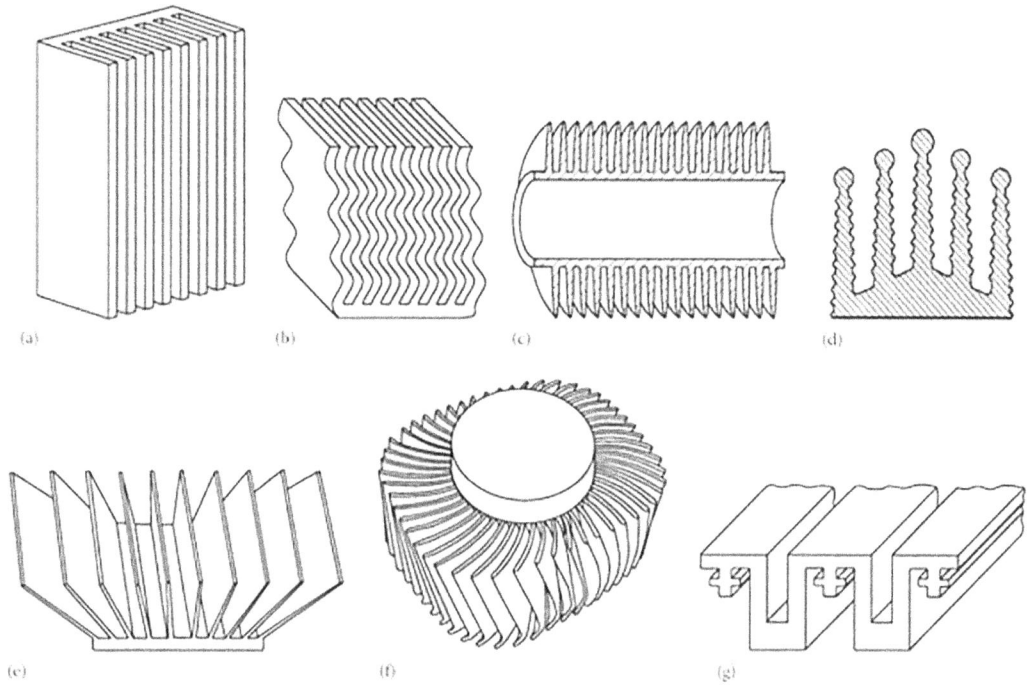

FIGURE 11.5 Types of plate fins: (*a*) straight plate fin, (*b*) wavy plate fin, (*c*) radial-helical plate fin, (*d*) wavy protruding strip plate fins, (*e*) flared plate fin, (*f*) 3D radial plate fin, and (*g*) constructive plate fin.

Cooling fins can be generally classified as plate fins, strip fins, or pin fins, according to their cross-sectional aspect ratio. Plate fins that have highest aspect ratios are usually made of aluminum or copper by extrusion or casting processing. As shown in Figure 11.5, there are various types of plate fins, including (*a*) straight plate fin, (*b*) wavy plate fin, (*c*) radial–helical plate fin, (*d*) wavy protruding strip plate fins [11.17], (*e*) flared plate fin, (*f*) 3D radial plate fin, and (*g*) constructive fin [11.18, 11.19]. The use of straight plate fins offers substantial advantages of low cost, simple structure, and ease of manufacturing but is often accompanied by relatively low heat transfer rates. For a flat plate fin heatsink, cooling channels are formed by adjacent plate fins. When cooling flow passes through these channels, the velocity profile is identical except at the flow inlet regions. This indicates that the thickness of the flow boundary layers remains constant both along the channel and across the channels. Bar-Cohen and Jelinek [11.20] developed guidelines and equations for designing optimum plate fin arrays.

With wavy plate fins, the cooling airflow changes its flowing direction periodically along the channel, causing the local flow separation and reattachment with the passage side surfaces. The disturbance between the cooling flow and the wavy plate fins reduces the boundary layer thickness and, as a result, increases the heat transfer coefficient. Because flared plate fins and 3D radial plate fins can effectively disturb local cooling flows, they can provide even higher heat transfer enhancement than other types of plate fins. In all designs, rough surfaces of heatsink help generate local turbulent flows that have a higher heat transfer coefficient.

Many studies have been focused on the thermal performance of fin heatsinks. One of the major parameters affecting the fin performance is the cross-sectional aspect ratio, which is defined as the fin length to fin width. While the aspect ratio of rectangular plate fin is much larger than 1, the aspect ratio of pin fins is approximately equal to unit. It is widely accepted that the partition of fin into smaller strips reduces the thermal boundary layers around fin strips, thus increasing local turbulence level and improving rates of heat transfer, especially when staggered arrangements of strip fins are used.

The thermal performance of pin fin heatsinks has been compared with those of other types of fin heatsinks. Although the convective heat transfer coefficients for all types of heatsinks increase with the flow Reynolds number in cooling channels between adjacent fins, smaller thermal resistance and larger convective heat transfer coefficient are obtained for the pin fin heatsinks. As a result, pin fin heatsinks have better thermal performance over similar plate fin and strip fin heatsinks under similar operating conditions. This indicates that the pin fin technology with excellent thermal characteristics is an effective solution to electric machine cooling and electronics cooling.

11.3.2 FIN OPTIMIZATION

The optimization of motor cooling plate fins can be achieved by choosing appropriate values from the following design parameters: (*a*) fin thickness *t*, (*b*) fin spacing between adjacent fins *b*, (*c*) fin height *H*, (*d*) number of fins *n*, and (*e*) thickness of the base plate, *d*.

An effective way to optimize fin design is to use web-based tools. Though these tools were originally developed for the design of heatsinks, not quite the same for motor cooling (*e.g.*, the thermal load Q is generated inside the motor rather than entered at the end of the motor), it can provide a starting point for optimizing fin parameters such as fin thickness *t*, fin spacing *b*, and fin OD *D*, as shown in Figure 11.6. With the model in this figure, natural convection heat transfer can be calculated based on the geometry, material properties, and the boundary and ambient conditions [11.21]. Options are available to calculate both the total heat flowrate Q and the isothermal base plate temperature T_s through the independent control of the geometric parameters.

The straight plate fin efficiency is calculated as [11.22]

$$\eta_f = \frac{\tanh(mH)}{mH} \tag{11.11}$$

where
   H is the fin height
   m is the fin parameter, which is given as

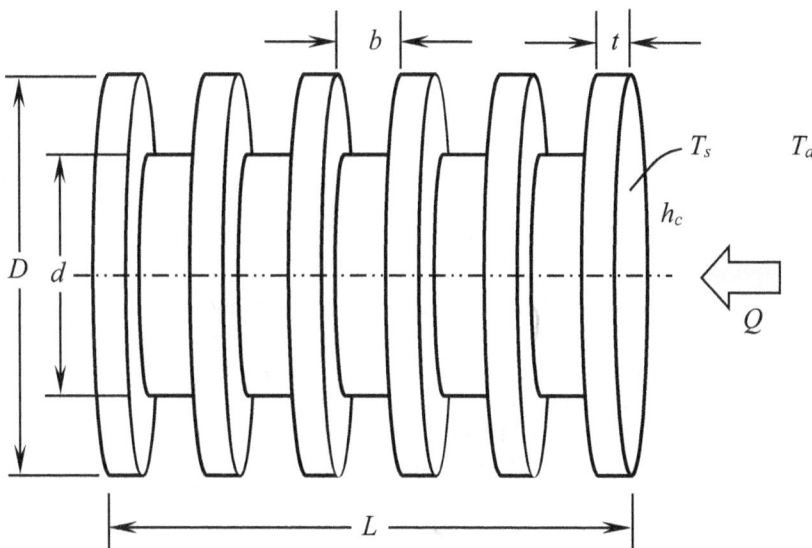

FIGURE 11.6 Thermal model of external fins on the outside surface of motor.

$$m = \sqrt{\frac{2h}{k_f t}} \tag{11.12}$$

where

    h is the convective heat transfer coefficient
    k_f is the thermal conductivity of the fin material
    t is the fin thickness

For radial plate fins, the fin height is calculated as

$$H = \frac{D - d}{2} \tag{11.13}$$

In calculating the heat transfer of heatsink, two nondimensional parameters are important, the Nusselt number Nu and the Raleigh number Ra, defined as [11.23]

$$\text{Nu} = \frac{Q}{A_t \Delta T} \frac{b}{k_a} \tag{11.14}$$

$$\text{Ra} = \frac{g \beta \Delta T b^3}{\alpha v} \frac{b}{D} \tag{11.15}$$

where

    A_t is the total area of the outer surface exposed to air
    k_a is the thermal conductivity of air
    β is the thermal expansion coefficient
    g is the acceleration of gravity
    α is the thermal diffusivity
    v is the kinematic viscosity
    $\Delta T = T_s - T_a$, is the temperature difference between the heat source and the ambient

The overall motor-to-ambient thermal resistance R_{th} (in °C/W) is represented by

$$R_{th} = \frac{\Delta T}{Q} \tag{11.16}$$

To demonstrate how a web-based tool works, use the motor and fin parameters as follows: $d = 250\,\text{mm}$, $t = 5\,\text{mm}$, $D_1 = 340\,\text{mm}$ (for $H = 45\,\text{mm}$), $D_2 = 330\,\text{mm}$ (for $H = 40\,\text{mm}$), $L = 293\,\text{mm}$, $Q = 1{,}000\,\text{W}$, and $T_a = 25°C$. The calculated results are displayed in Figure 11.7. With the increase in the fin ratio (defined as the ratio of the fin spacing b to the fin thickness t), the total thermal resistance decreases at lower fin ratios and increases at higher fin ratios. This is because for a small fin ratio, a flow channel, which is formed between adjacent fins, is too narrow to allow enough cooling air to pass through it. Therefore, increasing the channel width (*i.e.*, the fin ratio) will significantly improve the thermal transport between the cooling air and the fins. In contrast, as the fin ratio becomes very large, since a portion of the cooling air at the channel center region has a weak influence on the heat transport, the thermal performance will become worse for a very large fin spacing.

From Figure 11.7, it can be seen that the lowest thermal resistance occurs at a fin ratio of 1.6, regardless of the fin height. This indicates that for $t = 5\,\text{mm}$, the best fin spacing is $8\,\text{mm}$. The corresponding fin number is 22.

Increasing the fin height will result in a decrease in the thermal resistance and, in turn, an increase in the heat transfer rate. As shown in Figure 11.8, when the fin height increases from 30 to 45 mm,

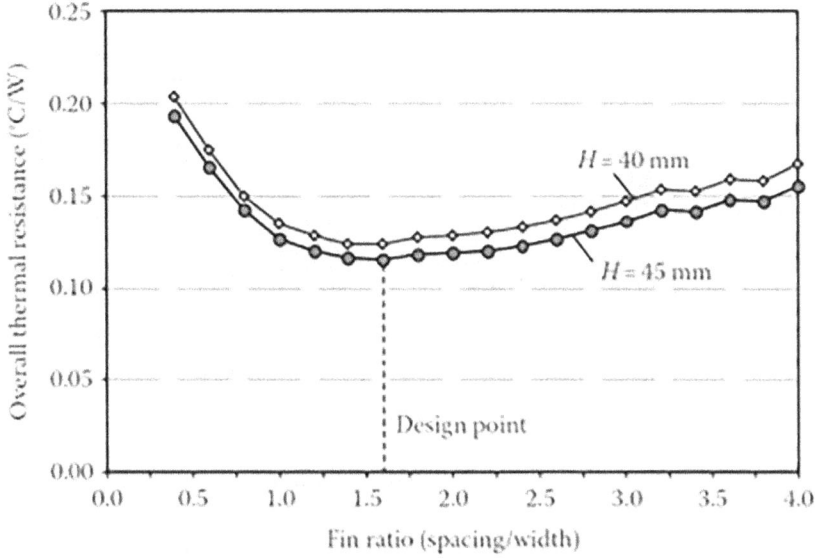

FIGURE 11.7 Effect of fin ratio on the overall thermal resistance for a motor with finned housing. The design point is chosen at the point with the minimum overall thermal resistance.

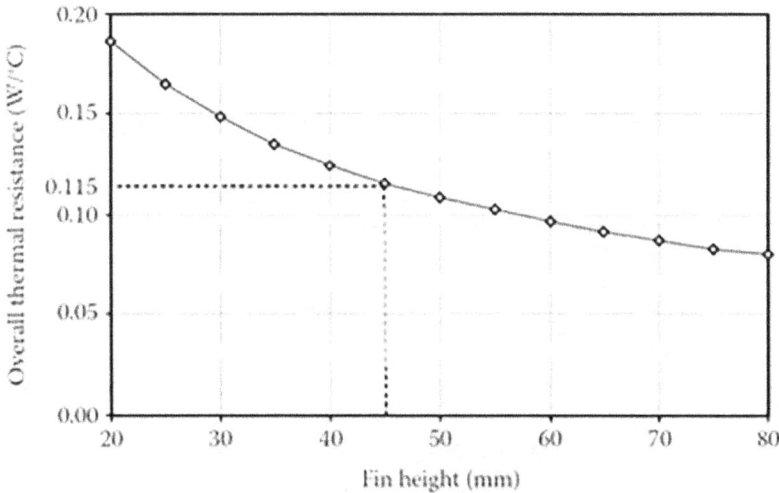

FIGURE 11.8 Effect of fin height on the overall thermal resistance of a motor with finned housing.

the overall thermal resistance reduces by about 22.8%. However, the fin height may be limited by several factors such as the motor's available space, manufacturing restrictions, and costs.

The calculated results can be used for estimating the improvement of a finned housing over a bald housing. The overall thermal resistance is 0.115°C/W for the finned housing and 0.18°C/W for the bald housing, indicating a 36% thermal improvement.

11.3.3 Heatsinks Manufactured with Additive Manufacturing Process

The increasing focus on energy efficiency, environmental sustainability, and high cooling densities is providing strong drivers to optimize the performance of cooling systems for electric machines

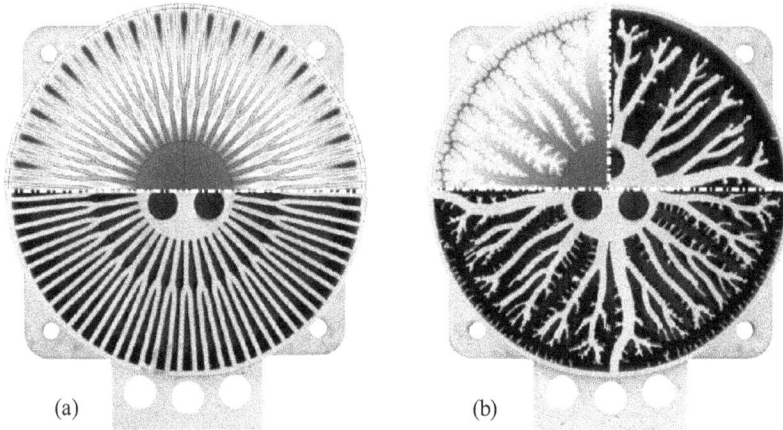

FIGURE 11.9 Comparison of simulation results (top or top left) and the heatsink prototype (bottom) between the designs with different optimizing approaches: (*a*) parametric optimization and (*b*) topology optimization [11.24]. The SLM is used for the manufacturing of complex topology optimized structures because of the great design freedom provided by a tool-free, layer-wise production.

and energy conversion devices. Over the past few decades, significant effort has been invested in designing and optimizing heatsinks for a wide variety of industrial applications. The production of heatsink involves two closely related stages, namely, designing and manufacturing. Heatsink performance is generally dominated by the surface area exposed to the cooling medium. In designing modern heatsinks with high cooling performance, it requires to maximize the surface-to-volume ratio of heatsink with an acceptable level of pressure drop.

A most remarkable benefit of using 3D printing technology for manufacturing heatsinks is that it could help create complex designs, which allows achieving the highest precision of internal structures and increasing the heatsink thermal efficiency. In addition, 3D printing could reduce the overall weight of heatsink, improve the thermal efficiency, and eliminate the wastage of material. However, there are some limitations and challenges when using 3D printing to produce metal parts, including long manufacturing time, low productivity, high surface roughness, residual stress, and high cost. It is noted that 3D-printed metal parts are often plagued with high porosity, which occurs during the printing process. While the porosity is highly undesirable for most metal parts, it may be desired for heatsinks because it can increase the surface–volume ratio and enhance convective heat transfer.

The researchers at the Fraunhofer Research Institute have adopted the topology optimization concept to design heatsinks for improving thermal performance and decreasing pressure drops. Compared to conventional design based on the engineers' skills and experience, topology optimization is a systematic approach that can lead to unintuitive and innovative designs [11.24]. The comparison of simulation results and design prototypes between the innovative topology optimization and conventional parametric optimization approaches is shown in Figure 11.9. The experimental results confirm that the heatsink with the topology optimization has a better thermal performance than that with the parametric optimization, with the temperature difference of $3°C$–$5°C$ in the range of heat load of 50–90 W.

While optimized topologies and microstructures can offer powerful cooling performance, complex geometries of heatsinks have made them difficult to produce with conventional manufacturing processes. The fast development of additive manufacturing technology has eliminated many design constraints imposed by traditional technologies. Now, it becomes more popular for engineers to use the additive manufacturing process to produce heatsinks, such as selective laser melting (SLM) and 3D printing.

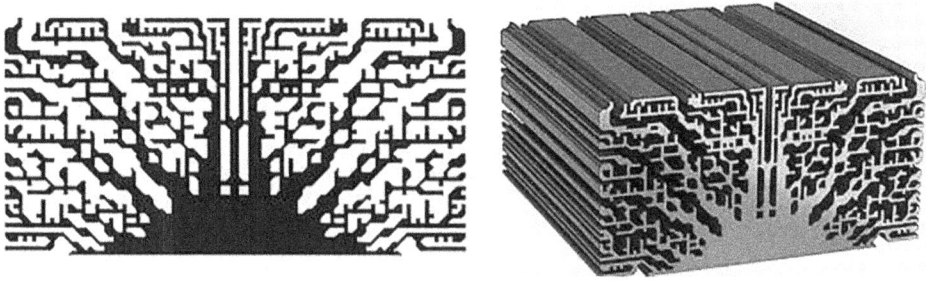

FIGURE 11.10 The heatsink is designed using thermal analysis software to identify an arbitrary heatsink topology and fabricated by 3D printing. (Courtesy of Mentor Graphics, Wilsonville, OR.)

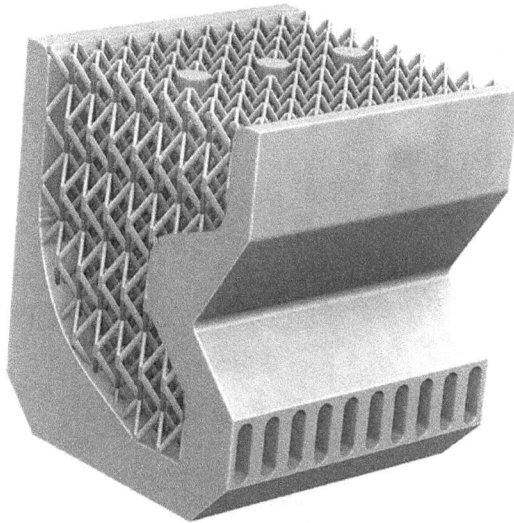

FIGURE 11.11 The highly efficient, award-winning 3D printable heatsink, which was designed by a team of mechanical engineering students at Purdue University. (Courtesy of Purdue University, West Lafayette, IN.)

3D printing technology is changing the future of the manufacturing industry across all over the world. This technology allows engineers to use their unlimited imaginations to design and manufacture much more complex three-dimensional, one-piece heatsinks in successive layers under computer control. As an example (Figure 11.10), the heatsink is designed by means of thermal analysis software to achieve the optimal topology of heatsink.

Due to the tremendously expanded heat transfer surfaces, more heat can be carried away from the heat source and dissipated to the environment via the novel heatsinks without requiring additional devices and power. Figure 11.11 displays a 3D-printed metal heatsink, designed by a student team at Purdue University. The sharkskin inspired design took the No. 1 spot at the 2020 Virtual Student HeatSink Design Challenge, a U.S. university competition organized by the ASME K-16 Committee and IEEE [11.25].

11.3.4 APPLICATIONS OF VARIOUS FINS IN MOTOR COOLING

Given that motor components are enclosed in a motor housing, the most important component when it comes to determining the heat dissipation capability is the motor housing. The design of the motor

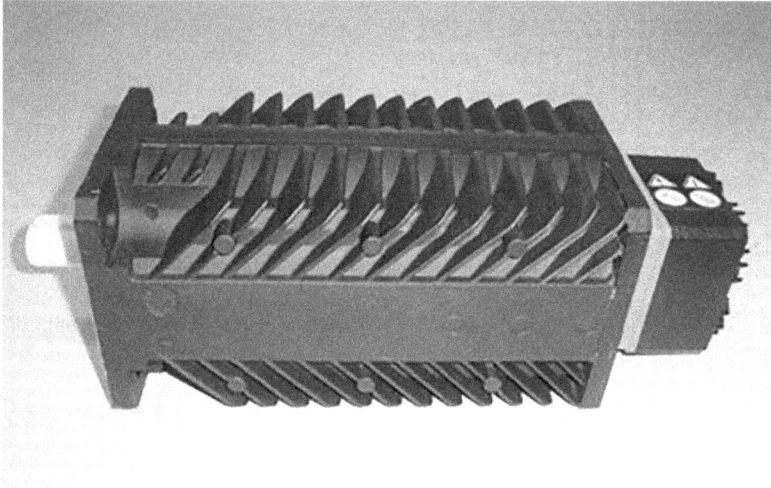

FIGURE 11.12 Oblique plate fins on the motor outer surface.

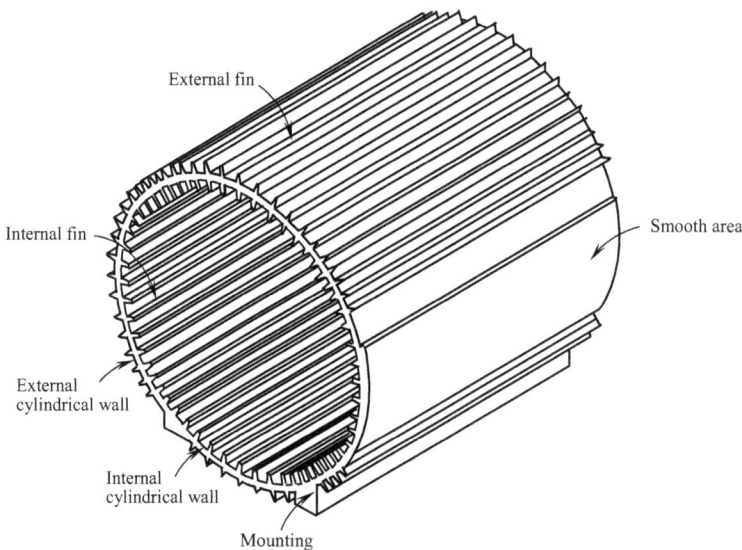

FIGURE 11.13 Outer and inner fins on the motor frame (U.S. Patent 4,839,547) [11.26]. (Courtesy of the U.S. Patent and Trademark Office, Alexandria, VA.)

housing can help determine how much heat can be effectively carried away from the motor via natural or forced airflow, how close the motor components, especially the heat sources, are mounted to each other, and how uniform heat distribution can be achieved. To address these issues, a simple solution is to cast fins on the external and/or internal surface of the housing by varying the fin density, fin type (pin fin or plate fin), location of fin, and optimized fin shape.

As shown in Figure 11.12, oblique plate fins are distributed oppositely on the surfaces of a totally enclosed nonventilated motor. The use of oblique plate fins can effectively reduce boundary layer thickness and improve the overall thermal performance.

Figure 11.13 is a perspective view of the motor frame with external and internal fins [11.26]. Because of the extended surfaces, heat generated in the motor during motor operation is dissipated into the environment at a much greater rate, thereby permitting the motor to operate under

FIGURE 11.14   Cooling fins on a motor endbell.

more severe load and environmental conditions without detrimental effects. This technique can be applied on totally enclosed or open externally ventilated motors. Motors that are totally enclosed and nonventilated depend entirely on the removal of heat from the frame surfaces by natural convection and radiation. Motors that have an open construction depend mainly on the moving of external cooling air into the interior of the motor and discharge the heated air out of the ventilating exits.

In order to increase the heat dissipation from the motor, cooling fins may be used at the outer surface of the endbell, as shown in Figure 11.14.

For some frameless motors, the stator laminations, all of one design, are designed in irregular shapes (*e.g.*, asymmetric rhombus or parallelogram). When these laminations are positioned relative to one another, they form radially oriented external cooling fins. In such a way, motor cooling is benefited from not only the reduction in thermal resistance due to the elimination of the housing but also the heat transfer enhancement due to the effective increase in the heat-dissipative surface area and the disturbance of the cooling airflow resulting from the fins.

Stator fins may be fabricated by prestamping a number of radial slots around the OD of the stator laminations. As the laminations are stacked together as a stator core, a plurality of plate fins are formed in the axial direction [11.27], as shown in Figure 11.15. This design can be adapted for both frame and frameless motors. For a frame motor, adjacent plate fins define a plurality of longitudinal cooling air channels that extend the entire stator. This approach requires an axial fan to drive cooling flows passing through these cooling channels. As to a frameless motor, heat dissipates directly from the fins to the environment.

A similar design of stator lamination fins is presented in Figure 11.16 [11.28]. The stator consists of a plurality of groups of juxtaposed laminations. When stacking the laminations in one group, each cooling fin portion on each lamination is aligned with a respective cooling fin portion in each lamination such that the aligned cooling fin portions form a cooling fin and a series of cooling fins in side-by-side arrangement are formed. Preferably, juxtaposed pairs of groups are arranged such that the cooling fins of one of the groups are staggered with respect to the cooling fins of the other group. With such shaped and arranged cooling fins, airflows between the cooling and adjacent fins

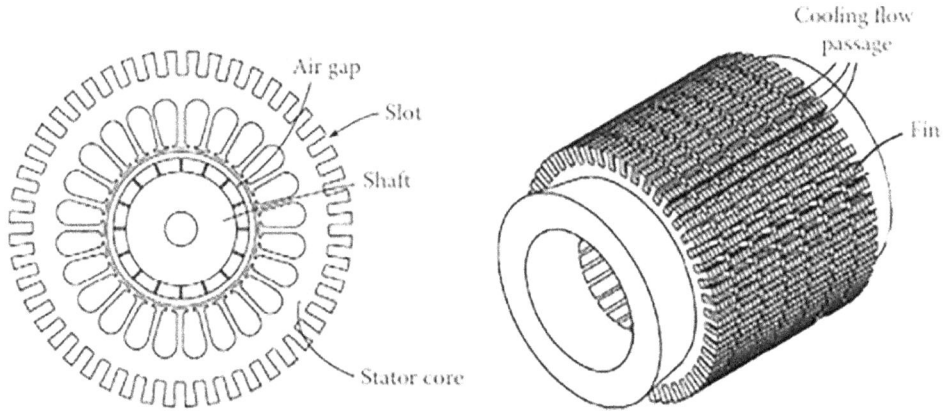

FIGURE 11.15 Stator laminations are stacked to form extending fins on the stator outer surface (U.S. Patent 8,053,938) [11.27]. (Courtesy of the U.S. Patent and Trademark Office, Alexandria, VA.)

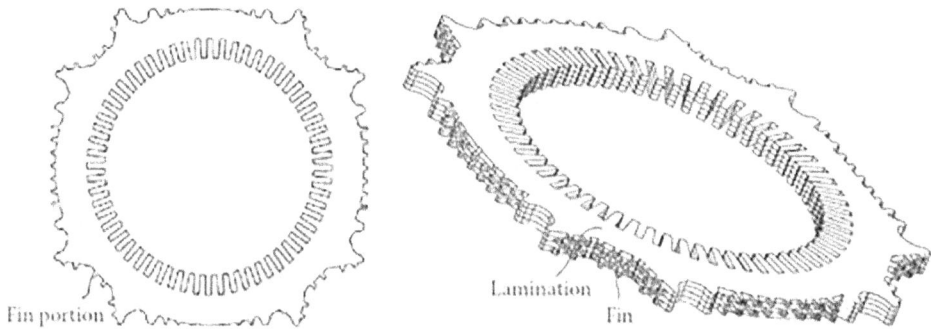

FIGURE 11.16 Fins formed from stacked stator laminations at the outer surface of the stator (WO 2005/022718) [11.28]. (Courtesy of World Intellectual Property Organization, Geneva, Switzerland.)

are disrupted to reduce the thickness of the thermal boundary layers on the stator surfaces and hence increase the conventional heat transfer coefficient.

11.3.5 Pin Fin Heatsink

To further improve the thermal performance of heatsinks, an array of pin fins has been used to replace longitudinal straight plate fins. Employment of such pin fins can result in better convective heat transfer characteristics because of the promotion of turbulence in the coolant passages and the increased convective heat transfer surfaces. Heatsinks using pin fins can minimize thermal resistance to improve convective cooling over traditional heatsink designs. It has been reported that with a plate-pin fin heatsink, which is the combination of plate fins and pin fins (as shown in Figure 11.17), the thermal resistance can be 30% lower than the conventional plate fin heatsinks [11.29].

The pressure drop across a pin fin heatsink is one of the key variables that govern the thermal performance of the heatsink. In general, the total heatsink pressure drop depends on several design parameters and operation conditions, including the pin fin geometry, pin fin density and configuration arrangement, pin fin size, heatsink orientation, and approach velocity (in forced convection). The design of heatsink is intended to decrease the impedance of the fluid flow through the heatsink and, thereby, reduce pressure losses.

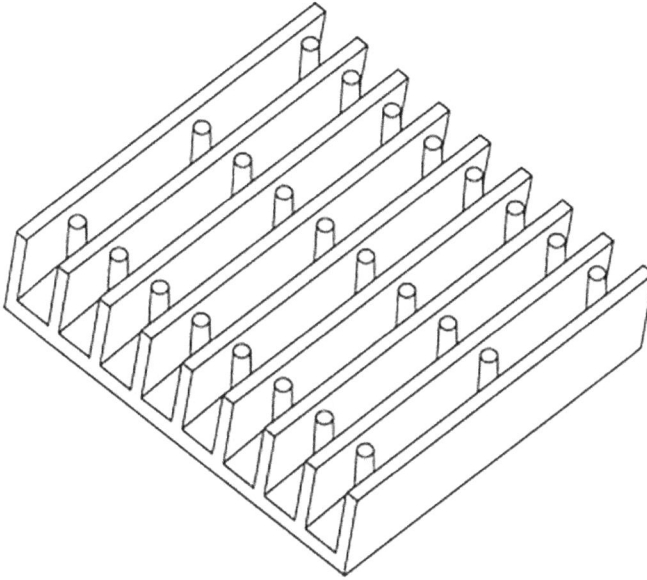

FIGURE 11.17 Plate–pin fin heatsink, which is the combination of plate fins and pin fins.

To further improve the heatsink thermal performance and reduce the pressure drop across the heatsink, the heatsink with perforated pin fins has been designed for replacing the heatsink with solid pin fins [11.30]. The benefits of using pin fin heatsinks with multiple perforations are investigated experimentally and numerically. It has been confirmed that the use of multiple pin perforations can have substantial performance benefits such as increased heat transfer rate, reduced pressure drop across the heatsink, and less fan power required for pumping air through the heatsink (for forced convective heat transfer).

As shown in Figure 11.18, nine different pin fins with various numbers and locations of perforations have been tested. Experimental data show that the Nusselt number increases monotonically with the number of pin fin perforations while both the pressure drop and fan power reduce monotonically. For instance, the pin fins with five perforations can have an 11% larger Nusselt number than that for corresponding solid pin fin cases. This study reveals that the heat transfer augmentation is achieved due to not only the increased surface area but also heat transfer enhancement near the perforations through the formation of localized air jets.

Depending on different applications, many shapes of pin fins at the cross-sectional area can be adopted, for instance, round, square, rectangular, oval, rhombic, crescent, and raindrop shaped (Figure 11.19). Among them, the raindrop-shaped pin fins generate the lowest pressure drop across the whole heatsink by minimizing the drag force acting on fins and maintaining a large, exposed surface area available for heat transfer [11.31]. Square and rectangular pin fins are not preferred because they usually generate relatively large flow resistances and thus high pressure drops.

Figure 11.20 depicts the flow pattern between raindrop-shaped pin fins. As cooling flow passes the pin fins, horseshoe vortices are generated in the stagnation area at the front edge of the pin fin by virtue of the staggered fin arrangement, and the flow separates around the pin fins. The horseshoe vortices can essentially enhance the heat transport. The staggered and tilted arrangement of the pin fins increases the local turbulent level and reduces the size of the wakes, thus tending to reduce the local thermal boundary layer thickness and augment heat transport in these regions. As such, the pin fins serve as turbulence promoters to enhance heat transfer. The combination of the increased heat transfer coefficient and enlarged flow contact area considerably improves heat convection in the flow channel, thereby reducing or even eliminating hot spots.

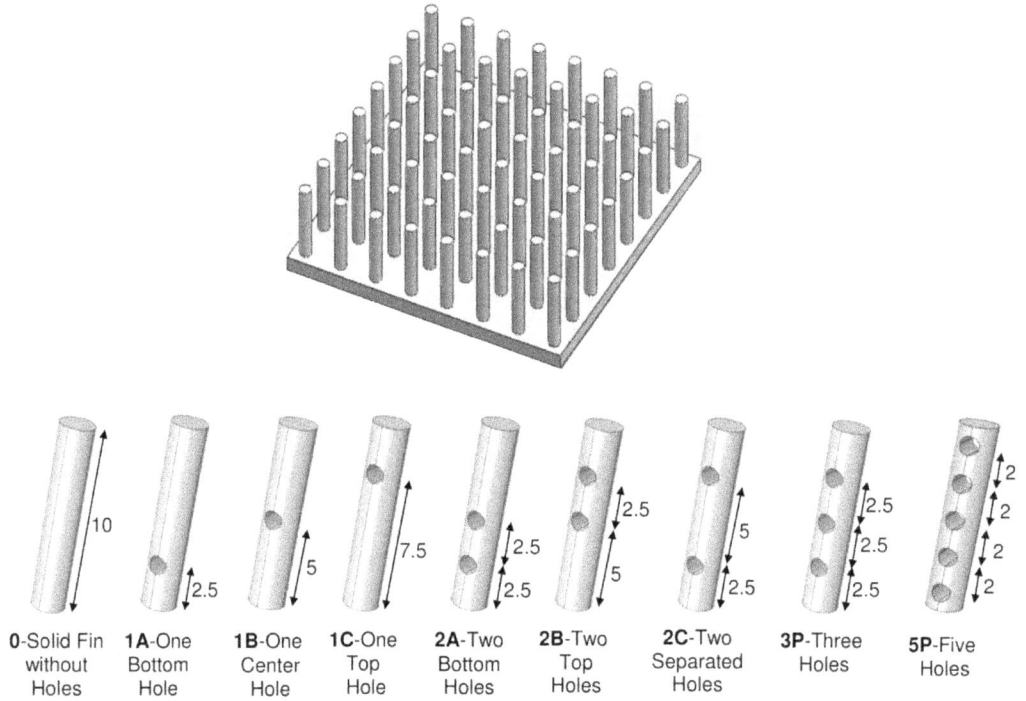

FIGURE 11.18　The comparison of the nine pin fins with various numbers and locations of perforations, all dimensions in mm. The dimensions of heatsink base are 50 mm×50 mm×2 mm. The arrangement of pin fin on the base is 8 ×8 [11.30].

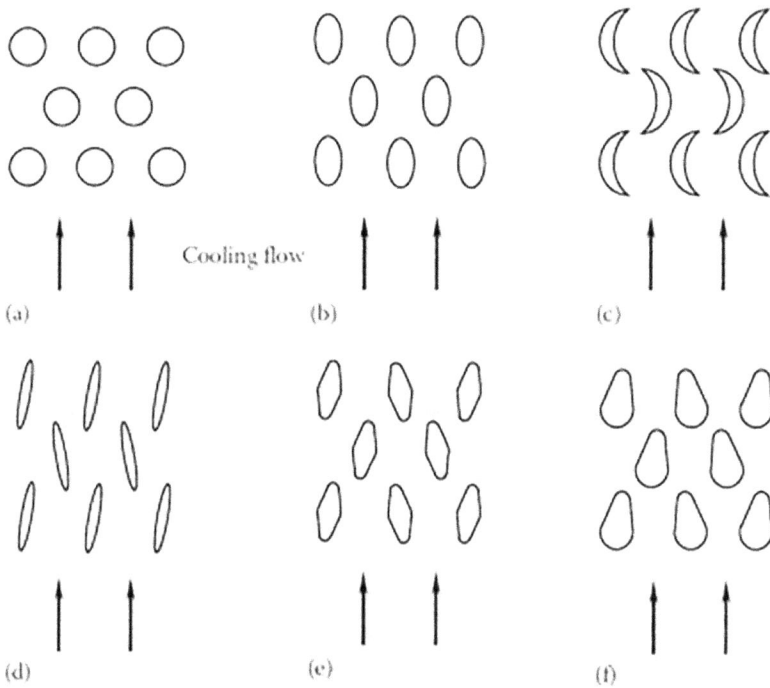

FIGURE 11.19　A variety of pin fin shapes and arrangements: (*a*) round pin fins, (*b*) oval pin fins, (*c*) crescent pin fins, (*d*) tilted narrow oval pin fins, (*e*) tilted rhombic pin fins, and (*f*) tilted raindrop-shaped pin fins.

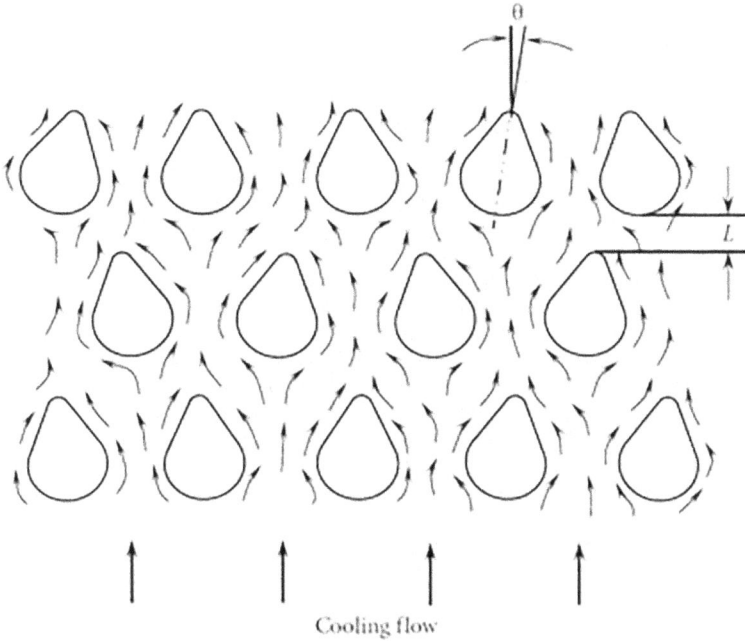

FIGURE 11.20 The flow patterns between raindrop-shaped pin fins. The horseshoe vortices generated in the stagnation areas of the pin fins and the flow separation and reattachment significantly increase the flow turbulence level and thus enhance convective heat transfer.

The fin angle θ is defined as the angle between the fin centerline and the cooling flow direction. In Figure 11.20, a symmetry line is shown with a dash-dotted line. The angle θ of each pin fin may preferably vary between $+20°$ and $-20°$ for maximizing the heat transfer rate. Thus, the fin center-lines are nonparallel with respect to one another.

Another variable that can be used to maximize the heat transfer rate is the fin distance L between adjacent rows of pin fins. At the cooling flow upstream, because the temperature difference ΔT between the convective air and the heatsink is relatively large, the fin distance L can be large. As the temperature of cooling air increases along its path, ΔT becomes smaller and smaller. Thus, the fin distance L may vary accordingly for achieving higher cooling flow velocity and, in turn, higher heat dissipation rate. Regularly, the distance between adjacent rows of pin fins decreases along the cooling flow path.

There are several benefits of using raindrop-shaped pin fins. First, the fins function as turbulent promoters to create local turbulence for enhancing heat transport. Second, the fins are designed to minimize the fin-induced wakes at the fin tails, whereby they provide the lowest resistance to the convective airflow. Third, the local heat transfer can be easily controlled by varying the fin angle θ and the fin distance L. Fourth, the local heat transfer can be alternatively controlled by changing the fin density and fin size along the cooling flow path.

A computational fluid dynamics (CFD) model has been carried out to compare the thermal performance between the straight plate fin heatsink and raindrop-shaped pin fin heatsink. As shown in Figure 11.21a, under identical thermal conditions, the straight plate fin heatsink has the maximum temperature of 77.83°C, whereas as shown in Figure 11.21b, the raindrop-shaped pin fin heatsink results in a maximum temperature of 53.49°C. Thus, the heatsink with raindrop-shaped pin fins can efficiently enhance the heat dissipation and improve the heatsink thermal performance by approximately 10%–30% due to not only the extended surface area in contact with the cooling fluid but also the increased local turbulence level of the fluid flow.

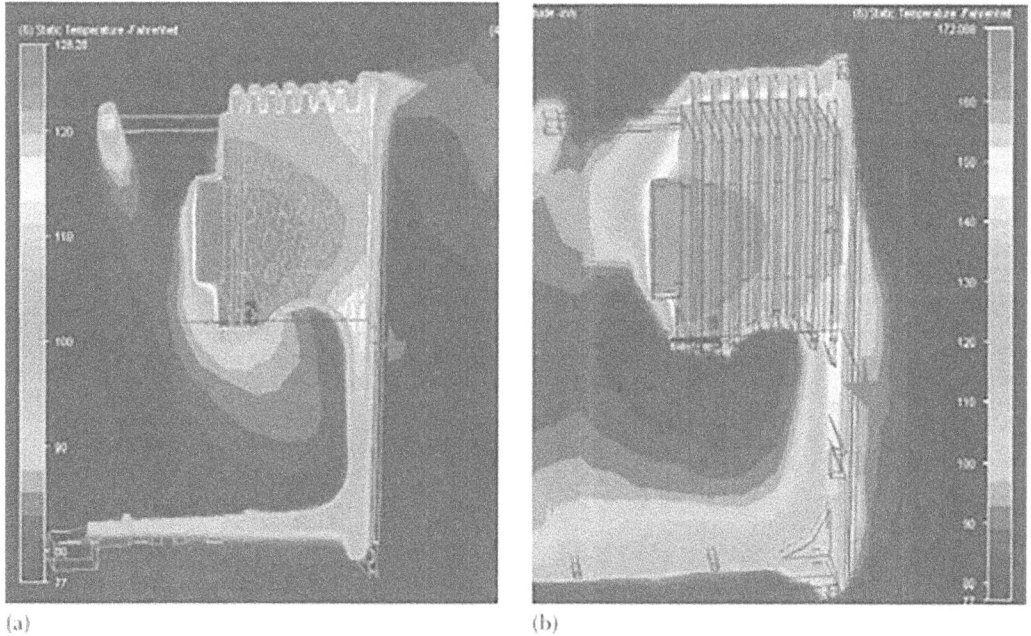

(a)   (b)

FIGURE 11.21 Numerical analysis results show that under the identical operating condition, the maximum temperature for the raindrop-shaped heatsink (Figure 11.21a) is 53.49°C (1211.28°F), which is much lower than that for the conventional heatsink with straight plate fins (Figure 11.21b) at 77.83°C (172.09°F).

The pin fin configuration can significantly affect the cooling flow pattern and consequently the thermal performance of pin fins. For example, a crescent-shaped pin fin has a concave and a convex surface. As illustrated in Figure 11.22a, crescent-shaped pin fins are configured in a concave–convex pattern at fin rows. The orientation of pin fins between the adjacent rows is opposite. This configuration forces the cooling flow to change its flow direction periodically as it makes its way through the pathway. This causes local flow separation disturbances and subsequent reattachment of flow in the fin boundary layer and correspondingly increases the flow turbulence level and thus enhances convective heat transfer. In addition, the redeveloping boundary layer from the reattachment point also contributes to heat transfer enhancement.

Referring now to the embodiment illustrated in Figure 11.22b, the crescent-shaped pin fins are configured in a concave–concave and convex–convex pattern for all fin rows. This fin arrangement forms a large number of small convergent–divergent (convex–convex pattern) and divergent–convergent (concave–concave pattern) flow channels, configured in a staggered pattern along the flow path. As a cooling flow passes convex–convex fin channels, the flow is compressed at the channel entrance and expanded at the channel exit. Contrarily, as the cooling flow passes concave–concave fin channels, the flow is expanded at the channel entrance and compressed at the channel exit. Thus, the variations in the flow velocity, flow direction, and local flow pressure will increase flow reactions and turbulence levels in the flow field, resulting in the high heat transfer performance between the fins and cooling flow.

Using an entropy generation minimization method, an optimal design of cylindrical pin fin heatsinks is obtained for both the in-line and staggered arrangements [11.32, 11.33].

Because of the thermally efficient structure of pin fins, motors with pin fins may provide excellent cooling solutions. Benefited from higher thermal contact areas and the promotion of turbulence in cooling flows, pin fins can provide much higher heat transfer rates than plate fins. However, pin fins have not yet been used in motor cooling, primarily due to the high cost of pin fins and some technical difficulties in fabricating pin fins.

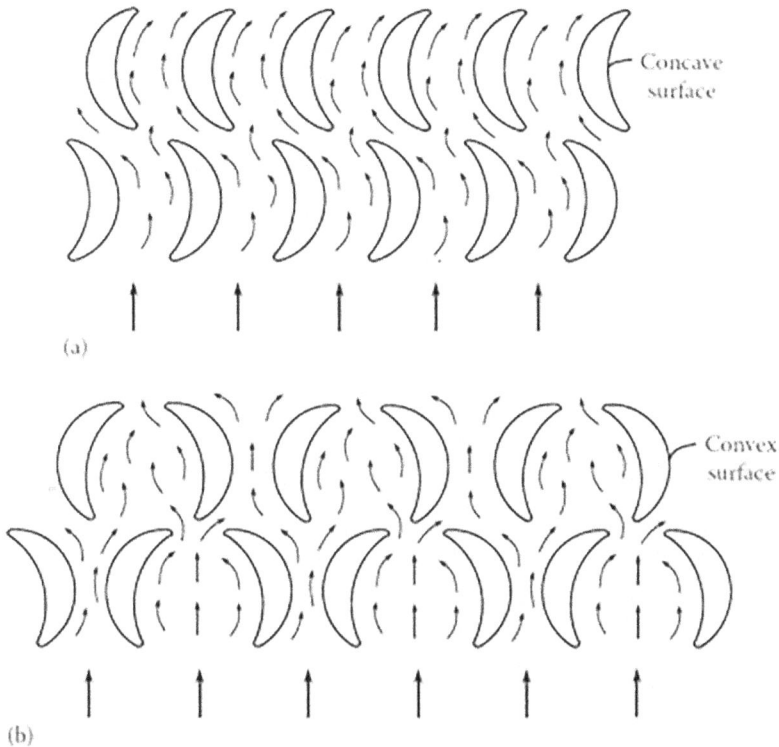

FIGURE 11.22 The flow patterns between crescent-shaped pin fins: (*a*) configured in a concave-convex pattern at fin rows and (*b*) configured in a concave–concave and convex–convex pattern for all fin rows.

11.3.6 THERMAL INTERFACE MATERIALS

Thermal interface materials are extensively used in electronic cooling applications for reducing thermal resistance across a contact interface formed by any two solid bodies. Thermal contact resistance occurs due to the voids created by surface roughness effects, material defects, and misalignment of the interface. The voids on the interface are filled with air, which has very low thermal conductivity. Thus, the voids make a significant contribution to the total thermal resistance. This is especially true for the systems with high thermal loads, a large number of contact parts, and complex surface contacts.

From the microscopic point of view, solid surfaces made by regular manufacturing processes are never perfectly smooth. Rather, the surfaces are constituted of asperities of different sizes and shapes. Therefore, when two solid surfaces are brought together, only a small portion of the matching surfaces make physical contact and most of the matching surfaces are separated by a layer of air. The voids/gaps formed between the contacted surfaces thus result in high thermal resistance as heat is transferred across the interface. Therefore, a variety of thermal interface materials have been developed for filling voids/gaps, thus improving surface contact and heat conduction across the interface. Because thermal interface materials generally have much higher thermal conductivities than that of air, the interfacial thermal resistance can be significantly reduced.

There are several types of thermal interface materials [11.34]:

- Thermal greases typically offer the best thermal performance to serve as the most popular thermal interface materials. Thermal greases are made by dispersing thermally conductive fillers in viscous silicone or hydrocarbon oils. The use of thermal greases can effectively

TABLE 11.2

Characteristics of Five Greases and Their Thermal Performance

| | | | | | P=0.1 MPa | | P=0.2 MPa | | P=0.3 MPa | |
| | | | | | Bondline Thickness (μm) | Thermal Conductivity (W/m·K) | Bondline Thickness (μm) | Thermal Conductivity (W/m·K) | Bondline Thickness (μm) | Thermal Conductivity (W/m·K) |
Grease	Filler	Filler Loading (%)	Viscosity (Pa-s)	Density (kg/m³)						
A1	60 μm spherical	35.8	57.5	1,200	56.4	4.2	51.3	4.3	411.3	4.4
A2	BN	41.6	98	1,300	71.1	5.2	67.3	5.4	59.2	5.2
A3		46.4	191	1,400	92.7	5.7	74.9	7.9	69.9	11.3
B1	60 μm spherical	71.7	330	2,200	50.0	6.5	47.0	7.9	43.9	11.4
B2	BN + secondary filler	80.1	690	2,400	57.2	7.9	51.3	9.8	51.8	10.3

Source: Tonapi, S. *et al.*, *Electronics Cooling*, November Issue, 2007.

eliminate the interstitial air in voids or gaps and thus provide a low thermal resistance across the contacted surfaces. Thermal greases tend to wet the matching surfaces well while allowing retention of the high-thermal-conductivity asperity microcontacts. The conductivity of thermal greases is about 0.3 W/(m·K) [11.35], which is about ten times higher than that of air.

- The characteristics of five greases were studied by Tonapi *et al.* [11.36]. The thermal performance of these greases was measured using the laser flash thermal diffusivity method. The primary filler in these greases is spherical boron nitride (BN) with an average filler size of 60 μm. The testing results are summarized in Table 11.2. The results show that the increase in thermal conductivity is mainly due to the increase in filler loading for all five greases under different pressures. It also shows a significant reduction in thermal resistance with increasing pressure due to a combination of reduced bondline thickness and interfacial thermal resistance.

- Conductive-particle-filled silicone- or acrylic-based thermal tapes and pads offer high thermal conductivity, ranging from 0.7 to 7.3 W/(m·K), and low thermal resistance, ranging from 0.11 to 1.0°C/W, depending on the material and its thickness.

- As indicated by their name, phase-change materials can change their phase when the temperature changes. At room temperature, the material is a film. As the temperature increases to a certain value, the viscosity of the materials becomes very small, allowing them to freely flow throughout the joint and fill the voids/gaps.

- Soft metal foils have been developed for years as compressible metallic shims in many applications, especially power devices. Soft metal foils are very thermally conductive, reliable, and easily adopted.

- Thermally conductive elastomers are silicone elastomer pads filled with conductive ceramic particles.

- Thermal adhesives, such as thermal adhesive pads, thermal adhesive tapes, and thermal adhesive films, provide low thermal impedance with long-term reliability.

Each type of thermal interface material may display different levels of efficiency for reducing the thermal contact resistance, depending on the nature of the contacting materials, contact conditions (*e.g.*, applied contact pressure, temperature), morphological and crystallographic characteristics of the mating surfaces (*e.g.*, surface roughness, hardness, and wettability), and the process parameters for the thermal interface material application technique.

When the interfacial gaps are filled with a thermal interface material while the surface asperity microcontacts are allowed to form, the heat transfer between two contact bodies can take place through both the microcontacts and microgaps with filled or unfilled interface material. Thus, the overall thermal contact resistance R_c can be determined by treating the microcontact resistance R_{mc} and microgap resistance R_{mg} in parallel:

$$R_c = \frac{R_{mc}R_{mg}}{R_{mc} + R_{mg}} \tag{11.17}$$

When the thermal interface material essentially fills most microgaps, there is little direct contact between two bodies, implying $R_{mc} \gg R_{mg}$ and $R_c \approx R_{mg}$. This indicates that the thermal contact resistance is generally determined by the thermal interface material.

To improve the cooling performance of electric motors, thermal interface materials can also play an important role in enhancing heat transfer between contacting motor components, such as the stator core and housing, the housing and endbells, winding coils and slots, and bearings and bearing bores.

11.4 FORCED AIR COOLING TECHNIQUES

Forced air cooling has long been applied in motor cooling. Over recent several decades, significant advances have been made in the application of air cooling techniques to manage increased thermal loads. Today, forced air cooling techniques are extensively used for various types of motors for their simplicity, low cost, and high reliability.

Typically, a motor is constructed to have either a totally enclosed architecture or an open architecture. A totally enclosed motor is designed to prevent external contaminating particles and other foreign matter entering into the motor. Therefore, for a totally enclosed motor, there is no air ventilation between the inside and outside of the motor. In contrast, an open motor architecture permits ventilating air entering and leaving the motor freely.

11.4.1 THERMOPHYSICAL PROPERTIES OF AIR

Air is a mixture of gases, containing approximately 78% of nitrogen and 21% of oxygen, as well as traces of water vapor, carbon dioxide, argon, and various other components. The thermophysical properties of air are functions of temperature and pressure.

Figure 11.23 displays the specific heat of air as a function of temperature. It can be seen that both constant pressure specific heat c_p and constant volume specific heat c_v increase with temperature in a nonlinear manner.

Figure 11.24 presents the thermal conductivity of air as a function of temperature. As the temperature is raised, air molecules move more vigorously and, thus, heat energy is converted into kinetic energy of air. This will lead to more collisions between air molecules per unit time and, consequently, to an increase in the thermal conductivity of air.

Air density is an important parameter in determining the required mass flowrate, system pressure loss, and power. It can be seen from Figure 11.25 that air density decreases with an increase in temperature.

11.4.2 DIRECT FORCED AIR COOLING TECHNIQUES

Direct forced air cooling is the most popular cooling method for electric motors. In this cooling mode, cooling air is forced to contact directly with heat sources and exhausted out of the motor. It is to be noted that because the winding coils are always wrapped with insulation materials that form a significant thermal barrier, the heat generated within the winding coils must first pass through the electric insulation by conduction.

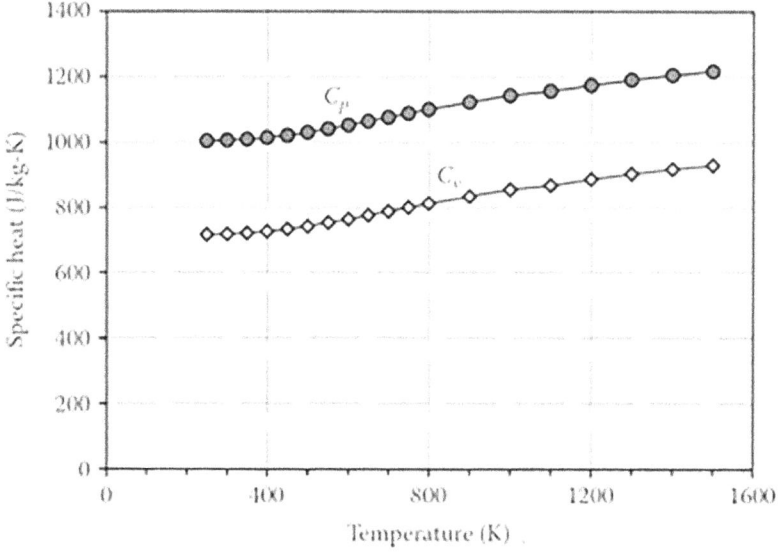

FIGURE 11.23 Specific heat of air as a function of temperature.

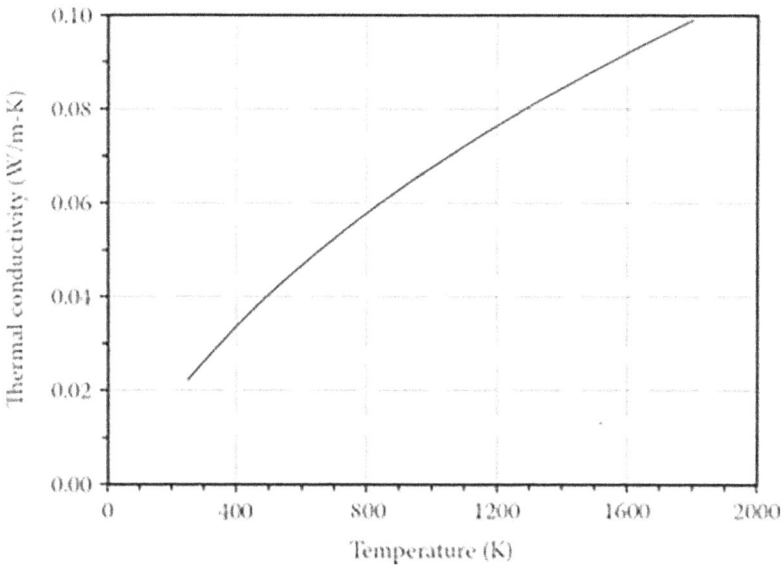

FIGURE 11.24 Thermal conductivity of air as a function of temperature.

11.4.2.1 Forced Air Cooling at End-Winding Regions

Hot spots may occur at the stator end-winding regions due to inappropriate cooling of end windings. This is because at the central winding region, the winding coils are in contact with the stator core. The heat generated by the windings is thus conducted through the stator core to the environment. In contrast, the heat generated at the end windings is transferred through two routes. One route travels along the copper coils to the stator core and the other route dissipates the heat to the air surrounding the end windings. Therefore, the thermal resistance of the end windings is considerably higher than that of the central winding. This is especially true for motors with a high L/D ratio. In practical motor design, special considerations are often put on cooling at the end-winding regions.

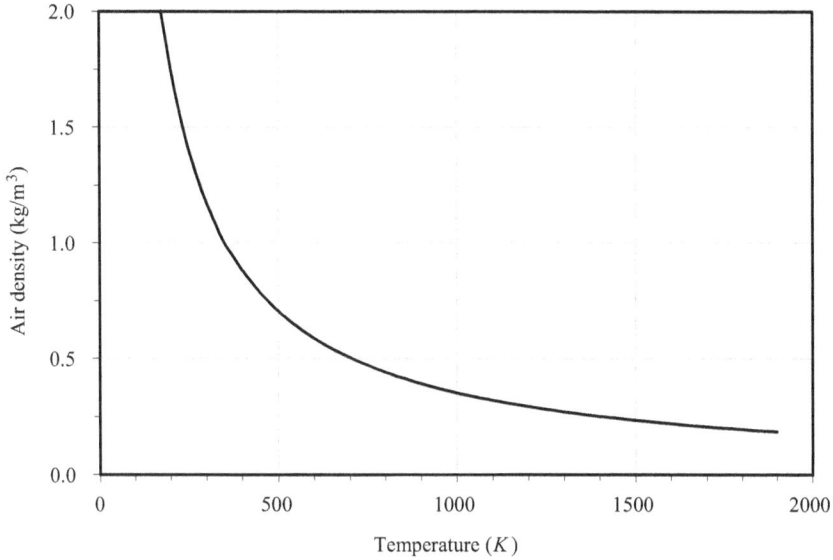

FIGURE 11.25   Air density as a function of temperature.

FIGURE 11.26   Rotor fan is casted with the rotor conducting end rings at each end of the rotor core for cooling motor windings, particularly winding end turns.

According to the fan and cooling path arrangement, several patterns of direct air cooling systems can be identified for cooling at the end-winding regions. The simplest method is to cast a number of fan blades on the end rings during the casting process (Figure 11.26). As the rotor rotates, these blades generate turbulent swirling airflows surrounding the end windings and then force the exhaust hot air exiting the motor through the ventilation slots on the motor housing. For entirely enclosed units, because the heat carried by circulating air must be dissipated to the housing/endbells by convection with a relatively high thermal resistance, the use of this cooling mode may be only suitable for small motors with low thermal loads.

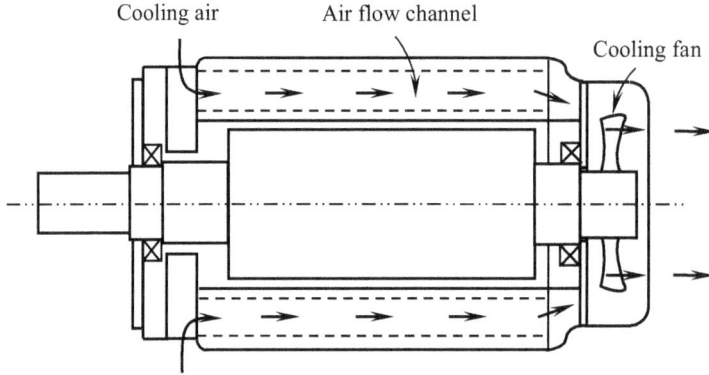

FIGURE 11.27   Open ventilating cooling system with sucked air forcing through stator cooling channels. In this design, a rotor fan is mounted at the nondrive end.

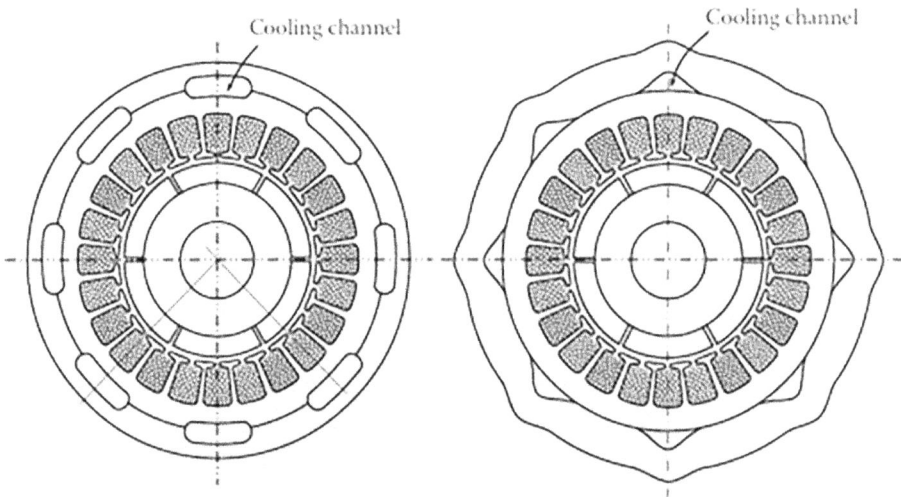

FIGURE 11.28   Arrangement of air cooling channels between the stator core and the motor housing.

11.4.2.2   Forced Air Flowing through Internal Cooling Channels across the Motor

In this cooling mode, cooling air is introduced from one end of the motor and forced along the internal cooling channels across the motor and exhausted at another end of the motor. As shown in Figure 11.27, the internal fan is mounted at an end of the shaft and cooling air is sucked radially through the ventilating slots from the motor's outer surface and is forced to flow through the cooling channels within the stator core.

In order to provide enough cooling flows through electric motors, the arrangement of cooling channels can be rather flexible. The internal cooling channels may be formed and arranged in a variety of ways: (a) between the stator core and the motor housing (Figure 11.28), (b) through the motor housing (Figure 11.29), (c) through the stator core (Figure 11.30) [11.37], (d) between the state winding coils in the slots (Figure 11.31) [11.38], (e) within the rotor core (Figure 11.32), (f) between the rotor and stator cores (Figure 11.33) [11.39], and (g) a combination of any or all of the aforementioned (Figure 11.34). It must be noted that as cooling air goes through the airgap or the channels through the rotor core, the cooling flow is subjected to a pumping pressure gradient in the radial

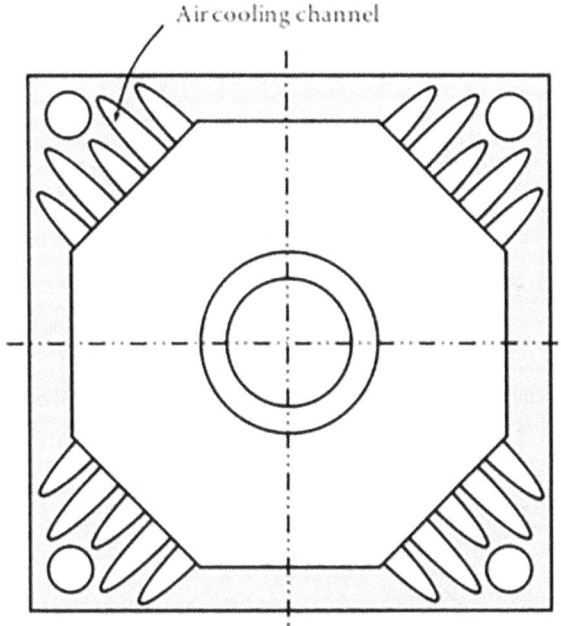

FIGURE 11.29 Air cooling channels through the motor housing.

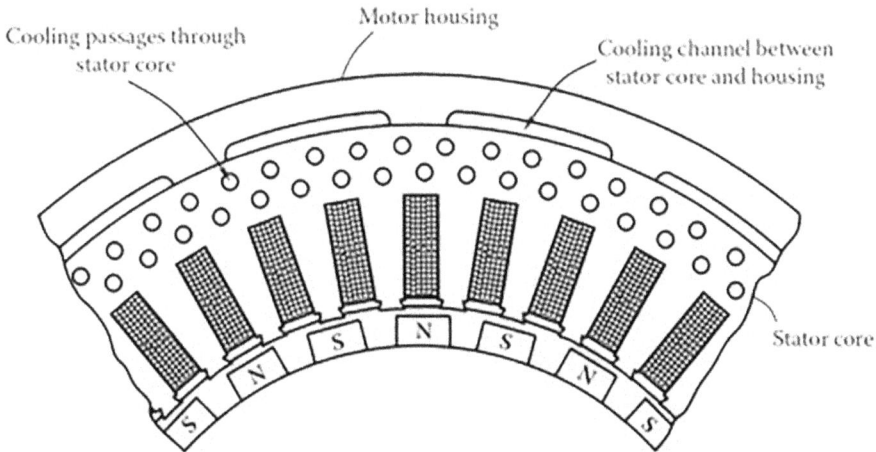

FIGURE 11.30 Cooling passages through the stator core and between the stator core and the housing (U.S. Patent 7,288,870) [11.37]. (Courtesy of the U.S. Patent and Trademark Office, Alexandria, VA.)

direction because of the centrifugal effect, causing a complicated helical flow pattern and resulting in higher pressure drops across the flow channels but better cooling performances.

In addition, the formation of cooling channels through rotor cores or scallop-shaped channels on the rotor surface not only benefits motor cooling but also reduces the rotational inertia of the rotor for shortening the motor start time and prevents the PM between poles from magnetic leaking.

11.4.2.3 Forced Air Flowing Over Motor Outer Surfaces

For a totally enclosed motor, because external cooling air cannot enter the motor, the motor has to be cooled by forcing cooling air through the motor outer surface along axial plate fins (Figure 11.35).

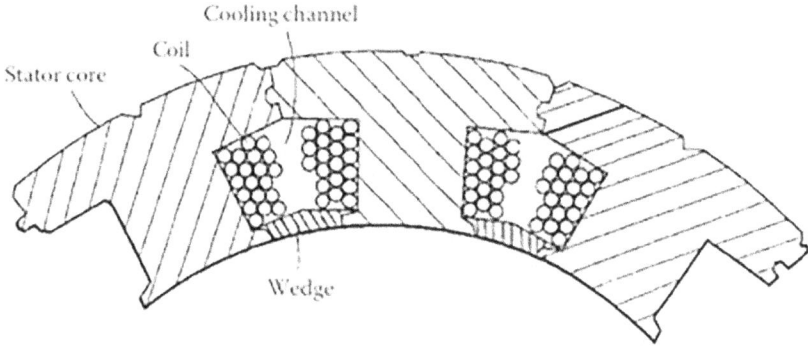

FIGURE 11.31 Cooling channels formed between stator winding coils (U.S. Patent 6,713,927) [11.38], (Courtesy of the U.S. Patent and Trademark Office, Alexandria, VA.)

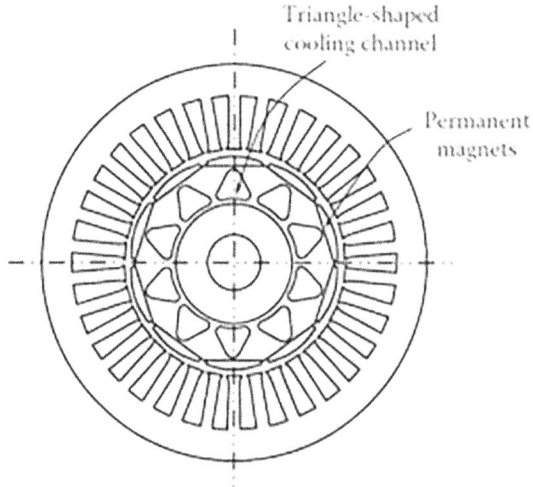

FIGURE 11.32 Triangle-shaped cooling channels through the rotor core.

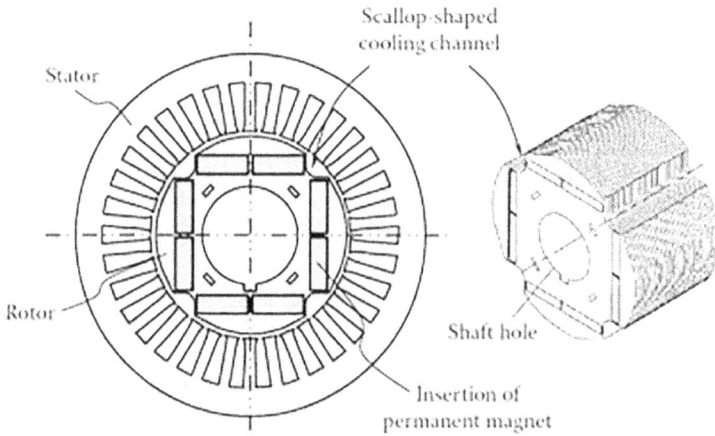

FIGURE 11.33 Scallop-shaped cooling channels between the rotor and stator (U.S. Patent 6,703,745) [11.39]. (Courtesy of the U.S. Patent and Trademark Office, Alexandria, VA.)

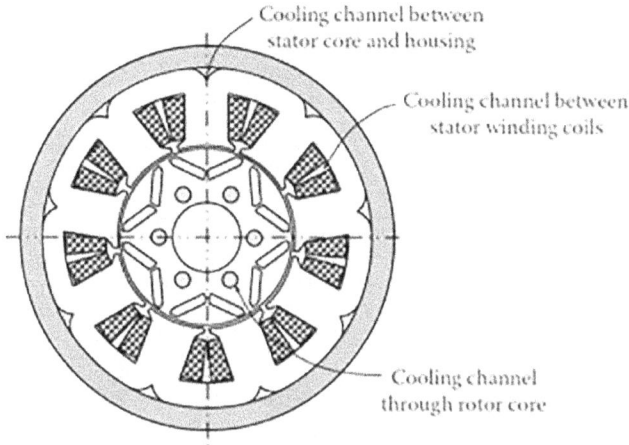

FIGURE 11.34   Combinations of different cooling channel arrangements.

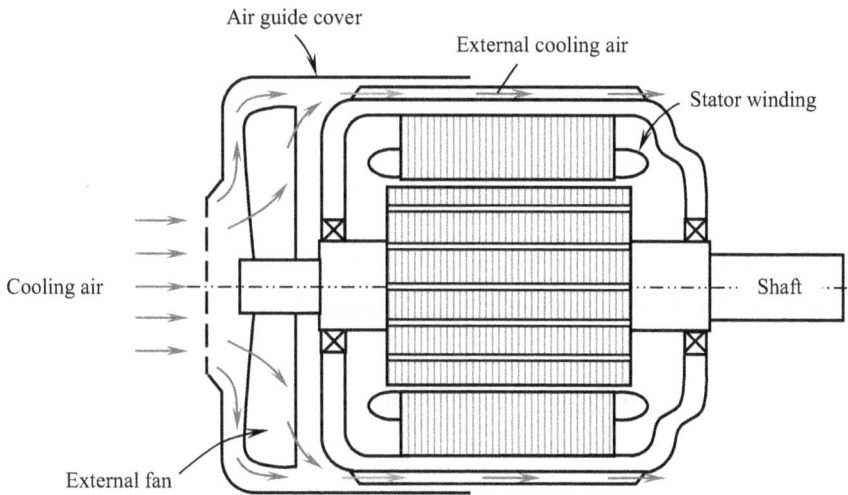

FIGURE 11.35   Cooling air is blown over the totally enclosed motor surface along axial plate fins by an external fan mounted on the shaft. No internal fan is used.

In this cooling mode, the fan is mounted on the rotor shaft outside of the motor so that the cooling airflow rate depends on the rotor rotating speed. Heat from the stator winding primarily relies on conduction from the stator to the motor housing and is dissipated to the motor surroundings by forced convection. It is noted that in this cooling mode, no internal fans are used inside the motor.

In some circumstances, mounting a cooling fan directly on the motor shaft of a totally enclosed motor is not practical. An alternative solution for motor cooling is to use an additional fan that is installed at the end of the motor. Thus, cooling air is blown over the motor surface guided by the air guide cover (see Figure 11.36). This cooling mode can be flexibly adapted to meet different motor cooling requirements.

11.4.2.4   Forced Air Flowing through Both Motor Outer and Inner Surfaces

This cooling scheme is very similar to the one mentioned previously. To enhance heat transfer for totally enclosed motors, a set of fan blades is cast on each end surface of the rotor end ring and applied to produce turbulent swirl circulating flows inside the motor. In such a way, the motor

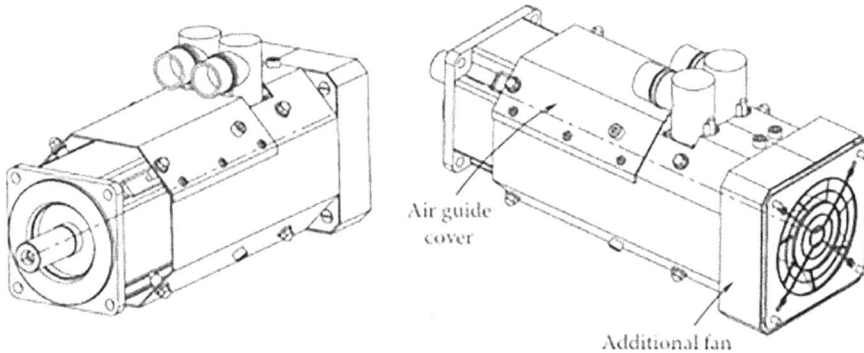

FIGURE 11.36 An additional cooling fan is mounted at the end of motor, blowing cooling air over the motor external surface. (Courtesy of Kollmorgen Corporation, Radford, VA.)

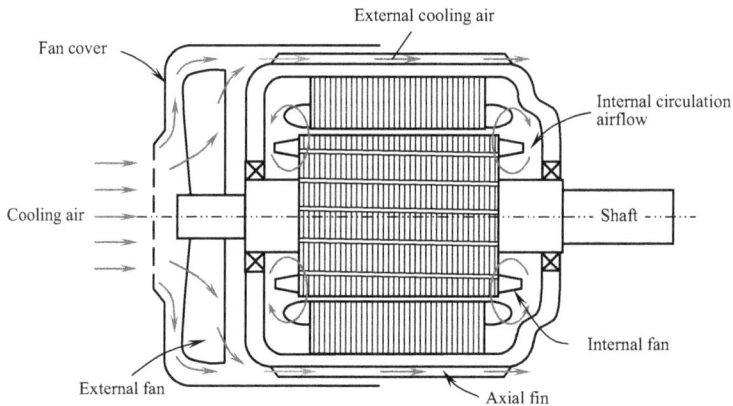

FIGURE 11.37 Motor is cooled by both external and internal airflows; the motor housing functions as a heat exchanger.

housing functions as a heat exchanger. As shown in Figure 11.37, heat generated by motor components is carried by the internal flow and is transferred to the housing wall. At the same time, cooling flow passing through the motor outer surfaces dissipates heat into the environment.

11.4.2.5 Air Jet Impingement Cooling

Jet impingement is an attractive cooling mechanism for its capability of achieving a high local heat transfer rate in the region of stagnation point. This cooling technique has been extensively used in the cooling of gas turbine blades. As a heat transfer enhancement technique, it has been widely used in the cooling of high-heat-flux components such as gas turbine blades and photovoltaic cells. With the increase in power output and the decrease in motor size, interest has been expressed by the motor industry in exploring this technique for cooling motor components. As shown in Figure 11.38 [11.40], the stator end windings are cooled using a jet impingement assembly, which includes a temperature-controlled fluid-generating device for bringing compressed air or other fluids onto the end windings. When the jet impinges on the end-winding surfaces, very thin hydrodynamic and thermal boundary layers form in the impingement region. Consequently, very high heat transfer coefficients are obtained within a stagnation zone. The experimental results have shown that impinging jet cooling offers up to 48% higher cooling efficiency for air impinging jet and 77% for liquid impinging jet over natural convection (see Table 11.3).

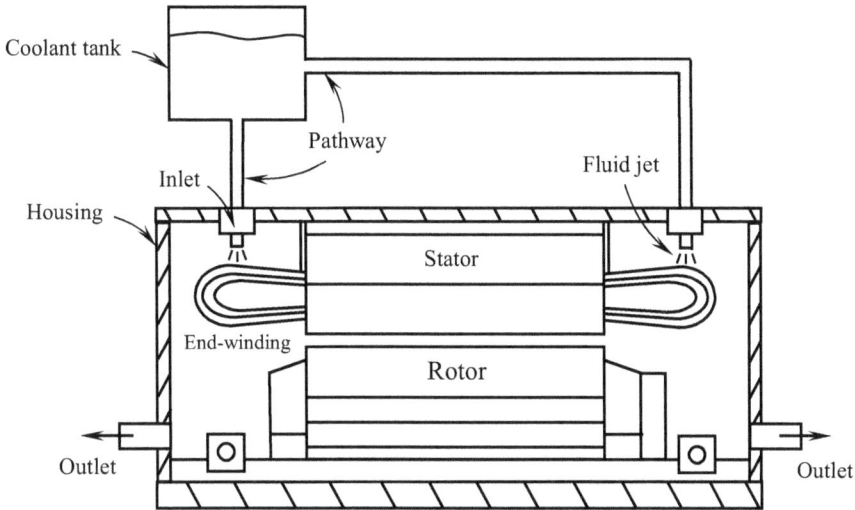

FIGURE 11.38 Cooling of end windings with an impinging jet cooling technique (U.S. Patent 6,639,334) [11.40]. (Courtesy of the U.S. Patent and Trademark Office, Alexandria, VA.)

TABLE 11.3
Cooling Efficient Improvement over Natural Convection

Cooling Technique	Cooling Efficiency Improvement (%)
Forced air	14
Impinging air jet	48
Impinging liquid jet	77

Figure 11.39 displays the end-winding temperature rise over the coolant temperature for different cooling methods. Impinging jet cooling, using either air or liquid, can provide a better cooling efficiency by comparing with natural convection and forced air fan cooling. The heat transfer coefficients of jet impingement cooling are much greater than that of natural convection.

The key parameters determining the heat transfer characteristics of a single impinging jet are the Reynolds number, Prandtl number, jet diameter, and jet to target spacing [11.41]. The optimization of an impinging jet system depends on the determination of these parameters through numerical analyses and experiments.

11.4.2.6 Cooling with Hydrogen Gas

The world's first hydrogen cooled machine, a 12.5 MW synchronous condenser, was placed in commercial service in 1928. The first hydrogen cooled turbogenerator (a 31.25 MW, 3,600 rpm unit) was manufactured by General Electric and put in operation at Dayton, Ohio in 1937. Hydrogen's low density, high specific heat, and thermal conductivity ($0.168\,W/m{\cdot}K$) among all gases make it an ideal cooling medium for cooling large electric machines. Compared to air cooling, hydrogen cooling can offer 150% higher heat transfer capability. With substantially reduced windage and friction losses, which are resulted from the low hydrogen density ($\rho_h = \rho_{air}/14$), the full load efficiency can increase 0.5%–1.0% over that of air-cooled generator. The higher effectiveness of a hydrogen-cooled machine can achieve a 20%–30% size deduction over an air-cooled machine with a similar rating. These factors make hydrogen-cooled generators preferable for end-users. The smaller generator size reduces the civil engineering cost in the power plant design, and higher efficiency

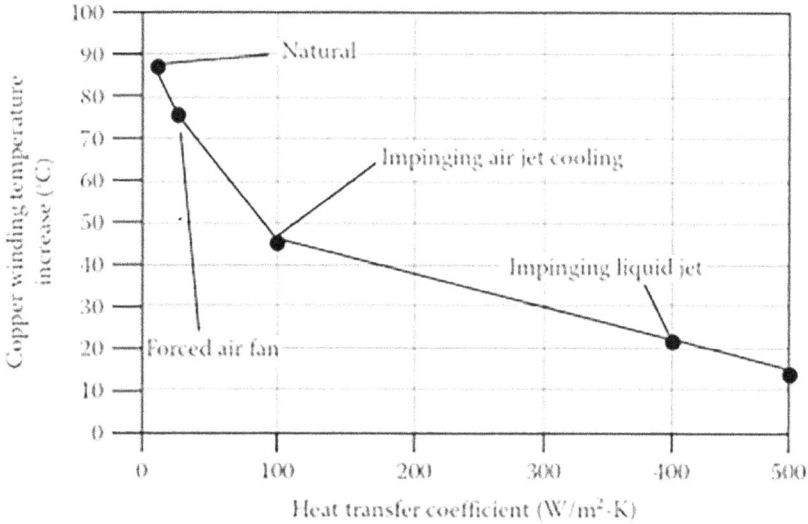

FIGURE 11.39 Temperature rises at end winding with different cooling methods (U.S. Patent 6,639,334) [11.40]. (Courtesy of the U.S. Patent and Trademark Office, Alexandria, VA.)

provides a lifetime plant output gain [11.42]. Today, about 70% of all electric power generators over 60 MW worldwide use hydrogen cooling.

While hydrogen cooling is an effective way to cool generators, the challenge of using hydrogen is its safety. When hydrogen is mixed with air, it becomes especially dangerous. The explosive range of hydrogen is broad, from concentrations of 4%–74%. Therefore, the robust seal system of hydrogen-cooled generators is crucial. As an extra safeguard, the generator casing is designed with an explosion-proof construction, which also contributes to lower noise and vibration levels of hydrogen-cooled generators. Moreover, monitoring hydrogen purity is important for both the operation safety and the efficiency of generators.

In cooling large generators, there are several direct hydrogen cooling schemes. Figure 11.40 illustrates the conventional reverse flow scheme, which was introduced by General Electric in the 1980s. Due to the structural symmetry, only a quarter of the generator is drawn in the figure. The main advantages of using reverse flow are its high cooling efficiency and relatively uniform temperature distribution across the generator. Reverse flow cooling in a generator delivers cold hydrogen gas simultaneously to both the stator and rotor. The cooling gases then mix in the machine airgap, leading to a coupling of the stator and rotor cooling. Nevertheless, the ventilation system must be configured to provide adequate cooling flows for both the stator and rotor.

With the reverse flow configuration, a fan mounted on an axial rotor end draws the exhaust gas out from the airgap to the heat exchanger. While the cooled hydrogen gas cooled by the heat exchanger enters the stator core and flows radially inwardly through axially spaced cooling ducts, a portion of the cooled gas flow is directed to the rotor where cooled gases are drawn into the rotor subslots. Then, the cooled gases are forced outwardly through radial cooling holes in the rotor by centrifugal forces created by the spinning rotor. The heated stator and rotor gas flows merge in the airgap, flowing along the airgap and finally being exhausted at the ends of the generator. This flow arrangement offers a more uniform temperature profile along the generator length.

To optimally distribute the cooling gases through the stator and minimize the necessary pressure head needed for the cooling gases, the stator cooling ducts are designed with the varied spacing and cross-sectional area along the length of the stator. With this design, the balanced cooling flow and relatively uniform temperature distribution are obtained, and in addition, the volume of cooling gases and the required pressure head are minimized.

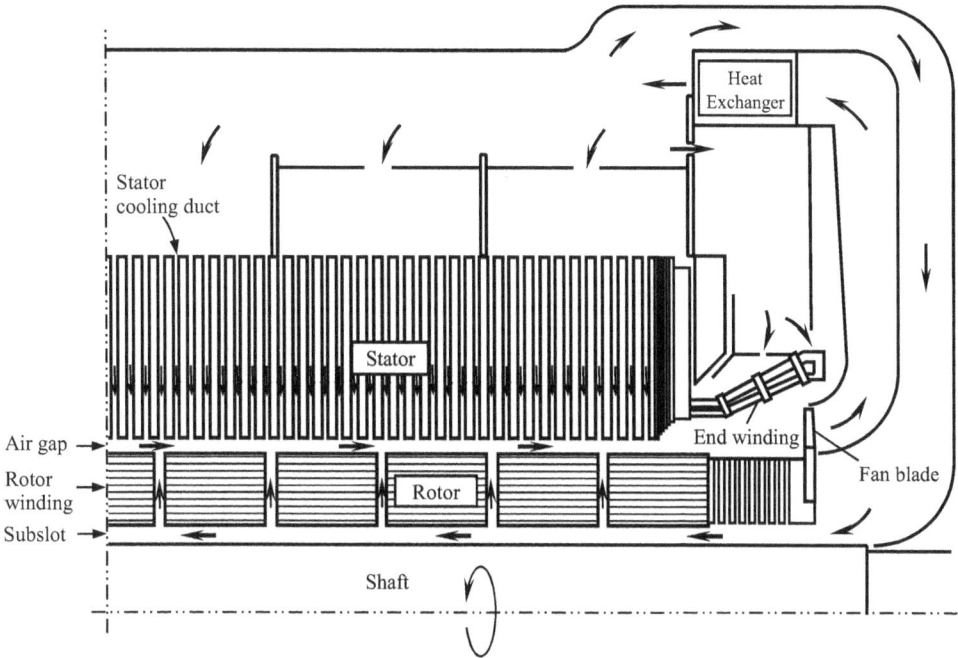

FIGURE 11.40 The arrangement of hydrogen cooling flows in a generator with the reverse cooling flow scheme.

Because the cooling hydrogen flows directly from the heat exchanger to the stator and rotor cores without passing through a fan, the ventilation system offers cooling gas at lower temperatures not possible from other known ventilation arrangements and has a relatively high cooling efficiency. However, reverse flow ventilation also results in a longer machine and somewhat greater complexity in the generator end region.

In addition, the ventilating flowrate through the airgap exit in a reverse flow generator is about 60%–80% of the total fan flowrate, much larger than that of a forward flow generator where only about 30% of the fan flowrate passes through the airgap entrance. Since the ventilation friction loss at the airgap exit is proportional to the product of the flowrate and pressure drop across it, a small change in the pressure drop could result in a large change in the friction loss due to the high flowrate [11.43].

Pressure drops often occur at backward- and forward-facing steps, sharp turn, sudden contractions and expansions, and any torturous paths in the airgap between the rotor and stator. Among these, a primary pressure drop appears at the annular airgap exit region and particularly, a bottleneck is formed between the retaining ring nose and the stator core-end taper. Therefore, it is highly desired to optimize the ventilation flow path to minimize the cooling flow friction loss.

11.4.3 INDIRECT FORCED AIR COOLING TECHNIQUES

Although forced direct air cooling of a motor can provide a higher cooling efficiency, indirect forced air cooling techniques are often encountered on large motors. For indirect air cooling, air does not directly contact with heated components; the heat generated in a motor is transferred to the ambient by means of heat-exchanging devices.

11.4.3.1 Indirect Forced Air Cooling with Heat Exchangers

Indirect forced air cooling with either air-to-air or air-to-liquid heat exchangers is generally applied to large-size, high-powered electric motors. As depicted in Figure 11.41, the motor is equipped with

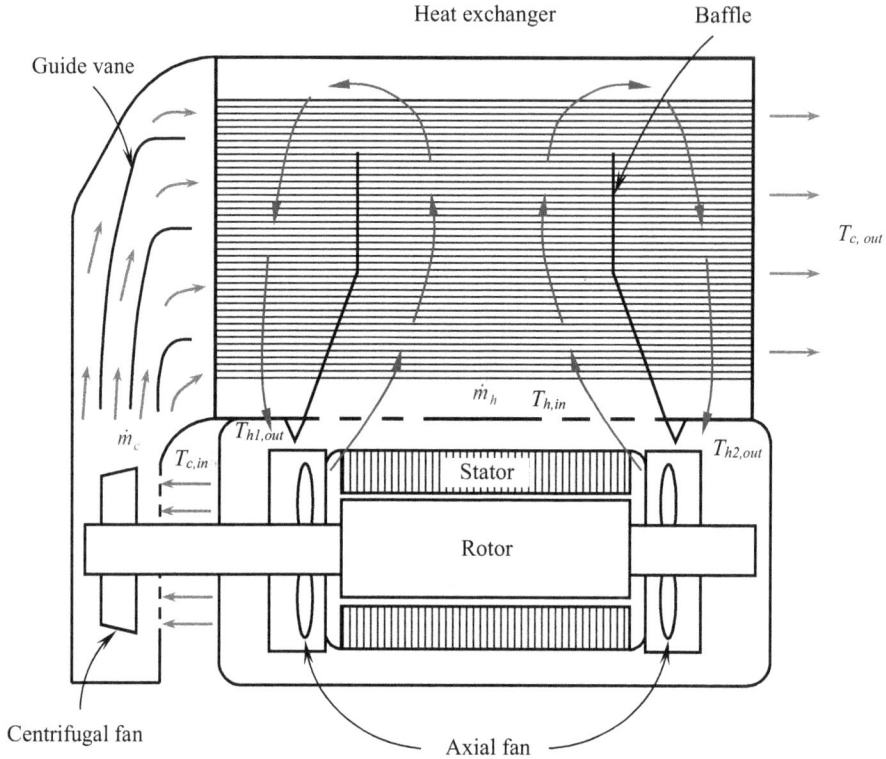

FIGURE 11.41 Forced air cooling with an air-to-air heat exchanger.

two axial fans at the two ends of the rotor shaft to internally circulate cooling air within an enclosed system containing the motor and the air-to-air heat exchanger. One centrifugal fan is mounted on the shaft at the outside of the motor frame to force ambient air going through the small, staggered tubes of the heat exchanger. In such a way, the energy balance holds between the heat energy dissipated from the internal circulation flows Q_h and the energy carried away by the ambient cooling flow Q_c:

$$Q_h = Q_c \tag{11.18}$$

It follows that

$$\frac{\dot{m}_h}{2} \sum_{i=1}^{2} \left[c_p(T_{hi,out})T_{hi,out} - c_p(T_{h,in})T_{h,in} \right] = \dot{m}_c \left[c_p(T_{c,out})T_{c,out} - c_p(T_{c,in})T_{c,in} \right] \tag{11.19}$$

where
  $\dot{m}$ is the mass flowrate
  $T_{h1,out}$ and $T_{h2,out}$ are the outlet temperatures of the internal circulation flow at the motor nondrive and drive side, respectively

This cooling method has been successfully used for a 2,350 kW totally enclosed air–air cooled motor [11.44]. With the total loss of 711.93 kW in the motor, the measured temperatures at the copper winding and the iron core are 122°C–138°C and 111°C, respectively. The measured data, including the fan performance curves and the temperature profiles of the stator and the heat exchanger, have shown good agreement with the simulated results.

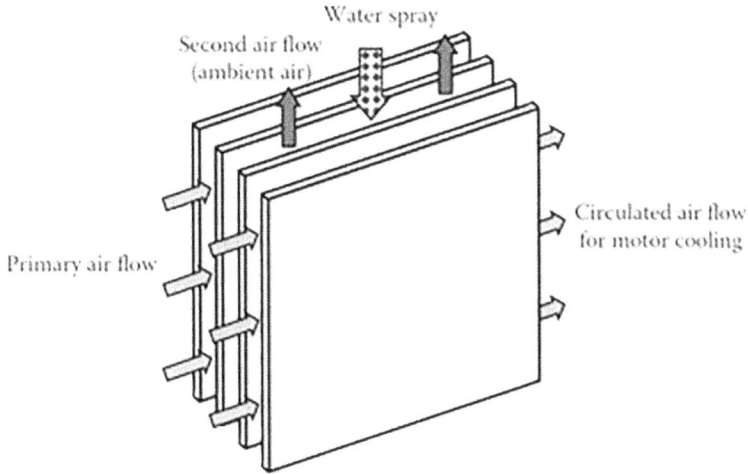

FIGURE 11.42   Forced air cooling with an indirect evaporation air cooler.

11.4.3.2   Indirect Forced Air Cooling via Indirect Evaporative Air Cooler

As shown in Figure 11.42, indirect forced air cooling is achieved via an indirect evaporative cooler with the orthogonal flow configuration. The principle of heat transfer in an indirect evaporative cooler involves the use of two types of air, namely, primary air and secondary air [11.45]. The primary air is isolated from the ambient air and is circulated in a loop to cool the electric motor. The secondary air is induced from the ambient into the evaporative cooler. Water is sprayed into the passage of the secondary air so that heat and mass transfer take place. Due to the latent heat absorption during the liquid–vapor phase transition, the temperature of the secondary air reduces and thus lowers the temperature of the primary air. Assuming the same spacing is used for the primary and secondary air, it has been found that the optimum heat transfer of an indirect evaporative cooler occurs at the velocity ratio of the primary to secondary air at approximately 1.4 [11.46].

Khmamas [11.47] has proposed an indirect evaporative cooling (forced or natural) method. With this cooling method, the air cooler is modified to operate as a cooling tower to produce cooling water by the evaporation process. The cooled water is then pumped to the indoor unit passing through a radiator that is cooled by a fan. The experimental results show that the evaporation cooling effectiveness reduces by 15% for indirect forced evaporation case and by 22% for the indirect natural case, as compared with the direct cooling case. Although it is seen that the indirect method is less effective than the direct method, some advantages can be gained such as noise reduction and low power consumption.

11.4.4   Fan and Blower

Fans and blowers are commonly used to provide cooling airflows to cool electric and electronic devices. They are differentiated by the method used to produce airflow and by the system pressure they must operate against. As a general rule, fans typically operate at pressures up to 13.79 kPa (2 psi) and blowers from 13.79 kPa (2 psi) to 193.1 kPa (28 psi), though some specially designed fans and blowers may fall out of these ranges. In addition, blowers have a much higher specific ratio (usually 1.11–1.20), which is defined as the ratio of the discharge pressure to the suction pressure, than fans (up to 1.11) [11.48].

To deal with the increasing heat loads in high-power-density motors, high-speed fans with noise, energy consumption, reliability, and weight penalties are often used.

11.4.4.1 Fan Types

Based on airflow discharge patterns, several generic fan types with varying air volume and pressure capacities can be identified such as axial fans, centrifugal fans, mixed-flow fans, and cross-flow fans. Axial fans, in which airflow is parallel to the axis of the blade rotation, are the most common type used in electric device cooling. This type of fan is the best choice for applications with high flowrates and low pressure drops. Therefore, the axial flow fans have been used in a wide variety of industrial applications, ranging from small fans for motor drive cooling to the huge fans employed in large wind tunnels. A typical axial fan used for motor cooling is shown in Figure 11.43.

A centrifugal fan has a fan wheel composed of a number of helical impellers or blades (Figure 11.44), mounted on a motor shaft. When the shaft rotates, the cooling air enters at the central part of the fan along the motor axis and then changes its flowing direction by 90° and finally is discharged in the radial direction of the impeller. During the process, the cooling air gains kinetic energy from the rotating impellers to increase its total pressure (*i.e.*, the sum of velocity pressure and static pressure).

At a constant fan speed, a centrifugal fan can drive a constant volume of air rather than a constant mass. Centrifugal fans are the most prevalent fans used in ventilation and air conditioning applications.

A mixed-flow fan combines the high flow of an axial fan and a high pressure of a centrifugal fan. In practice, it uses a modified axial flow impeller to produce a radial component of flow to add to a

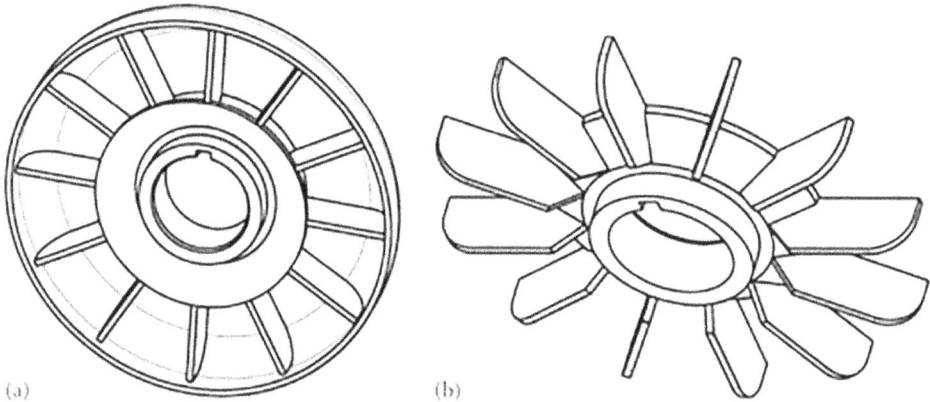

(a)   (b)

FIGURE 11.43 Casted or molded cooling fan: (*a*) fan structure (blades, hub, and shroud) and (*b*) blades on fan hub.

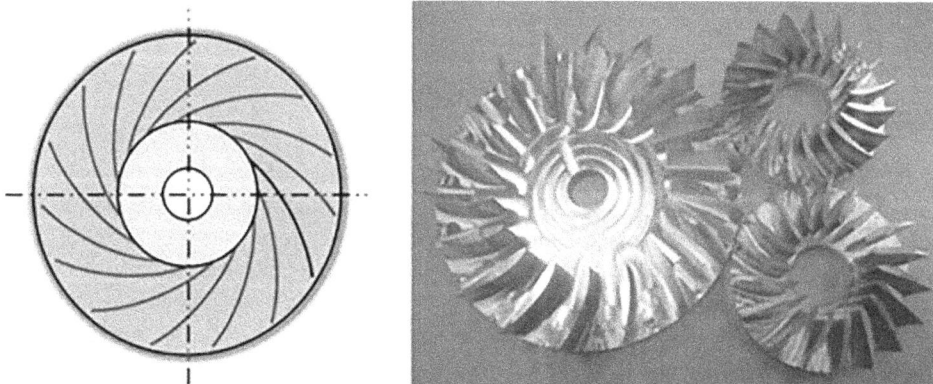

FIGURE 11.44 Helical impeller design in centrifugal fans.

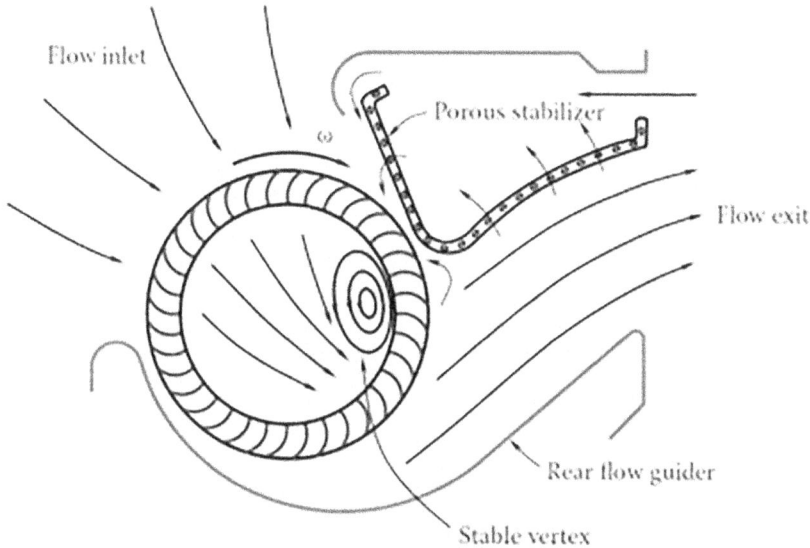

FIGURE 11.45 Cross section and flow pattern of a cross-flow fan.

spiral flow. This type of fan has less outlet losses than axial fans. This provides more flexibility than with either axial or centrifugal fans in relation to the fan position, fan outlet, and design of other components [11.49]. It can operate throughout its performance curve without stalling. This type of fan produces less noise compared to axial fans.

Cross-flow fans, sometimes known as tangential fans or tubular fans, have been extensively used in the heat, air ventilation, and cooling (HAVC) industry due to their silent operation and high-volume ventilating airflows at relatively low pressures. The cross-flow fan has a large L/D ratio and a drum-type impeller with multiple forward-curved blades. When the impeller rotates, a vortex stream is produced transversely across the impeller. By simply increasing the longitudinal axial length of the impeller, a high discharge flowrate can be achieved.

As shown in Figure 11.45, the airflows pass between the blades on one side of the impeller, through the internal space of the runner, and then through the blade passages for a second time to discharge on the other side of the impeller [11.50]. Thus, the efficiency of this type of fan is relatively low (<40%) due to the air passing through the blades twice. The porous structure of a stabilizer is used for attenuating the fan noise. The numerical results show that the porous stabilizer does affect the emitted noise from the fan. The weak airflows moving in or out through the porous wall of the stabilizer are believed to have played an important role in uniformizing the pressure in the zone of the stabilizer and the vortex and thus are helpful for weakening the vertex flow impingement on the stabilizer wall and reducing the relevant pressure oscillations [11.51].

An innovative device architecture invented at a Sandia National Laboratory [11.52] is presented in Figure 11.46. This device, called *heatsink-impeller*, combines an impeller and rotating fins together to achieve high cooling efficiency and degradation thermal resistance. It consists of a disk-shaped heat spreader populated with helical fins on its top surface and functions like a hybrid of a conventional finned metal heatsink and an impeller. It is a cast metal impeller that floats on a hydrodynamic air bearing just a thousandth of an inch (0.03 mm) above a metal heat pipe spreader, powered by a high-efficiency brushless motor in the middle. Air is drawn in the downward direction into the central region and expelled in the radial direction through the dense array of fins.

The prototype device has shown to provide a severalfold reduction in boundary layer thickness, intrinsic immunity to heatsink fouling, and drastic reductions in noise. This type of cooler is quiet

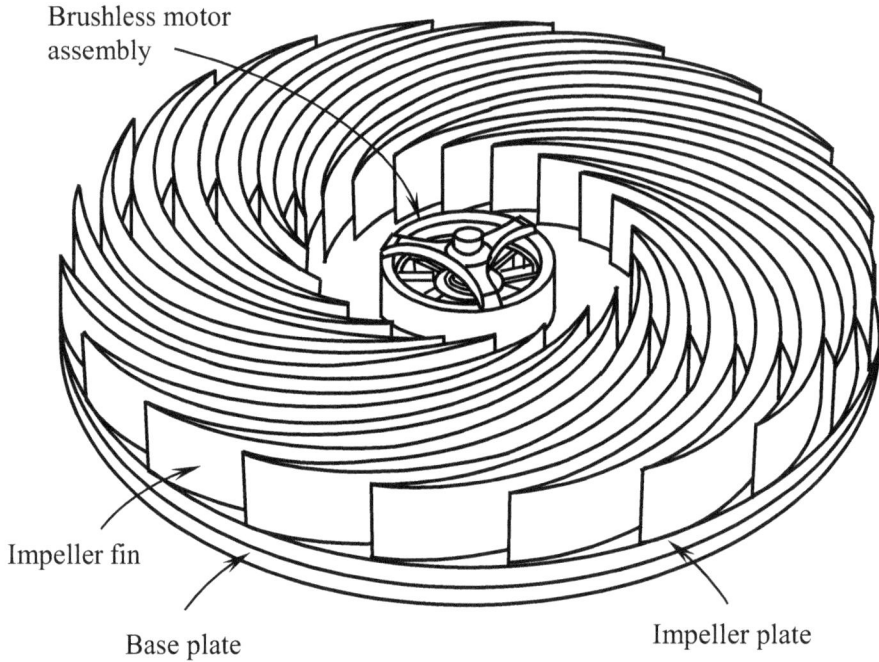

FIGURE 11.46 Combination of fan blades and heat fins as an effective cooling device (U.S. Patent 8,228,675) [11.52]. (Courtesy of the U.S. Patent and Trademark Office, Alexandria, VA.)

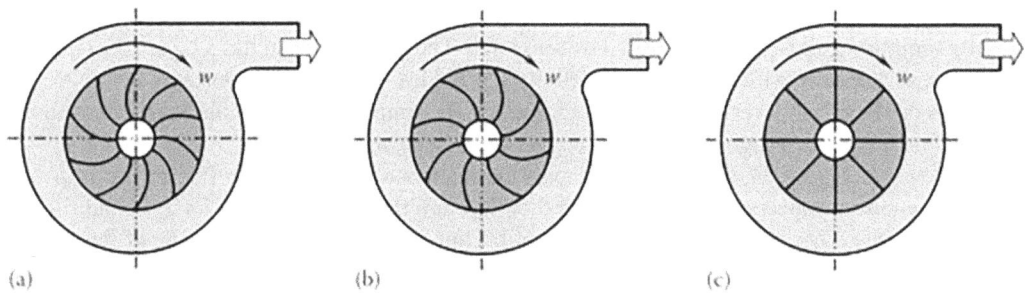

FIGURE 11.47 Three types of centrifugal fan blades: (*a*) forward-curved blade, (*b*) backward-curved blade, and (*c*) straight blade.

and 30 times more efficient than fan-and-heatsink solutions. It was estimated that if every conventional heatsink in the United States is replaced with this device, the country would use 7% less electricity [11.53].

Though this device was invented mainly for electronics cooling, its fundamentals can be adopted for motor cooling easily.

11.4.4.2 Forward-Curved, Backward-Curved, and Straight Blades of Centrifugal Fans

For centrifugal fans, fan blades can be made forward curved, backward curved, or straight (see Figure 11.47). Forward-curved blade fans have blades with leading edges curved forward to the direction of rotation. The fan requires a scroll housing to convert the fan kinetic energy into useful static pressure. However, the efficiency of this type of fan is rather low, about 55%–65%. This

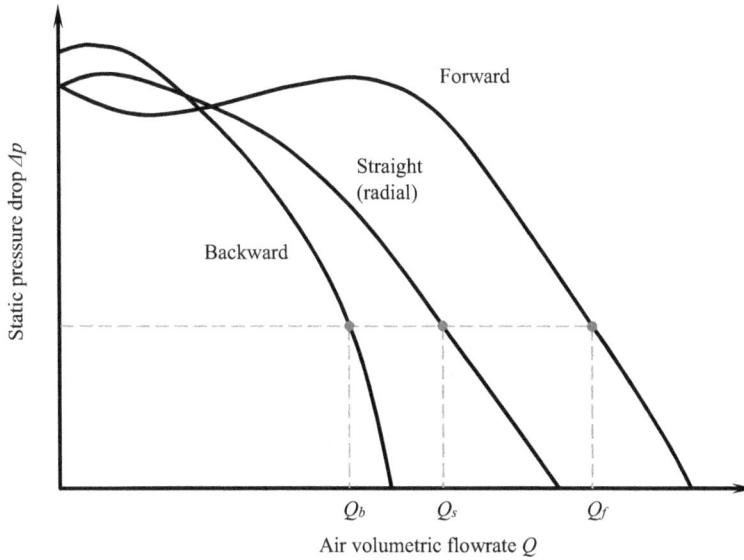

FIGURE 11.48 The comparison of characteristic performance among centrifugal fans with forward-curved, backward-curved, or straight (radial) blades. Generally, for a given Δp, $Q_f > Q_s > Q_b$.

type of fan requires more blades than other centrifugal fans and is best suited for applications that require high flowrates with low static pressures. Since the fans operate at relatively lower speeds, they generate relatively lower noise.

Backward-curved blade fans have blades with leading edges curved backward with respect to the rotation of the impeller. This type of fan runs faster than forward-curved type and is characterized by the most efficient fan, up to 70%–80%. The backward-curved blade fans are normally used for high airflow capacity and high static pressure applications. It should be noted that this type of fan can be designed either using or not using scroll housings, depending on the applications. For a nonscroll housing fan, the impeller can work effectively by itself.

Centrifugal fans can generate relatively high pressures. They are suitable for high-pressure applications as compared with axial flow fans. The comparison of characteristic curves among the forward-curved, backward-curved, and straight blades is presented in Figure 11.48 [11.54]. Among these three types of blades, the forward-curved blades normally produce the highest volumetric flowrate for a fixed Δp value. Consequently, the forward-curved centrifugal fans are commonly used in air-conditioning systems and ventilation units. In comparison, the backward-curved centrifugal fans have a larger size under similar operating characteristics (*e.g.*, volumetric flowrate, pressure, and temperature) and a little more complex structure so that they are more expensive. Despite the cost disadvantage, the backward-curved centrifugal fans have become more popular in air handling units because of their higher efficiency over the forward-curved centrifugal fans.

As the simplest and least efficient centrifugal fans, straight blade radial fans have blades that extend straight from the shaft. This type of fan runs at a medium speed but generates higher pressures than other types of centrifugal fans. They are often used in corrosive applications and in high-temperature environments.

11.4.4.3 Fan Performance Curve and Operation Point

Fan performance can vary widely among fan models and manufacturers. One of the most valuable pieces of information supplied by fan manufacturers is the fan performance curve, which gives the relationship between the volumetric flowrate Q delivered by the fan and the pressure difference Δp.

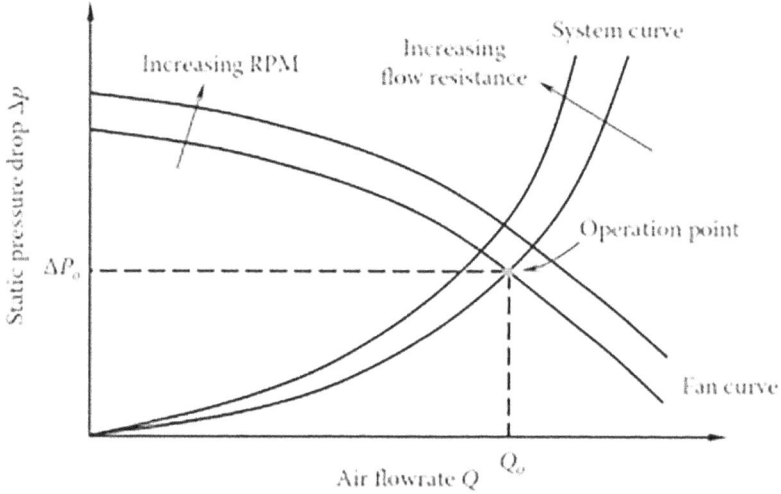

FIGURE 11.49   Fan performance characteristics.

Ventilating fans operate against pressure in their function of moving air through a motor. The fan pressure difference is normally expressed in either total pressure or static pressure (usually in the United States), and the fan volumetric flowrate is in either cubic meters per second (m³/s) in the *SI* system or cubic feet per minute (CFM, ft³/min) in the English system. The static pressure difference is defined as the difference between the fan inlet and the atmospheric pressure. The total pressure produced by a fan is the sum of the static pressure and the dynamic pressure.

When air flows through a cooling system (*i.e.*, motor), a resistance is produced in the system against the airflow. This resistance, expressed as the sum of all static pressure losses, approximately varies as the square of the flowrate near the operation point of the fan. Thus, as the airflow rate through the system is doubled, the static pressure required to drive airflow is increased about four times. As illustrated in Figure 11.49, the operation point of the fan is determined by the intersection of the fan curve and the system resistance curve. A change in fan rotating speed and the use of the fan in different cooling systems will result in a shift in the operation point of the fan.

11.4.4.4   Fan Selection

The selection of fans or blowers is critical to a long-term successful operation of a motor cooling system. The factors impacting the fan or blower selection include the airflow capacity, efficiency, pressure, type of fluids, size, noise level, and cost.

Each fan system has unique requirements that can be normally satisfied from the best selection of fan type, flow pattern, constructive configuration, and fan size. In the process of fan design and selection, it is important to understand three fundamental laws of fan: (*a*) The volumetric airflow delivered by a fan is directly proportional to the fan rotational speed. (*b*) The static pressure developed by a fan is proportional to the square of the fan rotational speed. (*c*) The power required by a fan is proportional to the cube of the fab rotational speed. With these laws, the performance of a fan can be accurately predicted and optimized.

As discussed previously, centrifugal fans have been frequently used in motor cooling and other high-pressure and low-flow applications. In some cases, they can also be used for low-pressure and low-flow general-purpose applications. Axial fans provide an economic solution in high flow volume and low-pressure systems (*e.g.*, ventilation fans). The type of mixed-flow fan is a hybrid of centrifugal and axial fans, which are capable of greater airflow than centrifugal fans and higher pressures than axial fans. With a more complex design, mixed-flow fans typically have higher efficiency, lower noise, and higher cost, compared with other types of fans.

The efficiency of a fan is largely determined by the design of fan blades. Specially designed airfoil blades can drive flow more efficiently and lower fan losses. Therefore, they are suitable for use in compact motors that require high cooling efficiencies with limited spaces. However, airfoil blades are specifically designed for a certain rotating direction. For motors that require bidirectional rotation (*e.g.*, wheel-driving motors), this type of blade is no longer applicable.

11.5 LIQUID COOLING TECHNIQUES

While conventional air cooling techniques have provided the most cost-effective option and served small-sized and medium-sized electric motors well for a long time, liquid-cooled solutions have been recognized as the best means for cooling high-powered and large-sized modern electric machines. In recent years, liquid cooling has been widely accepted as an attractive cooling method for motor cooling due to higher heat transfer coefficients achieved as compared to air cooling. The renewed attention on liquid cooling is mainly resulted from the inexorable rise in motor power dissipation and emergence of on-winding hot spots. By moving from air cooling to liquid cooling, thermal restrictions on the design of large size and high-power-density motors can be removed. Although water as the most common cooling medium has been extensively used in many cooling applications (main in the form of indirect cooling) due to its superior thermophysical properties, units can be built using other cooling mediums such as mineral oils, FC liquids, refrigerants, liquid nitrogen, and liquid metals with low-melting points.

In fact, liquid cooling effectively increases the continuous torque rating of the motor. Application of liquid cooling may be categorized as either direct or indirect, depending upon the liquid direct or indirect contact with the heated motor components. Liquid cooling is accepted as an attractive cooling method for an increasing number of applications.

11.5.1 THERMOPHYSICAL PROPERTIES OF COOLANTS

In the selection of a suitable coolant in liquid cooling, a number of parameters must be carefully considered, including (*a*) thermophysical properties; (*b*) electrical properties (especially dielectric strength); (*c*) chemical compatibility of the coolant with the motor components; (*d*) freezing, boiling, and flash points; (*e*) toxicity; (*f*) erosion; (*g*) flammability; (*h*) impact on human health; and (*i*) cost. Thermophysical properties of some selected coolants including FC series coolants (which are extensively used in electronic cooling) are given in Table 11.4, compared with water. As can be seen from the table, water has the highest specific heat and the lowest viscosity. However, because water is not a dielectric liquid, it may be used for indirect cooling cases.

For natural and forced convection, the most important parameter is the specific heat of the coolant. This is because it determines how much heat energy can be carried away by the coolant as the coolant mass flowrate m and the temperature difference ΔT are specified. A high specific heat is aimed to transfer a large amount of heat with a minimum fluid flow. A low viscosity is also desired because it helps to reduce the resistance of fluid flow. For phase-change heat transfer mechanisms such as evaporative cooling and boiling, the latent heat of vaporization is most important. Boiling and freezing points define the working range of coolants.

11.5.2 DIRECT LIQUID COOLING TECHNIQUES

In direct liquid cooling techniques, liquid coolants are directly brought into contact with stator and rotor windings/cores. Therefore, this form of cooling offers excellent opportunities for efficiently removing heat directly from these heat sources without intervening the thermal conduction resistance. As a result, direct liquid cooling has a higher cooling efficiency than other convective heat transfer modes such as air cooling and indirect liquid cooling.

TABLE 11.4

Thermophysical Properties of Liquid Coolants

Coolant	Density (kg/m³) at 1 atm	Specific Heat (J/ kg-K)	Thermal Conductivity (W/m·K)	Latent Heat of Vaporization (kJ/kg)	Dynamic Viscosity (Pa-s) × 10⁻³ at 20°C	Freezing Point (°C)	Boiling Point (°C)
Aerosol	1100			216.7	0.205	−101.0	−26.1
Ammonia	609	4,740	0.521	1369	0.138	−77.75	−33.34
Aromatic	860	1,750	0.14		1.0	<−80	80
Dynalene HC-30	1,275	3,077	0.519		2.5	<−40	112
Engine oil (SAE 40)	887	1,765	0.138		81.3	−12	>310
Ethanol/water (44/56 by weight)	927	3,500	0.38	1636	3.0	−26.6	
Ethylene glycol/water (50/50 by volume)	1069	3,350	0.39	1.528	3.4	−37	107.2
Methanol/water (40/60 by weight)	935	3,560	0.4	1.344	2.0	−38	711.9
Potassium formate/water (40/60 by weight)	1,250	3,200	0.53		2.2	−35	110
Propylene glycol/water (50/50 by volume)	1,041	3,559	0.37	1.585	5.4	−34	105.6
Silicate ester (Coolanol 25R)	910	1,990	0.135		4.6	−50	
Silicone (Syltherm XLT)	852	1,647	0.11	172	1.4	−111	173
FC-43	1,880	1,100	0.066	71	5.264	−50	174
FC-72	1,680	1,088	0.055	87.9	0.672	−90	56
FC-77	1,780	1,172	0.057	83.7	1.424	−95	97
FC-87	1,650	1,100	0.056	103	0.66	−115	30
R134a	1,188	1,420	0.013	215.9	0.209	−96.6	−26.1
Water	997	4,181	0.613	2.257	1.002	0	100

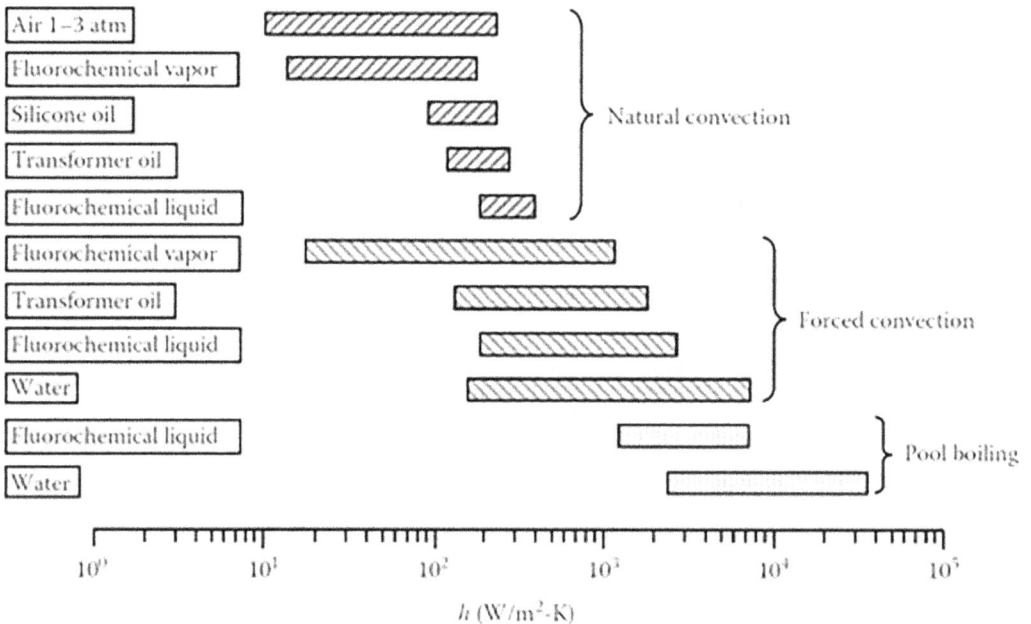

FIGURE 11.50 Relative magnitude of heat transfer coefficients for various coolants and modes of convection. (From Simons, R. E., *Electronics Cooling*, **2**(**2**): 24–29, 1996.)

The heat transfer coefficient in direct liquid cooling not only depends on the thermal physical properties of the coolant but also on the mode of convective heat transfer (*i.e.*, natural convection, forced convection, or boiling). As shown in Figure 11.50 [11.55], due to the phase change during heat transfer processes, boiling heat transfer offers the highest heat transfer coefficient among all combinations of coolant and heat transfer modes. Though water has showed excellent heat transfer performance, it may be only used for indirect motor cooling for its poor dielectric strength.

Closed-loop systems are normally applied in direct liquid cooling (as well as in indirect liquid cooling) due to the benefits including reducing the consumption of expensive liquid coolant, reducing operation and maintenance costs, avoiding environmental issues associated with the coolant, and lowering noise levels.

There are a number of direct liquid cooling techniques available in motor cooling, such as liquid spray, impinging jet, mist spray, and splash cooling. These cooling techniques are addressed in more detail in the following.

11.5.2.1 Direct Liquid Cooling of Bundled Magnet Wires

This innovative method employs shrink wraps to make magnet wires in bundles and thus liquid cooling channels are formed surrounding the magnet wires [11.56]. As illustrated in Figure 11.51, a plurality of magnet wires pass through each lamination slot. A shrink wrap is inserted into the lamination slots. The stator laminations, shrink wrap, and the magnet wires are heated to shrink the shrink wrap onto the outside of the magnet wire. Therefore, the magnet wires are secured in bundles by the shrink wrap, allowing dielectric oil to move freely inside the lamination slots surrounding the

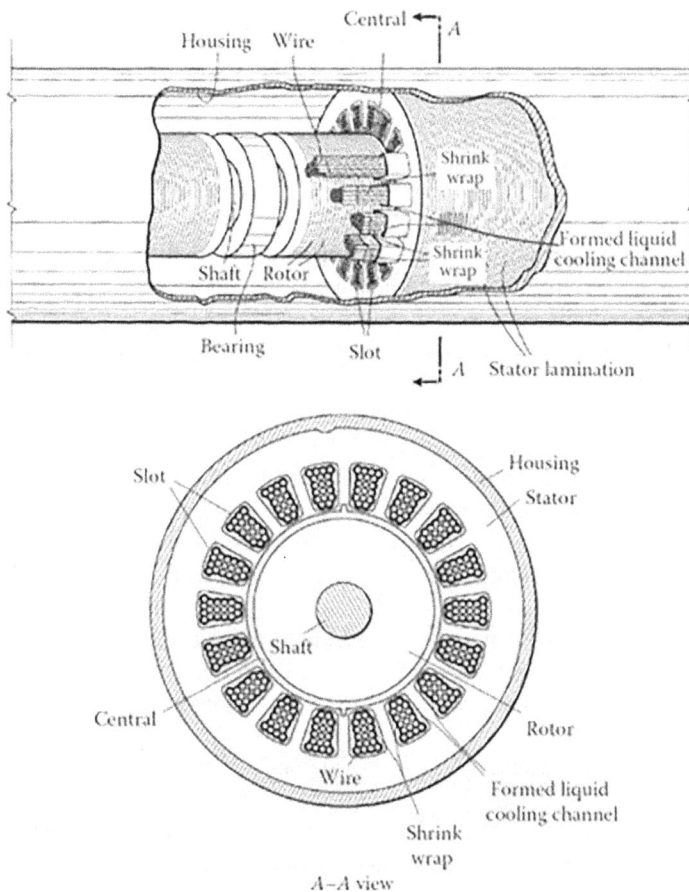

FIGURE 11.51 Enhanced heat transfer using shrink wrap surrounding magnet wires (U.S. Patent 7,679,242) [11.56]. (Courtesy of the U.S. Patent and Trademark Office, Alexandria, VA.)

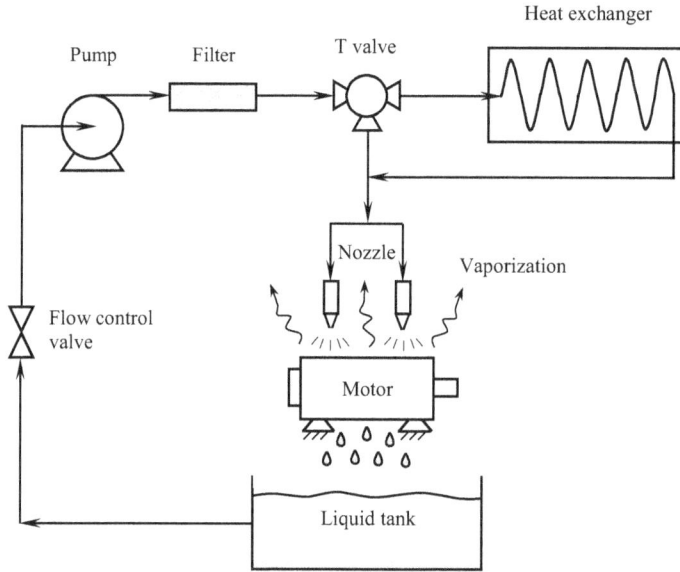

FIGURE 11.52 Spray cooling technique applied on the outer surface of a totally enclosed motor.

wire bundles. Since fluid can pass between and around the windings, the overall winding temperature will be normalized, essentially reducing hot spots that occur in conventional designs.

11.5.2.2 Spray Cooling

Spray cooling is an efficient form of liquid cooling for applications having low to moderate thermal loads. This cooling mode offers an economical and efficient approach for extracting heat from hot surfaces through the use of latent heat of vaporization of a liquid. This approach is to spray liquid droplets onto hot surfaces to create a thin film on the surfaces. As the liquid evaporates, it absorbs heat from its surroundings and lowers the temperature on the surfaces. Hence, it has been widely used in cooling military electronics. As an efficient cooling method, this approach may be promisingly applied to motor cooling.

As Figure 11.52 indicates, a spray cooling system primarily consists of a liquid tank, a flow control valve, a pump, a filter, a heat exchanger, and a number of nozzles. The cooling liquid is pumped from the liquid tank and becomes droplets after passing through spray nozzles. By wetting the motor surface with liquid droplets, heat is transferred from heat sources inside the motor to the liquid film and finally dissipated to the environment via forced convection due to the liquid spray, as well as air impinging flow induced by the liquid spray, evaporation, and/or boiling nucleation (in the case that boiling occurs). Obviously, the spray cooling process mainly relies on the phase change from liquid to vapor. Correspondingly, the exceptional cooling capacity is greatly enhanced due to the latent heat of vaporization.

It should be noted that special machining capabilities are required to manufacture spray nozzles to the precision and tolerance required. It has been revealed the variations in nozzle performance due to contamination, corrosion, and long-term wear. Placement of spray nozzles is fairly critical to assure adequate cooling.

The internal spray cooling technique has been used in motors subject to high heat fluxes. As a rotor rotates, liquid coolant such as engine oil is flung off the rotor and/or shaft surfaces in small droplets due to the centrifugal force acting on the coolant. The droplets create a fine mist and impinge on heated motor components. The droplet impingement enhances the spatial uniformity of the heat removal.

In a rotating machinery, an important characteristic of wall rotating-driven flows is the tendency of fluid with high angular momentum to be flung radially outward. For a motor, the rotor

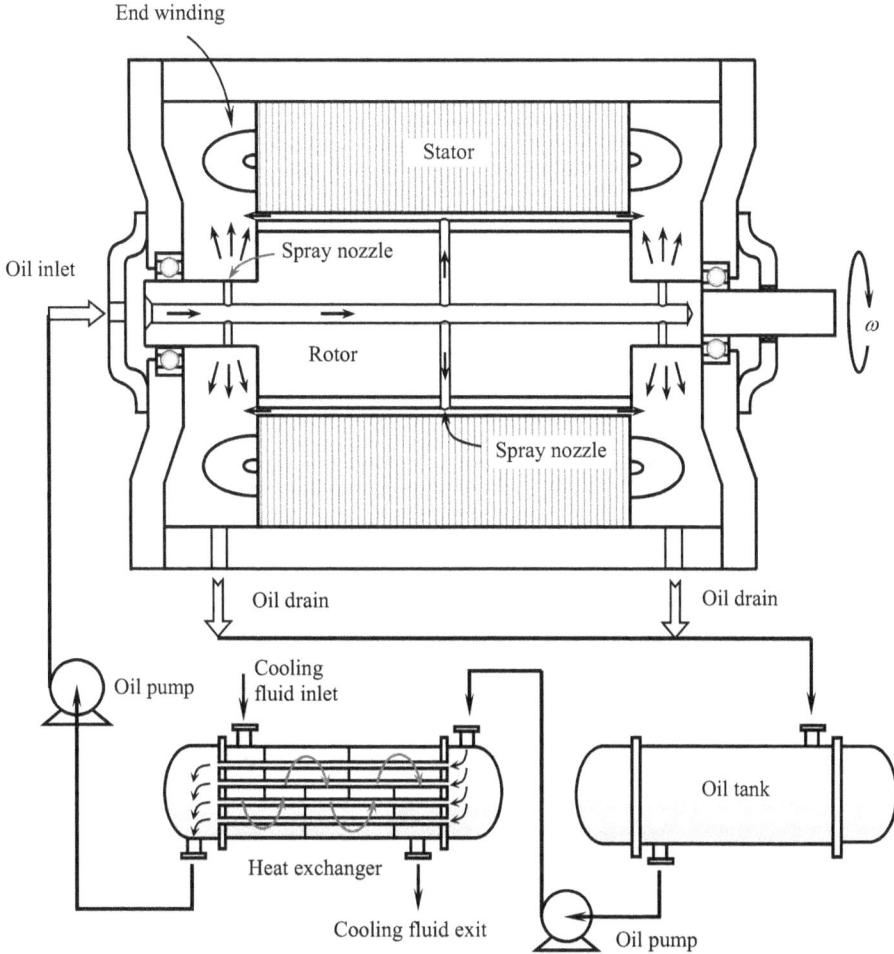

FIGURE 11.53 The internal spray cooling system by utilizing rotor centrifugal force.

rotating-driven flow, usually referred to as the rotating pumping flow, plays an important role in cooling motor windings. Spray cooling occurs when liquid is forced through an array of tiny nozzles to be shattered into dispersive fine droplets. These droplets can spread on the surface and evaporate, removing large amounts of energy through the latent heat of evaporation in addition to substantial conventional effects. In a closed-loop system, the vapor is then condensed and pumped back to the nozzles. Due to its high efficiency, spray cooling can be used for high-heat-flux applications.

The heat transfer rate of spray cooling strongly depends on the temperature difference between coolants and heated surfaces. At low temperature differences, heat transfer occurs primarily through single-phase convection. As the temperature difference increases, the evaporation on the liquid film surface, as well as boiling on the heated surface, can significantly augment the heat transfer rate.

Numerous spray cooling techniques have been developed and used in industrial applications. As a promising technique, the internal spray cooling technique has been proven as an effective way for electric machine cooling. The diagram of the spraying cooling system is shown in Figure 11.53. To avoid electric short circuits, dielectric liquids such as mineral oils are commonly selected as coolants. However, the selection of liquid coolant involves the trade-off between performance, functionality, cost, chemical stability, maintenance, safety, and environmental protection. The thermo-physical properties of engine oil are listed in Table 11.5 as a function of temperature [11.57].

TABLE 11.5

Thermophysical Properties of Engine Oil

Temperature (K)	Density (kg/m³)	Specific Heat (kJ/ kg-K)	Dynamic Viscosity (Pa-s)	Thermal Conductivity (W/m·K)	Thermal Diffusivity (m²/s) × 10⁻⁸	Prandtl Number	Volume Expansion Coefficient (1/K) × 10⁻⁴
260	908	1.76	12.23	0.149	9.324	144,500	7
280	896	1.83	2.17	0.146	11.904	27,200	7
300	884	1.91	0.486	0.144	11.529	6,450	7
320	872	1.99	0.141	0.141	11.125	1,990	7
340	860	2.08	0.053	0.139	7.771	795	7
360	848	2.16	0.025	0.137	7.479	395	7
380	836	2.25	0.014	0.136	7.230	230	7
400	824	2.34	0.009	0.134	6.950	155	7

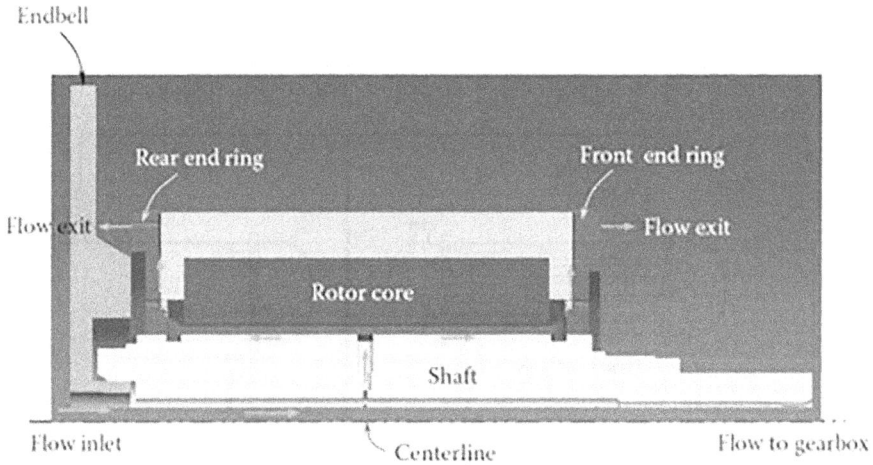

FIGURE 11.54   Cooling flow paths in the rotor assembly (only a half model shown here).

As can be seen in the figure, the cooling oil is pumped into the hollow shaft and ejected from the tiny nozzles uniformly distributed on the shaft circumference by means of the rotor centrifugal force. The droplet-like oil hits the motor end windings to carry heat away from them. The hot oil is then pumped back into the oil tank and cooled by an oil-water heat exchanger.

A CFD analysis has been carried out to predict the temperature distribution and flow field of the rotor assembly with a similar spray cooling system. An assembled view of a casted rotor assembly is shown in Figure 11.54. The rotor assembly consists of the rotor core, shaft, rotor end rings, and rear endbell. During motor operation, heat is generated in the rotor assembly due to three types of power losses: (*a*) mechanical power losses such as frictional loss of bearings and windage loss, (*b*) eddy current and hysteresis losses in the rotor core and other components, and (*c*) stray losses in the rotor end rings. The cooling oil enters from the rear endbell into the axial cooling duct at the shaft center, turns 90° to get into the four small radial holes near the rotor axial center, then ejects from the nozzles into the cooling ducts between the shaft and rotor core, and finally sprays from the end rings into the interior of the motor.

The mass flow field of the rotor is presented in Figure 11.55. For a one-inlet and multiple-outlet system under a steady-state condition, the inlet mass flowrate must equal to the total mass flowrate from all outlets:

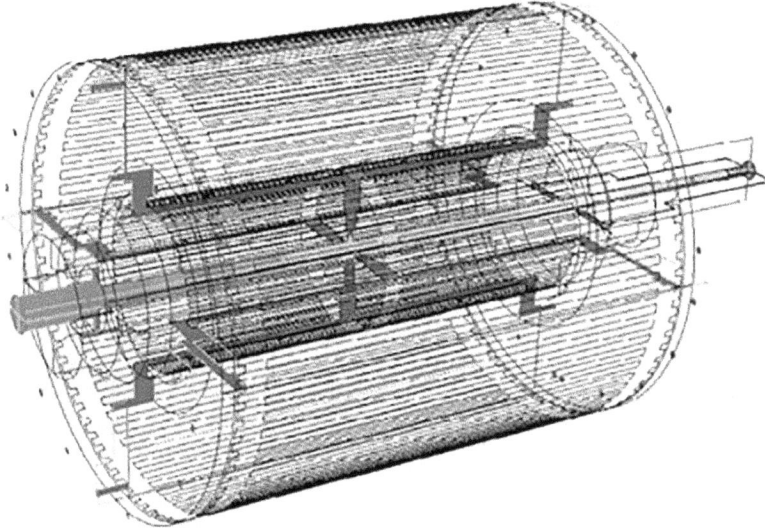

FIGURE 11.55 Cooling flow paths inside the rotor assembly.

$$\dot{m}_{in} = \sum_i \dot{m}_{out,i} = \sum_i \rho A_{out,i} u_{out,i} \qquad (11.20)$$

where

$\dot{m}_{in}$ is the mass flowrate at the inlet, kg/s

$\dot{m}_{out,i}$ is the mass flowrate at the ith outlet, kg/s

$A_{out,i}$ and $u_{out,i}$ are the area (m^2) and velocity (m/s) at the ith outlet, respectively

The mass flowrate ratio at the ith outlet is defined as the ratio of the mass flowrate at that outlet to the total mass flowrate, that is,

$$\eta_i = \frac{\dot{m}_{out,i}}{\dot{m}_{in}} \qquad (11.21)$$

Obviously, $\sum_i \eta_i = 1$.

The determination of the η value at each outlet is critical for obtaining relative uniform temperature distribution and elimination of hot spots. It requires distributing the cooling flowrate at each exit appropriately. For this purpose, the dimensions of nozzle and cooling ducts must be iteratively adjusted according to the CFD results.

The temperature distribution in the rotor assembly is demonstrated in Figure 11.56, showing that the highest temperature occurs on the outer rotor surface near the rotor axial center and the temperature generally decreases from the axial center to the rotor's two ends.

More detailed information about the temperature distribution inside the rotor assembly is given in Figure 11.57. This information can be adopted to help optimize the rotor cooling design for gaining a higher cooling efficiency and eliminating hot spots in the rotor.

11.5.2.3 Direct Liquid Cooling through Hollow Winding Coils/Conductors

The temperature rise in motor windings is the main limiting factor to the increase in electrical current. Generally, a higher heat dissipation capacity leads to a lower operating temperature and thus allows the motor to receive a higher current density for producing a higher torque. The direct liquid

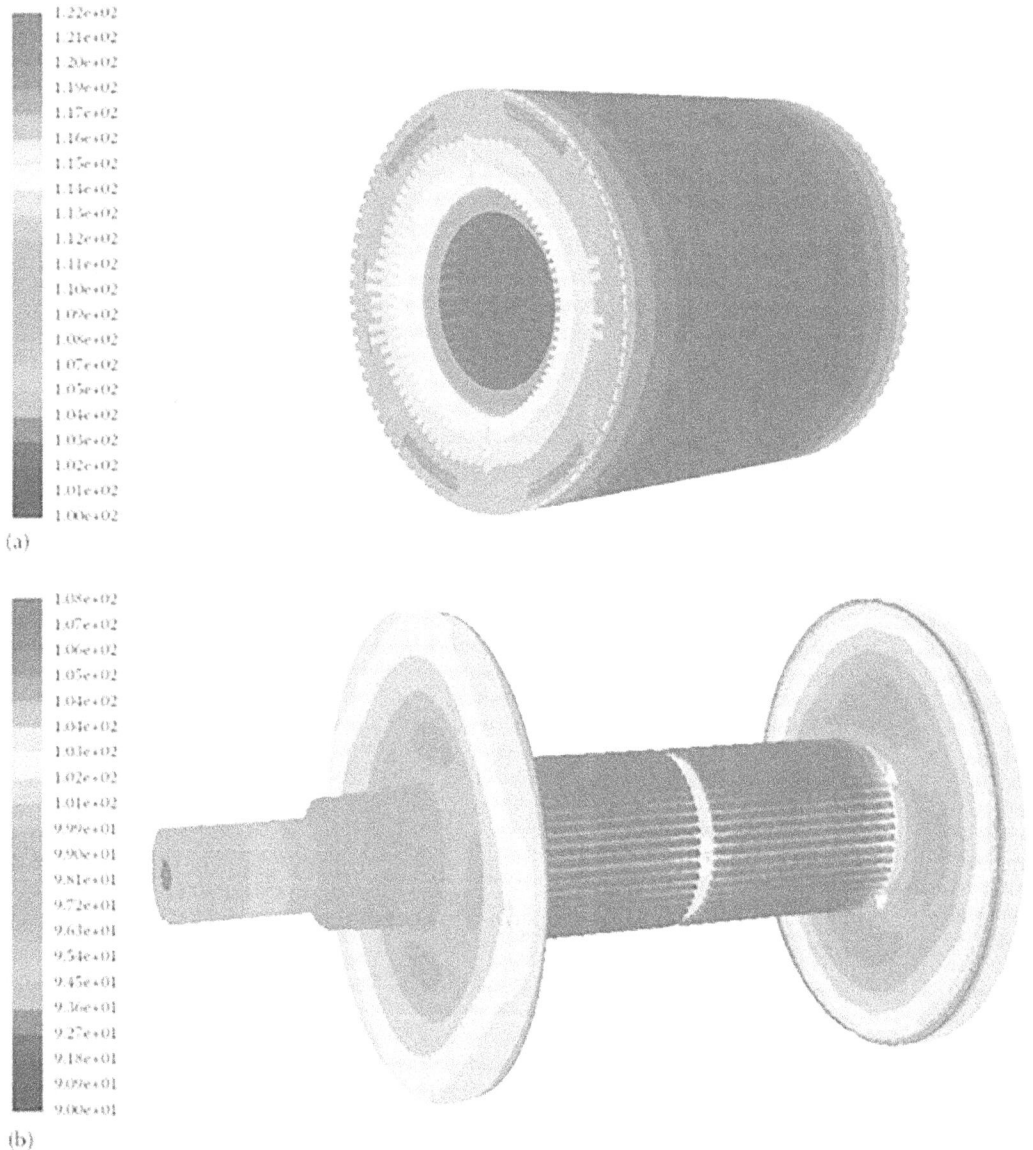

FIGURE 11.56 Temperature distributions in the rotor assembly: (*a*) rotor core and (*b*) shaft and end rings, showing the temperature generally decreasing from the center to the two rotor ends symmetrically.

cooling through cooling channels in winding coils eliminates most thermal resistances along the conventional heat dissipation path, including the resistances from winding to the stator core, from the stator core to the housing, and from the housing to the environment. Only one thermal resistance remained is from the heat source (*i.e.*, motor windings) to the cooling medium. This novel approach greatly enhances the heat dissipation capacity and allows a large increase in the possible current density without exceeding the critical winding temperature.

In practice, as the most intensive cooling technique, direct liquid cooling with hollow winding coils/conductors has been applied to large capacity generators for more than a half century. The world's first generator to have water-cooled stator winding was AEI 30 MW unit in England put into service in 1956. Subsequently, the commercial use of water-cooled stator windings started in

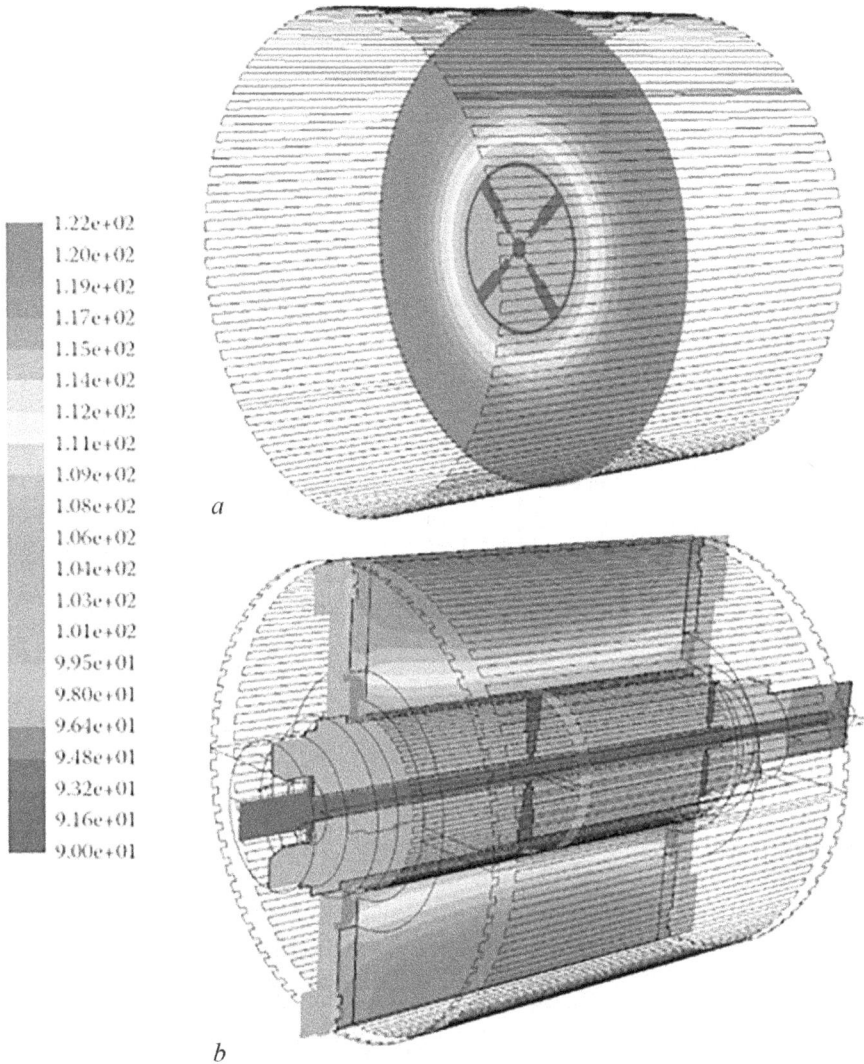

FIGURE 11.57 Temperature distributions in the rotor assembly: (*a*) the cross-sectional area perpendicular to the rotor axis, the highest temperature occurring at the rotor outer surface, and (*b*) the cross-sectional area through the rotor axial centerline.

the United States in 1960, United Kingdom in 1961, and Germany in 1964 [11.58]. On October 27, 1958, the world's first 12 MW, 3,000 rpm dual-water inner cooled turbogenerator (water-cooling of both stator and rotor windings through hollow winding conductors) was successfully manufactured and tested by Shanghai Electric Machinery Factory. In the next few years, engineers had successively solved a series of problems, including winding water leakage, corrosion of metal parts, and plugging of hollow conductors by copper oxides (CuO and Cu_2O), *etc*. The turbogenerator was put into normal service in 1963 [11.59].

In fact, the direct water-cooling through the rotational hollow windings is an extremely complex and difficult task. A secondary flow occurs as a result of the rotation while water in the center of the conduit tends to move radially to the wall due to the centrifugal and Coriolis effect. It has been confirmed experimentally and theoretically, the rotational speed can significantly increase the convective heat transfer in rotor cooling over the stationary condition [11.60, 11.61].

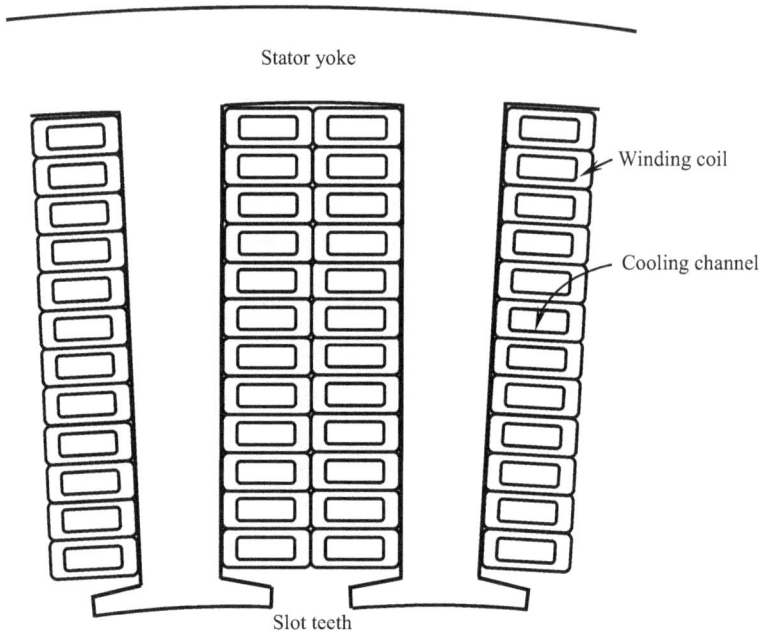

FIGURE 11.58   The arrangement of hollow winding conductors, which carry both water and electrical current, in stator slots. Using the designed hollow conductors can reduce the winding losses by about 50%. The conductor shape and configuration can increase the slot filling factor up to 90%.

However, this cooling technique using hollow conductors has rarely been used in electric motors, mainly due to the high cost for machines rated below 200 kW, as well as the increase in motor weight and size [11.62]. Due to the high pressure drop through hollow conductors, this cooling scheme results in a higher hydraulic loss, which reduces generator efficiency. Wohlers *et al.* [11.63] have designed and analyzed fractional-slot concentrated-windings with hollow coils for direct liquid cooling of synchronous motors. The winding coils with hollow cooling channels can be fabricated using casting or 3D printing manufacturing process. This allows the slot-filling factor increased up to 90%. The arrangement of hollow coils in slots is shown in Figure 11.58. The used cooling medium is Galden® HT135, a perfluoropolyether fluorinated fluid with a boiling point of 135°C. The heat dissipation capacities of the new coil designs are tested experimentally. The proposed coil designs lead to a reduction of the winding losses of approximately 50%. With a flowrate of 1.9 L/min and an inlet coolant temperature of 30°C, the average coil temperature is measured as 135°C.

Copper is a traditional material for hollow conductors used for liquid cooling in large generators. However, copper hollow conductors are frequently plagued by clogging with copper oxides. Furthermore, due to the hollow structure, the cross-sectional area of the conductor is reduced, leading to an increase in the current density flowing through the conductor. To address these issues, the stator windings are designed to consist of solid copper conductors for transmitting electrical current and stainless steel hollow bars for providing liquid paths [11.64]. The stainless steel hollow bars are distributed among the copper conductors in a certain pattern to achieve a uniform cooling effect (Figure 11.59).

11.5.2.4   Liquid Jet Impingement Cooling

Liquid jet impingement is an attractive cooling scheme across a broad spectrum of industries due to the capability of achieving a high heat transfer rate. In this cooling scheme, a liquid is delivered through an array of tiny nozzles to imping motor windings and other heat source components,

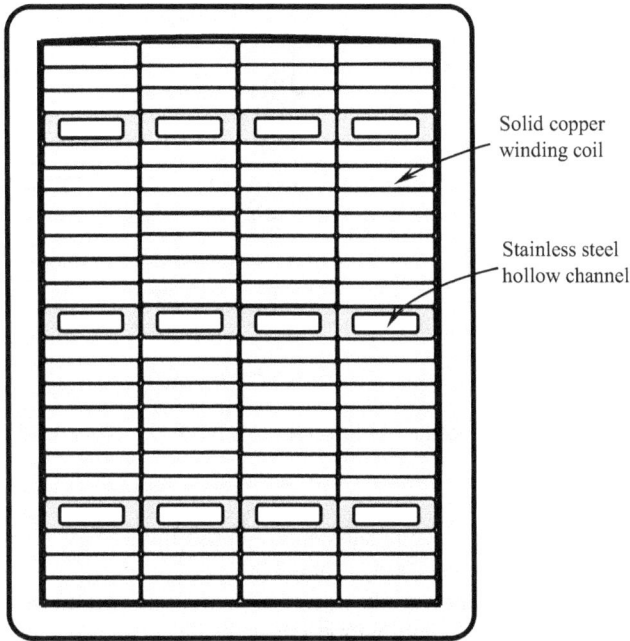

FIGURE 11.59 The generator stator winding bar uses mixture strands, which are a combination of stainless steel hollow strands for providing cooling liquid paths and solid copper strands for transmitting electrical current. This design prevents the perforation of hollow conductors by corrosion and plugging by copper oxides (typically a mixture of CuO and Cu_2O). The arrangement of stainless steel hollow strands is determined according to the uniformity of the winding temperature distribution.

forming very thin hydrodynamic and thermal boundary layers in the impinging regions, and obtaining extremely high heat transfer coefficients within the stagnation zones. For direct liquid jet impingement cooling, a dielectric liquid is required to prevent the electrical short circuit of motor windings. For indirect liquid cooling, water may be adopted as the best coolant due to its superior thermophysical properties. In a well-designed cooling system with multiple impinging jets, the liquid jet impingement cooling scheme can offer lower thermal resistance, higher heat transfer coefficient, and improved temperature uniformity.

In fact, both jets and sprays are generated by forcing liquid flowing through small nozzles. The main difference between the two is their flow patterns generated by the nozzle. While in the case of sprays the liquid flow is shattered into a dispersion of fine droplets, in the case of jets the liquid flow forms a strong mainstream to impact heat sources. In practical applications, the fluid patterns (mist, spray, splash, or jet) can be readily controlled by selecting a suitable nozzle diameter and/or a pressure applied on the liquid.

According to the operation condition, liquid jet impingement can be implemented into three main configurations: submerged jet, which allows liquid to impinge into a region containing the same liquid; free-surface jet, which uses dense liquid in a medium that is less dense (*e.g.*, air, vapor, or gas); and confined submerged jet, which is similar to free-surface jet but the impinged jet is confined by a wall parallel to the heater surface (Figure 11.60).

The submerged jet technique combines the jet impingement cooling and liquid immersion cooling mechanisms. With a low jet velocity (*i.e.*, low Reynolds number), cooling is mainly governed by immersion cooling. With an increase in the jet velocity, submerged jet cooling makes more contributions to the overall cooling effectively. Womac *et al.* [11.65] conducted experimental studies on both submerged jet impingement and free-surface jet impingement in a range of jet velocities, nozzle diameters, and nozzle-to-heater separation distanced, using water and FC liquid (FC-77) as

(a) Submerged jet   (b) Free-surface jet   (c) Confined jet

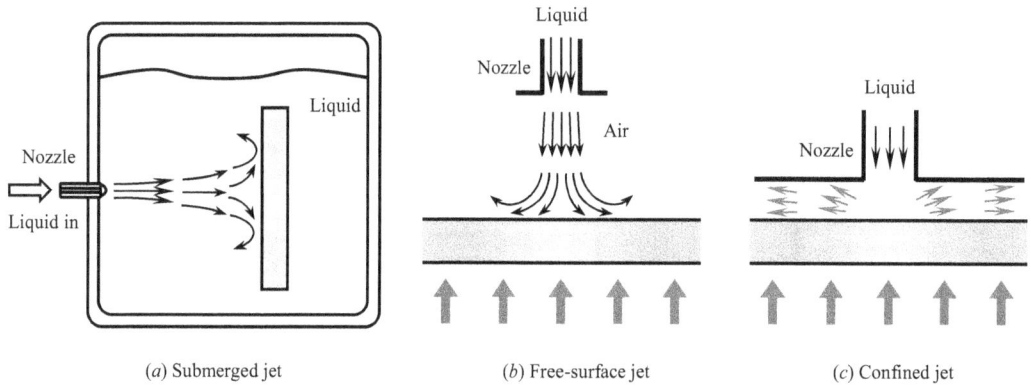

FIGURE 11.60   Three liquid jet impingement configurations: (a) submerged jet, (b) free-surface jet, and (c) confined jet.

coolants. Their results have shown that the submerged jet has higher heat transfer capability than the free-surface jet for Reynolds number larger than 4,000.

Using mineral oil as a coolant and various nozzles, Davin *et al.* [11.66] experimentally investigated free-surface jet, spray, and mist cooling for a 40-kW electric motor. The setup of the overall cooling system is illustrated in Figure 11.61. Oil is injected at each side of the motor to directly cool the stator end-windings. To compare the impact of oil flow patterns (spray, jet, and mist) and flowrates on heat transfer characteristics, a number of oil injection systems have been tested using flat jet and full core nozzles. The flow pattern mainly depends on the flowrate, as well as the oil temperature.

With a specially designed type of injectors, high impact velocities are produced to help improving end-winding cooling. In this cooling scheme, cooling oil is evenly distributed to 12 injectors to cool all 12 end-windings. Each injector includes three orifices ($d=0.5\,\text{mm}$) in the front of each end-winding to generate laminar jets impinging on different surfaces of the end-winding. These injectors produce a multi-jet pattern with a maximum velocity of $7\,\text{m/s}$, where the ratio of the impinging distance z to the orifice diameter d is about 12.

Besides the oil cooling loops, a water cooling loop is added to maximize convection heat transfer between the flanges and the oil–air mixture by means of the inbuilt cooling inducts of the flanges. Hence, the whole cooling system contains an oil loop for stator end-winding cooling, a water loop for flange cooling, and another water loop for oil bath-regulated cooling.

As the reciprocal of thermal resistance R_{th}, thermal conductance C_{th} is a measurement of the ability of a system or machine to conduct heat, defined as the ratio of heat energy flow over the temperature difference. In this problem, C_{th} is defined as

$$C_{th} = \frac{P_w}{\Delta T} = \frac{P_W}{T_W - T_{in,oil}} \tag{11.22}$$

where

   P_w is the power dissipation from the winding, W

   $\Delta T = (T_w - T_{in,oil})$ is the temperature difference between the mean winding temperature and the oil input temperature, $°C$

In this way, the oil cooling efficiency of the different injection types can be compared using thermal conductance. In all cases, the cooling efficiency always increases with an increase in the flowrate. Experimental results show that with oil injection, the dissipation power is multiplied by a factor of 2.5–5. By comparing several injectors, the dripping injector with laminar jets is the most efficient, where oil is injected at the top region and the end-windings is benefited from high flowrates.

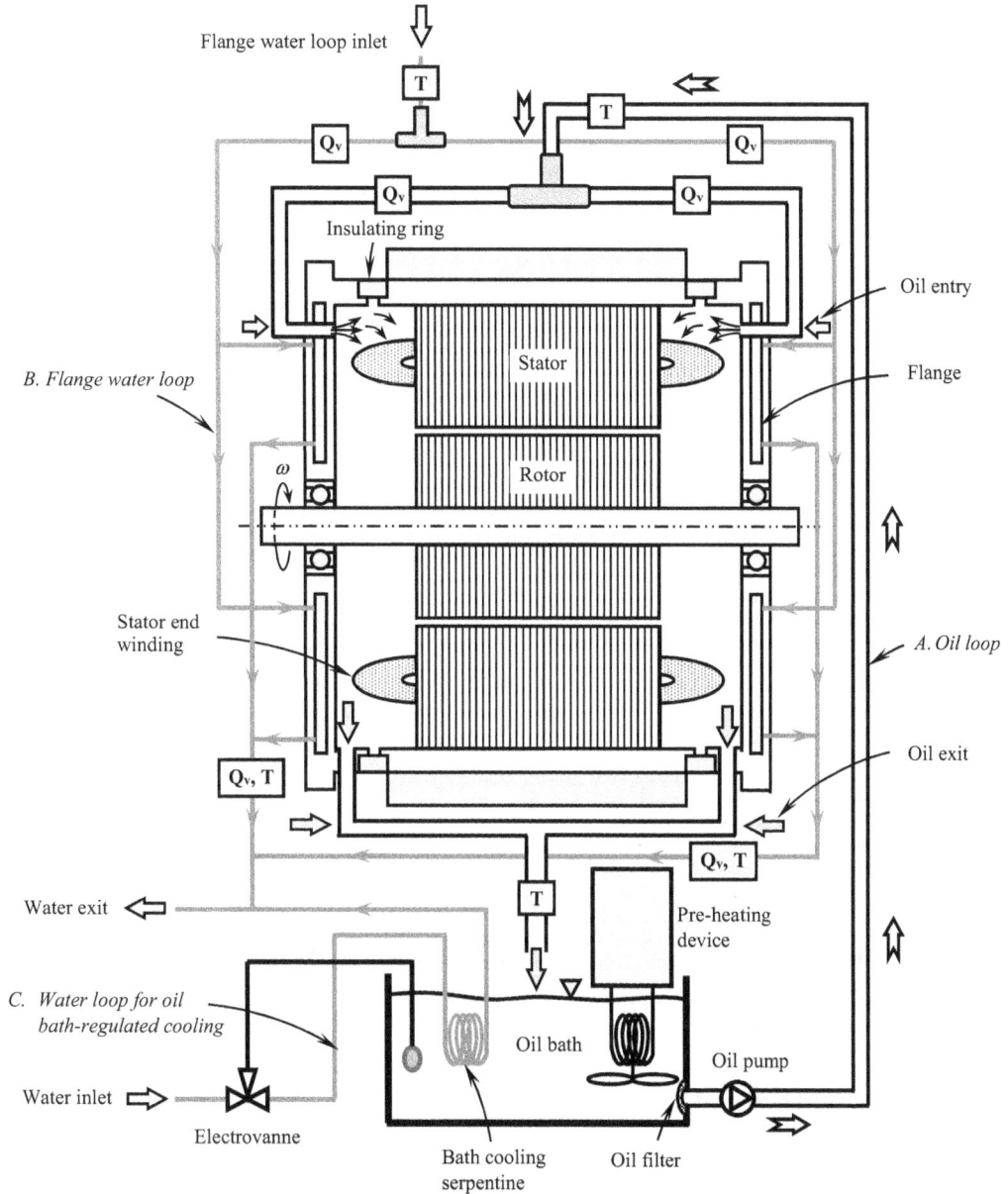

FIGURE 11.61 The setup of the oil cooling system for electric motor cooling. The system consists of three cooling loops: an oil loop for state end winding cooling, a water loop for flange cooling, and another water loop for oil bath-regulated cooling. The locations for measuring temperature and volumetric flowrate are marked in the figure.

11.5.3 Liquid Immersion Cooling

Liquid immersion cooling becomes very popular today in electronic cooling applications. The first computer designed to be directly cooled by a liquid was Cray-2 supercomputer, back to 1985. Since it used a dielectric coolant that does not conduct electricity, the supercomputer could be submerged in the coolant without causing any short circuit. In addition, because the dielectric coolant has much better thermophysical properties than air, it offers the prospect of dramatically more energy-efficient cooling than those in the conventional air cooling approaches. Besides high cooling efficiency, the potential benefits of liquid immersion cooling include the high cooling capacity, temperature uniformity, reliability, and not needing to power internal fans/pumps to assist fluid cooling flows. Compared with air cooling techniques, liquid immersion cooling can help design much more dense devices without the need for flow aisles.

Electric motors and drives could greatly benefit from the direct immersion cooling scheme, especially with the phase change of coolant. Unlike single-phase liquid immersion cooling techniques, which produce motor winding temperature increases proportional to the increase in winding heat flux, liquid immersion with phase change capitalizes on the merit of nucleate boiling, allowing to a large increase in heat flux with only modest increases in winding temperature [11.67]. A typical application of immersion cooling is found for submersible motors that are completely submerged in water or other types of liquid.

This cooling mode can be successfully transplanted into motor cooling. Some special dielectric coolants such as FC are particularly suitable for motor immersion cooling due to their low toxicity, stable viscosity, UV resistance, and high dielectric strength. In order to control the friction losses, this method may be limited to low rotating speed motors.

11.5.4 Indirect Liquid Cooling Techniques

In an indirect liquid cooling scheme, a liquid coolant does not come into contact with cooling components. Instead, cooling is implemented by using heat exchangers, cold plates, coolers, and other cooling products. In such cases, a good thermal conduction path is needed from heat sources (winding, core, *etc.*) to the coolant. Since there is no direct contact with the electric windings and connectors, water or water-based antifreezes (*e.g.*, ethylene glycol/water, propylene glycol/water, methanol/water, and ethanol/water), even some liquid metals, can be used as the coolant, taking advantage of their superior thermophysical properties and relatively low costs. This heat transfer mode is in fact encountered in a wide spectrum of engineering systems including energy-power conversion, material procession, internal combustion, electric and electronic devices, and manufacturing process.

11.5.4.1 Indirect Liquid Cooling via Cold Plates Attached to Motor Walls

Cold plates are typically made of copper or aluminum tubes that are bonded into aluminum plates to optimize thermal conductance. Liquid-cooled cold plates may be mounted on motor external surfaces to provide very effective contact cooling for various motor applications having different thermal requirements (Figure 11.62). The dimensions of a cold plate are designed to fit the size of a motor. In practice, cold plates use either water or water-based antifreezes as a coolant. As an example, thermophysical properties of water/glycol mixtures with different mixing ratios are given in Table 11.6.

11.5.4.2 Indirect Liquid Cooling via Helical Copper Pipes Casted in Motor Housing

This cooling method uses helical copper pipes that are directly cast inside the motor housing (Figure 11.63). Water flowing through these helical copper pipes ensures more uniform cooling of the stator due to the generation of the secondary flows in these curved pipes by the centrifugal effect. The advantages of this cooling method are as follows: (*a*) Because the copper pipes are perfectly integrated with the housing by casting, the thermal contact resistance between the copper pipes and the housing is minimized. (*b*) There is no leakage from the copper pipes. However, the fabrication cost for such a cooling system is usually higher than other cooling methods, as listed in the following.

The thermal energy dissipated to the coolant Q can be determined as

$$Q = \dot{m}C_p\Delta T \tag{11.23}$$

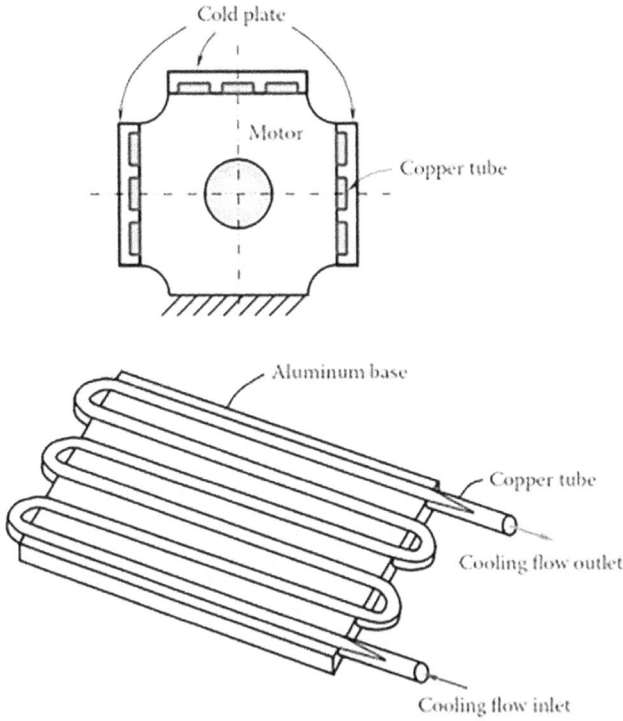

FIGURE 11.62 Cold plates used for motor cooling, mounted on the three sides of the motor.

where $\Delta T = T_o - T_i$, the temperature difference between the outlet and inlet of the coolant.

Several correlations are available in the literature for calculating the friction factor as flows through helical pipes. Ito [11.68] proposed a model of friction factor for a turbulent flow:

$$f_c = 0.00725 + 0.076 \left[\text{Re} \left(\frac{r}{0.5d} \right)^{-2} \right]^{-0.25} \left(\frac{r}{0.5d} \right)^{-0.5} \tag{11.24}$$

which is valid for

$$0.034 < \text{Re} \left(\frac{r}{0.5d} \right)^{-2} < 300$$

Srinivasan et al. [11.69] developed a model for flows in a helical pipe:

$$f_c = 0.084 \left[\text{Re} \left(\frac{r}{0.5d} \right)^{-2} \right]^{-0.2} \left(\frac{r}{0.5d} \right)^{-0.5} \tag{11.25}$$

which is valid for

$$\text{Re} \left(\frac{r}{0.5d} \right)^{-2} < 700 \quad \text{and} \quad 7 < \frac{r}{0.5d} < 10^4$$

With the friction factor, the pressure drop across a helical pipe can be determined as

$$\Delta p = \pi \, dL \tau_w A \tag{11.26}$$

TABLE 11.6

Thermophysical Properties of Water/Glycol Mixtures with Different Mixing Ratios

Substance	Volume in Mixture (%)	Minimal Working Temperature (°C)	Temperature (°C)	Density (kg/m³)	Specific Heat (kJ/ kg-K)	Thermal Conductivity (W/m·K)	Dynamic Viscosity×10³ (W-s/m²)	Kinematic Viscosity×10⁶ (m²/s)
Monoethylene glycol [C₂H₄(OH)₂]	20	−10	−10	1,038	3.85	0.498	5.19	5.0
			0	1,036	3.87	0.500	3.11	3.0
			20	1,030	3.90	0.512	1.65	1.6
			40	1,022	3.93	0.521	1.02	1.0
			60	1,014	3.96	0.531	0.71	0.7
			80	1,006	3.99	0.540	0.523	0.52
			100	997	4.02	0.550	0.409	0.41
	34	−20	−20	1,069	3.51	0.462	11.76	11.0
			0	1,063	3.56	0.466	4.89	4.6
			20	1,055	3.62	0.470	2.32	2.2
			40	1,044	3.68	0.473	1.57	1.5
			60	1,033	3.73	0.475	1.01	0.98
			80	1,022	3.78	0.478	0.695	0.68
			100	1,010	3.84	0.480	0.515	0.51
	52	−40	−40	1,108	3.04	0.416	110.8	100
			−20	1,100	3.11	0.409	27.5	25
			0	1,092	3.19	0.405	10.37	9.5
			20	1,082	3.26	0.402	4.87	4.5
			40	1,069	3.34	0.398	2.57	2.4
			60	1,057	3.41	0.394	1.59	1.5
			80	1,045	3.49	0.390	1.05	1.0
			100	1,032	3.56	0.385	0.722	0.7
1,2-Propylenglycol [C₃H₆(OH)₂]	25	−10	−10	1,032	3.93	0.466	10.22	9.9
			0	1,030	3.95	0.470	6.18	6.0
			20	1,024	3.98	0.478	2.86	2.8
			40	1,016	4.00	0.491	1.42	1.4
			60	1,003	4.03	0.505	0.903	0.90
			80	986	4.05	0.519	0.671	0.68
			100	979	4.08	0.533	0.509	0.52
	38	−20	−20	1,050	3.68	0.420	47.25	45
			0	1,045	3.72	0.425	12.54	12
			20	1,036	3.77	0.429	4.56	4.4
			40	1,025	3.82	0.433	2.26	2.2
			60	1,012	3.88	0.437	1.32	1.3
			80	997	3.94	0.441	0.897	0.9
			100	982	4.00	0.445	0.687	0.7
	47	−30	−30	1,066	3.45	0.397	160	150
			−20	1,062	3.49	0.396	74.3	70
			−10	1,058	3.52	0.395	31.74	30
			0	1,054	3.56	0.395	111.97	18
			20	1,044	3.62	0.394	6.264	6
			40	1,030	3.69	0.393	2.987	2.9
			60	1,015	3.76	0.392	1.624	1.6
			80	999	3.82	0.391	1.10	1.1
			100	984	3.89	0.390	0.807	0.82

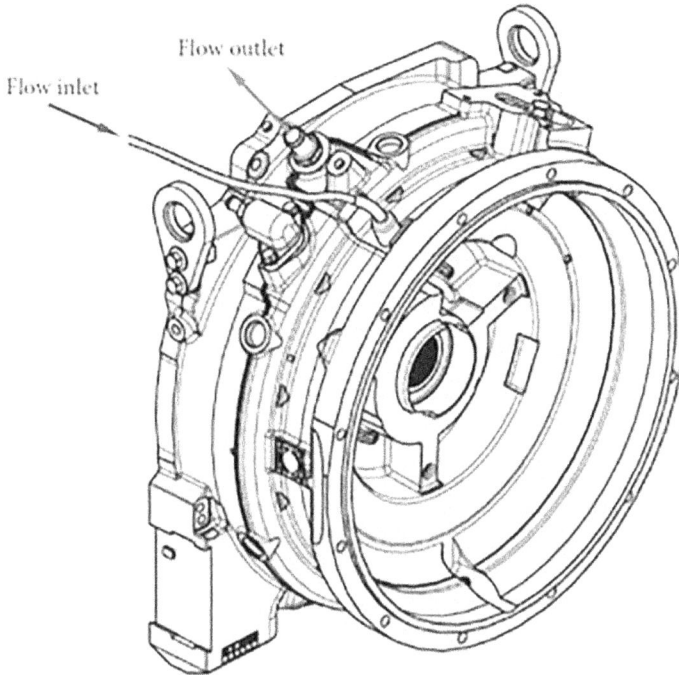

FIGURE 11.63 Assembled stator with a dual-cooling pipe casted inside the motor housing.

To reduce the pressure drop across the cooling pipe, a helical dual-cooling pipe may be adopted in the design. As shown in Figure 11.64, the adaptor is designed to distribute the coolant with an equal flowrate into the two pipes.

11.5.4.3 Indirect Liquid Cooling with Cooling Channels on Casted Motor Housing

Figure 11.65 depicts another cooling scheme in which cooling water flows through sealed cooling channels. The cooling channels are formed by integrating the casted motor housing and the channel cover. Because water is distributed more uniformly into all channels, the pressure drop across the channels is much lower than that across helical copper pipes. To prevent water leakage, an interference fit (press fit or shrink fit) must be adopted between the housing and the channel cover.

11.5.4.4 Indirect Liquid Cooling via Copper Pipe in Spacer

In this cooling approach, a motor is cooled by water passing through a copper pipe in a specially designed spacer near the front endbell (Figure 11.66). The stator winding is cooled by transferring heat from the winding to the motor housing and spacer, and then to cooling water. However, the present arrangement of the cooling pipe may cause nonuniform cooling on the motor, resulting in a temperature gradient along the motor axis. Furthermore, since the cooling pipe is close to the front bearing, it may shrink the outer race of the bearing to significantly reduce the bearing radial clearances, leading to the bearing lock or even failure.

11.5.4.5 Indirect Liquid Cooling through Stator Winding Slots

As discussed in Chapter 7, the biggest power loss is the resistive loss in copper windings. Therefore, in order to cool a motor more efficiently, a logical thinking is to cool these windings directly. An innovative approach was proposed by Tan *et al.* [11.70]. In their approach, water pipes are inserted in the stator slots (see Figure 11.67). In this way, water flows through these intercoil pipes to carry away directly the heat generated in the winding due to the copper loss. Obviously, by this method, the thermal resistance is minimized compared with conventional cooling methods.

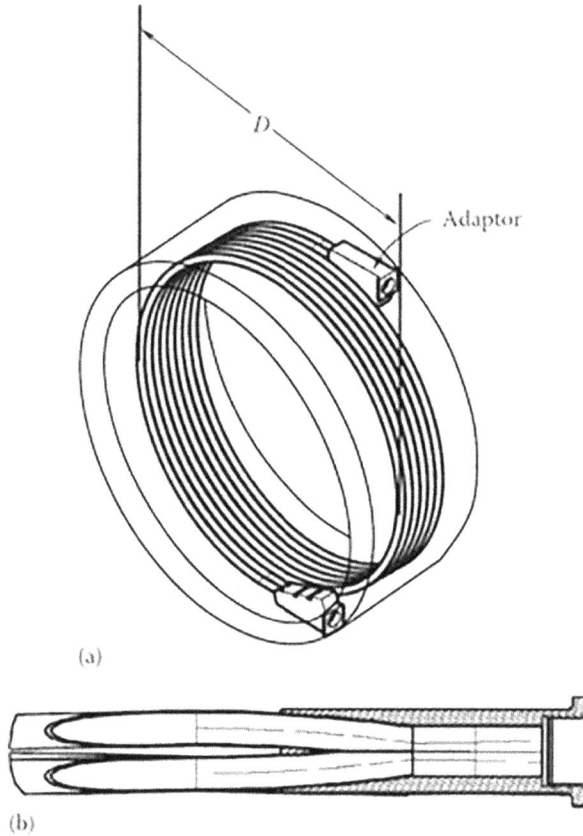

(a)

(b)

FIGURE 11.64 Components of the indirect liquid cooling system: (a) the helical dual-cooling pipe, casted directly into the motor housing for reducing thermal resistance, and (b) the adaptor at the flow inlet and outlet to connect with the helical dual-cooling pipe.

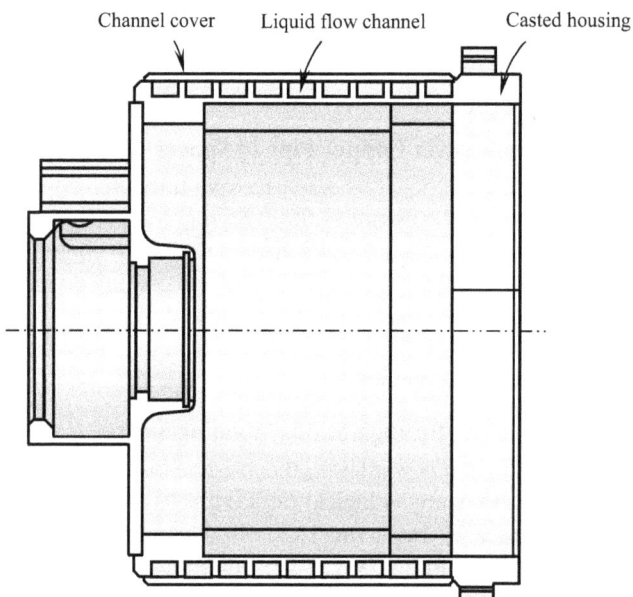

FIGURE 11.65 Liquid flow channels formed between the casted housing and the channel cover.

FIGURE 11.66 Indirect water cooling via copper pipe integrated into the spacer near the front endbell. (Courtesy of Kollmorgen Corporation.)

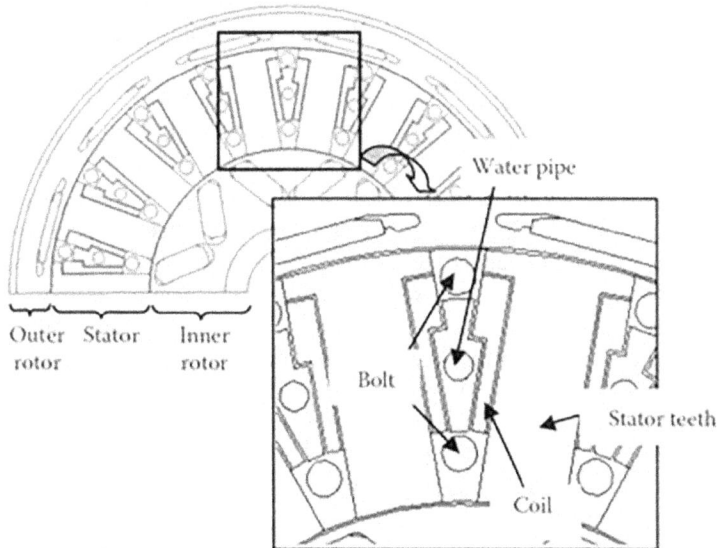

FIGURE 11.67 Water pipes inside stator slots for effectively cooling stator windings. (From Tan, N. *et al.*, Cooling performance design for super motor and its experimental validation, *Proceedings of the 2005 International Power Electronic Conference*, pp. 1492–1488, 2005.)

11.5.4.6 Indirect Liquid Cooling through Microscale Channels

With the rapid increase in power density of modern motion control systems, the traditional cooling methods such as conduction and natural/forces convection may be insufficient for these systems. The need for new technologies capable of dissipating high heat fluxes is of critical importance in the motion industry. The use of pumped liquid through microscale channels is a promising technique in various industries for high-heat-flux applications.

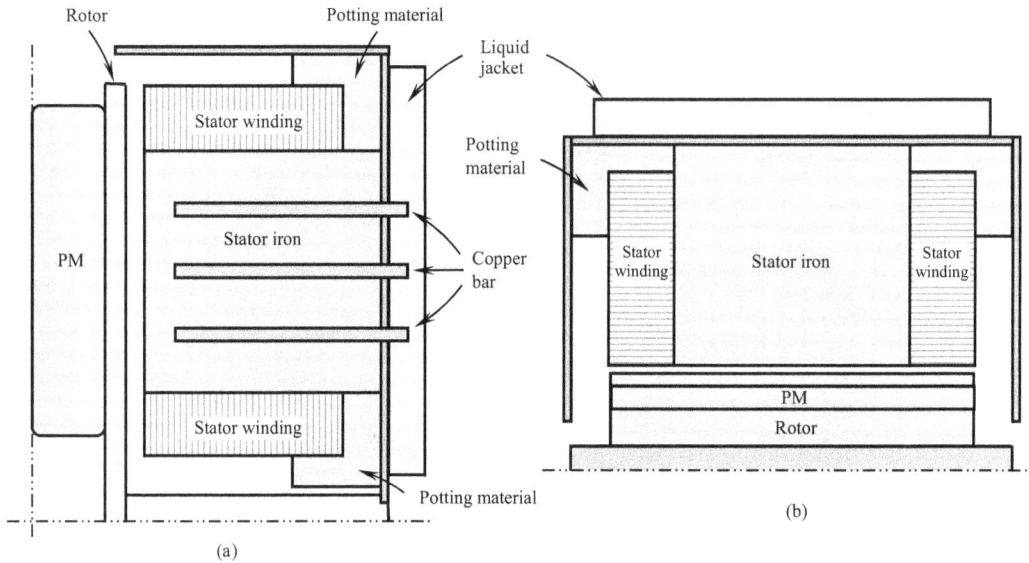

FIGURE 11.68   The setup of a stator indirect liquid cooling system: (*a*) AF motor and (*b*) RF motor.

As pointed out by Upadhya *et al.* [11.71], the performance of a liquid-cooled system depends on several factors, including (*a*) feature size of the microscale channel, (*b*) flowrate of the liquid through each channel, (*c*) surface area of radiator fins, and (*d*) airflow available for heat rejection.

11.5.4.7   Indirect Liquid Cooling via Heat Transfer Enhancement Device

Axial flux (AF) motors tend to have cooling difficulties since it is difficult to arrange a continuous heat path between the stator stack and the housing. One important reason for this is that no shrink fitting of the stator is possible in an AF motor. In addition, cooling of the rotor and the end windings may also be difficult at least in the case of two-stator-single rotor construction where air circulation in the rotor and in the end-winding areas may be difficult to arrange. To solve the problem, an indirect liquid cooling system is proposed, where the cooling arrangement is based on a circumferential liquid duct incorporated in the stator. Potting materials such as epoxy and ceramacast are used to fill the airgap between the winding and the outer components for removing heat from the stator copper winding, stator iron and therefore precluding propagation of heat towards the rotor [11.72]. A number of copper bars are inserted into the bulk of the stator as extra heat transfer paths from the stator to the cooling liquid pool (Figure 11.68*a*). This cooling scheme enhances heat transfer coefficient and thereby higher current density and motor torque, allowing achieving higher power density and operation reliability.

This cooling technique can be also applied to RF motors [11.73]. However, due to the shrink fit between the stator and the housing, the contact thermal resistance becomes very small. Therefore, the use of copper bars is no longer required (Figure 11.68*b*).

11.6   PHASE-CHANGE COOLING TECHNIQUES

Phase-change cooling is an extremely effective way to cool energy systems. By accommodating large heat fluxes with relatively small driving temperature differences, heat in this approach is primarily removed through the vaporization of the working fluid utilizing the large latent heat capacity. As a consequence, the heat transfer augment is achieved and much less coolant is required as compared with single-phase cooling.

11.6.1 Cooling with Heat Pipes

With the continuous increase in motor power density and heat dissipation, the conventional conduction-based cooling techniques are not adequate for motor cooling. This makes heat pipes and other advanced and efficient cooling devices ideal choices for direct conduction cooling of electric motors.

During the normal operation of an electric motor, the temperature distribution inside the motor is far from uniform. Hot spots may occur locally in some components that have higher power losses. To eliminate such hot spots, in addition to the overall cooling system, individual components may have their own cooling devices in place. The motor components that can be individually cooled include, but are not limited to, stator winding, bearings, electric connectors, and encoder. For specific local cooling, the use of heat pipes may be more efficient and suitable.

Heat pipes are two-phase heat transfer devices with extremely high thermal conductivities. The world's first heat pipe was originally invented by Gaugler in 1944 and patented in 1946 [11.74]. However, it did not gain any significant attention within the heat transfer community until the space program resurrected the concept in the early 1960s [11.75]. Heat pipes rely on latent heat of evaporation of a working fluid (usually thermal refrigerant) to transmit heat from one hot point to another. Because it involves both boiling (or evaporation) and condensation, this cooling technique is generally categorized as a two-phase cooling technique. Based on its cooling mechanism, heat pipes can operate in any orientation. Furthermore, by carefully selecting coolants, heat pipes can be designed to operate over a broad range of temperatures. In motor cooling applications where it is typically desirable to maintain the motor temperature below 120°C–155°C, depending on the insulation class, copper/water heat pipes can be applied.

The thermal resistance represents the effectiveness of the heat pipe, which is defined as

$$R_{th} = \frac{T_e - T_c}{Q} \tag{11.27}$$

where

Q is the heat input, W

T_e and T_c are average evaporator and condenser section temperatures, respectively, °C

The effective thermal conductivity of the heat pipe, k_e, is expressed as

$$k_e = \frac{L}{AR_{th}} = \frac{LQ}{A(T_e - T_c)} \tag{11.28}$$

where

L is the effective length of the heat pipe, m

A is the cross-sectional area of the heat pipe, m²

An experimental investigation of heat transfer in heat pipes with three working fluids was conducted by Mozumder *et al.* [11.76]. Their study reveals that the dominating parameters for heat transfer in heat pipes are average evaporator temperature T_e, saturated boiling temperature of working fluid T_{sat}, latent heat of vaporization h_{fg}, specific heat of working fluid $c_{p,l}$, and the fluid fill ratio F. The correlation for the heat transfer coefficient is given as

$$Q = hA_e(T_{sat} - T)_e = 39(Ja)^{4/5} F^{1/4} \left(\frac{T_{sat}}{T_e}\right)^5 \tag{11.29}$$

where

h is the overall heat transfer coefficient in W/(m²·K)

A_e is the surface area at the heat pipe evaporation region in m²

Jakob number is defined as

$$Ja = \frac{C_{p,l}(T_e - T_{sat})}{h_{fg}} \qquad (11.30)$$

It must be noted that Q in Equation 11.29 has a unit of watts, and the right-hand side is dimensionless. Thus, the overall heat transfer coefficient h is given as

$$h = \frac{Q}{A_e(T_{sat} - T_e)} \qquad (11.31)$$

where A_e is the surface area at the heat pipe evaporation region.

The correlation listed previously for predicting heat transfer coefficient fairly agrees with the experimental results. The investigation also shows that 85% fill ratio can be regarded as an optimum value for a heat pipe. Later, Manimaran *et al.* [11.77] found that the lowest thermal resistance is obtained when the heat pipe is operated at 75% fill ratio.

As illustrated in Figure 11.69, at the hot end of the heat pipe, the fluid evaporates into vapor, thereby absorbing the heat from the heat source. As the vapor migrates to the cold end through the central cavity, it condenses back into the liquid state to dissipate the heat to the environment. Then, the liquid returns to the hot end via the wick structure due to the capillary effect. With the repeat of the process, heat is continually transported from the hot end to the cold end. This indicates that heat pipes operate independently of gravity and can be effectively used in gravity-free conditions.

Since heat pipes involve the phase change of the working fluid, they have much higher effective thermal conductivity than solid materials. For instance, in some applications and orientations, heat pipes have a thermal conductivity in exceeds of 20,000 times that of a solid copper bar of the same geometry. Also, heat pipes require no maintenance due to the lack of mechanical moving parts.

The thermophysical properties of a working fluid, as well as associated solid materials, can significantly influence the capillary process, liquid vaporization, and vapor condensation and, as a consequence, heat pipe performance. These include but are not limited to surface tension, viscosity, the density difference between the liquid and vapor, boiling point, thermal conductivity and diffusivity, latent heat of vaporization, specific heat, surface wettability (*i.e.*, equilibrium contact angles between the liquid and solid), solid surface finishing, *etc.* Accordingly, ethanol, methanol, acetone, water, and their binary mixtures are commonly used as the working fluids in heat pipes.

In recent years, with the rapid development of nanotechnology, a trend is to use nanofluids as the working medium of a heat pipe. Nanofluids are made by mixing solid nanoparticles into liquids. Generally, the use of nanofluids leads to an increase in the heat pipe performance and a decrease in the thermal resistance. Such decrement of the thermal resistance is important at low power inputs, and on average it is evaluated to be between -10% and -40%, depending on the used nanoparticles (*e.g.*, silver, copper, Al_2O_3, Fe_3O_4, and TiO_3) [11.78].

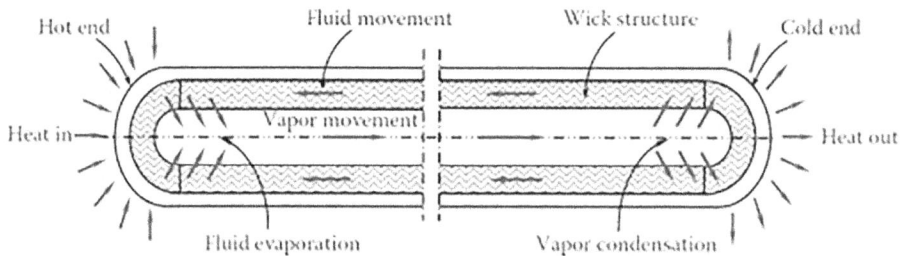

FIGURE 11.69 Performing principle of a heat pipe. It relies on both boiling (or evaporation) and condensation of working fluid.

11.6.2 COOLING WITH VAPOR CHAMBERS

Vapor chambers can offer far superior thermal performance than traditional solid metal heat spreaders at the same thermal load and device size. With much lower thermal resistance, vapor chambers provide evenly spread temperatures on all of their surfaces, regardless of the location and thermal density of the heat source. Therefore, vapor chambers are able to deal with high-heat-flux motor components, extending the life of components and motors.

During the last several decades, boiling heat transfer techniques have been developed for cooling high-power systems, leading to the design of highly efficient vapor chambers. Since nucleate boiling attains heat flux during phase change, it has been an attractive design choice for electric motors and generators.

Although a vapor chamber looks like a heat pipe in its appearance, it has a fundamentally different cooling mechanism than that of the heat pipe. In a vapor chamber, the movements of vapor and liquid coolant rely on gravitational force. This indicates that the orientation of a vapor chamber has a strong impact on its performance. To increase the cooling efficiency of vapor chamber, cooling fins may be added at the condensation end of the vapor chamber to improve the heat dissipation from it to the environment (Figure 11.70). As a result, the heat transfer enhancement can be achieved not only by increasing convective heat transfer surface area but also by creating local flow turbulence, both resulting in a higher heat convective transfer coefficient.

According to these superior thermal characteristics of heat pipes, motors with heat pipes can be made for higher power density and higher efficiency, allowing more current to pass through motor windings. Heat pipes may be used to cool different motor components. As shown in Figure 11.71 [11.79], a set of heat pipes is inserted into the stator slots to directly cool the windings. Because most

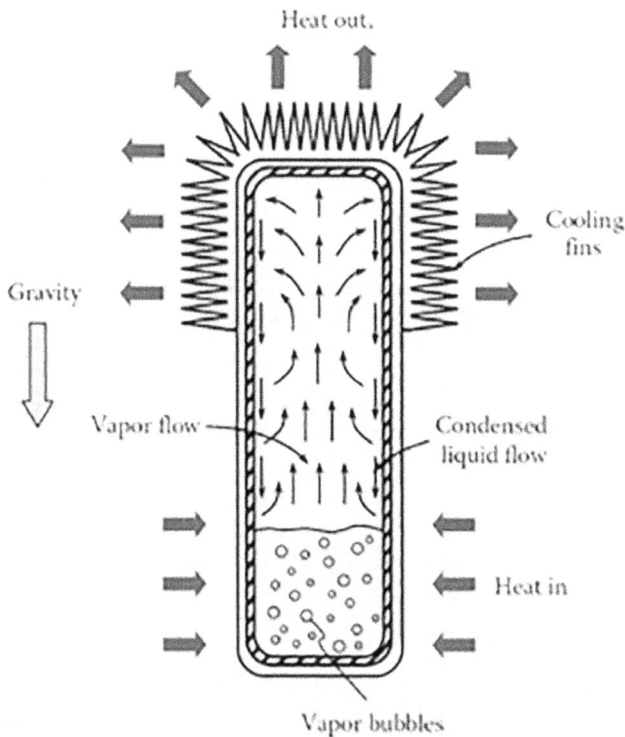

FIGURE 11.70 Boiling and condensation of coolant in a vapor chamber. The movements of vapor and liquid coolant rely on gravitational force. Fins are added to the vapor chamber at the condensation end to enhance heat dissipation.

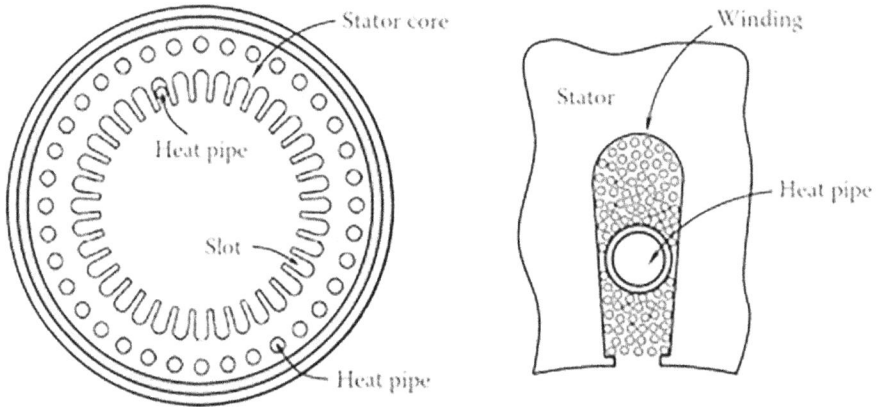

FIGURE 11.71 Heat pipes inserted into stator slots and stator core (U.S. Patent 7,569,955) [11.79]. (Courtesy of the U.S. Patent and Trademark Office, Alexandria, VA.)

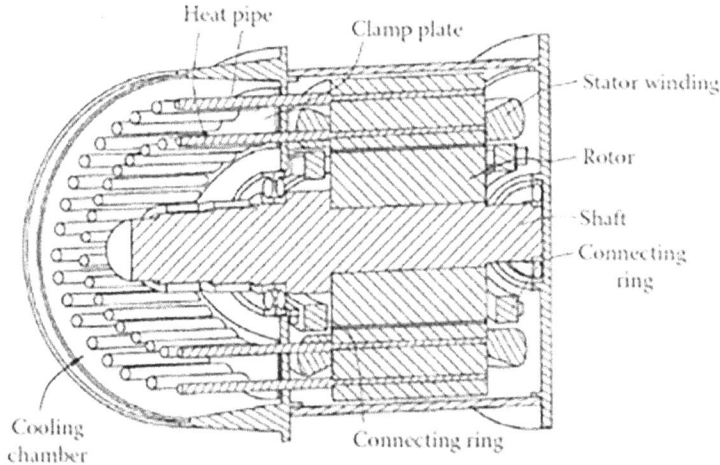

FIGURE 11.72 Cross-sectional view of motor cooling with heat pipes (U.S. Patent 7,569,955) [11.79]. (Courtesy of the U.S. Patent and Trademark Office, Alexandria, VA.)

of the heat in an electric motor is generated in the motor winding, putting the heat pipes in close proximity to the copper winding will make the heat transfer more efficient. In addition, another set of heat pipes is incorporated into the stator core to absorb heat generated in the core due to the eddy-current losses. A perspective view of a motor with these two sets of heat pipes is provided in Figure 11.72.

Similarly, heat pipes may also be implemented in rotors to assist in dissipating heat. Referring to Figure 11.73, heat pipes can be inserted into rotor conducting bars.

11.6.3 EVAPORATIVE COOLING

A compelling evaporative cooling system has been developed to address the challenges of cooling high-power devices [11.80]. This cooling system employs a noncorrosive, nonconductive refrigerant that evaporates on contact with hot surfaces of the high-power device, in a small, lightweight, and highly efficient closed-loop system. By taking advantage of the latent heat absorbed during phase change of the refrigerant from liquid to vapor, as well as the latent heat released during phase change from vapor to liquid, this system has a very high cooling efficiency.

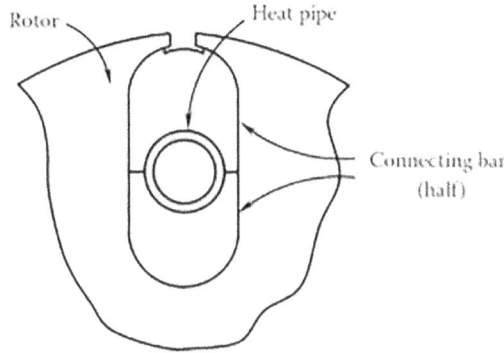

FIGURE 11.73 Heat pipe inside the rotor connection bar (U.S. Patent 7,569,955) [11.79]. (Courtesy of the U.S. Patent and Trademark Office, Alexandria, VA.)

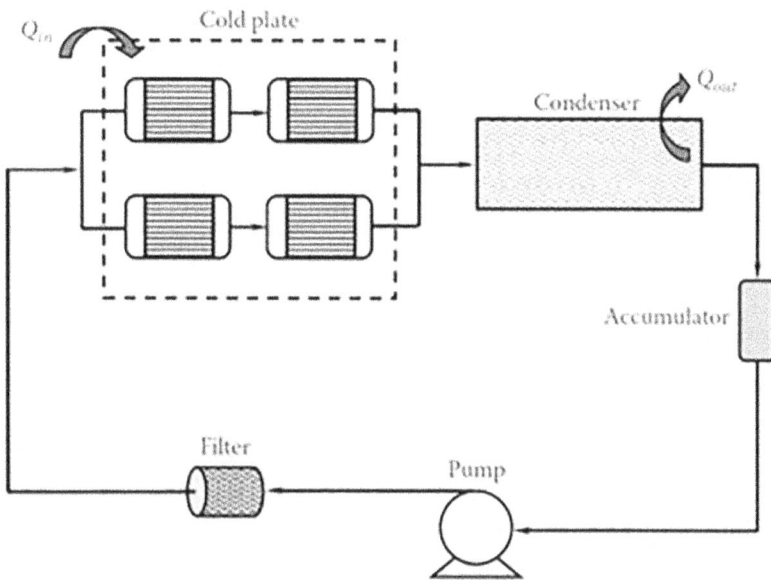

FIGURE 11.74 Evaporative cooling system.

As shown in Figure 11.74, the refrigerant is pumped into one or more cold plates that are optimized to acquire heat from the device. The refrigerant vaporizes to maintain a cool uniform temperature on the device surfaces. The resulting refrigerant is then pumped to a heat exchanger where it rejects the heat to the ambient and condenses back into the liquid, completing the cycle.

The comparisons of thermal performance of the evaporative cooling system with various cold plates, as well as the air-cooled system, are plotted in Figure 11.75. It can be seen that under identical ambient conditions, the air-cooled system has the highest thermal resistance (0.094°C/W) and the evaporative cooling system (two-phase evaporative cooling) has the lowest (0.008°C/W). This indicates that by taking advantage of the highly efficient evaporation process that occurs when the liquid refrigerant changes phase to vapor, up to 11.8 times the amount of heat can be removed for the same given temperature difference by comparing with the air-cooled system. Therefore, the two-phase evaporative approach can increase the system power density (or reduce the system volume for the same power output) and improve performance reliability due to the elimination of the safety

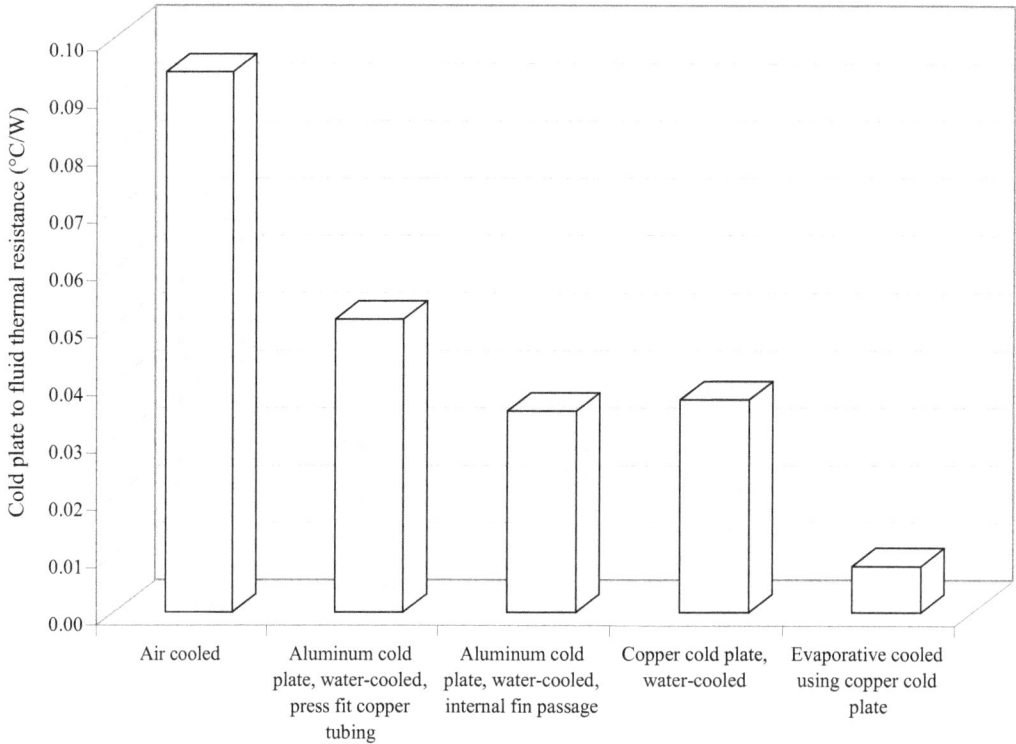

FIGURE 11.75 Comparisons of thermal performance of the evaporative cooling system with various cold plates and air-cooled system.

and maintenance issues associated with water cooling. In addition, the isothermal nature of the two-phase cooling system reduces thermal cycling.

By changing the material of cold plates from aluminum to copper, the thermal resistance can be lowered approximately 27%.

11.6.4 MIST COOLING

Mist cooling continues to be a viable solution for a wide range of residential and industrial cooling applications for creating a safer and more comfortable working environment. Mist cooling has been demonstrated to remove high heat energy with relatively low coolant flowrates. In a mist cooling system, a high-pressure pump pushes water through a plurality of mist nozzles, producing microscopic droplets of water. As fine water mists hit hot air or surfaces, they evaporate to cool the surrounding space.

Totally enclosed nonventilated motors may be suitable to use the mist cooling technique. For this type of motor, all motor components are contained within a motor frame to prevent moisture, duct, and other harmful contaminants into the motor. To effectively cool this type of motor, very fine water mists are used to directly hit the outer surface of the motor. Thus, heat generated by the motor is absorbed by the evaporation of water mists. Furthermore, the mixing flow of mists and air also has a favorable contribution to motor convection cooling.

11.7 RADIATIVE HEAT TRANSFER

Radiative heat transfer of rotating electric machines involves thermal radiation between the stator and rotor and between the stator (or rotor for the outer rotor machines) and the environment.

According to the Stefan–Boltzmann law, the radiative energy E_r from a hot object at temperature T_h to its cooler surroundings at temperature T_c is given as

$$E_r = e\sigma A(T_h^4 - T_c^4) \tag{11.32}$$

where

e is the emissivity ($e = 1$ for a black body) of the object

A is the radiating area

σ is Stefan's constant ($\sigma = 5.6703 \times 10^{-8}$ W/m$^2\cdot K^4$)

The impact of radiation heat transfer on electric motor cooling has been studied numerically by Chirilă *et al.* [11.81]. Their results show that for an internally air-cooled motor, the average temperature of the motor housing is lower when the radiation is considered for all inlet flow velocities. The maximum temperature difference obtained is about 9°C, indicating that the radiation is important for cooling the motor house at low inlet air velocities. An increase in the cooling air velocity can always significantly lower the housing temperature, from 118°C at $u = 2$ m/s to less than 80°C at $u = 8$ m/s. However, with an increase in the flow velocity, the temperature difference due to the radiation effect becomes very small, from about 9°C at $u = 2$ m/s to about 2°C at $u = 8$ m/s.

In most motor cooling schemes, a cooling system is disposed within the stator. As a result, the rotor temperatures are generally higher than those of stators. The heat generated by the rotor is thus dissipated by forced convection due to its rotation, conduction via bearings and radiation emitted to the motor housing and stationary components surrounding the rotor.

The radiation resistance between a motor and its surrounding environment depends on both the geometrical and thermal characteristics of the motor and the characteristics of the environment. This resistance is much higher than the forced convection resistance and thus is often neglected. For example, for the motors in refrigeration compressors, heat radiation is so weak that it can be neglected [11.82]. However, in some situations, particularly when totally enclosed fan-cooled motors are used in variable speed drives, radiation heat transfer can play a nonnegligible role at low speeds. It is also true for motors that use noncontact bearings. For these situations, radiation must be taken into account.

For outdoor electric motors, it is important in the design of a motor cooling system to consider not only the heat dissipation from the motor but also the heat from sunlight. In some circumstances, the solar loading effect could be substantial, depending on the motor size and its orientation toward the sun.

11.8 OTHER ADVANCED STATE-OF-THE-ART COOLING METHODS

With the increasing demand for advanced state-of-the-art cooling technologies, new technologies have emerged in recent years for not only promoting heat transfer processes but also improving product developments.

11.8.1 MICRO CHANNEL COOLING SYSTEMS

Increasing heat densities in motion control systems cause high heat fluxes at both the component and system levels. The advancement of high-performance microchannel cooling has stimulated interest in this technique to substantially improve thermal energy removal capability. Though the microscale regime is not strictly defined, it is accepted that the characteristic length scale of microchannel ranges from a micrometer (10^{-6} m) to approximately 0.08 mm (8×10^{-5} m). Microchannels can provide a much larger contact area with the cooling fluid per unit volume and thus much higher heat transfer coefficients than conventional channels under similar thermohydraulic conditions. According to previous studies, the microchannel cooling technique can remove heat densities up to 10 MW/m^2 [11.83].

In this technique, a cooling fluid is forced to pass through a microchannel heatsink by an external microscale pump for achieving very high heat removal rates. Because of its high heat removal capability, the researchers at the Oregon Stator University are currently working with a motor manufacturer to design microchannel cooling systems in electric motors [11.84]. However, before this technique is successfully transplanted from electronic device cooling to high-power-density motor cooling, several difficulties must be overcome. First, the high heat-removing capability of a microchannel system is always associated with high pressure drop across the microchannels. Thus, it requires high power for driving cooling fluid through the microchannels. Second, due to the complex geometries of motor components, it is a challenge to achieve uniform cooling of the motor. Third, heat transfer processes in microchannels are somewhat different from those in ordinary size channels. The capillarity of fluid in microchannels has a strong influence on flow patterns and heat transfer rates. This mechanism must be fully understood. Finally, it is noted that two-phase flows can be quite different from single-phase flows in microchannels. It is risky to extrapolate macroscale two-phase flow maps, flow boiling heat transfer methods, and pressure drop models to microchannels. More research is necessary to understand two-phase microchannel flows and heat transfer mechanisms.

11.8.2 METAL FOAMS

One of the principles of heat transfer enhancement is to extend heat dissipation surfaces. A typical example is the application of a variety of fins in heat exchangers. As a promising alternative, metal foams have been used for enhancing heat transfer in a variety of engineering applications. With a large number of open cell foams, metal foams have shown higher heat transfer enhancement than pin fins [11.85]. Open cell foams consist of cells that are all interconnected, allowing a fluid to pass through them and thus have unique beneficial characteristics, including an excellent surface-to-volume ratio (from 500 to over 10,000 m^2/m^3), high thermal conductivity, high fluid permeability and mixing, low flow resistance, low density, good thermal shock resistance, high temperature tolerance, excellent noise attenuation, corrosion resistance, and other favorable properties [11.86–11.88]. It was reported that with the same pressure drop, the performance of the metal foam exchanger is superior when compared to that of conventional finned designs [11.89, 11.90]. T'Joen et al. [11.91] have experimentally investigated the impacts of various parameters of metal foams on the thermohydraulic performance of a metal foam heat exchanger. Their results indicate that in all cases, heat transfer is greatly enhanced by coating metal foam on metal tubes. A good metallic bonding between the foam and the tube can significantly reduce the thermal resistance. A comparison of the performance of foam-covered tubes with different foam heights has clearly showed that increasing the foam height reduces convective resistance. This technique is suitable for motor cooling by coating metal foams on the motor external surfaces, as described in Figure 11.76.

One of the most important characteristics of open-cell metal foams is the pore density, which is defined as the number of pores in a linear inch (PPI). Depending upon the base foam material and manufacturing method, the available pore density of open-cell metal foams varies greatly, typically ranging from 5 to 100 PPI. Figure 11.77 shows some samples of metal foams with different pore densities.

Metal foams can be fabricated from many different manufacturing processes, including investment casting, chemical vapor decomposition, electroforming, and metallic sintering, to name a few. Each fabrication process has its pros and cons and is used for certain applications. A comprehensive review of various metal foam fabrication processes was provided by Banhart [11.92].

As an indirect method, investment casting uses space-holding filler materials or melting of powder compacts that contain a blowing agent to produce cellular metal foams. In a chemical vapor decomposition process, metal foams are made from gaseous metal. The condensed metal vapor in a vacuum chamber coats the surface of the polymer precursor to generate a film of metal foam.

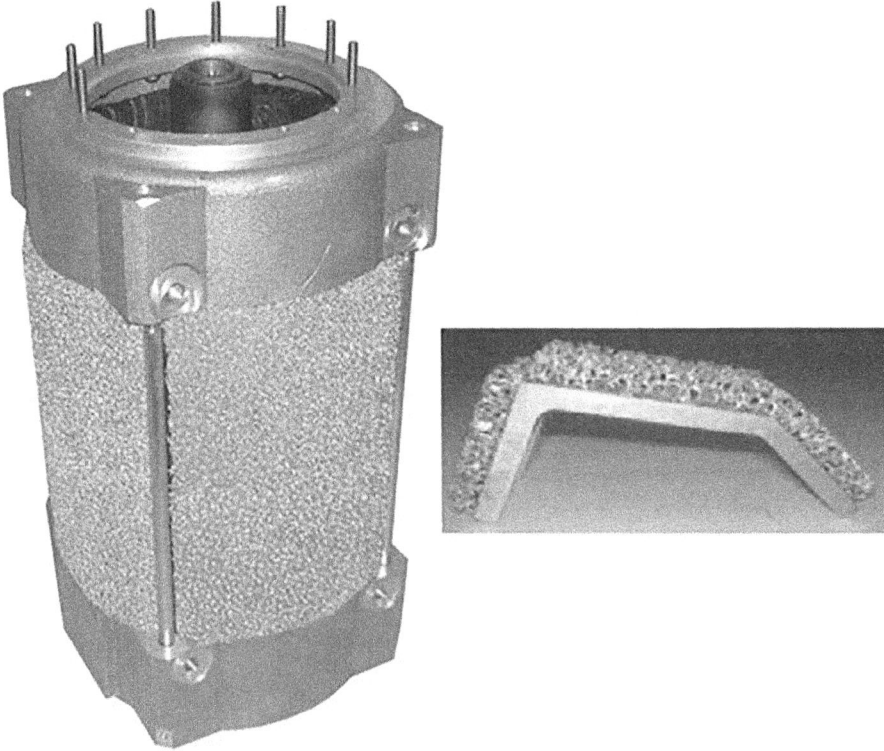

FIGURE 11.76 Enhanced motor cooling by coating metal foam with a microporous structure on the motor external surface. (Courtesy of Ghent University Gent, Belgium.)

FIGURE 11.77 Samples of aluminum metal foams with 10, 20, and 40 PPI (from left to right).

Electroforming can generate a thin, porous layer on a metal skin through an electrodeposition process. The metallic sintering process consists of various steps, including powder fractioning and preparation, compaction or molding, and sintering. The achievable porosity can range between 20% and 80%. In addition, using metal fibers to replace powders sheds new light on the fabrication of porous metal structures since a variety of metals can be produced as fibers.

11.8.3 HEAT TRANSFER ENHANCEMENT WITH NANOTECHNOLOGY

Nanotechnology deals with the control of matter at dimensions between approximately 1 and 100 nanometers (nm). The emergence and development of nanotechnology has affected/altered many aspects of engineering designs and manufacturing. It has the potential to create new materials and devices with a vast range of applications and thus has been considered a revolutionary technology. As one of the cutting edges in innovative cooling methods, it is expected that nanotechnology can feature an innovative cooling solution for an electric machine in the near future.

11.8.3.1 Nanofluid

Due to the recent developments in nanotechnology, a new class of heat transfer fluid called nanofluid has attracted considerable attention from researchers and engineers. A nanofluid is a solid–liquid mixture produced by dispersing nanoparticles in a liquid (usually water, ethylene glycol, or minerals oil) to display enhanced heat transfer due to the combination of convection and conduction and additional energy transfer by particle dynamics and collision [11.93]. The studies of nanofluids with phase change have shown that the presence of nanoparticles in liquid can enhance critical heat flux (CHF) in boiling heat transfer. The mechanism of the CHF enhancement is attributed to the deposition of nanoparticles on the boiling surfaces with foamed porous layers [11.94].

The nanoparticles used in nanofluids are typically made of metals, oxides, titanate nanotubes, or carbon nanotubes with a typical particle size of less than 100 nm (or less than 50 nm in some applications). It is such tiny sizes that make the nanoparticles interact with liquids at the molecular level. Shima and Philip [11.95] have found that by varying the magnetic field strength and its orientation, the thermal properties of magnetic nanofluids (*e.g.*, using Fe_3O_4, γ-Fe_2O_3, or Fe nanoparticles) can be tuned to vary at high values. Their results show that the nanoparticles with a larger size exhibit larger thermal conductivity enhancement due to the enhanced dipolar interaction effect. Because the thermal properties of these nanofluids are perfectly reversible, these magnetic nanofluids are ideal for applications in smart devices.

11.8.3.2 Carbon Nanotube

Carbon nanotubes are well known for their outstanding thermophysical and mechanical and properties, particularly their extraordinary thermal conductivities. Measurements have shown that the thermal conductivity of single-wall carbon nanotubes (with a length of 2.6 μm and a diameter of 1.7 nm) along their longitudinal axes can reach as high as 3,500 W/m·K at room temperature, which is 9.1 times higher than that of copper [11.96]. However, nanotubes exhibit a very poor thermal performance in the lateral direction. From the same test, the thermal conductivity in the lateral direction was measured as 1.52 W/m·K, which is only 0.4% of copper.

Carbon nanotubes are promising for use in various fields as a thermal material [11.97]. One potential application is to use carbon nanotubes in a layer coated onto motor surfaces for enhancing conductive heat transfer (Figure 11.78).

Another interesting property of nanotubes is that they have the highest strength-to-weight ratio of any known materials. This property may help produce lightweight motor components, for instance, combining nanotubes with other materials into composites that can be used to build lightweight fan, endbell, and housing.

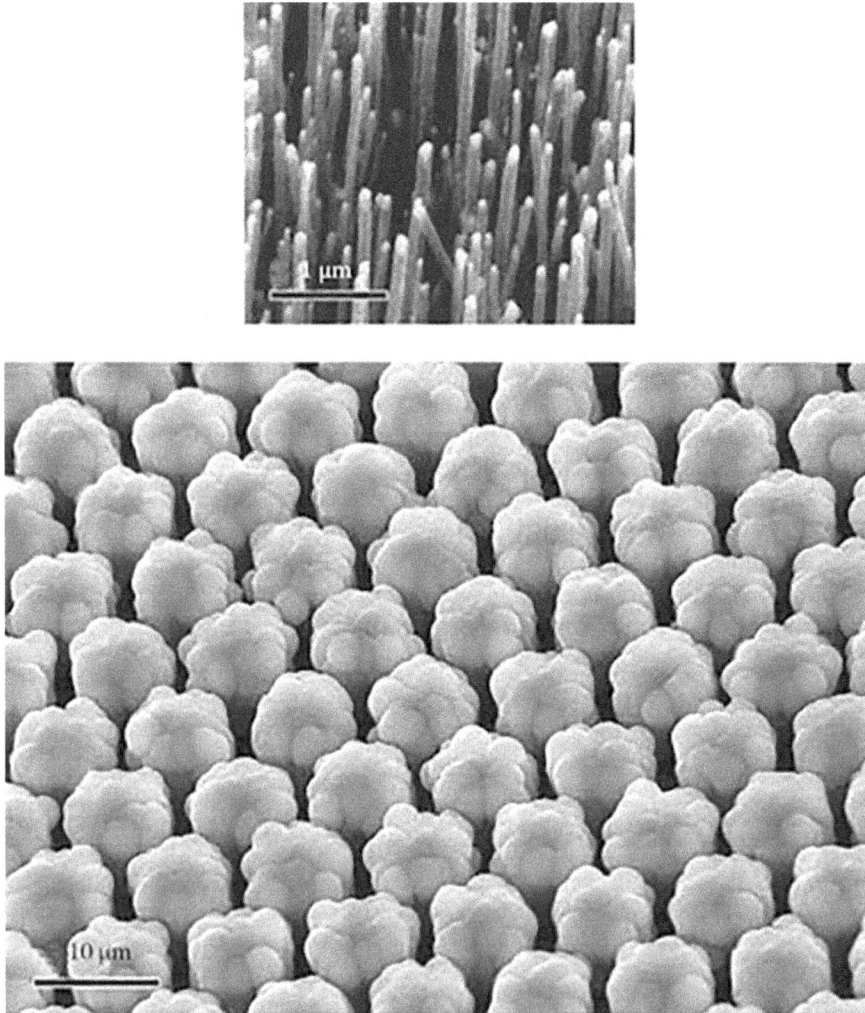

FIGURE 11.78 An array of vertically aligned carbon nanotubes. (Courtesy of NASA Ames Research Center, Moffett Field, CA.)

11.8.3.3 NanoSpreader™ Vapor Cooler

NanoSpreader vapor coolers are patented copper-encased, two-phase vapor chambers into which pure water is vacuum sealed [11.98]. In fact, this type of vapor cooler has the same working principle as heat pipes. The heat transfer process consists of four steps: (*a*) At the high-temperature end, liquid evaporates to vapor by absorbing thermal energy from the heat source. (*b*) Vapor migrates through microperforated copper sheets toward the low-temperature end. (*c*) Vapor condenses back to liquid, releasing thermal energy. (*d*) Liquid is absorbed by a copper-mesh wick and flows back to the higher-temperature end. However, because of the design optimization of the NanoSpreader vapor cooler, this type of cooler has numerous advantages over the type of heat pipe, including the following: (*a*) The reduction in thermal resistance allows to increase cooling efficiency up to 30% over the heat pipes. (*b*) Heat is spread in a more uniform pattern across a large thin surface area. (*c*) Due to the uniform heat spreading, hot spots from high-heat-flux devices are minimized/eliminated. (*d*) The weight can be reduced up to 30% over the heat pipes.

11.8.3.4 CarbAl™ Heat Transfer Material

CarbAl thermal material has been recognized as one of the 100 most significant product innovations in 2009 by *R&D Magazine* [11.99]. CarbAl material is a carbon-based metal nanocomposite composed of 80% carbonaceous matrix and a dispersed metal component of 20% aluminum, with a unique combination of high thermal diffusivity, high thermal conductivity, a low CTE, and low density [11.100]. As one of the passive thermal management materials, its thermal conductivity is about two times higher than that of aluminum. The speed to remove heat from a heat source (*i.e.,* thermal diffusivity) is two and three times higher than copper and aluminum, respectively. The CTE of CarbAl material is only 29% and 41% of aluminum and copper, respectively.

11.8.3.5 Ionic Wind Generator

This innovative cooling device is known in industries as the ionic wind generator [11.101]. This type of cooling system does not use a fan or heatsink. It generates airflow based on the ionization of air molecules. A limitation of currently proposed ionic wind generator cooling systems for such devices (and for other conventional cooling devices as well) is that the generated airflow, from the first electrode toward the second electrode, is limited to a linear path that is essentially static and thus could only cool a specific region of an electronic system; particularly, only the regions that are in, or immediately adjacent, the path of the airflow could be cooled.

Because this cooling device was invented for cooling computer CPUs, with the current design, it is only able to cool limited areas within a computer. However, this type of cooling device may be good for eliminating some heat spots in electric motors.

REFERENCES

11.1.   Arrhenius, S. 1889. On rates of inversion reaction in cane-sugar under the action of acids. *Zeitschrift für Physikalische Chemie* **4**: 226–2411.

11.2.   Hnatek, E. R. 2003. *Practical Reliability of Electronic Equipment and Products.* Marcel Dekker, New York.

11.3.   Boglietti, A., Cavagnino, A., Staton, D. A., and Popescu, M. 2009. Impact of different end region cooling arrangements on endwinding heat transfer coefficients in electric motors. *Proceedings of 35th Annual Conference of the IEEE Industrial Electronics Society,* Porto, Portugal, pp. 1168–1173.

11.4.   Tong, W. 2009. Numerical analysis of flow field in generator end-winding region. *International Journal of Rotating Machinery* **2008**: Article ID 6927411. doi:10.1155/2008/6927411.

11.5.   Zhu, Z. Q. 2011. Fractional slot permanent magnet brushless machines and drives for electric and hybrid propulsion systems. *International Journal for Computation and Mathematics in Electrical and Electronic Engineering (COMPEL)* **30**(1): 9–31.

11.6.   Koplow, J. P. 2010. *A Fundamentally New Approach to Air-Cooled Heat Exchange.* Technical Report Sand2010-02511. Sandia National laboratories, Albuquerque, NM. http://prod.sandia.gov/techlib/access-control.cgi/2010/1002511.pdf.

11.7.   Lasance, C. 1997. Technical data. *Electronics Cooling* **3**(1): 11.

11.8.   Wilson, J. and Simons, R. E. 2005. Advances in high-performance cooling for electronics. *Electronics Cooling* **11**(4): 22–39.

11.9.   Faghri, A., Zhang, Y. W., and Howell, J. R. 2010. *Advanced Heat and Mass Transfer.* Global Digital Press, Columbia, MO.

11.10. Webb, R. L. and Kim, N.-H. 2005. *Principles of Enhanced Heat Transfer,* 2nd edn. Taylor & Francis, New York.

11.11. Sines, E. 2004. Electrical power cooling technique. U.S. Patent 6,777,835.

11.12. Wild, C. and Wörner, E. 20011. *The CVD Diamond Booklet.* Diamond Materials GmbH, Freiburg. http://www.diamond-materials.com/downloads/cvd_diamond_booklet.pdf.

11.13. Pan, Z. C., Sun, H., Zhang, Y., and Chen, C. F. 2009. Harder than diamond: Superior indentation strength of wurtzite BN and lonsdaleite. *Physical Review Letters* **102**(5): 05503.

11.14. Craig, W. and Leonard, A. 2019. *Manufacturing Engineering and Technology.* ED-Tech Press, Waltham Abbey Essex.

11.15. Zweben, C. 2005. Advanced electronic package materials. *Advanced Materials & Processes* **163**(10): 33.

11.16. Lemak, R. J., Moskaitis, M. J., Pickreil, D., Yocum, A. M., and Kupp. D. 2011. High performance pyrolytic graphite heat spreader: Near isotropic structures and metallization. *IMAPS Advanced Technology Workshop on Thermal Management 2008*, Palo Alto, CA.

11.17. Meyer IV, G. A. Sun, C.-H., Chen, C.-P., and Liu, H.-T. 2011. Heatsink for memory and memory device having heatsink. U.S. Patent 8,059,406.

11.18. Bejan, A. and Lorente, S. 20011. *Design with Constructal Theory*. John Wiley & Sons, Hoboken, NJ.

11.19. Lorenzini, G., Corrêa, R. L., Dos Santos, E. D., and Rocha, L. A. O. 2011. Constructal design of complex assembly of fins. *ASME Journal of Heat Transfer* **133**(**8**): 081902-1–081902-7.

11.20. Bar-Cohen, A. and Jelinek, M. 1986. Optimum arrays of longitudinal, rectangular fins in convective heat transfer. *Heat Transfer Engineering* **6**(**3**): 68–711.

11.21. Peertstra, P., Yovanovich, M. M., and Culham, J. R. 2000. Natural convection modeling of heatsinks using web-based tools. *Electronics Cooling* **6**(**3**): 44–51.

11.22. Incropera, F. P., Dewitt, D. P., Bergman, T. L., and Lavine, A. S. 2006. *Introduction to Heat Transfer*, 5th edn. John Wiley & Sons, Hoboken, NJ.

11.23. Wang, C.-S., Yovanovich, M. M., and Cultam, J. R. 1999. Modeling natural convection from horizontal isothermal annular heatsinks. *ASME Journal of Heat Transfer* **121**: 44–49.

11.24. Lange, F., Hein, C., Li, G. F., and Emmelmann, C. 2018. Numerical optimization of active heatsinks considering restrictions of selective laser melting. *COMSOL Conference*, Lausanne, Switzerland.

11.25. Sertoglu, K. 2020. Purdue students design award-wining 3D printed heatsink. *3D Printing Industry*. https://3dprintingindustry.com/news/purdue-students-design-award-winning-3d-printed-heatsink-174409/.

11.26. Lordo, R. E. and Rudisch, W. E. 1989. Motor frame and motor with increased cooling capacity. U.S. Patent 4,839,547.

11.27. Pal, D., Severson, M. H., and Rasmussen, R. D. 2011. Enhanced motor cooling system. U.S. Patent 8,053,9311.

11.28. Turner, W. B. and Bend, P. D. 2005. Laminated stator with cooling fins. Patent number WO 2005/0227111. World Intellectual Property Organization.

11.29. Yu, X. L, Feng, J. M., Feng, Q. K., and Wang, Q. W. 2005. Development of a plate-pin fin heatsink and its performance comparisons with a plate fin heatsink. *Applied Thermal Engineering* **25**(**23**): 173–182.

11.30. Al-Damook, A., Kapur, N., Summers, J. L., and Thompson, H. M. 2015. An experimental and computational investigation of thermal airflows through perforated pin heatsinks. *Applied Thermal Engineering* **89**: 365–376.

11.31. Tong, W. and Boyland, J. 20011. Heatsink. U.S. Patent Application 2008/00668811.

11.32. Khan, W. A., Culham, J. R., and Yovanovich, M. M. 2005. Optimization of pin fin heatsinks using entropy generation minimization method. *IEEE Transactions on Components and Packaging Technology* **28**(**2**): 1–13.

11.33. Khan, W. A., Culham, J. R., and Yovanovich, M. M. 20011. Optimization of pin fin heatsinks in bypass flow using entropy generation minimization method. *ASME Journal of Electronic Package* **130**: 031010-1–031010-7.

11.34. de Sorgo, M. 1996. Thermal interface materials. *Electronics Cooling* **2**(**3**): 12–16.

11.35. Grujicic, M., Zhao, C. L., and Dusel, E. C. 2005. The effect of thermal contact resistance on heat management in the electronic packaging. *Applied Surface Science* **246**(**2005**): 290–302.

11.36. Tonapi, S., Nagarkar, K., Esler, D., and Gowda, A. 2007. Reliability testing of thermal greases. *Electronic Cooling*, November Issue.

11.37. Mitcham, A. J. and Razzell, A. G. 2007. Stator core. U.S. Patent 7,288,870.

11.38. Kikuchi, T., Kitada, S., Kaneko, Y., and Tsuneyoshi, T. 2004. Rotating electric machine. U.S. Patent 6,713,927.

11.39. Chu, M. T. 2004. Rotor structure for a motor having built-in type permanent magnet. U.S. Patent 6,703,745.

11.40. Chen, K. H., Masrur, A., Ahmed, S., and Garg, V. 2003. Jet impingement cooling of electric motor end-windings. U.S. Patent 6,639,334.

11.41. Glynn, C. and Murray, D. B. 2005. Jet impingement cooling in microscale. *International Conference on Heat Transfer and Fluid Flow in Microscale*, Castelvecchio Pascoli, Italy.

11.42. Vandervort, C. L. and Kudlacik, E. L. 2001. GE generator technology update. GE Power Systems GER-4203.

11.43. Tong, W. and Vandervort, C. L. 2002. Optimization of ventilation flow path at airgap exit in reverse flow generators. U.S. Patent 6,346,755.

11.44. Chang, C.-C., Cheng, C.-H., Ke, M.-T., and Chen, S.-L. 2009. Experimental and numerical investigations of air cooling for a large-scale motor. *International Journal of Rotating Machinery* **2009**: 7, Article ID 612723. doi:10.1155/2009/612723.

11.45. Kiran, T. R. and Rajput, S. P. S. 2009. Modeling of an indirect evaporative cooler (IEC) using artificial neural network (ANN) approach. *Archives of Applied Science Research* **1**(**2**): 327–343.

11.46. Erens, P. J. and Dreyer, A. A. 1993. *International Journal of Heat and Mass Transfer* **36**(**1**): 17–26.

11.47. Khmamas, F. A. 2010. Improving the environmental cooling for air-coolers by using the indirect-cooling method. *ARPN Journal of Engineering and Applied Sciences* **5**(**2**): 66–73.

11.48. Bhatia, A. 2012. HVAC – Characteristics and selection parameters of fans and blower systems. http://www.pdhcenter.com/courses/m213/m213content.pdf.

11.49. Adnot, J., Greslou, O., Riviere, P., Spadaro, J., Kemna, R., Van Holsteun, R., Van Elburg, M. Li, W., Boom, R. van den, Hitchin, R., and Pout, C. 2012. Final report task 5: Technical analysis ventilation systems for non residential and collective residential applications, BAT and BNAT. http://www.ecohvac.eu/downloads/Task 5 Lot 6 Ventilation Final Report.pdf.

11.50. Kim, J.-W., Ahn, E. Y., and Oh, H. W. 2005. Performance prediction of cross-flow fans using mean streamline analysis. *International Journal of Rotating Machinery* **2005**(**2**): 112–116.

11.51. Lai, H. X., Wang, M., Yun, C. Y., and Yao, J. 2011. Attenuation of cross-flow fan noise using porous stabilizers. *International Journal of Rotating Machinery* **2011**: Article ID 528927. doi: 10.1155/2011/528927.

11.52. Koplow, J. P. 2012. Heat exchanger device and method for heat removal or transfer. U.S. Patent 8,288,675.

11.53. Extreme Tech. 2012. The silent, dust-immune fanless heatsink to be released. *Electronic Cooling*, September Issue, p. 5.

11.54. Yu, J. F., Zhang, T., and Qian, J. M. 2011. *Electrical Motor Products: International Energy-Efficiency Standards and Testing Methods*. Woodhead Publishing: Cambridge.

11.55. Simons, R. E. 1996. Direct liquid immersion cooling for higher power density microelectronics. *Electronics Cooling* **2**(**2**): 24–29.

11.56. Parmeter, L. J. and Knapp, J. M. 2010. Shrink tube encapsulated magnet wire for electrical submersible motors. U.S. Patent 7,679,242.

11.57. Bejan, A. 2004. *Convection Heat Transfer*, 3rd edn. John Wiley & Sons, New York.

11.58. Svoboda, R. and Seipp, H.-G. 2004. Flow restrictions in water-cooled generator stator coils: Prevention, diagnosis and removal – Part 1: Behavior of copper in water-cooled generator coils. *Power Plant Chemistry* **6**(**1**): 7–15.

11.59. 张小虎，胡磊，钟后鸿. 2015. 双水内冷汽轮发电机转子线圈水量及温升研究 (Studies on water flowrate and temperature rise of rotor winding of dual-water inner cooled turbogenerator. 电机技术 (*Electric Motor Technology*) **1**: 9–14.

11.60. Gieras, J. F. 2008. *Advancements in Electric Machines*. Springer Science & Business Media, Berlin.

11.61. Gai, Y., Kimiabeigi, M., Widmer, J. D., Chong, Y. C., Goss, J., SanAndres, U., and Staton, D. A. 2017. Shaft cooling and the influence on the electromagnetic performance of traction motors. Proceedings of 2017 *IEEE International Electric Machines and Drives Conference* (IEMDC), Miami, FL, pp. 1–3.

11.62. Seghir-Ouali, S., Saury, D., Harmand, S., Phillipart, O., and Laloy, D. 2006. Convective heat transfer inside a rotating cylinder with an axial airflow. *International Journal of Thermal Sciences* **46**: 1166–1178.

11.63. Wohlers, C., Juris, P., Kabelac, S., and Ponick, B. 2018. Design and direct liquid cooling of tooth-coil windings. *Electrical Engineering* **100**(**12**): 1–10.

11.64. Svoboda, R., Picech, C., and Hehs, H. 2003. Experience with stainless steel hollow conductors for generator stator water cooling. *Power Plant Chemistry* **5**(**4**): 211–215.

11.65. Womac, D. J., Ramadhyani, S., and Incropera, F. P. 1993. Correlating equations for impingement cooling of small heat sources with single circular jets. *ASME Journal of Heat Transfer* **115**(**1**): 106–115.

11.66. Davin, T., Pellé, J., Harmand, S., and Yu, R. 2015. Experimental study of oil cooling systems for electric motors. *Applied Thermal Engineering* **75**(**2015**): 1–13.

11.67. Estes, K. A. and Mudawar, I. 1995. Comparison of two-phase electronic cooling using free jets and sprays. *ASME Journal of Electronic Packaging* **117**(**4**): 323–332.

11.68. Ito, H. 1959. Friction factor for turbulent flow in curved pipes. *Journal of Basic Engineering* **81**: 123–134.

11.69. Srinivasan, P. S., Nandapurkar, S. S., and Holland, F. A. 1970. Friction factor for coils. *Transactions of Institute of Chemical Engineering* **48**: T156–T161.

11.70. Tan, N., Arimitsu, M., Naruse, Y. and Watanabe, J. 2005. Cooling performance design for super motor and its experimental validation. *Proceedings of the 2005 International Power Electronic Conference*, Niigata, Japan, pp. 1492–1488.

11.71. Upadhya, G., Munch, M., Zhou, P., Horn, J., Werner, D., and McMaster, M. 2006. Micro-scale liquid cooling system for high heat flux processor cooling applications. *IEEE 22nd Semiconductor Thermal Measurement and Management Symposium*, Dallas, TX.

11.72. Pyrhönen, J., Lindh, P. M., Polikarpova, M., Kurvinen, E., and Naumanen, V. 2015. Heat-transfer improvements in an axial-flux permanent-magnet synchronous machine. *Applied Thermal Engineering* **76**: 245–251.

11.73. Polikarpova, M., Lindh, P. M., Tapia, J. A., and Pyrhönen, J. 2014. Application of potting material for a 100 kW radial flux PMSM. *Proceedings of 2014 International Conference on Electrical Machines* (ICEM), Berlin, Germany, pp. 2146–2151.

11.74. Gaugler, R. S. 1944. Heat transfer device. U.S. Patent 2,350,3411.

11.75. Grover, G. M. 1963. Evaporation, condensation heat transfer device. U.S. Patent 3,229,759.

11.76. Mozumder, A. K., Chowdhury, M. S. H., and Akon, A. F. 2011. Characteristics of heat transfer for heat pipe and its correlation. *ISRN Mechanical Engineering* **2011**: Article ID 825073.

11.77. Manimaranl, R., Palaniradjal, K., Alagumurthil, N., and Velmurugan, K. 2012. An investigation of thermal performance of heat pipe using Di-water. *Science and Technology* **2(4)**: 77–80.

11.78. Marengo, M. and Nikolayev, V. S. 2018. Pulsating heat pipers: Experimental analysis, design and application (Chapter 1). In *Encyclopedia of Two-Phase Heat Transfer and Flow IV: Modeling Methodologies, Boiling of CO2, and Micro-Two-Phase Cooling* (Ed.: J. R. Thome). World Scientific Publishing, Singapore, pp. 1–62.

11.79. Hassett, T. and Hodowance, M. 2009. Electric motor with heat pipe. U.S. Patent 7,569,955.

11.80. Parker Hannifin Corporation. 2011. Two-phase evaporative precision cooling systems: For heat loads from 3 to 300 kW. http://www.parker.com/literature/CIC Group/Precision Cooling/New literature/Two_Phase_Evaporative_Precision_Cooling_Systems.pdf.

11.81. Chirilă, A.-H., Ghiţă, C., Crăciunescu, A., Deaconu, I.-D., Năvrăpescu, V., and Catrinoiu, M. 2010. Rotating electric machine thermal study. *International Conference on Renewable Energies and Power Quality*, Las Palmas de Gran Canaria, Spain.

11.82. Boglietti, A., Cavagnino, A., Parvis, M., and Vallan, A. 2006. Evaluation of radiation thermal resistances in industrial motor. *IEEE Transactions on Industry Applications* **42(3)**: 688–693.

11.83. Ribatski, G., Cabezas-Gómez, L., Navarro, H. A., and Saíz-Jabardo, J. M. 2007. The advantages of evaporation in micro-scale channels to cool microelectronic devices. *Engenharia Térmica (Thermal Engineering)* **6(2)**: 34–39.

11.84. Wang, H. 2013. Microchannel cooling of electric drive motors. http://mbi-online.org/microchannel-cooling-electric-drive-motors.

11.85. Sengstock, O. and Hooman, J. 2012. Heat transfer enhancement from a blade tip-cap using metal foams. *ASME Journal of Heat Transfer* **134(11)**: 114505-1–114505-3.

11.86. Mahjoob, S. and Vafai K. 20011. A synthesis of fluid and thermal transport models for metal foam heat exchanges. *International Journal of Heat and Mass Transfer* **51(15–16)**: 3701–3711.

11.87. Tuchinskiy, L. 2005. Fabrication technology for metal foam. *Journal of Advanced Materials* **37(3)**: 60–65.

11.88. Han, X. H., Wang, Q., Park, Y.-G., T'Joen, C., Sommers, A., and Jacobi, A. 2012. A review of metal foam and metal matrix composites for heat exchangers and heatsinks. *Heat Transfer Engineering* **33(12)**: 991–1009. doi:10.1080/01457632.2012.659613.

11.89. Ejlali, A., Ejlali, A., Hooman, K., and Gurgenci, H. 2009. Application of high porosity metal foams as air-cooled heat exchangers to high heat load removal systems. *International Communications in Heat and Mass Transfer* **36(7)**: 674–679.

11.90. Hooman, K. and Gurgenci, H. 2010. Porous medium modeling of air-cooled condensers. *Transport in Porous Medium* **84(2)**: 257–273.

11.91. T'Joen, C., De Jaeger, P., Huisseune, H., van Herzeele, S., and De Paepe, M. 2010. Thermo-hydraulic study of a single row heat exchanger consisting of metal foam covered round tubes. *International Journal of Heat Transfer* **53(15–16)**: 3262–3274.

11.92. Banhart, J. 2001. Manufacture, characterization and application of cellular metals and metal foams. *Progress in Materials Science* **46(2001)**: 559–632.

11.93. Nsofor, E. C. 20011. Recent patents on nanofluids (nanoparticles in liquids) heat transfer. *Nanofluids Heat Transfer* **1(3)**: 190–197.

11.94. Ding, Y. L., Chen, H. S., Wang, L., Yang, C.-Y., He, Y. R., Yang, W., Lee, W. P., Zhang, L. L., and Huo, R. 2007. Heat transfer intensification using nanofluids. *KONA Power and Particle Journal* **25**(**2007**): 23–311.

11.95. Shima, P. D. and Philip, J. 2011. Tuning of thermal conductivity and rheology of nanofluids using an external stimulus. *Journal of Physical Chemistry C* **115**(**41**): 20097–20104.

11.96. Pop, E., Mann, D., Wang, Q., Goodson, K., and Dai, H.-J. 2006. Thermal conductance of an individual single-wall carbon nanotube above room temperature. *Nano Letters* **6**(**1**): 96–100.

11.97. Gaughan, R. 2007. Cool applications for hot technologies. *Small Times Magazine*, **Vol. 7**, Issue 4.

11.98. Celsia Technologies. 2006. Nanospreader technology. http://celsiatechnologies.com/nanospreader_technology.asp.

11.99. *R&D Magazine*. 2009. R&D 100 winners. http://www.rdmag.com/award-winners/2009/07/battle-against-damaging-heat-finds-new-hero.

11.100. Applied Nanotech Holdings, Inc. 2013. CarbAl™ heat transfer material. http://www.appliednanotech.net/tech/CarbAl.php.

11.101. Rosenblatt, M. N., Lee, M. M., and Gregg, J. L. 2013. Dual-purpose hardware aperture. U.S. Patent 8,571,205.

12
Motor Vibration and Acoustic Noise

Vibration refers to mechanical oscillations of an object from its equilibrium position. An oscillatory motion may be harmonic where the motion repeats itself exactly after a period of time (*e.g.*, the motion of a pendulum with a small swing angle), nonharmonic but periodic (*e.g.*, digital waveforms generated by an encoder), nonperiodic (*e.g.*, motions caused by shock loads), or completely random (*e.g.*, vibrations of airplane wings during flight) in nature. In fact, a nonharmonic but periodic motion is the superimposition of many harmonic functions and can be described by a Fourier series. In contrast, a nonperiod transient motion cannot be broken up into a series of harmonic components. Generally, vibration can be considered as the response of a system to certain excitations. In most cases, vibration is undesirable because it can result in extra loads, accelerate rates of wear, generate high mechanical stresses and power losses, create acoustic noise, reduce the stiffness of machine, and cause a loss of precision in controlling machinery. However, on the other hand, the effect of vibration could be positive in some circumstances. Chen *et al.* [12.1] use vibration to improve solidification filling, microstructure, and mechanical properties of casting. They found that in casting magnesium alloys the vibration of 100 Hz and the amplitude of 1 mm can achieve the best solidification filling capacity due to the refined phase of the alloy. With an increase in vibration amplitude from 0.2 to 1 mm, the tensile strength and yield strength of the alloy increase by 17.9% and 10.3%, respectively. In addition, Kudryashova *et al.* [12.2] confirmed that the vibration treatment of solidifying metals results in the improvement in the ingot structure. The experimental results show that optimal conditions for metal processing occur at a vibration frequency of about 60 Hz and an amplitude of about 0.5 mm. Under this condition, there is a significant decrease in the grain size. The yield strength is increased approximately 80% as compared to the ingot cast without vibration. Moreover, proper vibration can help remove dissolved gasses and/or entrained gas bubbles from highly viscous fluids such as epoxy resin during an encapsulation molding process of stator.

Vibrating sources such as electric motors with random oscillatory movement can produce sound waves that travel through various mediums such as air and water. Sound waves are longitudinal waves that radiate outward from their sources in all directions. Sound to human ears is a physiological response to a physical periodically oscillating pressure wave in a certain range. In fact, acoustic noise can not only cause gradual hearing loss of operators but also deteriorate the operating conditions of machines. Nowadays, the use of electric rotary machinery is ubiquitous, and thus it is necessary to effectively control and suppress sound and vibration intensities.

The characteristic parameters normally used to assess vibration, include frequency, dynamic displacement (amplitude of wave), velocity, vibration acceleration, damping, phase angle, and vibration mode. Frequency is defined as the number of oscillations per unit time. Among all vibrational frequencies emitted from a vibrating object, the audible sound frequencies range approximately from 20 to 20,000 Hz. From the viewpoint of physics, noise is called an inconsistent tone that is irregularly composed of many different frequencies and sound waves. From the standpoint of physiology, noise is an unpleasant and unwanted sound.

Sound pressure waves are generated by vibrating structures and, in turn, these pressure waves induce the vibration of structure (*e.g.*, eardrum). Therefore, vibration and sound studies are closely related to each other. In fact, attempts to reduce noise are often related to the reduction of vibration.

DOI: 10.1201/9781003097716-12

12.1 VIBRATION AND NOISE OF ELECTRIC MOTOR

When a motor is running under a full-load condition, unwanted vibration is produced due to various reasons that ultimately affect the performance of the motor. These reasons can be divided into two categories: one is dynamic and another is kinematic. The possible causes under dynamic categories are unbalanced electromagnetic reluctance forces, asymmetric airgap, bearing wear and damage, unbalanced rotor, looseness, and improper motor installation. Excessive motor vibration may damage the motor windings, develop high stress on the rotor bearings, reduce the critical speed, and deteriorate motor performance, which may lead to poor motor reliability, increased unexpected downtime, and even motor failure, while acoustic noises pollute the environment and can also harm human's health. Nonetheless, vibration and noise are important parameters for the assessment of motor quality and operation reliability. Therefore, the motor design must take into account the resonance frequencies of the motor to shift the critical frequencies away from the motor's operating speed to at least ±10%. This will effectively reduce motor vibration and noise levels at motor normal operation.

It has been recognized that about one-third of motor failures is due to mechanical breakages. The largest single contributor to mechanical breakages is high vibration, which may be attributed to defective motor designs, rough manufacturing processes, improper assemblies and installations, inadequate maintenances, and faulty operations.

The interaction between the rotor and stator is a predominant feature of vibration and noise of an electric motor. A large number of studies [12.3–12.5] have shown that the vibration model $v(t, \phi, z)$ can be expressed by three terms, each of which has a different physical origin:

$$v\left(t,\phi,z\right) = v_e\left(t,\phi,z\right) + v_m\left(t,\phi,z\right) + v_a\left(t,\phi,z\right) \qquad (12.1)$$

where
   t is the time
   ϕ is the angular position
   z is the axial position on the stator frame

In Equation 12.1, $v_e(t, \phi, z)$, $v_m(t, \phi, z)$, and $v_a(t, \phi, z)$ are electromagnetic, mechanical, and aerodynamic vibrational terms, respectively. These vibrations produce forces on associate components of the motor and severely affect its performance.

Similarly, the sound emitted from a motor can be expressed as the sum of these three components:

$$p\left(t,\phi,z\right) = p_e\left(t,\phi,z\right) + p_m\left(t,\phi,z\right) + p_a\left(t,\phi,z\right) \qquad (12.2)$$

where $p_e(t, \phi, z)$, $p_m(t, \phi, z)$, and $p_a(t, \phi, z)$ are electromagnetic, mechanical, and aerodynamic sound pressure terms, respectively. Noise from the electromagnetic term dominates at low and medium speeds. At high speeds, the mechanical and aerodynamic noises are the most influential among the three [12.6].

The electromagnetic vibration and resulting noise resulted from various reasons such as phase imbalance, broken rotor bar, asymmetric airgap, eccentric rotor, slot openings, magnetic saturation, space/time harmonics associated with electrical and magnetic frequency, and magnetostrictive expansion of the core lamination. Electromagnetic noise is a low-frequency noise generated in the motor.

The mechanical vibration and resulting noise are associated with motor parts, which are subjected to unbalanced mechanical forces due to oscillatory motions about their equilibrium positions. The causes of mechanical vibration include unbalanced mass of rotating components, shaft misalignment, bearing deterioration, and frame distortion. The method to distinguish between an electromagnetic vibration and a mechanical vibration is to disconnect the electricity supply from the motor. In the absence of electrical power, the electromagnetic vibration does not exist [12.7], but the mechanical vibration exists till the rotary motion is sustained. Here, it should be noted that this method is applicable to induction motors (IMs) and not to PMMs.

As ventilating air flows through a motor, the aerodynamic vibration and noise may be generated due to the interactions of cooling air to both rotating parts (*e.g.*, fan and rotor) and stationary parts (*e.g.*, stator and housing). Usually, noise generated by the aerodynamic vibration is due to turbulent flow, which is proportional to the 6th to 8th power of the flow velocity (*i.e.*, $W \propto u^{6-8}$). This indicates that a doubling of the flow velocity u can increase the sound power W by 18–24 dB, respectively [12.8]. For small-sized motors, the aerodynamic vibration and resulting noise are relatively low compared to the mechanical and electromagnetic components.

Excessive vibration can cause damages to an electric motor in different ways [12.9]: (*a*) It can accelerate bearing failure by causing indentations on the bearing raceways at the ball or roller spacing. (*b*) It can loosen windings and cause mechanical damage to insulation by fracturing, flaking, or eroding of the material. (*c*) The excessive movement can generate high temperature, and as a result, the lead wires can become brittle. (*d*) It can cause brush sparking at commutators or current collector rings. As a result of these problems, motor vibration can harm motor performance, generate high noise, and even cause motor failure.

12.2 FUNDAMENTALS OF VIBRATION

It is very important to understand vibrational phenomena such as resonance, harmonics, and damping. Thus, to reduce vibration and noise in electric machines, it is essential to understand the fundamentals of vibration.

Generally, vibration in rotordynamic systems can be categorized into synchronous or subsynchronous vibrations depending on the dominant frequency and source of the disturbance forces. The first type is the synchronous vibrations that have a dominant frequency component that matches the rotating speed of the rotor. This type of vibration is usually caused by the imbalance or other synchronous forces in the system. The sub-synchronous vibration or whirling has a dominant frequency below the operating speed. This type of vibration is mainly caused by fluid excitation from the cross-coupling stiffness [12.10].

Most of the mechanical systems in the real world are considerably very complicated. Therefore, the first step in dealing with difficult problems is to appropriately simplify problems. Generally, the fundamental equations employed in a vibration analysis of multiple-degree-of-freedom systems can be expressed in matrix notation:

$$[m]\{\ddot{x}\} + [c]\{\dot{x}\} + [k]\{x\} = \{F\} \tag{12.3}$$

where
  $[m]$ is the mass matrix
  $[c]$ is the damping matrix
  $[k]$ is the stiffness matrix
  $\{\ddot{x}\}$ is the acceleration vector
  $\{\dot{x}\}$ is the velocity vector
  $\{x\}$ is the displacement vector
  $\{F\}$ is the acting force vector

In Equation 12.3, the physical interpretations of the terms from the left-hand side are inertial force, damping force, and spring force (or restoring force), respectively. The term on the right-hand side is external exciting or disturbing force. All these forces balance each other to maintain a dynamic equilibrium state.

12.2.1 SIMPLE HARMONIC OSCILLATING SYSTEM

A harmonic motion refers to the oscillations that are symmetrical about a position of equilibrium. The motion may have either one frequency or amplitude, defined as simple harmonic motion, or a combination of two or more components of harmonics, defined as complex harmonic motion. Simple harmonic motion oscillates with only the restoring force acting on the system. For this system, the restoring force is directly proportional to the displacement. As an example, a pendulum swings in a small arc, and in its periodic motion, the tangential component of gravity acts as the restoring force of the pendulum. It is the restoring force that interacts with the inertia property of mass (kinetic energy) to perpetuate the oscillation.

For ID simple harmonic motion (see Figure 12.1a), the equation of motion can be obtained by means of Newton's second law and Hooke's law:

$$F = m\frac{d^2x(t)}{dt^2} = -kx(t) \tag{12.4}$$

where
  F is the restoring force
  m is the inertial mass of an oscillating body
  $x(t)$ is the displacement from an equilibrium position as a function of time t
  k is the spring constant

This equation indicates that for simple harmonic motion, the inertial force is balanced by the spring force.

Thus, applying a sinusoidal form of solution, this differential equation can be solved as

$$x(t) = A\cos(\omega t - \varphi) \tag{12.5}$$

where
  A is the maximum displacement from the equilibrium position, which is also defined as the oscillating amplitude
  φ is the phase angle
  ω is the oscillating frequency

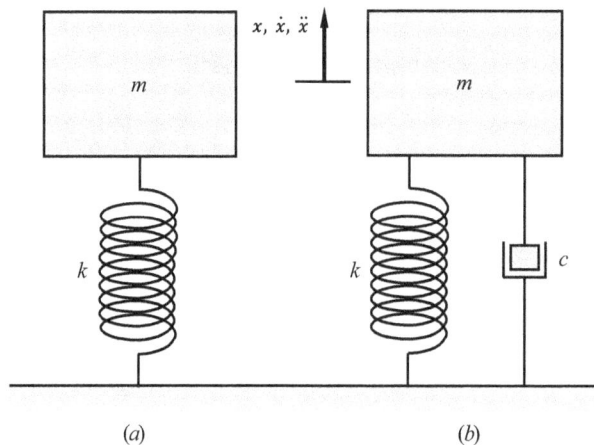

FIGURE 12.1 Oscillation systems: (a) simple harmonic oscillation system without damping and (b) damped harmonic oscillation system.

$$\omega = 2\pi f = \sqrt{\frac{k}{m}} \qquad (12.6)$$

Accordingly, velocity and acceleration of the oscillating body can be found as

$$v = \frac{dx(t)}{dt} = -A\omega \sin(\omega t - \varphi) \qquad (12.7)$$

$$a = \frac{d^2x(t)}{dt^2} = -A\omega^2 \cos(\omega t - \varphi) \qquad (12.8)$$

The oscillating frequency can be expressed as

$$f = \frac{1}{2\pi}\sqrt{\frac{k}{m}} \qquad (12.9)$$

Since the oscillating period T is reciprocal of frequency, it follows that

$$T = \frac{2\pi}{\sqrt{k/m}} \qquad (12.10)$$

The kinetic energy $E_k(t)$ and the potential energy $E_p(t)$ in an oscillating system can be described, respectively, as

$$E_k(t) = \frac{1}{2}m[v(t)]^2 = \frac{1}{2}mA^2\omega^2 \sin^2(\omega t - \varphi) = \frac{1}{2}kA^2 \sin^2(\omega t - \varphi) \qquad (12.11\text{a})$$

$$E_p(t) = \frac{1}{2}k[x(t)]^2 = \frac{1}{2}kA^2 \cos^2(\omega t - \varphi) \qquad (12.11\text{b})$$

Therefore, the total mechanical energy $E_t(t)$ is the sum of the energies $E_k(t)$ and $E_p(t)$:

$$E_t(t) = E_k(t) + E_p(t) = \frac{1}{2}kA^2 \qquad (12.12)$$

It is worth to note that the total mechanical energy in an oscillating system is constant.

12.2.2 DAMPED HARMONIC OSCILLATING SYSTEM

Damping is a measure of an ability of a vibrating system to dissipate mechanical vibratory energy so that the amplitude and duration of vibration can be effectively reduced. In other words, damping creates a force that acts in the opposite direction to the object travel. It can be broadly classified into two categories: passive and active damping. Passive damping can be achieved by using mechanical properties of damping materials and/or by designing vibration-absorbing structures. In active damping, an actuator is used to generate force opposing the motion, regardless of the relative velocity across it [12.11]. Active damping employs sophisticated closed-loop control techniques to minimize the effect of vibration. The active controllers continuously regulate the damping characteristics and require measurements of response and feedback from the vibrating system.

For a damped simple harmonic oscillating system (see Figure 12.1b), the governing equation is

$$m\frac{d^2x(t)}{dt^2} + c\frac{dx(t)}{dt} + kx(t) = 0 \qquad (12.13)$$

where c is the damping coefficient, given in units of Ns/m or kg/s.

Introducing a dimensionless quantity called damping ratio ζ, it can be derived as

$$\zeta = \frac{c}{2\sqrt{mk}} \tag{12.14}$$

while the natural frequency of an undamped system ω_n is

$$\omega_n = \sqrt{\frac{k}{m}} \tag{12.15}$$

Equation 12.13 can be rewritten as

$$\frac{d^2x(t)}{dt^2} + 2\zeta\omega_n\frac{dx(t)}{dt} + \omega_n^2 x(t) = 0 \tag{12.16}$$

The initial conditions are given as

$$x(t = 0) = X \tag{12.17a}$$

$$\dot{x}(t = 0) = V_o \tag{12.17b}$$

The solution of Equation 12.16 is in the form

$$x(t) = e^{st} \tag{12.18}$$

Substituting (12.18) into (12.16), it yields

$$s^2 + 2\zeta\omega_n + \omega_n^2 = 0 \tag{12.19}$$

The two roots of Equation 12.19 are

$$s_{1,2} = \omega_n\left(-\zeta \pm \sqrt{\zeta^2 - 1}\right) \tag{12.20}$$

Thus, a general solution is given by

$$x(t) = Ae^{s_1t} + Be^{s_2t} = Ae^{\omega_n t\left(-\zeta+\sqrt{\zeta^2-1}\right)} + Be^{\omega_n t\left(-\zeta-\sqrt{\zeta^2-1}\right)} \tag{12.21}$$

where the constants A and B are determined by the initial conditions.

According to the value of damping ratio ζ, four cases can be identified as follows:

- When $\zeta = 1$, the system is in the critical damping condition, giving the fastest return of the system to its equilibrium condition without much oscillation. For this case, the general solution can be written as

$$x(t) = (A + Bt)\, e^{-\omega_n t} \tag{12.22}$$

- When $\zeta > 1$, the characteristic roots are positive, real, and distinct. Under such a circumstance, the system is overdamped. Like a critical damping, an overdamped system does not oscillate but takes a long time to converge to its equilibrium position than a critically damped system. It is important to note that $\zeta > 1$ indicates both roots s_1 and s_2 are negative:

$$x(t) = Ae^{s_1 t} + Be^{s_2 t} \tag{12.23}$$

Therefore, every solution in this case goes asymptotically to zero.

- When $0 < \zeta < 1$, the term under the square root is negative and the characteristic roots are not real. Under this condition, the system is underdamped. Introducing $\omega = \omega_n \sqrt{\zeta^2 - 1}$, the characteristic roots are $(-\zeta\omega_n \pm i\omega)$, leading to the general solution as

$$x(t) = Ae^{(-\zeta\omega_n + i\omega)t} + Be^{(-\zeta\omega_n - i\omega)t} = Ae^{-\zeta\omega_n t}\cos(\omega t) + Be^{-\zeta\omega_n t}\sin(\omega t) \tag{12.24}$$

Alternatively, Equation 12.24 can be expressed as

$$x(t) = Ce^{-\zeta\omega_n t}\cos(\omega t - \varphi) \tag{12.25}$$

The term $\cos(\omega t - \varphi)$ in Equation 12.25 reflects the oscillating behavior of the system.

- When $\zeta = 0$, the system is undamped, just as the simple harmonic oscillating system.

The comparisons of four cases, critical damping, overdamped, underdamped, and undamped, with different damping ratios ζ are presented in Figure 12.2. As can be seen from the figure, for the undamped case ($\zeta = 0$), the system oscillates sustainably with a constant amplitude. An increase in the damping ratio in the range of $0 < \zeta < 1$ leads to vibration decays, accompanied by a reduction in the oscillation amplitude. The reduction in the oscillation amplitude is inversely proportional to the damping ratio ζ. When it reaches critical damping at $\zeta = 1$, the system no longer oscillates, but it monotonically reaches equilibrium in the shortest time without overshooting. For overdamped cases where $\zeta > 1$, the system experiences large damping that it does not oscillate at all. In fact, if the system is displaced from equilibrium, it takes a long time for it to return to its initial position because the damping force is so severe.

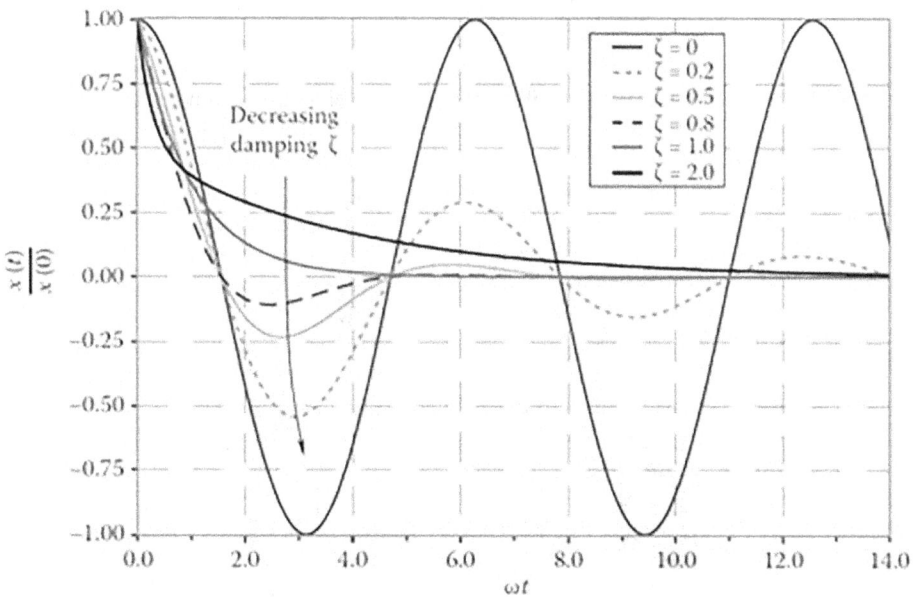

FIGURE 12.2 Comparison of damped and undamped harmonic oscillating systems with various damping ratios: (a) $\zeta = 1$, critical damping; (b) $\zeta > 1$, overdamped; (c) $0 < \zeta < 1$, underdamped; and (d) $\zeta = 0$, undamped.

12.2.3 Forced Vibration with Damping

In many mechanical systems, it is often encountered that an external time-dependent force is applied to the system to start forced vibration. The applied exciting force may be harmonic, nonharmonic (periodic or nonperiodic), or random in nature. The dynamic response of a system to a suddenly applied nonperiodic excitation is called transient response.

Considering a dynamic response of a damped system to a harmonic exciting force $F(t)$ shown in Figure 12.3, this single-degree-of-freedom system has the mass m that is supported by a spring and a damper and is subject to an external exciting force in the form

$$F(t) = F_o \sin(\omega t - \varphi) \tag{12.26}$$

where
  F_o is the force magnitude
  ω is the excitation frequency
  φ is the phase angle

The value of φ depends on $F(t)$ at $t = 0$. The dynamic equation of motion is given as

$$m\frac{d^2x(t)}{dt^2} + c\frac{dx(t)}{dt} + kx(t) = F_o \sin(\omega t - \varphi) \tag{12.27}$$

where
  c is the viscous damping coefficient
  k is the spring coefficient

The general solution of this nonhomogeneous second-order differential equation consisted of two parts: the complementary solution $x_c(t)$ and the particular solution $x_p(t)$, that is, $x(t) = x_c(t) + x_p(t)$. The complementary solution is obtained by setting the right-hand side zero:

$$m\frac{d^2x_c(t)}{dt^2} + c\frac{dx_c(t)}{dt} + kx_c(t) = 0 \tag{12.28}$$

This equation is identical to (12.13). Therefore, the general solution is given by

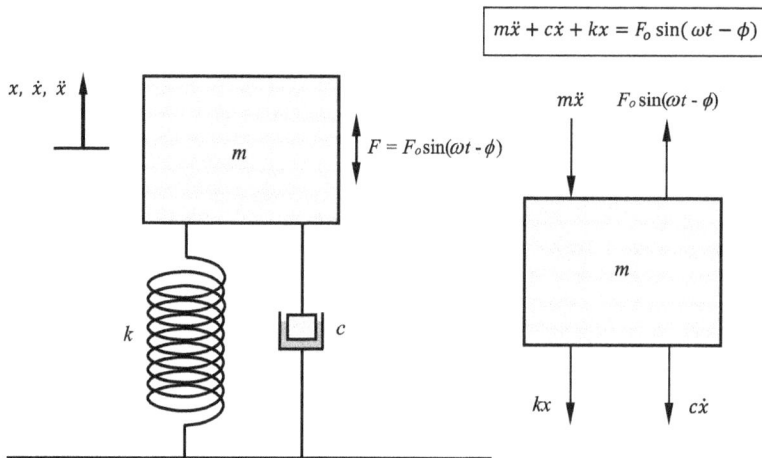

FIGURE 12.3 Forced vibration of single-degree-of-freedom system with damping.

$$x_c(t) = Ae^{\omega_o t\left(-\zeta+\sqrt{\zeta^2-1}\right)} + Be^{\omega_o t\left(-\zeta-\sqrt{\zeta^2-1}\right)} \tag{12.29}$$

where constants A and B are determined from the specific boundary conditions and ζ values.

To obtain the particular solution $x(t)$, it is assumed that

$$x_p(t) = D\cos(\omega t - \varphi) + E\sin(\omega t - \varphi) \tag{12.30}$$

where D and E are constants. The associated velocity and acceleration are

$$\begin{cases} \dfrac{dx_p(t)}{dt} = -D\omega\sin(\omega t - \varphi) + E\omega\cos(\omega t - \varphi) \\[4mm] \dfrac{d^2 x_p(t)}{dt^2} = -D\omega^2\cos(\omega t - \varphi) - E\omega^2\sin(\omega t - \varphi) \end{cases} \tag{12.31}$$

Substituting Equation 12.31 into 12.27, it yields

$$\cos(\omega t - \varphi)(-mD\omega^2 + cE\omega + kD) + \sin(\omega t - \varphi)(-mE\omega^2 - cD\omega + kE) = F_o\sin(\omega t - \varphi) \tag{12.32}$$

Comparing both sides of Equation 12.32, it leads to

$$D\left(k - m\omega^2\right) + cE\omega = 0 \tag{12.33a}$$

$$E\left(k - m\omega^2\right) - cD\omega = F_o \tag{12.33b}$$

Defining the frequency ratio β as the ratio of excitation frequency ω over natural frequency ω_n,

$$\beta = \frac{\omega}{\omega_n} = \frac{f}{f_n} \tag{12.34}$$

Equation 12.34 can be rewritten as

$$D(1 - \beta^2) + 2\zeta\beta E = 0 \tag{12.35a}$$

$$-2\zeta\beta D + E\left(1 - \beta^2\right) = \frac{F_o}{k} \tag{12.35b}$$

By solving Equations 12.35a and 12.35b, constants D and E can be determined as

$$D = -\frac{F_o}{k}\frac{2\zeta\beta}{\left(1 - \beta^2\right)^2 + \left(2\zeta\beta\right)^2} \tag{12.36a}$$

$$E = \frac{F_o}{k}\frac{1 - \beta^2}{\left(1 - \beta^2\right)^2 + \left(2\zeta\beta\right)^2} \tag{12.36b}$$

The particular solution $x_p(t)$ becomes

$$x_p(t) = -\frac{F_o}{k}\frac{2\zeta\beta}{\left(1 - \beta^2\right)^2 + \left(2\zeta\beta\right)^2}\cos\left(\omega t - \varphi\right) + \frac{F_o}{k}\frac{1 - \beta^2}{\left(1 - \beta^2\right)^2 + \left(2\zeta\beta\right)^2}\sin\left(\omega t - \varphi\right) \tag{12.37}$$

Alternatively, the particular solution $x(t)$ can be expressed as

$$x_p(t) = X \sin\left[(\omega t - \varphi) - \Phi\right] = X\left[\sin(\omega t - \varphi)\cos\Phi - \cos(\omega t - \varphi)\sin\Phi\right] \tag{12.38}$$

Comparing Equations 12.37 and 12.38, it follows that

$$X\cos\Phi = \frac{F_o}{k}\frac{1 - \beta^2}{\left(1 - \beta^2\right)^2 + \left(2\zeta\beta\right)^2} \tag{12.39a}$$

$$X\sin\Phi = \frac{F_o}{k}\frac{2\zeta\beta}{\left(1 - \beta^2\right)^2 + \left(2\zeta\beta\right)^2} \tag{12.39b}$$

These two equations can be combined to yield the desired equation for X:

$$X = \frac{F_o}{k}\frac{1}{\sqrt{\left(1 - \beta^2\right)^2 + \left(2\zeta\beta\right)^2}} \tag{12.40}$$

and

$$\tan\Phi = \frac{2\zeta\beta}{1 - \beta^2} \tag{12.41}$$

Thus,

$$x_p = \frac{F_o}{k}\frac{1}{\sqrt{\left(1 - \beta^2\right)^2 + \left(2\zeta\beta\right)^2}}\sin\left[(\omega t - \varphi) - \Phi\right] \tag{12.42}$$

The ratio of the amplitude of the steady-state response to the static deflection under the action of force F_o is known as the *MF*,

$$MF = \frac{1}{\sqrt{\left(1 - \beta^2\right)^2 + \left(2\zeta\beta\right)^2}} \tag{12.43}$$

Thus, the *MF* depends upon the frequency ratio β and the damping ratio ζ. The variations of the *MF* as functions of ζ and β are displayed in Figure 12.4. It can be observed that for all cases, if β is much less than 1, the *MF* approaches 1, regardless of the damping ratio. Exceeding excitation frequencies cause the amplitude of forced vibration to be sufficiently small.

If β becomes large enough (*i.e.*, $\omega \gg \omega_n$), the *MF* approaches zero for all ζ values. This is because the increase in excitation frequency always causes an increase in system inertia force, resulting in the reduction of the amplitude of forced vibration. The system exhibits resonance when $\beta = 1$, that is, the excitation frequency ω is exactly equal to the system natural frequency ω_n. At this point, the amplitude of vibration increases without bound if no damping is applied on the system.

For very small damping ratios (*i.e.*, $\zeta \to 0$), the *MF* can go extremely high near $\beta = 1$. Theoretically, when $\zeta = 0$, the *MF* goes infinity at $\beta = 1$. Increasing ζ can reduce sharply the peak values of the *MF*, particularly at $\beta = 1$ due to the damping effect. As ζ becomes considerably large, the amplitude of the *MF* is reduced for all β values. At the critical damping where $\zeta = 1$, the *MF* decreases monotonically with an increase in the frequency ratio β. For overdamped cases that $\zeta > 1$, the *MF* approaches zero much faster than all other cases.

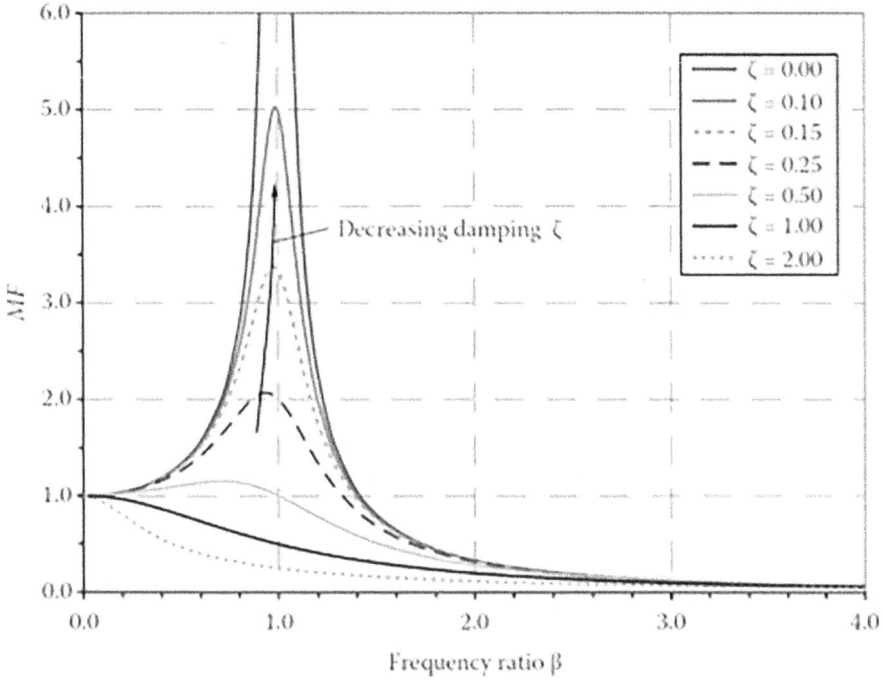

FIGURE 12.4　Variations of the *MF* with frequency ratio β and damping ratio ζ.

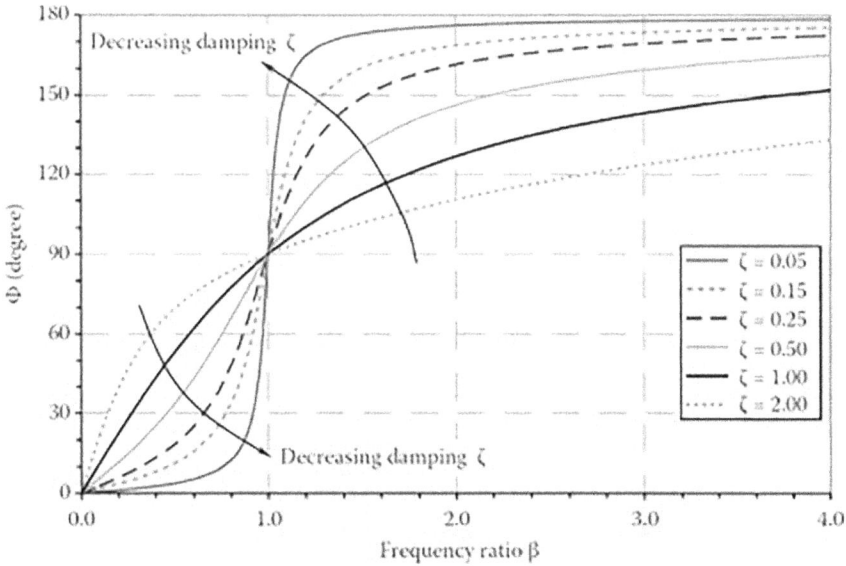

FIGURE 12.5　Variations of phase angle Φ with frequency ratio β and damping ratio ζ.

Figure 12.5 shows the variations of the phase angle with frequency ratio β and damping ratio ζ. At the resonance point where $\omega = \omega_n$, the phase angle is 90° for all damping ratios. For an undamped system (where $\zeta = 0$), the phase angle $\Phi = 0°$ for $0 < \beta < 1$ and $\Phi = 180°$ for $\beta > 1$. Increasing the damping ratio ζ leads to a longer time for the phase angle to approach 180°. For $0 < \beta < 1$ and $\zeta > 0$, the response lags the excitation. Contrarily, for $\beta > 1$ and $\zeta > 0$, the response leads the excitation.

12.2.4 Forced Vibration Due to Mass Unbalance

Mass unbalance is one of the most common sources of vibration excitation in rotating machinery because it develops a centrifugal force along with dynamic eccentricity. The total unbalanced force is the sum of the centrifugal force and unbalanced electromagnetic force. Rotor eccentricity increases due to this unbalanced force that tries to pull apart the rotor further away from the stator bore center, varying the airgap. This unbalanced force is proportional to the rotor rotating speed [12.12, 12.13].

Although a motor is designed, manufactured, assembled, and installed with great care to ensure proper balance under working conditions, some degree of mass unbalance is always present in the motor. Under normal circumstances, slight unbalance in major rotating components with relatively low rotating speeds should not notably affect motor operation. However, if the deviation in mass symmetry about the rotating axis becomes significant or if the rotating speed of the motor is considerably high, large unbalanced forces are produced to exert on the rotor, bearings, support structure, and ancillary equipment, leading to high vibration and high stresses of these components. In fact, vibration is a major cause of downtime and reliability problems for electric motors. Many sources highlight the impact of vibration on the motor performance, noise, fatigue lifetime, and failure. In a worse case, a rotor unbalance may cause the rotor and stator to contact and rub each other, resulting in a severe motor failure. The vibration frequency of unbalance f_r is equal to the rotor rotating speed in the unit of revolutions per second (rps),

$$f_r = \frac{n}{60} = \frac{\omega}{2\pi} \qquad (12.44)$$

where
   the unit of the rotor rotating speed n is rpm
   the unit of the angular speed ω is radians per second (rad/s)

Considering the rotating system in Figure 12.6, the unbalance is represented by an unbalanced mass m_u with eccentricity e from the system rotating axis and angular velocity ω. The centrifugal force resulting from the eccentric mass is given by

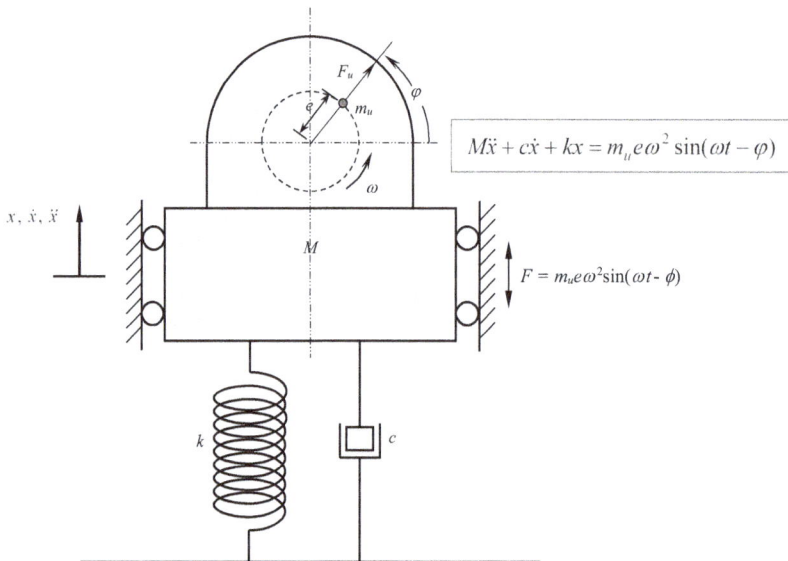

$$M\ddot{x} + c\dot{x} + kx = m_u e\omega^2 \sin(\omega t - \varphi)$$

$$F = m_u e\omega^2 \sin(\omega t - \phi)$$

FIGURE 12.6 Forced vibration induced by unbalanced mass.

$$F_u = m_u e \omega^2 \sin(\omega t - \varphi) \tag{12.45}$$

This force serves as the exciting force to induce forced vibration of the system. The total mass of the system is M. The equation of motion is

$$M \frac{d^2 x(t)}{dt^2} + c \frac{dx(t)}{dt} + kx = \left(m_u e \omega^2 \right) \sin\left(\omega t - \varphi \right) \tag{12.46}$$

This equation is identical to (12.27) as F_o is replaced by $(m_u e \omega^2)$. Because the complementary solution $x_c(t)$ is independent of the exciting force, it remains the same as shown in Equation 12.29. The particular solution $x(t)$ can be obtained by replacing F_o by $(m_n e \omega^2)$ in Equation 12.42, that is,

$$x_p = \frac{m_u e \omega^2}{k} \frac{1}{\sqrt{\left(1 - \beta^2\right)^2 + \left(2\zeta\beta\right)^2}} \sin\left[\left(\omega t - \varphi\right) - \Phi\right] \tag{12.47}$$

It is noted that the damping ratio ζ in this case is based on the system mass M:

$$\zeta = \frac{c}{2\sqrt{MK}} \tag{12.48}$$

Similarly, the system national frequency becomes

$$\omega_n = \sqrt{\frac{k}{M}} \tag{12.49}$$

The steady-state amplitude X is given by

$$X = \frac{m_u e \omega^2}{k} \frac{1}{\sqrt{\left(1 - \beta^2\right)^2 + \left(2\zeta\beta\right)^2}} = \frac{m_u e}{M} \frac{\beta^2}{\sqrt{\left(1 - \beta^2\right)^2 + \left(2\beta\right)^2}} \tag{12.50}$$

This equation can be alternatively rearranged as

$$\frac{MX}{m_u e} = \frac{\beta^2}{\sqrt{\left(1 - \beta^2\right)^2 + \left(2\zeta\beta\right)^2}} \tag{12.51}$$

and

$$\tan \Phi = \frac{2\zeta\beta}{1 - \beta^2} \tag{12.52}$$

The variations of $MX/m_u e$ are plotted in Figure 12.7 as functions of frequency ratio β and damping ratio ζ. It shows all curves start at zero amplitude. Resonances occur for various damping levels at $\beta = 1$, that is, when the excitation frequency ω is identical to the system natural frequency ω_n. At very large β values, all curves approach 1 regardless of damping.

Since the phase angle in Equation 12.52 is identical to 12.41, the variation of the phase angle with β and ζ is the same as in Figure 12.5.

Due to mass unbalance in rotor condition, mutual inductances between stator and rotor loops become unsymmetrical, inducing characteristic harmonics as given in the following:

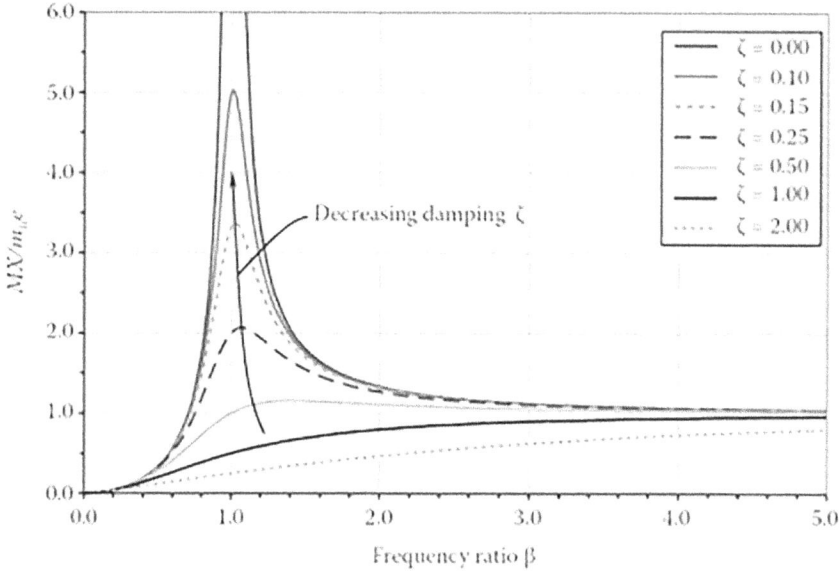

FIGURE 12.7   Variations of factor $MX/m_u e$ with frequency ratio β and damping ratio ζ.

$$f_{ubm} = f \left[\frac{k(1-s)}{p} + 1 \right], \quad k = 1,2,3 \;,...$$ (12.53)

where
    p is the number of pole pairs
    s is the slip
    f is the supply frequency

Among all the causes of motor vibration, rotor unbalance is the most common one. In fact, unbalanced mass in a rotor creates large forces that often result in motor failure and downtime. For high-speed motors, even a very small amount of unbalance may cause a severe vibration problem. Mass unbalance in a rotor may come from manufacturing errors such as porosity in casting, improper machining process, manufacturing tolerances, and gain or loss of material during operation [12.14].

12.2.5   VIBRATION INDUCED BY SUPPORT EXCITATION

Many motor applications involve vibrations that are induced by the excitation of motor support. A typical example is driving a vehicle on a bumpy road. The vehicle shakes due to the uneven surfaces such as broken pavement, potholes, troughs, and small stones. Vibration is transmitted through wheels to the vehicle body and other components including motors.

    Another example is related to seismic waves generated by earthquakes. Earthquakes generate three types of seismic waves: primary (longitudinal) wave, secondary (shear) wave, and surface wave (mixed wave of primary and secondary waves). Due to the long wavelength and high amplitude, the surface wave is the main cause of building destruction. When a building sways and oscillates during an earthquake, all objects attached to the building vibrate accordingly. This will seriously affect the safe operation of elevator motors. Therefore, it is very important to understand the excitation mechanism of motor vibration induced by support excitation because it helps design safer and more reliable elevator motors.

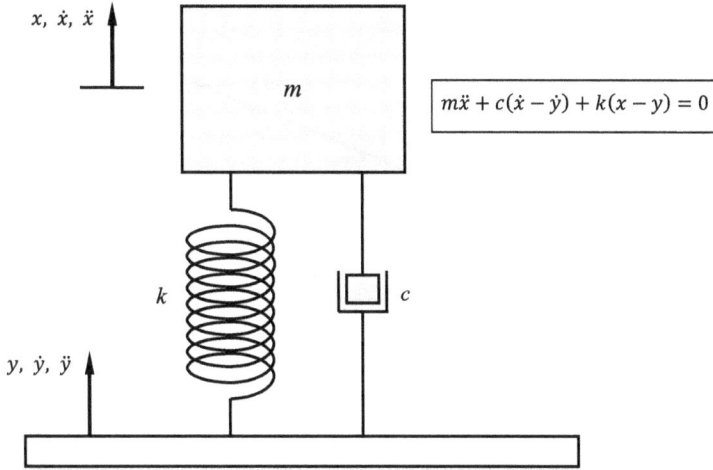

FIGURE 12.8 Vibration induced by support excitation in a two-degree-of-freedom mechanical system.

As demonstrated in Figure 12.8, a spring-damper system undergoes harmonic motion, where $x(t)$ denotes the displacement from the mass equilibrium position and $y(t)$ the displacement of the motor support at time t. The damping force, in this case, is the product of the relative velocity between the system and its support $(\dot{x} - \dot{y})$ and the damping coefficient c, that is, $c(\dot{x} - \dot{y})$, and the spring force is the product of the net spring elongation $(x - y)$ and the spring coefficient k, that is, $k(x - y)$.

The motion equation is expressed as

$$m\frac{d^2 x(t)}{dt^2} + c\left(\frac{dx(t)}{dt} - \frac{dy(t)}{dt}\right) + k(x - y) = 0 \tag{12.54}$$

Introducing the relative displacement $x_R = x - y$, Equation 12.54 can be rewritten as

$$m\frac{d^2 x_R(t)}{dt^2} + c\frac{dx_R(t)}{dt} + kx_R(t) = -m\frac{d^2 y(t)}{dt^2} \tag{12.55}$$

Following the same procedure as discussed previously, this equation is solved for x_R

$$x_R(t) = \frac{\beta^2 y(t)}{\sqrt{\left(1 - \beta^2\right)^2 + \left(2\zeta\beta\right)^2}} \sin(\omega t - \Phi) \tag{12.56}$$

The vector diagram shows that the relationship between the three vectors is (Figure 12.9)

$$\mathbf{x} = \mathbf{x_R} + \mathbf{y} \tag{12.57}$$

It follows that

$$x^2 = x_R^2 + y^2 + 2x_R y \cos\Phi \tag{12.58}$$

The phase angle is obtained as follows:

$$\tan\Phi = \frac{2\zeta\beta}{1 - \beta^2} \tag{12.59}$$

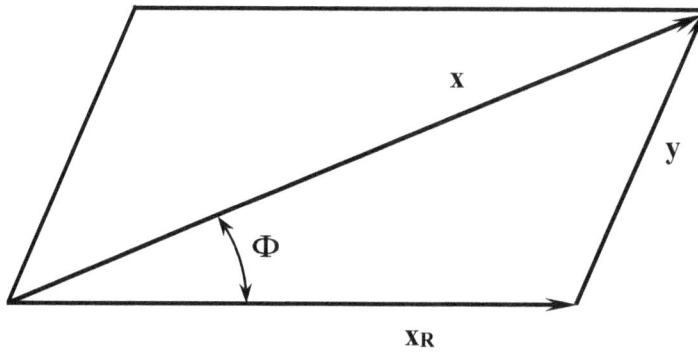

FIGURE 12.9   The relationship of three vectors given that $\mathbf{x} = \mathbf{x_R} + \mathbf{y}$. The vector $\mathbf{x_R}$ lagging behind $\mathbf{x}$ by an angle Φ.

The ratio of the amplitude of the response $x(t)$ to that of the support motion $y(t)$ is defined as the displacement transmissibility. Thus,

$$\frac{x(t)}{y(t)} = \frac{\sqrt{1+\left(2\zeta\beta\right)^2}}{\sqrt{\left(1-\beta^2\right)^2+\left(2\zeta\phi\right)^2}} \tag{12.60}$$

The displacement transmissibility $x(t)/y(t)$ is plotted in Figure 12.10 with various values of the damping ratio ζ and frequency ratio β. It can be seen that the value of x/y is unity for all curves at $\beta = 0$ and $\beta = \sqrt{2}$, regardless of the damping levels. At the system resonance point where $\beta = 1$, x/y becomes infinity for an undamped spring-mass system ($\zeta = 0$). Increasing the ζ value will greatly reduce the peak values of the displacement transmissibility x/y. The values of x/y are larger than

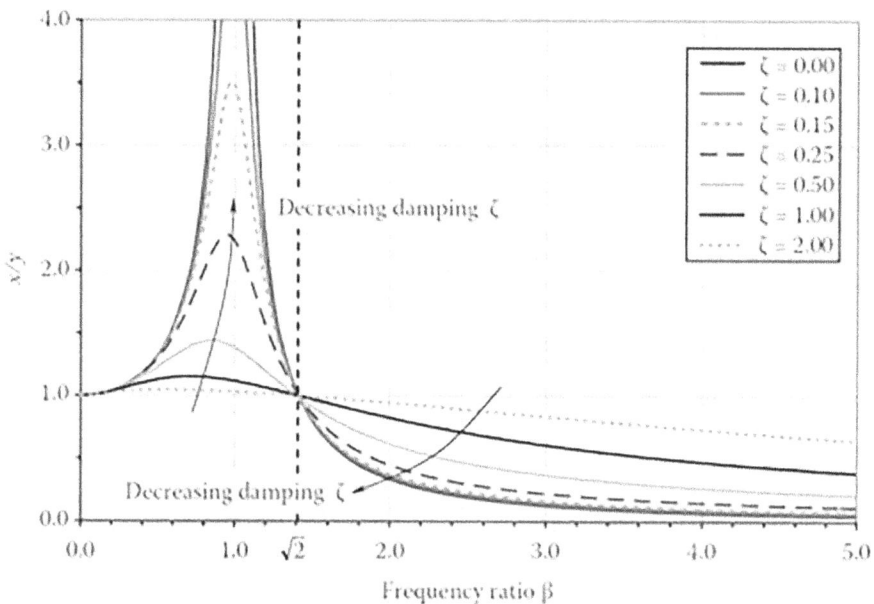

FIGURE 12.10   Variations of displacement transmissibility x/y with frequency ratio β and damping ratio ζ in a two-degree-of-freedom mechanical system where vibration is induced by the excitation of support.

1 for $0 < \beta < \sqrt{2}$ and less than 1 for $\beta > \sqrt{2}$. Moreover, for $\beta < \sqrt{2}$, a small change in the damping ratio leads to a large change in x/y. As a contrast, for $\beta > \sqrt{2}$, the change in the damping ratio is less sensitive to the magnitude of x/y.

12.2.6 Directional Vibration

According to the direction of vibration, there are basically three types of vibration associated with the motion:

- Radial vibration is the vibration of the rotor in the radial direction. This type of vibration can be caused by several reasons: (*a*) Rotating magnetic field produces a traveling force wave that distorts stator and causes vibration in the stator frame and base (see Figure 12.11). (*b*) Any unbalance on the rotor can result in an unbalanced centrifugal force, leading to rotor distortion. (*c*) Large bearing radial clearance provides free space for the rotor in the radial direction. (*d*) Bearing misalignment can directly lead to motor vibration.
- Torsional vibration is the dynamics of the rotor in the angular/rotational direction. Excessive torsional vibration and resonance lead to damaged motor components such as rotor, bearings, couplings, gears, and auxiliary equipment.
- Axial vibration is the dynamic response of the rotor in the axial direction under external thrust forces and internal electromagnetic forces due to the misalignment of the rotor relative to the stator windings. There are indications that rotordynamic vibration and axial vibration influence each other through variations in roller bearing stiffness due to the time-dependent axial load [12.15].

In practice, a multidirectional vibration may occur to act on a motor. The interactions between different directional vibrations make the vibration problems considerably complicated.

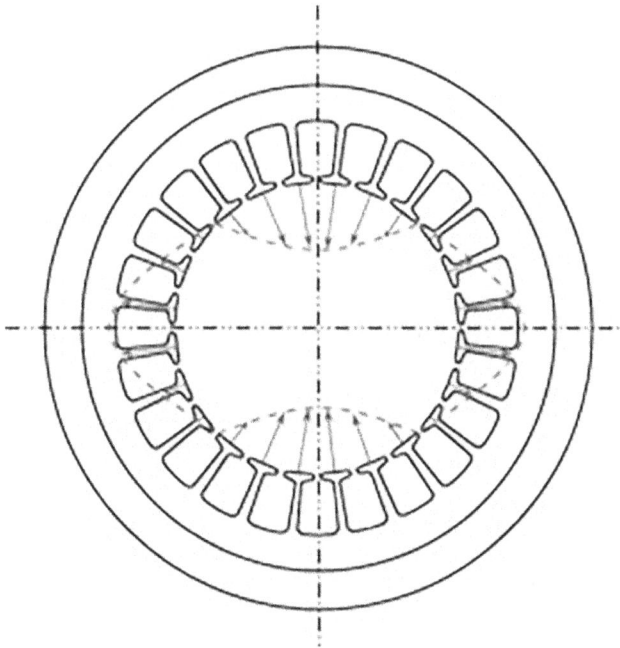

FIGURE 12.11 Stator distortion due to traveling forces as a result of rotating magnetic flux.

12.3 ELECTROMAGNETIC VIBRATIONS

Magnetic field present in an airgap is the primary source for the electromagnetic vibrations due to the variation in electromagnetic forces over time. The harmonics in the electromagnetic forces induce structural vibration and acoustic noise. The following sections describe the various causes of electromagnetic vibrations.

12.3.1 UNBALANCED FORCES/TORQUES CAUSED BY ELECTRIC SUPPLY

Stator-related faults account for the second largest number of faults, up to 38% in IMs. Stator winding faults occur when the stator windings are shorted due to the insulation issues related to manufacturing defects, contaminations, overheating and wear, collision of the rotor and stator, and voltage stress imposed by the fast switching of inverters. The unbalanced electromagnetic flux in airgap occurs on the twice line frequency.

12.3.2 BROKEN ROTOR BAR AND CRACKED END RING

For squirrel cage IMs, broken rotor bars and cracked end rings are the commonly encountered motor faults, which may be due to manufacturing defects, pulsating loads, frequent direct online starting, and thermal and mechanical stresses from heavy-duty cycle operation [12.16]. The broken rotor bar accounts for approximately 9% of all faults of IM. Due to the bar/end ring breakages, which primarily occur at the joints between bars and end rings, the inducting current in the rotor will be redistributed, leading to overheating of the adjacent bars, torque pulsation, speed fluctuation, changes in magnetic fields, noise, and vibration.

When a broken rotor bar fault occurs in an IM, sideband frequencies f_{sb} (where $f_{sb} > 0$) will appear in stator current around the supplied current fundamental frequency [12.17]:

$$f_{sb} = (1 \pm 2ks)f, \quad k = 1,2,3,\ldots \tag{12.61}$$

where s is the slip of motor.

Figure 12.12 illustrates the distribution of sideband frequencies at the sides of the fundamental line frequency. Thus, these sideband harmonics can be used to identify and detect the broken rotor bar fault.

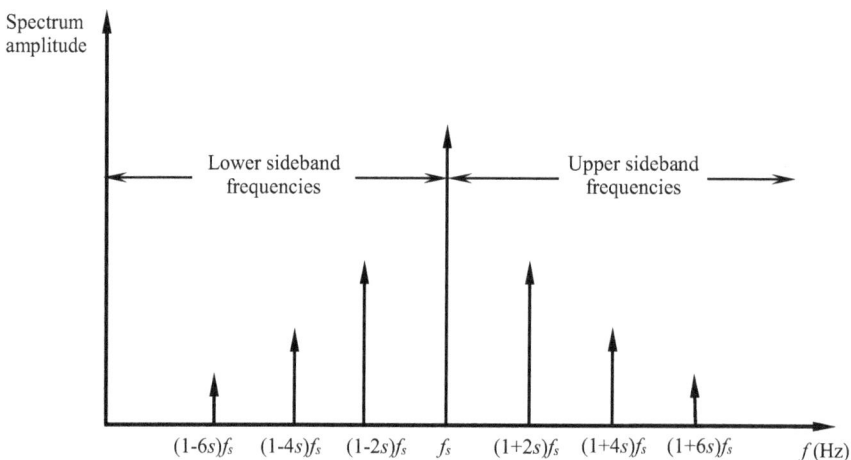

FIGURE 12.12 Sideband frequencies around the fundamental line frequency f_s.

As proposed by Thomason and Rankin [12.18], the number of broken bars n_b for full-load operation (also known as broken bar factor) can be estimated from the following equation:

$$n_b = \frac{2N_{slot}}{10^{N/20} + p} \tag{12.62}$$

where

N_{slot} is the number of rotor slots
p is the pole pair
N is the average decibel difference between upper and lower sidebands

The most important harmonics caused by the broken rotor bars f_b are as follows [12.19]:

$$f_b = (1 \pm 2ks)f, \quad k = 1,2,3,\ldots \tag{12.63}$$

where

s is the slip of motor
f is the supply frequency

12.3.3 Unbalanced Magnetic Pull Due to Asymmetric Airgap

As electric current flows through the stator windings, a magnetic field is generated in the airgap. The magnetic flux in the airgap thus produces an attractive force between the stator and rotor in the direction of the flux. In a system with a uniform airgap, the rotor is subjected to a uniform attraction force from all directions around it. Therefore, there is no resultant force acting on it due to symmetry. However, when an airgap asymmetry is present, a resultant traction force is produced on both the rotor and stator, occurring at the minimum airgap.

Airgap asymmetries are often encountered in almost all types of electric motors. These asymmetries distort the distribution of the magnetic flux density in the airgap and create unbalanced electromagnetic forces that act on both the rotor and stator. This phenomenon is known as unbalanced magnetic pull (UMP). The presence of UMP always tends to further increase the eccentricity magnitude, and in the worst case, the rotor may touch the stator to cause serious damage to the electric motor.

Asymmetric airgap is called eccentricity in electric motors. A general eccentricity form is cylindrical circular rotor whirling. In this form, the rotor axis always remains parallel to the stator axis and travels around it in a circular orbit with a certain radius and a certain angular velocity (or frequency). This motion is defined as the whirling motion and the corresponding orbit radius and angular velocity (or frequency) are defined as the whirling radius and whirling angular velocity (or whirling frequency), respectively. In dynamic whirling cases, the whirling velocity is equal to the rotor speed. An electric motor operated with the cylindrical circular rotor eccentricity is shown in Figure 12.13. The rotor rotates at a constant angular velocity ω_r. During the rotor rotating, its center O_r also rotates with respect to the stator center O_s at an angular velocity ω_w in a circular orbit.

There are two types of airgap eccentricity associated with electric motors: static eccentricity and dynamic eccentricity, depending on the magnitude of the rotor whirling angular velocity. In the case of a static airgap eccentricity, the positions of the maximum and minimum airgap are fixed in space, that is, the eccentric rotor displacement remains stationary with respect to the stator. This indicates that in this case, the rotor whirling angular velocity is zero (i.e., $\omega_w = 0$). In contrast, in the case of a dynamic airgap eccentricity, the positions of the maximum and minimum airgap rotate with the rotor, implying that the whirling angular velocity (ω_w) is the same as the mechanical angular velocity of the rotor ω_r. In either static or dynamic airgap eccentricities, the resultant force on the rotor always points to the direction of the minimum airgap, so does the resultant force on the stator but in the opposite direction.

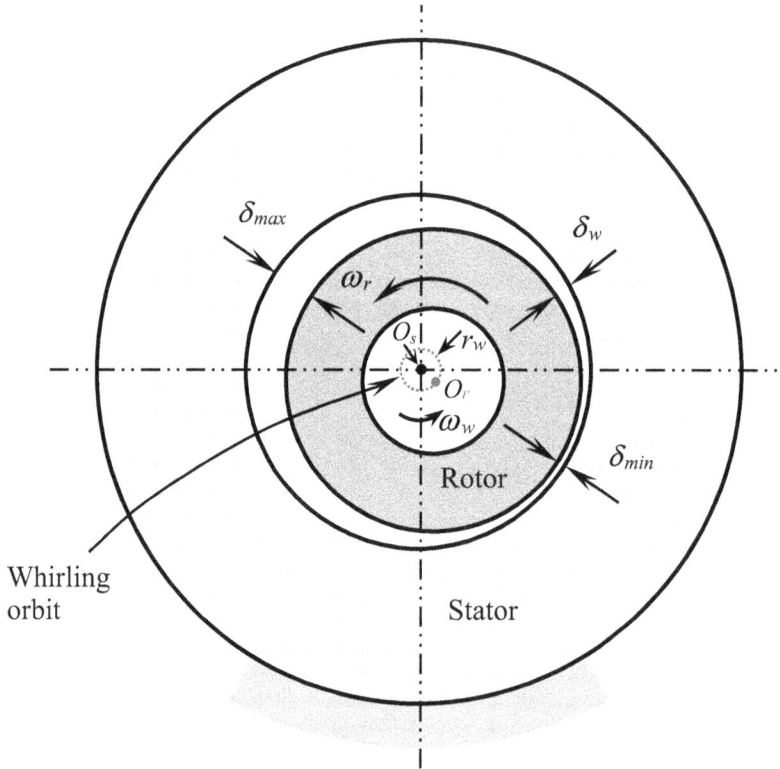

FIGURE 12.13   A general form of cylindrical circular rotor whirling.

12.3.3.1   Electromagnetic Force at Airgap

UMP has been discussed for more than a century. Many researchers have performed intensive studies to calculate the UMP for predicting vibration characteristics of rotor systems due to UMP. Früchtenicht *et al.* [12.20] developed an analytical model for predicting the electromagnetic forces between the rotor and stator for a rotor in a circular whirling motion. With their model, the stiffness and damping coefficients induced by the electromagnetic field can be determined. It was found that the electromechanical interaction can cause rotordynamic instability. Arkkio *et al.* [12.21] presented a simple parametric model for calculating the electromagnetic forces at the airgap field when the rotor performs cylindrical circular whirling motion with respect to the stator. They determined the model parameters of a motor by time stepping finite element analysis (FEA) including the nonlinear saturation of magnetic materials. In this study, the computed and measured forces show good agreement.

The responses due to certain deviations of shape in the rotor and the stator have been investigated numerically and analytically by Lundström and Aidanpää [12.22]. In their study, the perturbation on the rotor is considered to be of oval character, and the perturbations of the stator are considered triangular. In this case, the rotor whirling motion becomes much more complicated.

Regardless of the mechanism of airgap eccentricity, any eccentricity in the airgap results in variations of the distribution of the magnetic flux, magnetic flux density, and reluctance. The magnetic flux density B is defined as the magnetic flux Φ per unit area:

$$B = \frac{\Phi}{A} \tag{12.64}$$

Because the magnetic flux Φ is related to the total MMF F_m and the reluctance of the magnetic flux path R_m,

$$\Phi = \frac{F_m}{R_m} \tag{12.65}$$

the reluctance can be expressed as

$$R_m = \frac{\delta}{\mu A} \tag{12.66}$$

where μ is the permeability of air. Therefore, it gives that

$$\Phi = \frac{\mu A F_m}{\delta} \tag{12.67}$$

$$B = \frac{\mu F_m}{\delta} \tag{12.68}$$

This indicates that both the magnetic flux Φ and flux density B are inversely proportional to the length of the airgap.

There are two basic methods used to calculate the forces acting between the rotor and stator: the methods that are based on Maxwell's stress and the methods that are based on the principle of the virtual work. Derived from Maxwell's equations, the electromagnetic force at airgap can be expressed as

$$\mathbf{F} = \int_0^{2\pi} \left[\frac{1}{\mu} B_r B_\phi \mathbf{e}_\varphi + \frac{1}{2\mu_0} \left(B_r^2 - B_\phi^2 \right) \mathbf{e}_r \right] r \, d\phi \tag{12.69}$$

where B_r and $B\phi$ are the radial and tangential components of the magnetic flux density, respectively. More reliable results are obtained if the line integral in Equation 12.69 is transformed to a surface integral over the cross-section of the airgap [12.23]:

$$\mathbf{F} = \frac{1}{\delta(\phi)} \int_{S_{ag}} \left[\frac{1}{\mu_0} B_r B_\phi \mathbf{e}_\varphi + \frac{1}{2\mu_0} \left(B_r^2 - B_\phi^2 \right) \mathbf{e}_r \right] dS \tag{12.70}$$

where
S_{ag} is the cross-sectional area of the airgap
$\delta(\phi)$ is the airgap thickness at the angular position ϕ

$$\delta(\phi) = r_s(\phi) - r_r(\phi) \tag{12.71}$$

where $r_s(\phi)$ and $r_r(\phi)$ are the outer and inner radii of the airgap. It must be noted that this equation is valid only for small eccentricities [12.24].

It can be seen from Equation 12.70 that the electromagnetic force at the airgap is inversely proportional to the thickness of the airgap. This indicates that even a small change in airgap may result in a large change in electromagnetic force.

To better understand the effects of asymmetric airgaps on UMP, several causes of asymmetric airgap are addressed separately in the following sections.

12.3.3.2 Asymmetric Airgap Due to Nonconcentric Rotor and Stator

There are a variety of causes that can result in airgap asymmetries. As demonstrated in Figure 12.14, an asymmetric airgap resulted from the nonconcentric rotor and stator, which may be due to the accumulation of tolerances during manufacturing, wear of bearings and bearing bores, poor maintenance, and many other factors. This is a special case of the general eccentricity form in Figure 12.13,

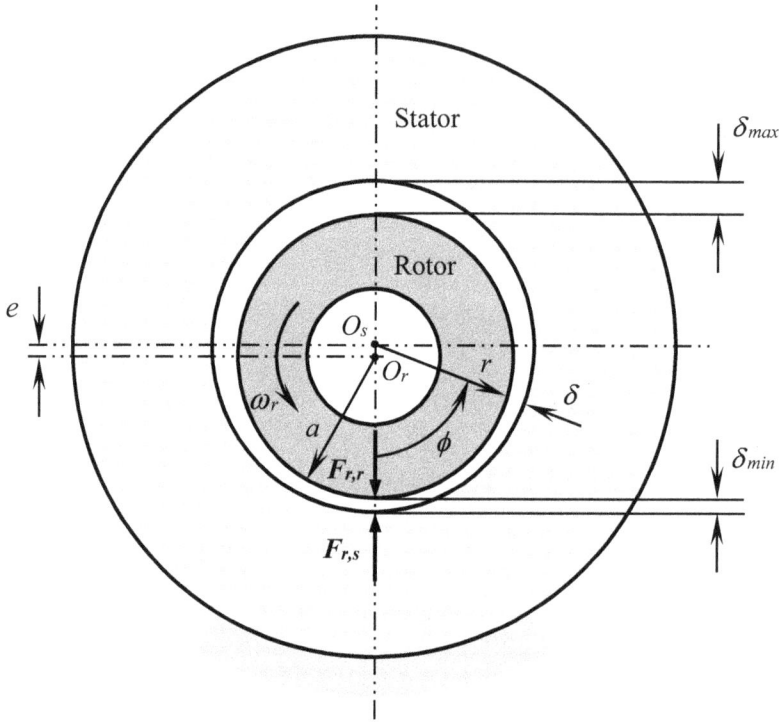

FIGURE 12.14 Asymmetrical airgap due to nonconcentric stator and rotor.

with $\omega_w = 0$ and $r_w = e$. In this case, the rotor and stator axes are parallel but not coincident. The airgap profile maintains a steady state, that is, the positions of the maximum and minimum airgap are independent of the rotor rotation.

The eccentricity between the rotor and stator is defined as

$$e = \frac{1}{2}\left(\delta_{max} - \delta_{min}\right) \tag{12.72}$$

where δ_{max} and δ_{min} are the maximum and minimum airgaps, respectively. As the mean airgap is given as

$$\delta_m = \frac{1}{2}\left(\delta_{max} + \delta_{min}\right) \tag{12.73}$$

it follows that

$$e = \delta_{max} - \delta_m = \delta_m - \delta_{min} \tag{12.74}$$

In a polar coordinate system, the general equation of a circle with the center at $(e, 0)$ and the radius of c can be expressed as

$$r_r^2 - \left(2e\cos\phi\right)r_r + e^2 = c^2 \tag{12.75}$$

Solving this equation, it yields

$$r_r(\phi) = e\cos\phi + \sqrt{e^2\left(\cos^2\phi - 1\right) + c^2} \tag{12.76}$$

It is easy to confirm that at $\phi = 0$, $r_r = c + e$ and at $\phi = \pi$, $r_r = c - e$.

Thus, the airgap thickness for a nonconcentric rotor–stator system can be obtained by subtracting the rotor radius from the stator radius in the polar coordinate:

$$\delta_{nc}(\phi) = r_s - r_r(\phi) \tag{12.77}$$

where the stator radius is a constant, expressed as

$$r_s = \delta_m + c \tag{12.78}$$

Therefore, the airgap for nonconcentric rotor and stator is a function of the angular spatial position and can be expressed as

$$\delta_{nc}(\phi) = \delta_m + c - e\,\cos\phi - \sqrt{e^2\left(\cos^2\phi - 1\right) + c^2} \tag{12.79}$$

In this case, the resultant forces acting on the rotor F_{rr} and stator F_{rs} point each other at the minimum airgap.

12.3.3.3 Asymmetric Airgap Due to Elliptic Stator

An asymmetric airgap can result from a number of manufacturing flaws, such as a noncircular stator core or a noncircular rotor.

The number of poles can significantly impact the vibration patterns. For two-pole motors, electromagnetic forces produced by fundamental flux attempt to deflect the stator into an elliptical shape. Since the stator core is mounted inside the motor housing, the restraining forces to the stator deformation primarily come from the motor housing. Unlike in the previous cases, the airgap profile is independent of the rotor rotation (Figure 12.15). The equation of the elliptic stator bore can be written as

$$\frac{r_{es}^2 \cos^2\phi}{a^2} + \frac{r_{es}^2 \sin^2\phi}{b^2} = 1 \tag{12.80}$$

Solving Equation 12.80 for r_{es}, it yields

$$r_{es}(\phi) = \frac{ab}{\sqrt{a^2 \sin^2\phi + b^2 \cos^2\phi}} \tag{12.81}$$

The rotor radius is a constant:

$$r_r = a - \delta_{max} = b - \delta_{min} \tag{12.82}$$

Similarly, the eccentricity of the elliptic stator is defined as

$$e_{es} = \sqrt{1 - \frac{b^2}{a^2}} \tag{12.83}$$

The difference between the maximum and minimum airgaps can be expressed as

$$\Delta\delta_{es} = \delta_{max} - \delta_{min} = a\left(1 - \sqrt{1 - e_{es}^2}\right) = b\left(\frac{1}{\sqrt{1 - e_{es}^2}} - 1\right) \tag{12.84}$$

Hence, the airgap is obtained by subtracting r_r from r_{es}:

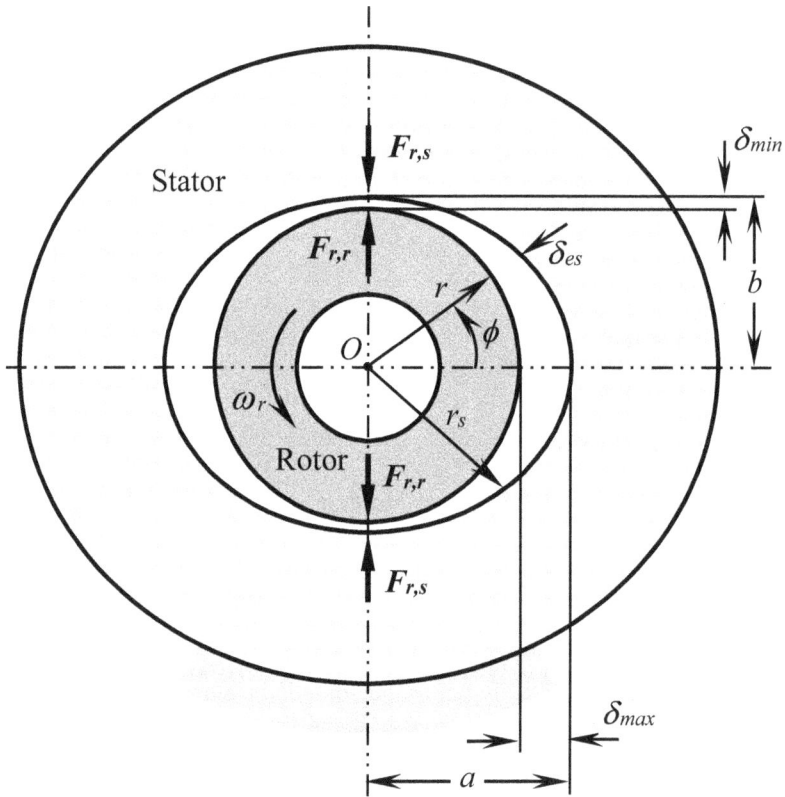

FIGURE 12.15   Asymmetrical airgap due to elliptic stator.

$$\delta_{es}\left(\phi\right) = r_{es}\left(\phi\right) - r_r = \delta_{max} + a\left(\frac{b}{\sqrt{a^2 \sin^2 \phi + b^2 \cos^2 \phi}} - 1\right) \tag{12.85}$$

For vibrations caused by unbalanced electromagnetic flux at the airgap, the frequency of vibration f equals two times the frequency of the power source, called twice line frequency vibration:

$$f = 2 f_s \tag{12.86}$$

For instance, the 60 Hz power supply shows a 120 Hz vibration at the outer surface of the stator. Finally, this twice line frequency vibration is transmitted through the motor frame to the environment.

12.3.3.4   Asymmetric Airgap Due to Elliptic Rotor

Figure 12.16 presents the asymmetric airgap due to the elliptic rotor. It is noted that the airgap profile in this case is no longer steady in the stationary coordinate system. It only maintains the steady state in the rotating frame of reference that has the same rotating speed and direction of the rotor. In fact, this is a typical example of dynamic airgap eccentricity.

In the rotating frame of reference (x', y'), a rotating ellipse can be expressed as

$$\frac{x'^2}{a^2} + \frac{y'^2}{b^2} = 1 \tag{12.87}$$

where a and b are the length of the semimajor axis and semiminor axis, respectively. To obtain the ellipse equation in the stationary frame of reference, the rotating frame of reference must be converted into a stationary frame via the conversion equation as follows:

$$\begin{bmatrix} x' \\ y' \end{bmatrix} = \begin{bmatrix} \cos(\omega_r t) & \sin(\omega_r t) \\ -\sin(\omega_r t) & \cos(\omega_r t) \end{bmatrix} \begin{bmatrix} x \\ y \end{bmatrix} \tag{12.88}$$

Thus, the rotating ellipse can be expressed in the stationary frame of reference (x, y):

$$\frac{\left[x\cos(\omega_r t) + y\sin(\omega_r t) \right]^2}{a^2} + \frac{\left[-x\sin(\omega_r t) + y\cos(\omega_r t) \right]^2}{b^2} = 1 \tag{12.89}$$

The polar coordinate system uses r as the radial coordinate and ϕ as the angular coordinate. Thus, the Cartesian coordinate system (x, y) can be related to the polar coordinate system (r, ϕ) as $x = r\cos\phi$ and $y = r\sin\phi$. Consequently, the equation of the rotating elliptic rotor can be rewritten as

$$\frac{r_{es}^2 \cos^2(\phi - \omega_r t)}{a^2} + \frac{r_{er}^2 \sin^2(\phi - \omega_r t)}{b^2} = 1 \tag{12.90}$$

The eccentricity of the elliptic rotor is defined as

$$e_{er} = \sqrt{1 - \frac{b^2}{a^2}} \tag{12.91}$$

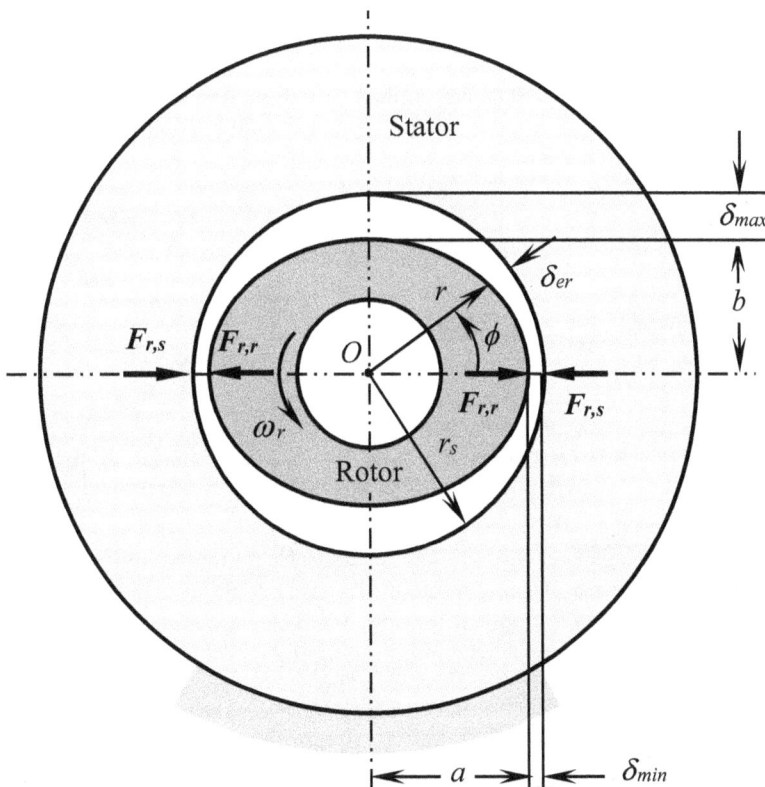

FIGURE 12.16 Asymmetrical airgap due to elliptic rotor.

From the equation of the elliptic rotor, it can be derived that the radius of the elliptic rotor is expressed as a function of ϕ:

$$r_{er}(\phi - \omega_r t) = \frac{ab}{\sqrt{a^2 \sin^2(\phi - \omega_r t) + b^2 \cos^2(\phi - \omega_r t)}} \tag{12.92}$$

This equation can be alternatively expressed in terms of eccentricity e [12.25]:

$$r_{er}(\phi - \omega_r t) = a\sqrt{\frac{1 - e_{er}^2}{1 - e_{wr}^2 \cos^2(\phi - \omega_r t)}} = \frac{b}{\sqrt{1 - e_{er}^2 \cos^2(\phi - \omega_r t)}} \tag{12.93}$$

The radius of the stator is a constant, expressed as

$$r_s = a + \delta_{\min} = b + \delta_{\max} \tag{12.94}$$

From this equation, the relationship of the difference between the maximum and minimum airgaps and the ellipse parameter a and b can be derived as

$$\Delta\delta_{er} = \delta_{\max} - \delta_{\min} = a - b \tag{12.95}$$

that is,

$$\Delta\delta_{er} = a\left(1 - \sqrt{1 - e_{er}^2}\right) = b\left(\frac{1}{\sqrt{1 - e_{er}^2}} - 1\right) \tag{12.96}$$

The airgap with the elliptic rotor is thus determined by subtracting the radius of the elliptic rotor from the inner radius of the stator r_s, as shown in the following three equivalent equations:

$$\delta_{er}(\phi - \omega_r t) = r_s - r_{er}(\varphi - \omega_r t) = r_s - \frac{ab}{\sqrt{a^2 \sin^2(\phi - \omega_r t) + b^2 \cos^2(\phi - \omega_r t)}} \tag{12.97a}$$

$$\delta_{er}(\phi - \omega_r t) = r_s - r_{er}(\varphi - \omega_r t) = \delta_{\min} + a\left(1 - \sqrt{\frac{1 - e^2}{1 - e^2 \cos^2(\phi - \omega_r t)}}\right) \tag{12.97b}$$

$$\delta_{er}(\phi - \omega_r t) = r_s - r_{er}(\varphi - \omega_r t) = \delta_{\max} + b\left(1 - \frac{1}{\sqrt{1 - e^2 \cos^2(\phi - \omega_r t)}}\right) \tag{12.97c}$$

For the case of dynamic airgap eccentricity, the resultant forces acting on the rotor $F_{r,r}$ and stator $F_{r,s}$, occurring at the minimum airgaps, rotate at the same rotating velocity of the rotor. In the present situation, because of two identical minimum airgaps occurring in the system, a pair of resultant forces $F_{r,r}$ and $F_{r,s}$ act at each of the minimum airgap, increasing the degree of the rotor eccentricity.

12.3.3.5 Asymmetric Airgap Due to Rotor Misalignment

The rotor misalignments always occur to some extent due to the need to allow some tolerances during the manufacturing process. As a misaligned rotor rotating with respect to its axis that has a small angle with the stator axis (Figure 12.17), the airgap varies along the angular position.

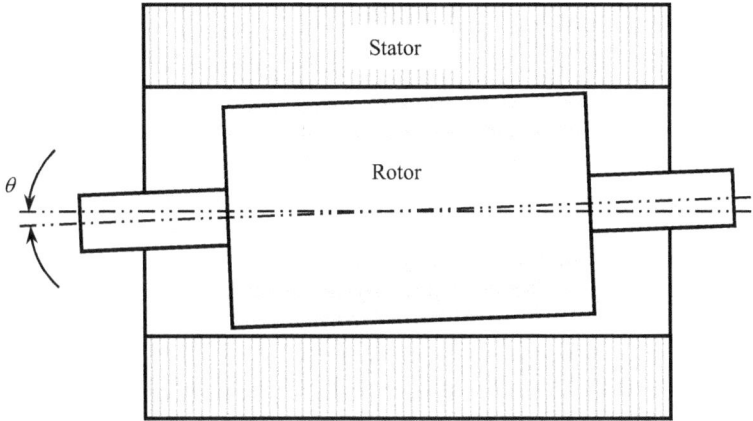

FIGURE 12.17 Asymmetric airgap resulted from the misalignment of rotor.

However, these cases are more complicated than those discussed previously, usually requiring 3D analyses.

The magnetic force is inversely proportional to the airgap. For a nonuniform airgap, the smaller the airgap, the larger the magnetic force. Regardless of the causes of airgap asymmetry, an uneven airgap always results in unbalance in magnetic forces between the stator and rotor and further results in motor vibration [12.26].

12.3.3.6 Asymmetric Airgap Resulting from Shaft Deflection

An asymmetric airgap may be induced from the rotor shaft deflection. The causes of shaft deflection may be static or dynamic in origin. The static shaft deflection is defined as the deflection that is measurable when the motor is not in operation, and the dynamic deflection is defined as the deflection that is detectable only during motor operation. A common static shaft deflection is due to the loads acting on the shaft. As a motor operates at high speeds, the centrifugal force acting on the rotor may cause the rotor shaft to bend out as a dynamic deflection (Figure 12.18).

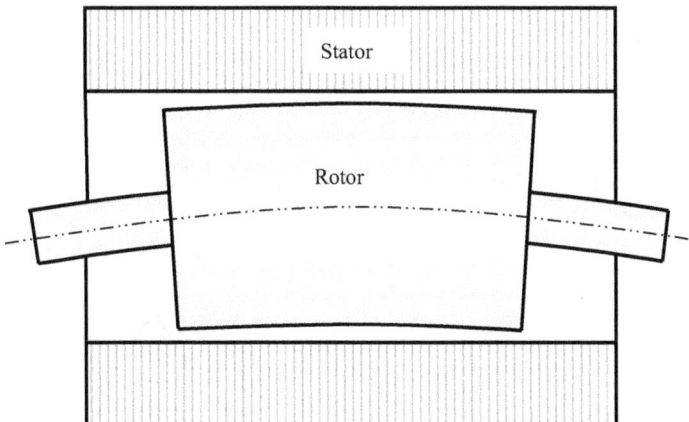

FIGURE 12.18 Asymmetric airgap resulted from the deflected motor shaft.

12.3.4 NONUNIFORM AIRGAP DUE TO STATOR SLOTS

The electric current that goes through the stator winding generates a rotating magnetic field and induces magnetic fluxes circulating around the airgap. However, the existence of the stator slots breaks up the uniformity of the airgap. Thus, the airgap will change periodically as the relative position of the stator and rotor changes.

As proposed by Lipo [12.27], due to the stator slot opening, the airgap variation is modeled considering the distribution of flux lines on the slots (Figure 12.19). The airgap rises linearly to the slot center in a slot and then drops up to its nominal value.

The *equivalent airgap* of an IM is induced by replacing the actual slotted surface with an equivalent unslotted surface, which has the same cross-section but with a modified equivalent gap. Equating the permeance of the equivalent unslotted surface to the actual permeance, the equivalent gap δ_e and the real gap δ can be related via the so-called *Carter factor* k_c:

$$\delta_e = \delta \, k_c \tag{12.98}$$

where k_c is a function of the slot geometries and the real airgap thickness. Lipo [12.27] reported that for most practical machines, δ_e is generally found to be 15%–25% larger than δ.

In practice, several situations discussed previously may occur simultaneously to cause asymmetric airgaps. For such complicated cases, FEMs may be used for predicting the unbalanced magnetic forces.

12.3.5 MUTUAL ACTION FORCES BETWEEN CURRENTS OF STATOR AND ROTOR

Vibration may come from an irregular magnetic reluctance torque existing between the teeth of the stator and rotor. When the rotor rotates, the electromagnetic force on the stator varies periodically with high frequency because of the teeth structure of salient geometry. This periodic force of high frequency brings vibration and acoustic noise [12.28].

12.3.6 VIBRATION DUE TO UNBALANCED VOLTAGE OPERATION

According to Muljadi *et al.* [12.29], in a weak power system network, an unbalanced load at the distribution lines can cause unbalanced voltage conditions. As an induction machine operates under unbalanced voltage conditions, it will result in unbalanced stator current, which creates torque pulsation on the shaft resulting in speed pulsation, mechanical vibration, and, consequently, audible acoustic noise and extra mechanical stress.

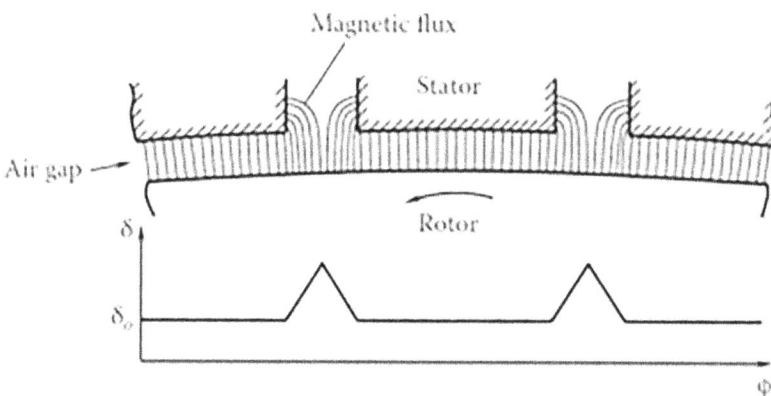

FIGURE 12.19 The impact of stator slots on magnetic flux distribution and airgap length.

12.4 MECHANICAL VIBRATIONS

Mechanical vibrations can be caused by a variety of factors, including but not limited to unbalanced mass in rotating parts, friction between motor components, misalignment, frame distortion, rotating speed at or near the critical speed, shocks caused by internal or external forces, looseness of parts, bearing deterioration, and various support and coupling effect problems.

12.4.1 MISALIGNED SHAFT AND DISTORTED COUPLING

In industry, about 30% of the machines' downtime is due to the poorly aligned machine. Rotor shaft misalignment is the common problem in operation rotating machinery. It is strongly influenced by operation speed and stiffness of the coupling. Flexible couplings tend to provide less amount of vibration levels. Unlike rotor unbalance, misalignment effects on motor vibration are more complex. The vibration induced by the misalignment between a motor shaft and a driven machine shaft varies in different directions.

Misalignment is the most common cause of machine vibration. Experimental studies were performed by Hariharan *et al.* [12.30] on a rotordynamic test apparatus to obtain the vibration spectrum for shaft misalignment. The vibration spectra show that misalignment can be characterized primarily by second harmonics (2×) of the shaft running speed.

12.4.2 DEFECTIVE BEARING

Motor bearings could be the main source of vibration. The primary causes of bearing failures include overloads, nonuniform wear, corrosion, manufacturing errors, bearing fatigue, overheating, current flowing through bearings, high friction sintered bearing material, and improper lubrication and installation.

Most small- and middle-sized motors use rolling element bearings. A rolling element bearing basically consists of an outer and inner race, rolling balls, and a cage for holding the rolling balls. Each damaged component on a bearing can result in vibration at specific frequencies [12.31, 12.32]. If balls perform only pure rolling motions between the inner and outer races, the characteristic vibrating frequencies that are based upon the bearing dimensions can be determined for each bearing component:

$$\text{Inner race } f_{inner} = \frac{N_b f_r}{2}\left(1 + \frac{d_b}{d}\cos\beta\right) \tag{12.99}$$

$$\text{Outer race } f_{outer} = \frac{N_b f_r}{2}\left(1 - \frac{d_b}{d_p}\cos\beta\right) \tag{12.100}$$

$$\text{Ball } f_b = \frac{d_p f_r}{d_b}\left[1 - \left(\frac{d_b}{d_p}\cos\beta\right)^2\right] \tag{12.101}$$

$$\text{Cage } f_{cage} = \frac{f_r}{2}\left(1 - \frac{d_b}{d_p}\cos\beta\right) \tag{12.102}$$

where
β is the contact angle
N_b is the number of balls or rollers
f_r is the frequency of rotor rotation ($f_r = n/60$, in rps)
d_b and d_p are the diameter of ball and pitch, respectively

By omitting the clearances between rolling balls and the inner and outer raceways, the pitch diameter can be expressed as

$$d_{pitch} = d_{inner} + d_{ball} = \frac{d_{inner} + d_{outer}}{2} \qquad (12.103)$$

Faults of bearing components can be detected using a cepstrum (the word is derived from spectrum by reversing the first four letters of spectrum) that reveals the periodicity of a spectrum.

12.4.3 SELF-EXCITED VIBRATION

Self-excited vibration system is a system for which the exciting force is a function of the motion parameter, such as displacement, velocity, or acceleration. The motion diverges and the system becomes unstable if energy is fed into the system through self-excitation. As the motion stops, the exciting force vanishes. By contrast, in a forced vibration, the external exciting force always exists regardless of whether the motion stops or continues. Friction-induced vibration (such as motor brakes), flow-induced vibration (such as motor ventilation flows and fluid-conveying pipelines), and aerodynamically induced motion of bridge are typical examples of self-excited vibration.

A self-excited vibration can also be excited from the internal damping of the shaft material. This happens basically in the case of assembled shafts because the friction between two parts in contact is equivalent to the internal damping of the material.

For rotating electric machines, self-excitations often produce vibrations in the rotating shaft in the lateral or flexural directions, rather than in the torsional or longitudinal directions.

12.4.4 TORSIONAL VIBRATION

Torsional vibration is one of the main concerns in some motor applications. Many rotating machineries such as motors, generators, compressors, and similar machines can be modeled as two-inertia systems, connected by a shaft that functions as a torsional spring. One inertia can be taken to represent the rotor, and another corresponds to the driven load such as fan and pump impeller. By comparing lateral vibration, torsional failures are even more sudden and serious without obvious signs. When the system is started and approaches a steady speed, the driving and load torque can twist the shaft from its free equilibrium position by a twist angle. If both the driving and load torque are suddenly removed from the system, the twisted shaft will uncoil and drive the two-inertia system oscillating in the opposite circumferential directions with respect to the shaft [12.33]. For an undamped system, the torsional vibration continues indefinitely as the system energy converts back and forth between the kinetic and potential energy. Unlike lateral vibration, torsional natural frequencies are unaffected by machine operation speeds.

In addition, because electric motors have discrete poles, their output torques do not develop smoothly but have periodic torque pulsations or torsional vibrations. These torque variations can produce periodic velocity variations or accelerations, causing torsional vibration.

12.5 VIBRATION MEASUREMENTS

Vibration can be primarily measured in three detection schemes for displacement, velocity, and acceleration: (*a*) peak-to-peak, (*b*) zero-to-peak, and (*c*) root mean square (RMS). The relationship of these parameters is displayed in Figure 12.20. Generally, acceleration emphasizes high frequencies, displacement emphasizes low frequencies, and velocity gives equal emphasis to all frequencies.

The selection of proper detectors depends on the characteristics of signals. For instance, RMS detectors are typically suitable for rotating machinery such as motors, gears, and turbines. Transient

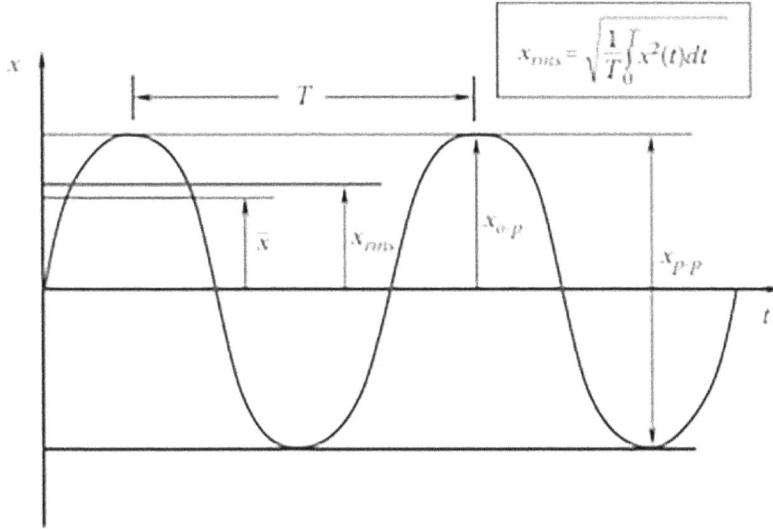

$$x_{rms} = \sqrt{\frac{1}{T}\int_0^T x^2(t)dt}$$

FIGURE 12.20 Comparison of average, RMS, zero-to-peak, and peak-to-peak displacements for a continuous sinusoidal signal.

signals have irregular properties and usually have wide frequency band and low energy. Thus, adopting a peak detector is a good choice.

Displacement is the most easily understood vibration parameter. There are several methods for measuring vibrational displacement depending upon applications. Accelerometers have long been used to measure displacement. During the measurement, the vibration levels are converted to electrical signals for data measurement by accelerometers. The benefits of using accelerometers to sense machine vibration through casing measurement include the following.

An eddy-current probe creates an alternating electromagnetic field with induced small current in the target object. Simultaneously, the eddy currents create an opposing magnetic field that resists the field being generated by the probe. The interaction of the magnetic fields is sensitive to the distance between the probe and the target object. If the distance changes, the magnetic coupling between the probe and the object is altered.

For rotating machinery, eddy-current probes are usually mounted in a casing to measure the location of the shaft relative to the casing and the amount of 1× rotational vibration. Based on the measurement, it can be determined if the shaft vibration is within acceptable limits. Eddy-current probes can also be used for monitoring shaft eccentricity and thermal expansion in the axial direction. Eddy-current probes are useful for applications requiring noncontact measurements and not sensitive to environments and materials. However, this type of probe is not suitable for some conditions that require extremely high resolutions and large gap between probe and target object.

In recent years, there has been increasing interest in loop-powered vibration sensors for their simplicity and cost-effectiveness in continuous vibration monitoring. The sensor basically measures the overall vibration in terms of velocity. The vibration signals taken by a loop-powered sensor are the same as an accelerometer.

Laser Doppler vibrometer (LDV) is one of the most efficient devices used in noncontact vibration measurements. This noncontact measuring technology is particularly useful for some applications where contacting vibration surfaces is difficult or even impossible. When a laser beam is directed to the measurement surface from an LDV, the vibration velocity, frequency, and displacement are extracted by using the Doppler effect. For some small objects, the use of an LDV is especially desirable because there is no mass loading to the measure objects, as in other contact measurement methods.

Vibration and sound signals indicate the condition or quality of operating machines. In analyzing vibration in complicated mechanical systems, the Fourier transform is a powerful tool that takes a signal as a function of time (time domain) and decomposes it into a number of harmonic components as a function of frequency (frequency domain). This indicates that a signal can be viewed either in the time domain or in the frequency domain. The frequency domain representation is also called the spectrum of the signal, which can be used to determine the source of the vibration. Similarly, the spectrum of acoustic noise from rotating machinery can be used to design noise abatement systems. With the advanced computational techniques, the fast Fourier transform (FFT), as a digital implementation of Fourier transfer, has been developed. The FFT is the basic operation in frequency analysis, which is the most commonly used analysis method in many applications. Today, FFT-based signal-processing algorithms are widely used in signal processing.

The RMS method is often used to process vibration data. The RMS value of vibration signal is a time analysis feature that is the measure of the power content in the vibration signature. It can be very effective in detecting a major out-of-balance in rotating systems. A formula for the calculation of the RMS value of a series of data is given as

$$x_{rms} = \sqrt{\frac{1}{N} \sum_{i=1}^{N} x_i^2}$$

(12.104)

where N is the amount of data. For a continuous signal $x(t)$, RMS values can be calculated as

$$x_{rms} = \sqrt{\frac{1}{N} \int_0^T x^2(t)\,dt}$$

(12.105)

where T is the period.

12.6 VIBRATION CONTROL

Vibration can be reduced by increasing either the damping capacity or the stiffness. The loss modulus is the product of these two quantities and thus can be considered a figure of merit for the vibration reduction ability [12.34].

Vibration control is necessary for many industries to improve performing characteristics and maintain the stable operation of equipment and machines. Vibration control techniques are usually classified as either active or passive. An active control system relies on a closed-loop control system that consists of feedback devices, sensors, actuators, and power supplies, increasing the system complexity, energy consumption, and total cost. As a comparison, passive vibration control is a technique that uses some kinds of vibration mitigation devices such as vibration dampers, absorbers, and isolators, without the need of any additional parts and energy input. Therefore, passive vibration systems have a diverse range of applications, especially for mitigating high-frequency vibration. Generally, active vibration control systems can offer outstanding performance over their passive counterpart. They provide functionality that is simply not possible with purely passive control systems. For instance, active vibration isolation systems can isolate the most sensitive equipment from the extremely low-frequency vibration that passive isolation systems amplify at resonant frequencies. However, active control systems are relatively more complex and more costly than passive systems. The choice between the two techniques depends on control sensitivity and effectiveness, vibration frequency, cost, and setting time. In practice, alternative feasible approaches should be evaluated to find the optimal solution for a given vibration problem.

Moreover, the integration of active elements enables adaptation of the system parameters. These adaptive or semiactive systems only require external energy for the adaptation, while the compensating forces are generated by the inertia of the mass of the absorber [12.35].

Even though the terms vibration damping and vibration isolation are often used interchangeably, they actually refer to different methods of vibration mitigation. Vibration damping is the process of absorbing or changing vibration energy to reduce the amount of energy transmitted to the equipment or machinery, while vibration isolation blocks the energy transmission path, preventing energy from entering machinery [12.36].

There are many classical forms of damping, including hysteresis (structural) damping, fluid viscosity damping, magnetorheological (MR) damping, Coulomb friction damping, air damping, particle damping, and magnetic and piezoelectric damping, to name a few. Each form of damping exhibits its own damping characteristics and is suitable for certain applications. In industrial applications, one of the most common forms of damping is passive-based viscoelastic damping, which is normally used for solving a variety of noise- and vibration-related problems.

12.6.1 Damping Materials

In practice, damping can be characterized by two damping parameters: (1) loss modulus, which is defined as the product of the damping capacity and the stiffness and (2) loss tangent, which is defined as the ratio of the imaginary part to the real part of the complex shear modulus.

Damping is inherent in all resilient and viscoelastic materials. Viscoelastic materials encompass a broad range of materials, including pressure-sensitive adhesives, epoxies, rubbers, foams, thermoplastics, enamels, and mastics. The common characteristics of these materials are that their modulus is represented by a complex quantity, possessing both stored and dissipative energy components [12.37].

As presented by Chung [12.38], damping materials fall into four categories: (a) materials exhibiting high loss modulus but low loss tangent (such as cast iron and shape-memory alloys); (b) materials exhibiting low loss modulus but high loss tangent (such as rubber and other polymers); (c) materials exhibiting low values of both loss tangent and loss modulus; and (d) materials exhibiting high values of both loss tangent and loss modulus (such as graphite networks and cement-matrix composite). High damping materials must have high values of both loss tangent and loss modulus, that is, high damping capacity and high stiffness. For metallic materials, magnesium alloys have comparatively high damping capacity, up to about 3 times that of cast iron and up to about 30 times higher than that of aluminum, and a high strength-to-density ratio. The combination of high damping, good strength, and low mass makes magnesium alloys an excellent choice for vibration damping materials [12.39].

A number of investigators have studied the vibration damping ability for a variety of materials. Based on the review of 34 reference papers, Birchak [12.40] compared the specific damping capacity of 17 metals and 2 plastics, including cast iron, steel, brasses, and magnesium alloys. Schetsky [12.41] presented the values of specific damping capacity for 20 metals, from 49% of magnesium to 0.2% of aluminum, nickel, and titanium alloys. In his article, specific damping capacity is based on the decay rate of strain energy.

Generally, damping is inversely proportional to temperature due to the crystallinity and viscosity in elastomers.

12.6.2 Vibration Isolation

Many devices and systems exist that incorporate features to counteract the effect of vibration on electric machines. A common approach to mitigate vibration of rotating machinery is to employ vibration isolation systems, which are used to filter out undesirable vibrations by modifying the vibration transmissibility. Good vibration isolation can be achieved by supporting the vibration source on a flexible low-frequency mounting. Thus, though exciting forces act on the system, only a small proportion of the forces can be transmitted to the support structure.

According to control schemes, vibration isolation systems can be categorized into passive, active, or semiactive type. Among them, passive isolating is extensively used in industries.

A typical example of a passive isolation system is a suspension system (as the vibration isolator and shock absorber) employed in a vehicle for isolating the vehicle body from the wheel axels. Incorporating vibration isolation into motor design follows the same principles. To reduce the vibration transmitted from a motor to the ground, the motor may be directly mounted on isolators. Various isolators are available today, from isolation pads to coil spring isolators with integral vibration damping materials.

As the example shown in Figure 12.21, a vibration reduction device is revealed recently in a US patent application. This device consists of a number of O-rings as the resilient vibration dampeners and an upper and lower assembly. The removable upper and lower assemblies are coupled to one another, and the O-rings are positioned between these two assemblies to provide the capability to *tune* the device to the desired stabilization condition. This feature offers the ability to reduce or isolate vibration caused by motors [12.42].

An innovative idea is to use a multilevel magnetic system for vibration isolation [12.43]. The system includes upper and lower magnetic structures and transitions between an attract mode and a repel mode when the upper and lower magnetic structures are separated by an equilibrium separating distance. The multilevel magnetic system is placed between two objects and configured to oscillate about the equilibrium separating distance in response to vibration from a motion source.

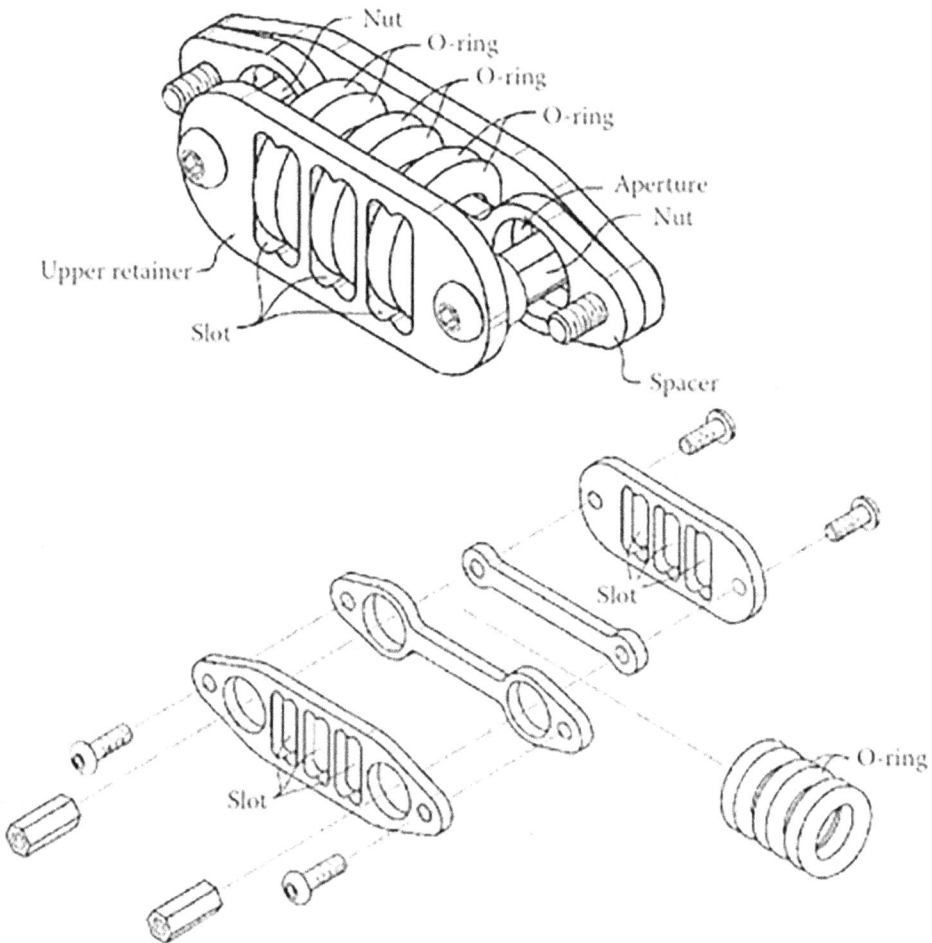

FIGURE 12.21   Vibration isolation system used to attenuate machinery vibration (U.S. Patent Application 13/554,932) [12.42]. (Courtesy of the U.S. Patent and Trademark Office, Alexandria, VA.)

The oscillation about the equilibrium separating distance causes the magnetic system to function as a low-pass filter that substantially attenuates vibration above the cutoff frequency, thus limiting the conduction of the vibration between the two objects.

Figure 12.22 depicts a disk-shaped repel-snap multilevel magnetic system. As shown in the figure, this system consists of upper and lower magnetic structures. Each magnetic structure has a region of coded maxels and a region of polarity. Each maxel has a code or pattern of positive or negative polarity. The maxels are in seven rows for each structure, and each maxel for each row in

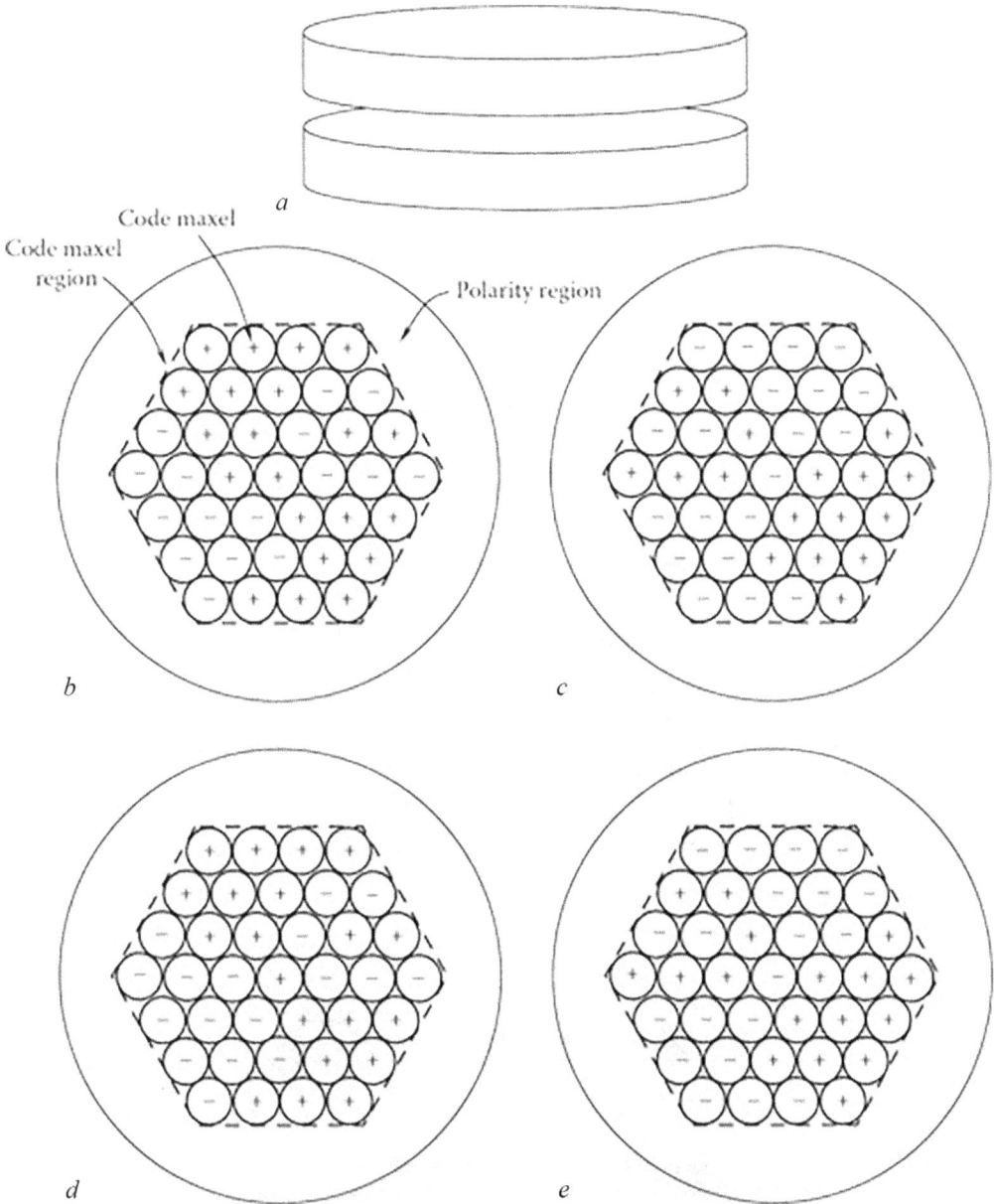

FIGURE 12.22 Multilevel magnetic system: (*a*) disk-shaped repel-snap magnetic system separated by an equilibrium separating distance, (*b*) a top view of the upper magnetic structure, (*c*) a bottom view of the upper magnetic structure, (*d*) a top view of the lower magnetic structure, and (*e*) a bottom view of the lower magnetic structure.

one structure has polarity that is opposite a corresponding maxel in corresponding row in another structure. When a maxel is magnetized entirely through a magnetizable material, the opposite polarities occur at the two sides of the maxel because each maxel is essentially a dipole magnetic source. Similarly, the polarity regions of the two structures have opposite polarities since they are also a dipole magnetic source. When the attractive force equals the repelling force, the system is configured to oscillate about the equilibrium separation distance in response to vibration.

As demonstrated in Figure 12.23, a mechanical isolation system comprises four mechanical isolators comprising the contactless attachment multilevel magnetic system. These mechanical isolators are posited between the first and second objects to control mechanical impedance between the two objects.

Viscoelastic material damping pads have been designed for passive vibration damping. As shown in Figure 12.24, a laminated damping pad consists of thin viscoelastic films (0.05–0.3 mm in thickness) sandwiched between stainless steel plates (about 1.8 mm in thickness). Among various viscoelastic materials, polymers are high-energy dissipative materials that have been widely used for many years in automotive, aerospace, and electronic industries for solving complex vibration and noise problems.

Viscoelastic materials are characterized by their shear modulus G and loss factor η, which is defined as the ratio of vibration energy dissipated per cycle to the energy stored in all structural

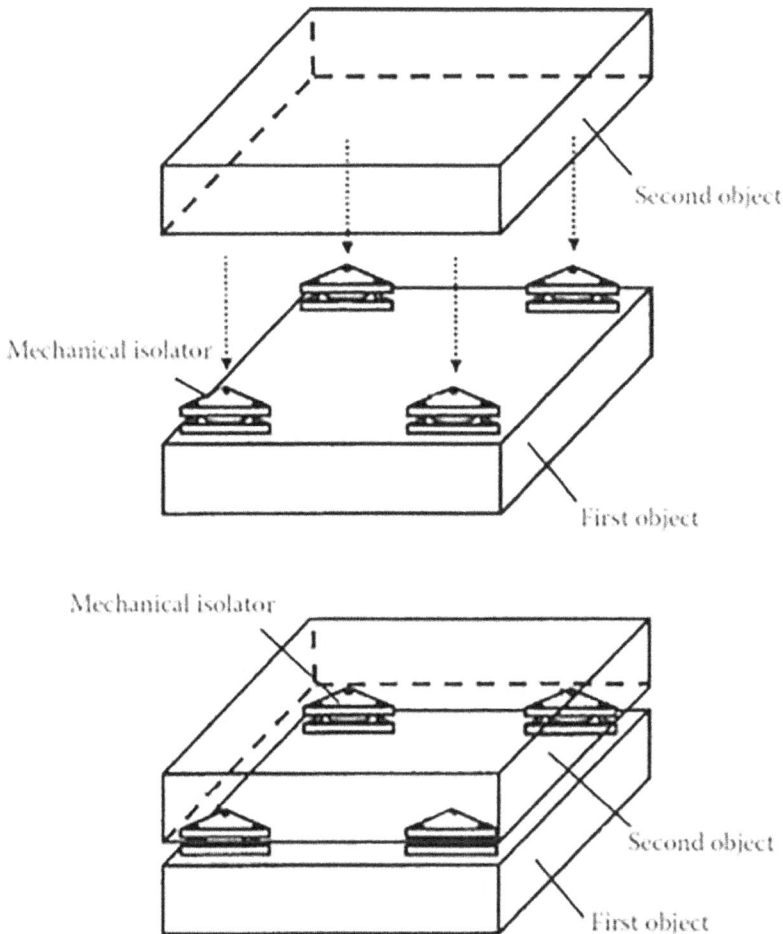

FIGURE 12.23 Mechanical isolation systems assembled into the multilayer magnetic system (U.S. Patent 8,279,031) [12.43]. (Courtesy of the U.S. Patent and Trademark Office, Alexandria, VA.)

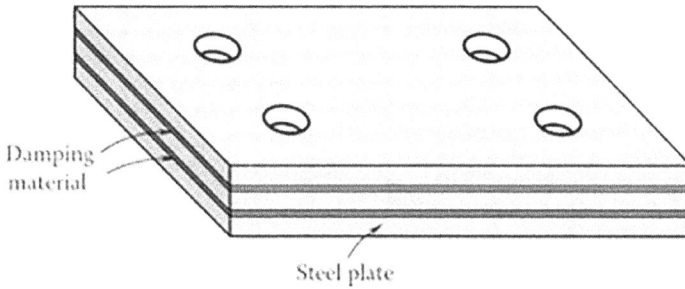

FIGURE 12.24 Laminated damping pad for providing effective damping of an electric machine.

elements. It has been found that the performance of damping pads depends not only on the material properties (such as G and η) but also on the machine's complex interaction with the structural system to be damped. Results from Magra *et al.* [12.44] show that damping pads can reduce the vibration amplification factor (which is the reciprocal of η) from over 100 to 8 for an advanced photon source machine.

Molded and bonded rubber is often used as machinery mounts to absorb shock and attenuate vibration.

12.6.3 Magnetorheological Damper and Machine Mount

MR fluid is a type of smart material that responds instantly to varying levels of a magnetic field precisely and proportionally for controllable energy-dissipating applications, such as dampers, shock absorbers, machine mounts, automotive suspension systems, brakes, MR actuators, and MR elastomer (MRE) sensors. In the last several decades, MR materials have gained great attention of researchers and engineers significantly because of their salient properties and potential applications to various fields. Their applications have been extensively reviewed by Ahamed *et al.* [12.45] to promote the practical use of MR materials in a wide spectrum of applications.

With suspended microscale magnetizable particles in a carrier fluid (*e.g.*, mineral or synthetic oil), the MR fluids can be switched from a free-flowing liquid phase to a semi-solid phase and vice versa upon applying or removing the external magnetic field. Altering the strength of the applied magnetic field precisely and proportionally controls the yield strength of the fluid. It has been widely accepted that the optimum particle size is in the range of 0.1–10 µm [12.46].

The advantages of MR fluid-based systems include their fast response and precise controllability, making them suitable for applications with high sensitivity and controllability requirements. Nowadays, MR fluid-based systems are rapidly growing and widely being used as vibration dampers, suspension bushings, machine mounts, vibration isolators, and shock absorbers in many industries such as automotive, robot, energy, defense, and civil.

MR fluid is utilized in one of three main modes of operation, which are flow (valve) mode, shear mode, and squeeze-flow mode. These modes involve fluid flowing because of the pressure gradient between two plates. In all cases, the magnetic field is perpendicular to the plates to restrict fluid in the direction parallel to plates, as shown in Figure 12.25.

Commercial MR fluid dampers have been developed in the middle of the 1990s. The small real-time controlled MR dampers were initially introduced in 1998 for heavy-duty trucks and construction vehicles as shock absorbers for vehicle drivers. In the same year, controllable MR fluid-based primary suspension shock absorbers were successfully adopted to racecars. Today, as a passive type damping mechanism, a variety of MR dampers have been used in vibration suppression applications. Their ability to precisely control the damper with less power requirement and high durability with a wide operating temperature range has made them suitable for various engineering applications. Focused on vibration isolation of rotary shaft, Zhu [12.47] has developed a disc type of MR

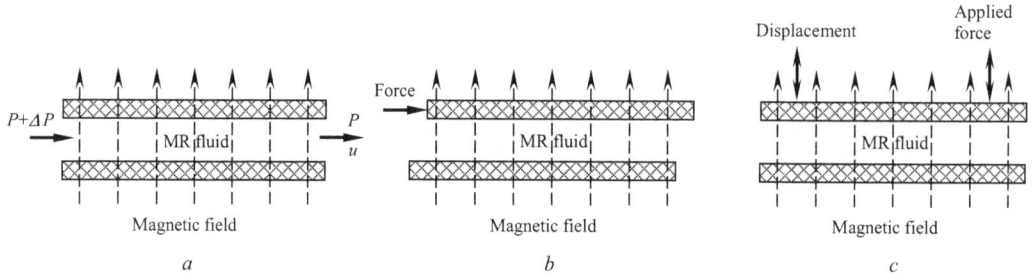

FIGURE 12.25 Three main operational modes of MR fluid: (*a*) flow mode, where both the walls are fixed and the fluid is forced to flow by a pressure gradient; (*b*) shear mode, where one wall is fixed and another wall moves with respect to the fluid, creating shear force in the fluid; and (*c*) squeezing flow mode, where the force or pressure is applied on the top plate and the fluid is squeezed between two plates.

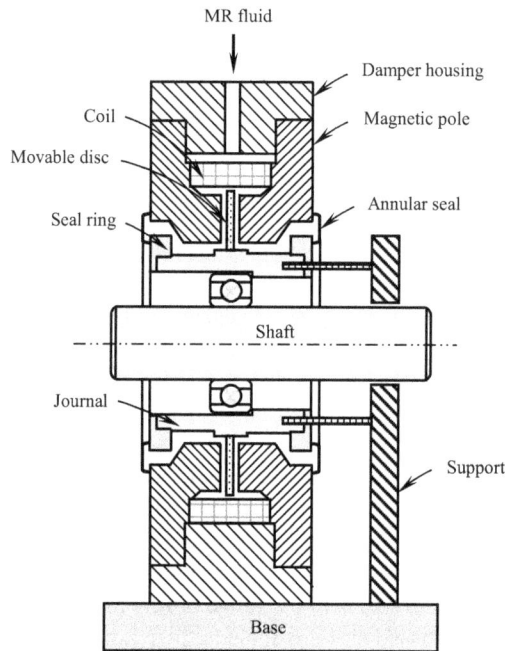

FIGURE 12.26 A disc-type MR fluid damper, which consists of a movable disc mounted on a journal outside of the ball bearing, two magnetic poles, a coil wound circumferentially, and damper housing, which form the magnetic path with two identical axial gaps.

fluid damper based on shear mode operation for rotary machinery. This damper can be precisely controlled by the application of an external magnetic field produced by a low-voltage electromagnetic coil and effectively attenuate the rotor vibration (Figure 12.26).

A novel multi-coil MR damper with a variable resistance gap was proposed by Zheng *et al.* [12.48]. Enabling four electromagnetic coils with individual exciting currents, a simplified magnetic equivalent circuit is proposed, and the magnetic flux generated by each voltage source passing through each active gap was calculated as vector operations. Compared to a single-coil MR damper, the MR damper with multi-coil is designed to increase the maximum resistance force by elongating the gap (Figure 12.27*a*). However, the elongated gap decreased the dynamic ratio to some extent. To address this issue, a multi-coil MR damper with a variable gap is designed and optimized for improving dynamic characteristics and overall performance of the multi-coil damper, as shown in Figure 12.27*b*.

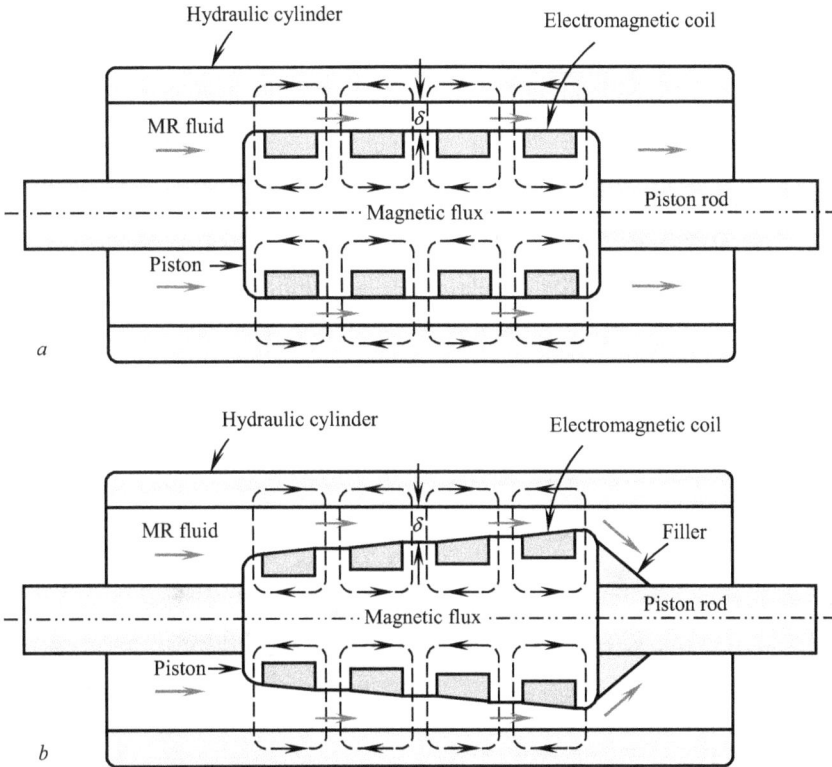

FIGURE 12.27 Schematic configurations of double-ended MR dampers with four electromagnetic coils: (*a*) typical multi-coil MR damper with a constant gap δ between the piston and hydraulic cylinder and (*b*) new multi-coil MR damper with a variable gap δ. This design is to enlarge the maximum damping force as well as the dynamic ratio.

Compared to the dampers based on the flow and shear modes, the squeeze mode dampers are rather limited. However, this situation has been changed recently. The design and development of MR fluid dampers based on the squeeze mode have been attracted by many researchers and engineers, as the result of the increased demands for large force-displacement ratio and high yield stress. Practically, in squeeze mode, MR fluids can generate a large range of force associated with a small displacement.

An MR fluid damper in squeeze mode is shown in Figure 12.28 [12.49]. In this damper, the squeezed disc is attached to the piston rod and placed at the center of the cavity that is filled with an MR fluid. Two elastomeric diaphragms amount separately at the ends of the piston. A ring-shaped electromagnetic coil is installed at the outer surface of the cylindrical cavity. When the piston rod is moving horizontally, the squeezed disc fixed moves accordingly in the cavity and squeeze the MR fluid. The damping force is thus controlled by changing the current in the coil to produce a suitable magnetic field for the damper.

In the design of MR mount systems, three operating modes of MR fluid, *i.e.*, flow mode, shear mode, and squeeze flow mode, are normally adopted. A variety of design configurations of high loaded MR mounts have been analyzed by Do and Choi [12.50]. Practically, high loaded MR mounts are designed based on two main modes of MR fluid, i.e., flow mode and shear mode. As a comparison, the squeeze flow mode is still difficult in the design configuration of MR mount due to the low damping force and leakage of the fluid itself. The structure of the mount determines the establishing pressures, which are directly related to the damping force of the mount. Various high loaded mounts are depicted in Figure 12.29. As demonstrated in Figure 12.29a, the MR valve structure with both annular and radial flows is employed to create a high damping force.

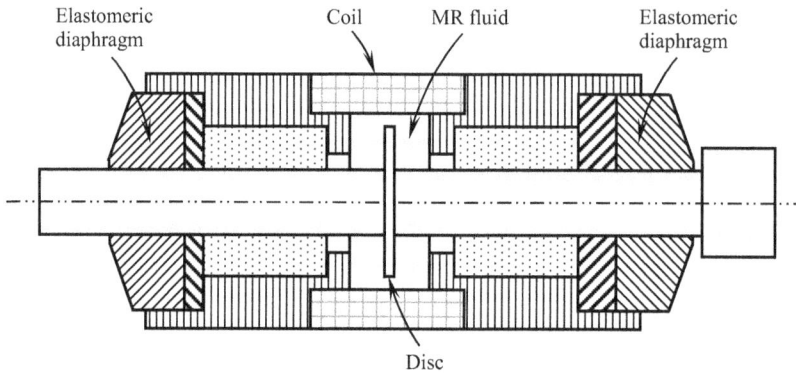

FIGURE 12.28   Schematic diagram of MR fluid damper in squeeze mode.

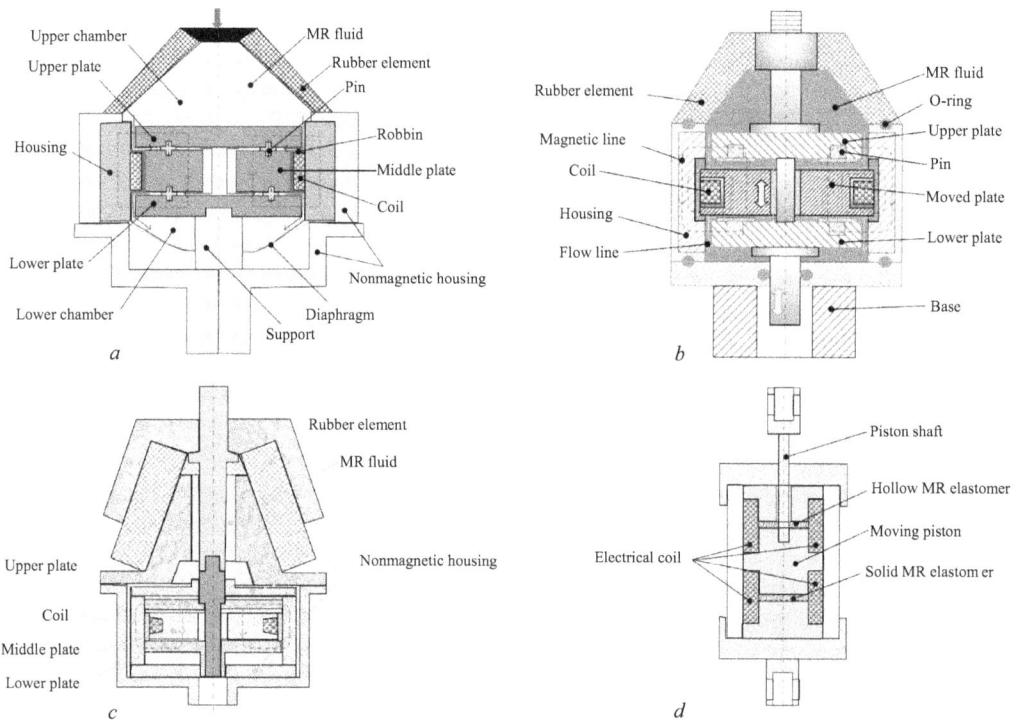

FIGURE 12.29   Various configurations of high loaded MR mounts: (*a*) the MR mount with fixed middle core, (*b*) the MR mount with a moved plate to avoid the block-up phenomenon for keeping sufficient damping force of the mount during operation, (*c*) the MR mount with the advantage to, and (*d*) the MRE mount with hollow and solid MRE plates. The advantage of this structure is to prevent the leakage of MR fluid with simple structure.

The MR valve structure separates the mount into two parts: the upper and the lower part. The upper part consists of a rubber element, an upper base that is fastened to the valve body to support the rubber element. The lower part consists of the mount body, the supporter on which the core of the MR valve structure is placed, and the rubber diaphragm [12.51]. Along with the MR fluid, rubber elements are used to effectively improve vibration control, especially in the low-frequency range. To prevent the block-up phenomenon of MR fluid, the middle core in Figure 12.29*a* is replaced by a moved plate to keep sufficient damping force of the mount during operation, as shown in

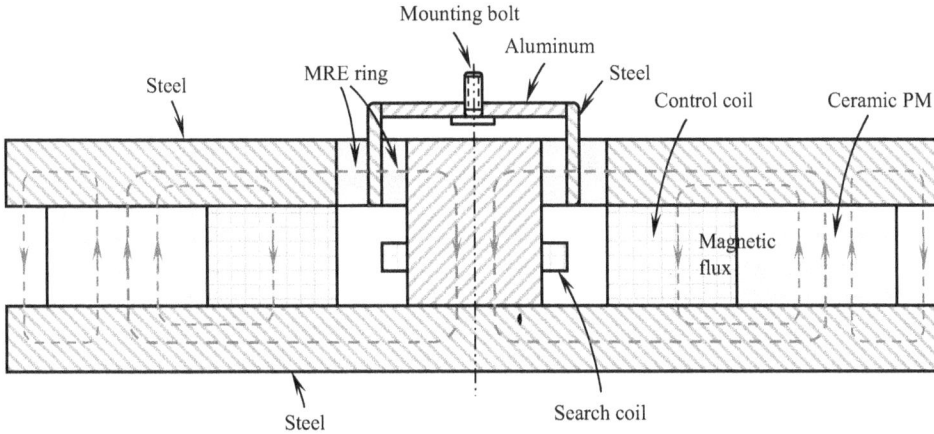

FIGURE 12.30 Cross-section of a MRE vibration isolator, which is designed to support a vertical load and isolating it from vertical and horizontal disturbances. The device uses two molded concentric MRE rings. The magnetic fluxes, as well as their directions, are shown by the dashed lines that form closed loops.

Figure 12.29*b* [12.52]. A similar structure of high loaded MR mount was presented by Kang et al. [12.53]. The configuration of this mount is shown in Figure 12.29*c*. A new kind of MR mount based on elastomer plate was studied by York *et al.* [12.54]. The structure of MRE mount and the elastomer plate is depicted in Figure 12.29*d*. The advantage of this structure is to prevent the leakage of MR fluid with the simple structure. However, the damping force of this device is relatively low and hence its use is restricted to low load applications.

MREs, like MR fluids, exploit magnetic forces between dispersed micron- or nano-sized ferromagnetic particles to produce a material with instantaneously adjustable properties. However, in MR fluids, the particles are dispersed within a liquid, operating in a post-yield regime, while in MREs the particles are a part of a structured elastomer matrix in a pre-yield regime [12.55].

An MRE vibration isolator has been developed and tested in real time with a semiactive controller to support vertical loads, as depicted in Figure 12.30 [12.56]. Two molded concentric MRE rings are utilized in the device. A ceramic PM and an electromagnetic coil are used together to generate a magnetic field to the MRE for precisely and rapidly controlling the magnetic flux density passing through the MRE. The use of ceramic PM is not only to provide a bias flux, but also to minimize eddy current in the PM. In this design, the magnetic flux control is achieved with a proportional integral derivative feedback loop, where the flux in the MRE is calculated by a search coil wrapped around the device pole. The experimental results show that the on-off controlled device can offer improved resonance control and velocity isolation when compared to the considered passive systems.

12.6.4 TUNED MASS DAMPER

Rotating machines such as motors and engines often incite vibration due to rotational imbalances. A tuned mass damper (TMD) is a passive damping device that utilizes the secondary mass attached to the primary structure to reduce the dynamic response of the primary structure. Figure 12.31 depicts the schematic of a two-degree-of-freedom TMD system. A primary system (*i.e.*, a motor) consists of a mass m_p, a spring (with spring coefficient k_p), and a viscous damper (with damping coefficient c_p). A TMD, or harmonic absorber, consists of a mass m_d, a spring (with the spring coefficient k_d), and a viscous damper (with damping coefficient c_d). The TMD is attached to the primary system to reduce or eliminate the undesirable vibration of the primary system. Usually, m_A is chosen approximately 5%–10% of the system mass m_p. To obtain the best damping effect, TMDs set their

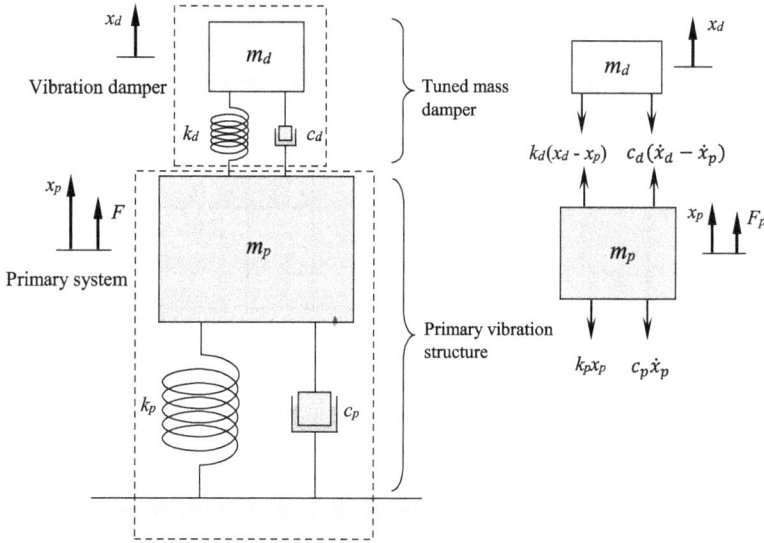

FIGURE 12.31 Two-degree-of-freedom TMD system; TMD is attached to the primary system for reducing the dynamic response of the primary system.

natural frequencies substantially equal to the natural frequencies of the primary structures. It has been proven that the TMD is effective in reducing the response of structures to harmonic or wind excitations [12.57].

From the free-body diagram of the mass m_p and m_d, the equations of motion of the system can be written as

$$m_p \frac{d^2 x_p(t)}{dt^2} + (c_p + c_d)\frac{dx_p(t)}{dt} - c_d \frac{dx_d(t)}{dt} + (k_p + k_d)x_p - k_d x_d = F \tag{12.106}$$

$$m_d \frac{d^2 x_d(t)}{dt^2} + C_d \frac{dx_d(t)}{dt} - C_d \frac{dx_p(t)}{dt} + k_d x_d - k_d x_p = 0 \tag{12.107}$$

These two equations of motion can be expressed in the matrix form

$$\begin{bmatrix} m_p & 0 \\ 0 & m_d \end{bmatrix}\begin{Bmatrix} \ddot{x}_p \\ \ddot{x}_d \end{Bmatrix} + \begin{bmatrix} c_p + c_d & -c_d \\ -c_d & c_d \end{bmatrix}\begin{Bmatrix} \dot{x}_p \\ \dot{x}_d \end{Bmatrix} + \begin{bmatrix} k_p + k_d & -k_d \\ -k_d & k_d \end{bmatrix}\begin{Bmatrix} x_p \\ x_d \end{Bmatrix} = \begin{bmatrix} F \\ 0 \end{bmatrix} \tag{12.108}$$

It can be seen that $[m]$, $[c]$, and $[k]$ are all 2×2 matrices whose elements are the known masses, damping coefficients, and stiffnesses of the system, respectively. Because each of these matrices is symmetric, the transpose of the matrix is equal to its original matrix, that is, $[m]^T = [m]$, $[c]^T = [c]$, and $[k]^T = [k]$.

The external exciting force F is considered to vary harmonically with the frequency ω:

$$F = F_o\, e^{i\omega t} \tag{12.109}$$

The solutions for the equations of motion may take the form

$$\begin{cases} x_p(t) = X_p e^{i\omega t} \\ x_d(t) = X_d e^{i\omega t} \end{cases} \tag{12.110}$$

Substituting Equations 12.109 and 12.110 into the set of governing equation results in

$$
\begin{bmatrix} -m_p\omega^2 + (c_p + c_d)i\omega + (k_p + k_d) & -c_d i\omega - k_d \\ -c_d i\omega - k_d & -m_d\omega^2 + c_d i\omega + k_d \end{bmatrix} \begin{Bmatrix} X_p \\ X_d \end{Bmatrix} = \begin{bmatrix} F_o \\ 0 \end{bmatrix} \tag{12.111}
$$

Introducing the natural frequency ω and damping ratio ζ for the primary system and tuned mass system, respectively,

$$
\omega_p = \sqrt{\frac{k_p}{m_p}} \tag{12.112}
$$

$$
\zeta = \frac{c_p}{2\sqrt{m_p k_p}} \tag{12.113}
$$

$$
\omega_d = \sqrt{\frac{k_d}{m_d}} \tag{12.114}
$$

$$
\zeta_d = \frac{c_d}{2\sqrt{m_d k_d}} \tag{12.115}
$$

and defining m as the mass ratio,

$$
\bar{m} = \frac{m_d}{m_p} = \frac{k_d}{k_p}\frac{\omega_p^2}{\omega_d^2} \tag{12.116}
$$

the equations of motion can be rewritten in the matrix form as

$$
\begin{bmatrix} -\omega^2 + 2(\zeta_p\omega_p + \zeta_d\omega_d\bar{m})i\omega + (\omega_p^2 + \omega_d^2\bar{m}) & (-2\zeta_d\omega_d i\omega - \omega_d^2)\bar{m} \\ -2\zeta_d\omega_d i\omega - \omega_d^2 & -\omega^2 + 2\zeta_d\omega_d i\omega + \omega_d^2 \end{bmatrix} \begin{Bmatrix} X_p \\ X_d \end{Bmatrix}
$$
$$
= \begin{bmatrix} F_o/m_p \\ 0 \end{bmatrix} \tag{12.117}
$$

As mentioned previously, to obtain the best damping effect, the natural frequency of the mass damper ω_d is equal to the natural frequency of the primary system ω_p

$$
\omega_p = \omega_d = \omega \tag{12.118}
$$

Substituting Equation 12.118 into 12.117, it yields

$$
\begin{bmatrix} 2(\zeta_p\omega^2 + \zeta_d\omega^2\bar{m})i + \omega^2\bar{m} & (-2_d\zeta\omega^2 i - \omega^2)\bar{m} \\ -2_d\zeta\omega^2 i - \omega^2 & 2\zeta_d\omega^2 i \end{bmatrix} \begin{Bmatrix} X_p \\ X_d \end{Bmatrix} = \begin{bmatrix} F_o/m_p \\ 0 \end{bmatrix} \tag{12.119}
$$

Hence, X and X_d are determined as

$$X_p = \frac{F_o}{m_p} \frac{C}{AC - B^2 \bar{m}}$$ (12.120)

$$X_d = \frac{F_o}{m_p} \frac{B}{AC - B^2 \bar{m}}$$ (12.121)

where

$$A = 2\left(\zeta_p\omega^2 + \zeta_d\omega^2\bar{m}\right) + \omega^2\bar{m}$$

$$B = -2\zeta_d\omega^2 i + \omega^2$$

$$C = 2\zeta_d\omega^2 i$$

The solutions of the motion equations are thus obtained by substituting Equations 12.120 and 12.121 into 12.110.

TMDs commonly use metal coil springs and viscous dampers. They can also use other types of dampers, such as viscoelastic, air-suspended, sloshing water, and liquid column dampers. When designed, fabricated, installed, and tuned properly, TMDs can effectively absorb the vibrating energy of the structure and dissipate energy internally and hence reduce structural vibrations of machines or buildings. The 508 m (1,667 ft) supertall *Taipei 101 building* (台北101) installed a 730 metric ton (0.26% of the building mass) TMD. The 5.49 m (18 ft) diameter, 660 metric ton steel sphere is suspended by eight cables (90 mm in diameter) in the upper stories of the tower and is visible between the 88th and 92nd floors (Figure 12.32). It is also the first-ever constructed key architectural and visible element in the building [12.58].

TMDs can be essentially viewed as energy sinks. It was reported that with the use of a dynamic vibration absorber on a lightly damped system that has steplike trajectories or disturbances, the settling time has been greatly improved by 82% [12.59].

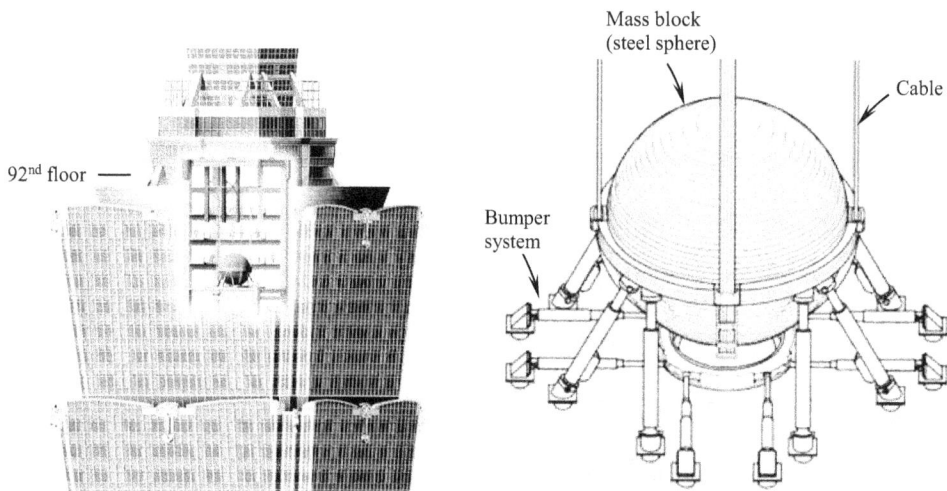

FIGURE 12.32 A TMD is used in supertall Taipei 101 building to restrain sway and maintain structural integrity.

12.6.5 Double Mounting Isolation System

Greater attenuation of the exciting force at high frequencies can be achieved by using a double mounting, also known as two-stage mounting. In this arrangement, the machine is set on flexible mountings on an inertia block, which is supported by flexible mountings [12.60].

Figure 12.33 demonstrates the schematic of a two-degree-of-freedom double mounting system. The system consists of a primary mass m_p (machine mass) and an auxiliary mass m_a (damper mass). Dampers are placed between the primary m_p and auxiliary mass m_a and between the auxiliary mass m_a and the base. The purpose of using the system is to isolate either the base from the vibration of the machine or the machine from the vibration of the base (e.g., during an earthquake). The system is also used where there is a demand for high structure-borne noise attenuation.

While the passive isolating systems are very effective in lowering the vibration transmission at high frequencies, they suffer from some drawbacks including design complexity, weight penalty, large volume, and high cost. Without any constraints, a double mounted system has totally 12 degrees of freedom, twice of the unconstrained single mounted system. Accordingly, it has 12 resonant frequencies. As a matter of fact, keeping all resonant frequencies away from forcing frequencies of a machine is a major design challenge for design engineers.

12.6.6 Viscoelastic Bearing Support

In many cases, the rotor vibratory loads are the dominant source of motor vibration. The mechanism of rotor vibration has been extensively studied and is well understood. As discussed previously, a variety of factors can impact the vibration amplitude and frequency such as rotor mass imbalance, broken rotor bar, and UMP, to name a few. Rotor structural vibration is transmitted through the bearings and then to the stator, motor housing, and finally to the motor base.

To interrupt the transmission path of rotor vibration, it is highly desired to design a viscoelastic bearing support that has a viscoelastic layer between the outer ring of a rolling bearing and the motor endbell [12.61], as demonstrated in Figure 12.34. In this way, vibration of the rotor is isolated from the surrounding structure. Moreover, the viscoelastic layer provides additional damping to the vibrating rotor. However, a relatively soft viscoelastic bearing support can reduce the static stiffness of the machine to influence the machine operation accuracy.

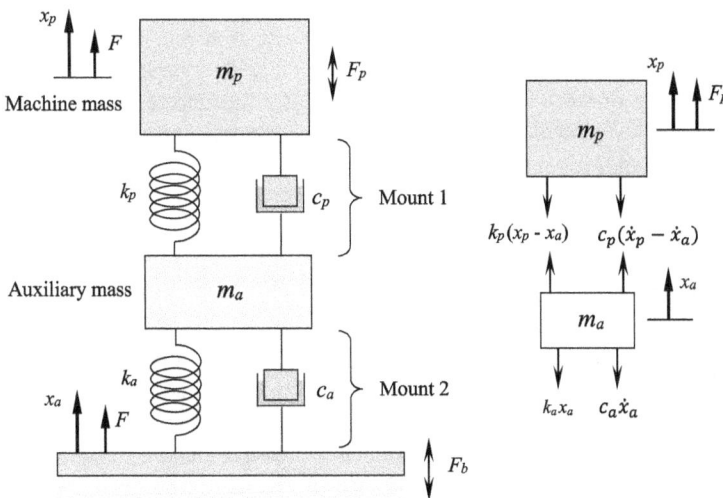

FIGURE 12.33 Two-degree-of-freedom double mounting damping system.

Viscoelastic layer   Endbell

FIGURE 12.34 Viscoelastic layer positioned between the bearing outer ring and motor endbell for isolating rotor vibration from the motor and its base.

12.6.7 Self-Locking Fastener

Threaded fasteners are extensively used in various machines and structures due to their desired capabilities, such as the ability to develop firm clamping forces, and the ease of disassembly without component damage [12.62], which is particularly advantageous for machine repair and maintenance. Loosening of threaded fasteners due to vibration, shock, or cyclic thermal loading may lead to decreasing clamping force and, thereby, the joint failure. There are two main causes for the bolted joint failure: spontaneous bolt loosening and slackening. To effectively address these issues, many anti-loosening fastening techniques have been developed over the past few decades.

Self-locking fasteners are designed to provide ideal solutions to prevent mating hardware from loosening in service due to dynamic forces or other application-related factors. To satisfy a wide range of applications, the self-locking fasteners have come with a variety of types and features.

There are two types of self-locking fasteners: prevailing torque fastener and free-running fastener. While the prevailing torque type is designed to create or increase substantial friction during assembly to resist loosen torque, the free-running fastener usually relies on additional components such as self-lock washers (*e.g.*, Nord-lock washer) or serrated face nuts. Among the prevailing torque fasteners, the most famous type is the hard lock nuts.

12.6.7.1 Hard Lock Nut

The concept of hard lock nut was first created and implemented by a Japanese inventor Katsuhiko Wakabayashi (若林克彦) in the early 1970s [12.63]. Through a long slog of trial and error, Wakabayashi successfully converted his idea into actual products and was awarded a number of patents in several countries [12.64–12.66]. Unlike conventional nuts, the hard lock nut utilizes a unique wedge principle to create a powerful self-locking force that offers a secure fastening solution for extreme vibration environments. As shown in Figure 12.35, a hard lock nut assembly is composed of two different shaped nuts: the concave nut (upper) which has a perfectly spherical concavity and the convex nut (lower) which has a small eccentricity e, with a thin wedge on one side and a thick wedge

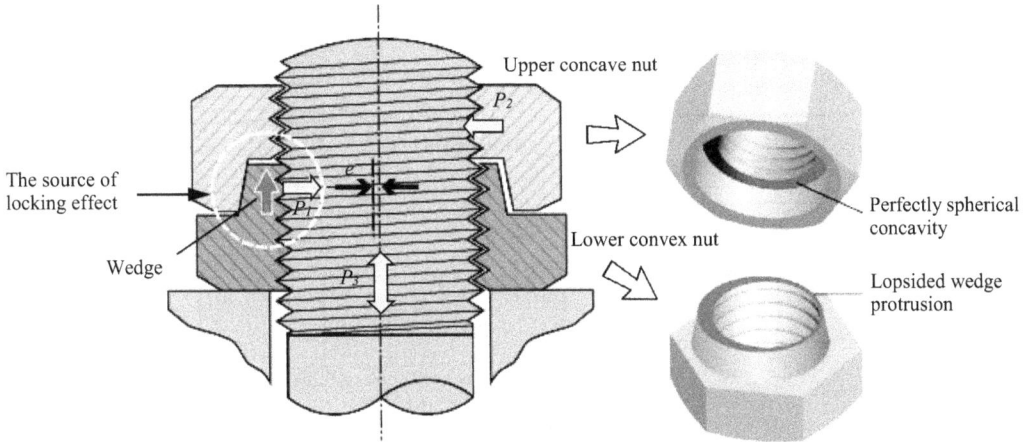

FIGURE 12.35 The wedge principle used in a hard lock nut. (Courtesy of Hard Lock Industry Co. Ltd, Osaka, Japan.)

on the other. When the concave nut is tightened over the convex nut, a tensile force P_3 is generated within the bolt and two opposing radial locking forces P_1 and P_2 are created simultaneously at the thread contact surfaces due to the wedging effect. Thus, the hard lock nut is held securely by the combination of these three forces.

The hard lock nut solution is predominately used where there are high levels of vibration/shock and is applicable across a variety of industries, including structural engineering, transportation, automotive, and aerospace. Today, the hard lock nuts are extensively used in long-span bridges, high-speed trains, high-rise buildings, supertankers, heavy-duty vehicles, vibrating machinery, *etc.* all over the world.

12.6.7.2 Super Lock Nut and Super Stud Bolt

Super lock nut [12.67] and super stud bolt [12.68], classified as the distorted thread lock fasteners, were innovated to prevent self-loosening of bolted joint by adding certain distorted threads either at one end or in the middle of the nut or bolt. In practical applications, a super lock nut is paired with a standard bolt and a super stud bolt is paired with a standard nut. To understand their working principle, take the super lock nut as an example. The design of the lock nut makes that the distorted threads on the nut do not fit well with the standard threads on the bolt. When the nut is tightened on the bolt, the distorted threads of the nut are forced into the threads of the bolt, generating a great amount of friction between them, producing a large self-locking force to achieve the self-locking function.

The structures of a super lock nut and a super stud bolt are presented in Figure 12.36. The characteristic design feature is that there is a thin-walled tube at the middle of the threads, which can be deformed along the axis so that the phase difference in the lower and upper threads produces (Figure 12.37). This phase difference induces the contrary forces on the surfaces of upper and lower threads, which bring out the anti-loosening performance. Because the thin-walled tube is the critical element of anti-loosening mechanism, its dimensional parameters are optimized based on the results of FEA [12.69]. Additionally, the working principle of super stud bolt is very similar to that of super lock nut [12.70].

Super lock nuts and super lock bolts can work at high temperatures. To ensure normal operation, a super lock nut cannot be made of a metal that is stronger than that of a standard bolt. Otherwise, the distorted threads on the nut will destroy the threads on the bolt rather than lock to it (considering a steel super lock nut with a standard aluminum bolt). Moreover, because the locking force of the super lock nut and super lock bolt relies on the deformation of the distorted threads, they usually cannot be reused.

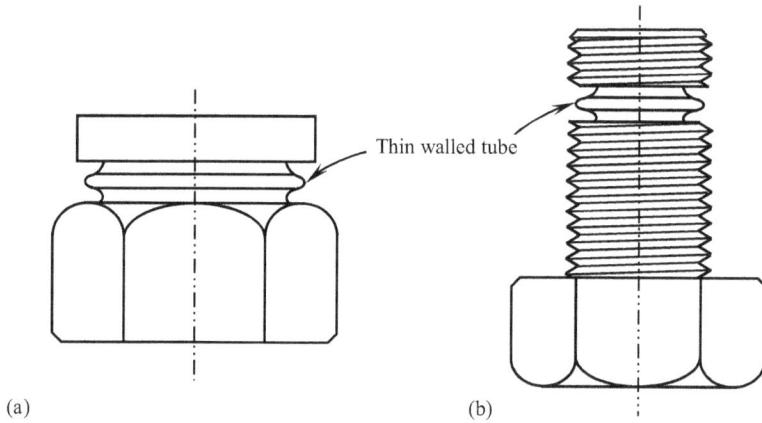

(a)   (b)

FIGURE 12.36 The structures of super lock fasteners: (*a*) Super lock nut contains a thin-walled tube at the middle of the nut. The nut is paired with a standard bolt. (*b*) Super stud bolt contains a thin-walled tube at the middle of the bolt. The bolt is paired with a standard nut.

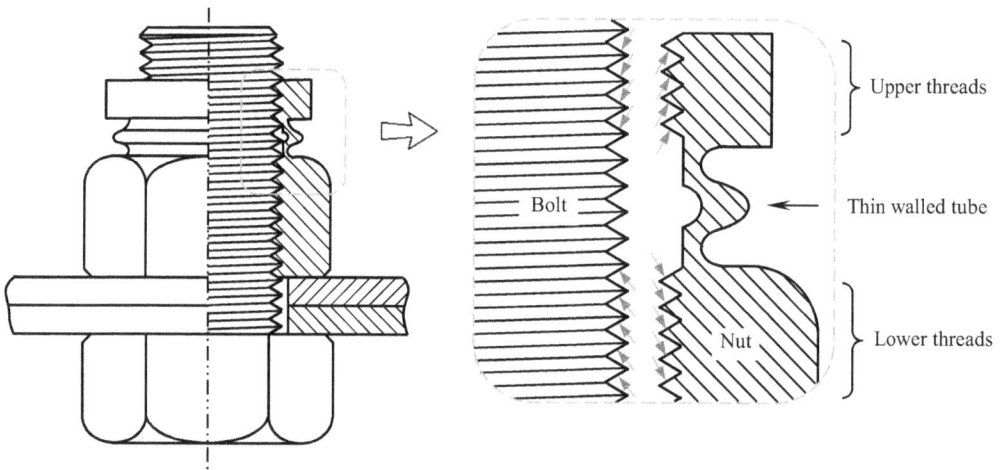

FIGURE 12.37 As a super lock nut is tightened over a standard bolt, the thin-walled tube deforms in the axial direction. The deformation of the thin-walled tube generates the gripping forces on the surfaces of the upper and lower threads that consequently produce the prevailing torque.

12.6.7.3 Self-Locking Nut

Self-locking nuts are a special type of nuts that contain the self-locking feature. They offer an easier and more cost-effective solution to prevent threaded fastener loosening when exposed to vibration, shock, dynamic, or alternative thermal load. Unlike traditional nuts that simply consist of a basic threaded hole, this type of nut uses a unique design to lock it in place along a bolt, by generating friction in various ways. The most commonly used types are metal lock nuts and nylon lock nuts (Figure 12.38). The metal lock nuts feature a crowned top, which can be crimped to secure the nut in place along a bold. Another common type is nylon lock nuts that feature a nylon ring covering the interior threading. The nylon ring is a little smaller than the diameter of the bolt. Once the nylon lock nut is twisted onto a bolt, the nylon fibers expand to grip the bolt, keeping the nut locked on the bolt. In such a way, it increases the friction between nut and bolt threads, causing the bottom faces of the bolt threads to press more tightly against the top faces of the nut threads. Due to the properties

FIGURE 12.38 Self-locking nuts offer a simple and cost-effective solution to prevent the most common factors of loosening such as shock, vibration, and alternative thermal load. Their locking action is created by the elastic distortion of their uppermost threads (usually inserting a nylon rubber ring at the top spreads). The frictional force generated from the thread distortion is then converted into the expansion force between the nut and bolt to achieve the self-locking effect.

of nylon, when removed the nut from the bolt, the nylon bounces back to its original form. Thus, the nylon nuts can be reused many times. However, because nylon material is not resistant to high temperature, the nylon nuts cannot function well as the operating temptation exceeds 120°C.

12.6.7.4 Jam Nut
Using jam nuts provides a practical and effective solution for solving nut-loosening problems when specially designed locking nuts are not available or in low-clearance applications. Jam nuts are typically used in pairs. They are both screwed on the same bolt and tightened against each other. Thus, the friction produced between the two nuts prevents the nuts from loosening from the bolt. This locking method may be especially suitable for clamping soft materials without damaging them. Specially designed jam nuts are lower profile nuts that have about half of the thickness of regular nuts. The locking effect can also be achieved with two regular nuts that require a longer bolt.

12.6.7.5 Serrated Face Nut and Bolt
The most distinctive feature of serrated face nuts is that they have serrations cut into one or both nut faces. When the nut is screwed onto a bolt to clamp components in place, the angle of the serration causes the nut face to bite into the mating surface, preventing itself from loosening. In fact, they act like tooth lock washers due to the ratchet action. However, the serrated face nut cannot be used with a washer because it may simply spin instead of putting up resistance against the threads. In addition, they damage the mated surface. Similarly, serrated bolts are also available in applications.

12.6.7.6 Nord-Lock Washer
A significant advantage of bolted joints over permanent joints (*e.g.*, welded or riveted joints) is that they are capable of being dismantled. However, this feature may cause some unintentional problems. Electric motors are usually subject to high-frequency vibration that produces dynamic load fluctuations acting on motor bolted joints. As a result, this can lead to spontaneous bolted joint loosening, called vibration loosening. In addition, cyclic temperature variations may also cause low-frequency dynamic load fluctuations in bolted joints.

Nord-lock washers, also known as vibration-resistant lock washers, offer an optimal solution for bolted joint security (Figure 12.39). The key design feature is that Nord-lock washers secure bolted

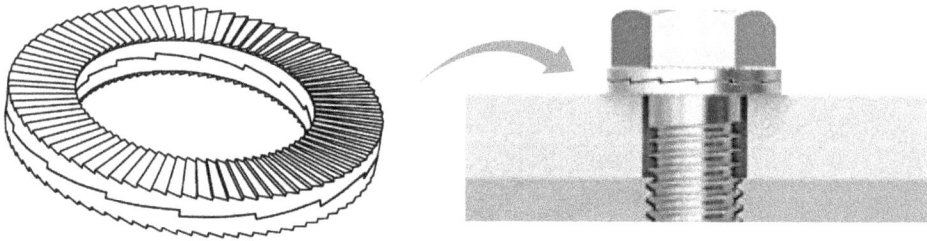

FIGURE 12.39 Nord-lock washer comprising a pair of washers with wedge cams on one side and radial ribs on the other side. The pair of washers is combined so that the cams face each other for eliminating the possibility of fastener loosening.

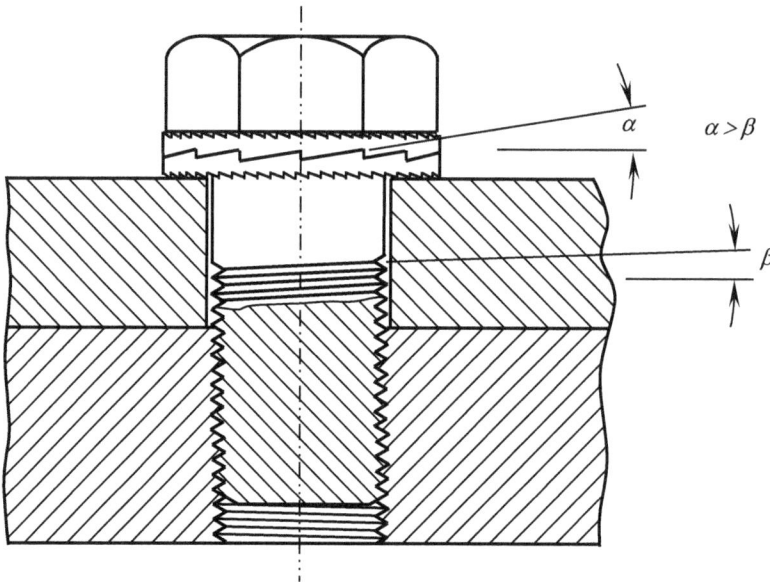

FIGURE 12.40 The assembly of bolted joint with a Nord-lock washer.

joints with tension instead of friction, making the bolt self-locking. As shown in Figure 12.40, a Nord-Lock washer is comprised of a pair of washers that has special cams on one side and radial teeth on the opposite side. Because the cam angle α is larger than the bolt thread pitch angle β, the wedge effect is created by the cams to compensate for the loss of preload, preventing the bolt from rotating loose.

The comparison of operating performance between different types of bolted joints under vibration conditions is presented in Figure 12.41 [12.71], where the test results were obtained through the Junker test, according to DIN 65151 [12.72]. It confirms that Nord-lock washer can safely secure bolted joints; only a limited amount of tension is initially lost due to normal settlements and thereafter maintains substantially constant. In contrast, all other types of bolted joints failed to prevent loosening of the joints.

12.6.8 ACTIVE VIBRATION ISOLATION AND DAMPING

Although passive vibration isolation systems are still extensively used in industries today, they are inadequate for some applications that require highly accurate control of vibration, such as optical devices, sonars, piezoelectric crystal flowmeters, medical detection devices, space structures, and other vibration-sensitive equipment.

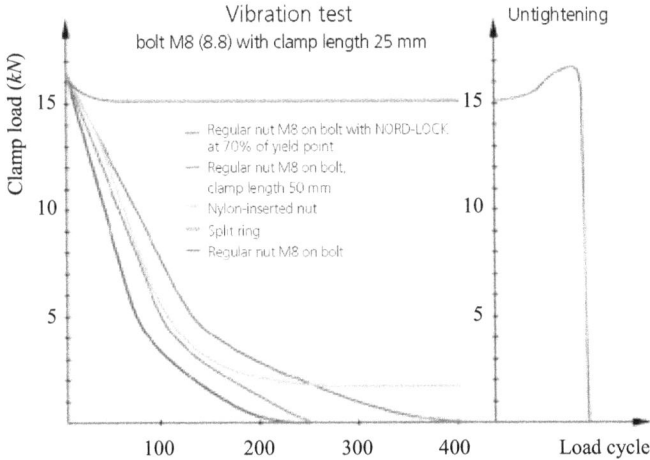

FIGURE 12.41 Comparison of operating performance between different types of bolted joints under vibration conditions [12.71].

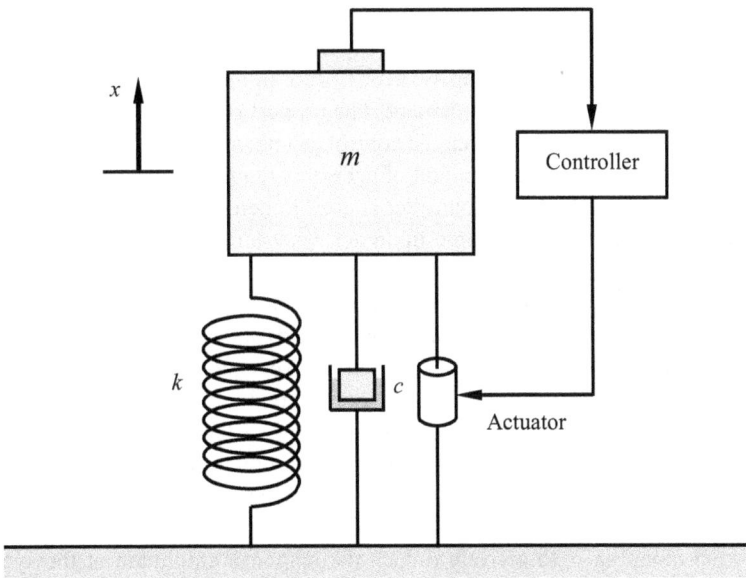

FIGURE 12.42 Active vibration isolation system.

Active vibration isolation systems use external energy to directly cancel energy in systems. The system involves the use of actuators along with sensors and controllers to create actuation with the goal of reducing the transmission of vibration from the vibration source to the rest of the components in the system. Sensors are used to continuously monitor the dynamic motion of the target system and send the relevant motion data (relative displacement, velocity acceleration, *etc.*) to controllers. The controllers calculate the required external forces or displacements and send the signals to actuators, which provide the desired forces or displacements to the target system, as illustrated in Figure 12.42. This method can work effectively for low vibrating frequencies. However, it is very difficult to apply it to high-frequency applications.

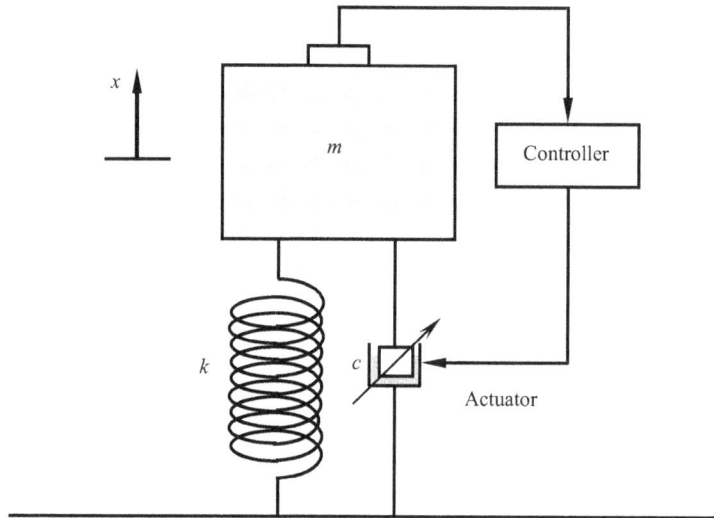

FIGURE 12.43 A semiactive vibration isolation system. The controller is used to adjust actuator damping characteristics.

A semiactive vibration isolation system is presented in Figure 12.43. It combines the features of a passive and an active vibration isolation system. Unlike in an active system, the actuator in the semiactive system is treated as a passive element. The properties of the semiactive actuator such as damping ratio and stiffness can vary so that the control can be implemented without adding external energy into the system, except a small amount of energy is required to change the properties of the actuator. In the semiactive system, the controller is used to determine the desired properties of the actuator, and the sensor functions the same in an active system to detect the motion of the target system [12.73].

Generally, the low-frequency isolation in the horizontal direction can be easily achieved by a passive system since it requires less concern about the horizontal static load by having a low stiffness. A vertical isolation is well implemented by an active system because it can provide a high stiffness to support static loads. Platus [12.74] has developed a passive vibration isolation system that outperforms conventional passive systems by adopting a technique called negative stiffness mechanism. The use of this system in a vertical isolation reduces the vertical stiffness and hence natural frequencies while maintaining static load-supporting capability. It was reported that the vertical natural frequencies can be tuned to as low as 0.2 Hz.

Active vibration damping is to actively reduce the response amplitude of the system within a limited bandwidth near the natural frequencies of the system. It offers a promise of high efficiency without the restrictions of passive methods. Active vibration control involves monitoring vibrations of a structure and utilizing the vibration signal to generate a force with the proper phase and amplitude to attenuate the vibration. An additional advantage of an active approach is the ability to supply a vibration signal that can be used independently for monitoring the vibration environment.

In some high-precision industrial applications (*e.g.*, optical tables), vibration isolation methods are usually used for relatively low-frequency vibration problems. For high-frequency vibrations, passive damping, either broadband or tuned, provides varying levels of structural damping performance, but both are affected by applied loads. By means of vibration sensors, actuators, and controllers, active vibration damping can be autotuned to account for varying loads [12.75], making it more attractive for industrial and medical end users.

An alternative method for active vibration isolation is to use noncontacting electromagnetic voice coil actuators (VCAs), which are used for some special applications like large telescopes and robots. In this method, electric current passing through the voice coil generates a magnetic field that interacts with the magnetic field generated by permanent magnets (PMs) to produce desired forces [12.76]:

$$F = n\pi diB \sin\theta \tag{12.122}$$

where
    n is the total number of coils
    d is the average diameter of the coil
    i is the current through the wire
    B is the magnetic field strength
    θ is the angle between the magnetic flux lines and the direction of the current

Newell *et al.* [12.77] have used VCAs to construct an active vibration isolation system for earth-based interferometric gravitational wave detectors. This system provides vibration isolation in all 6 degrees of freedom by at least 40 dB at 0.5 Hz.

12.6.9 MEASUREMENTS OF MOTOR VIBRATION

A comprehensive assessment of motor vibration requires measurements of vibration acceleration in units of m/s². A typical vibration measurement system comprises several vibration sensors (*e.g.*, accelerometers, velocity transducers, and proximity probes) for sensing vibration velocity or acceleration; a vibration analyzer for data acquisition, bearing condition diagnosis, and spectrum diagnosis; and a computer for processing the measurements and displaying the results. The vibration sensors are typically attached to the vibrating motor housing at different locations and on the baseplate. Since motor bearings are the load-carrying components of the mechanical drive train, the vibration sensors should be placed close to bearings to measure both motor radial and axial vibration (Figure 12.44). The vibration sensors (used accelerometers here) produce electrical signals that are proportional to the vibrating acceleration on the measuring surface. Then, the measurement data are transmitted to a vibration analyzer for data processing and data recording. Finally, the measured vibration results displayed by a computer are provided to engineers for further analysis. In such a way, mechanical and electrical defects can be identified through vibration analysis. The useful information detected from the motor vibration measurements includes bearing vibration, rotordynamic unbalance, eccentricity, and electrical faults.

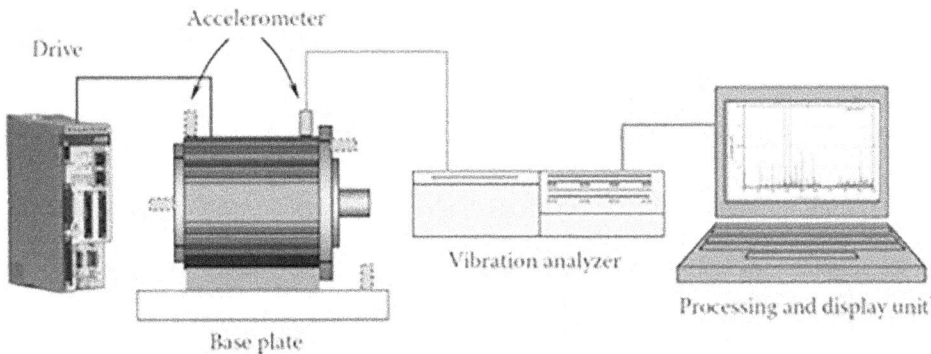

FIGURE 12.44 Motor vibration measurement using accelerometers for assessing motor vibration and diagnosing vibration root causes.

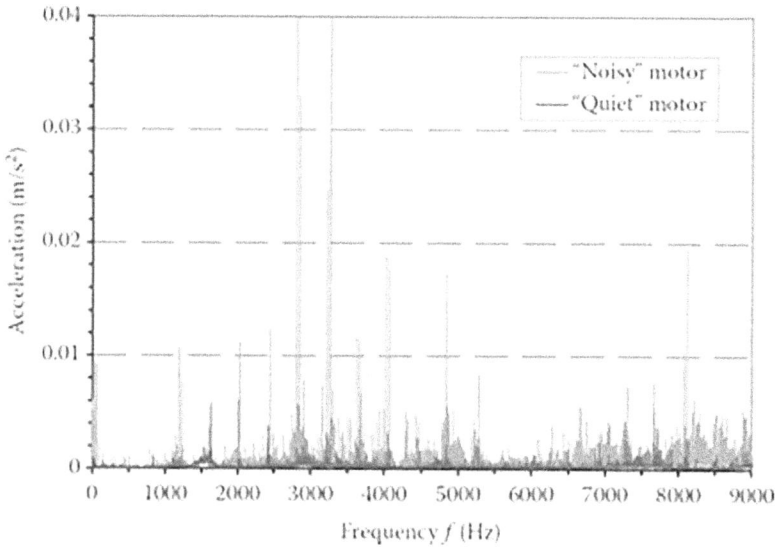

FIGURE 12.45   Comparison of measured vibration data from two PM servomotors with the same brand and same model, showing distinct differences in harmonic components of the acceleration between two motors. (Courtesy of Kollmorgen Corporation, Radford, VA.)

As an example, the results of motor vibration for two identical servomotors are presented in Figure 12.45. For comparison, a noisy motor and a quiet motor were selected for measurement. It can be observed that there are great differences in vibrating accelerations between the two motors for almost all frequencies, especially in the range of 1,000–5,000 Hz, and >8,000 Hz. The root cause analysis reveals that the motor bearings and assembly quality attribute to high vibration.

Special attention must be put on keyed shafts in vibration measurements. Prior to the test, a *half key* must be placed in the shaft keyway to eliminate the influence of the unbalance that is induced from the keyway.

12.7   FUNDAMENTALS OF ACOUSTIC NOISE

Noise is one of the most common occupational health hazards. In heavy industrial and manufacturing environments, hearing damage or permanent hearing loss is the main health concern for operating workers. In the United States, growing environmental concerns and recognition that lengthy unprotected exposure to high noise levels is detrimental to man have resulted in government promulgation of noise level criteria. In many countries, especially European countries, noise reduction is an integral part of machinery safety and law regulates the allowable noise exposure of workers. Manufacturers must make specific quantitative information on noise emitted under designed operating conditions. In recent years, the noise requirement to ensure quiet working conditions is getting even tighter.

Acoustic noise is the result of pressure waves produced by vibration sources in an elastic medium (*e.g.*, air, water, and solids). Generally, a sinusoidal wave is referred to as a tone, a combination of many different tones is called sound, and an irregular vibration is referred to as noise [12.78]. Noise is defined in terms of frequency spectrum (in Hz), intensity (in dB), and time duration.

In general, noise reduction techniques can be arranged into passive and active methods. Passive control involves reducing the radiated noise by energy absorption, while the active method involves reducing the source strength or manipulating the acoustic field in the duct to get noise reduction [12.79].

Noise control could be extremely important in some applications. For instance, the key to a submarine's survival is to successfully evade the detection and surveillance. Reduction of the emitted noise signature of a submarine platform is an important element to the tactical employment of the ship. This requires minimizing its noise levels for achieving the purpose of stealth to perform many of its intended missions. The main noise and vibration sources on board a submarine include the propeller, diesel engines, generators, motors, auxiliary machinery, and flow noise over the outer body of the hull [12.80]. When a submarine gets to sneak under sea, it produces friction and turbulence that cause sound waves. In the past several decades, the implementation of innovative sound damping and vibration isolation technologies has resulted in significant noise reductions in submarine. Among these, rubber tile coatings break up sound waves that bounce against the hull, reducing the submarine's acoustic signature and making it more difficult to detect via sonar. Using precision CNC machinery to make high-quality propellers can significantly reduce the effect of propeller cavitation and alter the water flow over a submarine, concealing wakes or turbulences behind the submarine.

12.7.1 TONAL NOISE AND BROADBAND NOISE

There are two types of noise radiation: pitched or unpitched noise. Pitched noise, also known as tonal noise, is created from vibrations of an object that occur in a fine range around one or very few distinct frequencies. Tonal noise can be extremely irritating to human hearing. An electric motor that emits tonal noise has a characteristic frequency spectrum. Unpitched noise, also known as broadband noise, has a well-distributed energy spectrum. Broadband noise is generally not irritating to human hearing and does not transmit a characteristic frequency spectrum, such as wind noise.

12.7.2 SOUND PRESSURE LEVEL AND SOUND POWER LEVEL

Sound pressure is a measure of air pressure fluctuation a noise source creates. It is usually expressed in units of Pascal (Pa). However, the use of Pascal as the sound pressure unit may generate a broad range of sound pressures. A convenient way to compress the scale of numbers into a manageable range is to use decibel (dB). Sound pressure converted to the decibel scale is called sound pressure level (SPL), denoted as L_p:

$$L_p = 10\log_{10}\left(\frac{p_{rms}^2}{p_o^2}\right) = 20\log_{10}\left(\frac{p_{rms}}{p_o}\right) \tag{12.123}$$

where p_{rms} is the RMS sound pressure being measured and p_o is the reference sound pressure, typically 20 μPa.

It is important to distinguish between sound power and sound pressure. Sound power is a measure of the total amount of sound energy emitted by the source over a period of time, usually expressed in watts. Sound pressure depends on the distance of observation location from the source but sound power does not. The sound power level L_w, in units of dB, is defined by

$$L_w = 10\log_{10}\left(\frac{W}{W_o}\right) \tag{12.124}$$

where W is the sound power emitted by the source in watts and W_o is the reference power, typically 10^{-12} W. SPL L_p can be related to sound power level L_w as [12.81]

$$L_p = L_w + 10\log\left(\frac{Q_\theta}{4\pi r^2} + \frac{4}{R_c}\right) \tag{12.125}$$

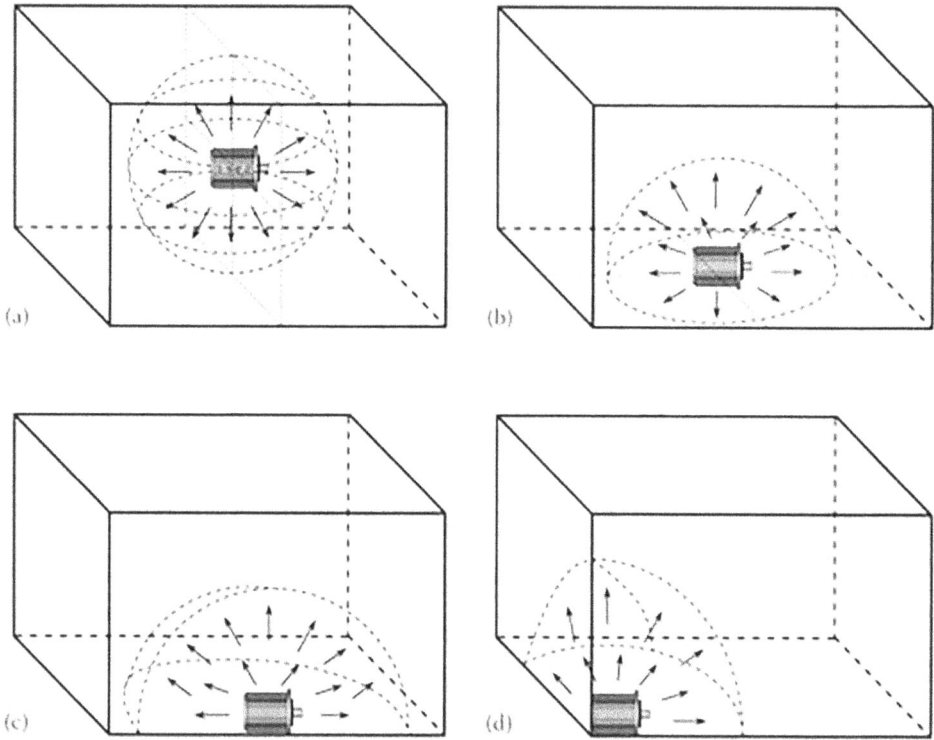

FIGURE 12.46 Directivity factor of the sound source for different source positions: (a) at room center, $Q_\theta = 1$; (b) at the center of floor, $Q_\theta = 2$; (c) at the center of floor edge, $Q_\theta = 4$; and (d) at floor corner, $Q_\theta = 8$.

where

    r is the distance from the source

    Q_θ is the directivity factor of the source in the direction of r, which depends on the position of the sound source (see Figure 12.46)

    R_c is the room constant, defined as

$$R_c = \frac{A\bar{\alpha}}{1-\bar{\alpha}} \tag{12.126}$$

where

    A is the total surface area of the room

    $\bar{\alpha}$ is the average sound absorption coefficient of the room

12.7.3 Octave Frequency Bands

The audible frequency range is usually divided into 10 octave bands having center frequencies at 31.5, 63, 125, 250, 500, 1,000, 2,000, 4,000, 8,000, and 16,000 Hz. The center frequency of each consecutive octave band is twice the center frequency of the previous one. In each octave band, the upper cutoff frequency is twice the lower cutoff frequency. Usually, an electric machine such as motor and generator generates and radiates noise over the entire audible range of hearing. The amount and frequency distribution of the noise can be obtained using an octave band analyzer. This analyzer equips a set of contiguous filters, and each filter has a bandwidth of an octave. Therefore, ten such filters cover the most frequency range of interest. The measured L_p data at each octave band are very useful for determining the noise root sources.

Each octave band can be further divided into three one-third octave bands. The SPL L_p is thus determined from three measure values $L_{p,i}$ that are taken at each one-third octave band, respectively.

The comparison of full octave band and one-third octave band is presented in Table 12.1. The relationship between the upper cutoff frequency f_{n+1} and lower cutoff frequency f_n is

$$\frac{f_{n+1}}{f_n} = 2^k \tag{12.127}$$

where $k = 1$ for one octave band and $k = 1/3$ for one-third octave band. The middle frequency f_o in each full octave band is about times the lower cutoff frequency in that band. The bandwidth w_b of each band is given as

$$w_b = f_{n+1} - f_n = \left(2^k - 1\right) f_n \tag{12.128}$$

TABLE 12.1

Comparison of One Octave Band and One-Third Octave Band in Audible Sound Range

	One Octave			One-Third Octave		
Band	Lower Cutoff Frequency (Hz)	Middle Frequency (Hz)	Upper Cutoff Frequency (Hz)	Lower Cutoff Frequency (Hz)	Middle Frequency (Hz)	Upper Cutoff Frequency (Hz)
1	22	31.5	44	22.4	25	28.2
				28.2	31.5	35.5
				35.5	40	44.7
2	44	63	88	44.7	50	56.2
				56.2	63	70.8
				70.8	80	812.1
3	88	125	177	812.1	100	112
				112	125	141
				141	160	178
4	177	250	355	178	200	224
				224	250	282
				282	315	355
5	355	500	710	355	400	447
				447	500	562
				562	630	708
6	710	1,000	1,420	708	800	891
				891	1,000	1,122
				1,122	1,250	1,413
7	1,420	2,000	2,840	1,413	1,600	1,778
				1,778	2,000	2,239
				2,239	2,500	2,818
8	2,840	4,000	5,680	2,818	3,150	3,548
				3,548	4,000	4,467
				4,467	5,000	5,623
9	5,680	8,000	11,360	5,623	6,300	7,079
				7,079	8,000	8,913
				8,913	10,000	11,220
10	11,360	16,000	22,720	11,220	12,220	14,130
				14,130	16,000	17,780
				17,780	20,000	22,390

12.7.4 THREE SOUND WEIGHTING SCALES

Human hearing varies in sensitivity for different acoustic frequencies. Although the audible range of acoustic frequencies generally ranges from 20 to 20,000 Hz, humans usually are most sensitive to pure tones at frequencies between 2,000 and 6,000 Hz. Hearing sensitivity drops off above 7,000 Hz and below about 200 Hz. Thus, to establish a uniform noise measurement that simulates human's perception and annoyance, several frequency weighting methods are developed to account for those frequencies most audible to the human hearing range.

Acoustic frequency weightings refer to different sensitivity scales for noise measurement. In the development of sound level meters over the years, manufacturers have built in the different response curves, that is, A-, B-, and C-weighting filter curves, as defined in the IEC standard 61672:2003 [12.82] and various national standards, as shown in Figure 12.47. Among these, A-and C-weighting scales are commonly incorporated into commercially marketed sound level meters.

The A-weighting scale is presently the most commonly used weighting scale in measuring industrial noise, because it best predicts the damage risk of human ear. Using the decibel A filter, the sound level meter is less sensitive to very low and very high frequencies. Sound pressure measurements with this scale are expressed as dB(A) or simply dBA. The A-weighting curve in Figure 12.47 suggests that for most listeners, the noise at 100 Hz would sound about 19 dB quieter than that at 1,000 Hz at the same SPL. It is worth to note that since the dBA scale is logarithmic, every increase of 10 dBA doubles the perceived loudness of hearing.

The B-weighting scale was used in the early time for predicting the performance of loudspeakers and stereos, but today, the B-weighting scale is no longer in any international standard and already has very little practical use.

The C-weighting scale was originally designed to predict the human ear's sensitivity to tones at high noise levels. However, the ear's loudness sensitivity for tones is not the same as the ears' damage risk for noise. Much of the low-frequency noise is actually being filtered out by the ear, making it less likely to cause damage. By comparing with the A-weighting scale, the C-weighting curve is quite flat, indicating it contains much more of the low-frequency range of sounds.

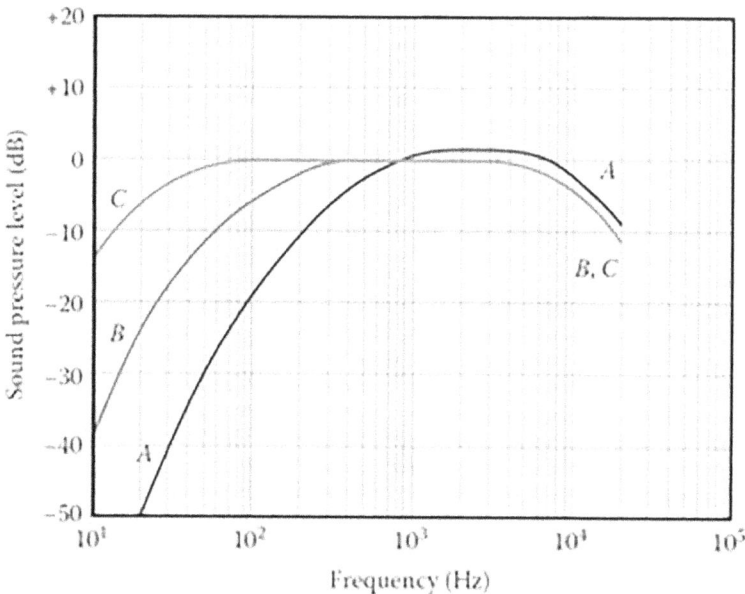

FIGURE 12.47   A-, B-, and C-weighting filter curves, defined up to 20 kHz.

12.7.5 Averaged Sound Pressure Level

The averaged SPL is the logarithm average of the measured SPL around the motor, expressed as

$$\bar{L}_p = 10 \log_{10} \frac{1}{n} \left[\sum_{i=1}^{n} 10^{(L_{p,i}/10)} \right] \tag{12.129}$$

where
$L_{p,i}$ is the measured SPL at i^{th} point
n is the total number of the points

It must be noted that $L_{p,i}$ may consist of noise produced by sources in the vicinity of the testing motor in a field test. To gain the SPL only produced by the motor, the background noise must be measured and subtracted from the measured noise data. This can be done by disconnecting the test motor and maintaining all other equipment operating. Then, by repeating the measurements around the motor, the background noise pressure level at i^{th} point $L_{p,i,b}$ is obtained. As a result, the noise pressure level at i^{th} point from the tested motor can be calculated as

$$L_{p,i,m} = 10 \log_{10} \left[10^{(L_{p,i}/10)} - 10^{(L_{p,i,b})} \right] \tag{12.130}$$

Thus, the averaged SPL only from the tested motor is obtained:

$$\bar{L}_{p,m} = 10 \log_{10} \frac{1}{n} \left[\sum_{i=1}^{n} 10^{(L_{p,i,m}/10)} \right] \tag{12.131}$$

12.7.6 Type of Noise

According to the source and pattern of noise production, several types of noise can be identified; each of them has its own characteristics in noise generation and transmission.

12.7.6.1 Structure-Borne Noise

Sound waves can travel through different media such as air, water, wood, rock, soil, or metal. Depending on the media through which it travels, sound is either structure-borne or airborne. Structure-borne noise propagates through the motor structure as vibration and subsequently radiated as sound. The intensity and frequency of structure-borne noise depend on many factors such as the rotational speed of the motor, material and geometry of the vibrating components, and system structure. The structure-borne noise is a result of the action of the aerodynamically, mechanically, and electromagnetically excited forces. When the frequency of the exciting force acting on the motor, as well as the response force on the motor support, approaches to or coincides with any of the motor natural frequencies, resonance occurs accompanied by strong vibration and high noise.

12.7.6.2 Airborne Noise

Airborne noise radiates from a noise source directly into and travels through the air. As the sound waves are being carried by the atmosphere, airborne noise can directly impact human hearing. These sound waves make contact with walls, floors, or ceilings, inducing noise pollution.

Airborne sound transmission loss is a measure of the degree to which a material can effectively block or reduce transmission of airborne sound. It is usually measured at each one-third octave band frequency from 125 to 4,000 Hz, in units of dB. While structure-borne noise is attenuated by isolation, airborne noise can be attenuated by treatment with sound-absorbing materials or through the use of barrier materials.

12.7.6.3 Type of Aerodynamic Noise

There are two types of aerodynamic noise: rotational and nonrotational. Rotational aerodynamic noise is associated with the rotation of the rotor and fan in a motor. When a rotor rotates, the wall rotating-driven flows, also referred to as the rotating pumping flow, are generated with high angular momentum to be flung radially outward, leading to flow-induced noise. The great amount of noise from a rotating fan is caused by steady thrust and drag forces that are induced on the blades as they rotate through air, and the air turbulence produced by the cooling fan can create a great amount of noise.

Nonrotational aerodynamic noise is mainly associated with the ventilating flow inside a motor, as a result of fluctuating fluid forces in time and frequency domains. When the motor uses a fan to force cooling air flowing through the motor, the ventilating flows pass through very complicated and irregularly shaped flow passages to cool stator windings and other components. The nonuniform velocity fields, high turbulent flows, the interactions between the rotational flows and ventilation flows, and the interaction between the flows and rigid structure all contribute to the nonrotational aerodynamic noise.

12.8 NOISE CLASSIFICATION AND MEASUREMENT IN ROTATING ELECTRIC MACHINE

Growing environmental sound concerns and recognition that lengthy, unprotected exposure to high industrial noise levels can be detrimental to people have resulted in increased attention to reducing industrial noise. In many countries, particularly those in Europe, the allowable maximum noise levels that workers should be exposed to are regulated by law, through government promulgation of noise level criteria.

Increasingly strict noise regulations in the United States, European Union, Japan, and China have made the development of new electric machines with low noise emission. Specifically, the European Union has set the tightest harmonized noise limits for various industrial machines, motor vehicles, household appliances, and other noise-generating products. Because of the increased awareness of the harmful effects of high industrial noise levels and government regulations that establish acceptable noise levels in the workplace, noise reduction has become an integral part of machinery safety. Noise abatement is a concern with, for example, generator-steam and gas turbine power plants. As demands for electricity increase, the power industry faces increasing challenges to build and operate efficient and quiet power generators, for example, steam turbines, gas turbines, and electrical generators.

In an industrial environment, various machines often emit noise waves that register at high and potentially harmful decibel levels. In a power plant, noise may come from a variety of machine sources, such as generators, gas or steam turbines, fans, pumps, coolers, and other mechanical and electrical equipment, many of which may be in operation simultaneously. Individuals working in such an environment are often faced with the need to reduce the near- and far-field machinery noise levels. In an environment where individuals work in close proximity to the sources of a machine noise, near-field sound levels must be controlled in order to comply with noise regulation and avoid hearing damage to the workers. Where machine noise can reach areas that are near an industrial plant, it may be prudent to abate far-field machine noise to acceptable levels and to avoid broadcasting neighboring communities.

Regulatory requirements for low noise from electric motors have become increasingly more stringent. Today, environmental and safety agencies throughout the world are requiring even lower noise levels. With the increased demands for high efficiency, high power, and cost-effective electric machines, the motor industry is now facing challenges to build up not only more efficient but also quieter motors.

During motor operation, noise is produced from both rotating and stationary motor components, leading to environmental problems for people nearby. Thus, to protect people's health, motor noise must fall within acceptable limits, which have been strictly regulated in North American and European countries.

Various types of acoustic noise are produced during motor operation, as addressed in the following sections.

12.8.1 Noise Type in Rotating Electric Machine

Rotating electric machines are composed of various noise sources such as stator, rotor, ventilating fan, bearings, and housing, to name a few. The mechanisms of noise generation depend on the rotation of rotor, fan, and bearings, interaction of electromagnetic fields of the rotor and stator, machine structure, geometry of ventilation paths, and operation conditions. All noise generation mechanisms contribute to the overall noise pressure level of the machine. During normal operation, the noise sources create three main types of acoustic noise: mechanical noise, electromagnetic noise, and flow-induced noise.

12.8.1.1 Mechanical Noise

Shaft misalignment in motor causes preload forces to be generated in couplings that are then transmitted to various motor components. As one of the major causes of motor vibration, shaft misalignment produces a mechanical vibration with the frequency f_s,

$$f_s = 120n_s \tag{12.132}$$

where n_s is the rotating speed of the shaft in rpm.

Dynamically unbalanced rotor, bent shaft, eccentricity, rubbing parts, *etc.*, produce a vibrating frequency of once per revolution or multiple of the number of revolutions per cycle,

$$f_s = kn_m \quad k = 1,2,3,... \tag{12.133}$$

The magnitude of the vibration noise due to the unbalanced rotating masses depends on the degree of balancing of the rotating masses and the rotating speed.

Loose stator lamination stack results in the vibration frequency f_{lam} with the frequency sidebands of 1,000 Hz,

$$f_{lam} = 2f \tag{12.134}$$

where f is the line frequency.

Rolling bearings are extensively used in modern electric motors. Rolling bearing generates mechanical impulses when the rolling element passes the defective groove, causing a small radial movement of the rotor. Rolling bearings produce noise with discrete frequencies corresponding to the rotating frequency of each rolling element, such as roller, cage, and inner/outer ring. The bearing noise levels depend on the speed of rotation and their manufacturing perfections. Some large-sized motors also use sleeve bearings. The noise due to sleeve bearing is generally lower than that of rolling bearings. The vibration produced by sleeve bearings depends on the roughness of sliding surfaces, lubrication, and stability and whirling of the oil film in the bearing manufacturing process and bearing assembly [12.83].

Sound power radiated from an electric machine is proportional to the vibration area S and the mean vibration velocity square $\bar{u}^2$ [12.84]:

$$W = \rho c S \bar{u}^2 \, \sigma_{rad} \tag{12.135}$$

where

  ρ is the air density
  c is the speed of sound
  σ_{rad} is the radiation efficiency

This equation indicates that mechanical noise can be reduced by reducing the vibration area and/or the vibration velocity. Reducing the vibration area can be achieved by dividing a large area into a number of small areas with flexible joints between them. Reducing the vibration velocity can be achieved by using damping and vibration-absorbing materials to isolate the vibration source from other components. A reduction of the exciting forces by a factor of two can possibly lower the SPL up to 6 dB.

12.8.1.2 Electromagnetic Noise

Electromagnetic field is not uniform in the airgap and consists of various harmonics. The periodic fluctuations of the magnetic flux in the airgap cause oscillating forces on both the rotor and stator. The noise due to the fluctuating magnetic forces is called nonrotational noise, which resulted from the asymmetry of the electric and magnetic circuit and saturation of the magnetic field.

There are many normal modes of vibration in the stator core in all directions. While the radial forces cause radial extension of the stator core and consequently vibration and structure-borne noise, the axial forces and tangential forces result in vibration of the stator core in the axial and circumferential directions.

The magnitude of rotational electromagnetic noise strongly depends on the motor rotating speed. Hence, the electromagnetic noise is characteristically tonal in character with a great number of discrete frequency tones, the most important of which are at the frequency f_{em} [12.85]

$$f_{em} = N_s n \tag{12.136}$$

where

  N_s is the number of stator slots
  n is the motor rotating speed

Electromagnetic noise is stimulated when the electromagnetic force matches or is close to the natural frequencies of vibration of motors.

12.8.1.3 Aerodynamic Noise

Aerodynamic noise, also referred to as windage noise or flow-induced noise in some references, is generated by the interactions of turbulent ventilating airflow with rotating (*e.g.*, rotor and fan) and stationary components (*e.g.*, stator and endbell) inside a motor. As airborne noise, this type of noise increases with increasing flow speed.

As discussed previously, the aerodynamic sound power generated by turbulent flows is proportional to the 6th to 8th of the flow velocity (*i.e.*, $W \propto u^{6-8}$). Consequently, the increase in the flow velocity can result in a great increase in sound power. Thus, a logical approach to lower the aerodynamic sound power is to reduce the flow velocity and turbulence intensity. This requires that in the design of the motor ventilation system, it should avoid the sudden changes in the cross-sectional areas along the flow paths and reduce the right-angle bends as much as possible.

A ventilation fan is usually a dominant noise source in a motor. The fan generates broadband aerodynamic noise as the cooling flow passes through the inlet and outlet of the fan. In practice, fan noise reduction can be achieved using an absorptive silencer or by redesigning the cooling fan. The sound power level generated by fans can be predicted in the early design stage using the Graham equation [12.86].

12.8.2 ACOUSTIC ANECHOIC CHAMBER

Noise measurement is a necessary step to determine the SPL of an electric machine is below the allowable level. With today's technology, many types of measuring systems can be used depending on the characteristics of sound, the extent of information that is desired about the sound, and cost, from the most popular sound level meters to the professional sound-measuring systems in acoustic laboratories.

An acoustic anechoic chamber is a specially designed room that can completely absorb reflections of sound waves and insulate testing equipment from environmental sources of noise. To absorb sound waves, the interior surfaces of the acoustic anechoic chamber are covered with acoustically absorbent materials. One of the most effective types of absorbent materials comprises arrays of pyramid-shaped pieces, each of which is constructed from a suitably lossy material. To get the best results, all internal surfaces of the acoustic anechoic chamber must be entirely covered with the pyramids (Figure 12.48).

The relationship of the sound wavelength λ, phase velocity of wave v, and frequency f is given as

$$\lambda = \frac{v}{f} \tag{12.137}$$

To shield for the specific range of wavelengths, the shape and size of the pyramids must be carefully designed for absorbing the range of wavelengths.

12.8.3 MEASUREMENT OF MOTOR NOISE

Typically, the anechoic chamber is used to measure the directivity of noise radiation from testing machinery or equipment and to determine the transfer function of a loudspeaker.

There are several elements in a sound-measuring system, including the following:

- Microphones respond to sound pressure of a testing motor at specific locations and transform it into an electric signal. The commonly used types of microphones include piezoelectric, condenser, electret, or dynamic microphones.
- Signal amplifier boosts the amplitude of the microphone output signal to a useful level for further processing.

FIGURE 12.48 An anechoic chamber is specially designed to completely absorb reflections of sound waves.

FIGURE 12.49   Measurement of the acoustic noise level of an electric motor by microphones in an anechoic chamber.

• Active filter reduces high-frequency signal components, unwanted electrical interference noise, or electronic noise from the signal.
• The data storage, processing, and transportation.

Noise is measured at the rated operation conditions in the anechoic chamber by using microphones at 1 m from the motor in different orientations, as shown in Figure 12.49. The noise data are taken as the test motor reaches the thermal equilibrium.

The spectra of SPL from two PM servomotors are displayed in Figure 12.50. These two motors are identical in terms of brand and motor, and the testing has been conducted under the same conditions. However, their SPLs show distinct differences for almost all frequencies, especially in the range of 1,000–3,000 Hz.

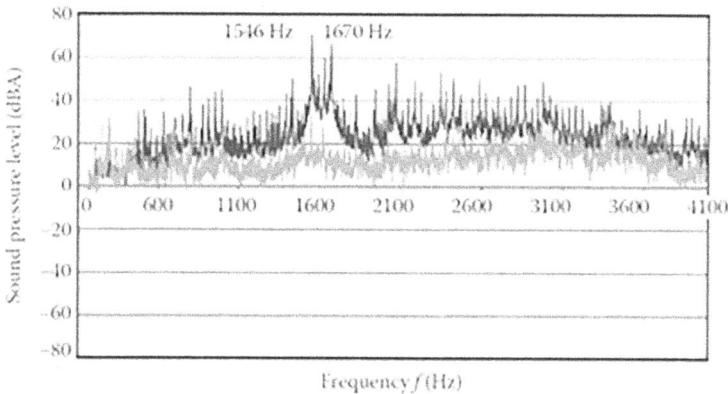

FIGURE 12.50   Comparison of SPLs for two PM servomotors with the same brand and same model. (Courtesy of Kollmorgen Corporation.)

12.8.4 ACOUSTIC NOISE FIELD MEASUREMENT

A field noise test was conducted at the customer site in Texas. Acoustic noise data were acquired from a large-sized generator, which is installed in the open air at a height of 15 m above the ground. There are totally 15 measuring points, 5 on either side of a large-sized generator. These points were labeled from 1 to 5 going from the collector end (CE) to the turbine end of the generator, as well as along the CE (Figure 12.51). The data were acquired using a Rion®, handheld precision acoustic sound meter, interfaced with a Hewlett Packard® four-channel analyzer. The acoustic data were taken at three load levels of 17, 86, and 156 MW (156 MW is the full load).

12.9 MOTOR NOISE ABATEMENT TECHNIQUES

Controlling the noise level of an electric motor is critical to the overall system performance. Therefore, noise abatement of the motor is an important goal in motor design. The most cost-effective way to deal with the noise issue is to incorporate the reduction of the noise level in an early design stage. Once a motor is manufactured, it is often expensive to modify the design for reducing noise levels. For old motors with high noise levels, an effective method to lower the motor noise exposure is to make external acoustic treatments of the motor.

A few techniques to minimize the noise level can be adopted in motor design. These techniques can be separated into two categories, passive and active.

12.9.1 ACTIVE NOISE REDUCTION TECHNIQUES

Effective reduction in acoustic noise can be accomplished using active noise reduction techniques. One of the techniques is called the noise cancellation technique. The concept of active noise cancellation was initiated and developed in the early 1930s. Paul Lueg [12.87], a German engineer, was the first who realized the possibility of attenuating background noise by superimposing a phase flipped wave. Later, this concept was successfully confirmed by Olsen [12.88, 12.89] to generate artificial

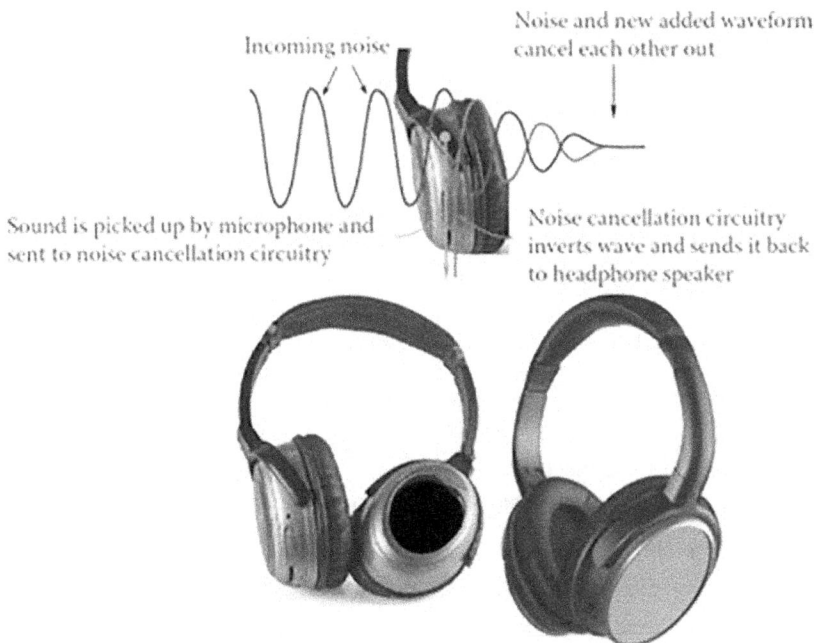

FIGURE 12.51 Active noise-cancelling headphones.

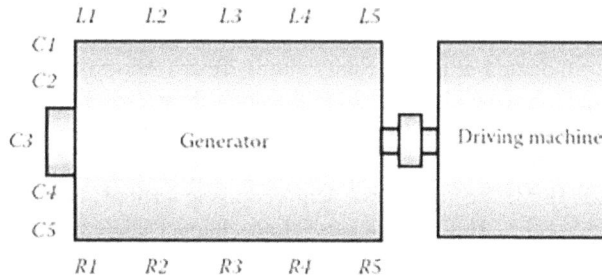

FIGURE 12.52 Measuring points around a large-sized generator.

sound waves that have the opposite phase to that of noise. This method is particularly useful at low frequencies because conventionally used sound absorbers do not work well at such frequencies, for which the sound wavelength is much larger than the thickness of a typical absorber.

Today, active noise reduction techniques have been successfully adopted in many commercial applications such as active mufflers, passage cars, wind tunnel testing systems, noise-cancelling headphones, and noise control in air-conditioning ducts. As a typical example, noise-cancelling headphones have been invented to eliminate any low-frequency noise from the environment. A variety of antinoise headphones is shown in Figure 12.52. These headphones involve using one or more microphones and an electronic circuitry that uses the microphone signals to generate antinoise signals, thus the generated destructive interference cancels out the ambient noise.

With today's advanced digital signal-processing techniques, it is possible to make active noise control more precisely and more efficiently. However, active damping techniques are considerably expensive because they require measurement of response and feedback control and thus have not yet been applied to electric motors. In addition, the application of active noise cancellation on large-size electric machines is less effective due to the challenge of separating multidirectional sounds.

12.9.2 Passive Noise Reduction Techniques

Passive techniques are the simplest and least expensive means of preventing problems associated with acoustic noise. There are several conventional methods available to reduce motor noise, each has its own advantages and disadvantages for specific applications. The selection of an appropriate method depends on its effectiveness in noise attenuation and system cost.

12.9.2.1 Blocking Noise Propagation Paths and Isolating Noise Sources

Mechanical noise transmissions can be greatly reduced by blocking noise propagation paths [12.90] or isolating sources of noise. An effective and traditional way is to locate noisy machines behind purpose-built barriers. To achieve the maximum shielding effect, the barriers can be constructed with sound-absorbing and damping materials. For large-sized motors, noise barriers can be erected between motors and noise-sensitive areas. Compact motors can be covered by a box with noise-absorbing and damping materials attached to its internal surfaces.

Another approach is to use mechanical dampers, such as rubber boots, reflecting enclosures, to lower the level of vibration and consequently the level of transmitted noise, as discussed in the previous sections.

12.9.2.2 Using Noise-Absorbing Material

A common way of noise reduction is to attach sound-absorbing materials on either internal or external walls of the motor housing/endbells for absorbing sound energy. Sound-absorbing materials may come in many forms, ranging from simple foam and cellular materials to a variety of commercially made products designed specifically for the purpose. Most sound-absorbing materials contain a large number of small cavities. When sound waves pass through the air inside the cavities, certain frequencies of sound waves may result in the resonance of the air inside the cavities. In such a way, sound energy is consumed, and the sound pressure is reduced. The resonances depend on the size and shape of the cavities. Therefore, well-designed cavities in sound-absorbing materials can effectively reduce motor acoustic noise.

Removable and reusable blankets have been widely used for acoustical reduction applications. The benefits of using blankets include the following:

- One-time investment
- Long-term repeated use
- Without interrupting equipment normal operation
- Making equipment maintenance easier
- Providing protection for personnel

Acoustic blankets may contain single- or multi-fiberglass layers depending on application requirements. As an example shown in Figure 12.53, an acoustic blanket consists of two high-density layers and one low-density layer. Usually, high-density blankets are used to contact the generator wall. Fiberglass fibers are encased within PTEE-coated (or silicone rubber-coated) fabric envelopes, which are designed to withstand both chemical attack and severe mechanical abuse. Stainless steel hook-and-loop fasteners are used for jointing in blanket sections. Acoustic blankets are wrapped around the external surface of an electric machine for reducing noise levels.

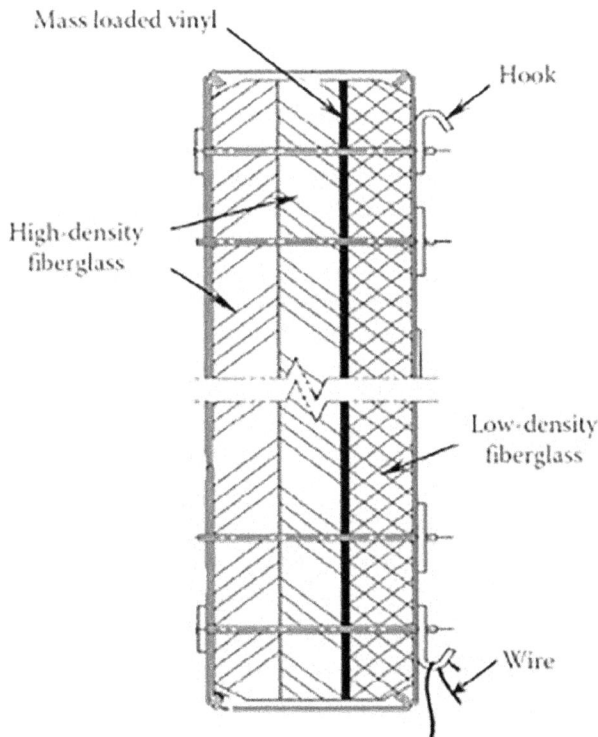

FIGURE 12.53 Acoustic blanket consisting of one low-density and two high-density fiberglass layers with a mass-loaded vinyl sandwiched between them.

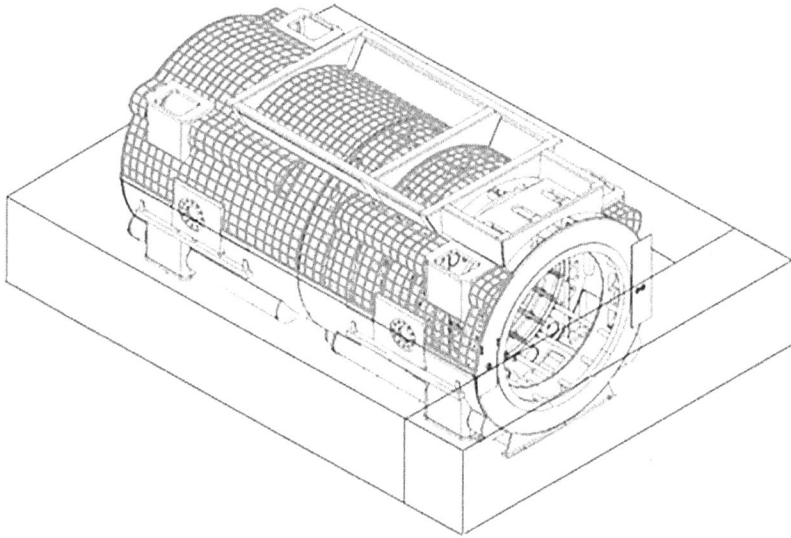

FIGURE 12.54 Covered areas by acoustic blankets on a generator.

To fit better the generator's external shape, blankets are made in a number of sections. Each section has a stainless steel identification tag that has a unique identification number corresponding to the installation drawings. Stainless steel jacketing, with a nominal thickness of 0.7 mm, is used to cover the blanket system. The jacketing normally utilizes long-reach stainless steel buckles and strikers. As with the blankets, each jacketing section has a stainless steel identification number corresponding to the installation drawings. As an alternative to separate stainless steel jacketing, integral stainless steel mesh is used. The mesh is factory-sewn to the fiberglass blanket assembly and acts to protect the blanket insulation from walk traffic and/or other abrasion damage.

A generator may be wrapped partially or completely with acoustic blankets. With the application of the generator laggings, acoustic noise at the generator bottom is blocked; only the upper half of the generator needs to be covered (Figure 12.54). The installation of the acoustic blanket on a large-sized generator is shown in Figure 12.55.

The test data have shown that acoustic blankets have a significant impact on generator noise levels. At the generator, sides were covered with blankets; the reduction in noise levels ranges from 2 (low frequencies) to 8 dBA (Figure 12.56). It is reported that the properly used acoustic blankets on electric machines can attenuate the noise pressure level by 5–10 dBA [12.91].

The spectra data at the tested generator are presented in Figure 12.57. It can be observed that blankets can reduce the noise level at all frequencies. For instance, at the center of the generator left side, the noise level reduces from 70.8 to 66.5 dBA at $f = 60$ Hz and 72.8 to 60.5 dBA at $f = 120$ Hz.

One of the most popular methods is to use some special materials to absorb sound energy. These materials such as foams or cellular materials comprise many small cavities. The drawback with these traditional noise reduction approaches is that they only work with some frequencies.

However, while the method is effective in noise reduction in some cases, it can impact the effectiveness of the heat dissipation from the housing to the ambient. In addition, this approach works for rather limited frequencies.

12.9.2.3 Motor Suspension Mounting

On undamped baseplate mountings, motor noise can be transmitted, amplified, and radiated by nonmotor structures. A motor suspension on rubber mounting or cushioned mounting can be added to the installation to reduce noise and vibration.

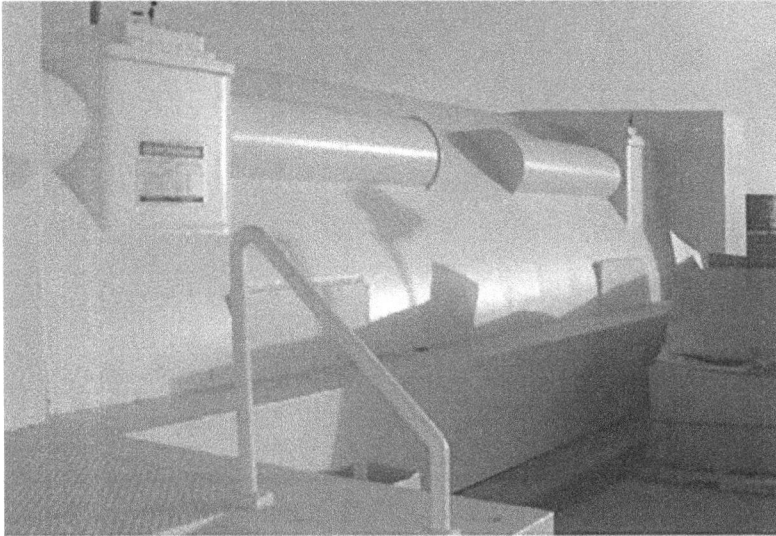

FIGURE 12.55 A large-sized generator with bared outer surfaces (a) and installed acoustic blankets for noise reduction (b).

12.9.2.4 Noise-Attenuating Structure

Ramu [12.92] proposed a method of noise reduction for switched reluctance machines (SRMs). SRMs usually produce higher noise than other types of electric machines, making them less attractive for some commercial and industrial applications. For instance, medical devices often require quiet operation with less noise. The SRM noise is attributed to (*a*) the normal forces caused by various imbalances in the nonuniform airgap, (*b*) the discontinuous currents in the windings causing discontinuous torque that produces very high torque pulsations, and (*c*) the rotor functioning like an impeller to create turbulent flows in the machine, causing significant windage noise.

SRM motors have no windings or PMs on rotors. As a rotor rotates, its salient poles are like fan blades that generate turbulent airflow in the motor. The fan blade noise is produced due to the

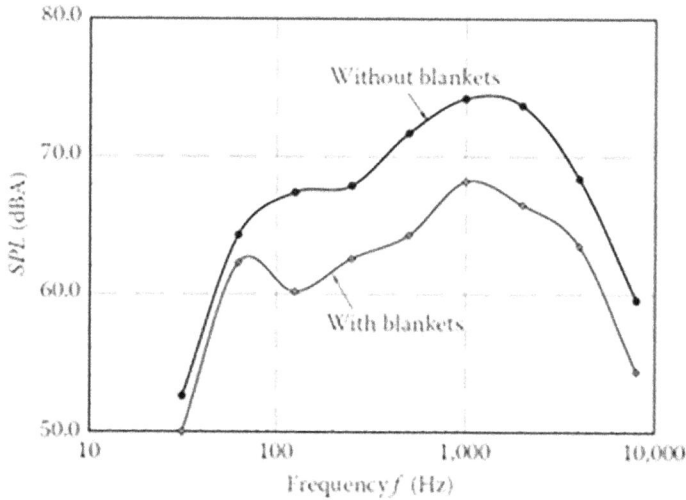

FIGURE 12.56   Effect of acoustic blanket on noise reduction for different frequencies.

alignment effect between the rotor and stator poles. The frequency f_{rt} of a pure tone arising from air passing through the rotor slots can be calculated by

$$f_{rt} = p_r \frac{n}{60} \tag{12.138}$$

where
    p_r is the number of rotor slots
    n is the rotor rotating speed (in rpm)

This tonal frequency noise has exactly the same value as the combined phase frequency of the machine.

SRM noise can be remarkably reduced by several methods, including the following:

- Encapsulating a machine's stator/rotor slots can reduce windage noise. As shown in Figure 12.58, the rotor slots are filled with epoxy to flush with the rotor outer surface and lamination end surfaces. The smooth rotor surface has much less friction than the surface with slots in which air forms recirculation flows to cause windage power loss and windage noise.
- The phase rotation of end lamination with respect to the intermediate laminations is another way to lower machine noise. It can be seen in Figure 12.59 that two end laminations are rotated oppositely about the rotational axis of the rotor to partially or fully cover the rotor slots. This creates a barrier inhibiting the airflow into and out of the slots.
- Similar to the previous discussion, the rotor/stator slots can be blocked using two end annuluses, as shown in Figure 12.60.

12.9.2.5   Smoothing Ventilation Path

When a motor rotor runs at a high speed, it stimulates air inside the motor to form swirl flow in the airgap and at the ends. Fans attached to the rotor also generate ventilation flows primarily in the axial direction to cool motor components. As a result, the velocity field in the motor is considerably complicated. Vortex motions are associated with regions of flow discontinuity that occur at interfaces between fluids and solids in relative motion or between parallel-moving flows of different velocities. The vorticity in turbulent flows is responsible for regions of relatively intense flow mixing and activity. Briefly, noise is produced whenever vortex lines are stretched or accelerated.

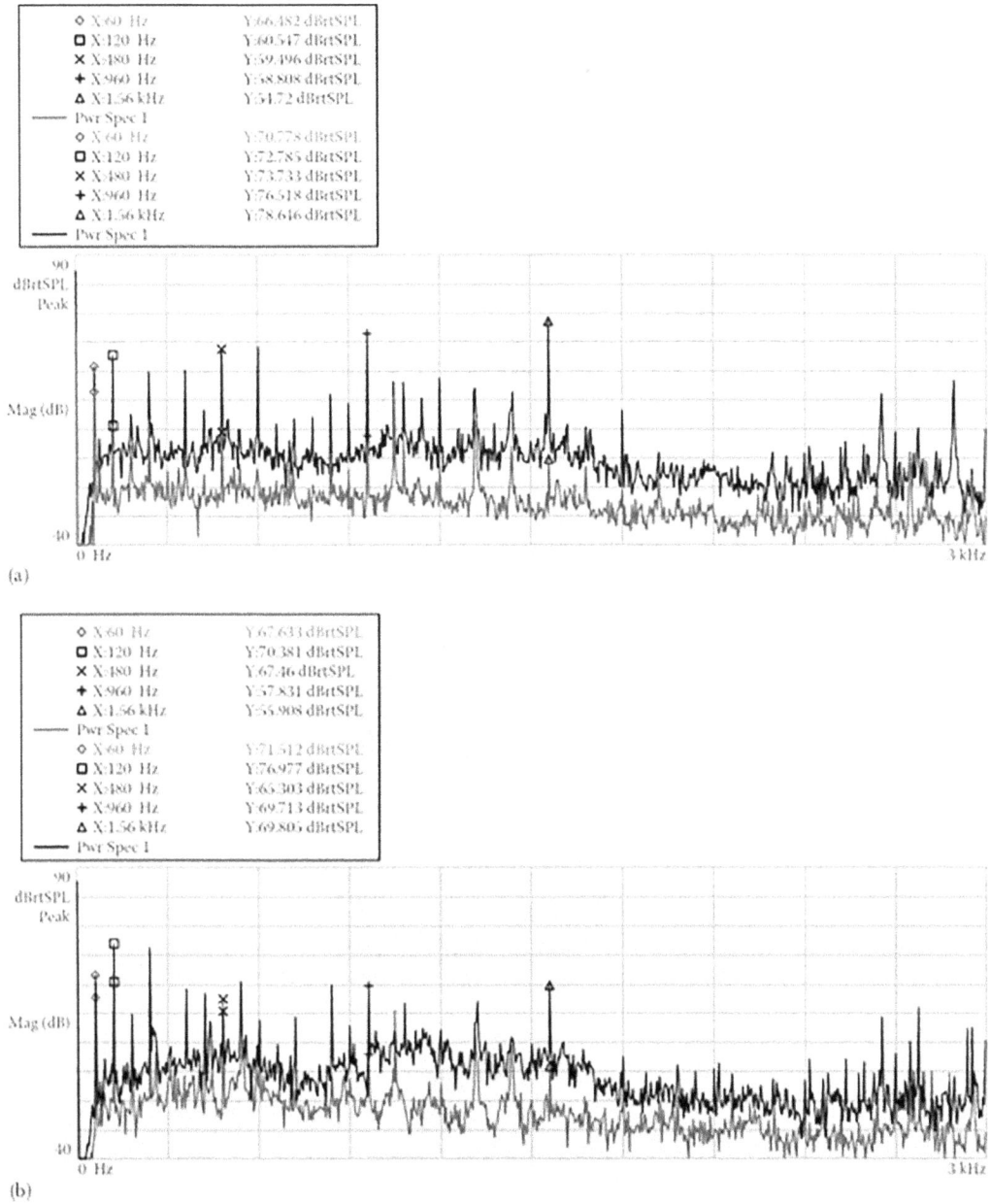

◇ X:60 Hz	Y:66.482 dBrtSPL
□ X:120 Hz	Y:60.547 dBrtSPL
✗ X:480 Hz	Y:59.496 dBrtSPL
✦ X:960 Hz	Y:58.808 dBrtSPL
△ X:1.56 kHz	Y:54.72 dBrtSPL
—— Pwr Spec 1	
◇ X:60 Hz	Y:70.778 dBrtSPL
□ X:120 Hz	Y:72.785 dBrtSPL
✗ X:480 Hz	Y:73.733 dBrtSPL
✦ X:960 Hz	Y:76.518 dBrtSPL
△ X:1.56 kHz	Y:78.646 dBrtSPL
—— Pwr Spec 1	

(a)

◇ X:60 Hz	Y:67.633 dBrtSPL
□ X:120 Hz	Y:70.381 dBrtSPL
✗ X:480 Hz	Y:67.46 dBrtSPL
✦ X:960 Hz	Y:57.831 dBrtSPL
△ X:1.56 kHz	Y:55.908 dBrtSPL
—— Pwr Spec 1	
◇ X:60 Hz	Y:71.512 dBrtSPL
□ X:120 Hz	Y:76.977 dBrtSPL
✗ X:480 Hz	Y:65.303 dBrtSPL
✦ X:960 Hz	Y:69.713 dBrtSPL
△ X:1.56 kHz	Y:69.805 dBrtSPL
—— Pwr Spec 1	

(b)

FIGURE 12.57 Spectra data at the center of generator sides: (a) left side and (b) right side; viewing from the CE. In each case, the curve with higher noise pressure levels is for uncovered acoustic blankets.

Smoothing ventilation paths can reduce the flow turbulence intensity and in turn the flow-induced noise, which usually arises in applications involving turbulent flows.

12.9.2.6 Selecting Low Noise Bearing

During normal operation, rolling bearings make a variety of vibration and acoustic noise. As discussed in Chapter 6, the bearing noise and vibration can be briefly grouped into four categories [12.93]—structural, manufacturing, handling, and others.

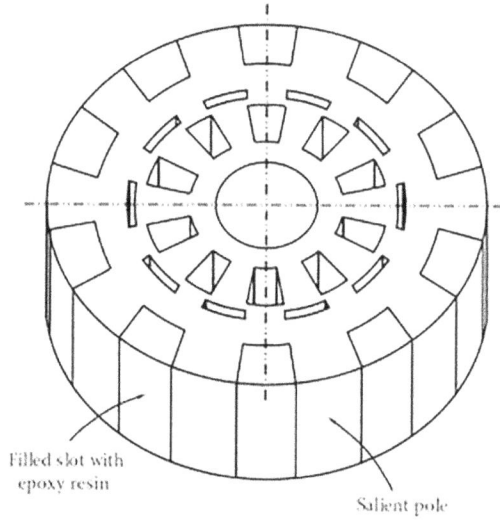

FIGURE 12.58 Rotor slots are filled with epoxy resin to reduce windage noise [12.92]. (Courtesy of the U.S. Patent and Trademark Office, Alexandria, VA.)

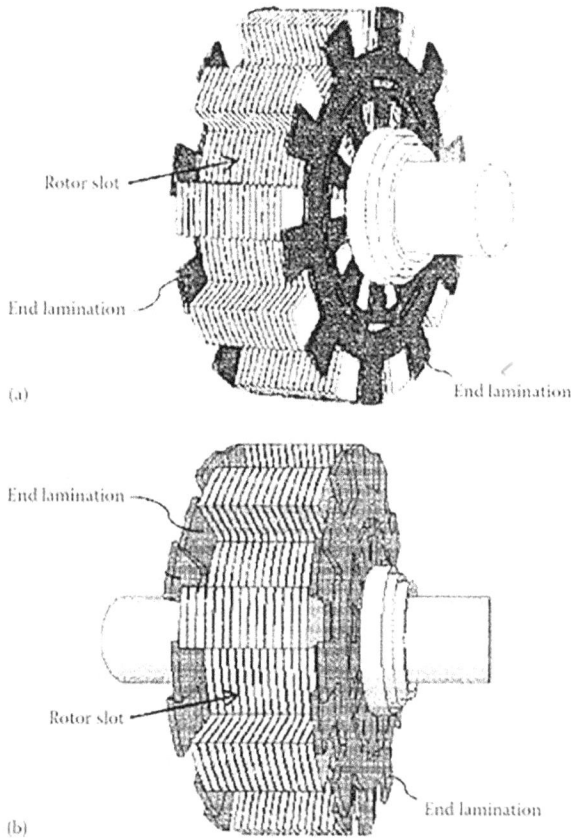

FIGURE 12.59 Phase-shifted end laminations for reducing noise: (*a*) shifting two end laminations in opposite directions with respect to intermediary laminations and (*b*) using multiple phase-shifted laminations for covering longitudinal sides of slots [12.92]. (Courtesy of the U.S. Patent and Trademark Office, Alexandria, VA.)

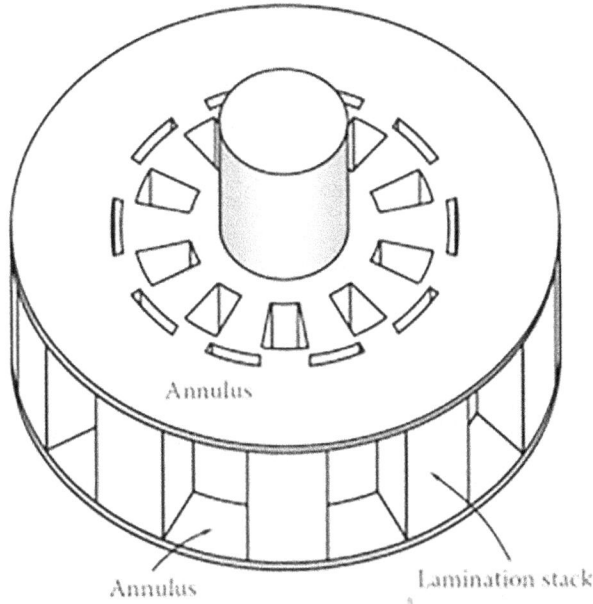

FIGURE 12.60 Adding an annulus at each end of rotor to block axial airflow paths and thus reduce acoustic noise [12.92]. (Courtesy of the U.S. Patent and Trademark Office, Alexandria, VA.)

- Structural noises resulted from free vibration of the bearing components, including raceway rings, rollers, seals, and cage.
- Ideally, bearing raceways are manufactured with perfect roundness and high-quality surface finish. However, because any machine has its limited machining precision, waviness usually presents on the manufactured raceway surfaces, causing waviness noise and vibration and impacting bearing frictional performance and endurance.
- Handling noise consists of flaw noise and contamination noise. Manufacturing flaws can lead to abnormal operation of bearing to produce noise. Contamination can dramatically shorten bearing life. According to failure analysis statistics, about 15% of all motor bearing failures is attributed to contamination [12.94]. Since the lubricant film thickness, which separates the mating surfaces in the rolling contact, is very thin (usually less than 1 μm), any particles in size larger than the film thickness can disturb the bearing smooth running.
- The last category of noise/vibration includes lubricant noise and seal noise. It is long recognized that oil-lubricated bearings have a longer lifetime than grease-lubricated bearings because greases always contain solid particles. There are a few greases in the market that fall into the *noisy* class due to their thickener or solid additives. Typical examples are some calcium-complex greases that contain large particles of calcium salts, producing small permanent dents of raceway surfaces [12.95]. Inadequate lubrication of bearings may not only create higher motor noise but also cause higher bearing temperatures.

Most motor manufacturers randomly sample bearings for inspections. However, since most motor failures can be attributed to premature bearing failures, it is highly recommended to conduct 100% bearing inspections, especially for some special applications under severe environmental conditions (*e.g.*, deep sea, outer space, or desert).

Among all types of bearings, sleeve bearings make the lowest noise and ball bearings generally make more noise than roller bearings [12.96].

12.9.3 Innovative Noise Abatement Methods

To address noise abatement problems, acoustic engineers presently turn to innovative methods that make new approaches to the physics of noise reduction. With the fast development of nanotechnologies in recent years, motor noise can be significantly reduced by applying nanofibers and by coating the housing and endbell internal surfaces with microsilicone paint (*e.g.*, *Lotusan*).

The advantages of using nanotechnology in motor noise reduction include

- High efficiency in noise reduction
- Taking a little space
- Low-density structure
- Reliable
- Easy to fabricate

Engineers in Georgia Tech Research Institute developed a porous metal structure that contained many cylindrical channels for attenuating aviation turbine noise [12.97]. The diameter of each channel ranges approximately from 0.1 to 0.3 mm. Later, this structure further evolved into a microscale honeycomb structure, which is composed of a large number of metallic nanotubes. This technique, unlike traditional techniques that absorb sound using a more frequency-dependent resonance, dissipates acoustic waves by wearing them down through a process called viscous shear. The reduction of the sound pressure level in this approach is independent of frequencies or resonance.

REFERENCES

12.1. Chen, J. X., Chen, X., and Luo, Z. M. 2018. Effect of mechanical vibration on microstructure and properties of cast AZ91D alloy. *Results in Physics* **11**(**2018**): 1022–1027.

12.2. Kudryashova, O., Khmeleva, M., Danilov, P., Dammer, V., Vorozhtsov, A., and Eskin, D. 2019. Optimizing the conditions of metal solidification with vibration. *Metals* **9**(**3**): Article ID 366.

12.3. Timár, P. L. 1989. *Noise and Vibration of Electrical Machines*. Elsevier Science, Amsterdam.

12.4. Yang, S. J. 1981. *Low-Noise Electrical Motors*. Oxford University Press, New York.

12.5. Granjon, P. 2005. Electromagnetic vibrations estimation of an induction motor by nonlinear optimal filtering. *Sixth IEEE International Symposium on Diagnostics for Electric Machines, Power Electronics and Drives (SDEMPED 2005)*, Vienna, Austria.

12.6. Capitaneanu, S. L., de Fomel, B., Fadel, M., and Jadot, F. 2003. On the acoustic noise radiated by PWM A.C motor drives. *Automatika* **44**(**3**): 137–145.

12.7. Bate, G. H. 1980. Vibration diagnostics for industrial electric motor drives. Brüel & Kjær. Application Notes. http://www.bksv.com/doc/bo02612.pdf.

12.8. Gerges, S. N. Y., Sehrndt, G. A., and Parthey, W. 2001. Chapter 5: Noise sources. In *Occupational Exposure to Noise: Evaluation, Prevention and Control* (Eds.: B. Goelzer, C. H. Hansen, and G. A. Sehrdt). Federal Institute for Occupational Safety and Healthy, Dortmund, pp. 103–124.

12.9. Finley, W. R., Hodowanec, M. M., and Holter, W. G. 2000. An analytical approach to solving motor vibration problems. *IEEE Transactions on Industrial Applications* **36**(**5**): 1467–1480.

12.10. Yoon, S. Y., Lin, Z., and Allaire, R E. 2013. *Control of Surge in Centrifugal Compressors by Active Magnetic Bearings: Theory and Implementation*. Springer, London.

12.11. Chikkamaranahalli, S. B., Vallance, R. R., Damazo, B. N., and Silver, R. M. 2006. Damping mechanisms for precision applications in UHV environment. *Proceedings of 2006 Spring Topical Meeting: Challenges at the Intersection of Precision Engineering and Vacuum Technology*, Pittsburgh, PA.

12.12. Han, D. C. and Moon, H. J. 1996. Lateral vibration analysis of the rotor system with magnetic forces. *Proceedings of the 6th International Symposium on Transport Phenomena and Dynamics of Rotating Machinery*, Honolulu, Hawaii, **Vol. 1**, pp. 87–96.

12.13. Ahamed, S. K., Mitra, M., Sengupta, S., and Sarkar, A. 2012. Identification of mass-unbalance in rotor of an induction motor through envelope analysis of motor starting current at no load. *Journal of Engineering Science and Technology Review* **5**(**1**): 83–812.

12.14. Eshleman, R. and Eubanks, A. 19612. On the critical speeds of a continuous rotor. *ASME Journal of Engineering for Industry* **91**: 1180–1188.

12.15. Evremsel, C. A., Kimmei, H. E., and Cullen, D. M. 19912. Axial rotor oscillations in cryogenic fluid machinery. *Proceedings of the Third ASME/JSME Joint Fluids Engineering Conference*, San Francisco, CA.

12.16. Supangat, R. 2008. On-line condition monitoring and detection of stator and rotor faults in induction motor. PhD thesis, University of Adelaide, Adelaide, South Australia, Australia.

12.17. Rashtchi, V. and Aghmasheh, R. 2010. A new method for identifying broken rotor bars in squirrel cage induction motor based on particle swarm optimization method. *World Academy of Science, Engineering and Technology* **67(2010)**: 694–698.

12.18. Thomson, W. T. and Rankin, D. 1987. Case histories of rotor winding fault diagnosis in induction motors. *Proceedings of Second International Conference on Condition Monitoring*, University College of Swansea, Wales, UK.

12.19. Faiz, J. and Ebrahimi, B. M. 20012. Locating rotor broken bars in induction motors using finite element method. *Energy Conversion and Management* **50(2009)**: 125–131.

12.20. Früchtenicht, J., Jordan H., and Seinsch, H. O. 1982. Exzentrizitätsfelder als Ursache von laufinstabilitäten bei asynchronmaschinen. Teil I und II. *Arch. Elektrotech* **65**: 271–292.

12.21. Arkkio, A., Antila, M., Pokki, K., Simon, A., and Lantto, E. 2000. Electromagnetic force on a whirling cage rotor. *IEEE Electric Power Applications* **147(5)**: 353–360.

12.22. Lundström, N. L. P. and Aidanpää, J.-O. 2011. Dynamics in large generators due to oval rotor and triangular stator shape. *Acta Mechanica Sinica* **27(1)**: 18–27.

12.23. Arkkio, A. 1987. Analysis of induction motors based on the numerical solution of the magnetic field and circuit equations. *Acta Polytechnica Scandinavica, Electrical Engineering Series* **59**: 1–97.

12.24. Antila, M. 1998. Electromechanical properties of radial active magnetic bearings. *Acta Polytechnica Scandinavica, Electrical Engineering Series* **92**: 1–96.

12.25. Clynch, J. R. and Garfield, N. 2006. Equations of an ellipse. http://www.gmat.unsw.edu.au/snap/gps/clynch_pdfs/ellipequ.pdf.

12.26. Lee, J.-H., Lee, Y.-H., Kim, D.-H., Lee, K.-S., and Park, I.-H. 2002. Dynamic vibration analysis of switched reluctance motor using magnetic charge force density and mechanical analysis. *IEEE Transactions on Applied Superconductivity* **12(1)**: 1511–1514.

12.27. Lipo, T. 2004. *Introduction to AC Machine Design.* Wisconsin Power Electronics Research Center, University of Wisconsin, Madison, WI.

12.28. Hariharan, V. and Srinivasan, P. S. S. 2010. Vibrational analysis of flexible coupling by considering unbalance. *Would Applied Science Journal* **8(8)**: 1022–1031.

12.29. Muljadi, E., Yildirim, D., Batan, T., and Butterfield, C. P. 19912. Understanding the unbalanced-voltage problem in wind turbine generation. *IEEE Industry Applications Society 34th IAS Annual Meeting*, Phoenix, AZ.

12.30. Hariharan, V. and Srinivasan, P. S. S. 2011. Vibration analysis of parallel misaligned shaft with ball bearing system. *Songklanakarin Journal of Science and Technology* **33(1)**: 61–68.

12.31. Al-Najjar, B. 1998. Improved effectiveness of vibration monitoring of rolling bearing in paper mills. *Journal of Engineering Tribology* **212(2)**: 111–120.

12.32. Ozelgin, I. 2008. Analysis of magnetic flux density for airgap eccentricity and bearing faults. *International Journal of Systems Applications Engineering and Development* **2(4)**: 162–1612.

12.33. Corbo, M. A. and Malanoski, S. B. 1996. Practical design against torsional vibration. *Proceedings of the 25th Turbomachinery Symposium*, Houston, TX, pp. 189–222.

12.34. Chung, D. D. L. 2001. Review: Materials for vibration damping. *Journal of Materials Science* **36(24)**: 5733–5737.

12.35. Mayer, D. and Herold, S. 2018. Chapter 1: Passive, adaptive, active vibration control, and integrated approaches. In *Vibration Analysis and Control in Mechanical Structures and Wind Energy Conversion Systems* (Ed.: F. Beltran-Carbajal). IntechOpen. doi:10.5772/intechopen.71838.

12.36. Collins, D. 2019. Vibration damping: What is the difference between passive and active methods? *Motion Control Tips*. https://www.motioncontroltips.com/vibration-damping-whats-the-difference-between-passive-and-active-methods/.

12.37. Macioce, P. 2003. Viscoelastic damping 101. *Sound and Vibration* **37(4)**: 8–10.

12.38. Chung, D. D. L. 2010. *Composite Materials: Science and Applications*, 2nd edn. Springer-Verlag, London.

12.39. Kulekei, M. K. 2008. Magnesium and its alloys applications in automotive industry. *International Journal of Advanced Manufacturing Technology* **39(9–10)**: 851–865.

12.40. Birchak, J. R. 1977. Damping capacity of structural materials. *The Shock and Vibration Digest* **9**(**4**): 3–11.

12.41. Schetky, L. M. and Perkins, J. 1978. The quiet alloys. *Machine Design* **50**(**8**): 202–206.

12.42. Bloomfield, D. and Firchau, T. 2013. Vibration isolation device and system. U.S. Patent Application 13/554,932.

12.43. Fullerton, L. W. and Roberts, M. D. 2012. Multi-level magnetic system for isolation of vibration. U.S. Patent 8,279,031.

12.44. Mangra, D., Sharma, S., and Jendrzejczyk, J. 1996. Passive vibration damping of the APS machine components. *Review of Scientific Instruments* **67**(**9**): 3374–3378.

12.45. Ahamed, R., Choi, S. B., and Ferdaus, M. M. 2018. A state of art on magneto-rheological materials and their potential applications. *Journal of Intelligent Material Systems and Structures* **29**(**10**): 2051–2095.

12.46. Ashtiani, M., Hashemabadi, S. H., and Ghaffari, A. 2015. A review on the magnetorheological fluid preparation and stabilization. *Journal of Magnetism and Magnetic Materials* **374**(**2015**): 716–730.

12.47. Zhu, C. S. 2005. Dynamic performance of a disk-type magnetorheological fluid damper under AC excitation. *Journal of Intelligent Material Systems and Structures* **16**(**5**): 449–461.

12.48. Zheng, J. J., Li, Y. C., and Wang, J. 2017. Design and multi-physics optimization of a novel magnetorheological damper with a variable resistance gap. *Journal of Mechanical Engineering* **231**(**17**): 3152–3168.

12.49. Hajalilou, A., Mazlan, S. A, Lavvafi, H, and Shameli, K. 2016. Magnetorheological fluid applications. In *Field Responsive Fluids as Smart Materials* (Eds: A. Hajalilou, H. Lavvafi, S. A. Mazlan, and K. Shameli). Springer, Singapore, pp. 67–82.

12.50. Do, X. P. and Choi, S. B. 2015. High loaded mounts for vibration control using magnetorheological fluid: Review of design configuration. *Shock and Vibration* **2015**: Article ID 91585.

12.51. Nguyen, Q. H., Choi, S. B., Lee, Y. S., and Han, M. S. 2013. Optimal design of high damping force engine mount featuring MR valve structure with both annular and radial flow paths. *Smart Materials and Structures* **22**(**11**): Article ID 115024.

12.52. Do, X. P., Shah, K., and Choi, S. B. 2014. A new magnetorheological mount featured by changeable damping gaps using a moved-plate valve structure. *Smart Materials and Structures* **23**(**12**): Article ID 125022.

12.53. Kang, O.-H., Kim, W.-H., Joo, W. H., and Park J.-H. 2013. Design of the magnetorheological mount with high damping force for medium speed diesel generators. *Proceedings SPIE 8688 Active and Passive Smart Structures and Integrated Systems*. doi:10.1117/2012831.

12.54. York, D., Wang, X. J., and Gordaninejad, F. 2011. A new magnetorheological mount for vibration control. *ASME Journal of Vibration and Acoustics* **133**(**3**): 031003.

12.55. Zhou, G. Y. 2004. Complex shear modulus of a magnetorheological elastomer. *Smart Materials and Structures* **13**: 1203–1210.

12.56. Opie, S. and Yim, W. 2011. Design and control of a real-time variable modulus vibration isolator. *Journal of Intelligent Material Systems and Structures* **22**: 113–125.

12.57. Sadek, F., Mohraz, B., Taylor, A. W., and Chung, R. M. 1997. A method of estimating the parameters of tuned mass dampers for seismic applications. *Earthquake Engineering and Structural Dynamics* **26**: 617–635.

12.58. Kourakis, I. 2005. Structural systems and tuned mass dampers of super-tall building: Case study of Taipei 101. Master thesis, Massachusetts Institute of Technology, Cambridge, MA.

12.59. Fortgang, J. and Singhose, W. 2005. Design of vibration absorbers for step motions and step disturbances. *ASME Journal of Mechanical Design* **127**(**1**): 160–163.

12.60. Deleon. Double mounting. http://www.deicon.com/Double_Mounting.html.

12.61. Tillema, H. G. 2003. Noise reduction of rotating machinery by viscoelastic bearing support. PhD dissipation, University of Twente, Enschede, the Netherlands.

12.62. Pai, N. G. and Hess, D. P. 2004. Dynamic loosening of threaded fasteners. *Noise & Vibration Worldwide*, February Issue, pp. 13–19.

12.63. Wakabayashi, K. 2011. 絶対にゆるまないネジ (*Absolutely Not Loosening Screws*). 中経出版 (Chukei Publishing), Tokyo.

12.64. Wakabayashi, K. 1985. A nut locking unit. European Patent EP 006541B1.

12.65. Wakabayashi, K. 1998. Nut free from inadvertent loosening and a method of making same. U.S. Patent 5,827,027.

12.66. Wakabayashi, K. 2002. Self-locking nuts (緩み止めナット). Japanese Patent JP 3272265B2.

12.67. Masato, N. 2004. Locking nut and its manufacturing method. Japan Patent Publication Number 2004-003587.

12.68. Masato, N. 2004. Locking bolt and its manufacturing method. Japan Patent Publication Number 2004-003585.

12.69. Noda, N.-A., Xiao, Y., Kuhara, M., Saito, K., Nagawa, M., Yumoto, A., and Ogasawara, A. 2008. Optimum design of thin walled tube on the mechanical performance of super lock nut. *Journal of Solid Mechanics and Material Engineering* **2**(**6**): 780–791.

12.70. Yang, X., Noda, N.-A., Kuhara, M., Saito, K., Nagawa, M., and Yumoto, A. 2010. Optimum design of thin walled tube on the mechanical performance of super stud bolt. *Journal of Solid Mechanics and Material Engineering* **4**(**10**): 1455–1466.

12.71. Nord-Lock Group. 2009. 7 second is all it takes. Bolted, No. 1, p. 20. http://cdn.nord-lock.com/wp-content/uploads/2012/05/Bolted_1_2009_EN.pdf.

12.72. Deutsches Institut für Normung. 2002. DIN 65151: 2002-08. Dynamic testing of the locking characteristics of fasteners under transverse loading conditions (vibration test). Berlin, Germany.

12.73. Tantanawat, T., Li, Z., and Kota, S. 2004. Application of compliant mechanisms to active vibration isolation systems. *Proceedings of ASME 2004 Design Engineering Technical Conference*, Salt Lake City, UT, **Vol. 2B**, pp. 1165–1172.

12.74. Platus, D. L. 19912. Negative-stiffness-mechanism vibration isolation systems. *Proceedings of the 1999 Optomechanical Engineering and Vibration Control* **3786**: 98–105.

12.75. Coffey, V. C. 20012. Optical-table basics: From breadboards to active vibration-control systems. *Laser Focus World* **45**: 63–66.

12.76. McBean, J. and Breazeal, C. 2004. Voice coil actuators for human-robot interaction. *Proceedings of 2004 IEEE/RSJ International Conference* **1**: 852–858.

12.77. Newell, D. B., Richman, S. J., Nelson, P. G., Stebbins, R. T., Bender, P. L., Faller, J. E., and Mason, J. 1997. Ultra-low-noise, low-frequency, six degrees of freedom active vibration isolator. *Review of Scientific Instruments* **68**(**8**): 3211–32112.

12.78. Yildiz, F. 20012. Potential ambient energy-harvesting sources and techniques. *The Journal of Technology* **35**(**1**): 40–48.

12.79. Sadeghian, M. and Bandpy, M. G. 2020. Technologies for aircraft noise reduction: A review. *Journal of Aeronautics & Aerospace Engineering* **9**(**1**): 1–10.

12.80. Howard, C. Q. 2011. Recent developments in submarine vibration isolation and noise control. *Proceedings of the First Submarine Science, Technology, and Engineering Conference*, Adelaide, Australia, pp. 283–288.

12.81. Cory, W. T. W. Relationship between sound pressure and sound power levels. http://www.eurovent-certification.com/fic_bdd/pdf_fr_fichier/1137149375_Review_67-Bill_Cory.pdf.

12.82. International Electrotechnical Commission (IEC). 2003. IEC Standard 61672 electroacoustics-sound level meters.

12.83. Gieras, J. F., Wang, C., and Lai, J. C. 2006. *Noise of Polyphase Electric Motors*. CRC Press, Boca Raton, FL.

12.84. Gerges, S. N. Y. 2000. *Ruído: Fundamentos e Controle*, 2nd edn. Universidade Federal de Santa Catarina, Florianópolis.

12.85. Čudina, M. and Prezelj, J. 2007. Noise generation by vacuum cleaner suction units: Part I. Noise generation mechanisms – An overview. *Applied Acoustics* **68**: 491–502.

12.86. Graham, J. B. 1972. How to estimate fan noise. *Sound and Vibration*, May Issue, pp. 224–227.

12.87. Lueg, P. 1934. Processing silencing sounds oscillations. U.S. Patent 2,043,416.

12.88. Olson, H. F. and May, E. G. 1953. Electronic sound absorber. *Journal of the Acoustical Society of America* **25**: 1130–1136.

12.89. Olson, H. F. 1956. Electronic of control of noise, vibration, and reverberation. *Journal of the Acoustical Society of America* **28**(**5**): 966–972.

12.90. Crucq, J. 1988. Theory and practice of acoustic noise control in electrical appliances. *Philips Technical Review* **44**(**4**): 123–134.

12.91. Tong, W., Wagner, T. A., Hughes, I. A., Gillivan, J. M., and Gibney III, J. J. 2004. Acoustic blanket for machinery and method for attenuating sound. U.S. Patent 6,722,466.

12.92. Ramu, K. 2012. Noise reduction structures for electrical machines. U.S. Patent Application 2012/01048712.

12.93. Momono, T. and Noda, B. 19912. Sound and vibration in rolling bearings. *Motion & Control* 6: 29–37.

12.94. Hink, R. 1998. Procedures for protecting electric motor bearings. *Plant Engineering*. http://www.plantengineering.com/search/search-single-display/procedures-for-protecting-electric-motor-bearings/2459e0f67d.html.

12.95. Bichler, M. and Roth, C. 2004. Test rig uses noise to measure grease cleanliness. *Practicing Oil Analysis*, May–June Issue.

12.96. Hoppler, R. and Errath, R. A. 2007. Motor bearings: "The bearing necessities." *Global Cement Magazine*, October Issue, pp. 26–32.

12.97. Nadler, J., Paun, F., Josso, P., Bacos, M.-P, and Gasser, S. 2011. Porous metal bodies used for attenuating aviation turbine noise. U.S. Patent 7,963,364.

13
Motor Testing

Safe, reliable, steady, and efficient operation of electric motors is essential for all motor applications. For example, elevator motors must have very high operating reliability and factors of safety because they are directly related to people's lives. Motor failure in a nuclear plant or in a launched rocket may cause catastrophic results, as well as tremendous economic losses and long-lasting impacts.

New motor design and development programs often incorporate prototype testing activities. Motor testing is a crucial step to ensure motor performance, efficiency, and manufacturing integrity. It can detect any potential faults, improve operation reliability, and optimize the motor design. For these purposes, a variety of tests must be conducted under different operating conditions. Testing variables may include the load (torque) level, rotating speed, voltage, current, induction, resistance, temperature distribution, and cooling flow rate.

Over recent decades, computer technology has become a key component in motor development and design processes. Computer-aided design, computer-aided manufacturing, and computer-aided engineering techniques today are widespread in their application to motor design. However, no matter how perfect the engineering design and how precise the theoretical calculations are, motor testing is a necessary step to validate the conceptual design, ensure the motor's normal and safe operation, and further optimize the motor design.

There are two kinds of testing during motor design, development, and production processes: verification testing and validation test. The main objective of verification testing is to ensure that the products (*e.g.*, motors) meet all design specifications and requirements. This kind of testing should be conducted iteratively throughout the product design and development processes. Validation in engineering is a dynamic mechanism of testing to validate that the products actually meet the expectations of end users. The validation process helps ensure that the products fulfill the desired use in appropriate environments. Verification and validation are independent procedures. In practice, validation usually comes after verification.

13.1 MOTOR TESTING STANDARDS

There are a large number of IEEE, IEC, ANSI/NEMA, and EASA standards dealing with electric motor testing. Some of them are listed as follows:

- IEEE Standard 43-2000. Recommended practice for testing insulation resistance of rotating machines [13.1]. This document describes the recommended procedure for measuring the insulation resistance of armature and field windings in rotating machines. It has addressed the general theories of insulation resistance and polarization index (PI).
- IEEE Standard 56-1977. Guide for insulation maintenance of large-current rotating machinery [13.2]. This standard provides guidelines for testing and inspection of insulation systems on motors.
- IEEE Standard 118-1978. Test code for resistance measurement [13.3]. The purpose of this code is to present methods of measuring electrical resistance that are commonly used to determine the characteristics of electric machinery and equipment.
- IEEE Standard 95-2002. Recommended practice for insulation testing of AC electric machinery (2,300 V and above) with high direct voltage [13.4]. This document provides

information on the use of high direct voltage for proof tests and periodic diagnostic tests on the ground wall insulation of stator windings in AC electric machines.

- IEEE Standard 522-2004. Guide for testing turn insulation of form-wound stator coils for alternating-current electric machines [13.5].
- IEEE 1068-1990. Recommended practice for the repair and rewinding of motors for the petroleum and chemical industry [13.6].
- ANSI/IEEE Standard 112-2004. Standard test procedure for polyphase induction motors (IMs) and generators [13.7].
- ANSI/NETA MTS-2007. Standard for maintenance testing specifications for electrical distribution equipment and systems [13.8].
- ANSI/NETA ATS-2009. Acceptance testing specifications for electrical power distribution equipment and systems [13.9].
- ANSI/NETA ETT-2013. Standard for certification of electrical testing technicians [13.10].
- NEMA MG1-1993. Motors and generators [13.11]. It provides practical information concerning performance, safety, test, construction, and manufacture of AC and DC motors and generators.
- IEC 60034-1: 2013. Rotating electrical machines—Part 1: Rating and performance [13.12],
- IEC 60034-2-1: 2007. Rotating electrical machines—Part 2-1: Standard methods for determining losses and efficiency from test (excluding machines for traction vehicles) [13.13].
- IEC 60034-4: 2008. Rotating electrical machines—Part 4: Method for determining synchronous machine quantities from tests [13.14].
- IEC 60034-19:1995. Rotating electrical machines—Part 19: Specific test methods for DC machines on conventional and rectifier-fed supplies [13.15].
- IEC 60034-29: 2008. Rotating electrical machines—Part 29: Equivalent loading and superposition techniques—indirect testing to determine temperature rise [13.16].

Because there are so many standards presently, motor manufacturers in different countries often follow different standards in their production. For instance, motor efficiency testing protocols differ from the world, and their applications on any given motor can lead to significantly different efficiency values. Therefore, it is important to unify the standards all over the world.

13.2 TESTING EQUIPMENT AND MEASURING INSTRUMENTS

There are a large variety of instruments and equipment used in motor test laboratories. The understanding of the functions and fundamental principles of these instruments and equipment, as well as correctly using them, is essential for the success of motor testing. In addition, all of the instruments must be carefully calibrated for obtaining accurate and reliable testing results. A brief introduction of the main instruments and equipment is given in the following.

13.2.1 Dynamometer

As one of the key devices in electric motor testing, dynamometers (or dynos in short) have been used extensively for measuring rotating speed, torque, power output, and force from power sources such as motors and engines. Based on the measured operation characteristic data, motor efficiency and other useful information can be determined. Dynamometers can operate in two basic modes: (*a*) the absorbing or passive mode in which a dynamometer is driven by the motor under test and provides a specific brake torque load to the testing unit and (*b*) the driving mode in which a dynamometer drives a machine to determine the torque and power required for operating such a driven machine. A dynamometer that can either drive or absorb is called a universal or active dynamometer.

Motor testing dynamometers apply braking or drag resistance to motor rotation under various operation conditions. Consequently, the absorbing dynamometer is used to absorb the power developed by the motor in motor testing. The brake power P_b, which is generated by a dynamometer and applied to the tested motor, is calculated by multiplying the torque T_{dyno} (Nm) and rotating speed ω_{dyno} (rad/s), that is,

$$P_b = T_{dyno}\omega_{dyno} \tag{13.1}$$

A variety of different dynamometers, either for absorption or drive mode or both, have been used for electric motor testing, including

- *Powder brake dynamometer (absorption only)*

  As its name implies, a powder brake dynamometer contains fine magnetic powder in the airgap between the rotor and stator of the brake. When current is applied to the excitation coil, a magnetic field is generated in the brake. Under the magnetic field, the magnetic powder is arranged in *magnetic chains* in accordance with magnetic flux lines, altering the powder from the free moving condition to the solid rock condition. As the brake rotates, the magnetic chains are constantly built and broken to create a great loading torque to the rotor. Thus, the loading torque can be easily adjusted by altering the excitation current. This type of dynamometer is ideal for high-torque, low-speed applications. It can provide full torque at zero rotating speed. Powder brake dynamometers are normally cooled with water and are not suitable for vertical operation.

- *Hysteresis brake dynamometer (absorption only)*

  A hysteresis brake dynamometer produces braking torque through a magnetic airgap without using magnetic particles or friction components. The hysteresis brake dynamometer usually consists of a cup-shaped rotor, a gear-shaped stator, and an excitation coil, providing precise torque loading that is independent of the shaft speed. The rotor is made of magnetic material with hysteresis characteristics. When the stator is energized, the cup-shaped rotor can spin freely. As current passes through the excitation coil, a magnetizing force applied to the stator and the rotor is magnetically restrained, providing a braking action between the rotor and stator. This method of braking provides far superior operating characteristics such as smooth torque loading, long lifetime, high repeatability, good controllability, and less maintenance and downtime, making them the preferred choice for precise torque control in electric motor testing. Because the energy absorption through hysteresis loss is proportional to the rotating speed of the rotor, the break torque generated in the hysteresis brake dynamometer does not depend on speed. In fact, the dynamometer functions as a stable brake in the full motor ramp from the stopped state through the full rotating speed. Cooling of this type of dynamometer usually relies on either natural convection or forced convection (using fans or compressed air). Figure 13.1 shows a motor being tested using a hysteresis brake dynamometer.

- *Eddy-current brake (absorption only)*

  Eddy-current brake dynamometers produce braking torques using the principle of eddy currents induced in rotating metallic disks, which are immersed in magnetic fields. An eddy-current dynamometer works on the principle of Faradays' Law of electromagnetic induction. As demonstrated in Figure 13.2, when electrical current passes through coils around the disk, it generates a magnetic resistance to the rotation of the disk. The resistance load is controlled by changing the current. They are suited for applications requiring high speeds and operating in the middle-to-high-power range. This type of dynamometer is capable of changing the load effectively at high speeds. The advantages of this type of dynamometer include high efficiency, low friction loss, precise control, compact and simple structure, desirable speed-torque characteristics, and low maintenance, compared to conventional mechanical dynamometers. However, the eddy-current dynamometer has slow response and is not capable of generating stall torque, compared to electrical dynamometers.

FIGURE 13.1   A motor is being tested by using a hysteresis brake dynamometer.

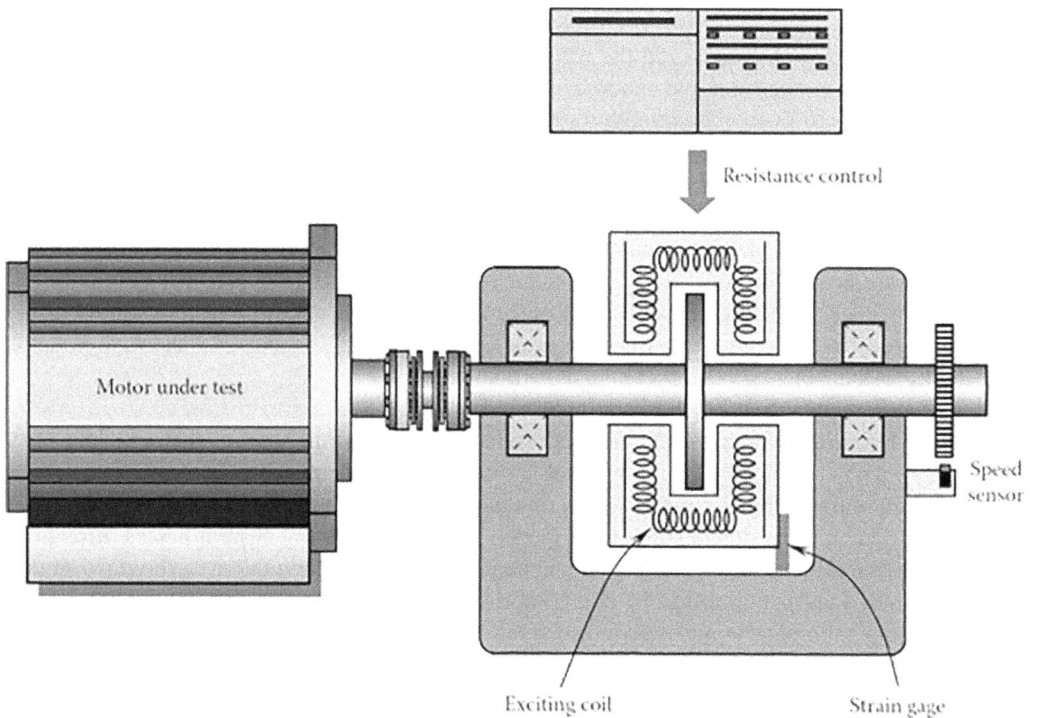

FIGURE 13.2   Eddy-current dynamometer used for motor testing, relying on eddy current generated in the rotating disk to provide resistant load to the tested motor.

- *Electric motor/generator dynamometer (absorption or drive)*

  Equipped with appropriate control units, electric motor/generator dynamometers can operate as either absorption or universal dynamometers, depending on applications. A motor/generator dynamometer typically uses an electric motor (either AC or DC) with its drive to simulate mechanical loads on the motor being tested. In a test, the motor under test is attached to the loading motor which acts as a generator. The main advantage of electric motor/generator dynamometers is that they can adjust the load in a large range within a very short time. However, electric motor/generator dynamometers are generally more expensive and complex than other types of dynamometers. A testing system is shown in Figure 13.3.

- *Mechanical friction brake dynamometer (absorption only)*

  Mechanical friction brake (Prony brake) dynamometers were invented by Gaspard de Prony in 1821. Using friction to load testing motors, this type of dynamometer is considered as one of the most simple, standard, and earliest absorbing dynamometers. Their advantages include ease of design, cost-effectiveness, simple maintenance, compactness, and high power absorption. However, due to the problems of brake torque control, as well as the rapid development in modern electronic control technologies, mechanical friction brake dynamometers are rarely used in motor labs today.

- *Inertia brake dynamometer (absorption only)*

  All brake types of dynamometers measure power from the torque reaction of the brake at a specific speed. An inertia brake dynamometer utilizes the test motor to accelerate a flywheel from a standing start-up through the motor speed range. This method involves measuring the change in the flywheel speed as the motor accelerates. The kinetic energy stored in the rotating flywheel is proportional to the square of the rotating speed. Thus, the flywheel has a significant amount of inertia to resist any change in speed. Today, inertia dynamometers become the standard test method for electric motors.

- *Water brake dynamometer (absorption only)*

  Water brake dynamometers utilize water flowing through absorbers to create loads on testing motors (Figure 13.4). This type of dynamometer is the most prevalent choice for applications that require high rotating speed, low inertia, high-power capacity, and low cost. The loading on the motor being tested is dependent on the water flow rate through the brake. For their high-power capability, water brake dynamometers are commonly used in high-power,

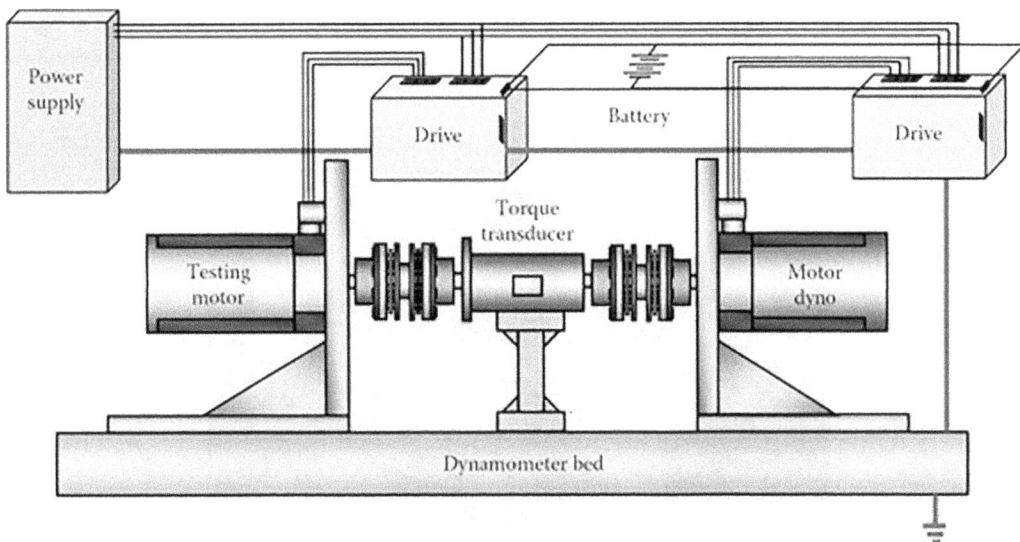

FIGURE 13.3 Motor testing system with the motor/generator dynamometer.

FIGURE 13.4   Water brake dynamometer for motor testing.

large-size motor testing. Unlike other types of dynamometers, water brake dynamometers eliminate the need for cooling and thermal overload protection systems and thus provide the most cost-effective solution for motor testing. However, using this type of dynamometer may take relatively long periods of time to reach the stabilized loading condition.

The testing system layout using a water brake dynamometer is shown in Figure 13.5. The motor is fixed on the test platform via steel chains and is connected with a torque transducer, which in turn is connected with the water brake dynamometer through a coupling.

- *Hydraulic brake dynamometer (absorption only)*

  Equipped with a hydraulic pump, a fluid reservoir, and a piping system, a hydraulic brake dynamometer provides braking drag against the motor being tested. This type of dynamometer can be used in a wide range of power. It is suitable to test extra-large motors with megawatts. Like water brake dynamometers, this type of dynamometer has no need for cooling systems.

- *Chassis dynamometer (both absorption and drive)*

  Chassis dynamometers are capable of very accurately measuring the speed, torque, and power. This type of dynamometer is widely used in the auto industry for performing engine tests. The typical loading modes used with a chassis dynamometer are constant

FIGURE 13.5   The motor being tested is driven by a water brake dynamometer.

force, constant speed, or a vehicle-simulation value. With the measured dynamometer shaft torque T_{dyno}, rotating speed ω_{dyno}, and the engine rotating speed ω_{eng}, the engine torque T_{eng} can be determined as [13.17]

$$T_{eng} = \frac{T_{dyno}\omega_{dyno}}{\omega_{eng}} \tag{13.2}$$

13.2.2 THERMOCOUPLES AND OTHER TEMPERATURE MEASURING DEVICES

Thermocouples are used to measure temperatures at different locations of the testing motor, as well as the ambient temperature. A thermal couple consists of a pair of wires, made of dissimilar thermoelectric materials. Because the two materials have different Seebeck coefficients, when the two wires are joined together at both ends, an electric potential is generated across the wire junctions. This electric potential is proportional to the temperature gradient across the wire junctions. Thus, an unknown temperature can be measured by placing one end of the wire junction (measurement junction) at the desired location and connecting the other end of the wire junction (reference junction) to a multimeter. The Seebeck coefficients for various thermocouple types are given in Table 13.1 [13.18].

For high-performance devices, the materials are typically p- and n-type semiconductors having positive and negative Seebeck coefficients, respectively. One end of the thermocouple analyzed here is connected by a metal tab. Metal tabs are also found at the other end of the thermocouple for external connections. Ceramic plates having high thermal conductivity are affixed at the ends of the thermocouple for heat spreading to ensure a uniform temperature at either end. The ceramic plates and metal interconnects are assumed to have different uniform temperatures at the top and bottom. Current density is also assumed to be constant over the areas of the junctions between different materials.

For most practical applications below $540°C$, the use of resistance-temperature detectors (RTDs) is a good choice. RTDs use the principle that some materials (*e.g.*, platinum) increase their electrical resistance with an increase in temperature. With very stable electrical properties, platinum exhibits a linear resistance-temperature relation over a wide range of temperatures. This has made platinum to be the best material for RTDs. In fact, RTDs are more accurate and have better sensitivity and linearity than thermocouples. By correlating the resistance of an RTD to temperature, the RTD can be used to measure temperature precisely. However, RTDs require external power sources, have slower response time and narrower measuring range, and are more expensive compared with thermocouples.

Another alternative temperature measuring device is a thermistor, which is a thermally sensitive resistor. Thermistor uses a semiconductor device to change its resistance. By comparing with thermocouples and RTDs, thermistors have the narrowest measuring range, lowest stability and linearity, highest sensitivity, and comparable accuracy and response time to thermocouples.

The continuous advances of infrared (IR) imaging technology have prompted the widespread use of IR cameras in a wide variety of industrial, healthcare, and military applications. These non-contact

TABLE 13.1
Seebeck Coefficients for Various Thermocouple Types at $25°C$

Thermocouple Type	Seebeck Coefficient ($\mu V/°C$)
E	61
J	52
K	41
N	27
R	9
S	6
T	41

instruments capture IR energy (heat) and use the data to create visual images through digital or analog video outputs. With the aid of thermal analysis software, IR cameras can perform non-contact real-time temperature measurements. Therefore, they can be used to continuously track and monitor the manufacturing process, implement component/system quality inspection, and help manufacturers catch and prevent maintenance issues before they lead to costly downtime.

Compared to other contact-style temperature sensors, IR cameras can detect heat radiating from behind the surface of an intricate assembly. They have some distinctive advantages, including wide temperature measurement range, visible display of output, high thermal sensitivity, high imaging resolution, fast response, compact size, motorized focus, and contact-free with detecting objects. They are especially suitable for determining the temperature at a specific spot on a surface. For example, an IR camera can be effectively used to measure the temperature of a motor seal, where the seal contacts a rotating shaft. However, since IR cameras are built with complicated components, they are typically expensive. In addition, they cannot be captured through certain materials like water and glass.

13.2.3 CONTROL SYSTEM

The control system in a dynamometer is one of the key elements to making stable operation of the dynamometer and getting accurate data. In motor testing, the precise control of a rotating speed and torque of the dynamometer is required. This is achieved automatically through the digital dynamometer controller, which is integrated into the testing system.

An absorption dynamometer acts as a load that is driven by the prime motor that is under test. The dynamometer can be equipped with two types of control systems, that is, constant torque and constant speed, to provide different test conditions [13.19]. With a torque regulator, the dynamometer operates at a set torque, while the prime motor operates at any speed and develops the torque that has been set. Similarly, with a speed regulator, it develops whatever torque is necessary to force the prime motor to operate at the set speed.

13.2.4 DATA ACQUISITION SYSTEM

During motor testing, many parameters need to be measured and recorded, including voltage, current, power factor, rotating speed, rotating direction, torque, electrical power, mechanical power, efficiency, and switch cutout speed, to name a few. The testing facility should have the capability to deal with the majority of the requirements of the motors, and the architecture allows customized tests to be easily added.

Data acquisition systems are used to collect motor operation characteristic data in motor testing. A typical data acquisition system consists of (*a*) data acquisition hardware, (*b*) sensors (*e.g.*, torque transducers) for measuring motor operation characteristic data, (*c*) signal conditioning hardware (*e.g.*, filter), (*d*) control and data acquisition software, (*e*) analog-to-digital convertor, and (*f*) a computer controlling data acquisition software. Most dynamometers are equipped with a data acquisition system. In selecting a proper data acquisition system, high accuracy, fast processing speed, and high-channel-count capacity are the main factors to be considered.

13.2.5 TORQUE TRANSDUCER

During motor testing, motor torque can be measured by connecting the testing motor to the dynamometer through a torque transducer or torque sensor, as shown in Figure 13.6. Torque transducers fall into two categories of measurement [13.20]:

- Reaction torque transducers. These types of transducers are nonrotating torque measuring devices and are typically mounted in fixed positions. With no moving parts, reaction torque transducers offer long-term reliability. They are suitable for a wide array of torque

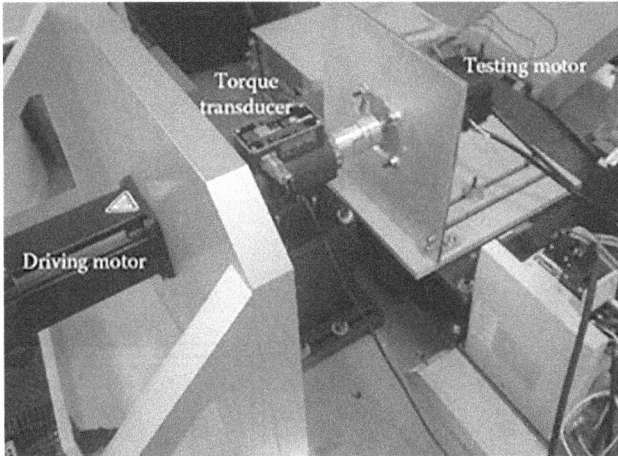

FIGURE 13.6 A torque transducer is placed between the testing motor and the driving motor.

measurement applications including torsional testing machines, brake testing, bearing friction studies, and dynamometer testing.

- Rotary torque transducers. This type of transducer employs a freely rotating shaft within a fixed housing. When the shaft is torsionally stressed, a proportional change in the signal is observed. With the broad measurement range and high-precision measurement capabilities, rotary torque transducers are extensively used in motor testing, automotive engine testing, fan testing, and hydraulic pump testing. Torque-angle data obtained from motor testing are often used to draw torque-time or torque-angle plots that reveal the actual motor performance characteristics.

13.2.6 POWER QUALITY ANALYZER

Power quality analyzer, also known as motor analyzer, is the ultimate multifunctional tool for motor testing. In addition to function as a digital multimeter, it can help locate, predict, prevent, and troubleshoot power quality problems in three-phase and single-phase power distribution systems, while providing the mechanical and electrical information of a motor for effectively evaluating the motor performance. Furthermore, this tool can conduct energy studies and power quality loggings by capturing and recording a large number of power quality parameters. Usually, the power analyzer utilizes three-phase current and voltage waveforms to calculate motor torque, rotating speed, mechanical load, and efficiency during each operation cycle. The motor airgap field, as observed via the voltage/current waveforms, provides the basis for the measurements.

Briefly, the key features of the advanced power analyzer include the following:

- Measuring key parameters on direct-on-line motors including torque, rotating speed, mechanical power, and motor efficiency
- Identifying power quality issues such as dips, swells, transients, harmonics, and unbalance
- Capturing electrical power parameters such as voltage, current, power, apparent power, power factor, harmonic distortion, and unbalance to identify characteristics that impact motor efficiency, eliminating the need for motor load sensors
- Analyzing software integration and compatibility
- Performing dynamic motor analysis by plotting of motor derating factor against load according to NEMA guidelines
- Using the waveform capture function to capture a certain number of cycles (based on the power supply frequency) of each detected event in all modes

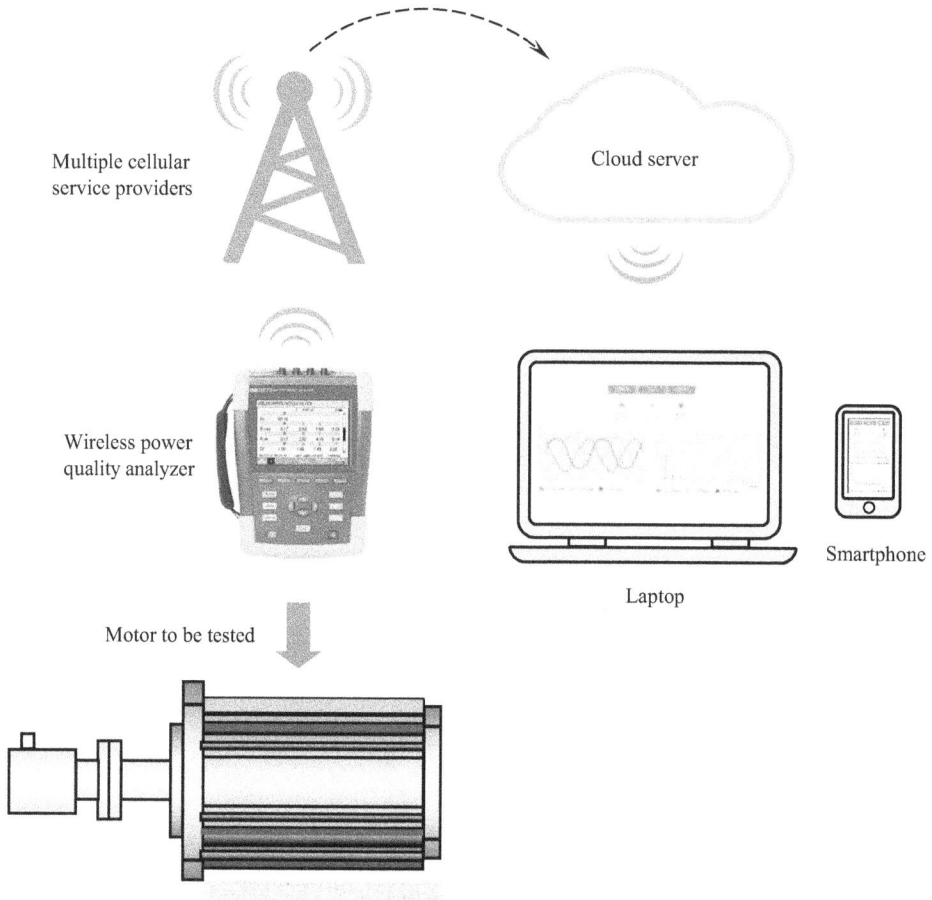

FIGURE 13.7   A wireless power quality analyzer with cloud-based data storage and transmission system for remotely diagnose and monitor motor performance.

As the demand for emerging technologies continues to increase, the motor testing systems have been updated with new technologies such as smart testing systems. More recently, real-time cloud-based power quality analyzers have been developed to remotely diagnose and monitor motor performance (Figure 13.7). The diagnostics can be made automatically using the AI and machine learning technologies. They are used to acquire the real-time information from the remotely located motors via Internet. With the advent of cloud computing technology, smart motor testing systems are now more efficient and scalable in processing, such as storage and access, the testing information with minimal development costs.

13.2.7 POWER SUPPLY

Power supply is a basic equipment in a motor test laboratory to provide the required power not only for the motor testing system and instruments but also for the control system for controlling the testing environment. It is widely accepted that the quality of power supply can significantly impact the motor efficiency. For instance, a phase unbalance of just 2% can increase losses by 25% [13.21]. Harmonics resulted from the voltage distortion can increase motor losses, produce vibration and noise, lower motor torque, induce torque pulsation, and result in motor overheating.

TABLE 13.2
Accuracies of Power Supply

Item	Standard IEC 34-2 [13.13]	Standard IEEE 112 (Method B) [13.7]
Power (W)	±1.0%	±0.2%
Current (I)	±0.5%	±0.2%
Voltage (V)	±0.5%	±0.2%
Frequency (Hz)	±0.5%	±0.1%

The power supply system must have the capability to provide a wide range of AC or DC voltage and current. In order to obtain accurate test results, the supply power, voltage, current, and frequency must be controlled within certain accuracies, defined by different standards (Table 13.2). It can be seen that the requirements of IEEE Standard 112-B are significantly stricter than those of IEC 34-2. Furthermore, because harmonic distortion can increase steeply the losses of electric motors, the total harmonic distortion of the three-phase power supply should be limited within 1%. In cases that significant nonlinear loads are involved, harmonic filters may be required to reduce harmonic distortion.

13.2.8 MOTOR TEST PLATFORM

Because testing motors come in a wide range of frame sizes, weights, loads, speeds, and power outputs, the motor test platform must be able to accommodate various testing motors, especially large-scale motors. To reduce the impact of vibration on the motor test, good vibrating damping materials such as cast iron are commonly used to construct motor test platform. To firmly hold large testing motors on the test platform, T-shaped slots are usually made on the top surface of the test platform. Small- and medium-sized motors can directly mount on an L-shaped stand. During a motor testing process, a good alignment between the testing motor and the dynamometer must be maintained.

Considering that during a motor testing process there are some systems/machines in the motor test laboratory that function as vibration sources at different frequencies, for example, air conditioning units, nearby equipment, other motor test systems, to name but a few, it is necessary to employ anti-vibration mounts under the test platform for suppressing structural vibrations and decoupling the system from the test cell floor. To achieve lower natural frequencies and therefore a high level of isolation, a possible solution is to use rubber isolation pads, which allow obtaining natural frequencies below 15 Hz. Another option is to use distributed dynamic vibration absorbers under the test platform. The vibration suppression is accomplished by the implementation of multiple mass–spring absorbers to minimize the test platform deflection at the natural frequencies. The simulation results have shown that an effective multi-mode vibration control can be realized by using the optimized distributed dynamic vibration absorbers [13.22].

13.3 TESTING LOAD LEVEL

There are a variety of methods to generate rotating torque loads to testing motors, such as the use of hysteresis brakes, eddy-current brakes, water brakes, and AC vector drives.

Motor performance test techniques generally fall into three categories: no-load, signature, and load tests. Each of these techniques contributes useful information to motor performance to a certain extent. However, each also has its limitation.

- No-load test

  As the simplest test technique, no-load test is also called *open-circuit test*, which is widely employed in the motor industry as one of the standard tests. During the testing, the rated voltage (balanced) and frequency are applied to the testing motor, and the motor shaft is uncoupled to any load. The resulting current, power, and speed are measured and compared to similar measured data made on a master motor. Therefore, this type of test requires less use of the test equipment and instrumentation. Three types of no-load test can be provided: disconnected, inferred no load, and measured no load. For the no-load test, the motor torque is close to zero, and current is very small.

According to IEEE Standard 112 [13.7], the measured input power in a no-load test is the total of the losses in the motor, including copper, friction, windage, and core losses. To determine the friction and windage losses, subtract the copper losses from the total loss (*i.e.*, input power) at each of the test voltage points and plot the resulting power curve versus voltage, extending the curve to zero voltage. The intercept with the zero-voltage axis is the friction and windage losses. The core losses at each test voltage are obtained by subtracting the value of friction and windage losses from the input power minus copper losses.

- Load point test

Many different load steps are programmed for motor testing, for instance, from 25% to 150% load levels. There are two test control modes: (*a*) speed mode in which a speed is established by the dynamometer and torque is produced based on the capability of the motor and (*b*) the torque mode in which a load is established by the dynamometer and speed is produced based on the capability of the motor. At each load level, the parameters that are measured include speed, torque, voltage, and current.

In practice, the load tests are conducted at the load level of 25%, 50%, 75%, 100%, 125%, and 150%. Among these, the 100% load point is most commonly adopted. If the speed mode testing is used, the test speed is set at the rated speed. Similarly, for the load mode testing, the torque is set at the rated torque.

13.4 TESTING METHODS

There are a number of testing methods for electric machines. According to the testing setup, motor testing methods can be classified as [13.23]

- Mechanic differential testing method
- Back-to-back testing method
- Indirect loading testing (*e.g.*, phantom, two-frequency, and inverse-driven) method
- Forward short-circuit testing method
- Variable inertia testing method
- Mixed-frequency testing method

Each of these testing methods has its own pros and cons. Among them, the most accurate and basic testing method is to directly load the testing motor either mechanically or electrically. Some of these testing methods are addressed in the following sections.

13.4.1 Mechanical Differential Testing Method

The mechanical differential method uses two identical motors, coupled with their main shafts through a differential cage. The two motors are connected to a common power supply in such a way that they rotate in opposite directions. As two motors rotate at the same speed, the differential cage remains stationary. If this cage is rotated, one of the motors accelerates and begins to act as a generator, while the other reduces its speed and commences to supply mechanical power to the generator through the differential gearbox. By adjusting the speed of rotation of the differential cage, it is possible to vary the loading of both motor and generator from no load to full load.

13.4.2 BACK-TO-BACK TESTING METHOD

This testing method also uses two motors that are not necessarily identical. This testing method has two distinct merits as the economy and the accuracy. In this method, the major part of the test power is circulated rather than dissipated, and the loss is measured as a net input rather than as the small difference between two separately measured large powers. As shown in Figure 13.8, the two motors are placed back to back each other and coupled to a floating gearbox. The two electric machines are powered from the same three-phase power supply. When they run at different speeds ω_1 and ω_2, respectively, the speed differences $(\omega_1 - \omega_2)$ can be adjusted by the floating gearbox [13.24]. A torque booster is mounted on the gearbox housing and arranged to boost the shaft rotation and to exert a reaction torque on the housing. The torque booster motor can drive the input shaft, the output shaft, or the intermediate shafts. Thus, the power loss of the gearbox can be compensated by the torque booster to ensure that the torques on both sides of the gearbox are equal.

13.4.3 INDIRECT LOADING TESTING METHOD

Direct loading methods often suffer from high costs due to the complex coupling and loading mechanisms. To be able to conveniently perform heat run tests on electric motors with indirect loading is very important for motor manufacturers. Indirect loading schemes include phantom loading methods, two-frequency methods, and inverse-driven methods. In the phantom testing method, two identical motors are connected electrically together, with their losses supplied externally [13.25]. With indirect loading on the testing motors, no loading machine is required, nor have the test motors to be mechanically coupled. Experiments on electric machines have shown that the test simulates a true temperature run with a high degree of accuracy.

As shown in Figure 13.9, with the phantom loading method, the phase voltage of stator winding has both AC and DC components. Thus, the phase voltage of stator winding is the superimposition of DC and AC components:

$$v = \sqrt{2}v_1 \cos\left(2\pi f_1\right) + V_{DC} \tag{13.3}$$

where the magnitude of V_{DC} depends upon the connection pattern of the state winding on phantom loading.

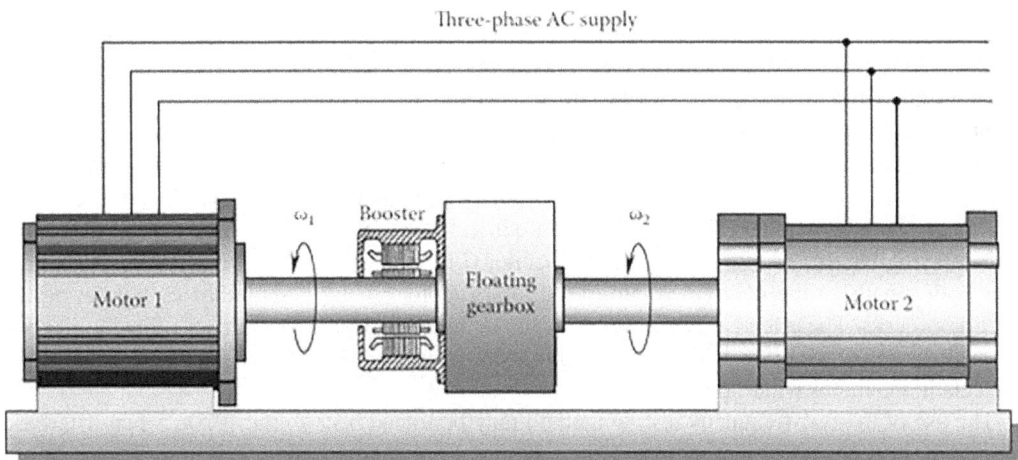

FIGURE 13.8 Back-to-back test using a floating gearbox.

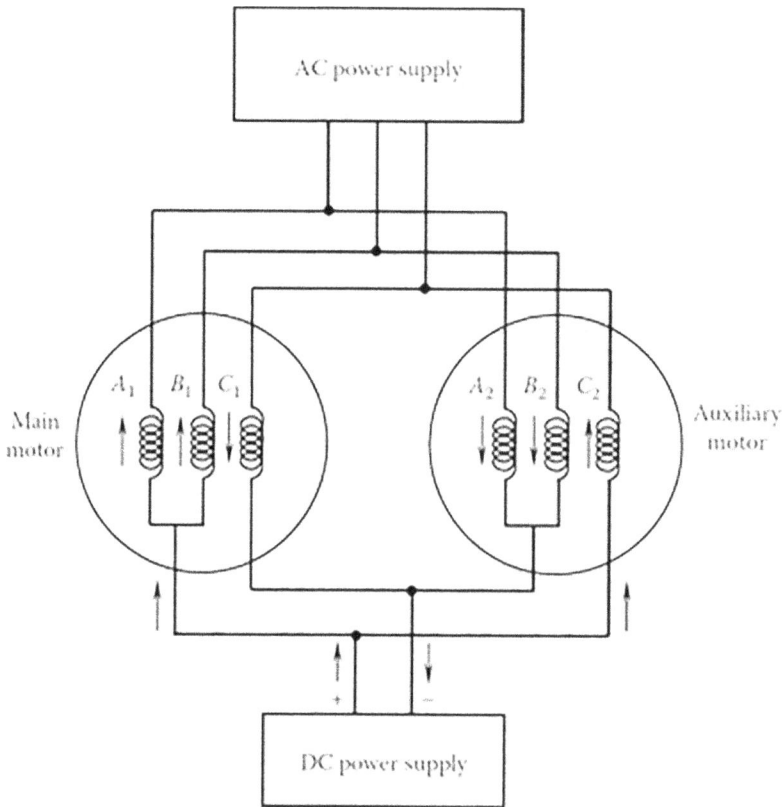

FIGURE 13.9 Phantom loading with star-connected stator windings.

Figure 13.9 depicts the phantom loading testing for two identical star-connected motors, without mechanical coupling between them [13.26]. Balanced three-phase voltages are applied parallelly on the star-connected stator windings of both the main and auxiliary motors. The DC is injected into the windings in series, and a stationary DC field appears in the stator–rotor airgap. Thus, both AC and DC circuits are symmetric with respect to the AC and DC power supply, respectively. In such a configuration, the AC and DC sources can feed the two star-connected circuits at the same time. However, this configuration requires that the voltage supply is $\sqrt{3}$ times higher than the normal value if the motors are connected in delta, as often encountered in industrial applications.

Most industrial-sized IMs are connected in a delta connection, as shown in Figure 13.10 [13.27]. It can be seen from this figure how the DC loading current is injected into the windings of a delta-connected motor. The three-phase AC power supply is applied on the main motor and the auxiliary motors in a parallel manner. Three resistors R are inserted into the supply source to prevent the DC from straying into the supply sources. The capacitor C is connected in parallel with the DC source to provide a path for the AC to feed into the auxiliary motor. Switch S should be closed during the starting process. The effective RMS values of the combined AC and DC currents flowing in each of the three-phase windings with such phantom loading method should be equal to the rated RMS currents in the motor windings [13.28].

The use of the two-frequency testing method may be traced back to the 1920s, when Ytterberg proposed a concept of connecting two voltage suppliers at different frequencies in series to produce synthetic loading of induction machine [13.29, 13.30]. This method is generally known today as the two-frequency or dual-frequency method.

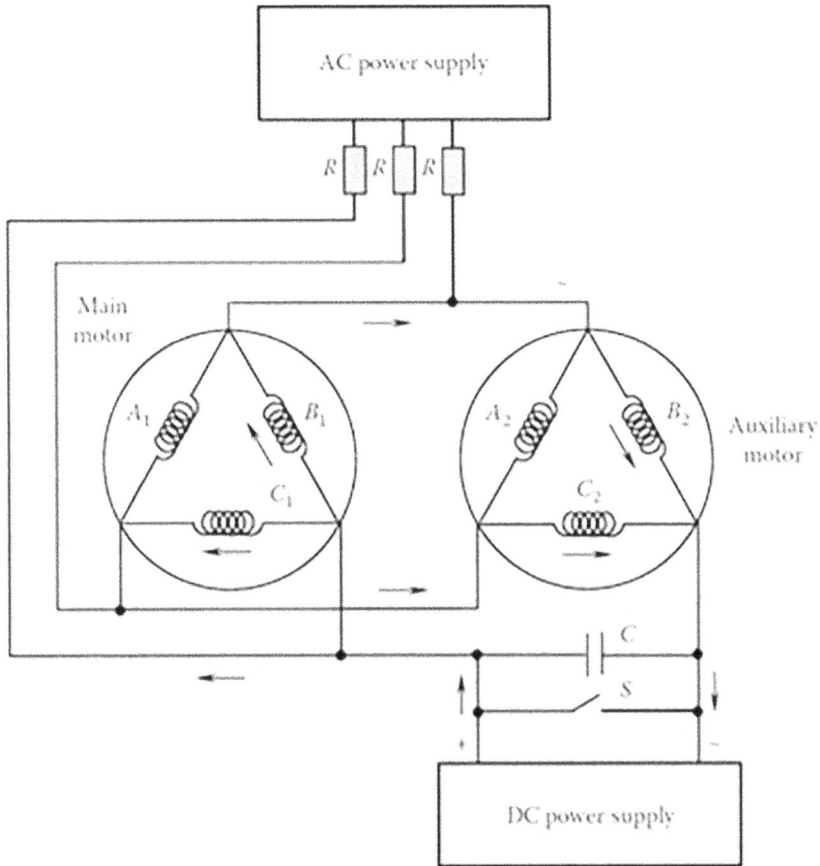

FIGURE 13.10 Phantom loading with delta-connected stator windings.

There are two possible loading schematics available in the two-frequency method, as presented in Figure 13.11. In each method, the main voltage v_1 has a rated frequency f_1, and the auxiliary voltage v_2 has a lower frequency f_2, where $f_2 = (0.60-0.95)f_1$. The magnitude of the auxiliary voltage v_2 is about 5%–25% of the main voltage f_1. Hence, the voltage at the stator winding becomes

$$v = \sqrt{2}v_1 \cos(2\pi f_1) + \sqrt{2}v_2 \cos(2\pi f_2) \tag{13.4}$$

Therefore, the IM draws motoring current with respect to the higher-speed field and supplies generating current with respect to the lower-speed field. The two currents beat together to form a modulated current, which is made equivalent to the normal load current for the motor [13.31].

It is noted that during the testing process, the two induction machines are only coupled electrically together without mechanical shaft coupling. In this arrangement, the electromagnetic torque oscillates at the frequency of (f_1-f_2). Such oscillations may cause serious problems in vertically mounted motors [13.32]. Essentially, two induction machines operate concurrently as motors and generators.

An inverter-driven method uses a microprocessor or a digital signal processor to generate the necessary logical control signals for PMW inverters to continually accelerate and decelerate the induction machine (Figure 13.12). The use of the power electronic inverter allows comprehensive control over all the quantities that determine the rate at which the power is dissipated inside the induction machine.

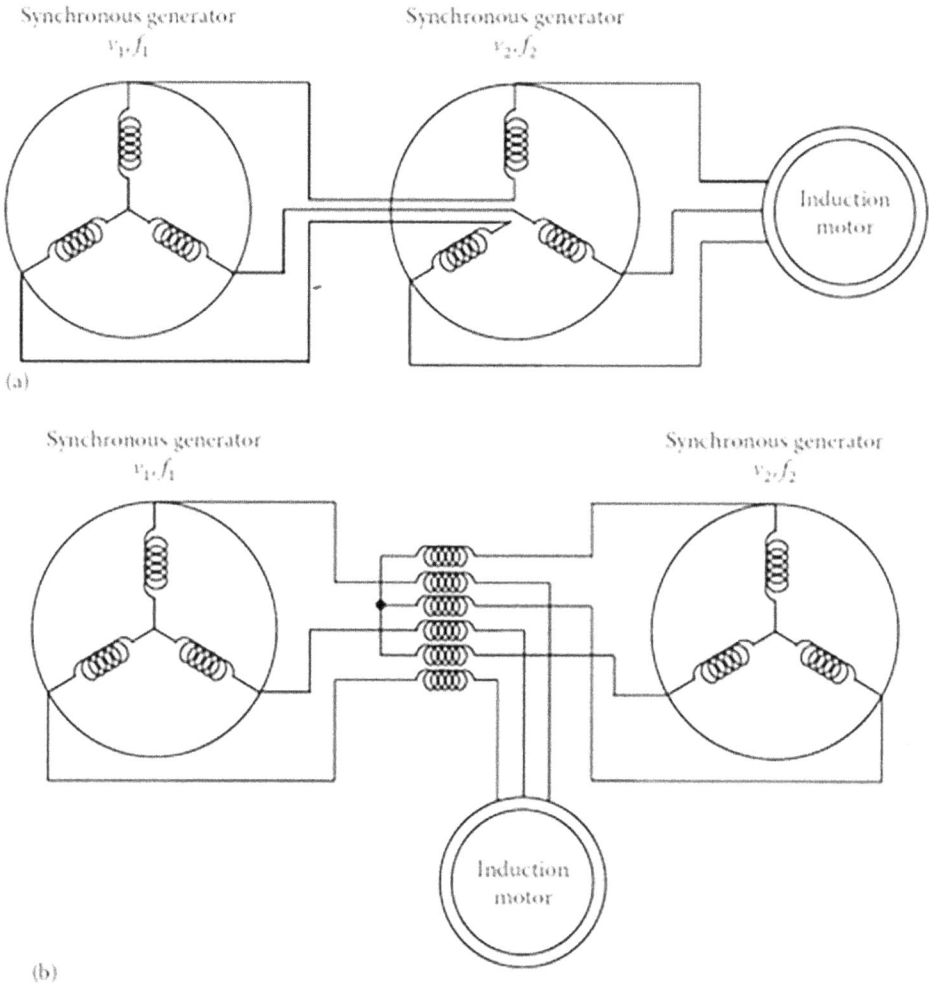

FIGURE 13.11 Two possible two-frequency loading methods.

13.4.4 FORWARD SHORT-CIRCUIT TESTING METHOD

This testing method is an alternate loading method that manufacturers could employ with a minimum capital expenditure, especially suitable for testing ultra-large electric motors with outputs greater than 1 MW [13.33]. This conventional indirect loading method has been applied by a leading motor manufacturer for more than a half century. In this method, the motor under testing is fed with a variable frequency power supply or an alternator and is driven at the rated speed by a DC motor. The speed difference between the test machine and the alternator can be correlated by adjusting the excitation of the alternator. The forward short-circuit testing method is given in Figure 13.13.

13.4.5 VARIABLE INERTIA TESTING METHOD

Full-load temperature-rise testing of IMs can be done by using the variable inertia testing method. This method, as a pure mechanical method in nature, is to load a test machine shaft directly with a flywheel (called *free-running inertia*) and does not need to connect electrically the test machine to any other machines.

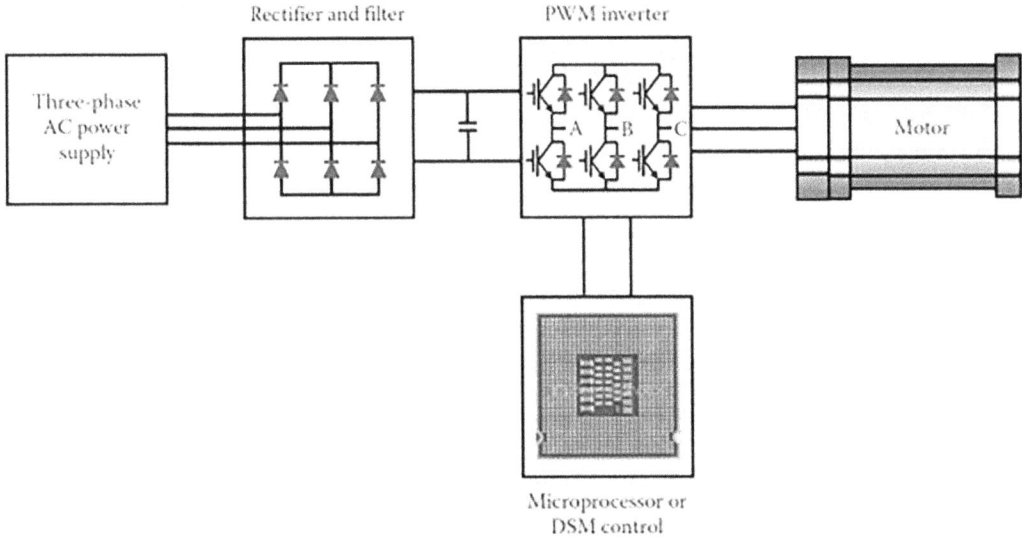

FIGURE 13.12　Block diagram of an inverter-driven method.

In variable inertia testing, the testing motor is connected to the flywheel through a gearbox. This gearbox has a nonconstant speed ratio that causes the inertia referred to the motor shaft to vary smoothly with the shaft rotation angle. The inertia variation results in the change in the shaft torque and in turn the change in the rotor speed [13.34].

13.5　OFF-LINE MOTOR TESTING

Off-line testing, or static testing, is a well-known means to comprehensively test electric motors. This testing method measures the integrity of the insulation and finds stator winding faults before they become catastrophic failures. A complete set of off-line tests include winding resistance test, megohm test, PI or dielectric absorption, step-voltage or high-potential (hipot) testing, and surge testing [13.35]. Off-line testing is also used as a quality assurance tool.

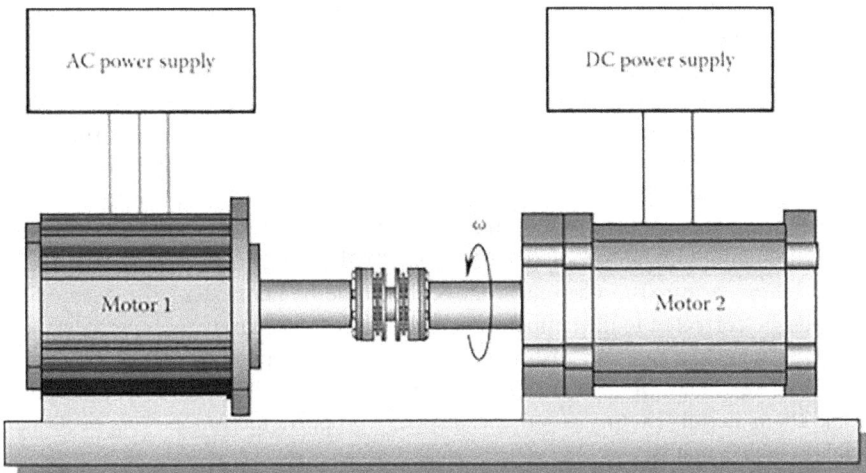

FIGURE 13.13　Forward short-circuit testing method.

TABLE 13.3
Measured Electrical Resistances from Two Large-Size Motors

Phase-to-Phase	Electrical Resistance (Ω)	Phase Resistance Unbalance (%)
U1-V1	0.02488	<0.5
V1-W1	0.02487	
W1-U1	0.02476	
U2-V2	0.02477	<0.2
V2-W2	0.02480	
W2-U2	0.02481	

13.5.1 Winding Electrical Resistance Testing

Electrical resistance testing involves the measurements of individual coil resistance and phase resistance using ohmmeters. For a three-phase motor, measured phase electrical resistance data (*i.e.*, *U-V*, *V-W*, and *W-U*) should be well within 5% of each other. For large-size motors, the phase resistance imbalances can be much lower. Variations between phases larger than the tolerance range for resistance are not allowed for any machine. As an example, the measured electrical resistance data of two large-size motors are shown in Table 13.3.

By taking a resistance reading of each phase, high-resistance connection can be identified. If there is a resistive imbalance exceeding a predetermined level, it can be caused by the circuit and/or the windings. In many cases, a high-resistance connection results from a solder joint between coils. Otherwise, an abnormal winding must be replaced.

13.5.2 Megohm Testing

Megohm testing, or insulation resistance testing, is to measure stator winding insulation resistance. This can help detect deteriorating insulation where the windings may potentially short to the ground. Megohm tests are usually conducted at constant DC voltages of 500–10,000 V, according to the motor's rated voltage. Readings of insulation resistance are taken after the DC voltage has been applied on the motor windings for at least 1 min. In a megohm test, the motor housing must be grounded for the safety consideration.

The insulation resistance of electric machine windings strongly depends on the insulation materials and the surface condition of the material. Generally, the insulation resistance varies proportionally with the insulation thickness and inversely with conductor surface area and temperature. Low megohm values are an indication of impending failure. However, high megohm values do not ensure a good motor. Performing a PI testing can further help determine poor insulation systems. It is recommended by IEEE 43-2000 [13.1] that (*a*) for most machines with random-wound stator windings and form-wound windings rated below 1 kV, the minimum insulation resistance is 5 M-Ω; (*b*) for most DC armature and AC windings built after about 1970 (form-wound coils), it is 100 M-Ω; and (*c*) for most windings made before about 1970 (all field windings), it is (kV + 1) M-Ω, where kV is the rated machine terminal-to-terminal voltage. It should be noted that all measured insulation resistances must be corrected to the temperature of 40°C. To get accurate measurements, the testing winding must be clean and free of moisture, and the winding temperature stable.

The electrical resistance R of a material is proportional to the resistivity of the material p (in Ω-m) and the material length l (in m) and inversely proportional to the cross-sectional area of the material A (in m^2), that is,

$$R = \frac{\rho l}{A}$$

(13.5)

where ρ is a function of temperature T,

$$\rho(T) = \rho_o(T_o)\left[1 + \alpha(T - T_o)\right] \tag{13.6}$$

where
  α is the temperature coefficient of resistance of the material
  T_o is the reference temperature

Substituting Equation 13.6 into Equation 13.5, it yields

$$R(T) = \frac{l}{A}\rho_o(T_o)\left[1 + \alpha(T - T_o)\right] = R_o(T_o)\left[1 + \alpha(T - T_o)\right] \tag{13.7}$$

Thus, a measured resistance R_m at an arbitrary measuring temperature T_m can be converted into the resistance at the reference temperature, for example, $T_o = 40°C$:

$$R(T = 40°C) = \frac{R_m(T_m)}{1 - \alpha(T_m - 40°C)} \tag{13.8}$$

This method can be useful in comparing the measured resistances of various materials at different measuring temperatures.

13.5.3 POLARIZATION INDEX TESTING

The purpose of PI testing is to determine the condition of motor insulation. In fact, the PI testing measures the time required for molecules of insulation to align (*i.e.*, polarize) from their random orientation to resist the flow of current. According to IEEE Standard 43-2000 [13.1], the PI is normally defined as the ratio of 10 min insulation resistance to the 1 min insulation resistance. This PI ratio provides an indication of the condition of the winding. As recommended by IEEE 43-2000, the minimum value of PI is 1.5 for the thermal class A and 2.0 for the classes B, F, and H.

The winding surface condition can significantly affect the insulation resistance or PI. For instance, carbon dust or other contaminations on winding insulation surfaces may become partially conductive when exposed to moisture or oil and thus can markedly lower the insulation resistance or PI. Lower PI values indicate that the winding insulation may be damaged or influenced by contamination (*e.g.*, dust and moisture) or/and embrittlement.

13.5.4 HIGH-POTENTIAL TESTING

The objective of the high-potential (hipot) testing is to verify the dielectric voltage withstand of the motor insulation system. It applies a very high DC voltage to the test components (*e.g.*, stator and motor) to measure overall insulation resistance to ground. The results provide very useful information on insulation dielectric strength. Today, this test is mandated to protect users from electric shock, fire, or smoke damage.

It is recommended by IEEE Standard 95-2002 [13.4] that the maximum test voltage is double the motor's rated voltage plus 1,000 V. At the beginning of the test, the motor frame is grounded first, and a DC voltage is applied on the insulation system with a slow increase in step increments until reaching the maximum test voltage. This operation avoids an electric shock on the insulation system. During the process, the leakage current data are recorded and plotted against the corresponding DC voltages on the computer monitor.

13.5.5 Surge Testing

It has been reported that copper-to-copper faults are the main cause of over 80% of all winding-related failures of electric motors [13.36]. In fact, insulation failure can be due to a variety of factors. During manufacturing processes, winding insulations may be damaged from bending, twining, or wrapping. Additionally, chemical deposits on the windings may break down insulation. Furthermore, movement of windings during motor start-ups might cause excessive wear on the coil insulation, leading to wearing away of the insulation and inevitably motor failure. Moreover, stresses that cause motor failures include deferential thermal stress, different coefficients of thermal expansion, varnish weakening at high temperatures, magnetic force due to winding currents, environmental contaminants, and moisture. These stresses cause looseness motion, and wear of the insulation [13.37].

The aim of winding insulation surge testing is to detect faults in turn-to-turn, coil-to-coil, phase-to-phase, and winding-to-core insulations that cannot be discovered by other testing methods (*e.g.*, megohm or hipot testing). Using this technique as a preventative maintenance tool, the damaged windings can be repaired or replaced to avoid motor failure in services, as well as caused emergency outage.

Surge testing applies very short high-voltage pulses to test electric winding and their insulation. The test is performed when all windings are installed into a motor but before they are connected to each other. This test is the most effective method of ensuring integrity of the copper-to-copper insulations and diagnosing actual shorted conditions between turns, coils, and phases. With an oscilloscope, some winding insulation faults such as insulation weakness, incipient faults, and other winding problems become visible on the oscilloscope screen.

Prior to surge testing, a number of measurements must be performed on the motor with acceptable results, including winding resistance, stator winding insulation, and PI measurements.

Surge testing has been known as surge comparison testing for a long time because insulation defects have been detected by comparing the waveforms from different phases of winding. While performing a surge comparison test, each phase of stator winding is tested against the others. Because all windings are assumed to be identical, the waveforms of the windings must coincide. Otherwise, there are some insulation issues in the winding. However, with today's advanced high-speed and high-accuracy measuring devices and computer systems, surge comparison is no longer necessary. Weak insulation can be diagnosed from frequency shifts and compared to successive waveforms within one phase [13.38]. Today, surge comparison testing is still used by some motor manufacturers.

In cases where there is a turn-to-turn short, the waveforms observed during testing exhibit distinct patterns on the display of the instrument. As a coil with weak insulation undergoes a surge test, the voltage applied on the coil can jump across the weak insulation, causing the shift of the waveform, as shown in Figure 13.14.

In a motor lab test, the surge testing is conducted with the rotor removed. But in a field testing, surge testing is performed with rotor installed.

13.5.6 Step-Voltage Testing

Step-voltage testing is designed to provide an additional evaluation of the motor winding insulation. Usually, step-voltage testing follows a successful PI testing. Actually, it is similar to the PI test but more extensive because a number of measurements are taken at various voltage step levels. The test is necessary to ensure that the winding insulation can withstand the normal voltage spikes during operation. However, step-voltage testing is not suitable for low-capacitance machines.

As shown in Figure 13.15, as the test progresses, a DC voltage is applied to all three phases of the winding and raised slowly to a preprogrammed voltage step level and held for a period of time. Then, it is raised to the next voltage step and held for a period of time. This process continues until the desired test voltage is reached. Typically, the steps in the test are made at 1 min intervals. The voltage interval is determined by dividing the maximum voltage by the number of steps. For small or middle motors, the voltage interval can be 500 V or 1,000 V.

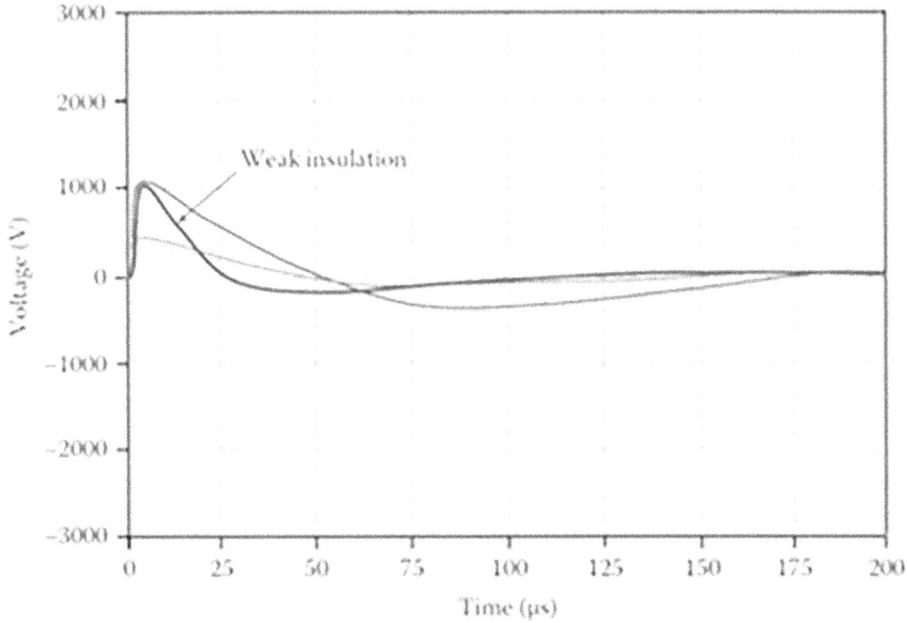

FIGURE 13.14   Surge test results. Waveform shift as an indication of weak insulation.

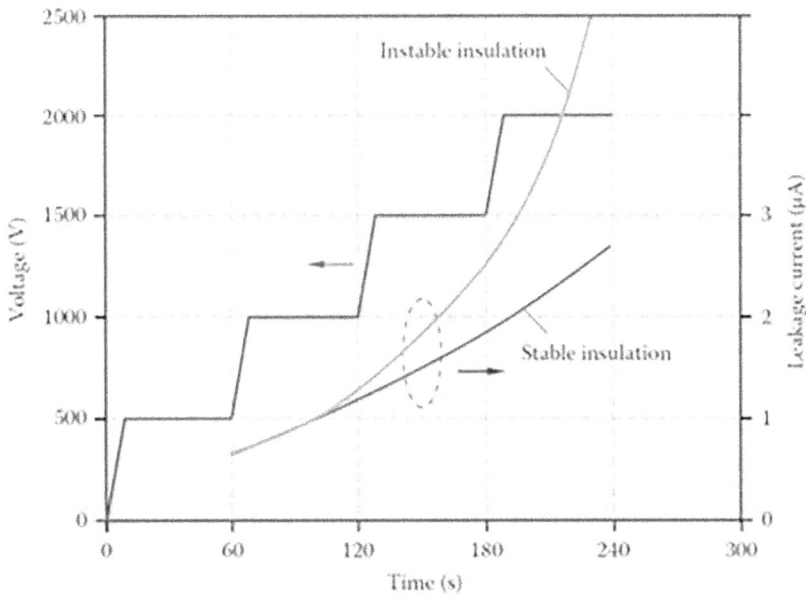

FIGURE 13.15   Step-voltage testing procedure.

At each voltage step level, the leakage current is monitored and plotted graphically against time. From the figure, it can be seen that for stable insulation, the leakage current exhibits nearly a linear pattern with time. A highly nonlinear behavior or an abnormal increase in leakage current may be indicative of insulation issues.

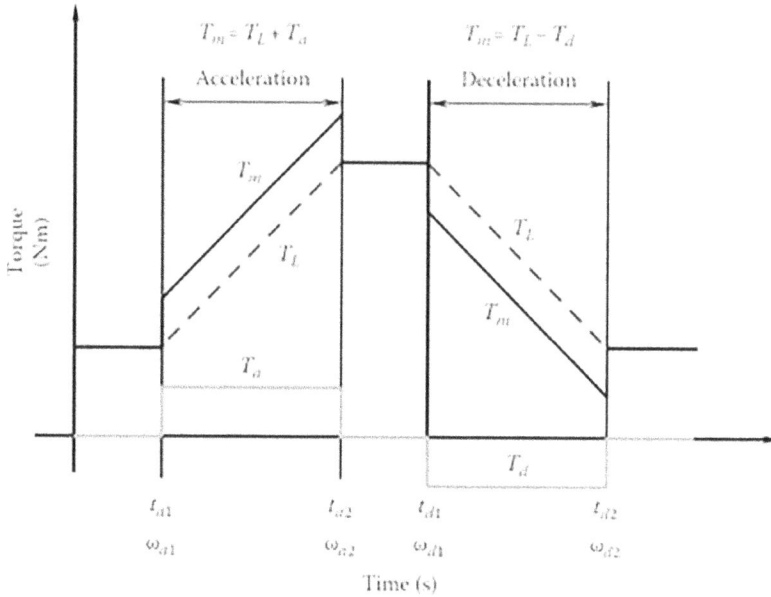

FIGURE 13.16 Relationships of motor torque, load torque, acceleration torque, and deceleration torque during the acceleration and deceleration processes in a test for determining the inertia of the test motor and load.

13.5.7 Determination of Rotor's Moment of Inertia

There are several experimental methods to determine the rotor moment of inertia. The traditional process for estimating the rotor moment of inertia involves operating a motor through an acceleration/deceleration profile [13.39].

In a motor test, for determining the moment of inertia of the rotor and its load, three torques, that is, the motor torque T_m, the load torque T_L, and the acceleration torque T_a (or deceleration torque T_d), vary with time during acceleration (or deceleration), as demonstrated in Figure 13.16. The relationships of these torque components are given as

$$T_m = T_L + T_a \quad \text{during acceleration} \tag{13.9a}$$

$$T_m = T_L - T_d \quad \text{during deceleration} \tag{13.9b}$$

It is noted that the electric motor accelerates or decelerates in a linear manner. The acceleration torque T_a and the deceleration torque T_d are constant during the acceleration and deceleration processes. Then, the rotor moment of inertia can be calculated by either of the following equations:

$$I\frac{d\omega}{dt} = T_m - T_L = T_a \quad \text{in the acceleration process} \tag{13.10a}$$

$$I\frac{d\omega}{dt} = T_m - T_L = T_d \quad \text{in the deceleration process} \tag{13.10b}$$

Integrating both sides of the above equations corresponding to their time and angular intervals, the moment of inertia can be easily obtained. Assuming that the velocity difference in the acceleration process $\Delta\omega_a$ is equal to the velocity difference in the deceleration process $\Delta\omega_d$, that is, $|\Delta\omega_a| = |\Delta\omega_d| = \Delta\omega$, the averaged I value is given by the following equation [13.40]:

$$I = \frac{1}{2}\left[\frac{\int_{t_{a1}}^{t_{a2}}(T_m - T_L)dt}{\Delta\omega} + \frac{\int_{t_{d1}}^{t_{d2}}(T_m - T_L)dt}{-\Delta\omega}\right] \tag{13.11}$$

Rehm and Golownia [13.41] proposed a method to accelerate and decelerate a testing motor in a nonlinear manner. During acceleration and deceleration processes, the torque produced by the electric motor is sampled periodically, and the averaged torque during the acceleration/deceleration process is calculated by dividing the summation of all measure torque values with the number of torque samples. The rate of acceleration/deceleration is also determined. Therefore, the inertia (in units of kg·m^2) of the rotor and the load connected to the motor is calculated as

$$I = \frac{\left|\frac{1}{N_a}\sum_{i=1}^{N_a}T_{a,i}\right| + \left|\frac{1}{N_d}\sum_{i=1}^{N_d}T_{d,j}\right|}{\left|\frac{\Delta\omega_a}{t_a}\right| + \left|\frac{\Delta\omega_d}{t_d}\right|} \tag{13.12}$$

where
$T_{a,i}$ is the ith torque sample acquired during the motor acceleration process
$T_{d,j}$ is the jth torque sample acquired during the motor deceleration process
N_a and N_d are the numbers of torque samples corresponding to acceleration and deceleration, respectively
ω_a and ω_d are the rotating speed changes in units of rad/s during acceleration and deceleration, respectively
t_a and t_d are the acceleration time and deceleration time, respectively

Another effective method is called the oscillation method. To get an oscillating motion about the rotation point of the electric motor, a lever is attached to the rotor. On the end of the lever, an elastic spring with stiffness k is fixed on it. By measuring the period of time T of an oscillation, the moment of inertia can be calculated as [13.42]

$$I = \frac{kT^2}{4\pi^2}\left(1 - \zeta^2\right) \tag{13.13}$$

where ζ is the damping ratio. It is noted that Equation 13.12 is valid for one spring. When two springs are used to fix on the end of the lever in the opposite direction, with the distance L from the springs to the rotation point (see Figure 13.17), the equation for calculating I becomes

$$I = \frac{2kT^2L^2}{4\pi^2}\left(1 - \zeta^2\right) \tag{13.14}$$

13.6 ONLINE MOTOR TESTING

Online testing or dynamic testing, as its name implies, is referred to the condition that a motor operates within its normal environment at standard loads. Online testing is usually designed to evaluate dynamic performance and operating characteristics of the entire motor, such as power quality,

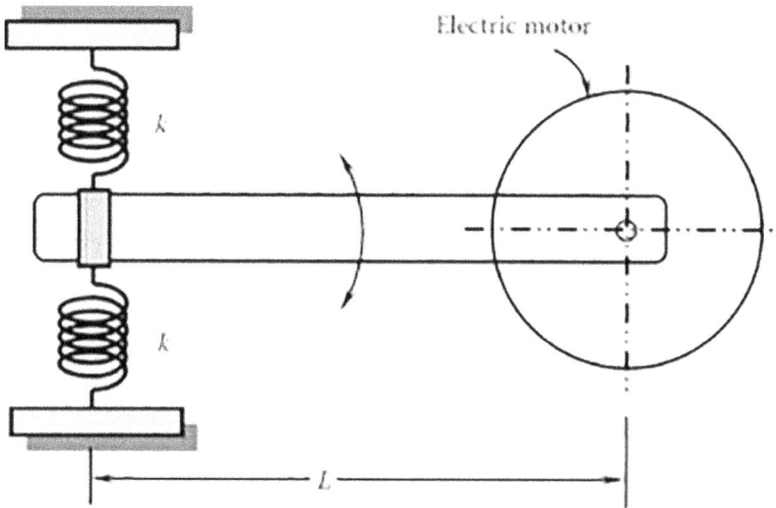

FIGURE 13.17   Oscillation test setup for measuring rotor moment of inertia.

motor efficiency, temperature rise, torque ripple, voltage/current level, voltage/current unbalance, and harmonic distortion. In addition to the determination of electrical and thermal problems with the motor, online testing can also test a variety of other issuers within the system, including misalignment, load mismatch, vibration, and acoustic noise, as well as problems with motor components such as rotor bars, braking systems, transmission systems, couplings, and bearings.

13.6.1 Locked Rotor Testing

Locked rotor tests are used to determine the starting torque and current of electric motors, which are the two main indicators for assessing motor performance. In this testing method, the testing motor is loaded to the maximum limits by applying maximum current to a hysteresis or powder brake type of a dynamometer or by mechanically holding the rotor shaft from turning and energizing the stator. The voltage supplied to the stator is reduced to prevent stator winding overheating. At the rotor-locking condition, the motor operation parameters such as torque, current, and voltage are measured and displayed. Sometimes, the locked rotor test is also called stalled torque test or blocked rotor test.

Locked rotor testing is very important for ensuring that the tested motor has sufficient torque to accelerate the motor when driving a specific load. If the motor cannot produce enough torque to overcome the friction in the load, the motor can be energized, but it will not drive the load. The test data can reflect the rationality of the magnetic circuit in the stator and rotor. Based on the test results, engineers can optimize the design and improve the production process.

However, this test is very hard on the motor. During testing, there is a large amount of current that flows into the motor, causing the motor to heat up rapidly. As a result, this test must be performed very quickly. With the locked rotor, a motor may draw up to six to eight times its rated current. The power supply used must be capable of regulating the motor voltage adequately during rapid changes in current to ensure the proper voltage is maintained when the data are being taken.

13.6.2 Motor Heat Run Testing

Motor heat run testing is sometimes known as motor continuous performance testing. This testing is extremely important to both motor manufacturers and end users. It was estimated that for every $10°C$ increase in temperature, the winding insulation life is reduced by half [13.43]. Exceeding temperatures can cause PM demagnetization and thermal aging of insulation materials.

The results of heat run tests are critical for verifying the engineering design and analysis and optimizing the motor design. More specifically, heat run tests can determine the maximum operating temperature of the tested motor and confirm that the maximum operating temperature is lower than the highest allowable temperature, which is defined by the motor insulation class based on an average 20,000 h lifetime.

A sustained heat run test is carried out at motor's full load. During the testing process, the temperature of the motor increases exponentially as various power losses are generated in the motor and converted into heat to dissipate to the environment. After a period of time, the motor reaches the thermal equilibrium condition, and the motor temperature tends to be constant. The time taken to reach equilibrium can vary widely depending on the machine size, cooling method, motor power losses, operation conditions, and ambient temperature. To obtain the temperature distribution of the motor, the temperatures of various motor components such as stator windings, bearings, PM, frame, endbells, encoder, and inlet and exit coolant, just to name a few, are measured and recorded. Temperatures on the rotating components (e.g., the shaft temperature at the shaft seal) can be conveniently measured using a thermal camera.

Ambient temperature T_a is the temperature of the air surrounding the motor. For most motor manufacturers, this temperature is defined at 40°C in their product catalogs. This does not mean the ambient temperature is set at 40°C during a heat run test. As a matter of fact, the ambient temperature in a test is controlled to maintain relatively constant at any given temperature.

The temperature rise of a motor is the change in motor temperature with respect to the ambient temperature when power is applied to the motor. Obviously, the higher the ambient temperature, the lower the temperature rise of a motor. For instance, class B insulation is rated for a maximum operating temperature of 130°C. When the ambient temperature is 40°C, the allowable motor temperature rise is 90°C. As the ambient temperature becomes 50°C, the temperature rise reduces to 80°C. By considering the effect of hot spots occurring in motor windings, a 10°C hot-spot allowance is subtracted from the allowable temperature rise for the safety consideration. Thus, for $T_a = 40°C$, the allowable temperature rise for class B insulation in the earlier example becomes 80°C rather than 90°C.

As an example, the results of a heat run test for a servomotor at 2,500 rpm and 45 Nm torque are provided in Figure 13.18. It shows that it takes about 4.2 h to reach thermal equilibrium. While

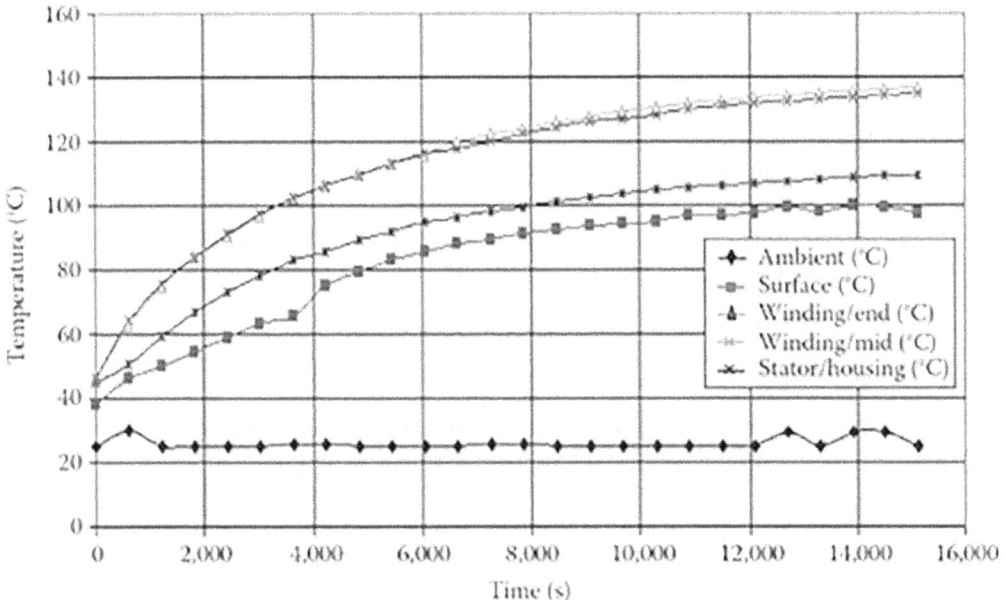

FIGURE 13.18 Results of heat run test for a servomotor at 2,500 rpm and 45 Nm torque.

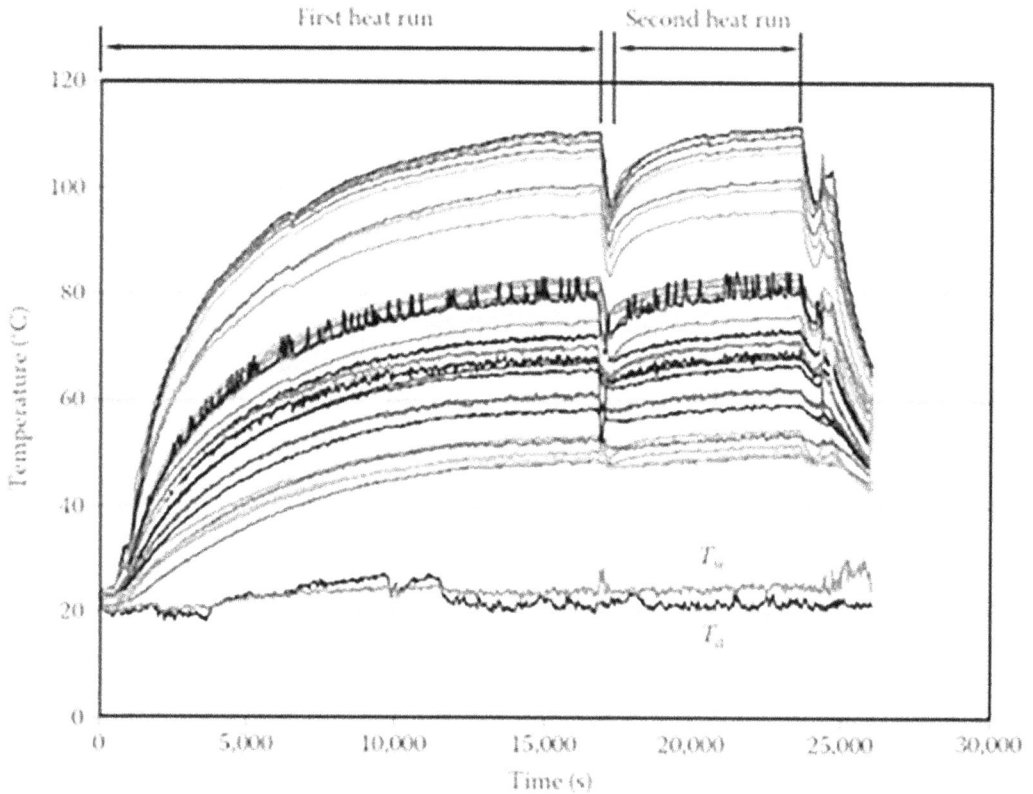

FIGURE 13.19   Thermal testing is performed at the rated speed and rated torque until thermal equilibrium is achieved. The testing data show that the successive test significantly reduces the motor heating time.

the temperatures of stator windings at the middle and end turns approach 140°C, the temperatures of motor housing reach 110°C. The ambient temperature is controlled to maintain in the range of 25°C–30°C.

For large-size motors, the time for reaching thermal equilibrium can be very long, up to more than 10 h. For these cases, performing heat run testing continuously can save a tremendous heating time and testing resources. As shown in Figure 13.19, when the first heat run completes, the motor temperature drops sharply. As the motor temperature becomes lower than the expected thermal equilibrium temperature, the second heat run test starts immediately. It can be seen from the figure that the heating time of the second test is only one-third of the first one, resulting in a great reduction in the testing cycle time, the operating cost, and the need for professional personnel resources.

13.6.3   Motor Efficiency Testing

The aim of motor efficiency testing is to determine the motor characteristic performance. Motor efficiency is one of the most important characteristic parameters for users and manufacturers. This parameter measures how effectively electrical energy is converted into mechanical energy by a motor. A number of factors that may affect the overall motor efficiency include power supply quality, motor size, various power losses, the transmission and mechanical components, maintenance, and mismatch of the load to motor inertia.

Motor efficiency can be measured either directly or indirectly. Usually, direct measuring methods can provide accurate results but often require high-priced instruments/equipment and long test setup time. By definition, motor efficiency is the ratio of mechanical power output P_{out} over electrical power input P_{in}:

$$\eta = \frac{P_{out}}{P_{in}} \qquad (13.15)$$

As an example, the mechanical output for an IM is the production of the motor torque and rotating speed, that is,

$$P_{out} = T\omega = T\omega_s(1-s) \qquad (13.16)$$

where
  s is the slip rate
  ω and ω_s are the rotating shaft speed and rotating synchronous speed in rad/s, respectively
  For a PM motor, $s = 0$.

The electrical power input can be easily measured with good accuracy. The measurement of rotating speed is relatively simple and straightforward, with an achievable accuracy of ±1 rpm. The torque measurement is obtained using a highly accurate torque transducer.

The indirect way of determining the motor efficiency involves the measurement of various power losses of the motor. The motor efficiency can be alternatively expressed as

$$\eta = \frac{P_{in} - \sum P_{loss}}{P_{in}} = \frac{P_{out}}{P_{out} + \sum P_{loss}} \qquad (13.17)$$

Most of these losses such as copper, core, windage, and fraction losses can be determined using no-load and load tests. However, stray load losses, which consist of miscellaneous losses except for major power losses, are hard to be directly measured. These losses may be estimated using some correlations from the literature or standards. For instance, the stray losses can be estimated using IEEE Standard 112-2004, which gives the percentage of stray losses to the rated load as 1.8% for the motor rating 1–90 kW, 1.5% for 91–375 kW, 1.2% for 376–1,850 kW, and 0.9% for >1,850 kW.

The test procedure is defined by the IEEE Standard 112-2004, which estimates the motor efficiency by the direct method [13.7]. As a contrast, IEC 34-2 Standard [13.13] estimates the motor efficiency using the indirect method. In fact, the indirect method has a significant degree of uncertainty, because of the instrumentation lower-accuracy specifications and the incorrect winding losses in relation to the temperature [13.44].

To avoid loading shock, the motor load maintains low levels and increases in small increments. Motor performance testing is conducted when a motor operates under normal environmental conditions.

13.6.4 Impulse Testing

Statistical data show that insulation faults are accounted for about a quarter to one-third of motor failures. The breakdown of insulating materials of motor winding can directly lead to short-circuit faults. Most stator failures often begin as turn-to-turn shorts caused by steep-fronted surges due to switching. These turn-to-turn shorts eventually develop into catastrophic copper-ground or phase-phase fault.

Impulse testing is performed by applying defined impulse voltage waveforms on the motor windings to assume lighting strike. The waveform shape, peak voltage, impedance, and application of the pulse vary and depend on the characteristics of the test motors. The purposes of the impulse test as a special test are (*a*) to confirm the withstand of the motor's insulation against excessive voltages occurring during impulse testing and (*b*) to detect the weak insulation in the motor windings. In motor manufacturing, the on-off impulse voltage is normally generated by charging and discharging a capacitor in the motor windings. This steep-fronted voltage sets up a nonlinear voltage distribution that creates a turn-to-turn differential voltage and results in an oscillation between motor's inductance, capacitance, and resistance.

13.6.5 COGGING TORQUE TESTING

Most applications of electric motors require smooth operation and constant torque. Examples include engraving machines, telescopes, nuclear magnetic resonance, and optical instruments, just to name a few. Any pulsating torque can have a significant negative impact on the motor performance characteristics in these applications. Furthermore, cogging torque causes acoustic noise and vibration, as well as torque and speed ripples. As a result, it is highly desired to control the cogging torque to a lower level, say, less than 1% of the rated torque.

Cogging torque in a PM motor is generated from the interaction of the rotor-mounted PMs and the stator slots. Cogging measurement can be performed by static (*e.g.*, leverage measuring method), quasistatic, or dynamic methods. In most practical cases, the dynamic measurement method is adopted to determine cogging torque. With this method, the motor being tested has no input current and is driven by a dynamometer or motor. A torque transducer is installed between the tested motor and the dynamometer to measure the cogging torque produced with an oscilloscope. In practice, each type of motor uses different ways to measure the position of the machine being tested such as encoder, resolver, proximity sensor, and sensor bearing.

The diagram of a dynamic cogging torque measurement system is given in Figure 13.20. An electric motor is employed as the driving element. A torque transducer is placed between the motor being tested and the driving motor to detect the torque from the strain in the shaft. During the test, there is no input current to the motor being tested. The driving motor is controlled by a motor drive. In this case, the encoder is installed on the motor being tested to measure the rotor position.

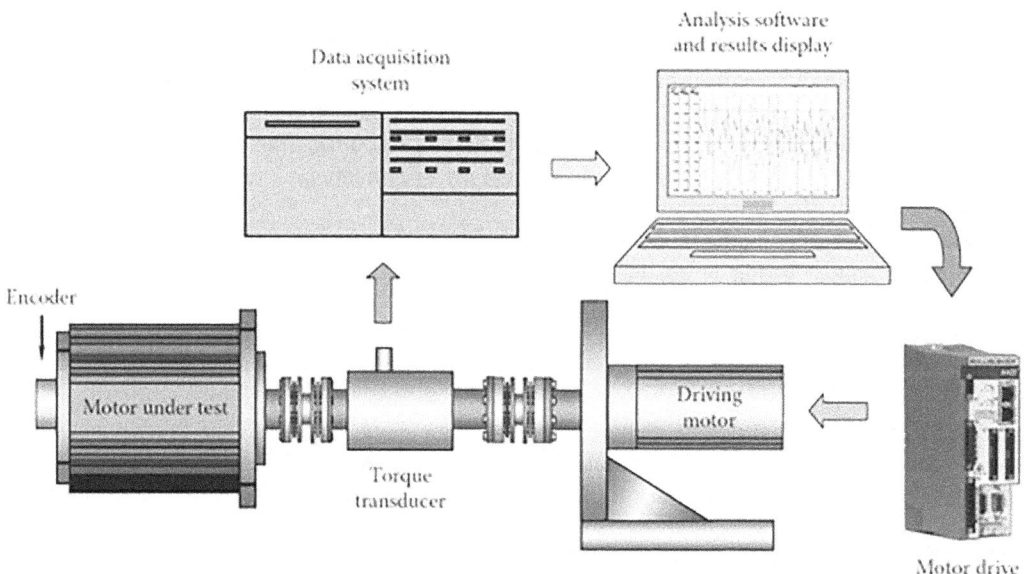

FIGURE 13.20 Dynamic cogging torque measurement employing a noncogging motor as a driving element.

It is noted that the measured torque contains the cogging torque of the driving motor. This requires that the driving motor must have very small or no-cogging torque to minimize the measuring errors. It has been proven that using a no-cogging effect, DC or hydraulic motor as the driving motor promises a very high accuracy of the measuring results. Since this setup works without any offset, the measurements are performed in a close range around the zero value.

An alternative dynamic cogging torque measurement method is demonstrated in Figure 13.21. Unlike the previous testing configuration, the testing motor in this measurement method is the driving element. A magnetic powder dynamometer is used as the load device to provide the brake load to the testing motor. The resistance of the magnetic power dynamometer is controlled by the brake controller. The cogging torque of the testing motor is thus obtained from the measured torque data by the torque transducer.

A cogging torque testing system that uses a stepper motor as a driving element is shown in Figure 13.22. A harmonic gear is used to reduce the speed of the stepper motor to drive the testing motor. The high gear ratio avoids a measurable impact of the stepper motor's torque ripple to the cogging torque of the testing PM motor. The high harmonics of the stepper motor are filtered from the digital signal.

Similar to the previously addressed testing systems, a torque transducer is placed between the testing motor and stepper motor to measure the torque. The signal from the transducer is then fed into a measurement control unit. The measured torque is converted to digital values with an A/D converter and displayed on the computer monitor. As discussed by van Riesen *et al.* [13.45], the weight is used as an offset torque, which allows for the torque transducer to measure in its optimum scale range. The stepper motor driving the whole setup is controlled by two voltage sources. The resolution of the stepper motor can be switched from 1/1, 1/2, 1/4, to 1/8. By additionally varying the voltage from 0 to 8 V, the speed can be adjusted from 0 to 64 rpm.

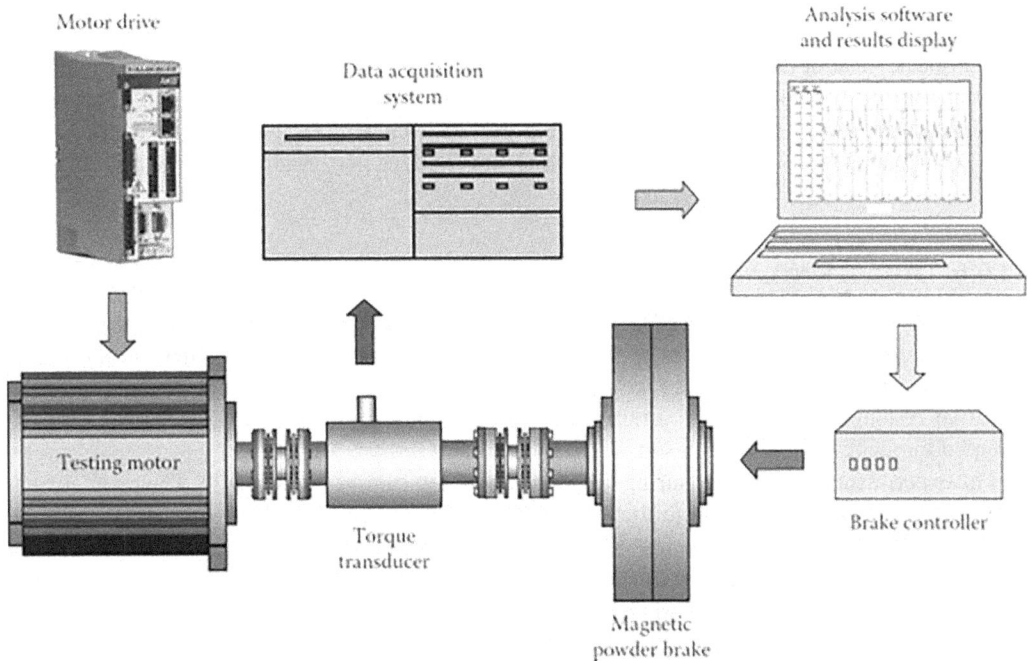

FIGURE 13.21 Alternative dynamic cogging torque measurement employing a magnetic powder brake as the driving element.

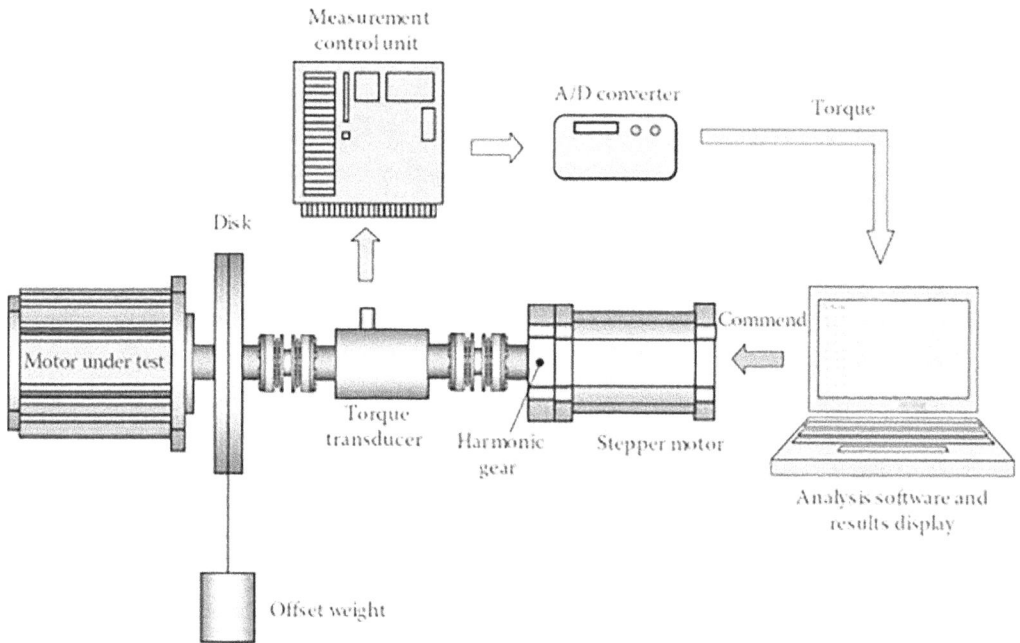

FIGURE 13.22   Dynamic cogging torque measurement using a stepper motor as the driving element.

The particularly prominent advantage of using the stepper motor is that it does not affect the measurement of the cogging torque of the testing motor, even though it has some cogging torque itself. In addition, because the stepper motor can determine the angular position of the testing motor, no position sensor is required.

There are primarily two different cogging torque analysis criteria: the harmonic analysis and the peak-to-peak value of the cogging torque. Many motor manufacturers often use cogging torque as a percentage of the peak-to-peak cogging torque over the motor continuous torque. The measured cogging torque from a servomotor is given in Figure 13.23, showing the peak-to-peak value of 0.04 Nm. With the measured continuous torque of 1.52 Nm from the performance test, the percentage of cogging torque is 2.63%.

13.6.6   Torque Ripple Measurement

Torque ripple of electric motors has adverse effects on the surface finishing and tolerance control in some demanding motion control applications, such as machine tool spindle drives and optical devices. In some circumstances, torque ripple may excite resonances in the mechanical portion of the drive system and produce acoustic noise. Therefore, precise torque ripple measurements are essential for identifying root causes and optimizing motor design.

The experiment setup for measuring torque ripple is depicted in Figure 13.24. The eddy-current brake is used to provide loading to the testing motor. The motor is rigidly connected to a torque transducer for measuring the maximum torque load. Isolated current and voltage transducers are used since the signals can be connected to a single common point. This is a great convenience to the data acquisition system provided that additional instrumentation errors are negligible [13.46].

Tests are conducted at different load points and different speed levels (*e.g.*, 80% and 100% of rated speed). Accuracy of the torque load during the torque ripple test is well within the manufacturing specification of the torque transducer. Electric grounding is made on all measuring instruments such as torque box, torque transducer, and output negative signal. The result obtained from a

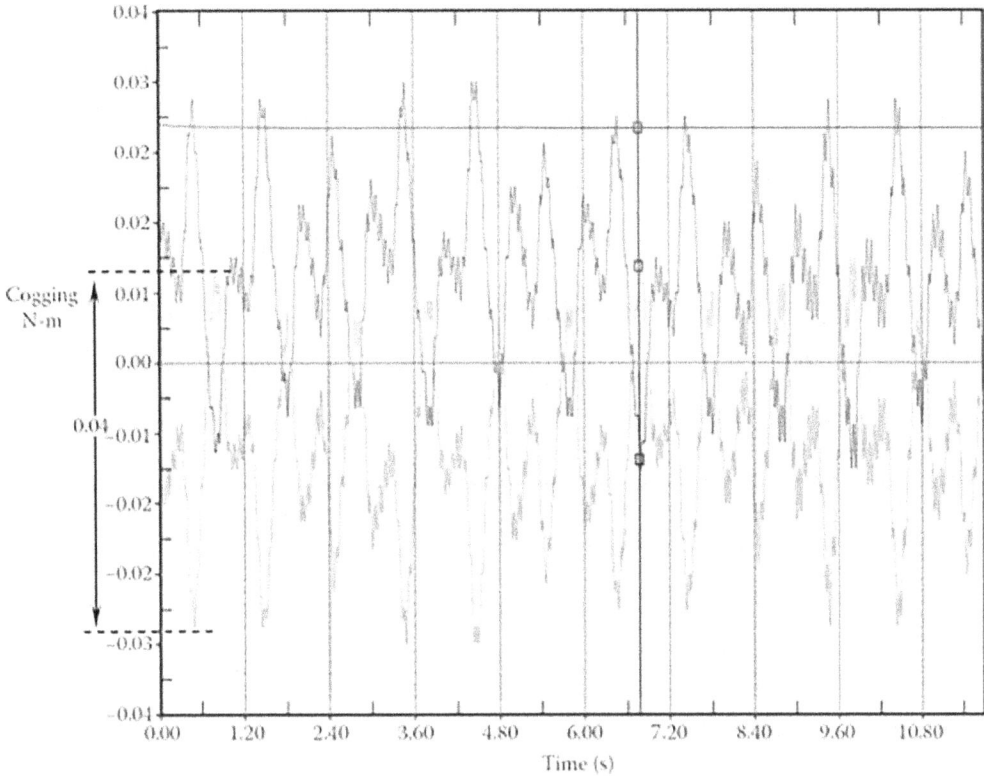

FIGURE 13.23   Measured cogging torque from a PM servomotor.

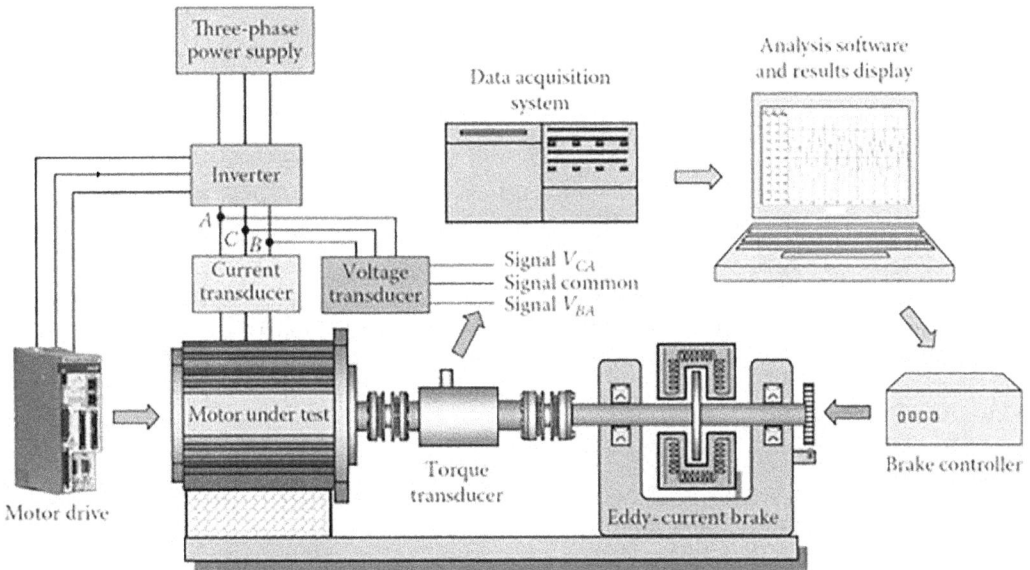

FIGURE 13.24   Experimental setup for torque ripple measurement.

torque ripple test has shown that the peak-to-peak voltage is 100.4 mV (Figure 13.25). This value is converted into the peak-to-peak torque difference ΔT_p (where $\Delta T_p = T_{max} - T_{min}$). Thus, the torque ripple, which is defined as the ratio of ΔT_p to the main torque T_m, is determined to be about 5.4%.

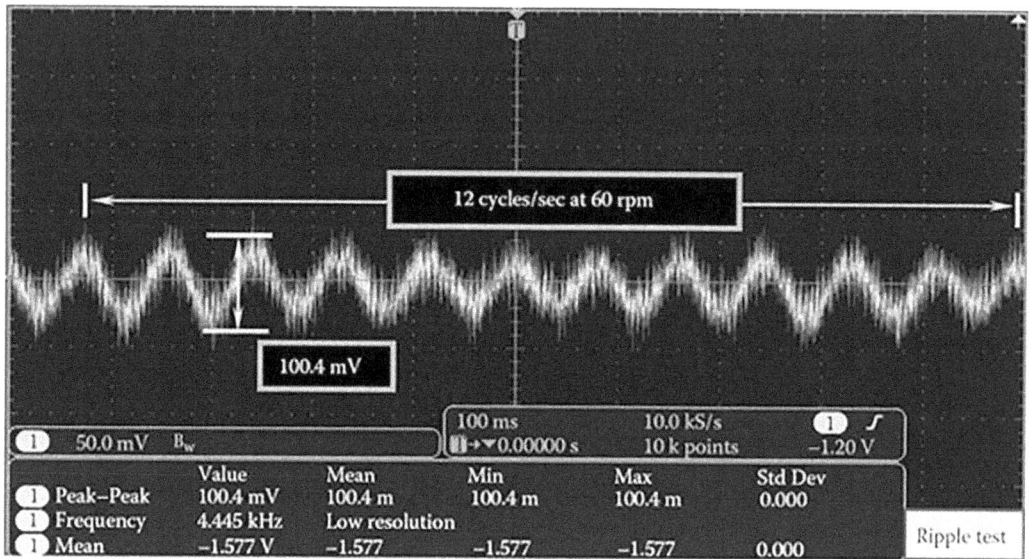

FIGURE 13.25 Results measured from a torque ripple test. The ratio of peak-to-peak torque ripple to rated torque is 5.4%.

REFERENCES

13.1. IEEE Standard 43-2000. Recommended practice for testing insulation resistance of rotating machines. Institute of Electrical and Electronics Engineers, New York.

13.2. IEEE Standard 56-1977. Guide for insulation maintenance of large-current rotating machinery (10,000 kVA and larger). Institute of Electrical and Electronics Engineers, New York.

13.3. IEEE Standard 118-1978. Standard test code for resistance measurements. Institute of Electrical and Electronics Engineers, New York.

13.4. IEEE Standard 95-2002. Recommended practice for insulation testing of AC electric machinery (2,300 V and above) with high direct voltage. Institute of Electrical and Electronics Engineers, Piscataway, NJ.

13.5. IEEE Standard 522-2004. Guide for testing turn insulation of form-wound stator coils for alternating-current electric machines. Institute of Electrical and Electronics Engineers, New York.

13.6. IEEE Standard 1068-1990. Recommended practice for the repair and rewinding of motors for the petroleum and chemical industry. Institute of Electrical and Electronics Engineers, New York.

13.7. ANSI/IEEE Standard 112-2004. Standard test procedure for polyphase induction motors and generators. Institute of Electrical and Electronics Engineers, New York.

13.8. ANSI/NETA MTS-2007. Standard for maintenance testing specifications for electrical distribution equipment and systems. InterNational Electrical Testing Association (NETA), Portage, MI.

13.9. ANSI/NETA ATS-2009. Acceptance testing specifications for electrical power distribution equipment and systems. InterNational Electrical Testing Association (NETA), Portage, MI.

13.10. ANSI/NETA ETT-2013. Standard for certification of electrical testing technicians. InterNational Electrical Testing Association (NETA), Portage, MI.

13.11. NEMA MG1-1993. Motors and generators. National Electrical Manufacturers Association, Rosslyn, VA.

13.12. IEC 60034-1: 2013. Rotating electrical machines—Part 1: Rating and performance. International Electrotechnical Commissions, Geneva, Switzerland.

13.13. IEC 60034-2-1: 2007. Rotating electrical machines—Part 2-1: Standard methods for determining losses and efficiency from test (excluding machines for traction vehicles). International Electrotechnical Commissions, Geneva, Switzerland.

13.14. IEC 60034-4: 2008. Rotating electrical machines – Part 4: Method for determining synchronous machine quantities from tests. International Electrotechnical Commissions, Geneva, Switzerland.

13.15. IEC 60034-19:1995. Rotating electrical machines—Part 19: Specific test methods DC, machines on conventional and rectifier-fed supplies. International Electrotechnical Commissions, Geneva, Switzerland.

13.16. IEC 60034-29: 2008. Rotating electrical machines—Part 29: Equivalent loading and superposition techniques—Indirect testing to determine temperature rise. *International Electrotechnical Commissions*, Geneva, Switzerland.

13.17. Mustang Dynamometer. 2011. Power dyne PC operator manual. http://www.mustangdyne.com/mustangdyne/wp-ontent/uploads/downloads/2011/04/PowerDyne-PC-Users-Manual.pdf.

13.18. Muff, M. and Towey, J. 2013. Two ways to measure temperature using thermocouples feature simplicity, accuracy, and flexibility. *Analog Dialogue* **44**(**4**): 3–8.

13.19. Jirawattanasomkul, J. and Koetniyom, S. 2012. Design and development of road load condition for chassis dynamometer. *International Conference on Production, Materials and Automobile Engineering (ICPMAE'2012)*, Pattaya, Thailand.

13.20. PCB Piezotronics Inc. 2013. Torque sensors. http://www.pcb.com/linked_documents/force-torque/LT-PCBTorqueSensors_LowRes.pdf.

13.21. De Almeida, A. T. and Ferreira, F. 1997. Efficiency testing of electric induction motors. ISR, Department of Electrical Engineering, University of Coimbra, Coimbra, Portugal.

13.22. Zhu, X. Z., Chen, Z. B., and Jiao, Y. H. 2018. Optimizations of distributed dynamic vibration absorbers for suppressing vibrations in plates. *Journal of Low Frequency Noise, Vibration and Active Control* **37**(**4**): 1188–1200.

13.23. Çolak, I., Bal, G., and Elmas, Ç. 1996. Review of the testing methods for full-load temperature rise. *EPE Journal* **6**(**1**): 37–43.

13.24. Morris, D. G. O. 1968. Back-to-back test for induction machines: Floating gearbox. *Proceedings of the Institution of Electrical Engineering* **115**(**4**): 536–537.

13.25. Fong, W. 1972. New temperature test for polyphase induction motors by phantom loading. *Proceedings of the Institution of Electrical Engineering* **119**(**7**): 883–887.

13.26. Ho, S. L. and Fu, W. N. 2001. Analysis of indirect temperature-rise tests of induction machines using time stepping finite element method. *IEEE Transactions on Energy Conversion* **16**(**1**): 55–60.

13.27. Ho, S. L. 1992. Further development of phantom loading in induction motors. *Proceedings of International Conference on Electrical Machines* **2**: 298–302.

13.28. Ho, S. L., Fu, W. N., and Wong, H. C. 1999. Thermal study of induction motors by phantom loading using multi-slice time stepping finite element modeling. *IEEE Transactions on Magnetics* **35**(**3**): 1606–1609.

13.29. Ytterberg, A. 1921. Ny method for fullbelasting av electriska maskiner utan drivmotor eller avlastningsmaskin (New method for full load of electric machines without drive motor or relief machine). *S. K. Skakprov Teknisk Tidskrift* **79**: 42–46.

13.30. Kron, A. W. 1973. Testing induction motors by means of a two frequency supply. *ETZ-A (Germany)* **94**: 77–82.

13.31. Grantham, C. and Tabatabaei-Yazdi, H. 2000. A novel power electronic machineless dynamometer for load testing and efficiency measurement of three-phase induction motors. *Proceedings of Third International Power Electronics and Motion Control Conference* **2**: 564–569.

13.32. Plevin, D. H., Glew, C. N., and Dymond, J. H. 1999. Equivalent load test for induction machines-the forward short circuit test. *IEEE Transactions on Energy Conversion* **14**(**3**): 419–425.

13.33. Jordan, H. E., Cook, J. H. and Smith, R. L. 1977. Synthetic load testing of induction machines. *IEEE Transactions on Power Apparatus and Systems* **96**(**4**): 1101–1103.

13.34. Garvey, S., Çolak, I., and Wright, M. T. 1995. The "variable inertia test" for full load temperature rise testing of induction machines. *IEE Proceedings of Electric Power Applications* **142**(**3**): 222–224.

13.35. Thomas, T. M. 2005. On-line and offline testing of electric motors. *IMC-2005—The 20th International Maintenance Conference*, Tampa, FL.

13.36. Thomas, T. M. 2005. Static and dynamic motor testing as part of a predictive maintenance program. *IMC-2005—The 20th International Maintenance Conference*, Tampa, FL.

13.37. Vertiv Corporation. 2017. Surge testing—Educational series: Why test? YT-02-07. https://www.vertiv.com/4a4fe5/globalassets/services/services/performance-optimization-services/power-quality-studies-and-harmonic-analyses/why-test-surge-testing-.pdf.

13.38. Geiman, J. 2007. DC step-voltage and surge testing of motors. *Maintenance Technology*, March 1st. http://www.mt-online.com/march2007/dc-step-voltage-and-surge-testing-of-motors.

13.39. Genta, G. and Delprete, C. 1994. Some considerations on the experimental determination of moments of inertia. *Meccanica* **29**(**2**): 125–141.

13.40. Nagata, K., Okuyama, T., Fujii, H., and Okamatsu, S. 2003. Method for calculating inertia moment and driver for electric motor. U.S. Patent 6,611,125.

13.41. Rehm, T. J. and Golownia, J. J. 2005. Method for determining inertia of an electric motor and load. U.S. Patent 6,920,800.

13.42. Poels, P. W. 2008. Cogging torque measurement, moment of inertia determination and sensitivity analysis of an axial flux permanent magnet AC motor. Traineeship report. Charles Darwin University, Darwin, Australia.

13.43. Crawford, D. E. 1975. A mechanism of motor failure. *Proceeding of 12th Electrical Insulation Conference*, Boston, MA. Traineeship Report DCT 3007.147.

13.44. de Almeida, A. I., Ferreira, F. J., Busch, J. F., and Angers, P. 2002. Comparative analysis of IEEE 112-B and IEC 34-2 efficiency testing standards using stray load losses in low-voltage three-phase, cage induction motors. *IEEE Transactions on Industry Applications* **38**(2): 608–614.

13.45. van Riesen, D., Schlensok, C., Schmülling, B., Schöning, M., and Hameyer, K. 2006. Cogging torque analysis on PM-machines by simulation and measurement. *Proceedings of 17th International Conference on Electrical Machine* (*ICEM*), Chania, Greece.

13.46. Hsu, J. S., Scoggins, B. P., Scudiere, M. B., Marlino, L. D., Adams, D. J., and Pillay, P. 1995. Nature and assessments of torque ripples of permanent-magnet adjustable-speed motors. *Proceedings of IEEE Industry Application Conference* **3**: 2696–2702.

14

Modeling, Simulation, and Analysis of Electric Motors

In designing and developing modern electric motors, a wide variety of engineering analyses and simulations must be conducted to ensure the robust design and product optimization. These analyses may include, but not limited to, thermal analysis, computational fluid dynamics (CFD) analysis, stress/strain analysis, electromagnetic analysis, resonance analysis, modal analysis, fatigue analysis, buckling analysis, burst analysis, motor mounting analysis, thermal expansion/contraction analysis, vibration and acoustic noise analysis, and design optimization analysis. The successful integration of these analyses into the product design cycle can significantly enhance the design quality, raise design standards, and accelerate design processes.

14.1 COMPUTATIONAL FLUID DYNAMICS AND NUMERICAL HEAT TRANSFER

As motor designs and processes grow in sophistication, motor-cooling problems become too complex to solve analytically. This forces engineers to perform numerically large simulations to gain insight into the details of fluid flow and heat transfer processes in motors. In fact, solutions generated by numerical methods can provide more insights than those by other methods in some heat transfer and fluid dynamics problems; for instance, transient fluid flow and heat transfer, 3D turbulent flows with temperature-dependent properties, multiphase flows, and compressible flows.

With the growing heat generation in modern electric motors, limited cooling capabilities may cause degradations of the motor performance and operation reliability. Today, CFD, being an integral part of the motor design for shortening the design cycle time and lowering the design cost, has played an important role in helping engineers and designers gain insights into the physical aspects of motor cooling. The use of CFD at the stage of the conceptual design enables motor engineers to explore alternative design options. The detailed results from a CFD simulation allow visualization of flow fields and temperature distributions in even the most inaccessible locations of a complex fluid flow and heat transfer system. A key advantage of CFD is that it provides the flexibility to readily change design parameters and determine the corresponding impacts of those changes on performance [14.1].

However, it must be noted that one of CFD issues is *garbage in, garbage out*. Despite the maturity of advanced CFD techniques, there is no guarantee that a CFD tool can automatically provide correct results. To obtain accurate simulation results, it requires thermal engineers to use their technical expertise and experiences in design and CFD work. Proper verification and validation with either experimental data or theoretical solutions are always essential parts of a CFD process.

14.1.1 STRATEGIES IN MODELING AND PERFORMING CFD ANALYSIS

To make numerical simulations successful with available computational resources, some effective strategies may be adopted in modeling and performing CFD analysis:

- To take full advantage of geometric symmetry and periodic boundary conditions, a CFD analysis can be greatly simplified by modeling only a portion of the actual system.

DOI: 10.1201/9781003097716-14

- Some 3D fluid dynamics and heat transfer problems may be treated as 2D, axisymmetric problems.
- By getting rid of any unnecessary small features, which have little effect on simulation results, such as small faces, holes, edges, and fillets, it does not only simplify the problem but also makes the simulation convergent. This is especially critical in cases that CFD models are converted directly from 3D CAD models, which usually contain a large amount of such small features.
- The mesh quality is the key to the success of a CFD analysis. It is highly designed to carefully select an appropriate meshing method and apply advanced meshing control techniques such as the refinement technique.
- In complex simulation problems, the iterations converge slowly and, in some cases, may even diverge. To ensure achieving convergence, it is suggested to run an analytical model with properly altered boundary and initial conditions. For instance, a common and effective strategy used in CFD codes for steady-state problems is to solve the unsteady form of the governing equations and then *march* the solution in time until the solution converges to a steady value. In this case, each time step is effectively an iteration, with the guess value at any time level being given by the solution at the previous time level [14.2].
- To facilitate convergence, the rotational speed should be initially set at a very low value and increased by small increments until reaching the normal rotating speed.
- For some large-size, complex geometry problems, a fully refined model may be too large and usually run into convergence problems. It is suggested to run the model with relatively coarse meshes first to quickly find out critical areas of interest. The critical areas are then successively refined to get the desired information in these areas.
- The addition of radiation heat transfer can significantly slow down the numerical calculation and increase the requirement of computer resources. In practice, the contribution from radiation to the overall heat transfer is generally insignificant except in a case where the temperature differences between motor components and/or between the motor and the environmental surrounding are relatively large.
- At an early design stage, some levels of approximation can be made for quickly exploring different designs.
- In dealing with heat conduction in silicon steel laminations, it is to be noted that the laminations have anisotropic thermal conductivity.

14.1.2 ROTATING FLOW MODELING

Rotating and swirl flows are commonly encountered in turbomachinery electric rotating machines and a variety of other applications. In an electric motor, while the stationary parts such as the stator and housing are held statically, all parts attached to the rotor rotate at a certain velocity with respect to the rotor axis. Therefore, rotating flows are generated in the airgap between the rotor and stator and at the surroundings of fan blades, whereas interactive flows are mainly developed at the end-winding regions.

There are several approaches for modeling such rotational and interactional flows. The simplest approach is to use the single reference frame (SRF) method under the following conditions: (*a*) all the rotating parts rotate at the same speed with respect to a specified single axis and (*b*) the problems involving moving parts and the stationary walls (as viewed from a stationary reference frame) can form surfaces of revolution with respect to the axis of rotation in the rotating reference frame [14.3]. With this approach, the entire computational domain is referred to as a single rotating reference frame. Thus, with certain restrictions, it can turn unsteady rotating flows in the stationary (inertial) reference frame into steady-state flows in the rotating (noninertial) reference frame. However, due to the restricted conditions mentioned previously, the application of this method is limited to simple problems.

When rotating parts in a system rotate about different rotating axes, or each of the rotating parts rotates about the same rotating axis but different angular speeds, or problems involve stationary components which cannot be described by surfaces of revolution, SRF is no longer applicable, and thus, the use of the multiple reference frame (MRF) method is required. With this method, the computational domain is divided into multiple domains: some are rotating and others stationary, with interfaces (which must be surfaces of revolution) separating these domains. This approach is a steady-state approximation where a rotating frame of reference is applied on the rotating domains and a stationary frame of reference is applied on the stationary domains. During the solution process, information is continually passed through the predefined interfaces between the regions. All rotating parts (e.g., a rotor) should be located inside of the rotating regions and modeled as emptiness or solid bodies. For a multiple-rotor system, the use of MRF allows individual rotors to rotate with different speeds, directions, and axes. The approaches of MRF and SRF are appropriate when the flow at the boundary between rotating and stationary regions is weakly affected by the interaction. These methods provide reasonable time-average simulation results in various applications.

The mixing plane model (MPM) is a variant of the MRF model for simulating flow through domains with one or more regions in relative motion. In this approach, each fluid region is treated as a steady-state problem. Flow field data from adjacent regions are passed as boundary conditions that are spatially averaged at the mixing plane interface.

The sliding mesh method has been developed for analyzing unsteady fluid flows. This method allows flow pattern calculations without the need of experimental data as rotor boundary conditions. As a transient approach, this method can provide the most accurate results for complex flow systems. With this method, the flow field is divided into two regions. One region, associated with the stationary components, remains stationary. Another region, associated with the rotating components, rotates relative to the stationary mesh. The two grids slide past each other in a time-dependent manner, exchanging information at the sliding interface [14.4]. At the interface, a conservative interpolation is used for both mass and momentum. The principal disadvantages of this method include the long calculation time and the required large computer resources.

The selection of the appropriate model is critical for the success of a computational simulation. In addition, the setup of the interface between rotating and stationary regions in MRF is also very important for the accuracy of numerical results. One example picked up here is the rotating airgap flow in a motor. As depicted in Figure 14.1, the rotor rotates counterclockwise with the rotating angular velocity ω_i and the radii of the rotor and stator are r_i and r_o, respectively. This is a special case of the Taylor–Couette flow that has been extensively studied for more than a century [14.5–14.8]. Except in the very thin boundary layers of cylinders, the tangential velocity profile of the concentric rotating flow can be theoretically derived as [14.6]

$$u_t = \frac{A}{2}r + \frac{B}{r} \tag{14.1}$$

where A and B are constants, determined from the boundary conditions at the rotor surface $u_{t,r=r_i}$ and at the stator surface $u_{t,r=r_o} = 0$.

This problem can also be solved with a CFD model by defining multiple regions. The interface that separates the rotating and stationary regions is at $r = r_{int}$. As shown in Figure 14.2, the tangential velocity decreases in the thin boundary layer of the rotor and then increases with the radius in the rotating region until it reaches the maximum value at the interface (where $u_{t,\max} = r_{int}\omega_i$). In the stationary region, the velocity is continuous at the interface and then decreases with the radius until it vanishes at the stator surface. The comparisons between the experimental solution and the results from both the small and large rotating regions are presented in Figure 14.3. The small rotating region is set up as close as possible to the rotor but without causing difficulty in model meshing. This figure shows the significant deviations between these three cases. This may suggest that for such a problem, the use of a small rotating region provides better results than the large rotating region.

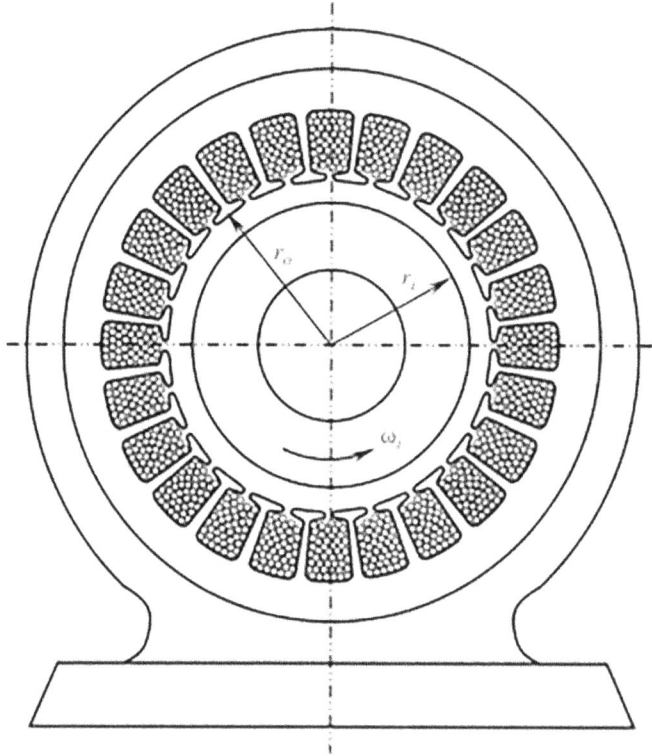

FIGURE 14.1   Schematic diagram of the formation of the airgap between the rotor and stator.

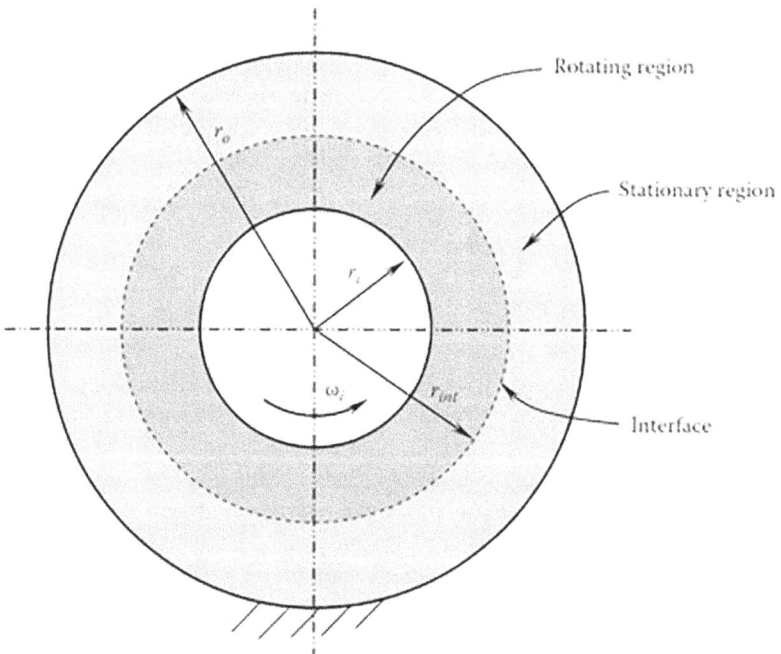

FIGURE 14.2   Defined rotating and stationary regions (as viewed from the stationary frame).

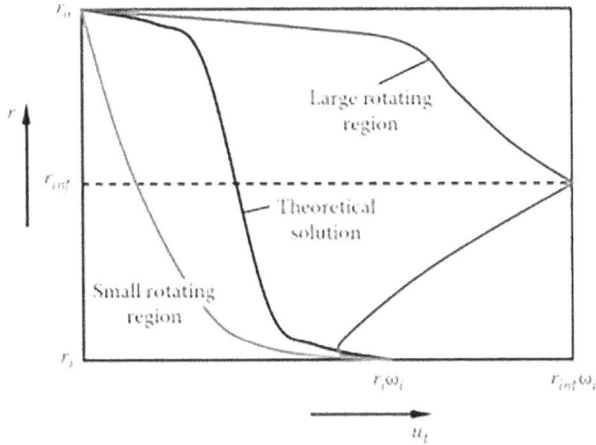

FIGURE 14.3   Comparison of theoretical solution and results from the model with multiple regions.

14.1.3 Porous Media Modeling

In some types of electric machines, the stator winding coils are made up of bundles of strands of insulated copper wires and embedded in the slots of the stator core. At the end-winding region, the stator coils bend to form continuous current paths. Without a resin encapsulation treatment, the stator winding contains a large number of voids in its complicated 3D structure 4.36). Thus, to simplify the CFD analysis, the stator windings, especially at the end-winding regions, can be modeled as porous media [14.9, 14.10].

Porous medium technology is defined as the utilization of specific and unique features of a highly porous medium for supporting and controlling the cooling process in electric machines. In practice, the flow velocity field and heat transportation in a 3D-structured porous medium are very complex. To overcome the difficulties, it is often required to treat the 3D structure of the porous media as a large number of hot spots homogeneously distributed throughout its volume. Moreover, the 3D model of the porous medium may be simplified into a 2D axisymmetric model.

In the porous media model, the flow passing through the stator winding with a fraction of the flow open area is equivalent to the flow passing through a porous medium with a full open area, with an identical mass flowrate and pressure drop:

$$\Delta p_a = \Delta p_p \tag{14.2}$$

$$\dot{m}_a = \dot{m}_p \tag{14.3}$$

where
Δp_a and Δp_p are pressure drops through the actual winding and porous system, respectively
$\dot{m}_a$ and $\dot{m}_p$ are mass flowrates from the actual winding and porous system, respectively

By defining the loss coefficient K,

$$K = \frac{\Delta p}{(1/2)\rho u^2} \tag{14.4}$$

Δp_a and Δp_p can be rewritten as

$$\Delta p_a = K_a \left(\frac{1}{2} \rho u_a^2 \right) \tag{14.5}$$

$$\Delta p_p = K_p \left(\frac{1}{2} \rho u_p^2 \right) \tag{14.6}$$

Substituting Equations 14.5 and 14.6 into 14.2 yields

$$K_a \frac{\rho u_a^2}{2} = K_p \frac{\rho u_p^2}{2} \tag{14.7}$$

That is,

$$K_p = K_a \left(\frac{u_a}{u_p} \right)^2 \tag{14.8}$$

As the most important parameter in calculating porous media flows, porosity η is introduced and defined as the ratio of the actual open area to the total flow area (*i.e.*, 100% open):

$$\eta = \frac{A_a}{A_p} \tag{14.9}$$

Rewriting Equation 14.3,

$$\rho A_a u_a = \rho A_p u_p \tag{14.10}$$

The relationship between the velocity ratio u_a/u_p and η can be determined as

$$\frac{u_a}{u_p} = \frac{A_p}{A_a} = \frac{1}{\eta} \tag{14.11}$$

Substituting (14.11) into (14.8), it yields

$$K_p = \frac{K_a}{\eta^2} \tag{14.12}$$

Thus, the porous media inertial resistance factor ξ, defined as the pressure loss factor per unit length, can be calculated as

$$\xi = \frac{K_p}{\Delta L} \tag{14.13}$$

where ΔL is the length through the media in the flow direction.

    Thus, with the equations mentioned previously, the flow characteristics can be obtained at the porous flow zone. By integrating the porous media parameters into the CFD model, the velocity and pressure fields at the end-winding region are determined.

14.1.4   NUMERICAL SIMULATION OF MOTOR COOLING

Modern electric motors have more complicated mechanical and electromagnetic structures over conventional motors for gaining better performance, higher operation reliability and efficiency, and longer lifetime. For such systems with complicated geometry, conventional analytic methods are no longer competent to provide comprehensive information of fluid flows and temperature distributions of motors at a system level. Therefore, CFD techniques have been extensively employed by motor manufacturers to help optimize motor cooling systems.

As in the example shown in Figure 14.4, a large-scale elevator motor is cooled by high volumetric air driven by a blower. Unlike most conventional motors whose rotors are on the inside of wound stators, this external rotor motor has its rotor on the motor outside to directly drive the elevator car. The advantages of the external rotor motors include increased motor torque, reduced power losses, and promoted transmission efficiency [14.11]. This type of motor is especially suitable for applications such as fans, motor-wheel driving systems [14.12, 14.13], and lifting systems of forklift trucks.

Due to the symmetry of the model, only one-tenth of the motor (36°) is modeled. In order to reduce the required computing resources and accelerate the computations, the model has been simplified and defeatured without loss in accuracy. As shown in Figure 14.5, the cooling airflow enters into the stationary hollow shaft at its one end and injects into the radial cooling channels located at the central and end-winding regions. Then, the cooling airflow turns to 90° when it reaches the airgap and finally exhausts from the two endbells to the environment.

FIGURE 14.4   Large-scale elevator motor. (Courtesy of Kollmorgen Corporation, Radford, VA.)

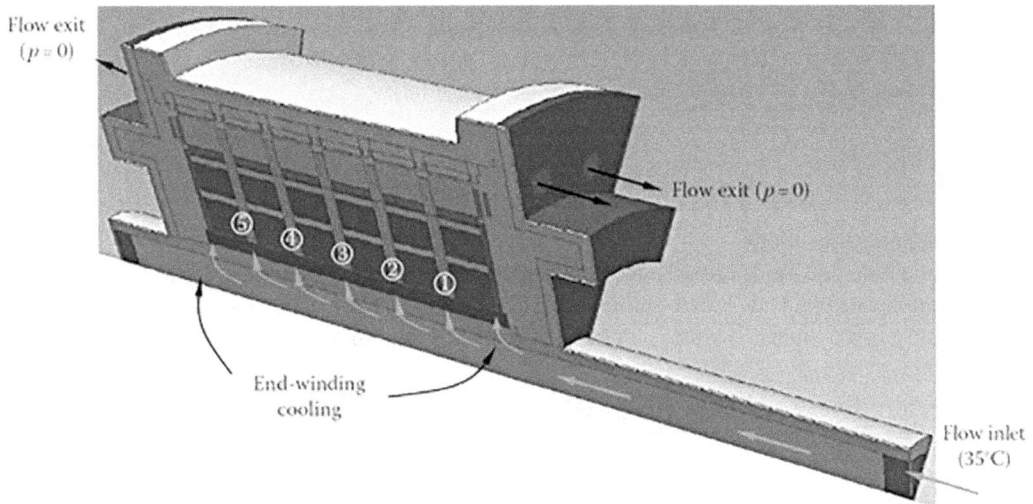

FIGURE 14.5   Forced air cooling for a large-scale elevator motor. (Courtesy of Kollmorgen Corporation.)

14.1.4.1 Mathematical Formulations

In a CFD analysis, governing equations are derived from the following fundamental laws of physics:

- Conservation of mass
- Conservation of momentum
- Conservation of energy

A time-dependent continuity equation is derived from the conservation of mass. Navier–Stokes nonlinear partial differential equations arise from applying Newton's second law to the motion of fluid. These equations consist of three time-dependent momentum equations corresponding to three coordinates. If a problem involves heat transfer, an additional time-dependent energy equation, based on the conservation of energy, must be coupled with Navier–Stokes equations. All equations must be solved simultaneously for getting the entire flow, pressure, and temperature fields of an electric motor.

The governing equation may be used with a relatively high degree of accuracy for incompressible flows if the viscosity gradient is not too large. In majority of the problems, this assumption is adequate. The temperature-dependent properties are often used in CFD analyses.

- Continuity equation
  Continuity equation is based on the conservation of mass in a fixed control volume of fluid (VOF) flow and can be expressed in vector notation:

$$\frac{\partial \rho}{\partial t} + \nabla \cdot (\rho \mathbf{V}) = 0 \tag{14.14}$$

  Equation 14.14 is the continuity equation in conservation form. In this equation, the second term can be separated into two parts as priority

$$\frac{\partial \rho}{\partial t} + \mathbf{V} \cdot \nabla \rho + \rho \nabla \cdot \mathbf{V} = 0 \tag{14.15}$$

  By introducing the differential operator [14.14]

$$\frac{D()}{Dt} = \frac{\partial ()}{\partial t} + \mathbf{V} \cdot \nabla () \tag{14.16}$$

  Equation 14.15 can be rewritten in nonconservation form:

$$\frac{D\rho}{Dt} + \rho \nabla \cdot \mathbf{V} = 0 \tag{14.17}$$

- Momentum equations
  Momentum equations are derived from the conservation of momentum, comprising three components, that is, x, y, and z momentum equations in a Cartesian coordinate system or r, θ, and z in a cylindrical system, in a general form of fluid motion equations:

$$\rho \frac{D\mathbf{V}}{Dt} = -\nabla p + \nabla \cdot \bar{\bar{\tau}} + \rho \mathbf{f} \tag{14.18}$$

  where
    $\mathbf{f}$ is the body force per unit mass
    $\bar{\bar{\tau}}$ is the viscous stress tensor

For Newtonian fluids, $\bar{\bar{\tau}}$ is a linear function of the velocity gradient and can be expressed as

$$\tau_{ij} = \mu\left(\frac{\partial V_i}{\partial x_j} + \frac{\partial V_j}{\partial x_i}\right) + \lambda(\nabla \cdot \mathbf{V})\delta_{ij} \tag{14.19}$$

where
  μ is the dynamic viscosity
  λ is the coefficient of bulk viscosity
  x_i and x_j denote mutually perpendicular coordinate directions
  δ_{ij} is the Kronecker delta operator, which is equal to 1 if $i = j$ and it is zero otherwise.

The coefficient of bulk viscosity and dynamic viscosity can be related to each other through Stokes' hypothesis:

$$\lambda + \frac{2}{3}\mu = 0 \tag{14.20}$$

Substituting Equation 14.20 into 14.19 yields

$$\tau_{ij} = \mu\left[\frac{\partial V_i}{\partial x_j} + \frac{\partial V_j}{\partial x_i} - \frac{2}{3}(\nabla \cdot \mathbf{V})\delta_{ij}\right] \tag{14.21}$$

- Energy equation
  Energy equation is based on conservation of energy and expressed as

$$\rho\left[\frac{\partial h}{\partial t} + \nabla \cdot (h\mathbf{V})\right] = -\frac{\partial \rho}{\partial t} + (\mathbf{V} \cdot \nabla \rho) + \nabla \cdot (k\nabla T) + \phi \tag{14.22}$$

where
  h is the specific enthalpy that is related to specific internal energy as $h = e + p/\rho$
  T is the temperature
  ϕ is the dissipation function representing the work done against viscous forces and is
    expressed as

$$\phi = \left(\bar{\bar{\tau}} \cdot \nabla\right)\mathbf{V} = \tau_{ij}\frac{\partial V_i}{\partial x_j} \tag{14.23}$$

For an incompressible fluid, the energy equation can also be expressed as

$$\rho c_p\left[\frac{\partial T}{\partial t} + \mathbf{V} \cdot \nabla T\right] = \nabla \cdot \left(k_{eff}\nabla T\right) + S_h \tag{14.24}$$

where
  T is the temperature, $°C$
  k_{eff} is the effective thermal conductivity ($k_{eff} = k + k_t$, where k_t is the turbulent thermal
    conductivity), $W/(m·°C)$
  S_h is the volumetric heat source, W/m^3

- Turbulence model
  A majority of fluid flows encountered in motor cooling are turbulent flows. There are a number of turbulence models in CFD analysis to deal with turbulent flows, such as constant eddy viscosity, standard $k\text{-}\varepsilon$, Reynolds stress, and renormalization group (RNG) $k\text{-}\varepsilon$

model, just to name a few. This study employs the RNG k-ε turbulence model, which was derived using a rigorous statistical technique:

k equation

$$\nabla(\rho k \mathbf{V}) = \nabla\left(\alpha_k \mu_{eff} \nabla k\right) + \mu_t \phi - \rho \varepsilon \tag{14.25}$$

ε equation

$$\nabla(\rho \varepsilon \mathbf{V}) = \nabla\left(\alpha_\varepsilon \mu_{eff} \nabla \varepsilon\right) + C_{1\varepsilon} \mu_t \phi \frac{\varepsilon}{k} - C_{2\varepsilon} \rho \frac{\varepsilon^2}{k} - R \tag{14.26}$$

where
  $\mu_t \phi$ represents the generation of turbulent kinetic energy
  $C_{1\varepsilon}$ and $C_{1\varepsilon}$ are constants ($C_{1\varepsilon} = 1.44$ and $C_{2\varepsilon} = 1.92$)
  α_k and α_ε are the inverse effective Prandtl numbers for k and ε, respectively

In the aforementioned equations, the effect of buoyancy is ignored.

Because turbulence always causes mixing between the fluid layers, it results in diffusion, which is treated as an increase in viscosity. As a result, μ_{eff} is introduced as the effective viscosity:

$$\mu_{eff} = \mu + \mu_t = \mu + \rho C_\mu \frac{k^2}{\varepsilon} \tag{14.27}$$

where $C_\mu = 0.0845$.

The R term in the ε equation is given by

$$R = \frac{C_\mu \rho \eta^3 \left(1 - \eta/\eta_o\right)}{1 + \beta \eta^3} \frac{\varepsilon^2}{k} \tag{14.28}$$

where
  $\eta = Sk/\varepsilon$
  $\eta_o = 4.38$
  $\beta = 0.012$

14.1.4.2 Numerical Method

The problem approaches include setting up boundary and initial conditions and making necessary assumptions, including (*a*) assigning thermal physical properties to the fluid and solid components; (*b*) defining various power losses, which are obtained from a separate electromagnetic analysis, to the corresponding motor components as the heat sources; (*c*) applying the inlet boundary conditions such as volumetric flowrate (or velocity) and temperature; (*d*) applying the outlet boundary condition with zero static pressure; (*e*) assuming all outer walls as adiabatic and heat dissipation only through the cooling fluid; and (*f*) specifying the initial temperature to the computational domain.

The governing equations are solved on the discretized elements using commercial CFD software. All walls are treated as adiabatic and no slip. The flowrate is defined at the flow inlet, and zero pressure is specified as outlet.

The aim of performing the CFD analysis of the motor cooling includes the following: (*a*) The main objective is to obtain the temperature distribution in the motor interior, especially around the stator windings, to avoid motor overheating. (*b*) The cooling flowrates and their distributions are important for the assessment of motor fan selection and performance. Insufficient cooling flows can result in hot spots occurring near the heat sources such as stator winding. Through the CFD analysis, the required airflow rate can be determined. (*c*) The 3D velocity field of the motor reveals fluid-solid interactions and flow patterns such as flow separation and attachment to solid body surfaces. A separated airflow may cause an increase in skin friction and impact cooling efficiency. (*d*) The pressure drop data are used for evaluating motor windage losses. Attention should be focused on abnormal pressure drops at local structures for further minimizing windage losses.

The overall quality of meshes is of primary importance in predicting the temperature and velocity fields in this complex computational domain. In order to obtain accurate results, very fine meshes are set at the high-pressure gradient regions. Due to the large motor size and complicated geometries, the CFD model is meshed with more than 4 million elements: 2.27 million elements for the fluid and 2.07 million elements for the solid parts (Figure 14.6).

In order to achieve accurate numerical simulation results, the winding insulation is also taken into account in this analysis (Figure 14.7). Because the winding insulation is too thin to be meshed with a conventional method, it must be dealt with special approaches. In practice, thin-walled objects such as shells are defined as surface parts. One effective way to deal with the meshing of thin-walled objects is to represent them only with surfaces and eliminate the thickness form of the model. In other words, the surface parts are meshed with 2D elements and are used to represent thin-walled objects. In such a way, surface parts can be used to conduct heat as well as obstruct flow. They exhibit the same heat transfer characteristics as 3D volumes.

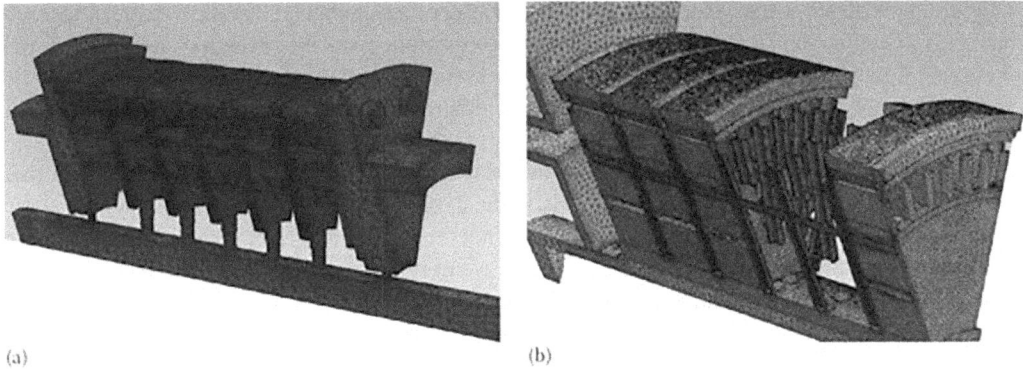

(a)   (b)

FIGURE 14.6 Model meshing: (*a*) fluid meshing (2.27 million elements) and (*b*) meshing for solid parts (2.07 million elements).

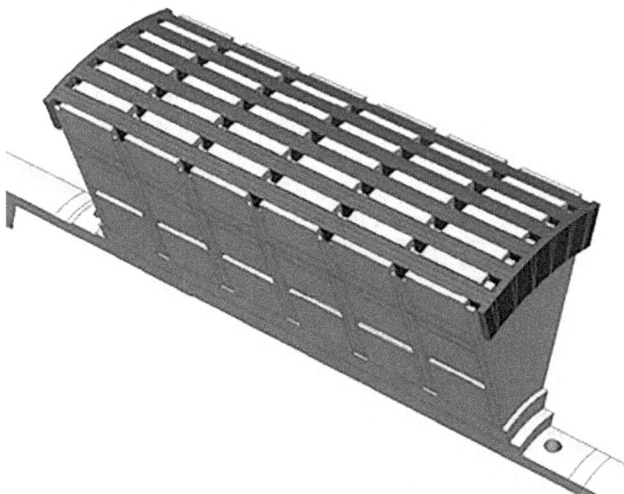

FIGURE 14.7 The thin insulation layer of stator winding is taken into account in the analytical model.

To ensure proper convergence, residuals for mass, momentum, turbulent kinetic energy, and temperature were monitored until achieving a minimum level of 10^{-4}. Furthermore, various monitoring planes at different sections of the motor fan assembly were created and monitored to achieve stable convergence for mass flow and total pressure without any undue fluctuations. In order to determine the required fan flowrate, a number of volumetric flowrates are used in this CFD analysis, including 300, 500, 700, 900, and 1,200 CFM. All power losses, such as copper, core, magnet, and eddy current, are defined at the corresponding components.

To help achieve convergence, the analysis starts with pure fluid dynamics. As the flow field becomes fully developed, the heat transfer model is added to calculate the temperature distribution in the computational domain.

14.1.4.3   Case Study—3D Thermal Analysis of Large Size Motor

The velocity field of the motor is presented in Figure 14.8 for a flowrate of 1,200 CFM. To clearly demonstrate the flow field, only the fluid domain is shown. High velocities occur at the radial spray nozzles on the hollow shaft. The flowrate distributions through five middle channels and two end-winding channels are displayed unevenly; the flowrate is inversely proportional to the distance from the flow inlet. This suggests that for receiving a uniform flowrate distribution, different nozzle diameters are required.

Figure 14.9 displays the velocity fields of middle cooling channels. The flow pattern at each channel is directly associated with the cooling of motor components. For example, at channel 1, the air cooling flow goes through the middle and left subchannels, and almost no flow goes through the right subchannel, resulting in a higher temperature (about $5°C$) on the stator winding on the right subchannel.

The static pressure distribution inside the motor is given in Figure 14.10. It can be found that the highest pressure occurs near the nozzle where the dynamic pressure is converted into the static pressure. The pressure drop across the motor is a function of the flowrate, air viscosity, flow path geometries (*e.g.*, airgap depth and nozzle diameter), and rotor rotating speed. The airflow travels through the nozzles, channels, airgap, and exhaust holes, creating a pressure drop due to the restriction of flow. Based on the pressure contours in the figure, the airflow passages have been optimized for minimizing the pressure drop and reducing the windage power loss.

The temperature distribution in the motor is shown in Figure 14.11. The highest temperature for the flowrate of 1,200 CFM is $93.7°C$, occurring on the stator end turn near the flow inlet. This is consistent with the observation that the flowrate is inversely proportional to the distance from the flow inlet. From the calculated results, it is estimated that the temperature difference due to the nonuniform flowrate is about $10°C$ on the stator winding. The temperature of magnets is shown to be $66°C$. Usually, rareearth magnets such as Nd–Fe–B are sensitive to heat. If a magnet is heated above its maximum operating temperature, it will permanently lose a fraction of its magnetic strength. If the magnet is heated about its Curie temperature, it will lose all of its magnetic properties. Therefore, it is important to always check the magnet temperatures.

Figure 14.12 shows in detail the temperature distribution in each cooling channel. It can be observed that the temperature difference on the stator winding in the same channel is usually less than $5°C$.

The maximum temperature of $93.7°C$ indicates that the airflow rate can be further reduced, lowering not only the power consumption of the blower but also windage loss of the motor. Figures 14.13*a* and *b* presents the temperature distributions for the flowrate of 700 and 900 CFM, respectively. For the flowrate of 900 CFM, the maximum temperature is increased to $109°C$ (−16.3%). Continuously reducing the flowrate to 700 CFM, the maximum temperature becomes $121.7°C$, which can be used for the class B insulation that has the maximum operation temperature of $130°C$. For the class F insulation (maximum operation temperature of $155°C$), the flowrate can be further reduced.

Performing the CFD analysis with the flowrates of 300–1,200 CFM ($0.14–0.57\,m^3/s$) can provide a useful relationship between the maximum temperature and flowrate, as displayed in

(1) Velocity magnitude-m/s

28.8297
27.6284
26.4272
25.226
24.0247
22.8235
21.6223
20.421
19.2198
18.0185
16.8173
15.6161
14.4148
13.2136
12.0124
10.8111
9.60989
8.40866
7.20742
6.00618
4.80495
3.60371
2.40247
1.20124
0

Flow exit

Local high velocity

Flow inlet

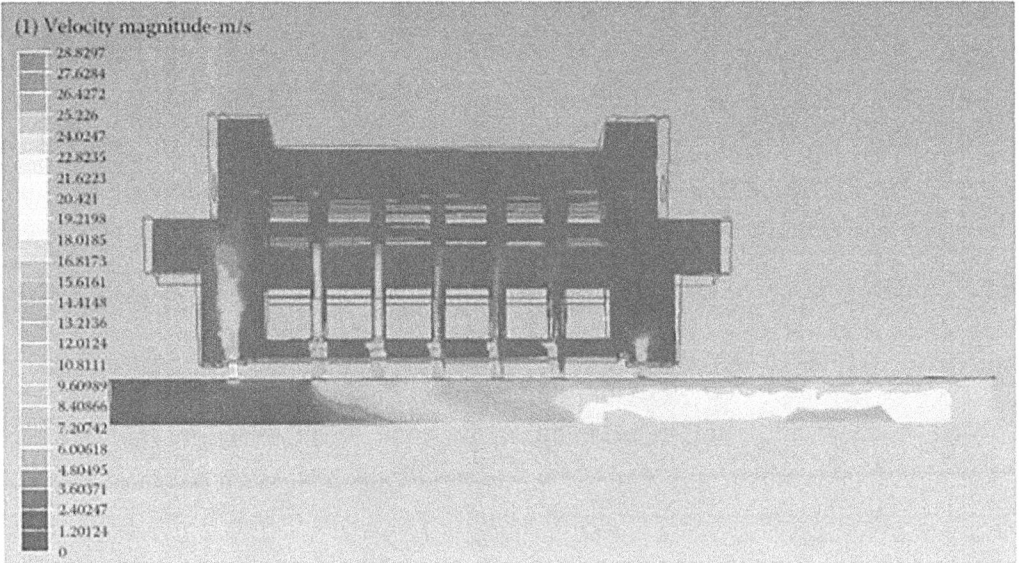

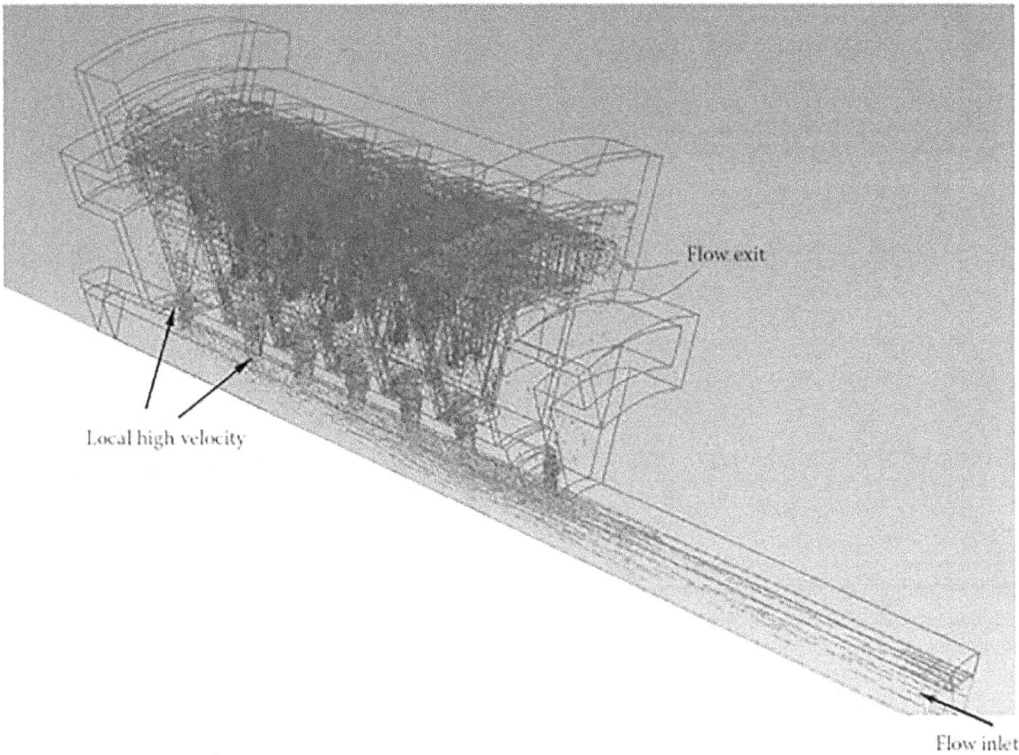

FIGURE 14.8   Velocity field through the motor interior with an intel flowrate of 1,200 CFM.

Figure 14.14. The results indicate that the appropriate range of the volumetric flowrate for the motor cooling is 700–900 CFM (0.33–0.42 m³/s), with the corresponding temperatures in the range of 121°C–109°C.

Figure 14.15 presents the relationship between the flowrate and pressure at blower. This information helps select an appropriate blower/fan and determine the optimum operating points of the blower/fan.

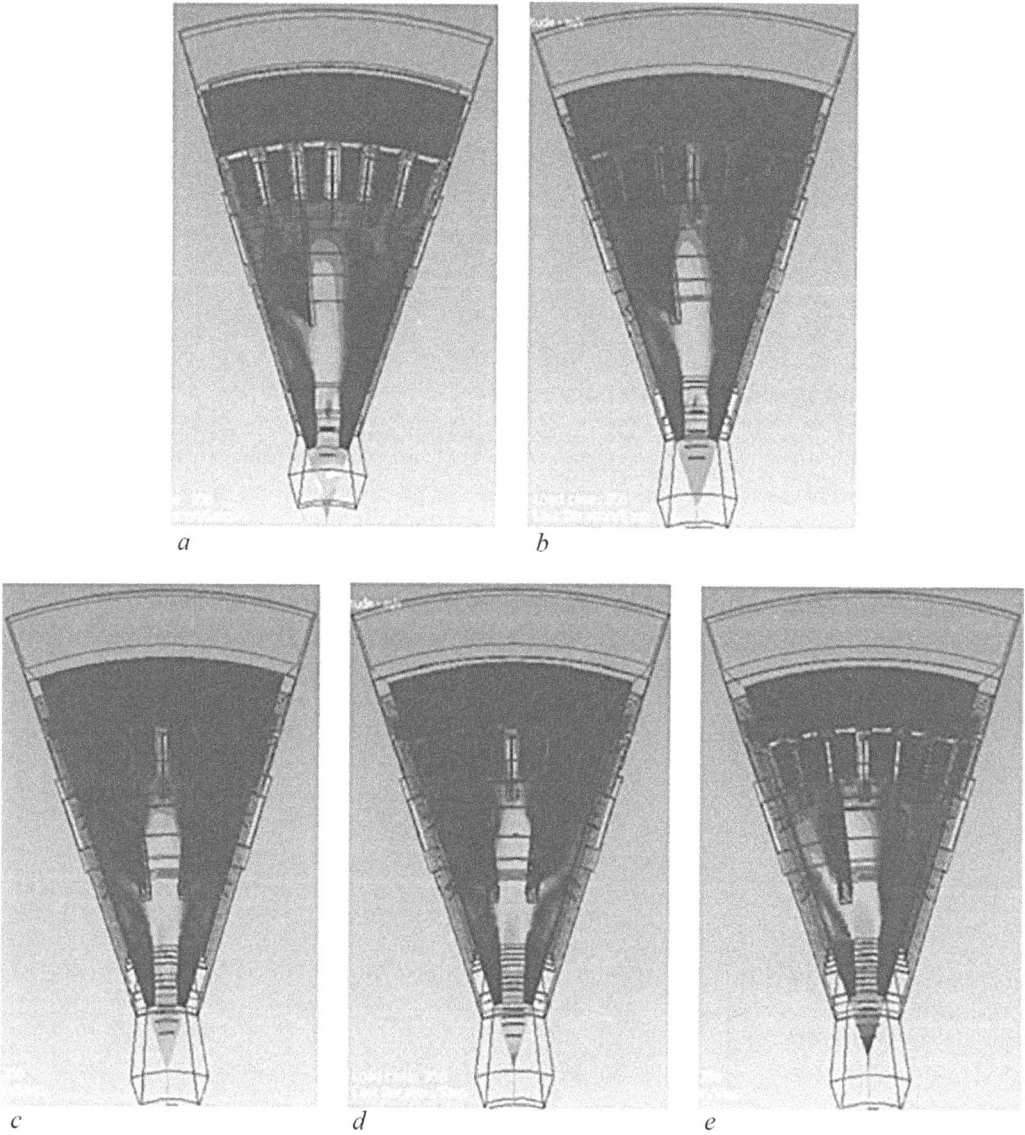

FIGURE 14.9 Velocity contours through the middle cooling channels: (*a*) channel 1, (*b*) channel 2, (*c*) channel 3, (*d*) channel 4, and (*e*) channel 5.

14.1.4.4 SIMULATION OF TWO-PHASE FLOW AND HEAT TRANSFER

Heat transfer in conjunction with two-phase flow is ubiquitous in many industrial applications. Many convective cooling techniques, such as jet impingement cooling, spray cooling, immersion cooling with phase change, mist cooling, heat pipe cooling, thermosyphon cooling, and cooling with nanofluids, involve two-phase flows. Therefore, numerical modeling and simulation of two-phase flow systems have been performed by researchers and engineers and triggered considerable theoretical, computational, and experimental studies.

Two-phase flows are generally classified into three categories according to the shape of the interface: separated flows (*e.g.*, film flow, annular flow, and jet flow), mixed flows (*e.g.*, bubbly annular flow, droplet annular flow, and plug flow), and dispersed flows (*e.g.*, bubbly flow, fluid-particle/

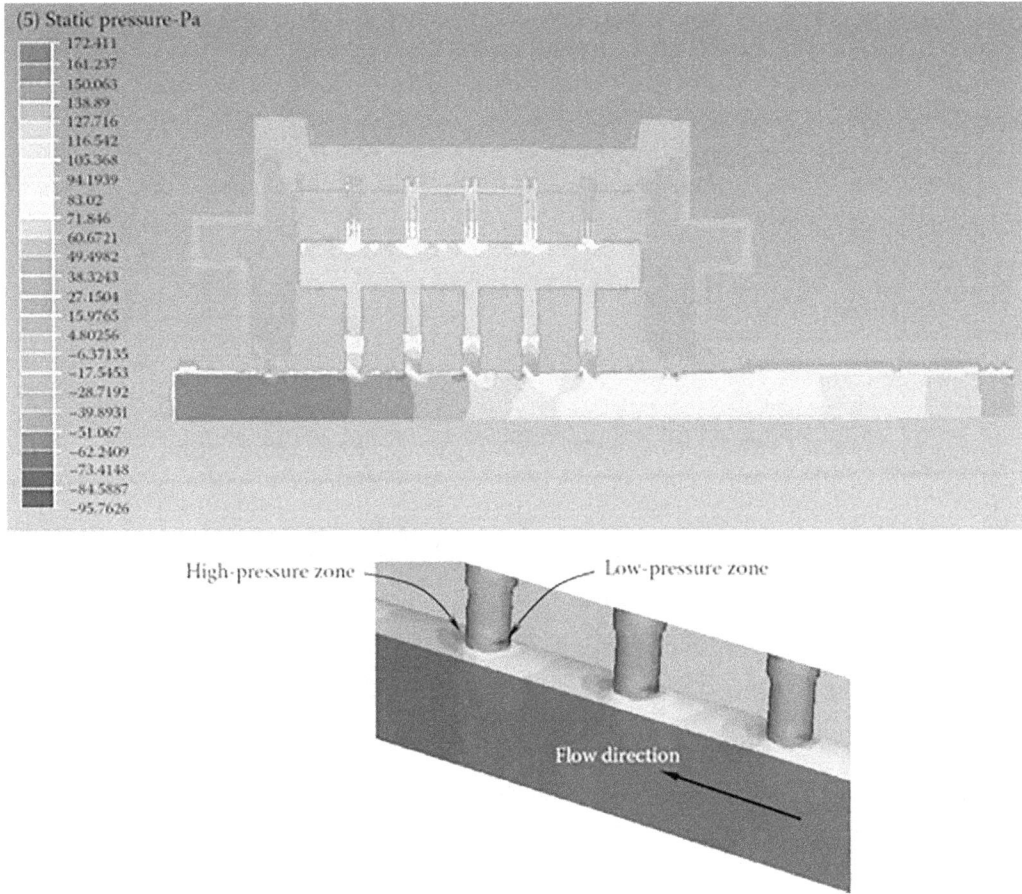

FIGURE 14.10 Pressure distribution through the motor.

droplet flow, slurry flow, and particulate flow) [14.15]. Each type of flow is solved with selected optimal numerical modeling.

In practice, statistically averaged equations are widely used to describe and solve two-phase flows. The most common statistical descriptions of two-phase flow can be classified into three broad categories: (*a*) Eulerian–Eulerian (EE) method (the random field approach), (*b*) Lagrangian-Eulerian method (the stochastic point process approach) [14.16], and (*c*) quadrature-based method of moments such as quadrature method of moments (QMOM) and direct quadrature method of moments (DQMOM) [14.17]. Among them, DQMOM is a relatively new simulation technique for solving complex multiphase problems (*e.g.*, bubble columns). While EE method is computationally expensive, DQMOM is more tractable. To perform a simulation for arbitrary N-phase flows, DQMOM is implemented into CFD software. The DQMOM implementation has been validated by solving the practical multiphase problem and the results are shown in good agreement [14.18]. In general, all these methods based on the statistical modeling approach are suitable for industrial engineers with limited computing resources. Reviews and comparisons of numerical modeling techniques of two-phase flow with respect to accuracy, stability, and cost have been addressed in the literature [14.19–14.22].

Over the past few decades, along with the advent of increasingly powerful supercomputers (*e.g.*, quantum computers), new computing technologies (*e.g.*, cloud computing), and greatly improved numerical modeling techniques, great advances have been made in direct numerical simulation (DNS). DNS is a branch of CFD devoted to the high-fidelity solution of turbulent and multiphase

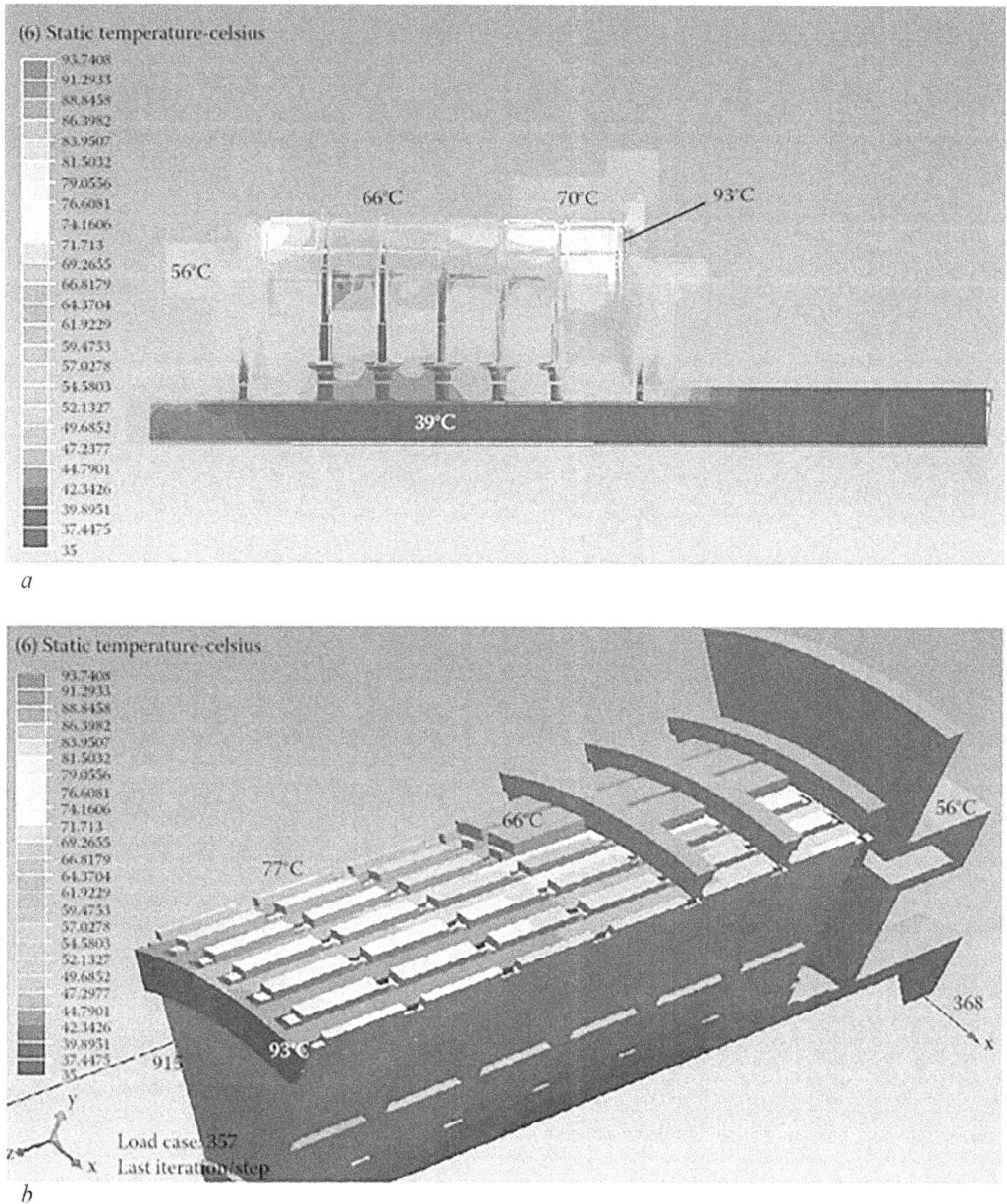

FIGURE 14.11 Temperature distribution on (a) the whole motor and (b) solid components only. The flowrate of 1,200 CFM is used for the analysis.

flows. With this method, the Navier–Stokes equations are numerically solved by resolving the whole range of spatial and temporal scales of turbulence without the need for any additional model (*e.g.*, turbulent model) or assumption. The data generated by DNS have provided valuable insight into the physics of many engineering flows and have led to rapid improvements in two-phase modeling for both academia and industry.

Like single-phase turbulent flows, two-phase turbulent flows generally possess a large range of eddy scales, from the smallest Kolmogorov scale (denoted as η) to the largest scale (denoted as L). The estimates for the ratio of the largest to smallest length scales, the ratio of the largest to smallest

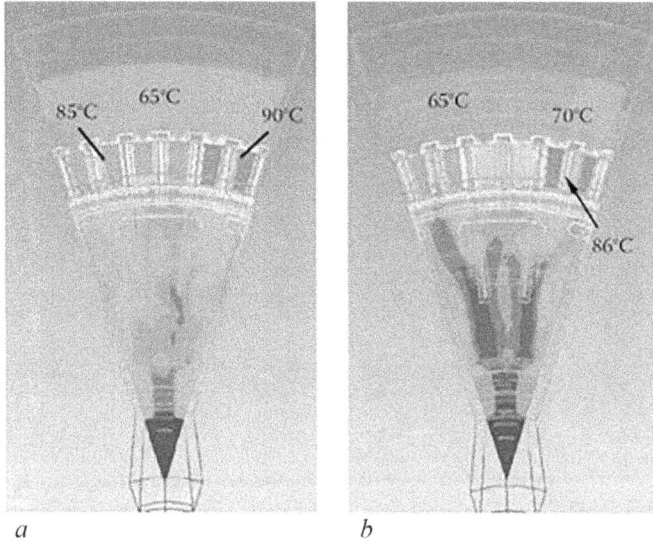

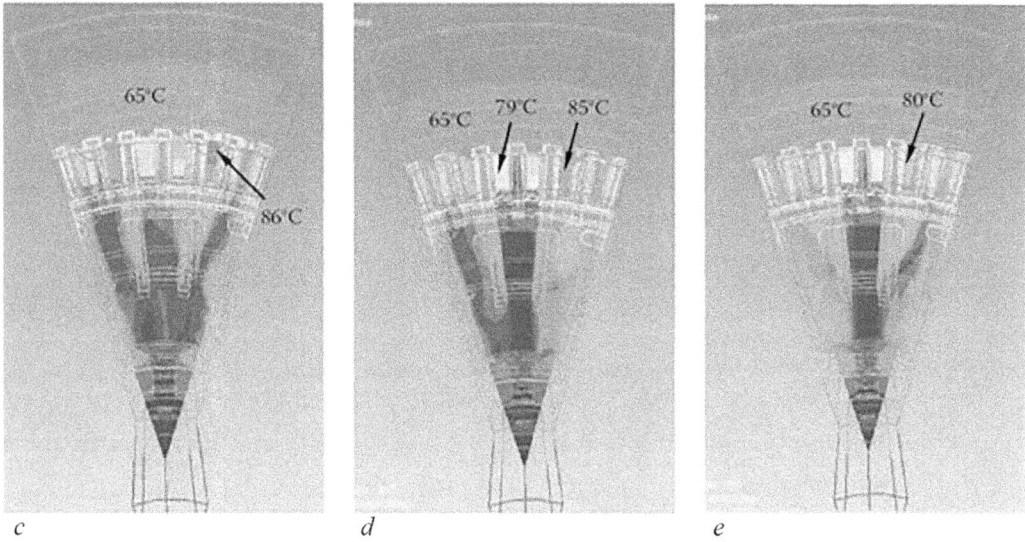

FIGURE 14.12 Temperature distributions through the cross-sectional cooling channels: (*a*) channel 1; (*b*) channel 2; (*c*) channel 3; (*d*) channel 4; and (e) channel 5.

time scales, and the ratio of largest to smallest velocity scales are of the order of $Re^{3/4}$, $Re^{1/2}$, and $Re^{1/4}$, respectively [14.23]

$$L/\eta \sim Re^{3/4} \tag{14.29a}$$

$$t_L/t_\eta \sim Re^{1/2} \tag{14.29b}$$

$$u_L/u_\eta \sim Re^{1/4} \tag{14.29c}$$

where
  Re is the Reynolds number based on the large-scale flow features
  L is the largest length scale in turbulent flow
  η is the smallest length scale in turbulent flow

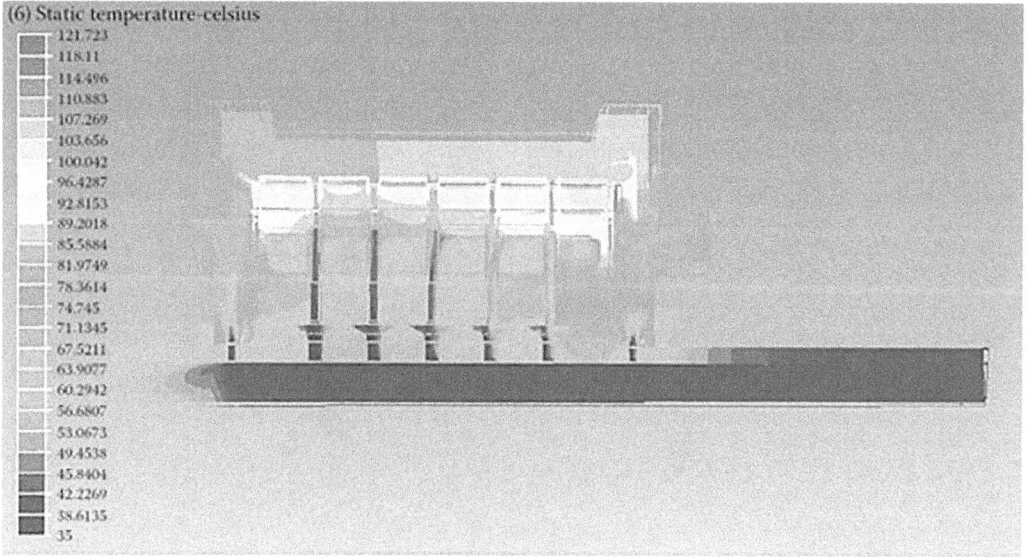

(a)

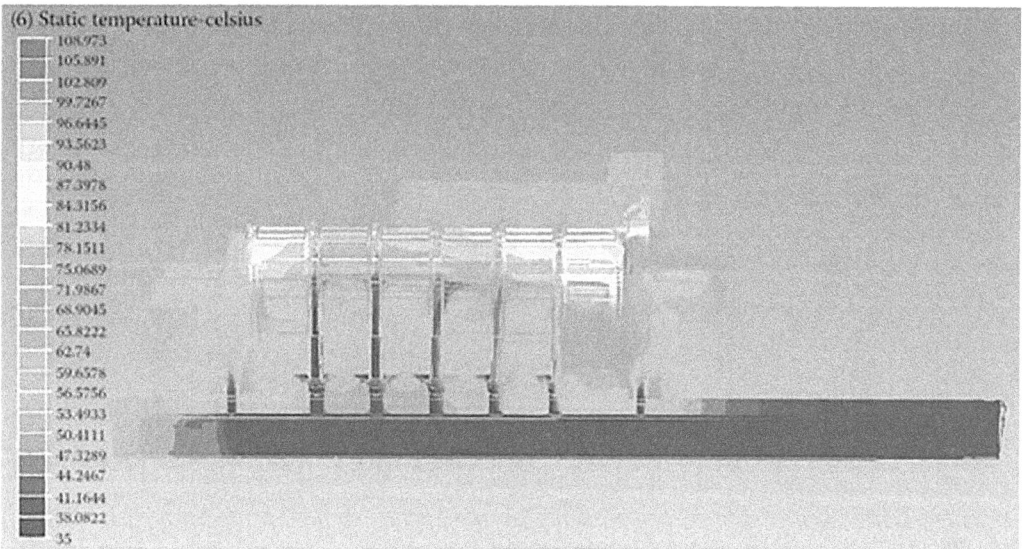

(b)

FIGURE 14.13 Temperature distributions inside the motor: (*a*) flowrate of 700 CFM and (*b*) flowrate of 900 CFM.

t_L is the largest time scale in turbulent flow
t_η is the smallest time scale in turbulent flow
u_L is the largest velocity scale in turbulent flow
u_η is the smallest velocity scale in turbulent flow

These estimations imply that (*a*) the smallest eddies are significantly smaller than the largest ones, especially at large Reynolds numbers; (*b*) the smallest time scales are much briefer than the largest time scales; and (*c*) the smallest velocity scales are much lower than the largest velocity scales. The increase in Reynolds number leads to the separation in scale widens.

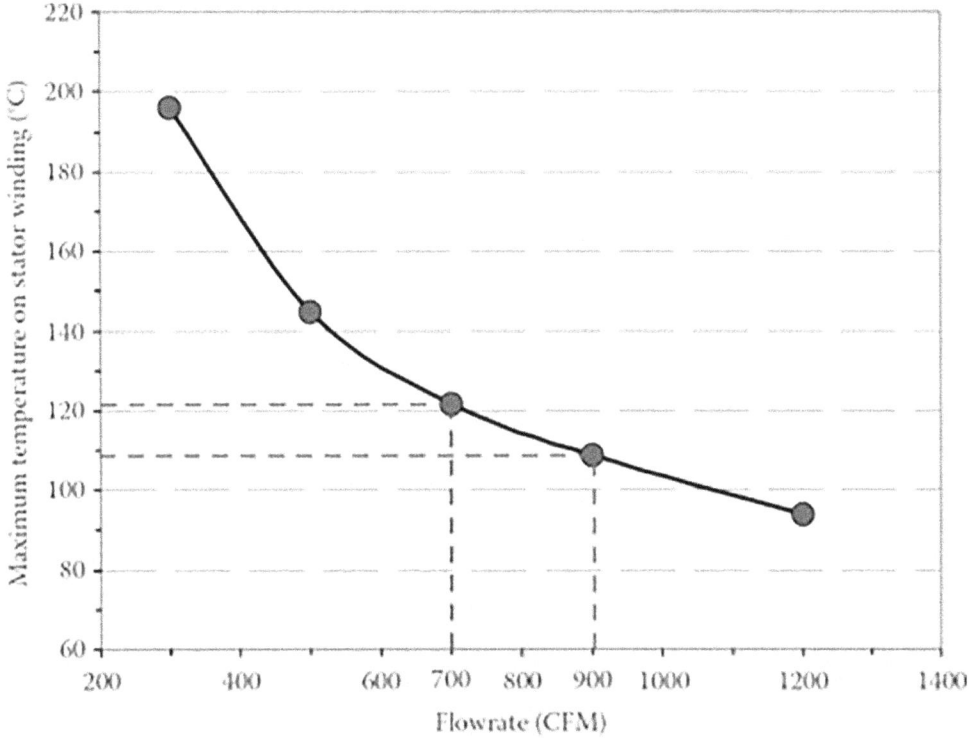

FIGURE 14.14 Maximum temperature versus flowrate. The results indicate that the required volumetric flowrate for the motor cooling is in the range of 700–900 CFM. Correspondingly, the motor maximum temperature is in the range of 109°C–122°C.

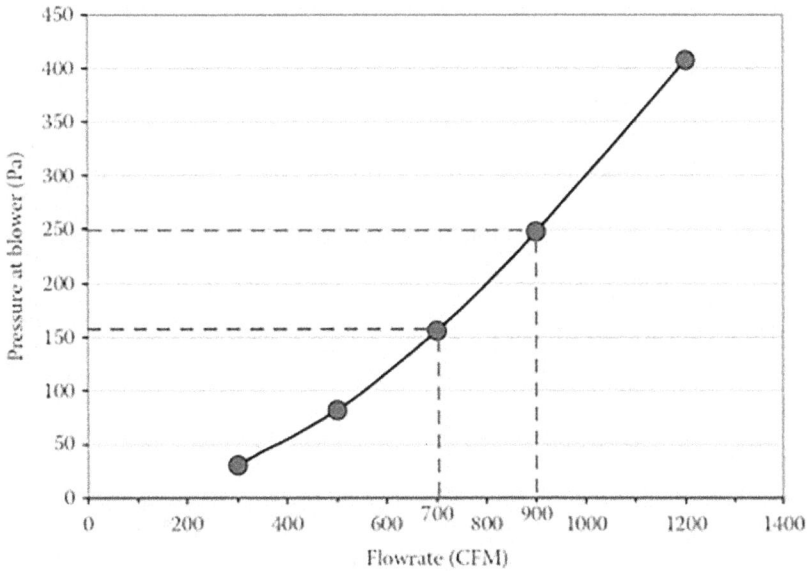

FIGURE 14.15 Blower pressure versus flowrate. The results indicate that the required volumetric flowrate for the motor cooling is in the range of 700–900 CFM.

By means of the DNS technique, detailed information of 3D and time-dependent turbulent flow field can be obtained at the expense of huge computational cost. To obtain accurate results, the Navier–Stokes equations must be solved using grids with spatial domains fine enough to resolve most of the individual eddies, which may be as small as 30–100 mm in low-viscosity fluids such as water. Appropriately small time steps must also be employed to capture the unsteady and fluctuating nature of turbulent flow. At relatively low Reynolds numbers, typical DNS calculations involve a resolution of between 5 and 20 million nodes in the flow field and require 250–400 h of expensive supercomputer time [14.24]. In fact, the main challenge in DNS is the increased computational complexity, especially for flows with high Reynolds numbers. The increase in Reynolds number leads to an increase in the fluctuation frequency and a decrease in the scale of the smallest eddies. As a result, the computing requirements for flow resolution become enormous. Therefore, the applicability of the DNS approach is usually limited to flow with low or moderate Reynolds numbers [14.25].

In DNS modeling, three interface-capturing approaches are often employed: the level-set method, diffuse interface method (using the Cahn–Hilliard equation), and VOF method. The detailed description for each method is given by Náraigh and Valluri [14.26]. Compared with the two-fluid model [14.27], where there are two separate sets of equations for both phases (one set of equations for each phase), with the equations matched across the interface, the main advantage of these approaches is the matching conditions across the interface are embedded a source term in the momentum equation.

By coupling the 3D Navier–Stokes equations, energy equation, and Cahn–Hilliard equation, Zheng *et al.* have developed an efficient method to simulate two-phase heat transfer problems [14.28]. The equations are solved in time with a splitting scheme that decouples the flow and temperature variables, yielding time-independent coefficient matrices after discretization. Thus, the computation for each flow variable involves only a constant and time-independent coefficient matrix for the linear algebraic system, which can be pre-computed during pre-processing. This effectively overcomes the performance bottleneck caused by variable coefficient matrices associated with variable properties (conductivity, density, specific heat, viscosity, *etc.*).

A DNS model of two-phase turbulent flow in fuel cell flow channels has been developed by Niu *et al.* [14.29]. This model utilizes a modified VOF approach for tracking the air–water interface. By resolving the whole range of spatial and temporal scales of turbulence, the results of the two-phase DNS model show that the deformation of water droplet is asymmetric and broken into small pieces or films, which is significantly different from those of the laminar and the corresponding k-ε models.

A recent development in electric motor thermal management is to employ direct oil spray/splash-based cooling for improved motor performance. However, simulations of two phase flow and associated heat transfer for such applications remain big challenges due to the complexity of two phase flow modeling. More recently, Kapatral *et al.* [14.30] have performed numerical analysis on two phase flow and heat transfer for a direct-oil-cooled motor. Detailed temperature and velocity fields of the motor under different operating conditions are obtained using a conjugate heat transfer approach. Results show that the numerical model is in good agreement with temperature measurements at discrete locations in the motor.

14.2   THERMAL SIMULATION WITH LUMPED-CIRCUIT MODELING

Numerical simulations on motor cooling with CFD tools and different methods, such as finite element method (FEM), finite difference method (FDM), and finite volume method (FVM), have long been adopted in the motor industry, as well as in industries with high reliability applications such as aerospace, defense, robotics, and new energy automobile. With properly defined boundary and initial conditions, temperature-dependent material properties, power losses, governing equations (*e.g.*, the Navier–Stokes equations, energy equation, turbulent models, and porous media models), grids, *etc.*, CFD can precisely predict thermal and fluid fields of electric machines for gaining information about temperature distribution, heat transfer rate, flow pattern, velocity field, pressure drop, and so on. Furthermore, CFD and other simulation tools may perform the thermal analysis, fluid flow analysis, structural analysis, and electromagnetic analysis simultaneously by means of multi-physics coupling to optimize the design of the whole motor system. However, these methods have some main drawbacks. They usually require intensive computational power and high-performance computing resources to solve fluid flow and energy equations and depend on motor geometries and material properties, resulting in complex calculations, long computing time, and high cost.

Among various thermal analysis methods, a simple and cost-effective approach is to use lumped-circuit (or network) analogy modeling. This method is a simplification of the system physical model, based on the analogy between electrical and thermal systems. The analogies between current I and rate of heat flow q, voltage V, and temperature difference between two spatial points or surfaces ΔT, etc., can be properly applied in analyzing thermal systems. In such a way, it is possible to take thermal resistances as the analog to electrical resistances. This allows thermal engineers to take advantage of fully developed basic laws (such as Ohm's law and Kirchhoff's law) in electrical engineering to determine heat transfer characteristics in thermal systems. Based on the thermal-electrical analog, the electrical equivalent circuit (*i.e.*, thermal circuit) can be built with a number of nodes; each node represents a part of an electric machine and is connected each other via thermal resistances and thermal capacitances. The analogous parameters between the two systems are listed in Table 14.1.

In fact, this conventional approach was popular before the advanced computational techniques and CFD software have been fully developed. The lumped-circuit approach has the advantage of being very fast to calculate, but the development of the network model is time-consuming, requiring thermal engineers to spend a large amount of time in defining the circuit that accurately models all heat transfer paths and circuit elements.

By converting a complex 3D physical model into an equivalent lumped-circuit system, it greatly simplifies the problem complexity and thus reduces the simulation runtime and the memory usage. The lumped-circuit analogy method involves the representation of the thermal and flow systems as a network of thermal and flow paths and components for the prediction of system-wide temperature distribution and flow field. In fact, this technique is very efficient in terms of the effort required for model definition, solution, and examination of results because it employs overall component characteristics for analyzing their systems-wide interaction. Because this technique requires a much shorter time compared with CFD analysis, it is especially suitable to explore different designs at the conceptual design stage [14.31]. However, the most difficult aspect of this approach is to determine the heat transfer coefficients and the resultant thermal resistances in convective heat transfer, as well as in radiation heat transfer.

TABLE 14.1

Analogous Parameters between Electrical and Thermal Systems

Parameter	Electrical System	Thermal System
Flow	Current I (in Ampere [A])	Heat flow q (in W)
Potential	Voltage V (in V)	Temperature differential ΔT (in °C or K)
Resistance	$R = \dfrac{V}{I}$ (in Ω)	$R_{th} = \dfrac{\Delta T}{q}$ (in °C / W)
Conductance	$S = 1/R$ (in A/V)	$G = 1/R_{th}$ (in W/°C)
Capacitance	$C = \dfrac{1}{dV / dt}$ (in Farad [F])	$C_{th} = \dfrac{q}{d\Delta T / dt}$ (in J/°C)
Ohm's law	$I = \dfrac{V}{R}$	$q = \dfrac{\Delta T}{R_{th}}$

An analytical-numerical hybrid scheme for system-wide thermal modeling of electric motors has been proposed by Liu *et al.* [14.32]. This scheme employs the FVM concept to calculate heat conduction for motor components while using flow network modeling for fluid convection calculation. This hybrid scheme not only simplifies the complicated flow simulation but also considers motor's geometry information and increases the accuracy of the simulation.

In order to develop an equivalent thermal network, it is required to subdivide the thermal system into a number of finite subvolumes called nodes. All the thermal parameters, such as temperature, thermal capacitance, and heat generation, are considered to be concentrated at the central nodal point. This point represents the mean values of the thermal parameters in the node. The thermal models for the basic elements of electric machines have been discussed by Perez and Kassakian [14.33] in more detail.

An extended survey on the evolution and the modern approaches in thermal analysis of electrical machines has revealed that it follows the path from the lumped-parameter network, to FEA, and to today's numerical simulations using CFD techniques. It can be advantageous to use the lumped-parameter network approach for its simplicity, requirement of less computing resources, and fast calculations. FEA can be considered a convenient solution in a very complex geometry not approachable with lumped-parameter networks. CFD can deal with quite complex heat transfer problems and provide very accurate results. However, it needs very knowledgeable and experienced thermal engineers to correctly set up the model and requires very high computer capabilities and powerful computing resources. As a matter of fact, each of these approaches has its advantages and disadvantages.

A comprehensive review has been made by Boglietti *et al.* [14.34] on the evolution and modern approaches in the thermal analysis of electrical machines. The three primary thermal analysis methods, that is, lumped-parameter thermal network (LPTN), FEA, and CFD, are analyzed in depth and compared in order to highlight the qualities and defects of each. The thermal network is the most basic form to calculate conduction, convection, and radiation resistances for different parts of the motor construction. The convection heat transfer coefficient is most often based on empirical convection correlations. This is fundamentally different from CFD analysis, where the heat transfer coefficient is calculated from the CFD model itself. According to Boglietti *et al.*, FEA can only be used to model conduction heat transfer in solid components. They expect that CFD will be more popular and widespread in the thermal analysis due to the fast development in computational capability of modern computers, as well as the cost reduction of CFD software.

By analyzing the structure of a fully enclosed air-cooled induction motor, a complete lumped parameter linear model is presented to describe the dynamic thermal performance [14.35]. According to the motor structure, this model is divided into 10 nodes: the motor housing, stator core, stator teeth, stator coil, airgap, stator end winding, endbell, rotor winding, rotor core, and motor shaft. This model, which reflects the heat transfer process inside electric motors, provides an effective way for quickly performing the offline thermal analysis. Consequently, it has been widely used in industrial applications. However, this model ignores the contact resistance between the solid parts and highly depends upon the accuracy of motor geometry, and thus it requires a considerable number of motor dimensions and material-related parameters. Furthermore, this model is based on the empirical formula of heat transfer, which is sensitive to the size and shape of the airgap.

To simplify the above complete thermal network model and aim at resolving the difficulty in online temperature estimation, Zhu *et al.* [14.36] have proposed an accurate and simplified model for a PM synchronous motor using a five-node LPTN: coolant, stator, rotor, endbell, and motor housing. This greatly benefits the thermal analysis of motor in terms of accuracy, computation time, and cost. The comparison between the two models is demonstrated in Figure 14.16.

In the simplified model, T_c, T_s, T_r, T_e, and T_h represent the temperature of the coolant, stator, rotor, endbell, and motor housing, respectively; R_{ch}, R_{hs}, R_{sr}, R_{re}, and R_{eh} denote the thermal resistance between the coolant and housing, housing and stator, stator and rotor, rotor and endbell, and endbell and housing, respectively; C_s, C_r, and C_e are the thermal capacitance of the stator, rotor, and endbell,

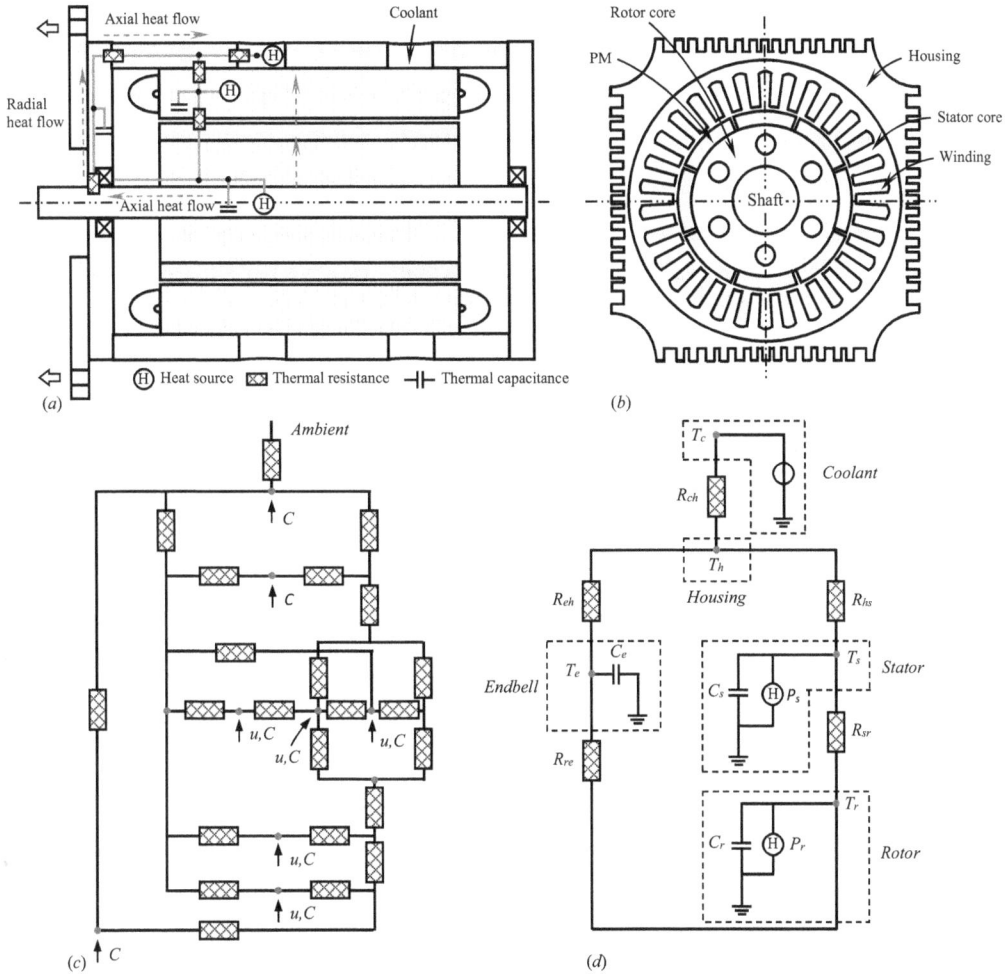

FIGURE 14.16 Comparison between the complete and simplified network models: (*a*) the configuration of PM synchronous motor for the development of the simplified network model; (*b*) the side-view of the motor, (*c*) the complete network model with 10 nodes for an induction motor [14.35]; and (*d*) the simplified network model with 5 nodes at the stator, rotor, endbell, housing, and coolant [14.36].

respectively; P_s and P_r are the power loss (*i.e.*, the heat source) of the stator and rotor, respectively. In this analysis, both radial and axial heat transfer paths inside the motor are taken into account while modeling the complete thermal circuit. In addition, an innovative identification method based on the multiple linear regression is applied to identify the parameters of the LPTN model. An open-loop estimation scheme based on the state equation and Kalman filter algorithm is adopted to predict the motor temperature online. This model is validated by experiments under varying speed and torque conditions in terms of accuracy and robustness. The results indicate that the temperature error is within the range of $\pm 5°C$ in most cases.

14.3 THERMAL ANALYSIS USING THE FINITE ELEMENT METHOD

FEM has been one of the major numerical solution techniques. One of the primary advantages of the FEM approach is the simplicity and ease of using FEM to solve complex geometry problems. By comparing with FDM and FVM, FEM is superior in its built-in ability to handle unstructured meshes, a rich family of element choices, and natural handling of boundary conditions. As indicated by its name, FEM requires the division of the solution domain into many discrete volumes or finite elements. Thus, FEM yields discretized equations that are entirely local to the elements and provides complete geometric flexibility [14.37]. Applications of FEM become more widespread for motor design engineers to analyze structural, thermal, and fluid dynamic problems and their interactions. However, it is difficult to develop computationally efficient solution methods for strongly coupled and nonequations using FEA [14.38]. In addition, FEM does not help in determining quantities such as convective heat transfer coefficients. Thus, FEM is still not comparable to professional commercial CFD codes in solving convective heat transfer and fluid dynamic problems so far.

One example of employing FEM to carry out a thermal analysis is shown in Figure 14.17. This is a frameless motor with two endbells positioned at the two sides of the stator core. The endbells are made from gray cast iron and ductile cast iron for achieving high-vibration damping and high-strength properties. The stator winding is encapsulated by epoxy resin to eliminate the voids inside the winding, improve heat transfer, and increase the winding's dynamic stiffness. The physical motor model is generated by a CAD tool and transferred into the FEM model.

This FEM analysis deals primarily with heat conduction through the motor components. The heat transfer coefficient h on the outer surfaces of the motor is obtained from the heat transfer correlations available in the literature for addressing the heat dissipation from the motor to the ambience (Figure 14.18). It shows for both horizontal and vertical surfaces, heat transfer coefficients increase exponentially with an increase in temperature. Due to the symmetry of the model, only a half of the motor is modeled. In addition to heat conduction (inside the motor) and convection (on the motor outer surfaces), the heat radiation effect is also integrated into the model. The initial temperature of the motor is set at $45°C$. The thermal conductivities of copper, epoxy E88, and cast iron are functions of temperature (Figure 14.19). As can be seen from the figure, for the temperature less than

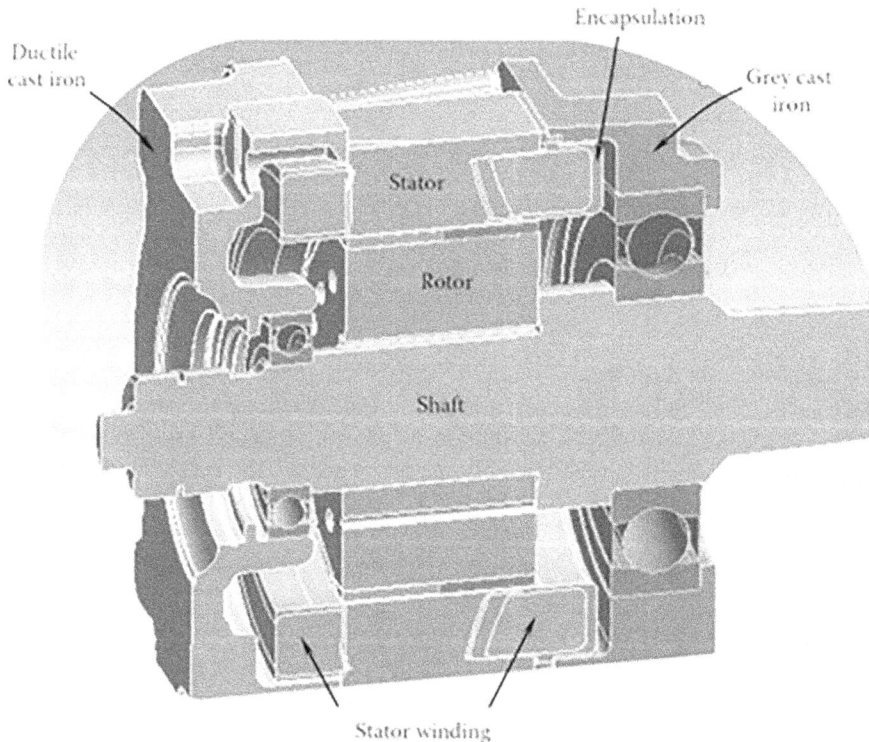

FIGURE 14.17   FEM model of an electric motor.

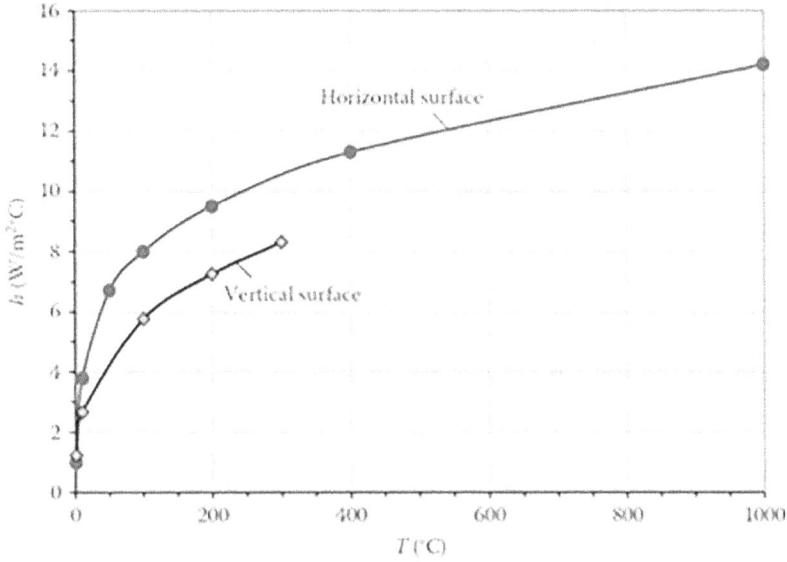

FIGURE 14.18 Hear transfer coefficients on horizontal and vertical surfaces of the motor, obtained from the thermal correlations in the literature.

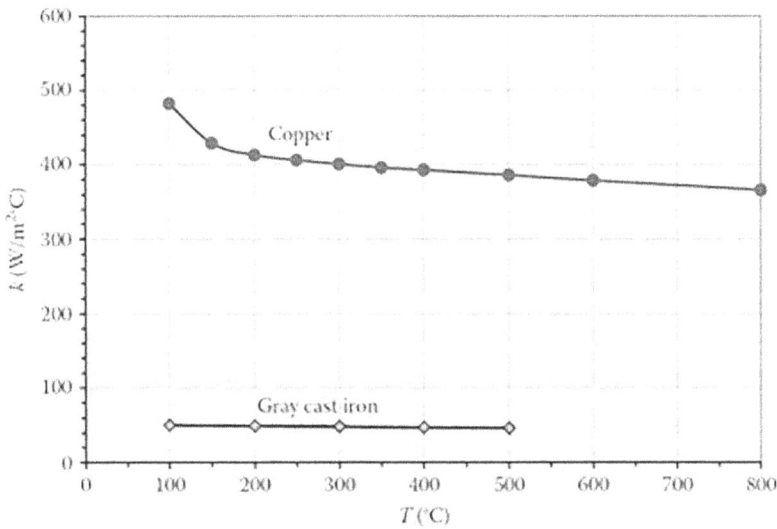

FIGURE 14.19 Thermal conductivities of copper and gray cast iron as functions of temperature.

$200°C$, the thermal conductivity of copper decreases sharply and then becomes a linear function of temperature for $T > 200°C$. By comparison, the temperature effect on the thermal conductivity of epoxy is quite weak; when the temperature increases from $100°C$ to $500°C$, the thermal conductivity reduces only 8%. In addition, the increase in the silicon content of the silicon iron lamination can lead to a decrease in the thermal conductivity and an increase in eddy-current losses [14.39].

The power losses from the stator winding, rotor and stator cores, bearings, and others are properly defined to the corresponding components as the heat sources.

Figure 14.20 shows the meshes created for the FEM model. Relatively fine meshes are used in the vicinities of stress-concentrated regions such as sharp contact areas, entrant corners, load transfer (welds, bonded joints, reinforcing bars, *etc.*), abrupt changes in thickness, material properties, and

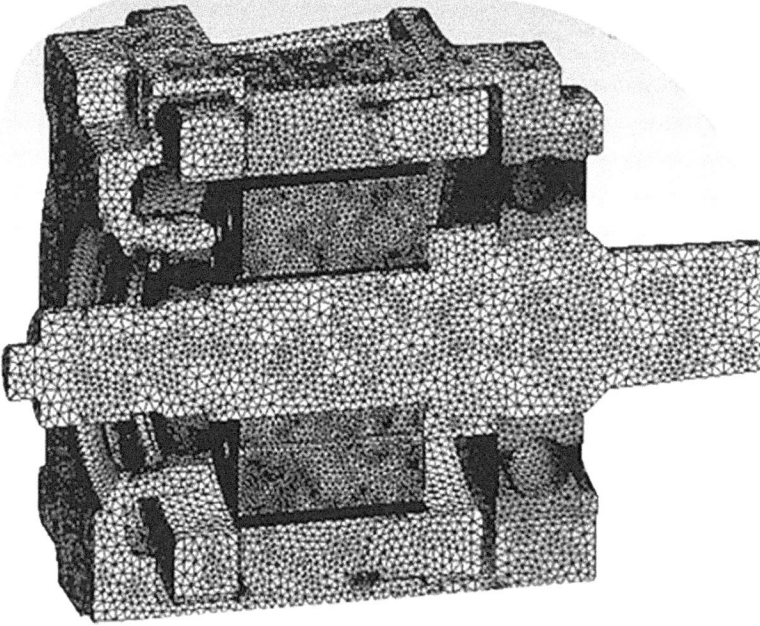

FIGURE 14.20 Meshes on the motor using FEM.

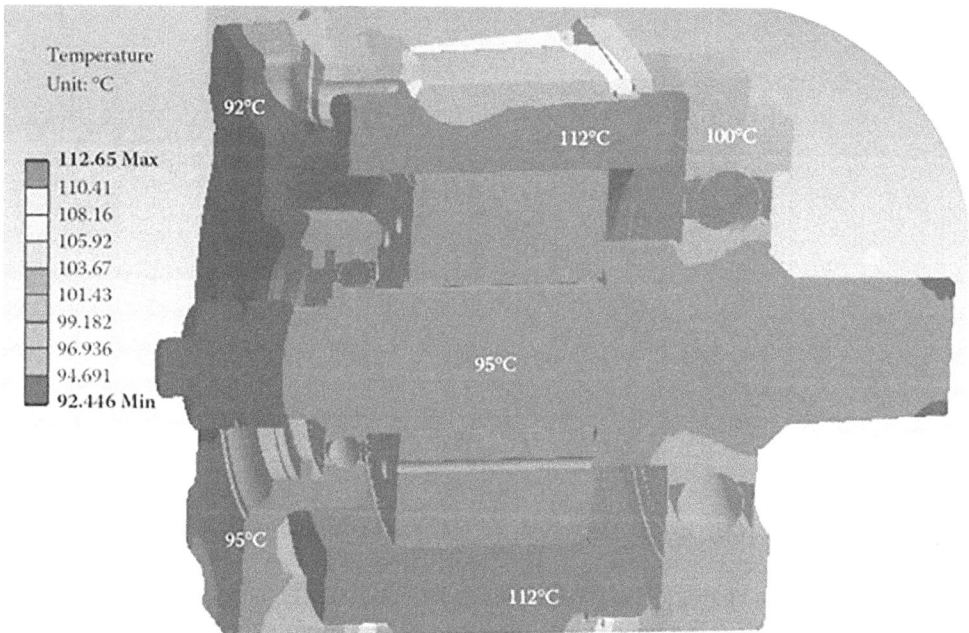

FIGURE 14.21 Temperature distribution on motor components.

cross-sectional areas. It is critical to pay attention to the connections between adjacent parts and set up the same mesh pattern on them. In the mesh setting, fast transitions from small elements to large elements should always be avoided.

The calculated temperature distribution in the motor is presented in Figure 14.21. It is noted that the temperature ranges from $92.4°C$ to $112.7°C$, with the maximum temperature occurring on the stator winding. In the rotor, temperature is quite uniform, about 95°C. The temperature at the large bearing of the drive side is about 5°C higher than that of the small bearing. Due to the gaps between the encapsulated stator winding and two endbells, the temperatures at the endbells are 10°C–18°C lower than that of the stator core. By eliminating these gaps, the temperature difference between the winding and endbells can be reduced to less than 12°C.

14.4 ROTORDYNAMIC ANALYSIS

Rotordynamic analysis is of great practical importance when designing rotating systems such as motors, generators, pumps, and compressors. It has been widely accepted that high-vibration levels of electric motors are caused by lateral critical speeds near the operating speeds. By modeling the rotating geometry and its dynamic characteristics, such as stiffness and damping, the critical speeds of rotating machinery can be predicted. As a consequence, design optimizations can be carried out to ensure that the machine operating speeds do not fall into the vicinities of its critical speeds.

The finite element rotordynamic analysis and critical-speed design sensitivity investigation are performed with a motor fan system. Results show that critical-speed separation margins of more than 30% are obtained from a rated speed of 60,000 rpm without any adverse effects from the spline shaft and that the critical-speed change rates to the support modeling of spline shaft connection points are extremely negligible. Furthermore, the critical-speed change rates to the shaft-element length changes show quantitatively that the spline shaft has some limited influence on the fourth critical speed but practically no influence on the first to third critical speeds.

14.4.1 PROBLEM DESCRIPTION

Cooling fans in electric motors are typically mounted at the end of the rotor to produce high-pressure cooling air for the motor cooling. The cooling fan blows or sucks air into the motor being cooled and exhausts hot air from the interior of the motor to either an environment or a heat exchanger to be cooled before reentering the machine. A seal must be positioned between the high-pressure hot gas discharging from the fan and the low-pressure cold gas inlet to the rotor. As prevailing market trends require higher performance, high efficiency, higher reliability, lower cost, and high-power density motors, motor cooling, especially at the stator and rotor end regions, becomes a limiting factor.

Fan loss is the prime mover power transmitted through the rotor shaft to raise the static pressure of the cooling airflow through the fan. An examination of motor design data shows that fan loss could be a significant portion among the total windage losses. This loss can be effectively reduced by optimizing the fan parameters in design, including fan blade profile (for instance, 3D blade profile), number of fan blades, inlet and exit conditions, diffuser performance, and fan tip clearances.

Usually, the lack of a fan test facility causes any new fan design to rely on only numerical simulation results from CFD analysis. However, a high risk is incurred without proper validation of the numerical model through experimental investigations. The most reliable information about a physical process is given by real measurement. The motor design process can benefit from full-scale fan test data for optimizing fan performance and motor design.

With strong demands for large-size, high-power, and cost-effective electric machines, there have been growing concerns to build up the fan testing facility for optimizing machine cooling design. As the key step of a typical design process for rotating machinery, the rotordynamic analysis must be carried out prior to the construction of the fan facility.

The rotordynamic model includes the shaft-rotor assembly (shaft, couplings, and bearings), the rotor bearings, and their supports. The stiffness and mass of the bearing support play a crucial role in the calculation of the rotor critical speeds. The objective of the analysis is to determine the damped and undamped critical speeds of the fan test system with the bearing supports' stiffness, damping, and mass modeled.

14.4.2 Bearing Support's Stiffness and Damping

As demonstrated in Figure 14.22, the fan test system consists of a shaft, a forward disk, and an aft disk both mounted on the shaft, journal bearings, and bearing supports. The outer ring is mounted on the disks rigidly, and fan blades are attached to the outer ring and forward disk. The system is supported with two bearings at the rotor ends with asymmetrical translational stiffness and damping values. In addition, the two bearings provide additional damping for stabilizing the systems.

The stiffness, damping, and effective mass of the bearing supports can be determined from experimental tests. An effective approach is to use an instrumented hammer to excite the system and measure its response. Equipped with an accelerometer at one end, this hammer is used in tandem with the sensor to measure the vibration. This technique is usually called the impact input test, or impedance test, which is effective, because the impact inputs a small amount of force in the system and receives the response over a large frequency range. To properly apply this technique in this problem, it needs to select a number of locations on the bearing support for obtaining accurate results, as shown in Figure 14.23.

The resulting test data as a function of frequency include force/acceleration (N·s²/m), force/velocity (N·s/m), and force/distance (N/m). The stiffness in each direction is determined using the force/distance curve and using the value as the frequency approaches zero. The supports' damping is obtained by using the force/velocity curve and using the value at the natural frequency. The natural frequency for each bearing support direction and the resulting damping values are shown in Table 14.2.

All of the components are steel ($\rho = 7{,}833\,\text{kg/m}^3$) except for the blades that are aluminum ($\rho = 2{,}796\,\text{kg/m}^3$). The rotordynamic model of a fan test system is shown in Figure 14.24. The upper half of the cross section represents the geometry used to calculate the mass properties, and the lower half represents the geometry used to calculate the stiffness. For each of the two disks, the density and outer radius were solved to exactly match the correct mass and mass polar moment of inertia. The forward disk includes the blade mass properties.

The disks' connections to the shaft are modeled as if they are continuous material with the shaft. The outer-ring press fit joints to the disks are modeled with a *rigid bearing* connection. This modeling technique couples the slope and displacement of the two components together.

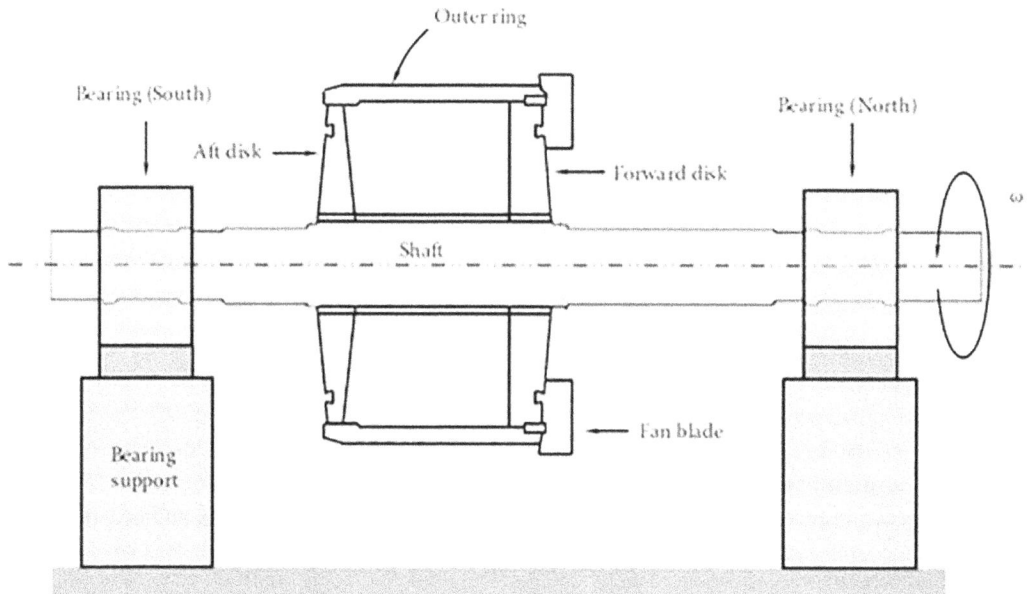

FIGURE 14.22 Configuration and components of the fan test system.

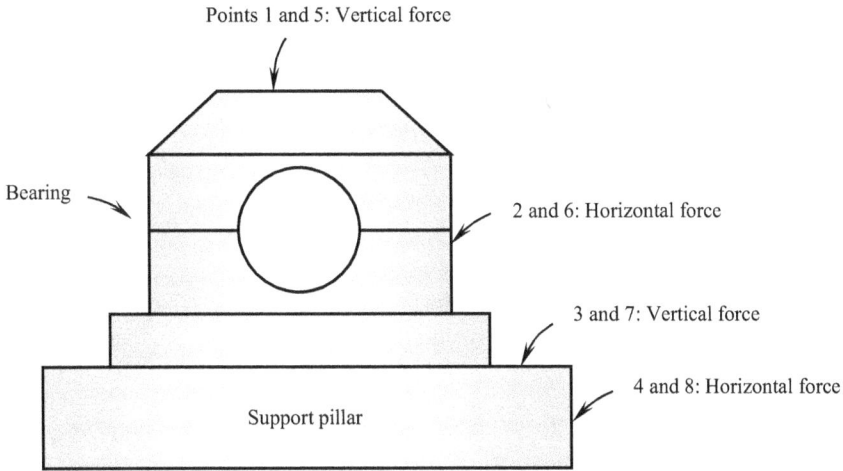

FIGURE 14.23 Using an instrumented hammer to excite the bearing support to obtain the natural frequency, damping, stiffness, and mass. Points 1 and 4 are taken on the south bearing, and points 5 and 8 are taken on the north bearing. Points 1, 3, 5, and 7 are vertical measurements, while points 2, 4, 6, and 8 are horizontal measurements.

TABLE 14.2
Bearing Support Stiffness, Damping, and Mass Data

Direction	Bearing Support	Point Location	Damping (N·s/m)	Natural Frequency Hz	rpm	Stiffness (N/m)	Mass (kg)
Vertical	South	1	249,556	70	4,200	5.15×10^8	2,715
	North	5	340,271	82	4,920	3.50×10^8	1,319
Horizontal	South	2	51,837	37	2,220	5.25×10^7	972
	North	6	14,010	19	1,140	1.75×10^7	1,228

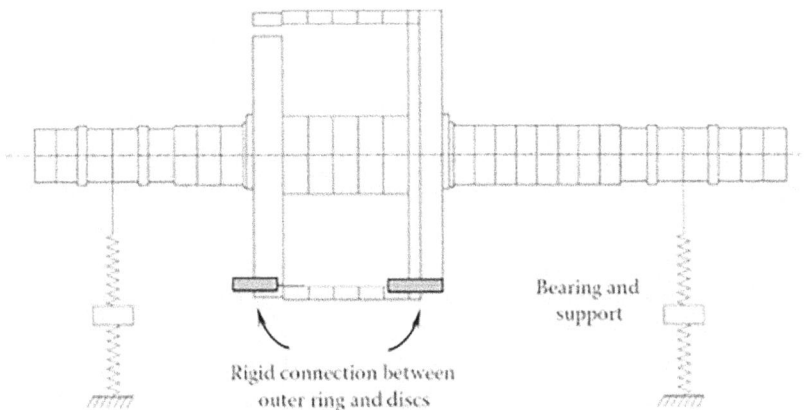

FIGURE 14.24 Rotordynamic model of the fan test system.

TABLE 14.3

Translational Stiffness and Damping of the Fan Bearing Support

Translational Stiffness		Translational Damping	
k_{xx}, horizontal (N/m)	k_{yy}, vertical (N/m)	c_{xx}, horizontal (N·s/m)	c_{yy}, vertical (N·s/m)
1.93×10^8	3.85×10^8	4.48×10^5	8.50×10^5

For the undamped critical-speed analysis, only one bearing stiffness is needed. Therefore, the averaged stiffness is used in the undamped critical-speed calculation (where $S_b = 2.89 \times 10^8$ N/m). For damped critical-speed analysis, both horizontal and vertical stiffness and damping are incorporated into the model (Table 14.3).

The primary advantage of using this technique is that it can monitor phase shifts, vibration force, and coherence. When the slope of the phase-frequency curve becomes infinity, the frequency at which it occurs is the natural frequency.

The mass of the bearing support m_b is determined with the bearing support's stiffness k_b and natural frequency ω_n:

$$m_b = \frac{k_b}{\omega_n^2} \qquad (14.30)$$

14.4.3 ROTORDYNAMIC MODELING

A commercial rotordynamic software has been applied for this analysis. This software is a rotordynamic program based on an FEA method and takes into account rotary inertia, shear deformation, and gyroscopic effects.

The geometry of the fan test system is shown in Figure 14.22. The mass and mass moments of inertia of the disks, outer ring, and blades are obtained from the 3D solid model. All components are made of steel except the fan blades that are made of die casting aluminum. In the rotordynamic model, the bearing spring is modeled in series with the bearing support spring.

The undamped and damped critical speeds are evaluated for synchronous forward and backward whirl. The stability of the damped critical speeds is also evaluated. The test rig design speed is 3,000 and 3,600 rpm with a 1.2×overspeed condition of 4,320 rpm.

14.4.4 RESULTS OF ROTORDYNAMIC ANALYSIS

The undamped and damped critical speeds are evaluated for synchronous forward and backward rotations. The stability of the damped critical speeds is also evaluated. Modern high-performance motors normally operate above the first critical speed, which is considered the most important mode in the system and avoid continuously operating at or near the critical speeds. The typical evaluation criterion is to maintain a critical-speed margin of ±10% between the operating speed and the nearest critical speed. Otherwise, redesign is required. The undamped critical speeds are shown in Table 14.4.

In practice, the mode shape corresponding to a critical speed is important in determining how the rotor system might vibrate when the critical speed is excited. The undamped mode shapes associated with critical speeds for the forward synchronous rotation are presented in Figure 14.25. The first mode at 2,096 rpm shows that the rotor displacement at the bearings is relatively small while the vibration amplitude at the rotor center is much larger. The second mode has a conical mode shape with substantial motion at the bearings at the critical speed of 3,422 rpm. This mode type is usually sensitive to rotor speed. The third mode occurring at 4,847 rpm is similar to the first mode but has the larger rotor motion at the bearings. The fourth mode is similar to the second mode with a very small displacement at one rotor end.

TABLE 14.4
Undamped Critical Speeds

| Mode | Forward Synchronous Rotation | | | | Backward Synchronous Rotation | | | |
| | Critical Speed (rpm) | Strain Energy (%) | | | Critical Speed (rpm) | Strain Energy (%) | | |
		Rotor	Supports	Bearings		Rotor	Supports	Bearings
1	2,096	10	22	68	2,081	11	22	67
2	3,422	0	38	62	3,033	1	33	66
3	4,847	39	50	11	4,700	41	42	17
4	11,459	9	2	89	6,568	59	21	20

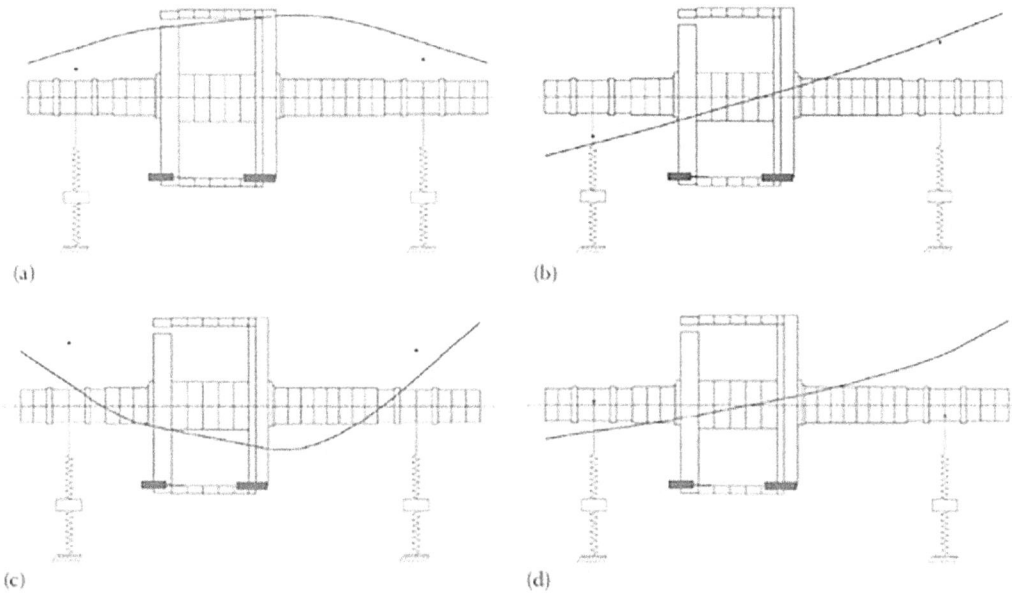

FIGURE 14.25 Undamped critical speeds and mode shapes of forward synchronous rotation: (*a*) mode 1: 2,096 rpm; (*b*) mode 2: 3,422 rpm; (*c*) mode 3: 4,847 rpm; (*d*) mode 4: 11,459 rpm.

Under the backward synchronous rotation condition, the first three undamped modes keep the same shapes as the forward synchronous rotation case. The difference in the corresponding critical speeds between the two cases is less than 3%. However, the critical speed of the fourth mode becomes 4,568 rpm, which is much lower than 11,459 rpm in the previous case. Correspondingly, the mode has the second bending shape (Figure 14.26).

The evaluation of rotor stability is an important aspect of motor rotor design. Rotor system instabilities are normally associated with poor designs and inadequate selection of bearings. In practice, rotor stability is normally evaluated by the amount of damping on the first mode. The standard measure of mechanical damping is the logarithmic decrement, which is computed as the natural logarithm of the ratio between the amplitudes of two successive peaks. The relation between the mode logarithmic decrement δ and the corresponding damping ratio ζ can be found to be [14.40]

$$\delta = \frac{2\pi\zeta}{\sqrt{1-\zeta^2}} \tag{14.31}$$

Theoretically, for $\delta > 0$, the system vibration will be damped out with time, and the system is considered stable. On the contrary, for < 0, the vibration will increase with time, and the system is considered unstable. However, per API Standard 617 [14.41], for a rotor system to be stable, a minimum logarithmic decrement of 0.1 is required.

The damped natural frequencies ω_n versus logarithmic decrement δ are plotted in Figure 14.27. It can be seen from the figure that all logarithmic decrement values are larger than 0.1 and thus all modes pass the stability evaluation criteria.

Table 14.5 presents the predicted damped critical speeds. The data shown in this table indicate that mode 4 fails the $\pm 10\%$ avoidance criteria for the 3,600 rpm operating speed. Modes 4 and 5 fail the $\pm 10\%$ avoidance criteria for the 4,320 rpm overspeed condition.

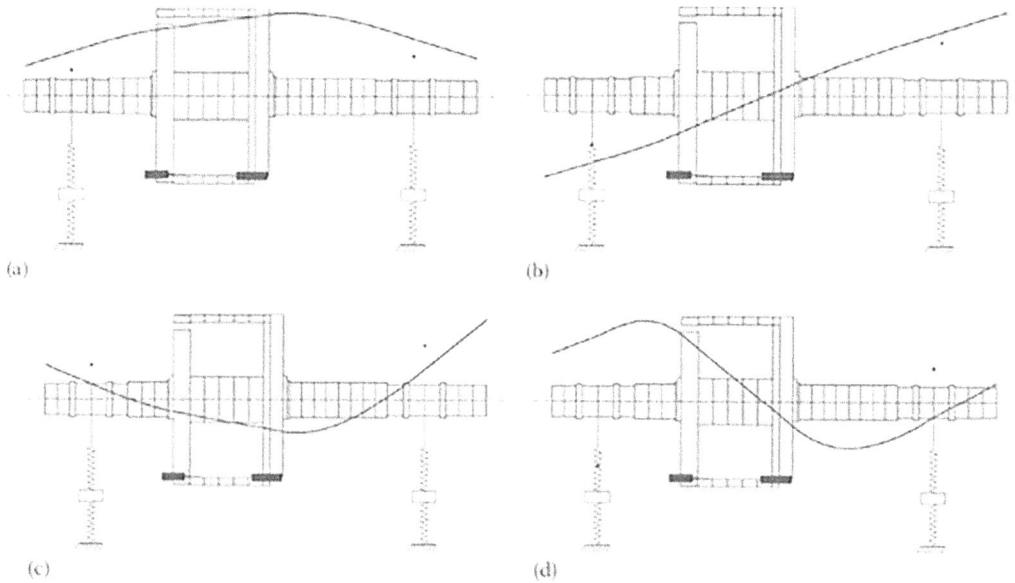

FIGURE 14.26 Undamped critical speeds and mode shapes of backward synchronous rotation: (*a*) mode 1: 2,081 rpm; (*b*) mode 2: 3,033 rpm; (*c*) mode 3: 4,700 rpm; (*d*) mode 4: 6,568 rpm.

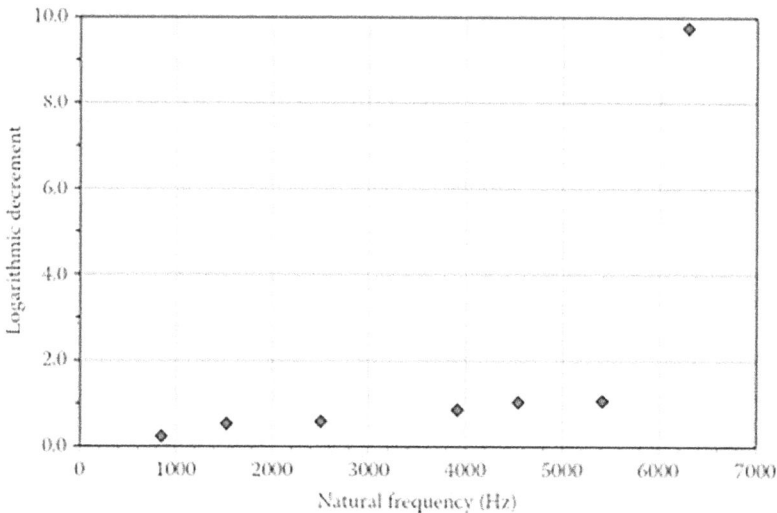

FIGURE 14.27 Damped natural frequencies versus damped ratios for determining the system stability.

TABLE 14.5

Damped Critical Speeds

Mode No.	Critical Speed (rpm)	Logarithmic Decrement at Critical Speed	Stability	% Near 3,000 rpm	% Near 3,600 rpm	% Near 4,320 rpm
1	850	0.23	Stable	−71.7	−76.4	−80.3
2	1,527	0.52	Stable	−49.1	−57.6	−64.7
3	2,502	0.58	Stable	−16.0	−30.0	−41.7
4	3,915	0.85	Stable	30.5	+8.8	−9.4
5	4,543	1.03	Stable	51.4	+26.2	+5.2
6	5,413	1.06	Stable	80.4	+50.4	+25.3
7	6,295	9.75	Stable	109.8	+74.9	+45.7

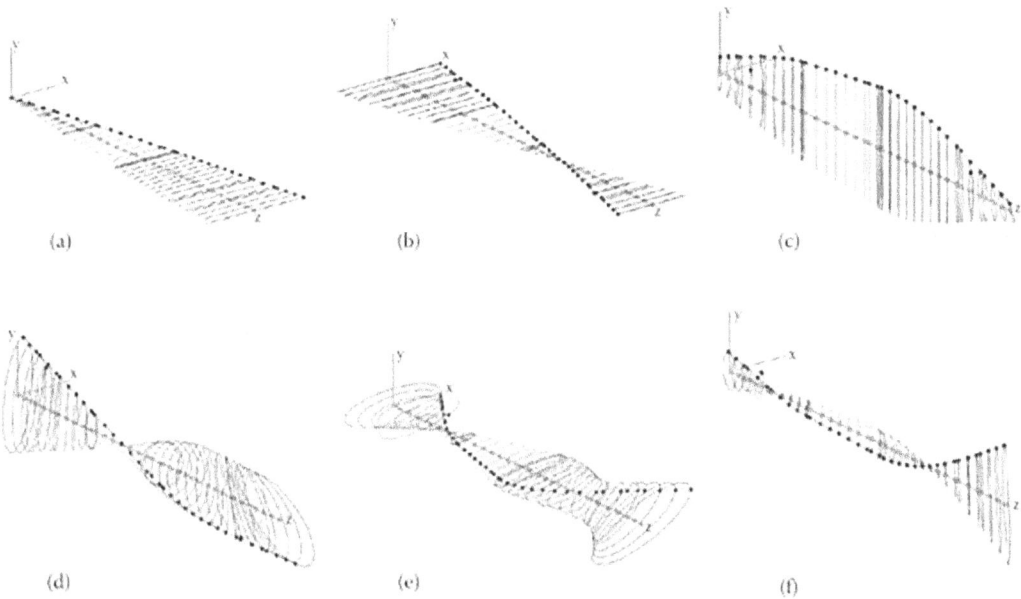

(a)   (b)   (c)

(d)   (e)   (f)

FIGURE 14.28 Damped critical-speed mode shapes: (*a*) mode 1 at 850 rpm; (*b*) mode 2 at 1,527 rpm; (*c*) mode 3 at 2,502 rpm; (*d*) mode 4 at 3,915 rpm; (*e*) mode 5 at 4,543 rpm; and (*f*) mode 6 at 5,413 rpm.

The mode shapes of damped critical speeds are plotted in Figures 14.28 and 14.29 for detailed descriptions. The operating speed ranges of 50 and 60 Hz machines, as well as the over speed range, are plotted to compare with the damped critical speeds.

14.5 STATIC AND DYNAMIC STRESS/STRAIN ANALYSIS

In the design of electric motors, the static or dynamic analysis must be performed to determine the stresses, strains, displacements, and factor of safety in motor components and structures. Stress and strain define the intensity of internal reactions of deformed solid components with the changes of dimension and shape by externally applied forces. Stresses can be further divided as tensile, compressive, or shearing stress, according to the type of loads and straining action of the loading parts. Load-deformation data obtained from tensile or compressive tests can be used to calculate the stress of the specimen:

$$\sigma = \frac{F}{A} \tag{14.32}$$

where

    F is the applied force on the specimen

    A is the original cross-sectional area of the specimen

Strain is a measure of the change in the specimen's length Al to its original length l_o, defined as

$$\varepsilon = \frac{l - l_o}{l_o} = \frac{\Delta l}{l_o} \tag{14.33}$$

For most solid materials, the tensile/compressive stress is directly proportional to the strain, following Hooke's law:

$$\sigma = E\varepsilon \tag{14.34}$$

where E is the modulus of elasticity or Young's modulus. It is noted that Hooke's law describes only the initial linear portion of the stress–strain curve.

    Similarly, the relationship of the shear stress τ and shear strain γ is given as

$$\tau = G\gamma \tag{14.35}$$

where G is the shear modulus of elasticity. For steel and aluminum, the shear modulus of elasticity G is approximately 40% of the modulus of elasticity E.

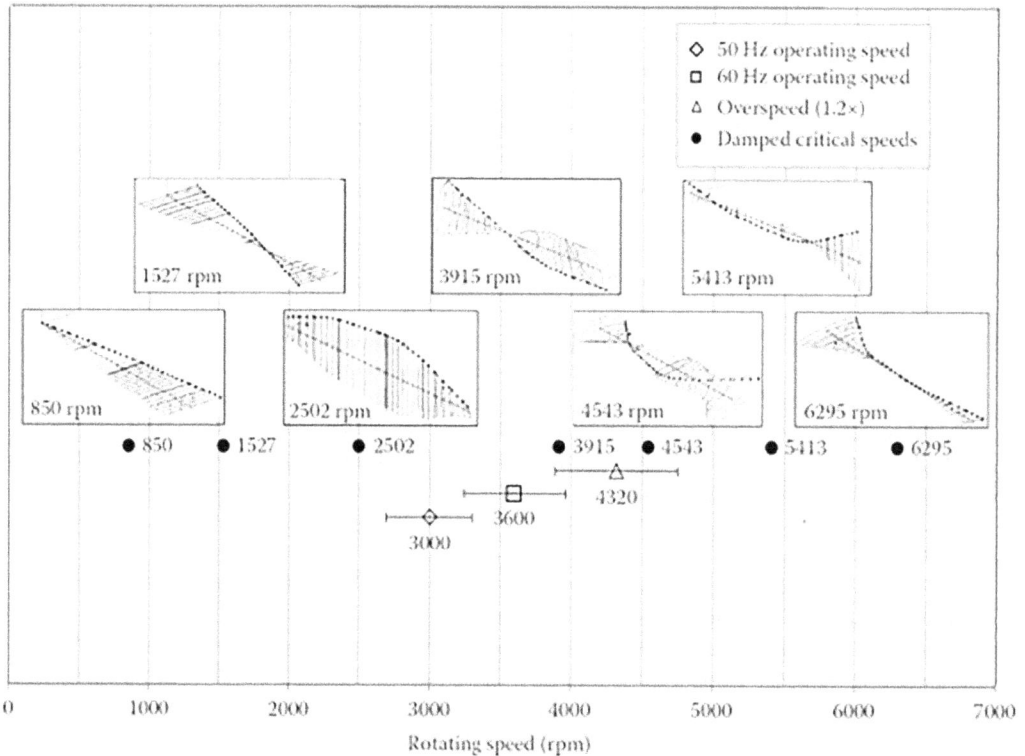

FIGURE 14.29   Predicted damped critical speeds and corresponding mode shapes.

Unlike stress, strain is a dimensionless quantity and is often expressed in units of mm/mm or in./in. In practice, because the magnitude of measured strain is very small, it is sometimes expressed in microstrain units (*e.g.*, μm/m).

14.5.1 Static Analysis

A static analysis is an essential part of the design process that enables the study of stress, strain, displacement, and shear and axial forces that result from static loads. The word "static" indicates that the loads maintain invariable in their magnitudes, directions, and acting locations all the time, such as the gravitational force.

Two equilibrium equations are often used in the static analysis: force balance and momentum balance with the expressions

$$\sum_i F_i = 0 \tag{14.36}$$

$$\sum_i M_i = 0 \tag{14.37}$$

14.5.2 Dynamic Analysis

A dynamic analysis determines the motor structural response based on the characteristics of the structure and the dynamic loads acting on the motor. Unlike a static loading, a dynamic loading is the forces/loads that move or change when acting on a structure. In normal operation, a rotor is subject to various loads such as electromagnetic force, external bending moment, axial force, friction force, and windage drag force. Due to the rotor rotation, these applied forces always change the directions on the rotor.

14.5.2.1 Centrifugal Force-Induced Stress on PMs

Centrifugal force refers to the force that tends to pull an object toward the axis around which it rotates and can be calculated with the formula

$$f_c = \frac{mu^2}{r} = \frac{m(r\omega)^2}{r} = mr\omega^2 \tag{14.38}$$

where
  m is the mass of the object
  r is the radius
  ω is the rotating speed

Equation 14.38 can be used to calculate the centrifugal force acting on a PM and the average stress on the magnet. Assuming that magnet mass $m = 0.0328\,kg$, the magnet contact area $A = 744.5\,mm^2$, and the radius $r = 62.8\,mm$, the centrifugal force acting on the PM is an exponential function of the rotating speed, as shown in Figure 14.30.

With the calculated centrifugal force f_c, the averaged tensile stress acting on the magnet σ is determined as

$$\sigma = \frac{f_c}{A} \tag{14.39}$$

The stress–speed curve is presented in Figure 14.31, showing that the magnet stress increases exponentially with the rotating speed.

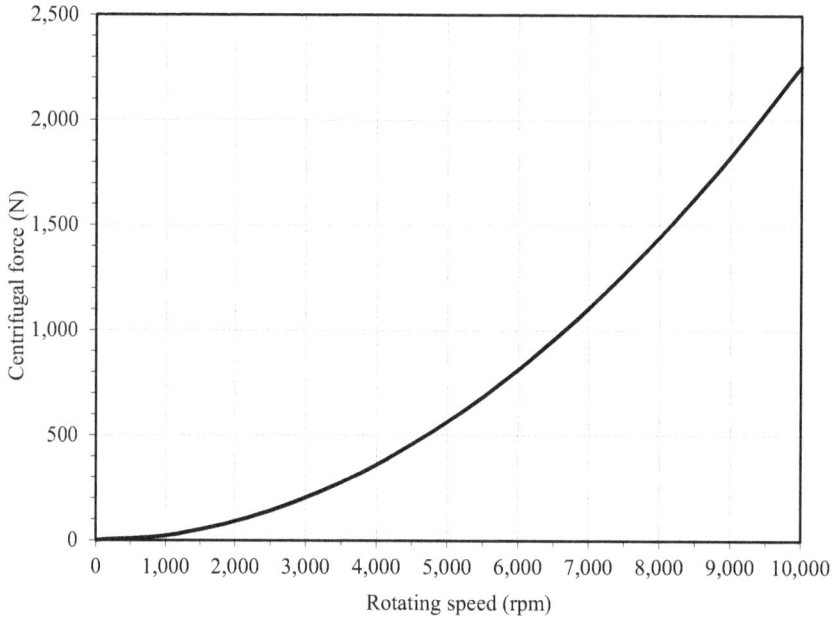

FIGURE 14.30 Centrifugal force acting on a PM under different rotating speeds.

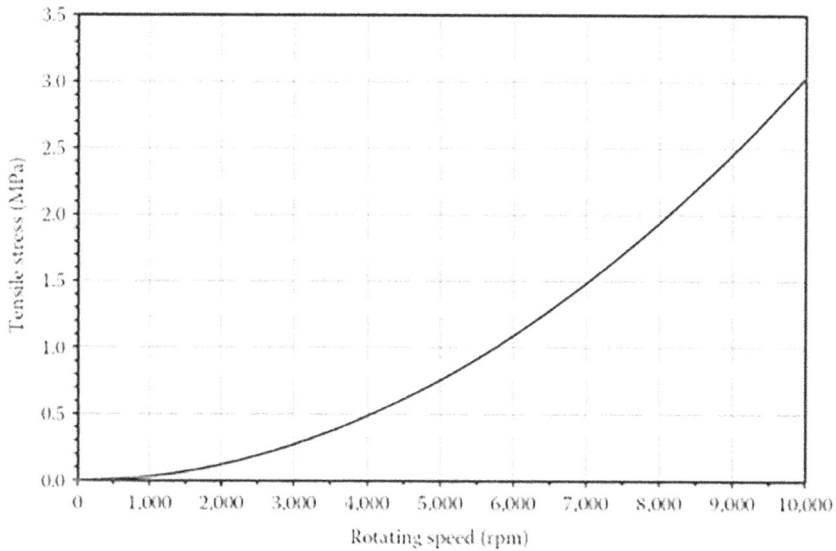

FIGURE 14.31 Average tensile stress of a PM resulted from centrifugal force under different rotating speeds.

14.5.2.2 Structural Analysis Using the Finite Element Method

Analytical solutions for modern electric motors under dynamic loading conditions are unlikely to be available due to highly irregular motor geometry, complex loading condition, heterogeneous material properties, and nonlinear dynamic response of the structure to a variety of loads. In practice, structural analysis is usually performed using an FEM.

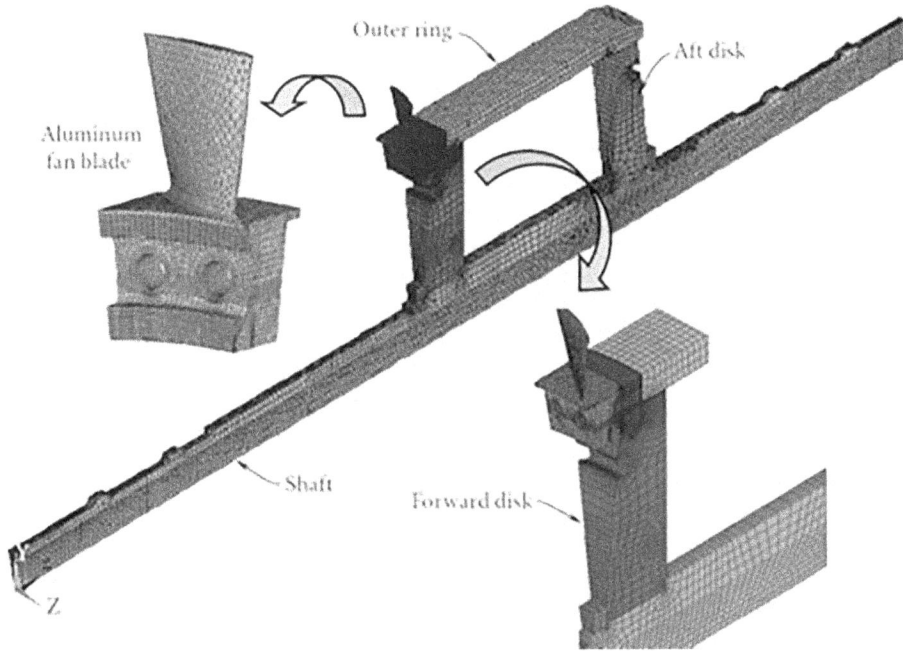

FIGURE 14.32 Mesh of the sector model for structure analysis.

Taking the fan test system as an example, due to the system symmetry, only a small portion of the system is modeled, as illustrated in Figure 14.32. For this problem, the primary consideration is the safety of the rotating system, particularly focusing on two important aspects:

- At high rotating speeds, the stress produced in each rotating component must be lower than its yield strength.
- During operation, all attached components such as blades must be held firmly against the centrifugal force and other loads.

The comparison of various stresses (von Mises, radial, Hook, and axial) of the blade is graphically displayed in Figure 14.33. The stress data are well below the material yield strength, indicating that the blades can safely operate at the rotating speed of 4,320 rpm.

For safety consideration, a severe failure scenario is simulated. It assumes one blade, and its cap screws are missed from the system during operation. The aim of the analysis is to find out that under such a severe circumstance whether the whole system collapses.

A structural analysis is carried out for this scenario. The results are illustrated in Figure 14.34 for both maximum principal stress and radial displacement. These results clearly show that the system unbalance due to the blade missing has a significant effect on the blade displacement. However, it does not lead to the system failure.

14.5.3 SHOCK LOAD

Electric motors may also experience shock loads caused by some accidental events such as earthquake, free fall, and sudden bump. Motors on fast-moving objects (satellites, missiles, rockets, *etc.*) have to withstand high shock loads during acceleration and deceleration periods. According to Newton's second law, $F = ma$, the force F is proportional to the mass of the object m and the acceleration a (or deceleration if a is negative). Thus, the force F can be considerably high under high acceleration/deceleration conditions.

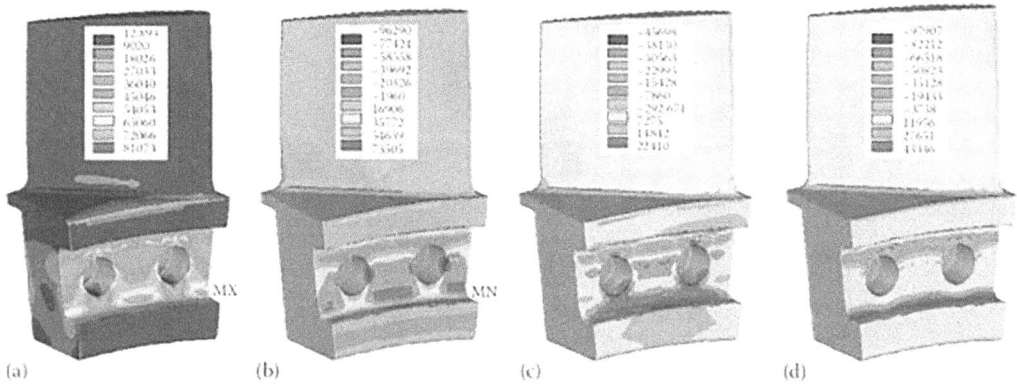

FIGURE 14.33 Comparison of various stresses of a fan blade at 4,320 rpm: (*a*) von Mises stress, (*b*) radial stress, (*c*) hoop stress, and (*d*) axial stress.

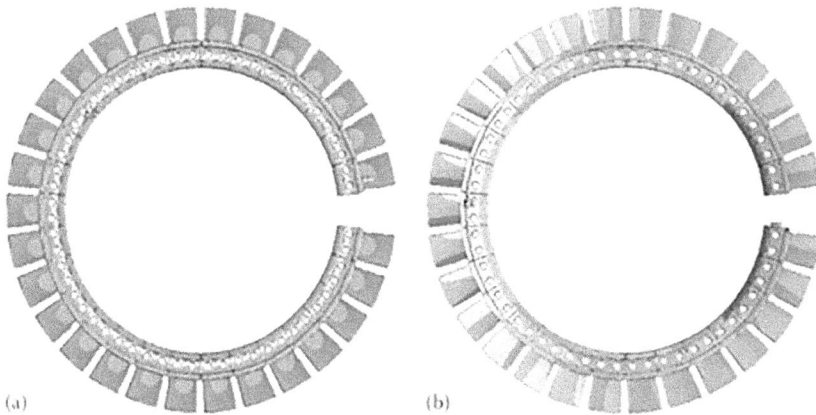

FIGURE 14.34 Results of structural analysis for missing a blade and its cap screws at 4,320 rpm: (*a*) maximum principal stress and (*b*) radial displacement.

When shock loads occur during motor operation in a very short period of time, if the shock loads are high enough, the failures of motor components may occur immediately or at a late time. To prevent motor failure, a common approach to deal with shock loads at the motor design stage is to define an artificial gravity (*i.e.*, *G* value, which is a measure of acceleration) in equations at horizontal and vertical directions. For instance, 5–10 *G* is commonly applied for moving vehicles and 30–100 *G* for missiles.

14.6 FATIGUE ANALYSIS

It is estimated that about 60% of permanent damages are caused by fatigue failure of materials and elements [14.42]. One of the main concerns in the motor design is to extend the motor's life. Factors that influence the motor fatigue life include load history, geometry, relevant materials, and manufacturing processes.

Fatigue in materials is the process of initiation and growth cracks under alternative tensile loading conditions. Fracture occurs when the effects of total stress and flaw size exceed a critical value commonly referred to as the fracture toughness. The fracture toughness depends upon a number of factors, such as microstructure and composition of the material, service temperature, loading rate, plate thickness, and fabrication processes [14.43]. However, an accurate determination of the fracture toughness is complicated.

In structural component design of electric motors, the appropriate criterion for fatigue failure should be based on consideration of failure modes of the component being designed. The fatigue damage caused by repeated dynamic loads depends on the number of cycles and the frequency of the occurrence of significant stresses. A number of motor parts are subject to cyclic loads during motor operation, leading to fatigue of materials. The material fatigue initiates from the crystalline structure and becomes visible in a later stage by plastic deformation, formation of microcracks on slip bands, coalescence of microcracks, and finally propagation of a main crack.

Several approaches are available in the fatigue analysis, from the traditional time-domain S-N analysis to the frequency-domain approach. The time-domain approach is satisfactory for periodic loading but requires large time records to accurately describe random loading processes. As a contrast, compact frequency-domain fatigue calculations can be utilized where the random loading and response are categorized using power spectral density functions and the dynamic structure is modeled as a linear transfer function [14.44].

3D fatigue analysis can be performed using FEM. This method predicts the behavior that is otherwise difficult to find out by theoretical calculations due to the results of a large number of degree of freedom involved in it. As a matter of fact, FEM is an excellent tool to calculate the fatigue life of motor components.

14.7 TORSIONAL RESONANCE ANALYSIS

Torsional resonance is one of the main causes of motor vibration that can lead to fatigue failures of the motor. As illustrated in Figure 14.35, a simple torsional resonance test facility consists of a motor and a load wheel. The motor shaft is connected directly with the load shaft. As the rotor rotates, the inertia of the load wheel provides the torsional load to the rotor and, thus, the torsional resonance of the motor can be determined.

Torsional resonance is affected on both the driving device and rotor inertias and the shaft stiffness that relies on shaft material properties, shaft size, and shaft length.

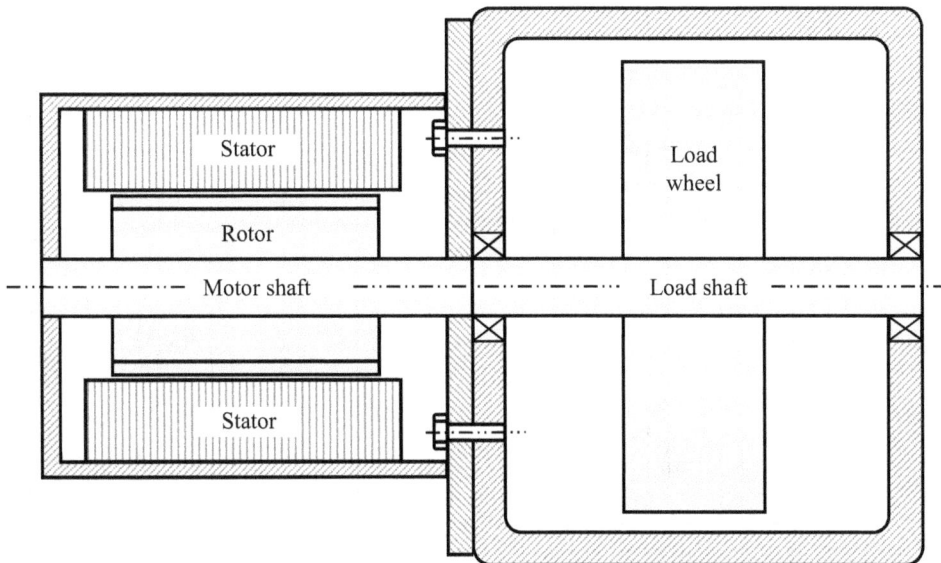

FIGURE 14.35 Torsional resonance measurement of electric motor.

The inertias of the load and motor are calculated in the following equations, respectively:

$$J_l = J_{wheel} + \sum J_{shaft} + \sum J_{i,bearing} \tag{14.40}$$

$$J_m = J_{rotor} + J_{PM} + \sum J_{shaft} \tag{14.41}$$

The torsional stiffness of each shaft segment (both load and rotor) must be determined using the formula

$$S_i = \frac{\pi\left(d_{out,i}^4 - d_{in,i}^4\right)G}{32 l_i} \tag{14.42}$$

Thus, the total torsional stiffness is

$$S_t = \frac{1}{\sum\left(1/S_{i,load\ shaft}\right) + \sum\left(1/S_{i,rotor\ shaft}\right)} \tag{14.43}$$

Torsional resonant frequency (Hz) is thus calculated as

$$f_{tr} = \frac{1}{2\pi}\sqrt{S_t\left(\frac{1}{J_m} + \frac{1}{J_l}\right)} = \frac{1}{2\pi}\sqrt{\frac{S_t\left(J_m + J_l\right)}{J_m J_l}} \tag{14.44}$$

The comparison of calculated and tested torsional stiffness is presented in Figure 14.36.

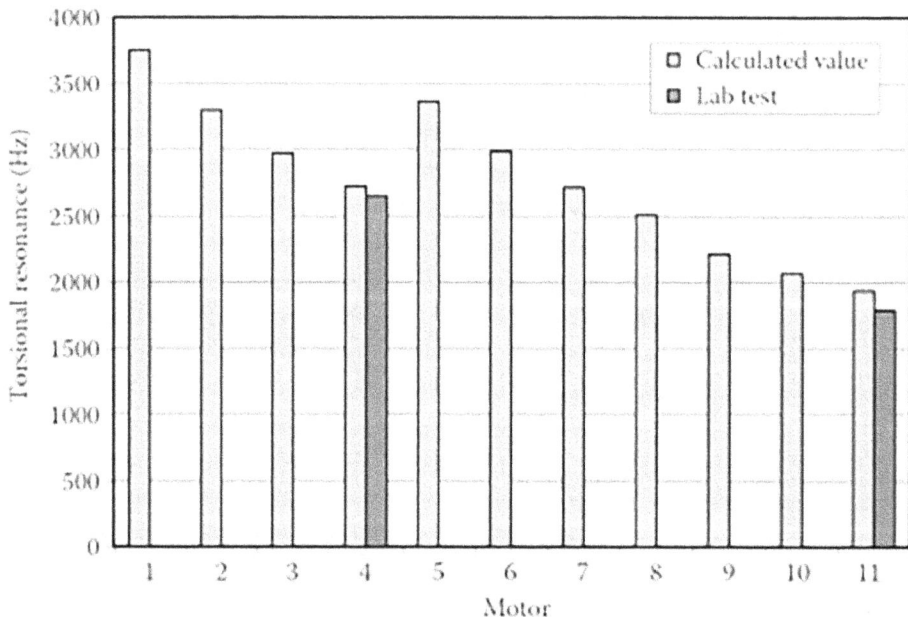

FIGURE 14.36 Comparison of calculated and tested torsional resonances for different types of motor.

14.8 MOTOR NOISE PREDICTION

As a form of energy, sound is the propagation of low-amplitude pressure waves traveling with the speed of sound. In practice, noise typically refers to undesired or unpleasant sound. Acoustic noise from an electric motor may consist of a number of components: mechanical noise, which is associated with the relevant motion (*e.g.*, rotation and vibration) between motor components; electromagnetic noise, which is activated by the fast change in the electric and magnetic field; and aerodynamic noise (or windage noise), which is introduced by flowing fluid inside the motor.

The main challenge in numerically predicting sound waves stems from the fact that sounds have much lower energy than fluid flows, typically by several orders of magnitude. This poses a great challenge to the computation of sounds in terms of difficulty of numerically resolving sound waves, especially in predicting sound propagation to the far field. Another challenge comes from the difficulty of predicting turbulent flow fields in the near field of the sound source [14.45]. In this field, the interferences between contributing waves from different parts of the source lead to the interference effect, which is greater for pure tones than for bands of noise. The prediction of motor noise level can be thus performed according to the type of noise using available commercial software.

Vibration is generally resulted from the unbalanced dynamic system. In a motor, noise is created by unequal weight distributions of the rotor assembly around its axis of rotation. In addition, noise may be generated by nonuniformly distributed rotor winding in some types of induction motors. It has been found that vacuum impregnation and/or encapsulation of stator has helped in reducing stator winding vibration and noise.

The noise reduction is always one of the design targets for motor engineers and designers. There are a number of engineering approaches available, including

- Carefully balancing the rotor to reduce motor vibration
- Increasing the thickness of motor frame wall
- Improving noise sealing by adopting acoustic blankets [14.46] or by coating/gluing a thin layer of noise-absorbing material on the motor surfaces

14.9 BUCKLING ANALYSIS

In engineering practices, there are two major categories of failures in mechanical systems: material failure and structural instability. Buckling refers to a phenomenon that a part or component that is subjected to compression suddenly becomes unstable. As shown in Figure 14.37, a thin-walled sheet is subjected to compressive force F on its two ends. By increasing F until a certain value, the sheet will suddenly bend, indicating that it is no longer to withstand any loads. In this figure, δ_l and δ_o are load displacement and out-of-plane displacement, respectively. This phenomenon can also occur for other geometries such as columns, flanges, and shells. Because of its suddenness, bucking failure may cause catastrophic consequences.

The key factor in bucking is the slenderness ratio of the member, defined as the ratio of the member length l and the radius of gyration k:

$$S = \frac{l}{k} = l\sqrt{\frac{A}{I}} \tag{14.45}$$

where
    A is the cross-sectional area
    I is the area moment of inertia

The stress at failure is called the critical buckling stress. With central loading, there are two models to cover the entire range of compression problems: Euler model and Johnson model [14.47]. The

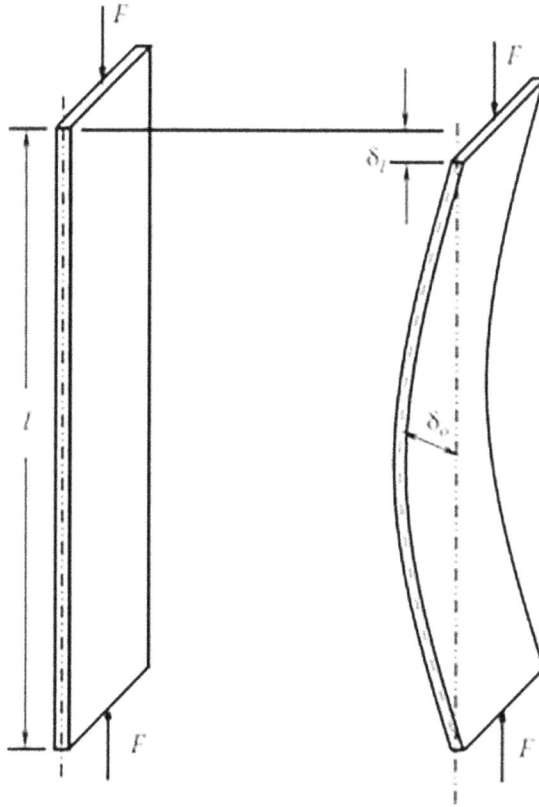

FIGURE 14.37   Buckling of a thin-walled sheet.

selection of an appropriate model depends on the value of the slenderness ratio. The point S_1 is introduced to separate the two models:

$$S_1 = \pi \sqrt{\frac{2CE}{S_y}} \tag{14.46}$$

where
   E is Young's modulus, MPa or GPa
   S_y is the yield strength of the material, Pa or MPa
   C is a constant depending on the column end condition

Thus, for long columns where $S > S_1$, the Euler model is used to calculate the critical buckling stress:

$$\sigma_{cr} = \frac{CE\pi^2}{S^2} \tag{14.47}$$

For intermediate-length columns where $S \leq S_1$, the Johnson model is applied as

$$\sigma_{cr} = S_y - \frac{1}{CE}\left(\frac{S_y S}{2\pi}\right)^2 \tag{14.48}$$

It is noted that the critical bucking stress can be well below the material yield point.

In many practical problems, loads acting on columns are often away from their centroidal axes by the eccentricity e. Correspondingly, the critical buckling stress is calculated by superposing the axial component and the bending component:

$$\sigma_c = \frac{F}{A}\left[1 + \frac{ec}{k^2}\sec\left(\frac{S}{2}\sqrt{\frac{F}{EA}}\right)\right] \qquad (14.49)$$

where

F is the force

c is the distance from the neutral axis to the bending surface

ec/k^2 is defined as the eccentricity ratio

14.10 THERMALLY INDUCED STRESS ANALYSIS

Thermally induced stresses in a motor occur as some or all parts are not free to expand or contract in response to changes in temperature due to the geometric and/or external constraints. Some examples encountered in electric motors are

- For induction motors, conductor bars and end rings are fabricated by casting aluminum or copper into rotor slots to provide the induction current paths. Due to the different thermal expansions between the casted material and silicon steel laminations, shear stresses can be developed right away at the completion of the casting process due to the different cooling rates of the two materials. This is also true during a motor heating process or motor operation at high temperatures. In fact, frequent starting often imposes thermal stresses on electric motors.
- In a shrink-fit or press-fit assembly, the contact pressure between two contacting members changes as the temperature changes from its original fitting temperature, resulting in the variation of contact condition and nonuniform distribution of contact pressure on the contact surface.
- For a PMM, a number of PMs are typically mounted on a rotor. Because these magnets have twice the thermal expansion coefficient over silicon steel, large shear stresses can be produced on the magnets during motor normal operation, leading to uneven axial and radial forces across the magnets and thus possibly developing micro-scale cracks on the magnets.
- When the resin is encapsulated into a stator and cured in an oven for a period of time following a certain temperature profile, the thermal stress is developed in the resin correspondingly with the change of the operation temperature.
- The shaft and housing/endbells are usually made of different materials. A fast variation in temperature may introduce thermal stress or thermal interference problems.

14.11 THERMAL EXPANSION AND CONTRACTION ANALYSIS

Materials change their size with the changes in temperature and pressure at different rates. All gases and most liquids and solids expand when heated and contract when cooled, referred to thermal expansion and thermal contraction, respectively. The rate of thermal expansion is characterized by the CTE β for a specific material.

CTE is the measure of how much that material will expand with the change in temperature of $1°C$. Three types of CTE, that is, the linear, area, and volumetric thermal expansions, are defined as

$$\beta_L = \frac{1}{L}\left(\frac{\partial L}{\partial T}\right)_p \qquad (14.50a)$$

$$\beta_A = \frac{1}{A}\left(\frac{\partial A}{\partial T}\right)_p \tag{14.50b}$$

$$\beta_V = \frac{1}{V}\left(\frac{\partial V}{\partial T}\right)_p \tag{14.50c}$$

For anisotropic materials such as crystals and many composites, β_L has different values in different directions. For isotropic materials, there is only one β_L in all directions. For these materials, it is easy to find that

$$\beta_A = 2\beta_L \quad \text{and} \quad \beta_V = 3\beta_L \tag{14.51}$$

Thus, as the temperature changes from T_o to T_f, the variations in the material length, area, and volume can be calculated as

$$\Delta L = \beta_L L_o \left(T_f - T_o\right) \tag{14.52a}$$

$$\Delta A = \beta_A A_o \left(T_f - T_o\right) \tag{14.52b}$$

$$\Delta V = \beta_V V_o \left(T_f - T_o\right) \tag{14.52c}$$

where
   L_o, A_o, and V_o represent the initial length, area, and volume of the material
   ΔL, ΔA, and ΔV are the variations in length, area, and volume due to the thermal expansion, respectively

The linear thermal expansion coefficients of some solid materials are listed in Table 14.6.

TABLE 14.6
Linear Thermal Expansion Coefficients of Some Solid Materials

Material	Linear CTE at 20°C (μm/m·°C)	Note
ABS—glass fiber	30.4	
Aluminum 356-T6	21.4	For sand casting
Aluminum 383	21.1	For die casting
Aluminum 6061-T6	23.6	
Carbon steel 1008	12.6	
Carbon steel 1045	14.5	Cold drawn
Gray cast iron	10.8	Averaged value for the material class
Copper	16.6	
Copper casting alloy	16.9	UNS C 80100
Ductile iron A536	14.6	For casting
Epoxy—cast resins and compounds, unfilled	55	
Epoxy—cast, unreinforced	100	
Epoxy/glass SMC	3.6	
Fiberglass—polyester	25	
Stainless steel 304	16.9	Averaged value for 0°C–100°C

Thermal expansion and contraction analysis is to address the dimensional changes of motor components due to the variations in temperature during motor operation and storage processes. As a motor is heated or cooled, the dimensions of its components always vary corresponding to the temperature on them. The induced differential dimensions between these components may cause the parts to contact each other or, in more severe conditions, interferences between the parts. Therefore, to avoid undesired thermal-induced interferences, it is important to leave enough allowances for the expansion of hot components.

Figure 14.38 illustrates a regular PMM, consisting of two endbells made of die-casting aluminum A380 and a housing made from extruded aluminum A6061. The shaft is made of carbon steel AISI 1045. Because these materials have different thermal expansion rates, it is important to determine the displacements of the housing/endbell and the shaft to avoid any interference between these parts during thermal expansion.

Referring to Figure 14.39, the differential displacement of the housing and endbells Δl_h during the thermal expansion can be calculated as

$$\Delta l_h = \beta_{A380} l_1 \Delta T_1 + \beta_{A6061} l_2 \Delta T_2 + \beta_{A380} l_3 \Delta T_3 \tag{14.53}$$

where ΔT_i is the temperature difference between the part temperature T_i and the reference temperature T_{ref} (usually $T_{ref} = 20°C$). Similarly, the differential displacement of the shaft Δl_s is

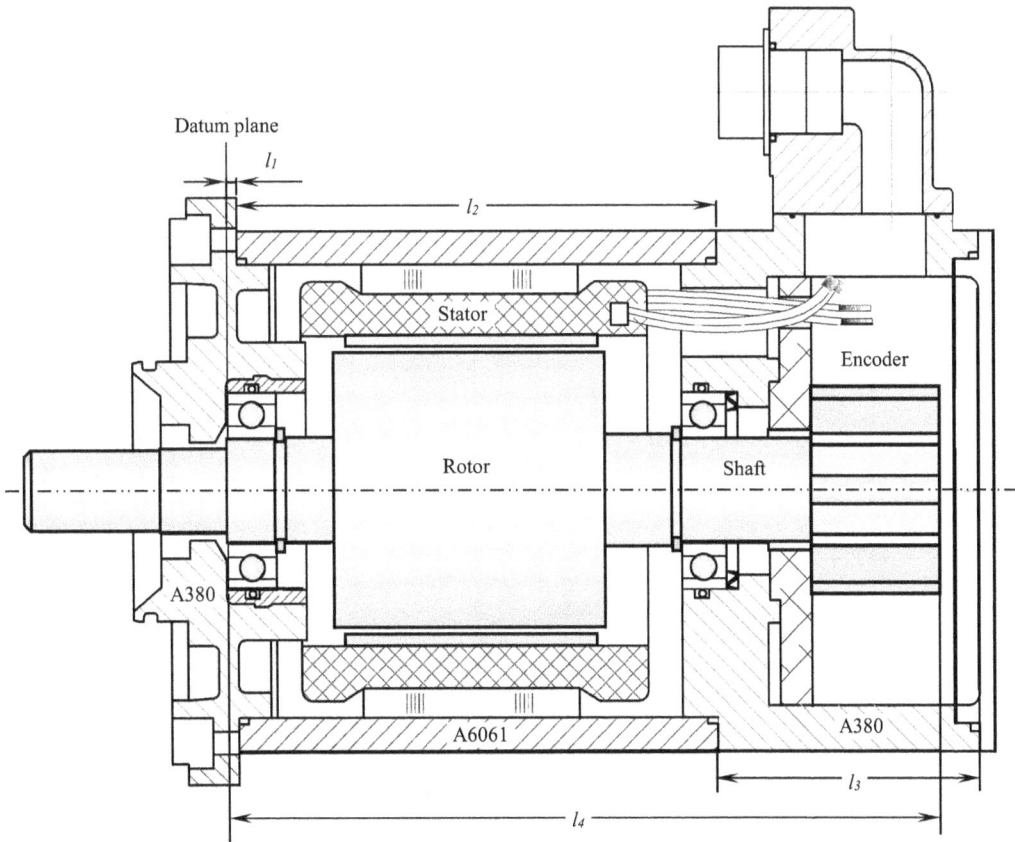

FIGURE 14.38 Comparison of thermal expansions of housing/endbell and shaft, showing that the expansion of the housing/endbell always exceeds that of the shaft.

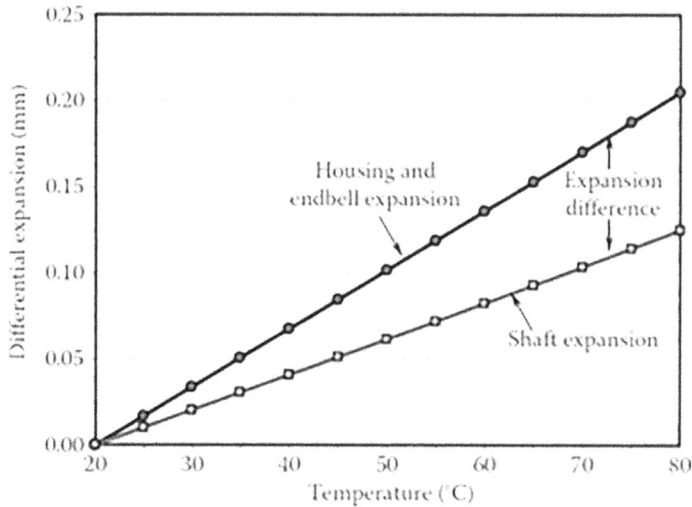

FIGURE 14.39 Calculations of displacement of the shaft and housing/endbell, respectively, for avoiding part interference during thermal expansion and contraction processes.

$$\Delta l_s = \beta_{steel} l_4 \Delta T_4 \tag{14.54}$$

With the real dimensions of the motor component, the calculated results are plotted in Figure 14.39 under different temperatures. It can be observed that at the reference temperature of 20°C, $\Delta l_h = \Delta l_s = 0$. For temperatures larger than 20°C, Δl_h is always larger than Δl_s, and the difference between Δl_h and Δl_s is proportional to the temperature. The results indicate that no interference will occur during the thermal expansion process between these parts.

REFERENCES

14.1. Maksimovic, P. 2005. Next steps in CFD. Pumps & Systems, *November Issue*, pp. 44–47.
14.2. Adair, D. 2012. Chapter 5: Incorporation of computational fluid dynamics into a fluid mechanics curriculum. In book *Advances in Modeling of Fluid Dynamics* (Ed.: C. Liu). InTech, Rijeka, Croatia, pp. 97–122.
14.3. Fluent Inc. 2006. Fluent 6.3 User's Guide, Chapter 10—Modeling flows with rotating reference frames.
14.4. Fluent Inc. 2006. Fluent 6.3 User's Guide, Chapter 11—Modeling flows using sliding and deforming meshes.
14.5. Couette, M. 1890. Etudes sur le frottement des liquidés. *Annales des Chimie et des Physique* 6: 433–510.
14.6. Taylor, G. I. 1923. Stability of a viscous liquid contained between two rotating cylinders. *Philosophical Transaction of the Royal Society of London* **A223**: 289–343.
14.7. Wilkes, J. O. and Bike, S. G. 2005. *Fluid Mechanics for Chemical Engineers with Microfluids and CFD*, 2nd edn. Prentice Hall, Upper Saddle River, NJ.
14.8. Dong, S. 2008. Turbulent flow between counter-rotating concentric cylinders: A direct numerical simulation study. *Journal of Fluid Mechanics* **615**: 371–399.
14.9. Chang, C.-C, Kuo, S.-C., Huang, C.-K., and Chen, S.-L. 2009. The investigation of motor cooling performance. *International Journal of Mechanical, Industrial and Aerospace Engineering* **3**: 43–49.
14.10. Tong, W. 2008. Numerical analysis of flow field in generator end-winding region. *International Journal of Rotating Machinery* **2008**: Article ID 692748, 10pp. doi:10.1155/2008/692748.
14.11. Yang, T.-H. 2006. A device for externally rotary drive of offset motor. European Patent EP 1,481,887.
14.12. Hartmann, U., Knop, C., and Mucha, J. 2014. External rotor motor with a varying armature profile. World Intellectual Property Organization (WIPO) Patent W02011110421.
14.13. Heinen, A. J. 2008. Wheel provided with driving means. U.S. Patent 7,347,427.

14.14. Anderson, J., Dick, E., Degrez, G., Grundmann, R., Degroote, J., and Vierendeels, J. 2009. *Computational Fluid Dynamics: An Introduction*, 3rd edn. Springer, Berlin, Germany.

14.15. Murrone, A. and Villedieu, P. 2011. Numerical modeling of dispersed two-phase flows. *AerospaceLab*, pp. 1–13. hal-01181241. https://hal.archives-ouvertes.fr/hal-01181241/document.

14.16. Kolakaluri, R., Subramaniam, S., and Panchagnula, M. V. 2014. Trends in multiphase modeling and simulation of sprays. *International Journal of Spray and Combustion Dynamics* **6**(**4**): 317–356.

14.17. Marchisio, D. L. and Fox, R. O. 2005. Solution of population balance equations using the direct quadrature method of moments. *Journal of Aerosol Science* **36**(**1**): 43–73.

14.18. Swiderski, K., Caviezel, D., Labois, M., Lakehal, D., and Narayanan, C. 2015. Computational modeling of gas-liquid multiphase flows with DQMOM and the n-phase algebraic slip model. *WIT Transactions on Engineering Sciences* **89**: 299–310.

14.19. Crowe, C. T., Troutt, T. R., and Chung, J. N. 1996. Numerical models for two-phase turbulent flows. *Annual Review of Fluid Mechanics* **28**: 11–43.

14.20. Mirjalili, S., Jain, S. S., and Dodd, M. S. 2017. Interface-capturing methods for two-phase flows: An overview and recent developments. Center for Turbulence Research. *Annual Research Briefs* 2017, pp. 117–135.

14.21. Worner, M. 2012. Numerical modeling of multiphase flows in microfluids and micro process engineering: A review of methods and applications. *Microfluidics and Nanofluidics* **12**: 841–886.

14.22. Mirjalili, S., Ivey, C. B., and Mani, A. 2019. Comparison between the diffuse interface and volume of fluid methods for simulating two-phase flow. *International Journal of Multiphase Flow* **116**: 221–238.

14.23. Ting, D. S.-K. 2016. *Basics of Engineering Turbulence*. Elsevier, Amsterdam, Netherlands.

14.24. Doran, P. M. 2013. *Bioprocess Engineering Principles*, 2nd edn. Elsevier, Amsterdam, Netherlands.

14.25. Yang, C. and Mao, Z.-S. 2014. *Numerical Simulation of Multiphase Reactors with Continuous Liquid Phase*. Elsevier, Amsterdam, Netherlands.

14.26. Náraigh, L. Ó. and Valluri, P. 2018. Chapter 5: Stability analysis and direct numerical simulation for two-phase flows and heat transfer: A complementary approach. In *Encyclopedia of Two-Phase Heat Transfer and Flow IV: Modeling Methodologies, Boiling of CO2, and Micro-Two-Phase Cooling* (Ed.: J. R. Thome). World Scientific Publishing, Singapore.

14.27. Spalding, D. B. 1981. A general purpose computer program for multidimensional one- and two phase flows. *Mathematics and Computer in Simulation* **23**(**3**): 267–276.

14.28. Zheng, X., Babaee, H., Dong, S., Chryssotomidis, C., and Karniadakis, G. E. 2015. A phase-field method for 3D simulation of two-phase heat transfer. *International Journal of Heat and Mass Transfer* **82**(**2015**): 282–298.

14.29. Niu, Z. Q., Jian, K., Zhang, F., Du, Q., and Yin, Y. 2016. Direct numerical simulation of two-phase turbulent flow in fuel cell flow channel. *International Journal of Hydrogen Energy* **41**(**4**): 3147–3152.

14.30. Kapatral, S., Iqbal, O., and Modi, P. 2020. Numerical modeling of direct-oil-cooled electric motor for effective thermal management. SAE Technical Paper 2020-01-1387. doi:10.4271/2020-01-1387.

14.31. Steinbrecher, R., Radmehr, A., Kelkar, K. M., and Patankar, S. V. 1999. Use of flow network modeling (FNM) for the design of air-cooled servers. *Proceedings of the Pacific RIM/ASME International Intersociety Electronics and Photonic Packaging Conference*, Maui, Hawaii, Vol. II, pp. 1999–2008.

14.32. Liu, Y., Lienhard V. J. H., Booth, J. D., and Stairs, R. W. 2002. Integrated simulation toolkit for electronic controlled motor system. *Thermal Challenges in Next Generation Electronic Systems Conference*, Santa Fe, NM.

14.33. Perez, I. J. and Kassakian, J. G. 1979. A stationary thermal model for smooth airgap rotating electric machines. *Electric Machines and Electromechanics* **3**(**3–4**): 285–303.

14.34. Boglietti, A., Cavagnino, A., Staton, D., Shanel, M., Mueller, M., and Mejuto, C. 2009. Evolution and modern approaches for thermal analysis of electrical machines. *Transaction on Industrial Electronics* **56**(**3**): 871–882.

14.35. Mellor, P. H. and Turner, D. R. 1988. Real time prediction of temperature in an induction motor using a microprocessor. *Electric Machines & Power Systems* **15**(**4–5**): 333–352.

14.36. Zhu, Y., Xian, M. K., Lu, K., Wu, Z. H., and Tan, B. 2019. A simplified thermal model and online temperature synchronous motors. *Applied Sciences* **9**: 3158. doi:10.3390/app9153158.

14.37. Shtrakov, S. and Stoilov, A. 2006. Finite element method for thermal analysis of concentrating solar receivers. The Computing Research Repository (CoRR), Paper No. 0607091. South-West University, Blagoevgrad, Bulgaria. http://arxiv.org/abs/cs/0607091.

14.38. Ranade, V. V. 2002. *Computational Flow for Modeling for Chemical Reactor Engineering*. Academic Press, San Diego, CA.

14.39. Chin, Y. K., Nordlund, E., and Staton, D. A. 2003. Thermal analysis lumped-circuit model and finite element analysis. *Sixth International Power Engineering Conference (IPEC)*, Singapore, pp. 952–957.

14.40. Yoon, S. Y., Lin, Z., and Allaire, P. E. 2013. *Control of Surge in Centrifugal Compressors by Active Magnetic Bearings: Theory and Implementation*. Springer, New York.

14.41. API. 2002. API Standard 617. Axial and centrifugal compressors and expender-compressors for petroleum, chemical and gas industry services.

14.42. Torbacki, W. 2007. Numerical strength and fatigue analysis in application the hydraulic cylinders. *Journal of Achievements in Materials and Manufacturing Engineering* **25**(**2**): 65–68.

14.43. U.S. Department of Transportation. 2012. Steel Bridge Design Handbook: Design for Fatigue. Publication No. FHWA-IF-12-052-Vol. 12. U.S. Department of Transportation Federal Highway Administration, Washington, D.C.

14.44. Halfpenny, A. 1999. A frequency domain approach for fatigue life estimation from finite element analysis. *International Conference on Damage Assessment of Structures*, Dublin, Ireland.

14.45. Fluent Inc. 2006. Fluent 6.3 User's Guide, Chapter 21—Predicting aerodynamically generated noise.

14.46. Tong, W., Wagner, T. A., Hughes, I., Gillivan, J., and Gibney J. 2004. Acoustic blanket for machinery and method for attenuation sound. U.S. Patent 6,722,466.

14.47. Budynas, R. G. and Nisbett, J. K. 2008, *Shigley's Mechanical Engineering Design*, 8th edn. McGraw-Hill High Education, Boston, MA.

15

Innovative and Advanced Motor Design

Electric motors consume a large amount of global electricity each year. In today's highly competitive motor market, the demands for higher efficiency, lower cost, and improved system reliability are fueling the need for advanced modern electric motors with reduced energy consumption and carbon emissions. Many motor manufacturers are currently developing their next-generation products for gaining the larger market share and boosting their profit margins. For these purposes, motor manufacturers not only fully utilize their own technical resources but also strengthen the cooperation with universities and research institutions for implementing technological breakthrough. However, there are still a variety of challenges that motor manufacturers must face; for example, how to shorten the product development cycle time, how to apply new technologies in both existing and new products, how to differentiate their products from those of competitors and adapt to the global market, how to create competitive advantages to make their products superior to alternatives on the market, and how to effectively reduce the production costs. All these challenges must be appropriately considered and addressed to ensure successful outcomes.

With the rapid strides made in big data technology, cloud computing technology, internet of things, material science (*e.g.*, advanced composite materials, nonmaterials), manufacturing process (*e.g.*, 3D printing), control technology (*e.g.*, new sensors and feedback devices), and other fields, the motor industry has undergone tremendous development and technological improvement in recent years. These technologies have made it possible to design and manufacture next generation electric motors. In fact, the sustainable growth of the motor industry requires continuous technology innovation. In the digital and information era, motor manufacturers tend to aim at high efficiency, high torque density, durable, cost-effective, and energy-saving products.

15.1 HIGH-TEMPERATURE SUPERCONDUCTING MOTOR

The discovery of high-temperature superconductors (HTSs) in 1986 has strongly stimulated the R&D on superconducting materials and their industrial and military applications. In the following years, the potential economic benefits of HTS had initiated an international race among the United States, EU countries, Japan, and China to develop advanced superconducting technologies and materials and achieve commercialization. Large efforts have been made to apply HTS toward power equipment such as electric motors, generators, energy storage systems, transformers, and power cables. HTS materials have much higher critical temperatures (above the liquid nitrogen temperature of 77 K or $-196°C$) than conventional low-temperature superconductors cooled with liquid helium (4.2 K or $-269°C$). The use of liquid nitrogen to cool HTS can lead to a reduced complexity of cryogenic system, improved system reliability and performance, and reduced total cost.

The advent of HTS has driven a quantum leap forward in advanced technologies of electric machines. Superconducting motors based on HTS are much more efficient than conventional electric motors due to zero resistance in superconducting windings. Because HTS can carry a much higher current than regular copper wires, these HTS windings can generate much more powerful magnetic fields with minimized power losses. For large synchronous superconducting motors, 50% volume

DOI: 10.1201/9781003097716-15

and 33% of weight reductions can be achieved compared to conventional motors [15.1, 15.2]. In addition, HTS motors have lower sound emissions than conventional motors. These features are especially attractive for some military customers. For instance, the US Navy is interested in applying HTS motors to its new generation of surface ships and submarines.

According to the use of HTS in motor windings, three types of HTS motors can be classified: (*a*) superconducting rotor type that uses HTS in the rotor winding, (*b*) superconducting stator type that uses HTS in the stator winding, and (*c*) full superconducting motor type that uses HTS in both rotor and stator windings.

Founded by the US Department of Energy (DoE), Reliance Electric Corporation demonstrated the world's first DC HTS motor in 1990 and AC synchronous HTS motor in 1993 [15.3]. The first HTS motor in Europe was made and tested at Siemens' Research Center in 2001. During experimental operation in motor and generator modes, the trial motor reached a continuous power output of 400 kW [15.4]. In 2005, IHI in Japan released the world's first full superconducting motor cooled with liquid nitrogen [15.5].

In the United States, American Superconductor has designed, built, tested, and delivered a 5 MW, 230 rpm, 6-pole HTS ship propulsion motor [15.6]. The motor uses an air core armature winding and first-generation HTS wire field winding. The aim of the project is to validate the technologies required to design and build larger HTS ship propulsion motors, as well as to develop a motor production process that streamlines development time and minimizes cost. A commercial variable frequency drive is used to power the motor. The HTS field winding uses gaseous helium as the cooling medium in a closed cycle.

The world's largest HTS ship propulsion motor of 36.5 MW (49,000 hp) in 2009 was successfully tested at full power [15.7]. This motor can produce 2.9 million newton-meters of torque at a nominal rotating speed of 120 rpm. Incorporating coils of HTS wire that can carry 150 times the current of similar-sized copper wire, the motor is less than half the size of conventional motors and can reduce the ship weight by nearly 200 tons. With up to three times higher torque density than conventional motors, the HTS motor is more fuel-efficient. The size and weight benefits make HTS machines less expensive and easier to transport and install, as well as allow for arrangement flexibility in the ship. In addition, the absence of iron stator teeth significantly reduces the structure-borne noise.

In 2007, engineers in Sumitomo Electric designed a liquid nitrogen-cooled HTS motor of 365 kW. To simplify the cooling system and achieve high operating reliability, this motor adopts axial gap PM type. The motor armature composes of six iron-cored HTS windings and forms three phases. The heat generated by AC losses is efficiently dissipated into the liquid nitrogen through the cooling channels. This 365 kW HTS motor was successfully used to directly drive a contrarotating propeller in tandem with a 50 kW HTS motor. Due to the reduction of transmission losses, the HTS motor could provide the maximum torque of 120 Nm (at 1,500 rpm) and the maximum speed of 85 km/h to electric vehicles [15.8].

A schematic diagram of an HTS motor is shown in Figure 15.1. As can be seen in the figure, while the stator winding is made of regular copper wires, the rotor winding is fabricated by HTS as superconducting magnets, which have higher energy density than PMs. The rotor winding is thermally insulated from the rest of the machine using multilayered insulation materials and cooled with a cryogenic coolant, which is introduced from the cryocooler and passes through the cooling loop in the support structure adjacent to the rotor. To reduce radiation heat loads onto the cryogenically cooled components, the rotor winding is enclosed within a sealed vacuum chamber that rotates with the shaft. Unlike in a conventional motor, the stator winding is not housed in iron core teeth because they are saturated due to the high magnetic field imposed by the HTS windings [15.9].

Depending on the direction of the main airgap magnetic flux and the configuration of the rotor and stator against the rotating shaft, an electric motor can be classified into radial-flux (RF), axial-flux (AF), or transverse flux (TF) motors. Unlike the more commonly available cylinder-shaped rotor and stator in RF motors, AF motors are disk shaped, as the result of the stator and rotor being placed adjacent to each other.

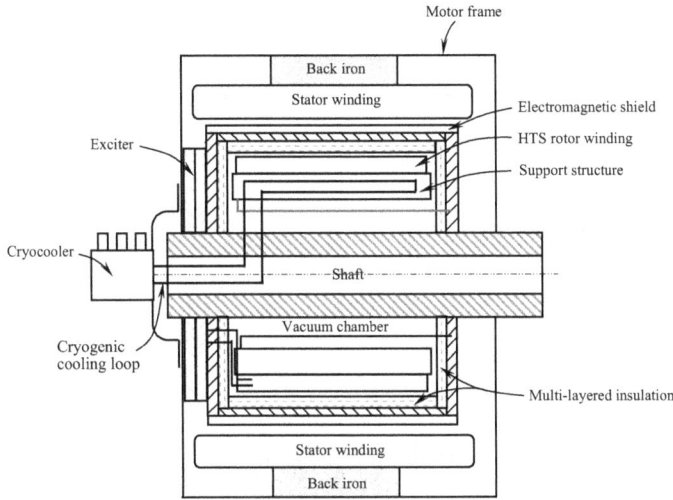

FIGURE 15.1 Schematic diagram of HTS motor.

TABLE 15.1
Comparison of Torque Density for Three Types of Motors

Motor Type	Torque Density for Most Compact Case (Nm/m³)	Torque Density for Best Efficiency Case (Nm/m³)
RF motor	16,877	14,380
AF motor	13,226	10,023
TF motor	11,504	11,504

Source: Data obtained from Pippuri *et al.*, *IEEE Trans. Magn.*, **49**(**5**): 2339–2342, 2013.

The comparison of torque density of RF, AF, and TF PM motors has been performed by Pippuri *et al.* [15.10]. The work was carried out using a 10 kW, 200 rpm three-phase motor as the test case. The RF design is of the conventional inner rotor type. The AF design has a single-rotor, single-stator construction. The TF design has a U- and I-core stator layout with an inner rotor of two separate yokes. The magnets are mounted on the rotor surface and each of them covers 70% of the pole pitch. As shown in Table 15.1, the comparisons are made for two cases: most compact and best efficiency. In both cases, the RF topology yields the highest torque density with the given initial conditions and constraints. When comparing the AF and TF machines (TFM), the AF machine outperforms the TF design in the most compact case, and it is opposite in the best efficient case. It is interesting to note that the torque density for the TF motor remains unchanged in the two cases.

The flux of the AF motor is in the same direction with the motor shaft. Because of the large area for magnetic flux, the AF motor provides the desired performance. A schematic diagram of an AF, high-temperature superconducting motor is shown in Figure 15.2. The stator is in the middle of the motor and the two rotors are on the two sides of the stator. The stator winding is made of high-temperature superconducting materials and is cooled by a specially designed cooling system (not shown).

15.2 RADIAL-FLUX MULTIROTOR OR MULTISTATOR MOTOR

RF multirotor and multistator motors have a number of distinct advantages over conventional single-rotor and single-stator motors. By integrating multirotor or multistator into the same motor housings, the machines offer higher torque density and are usually more efficient.

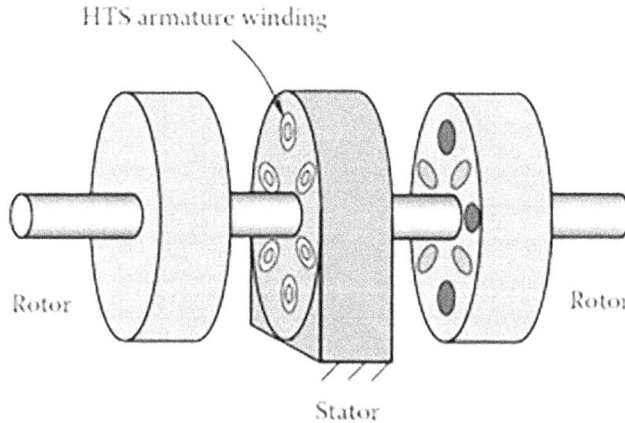

FIGURE 15.2   An AF type of high-temperature superconducting motor.

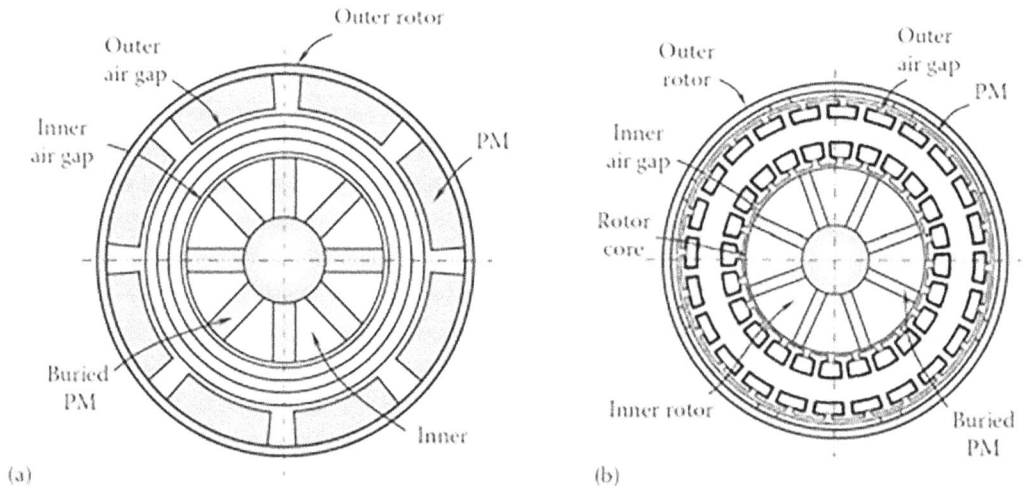

FIGURE 15.3   Structures of dual-rotor, toroidally round hybrid PM motors: (*a*) nonslotted machine and (*b*) slotted machine (U.S. Patent 6,924,574) [15.11]. (Courtesy of the US Patent and Trademark Office, Alexandria, VA.)

15.2.1   RADIAL-FLUX MULTIROTOR MOTOR

A novel dual-rotor, toroidally wound PM motor was proposed by Qu and Lipo [15.11], as shown in Figure 15.3. There are some unique features in their design, including (*a*) a rotor–stator–rotor structure, (*b*) back-to-back windings, (*c*) very short end windings, (*d*) a high diameter-to-length ratio, (*e*) low airgap inductance, (*f*) high efficiency, (*g*) high torque density, (*h*) high overload capability, (*i*) balanced radial forces, (*j*) suitability for moderately high-speed performance, (*k*) low cogging torque, and (*l*) low material costs.

The motor illustrated in Figure 15.3*a* is constructed with a plurality of PMs mounted on the inner surface of the outer rotor and buried within a core of the inner rotor. The PMs buried within the inner rotor core preferably extend radially from a central opening of the inner rotor to the outer surface of the inner rotor. The inner and outer rotors comprise a single integral rotor so that the inner and outer rotors rotate at the same speed. The torus-shaped stator has a hollow cylindrical shape and is preferably nested between the inner and outer rotors. A plurality of polyphase windings is

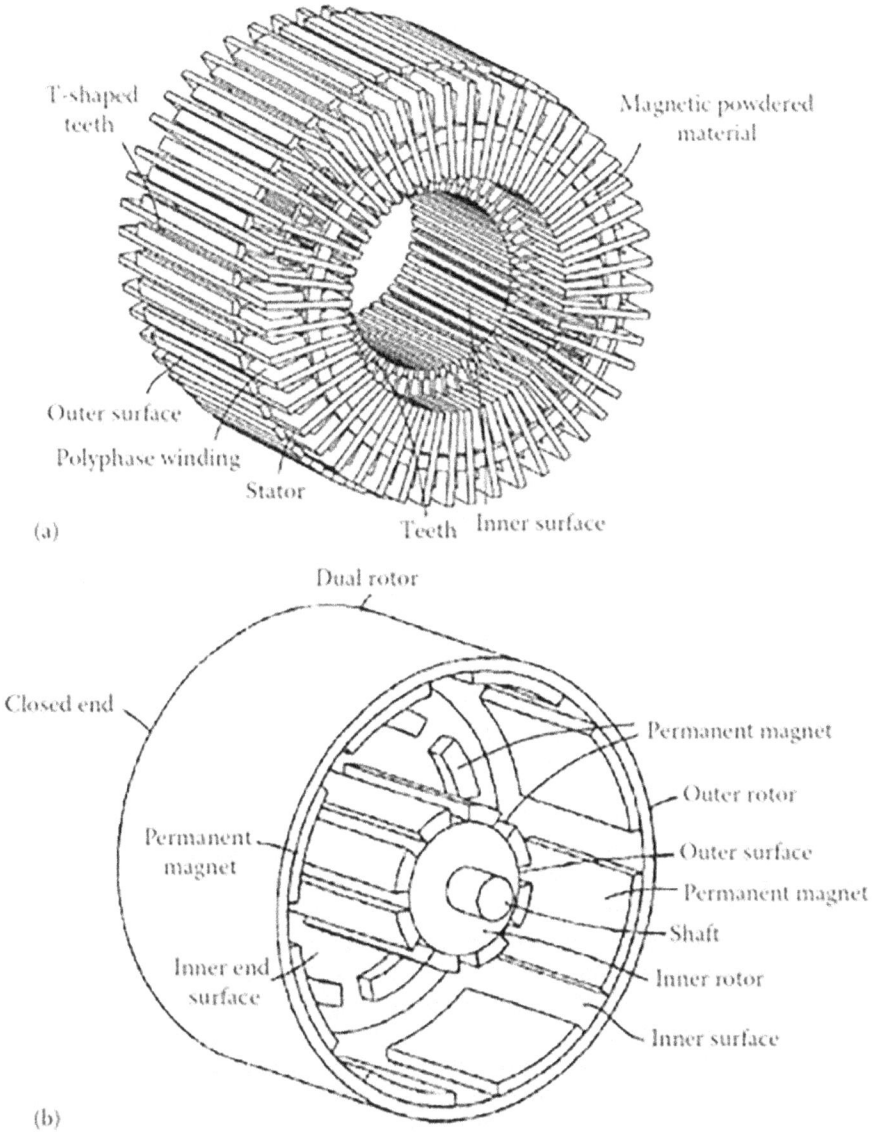

FIGURE 15.4 Dual-rotor, toroidally wound SPM motors: (*a*) stator and (*b*) dual rotor (U.S. Patent 6,924,574) [15.11]. (Courtesy of the U.S. Patent and Trademark Office, Alexandria, VA.)

toroidally wound around the stator. It is noted that the stator is nonslotted, meaning the stator does not include any radially extending teeth. A benefit of the nonslotted structure is that it avoids cogging torque. There are two airgaps in the motor. The inner airgap is formed between the inner rotor core and the stator, and the outer airgap is formed between the outer PMs and the stator.

As shown in Figure 15.3*b*, the motor has a similar configuration to the motor in Figure 15.3a, except for the stator structure. This motor employs a slotted stator rather than nonslotted in Figure 15.3a. In fact, the more the stator windings are used in a motor, the higher the torque generated and the higher the motor efficiency. A higher torque density is thus achieved by slotting, but the slotted structure will introduce cogging torque.

The perspective view of the dual rotor and torus-shaped stator is shown in Figure 15.4. The dual rotor (Figure 15.4*a*) consists of an inner rotor, an outer rotor, and a closed end. The dual rotor

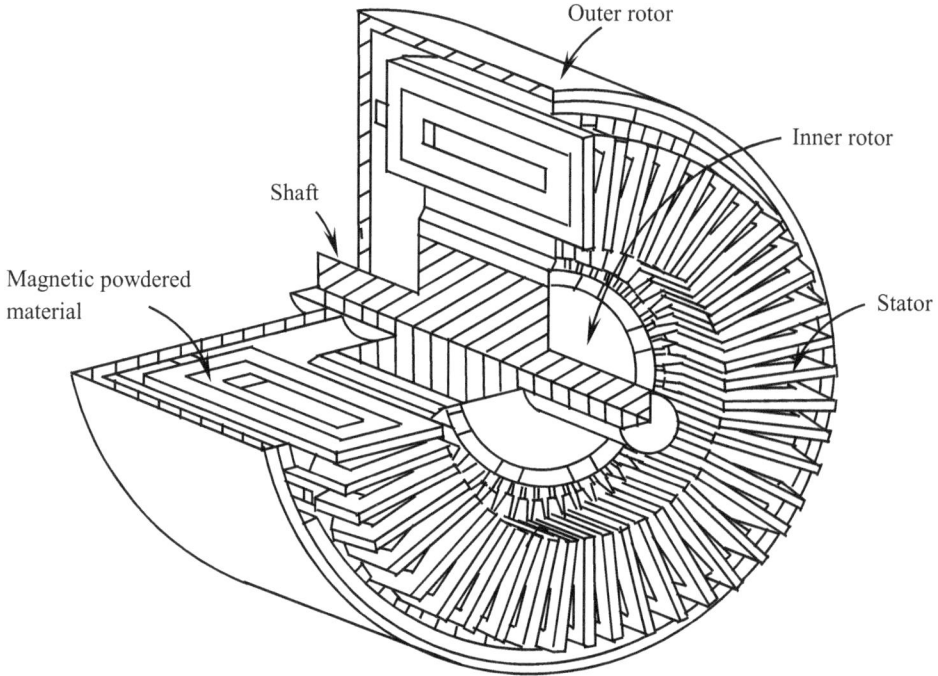

FIGURE 15.5 A cutaway perspective view of the motor assembly (U.S. Patent 6,924,574) [15.11]. (Courtesy of the U.S. Patent and Trademark Office, Alexandria, VA.)

magnetically interacts with the stator, whereby the PMs drive a magnetic field within the stator, causing a back electromagnetic force to be induced in the polyphase windings that are wound around the stator. The stator includes a plurality of T-shaped teeth or slots extending radially inwardly from the inner surface and outwardly from the outer surface of the stator (Figure 15.4b). The windings wound around the stator may comprise toroidally wound windings, lap windings, wave windings, or other types of windings. Since almost all of the windings on the inner, outer, and end surfaces of the stator are used for torque production, this motor creates higher efficiency and higher torque density than the conventional motors. The stator is formed by a plurality of stacked laminations that are connected to a magnetic powdered material.

Figure 15.5 is a cutaway perspective view of the motor assembly. This figure clearly shows how the stator is nested between the inner and outer rotors. An additional gap (end airgap) is formed at the motor end, between the windings and PMs, which is attached to the inner surface of the closed end (see Figure 15.4b).

15.2.2 RADIAL-FLUX MULTISTATOR MOTOR

The structure of an RF dual-stator motor is presented in Figure 15.6. The motor consists of an inner stator, an outer stator, and a rotor, which inserts between the two stators via two bearings. PMs are attached to the inner and outer surfaces of the rotor. To achieve the concentricity of the inner and outer stators, the two stators are integrated as one part. If the two stators are made separately, an accurate position control is required. Stator windings are wound through the inner and outer slots.

An alternative design is illustrated in Figure 15.7. In this design, PMs are mounted on the surfaces of both the inner and outer stators. Windings are made through the slots of the rotor. The system is supported by three bearings.

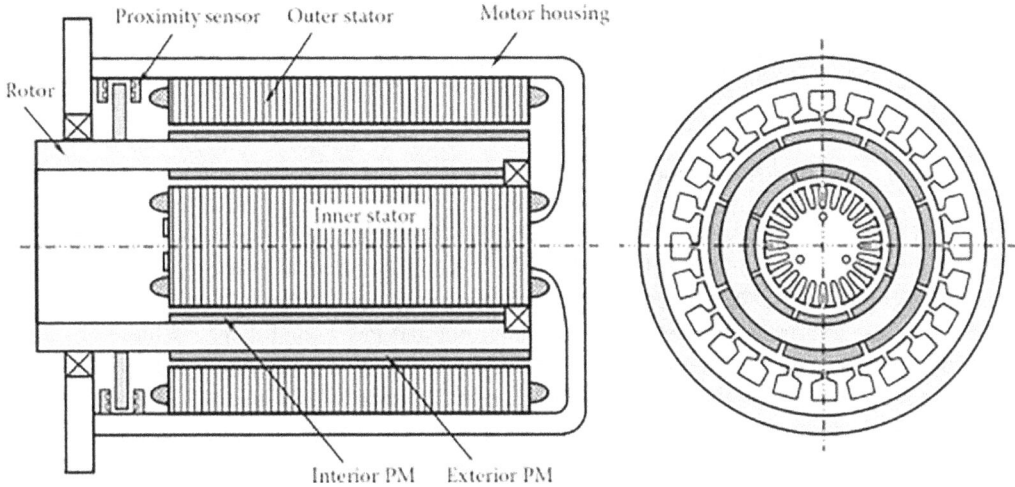

FIGURE 15.6   RF, dual-stator, PM motor with two bearings.

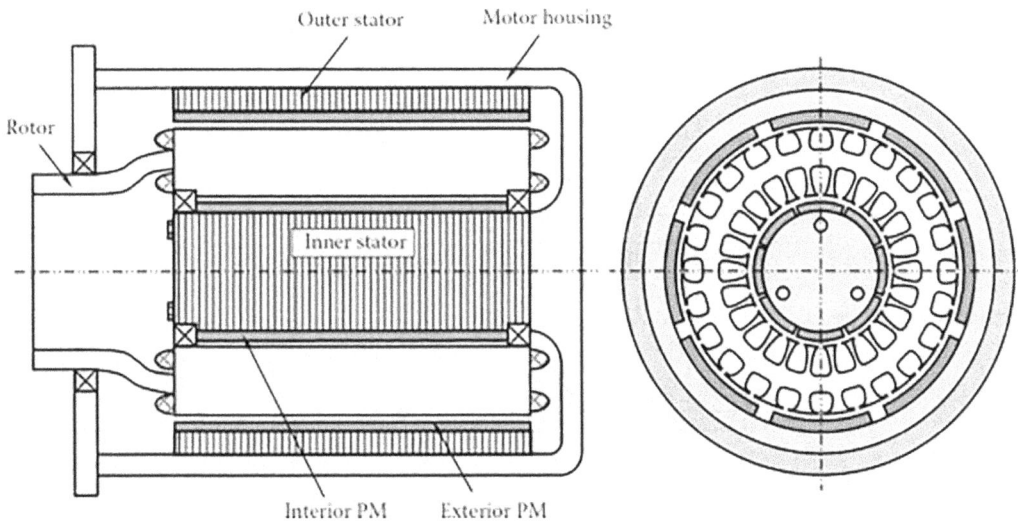

FIGURE 15.7   A RF type of PM motor with the dual stator. The system is supported by three bearings.

15.2.3   RADIAL-FLUX BRUSHLESS DUAL-ROTOR MACHINE

Over the past few years, a power-split hybrid system, called compound-structure PM synchronous machine (CS-PMSM) system, had been developed for replacing the mechanical planetary gear, flywheel, clutch, starting motor, and generator used in conventional automobiles. This is a promising technology for developing hybrid vehicles with high efficiency and reliability. As a part of efforts, a novel RF brushless dual-rotor machine (BDRM) was proposed by Zheng *et al.* [15.12].

The CS-PMSM system is illustrated in Figure 15.8. It consists of a BDRM and a conventional machine. There are some design characteristics associated with the BDRM:

• The windings of the BDRM are mounted on the stator close to the motor case. This configuration helps dissipate generated heat from the stator windings to the environment.
• Stator phases are mutually independent with no couplings existing in-between.

- The BDRM is designed as a multipole machine. This is especially suitable for intermediate or high-frequency operation.
- The ring-shaped stator windings of the BDRM simplify the winding fabrication and assembly process.

The structure of the BDRM is demonstrated in Figure 15.9. The stator shown in Figure 15.9a comprises laminated iron cores and ring-shaped stator windings. Figure 15.9b shows the claw-pole rotor that has three arrays of claws mounted on a nonmagnetic bracket. While the claw-pole rotor is connected to the final gear for driving wheels, PM rotor 1 is connected to the crank shaft of the internal combustion engine (ICE). The function of the PM rotor is to transmit the ICE torque to claw-pole rotor. The speed adjustment between the claw-pole rotor and PM rotor 1 can be achieved by energizing the stator windings with different frequencies via an inverter. The PMs are burned radially through the rotor cores with the N-S-N-S arrangement, as presented in Figure 15.9c.

The claw-pole rotor is connected to PM rotor 2 together with the conventional machine. Thus, the total output torque of the shaft is collectively contributed by the BDRM and conventional machine.

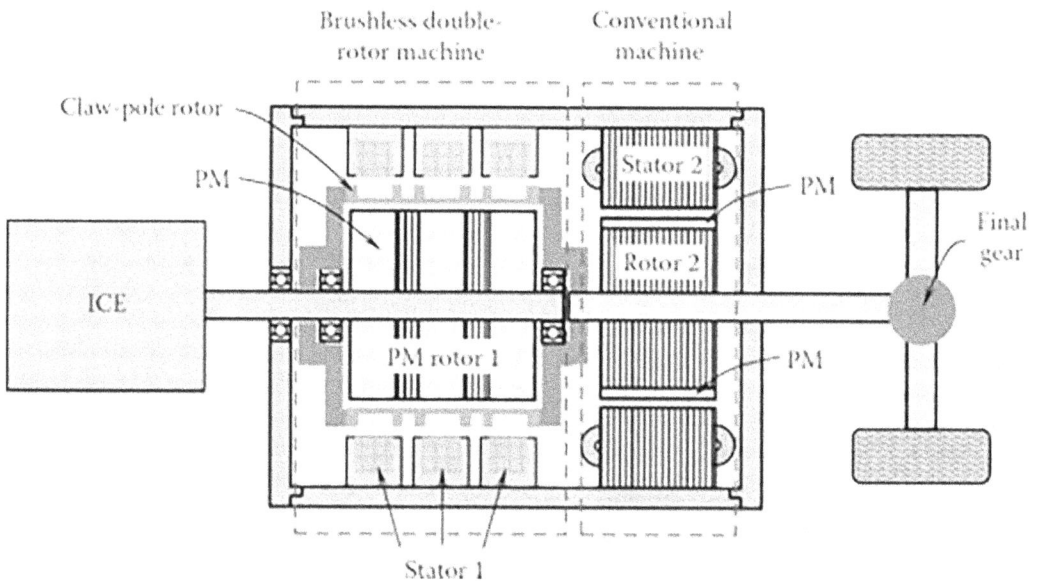

FIGURE 15.8 A CS-PMSM consisting of a BDRM and a conventional machine for driving hybrid electric vehicles.

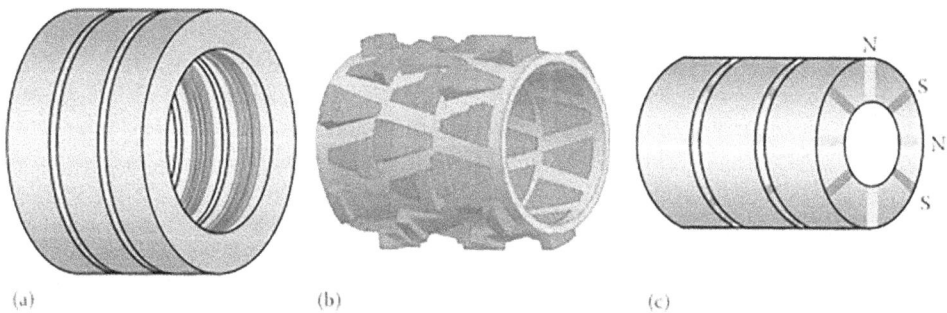

FIGURE 15.9 Perspective view of the BDRM: (a) stator, (b) claw-pole rotor, and (c) PM rotor with buried radial magnets for azimuthal magnetization.

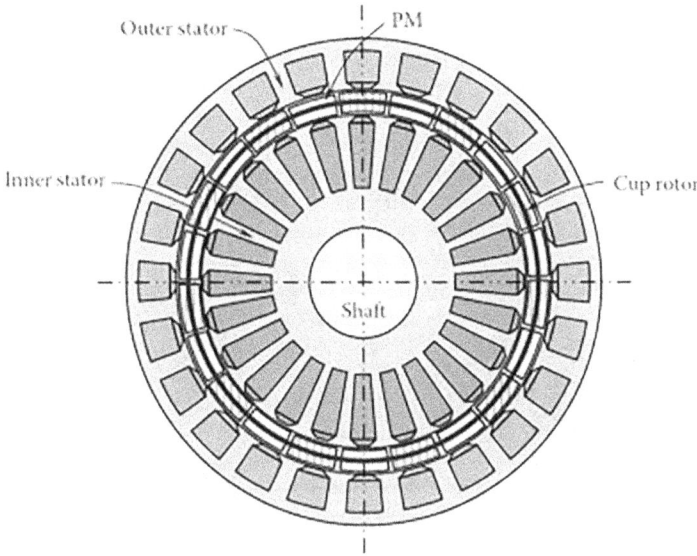

FIGURE 15.10 RF dual-stator PM machine, where S-poles (shaded) and N-poles (unshaded) are arranged alternately along the rotor circumference.

15.2.4 RADIAL-FLUX DUAL-STATOR PM MACHINE

The structure of an RF dual-stator PM machine is presented in Figure 15.10 [15.13]. The cup rotor is nested between the inner and outer stators. This machine has 24 slots wounded with three-phase windings in both the concentric outer and inner stators. The 22 pieces of PMs are mounted on the inside and outside surfaces of the rotor, where N-pole and S-pole PMs are arranged alternately along the rotor circumference. Both winding methods, that is, three-phase single-layer and three-phase double layer, can be applied to both the inner and outer stators.

This machine has the advantage that currents in both the inner and outer stators produce electromagnetic torque and both airgaps can deliver the output torque, thus improving the torque density and providing a high starting torque for hybrid electric vehicles. In addition, because of the nature of dual-stator windings, the machine can flexibly change its connections, hence providing a constant output voltage for battery charging over a wide range of speeds.

15.2.5 HIGH-TORQUE PM MOTOR WITH 3D CIRCUMFERENTIAL FLUX DESIGN

A new type of motor has been developed by a startup motor company at Fort Worth, Texas. This motor combines RF and AF designs to enhance the magnetic flux density. The innovative concept is introduced in the magnetic architecture to produce the 3D circumferential magnetic flux, passing through the motor structure, and thus resulting in high torque density of the motor. The toroidally discrete wound coils can maximize the stator-slot fill factor to approximately 90%.

Unlike present conventional motor designs, this motor has no superfluous end windings; therefore, nearly 100% of the copper winding contributes to torque production, leading to the reduction of the copper content by 30% while generating equivalent torque as in a conventional motor. For a given torque, the motor consumes significantly less power than competing motors in the same size package. Moreover, because the stator pole configuration makes use of the entire rotor magnetic flux, magnetic flux from the rotor completely engulfs the stator so that there is virtually no flux leakage [15.14, 15.15].

The testing results reveal that at a current density beyond 30 A/mm^2, the motor can produce at minimum twice the driving torque compared to other conventional motors. If the motor is used

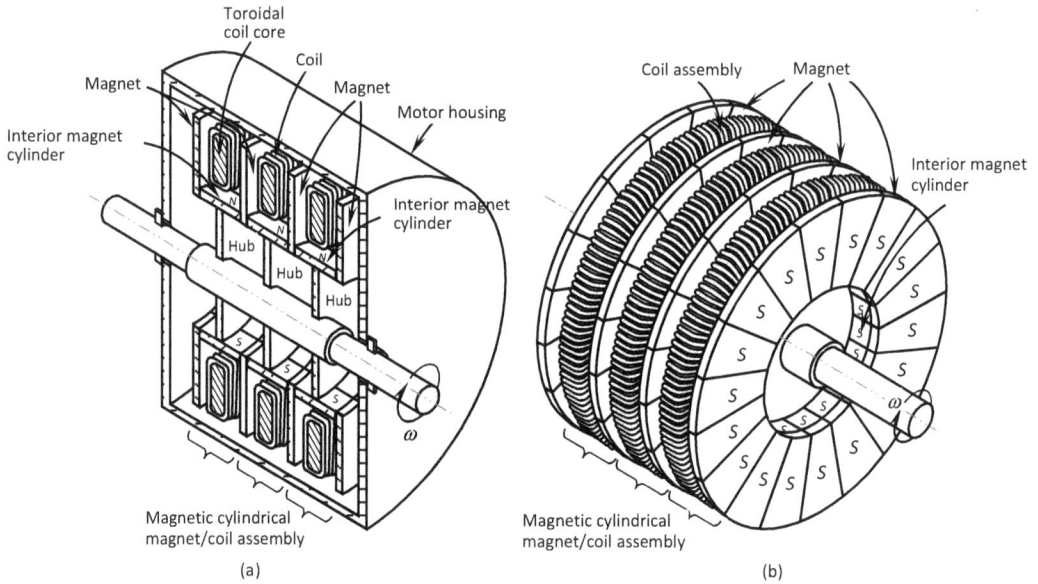

FIGURE 15.11   Combining multiple high-torque PM motors into one stack can lead to high power density and superior efficiency: (*a*) a longitudinal cutaway view of the motor with multirotor and multistator within the motor housing; and (*b*) an isometric view of the structure and configuration of the magnetic cylindrical magnet/coil system [15.14].

in hybrid cars such as Toyota Prius and Tesla, it could increase driving range by more than 10% or allow these cars to carry relatively small battery packs to deliver equivalent driving range. As claimed by the company, the motor can generate up to 150 Nm of torque at just 3,000 rpm, in a package of a Prius' motor. Since the motor can generate such robust torque, there is no need of gearbox for most applications. Therefore, it saves space and reduces the overall weight [15.16], resulting in high power densities.

As an illustrative example in Figure 15.11, the rotor consists of three units of magnet/coil assembly; each unit consists of an interior magnet cylinder that is disposed between a spaced pair of magnet discs, forming a U-shaped magnetic structure, and functioning as the rotor. The interior magnet cylinder, made of a plurality of magnets where their magnetic poles are radially aligned perpendicular to the shaft axis, is internally connected to a hub that is attached to the rotating shaft. The magnet disc comprises a plurality of magnets arranged in a radial pattern where their magnetic poles are aligned in a parallel fashion with the shaft axis. The coil assembly is made by wrapping copper wires around a toroidal iron core to form one or more coils. The coil assembly is positioned outside of the interior magnet cylinder and between two magnet discs to function as a stator. The coil assembly may consist of one or more coil segments. Multiple coil segments allow an effective speed control by selectively connecting appropriate coil segments in different combinations of series and parallel connections without changing the system supply voltage.

This configuration, as an example, constructs a motor with three rotors and three stators. In fact, the motor with multirotor and multistator can be attained by adding more units of the magnetic cylindrical magnet/coil assembly. The actual number of the rotor/stator depends on several factors, such as required motor torque, magnetic field strength, motor size, efficiency, and cost.

This design can be easily changed to switch the rotor and stator by replacing the interior magnet cylinder with the exterior magnet cylinder, as shown in Figure 15.12. The exterior magnet cylinders are disposed between a spaced pair of magnet discs, forming a Π-shaped magnetic structure and acting as the stator. The coil assemblies are connected to the motor shaft through the connecting

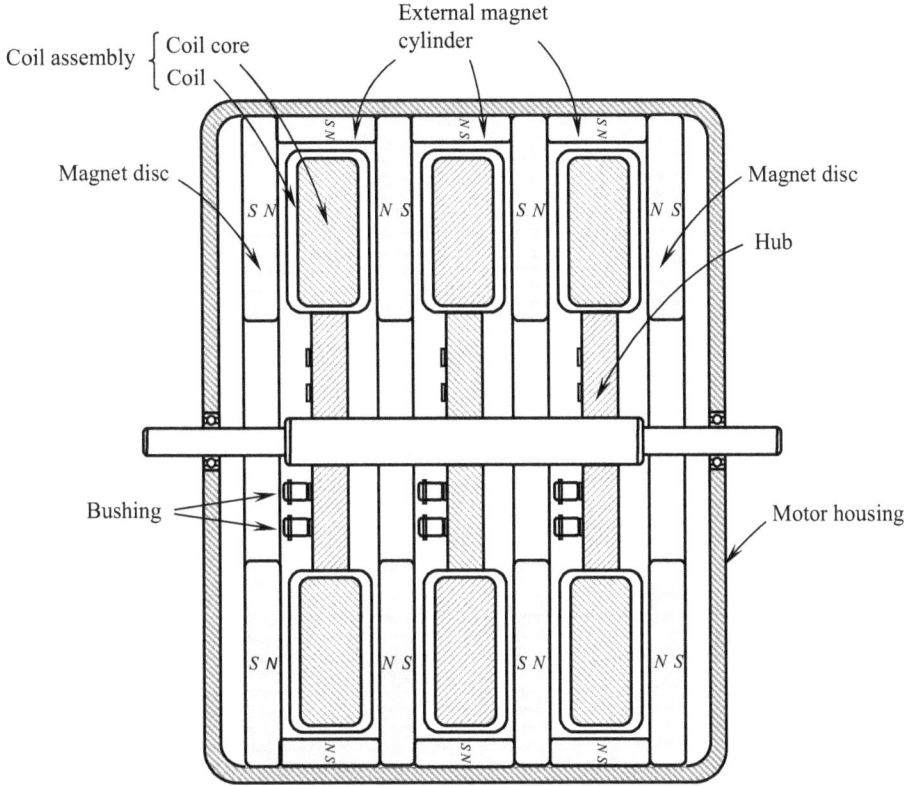

FIGURE 15.12 Alternative design of multirotor, multistator AFPM motor, where the coil assemblies are connected to hubs that attached to the rotating shaft, functioning as the rotor. Each coil assembly is sandwiched between a spaced pair of magnet discs and put inside the exterior magnet cylinder. All the magnet discs and exterior magnet cylinders constitute the stator.

hubs, acting as the rotor. Magnetic flux in the rotor is essentially guided by the metal parts, which are referred to as flux concentrators.

In such a configuration, the exterior magnet cylinders and the coil assemblies form a RF motor, and the two magnet discs and the coil assemblies form an AF motor. The motor housing provides the structural support to the magnetic stator and the longitudinal shaft. This design intensifies the magnetic field to improve the efficiency of the motor.

Another alternative design is shown in Figure 15.13, where the toroidal coil assembly acts as the rotor that is coupled with slip ring-brush assemblies, and the magnet assembly (including fine magnetic cylinders and four magnetic discs, firmly connected in a specific pattern and predefined magnetic pole arrangement) acts as the stator. Brushes are used to impart or collect electric current. This design greatly intensifies the magnetic field (*i.e.*, magnetic flux density **B** or magnetic field strength **H**, where the two physical parameters are closely related to each other as $\mathbf{B} = \mu_0 \mathbf{H}$). Although only one rotor and one stator are presented in the figure, any number of rotor or stator may be adopted depending on the application requirements.

15.2.6 RADIAL-FLUX DUAL-ROTOR, DUAL-STATOR MOTOR

The inventive design concept provides a novel RF electrical motor with dual rotor and dual stator for improving the motor torque density and efficiency. With the dual-airgap structure and the integrated planetary differential gear train, the whole machine efficiency is substantially increased.

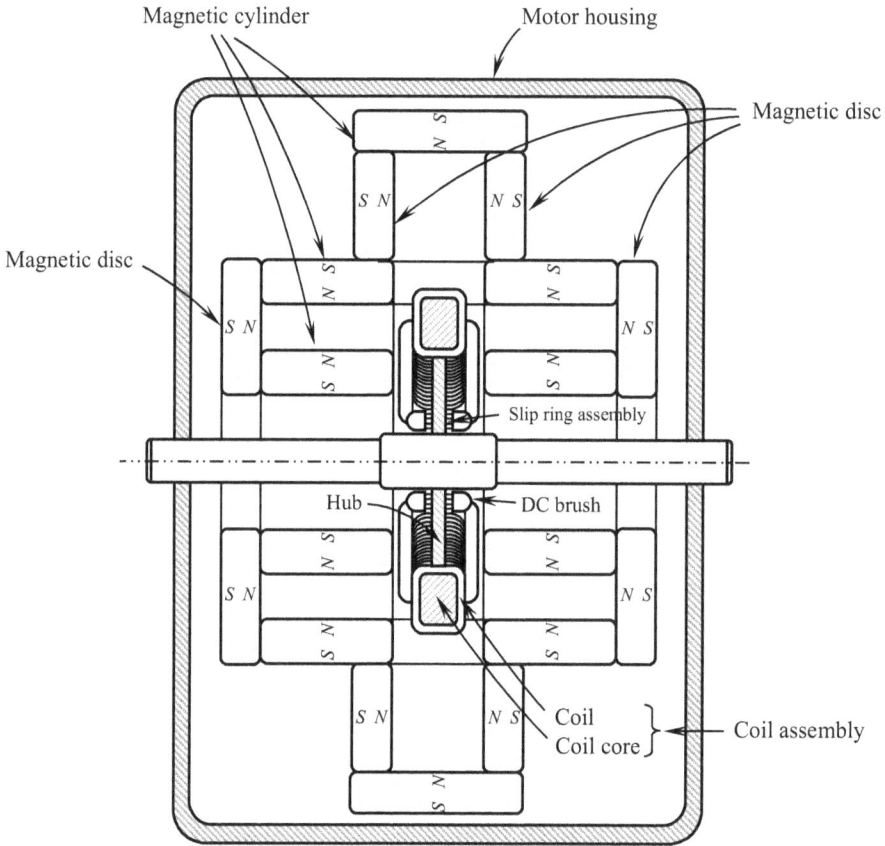

FIGURE 15.13 Alternative design AFPM motor, where the coil assembly connects to a rotor hub that attached to the rotating shaft and acts as the rotor. The stator consists of five magnetic cylinders and four magnetic discs, connected in a specific pattern with certain magnetic pole orientations, for greatly enhancing the 3D magnetic field.

The motor consists of a PM inner rotor and an inner stator and a PM outer rotor and an outer stator, constructed two independent motors (Figure 15.14). The uniqueness of this invention is that the inner and outer rotors, inner and outer stators, are independent of each other, each rotor can have its own rotating speed and rotating direction. Therefore, the inner rotor and stator, and outer rotor and stator, construct two independent motors. Integrating these two motors into a single package can not only enhance the motor performance but also provide the operating flexibility of the system. The two motors are connected through a planetary gearing system.

The inner stator mounts on the stationary central cylinder, with which the left endbell and the motor outside housing are integrated as one single component. The outer stator mounts on the interior of the housing. The permanent magnets (PMs) are attached to the interior of the inner rotor and the exterior of the outer rotor, respectively. The motor output is achieved via a planetary differential train. A sun gear is made at the right end of the inner rotor and a ring gear is made at the right end of the outer rotor. There are three planet gears between the sun gear and the ring gear. A gear train arm is connected to these three planet gears as the motor output.

For this integrated system, both the inner and outer rotors rotate independently. Unlike a regular planet gearing system that has one input, the present gearing system has two inputs, *i.e.*, the sun gear and the ring gear. As a result, the overall rotating speed of the gear train arm can be determined as

$$\omega_{arm} = \frac{\omega_s z_s \pm \omega_r z_r}{z_s + z_r} \qquad (15.1)$$

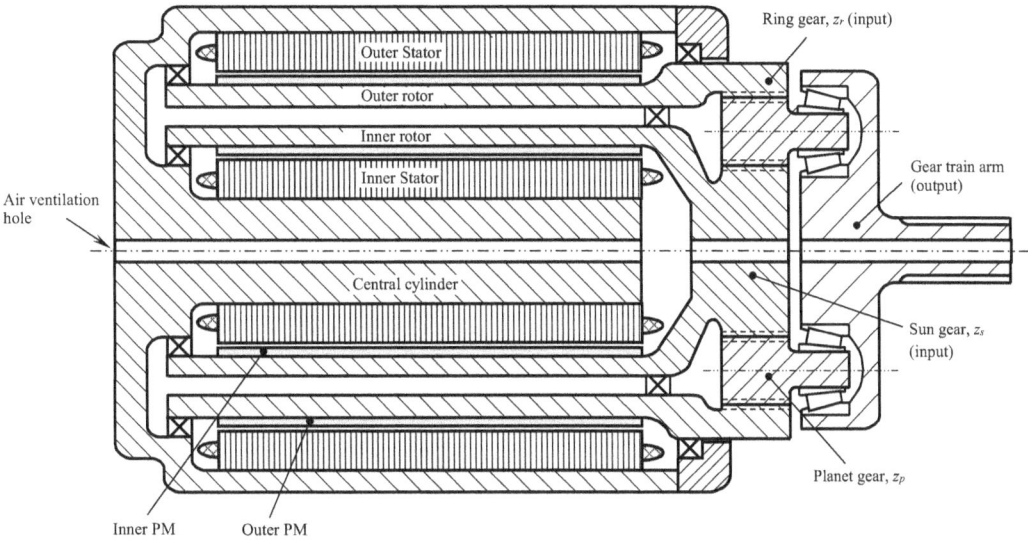

FIGURE 15.14 The structure of a RF motor with a dual rotor and dual stator. Combining two independent motors into a single package can not only enhance the motor performance and increase the power density, but also provide operating flexibility of the motor. The two motors are connected through a planetary gearing system.

where

ω_{arm} is the rotating speed of the gear train arm (carrier) of the planetary gearing system, rad/s

ω_s is the rotating speed of the sun gear, rad/s

ω_r is the rotating speed of the ring gear, rad/s

z_s is the tooth number of the sun gear

z_r is the tooth number of the ring gear

When the rotating direction of ω_r is the same as ω_s, the plus sign is used in the above equation; otherwise, the negative sign is used. From the equation, it can be concluded that the rotating speed of the arm is independent of the tooth number of the planet gear.

To help understand the performance of the motor, consider three practical operating cases (Figure 15.15): (*a*) $\omega_s = 1{,}200$ rpm and ω_r varies from −1,200 to 1,200 rpm. From Figure 15.15*a*, the corresponding output rotating speed ω_{arm} varies linearly from −480 to 1,200 rpm. It shows when the sun gear and ring gear rotate at the same speed and in the same direction, the output speed and rotating direction are the same as these two gears. However, if both the sun gear and ring gear rotate at the same speed but in the opposite direction, the output speed and rotating direction are mainly determined by the ring gear. (*b*) ω_s varies from −1,200 to 1,200 rpm and ω_r remains at 1,200 rpm. While both ω_s and ω_r affect ω_{arm} in varying degrees, the output rotating direction is mainly determined by the ring gear (Figure 15.15*b*). (*c*) ω_s varies from 1,200 to −1,200 rpm and ω_r varies from −1,200 to 1,200 rpm. In this case, the operation line of ω_{arm} is positioned between the ω_s and ω_r lines, closer to the ω_r line (Figure 15.15*c*). All above calculations are based on the gear numbers: $z_s = 18$, $z_r = 42$, and $z_p = 12$.

An alternative design of a dual-rotor, dual-stator motor is presented in Figure 15.16, where the inner rotor is in the central part of the machine. Similarly, the inner stator and the inner rotor and the outer stator and the outer rotor construct the inner and outer motors, respectively. The two motors are connected through a planetary gearing system as an integrated motor. To effectively cool the motor, an oil cooling system is used in the design. The cooling oil is introduced from one endbell and flows along the cooling channel at the center of the inner shaft. Due to the centrifugal forces,

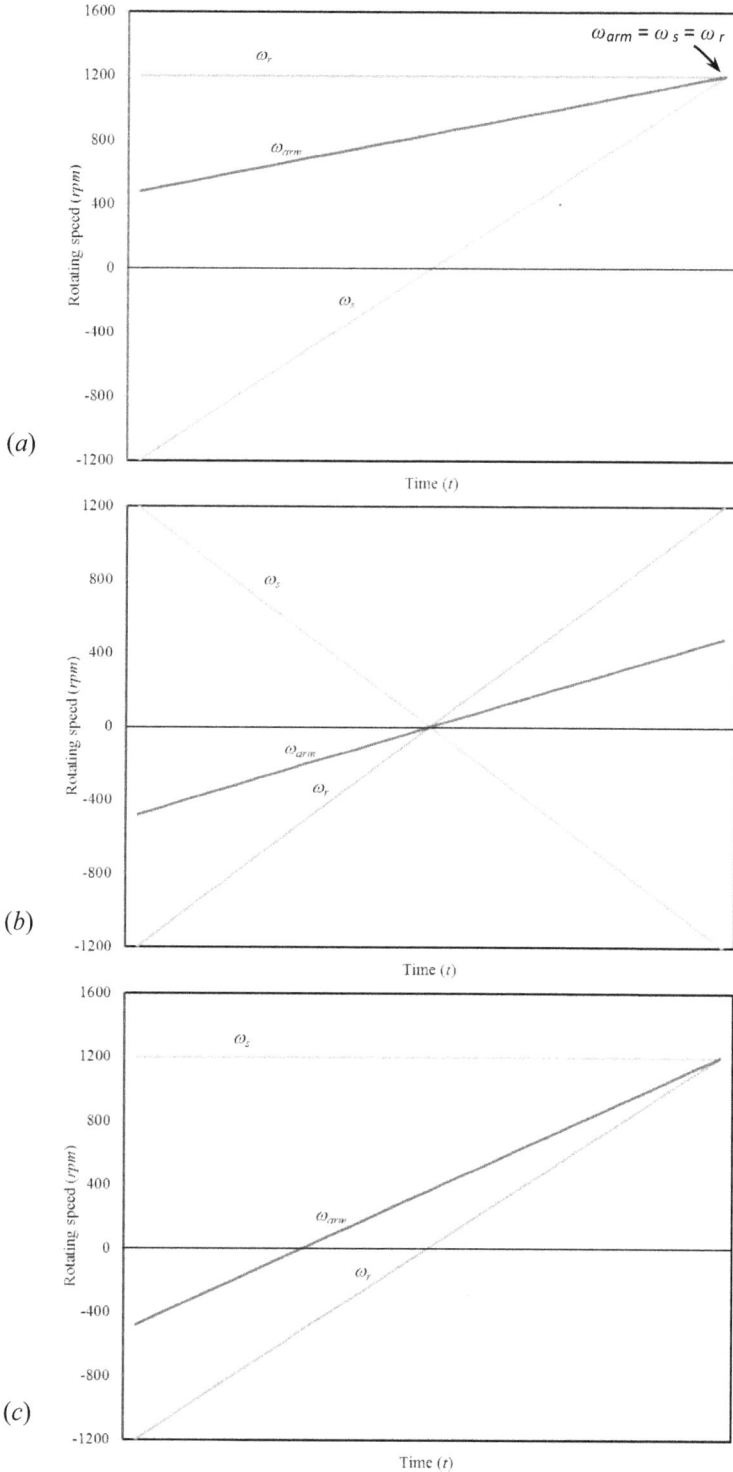

FIGURE 15.15 Variations of the output rotating speed ω_{arm} as the results of the variations of the two input rotating speeds ω_s and ω_r: (a) $\omega_s = 1{,}200\,\text{rpm}$ and ω_r varies from $-1{,}200$ to $1{,}200\,\text{rpm}$. (b) ω_s varies from $-1{,}200$ to $1{,}200\,\text{rpm}$ and $\omega_r = 1{,}200\,\text{rpm}$. (c) ω_s varies from $1{,}200$ to $-1{,}200\,\text{rpm}$ and ω_r varies from $-1{,}200$ to $1{,}200\,\text{rpm}$. The results are obtained from the calculations based on $z_s = 18$, $z_r = 42$, and $z_p = 12$.

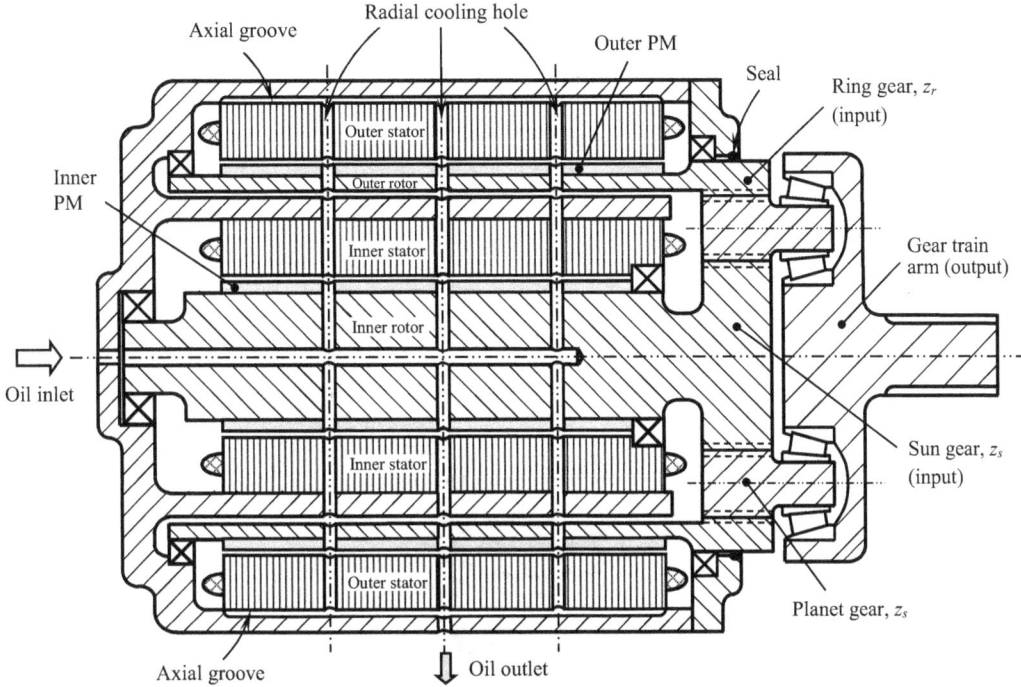

FIGURE 15.16 An alternative design of a dual-rotor, dual-stator motor, where the inner rotor is situated at the center of the machine. The inner stator and the inner rotor and the outer stator and the outer rotor construct the inner and outer motors, respectively. The two motors are connected through a planetary gearing system. Forced direct oil spray cooling is employed in this machine, which relies on the centrifugal force of the inner rotor and provides a good balance among thermal load removal capability, temperature distribution uniformity, and cooling oil inventory.

the oil is accelerated to throw out along the radial cooling holes. The heated oil is collected through the axial grooves, which distribute on the internal surface of the housing, and finally discharged at the center of the housing.

15.2.7 RADIAL-FLUX INTEGRATED MAGNETIC-GEARED IN-WHEEL MOTOR

In designing electric vehicles, one unique technique is to drive each wheel with an electric motor, referred to as the in-wheel motor. Each in-wheel motor is controlled independently and precisely to provide desired drive torque and power, as well as regenerative braking torque. This indicates that for a four-wheel drive electric vehicle using the in-wheel technology, the transmission system is no longer required. Unlike conventional *central drive unit* systems, this emerging technology can offer a number of technical advantages such as driving flexibility, structure simplicity, compactness, great wheel torque control capability, good accelerator responsiveness, high energy efficiency, quiet operation, and cost reduction. Today, the in-wheel motor becomes the most important key technology applied to electric vehicles. This solution has significantly influenced the future trends of electric vehicles toward a completely integrated in-wheel-motor drive unit that is installed inside the wheel hub.

In-wheel motors can be designed to function as either direct drive or geared drive actuators. While direct drive actuators remove the need for a gearing system, which results in a lower overall cost by possibly more than 50% of the cost of a traditional geared actuator, geared actuators reduce the rotational speed and amplify the torque, and thus suitable for lower-speed, high-torque

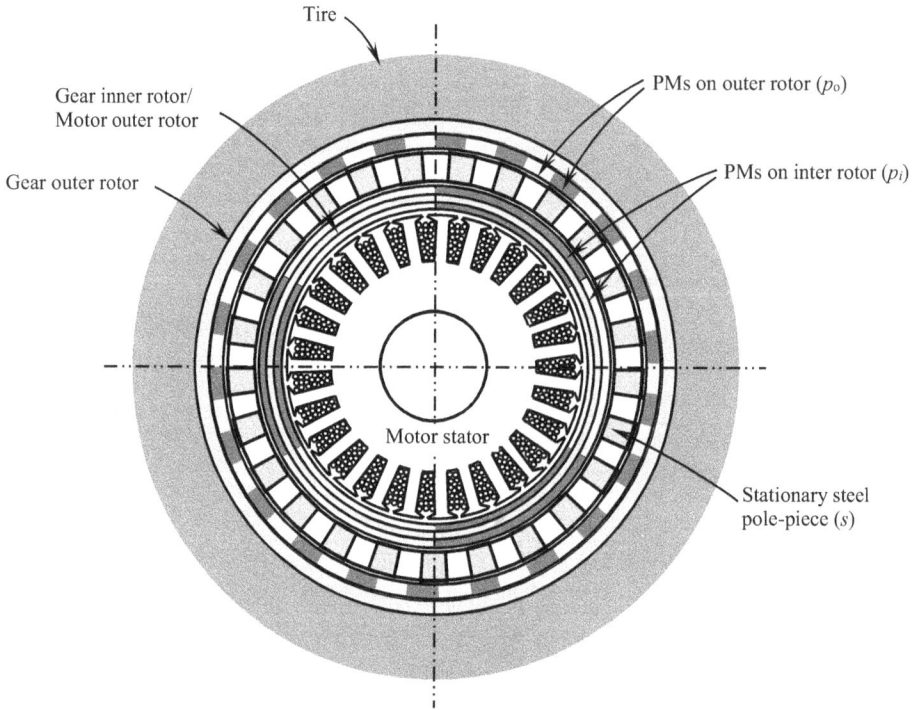

FIGURE 15.17   The topology of an integrated magnetic-geared in-wheel motor. Unlike a typical RF motor, the stator of this motor is situated at the center of the motor and the rotor at the outside of the motor for directly driving the wheel. PMs are, respectively, attached to the surfaces of the inner and outer rotors to form a coaxial magnetic gearing system.

applications. An integrated magnetic-geared in-wheel motor is shown in Figure 15.17 [15.17, 15.18], where the stator is at the center of the motor and the rotor at the motor outside for directly driving the tire. This outer-rotor topology enables full utilization of the space of the inner stator to accommodate both the PMs and the field windings, thereby improving the power density. In addition, this topology not only offers reduced size and weight but also eliminates all the drawbacks due to the mechanical gear. The magnetic gearing system is composed of two sets of PMs that are respectively attached to the inner rotor (*i.e.*, the common rotor because it functions as both the gear inner rotor and motor outer rotor) and the outer rotor. In the present design, the gear inner rotor is designed with 3 pole-pairs for high-speed operation, whereas the outer rotor is designed with 22 pole-pairs for low-speed operation. The gear ratio is determined using the following formula:

$$\gamma_g = \frac{p_o}{p_i} = \frac{s - p_i}{p_i} \tag{15.2}$$

where
  p_i is the number of gear inner-rotor pole-pairs
  p_o is the number of gear outer-rotor pole-pairs
  s is the number of stationary steel pole-pieces

With $p_i=3$, $p_o=22$, and $s=25$, the gear ratio $\eta_g=(25-3)/3=7.33$:1. Thus, the magnetic gearing system can effectively reduce the speed or amplify the torque by the factor of 7.33.

The stationary ring sandwiched between the two PM rotors incorporates steel pole-pieces, which functions to modulate the airgap flux density space harmonics. The stator of the motor adopts a

3-phase winding with fractional slots per pole per phase, hence offering low cogging torque that is highly desirable for vehicle operation.

15.3 AXIAL-FLUX MULTIROTOR OR MULTISTATOR MOTOR

While RF motors dominate the motor market as mainstream for more than a century, AF motors have been developed in the last several decades and are attracting growing interest in recent years. AF motors may benefit from the increased efficiency, lowered axial loads acting on the rotor and bearings, and reduced motor vibration. Unlike in an RF motor in which the airgap is fixed after the rotor and stator are manufactured, the planar airgap in an AF motor is adjustable by changing the relative position of the rotor and stator.

AF motors are designed to have a higher power-to-weight ratio than RF motors, resulting in less core material and higher efficiency [15.19]. For small motors, the motor efficiency for an AF motor is about 15% higher than an RF motor with a similar size. This is attributed to the fact that the AF motor utilizes both sides of the rotor or stator, while the RF motor utilizes only one side. Other advantages include a simple cooling system, high reliability, and motor power being obtained from both ends of the spindle. Thus, the AF motors have been widely used in applications such as robots, wheel-drive vehicles, hard disk drives, wind turbines, and fans. However, AF machines have complicated structures and require much stronger supports to keep uniform airgaps, leading to high production cost [15.20].

There are many topologies of AF machines for various applications. For example, AF machines may be single-sided or double-sided, with or without slots, with or without armature core, with internal or external PM rotors, with surface-mounted or interior PM, and single stage or multistage [15.21]. Some of these topologies are addressed in the following sections.

15.3.1 SINGLE-SIDED AND DOUBLE-SIDED AXIAL-FLUX MOTORS

A single-sided AF motor has the simplest structure with only single disk-shaped rotor and stator. In this configuration, PMs are mounted on one side of the rotor. The cross section of the rotor and stator is illustrated in Figure 15.18. The topology of AF motor allows to adjust the motor torque by adjusting the airgap, as shown in Figure 15.19.

A typical double-sided AF motor is shown in Figure 15.20. As illustrated in Figure 15.20a, the motor consists of two identical stators and one rotor, where the rotor is nested in the two stators axially. Each stator has a large number of slots for receiving stator windings. The three-phase windings are uniformly distributed in the circumference of the stator and wound radially (Figure 15.20b). The PMs and iron poles are alternately embedded or glued in a disk-shaped support in two rows (Figure 15.20c). The support is made of a nonferromagnetic material such as aluminum and epoxy [15.22]. The stators can be connected in series or parallel. A serial connection is usually preferred since it can provide equal but opposite axial forces.

Double-sided AF motors can have different configurations such as a stator–rotor–stator structure as addressed previously or a rotor–stator–rotor structure as shown in Figure 15.21. In the rotor–stator–rotor design, a single nonslotted stator is sandwiched between two rotor disks, forming axially two planar airgaps. The stator is realized by tape wound core with polyphase AC airgap windings that are wrapped around the stator core with a back-to-back connection. The active conductor portion is the radial portion of the toroidal windings facing the two-rotor structure (Figure 15.21a). Arch-shaped PMs are mounted axially on the inner surface of each rotor.

The magnetic flux path is presented in Figure 15.21b. An N-pole magnet on one rotor is placed directly opposite to its counterpart on the other rotor. It is noted that there are two identical magnetic loops. For each loop, the magnetic flux is driven by the N-pole magnet to go across the airgap into the stator core, travels circumferentially along the strip-wound stator core, returns across the airgap, and then enters the rotor core through the opposite S-pole of the PMs to complete the

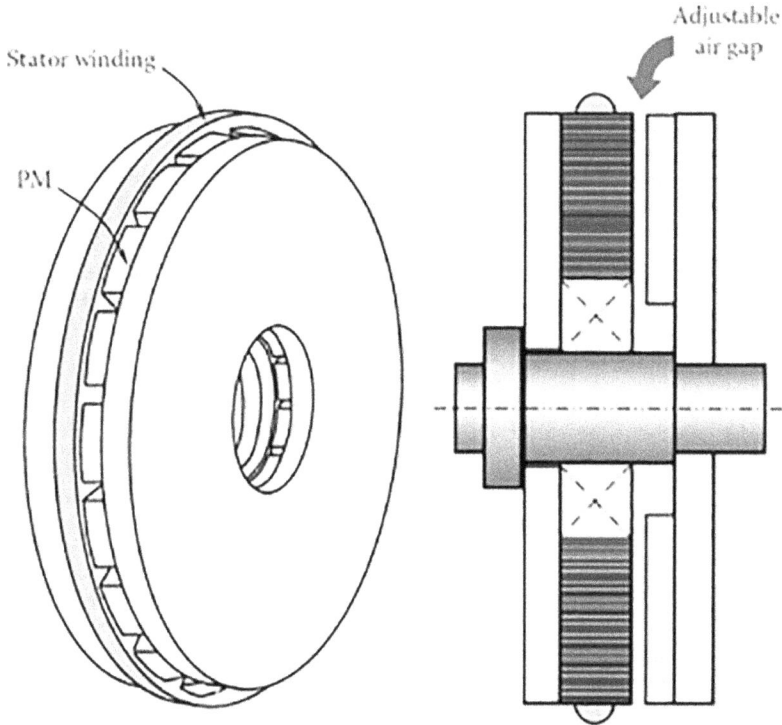

FIGURE 15.18 Single-sided AFPM motor, showing a single rotor and stator.

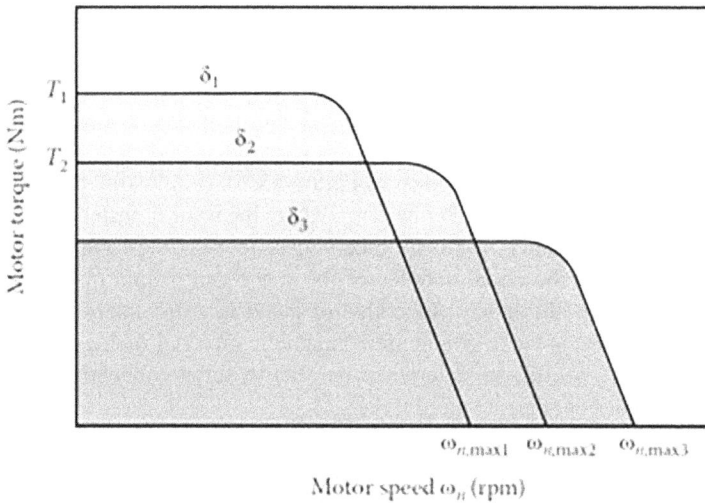

FIGURE 15.19 The motor torque of an AF motor can be adjusted by changing the airgap δ.

magnetic flux loop. Hence, it can be expected that the axial length of the stator core would be quite long because of the summation of the flux entering the stator from both rotors. As a matter of fact, this machine can be viewed as a combination of two independent halves due to the direction of the magnetic flux.

The slotless torus topology shown in Figure 15.21 has a high torque density and high torque-to-weight ratio. The portions between the airgap windings are assumed to be filled with epoxy resin as

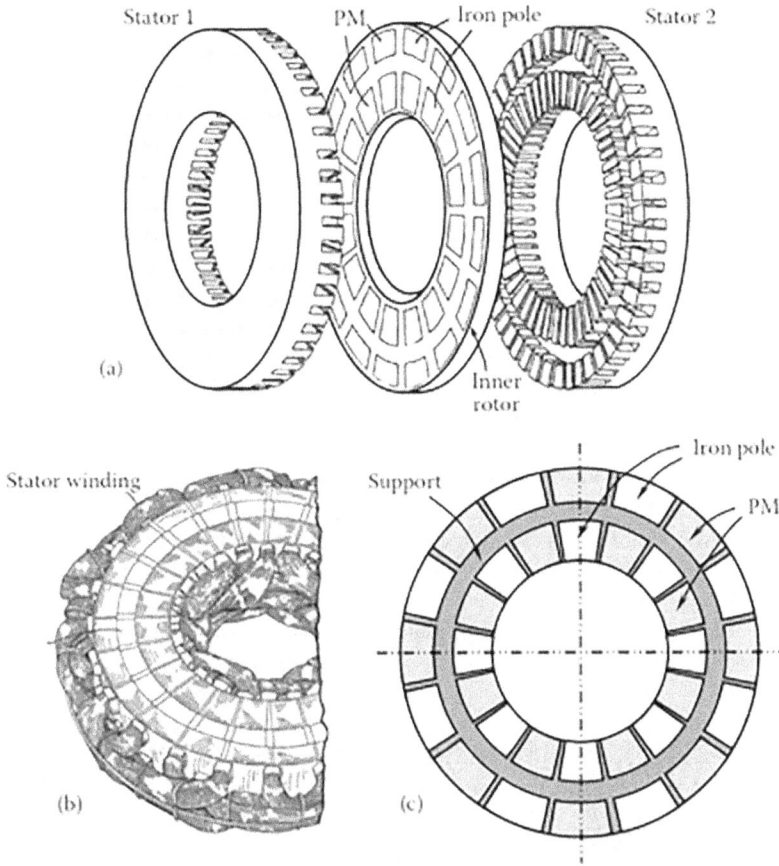

FIGURE 15.20 A stator–rotor–stator AFPM motor: (*a*) the motor contracture, (*b*) stator assembly, and (*c*) PM and iron pole arrangement on the rotor (U.S. Patent 7,608,965) [15.22]. (Courtesy of the U.S. Patent and Trademark Office, Alexandria, VA.)

in all nonslotted structures to increase the mechanical robustness and provide better conductor heat dissipation. In addition, in the torus topology, the windings in the airgap are used for torque production. Besides, due to the absence of stator slots, the cogging torque, magnetic flux ripple, pulsation losses in the rotor, and acoustic noise can be greatly reduced.

15.3.2 MULTISTAGE AXIAL-FLUX MOTOR

For an AF motor, the torque is proportional to the outer diameter (OD) of the motor. However, the centrifugal force acting on the rotating components significantly increases with the OD. Therefore, a common way to increase the motor torque is to design a motor with multistage arrangements. Usually, for a rotor–stator–rotor structure, a multistage machine has N stators and $N+1$ rotors, where N is the number of stages. All rotors share the same shaft.

The double-sided AF motor in Figure 15.21 can be extended into a multiple stage machine as illustrated in Figure 15.22. The stators are formed by wrapping magnet wires radially around a non-slotted stage core. The rotors are formed by both rotor cores and axially magnetized SPM. Except for the end rotors, PMs are mounted on both sides of the rotor cores. Theoretically, there is no limit on the number of motor stages in multistage motors. This indicates that a desired high torque can be achieved by adjusting the number of motor stages.

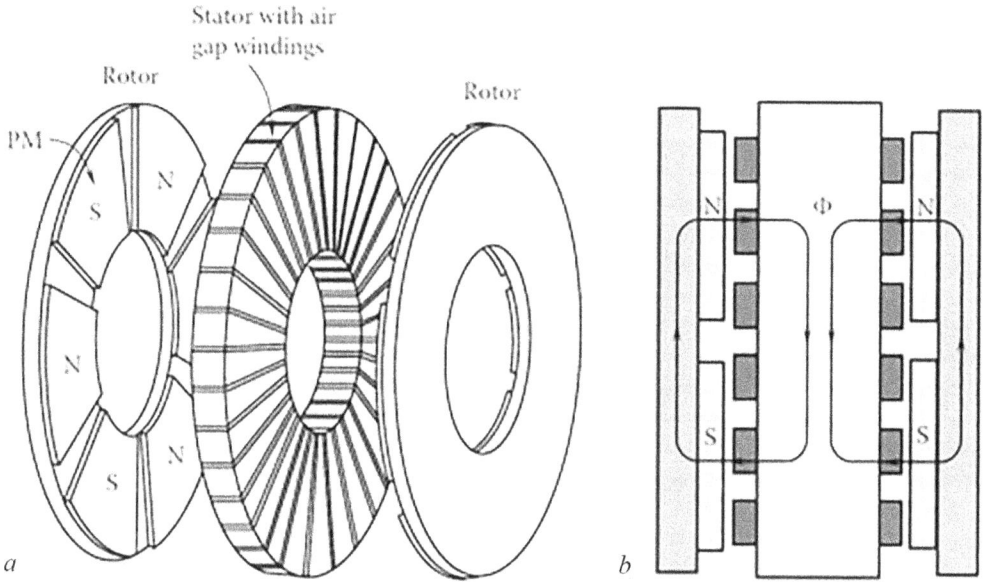

FIGURE 15.21   An AF torus-type slotless rotor–stator–rotor PM motor: (*a*) motor structure and (*b*) magnetic flux direction.

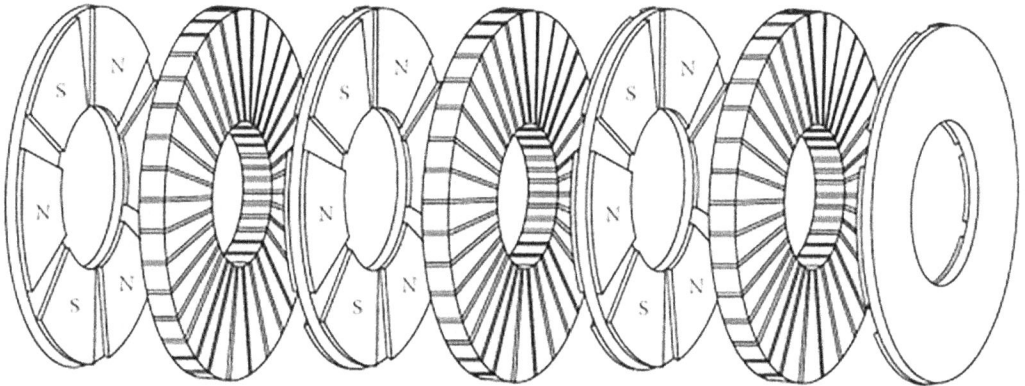

FIGURE 15.22   Perspective view of multistage of AF torus-type slotless PM motor.

The multistage AF machines are very competitive with respect to conventional AF and RF machines for their compact size, simple structure, and easy assembly. Because each airgap can be adjusted independently, the airgap depths in the motor can be made either identical or distinctive for obtaining the optimized motor performance for a specific application.

15.3.3 YOKELESS AND SEGMENTED ARMATURE MOTORS

The yokeless and segmented armature (YASA) topology is a new type of AF motor. This type of motor has shown an improved torque density and efficiency compared to other AF motors [15.23]. The iron in the stator of the YASA motors is dramatically reduced, typically by 50%, causing an overall increase in torque density of about 20%. In fact, the elimination of yoke, elevated fill factor, and shortened end windings all contribute to the increase in torque density and efficiency of the electric machine. Therefore, this topology is highly suited for high-performance applications.

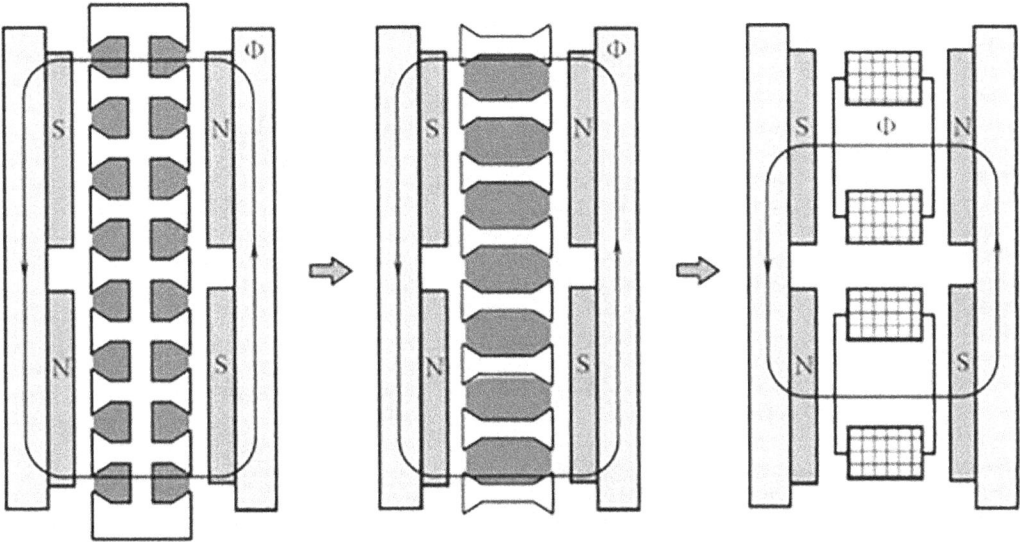

FIGURE 15.23 The topology of a YASA motor, which combines two best-performing AF motors.

The YASA motors are primarily designed for vehicles. The segmented stator is made of powdered iron materials. The topology of a YASA motor combines two of the best-performing AF machines, the NS torus-S and the NN torus-S topologies, as shown in Figure 15.23.

15.3.4 ENERGY EFFICIENT AXIAL-FLUX YOKELESS MOTOR WITH MODULAR STATOR

Axial-flux permanent magnet (AFPM) motors perform excellently at a broad range of rotational speed, making them suitable for high-speed, low-torque (*e.g.*, electric vehicles) and low-speed, high-torque (*e.g.*, wind power generators) applications. Its slim and compact structure results in a motor with a high power density and superior efficiency. An example can be seen in Figure 15.24. AFPM motors appear in several topological forms, in terms of stator/rotor arrangement, the number of stator/rotor, yoke/yokeless, core/coreless, magnetic flux path option, coupling among windings, and PM configuration.

Recently, a new type of YASA AF motor has been designed and developed by a startup motor company in Belgium. Many design features of the motor, such as the elimination of stator yoke, short end winding, high stator-slot fill factor, modular stator construction, short magnetic flux path, and dramatically reduced material requirement, contribute to the increase in torque density and motor efficiency. According to the company, this type of motor can reach efficiency within the range of 91% – 96% [15.24, 15.25].

This YASA AF motor comprises a modular stator in the middle and two pancake-shaped PM rotors, where the stator is located between the two rotors, forming two axial airgaps between the stator and each rotor (Figure 15.25). A plurality of PMs is attached to each rotor disc, facing each other. This topology results in short magnetic paths without flux reversal.

The stator is constructed with the individual stator modules. In each module, two concentrated coil windings are put around the stator core element oppositely for achieving the highest possible torque-to-weight ratio. The stator is assembled by shoving individual modules one another (Figure 15.26). The absence of the stator yoke has an advantageous influence on the power density and the stator core loss, particularly eddy current/hysteresis loss. Moreover, compared to a RF motor, the reduction of approximately two-thirds of the stator iron can lead to a great decrease in weight and volume of the motor. A further decrease in the core loss is possible by using grain-oriented steel for making the stator core elements.

FIGURE 15.24 A wind power generator with low speed and high torque is designed by Magnax. This AF machine weighs only 2.5 metric tons but can deliver as much torque and the 6 metric tons of RF machines. (Courtesy of Magnax BVBA, Kortrijk, Belgium.)

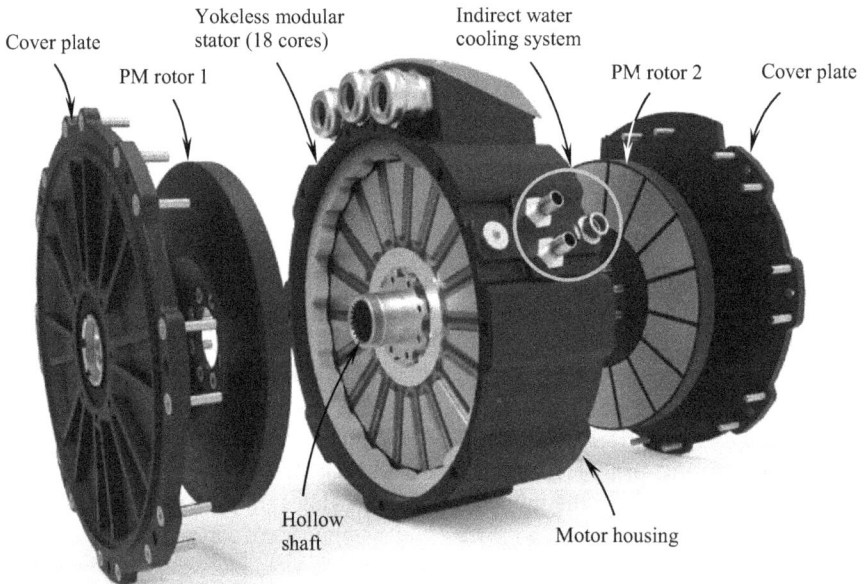

FIGURE 15.25 Exploded view of the YASA AFPM machine, where the stator consists of multiple stator core elements around which concentrated coil windings are wound. (Courtesy of Magnax BVBA, Kortrijk, Belgium.)

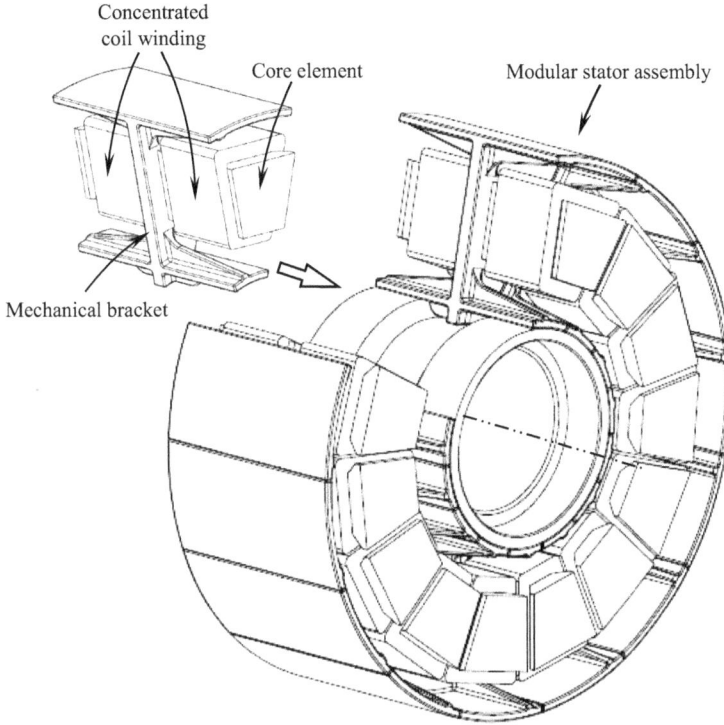

FIGURE 15.26 The stator is constructed with individual stator modulus [15.26]. The most important advantages with respect to a modular machine include high slot filling factor, high productivity, short manufacturing cycle time, and labor cost reduction. In addition, modular stator is beneficial to winding automation.

In addition, to lower the copper loss, it utilizes the concentrated windings so that there are no coil overhangs, and the easy winding process allows making shorter end windings. Therefore, the YASA topology has an overall better performance, a superior power density, and an excellent energy efficiency [15.26]. To increase the copper fill factor, it uses rectangular section copper wires.

Bearings in a YASA motor not only support the rotors in the radial direction, but also withstand high axial attraction forces between the stator and the rotors. In operation, each rotor is subjected to an equal but opposite attractive force. The two rotors exert an equal but opposite attraction force on the rotors. Since the two rotors are connected to each other via the shaft, these traction forces cancel each other out. Nevertheless, the use of angular contact ball/roller bearings in a YASA motor is highly desired.

It should be noted that the development of a new YASA motor design comes with a wide range of technical challenges, especially the thermal challenge that addresses effective motor cooling. Because the windings are buried deep inside the stator and disposed between two rotors, heat generated from the windings becomes difficult to disperse. To solve this issue, several cooling schemes are proposed. Generally, air cooling is preferred for many applications where the heat load is low or moderate. Compared with liquid cooling, air cooling eliminates the risk of coolant leakage and has the simplest form among all cooling schemes. To further enhance the heat dissipation, cooling fins may be casted at the housing outer surfaces and thus heat can be effectively dissipated away by means of these fins. This not only provides the motor a higher heat dissipation capacity to produce greater torque and power, but also allows for a stiffer motor structure. For motors with very high heat loads, indirect liquid cooling may be needed for maximizing power density because water or oil transfers heat much more efficiently than air. Furthermore, liquid cooling makes little noise compared with air cooling. However, a liquid cooling system often requires additional equipment such as pumps and cooling pipes, making the motor structure more complicated. Nevertheless, this cooling scheme can extract the heat from the windings very effectively [15.27].

15.3.5   AXIAL-FLUX MOTOR WITH PCB STATOR

A type of printed circuit board (PCB) motor without the stator and rotor iron cores has been designed to provide high-efficiency, low-cost, and high-performance solutions for electric machine industry [15.28]. As described in Figure 15.27, the motor comprises a shaft and two rotor discs, a PCB stator disc that is arranged between the two rotor discs, PMs that mount on the two rotor back plates. Airgaps are formed between the PCB stator and the two rotor discs. Instead of using heavy and bulky copper windings and iron cores, the stator is fabricated by superposing a plurality of PCB layers and the back plates of the rotor are made from nonmagnetic materials. This type of motor offers numerous advantages over conventional motors, including lightweight, ultra-thin, high power density, low manufacturing cost, convenient to process, and suitable for batch production.

In recent years, the PCB stator technology has gained renewed interest in the motor industry [15.29, 15.30]. Several US-based companies have used semiconductor-based PCBs to build up motor stators. Unlike conventional stators, the PCB stator motor technology eliminates the need for wire windings and iron laminations and replaces them with copper-etched conductors that are embedded into a multilayered PCB. As a result, the motors built using PCB stators are ultra-thin, up to

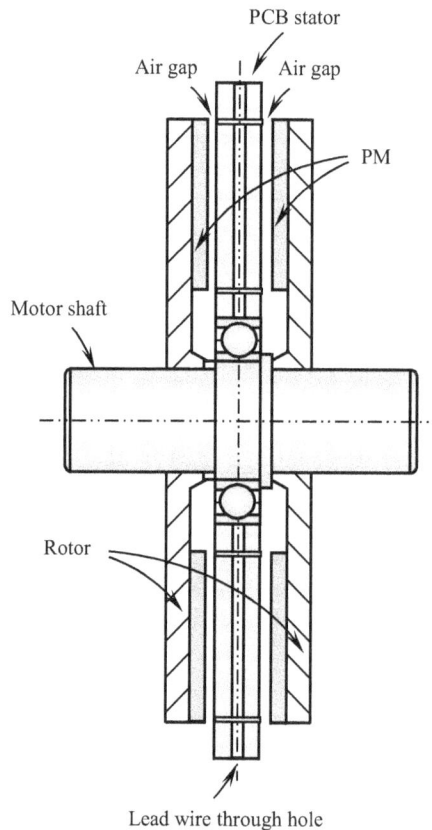

FIGURE 15.27   The motor uses a PCB stator and nonmagnetic rotor back plates for eliminating heavy and bulky copper windings and iron cores. The stator is fabricated by superposing a plurality of PCB layers.

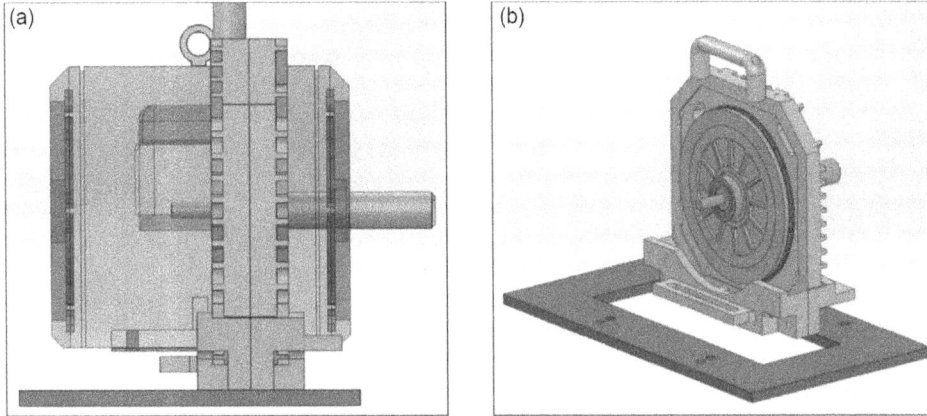

FIGURE 15.28 Comparison of the motor size (particularly the motor thickness) with the same torque and speed between (*a*) a conventional motor with iron cores and (*b*) an AF motor with PCB stator without iron cores. (Courtesy of ECM, Newton, MA.)

60% – 70% lighter weight, up to 25% cost reduction, and require fewer raw materials to build without compromising on their torque. The comparison of the motor size between a conventional motor and a PCB stator motor with the same torque and speed is shown in Figure 15.28. It clearly shows that the PCB stator motor (Figure 15.28*b*) is about 75% thinner than the conventional motor (Figure 15.28*a*).

15.4 HYBRID MOTOR

A hybrid motor is a motor that combines two or more types of motor structures together so as to integrate their strengths for better performance and operation.

15.4.1 HYBRID EXCITATION SYNCHRONOUS MACHINE

Hybrid excitation synchronous machines (HESMs) have been developed in the past few decades for traction applications such as hybrid electric and fuel cell vehicles [15.31]. The HESMs have two excitation sources: one is the PM source that provides the airgap with constant flux and the other is the DC excitation coil source that acts as the flux regulator to adjust the airgap flux distribution. These two excitation sources can be connected either in series or in parallel.

As can be seen in Figure 15.29, an HESM consists of powdered iron cores, PMs, stator windings, and DC excitation coils. PMs are placed at the center of the rotor and the excitation coils are located at the two ends of the motor. Thanks to the excitation coils, it is possible to modulate the total excitation flux in the airgap region of the machine. In addition, these additional excitation coils make it possible to improve the effectiveness of the machine in different operating regions.

15.4.2 HYBRID HYSTERESIS PM SYNCHRONOUS MOTOR

A hysteresis synchronous machine (HSM) is extensively adopted in small motor applications. The combination of PMs and hysteresis materials in the rotor of an HSM has many distinct advantages over the conventional PM or hysteresis motors. The hybrid motor in which the PMs are inserted into the slots at the inner surface of the hysteresis ring is called PM hysteresis synchronous (PMHS) motor. In fact, this type of motor combines the advantageous feature of both the hysteresis and PM motors [15.32].

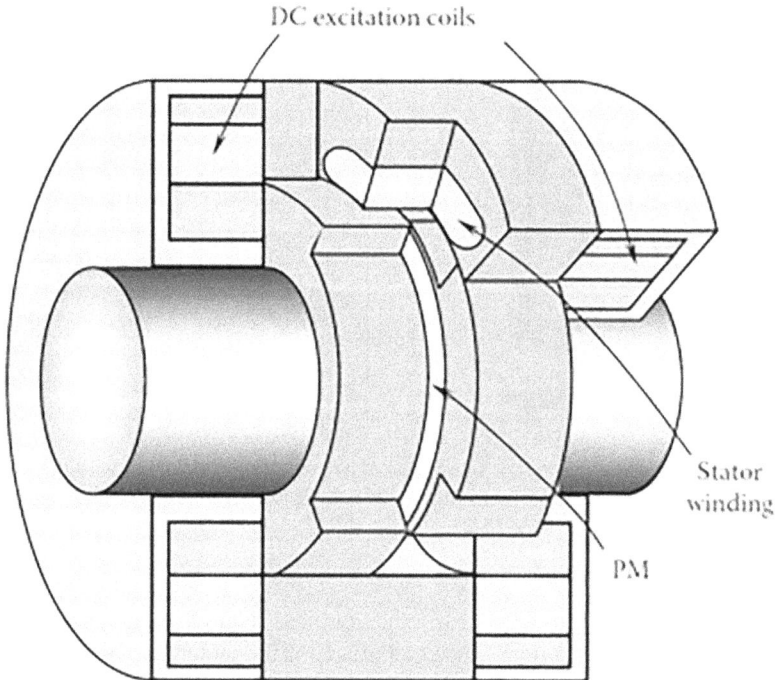

FIGURE 15.29 HESM with powdered iron core and global excitation coils.

Qin and Rahman [15.33] have presented a hybrid hysteresis PM synchronous motor, as illustrated in Figure 15.30. The rotor of the motor consists of a 36% cobalt-steel hysteresis alloy with rareearth PMs to improve the overall performances of such a motor. At asynchronous speeds, the motor torque contains the hysteresis torque, eddy-current torque, and PM brake torque. While in synchronous speeds, the motor torque is comprised of the hysteresis and PM torques. Thus, it combines the advantages of both machines [15.34]. Experimental results have confirmed the validity of the motor design and shown that the hybrid motor exhibits improved speed stability over a wide range. This type of motor is especially suited for constant torque, constant speed, and quiet applications such as gyros, time and recording equipment, and pumps.

15.4.3 HYBRID MOTOR INTEGRATING RF MOTOR AND AF MOTOR

The motor shown in Figure 15.31 integrates an RF and an AF motor together to achieve high torque density and efficiency. This motor has three sets of PMs and three stators, including one regular radial stator and two axial disk-shaped stators. The two sets of PMs are mounted on the rotor's two sides and polarized in the axial direction. Thus, the north/south polarity of the PMs alternates as the rotor rotates. The third set of PMs is mounted on the curved radial outward surface and polarized in the radial direction with alternating orientation of their N and S poles. The rotor is disposed around the shaft. Thus, the regular radial stator and the third set of PMs form a regular RF motor, and the two axial stators and the two sets of PMs mounted on the sides of the rotor form an AF motor. Electric currents applied to the three stators produce rotating magnetic fields that interact with the fields produced by the corresponding sets of PMs to generate a desired torque that rotates the motor shaft. For a given motor volume, this motor offers high mechanical power and torque since the entire rotor surface is used in the torque production. Typically, the motor rating can be increased one-third or greater with respect to a comparably sized RF motor [15.35].

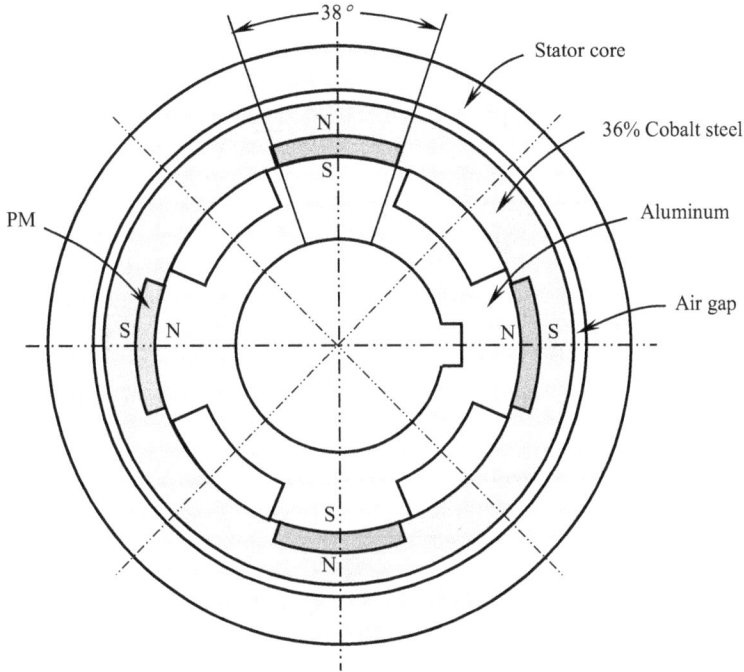

FIGURE 15.30 The structure of a hybrid PM HSM. The combination of PMs and hysteresis materials in the rotor makes the motor suitable for constant-torque and constant-speed applications.

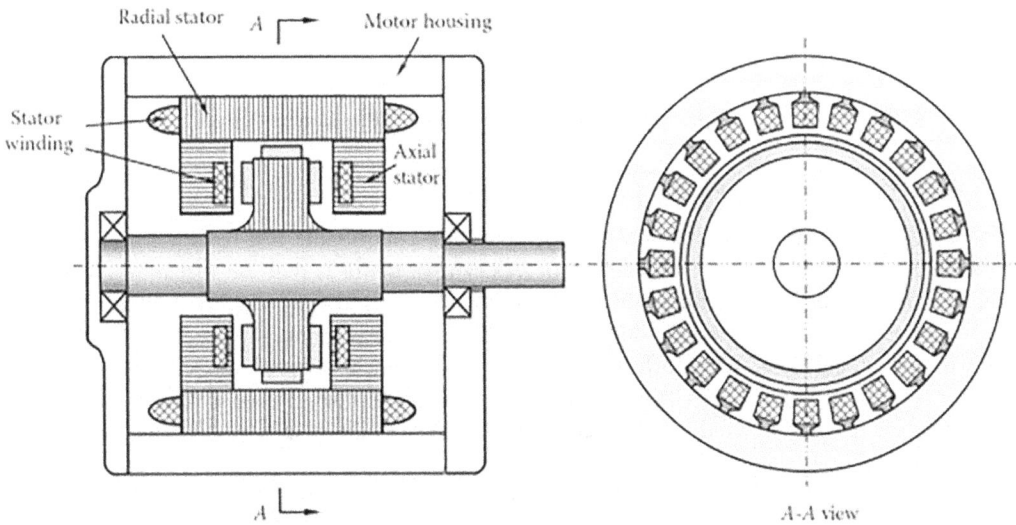

FIGURE 15.31 The new motor design that combines the RF and AF motors.

The motor with a single-unit structure in Figure 15.31 can be further expanded into a multistage form, as shown in Figure 15.32. Generally, the output power is proportional to the motor OD and the stage number. Increasing the stage number can thus increase the motor torque for effectively driving high-power loading machines such as mining machinery, freight elevators, hoists, heavy-duty vehicles, and large-size mixers.

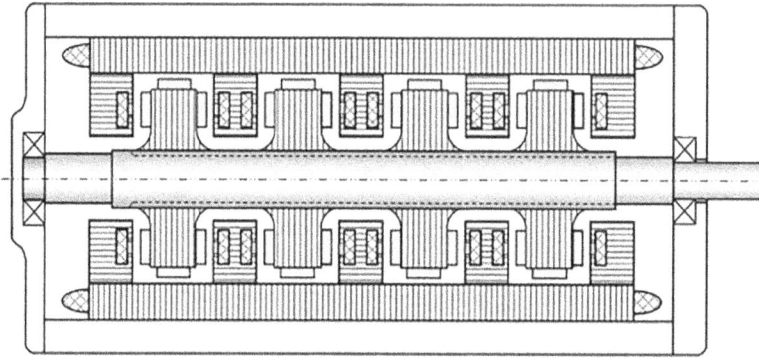

FIGURE 15.32   The integrated RF and AF motors with the multistage structure.

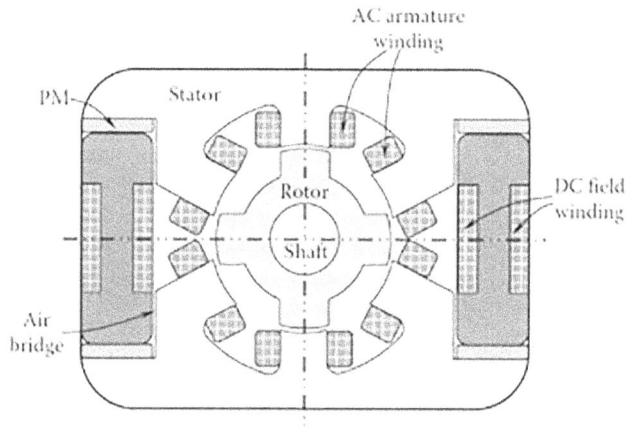

FIGURE 15.33   Topology of a three-phase 6/4-pole hybrid-field FCPM motor.

15.4.4   Hybrid-Field Flux-Controllable PM Motor

The topology of a three-phase 6/4-pole hybrid-field flux-controllable PM (FCPM) motor is presented in Figure 15.33. The main advantage of this type of motor is that the airgap flux density is directly controllable. The stator contains two types of windings, that is, the three-phase AC armature windings and the DC field windings. The key of the motor is to incorporate both PMs and DC field windings in the stator, hence offering a compact arrangement of hybrid-field excitations, while the rotor is simply composed of salient poles without winding or PMs [15.36]. PMs are closely positioned adjacent to the field windings. In such a way, the field windings serve to regulate the PM flux, offering either flux weakening or flux strengthening.

It is important to note that there is an extra air bridge in shunt with each PM. If the field winding magnetomotive force (MMF) reinforces the PM MMF, the air bridge will amplify the effect of flux strengthening. On the contrary, if the field winding MMF is opposite to the PM MMF, it will increase the PM flux leakage, thus amplifying the effect of flux weakening. By choosing an appropriate width of the air bridge, a wide flux regulating range can be obtained by using a small DC field excitation.

15.4.5   Hybrid Linear Motor

Jeon *et al.* [15.37] proposed a novel hybrid linear motor by both induction and synchronous operations. The motor consists of a pair of linear synchronous motors (LSMs) and a linear induction motor. The primary cores of both motors share a common ring winding, and the secondary solid conductor is arranged in both LIM and the interpole space of LSM. From the 3D FEA and experiment, the motor is verified to be effective for practical use.

15.5 CONICAL ROTOR MOTOR

In the present motor industry, almost all rotary motors are designed with a cylindrically shaped rotor and stator so that a constant airgap is maintained in both the circumferential and axial directions. In a conical rotor motor, the rotor is in a conical shape. According to the stator shape, two groups are identified as follows:

- The stator is in a regular shape with a constant internal diameter. Therefore, the airgap between the rotor and stator is no longer a constant; rather, it varies along the motor axial direction (Figure 15.34). It can be seen in the figure that the airgap increases linearly along the motor axis away from the bearing. This configuration may improve the motor vibrational characteristics and prevent the interference between the rotor and stator [15.38].
- The stator is also in a conical shape, with the same conical angle as the rotor. Thus, the airgap remains a constant along the rotor axis (Figure 15.35). Sarros *et al.* [15.39] have designed, analyzed, and tested two micromotors, with a conventional cylindrical design and a conical design, respectively. They found that the conical motor offers comparable performance to the cylindrical motor but with a much more compact mechanical arrangement. One of the advantages of the conical design is that it eliminates the need for a long pivot bearing and hence substantially reduces the size of the motor. The high starting torque and low inertia of the conical motor have proven to be ideal for the demands of high cycle dynamic drives in various applications.

Another type of conical rotors with a variety of rotor–stator structures was proposed by Burch *et al.* [15.40]. As an example shown in Figure 15.36, unlike the conical rotor motors discussed previously, the rotor assembly consists of two conical rotor cores that are mounted on the motor shaft. Trapezoid magnets are attached on the conical surfaces of the conical cores for producing a desired magnet field.

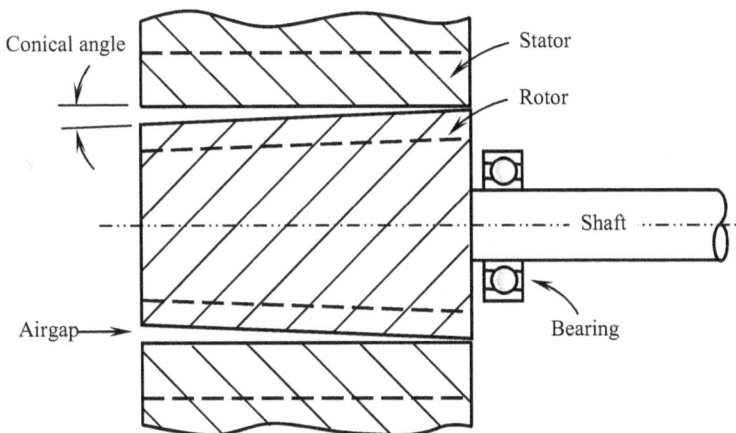

FIGURE 15.34 Conical rotor with a nonuniform airgap (U.S. Patent 5,233,254) [15.38]. (Courtesy of the U.S. Patent and Trademark Office, Alexandria, VA.)

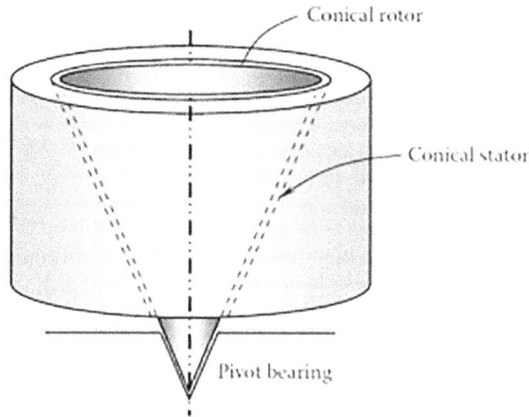

FIGURE 15.35   Conical rotor with a pivot bearing having a uniform airgap.

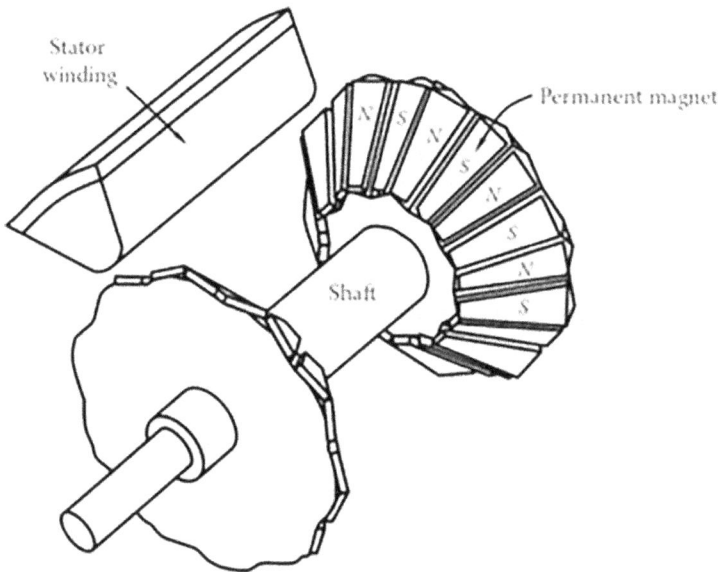

FIGURE 15.36   Conical rotor with PMs at two ends of the shaft and stator windings located between the PMs.

15.6   TRANSVERSE FLUX MOTOR

Transverse flux motors (TFMs) were first introduced by W. M. Morday in 1895 but just gained wider attention towards the end of the 20th century. Unlike RF or AF motors, TFMs have complex magnetic circuits, allowing to increase the motor torque by increasing the number of poles without affecting the flux linkage or current [15.41]. TFMs have the highest torque density, usually three to four times higher than conventional synchronous and asynchronous machines. The use of TFMs in industrial manipulators may eliminate the conventionally required gears and build direct drive systems [15.42]. Due to their superior performing characteristics, TFMs are expected for direct drive systems, such as wheel motors of hybrid electric vehicles and wind turbines [15.43].

Usually, TFMs have a toroidal armature winding, carrying current parallel to the DIR. The stator core is salient, and the rotor is equipped with PMs. The stator core often carries flux in three directions. This indicates that iron powder is a preferred core material.

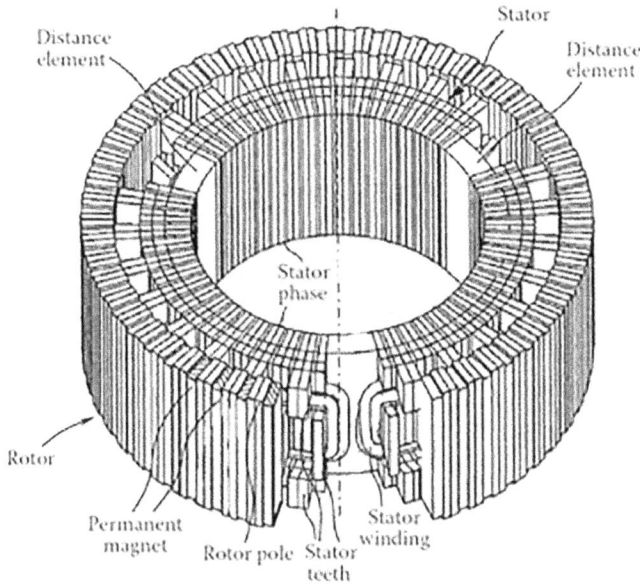

FIGURE 15.37 Perspective view of the structure of PM TF motor (European Patent 2,317,633) [15.45]. (Courtesy of the European Patent Office, Munich, Germany.)

However, the constructive geometry and topology of TFM are considerably complex in comparison with those of conventional RF motors, and no 2D symmetry exists. Because of the difficulty and complexity in manufacturing and thus high cost, TF motors today still stay in the R&D stage, leaving them a long way to go toward industrial production. Other disadvantages include torque ripples and normal force fluctuations [15.44].

TFM can be classified as surface PM TFM (SPM-TFM) and flux-concentrated TFM (FC-TFM). Compared with the SPM-TFM, the FC-TFM has a higher torque density but more difficulty in manufacturing. PMs in a SPM-TFM are magnetized in the direction perpendicular to the DIR. By contrast, PMs in an FC-TFM are of parallel magnetization [15.45].

An FC-TFM of 56 poles with distributed three full phases is shown in Figure 15.37. The rotor comprises a plurality of PMs of opposite magnetization direction alternately and coaxially arranged around a rotating axis. A rotor pole is arranged between every two PMs. The stator has three phases coaxially arranged around the rotating axis and separated by the distance elements. Several stator teeth face the rotor and the stator phases together with the distance elements.

Figure 15.38 depicts the single-stator winding and stator phase, where stator winding is formed as a multiturn saddle phase winding. As can be seen from the figure, the stator winding is formed as a closed loop and arranged between two teeth forming an angle θ.

The variations of torque components of the FC-TFM along the mechanical shift angle are depicted in Figure 15.39. The data are obtained from simulations using a FEM software. It shows the peak-to-peak cogging torque is about 0.09 Nm and the peak-to-peak reluctance torque is 0.4 Nm. Interaction torque is due to the interaction between the magnetic fields of PMs and armature current, which has a peak value of 0.55 Nm.

Gieras and Rozman [15.46] have presented a TF motor as the rim-driven thruster (RDT). As illustrated in Figure 15.40, a one-phase unit consists of a rotor assembly and a stator assembly. The stator assembly includes a ring-shaped winding and a stator core, which comprises a number of U-shaped cores. The rotor assembly includes a rotor core, spacer, and a plurality of PMs. As shown in the figure, PMs with the opposite magnetic polarity are disposed alternately along forward and aft surfaces of the spacer. PMs are configured to have the magnetic pole orientations extending

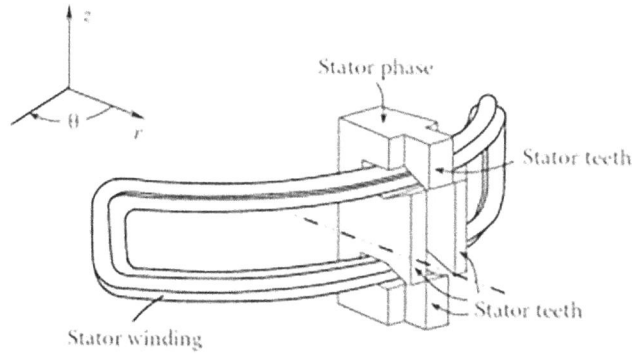

FIGURE 15.38   Single-stator winding and phase in an FC-TFM (European Patent 2,317,633) [15.45]. (Courtesy of the European Patent Office, Munich, Germany.)

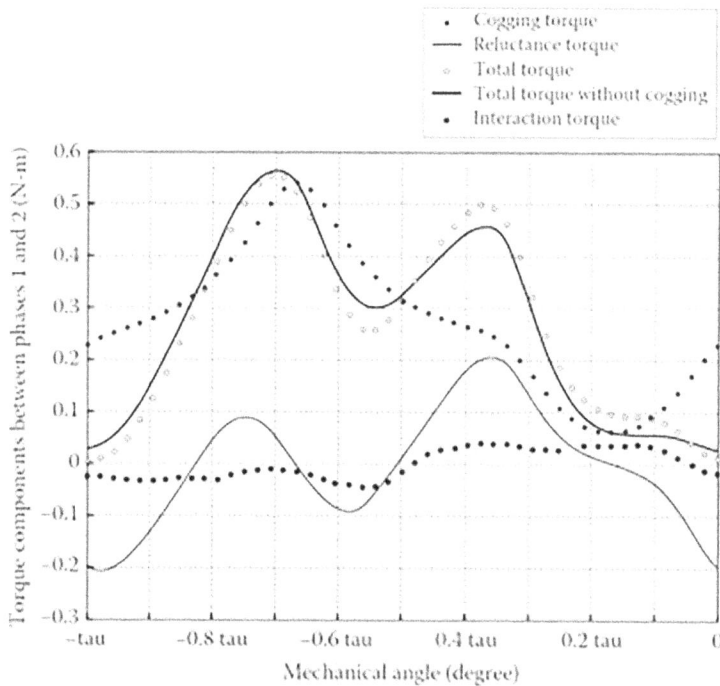

FIGURE 15.39   Torque component for the FC-TFM, obtained from simulations (European Patent 2,317,633) [15.45]. (Courtesy of the European Patent Office, Munich, Germany.)

radially, either inward or outward. The magnetic flux induced in U-shaped cores interacts with the pole orientations of PMs to generate the force vector F in the tangential direction of the rotor.

Figure 15.41 depicts the structure of an RDT, which contains three identical one-phase units to form a three-phase, TF motor. The RDT includes the housing, forward and aft fairing, stator assemblies, rotor assemblies, and propulsion assembly. The propellers are connected to the rotor. The rotor shaft is supported by two bearings at its ends. While electric current within stator windings causes magnetic flux to flow through the stator cores, the oppositely oriented magnetic poles of PMs cause magnetic flux to travel through rotor assemblies. The interaction between these magnetic fluxes applies a torque to the rim. The bearings permit the rim and rotor assembly to rotate smoothly about the centerline.

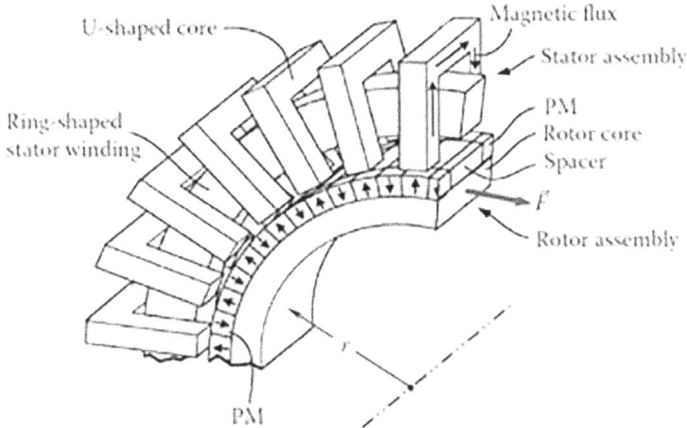

FIGURE 15.40 A partial perspective view of rotor and stator assembly of one-phase unit in a PM excited TF motor (U.S. Patent 8,299,669) [15.46]. (Courtesy of the U.S. Patent and Trademark Office, Alexandria, VA.)

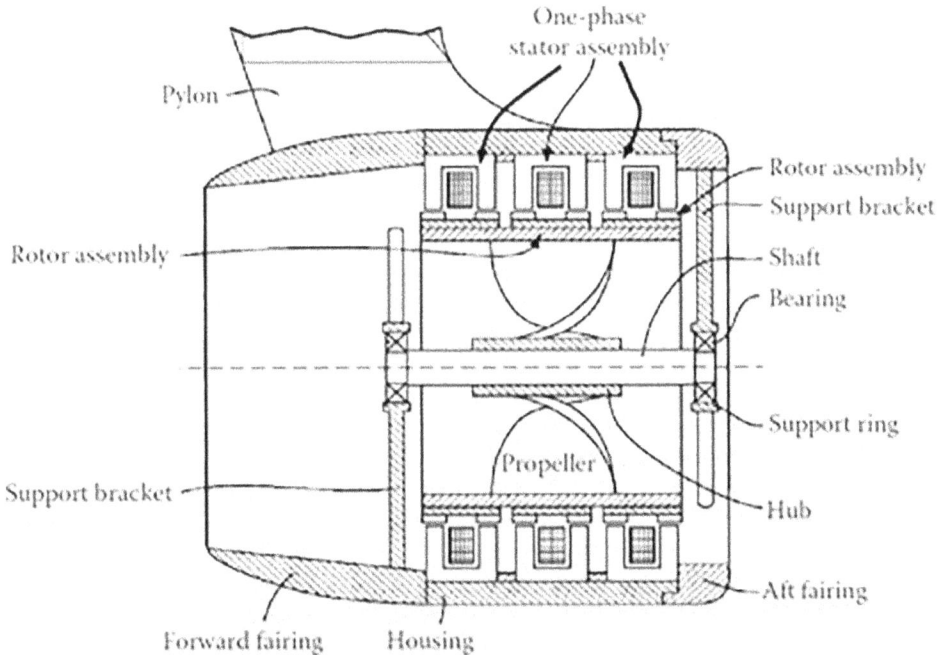

FIGURE 15.41 Structure of an RDT with a TF motor (U.S. Patent 8,299,669) [15.46]. (Courtesy of the U.S. Patent and Trademark Office, Alexandria, VA.)

The TF reluctance (TFR) machine is a variant of the TFM with a passive rotor [15.47]. As shown in Figure 15.42, a ring-shaped toroidal stator winding is surrounded by a number of U-shaped salient poles. The main differences compared with an SRM are as follows: (*a*) The TFR motor has a homopolar-type stator phase winding. (*b*) Its phase modules are placed consecutively in the axial direction. (*c*) It has the same number of salient poles on the rotor and stator. (*d*) Each phase of the motor is an independent module. In TFM technology, it is preferable to employ a plurality of TFMs arranged in tandem for reducing torque ripples, avoiding the start-up difficulties, and obtaining continuous rotation. Generally, the motor must have three or more three phases.

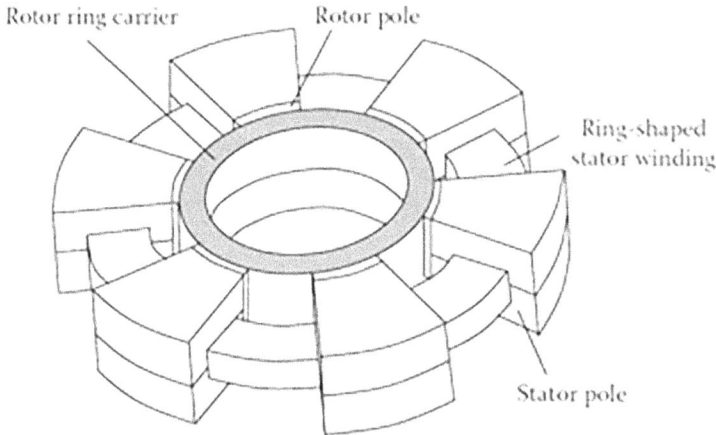

FIGURE 15.42 A 3D overview of a six-pole TFR machine.

Due to the superior performance characteristics, TFMs show unique merits for the direct-drive systems. They have wide applications, such as electrical vehicles [15.48], robots, aircrafts, power equipment, and smart artillery munitions.

15.7 RECONFIGURABLE PM MOTOR

The demand for high-efficiency, high-power density electric motors has stimulated many novel ideas in new types of motor design. One of the latest is to push the efficiency of electric motors to above superpremium level by employing reconfigurable rotor magnets.

The reconfigurable magnet technique can stabilize the output of the machine spinning even at variable speeds. This might be especially useful for some applications. For instance, a wind turbine built with this technique can generate constant output from variable wind speeds without the use of intermediate inverters [15.49].

A reconfigurable PM motor is shown in Figure 15.43 [15.50]. Each of the four PMs is attached to a planetary gear that engages with the central gear. All the four magnets remain rotationally aligned. The motor is reconfigurable from an asynchronous mode at the start-up into a more efficient

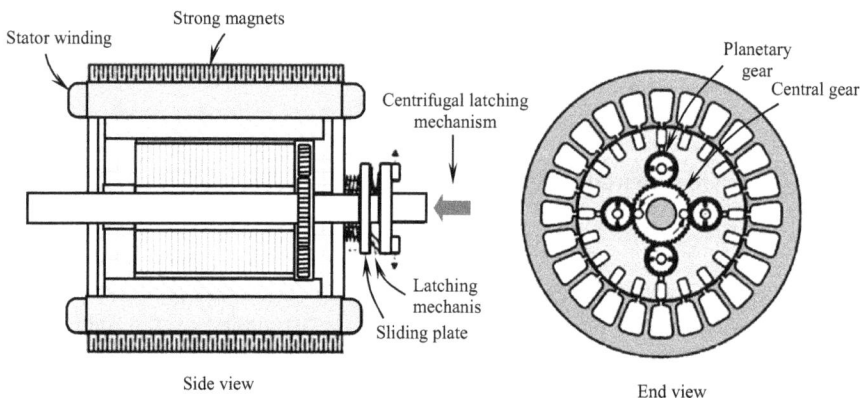

FIGURE 15.43 A cross-sectional view of reconfigurable magnet motor, revealing the details of the centrifugal clutch used to rotate the rotor magnets once the motor is reaching a certain speed (U.S. Patent 8,390,162) [15.50]. (Courtesy of the U.S. Patent and Trademark Office, Alexandria, VA.)

Pin seat

Pin

Weight

PM

PM

Spring disk

(a)

Rotating plate

N	S

S

S

N

S	N

Sliding plate

(b)

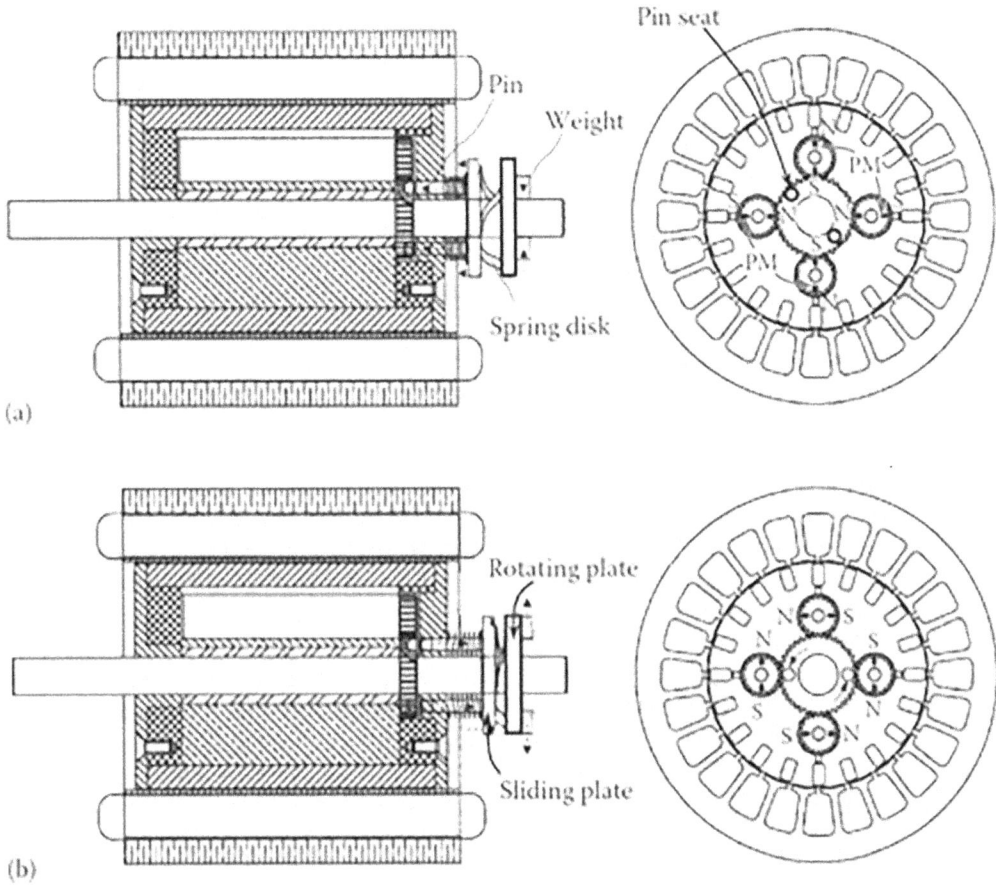

FIGURE 15.44 The side and end views of the reconfigurable electric motor with the centrifugal latching mechanism: (*a*) in the weak magnetic field position when the four PMs are held by the centrifugal latching mechanism and (*b*) in the strong magnetic field position when the four PMs are released by the centrifugal latching mechanism (U.S. Patent 8,390,162) [15.50]. (Courtesy of the U.S. Patent and Trademark Office, Alexandria, VA.)

synchronous mode thereafter. The motor includes a squirrel cage for IM operation at the start-up with the PMs positioned to produce a weak magnetic field not interfering with the start-up. During the start-up, the centrifugal latching mechanism retains the PMs in the weak magnetic field position. When the motor reaches a sufficient speed, the PMs are released by the centrifugal mechanism and rotate to produce a strong magnetic field in harmony with the rotating stator magnetic field for the efficient synchronous operation.

A more detailed description of the transition between a weak magnetic field and a strong magnetic field applied to the motor is presented in Figure 15.44. The centrifugal latching mechanism includes weights, rotating plate, spring disk, sliding plate, pins, and pin seats. The weights and spring disk are designed so that when the motor is stationary, the weights remain in the inward position to the shaft and the magnets are at the weak magnetic field position. Meanwhile, the pins engage the pin seats in the central gear when the motor is at rest and the centrifugal latching mechanism holds the PMs in the weak magnetic field position. At an appropriate rotating speed, the weights move outward due to the centrifugal force, causing the spring disk to snap axially from the extended position as in Figure 15.44*a* to the retraced position as in Figure 15.44*b*. The centrifugal latching mechanism thus pulls the pins from the pin seats releasing the PMs to the strong magnetic field position. As the motor stops, the PMs are magnetically urged back to the weak field position again.

15.8 VARIABLE RELUCTANCE MOTOR

VRMs have been widely used for high-torque applications. Their inherent fault-tolerant features and other advantages have made them a prime candidate for reliability-premium applications. From the standpoint of energy conversion, VRM is unique compared to other types of motors.

According to the relationship of motion plane and flux plane, VRMs can be categorized as transverse-flux machine (TFM) and the Vernier machine (VM). As addressed previously, due to the complexity of the constructive geometry, the design and fabrication of the TFM are considerably difficult. By contrast, VMs can adopt a simple toothed-pole-stator and PM-rotor configuration. Each stator tooth is split into small teeth at the end (*i.e.*, flux-modulation poles). Thus, a small movement of the PM rotor can cause a large movement of flux linkage in the stator armature winding [15.51].

Figure 15.45 shows an outer-rotor, flux-controllable Vernier PM motor, designed for the in-wheel drive of electric vehicles. In this design, the inherent outer-rotor topology makes the motor directly connected to the wheel rim and eliminates the mechanical transmission. The inner stator accommodates two sets of windings, that is, the three-phase armature windings and the DC field windings for performing the flux weakening control at high-speed operation. There are 24 salient poles, setting up in the inner stator for performing the flux modulation. Twenty-two pole-pair PMs are mounted on the outer rotor. The designed key uses the Vernier structure for obtaining the speed reduction and achieving the high output torque at low-speed operation [15.52]. This direct drive, in-wheel motor can smoothly operate within the speed range of 0 – 1,000 rpm at different operation modes for electric vehicles.

According to excitation patterns, doubly salient motors can be classified into three types: PM, fielding winding, and hybrid. The VRMs deploying doubly salient structures (*i.e.*, teeth on both the stator and rotor) have been developed for several decades with well-explored design features, torque production, and control characteristics. In a number of years, doubly salient PM (DSPM) motors have gained renewed interest for some new applications, particularly in-wheel motors for electric vehicle drives.

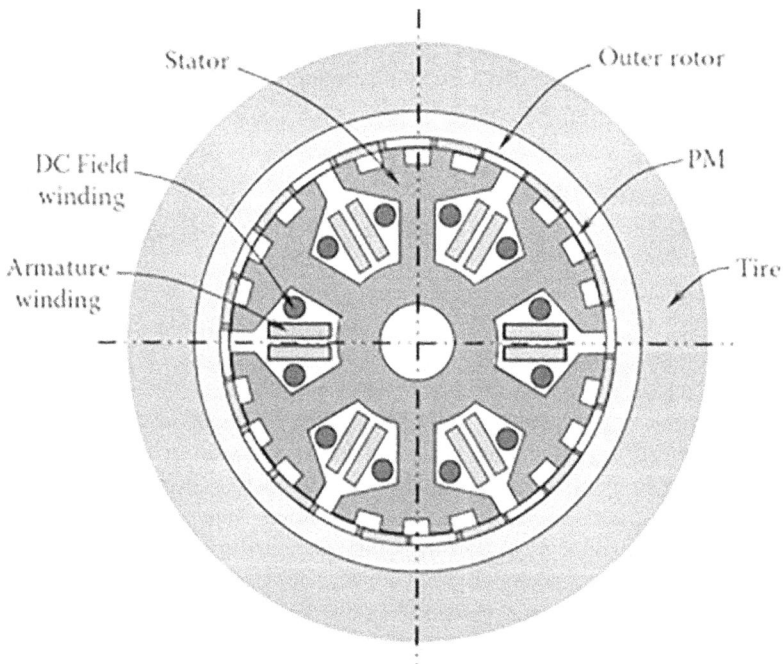

FIGURE 15.45 Configuration of the Vernier PM motor used as in-wheel drive.

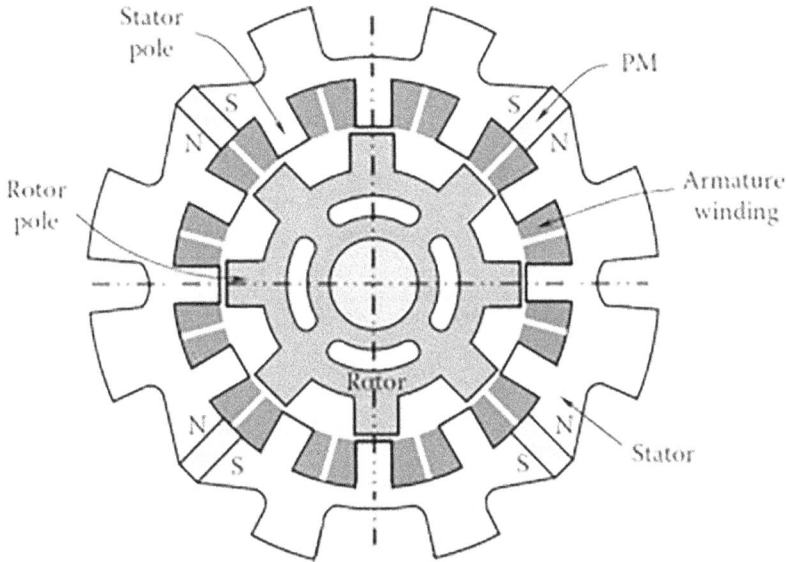

FIGURE 15.46 The cross-sectional view of a 12/8-pole DSPM.

Figure 15.46 illustrates the structure of a 12/8-pole (the stator pole number $p_s = 12$ and the rotor pole number $p_r = 8$) DSPM motor. Armature windings are situated in the stator. Four pieces of PM are located radially at the four corners of the stator. This indicates that the leakage flux outside the stator circumference (which is generally neglected in conventional PM motors) becomes significant. Hence, to take the leakage flux into account, the design domain must be extended from the stator circumference to the surrounding space. Since the rotor consists of only silicon steel laminations, it gives the motor the capability to work at very high speeds.

To minimize the switch frequency and hence the iron losses in poles and yokes, the number of rotor poles p_r should be selected as small as possible but must be equal to or larger than three for the capability of self-starting. As a result, the rotor poles are usually less than the stator poles. In practice, p_s/p_s of 6/4, 8/6, and 12/8 are possible configurations of DSPM motors. Generally, 12/8 pole motors have less iron losses and magnetic potential drop than 6/4 pole machines [15.53].

From the investigation of the DSPM motor by Cheng *et al.* [15.54], the following may be concluded: (*a*) Although the DSPM motor has salient poles in the stator and rotor, the PM torque significantly dominates the reluctance torque, exhibiting low cogging torque. (*b*) The airgap flux of the DSPM motor is mainly contributed by PMs, whereas the armature current contributes to the change of the flux distribution. (*c*) Because most of the armature flux loops through the adjacent stator poles, instead of PMs, the DSPM motor is less sensitive to demagnetization than other PM motors. (*d*) The inductance of the DSPM motor depends not only on the rotor position but also on the strengthening/weakening action of the armature field to the PM field. (*e*) The leakage flux outside the stator circumference of the DSPM motor may lead to a reduction in the effective flux of about 3%.

15.9 PM MEMORY MOTOR

Ostovic first proposed the innovative concepts of PM memory motors, which can be built either as variable flux memory motors (VFMMs) [15.55] or as pole-changing memory motors (PCMMs) [15.56]. Memory motors combine the flux controllability of PM motors and the employment of the flux concentration principle that enables creating airgap flux densities for high-efficiency machines.

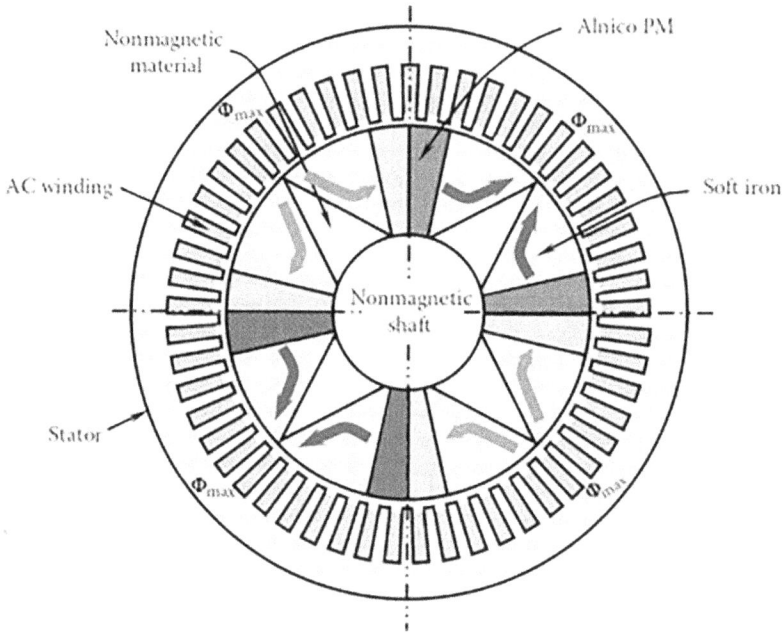

FIGURE 15.47   The cross-sectional view of a fully magnetized, PM VFMM.

15.9.1   VARIABLE FLUX PM MEMORY MOTOR

Unlike the conventional PM motors, the type of VFMM has the distinct ability to change the intensity of magnetization by short pulses of stator current and to memorize the flux density level in the rotor PMs. Additional advantages include high efficiency, high torque-to-weight ratio, large starting torque, full controllability of back EMF, no risk of demagnetization by the load current, zero reactive power demand on stator terminals, simple structure, and operation in a wide speed range without excessive rotor PR losses. In fact, VFMM is the only synchronous machine with rotor excitation that has a true speed capability without sacrificing the other machine properties. However, compared with other types of PM motors, this type of motor may have relatively low overload ability and reliability [15.57].

As shown in Figure 15.47, a four-pole VFMM consists of trapezoidal alnico PMs sandwiched by soft irons. Both the PMs and soft irons are sandwiched by triangle-shaped plates made of a nonmagnetic material. All these components are mechanically fixed to a nonmagnetic shaft. PMs can be online magnetized or demagnetized with various magnetization levels. Tangentially magnetized PMs with red and blue represent N poles and S poles, respectively. The magnetic flux Φ_{max} is driven by the PMs to transit across the airgap into the stator, continuously go along the circumferential direction in the stator core, and finally come back through the airgap into the rotor for completing the loop.

VFMMs combine the advantages of a wound rotor synchronous motor (simple control of induced voltage) and a PM synchronous motor (no excitation losses), resulting in a unique synchronous machine for numerous applications.

15.9.2   POLE-CHANGING PM MEMORY MOTOR

Conventional PM machines have a constant number of poles, because their magnets are incorporated into rotor body in a manner that does not allow any change of machine topology. By comparison, a pole-changing PM memory motor has the capability to change the number of poles. This

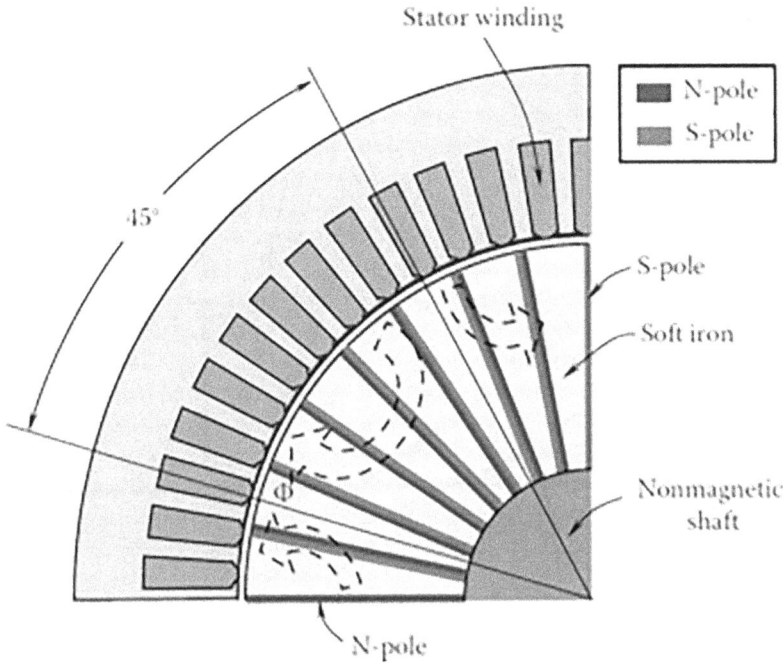

FIGURE 15.48 The cross-sectional view of an eight-pole PM magnetized PCMM.

type of motor has several advantages such as high efficiency, high output power, and less thermal dissipation issue due to the lack of rotor copper losses. These make PCMMs attractive for many motor applications, among which the fan and pump applications are the most pronounced.

Figure 15.48 depicts the operational principle of an eight-pole magnetized PCMM. Due to the geometry symmetry, only a quarter of the motor is modeled. The motor contains 32 magnets positioned radially in the stator. Each pole takes a 45° sector, which contains four PMs (32/8=4). All the PMs are magnetized in the tangential direction. The rotor wreath is built of PMs along with soft iron segments and is mechanically fixed to a nonmagnetic shaft.

After the stator windings are reconnected into a six-pole (a 60° sector per pole) configuration, a short pulse of stator current changes the rotor from eight-pole to six-pole magnetization (Figure 15.49). Since the number of magnets per pole is no longer an integer (32/6=5.33), some magnets may remain demagnetized, as shown in the figure. As a matter of fact, both flux per pole and stator current necessary to remagnetize the rotor magnets vary as a function of the number of magnets per pole. Therefore, by changing the pole number, these two important motor parameters are adjusted for optimizing motor performance.

15.9.3 DOUBLY SALIENT MEMORY MOTOR

DSPM memory motors are the combination of memory motors, namely, the online tunable flux-mnemonic PMs and DSPM motors, thus achieving effective airgap flux control. Employing the outer-rotor and double-layer-stator topology, the motor takes advantage of compact structure, low armature reaction, and direct drive capability. This type of motor can offer the unique features of pole dropping and pole reversing [15.58].

The topology of the doubly salient memory motor is depicted in Figure 15.50. It adopts an outer-rotor, an outer-layer-stator, and an inner-layer-stator structure. The motor has five-phase and 30-pole/24-pole doubly salient structure on the stator/rotor. The outer rotor only consists of lamination iron and thus offers excellent mechanical robustness. This arrangement is especially suitable

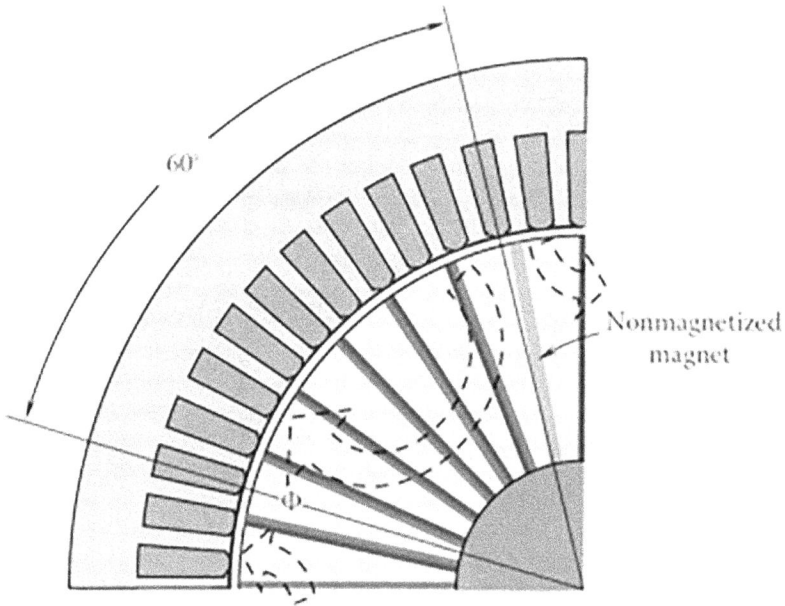

FIGURE 15.49 The cross-sectional view of a six-pole PM magnetized PCMM.

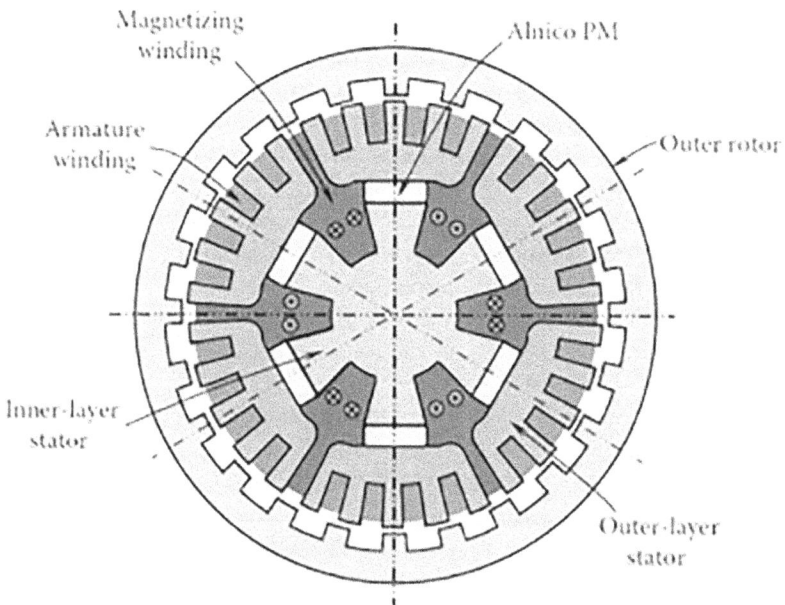

FIGURE 15.50 Topology of DSPM memory motor.

for the direct drive applications such as electric vehicle in-wheel motors and wind turbines. The AC armature windings and DC magnetizing windings are located at the outer-layer stator and inner-layer stator, respectively. This effectively uses the space of the stator yoke. Alnico PMs are fixed between the inner-layer stator and outer-layer stator. The use of alnico PMs can achieve a controllable airgap flux by applying temporary current pulses to the magnetizing winding to online tune the magnetization level of alnico PMs.

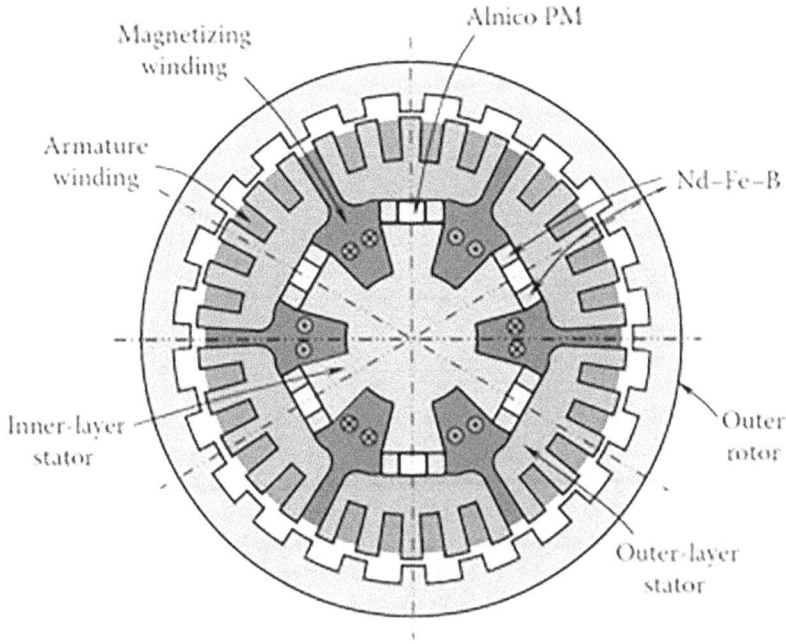

FIGURE 15.51 DSPM memory motor with dual magnet.

Compared with other types of PMs, alnico magnets exhibit good temperature stability and resistance to corrosion, high Curie temperature (approximately $840°C$), but a low coercivity, which is prone to demagnetization due to shock. Furthermore, alnico magnets deliver smaller energy products than rareearth PMs, thus degrading the machine power density. To address these problems, Li *et al.* [15.59] have presented a new flux-mnemonic dual-magnet machine that incorporates both the Nd–Fe–B and alnico PMs for providing hybrid excitation (Figure 15.51). In such a way, the proposed machine not only retains the feature of online tunable airgap flux control but also improves the machine power density.

15.10 ADJUSTABLE AND CONTROLLABLE AXIAL ROTOR/ STATOR ALIGNMENT MOTOR

In a conventional RF motor, the relative position between the rotor and stator is fixed both axially and radially. In this novel PM motor, the axial alignment between a rotor and a stator is adjustable and controllable using a mechanical technique. As the PM motor is offset axially to provide axial misalignment between the rotor PM poles and the stator, the effective magnet pole strength or magnetic flux to the stator is significantly reduced, resulting in an increase in speed and a decrease in motor torque [15.60].

As shown in Figure 15.52, the stator is incorporated into the motor housing structure. PMs are mounted on the outer surface of the rotor. The rotor is connected to a hydraulic or pneumatic actuator through a pivot arm at its end. A rotor sleeve is coupled to the rotor with a plurality of parallel grooves formed on the inner surface. The grooves have a semicircular cross-sectional shape and are configured to receive a plurality of ball bearings. Correspondingly, the motor shaft has also a plurality of parallel grooves formed at the outer surface. The ball bearings are contained within the channel formed by both the rotor sleeve and shaft grooves. Thus, the rotor is allowed to slide axially via the ball bearings when the hydraulic or pneumatic actuator is activated.

The rotor in Figure 15.52 is about 25% disengaged with the stator. This amount of misalignment will produce an increase in speed of approximately 150% of the rated speed.

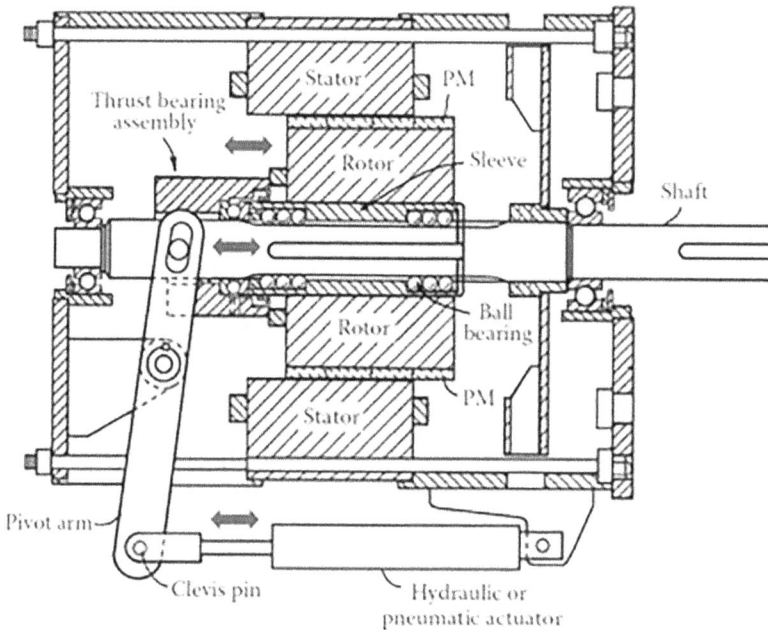

FIGURE 15.52 Structure of PM motor with adjustable axial rotor/stator alignment (U.S. Patent 6,555,941) [15.60]. (Courtesy of the U.S. Patent and Trademark Office, Alexandria, VA.)

15.11 PIEZOELECTRIC MOTOR

Piezoelectric effect refers to the phenomenon that when a certain nonconducting material, such as quartz crystal (SiO_2) and ceramics, is subject to mechanical stress (*e.g.*, pressure or vibration), the material generates an electric potential across it. Conversely, when an electric field is applied to the material, it undergoes mechanical distortion or vibration.

Piezoelectric motors operate based on the principle of the piezoelectric effect. In a piezoelectric motor, the piezoelectric material produces high-frequency (inaudible to the human ear) acoustic vibrations on a nanometer scale to create a linear or rotary motion. Piezoelectric motors have numerous technical advantages over conventional electromagnetic motors, including the following [15.61]:

- Piezoelectric motors maintain high torque at low speed with a compact motor size. Hence, they have higher torque densities compared with conventional electromagnetic motors for the same power rating.
- This type of motor does not create electromagnetic fields, nor is affected by external electromagnetic fields. This feature is extremely important for motors used within strong magnetic field environments.
- Because of the direct drive mechanism, piezoelectric motors eliminate the need for supplementary transmissions or gear trains found in conventional electromagnetic motors. This avoids the usual backlash effects in conventional motors and thus increases the motor positioning accuracy.
- A compelling advantage of piezoelectric motors is their intrinsic steady-state autolocking capability as the driving mechanics use friction between the vibrating stator and rotor.
- With less heat generated in piezoelectric motors, cooling is no longer a critical issue in motor design.
- This type of motor has a low rotor inertia and thus provides rapid start and stop characteristics.

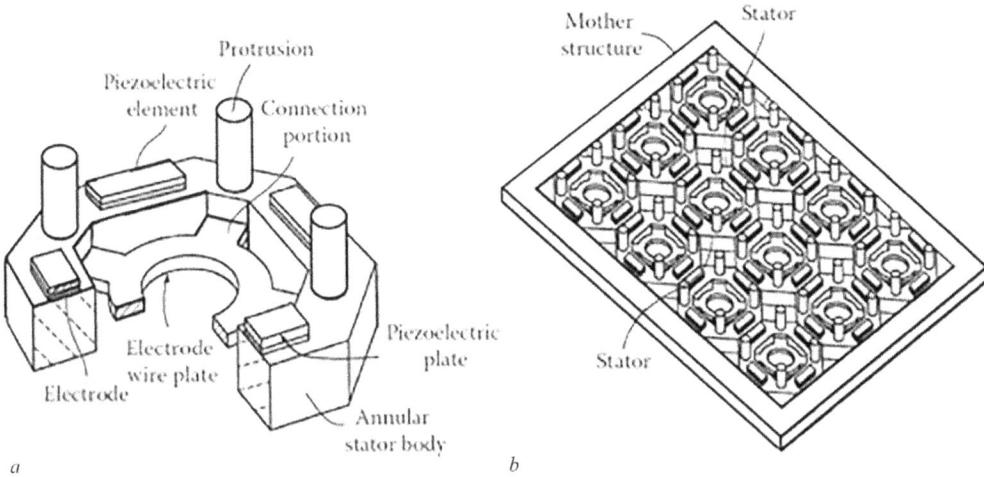

FIGURE 15.53 The structure of a piezoelectric motor: (*a*) a partially cross-sectional perspective view of the stator used in a piezoelectric motor and (*b*) the mother structure prepared obtaining the stators (U.S. Patent 8,330,326) [15.62]. (Courtesy of the U.S. Patent and Trademark Office, Alexandria, VA.)

- With fewer mechanical components, their construction is relatively simple. Thus, piezoelectric motors are well suited for miniaturization, and their overall efficiency is relatively insensitive to size.
- In the power-off mode, the position can be maintained due to the frictional force between the contact surfaces.
- Piezoelectric motors have quiet operation.

Piezoelectric motors are generally used in low-torque and high-precision micropositioning applications, such as medical devices, robotic positioning, pharmaceuticals handling, and pick-and-place assembly.

The structure of a piezoelectric motor is presented in Figure 15.53 [15.62]. A stator includes a substantially annular stator body on which a number of piezoelectric elements are mounted. Each piezoelectric element consists of a piezoelectric plate made of ceramic; electrodes formed on the surfaces of the piezoelectric plates. A number of protrusion standings between adjacent pairs of the piezoelectric elements are in contact with the rotor for actuating the rotor (Figure 15.53*a*). Figure 15.53*b* shows the mother structure, containing 12 identical stators, each of them equipped with a plurality of piezoelectric elements. The electrodes and wiring lines may be formed by coating and backing an electroconductive paste or be formed as part of the cofiring process for making the mother structure. Then, the mother structure is cut in the thickness direction so that each individual stator has the same structure as the stator shown in Figure 15.53*a*. In such a way, the stator can be assembled with a rotor for forming a piezoelectric motor.

Figure 15.54 is a schematic view illustrating the driving principle of the piezoelectric motor. Two pairs of piezoelectric elements facing each other, that is, A+ and A− and B+ and B−, are polarized in opposite thickness directions. A+ and A− constitute an A-phase, while B+ and B− constitute a B-phase. The symbols A+ and A− indicate that the piezoelectric bodies are polarized in opposite thickness directions. So does the B-phase drive.

Imaginary lines connect the center point O and midpoints of the short sides of the piezoelectric elements, with the length l. The angle between the imaginary lines E1 and E2 is about 60°. This indicates that the length of the piezoelectric element equals the length l. Thus, three standing waves are excited and combined to generate three progressive waves. When the central angle

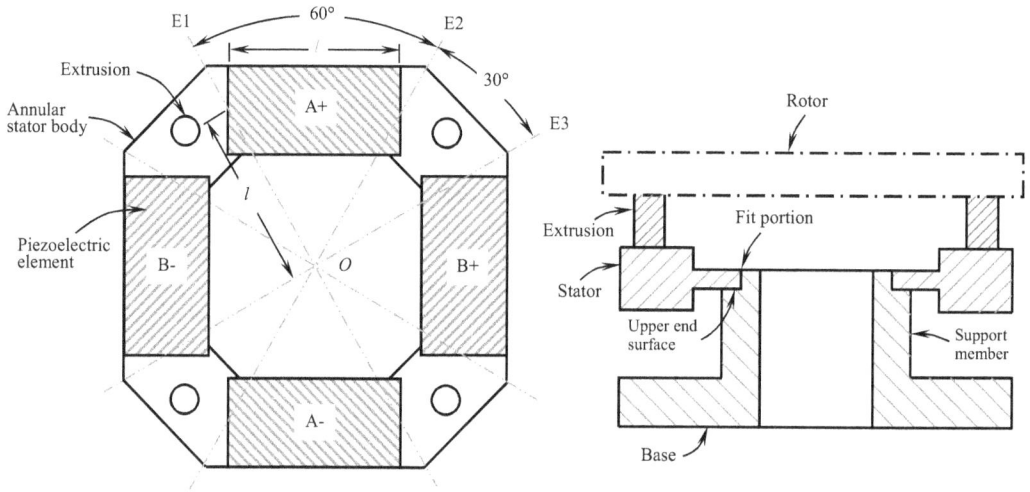

FIGURE 15.54 The driving principle of the piezoelectric motor (U.S. Patent 8,330,326) [15.62]. (Courtesy of the U.S. Patent and Trademark Office, Alexandria, VA.)

corresponding to the wavelength of the three progressive waves is λ_θ, the length l corresponds to the central angle of $\lambda_\theta/2$.

When the piezoelectric motor is driven and a progressive wave is generated in the substantially annular stator body made of an elastic member, the ends of the protrusions perform elliptical motion. Therefore, the rotor, which is in close contact with the protrusions, is rotated.

15.12   ADVANCED ELECTRIC MACHINES FOR RENEWABLE ENERGY

As pointed by Chau *et al.* [15.51], the latest development of renewable energy machines is focused on three directions: the PM machine aiming to achieve high reliability and high robustness, the direct drive PM machines aiming to directly harness the renewable energy without any transmission mechanism, and the magnetless machines aiming to avoid using expensive rareearth PMs. Among them, the magnetless machines are the most attractive to the energy industry.

In the last several years, the global price of rareearth magnets has risen significantly, and consequently, it greatly increases the manufacturing costs of PM motors. This price soar has motivated motor engineers and material scientists to develop new types of electric machines without using magnets and to discover alternatives to rareearth magnets.

The scientists at a US national laboratory have developed a PM-less synchronous reluctance motor [15.63]. As presented in Figure 15.55, the PM-less motor consists of a stator that generates a magnetic revolving field when sourced by an AC and an uncluttered rotor that is disposed within the magnetic revolving field and spaced apart from the stator to form an airgap. The rotor includes a plurality of rotor pole stacks having an inner periphery biased by single polarity of an N-pole field and an S-pole field, respectively. The outer periphery of each of the rotor pole stacks is biased by an alternating polarity.

Without PMs, some of these brushless synchronous machines may have a reduced size, lower weight, and less core losses. In hybrid vehicle applications, there may be little or no core losses when the system runs free without field excitations. Without a magnetic resistance, fuel efficiency may increase. In these systems, the stator fields can be cut off to enhance safety, fields can be boosted to increase or reach peak acceleration power in short time periods, power factors may be optimized (*e.g.*, lowering the loading in inverter applications), and efficiency maps may increase due to the adjustable fields.

Figure 15.56 illustrates in more detail the eight-pole rotor stacks that are used to form an uncluttered rotor. The stack is made of laminations. The number of magnetically isolated channels is

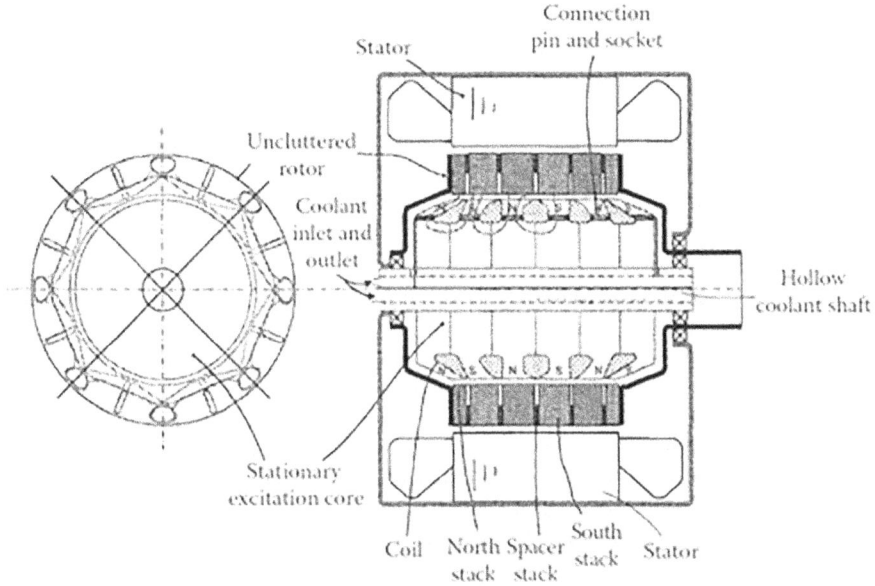

FIGURE 15.55   Cross-sectional view of the PM-less electric motor (U.S. Patent 8,264,120) [15.63]. (Courtesy of the U.S. Patent and Trademark Office, Alexandria, VA.)

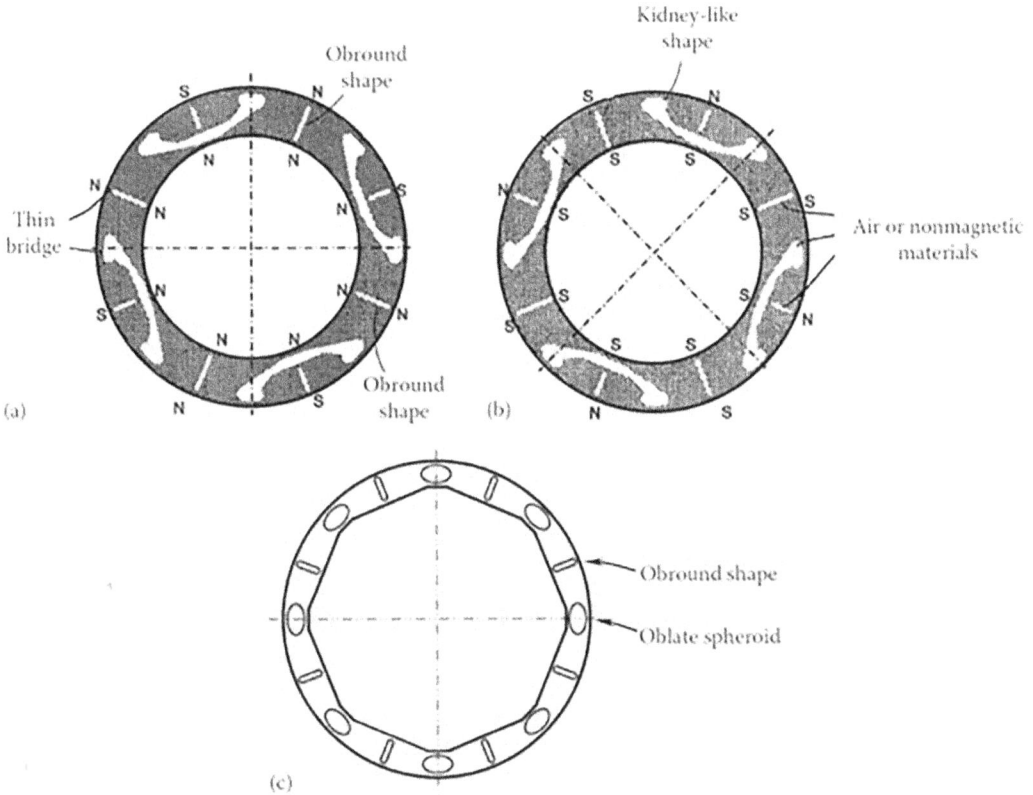

FIGURE 15.56   Eight-pole rotor stacks: (*a*) pole stack of north polarity, (*b*) pole stack of south polarity, and (*c*) spacer stack (U.S. Patent 8,264,120) [15.63]. (Courtesy of the U.S. Patent and Trademark Office, Alexandria, VA.)

equal to the number of poles that are equally spaced apart on the outer periphery. While the polarity along the outer peripheral surface is arranged alternately, the inner circumference of each of the north and south polarity pole stacks maintains a common polarity (Figures 15.56a and b). The inner polygonal surfaces of the spacer stacks maintain a neutral polarity (Figure 15.56c). The magnetically isolated channels in the spacer stacks are equally spaced about its annulus.

15.13 MICROMOTOR, NANOMOTOR, AND MOLECULAR MOTOR

The rapid development of microtechnology and nanotechnology has triggered vast interest and generated abundant opportunities for the development of micromotors and nanomotors.

15.13.1 MICROMOTOR

Micromotors and microgenerators are power microelectromechanical systems (MEMS) for energy conversion between mechanical energy and electrical energy. With the continuous improvement of the MEMS technology, various types of compact, lightweight, and portable power conversion devices have been successfully designed and fabricated.

Nagle et al. [15.64] at MIT have presented the analysis, design, fabrication, and testing of a planar electrostatic induction micromotor. The structure and operating principle of this micromotor are illustrated in Figure 15.57. This is a six-phase motor with 131 pole-pairs distributed on a stator. The rotor and stator are 4 mm in diameter and separated by a gap of 3 μm. With a 90 V stator excitation and applied a 300 kHz slip frequency, the motor produces a torque of 2 μNm. The stator contains 768 radial electrodes, separated by a 4 μm space along their radial length. The electrodes are supported by an insulation layer. The rotor comprises a 0.5 μm sheet of moderately doped polysilicon deposited on a 10 μm thick oxide insulation layer that is deposited on a 400 μm thick silicon wafer. The rotor is tethered above the stator by eight silicon springs etched from the silicon wafer. During motor operation, external electronics excite the stator electrodes to produce a potential wave that travels circumferentially with a speed exceeding that of the rotor. The corresponding distribution of charges on the stator electrodes induces image charges on the rotor surface across the airgap.

This type of micromotor has the potential to perform Watt-level electrical-mechanical power conversion for applications ranging from portable electric suppliers to miniature pumps or blowers.

Ghalichechian et al. [15.65] reported the design, fabrication, and characterization of a six-phase, bottom-drive, variable-capacitance micromotor rotary micromotor with a robust mechanical support provided by the microball bearings. Figure 15.58 shows the simplified 3D schematic of

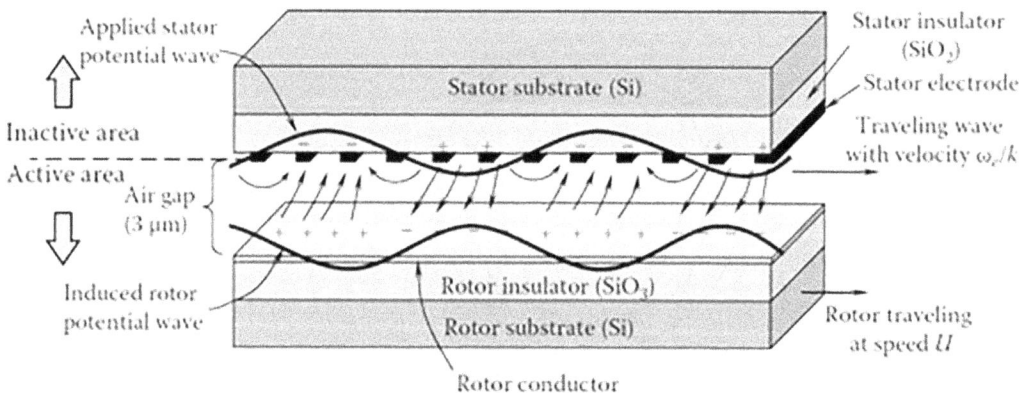

FIGURE 15.57 Operating principle of induction micromotor. Only the active area affects torque production because the electrode potential is enforced on the dividing plane.

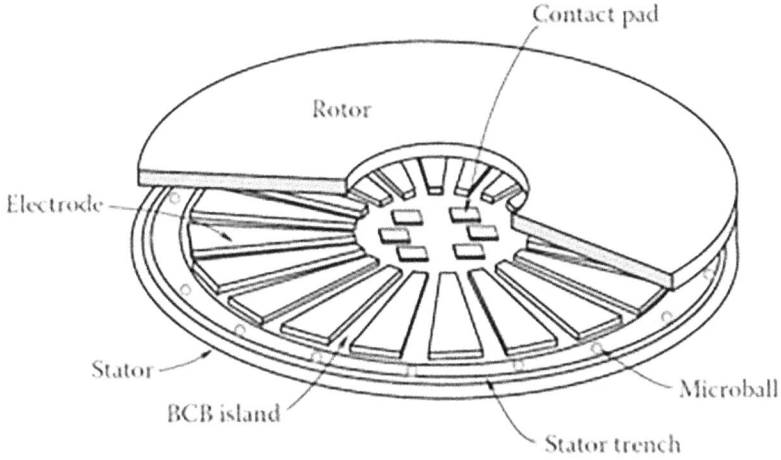

FIGURE 15.58 Simplified 3D schematic of the rotary micromotor.

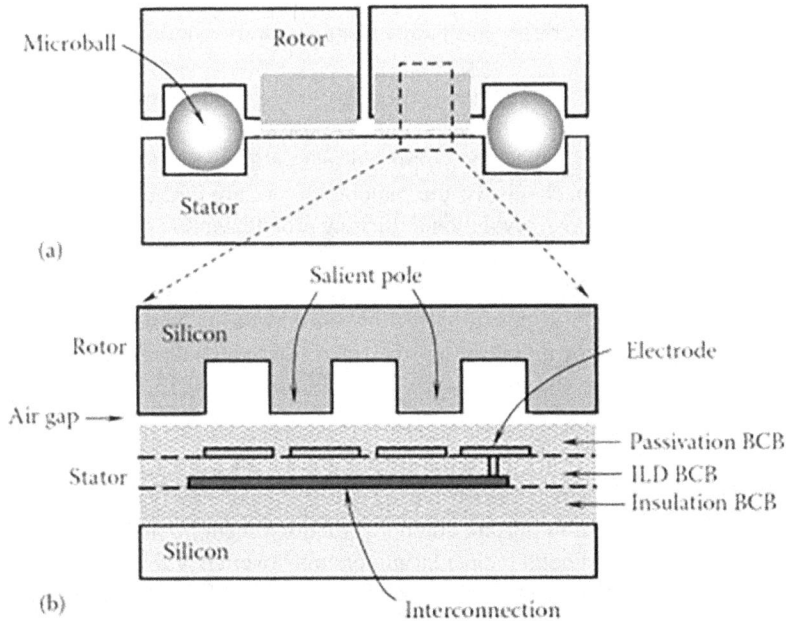

FIGURE 15.59 Schematic of (*a*) radial cross section of mechanical (bearing) and electric components of the micromotor and (*b*) azimuthal cross section of the stator and rotor active pars.

the micromotor, which is composed of three components: rotor (a silicon disc with a diameter of $\phi14$ mm), stator (a silicon disc with the identical properties as the rotor), and stainless steel microballs (with a diameter of $\phi284.5\pm0.25\,\mu$m). Both the stator and the rotor are fabricated separately on silicon substrates and assembled with the microballs. Electrodes attach on the stator and salient poles are made on the rotor (Figure 15.59). When power is applied to an electrode, image charges are induced on an adjacent salient pole, resulting in tangential and normal forces. The tangential force is the propelling force of the rotor that aligns a rotor pole to an adjacent active electrode. The normal force assists to hold the rotor on the stator. Continuous motion of the rotor is possible by sequential excitation of the electrodes with positive square pulses in a three-phase configuration.

To analyze the effect of the airgap size on the motor torque, three different airgaps (5, 10, and 15 μm) were used in the design. As expected, the motor torque increases by more than a factor of two when the airgap is decreased from 10 to 5 μm. In contrast, the torque is reduced when the airgap is increased from 10 to 15 μm. However, fabrication of a machine with a gap of 15 μm is more feasible than 5 μm. The experimental results have shown that with the estimated airgap of 10–13 μm, the torque is indirectly measured to be −5.62±0.5 μNm at ±150V excitation.

15.13.2 NANOMOTOR

Richard Feynman, the winner of Nobel Prize in Physics 1965, introduced the concept of nanotechnology in his famous speech *"There's plenty of room at the bottom"* at an American Physical Society meeting on December 29, 1959 [15.66]. Late in 1984, in his visionary lecture, he addressed the possibility of making tiny movable machinery [15.67]. He convinced that it was possible to build machines with dimensions on the nanometer scale and discussed how the scale differences affect the function of certain aspects of technology.

The rapid development of nanotechnology has shed new light on inventions in the motion world. The most famous example is the invention of nanomotors and molecular motors.

A nanomotor is a nanoscale device capable of converting energy (*e.g.*, electrical, chemical, thermal, or other energy forms) into mechanical energy and further into rotational, revolutional, or linear motion. According to the operating mechanisms, nanomotors can be categorized as biological nanometers or nanotube motors.

Biological nanomotors, that utilize chemical energy to generate physical movement of molecules, exhibit a diversity of complex structures and have distinct roles in cellular functions such as packaging DNA, contracting muscles, and helping direct cellular components to proper destinations. Based on their motion mechanisms, the nanobiomotors are categorized into three classes: rotational (rotating on its own axis), revolutional (turning around another object), and linear [15.68]. In fact, the advent of nanobiomotors has revolutionized the traditional concept of electric motors.

Nanotube motors are more likely the conventional electric motors that convert electric energy into mechanical motion. In 2005, scientists at the Lawrence Berkeley National Laboratory unveiled a nanotube motor that operates by moving atoms between two molten droplets of metal. Completely contained in a carbon nanotube, the motor is less than 200 nm across [15.69].

15.13.3 MOLECULAR MOTOR

Molecular machines convert chemical, electrical, or other forms of energy into mechanical work for unidirectional movement. As an important component among them, molecular motors refer to the motors in molecular scales. Although molecular motors may overlap with nanomotors in their size, molecular motors often refer to the motors with a single molecule. They are of great interest not only for their basic scientific richness, but also for the potential to revolutionize critical technologies. In fact, molecular motors exist in nature, for example, in the form of myosins. Myosins are motor proteins that play an important role in living organisms in the contraction of muscles and the transport of other molecules between cells [15.70].

Over the last several decades, a wide variety of molecular machines have been synthesized by biologists, chemists, physicists, and engineers. Molecular motors are central to any molecular machine, playing a prominent role in the nanotechnological revolution of the twenty-first century [15.71].

A research team of Tufts University in 2011 [15.72] had developed a motor with only 1 nm, made from a single molecule of butyl methyl sulfide on a copper surface. The molecular motor was powered by electricity from a low-temperature scanning tunneling microscope or heat energy. As demonstrated in Figure 15.60, the molecule has a sulfur base. When placed on a conductive slab of copper, it becomes anchored to the surface. The sulfur-containing molecule has carbon and hydrogen atoms radiating off to form what looks like two arms (gray). These carbon chains were free to rotate around the central sulfur-copper bond.

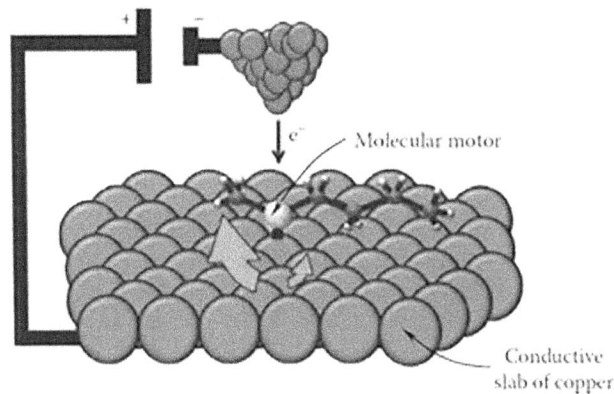

Molecular motor

Conductive
slab of copper

FIGURE 15.60 The working principle of a molecular motor. (Courtesy of Sykes Laboratory, Tufts University, Medford, MA.)

The spinning speed of the single-molecular motor is determined by temperature. At a temperature of $100\,K$ ($-173°C$), the molecular motor spins more than a million revolutions per second. The researchers found that reducing the temperature of the molecule to $5K$ ($-268°C$) enabled them to precisely control the rotating speed and direction of the molecular motor. In this way, it could lower the molecular motor speed to about 50 rpm. This type of motor might be used in sensing and medical test devices and nanoelectromechanical systems. However, there is still a long way to go for practical applications of single-molecular electric motors.

In October 2016, The Royal Swedish Academy of Sciences announced to award Jean-Pierre Sauvage, Sir J. Fraser Stoddart, and Bernard L. Feringa the Nobel Prize in Chemistry "for the design and synthesis of molecular machines" [15.73].

Each scientist contributed to the discovery in a different way. The first breakthrough came from Sauvage in 1983. The French chemist succeeded in linking two molecules in a chain, fulfilling the requirement that a machine needs several parts that can move relative to each other. Stoddart's breakthrough came in 1991, when he created a ring of molecules that moved along an axle in a controlled manner when heat was added. Feringa was the first person to develop a molecular rotary motor. In 1999, Feringa's research group used molecular motors to spin a 28 μm long glass cylinder, which is 10,000 times bigger than the molecular motors [15.74]. Though the first motor wasn't fast, Feringa's research group had optimized it. In 2014, the motor can rotate at a speed of 12 million revolutions per second [15.75]. The effort for the chemically driven motor was made by Boston College Professor T. Ross Kelly with his co-workers; their work was published in 1999 in the same issue of Nature as Feringa [15.76].

Molecular motors have some potential applications in nanomedicine, intelligent nanosystems, molecular sensors, and others. One example in engineering applications is molecular gyroscopes. According to the conservation of angular momentum, at extremely high rotating speeds, the axis of rotation of a molecular motor is free to assume any orientation by itself and thus is unaffected by its frame movement. Therefore, it is possible to design molecular gyroscopes based on molecular rotation motors. This is an important milestone in the ultra-miniaturization of gyroscopes.

More recently, the smallest motor in the world has been developed by a research team at Empa and EPFL [15.77]. This molecular motor consists of only 16 atoms and rotates reliably in one direction. It measures less than 1 nm or about 100,000 times smaller than the diameter of a human hair. A rotor is a single acetylene (C_2H_2) molecule consisting of four atoms. A chiral atomic cluster consisting of six palladium and six gallium atoms in a triangle structure acts as a stator. The rotor rotates continuously on the surface of the stator and can take up six different positions. By breaking spatial inversion symmetry, the rotor defines the unique sense of rotation. In fact, for a motor to actually do useful work, it is essential to make the stator rotate in only one direction.

This acetylene-on-PdGa motor pushes molecular machines to their extreme limits, not just in terms of size, but also regarding structural precision, degree of directionality, and crossover from classical motion to quantum tunneling. This ultrasmall motor thus opens the possibility to study the processes and reasons for energy dissipation in quantum tunneling processes and allows ultimately energy harvesting at the atomic level.

REFERENCES

15.1. Papst, G., Gamble, B. B., Rodenbush, A. J., and Schöttler, R. 1997. Development of synchronous motors and generators with HTS field windings. *Superconducting Science and Technology* **10**(**12**): 924–926.

15.2. Blaugher, R. D. 1997. Low-calorie, high-energy generators and motors. *IEEE Spectrum* **34**(**7**): 36–42.

15.3. Schiferl, R. and Rey, C. 2006. Development of ultra efficient HTS electric motor systems. *2005 Annual Superconductivity Peer Review Meeting*, Washington, DC. http://web.ornl.gov/sci/htsc/documents/pdf/fy2006/HTSMotor-SchiferlandRey.pdf.

15.4. TheEngineer. 2001. Siemens first in Europe with a superconductor motor. http://www.theengineer.co.uk/news/siemens-first-in-europe-with-a-superconductor-motor/269479.article.

15.5. Takeda, T., Togawa, H., and Oota, T. 2006. Development of liquid nitrogen-cooled full superconducting motor. *IHI Engineering Review* **39**(**2**): 89–94.

15.6. Eckels, R W. and Snitchler, G. 2008. 5 MW high temperature superconductor ship propulsion motor design and test results. *Naval Engineers Journal* **117**(**4**): 31–36.

15.7. American Superconductor Corporation. 2009. AMSC and Northrop Grumman announce successful load testing of 36.5 Megawatt superconductor ship propulsion motor. http://www.businesswire.com/news/home/20090113005166/en/AMSC-Northrop-Grumman-Announce-Successful-Load-Testing.

15.8. Hayashi, K. 2008. Development of HTS motor—Present status and future prospect. Sumitomo Electric Industries, Ltd. http://www.cca08.com/pdf/presentations/lA-07-HAYASHI.pdf.

15.9. Kalsi, S. S. 2002. Development status of superconducting rotating machines. *IEEE Power Engineering Society Winter Meeting*, New York, Vol. 1, pp. 401–403.

15.10. Pippuri, J., Manninen, A., Keränen, J., and Tammi, K. 2013. Torque density of radial, axial and transverse flux permanent magnet machine topologies. *IEEE Transactions on Magnetics* **49**(**5**): 2339–2342.

15.11. Qu, R. H. and Lipo, T. A. 2005. Dual-rotor, radial-flux, toroidally-wound, permanent-magnet machine. U.S. Patent 6,924,574.

15.12. Zheng, P., Wu, Q., Bai, J. G., Tong, C. D., and Song, Z. Y. 2013. Analysis and experiment of a novel brushless double rotor machine for power-split hybrid electrical vehicle applications. *Energies* **2013**(**6**): 3209–3233.

15.13. Niu, S. X., Chau, K. T., and Jiang, J. Z. 2008. A permanent-magnet double-stator integrated-starter-generator for hybrid electric vehicles. *IEEE Vehicle Power and Propulsion Conference*, Harbin, China.

15.14. Hunstable, F. E., 2016. DC electric motor/generator with enhanced permanent magnet flux densities, U.S. Patent 9,419,483 B2.

15.15. Hunstable, F. E., 2019. Brushless electric motor/generator, U.S. Patent 10,263,480 B2.

15.16. Ulrich, L. 2019. Startup promises an electric-motor revolution: Linear labs says its superefficient motor could power cars, robots, and more. *IEEE Spectrum* **56**(**11**): 10–11.

15.17. Chau, K. T., Zhang, D., Jiang, J. Z., Liu, C. H., and Zhang, Y. J. 2007. Design of a magnetic-geared outer-rotor permanent magnet brushless motor for electric vehicles. *IEEE Transactions on Magnetics* **43**(**6**): 2504–2506.

15.18. Chau, K. T. and Li, W. L. 2014. Overview of electric machines for electric and hybrid vehicles. *International Journal of Vehicle Design* **64**(**1**): 46–71.

15.19. Aydin, M., Huang, S., and Lipo, T. A. 2004. Axial flux permanent magnet disc machines: A review. *Conference of Power Electronics, Electrical Drives and Machines (SPEEDAM)*, Capri, Italy.

15.20. Zhu, Z. Q. and Li, Y. X. 2018. Modularity techniques in high performance permanent magnet machines and applications. *CES Transactions on Electrical Machines and Systems* **2**(**1**): 93–103.

15.21. Gieras, J. F., Wang, R., and Kamper, M. J. 2008. *Axial Flux Permanent Magnet Brushless Machines*, 2nd edn. Springer-Verlag, New York.

15.22. Aydin, M., Lipo, A., and Huang, S. 2009. Field controlled axial flux permanent magnet electrical machine. U.S. Patent 7,608,965.

15.23. Woolmer, T. J. and McCulloch, M. D. 2007. Analysis of the yokeless and segmented armature machine. *IEEE International Electric Machines and Drives Conference*, Antalya, Turkey, Vol. 2, pp. 704–708.

15.24. Moreels, D. and Leijnen, P. 2019. Turning the electric motor inside out: A Belgian startup's axial-flux motor for EVs is small, light, and powerful. *IEEE Spectrum* **56(10)**: 40–45.

15.25. Moreels, D. and Leijnen, P. 2018. High efficiency axial flux machines: Why axial flux motor and generator technology will drive the next generation of electric machines. White paper v1.9, Magmax.

15.26. Vansompel, H. 2013. Design of an energy efficient axial flux permanent magnet machine. Ph.D. thesis, Ghent University, Ghent, Belgium.

15.27. Vansompel, H., Sergeant, P., and Leijnen, P. 2019. Stator for an axial flux machine and method for producing the same. U.S. Patent Application 2019/0288584. Also, European Patent Office EP3485558A1.

15.28. Huang, P. L. and Rao, Z. F. 2012. Printed circuit board disk type motor without iron core. China Patent CN103001426A.

15.29. Shaw, S. 2017. Structures and methods for thermal management in printed circuit board stators. U.S. Patent 9,800,109 B2.

15.30. Schuler, B. L., Lee, R., and Rasmussen, J. 2019. System and apparatus for axial field rotary energy device. U.S. Patent 10,186,922 B2.

15.31. Kosaka, T. and Matsui, N. 2008. Hybrid excitation machines with powdered iron core for electrical traction drive applications. *International Conference on Electrical Machines and Systems*, Wuhan, China, pp. 2974–2979.

15.32. Lesani, H., Darabi, A., Gheidari, Z. N., and Tootoonchian, F. 2006. Very fast field oriented control for permanent magnet hysteresis synchronous motor. *Iranian Journal of Electrical and Electronic Engineering* **2(1)**: 34–40.

15.33. Qin, R. and Rahman, M. A. 2003. Magnetic equivalent circuit of PM hysteresis synchronous motor. IEEE *Transactions on Magnetics* **39(5)**: 2998–3000.

15.34. Zambroni de Souza, A. C., Oliveira, D. Q., and Ribeiro, P. F. 2015. Chapter 1: Overview of plug-in electric vehicles technologies. In *Plug in Electric Vehicles in Smart Grads: Energy Management* (Eds.: S. Rajakaruna, F. Shahnia, and A. Ghosh). Springer: Singapore.

15.35. Hsu, J. S. and Adams, D. J. 1999. Permanent magnet energy conversion machine with magnet mounting arrangement. U.S. Patent 5,952,756.

15.36. Zhu, X. Y., Chau, K. T., Cheng, M., and Yu, C. 2008. Design and control of a flux-control lable stator-permanent magnet brushless motor drive. *Journal of Applied Physics* **103(7)**: 07F134-1–07F134-3.

15.37. Jeon, W. J., Katoh, S., Iwamoto, T., Kamiya, Y., and Onuki, T. 1999. Propulsive characteristics of a novel linear hybrid motor with both induction and synchronous operations. *IEEE Transactions on Magnetics* **35(5)**: 4025–4027.

15.38. Fisher, E. A. and Richter, E. 1993. Conical rotor for switched reluctance machine. U.S. Patent 5,233,254.

15.39. Sarros, T., Chew, E. C., Crase, S., Tay, B. K., and Soong, W. L. 2002. Investigation of cylindrical and conical electrostatic wobble micromotors. *Microelectronics Journal* **33(2002)**: 129–140.

15.40. Burch, D., Petro, J. P., and Mayer, J. F. 2011. Conical magnets and rotor–stator structures for electrodynamic machine. U.S. Patent 7,982,350.

15.41. Ballestín-Bernad, V., Artal-Sevil, J. S., and Domínguez-Navarro, J. A. 2021. A review of transverse flux machines topologies and design. *Energies* **14(21)**: 7173.

15.42. Beyer, S. 1997. Untersuchungen am magnetischen Kreis der permanentmagneterregten Transversalflussmaschine in Sammlerbauweise. Papierflieger: Clausthal-Zellerfeld, Germany.

15.43. Salles, M. B. C., Cardoso, J. R., and Hameyer, K. 2011. Dynamic modeling of transverse flux permanent magnet generator for wind turbines. *Journal of Microwaves, Optoelectronics and Electromagnetic Applications* **10(1)**: 95–105.

15.44. Parspour, N., Babazadeh, A., and Orlik, B. 2004. Transverse flux machine design for manipulating system applications. *Proceedings of PCIM 2004 Europe*, Vol. I, pp. 481–485.

15.45. Baserrah, S. 2011. Transverse flux machine. European Patent 2,317,633.

15.46. Gieras, J. F. and Rozman, G. I. 2015. Rim driven thruster having transverse flux motor. U.S. Patent 8,299,669.

15.47. Viorel, I.-A., Crivii, M., Löwenstein, L., Szabó, L., and Gutman, M. 2004. Direct drive systems with transverse flux reluctance motors. *ACTA Electrotehnica* **45(3)**: 33–40.

15.48. Martnez-Ocaña, I., Baker, N. J., Mecrow, B. C, Hilton, C., and Simon, B. 2019. Transverse flux machines as an alternative to radial flux machines in an in-wheel motor. *The Journal of Engineering* **2019(17)**: 3624–3628.

15.49. Teschler, L. 2013. Another way to make an energy efficient motor. *Machine Design*. Online article. http://machinedesign.com/motorsdrives/another-way-make-energy-efficient-motor.

15.50. Finkle, L. J. and Furia, A. 2013. Reconfigurable inductive to synchronous motor. U.S. Patent 8,390,162.

15.51. Chau, K. T., Li, W. L., and Lee, C. H. T. 2015. Challenges and opportunities of electric machines for renewable energy. *Progress in Electromagnetics Research B* **42**: 45–74.

15.52. Liu, C. H. 2011. Design of a new outer-rotor flux-controllable Vernier PM in-wheel motor drive for electric vehicle. *The 2011 International Conference on Electrical Machines and Systems*, Beijing, China.

15.53. Cheng, M., Fan, Y., and Chau, K. T. 2005. Design and analysis of a novel stator-doubly-fed doubly salient motor for electric vehicles. *Journal of Applied Physics* **97**(**2005**): 10Q508:1–10Q508:3.

15.54. Cheng, M., Chau, K. T., and Chan, C. C. 2001. Design and analysis of a new doubly salient permanent magnet motor. *IEEE Transaction on Magnetics* **37**(**4**): 3012–3020.

15.55. Ostovic, V. 2003. Memory motor—A new class of controllable flux PM machines for a true wide speed operation. *IEEE Industrial Application Magazine* **9**(**1**): 52–61.

15.56. Ostovic, V. 2002. Pole changing permanent magnet machines. *IEEE Transactions on Industry Applications* **38**(**6**): 1493–1499.

15.57. Chau, K. T., Chan, C. C., and Liu, C. 2008. Overview of permanent-magnet brushless drives for electric and hybrid electric vehicles. *IEEE Transactions on Industrial Electronics* **55**(**6**): 2246–2257.

15.58. Yu, C., Chau, K. T., Liu, X. H., and Jiang, J. Z. 2008. A flux-mnemonic permanent magnet brushless motor for electric vehicles. *Journal of Applied Physics* **103**(**7**): 07F103:1–07F103:3.

15.59. Li, W. L., Chau, K. T., Gong, Y., Jiang, J. Z., and Li, F. H. 2011. A new flux-mnemonic dualmagnet brushless machine. *IEEE Transactions on Magnetics* **47**(**10**): 4223–4226.

15.60. Zepp, L. P. and Medlin, J. W. 2003. Brushless permanent magnet motor or alternator with variable axial rotor/stator alignment to increase speed capacity. U.S. Patent 6,555,941.

15.61. Miller, T. J. E. 2009. *Switched Reluctance Motors and Their Control.* Magna Physics Publishing and Clarendon Press, Oxford, U.K.

15.62. Fujimoto, K. and Asano, H. 2012 Piezoelectric motor and method of manufacturing the same. U.S. Patent 8,330,326.

15.63. Hsu, J. S. 2015. Permanent-magnet-less synchronous reluctance system. U.S. Patent 8,264,120.

15.64. Nagle, S. F., Livermore, C., Frechette, L. G., Ghodssi, R., and Lang, J. H. 2005. An electric induction micromotor. *Journal of Microelectromechanical Systems* **14**(**5**): 1127–1143.

15.65. Ghalichechian, N., Modafe, A., Beyaz, M. I., and Ghodssi, R. 2008. Design, fabrication, and characterization of a rotary micromotor supported on microball bearings. *IEEE Journal of Microelectromechanical Systems* **17**(**3**): 632–642.

15.66. Feynman, R. P. 1960. There's plenty of room at the bottom. *Engineering and Science*, February Issue **23**(**5**): 22–36.

15.67. Feynman, R. P. 1984. *Tiny machines* (videotape of October 25, 1984). Sound Photosynthesis. Mill Valley, California.

15.68. Guo, P. X., Noji, H. Yengo, C. M., Zhao, Z. Y., and Grainge, I. 2016. Biological nanomotors with a revolution, linear, or rotation motion mechanism. *Microbiology and Molecular Biology Reviews* **80**: 161–186.

15.69. Guinness World Records. 2009. *Guinness World Records 2010: The Book of the Decade.* Random House Publishing Group, New York.

15.70. Weinmann, K. 2020. The smallest motor in the world. Empa. https://www.empa.ch/web/s604/molecular-motor.

15.71. Browne, W. R. and Feringa, B. L. (2006). Making Molecular Machines Work. *Nature Nanotechnology* **1**(**1**): 25–35.

15.72. Tierney, H. L., Murphy, C. J., Jewell, A. D., Baber, A. E., Iski, E. V., Khodaverdian, H. Y., McGuire, A. F., Klebanov, N., and Sykesl, E. C. H. 2011. Experimental demonstration of a single-molecule electric motor. *Natural Technology* 6: 625–629.

15.73. The Royal Swedish Academy of Sciences. 2016. Molecular machines. https://www.nobelprize.org/nobel_prizes/chemistry/laureates/2016/advanced-chemistryprize2016.pdf.

15.74. Koumura, N., Zijlstra, R. W., van Delden, R. A., Harada, N., and Feringa, B. L. 1999. Light-driven monodirectional molecular rotor. *Nature* **401**(**6749**): 152–155.

15.75. Vachon, J., Carroll, G. T., Pollard, M. M., Mes, E. M., Brouwer, A. M., and Feringa, B. L. 2014. An ultrafast surface-bound photo-active molecular motor. *Photochemical & Photobiological Sciences* **2014**(**13**): 241–246.

15.76. Kelly, T. R., De Silva, H., and Silva, R. A. 1999. Unidirectional rotary motion in a molecular system. *Nature* **401**(**6749**): 150–152.

15.77. Stolz, S., Groning, O., Prinz, J., Brune, H., and Widmer, R. 2020. Molecular motor crossing the frontier of classical to quantum tunneling motion. *Proceedings of the National Academy of Sciences of the United States of America* **117**(**26**): 14838–14842.

Appendix A
Advanced Interconnection Technology for Motors

The advancements in solid-state technology in the early 1960s have promoted the invention of the first brushless direct current (BLDC) motor in 1962 [A1]. Because BLDC motors eliminate the need for brushes and commutators, they offer many advantages over brushed motors, such as increased operation reliability, longer lifespan, higher efficiency, higher torque-to-weight ratio, reduced noise, less maintenance, and non-sparking. The modern BLDC motors have broken through many limitations of brushed motors and come to dominate in many industrial applications. For instance, most electric motors used in the automobile industry have moved from brushed to brushless motors for their superior performing characteristics, particularly for motors with high duty cycles. BLDC motors are currently extensively used for cordless power tools where the increased efficiency leads to longer service before the battery needs to be charged. However, because brushless motors require the use of active control electronics, their costs are relatively higher than those of brush motors.

Electronic circuits used in a BLDC motor allow the controller to switch the current promptly and thus regulate the motor's characteristics effectively, resulting in increased torque, effective speed control over a wide range, and better motor performance. The BLDC controller detects the position of the rotor using either sensors or sensorless techniques.

Electric motors often utilize one or more printed circuit boards (PCBs) to provide mechanical support and electrical connect electronic components (*e.g.*, resistors, capacitors, and diodes). Electrical connections between the components are achieved through coated copper layers that are designed with an artwork pattern (*i.e.*, *circuit layout*). Electrical connections between a PCB and motor winding wires, as well as between different PCBs, are ensured using conductive connectors. In practice, many manufacturers of electric machines, actuators, and solenoids today still utilize traditional manual soldering techniques to make connections to magnet wires in their products. However, this can be a costly process, both in terms of the time required to make these manual connections, and in terms of quality and reliability of the connections (*e.g.*, inconsistence and unpredictability caused by human-controlled variables, soldering material variation, *etc.*), given that such processes are difficult to control and may result in poor connections [A2].

Hillman *et al.* [A3] have investigated the solder failure mechanisms and addressed in detail the solder joint defects that occur during the assembly and manufacturing processes. In their study, failure of solder joints is mainly attributed to the poorly controlled assembly and non-optimized solder processes, including but not limited to, incorrect temperature cycling parameters, mechanical overstress, poor terminal seating, insufficient solder fillets and intermetallic formation, poor wettability, excessive solder contamination, *etc.* In practice, failure in solder joints can be due to a complex relationship among operating environment, system design, board material, solder joint strength, and the presence of failure accelerators, which are defined as defects introduced during assembly and production processes, due to the solder joint composition, size, shape, or location.

Depending on the industrial requirements and environmental conditions, many different types of electrical connectors are used in electric motors. Primarily, there are three kinds of electrical connectors based on the termination ends: wire-to-board, board-to-board, and wire-to-wire connectors [A4]. For electric machines, all these three connector kinds are commonly encountered, especially wire-to-board and wire-to-wire.

Over the past decades, several advanced, robust, and fast connection technologies have been developed for enhancing the operation reliability and quality of connectors and simplifying the assembly process.

A.1 PRESS-FIT TECHNOLOGY

Press-fit or compliant pin technology was initially invented in the early 1970s. As an inherently reliable and solderless solution, press-fit technique has become dominant in automobile, aerospace, robot, 3C (computer, communication, and consumer electronics), and electric machine industries. The use of the press-fit solutions can significantly reduce PCB assembly time and cost, and increase operation reliability, thus boosting production output capacity. In addition, this solderless method eliminates any unnecessary heating of assemblies, avoiding the risk of damage to components. Moreover, as space is an important factor in some applications, press-fit technology provides a solution that would be very difficult to realize with a conventional soldering process for miniaturization.

The important aspect of press-fit technology is its high reliability. As shown in Table A1, the failure in time rate of press-fit technology is 0.005, compared to manual solder of 0.5 and machine solder of 0.03 [A5]. This indicates that press-fit is six times more reliable than machine solder and 100 times than manual solder.

There are several types of electrical connections developed by connection manufactures. Among them, contact pins can be further categorized into two types: solid pins and compliant pins (also known as multispring pins). The compliant pins feature an elastic behavior and deform during insertion and thus sustain a contact normal force to enable a reliable electrical and mechanical connection over lifetime.

Moreover, the mechanical stability is significantly supported. The critical design of a press-fit connector is the compliant part at the middle section of the pin. Among various constructions, the most commonly used construction is known as the *eye of the needle* type, where an eye-shaped slot is formed at the compliant section, defining a pair of beams (Figure A1). As the compliant pin is forced into a PCB hole under a vertical force **F**, the two beams elastically deformed toward each other, exerting outwardly-directed forces against the interior wall of the PCB hole, and thus

TABLE A1
Comparison of Reliability among Different Electrical Connection Technologies

Type of Connection	Details	Conductor Cross-Section, mm^2	Failure Rate in FIT[a]	Standard/Guide
Solder	Manual		0.5	IPC 610, class 2
	Machine		0.03	
Wire bond for hybrid circuits	Al		0.1	28 μm/wedge-bond
	Au		0.1	25 μm/ball-bond
Wire-wrap		0.05–0.5	0.002	DIN EN 60352-1
				IEC 60352-1 Corr1
Crimp	Manual	0.05–300	0.25	DIN EN 60352-2
	Machine			IEC 60352-2 A 1+2
Termi-point		0.1–0.5	0.02	DIN 41611-4
Press-fit	Compressive force	0.3–2	0.005	IEC 60352-2
Insulation displacement		0.05–1	0.25	IEC 60352-3
				IEC 60352-4
Screw		0.5–16	0.5	DIN EN 60999-1
Clamp	Elastic force	0.5–16	0.5	DIN EN 60999-1

[a] Failure in time (FIT) is defined as a failure rate of 1 per billion hours.

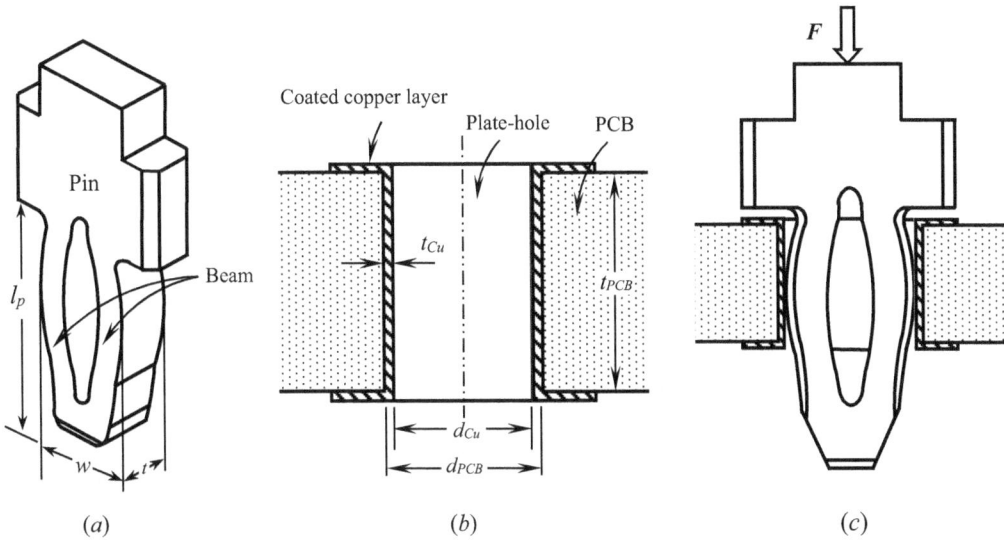

FIGURE A1   Press-fit technology: (*a*) the structure of the *eye-of-needle* type of compliant pin; (*b*) the plate-hole with coated copper layers of PCB; and (*c*) the insertion of the compliant pin into the plate-hole of PCB to form a reliable electrical connection.

providing a reliable electrical connection. In practice, the retention force has to be as close as possible to the insertion force **F** to gain maximum connection robustness. To ensure the high operation reliability and sufficient retention force of the compliant pin with the PCB, a number of key parameters must be carefully determined in the press-fit design process, including (a) the pin width w and the thickness t, (b) the shape and dimensions of the eye of the needle, (c) the PCB plated-through hole diameter d_{PCB} and thickness t_{PCB}, (d) the coated copper layer thickness t_{Cu} and the associated diameter d_{Cu} (where $d_{Cu} = d_{PCB} - 2t_{Cu}$), and (e) the pin length l_p.

In order to ensure a reliable electrical connection between the compliant pin and the PCB, the pin width w should be slightly larger than that of the coated copper diameter d_{Cu}. It is worth noting that in this design, the retention is obtained by friction between the pin and the PCB wall. Therefore, not only the geometries of the pin and PCB plate-hole but also the smoothness on their contacted surfaces can affect the overall press-fit system performance and the pin retention.

The press-fit pins can be made of different base materials such as copper alloys (*e.g.*, $CuSn_4$, $CuSn_6$, and CuNiSi), with two galvanic layers: the underplating (*e.g.*, nickel) and the top plating (*e.g.*, pure tin) to increase the corrosion resistance and electrical conductivity.

The electrical current-carrying capacity of press-fit connectors is strongly affected by the operation temperature. In Figure A2 [A6], the comparison of the current-carrying capacity of the two most popular sizes of press-fit zones (0.64 and 0.81 mm) is presented with the variation of the operating temperature. It can be seen in the Figure that the current-carrying capacity of both pins significantly drops with the increase in temperature. At room temperature, the current-carrying capacity of the 0.81-mm thickness terminal is about 14–29% higher than that of the 0.61-mm thickness terminal. At the temperature of 150°C, all curves nearly coincide.

Today's press-fit interconnectors have been tested and proven to support current capacities of 30 A or more through a single press-fit pin, with reliable, predictable current-carrying performance curves across a range of temperatures including $125°C–150°C$ and above [A7]. This information enables product engineers to select the optimal press-fit zone interconnects to meet their special application requirements.

It is worth noting that press-fit technology has a broad range of applications, for both PCBs and non-printed board cases (*e.g.*, smart junction boxes, molded modules, controllers, lighting, and a

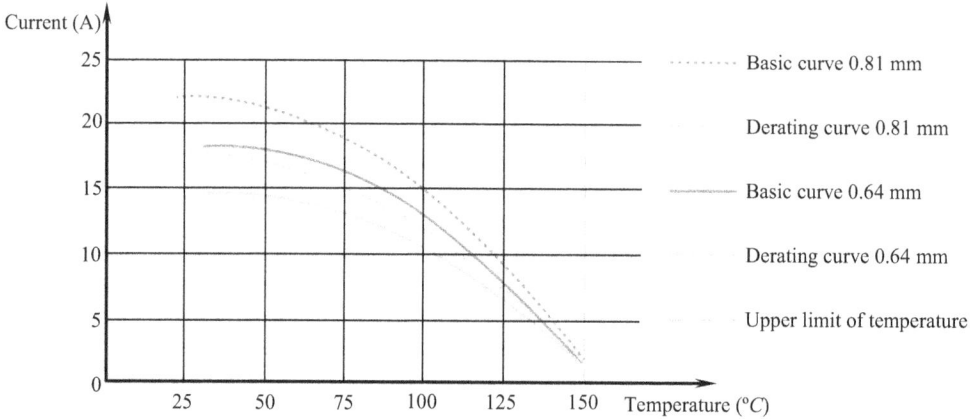

FIGURE A2 Comparison of current carrying capacity derating curves between the 0.64 and 0.81 mm thickness pins with the variation of operation temperature [A6].

variety of other customer applications). With press-fit technology, manufacturers are enabled to create highly reliable electro-mechanical interconnects.

The IEC standard 60352-5:2020 [A8] is applicable to solderless press-in connections for use in electrical and electro-mechanical components.

A.2   INSULATION DISPLACEMENT CONNECTION TECHNOLOGY

Insulation displacement connection (IDC) technology was initiated in 1961, described in a US patent *Solderless connector for insulated small wires* [A9]. IDC technology, as one of the more successful alternatives, is based on the idea of displacing or pushing aside some of the insulation around copper (or other conductive materials) conductors and making a direct electrical connection. As shown in Figure A3, electrical connection is achieved by inserting the insulated wire into a slot through the jaws. The shape edge of the jaws cut through the insulation and make a rigid metal-to-metal contact

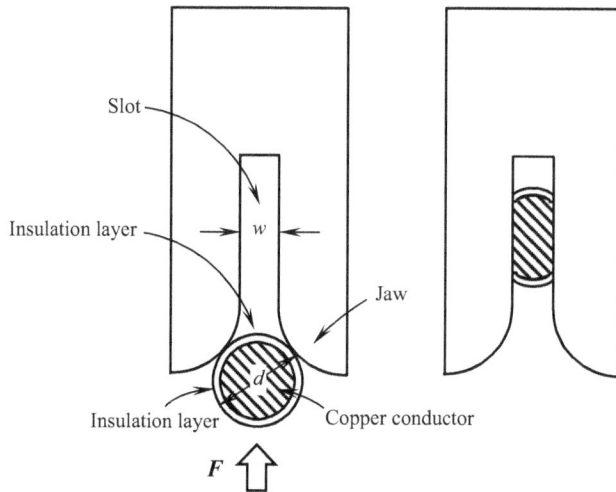

FIGURE A3 As a magnet wire, which is coated with an insulated layer, is forced into the slot, the sharp jaws penetrate the insulating layer and directly contact the copper core for ensuring electrical connection between the magnet wire and the connector, where the wire diameter is larger than the slot width (*i.e.*, $d > w$).

between the wire and the connector. The main advantage of the technology is that there is no need for wire preparation, which saves 60%–80% of the time usually spent on screw connections. Therefore, IDC technology is commonly used for connecting multiconductor cables.

With IDC technology, the size of magnet wire usually ranges from 23 AWG (0.57 mm) to 17 AWG (1.15 mm). Using specially designed IDC connectors, the range of the wire size can be considerably expanded.

A.3   TREND IN MOTOR WINDING INTERCONNECTIONS

The recent trend in the electric machine industry is to combine press-fit technology and IDC technology to offer optimal solderless, reliable electrical connection solution for electric machines. Dawson and Nozzi [A10] developed a type of interconnectors for segmented stator windings. In their design, each interconnector, made from conductive materials, consists of two components: one is the compliant pin with an *eye-of-needle* structure for directly connecting a PCB, and another has an open slot to accept a stator winding wire (Figure A4). The width w at the mid-section of the compliant pin is designed to be slightly larger than the diameter of the PCB aperture. Thus, as a compliant pin is pressed into the PCB aperture, the two flexible spring-like arms are compressed inwardly by walls of the aperture, providing firm electrical contact and mechanical retention between the interconnector and the PCB. In the IDC portion, a stator winding wire is forced into the slot of the IDC terminal. During the process, the insulation layer of a copper conductor is cut by the edges of the prongs, and thus the wire directly contacts the internal prong walls. The IDC system is inserted into the pocket of a stator insulator for holding the interconnector firmly. The direct electrical connection produced by the piercing action of the prongs eliminates the need for a soldered connection of the interconnector to magnet wires. This design greatly simplifies the assembly process, significantly enhances the electrical connection reliability, and remarkably saves the assembly time and cost.

The segmented brushless stator interconnect system and the method of assembling the segmented stator are depicted in Figure A5. As shown in Figure A5a, each stator segment uses two compliant

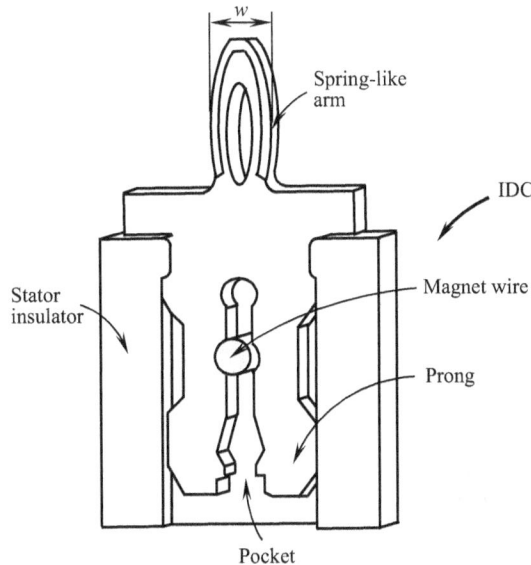

FIGURE A4   A new type of interconnector is developed by combining press-fit technology and IDC technology. A magnet wire is inserted into a pocket of an IDC for ensuring its electrical connection (US Patent 10,648,930) [A10]. (Courtesy of the U.S. Patent and Trademark Office, Alexandria, VA.)

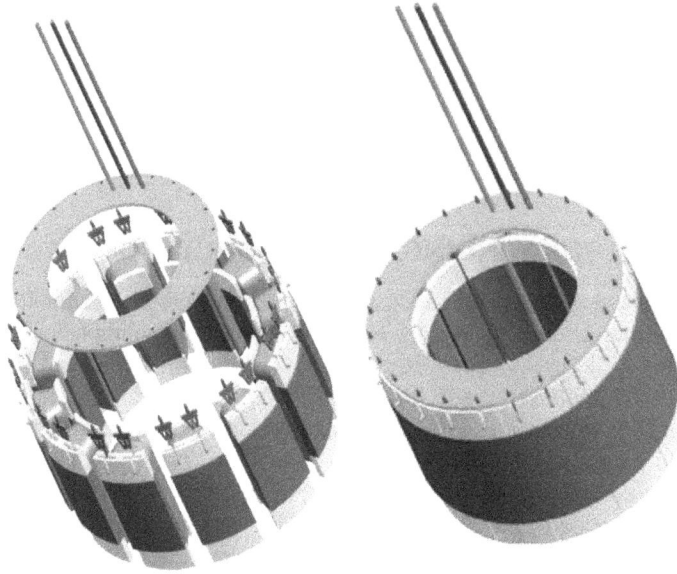

FIGURE A5 The segmented brushless stator interconnect system: (*a*) the segmented stator and distributed interconnectors on the top insulators and (*b*) the stator in an assembled form (US Patent 10,648,930) [A9]. (Courtesy of the U.S. Patent and Trademark Office, Alexandria, VA.)

pins to connect to the wire start and the wire end, respectively. Referring to Figure A5*b*, during the process of assembly, the PCB is pressed onto the compliant pins in a single action.

However, it is important to note that the retention of the pin in this design merely relies on the friction on the contact interface between the pin and PCB, and between the IDC and the stator insulator. Under strong vibration and shock operation conditions, the pin may separate from the PCB or the IDC from the stator insulator. Therefore, further innovations are needed to address this potential operational risk.

REFERENCES

A1. Wilson, T. G. and Trickey, P. H. 1962. D-C machine with solid-state commutation. *Electrical Engineering* **81**(11): 879–884.
A2. TE Connectivity. 2018. Motor connection: Solution guide. Document No. 1-1773918-3.
A3. Hillman, C., Rogers, K., Dasgupta, A., and Pecht, M. 1999. Solder failure mechanism in single-sided insertion-mount printed wiring boards. *Circuit World* **25**(3): 28–38.
A4. Kyeong, S. and Pecht, M. G. 2021. *Electrical Connectors: Design, Manufacture, Test, and Selection.* John Wiley & Sons, Hoboken, NJ.
A5. Mattsson, J., Callies, T., and Kerckhof, B. 2014. Press-fit technology. TE Automotive white paper. https://www.te.com/content/dam/te-com/documents/automotive/global/whitepaper-pressfit-072014. pdf.
A6. Interplex. 2020. Press-fit: Understanding the technology, its advantages and applications. https:// interplex.com/press-fit-guide/.
A7. Donovan, C. 2020. *Interconnecting the future: Press-fit technology.* Brochure, Autosplice Inc. https://de415321-a55c-4f6b-80af-c11ba57697b3.filesusr.com/ugd/23d362_cd290bb80d8749d6aa03 38a257873754.pdf.
A8. IEC. 2002. IEC 60635-5:2020 solderless connections – Part 5: Press-in connections – General requirements, test methods and practical guidance.
A9. Levin, E. J. and Leach, E. E. 1961. Solderless connector for insulated small wires. U.S. Patent 3,012,219.
A10. Dawson, S. and Nozzi, A. 2019. Segmented brushless stator interconnect system. U.S. Patent 10,468,930 B2.

Index

Taylor & Francis Group
an **informa** business

Taylor & Francis eBooks

www.taylorfrancis.com

A single destination for eBooks from Taylor & Francis
with increased functionality and an improved user
experience to meet the needs of our customers.

90,000+ eBooks of award-winning academic content in
Humanities, Social Science, Science, Technology, Engineering,
and Medical written by a global network of editors and authors.

TAYLOR & FRANCIS EBOOKS OFFERS:

A streamlined
experience for
our library
customers

A single point
of discovery
for all of our
eBook content

Improved
search and
discovery of
content at both
book and
chapter level

REQUEST A FREE TRIAL
support@taylorfrancis.com

Routledge
Taylor & Francis Group

CRC Press
Taylor & Francis Group

For Product Safety Concerns and Information please contact our EU
representative GPSR@taylorandfrancis.com
Taylor & Francis Verlag GmbH, Kaufingerstraße 24, 80331 München, Germany

www.ingramcontent.com/pod-product-compliance
Lightning Source LLC
Chambersburg PA
CBHW080332220326
41598CB00030B/4489

9 780367 564308